Catalysis and Reaction Engineering Division 2017

AA009374

Core Programming Area at the 2017 AIChE Annual Meeting

Minneapolis, Minnesota, USA
29 October – 3 November 2017

Volume 1 of 2

ISBN: 978-1-5108-5806-0

Printed from e-media with permission by:

Curran Associates, Inc.
57 Morehouse Lane
Red Hook, NY 12571

Some format issues inherent in the e-media version may also appear in this print version.

Printed by Curran Associates, Inc. (2018)

For permission requests, please contact AIChE
at the address below.

AIChE
120 Wall Street, FL 23
New York, NY 10005-4020

Phone: (800) 242-4363
Fax: (203) 775-5177

www.aiche.org

Additional copies of this publication are available from:

Curran Associates, Inc.
57 Morehouse Lane
Red Hook, NY 12571 USA
Phone: 845-758-0400
Fax: 845-758-2633
Email: curran@proceedings.com
Web: www.proceedings.com

TABLE OF CONTENTS

Volume 1

Volume 2

Some abstracts in this book refer to images that were not made available by the publisher

485774 Insight into the Formation Mechanism of an Industrially Relevant Ziegler-Natta Catalyst

Abstract:

Abstract for AIChE 2017 Antoine Klaue

Insight into the formation mechanism of an industrially relevant Ziegler-Natta catalyst

Antoine Klaue[1], Hua Wu[1], Massimo Morbidelli[1], Matthias Kruck[2], Nic Friederichs[2], Francesco Bertola[2]

[1]Institute for Chemical and Bioengineering, department of Chemistry and Applied Biosciences, ETH Zürich, 8093 Zürich, Switzerland

[2]SABIC, Technology & Innovation, P.O. Box 319, 6160 AH Geleen, The Netherlands

Ziegler-Natta (ZN) type polymerization processes have become very mature in terms of engineering and optimization [1]. However, few studies in the open literature exist dealing with the detailed chemistry during the formation of the ZN catalyst particles [2], of which knowledge is important in controlling the size and morphology of the ZN catalyst particles, thus those of the final polymer beads. In this work, we take a typical industrial ZN recipe to monitor the formation mechanism of the ZN catalyst particle. The recipe is a simple one-pot process [3], where an excess amount of aluminum alkyl dichloride is added into a precursor composed of magnesium and titanium alkoxides dissolved in hexane under agitation, leading to precipitation of the ZN catalyst particles. Through a combined technique based on ICP-OES, we have monitored the partition of the key elements in the liquid and solid phases along the catalyst formation. It is found that the first main reaction is the rapid precipitation of alkoxymagnesium chloride ($MgCl_x(OR)_{2-x}$): $Mg(OR)_2 + AlRCl_2$ à $MgCl_x(OR)_{2-x}+AlRCl_{2-x}(OR)_x$ Although the precipitation of alkoxytitanium chloride species ($TiCl_y(OR)_{4-y}$) also starts at the very beginning, $Ti(IV)(OR)_4 + AlRCl_2$ à $Ti(IV)Cl_y(OR)_{4-y}+AlRCl_{2-y}(OR)_y$, it proceeds at a much slower pace, such that the precipitation of $MgCl_x(OR)_{2-x}$ completes much earlier than that of $TiCl_y(OR)_{4-y}$. In fact, according to SEM-EDX measurements, $TiCl_y(OR)_{4-y}$ precipitates mainly on the surface of $MgCl_x(OR)_{2-x}$. We have also observed that at the later stage, further addition of aluminum alkyl dichloride leads to the formation of soluble bimetallic complexes of titanium and aluminum, leading to partial dissolution of titanium back to liquid phase, as shown in Figure 1. However, both the dissolved and precipitated complexes, once heated, lead to Ti(III)-type precipitation in the form of $Ti(OR)_{2-x}Cl_{x+1}$.

Figure 1 Composition measurement of the liquid phase during the particle formation process by ICP-OES. The dashed

line symbolizes the feed time.

[1] Böhm, L. L., Angew. Chem. Int. Ed. 2003, 42, 5010-2030

[2] Chien, J. C. W., J. of Polym. Sci., 1982, 20, 2019-2032

[3] Berger E., Derroitte, J.-L. (Solvay Cie, Brussels, Belgium). Polymerization of olefins, U.S. Patent, August 26, 1975.

499192 Use of Sacrificial Agent to Enhance Gas Transport through the Washcoat

Abstract:

Use of Sacrificial Agent to Enhance Gas Transport through the Washcoat

Pritpal Singh Dhillon, Michael P. Harold*

Di Wang[2], Ashok Kumar[2]

[1]Dept. of Chemical & Biomolecular Engineering, University of Houston, Houston, Texas

[2]Cummins Inc., Columbus, Indiana

* corresponding author e-mail: mpharold@central.uh.edu

Introduction

The Ammonia Slip Catalyst (ASC) is utilized to convert unreacted ammonia selectively into N_2 while minimizing undesired products NO, NO_2 and N_2O. The dual-layer ASC comprises a base layer of an oxidation catalyst (Pt/Al_2O_3) and a top layer of a SCR catalyst such as Cu-exchanged zeolite. The dual-layer ASC is known for higher nitrogen selectivity, but it adversely affects ammonia conversion due to mass transfer limitations. Increasing the top (SCR) layer loading improves nitrogen selectivity, simultaneously it reduces ammonia conversion. To increase nitrogen selectivity without affecting conversion, we prepared modified dual-layer ASC materials with enhanced top washcoat macro-porosity. Sacrificial agents including yeast and poly-(tert-butyl acrylate) were used to create voidage inside the top SCR layer with the intent of reduced the mass transfer limitations for ammonia [1]. The impact of the modified washcoat on the ammonia conversion and N_2 selectivity is the focus of the study. Reaction experiments were conducted using the modified dual-layer ASC showed significant increase in nitrogen selectivity and improvement in ammonia conversion simultaneously.

Materials and Methods

Bottom Layer: The base layer of a dual-layer ASC comprising Pt/Al_2O_3 was prepared through a sequence of steps. Catalyst powder was synthesized by incipient wetness impregnation using chloroplatinic acid hexahydrate as the Pt precursor. The Pt/Al_2O_3 slurry was prepared by adding deionized water and boehmite solution (AlOOH: 20wt %) to Pt/Al_2O_3 powder (mass ratio Pt/Al_2O_3: water: boehmite = 8:15:2) and ball milled for 20 h. The monolith was dipped into the Pt/Al_2O_3 slurry for 30 s and excess slurry was blown for 10 s. Subsequently, the monolith was dried in an oven at 120 °C for 2 h and the same dipping procedure was repeated to get the required loading. Finally, the monolith was dried overnight in an oven at 120 °C, followed by calcination at 550 °C for 5 h with a temperature ramp rate of 0.5°C/min. The freshly prepared monolith was then treated with 2% H_2 and balance Ar for 1 h at 500 °C, followed by degreening with 5% O_2 and balance Ar at same temperature for 2 h.

Conventional Dual-Layer: The dual-layer ASC comprising a Cu/SSZ-13 top layer and Pt/Al_2O_3 bottom layer was prepared as follows. Cu/SSZ-13 containing 3% Cu slurry was prepared by adding deionized water and boehmite solution (ALOOH: 20wt %) to Cu/SSZ-13 powder (mass ratio Cu/SSZ-13: water: boehmite = 4:8:5) and ball milled for 20 h. The monolith which was already coated with Pt base layer was dipped into the Cu/SSZ-13 slurry for 30 s and excess slurry was blown for 10 s. Subsequently, the monolith was dried in an oven at 120 °C for 2 h and the same dipping procedure was repeated to get the required loading. Finally, the monolith was dried overnight in an oven at 120 °C, followed by calcination at 550 °C for 5 h with a temperature ramp rate of 0.5°C/min. The type-I dual layer ASC was prepared (Fig 1).

Modified Dual-Layer: For the modified dual-layer ASC, the Cu/SSZ-13 (Cu/SSZ-13 containing 5% Cu) slurry was prepared by similar manner as above and ball milled for 20 h. Then, a calculated amount of sacrificial agent (yeast/polymer) was added to the slurry, intended to increase the porosity of top layer roughly by 20%. The monolith which was already coated with Pt base layer was dipped into the Cu/SSZ-13 slurry for 30 s and excess slurry was blown for 10 s. Subsequently, the monolith was dried at room temperature for 4 h and the same dipping procedure was repeated to get the required loading. Finally, the monolith was dried overnight at room temperature, followed by calcination in flow reactor (1000 sccm, 5% O_2 & balance Argon) at 550 °C for 5 h with a temperature ramp rate of 1°C/min. The type-II dual layer ASC was prepared (Fig 2).

Picture1.gif

Figure 1: (Type-I: Dual Layer ASC) Figure 2: (Type-II: Modified dual layer ASC)

Results and Discussion

The dual layer samples (type I) with varied top layer loadings were tested for NH_3 oxidation. The sample with the single Pt/Al_2O_3 layer achieved the highest NH_3 conversion approaching 95% but had a poor N_2 selectivity (Fig 3). The dual-layer sample with Cu/SSZ-13 loading of 3 g/in^3 gave the highest N_2 selectivity but had a lower NH_3 conversion than the single layer sample (Fig 4). The results expectedly showed that the increase in top zeolite layer loading increases the mass transfer limitations by restricting the amount of NH_3 reaching the base Pt layer [2]. To resolve the problem, modified dual-layer ASC is introduced.

Figure 3: NH_3 conversion vs temperature Figure 4: N_2 selectivity vs temperature

Two samples of almost same loadings were prepared, a dual-layer ASC {Pt(10)CuZ(2.6)} and modified dual-layer ASC {Pt(10)CuZ(2.7)} using the above-mentioned steps. The modified dual-layer showed a notably higher N_2 selectivity and somewhat improved NH_3 conversion at high temperature (Fig 5 & 6).

Figure 5: NH_3 conversion vs temperature

Figure 6: N_2 selectivity vs temperature

This is an ongoing study. The results we get so far are quite encouraging and more experiments are being done to check the consistency of reaction results. Most of the yeast or poly-(tert-butyl acrylate) burn away at 550 ^oC (TGA data not shown here for brevity). The catalyst characterization will be done to find if there is any change in catalyst morphology due to yeast/polymer combustion.

References:

1. Václavík, M., Dudák, M., Novák, V., Medlín, R., Štepánek, F., Marek, M., & Kocí, P. (2014). Yeast cells as macropore bio-templates enhancing transport properties and conversions in coated catalyst layers for exhaust gas oxidation. *Chemical Engineering Science*.

2. S. Shrestha, M. P. Harold, K. Kamasamudram, A. Kumar, L. Olsson and K. Leistner, "Selective oxidation of ammonia to nitrogen on bi-functional Cu-SSZ-13 and Pt/Al2O3 monolith catalyst," *Catalysis Today,* 2016.

497929 Bifurcation Analysis of Gas Phase Methane Oxidative Coupling

Abstract:

Abstract:

We present a detailed bifurcation analysis of methane oxidative coupling in the gas phase (without catalyst) using a global kinetic model with various oxidation, reforming and dehydrogenation reactions. The main goal is to determine the methane conversion (X) and C_2+ products yield (Y) in different types of large scale reactors under various feed and operating conditions. The global kinetic model used satisfies the thermodynamic constraints and is fitted to literature data as well as new data obtained under nearly isothermal conditions.

Hysteresis (ignition and extinction) behavior of the system is investigated as a function of the CH_4/O_2 ratio as well as feed temperature and space time. Hysteresis may exist at ambient feed temperature (300K), for low CH_4/O_2 ratios (3 or lower) with space times ~ 1s. When hysteresis exists, the best yield of ethylene can be obtained close to the extinction point on the ignited branches. Typically, reactions forming C_2+ products are dominating in the reactor when the temperature on the ignited branch is above 1200K and the space time is very short. Acetylene becomes the main C_2+ product, together with much excess of hydrogen and carbon monoxide than water and carbon dioxide, as reactor temperature or space time are increased beyond ignition.

The impact of axial backmixing of species and energy is also studied using axial dispersion model and single phase 2-mode model. It is found that when the thermal backmixing is highest and species backmixing is lowest, the extinction point can be pushed to lower values of feed temperatures and/or to lower space times with the yield of intermediate product increased.

491594 Nanoengineering the Environment Around Catalytic Active Sites

Abstract:

Finely engineered environments around catalytic active sites are a hallmark of enzymatic systems where functional groups are precisely placed, electronic charge is modulated, and hydrophobicity/hydrophilicity is controlled in the active site environment. Translating this precise control around an active site to a heterogeneous catalytic system is a lofty goal in heterogeneous catalysis.

The work presented here will highlight progress made in achieving this goal. One example uses immobilized Rh complexes inside of SBA-15 mesopores where for 1-octene hydroformylation, a rate enhancement was observed for the Rh confined inside of the mesoporous versus the activity of the homogeneous Rh catalyst. [1] . This is an example of using the pores of a support material to influence the environment around an active site. In another example, the environment around catalytically active gold nanoparticles in solution is modulated through different organic ligands bound to the surface of the gold nanoparticles. Using 4-nitrophenol reduction as a test reaction, our work demonstrates that the activity of the catalyst can be altered with different ligands bound to the surface of the gold nanoparticles. The results give additional insight into the active site requirements for 4-nitrophenol reduction when using gold nanoparticles as catalysts. [2, 3]

Further work in the synthesis of small gold clusters with tunable environments will be detailed with 11 atom gold clusters stabilized with organic ligands. The stability, characterization, and catalytic activity of these clusters will be assessed and presented.

1. Marras, F.; Wang, J.; Coppens, M.-O.; Reek, J.N.H.; *Chem. Commun. 2010*, **46**, 6587.

2. Nigra, M.M.; Ha, J.-M.; Katz, A.; *Catal. Sci. Technol.* **2013**, *3*, 2976.

3. Nigra, M.M.; Coppens, M.-O. **2017**, in preparation.

495224 Fabrication of Ultra-Thin PVP-Metal Oxide Nanofibers Via Electrospinning for Applications in Catalysis and Adsorption

Abstract:

Electrospinning has been recognized as an efficient technique for the synthesis of polymer-based nanofibers and has recently gained accrued attention due in part to a surging interest in nanotechnology. Electrospun nanofibers exhibit exceptionally high surface areas, large length to diameter ratios and effective electronic properties, which makes them promising candidates for catalytic, photocatalytic and sorption applications. In the electrospinning process, a high voltage is used to create an electrically charged jet of a polymer containing solution. Before reaching the collecting plate, the solution jet evaporates and small solid fibers in the micro-, sub-micro-, or nano-range are collected. Although numerous synthesis techniques of different inorganic materials have been reported and published over the past decade, little work has been done to understand the relationship between the electrospinning conditions and the microstructure and properties of the formed fibers.

In this work, sol-gel processing and electrospinning was used to fabricate 1-D inorganic-organic composite nanofibers from solutions containing polyvinylpyrrolidone (PVP) and suitable aqueous precursors of nickel, copper, calcium, magnesium, zinc, and aluminum nitrates or a combination thereof. The primary objective was to synthesize nanofibers that had diameters that were consistent and controllable, defect-free or defect-controllable, and continuous. To achieve this objective, the parameters influencing the transformation of the polymer-containing solution into fibers were investigated, which include (1) the polymer(s) properties such as, molecular weight and solubility; (2) solvent(s) properties such as, viscosity, conductivity, and surface tension; (3) the concentration of polymer(s) and metal precursor(s) in the solution; (4) the ratio of polymer(s) to metal(s) in the solution; (4) voltage and electric potential at the capillary tip; (4) distance between the tip and collecting plate; (5) extrusion rate; and (6) ambient parameters such as, solution temperature, humidity, and air velocity in the electrospinning chamber.

In this work, the electrospun solutions consisted of dissolving different amounts of PVP in ethanol/methanol/water/DMF mixtures and adding different ratios of polymer to metal salt. Electrospinning was conducted at wide range of operational conditions: 10-35 kV, 0.1-5 mL/hr extrusion rate, and 4-20 inch distance between the tip of the needle and collector plate. The formed fibers had diameters that ranged from 50µm to 10nm. The spun fibers were calcined at 500°C for 2 hours at a ramping rate of 2°C/min to completely remove the PVP matrix and convert the as-spun fibers to inorganic oxides for further characterization. The composition, physio-chemical properties, and structural morphologies of the PVP-metal nanofibers—before and after calcination—were examined by X-Ray diffraction (XRD), N2-Physisoprition (BET), thermogravimetric analysis (TGA), scanning electron microscopy (SEM) with electron dispersive spectroscopy (EDS), transmission electron microscopy (TEM), X-ray photoelectron spectroscopy (XPS), and atomic force microscopy (AFM) to elucidate information about the influence of synthesis parameters on the properties of the electrospun nanofibers.

The obtained results revealed that the concentration of PVP, ratio of PVP to metal in the polymer solution, and ambient conditions played a pivotal role in influencing the shape and diameter of the nanofibers as well as the formation of beads. For instance, ribbon-like structures were observed at high PVP concentrations while the onset of beads formation was observed at lower PVP content. The effects of extrusion rate, distance between the tip and collecting plate, and voltage were less pronounced, but seemed to strongly affect the range and distribution of the nanofibers' diameters. When the synthesis parameters and experimental conditions were optimized to yield smooth fibers with diameters that were 10-100nm, several advantageous catalytic and mechanical characteristics were observed which include: (1) a significant increase in the surface area to volume ratio, (2) flexibility in surface functionalities, and (3) superior mechanical performance (e.g., stiffness and tensile strength) compared with other forms of the material.

The synthesized fibers were tested in two applications. In the first, they were tested as sorbents for natural gas purification. Natural gas and oil typically contain a wide range of contaminants and require purification before processing because these contaminants damage process equipment and catalysts and are hazardous to humans and the environment. Metal oxides are extensively used to remove sulfur compounds via chemical reaction; however, their use is limited by incomplete conversion and mass-transfer limited sorption kinetics. Electrospun nanofibers are an attractive class of materials for reactive sorption of sulfur compounds because of their small reactive domains and large intra-fiber void spaces. This work studies the effects of synthesis parameters on the physical properties and reactivity of electrospun nanofibers. We investigate these properties using a combination of advanced characterization techniques and fixed bed sorption experiments.

In another application, the synthesized nanofibers were probed as hybrid catalyst-sorbents for sorption-enhanced steam methane reforming. Supplying distributed hydrogen at a price that is cost competitive with gasoline is the largest barrier to the widespread implementation of hydrogen fuel cell vehicles. While steam reforming of carbon-based feedstocks remains the most economical way to produce hydrogen, the economics of small scale reforming for hydrogen production are poor. An attractive alternative to conventional reforming is sorbent-enhanced steam reforming (SE-SR) using a catalyst-sorbent material that simultaneously adsorbs CO_2 during reforming reactions. Typical SE-SR processes use admixtures of reforming catalysts with mineral oxide powder based adsorbents. The limitations of these materials are incomplete conversion, loss of initial sorbent capacity, and mass-transfer limited sorption rates. Metal-oxide nanofibers, as hybrid catalyst-sorbent materials that place catalytic sites within molecular proximity of CO_2 sorption sites, allow for easy access of CO_2 to sorption sites, and thus these materials have the potential to achieve stoichiometric CO_2capacities without mass transfer restricted sorption rates. Furthermore, metal oxide nanofibers have the potential to be more active and stable catalysts for reforming compared to supported metal clusters because the nanofiber topology will prevent agglomeration of active metal sites and the deposition of carbon, which poisons catalytic sites.

This research effort provides shrewd insight into the design of fine-tuned and defect-free or defect-controllable nanofibers, which could be used in various applications—in particular as hybrid catalyst-sorbents for sorption-enhanced steam methane reforming and as sorbents in natural gas purification.

493379 Microfluidic Studies of Room-Temperature Synthesized Perovskite Nanocrystals

Abstract:

Emergence of organic/inorganic metal-halide perovskites as a new class of photovoltaic material with excellent electron conducting and light harvesting properties, have enabled breakthrough applications at device scales in light emitting diodes (LEDs) and solar cells.[1, 2] Over the past 7 years, efficiency of perovskite-based solar cells has increased significantly from 3.8% [3] to 21.1%,[4] illustrating their potential as a cost and energy-effective substitute to conventional silicon-based solar cells. Recent advancements in solution-phase processing of perovskite nanocrystals (*i.e.*, quantum dots)[5, 6] have stimulated research efforts towards continuous nano-manufacturing of high-quality perovskite nanocrystals. Precise band-gap engineering of perovskite nanocrystals, similar to well-studied II-VI and IV-VI semiconductor nanocrystals[7] would facilitate their utilization in photovoltaics and LEDs. However, the lack of fundamental understanding of different mechanisms involved during the colloidal synthesis of perovskite nanoparticles has limited their large-scale nanomanufacturing. Further integration of perovskite nanocrystals in solar cells and LED sectors would strongly benefit from a better understanding of different mechanisms involved during the early stage formation and growth kinetics of these nanocrystals.

In this work, we designed and developed a modular microfluidic platform for fundamental studies of the synthesis of colloidal cesium lead halide perovskite nanocrystals using *in-situ* absorption/fluorescence spectroscopy. The developed microfluidic platform comprised of a tubular microreactor (fluorinated ethylene propylene with an inner diameter of 750 Î¼m), 3 computer-controlled syringe pumps, and a custom-designed 3-port flowcell, enabling real-time spectral characterization of in-flow synthesized perovskite nanocrystals. The separation of the spectral characterization tool (*i.e.*, the portable flowcell) and the temperature-controlled microreactor enabled a systematic study of the effect of mixing timescale on the nucleation and growth of perovskite nanocrystals. The developed microfluidic strategy enabled studies of 4 orders of magnitude change in the reaction timescales (15 ms to 5 min) on the quantum yield and size distribution of the synthesized perovskite nanocrystals. Furthermore, the automated microfluidic setup was utilized for the high-throughput screening of the experimental parameter space associated with the synthesis of colloidal perovskite nanocrystals (2000 spectral data were obtained within 20 min). Furthermore, utilizing the developed microreactor, the effects of the solvent and surface ligands on the intermediate growth pathways, final size, quantum yield, and size distribution of the resulting nanocrystals were studied. The obtained characteristic kinetic information of different mechanisms involved in the nucleation and growth stages of perovskite nanocrystals will be utilized to develop the design principles and demonstrate the technology for continuous large-scale manufacturing of engineered perovskite nanocrystals for applications in photovoltaics and LEDs.

References

1. D. Liu and T. L. Kelly, *Nat Photon*, 2014, **8**, 133-138.

2. B. R. Sutherland, S. Hoogland, M. M. Adachi, P. Kanjanaboos, C. T. O. Wong, J. J. McDowell, J. Xu, O. Voznyy, Z. Ning, A. J. Houtepen and E. H. Sargent, *Advanced Materials*, 2015, **27**, 53-58.

3. A. Kojima, K. Teshima, Y. Shirai and T. Miyasaka, *Journal of the American Chemical Society*, 2009, **131**, 6050-6051.

4. M. Saliba, T. Matsui, J.-Y. Seo, K. Domanski, J.-P. Correa-Baena, M. K. Nazeeruddin, S. M. Zakeeruddin, W. Tress, A. Abate, A. Hagfeldt and M. Gratzel, *Energy & Environmental Science*, 2016, **9**, 1989-1997.

5. L. Protesescu, S. Yakunin, M. I. Bodnarchuk, F. Krieg, R. Caputo, C. H. Hendon, R. X. Yang, A. Walsh and M. V. Kovalenko, *Nano Lett.*, 2015, **15**, 3692-3696.

6. L. C. Schmidt, A. Pertegás, S. González-Carrero, O. Malinkiewicz, S. Agouram, G. Mínguez Espallargas, H. J. Bolink, R. E. Galian and J. Pérez-Prieto, *J. Am. Chem. Soc.*, 2014, **136**, 850-853.

7. M. Abolhasani, C. W. Coley, L. Xie, O. Chen, M. G. Bawendi and K. F. Jensen, *Chem. Mater.*, 2015, **27**, 6131-6138.

493468 Ultrafast Synthesis of Titanium Silica(TS-1) By Continuous Flow

Abstract:

For the preparation of porous crystalline materials, the characteristic of the higher energy consumption and long crystallization time and instability of products existing in traditional hydrothermal technology with a batch reactor is becoming a challenge to meet the requirements of green chemical industry. Recently, the method of fast flow crystallization using tubular reactors has attacked great interests from many researchers, due to its advantages of fast, controllable, safety, low energy consumption, and reliable scaling up. Herein, we report a rapid continuous hydrothermal synthesis method with a tubular reactor to get TS-1, and the nanoparticles of MFI structures have been successfully synthesized after ten-minutes synthesis in the tubes. The resulting samples are characterized by X-ray diffraction, UV-vis spectroscopy, SEM and N_2 adsorption to evaluate the textural properties. The obtained TS-1 is about 100 nm particles, and for UV-Vis results, only one peak exists in 210 nm, so the results also show that faster nucleation rate has advantages in titanium entering into the silicon skeleton.

491710 Nitrogen Recovery and Biocrude Formation from Hydrothermal Liquefaction of Protein

Abstract:

Nitrogen Recovery and Biocrude Formation from the Hydrothermal Liquefaction of Protein

James D. Sheehan and Phillip E. Savage

Hydrothermal liquefaction is a thermochemical process that uses water at elevated temperatures (200-370°C) and pressures (10-22 MPa) to transform wet biomass into an energy dense "biocrude" oil and aqueous-phase nutrient coproducts.[1–3] The biocrude can be upgraded and co-refined with petroleum crudes into liquid fuels while the N- and P-containing nutrient coproducts can be recovered and recycled to cultivate additional biomass (e.g., microalgae). N is an essential element for cultivating microalgae and its demand can be met in part by recycling recovered N-containing nutrients from HTL, which are mostly derived from microalgal proteins. Proteins are major biochemical constituents of microalgae and contain approximately 90% of microalgal N.[4]The N atoms in the biomass proteins are ideally recovered as nutrients, however N-containing compounds can also partition into biocrude, which lowers the fuel quality and reduces the heating value of biocrude. Thus, the sustainability of HTL can be highly dependent on proteins and their transformation into valuable nutrient coproducts or detrimental N-containing compounds that partition into biocrude.

We performed HTL of bovine serum albumin (BSA), a well-defined model protein, and observed its transformation into valuable nutrient coproducts (e.g. amino acids, ammonia) and biocrude. We used temperatures ranging from 200 °C to 400 °C and holding times between 30 s to 1 hour. BSA readily decomposed during HTL into smaller peptides. The presence of peptides with molecular weights exceeding 5 kDa was no longer observable at temperatures at 250°C or higher. At 200°C, BSA aggregated into aqueous- and organic solvent-insoluble solids, in yields up to 80 wt%, that eventually decomposed into peptides which persevered in the hydrothermal media. The yields of amino acids were less than 6 wt% under all conditions, however, at temperatures of 350°C or higher the yields of primary and secondary amines in the aqueous-soluble products were upwards of 50 wt% and over 40% of N was recovered as ammonia, a preferred N-source for cultivating microalgae. The highest yields of biocrude were between 17 and 24 wt% and were achieved at temperatures ranging 300 to 400°C and holding times of 1 to 10 min. Mild (200°C) and severe (400°C, >10 min) conditions resulted with biocrude yields less than 5 wt%. This presentation will also include a discussion of the reaction pathways, kinetics, and mechanisms and thereby breaks new ground by connecting operationally defined lumped products (e.g., biocrude) with molecular-level chemistry for HTL of proteins.

References

(1) Peterson, A. A.; Vogel, F.; Lachance, R. P.; Fröling, M.; Antal, Jr., M. J.; Tester, J. W. Thermochemical Biofuel Production in Hydrothermal Media: A Review of Sub- and Supercritical Water Technologies. *Energy Environ. Sci.* **2008**, *1*, 32–65.

(2) Toor, S. S.; Rosendahl, L.; Rudolf, A. Hydrothermal Liquefaction of Biomass: A Review of Subcritical Water Technologies. *Energy* **2011**, *36*, 2328–2342.

(3) López Barreiro, D.; Prins, W.; Ronsse, F.; Brilman, W. Hydrothermal Liquefaction (HTL) of Microalgae for Biofuel Production: State of the Art Review and Future Prospects. *Biomass and Bioenergy* **2013**, *53*, 113–127.

(4) Becker, E. *Microalgae Biotechnology and Microbiology*; Cambridge University Press: New York, 1994.

491087 Bio-Succinic Acid Production from Tartaric Acid

Abstract:

Fossil fuels are currently the primary source of energy and chemicals because of their high availability and energy content.[1] However, the depletion of fossil fuels and environmental problems such as greenhouse gas emissions have triggered the development of biomass-based technologies, as biomass is the only source of renewable carbon.[2] Those new technologies also bring novel value added chemicals by utilizing the functional groups which are already present in biomass molecules.[3] Succinic acid ($C_4H_6O_4$), a four-carbon dicarboxylic acid, is recognized by the Department of Energy (DOE) as one of the top biomass-derived platform chemicals.[4] Its main applications include replacement of maleic anhydride as the C4 building block and synthesis of biodegradable and biobased polymers (Scheme 1).[4] The traditional method for succinic acid production relies on fossil fuels. Fermentation has been proposed as a renewable alternative, yet low product concentration and high concentration of organic acid are still drawbacks.[5,6]

Scheme 1. Succinic acid synthesis and applications

We developed a novel one-step catalytic route to produce bio-succinic acid using biomass-derived tartaric acid (Scheme 1), in high yields. Tartaric acid, which is naturally produced in grapes, can be recovered from winery waste streams as a byproduct of wine industry.[7] A liquid-phase system under hydrogen atmosphere comprised of molybdenum oxide catalyst supported on carbon black (MoO_x/BC) is effective in catalyzing C-O bond cleavage of tartaric acid. The role of solvent was investigated to gain a better understanding of the reaction system. The MoO_x/BC performance under different pre-reduction temperatures combined with characterization data such as temperature programmed reduction (TPR), X-ray powder diffraction (XRD), X-ray photoelectron spectroscopy (XPS) reveals that the high catalyst activity is correlated with the formation of specific Mo oxidation states. Finally, recyclability studies and structural characterization of the catalyst after reaction indicate that MoO_x/BC remains active upon reuse.

References

(1) Beerthuis, R.; Rothenberg, G.; Shiju, N. R. *Green Chem.* **2014**, *17*, 1341–1361.

(2) Ruppert, A. M.; Weinberg, K.; Palkovits, R. *Angew. Chemie - Int. Ed.* **2012**, *51* (11), 2564–2601.

(3) Corma, A.; Iborra, S.; Velty, A. *Chem. Rev.* **2007**, *107* (6), 2411–2502.

(4) Carlson, A.; Coggio, B.; Lau, K.; Mercogliano, C.; Millis, J. *Chemicals and Fuels from Bio-Based Building Blocks*, **2016**, 173–190.

(5) Cheng, K. K.; Zhao, X. B.; Zeng, J.; Wu, R. C.; Xu, Y. Z.; Liu, D. H.; Zhang, J. A. *Appl. Microbiol. Biotechnol.* **2012**, *95* (4), 841–850.

(6) Bechthold, I.; Bretz, K.; Kabasci, S.; Kopitzky, R.; Springer, A. *Chem. Eng. Technol.* **2008**, *31* (5), 647–654.

(7) Oliveira, M.; Duarte, E. **2016**, *10* (1), 168–176.

495881 1,6-Hexanediol Synthesis from Cellulose

Abstract:

1,6-hexanediol (HDO) is a high value commodity chemicals ($4,400/ton) used as a polyester or polyurethane component traditionally produced from adipic acid. In this presentation we discuss a three step catalytic route to produce HDO from cellulose[1] that includes: i) hydrolysis/dehydration of cellulose into LGO and HMF with sulfuric acid, ii) hydrogenation/hydrogenolysis of LGO and HMF into THFDM over Pd/Si-Al_2O_3 [2] and iii) hydrogenolysis of THFDM into HDO over a bifunctional Pt-acid catalyst. The overall HDO yield is 50%. In this presentation we will discuss the catalytic chemistry of each of these steps and elucidate the relationship between the catalyst structure and performance. We use cellulose, white birch and corn stover as feedstocks.

The THFDM is converted into 1,2,6 hexanetriol and then HDO in the final step through a dehydration-hydrogenation mechanism on a bifunctional Pt-based catalysts in over 90% HDO selectivity (70% HDO yield). We will show that the active catalytic acid phase for the dehydration, a BrÃ¸nsted acid site, is generated on the oxophilic support through hydrogen spillover from the Pt.

Â

Reference:

1 J. He, K. Huang, K. J. Barnett, S. Krishna, D. M. Alonso, Z. J. Brentzel, S. P. Burt, T. Walker, W. F. Banholzer, C. T. Maravelias, I. Hermans, J. A. Dumesic and G. W. Huber, *Faraday Discuss.*, 2017.

2 S. H. Krishna, D. J. McClelland, Q. A. Rashke, J. A. Dumesic and G. W. Huber, *Green Chem.*, 2017, **19**, 1278â€“1285.

497020 High-Efficiency Photo-Electrochemical Chloro-Alkali Production

Abstract:

The increased penetration of photovoltaic (PV) technologies has led to significant challenges on the management of the electrical grid. This has made evident the need for energy storage technologies that can bridge the temporal mismatch between electricity production and demand imposed by intermittent solar power sources. Alternatively, the implementation of PVs for large-scale off-grid chemical processes could further accelerate the deployment of clean energy technologies. Electrochemical processes, such as the ones implemented for chlor-alkali production, account for more than 1% of the global energy consumption and could be adapted to directly operate with solar energy sources. In this study, we demonstrate a solar chloro-alkali reactor that operates at unprecedented solar-to-chemical efficiencies as high as 25.1%. The reactor implements high efficiency multi-junction InGaP/GaAs/InGaAsNSb solar cells within a novel planar solar concentrator which maintains a $\pm 40^o$ acceptance angle with minimal tracking requirements. The concentrated PV system provides the required potential ($V_{OC} = 3.16$ V) to drive a customized electrochemical cell fabricated via additive manufacturing. The electrochemical cell is composed of a Nickel cathode and a dimensionally stable anode based on a Titanium substrate coated with an Iridium and Ruthenium oxide blend. The two electrodes are separated by a Nafion® 117 membrane which allows for the collection of nearly pure H_2 and Cl_2 gas streams as well as a NaOH enriched aqueous stream in the cathode side. Under operation conditions the faradaic efficiency towards chlorine production of the anode was measured to be between 96-99%. The demonstration presented here provides insights and foundational design principles for the implementation of solar chemical processes. Related developments for the production of solar-Hydrogen and solar-textiles will be discussed.

500094 Gold Nanoparticle Catalysis: Colloidal Versus Supported Heterogeneous Catalysis and Methods for Colloidal Nanoparticle Recovery

Abstract:

Gold nanoparticles (AuNPs) have attracted enormous attention due to their unique catalytic activities. Potential advantages of colloidal AuNPs over supported AuNPs is the greater catalytic activity and selectivity. Colloidal catalysts also catalyze reactions under mild conditions and are very effective for chiral catalysis. Recovery of colloidal catalysts involve tedious methods like pH or temperature induced aggregation, solvent extraction, etc. The advantage of supported catalyst lies in its reusability and recovery by simple methods like centrifugation, filtration, sedimentation, etc. Supported catalysts are preferred over colloidal catalyst in industries because they can be used in a continuous reactor system. For colloidal systems, surface functionalization with ligands is a prerequisite to prevent aggregation. Surface passivation causes significant reduction in catalytic activity and selectivity. For supported catalysts on metal oxide supports, a certain portion of the active surface area of the catalyst is lost due to the immobilization process and simultaneously new edges at metal-metal oxide interfaces are created. In this work, we have attempted to characterize the catalytic activity and active sites of both colloidal and supported catalysts systems by employing a ligand adsorption based method in aqueous media. An in-depth study of the effect of nanoparticle stabilizing ligands such as thiolated polyethylene glycol (PEG) and 11-mercaptoundecanoic acid (MUA) on the nanoparticle catalytic activity has been carried out and we have quantified the loss of active surface sites due to the ligand stabilization. Decreasing PEG chain length and increasing surface coverage of PEG on AuNP reduces the catalytic activity. Moreover, the functionalization of AuNP with MUA completely passivates the surface and the AuNPs show no catalytic activity. Although, high MUA coverage on the AuNP are detrimental to the catalytic activity, it renders the AuNPs easily recoverable from the solution by pH induced aggregation and redispersion. To maintain both catalytic properties and reversible pH induced recovery, we have considered partial surface coverages of the AuNP with MUA. This work also reports the synthesis of thiolated poly (acrylic acid) (PAA) functionalized AuNPs and explored its application as a recoverable catalyst where the PAA provides pH responsive dispersibility in aqueous media. Thus, PAA-AuNPs are easily and completely recovered from the reaction mixture and reused in subsequent reactions. It was found that the AuNP-PAA catalyst was highly active and reusable with modest reductions in the reaction rate with up to four catalyst recycles. We have also supported â€˜bareâ€™ colloidal AuNPs (trisodium citrate stabilized) on metal oxides and quantified the loss of active surface area. We report the comparison of loss of surface area due to ligand stabilization and metal oxide supports and compare the effectiveness and recoverability of both system of catalysts. A direct consequence of being able to characterize active surface area of both colloidal and supported system simultaneously is the ability to measure the effect of the support on the catalytic activity of the AuNPs. We have quantified the metal oxide support effect on the catalytic activity of AuNPs for various metal oxide-AuNP catalysts.

500274 Isosorbide Production from Aqueous Sorbitol Solution over Silica-Aluminas

Abstract:

Isosorbide is an important chemical for the production of fuels, plastics, and pharmaceuticals. Current isosorbide production is from plant biomass. Most current studies have been conducted in non-aqueous systems or pure sorbitol. However, processing plant biomass usually involves water. Thus, isosorbide production in aqueous systems is more practical. In the present study, we investigated the conversion of aqueous sorbitol to isosorbide using amorphous silica-alumina (ASA) with various SiO2/Al2O3 ratios. The isosorbide yield of ~ 56% was achieved at 220ËšC and 22 h. Surface area/pore volume, densities of acid sites, and total acidity of ASA from Brunauer-Emmett-Teller (BET), Fourier transform infrared spectroscopy (FT-IR)-pyridine adsorption and NH3-Temperature Programmed Desorption (TPD), respectively, were found to be responsible for this high isosorbide selectivity in the aqueous system.

498148 Simple One-Step Reaction to Achieve Cross-Linked Polymer and Polymer Composites: Models for Rubber Tires That Can Recycled Multiple Times with Robust Recovery of Properties Related to Cross-Link Density

Abstract:

Conventional polymer networks cannot be recycled into high-value products because of permanent, covalent crosslinks. Rubber tires illustrate well the issues ranging from major economic loss to environmental problems that arise with spent, crosslinked polymer. Here, we will discuss a simple, one-step reaction strategy using nitroxide-mediated polymerization (NMP) that we developed to synthesize recyclable crosslinked polymer with dynamic covalent bonds from any monomers or polymers that contain carbon-carbon double bonds that are amenable to free radical polymerization. (See Advanced Materials 2016, 28, 6746.) Using this reaction strategy, we have designed network polymers using polybutadiene and styrene monomer as models for recyclable tires, yielding full property recovery after multiple melt-reprocessing steps. No previous study has achieved direct formation of recyclable networks from commercial monomer/polymer without presynthesis or prefunctionalization and demonstrated full recovery of crosslinks after multiple reprocessing steps. We have likewise demonstrated the effectiveness of this strategy with commercial natural (uncrosslinked) rubber, leading to crosslinked natural rubber that can be effectively recycled and reprocessed multiple times with full recovery of properties associated with crosslink density. Thus, the reaction strategy can be employed with exactly the same polymers currently used in the tire industry and with reaction temperatures comparable to those employed in tire molding operations. As a result, this strategy provides for green reaction engineering not with the use on biobased feedstock but instead by effective, high-value recycling of synthetic materials that have been highly engineered for demanding uses and that are unlikely to be replaced by biobased feedstock. Importantly, we have also shown that this strategy can work well with polymer composites, e.g. networks incorporating carbon black filler exactly of the type used in rubber tires. Additionally, we have also used one-step NMP to produce polymer networks from monomers with very good reprocessabiltiy and relative uniformity of chains between crosslinks. Examples of other novel material development using simple, one-step NMP will also be discussed, time permitting.

484667 Selective Liquid Phase Hydrogenation of Chloronitrobenzene over Aupd Nanoclusters on TiO_2 Catalysts

Abstract:

Application of heterogeneous catalytic reactions in the production of fine chemicals is of increasing interest. Such reactions are often carried out in batch and semi-batch reactors. In the case of three-phase heterogeneous catalytic reactions, the rate of process and its selectivity can be determined either by intrinsic reaction kinetics or by external diffusion as well as by internal diffusion through the catalyst pores. Careful analysis of mass transfer is important for the elucidation of intrinsic catalytic properties, for the design of catalysts, and for the scale-up of processes. Nevertheless, in scale-up the operating conditions are selected in a way that the process is shifted from the kinetic region to a diffusion-controlled one. That is why it is important to know how the shift will affect not only the conversion but also the selectivity.

Palladium catalyst has been well-known to have high activity for liquid phase hydrogenation reaction. However, it is too active to have high selective for intermediate compounds. How to increase selectivity of intermediate without losing overall activity is the key issue in catalyst design. Using bimetallic catalyst is an option to achieve this goal.

In this study, bimetallic AuPd was loaded on TiO_2 support. TiO_2 support was chosen because its isoelectric point is ~7, therefore it is easy to deposit Au and Pd on it by using deposition-precipitation technique with $HAuCl_4$ as the starting material. After preparation, the bimetallic catalyst was reduced by $NaBH_4$ at room temperature. Various Au/Pd atomic ratios were used. The catalysts were characterized by various techniques such as X-ray diffraction (XRD), transmission electron microscopy (TEM), high resolution transmission electron microscopy (HR-TEM), X-ray photoelectron spectroscopy (XPS) and temperature programmed reduction (TPR). Selective hydrogenation of chloronitrobenzene to chloroaniline was carried out in a Parr batch reactor. The conditions for hydrogenation reaction were 0.55 MPa as H_2 pressure, room temperature and 300 rpm. Methanol was used as the solvent, the concentration of *p*-CNB was 0.2 M (2.54 g *p*-CNB in 80 ml methanol) and the amount of metal-based catalyst was 0.5 g.

All of PdAu catalysts could reach 100% conversion of *p*-CNB on the same reaction conditions in short times. The reaction could take place at room temperature and low hydrogen pressure. Only very small amount of palladium could improve the reaction activity and selectivity. Adding palladium could reduce gold-valence state, increase reaction active sites. Therefore, gold and palladium catalysts could promote conversion and selectivity for hydrogenation reaction. While Pd and Au form nanoally, Au became surface-enriched. The presence of small amount of Au could suppress dechlorination significantly. Au and Pd form semispherical shape and have very strong interaction with TiO_2 support. H_2 is dissociatively adsorbed on Pd-TiO_2 interface. Since the reaction can be operated at room temperature, no extra heating energy is needed. The operation cost is decreased and it is very easy to operate. H_2 partial pressure is also low (4 atm). For hydrogenation of p-CNB, TiO_2 is a good support. TiO_2 could donate electrons to Au, made Au

become electron-rich and more active. AuPd/TiO_2 showed high activity and selectivity. Adding small amount of Pd could increase activity of Au catalyst significantly. AuPd/TiO_2 catalyst had a very high activity and selectivity at room temperature. AuPd form nanoalloy particles. Particle size < 4 nm. Pd was enriched on the surface of nanoalloy. The optimum ratio Pd/Au= 1/6.

491326 Process Intensification for the Aqueous Phase Dehydration of Biomass-Derived Carbohydrates

Abstract:

Biomass is an abundant renewable feedstock for energy conversion and an alternative source for the production of fuels and chemicals. Furanic compounds, such as 5-hydroxymethyl furfural (HMF), have been identified as key platform chemicals produced from the acid-catalyzed dehydration of biomass-derived carbohydrates. Micro-technology offers an economically sustainable route for the production of HMF through process intensification. In particular, the small length scale and high surface-to-volume ratio of continuous flow microreactors result into heat and mass transfer rates that are orders of magnitude greater than conventional macroscale systems. In addition, microreactors allow operation at reaction times from minutes to fractions of seconds under precisely controlled conditions for improved yields and selectivities.

In this work, we characterize the microreactor to assess its transport and mixing characteristics. We performed experiments for various carbohydrates and using various catalysts. As an example, we will present a kinetic study of the aqueous phase Brønsted acid-catalyzed dehydration of fructose to HMF in an in-house built microreactor using an HCl and KCl buffer solution as the catalyst. We also used a hybrid first-principles and data-driven kinetic model of the reaction network to determine the optimal reaction and operation conditions that maximize HMF yield. Conditions for optimal HMF yield were identified.

493462 Organocatalyzed Beckmann Rearrangement of Cyclohexanone Oxime in a Microreactor: Kinetic Model and Product Inhibition

Abstract:

Îµ-Caprolactam (CPL) is the monomer for nylon-6 with wide applications in textiles, automobiles, electronics, and mechanicals. Here, we present the kinetics study of Beckmann rearrangement of cyclohexanone oxime (COX) catalyzed by trifluoroacetic acid (TFA) and acetonitrile in microreactors. The effect of temperature, ratio of TFA to acetonitrile, and concentration of COX were investigated. The inhibition effect of CPL in this reaction system was firstly reported. Based on the results, a comprehensive mathematic kinetic model considering the product inhibition effect of CPL has been developed in the temperature range of 368 K to 391 K, which is in good agreement with experimental results across a broad experimental parameter space. In addition, kinetic study indicates that the esterification reaction of COX and the transposition reaction of the intermediate are both supposed to be the rate-determining steps, and in the catalyst system, the ratio of TFA and acetonitrile mainly influences the reaction rate and the activation energy of the transposition step. The developed model could provide much reliable knowledge for industrial application.

496082 Modeling the Influence of Dean Vortices in Improving the Performance of Catalytic Membrane Microreactors

Abstract:

Porous membrane microchannels are used for controlling heterogeneous reactions over catalyst embedded in the annular region of membrane microchannel. In this work, a mathematical model is developed for a heterogeneously catalysed reaction in a porous membrane micro-reactor. Hydrogenation of an aqueous nitrite contaminant in water is analysed as a model reaction. For a straight micro-reactor a numerical algorithm is developed to predict the behaviour for non-linear kinetics. This is benchmarked with an analytical solution and later compared with the experimental observations reported in the literature. Results depict that the straight reactor is inefficient and catalyst utilization is poor when higher throughputs are required because the system is diffusion limited.

A modified design is proposed which incorporates curving of the microchannel to induce passive mixing by generating â€œDean Vorticesâ€. The velocity fields are obtained using a perturbation analysis. This is incorporated in solving the convection-diffusion-reaction system which governs the species transport. The non-linear system of equations in core and the membrane is solved using an operator splitting technique. An effective regularization treatment is devised to numerically deal with the singularity at the center for non-axisymmetric systems. Reactor intensification is characterized using conversion, effective length of the reactor and utilization of catalyst inside the membrane. The non-monotonic enhancement in Sherwood number is interpreted by analysing the dynamics of the concentration depletion layer.

498213 Enhancing Reaction Rates By Automated *in silico* Solvent Screening

Abstract:

Solvents have a large impact on the overall performance of chemical processes. In particular, solvents can significantly enhance chemical reactions by increasing reaction rates and selectivities [1]. Thus, solvents need to be carefully selected. Selecting suitable solvents is a challenging task since experimental data for reaction rates and selectivities are often scarce in early process design. Thus, solvents are usually selected from a limited set of known solvents based on heuristic guidelines and expert knowledge. However, this conventional approach does not exploit the potential of the large molecular design space of all possible solvents and likely results in suboptimal choices [2].

To overcome this limitation, we propose an automated solvent screening approach to select solvents *in silico*for enhancing reaction rates. The key for this approach is the prediction of liquid phase reaction rates. For this purpose, we calculate gas phase reaction rates and solvation effects using quantum chemistry [3,4]. To reduce the computational burden of quantum chemistry to a level enabling efficient large-scale solvent screening, computationally expensive quantum mechanical calculations of gas phase reaction rates are performed only once per reaction. In contrast, solvation effects in the different solvents are computed with COSMO-RS [5] at low computational cost. Thus, reaction rates in different solvents are predicted efficiently and the resulting solvent screening allows for the evaluation of thousands of solvents within few days. Thereby, we can extend our recent solvent screening framework from equilibrium properties [6] to kinetic properties.

We apply our approach to a Carbamate cleavage reaction and identify solvents which enhance reaction rates compared to the benchmark solvent. The resulting framework is generally applicable to liquid phase reactions and allows the efficient screening of solvents for optimal reaction rates.

Literature

[1] Struebing, H., Ganase, Z., Karamertzanis, P. G., Siougkrou, E., Haycock, P., Piccione, P. M., Armstrong, A., Galindo, A. and Adjiman, C. S.: Computer-aided molecular design of solvents for accelerated reaction kinetics. Nature Chemistry 5 (2013), 952–957.

[2] Adjiman, C. S., Galindo, A. and Jackson, G.: Molecules Matter: the Expanding Envelope of Process Design. Computer-Aided Chemical Engineering 34 (2014), 55-64.

[3] Deglmann, P., Müller, I., Becker, F., Schäfer, A., Hungenberg, K.-D. and Weiß, H.: Prediction of Propagation Rate Coefficients in Free Radical Solution Polymerization Based on Accurate Quantum Chemical Methods: Vinylic and Related Monomers, Including Acrylates and Acrylic Acid. Macromolecular Reaction Engineering 3 (2009), 496-515.

[4] Kröger, L. C., Kopp, W. A. and Leonhard, K.: Prediction of Chain Propagation Rate Constants of Polymerization Reactions in Aqueous NIPAM/BIS and VCL/BIS Systems. The Journal of Physical Chemistry B 121 (2017), 2887-2895.

[5] Klamt, A., Eckert, F., Arlt, W.: COSMO-RS: an alternative to simulation for calculating thermodynamic properties of liquid mixtures. Annual Reviews of Chemical and Biomolecular Engineering 1 (2010), 101–122.

[6] Scheffczyk, J., Redepenning, C., Jens, C., Winter, B., Leonhard, K., Marquardt, W. and Bardow, A.: Massive, Automated Solvent Screening for Minimum Energy Demand in Hybrid Extraction-Distillation Using COSMO-RS. Chemical Engineering Research & Design 115 (2016), 433-442.

501831 Enhancement of the Activity and Selectivity of the Fischer-Tropsch Synthesis with Water/Oil Emulsions

Abstract:

Liquid phase reactions at the water/oil interface of Pickering emulsions have shown great advantages in heterogeneous catalysis. For example, solubility differences can drive the migration of desirable products to a different phase, thus facilitating the separation/removal of products and avoiding undesirable reactions. In recent work, we have used metal-doped nanohybrids (carbon nanotubes â€“ metal oxide) to stabilize emulsions during Fischer-Tropsch synthesis catalysis. We observed that the emulsion system extends the life of the catalyst and improves the C_1/C_{5+} product balance. Also, Aliphatic hydrocarbons migrate preferentially to the oil phase while short alcohols remain in the water phase, which facilitates the separation of products. In this contribution, we systematically studied the effects of support hydrophobicity and emulsion properties on the activity and selectivity enhancements observed during Fischer-Tropsch synthesis. Carbon nanotubes with varying hydrophilicity are decorated with Ru metal and tested for liquid-phase Fischer-Tropsch synthesis using a batch reactor (H_2:CO = 4:1, 220 °C, 55 Bar). Different ratios of water/oil are also used and the emulsion properties characterized. Interestingly, emulsions stabilized with a more hydrophobic support (pristine carbon nanotubes) show CO consumption rates nearly twice those observed in single water phase. The results of catalytic activity and selectivity are discussed in terms of the relationship between emulsion properties and the interaction of the catalyst particles with the water and oil phases at the interface of the emulsion.

502119 Comparison between Thermal- and Electro-Catalytic Upgrading of Furanic Compounds in the Liquid Phase: A Combined Experimental and DFT Study

Abstract:

The ability to control and manipulate selectivities in catalytic upgrading of biomass is desirable. In this work, we show that selectivity of liquid-phase hydrogenation of furfural, which is an important platform chemical produced from thermal treatment of biomass, varies significantly as compared to gas-phase reactions using palladium and copper catalysts. Presence of the liquid phase affects the adsorption of reactants, intermediates and products as well as reaction pathways leading to the varied selectivities. In addition, using an electrochemical cell, we show that conversion of furfural to methylfuran over a copper catalyst is the preferred route, while it has been shown that furfuryl alcohol is the dominant product in the thermal-catalytic reactions. This different selectivites indicate that the reaction mechanism is varied under these two reaction conditions. This work thus suggests that one can control the selectivity of furfural upgrading by using both liquid phase catalysis and electrocatalysis, which provides additional degrees of freedom for tailoring the chemical reactions.

Acknowledgements

The calculations were performed at NERSC and OU Supercomputing Center for Education and Research (OSCER). We appreciate support from the U.S. DOE through a DOE/EPSCOR grant (DESC0004600).

499902 First-Principles Multiscale Modeling of CO Oxidation on Polycrystalline RuO_2 in a Fixed Bed Reactor

Abstract:

CO oxidation is a widely employed prototype reaction for studying reactivity differences between different metal and metal oxide catalysts, and improved methods for simulating catalytic processes. Ru-based catalysts have been identified as being particularly active for CO oxidation at atmospheric pressures, despite their low activity under ultra-high vacuum conditions compared to similar catalysts.[1] There is still disagreement about the active phase under working conditions, despite significant experimental effort[2-4] Theoretical methods, in particular density functional theory (DFT) and kinetic Monte Carlo (KMC), have also been employed in an effort to elucidate the active phase under working conditions.[5, 6] Such efforts are hampered by the fact that neither DFT nor KMC incorporate explicit reactor-level effects that may induce emergent behavior, and reactor-level models incorporating KMC models are rare due to the challenges of coupling KMC simulations with reactor-level models.[7]

In this work we demonstrate a method for coupling multiple DFT-based KMC models to a computation fluid dynamics (CFD) model of a fixed bed reactor. We then use this multiscale model to investigate the effect of operating conditions and catalyst composition on the rate of CO oxidation, and we compare our simulation results to steady state experimental data. We find significant differences between the model and experimental apparent activation energies (~100 kJ/mol). Reaction orders are in better agreement. Our model also shows that the apical (111) facet of RuO_2 is several orders of magnitude more active than the (110) lateral facet of RuO_2 at the investigated conditions, in line with experimental evidence.[4] Overall, the nature of the active phase for Ru/RuO_2 catalysts under working conditions is still an open question, and more work must be done to develop improved kinetic models. The method used in this work provides one thrust to be followed for better understanding of Co oxidation over Ru/RuO2, as well as for other catalytic reactions where coupling of DFT calculated transition states to reactor-scale simulations is desirable.

1. Peden, C.H.F. and D.W. Goodman, *Kinetics of carbon monoxide oxidation over ruthenium(0001).* The Journal of Physical Chemistry, 1986. **90**(7): p. 1360-1365.

2. Gao, F., et al., *CO oxidation over Ru(0001) at near-atmospheric pressures: From chemisorbed oxygen to RuO2.* Surface Science, 2009. **603**(8): p. 1126-1134.

3. Assmann, J., et al., *Heterogeneous oxidation catalysis on ruthenium: bridging the pressure and materials gaps and beyond.* Journal of Physics-Condensed Matter, 2008. **20**(18).

4. Rosenthal, D., et al., *On the CO-Oxidation over Oxygenated Ruthenium.* Zeitschrift Fur Physikalische Chemie-International Journal of Research in Physical Chemistry & Chemical Physics, 2009. **223**(1-2): p. 183-207.

5. Reuter, K. and M. Scheffler, *First-principles kinetic Monte Carlo simulations for heterogeneous catalysis: Application to the CO oxidation at RuO2(110).* Physical Review B, 2006. **73**(4).

6. Hess, F., et al., *"First-Principles" kinetic monte carlo simulations revisited: CO oxidation over RuO2(110).* Journal of Computational Chemistry, 2012. **33**(7): p. 757-766.

7. Matera, S., et al., *Predictive-Quality Surface Reaction Chemistry in Real Reactor Models: Integrating First-Principles Kinetic Monte Carlo Simulations into Computational Fluid Dynamics.* Acs Catalysis, 2014. **4**(11): p. 4081-4092.

500910 Feature Engineering of Machine-Learning Models for Metal Oxides

Abstract:

Feature Engineering of Machine-Learning Models for Metal Oxides

Zheng Li, and Hongliang Xin*

Department of Chemical Engineering, Virginia Tech, Blacksburg, VA, 24061

*hxin@vt.edu

Session: Catalysis and Reaction Engineering Division – Poster Sessions

Perovskites with the ABO_3 type structure have attracted enormous research efforts in recent years due to their high activity for catalyzing oxygen evolution in (photo)-electrochemical systems. The thermodynamically and kinetically unfavorable oxygen evolution reaction (OER) is considered as a bottleneck for many energy conversion processes such as metal-air batteries and photocatalytic water splitting cells [1]. We aim to design cost-effective metal oxides for catalyzing the sluggish OER. It is, however, very time-consuming and costly to search for highly optimized metal oxides by high-throughput experiments and/or quantum-chemical calculations due to the vast materials space of composition and geometry. In this poster, we extend our previous study [2-3] on metal alloy catalysts to metal oxides by developing a machine-learning-augmented chemisorption model that captures adsorption properties of active sites on metal oxides.

As the chosen of appropriate data representation method is the most critical ingredient for the success of machine learning models, we applied and analyzed various types of data representation methods to describe the metal oxide catalyst surfaces. For instance, we explore electronic based features such as the B-site metal e_g band filling, d-band upper edge, oxidation state, d-orbital splitting, ligand strength, and easily accessible structural based features such as the coordination number, together with the intrinsic properties of metal atoms. By using both electronic based features and structural features, the machine-learning models (e.g., artificial neural network, kernel ridge regression) optimized with available *ab initio* adsorption energies on perovskite (001) surfaces can capture adsorption energies of *O, *OH, and *OOH intermediates with the root mean squared errors (RMSE) smaller than the DFT-GGA calculation error of 0.2 eV. To shed light on the underlying factors that govern the adsorbate/catalyst interaction, we also employ a feature importance analysis (e.g., exploratory analysis, principle component analysis) to the feature datasets. Compared with the traditional high-throughput computational and experimental trial-and-error approach, the machine-learning chemisorption models exhibit great potential in accelerating the discovery of high-performance oxide materials for electrocatalysis.

[1] J. Suntivich, K.J. May, H.A. Gasteiger, J.B. Goodenough, and Y. Shao-Horn, A perovskite oxide

optimized for oxygen evolution catalysis from molecular orbital principles, Science. 334 (2011) 1383–1385.

[2] X. Ma, Z. Li, L.E.K. Achenie, and H. Xin, Machine-Learning-Augmented Chemisorption Model for CO2 Electroreduction Catalyst Screening, J. Phys. Chem. Lett. 6 (2015) 3528–3533.

[3] Z. Li, X. Ma, and H. Xin, Feature engineering of machine-learning chemisorption models for catalyst design, Catalysis Today. 280, Part 2 (2017) 232–238.

491017 Combined Quantum Mechanical and Molecular Mechanical Method for Catalyst Design on the Nu-1000 Metal-Organic Framework

Abstract:

Metal-organic frameworks (MOFs) have high internal surface area and tailorable nodes and organic ligands, and for these reasons they are very promising for heterogeneous catalysis, hydrogen storage, and gas separation. Fully quantum chemical investigations of these systems are usually quite demanding in terms of time and computational resources; therefore we have developed a combined quantum mechanical and molecular mechanical (QM/MM) method to investigate these systems.

Our method is illustrated by calculations on MOF NU-1000 [Zr$_6$(Î¼$_3$-O)$_8$(O)$_8$(H)$_{16}$(TBAPy)$_2$], which is a superb candidate for use as a catalytic support for various reactions (e.g., alkane metathesis and ethylene dimerization). We study the relative energetics for various proton topologies of the NU-1000 node. Because of the high partial charges on the atoms of the highly polar bonds in this structure, this provides a challenging test of the adequacy of the QM/MM boundary treatment. In our QM/MM calculations, we use a cluster model of the node [Zr$_6$(Î¼$_3$-O)$_8$(O)$_8$(H)$_{16}$(PhCOO)$_8$] with eight phenyl rings (88 atoms) treated classically while the other atoms (62 atoms) are treated quantum mechanically. Two kinds of link atoms (in particular H atoms and tuned F atoms) are used to cap the dangling bonds in the QM subsystem, and six kinds of charge modification schemes are considered for the treatment of boundary charge. By comparing with full QM calculations, we found that the QM/MM method can give remarkably accurate results with a mean unsigned error of only 2 kJ mol^{-1} in the relative energies of seven proton topologies.

To facilitate the design of MOF-based catalysts, it can be desirable to further reduce the number of atoms in the QM subsystem. To do this for NU-1000 we have to cut through two types of polar bonds (Zr-O and O-Zr) in the same calculation. We are developing a new method based on a tuned F link atom to treat this situation.

Multi-Scale CO_2 Conversion Modelling on Cu-Based Catalysts

Matej Huš, Drejc Kopač, Neja Strah Štefančič, Damjan Lašič Jurković, Venkata Dasireddy and Blaž Likozar

National Institute of Chemistry
Ljubljana, Slovenia

2017 AIChE Annual Meeting

October 29, 2017
Minneapolis, Minnesota

Outline

1. **Background & Motivation**
2. **Experimental**
3. **Multi-Scale Approach**
   3.1 Atomic level (density functional theory)
   3.2 Mesoscale (kinetic Monte Carlo)
   3.3 Continuum (micro-kinetic modelling)
4. **Results**
5. **Conclusion**
6. **Advantages and Limitations**
7. **Our other endeavours**
8. **Team & Acknowledgements**

NATIONAL INSTITUTE OF CHEMISTRY

Background & Motivation

Rising concentration of CO_2 in the atmosphere

Periods of surplus energy

Solution: catalytic hydrogenation of CO_2 to methanol

Challenge: design of an efficient **catalyst**

MefCO2 – methanol from CO_2

NATIONAL INSTITUTE OF CHEMISTRY

Experimental

Synthesis:

Cu/M/Al catalysts (Cu : M : Al = 50:30:20, M = Zn, Mg, Cr, Fe)
Co-precipitation method
Calcination in air at 300 °C for 4 h
Preconditioning in hydrogen atmosphere

Catalytic testing:

GHSV = 6000 h^{-1}
$CO_2 : H_2 = 1 : 3$
Pressure = 20, 30, 40, 50 bar
Temperature = 200, 220, 240, 260, 280, 300 °C

NATIONAL INSTITUTE OF CHEMISTRY

Multi-Scale Approach to Modelling

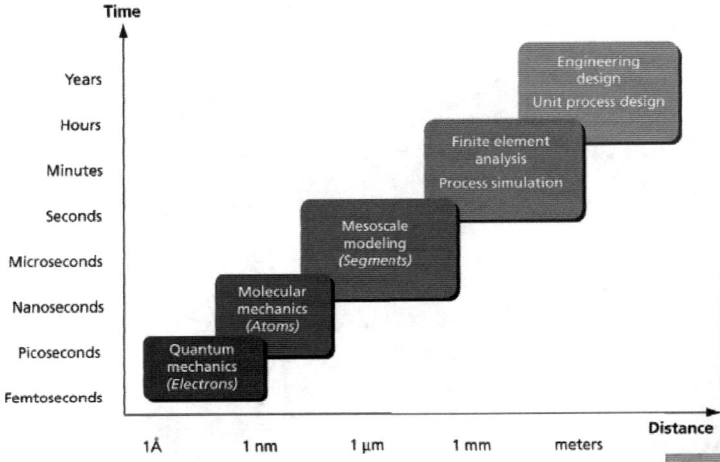

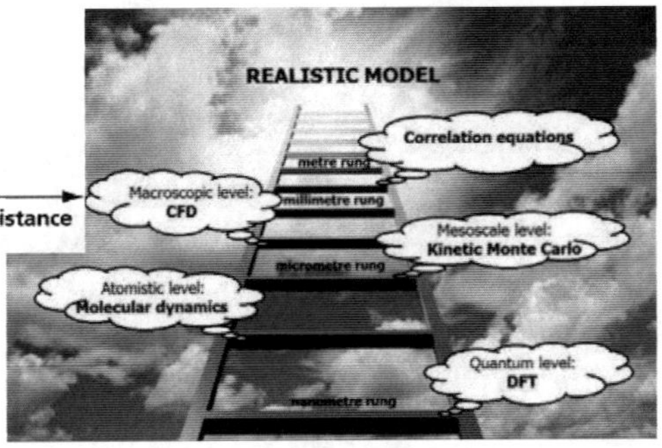

NATIONAL INSTITUTE OF CHEMISTRY

Atomic level (density functional theory)

DFT: best compromise between accuracy and computational speed for system with ~100 atoms

Plane-wave approach: periodic boundary conditions to model crystals and slabs

Ultrasoft pseudopotentials: only valence electrons matter

Transition state theory: identify TS along with E_A to obtain ΔG and k

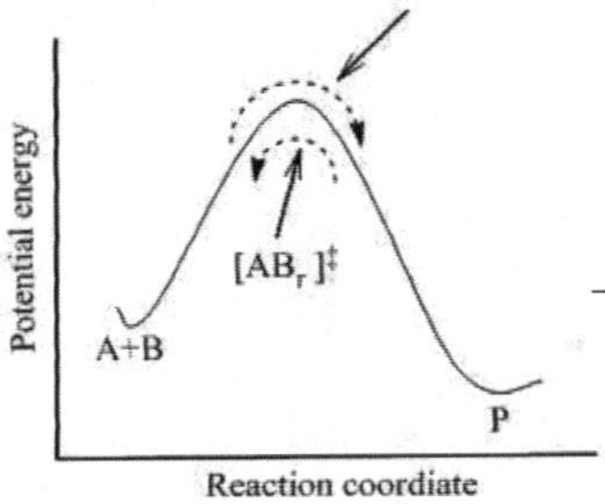

In reality multi-dimensional surface

Locating TS is difficult!

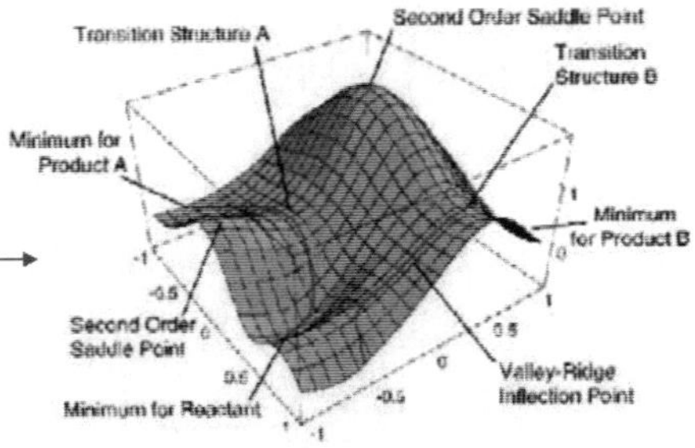

NATIONAL INSTITUTE OF CHEMISTRY

DFT model

We must account for metallic copper and metal oxide interface

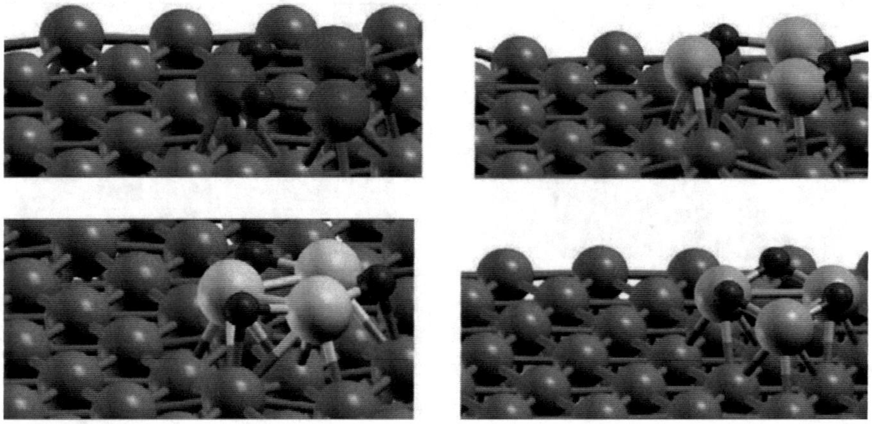

Cu(111) plane with M_3O_3 cluster (M=Zn, Cr, Fe, Mg)

Details: PBE (300 eV cutoff), MK K-mesh 4x4x1, 4 layers of Cu(111), bottom two fixed, unit cell 3x3, ultrasoft PPs, semiempirical Grimme (D2) corrections for vdW interactions, CI-NEB to search for TS, vibrational analysis of adsorbates

NATIONAL INSTITUTE OF CHEMISTRY

Mesoscale (kinetic Monte Carlo)

Using graph-theoretical KMC framework as integrated in **Zacros**
Periodic cell lattice with size of 48 x 48 cells
Unit vectors: 4.78 Å, α = 60°, corresponding to Cu(111)
Two distinct surface types (Cu and metal oxide)
Realistic denticity (number of O atoms = number of covered sites)
Full pathway
Molar ratio $H_2 : CO_2 = 3 : 1$

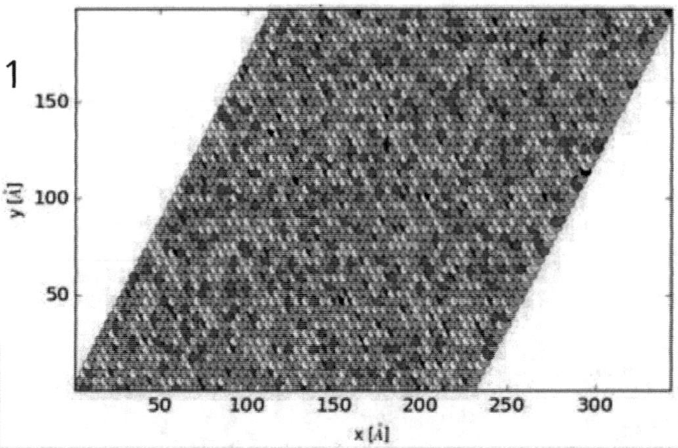

NATIONAL INSTITUTE OF CHEMISTRY

Continuum (micro-kinetic modelling)

Microkinetic modelling of a CSTR

Surface reaction rates: $r_i = k_i \cdot \theta_j \cdot \theta_m \dots - k_{-i} \cdot \theta_n \cdot \theta_o \dots$

Mass balances:

$$\frac{d\theta_j}{dt} = \sum_{i=1}^{n} S_{ij}\, r_i \qquad \frac{dC_j}{dt} = \frac{M_{sites}}{\varepsilon} \sum_{i=1}^{n} S_{ij}\, r_i + \frac{F}{\varepsilon\, V_R}\left(C_{j,inlet} - C_j\right)$$

Surface species   Gaseous species

$$CO_2\,\text{conversion [\%]} = \frac{n(CO_{2,in}) - n(CO_{2,out})}{n(CO_{2,in})} \cdot 100\ \%$$

$$S_{CH_3OH}\ [\%] = \frac{n(CH_3OH)}{n(CH_3OH) + n(CO) + n(CH_2O) + n(HCOOH)} \cdot 100\ \%$$

NATIONAL INSTITUTE OF CHEMISTRY

Results: DFT

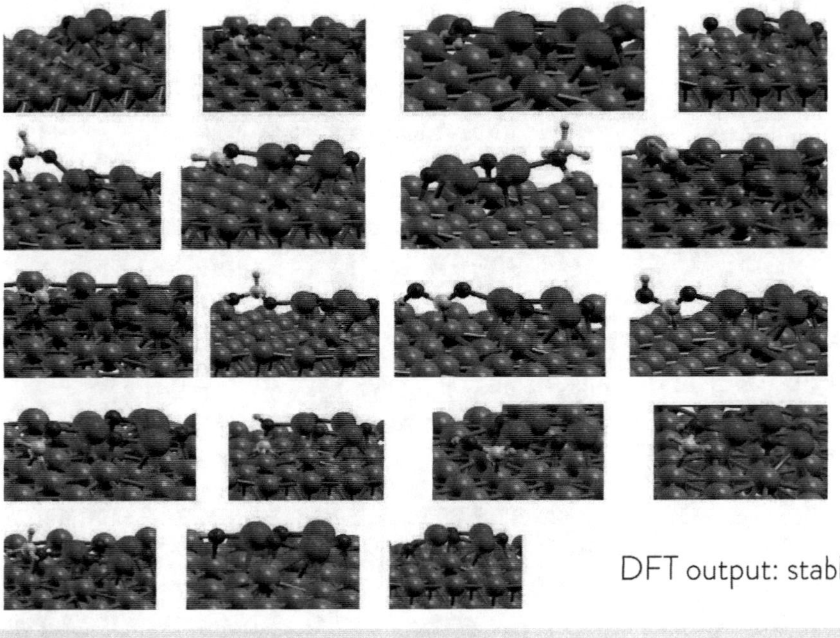

DFT output: stable geometries

NATIONAL INSTITUTE OF CHEMISTRY

Results: DFT

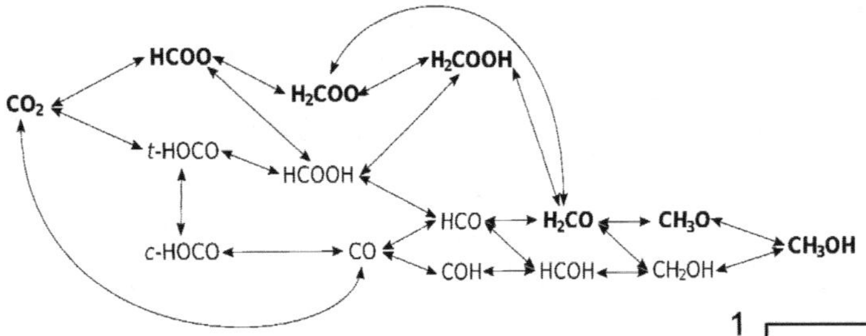

Reaction scheme with all possible elementary steps (H and OH omitted)

Most likely pathways (potential energy surface)

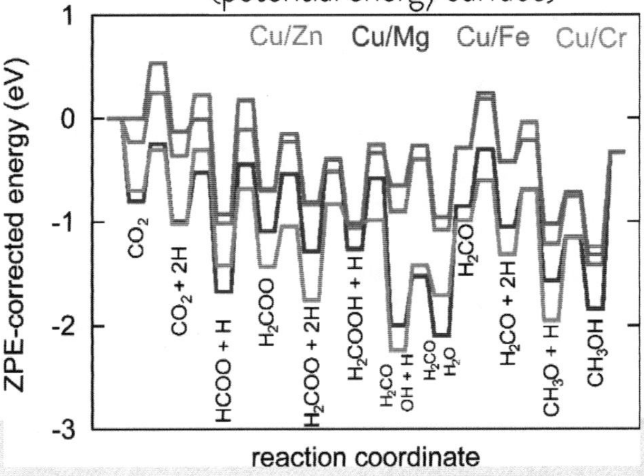

NATIONAL INSTITUTE OF CHEMISTRY

Results: KMC

Relative frequencies of elementary steps at 500 K and 40 bar.

Allow for determination of rate-limiting steps.

Show the most likely pathway = formate (HCOO) pathway

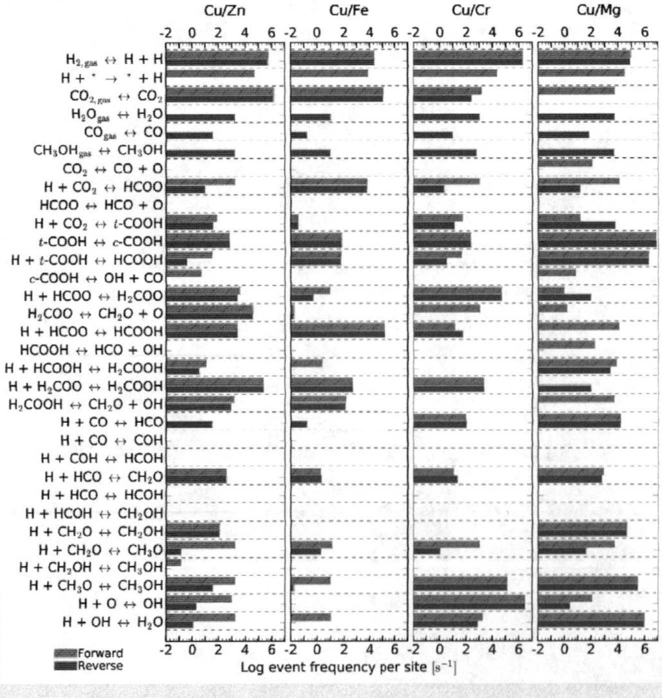

NATIONAL INSTITUTE OF CHEMISTRY

Results: KMC

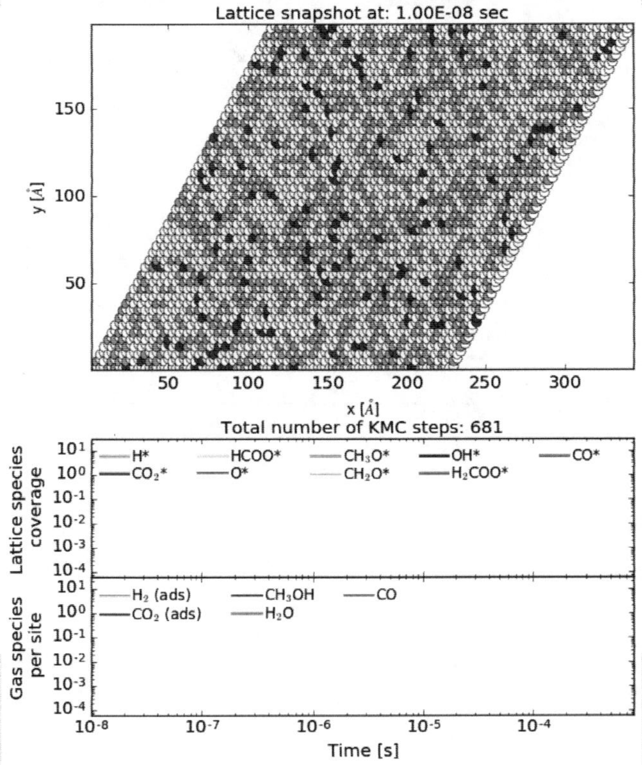

Lattice coverage snapshots at temporal evolution of lattice species and gaseous phase composition for Cu/Cr catalyst

Results: KMC

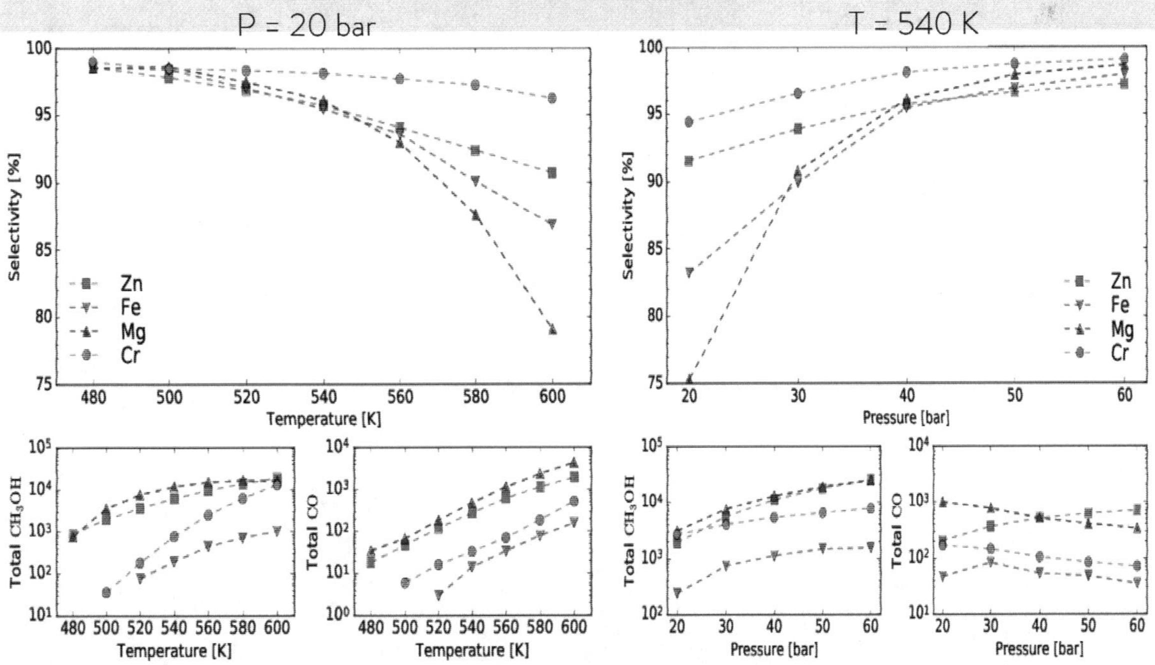

Results: micro-kinetic modelling

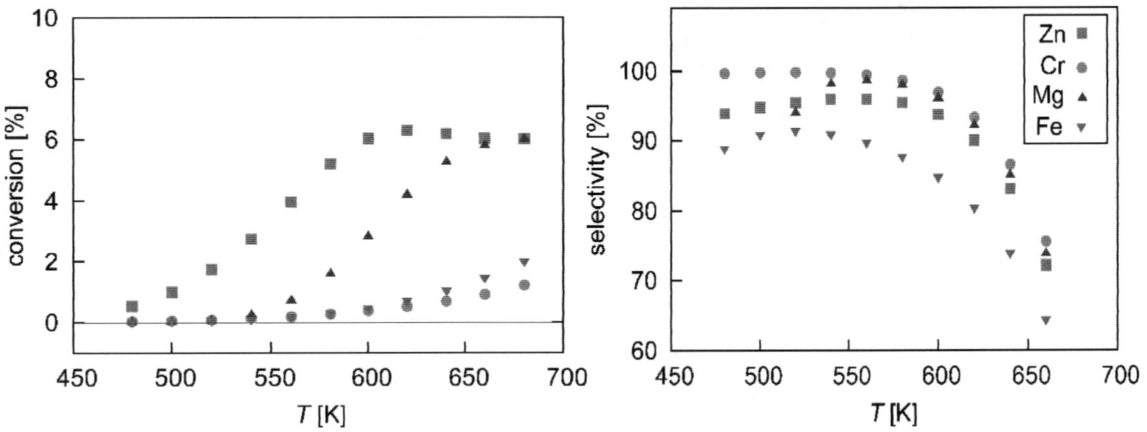

Temperature dependence of selectivity towards CH_3OH and conversion of CO_2 for all four investigated catalysts from micro-kinetic modelling at p = 40 bar and GHSV = 6000 h^{-1}.

NATIONAL INSTITUTE OF CHEMISTRY

Conclusion

- Superior performance of Cu/Zn catalyst, modest Cu/Mg
- Cu/Cr and Cu/Fe inappropriate
- Agreement between MK and KMC (both with DFT input) and experimental data
- Mechanistic insight into the reactions (reaction frequencies, surface coverages, operating conditions dependence)

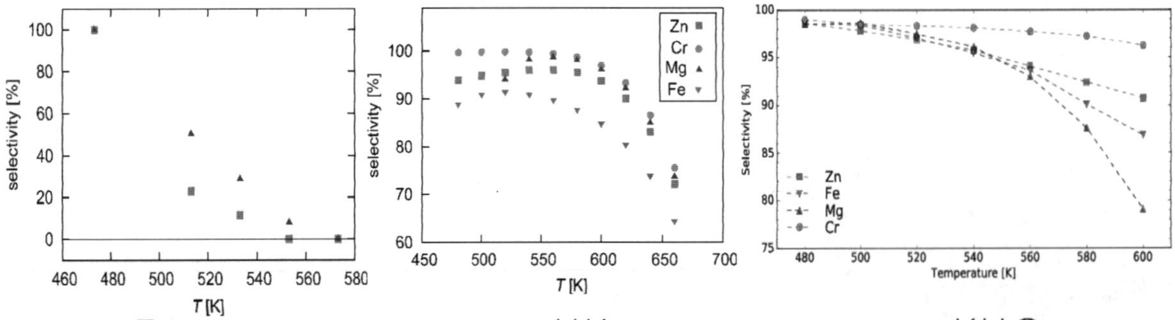

Experiment MK KMC

NATIONAL INSTITUTE OF CHEMISTRY

Advantages and limitations

- Long initial process of model set-up
+ After that, fast sweep over reaction conditions
+ Gives information that cannot be obtained experimentally
- Accuracy of current DFT techniques, approximations of KMC and micro-kinetics
+ Explains mechanistically why catalysts perform differently
- Rather sensitive to input parameters (high precision DFT calculations and KMC simulations required)
+ Allows further coupling with CFD to design reactors
+ Possible to perform a quick search of promising catalysts

+ Looking for a green solution in a green way – *in silico* instead of trial-and-error!

NATIONAL INSTITUTE OF CHEMISTRY

Our other endeavours

MefCO2 – methanol from CO_2

ADREM

Adaptable Reactors for Resource- and Energy-Efficient Methane Valorisation

next BioPharm DSP

Downstream processes unlocking biotechnological transformations

BioPharm.Si

Development of new generation of biological drugs

MarineBiotech

Bio-refinery and biotechnological exploitation of marine biomasses

FReSMe

From residual steel gases to methanol

CEL.KROG

Zavrženi potenciali biomase

Biomass valorization

NATIONAL INSTITUTE OF CHEMISTRY

Acknowledgements

The Team

Funding

European Union
Horizon 2020 – Grant N 637016

Republic of Slovenia
Research Agency – Core Funding P2-0152

NATIONAL INSTITUTE
OF CHEMISTRY

493759 Evaluation of Multi-Scale Models for Steam Methane Reforming in a Fixed Bed Reactor

Abstract:

Evaluation of Multi-scale Models for Steam Methane Reforming in a Fixed Bed Reactor

Anthony G. Dixon

Department of Chemical Engineering, Worcester Polytechnic Institute,

Worcester, MA, USA, 01609

Models of chemical reactions in fixed bed reactors, even at steady-state, must take into account phenomena occurring over a wide range of scales, from the catalytic site, to the pellet pore, the catalyst particle and finally to the reactor tube scale. In reactor tubes of low tube-to-particle diameter ratio (N) the latter can be broken down further into the tube radius scale and the tube length scale. Conventional pseudo-homogeneous effective (porous) medium fixed bed models take into consideration one or both tube scales, while the particle scale is averaged by use of the effectiveness factor. Heterogeneous effective medium models which solve explicit equations for diffusion and reaction in the catalyst pellet also include the pellet scale. Even in these multi-scale models, the transport parameters must lump several phenomena at various length scales. The question of whether the models are adequate to predict reactor performance is usually addressed by comparison to experimental data, an approach which is compromised by uncertainties in the reaction kinetics and in the correlations used for the effective transport parameters.

In this work a different method for the evaluation of two-dimensional heterogeneous fixed bed reactor models is presented, using 3D CFD resolved particle simulations to generate a detailed benchmark solution at both the tube and particle scales. The CFD simulation is run for known kinetics and fluid/particle properties, which then can be used in the effective medium models, eliminating the uncertainty about whether the chosen kinetics truly represent the experimental reacting system. The effective transport parameters can be obtained from CFD runs in the same bed of particles without reaction, under conditions which mimic the usual experimental set-up to determine these parameters, for example flow through a heated-wall tube for the effective radial thermal conductivity, or mass transfer from coated particles for the particle-fluid mass transfer coefficient. In this way, transport properties which represent the fixed bed being modeled are used, instead of introducing a large degree of uncertainty by choosing one of the many literature correlations.

The methodology described in the preceding paragraphs is illustrated in this talk for steam methane reforming in a packed bed of 807 spheres at N = 5.96, under conditions typical of the middle of a steam reformer tube, with specified wall heat flux. The 3D CFD simulations show a detailed picture of temperature, species and reaction rates at the level of a catalyst particle or lower (See Fig. 1 for contours of the water-gas shift reaction rate).

From the CFD simulations, it may be determined what level of detail is necessary for the 2D + 1D model to represent chosen features of the more detailed 3D model. The results show that the cross-sectional average profiles require detailed modeling of the reaction rate distribution (via the radial void fraction) as well as a radially-distributed axial velocity profile, but not a radially-varying effective radial thermal conductivity (see Fig. 2 for a comparison for fluid temperature).

These models which take account of local scale structure and flow, do not reproduce the radial profiles well. On the other hand, if a model for the local radial variation of the effective radial thermal conductivity is used, then more promising results are obtained (see Fig. 3).

501162 Parametric Studies of Steam Methane Reforming Using a Multiscale Reactor Model

Abstract:

In this work, a parametric study to verify the influence of a porous catalystâ€™s structural parameters on a packed bed reactor is performed through the application of a multiscale model. The set of constitutive equations for the catalytic pellet and the packed bed reactor are derived using the Reynolds Transport Theorem and applicable geometric parameters. Diffusive fluxes in the micro-scale (catalytic pellet) and the macro-scale (reactor) porous domains are calculated by using the rigorous Dusty Gas Model (DGM) and the Stefan-Maxwell equations respectively, while Chapman-Enskog theory is applied to estimate diffusion coefficients and viscosity. Simulations using the finite element method are carried out for a case study considering hydrogen production using steam methane reforming. The steam methane reforming processâ€™ effectiveness factors are quantified considering different structural characteristics of the porous catalyst, such as volumetric porosities, tortuosity factors, solid phase conductivity, and mean pore diameter.

490780 Hydrothermal Upgrading of Algal Bio-Oil By Supercritical Water

Abstract:

Global warming is a significant problem the world is currently facing as result of greenhouse gasses being emitted from the continual usage of fossil fuels. One particular solution that has been examined to reduce the rate of global warming is using bio-crude as a fuel replacement. Bio-crude is a more desired renewable fuel because liquid fuels are easier to transport and have higher energy densities than solid and gaseous fuels. Bio-crude is produced by the thermochemical or biochemical conversion of biomass through methods such as liquefaction, pyrolysis, solvent extraction, etc. Just like crude oil, bio-crude needs to be refined before it can be used; this is to remove oxygen and other heteroatoms that lower its fuel quality. The most common method of upgrading crude oil is hydro-deoxygenation which involves treating crude with hydrogen at high temperatures. Hydrogen is predominantly produced by natural gas reforming, thus it would be counter intuitive to use a fossil fuel in order to improve a renewable energy technology.

Supercritical water (SCW) treatment is examined as an alternative method to upgrade crude in a more sustainable manner. Water at supercritical conditions becomes a single phase and exhibits compressibility and density similar to that of gasses and liquids, respectively. Furthermore, its dielectric constant significantly decreases with increasing temperature, causing it to behave like a nonpolar solvent. SCW is also considered an environmentally-friendly solvent as it turns to regular water at ambient temperatures. These ideal solvent properties are proposed to be useful in crude upgrading as it is expected that SCW breaks down longer chain hydrocarbons while promoting oxidation reactions to remove heteroatoms from the crude.

This goal of this study was to upgrade bio-crudes produced from hydrothermal liquefaction (HTL) of microalgae (*N. salina*) with SCW in order to remove impurities, decrease oxygen content, reduce viscosity, and produce desired low carbon (C_6-C_{14}) alkane products. The bio-crude was treated with and without Pd/C catalyst at temperatures of 380, 400, and 420 °C at a fixed SCW density of 278 kg m^{-3}, thus allowing pressure to be monitored than controlled. The upgraded bio-crude quality was assessed by evaluating energy content, ultimate analysis, GCMS, FTIR, NMR, viscosity, and density. The energy content of the treated bio-oil increased with reaction temperature and catalyst loading as a result of the decrease in oxygen and sulfur content. Viscosity and density decreased due to the decomposition of large molecules, creating shorter straight chain hydrocarbons.

500381 Model Aldehyde Chemistry Using Cezrox Catalyst for Hydrothermal Liquefaction Upgrading

Abstract:

Green processes that upgrade biomass feedstocks and recycled bio-waste can both help meet our growing energy demands as well as reduce our overall carbon footprint. One promising technology is hydrothermal liquefaction (HTL), which breaks down wet organic solids at moderate temperatures and high pressures to form a carbon rich oil phase. While the bio-oil formed has a high heating value, the process often produces other byproducts such as solid biochar, carbon dioxide and wide distribution of organics that are water soluble. To design an economical HTL process, the water soluble organic byproducts must be minimized to increase oil product yields and reduce wastewater generation. Metal oxide catalysts are a promising option for HTL upgrading chemistry due to their catalytic activity as well as stability in hot liquid water conditions. $CeZrO_x$, in particular, is active for acid coupling reactions which simultaneously increase the carbon number of the product and decrease its oxygen to carbon ratio, both of which act to increase the oil yield and decrease the carbon loss to the aqueous phase. In our study, we sought to understand the key reactions that take place in the HTL process when using $CeZrO_x$ as a catalyst.

In preliminary runs, use of $CeZrO_x$ increased oil yields relative to other commonly used catalysts in the HTL conversion of food waste at 300 Â°C and 20.7 MPa. Subsequently, tests were performed to expose $CeZrO_x$ to hot liquid water conditions at 300 Â°C and 20.7 MPa for up to 165 hours. Post-run characterization was performed using both XRD and XPS to verify the hydrothermal stability of $CeZrO_x$, Model HTL reactions using $CeZrO_x$ were also performed with water soluble alcohols, aldehydes, ketones and carboxylic acids as reactants. $CeZrO_x$ had minimal activity for catalyzing reactions involving alcohol-alcohol, carboxylic acid-carboxylic acid or ketone-ketone coupling. In contrast, $CeZrO_x$ was highly active â€“ and selective â€“ for aldehyde coupling reactions, with the primary product being alkenes. For example, pentanal self-condensation reactions under HTL conditions produced a decene product that phase separated into an oil phase at room conditions. On-going work is focused on optimizing the catalyst properties and catalytic reactor conditions.

495412 Hydrothermal Liquefaction of Wild-Harvested Cyanobacteria from a Hypereutrophic Lake in Wisconsin

Abstract:

Processing by hydrothermal liquefaction (HTL) of wild-harvested cyanobacteria could have the potential to provide a sustainable, domestic fuel supply and nutrients for use in agriculture while removing a source of hypoxia from over-burdened waterways. Lab-scale cultured strains of microalgae has shown freshwater cyanobacteria as a promising feedstock with low ash content, leading to high biocrude yields. However, HTL of wild-harvested microalgal strains has yet to be fully investigated. Additionally, the intermittent nature of cyanobacteria blooms raises questions about the need to store harvested microalgae and the effect of storage time on the products.

Wild-harvested microalgae sampled from Lake Tainter, a hypereutrophic lake in Northwest Wisconsin, was collected and prepared in two ways: dried while fresh and stored long-term, and stored wet long term then dried before HTL processing. Fractions of the feedstock, mostly consisting of Microcystis Aeruginosa, were treated via HTL at a range of temperatures (280-300°C), solid loading ratios (10-25% w/w), and reaction holding times (30-60 min) in 30 mL batch reactors. The percent yield of the products and chemical composition of the biocrude and aqueous fraction (PHWW) was characterized by GC-MS, ICP-MS, and elemental analyzer.

496834 Multi-Scale Modeling of Heavy Oil Upgrading in Near-Critical/Supercritical Water

Abstract:

The upgrading of heavy oil using near-critical or supercritical water can be pushed towards better outcomes (higher light liquid product yield and lower coke formation) by suppressing the condensation reactions of light hydrocarbon species to heavier hydrocarbons and coke precursors. A potential method to achieve this is to ensure a much lower residence time for the light hydrocarbon species in the high-temperature reactor zone once they are formed by thermal cracking reactions of the heavy hydrocarbons.

This difference in residence times between the light and heavy species in the oil can in turn be realized in practice in a reactor configuration like the extractive upflow reactor which exploits gravity effects and the higher solubility of light hydrocarbons in the water phase. The oil phase being 2-8 times heavier than the water phase (depending on the operating pressure and temperature) circulates at the bottom of the reactor, while the water phase injected from the bottom flows through the bed of oil and exits the reactor at the top, in the process, extracting the light hydrocarbons soluble in the water phase. While experiments demonstrating the efficacy of the extractive upgrading process of heavy oil can be found in the literature, there remains a need to model and understand the complex coupling between the chemical reactions of interest and the interfacial mass transfer of hydrocarbon species and evaluate their effect on the final product yields.

With a view to address this need, we employ a two-phase well-stirred reactor (WSR) model with lumped chemical kinetics to study the overall coupling of the chemical reactions and the oil-water phase equilibrium in an extractive upflow reactor at P = 30MPa and T = 663,683 and 703K. Using this low-fidelity model, we demonstrate the improved yields of the upgrading process at higher inlet flow rates of water directly corresponding to higher rates of extraction of the light species from the oil to the water phase. This is in qualitative agreement with experimental demonstrations of the extractive upgrading configuration in literature. The WSR model also shows that the total interfacial species mass transfer rates in the reactor must be higher than the rates of retrograde reactions (condensation and coking reactions) for the best outcomes. This necessitates the need to develop methods to obtain realistic estimates for the interfacial species mass transfer rates in a realistic extractive upgrading reactor configuration.

In order to obtain a realistic range of values for the interfacial species mass transfer rates for different operating conditions of interest we use a multi-scale modeling approach, wherein the total interfacial species mass transfer rates in the reactor are estimated as the product of an average total interfacial area within the reactor and an averaged interfacial mass flux per unit area of the light hydrocarbon species.

The total average oil-water interfacial area in the reactor is calculated by simulating the macro-scale mixing process in a realistic lab-scale reactor geometry (diameter=30mm, height=100mm and 20g of oil feed) using two-phase flow simulations in 2-D with the Volume of Fluid (VoF) method with fluid properties like density, viscosity and interfacial tension estimated at the different given P, T conditions. The total average interfacial area in the reactor scales as the square root of the Weber number for a range of different operating conditions (pressure, temperature and inlet velocities). This is a useful intermediate result which can help engineers design extractive upgrading reactors for specified target light species extraction rates.

The average interfacial mass flux of light hydrocarbon species is estimated using 1-D droplet mixing simulations in spherically symmetric co-ordinates which account for the detailed oil-water phase equilibrium and diffusive fluxes at the interface. The heavy oil is modeled as a mixture of two pseudo-components, one corresponding to the light fraction and another corresponding to the heavy fraction in the mass ratio 1:1. The average interfacial mass flux of the light pseudo-component at P=30MPa increases from T=623K to 703K by more than an order of magnitude which is consistent with the rapidly increasing solubility of the light oil fraction in water with increasing temperature.

The total interfacial mass transfer rates estimated at P=30MPa and T=663-703K for inlet water mass flow rates in the range of 2.5-10 g/min and 20g of initial oil feed are an order of magnitude higher than the rates of the lumped cracking and coking reactions at these temperatures. This implies that the extractive upgrading reactor has the potential to effectively remove the light hydrocarbons from the oil phase as soon as they are formed and transport them safely out of the high temperature reaction zone. Although, this medium-fidelity approach to estimating the interfacial species mass transfer rates does not account for complex effects like the enhancement of the species fluxes due to sharpening of gradients in a flow field or the reduction in interfacial area due to interphase mass transfer, it nevertheless yields a reasonable first-order estimate which can be used to aid the design process of extractive upgrading reactors.

493158 Effects of Process Conditions on Metal Content in Biocrude from Hydrothermal Liquefaction of Microalgae

Abstract:

Hydrothermal liquefaction (HTL) converts wet biomass into a crude bio-oil that could be further treated to be acceptable for processing in petroleum refineries. HTL takes advantage of the properties of hot, compressed water near its critical point to break down biomacromolecules in the feedstock into smaller molecules. Several issues presently prevent the substitution of HTL biocrude for petroleum and one is the presence of metals in the crude bio-oil. Iron is found to plug the head of the catalyst beds that are used to hydrotreat algal biocrude oil [1]. Metal impurities in crude bio-oil can also cause catalyst deactivation, coke formation, corrosion, and deposition. Thus, further studies are required to understand how the HTL conditions influence metal content and to develop methods to reduce the metal content in algal biocrude for downstream processing operations.

We hydrothermally treated microalgae using different processing conditions (isothermal reaction vs. rapid heating, different temperatures and reaction times, different reactor loadings) and determined the amounts of thirteen different metals in the biocrude oil as well as in the other product phases. The metal content in biocrude oil undergoes dramatic change within the first 5 minutes of HTL reaction. Generally, lighter elements are transferred into bio-oil faster than heavier metals. Iron in bio-oil behaves differently than many of the other metals, and is the target metal to be removed due to its abundance. As the reaction conditions become more severe, less iron resides in bio-oil phase. Optimum reaction conditions could be identified in terms of a high bio-oil yield and low metal concentration in the bio-oil. The results from this study provide the first glimpse into how and whether HTL process conditions can be used to manipulate metal content in algae biocrude.

Reference

[1] Jarvis, Jacqueline M., et al. "Impact of iron porphyrin complexes when hydroprocessing algal HTL biocrude." *Fuel* 182 (2016): 411-418.

498900 Bio-Crude Quality and Composition from Algal Solids Cultivated in Varying Media, Growth Stage, and Solids Loading

Abstract:

Hydrothermal liquefaction (HTL) has become a promising technique for converting algal biomass to high-value products because it utilizes algaeâ€™s natural, wet environment for the conversion making it both energy and cost-effective. Furthermore, the necessary water and nutrients to cultivate algal biomass can be provided in a sustainable and inexpensive manner with the use of wastewater (WW) effluent. Algal biomass harvested from WW can also contain large amounts of inorganics, primarily calcium, depending on the treatment operation at the plant. During the HTL reaction these inorganics produce an abundant, solid co-product. X-ray Diffraction (XRD) has confirmed these solids to be one of three calcium materials: hydroxyapatite (HAp), $Ca_5(PO_4)_3OH$, tricalcium phosphate (TCP), $Ca3(PO4)_2$, or calcium carbonate ($CaCO_3$). The purpose of this presentation is two-fold. First, to demonstrate the impact media and growth stage have on biocrude composition. Second, to understand the role these calcium solids have on biocrude quality.

Conventional HTL reactions at 350°C for 60 minutes were performed on various WW and raceway-cultivated algal solids. To simulate seasonal and regional differences of WW, algae were grown in two separate raceway ponds: one raceway that had N-limited media and the other P-limited. Algal solids were then harvested at different growth stages thus altering their biomolecular content and ultimately varying the biocrude composition. To regulate inorganic content of the algal solids, the algae were harvested via auto-induced flocculation that occurred from a manual increase in media pH. At each growth stage the total Ca:P molar ratio of the algal solids were manipulated to range between 1 and 3 with the addition of lime ($Ca(OH)_2$), a common chemical used to balance the alkalinity and pH in WW treatment.

GC-MS results displayed dramatic change in oil composition dependent upon growth media, growth stage, and solids formed. Elemental CHN/O analysis indicated little change in N/C or O/C ratio between growth stages and media. However, the H/C ratio did change with the HTL solids produced from the reaction. Thus, further characterization of HTL solids were measured for their catalytic potential. In addition, fundamental reactions of model inorganic and algal compounds were also performed to better understand the mechanism of solid synthesis and biocrude upgrading.

Thermochemistry of Alkylaromatics Reconsidered

Presentation at the AIChE National Meeting

Lawrence Lai[1]

William H. Green[1]

1 – Massachusetts Institute of Technology, Cambridge, MA

1

Increase in Oil Demand

OPEC World Oil Outlook (2016)

Global oil demand growth in the long-term

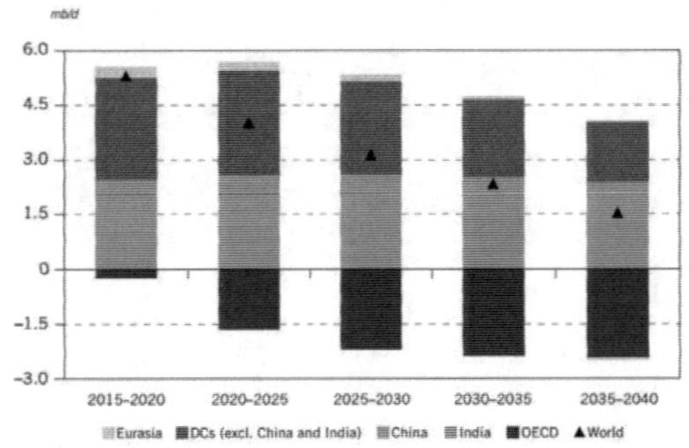

- Increase in global oil demand
- Demand to be met by heavy oil & bitumen
  - Difficult to vaporize - pyrolysis takes place
  - Not as clean combustion
  - High viscosity
  - Higher sulfur content

Caniaz and Erkey, *Process Intensification for Heavy Oil Upgrading Using Supercritical Water*

2

Supercritical Water Upgrading

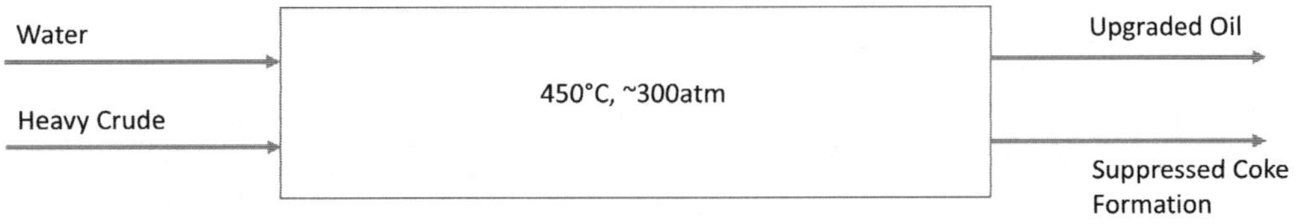

- Shown to suppress coke formation

With Water

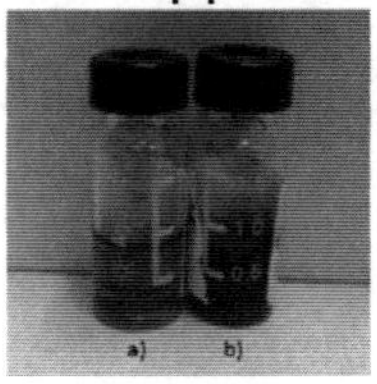

Without Water

aramco

Choi, *Supercritical Water Process to Upgrade Petroleum, WO2013066852 A1, 2013*

3

Hexylbenzene

- Focus of study to represent alkylbenzenes in crude oil
- Moderate size allows computational analysis using Reaction Mechanism Generator (RMG)
- Long alkyl chain to study cleavage chemistry
- Falls within GC-detection range

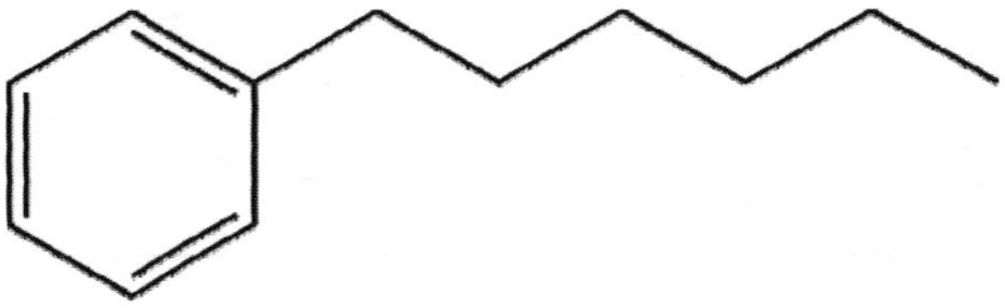

4

Hexylbenzene

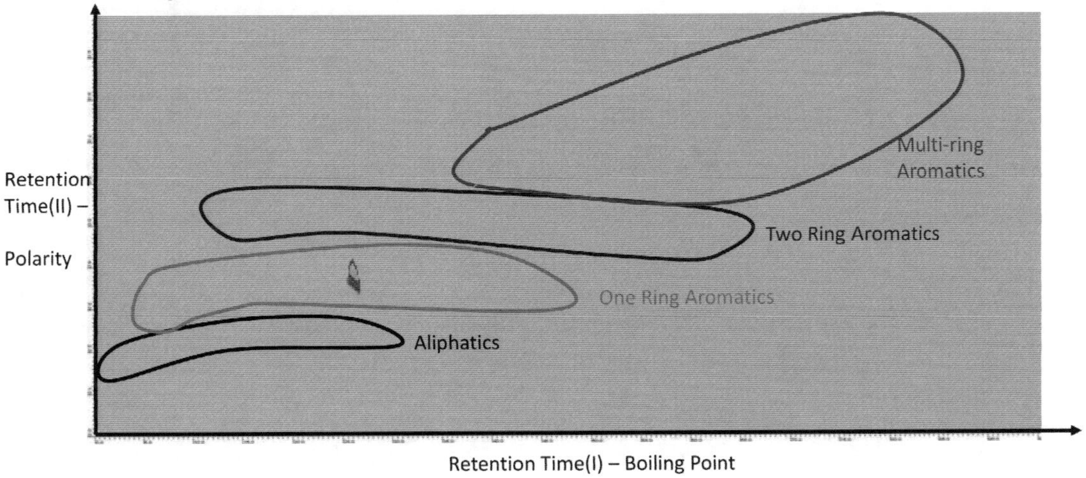

5

Supercritical Water Treated Hexylbenzene, 450°C, 45 minutes

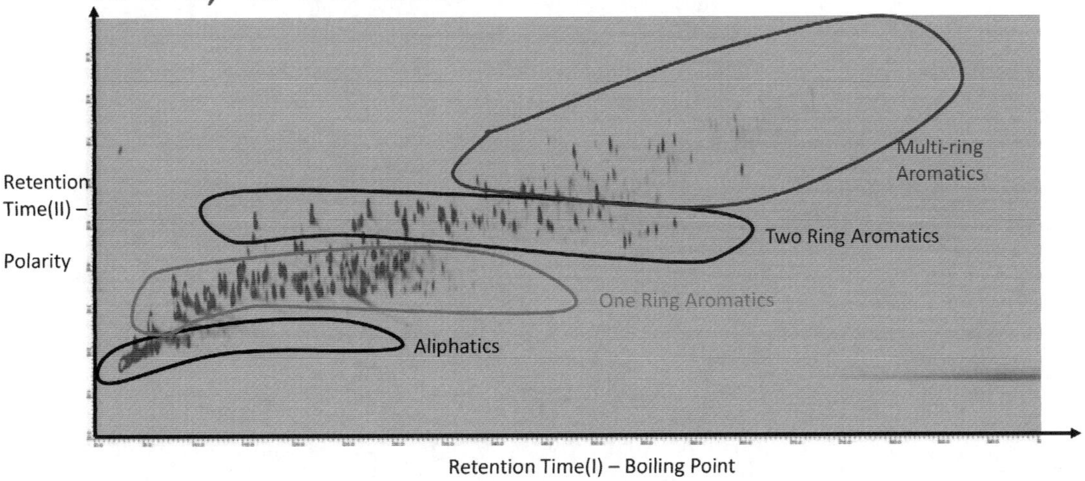

6

Intractable Experiment

- One reactant + water led to >500 Species
- Each species could be formed through multiple different reactions
- Radicals and other unstable intermediates are not observed on GCxGC
- Cannot use experimental means to uncover chemistry

Use the Green Group Reaction Mechanism Generator (RMG) to better understand the system

RMG does not show differences in presence of water

7

Reaction Mechanism Generator (RMG)

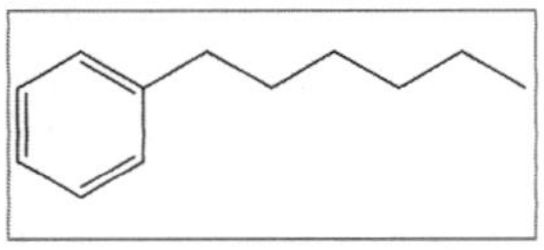

Core

8

Reaction Mechanism Generator (RMG)

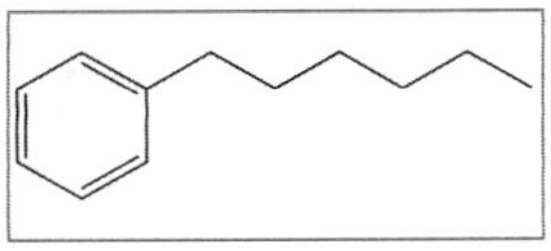

Core

Data Base of homolytic scission

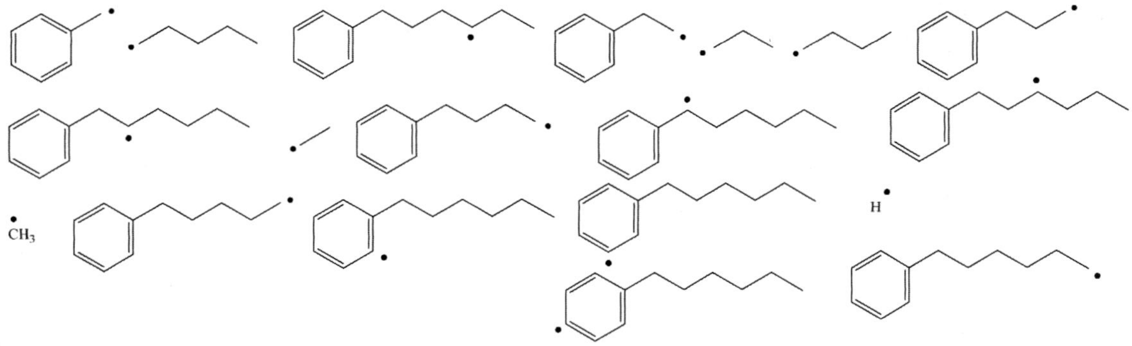

9

Reaction Mechanism Generator (RMG)

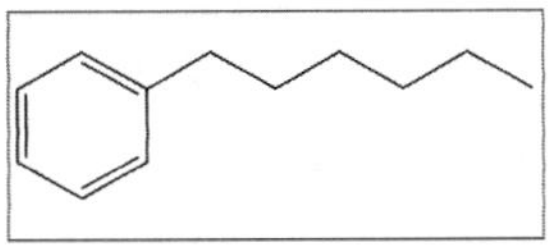

Core

Highest Flux Products

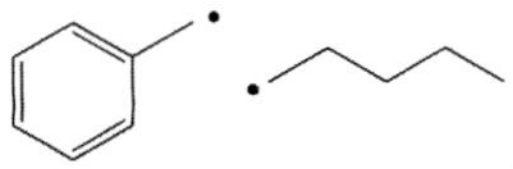

10

Reaction Mechanism Generator (RMG)

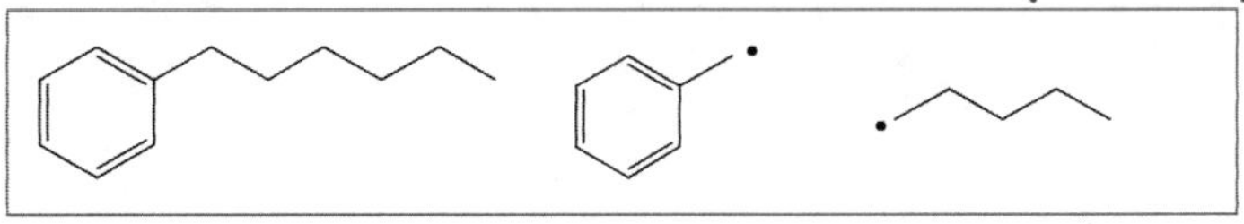

Core

11

Reaction Mechanism Generator (RMG)

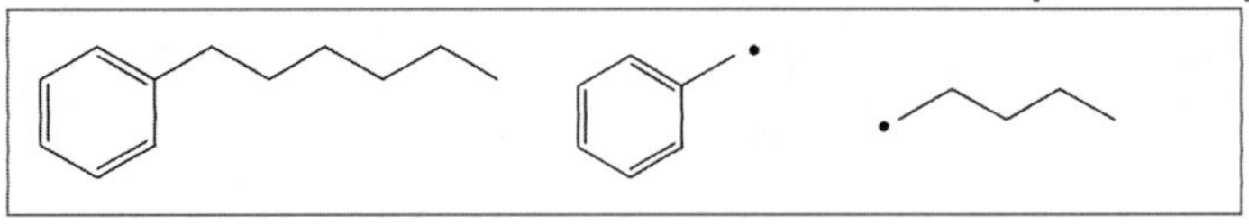

Core

New types of reactions can now take place:

Hydrogen Abstraction

Beta Scission

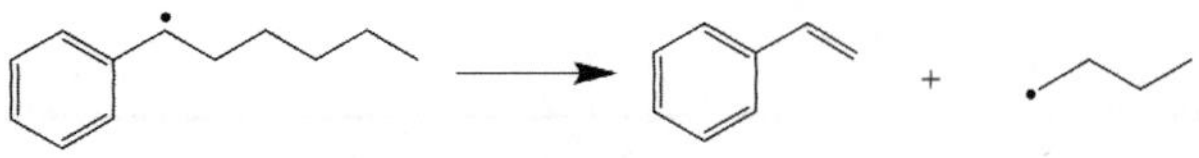

12

Overview of RMG algorithm

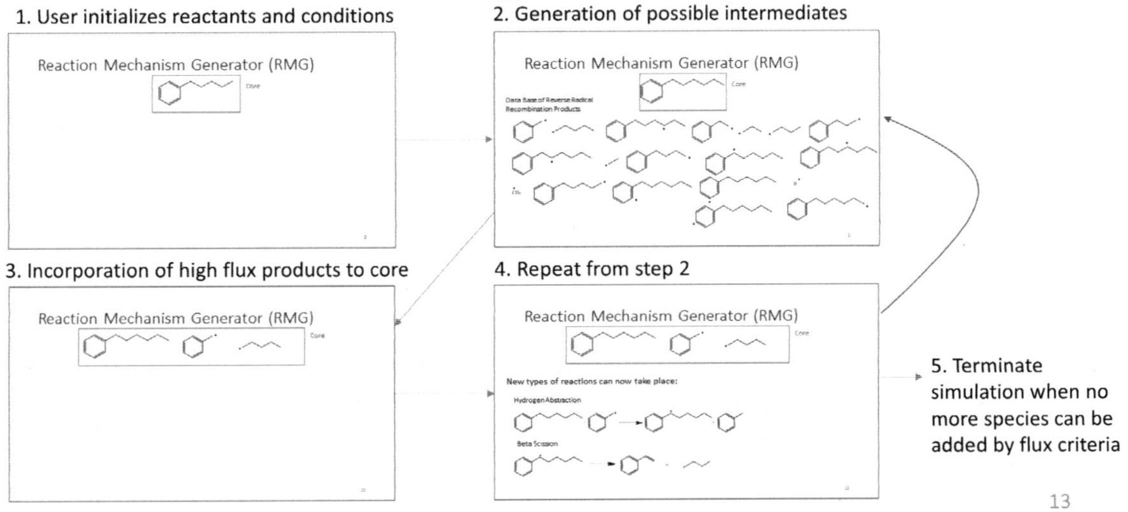

13

Model Conversion – Carr et al.[1]

- Model conversion factor of 2 off
- Inaccurate conversion leads to inaccurate product yields

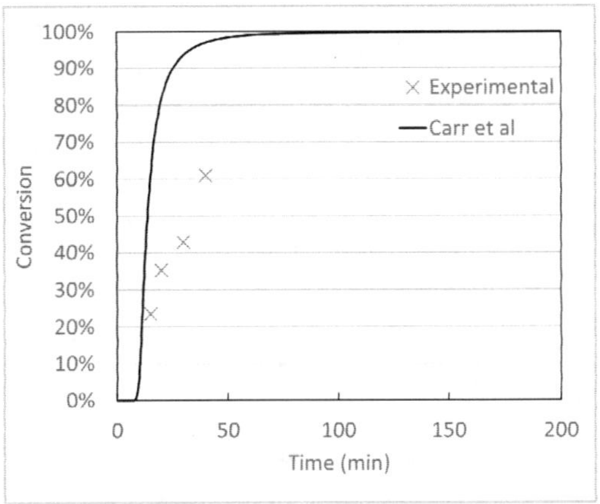

[1] Carr et al., *Energy and Fuels,* 2015.

14

"Retro-ene" Reaction

- "Retro-ene" Reaction thought to be important

(1)

	A (s^{-1})	n	E_a (kcal/mol)	Rate Constant at (723 K)
Burklé-Vitzthum[1] (Used in model)	$6.8 \times 10^{12}\ s^{-1}$	0	54.4	2.44×10^{-4}
CBS-QB3	$1.63 \times 10^{7}\ s^{-1}$	1.69	73.9	5.07×10^{-11}

- Rate used in Carr et al. was 7 orders of magnitude faster than reality
  - Highly accelerates hexylbenzene conversion

[1] V. Burklé-Vitzthum and R. Michels, *Ind. Eng. Chem. Res.*, 2005.

15

Model Conversion – Removed Retroene

- Conversion too slow
- Inspected model details

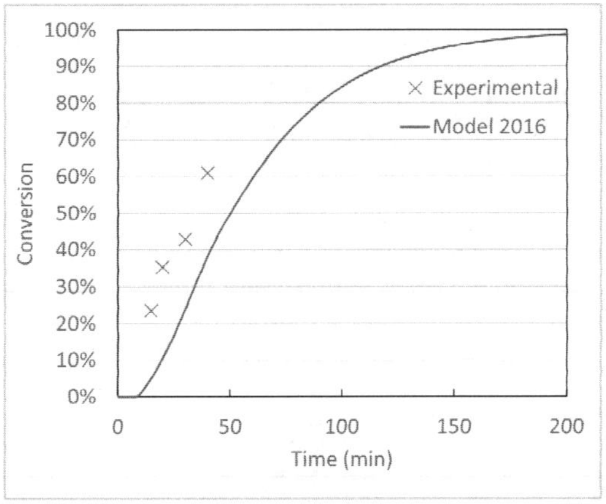

16

Hexylbenzene Model Summary

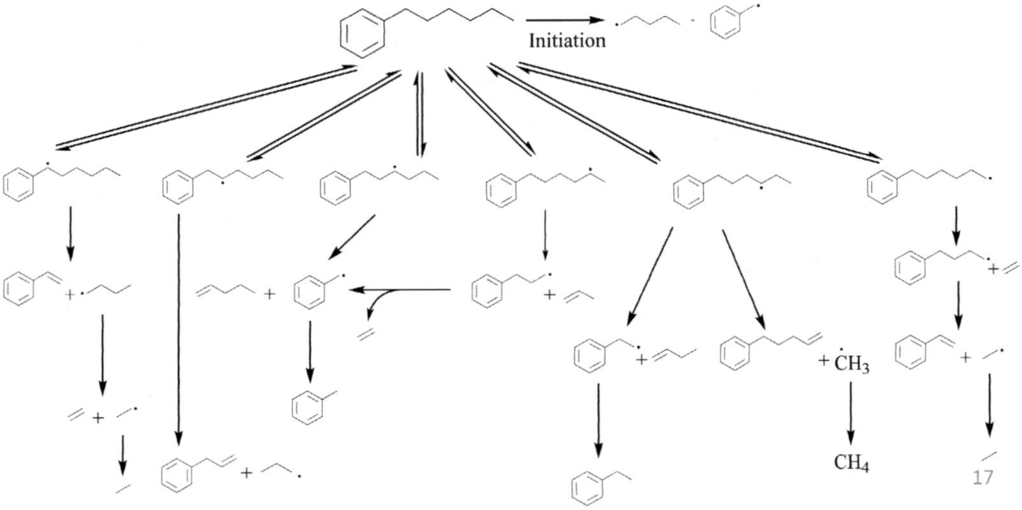

17

Initiation Step

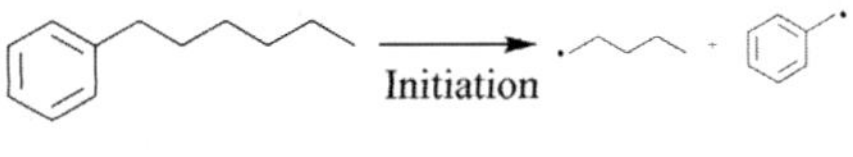

18

Hexylbenzene Radical Hydrogen Abstraction

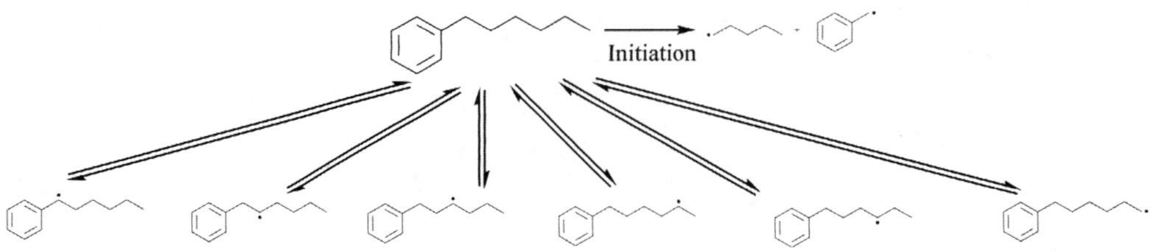

19

Main Beta Scission Pathway

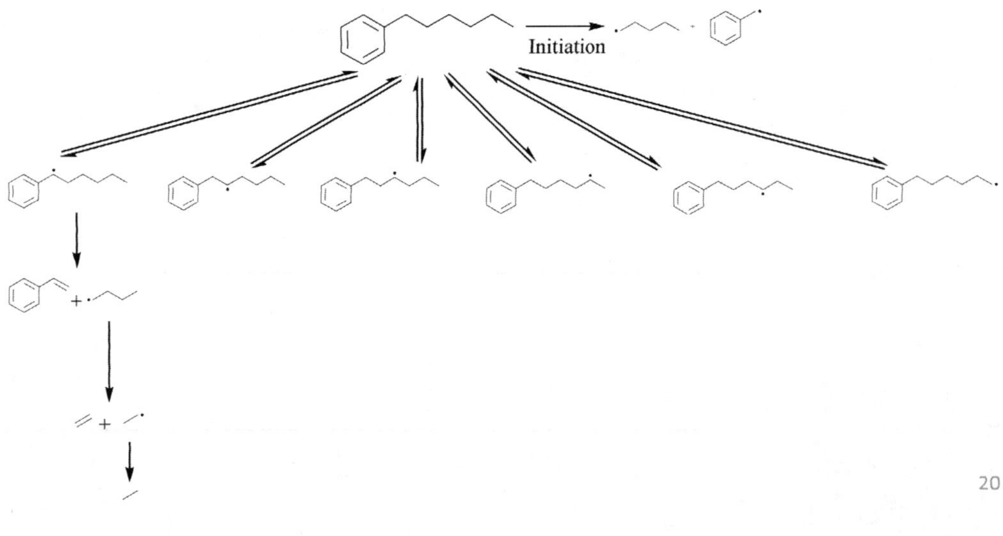

20

Main Product Flux

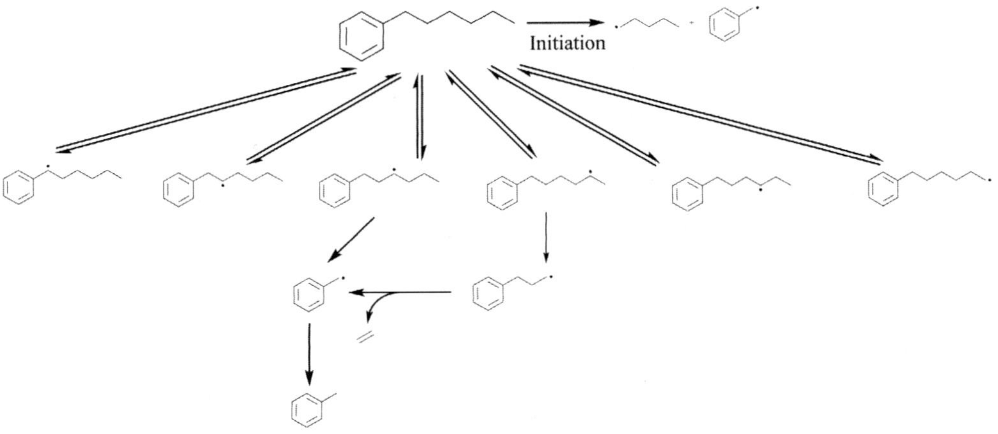

21

Hexylbenzene Model Summary

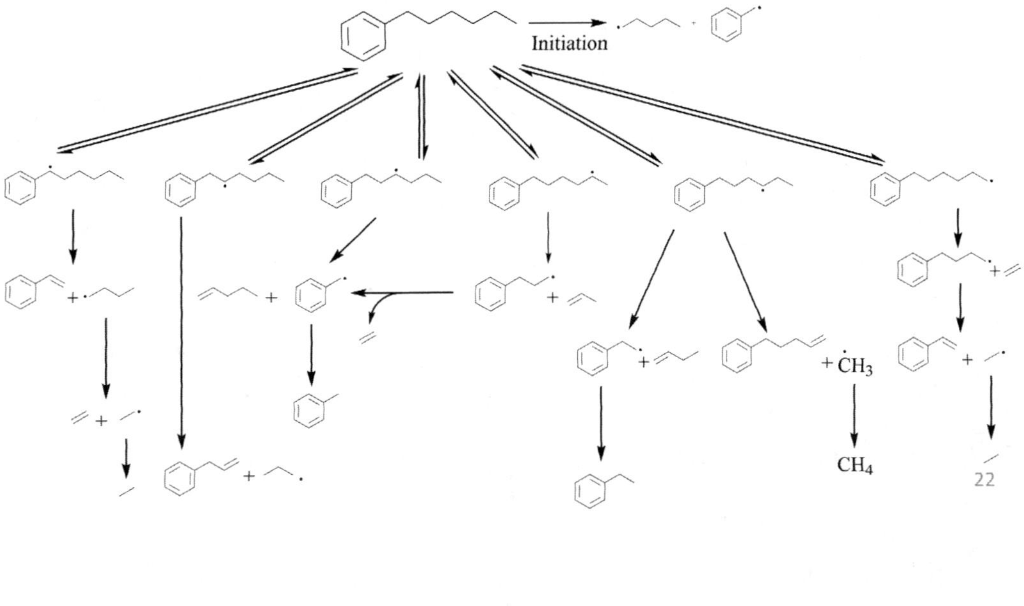

22

Sensitivity Analysis

$$s_i = \frac{\mathrm{d}\ln c_i}{d\ln k_j}$$

Where s is sensitivity
c is concentration
K is the rate constant
Subscript i indicates a species
Subscript j indicates a reaction

- Reaction (2)

- Reaction (3)

Discrepancy too large to be due to uncertainty in rate coefficients

New Rate Calculated

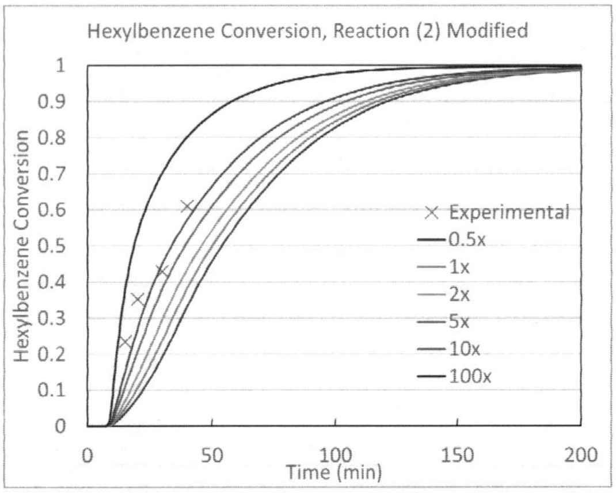

$k(T) = 2.1 \times 10^{13} \exp(-0.13 \text{ kcal mol}^{-1} /RT) \text{ cm}^3 \text{ mol}^{-1} \text{ s}^{-1}$

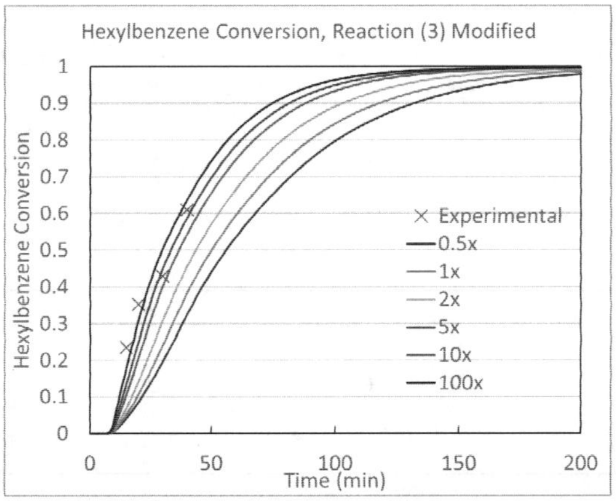

$k(T) = 2.4 \times 10^{2}\, T^{2.8} \exp(3.7 \text{ kcal mol}^{-1} /RT) \text{ cm}^3 \text{ mol}^{-1} \text{ s}^{-1}$

High Sensitivity to Thermochemistry

- Thermochemistry (Gibbs free energy) changes equilibria of reactions
- Small change in G leads to huge change in conversion

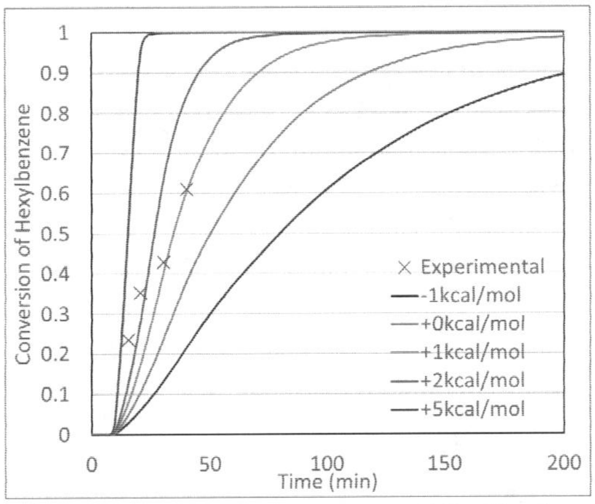

(Hexylbenzene Gibbs Free Energy = 118kcal/mol)

25

Thermochemistry Update

- Calculations done in CBS-QB3 level of theory; all hindered rotors included
- Species of interest:
  - Hexylbenzene and all aliphatic hexylbenzene radicals
  - All alkylbenzenes from toluene to pentylbenzene
  - All alkylbenzyl radicals adjacent to aromatic ring from benzyl to pentylbenzyl
  - Species involved in reactions (2) and (3): butyl, pentyl radical, and styrene

26

Updating Thermochemistry Resolves Discrepancy in Conversion

- Model matches experiments within 30% difference
- Updated thermochemistry available in RMG's thermochemistry database
  - Library named Lai_Hexylbenzene

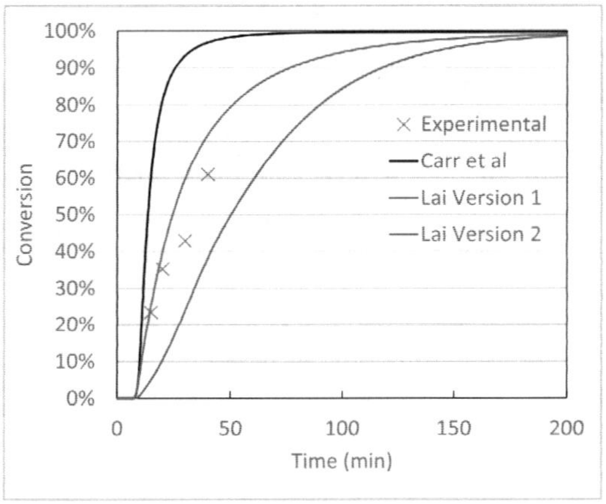

27

Individual Species Yield

- Species yield match at early times
- Model does not correctly depict chemistry at high conversion

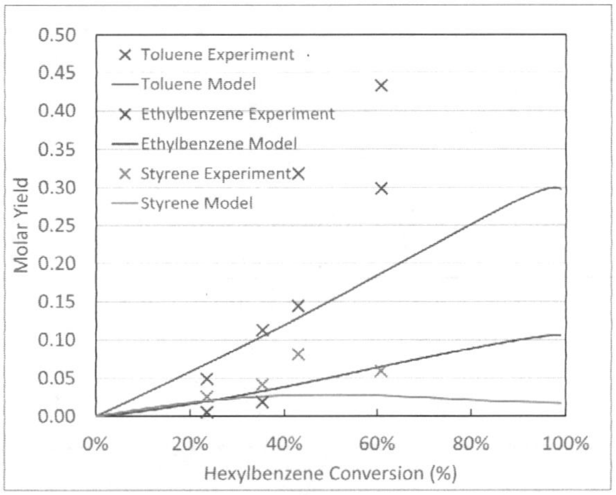

28

Error Analysis

- CBS-QB3 has an error of $\pm$2.4kcal/mol [1]
- With error bounds introduced, conversion of model is at an appropriate match with experiments

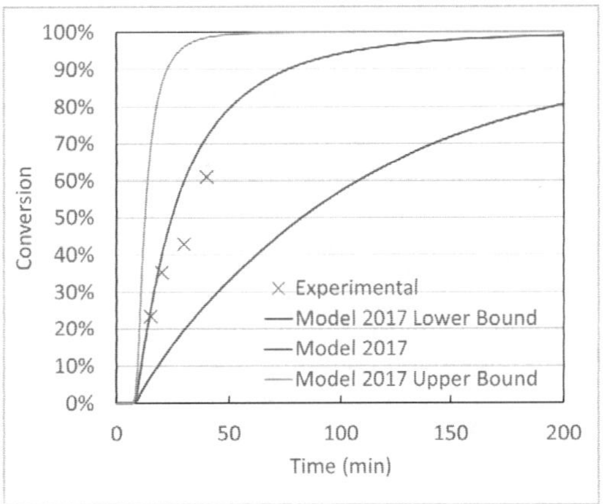

[1] K. P. Somers and J. M. Simmie, *The Journal of Physical Chemistry,* 2015.

29

Conclusion

- RMG was used to model pyrolysis of hexylbenzene
  - System is intractable through experimentation – use of RMG neccesary
- Thermochemistry affects conversion of reactant much more than kinetics
  - Pyrolysis at 723K
  - 2kcal/mol change in Gibbs Free Energy can affect conversion dramatically
  - 10x change in rate constant for initiation step required for similar change
- CBS-QB3 level of theory not precise
  - Model can be further improved using methods which compute/determine thermochemistry with tighter uncertainty ranges.

30

Acknowledgments

- Green Group
  - W. H. Green
  - Soumya Gudiyella
  - Max Liu
  - Zachary Buras
  - Kehang Han
- WPI
  - Mike Timko
- Funding

aramco

Founding member of the MIT Energy Initiative

31

Questions

- Lawrence Lai
- laitcl@mit.edu

32

References

- A. G. Carr, C. A. Class, L. Lai, Y. Kida, T. Monrose and W. H. Green, "Supercritical Water Treatment of Crude Oil and Hexylbenzene: An Experimental and Mechanistic Study on Alkylbenzene Decomposition," *Energy & Fuels,* vol. 29, pp. 5290-5302, 2015.

- V. Burklé-Vitzthum and R. Michels, "Experimental Study and Modeling of the ROle of Hydronaphthalenics on the Thermal Stability of Hydrocarbons under Laboratory and Geological Conditions," *Ind. Eng. Chem. Res.,* vol. 44, pp. 8972-8987, 2005.

- K. P. Somers and J. M. Simmie, "Benchmarking Compound Methods (CBS-QB3, CBS-APNO, G3, G4, W1BD) against the Active Thermochemical Tables: Formation Enthalpies of Radicals," *J. Phys. Chem. A,* vol. 119, pp. 8922-8933, 2015.

Backup Slides

Hexylbenzene Experimentation

Modeling
- Reaction Mechanism Generator (RMG)
- Species susceptible of formation included in initial conditions

Loading
- Hexylbenzene
- Naphthalene Standard
- SCW or Pyrolysis

Reaction
- Heating by sand bath
- Heat up time of 10 minutes - 450°C
- Pressure monitored
- 300 or 75 bar (SCW or Pyro)

Quenching
- Cooling by water
- Cool down time of 2-3 minutes

Analysis
- GC-MS Liquids
- GCxGC-FID for diagnosis
- GC-FID for gases

35

495612 Tailoring Zeolite ZSM-5 Crystal Morphology and Spatial Distribution of Acid Sites

Abstract:

The micropores of large zeolite crystals often impose severe mass transport limitations that adversely affect catalyst activity.[1] One of the challenges in zeolite design is to predictively control crystal size and habit to improve molecular diffusion. Here, we present a novel approach to tailor ZSM-5 crystallization using zeolite growth modifiers (ZGMs), which are molecules that bind to specific crystal surfaces and alter the anisotropic growth rates as a means of tailoring crystal habit.[2,3] We have observed that subtle changes in synthesis conditions (e.g., pH) can alter ZGM binding specificity and produce crystals with different morphology. This is due to the fact that modifier speciation can be controlled by adjusting alkalinity, which could prove to be a facile and highly effective method of tailoring zeolite growth.[4]

The unique properties of ZSM-5 catalysts, such as hydrothermal stability and shape-selectivity, are ideal for many reactions. Catalyst performance is dependent upon its chemical composition. A phenomenon known as *aluminum zoning* in ZSM-5 crystals has been reported in many studies; however, its origin remains elusive.[5] To address this topic, we have performed parametric investigations to assess the role of synthesis conditions on Al siting. We show that ZSM-5 crystals can be prepared with either Si- or Al-rich exteriors, which impacts their hydrothermal stability and product selectivity in catalytic reactions.[6]

The ability to design ZSM-5 catalysts with tunable morphology and Al distribution opens new avenues for tailoring catalyst performance, and understanding property-performance relationships. ZSM-5 properties can be selectively optimized for a wide range of applications. To this end, this study provides a general platform for zeolite design and optimization that could potentially be applied to other microporous framework types.

(1) Pérez-Ramirez, J. et al.; *Chem. Soc. Rev.* 37 **(2008)** 2530-2542

(2) Lupulescu, A. I.; Rimer, J. D.; *Angew. Chem. Int. Ed.* 51 **(2012)** 3345-3349

(3) Lupulescu, A. I.; Qin, W.; Rimer, J. D.; *Langmuir* 32 **(2016)** 11888-11898

(4) Qin, W.; Clark, R.J.; Palmer, J.C.; Rimer, J.D. (in preparation)

(5) Danillina, N. et al.; *J. Phys. Chem. C* 114 **(2010)** 6640-6645

(6) Qin, W.; Patton, M.D.; Rimer, J.D. (in preparation)

488629 Carbon Nanotube-Supported Catalysts Prepared By a Modified Photo-Fenton Process for Fischer-Tropsch Synthesis

Abstract:

We demonstrate use of a facile, scalable, modified photo-Fenton process for the preparation of carbon nanotube (CNT)-supported Fischer-Tropsch catalysts. Use of a green catalyst precursor and an oxidant in photo-Fenton process enhances the formation of small and well-dispersed catalyst particles on CNTs. Catalysts prepared by photo-Fenton process do not require a calcination step due to the absence of foreign components (nitrate, chloride, and acetate) that have to be removed during post-synthesis treatment. The synthesized catalysts were characterized by X-ray diffraction (XRD), BET, hydrogen temperature-programmed reduction (H_2-TPR), Raman spectroscopy, transmission electron microscopy (TEM), and inductively coupled plasma (ICP). H_2-TPR results show a high dispersion percentage and smaller catalyst particles with narrow particle size distributions. The performance of catalysts (CO conversion, C5+ selectivity, and catalyst lifetime) prepared by photo-Fenton process was evaluated during FTS and compared to CNT- and SiO_2-supported catalysts synthesized via conventional method (incipient wetness impregnation). In general, CNT-based catalysts prepared by photo-Fenton process (Co or Fe) show a higher performance at lower reaction temperature (200°C for Co and 250°C for Fe) in comparison to conventional catalysts. In particular, Co catalysts obtained from photo-Fenton process show outstanding selectivity for liquid hydrocarbons (C5+) ~ 70% and CO conversion ~80%.

491282 Synthesis Gas Conversion over Rh-Based Catalysts Promoted By Fe and Mn

Abstract:

The direct conversion of synthesis gas to C_{2+} oxygenates has attracted considerable interest owing to their wide range of application in chemical, fuel and polymer industries.[1,2] Rhodium has been reported to be the best metal for the selective production of oxygenates,[3,4] and transition metals have been shown to be promoters to increase the selectivity for C_{2+} oxygenates. To date, experimental studies addressing the interaction among promoters and the resulting impacts on synthesis gas conversion are limited.

Here, we present our results on synthesis gas conversion to oxygenates and hydrocarbons over Rh-based catalysts promoted by Fe and Mn. We show that selective deposition of Fe and Mn species on Rh nanoparticles can be achieved by controlled surface reactions. By carrying out reactions at 523 K, 580 psi and CO/H_2=1/1, the catalytic results show that the interaction between Rh and Fe promotes the selective production of ethanol through hydrogenation of acetaldehyde and enhances the selectivity towards C_2 oxygenates, which include ethanol and acetaldehyde. Mn promotes the overall reaction rate and the selectivity towards C_{2+} hydrocarbons, primarily alkenes. Trimetallic Rh-Fe-Mn catalysts surpass the performance of their bimetallic counterparts. The highest selectivities towards ethanol (36.9%) and C_2 oxygenates (39.6%) were achieved over the Rh-Fe-Mn ternary system with a molar ratio of 1:0.15:0.10, as opposed to the selectivities obtained over Rh/SiO_2 which were 3.5% and 20.4%, respectively. Moreover, 55% of the total products are industrially valuable oxygenates and C_{2+} hydrocarbons. To understand the nature of Fe and Mn species, measurements obtained from X-ray photoelectron spectroscopy suggest that Fe and Mn exist as metallic iron and manganese oxides on Rh surface upon reduction. Through density functional theory (DFT) calculations, we reveal that manganese oxide is more stable on the Rh (211) step-edges whereas metallic iron has no preference over Rh (111) and (211). Fourier transform infrared spectroscopy of adsorbed CO and DFT calculations suggest that the adsorption of CO on Fe and/or MnO_x modified Rh (211) step-edges is altered compared to the binding of CO to clean Rh (211) surface. Accordingly, MnO_x species and Fe alter the bonding of CO on Rh step-edges and shift their selectivity away from CH_4, while reduced Fe species also alter the reactivity of Rh terrace sites.

1. Angelici, C.; Weckhuysen, B. M.; Bruijnincx, P. C. A. *ChemSusChem* **2013**, *6*, 1595â€"1614.

2. Goldemberg, J.; Goldemberg, J. *Sci.* **2007**, *315*, 808â€"810.

3. Ichikawa, M. *J. Chem. Soc. Chem. Commun.* **1978**, *1500*, 566.

4. Van der Lee, G.; Schuller, B.; Post, H.; Favre, T. L. F.; Ponec, V. *J. Catal.* **1986**, *98*, 522â€"529.

493282 Tailored CeO_2 Supported Ni-Based Catalysts Prepared By Solution Combustion Synthesis for Hydrogen Generation from Hydrous Hydrazine

Abstract:

Hydrous hydrazine is a promising hydrogen carrier due to its high hydrogen content (8.0 wt.% for hydrazine monohydrate) and benign byproduct (only nitrogen) via complete decomposition. For its practical use, the development of cost-effective catalysts with high activity and 100% selectivity for hydrogen generation under mild conditions is required and is a significant challenge. In this work, CeO_2-supported non-noble metal containing Ni-based catalysts were prepared by solution combustion synthesis (SCS), varying the synthesis parameters in terms of ratio of metal oxidizers, fuel type and fuel-to-oxidizer ratio. The prepared catalysts were characterized and tested to understand the effects of the synthesis parameters on the SCS process and their correlation with the physical and catalytic properties. The experimental results demonstrate that the tailored catalyst exhibits high activity and 100% H_2 selectivity under PEM fuel cell conditions for vehicle applications.

492960 Synthesis of Rhodium Nanoclusters on Graphene/Cu(111) from Organometallic Precursors: A Combined DFT and STM Study

Abstract:

This paper aims to achieve fundamental understanding of surface reaction chemistry and catalysis by formation of rhodium metal nanoclusters from organometallic compounds ($Rh(CO)_2$(acac) and $Rh_6(CO)_{16}$) on an epitaxially grown graphene layer over Cu(111). This model catalyst was studied with ultrahigh vacuum scanning tunneling microscopy (UHV-STM) and Density Functional Theory (DFT) calculations. Several moiré patterns with different periodicities and different rotational alignments were observed in STM images, which indicate the weak interaction between graphene monolayer and Cu(111) substrate. DFT calculations show that Rh_9 cluster has the strongest bond to the graphene/Cu(111) surface with an adsorption energy of -4.16 eV and Rh-C bond of 2.11 Å. DFT calculations also shows that organometallic compounds that are introduced to the sample should first be dissociated on the surface so that rhodium nanoclusters can be formed on graphene hollow sites as oxygen weakly interacts with the surface (adsorption energies of -0.19 and -0.37 eV for RhCO and Rh(CO)2, respectively). These results explain the weak interaction of $Rh_6(CO)_{16}$ on the surface, which is attributable to the repulsive effect by oxygen atoms. Both STM and DFT results confirm that rhodium nanoclusters have preferential nucleation sites on graphene moireǐ pattern on Cu(111).

496587 Adsorption of Transition Metal Precursors on Reducible Metal Oxide Supports: Toward Rational Synthesis of Single-Site Catalysts

Abstract:

The ability of metal oxide supports to enhance the dispersion of the active metal on their surface and control their morphology and sintering kinetics is fundamentally related to the nature and strength of the metal–support interaction which is determined at the time of adsorption at the solid-liquid interface. Ceria is a reducible oxide support for transition metal catalysts well-known for its high "oxygen storage capacity" that allows it to successfully stabilize noble metals, inhibit sintering and maintain small sized nanoparticles on its surface compared to other oxide supports. However, fundamental molecular level of metal–ceria interface interactions lack substantial information on the energetics of binding and their influence on particle size and growth.

In this work, we study the adsorption of Pt and Pd based precursors, chloroplatinic acid, tetraammineplatinum (II) nitrate and ammonium tetrachloropalladate (II) on ceria and alumina under strong electrostatic solution conditions and by incipient wetness impregnation. Equilibrium adsorption isotherms aid in quantifying the amount of metal adsorbed on the support surface and by varying choice and weight loading of the precursors, we could identify that chloride ligand speciation chemistry around main metal center strongly influenced metal uptake. The binding energy of adsorption of metal precursors on the ceria surface is determined through the use of isothermal titration calorimetry (ITC) to experimentally evaluate the equilibrium binding constants, binding stoichiometry, and enthalpies of adsorption. Experimentally measured free energy of adsorption is found to be higher than theoretical columbic free energy under strong electrostatic conditions, thus substantiating the importance of solvation and ligand chemistry effects. The formation of atomically dispersed metal nanoparticles through formation of an M-O-Ce bond under strong oxidative conditions is examined using XPS. Support of high dispersion of ceria is provided via TEM imaging that shows Pt nanoparticles on ceria are under 2 nm, even at high metal weight loadings whereas they sinter to form clusters up to 10 nm in size on alumina. Based on these observations, we synthesized faceted nano-crystals of cerium oxide and subsequently deposited Pt precursors over a range of metal weight loading to optimize the support vacancy usage to form single atom catalysts. These catalysts are tested on methane oxidation reaction to study the effect of the transition from single atoms to nano-clusters on the activity of the reaction.

502563 Nanoparticle Synthesis Via Electrostatic Adsorption Using Incipient Wetness Impregnation

Abstract:

Nanoparticle Synthesis via Electrostatic Adsorption using Incipient Wetness Impregnation

Sonia Eskandari and John R.Regalbuto

The University of South Carolina (USA)

Department of Chemical Engineering, 541 Main Street, Columbia, SC 29208 (USA)

Incipient wetness impregnation is commonly used for supported metal catalyst synthesis given its simplicity. The disadvantage, however, is that most often the metal precursors do not interact with the support surface and form large agglomerates during the drying process, leading to poor metal utilization. In Strong Electrostatic Adsorption (SEA) [1], coulombic metal precursor-support interactions are induced between a protonated, positively charged surface and an anionic metal precursor complex such as platinum hexachloride, $[PtCl_6]^{2-}$, or a deprotonated, negatively charged surface and a cationic metal precursor such as platinum tetraammine, $[(NH_3)_4Pt]^{2+}$. The strong interaction of the precursors give high dispersion of the metal after drying and reduction [1]. SEA is normally employed with excess liquid solution, such that an additional filtering step is required vis-à-vis incipient wetness impregnation.

In this talk we demonstrate the extension of electrostatic adsorption, which gives small particle sizes, to the incipient wetness procedure; we call this method "Charge-Enhanced Dry Impregnation (CEDI) [2]. This is done by intensely acidifying or basifying the small quantities of solution needed for incipient wetness, to achieve a strong charge on the support surface. We extend our preliminary studies of CEDI [2,3] to a wider variety of noble (Pt, Pd) and base (Co, Ni, Cu) metals. We employed tetra- and hexaammine precursors over a high surface area silica support (Aerosil 300, 280 m^2/g). Using precursors with OH^- counterions, such as $(NH_3)_4Pt(OH)_2$, small particle size was immediately achieved, and when precursors with chloride counterions such as $(NH_3)_4PtCl_2$were used, particles were much larger. However, particle size could be reduced without much loss of metal by an appropriate washing step. We further demonstrate that the residual counterions can be used to control the particle size.

References

[1] J.R. Regalbuto, Strong Electrostatic Adsorption of Metals onto Catalyst Supports, in Catalyst Preparation: Science and Engineering, J.R. Regalbuto, ed., Taylor and Francis/CRC Press, 2007.

[2] X.R. Zhu, H.R. Cho, M. Pasupong, J.R. Regalbuto, ACS Catal. 3 (2013) 625–630.

[3] D,Souza, L., Barnes, S., and Regalbuto, J.R., Catalysts 2016, 6, 72.

496028 Escaping Linear Scaling Relations: Catalysis Beyond Constraints on Single Atom Alloys

Abstract:

A single atom alloy (SAA) is manufactured by combining a host metal with a dopant metal at such high dilutions that the dopant remains atomically dispersed. The dopant atoms in such alloys are subject to electronic and geometric effects, giving rise to catalytic sites with unique properties, not exhibited by either component of the alloy in the monometallic state. This behavior in turn generates novel opportunities for designing catalysts for industrially relevant chemistries such as methane activation, ammonia dissociation or hydrogenation reactions. A fundamental understanding of the properties of SAAs would be indispensable in formulating principles that would aid catalyst design efforts.

Motivated by these opportunities, we discuss our latest research that aims at advancing our theoretical understanding of SAAs. By calculating the energies of adsorption and bond activation for key species participating in small-molecule chemistries, we investigate pertinent reactivity trends for dissociation reactions such as C-H, C-O, N-H, and H-H, on SAAs. These materials are thus found to escape traditional Brønsted-Evans-Polanyi relations, an effect caused by the decoupling of the energies of the transition state versus the initial/final states. Our results are discussed in the context of experimental findings on C-H and H-H bond activation, showcasing the importance and benefits of these materials.

500853 Mechanistic Studies of Selective Hydrogenation Reactions with Palladium Gold Alloy Catalysts

Abstract:

Selective hydrogenation of alkynes to alkenes is an important process to produce alkyne-free feedstocks for the polymer industry. Pd with lead additives has shown good reactivity and selectivity in partial hydrogenation of alkynes and is widely used in industry. To further improve the catalytic performance of Pd catalysts and avoid the use of lead, different additives including Ag, Rh, Au, Cu, Zn, Cr, Bi and V have been reported. One promising approach to achieve the optimal material efficiency and selectivity of Pd catalysts is to form the alloy of Pd and Au at the single atom limit, which we call single atom alloy (SAA). Previous reports have demonstrated that PdCu and PdAu SAA catalysts are highly selective in partial hydrogenation of phenylacetylene[1] and 1-hexyne[2], respectively.

We have investigated the selective hydrogenation of 1-hexyne over PdAu alloys with different Pd ensembles, namely single atoms, clusters and nanoparticles. The surface structures of Pd species were studied by ATR-IR of chemisorbed CO. Kinetics studies indicate the hydrogenation reaction pathway and rate limiting step on PdAu SAA is same as that of Au nanoparticles. *In situ* ATR-IR tests were run to follow the evolution of the reactants and intermediates over the course of the hydrogenation reaction. The structure/ function relationships of different Pd ensembles in hydrogenation reactions were elucidated with these studies. The reactivity and characterization of the PdAu alloy nanoparticles with ATR-IR, STEM and XAS will be presented.

Acknowledgments: This work was supported as part of the Integrated Mesoscale Architectures for Sustainable Catalysis, an Energy Frontier Research Center funded by the U.S. Department of Energy, Office of Science, Basic Energy Sciences, under award #DESC0012573.

References

1. M. B. Boucher, B. Zugic, G. Cladaras, J. Kammert, M. D. Marcinkowski, T. J. Lawton, E. C. H. Sykes and M. Flytzani-Stephanopoulos, *Phys. Chem. Chem. Phys.*, 2013, **15**, 12187â€“12196.
2. J. Liu, J. Shan, Felicia R. Lucci, S. Cao, E. C. H. Sykes, M. Flytzani-Stephanopoulos, 2017, In Submission.

493593 Thermally Stable Single Atom Pt/m-Al_2O_3 for CO Oxidation and the Selective Hydrogenation of 1,3-Butadiene

Abstract:

Element sustainability has become a global issue. Maintaining current consumption rate, there are 22 elements facing depletion within the coming 50 years, including almost all Platinum Group metals that are crucial catalyst component[1]. Despite the critical role of Pt in catalysis[2], it is rare, in short supply in recent years, and has no adequate alternatives. In this regard, single-atom catalysts (or atomically dispersed catalysts), in which all the metal atoms are exposed on the support available for catalytic reactions, could help to address the problem[3]. The electronic properties of isolated metal atoms may be distinctly different from the active sites in bulk materials and nanoparticles, potentially triggering innovative applications and enabling more effective usage of noble metal elements[4].

A challenge in the development of single-atom catalysts is the increasing difficulty to stabilize single-atom species under drastic reaction conditions. Many industrially important catalytic processes involving Pt catalysts, such as reforming of hydrocarbons in petroleum refineries, are operated at several hundred degrees under oxidative or reductive atmosphere[5]. In this context, single-atom catalysts that are able to withstand harsh reaction conditions are highly desirable. Recent studies highlight the importance of manipulating the interactions between metal atoms and the host support to achieve high stability without losing reactivity in single-atom catalysis.

Figure 1 | Schematic illustration of the 0.2Pt/m-Al_2O_3-H_2 synthesis process.

Herein, we report a synthetic strategy for Pt single-atom catalysts with outstanding stability in a series of reactions under demanding conditions. The Pt-Al-O system is chosen because Al_2O_3 is a common support for Pt in industrial and environmental applications. The catalyst was prepared by a modified sol-gel solvent vaporization self-assembly method[6], followed by calcination in air and reduction with H_2 (Fig. 1).The Pt atoms are firmly anchored in the internal surface of mesoporous Al_2O_3 enriched with coordinatively unsaturated pentahedral Al^{3+} centers. Activity was fully maintained in CO oxidation after 60 cycles between 100 °C and 400 °C over a one-month period without any metal aggregation or support deterioration. The catalyst also kept its structural integrity and excellent performance for the selective hydrogenation of 1,3-butadiene after exposure to a reductive atmosphere at 200 ¡ãC for 24 h. Compared to commercial Pt nanoparticle catalyst on Al_2O_3 and control samples, this system also exhibited significantly enhanced stability and performance for *n*-hexane hydro-reforming at 550 ¡ãC for 48 h.

[1] A. J. Hunt, A. S. Matharu, A. H. King, J. H. Clark, *Green. Chem.* **2015**, *17*, 1949-1950.

[2] *Forecast of Platinum SUPPLY & DEMAND IN 2015*, Johnson Matthey, **2015**.

[3] a) X. F. Yang, A. Wang, B. Qiao, J. Li, J. Liu, T. Zhang, *Acc. Chem. Res.* **2013**, *46*, 1740-1748; b) J. M. Thomas, *Nature* **2015**, *525*, 325-326; c) M. Flytzani-Stephanopoulos, B. C. Gates, *Annu. Rev. Chem. Biomol. Eng.* **2012**, *3*, 545-574.

[4] a) F. Dvorak, M. Farnesi Camellone, A. Tovt, N. D. Tran, F. R. Negreiros, M. Vorokhta, T. Skala, I. Matolinova, J. Myslivecek, V. Matolin, S. Fabris, *Nat. Commun.* **2016**, *7*, 10801; b) P. Liu, Y. Zhao, R. Qin, S. Mo, G. Chen, L. Gu, D. M. Chevrier, P. Zhang, Q. Guo, D. Zang, B. Wu, G. Fu, N. Zheng, *Science* **2016**, *352*, 797-801; c) J. Liu, F. R. Lucci, M. Yang, S. Lee, M. D. Marcinkowski, A. J. Therrien, C. T. Williams, E. C. Sykes, M. Flytzani-Stephanopoulos, *J. Am. Chem. Soc.* **2016**, *138*, 6396-6399.

[5] S. H. Joo, J. Y. Park, C. K. Tsung, Y. Yamada, P. Yang, G. A. Somorjai, *Nat. Mater.* **2009**, *8*, 126-131.

[6] Q. Yuan, A. X. Yin, C. Luo, L. D. Sun, Y. W. Zhang, W. T. Duan, H. C. Liu, C. H. Yan, *J. Am. Chem. Soc.* **2008**, *130*, 3465-3472.

497895 Interactions of Water and Chiral Properties of a Single Site Pt "29" Cu_xo/Cu(111) Catalyst

Abstract:

Interactions of Water and Chiral Properties of a Single Site Pt "29" Cu_xO/Cu(111) Catalyst

Kyle Groden, Alyssa Hensley, Alex C. Schilling, Andrew Therrien, Renqin Zhang, E. Charles H. Sykes, Jean-Sabin McEwen

Figure 1. Illustration of the excellent agreement between the simulated STM image of the "29" structural oxide model and the experimental STM image. The experimental and the simulated STM image were taken at -0.5 V. The images are 3.5× 3.5 nm^2. The simulated image was done using the Tersoff-Hamann approach.

Single-site catalysts provide a promising means of using precious catalytic metals to a high degree of efficiency. Their design greatly reduces the metal loading and allows the atomically dispersed metal to actively participate in the underlying reaction of interest. Although the catalytic activity of single atoms has been under scrutiny[1], recent studies unambiguously show that individual Pt atoms on a "29" copper oxide are able to oxidize CO at low temperatures[2]. The model system for the "29" copper oxide is shown in Figure 1, which was obtained through an extensive collaboration between experiment and theory[3,4]. To further investigate the activity of this system, its interactions with water were examined from first principles. This is in response to the experimental work, which demonstrates that there is interchange of isotopically labelled surface oxygen and that of gas phase water. To explain the experimental results, a reaction mechanism was proposed and a corresponding microkinetic model was constructed to model the desorption kinetics.

We also examined the chiral properties of the "29" copper oxide support used for the previously mentioned work. Imaging has determined 6 different chiral configurations of the surface Cu_2O rings, each of which may behave in a chemically unique manner. This behavior may restrict the interactions of enantiomeric reactants to specific surface configurations, potentially leading to heterogeneous selectivity. As most reactions requiring any degree of enantioselectivity are typically performed homogeneously, this possible heterogeneous selectivity due to the surface orientation would be a powerful characteristic of this catalytic system. Our investigation is performed again using density function theory with propylene oxide as the enantiomeric probe molecule.

References:

1. Ding, K.; Gulec, A; Johnson, A. M.; Schweitzer, N. M.; Stucky, G. D.; Marks, L. D. Stair, P. C. Science 2015, 350, 189–192.

2. Therrien, A. J.; Hensley, A. J. R; Marcinkowski, M. D.; Lucci, F. R.; Coughlin, B.; Schilling, A.; McEwen, J.-S.; Sykes, E. C. H. (in preparation).

3. Hensley, A. J. R.; Therrien, A. J.; Zhang, R.; Marcinkowski, M. D.; Lucci, F. R.; Sykes, E. C. H.; McEwen, J.-S. *The Journal of Physical Chemistry C* 2016, *120* (44), 25387–25394.

4. Therrien, A. J.; Zhang, R.; Lucci, F. R.; Marcinkowski, M. D.; Hensley, A.; McEwen, J.-S.; Sykes, E. C. H. *The Journal of Physical Chemistry C* 2016, *120* (20), 10879–10886.

500136 Spectroscopic Signatures and Reactivity of CO Adsorbed to Atomically Dispersed Pt Atoms, Pt Oxide Clusters, and Metallic Pt Clusters on Anatase TiO_2

Abstract:

Due to their broad application in industrial chemical conversion processes and low abundance in the earths crust, significant efforts have been dedicated to developing efficient approaches for utilizing Pt-group metals as catalysts. The primary approach for optimizing Pt-group metal utilization in catalysis is synthesizing metal active sites that exhibit 100% dispersion - all metal atoms exposed to catalytic environment. However, there are significant challenges associated with the controlled synthesis and structural characterization of perfectly dispersed (this could be single metal atoms on the support, or sub-nanometer metal clusters) Pt-group metal species, which has led to a variety of inconsistent conclusions about their reactivity.

In this talk we will discuss a rigorous comparison of the reactivity of single Pt atoms, sub-nanometer PtO_x clusters, and sub-nanometer Pt^0 clusters supported on anatase TiO_2 for CO adsorption and catalytic CO oxidation. Control over the synthesized structure was achieved through the use of strong electrostatic adsorption synthesis protocols on extremely high surface area TiO_2 (290 m^2/g) at dilute synthesis conditions and varied Pt weight loadings (0.025-1%), combined with controlled oxidative or reductive pre-treatments. Using correlated IR spectroscopy and scanning transmission electron microscopy (STEM) we identify characteristics signatures of CO bound to cationic single Pt atoms, PtO_x clusters and Pt^0 clusters. Interestingly, it was identified that while single Pt atoms on TiO_2 and PtO_x clusters both exhibit similar electronic and local coordinative environments, they exhibit drastic variation in CO adsorption energies, with single Pt atoms adsorbing CO weakly and PtO_x clusters adsorbing CO even more strongly than Pt^0 clusters. Using the correlated STEM-IR characterization we compare the CO oxidation reactivity of single Pt atoms and metallic clusters under conditions of strict kinetic control where we find that both types of active sites exhibit an identical reaction mechanism, while single Pt atoms are more active on Pt mass and Pt surface area bases. We discuss the ramification of these findings in the context of designing stable catalysts with optimal Pt-group metal utilization.

500029 Reversible Transformation from Pt Single Atoms to Sub-Nanometer Particles for Low Temperature CO Oxidation

Abstract:

Reversible Transformation from Pt Single Atoms to Sub-Nanometer Particles for Low Temperature CO Oxidation

Xavier Isidro Pereira Hernández[1+], Andrew DeLaRiva[2+], Haifeng Xiong[2], Eric J. Peterson[2], Deepak Kunwar[2], Berlin Sudduth[1], Yong Wang[1,3,*] and Abhaya K. Datye[2,*]

[1] Voiland School of Chemical Engineering and Bioengineering, Washington State University, Pullman, WA 99164.

[2] Department of Chemical and Biological Engineering and Center for Micro-Engineered Materials, University of New Mexico, Albuquerque, New Mexico 87131, USA.

[3] Institute for Integrated Catalysis, Pacific Northwest National Laboratory, Richland, WA 99354.

*Corresponding authors: Prof. A. K. Datye (datye@unm.edu) and Prof. Y. Wang (yong.wang@pnnl.gov)

[+] These authors contributed equally to this work

Abstract: While single atom catalysts can provide high catalytic activity and selectivity, application in industrial catalysts demands long term performance and the ability to regenerate the catalyst. We have investigated factors that lead to improved catalytic reactivity of Pt for low temperature CO oxidation. Single atom Pt/CeO_2 becomes active for CO oxidation under lean condition only at elevated temperatures, because CO is strongly bound to ionic Pt sites. Activating the catalyst with a reduction treatment, even under mild conditions, leads to onset of CO oxidation activity even at room temperature. This high activity state involves the transformation of mononuclear Pt species to sub-nanometer sized Pt particles. Under oxidizing conditions, the Pt can be restored to its stable, single atom state. The key to facile regeneration is the ability to create mobile Pt species and suitable trapping sites on the support, making this a prototypical catalyst system for industrial application of single atom catalysis.

1. Introduction

One of the main objectives during the synthesis of heterogeneous catalysts is to have the active transition metal in an atomically dispersed form, in order to maximize atom efficiency. This can be achieved either through trapping the atomic species [1] or by limiting the transition metals weight loading [2], [3]. However, in most cases, increases in temperature under reaction conditions leads to sintering and loss of this dispersed form. Therefore, heterogeneous catalysts are based on catalyst supports that aid in the reaction and/or provide stability to the transition metal. Ceria is just one of the numerous supports that are used for Platinum Group Metals (PGM). Ceria is a commonly used component in automotive exhaust catalysts, often used as an additive for oxygen storage and in lean NOx traps. The unique ability of Ceria to change from Ce^{+4} to Ce^{+3} and the facile synthesis methods that can produce various morphologies make it ideal as a support for Pt. We show here a ceria-supported Pt single-atom catalyst with great stability even at temperatures as high as 800°C, as previously reported [4], and its activation (and structural changes) to achieve high activity for CO oxidation (low CO oxidation light-off temperatures) while maintaining its stability. Furthermore, we show that the catalyst can be restored to its initial state by high-temperature treatment under oxidative conditions, granting it flexibility to operate under the desired conditions. We propose that the Pt/Ceria catalyst be considered as a prototypical system that can toggle between an atomically dispersed form and self-assemble into a high activity state, providing the design elements for a thermally stable catalyst for industrial applications.

2. Experimental

The catalysts were synthesized via Dry Impregnation (DI) [5]. The activity measurements were performed using a stainless-steel reactor tube and monitored by a micro Gas Chromatograph (GC). The catalyst was characterized by means of High-Angle Annular Dark-Field Scanning Transmission Electron Microscopy (HAADF-STEM), Diffuse Reflectance Infrared Fourier Transform Spectroscopy (DRIFTS), X-Ray Photoelectron Spectroscopy (XPS), Raman Spectroscopy and X-Ray Powdered Diffraction (XRD).

3. Results and Discussion

The activity measurements for the As Synthesized catalyst (prior to activation) and the activated catalysts are observed on Figure 1. The activation of the catalyst is performed at 275°C under CO flow. It is seen that when the Pt/CeO_2 catalyst is activated, the light-off temperature decreases drastically (approximately 200°C).

Figure 1. CO oxidation reaction on a 1wt.%Pt/CeO_2 catalyst, As Synthesized and Activated.

DRIFTS measurements were performed on the catalyst As Synthesized, Activated and Regenerated and are observed on Figure 2. The catalyst as synthesized shows a very well defined peak at 2095 cm^{-1}, assigned to CO adsorbed on ionic Pt. Other authors have reported the adsorption of CO on ionic Pt at similarly high wavenumbers [6], [7]. After being activated, the presence of two peaks at 2086 cm^{-1} and 2073 cm^{-1} is observed. These two peaks are assigned to CO adsorbed linearly on Pt nanoparticles (NPs). At the same time, the Activated catalyst shows a peak at 2110 cm^{-1}, assigned to CO adsorbed on ionic Pt. The reason for the blueshift between the As Synthesized and Activated catalysts is associated to the presence of CO dipole-dipole interactions in the latter one, due to the existence of Pt NPs near the Pt single atoms. After performing a regeneration at 800°C under O_2 flow, the catalyst shows the same behavior as the As Synthesized catalyst, suggesting it can be restored to the initial state.

Figure 2. CO oxidation reaction on a 1wt.%Pt/CeO_2 catalyst, As Synthesized, Activated and Regenerated.

To further study the state of the catalyst, XPS measurements were performed and can be observed on Figure 3. The Ce 3d region shows that changes on the support start occurring during activation at temperatures as low as 125°C, in which Ce^{4+} is reduced to Ce^{3+}. Very little difference is observed between activation at 125°C and 275°C. Once regenerated, all the Ce^{3+} is converted to Ce^{4+}, although, this behavior is expected since CeO_2 can be oxidized relative easily. Analyzing the Pt 4f region, it can be observed that unlike on CeO_2, activation at 125°C only reduces partially the Pt^{2+} species, which are present in the catalyst As Synthesized. Activating the catalyst at 275°C shows a more well defined peak at 71.2 eV, associated to Pt^0. Once again, it is seen that the regeneration step restores the Pt species to their original state.

Figure 3. XPS spectra of the 1wt.% Pt/CeO2 catalyst. Left: Ce3d region. Right: Pt4f region. As Synthesized (Black), Activated at 125°C (Red), Activated at 275°C (Blue), Regenerated at 600°C (Green).

To confirm the results obtained by the previous characterization techniques, HAADF-STEM images on the Activated and Regenerated catalyst are presented on Figure 4. The Activated Catalyst (left) shows the presence of Pt species of 1 nm or smaller size (bright spots). These sub-nanometer particles are not observed anymore after regeneration (right) due to the redispersion of the Pt atoms.

Figure 4. HAADF-STEM images of the 1wt.% Pt/CeO_2 catalyst. Left: Activated catalyst. Right: Regenerated catalyst.

Conclusions

We presented a pathway to synthesize a Pt single-atom catalyst that is stable at temperatures as high as 800°C. This catalyst can be activated by a reduction treatment, leading to high activity for CO oxidation at low temperatures. Furthermore, the catalyst shows the capability to be restored to its initial state by regenerating it under an oxidation treatment at high temperatures, granting it flexibility to operate under the desired conditions in an industrial environment.

References

[1] X. Li *et al.*, "Single-Atom Pt as Co-Catalyst for Enhanced Photocatalytic H2 Evolution," *Adv. Mater.*, vol. 28, no. 12, pp. 2427–2431, 2016.

[2] M. Moses-Debusk *et al.*, "CO Oxidation on Supported Single Pt Atoms: Experimental and ab Initio Density

Functional Studies of CO Interaction with Pt Atom on θ-Al2O3(010) surface," *J. Am. Chem. Soc.*, vol. 135, no. 34, pp. 12634–12645, 2013.

[3] B. Qiao *et al.*, "Single-atom catalysis of CO oxidation using Pt1/FeOx," *Nat. Chem.*, vol. 3, no. 8, pp. 634–641, 2011.

[4] J. Jones *et al.*, "Thermally stable single-atom platinum-on-ceria catalysts via atom trapping," *Science (80-.).*, vol. 353, no. 6295, pp. 150–154, 2016.

[5] K. De Jong, *Synthesis of Solid Catalysts*. Wiley-VCH Verlag GmbH & Co. KGaA, 2009.

[6] K. Ding *et al.*, "Identification of Active Sites in CO Oxidation and Water-Gas Shift over Supported Pt Catalysts," *Science (80-.).*, vol. 350, no. 6257, pp. 189–192, 2015.

[7] S. C. Ammal and A. Heyden, "Water-Gas Shift Activity of Atomically Dispersed Cationic Platinum versus Metallic Platinum Clusters on Titania Supports," *ACS Catal.*, vol. 7, no. 1, pp. 301–309, 2017.

487347 Hydrodeoxygenation of Anisole As Bio-Oil Model Compound over Supported Non-Sulphided CoMo Catalysts: Effect of Co/Mo Ratio and Support

Abstract:

Hydrodeoxygenation of anisole as bio-oil model compound over supported non-sulphided CoMo catalysts: effect of Co/Mo ratio and support

Chanakya Ranga[a], Rune Lødeng[b], Vaios I. Alexiadis[a], Joris W. Thybaut[a*]

[a]Laboratory for Chemical Technology, Ghent University, Technologiepark 914, B-9052 Ghent, Belgium.

[b]SINTEF Materials & Chemistry, Department of Kinetics and Catalysis, N-7465 Trondheim, Norway.

*Corresponding Author: Joris W. Thybaut, E-mail: Joris.Thybaut@UGent.be

Introduction

Fast pyrolysis is considered as one of the prominent approaches to obtain liquid organic products with high yield from lignocellulosic biomass. However, pyrolysis oils (bio-oil) cannot be directly used as a fuel or fuel additive due to the high oxygen content [1], which results in undesirable properties such as low thermal and chemical stability, low heating value, immiscibility with conventional, fossil fuels and high acidity. These oils, hence, require upgrading via oxygen removal and increase in hydrogen content. Catalytic hydrodeoxygenation (HDO) is one of the most promising routes to upgrade pyrolysis oils for producing liquid transportation fuels [2].

The complexity of fast pyrolysis oil has prompted the use of model compounds such as phenolics, furans, ethers, acids etc. to study the intricacies of hydrodeoxygenation [1, 3]. Anisole, because of its methoxy group, has already been widely investigated as a model compound for lignin derived bio-oil [4]. The low sulfur content in bio-oil renders the use of traditional sulfided catalysts (NiMoS and CoMoS) for hydroprocessing less suited, as they require the presence of sulfur to retain their activity and, hence, result in contaminated end products. To mitigate this issue, non sulfided transition metal (Ni, Co, Mo) catalysts on various supports have been investigated for hydrotreatment of bio-oil model compounds [5-8], yet significant challenges remain in improving the catalyst activity, selectivity and stability. Our previous work on MoO_3/ZrO_2 catalysts provided insight in the effect of catalyst structural composition with varying preparation conditions on catalyst performance [7-8]. The present work explores the intricacies of anisole HDO reaction pathways upon addition of Co to the Mo catalysts by assessing their intrinsic kinetics performance in non-sulphided form. The catalyst performance is correlated with properties such as reducibility, active site dispersion, crystallite size, metal-support interaction.

Results

A series of supported CoMo catalysts were prepared with varying Co/Mo ratio (0.25, 0.58, and 1.07) while keeping the Mo loading between 8.3 – 10.2 wt%. A sequential incipient wetness impregnation (with Mo being introduced first) method using aqueous solutions of the corresponding precursor salts was employed during the synthesis procedure. Two different supports, i.e., Al_2O_3 and ZrO_2, were used. Material physicochemical characteristics were evaluated through ICP-OES, BET, (in-situ) XRD, H_2-TPR, NH_3-TPD, CO chemisorption, (S)TEM, and XPS techniques. The performance of these CoMo catalysts was tested for anisole HDO at gas phase conditions in a fixed bed tubular reactor in plug flow regime. A high-throughput kinetics screening set-up (HTK-S) [9] comprising 16 plug flow reactors with an internal diameter of 0.00211 m and a length of 0.85 m was used for the acquisition of intrinsic kinetic data.

CoMo/Al_2O_3 catalysts promote the transalkylation and isomerization reactions whereas, ZrO_2 supported ones promote mainly hydrodeoxygenation and transalkylation reactions. Catalyst stability as well as anisole conversion increased with Co/Mo ratio. Figure 1 displays the anisole conversion and product selectivity for CoMo/ZrO_2 and CoMo/Al_2O_3 catalysts with Co/Mo = 1.07 during a stability test (TOS of 50 h). Al_2O_3 supported catalysts displayed higher anisole conversions compared to ZrO_2 ones at the tested operating conditions. Further reaction pathway elucidation is done with the use of intrinsic kinetic measurements varying space time (50 – 250 kg_{cat} s mol^{-1}), temperature (523 - 623 K) and H_2/anisole inlet molar ratio (50-150 mol mol^{-1}) with best performing catalysts. Catalyst deactivation causes will be explored through characterization (O_2-TPO, XRD, XPS, (S)TEM) of spent catalyst samples.

Addition of Co improved the catalyst stability compared to the MoO_3 catalysts in our previous study [7-8]. Catalyst activity-structure correlations drawn from present study can serve significantly in guiding the synthesis of next generation catalysts with better catalyst stability and activity during anisole HDO.

(a)

(b)

Figure 1 Time on stream (TOS) experiment with (a) CoMo/ZrO_2 and (b) CoMo/Al_2O_3 at T = 573 K, P_T = 0.5 MPa, H_2/anisole = 50 mol mol^{-1}, space-time = 220 kg_{cat} s $mol^{-1}{}_{anisole}$. Co/Mo = 1.07 for both catalysts.

Conclusions

Supported CoMo catalysts observed to be relatively stable than the MoO_3 catalysts at the tested operating conditions during anisole HDO over 50 h TOS. Al_2O_3 supported catalysts exhibited higher anisole conversions and relatively lesser stability compared to ZrO_2 ones at the tested operating conditions. Catalysts stability as well as HDO activity observed during an extensive experimental campaign are correlated to the key properties such as active site dispersion, reducibility, crystallite size, and metal-support interaction. This will further enhance our understanding of the catalyst activity, stability and selectivity towards anisole HDO and provide valuable insights into the corresponding reaction pathways.

Acknowledgments

This research work is supported by the Innovative Catalyst Design for large-scale sustainable processes (*i*-CaD) project, which is a European Research Council (ERC) consolidator grant, by the European Commission in the 7th Framework Programme (GA n° 615456). It also fits in the framework of the FAST industrialization by Catalyst Research and Development (FASTCARD) project, which is a large scale collaborative project supported by European Commission in the 7th Framework Programme (GA n° 604277).

Keywords:

hydrodeoxygenation, reaction pathways, catalyst stability

References

1. Ruddy, D. A.; Schaidle, J. A.; Ferrell III, J. R.; Wang, J.; Moens, L.; Hensley, J. E., *Green Chemistry* **2014,** *16* (2), 454-490.

2. Li, C.; Zhao, X.; Wang, A.; Huber, G. W.; Zhang, T., *Chem. Rev* **2015,** *115* (21), 11559-11624.

3. Furimsky, E., *Applied Catalysis A: General* **2000,** *199*, 147–190.

4. He, Z.; Wang, X., *Catalysis for Sustainable Energy* **2013,** *1* (1), 28–52.

5. Sankaranarayanan, T. M.; Berenguer, A.; Ochoa-Hernandez, C.; Moreno, I.; Jana, P.; Coronado, J. M.; Serrano, D. P.; Pizarro, P., *Catalysis Today* **2015,** *243*, 163-172.

6. Zuo, H.; Liu, Q.; Wang, T.; Ma, L.; Zhang, Q.; Zhang, Q., *Energy & Fuels* **2012,** *26* (6), 3747-3755.

7. Lødeng, R.; Ranga, C.; Rajkhowa, T.; Alexiadis, V. I.; Bjørkan, H.; Chytil, S.; Svenum, I. H.; Walmsley, J.; Thybaut, J. W., *Biomass Conversion and Biorefinery* **2017**, 1-17.

8. Ranga, C.; Lødeng, R.; Alexiadis, V. I.; Rajkhowa, T.; Bjørkan, H.; Chytil, S.; Svenum, I. H.; Walmsley, J.; Detavernier, C.; Poelman, H.; Van Der Voort, P.; Thybaut, J. W., *Chemical Engineering Journal* **2017,** (submitted).

9. Van der Borght, K.; Toch, K.; Galvita, V. V.; Thybaut, J. W.; Marin, G. B., *Catalysts* **2015,** *5*, 1948-1968.

497051 Sulfur-Tolerant Molybdenum Carbide Catalysts Enabling Low-Temperature Stabilization of Fast Pyrolysis Bio-Oil

Abstract:

Fast pyrolysis is a cost effective lignocellulosic biomass liquefaction technology. However, the liquid product, bio-oil, is not suitable for direct use as transportation fuels or as a crude oil substitute due to undesirable qualities such as low heating value, corrosiveness, and instability directly related to its high oxygen and water content. The quality of bio-oils can be improved by catalytic upgrading (e.g., hydroprocessing, cracking) to eliminate oxygenated functionalities.

Developing catalytic bio-oil upgrading technology has proven challenging due in large part to short catalyst lifetime. Bio-oil contains a large quantity of carbonyl compounds which, due to high tendency toward polymerization reactions, lead to rapid catalyst fouling and bed plugging. Stabilization of bio-oil by converting reactive carbonyls prior to higher temperature upgrading is therefore critical. Low-temperature hydrogenation of carbonyl fractions has been recognized as a promising approach to bio-oil stabilization. While reduced precious metal catalysts such as ruthenium are effective in low-temperature hydrogenation reactions, they are highly sensitive to sulfur species present in biomass-derived feedstocks. Therefore, to address rapid performance loss due to S-poisoning of active sites, sulfur management strategies need to be implemented (e.g., sulfur guard, catalyst regeneration) which entails increased process cost and complexity.

Molybdenum carbides have shown potential in various biomass conversion reactions especially with high selectivity towards C-O bond cleavage. In this work, we have assessed Mo carbides in low-temperature hydrogenation of bio-oil with particular emphasis on understanding their sulfur tolerance. Our study revealed that molybdenum carbides are effective in low-temperature conversion of carbonyl compounds. Furthermore, carbides showed excellent sulfur tolerance, and structural robustness with respect to morphology, crystallography, and bulk composition. These properties enabled carbides to maintain catalytic performance in both aqueous-phase furfural (380 ppm S, 150 °C, 120 bar, batch reactor) and real bio-oil (52 ppm S, 140 °C, 124 bar, flow reactor, 100 h TOS) reactor feeds. Due to surface bifunctionality (metal-acid), carbides were able to catalyze both hydrogenation and C-C coupling reactions retaining most of carbon atoms in liquid products as more stable and higher molecular weight compounds. These findings (sulfur tolerance, durability, H and C atom economy) indicate that molybdenum carbides can provide a catalyst solution for the development of commercially viable bio-oil stabilization technology.

492815 Controlling Selectivity of Bio-Oil Model Compound Upgrading with Metal Promoters on Molybdenum Carbide

Abstract:

The upgrading of thermochemical biofuels could lead to cleaner, more homogeneous fuels[1], however, the processes and reactions are not very selective. For research described in this paper, we examine the use of various promoters on molybdenum carbide to improve the reaction selectivities. Crotonaldehyde is used as a model compound because it possesses multiple functionalities commonly found in pyrolysis bio-oils, while also allowing us to probe how changes to the catalyst affect selectivity to several different potential products[2].

In previous work, we have shown that potassium addition to Mo_2C catalyst increased the base site density and decreased the acid site density, and consequently shifted the crotonaldehyde and acetic acid conversion selectivities in the presence of H_2. The results suggested that the increase in base sites was responsible for the increase in selectivities to 3-butenal and acetone, respectively. In this paper, we report on how other metal promoters, including Fe, shift the crotonaldehyde upgrading selectivity relative to Mo_2C.

The Mo_2C catalysts were synthesized by a temperature-programmed reaction method as described previously[3]. Metal salt was added to the native (unpassivated) Mo_2C catalyst by incipient wetness in an oxygen-free, inert atmosphere. The catalysts were dried in H_2 for 2 hrs at 110‹C, reduced in H_2 for 4 hrs at 450‹C, and then passivated in 1% O_2/He for at least 6 hrs at room temperature.
All of the catalysts were characterized using x-ray powder diffraction, N_2 physisorption, and inductively coupled plasma optical emission spectrometry. Base and acid site concentrations were measured via temperature programmed desorption of CO_2 and NH_3. The activities and selectivities were measured at atmospheric pressure and 275-350‹C in a flow reactor with a H_2/crotonaldehyde ratio of 6. The ratio was selected because it is twice the stoichiometric amount of H_2 required for complete saturation and deoxygenation of crotonaldehyde to butane.

While K shifted the selectivity to 3-butenal, Fe on Mo_2C favored the formation of butene and suppressed formation of butyraldehyde at high temperatures. The model compound results presented in this paper will demonstrate the ability to tune the product selectivity of molybdenum carbide catalysts for specific desired reaction pathways by adding various metal promoters. Combining these results to promote Mo_2C with a mixture of these promoters could offer a powerful way to tune the selectivity of Mo_2C for bio-oil upgrading, depending on the desired character of the upgraded product.

[1] (a) Czernik, S.; Bridgwater, A.V.; *Energy & Fuels*, **2004**, *18*, 590-598; (b) Zhang, Q.; Chang, J.; Tiejun, W.; Yung, X. *Energy Conversion and Management*, **2007**, *48*, 87-92.

[2] Mullen, C.A.; Strahan, G.D.; Boateng, A.A. *Energy & Fuels*, **2009**, *23*, 2707-2718.

[3] Schaidle, J.A.; Schweitzer, N.M.; Ajenifujah, O.T.; Thompson, L.T. *J. Catalysis*, **2012**, *289*, 210-217.

490490 Reductive Conversion of Lignin with Copper-Doped Catalysts

Abstract:

The lignin component of biomass has potential as a renewable source for industrially useful aromatic chemicals. While technologies for the selective conversion of the carbohydrate components have been successful, lignin is generally treated as waste and burned for low grade heat. We have developed a catalytic system, based on a copper-doped porous oxide, which can reductively disassemble lignin in supercritical methanol with little to no char formation. While the copper-doped porous oxide catalyzes C-O hydrogenolysis of aryl-ether bonds linking lignin monomers, it also catalyzed ring methylation and hydrogenation, leading to lignin disassembly product proliferation. We have found that adding dimethyl carbonate can significantly suppress the hydrogenation of the phenolic intermediates responsible for much of the undesirable product diversity by methylation of phenolic hydroxyl groups to form a more stable aryl ether species. Using the methanol and dimethyl carbonate solvent system with the copper-doped porous oxide catalyst, O-methylation of phenolic hydroxyl groups was effective at increasing aromatic yields from a number of lignin model polymer systems and poplar organosolv lignin. These results demonstrate the promise of our copper-doped porous oxide catalyst when used in the presence of methanol and dimethyl carbonate to facilitate aromatic compound production from lignin.

498094 Hierarchical Nickel Incorporated USY Zeolites for Hydrodeoxygenation of Lignin-Derived Pyrolysis Oil Model Compounds

Abstract:

Pyrolysis of biomass and agricultural wastes is a readily achieved method for production of sustainable fuels and platform chemicals. While pyrolysis oil yields of up to 75 wt.% of the original biomass have been demonstrated,[1] raw pyrolysis oil is not suitable for combustion fuel applications due to poor stability and low heating value. Hydrodeoxygenation (HDO) of the oil fraction is a common method for removal of pyrolysis oil oxygen, which is eliminated in the form of water, thus preserving atomic carbon in the pyrolysis oil. Traditional HDO catalysts include NiMo, CoMo and precious metals, however the former require sulfidation and the latter are costly. Here, we propose Ultra-stable Y (USY) zeolite supported Ni catalysts for liquid-phase HDO of anisole, a common product formed from lignin pyrolysis.

Traditional zeolite catalysts are highly microporous, and as a result may limit diffusion of reactants and products to or from catalyst acid sites. Moreover, the low external surface area may result in increased micropore blockage after Ni impregnation. A variety of mesoporous USY zeolites were created using top-down alkaline desilication (DS), desilication with a pore-directing agent (DS-PDA)[2] and surfactant-mediated desilication (SA)[3]. As shown in Figure 1(A), DS formed a broad pore-size distribution with very little micropore volume, whereas DS-PDA preserved some microporosity and produced a narrower range of mesopores. The SA method preserved about 90% of the micropore volume and produced a narrow range of mesopore size at about 30 Å. Material acidity and crystallinity decreased after all treatments, and a clear correlation between material crystallinity, acidity and micropore volume was observed. Figure 1(B) shows a visualization of the types of pores formed from each treatment. Alkaline treatment strength and PDA concentration were varied in order to determine the effect of each variable on zeolite properties.

Ni was impregnated by the incipient wetness method, and NiO crystal size was approximately 30 nm, determined by application of the Scherrer equation to X-ray diffraction (XRD) patterns. After reduction, the Ni crystal size decreased significantly on the more microporous materials, but remained high on the larger-pore materials. Larger crystal sizes and greater interaction with the DS and DS-PDA zeolites was observed from temperature programmed reduction (TPR).

Each material was evaluated as a catalyst for the liquid-phase HDO of anisole at 200 °C and 750 psi. Overall reaction rate increased in the order Ni-SA-USY<Ni-USY<Ni-DS-USY<Ni-DS-USY-PDA, shown in Figure 1(C). A simplified kinetic model of the anisole reaction pathways was proposed, and kinetic parameters were fit to the data and normalized to the number of Ni and acid sites. Analysis of the fitted kinetic parameters showed that the larger mesoporous zeolites enhanced access of reactant anisole to the Ni sites for more efficient hydrogenation of the aromatic ring, whereas acid-catalyzed deoxygenation of the hydrogenated intermediate was about an order of magnitude faster. The Ni-DS-USY-PDA was the most effective catalyst overall, as hydrogenation of the aromatic ring was shown to be the rate determining step in this reaction system.

Figure 1. (A) BJH pore size distribution of mesoporous USY materials. (B) Visualization of each mesoporous USY material, white space indicates pore area. (C) Reaction rates for anisole hydrodeoxygenation (bars correspond to left axis), in addition to final anisole conversion and mass balance (corresponding to right axis).

References

[1] O. Onay, Fuel 86 (2007) 1452–1460.

[2] D. Verboekend, G. Vile, J. Perez-Ramirez, Cryst. Growth Des. 12 (2012) 3123–3132.

[3] J. García-Martínez, M. Johnson, J. Valla, K. Li, J.Y. Ying, Catal. Sci. Technol. 2 (2012) 987.

500416 *Operando* Solid-State NMR Study of Hydrogenolysis of Lignin Ether Linkages

Abstract:

Lignin is the most abundant and readily available feedstocks for bio-derived aromatics. Under mostly reductive conditions, transition metals like Ni and Pd catalyze the hydrogenolysis of lignin ether linkages in the liquid phase. However, the conversion chemistry is lack of molecular understanding especially at the solid-liquid interface, mostly owing to the demanding reaction conditions including high temperature/pressure. Customized high temperature/pressure solid-state NMR rotors were adapted, allowing *operando* solid-state NMR measurements of conversion of lignin catalyzed with supported metal catalysts. For example, benzylphenylether (BPE) as the a-O-4 model compound was studied with supported Ni catalysts at 120-225 °C. Kinetic and mechanistic study was performed with selective isotope labeling (^{13}C, ^{2}H) of both reactants and solvent. The pressure of 50 bar H_2 allows the simultaneous study of the further hydrogenation of phenol to cyclohexanol while toluene showed much higher stability with almost no conversion to methylcyclohexane. When no extra hydrogen gas is charged in the rotor, solvent 2-proponal can also serve as the hydrogen source and Ni can catalyze the transfer hydrogenation reaction with acetone formed and detected *in situ*.

490499 in-Situ Spectroscopy of Lignin Disassembly

Abstract:

Lignin (an aromatic macromolecule in biomass) is considered as waste and under-utilized. A more appealing and value-added use of lignin is its disassembly into aromatic platform chemicals, which can be achieved by appropriate combinations of catalyst, solvent, and reaction conditions. Mechanistic reaction network studies of lignin disassembly are challenging, because the reactions require relatively high temperatures (above 200 °C) and pressures (above 80 bar), as well as the co-existence of solid, liquid, and gas phases. We have used *in-situ* magic-angle-spinning (MAS) solid-state nuclear magnetic resonance (NMR) and small angle neutron scattering (SANS) investigations to study catalytic disassembly of lignin at multiple length scales and in real-time. Our project leverages unique high-temperature and -pressure MAS NMR technology at the Pacific Northwest National Laboratory (PNNL) and a pressure reaction cell for collecting time-resolved neutron scattering data during thermochemical treatment of biomass at Oak Ridge National Laboratory.

501759 UnderstandingÂ Catalyst Selectivity forÂ CO_2 Electro-Reduction

Abstract:

Developing new processes that can employ renewable electricity to produce fuels and chemicals using water and CO_2 as inputs could play an important role in a sustainable future. New catalysts are needed for these processes, particularly those that are selective to desired products, one of the principal challenges in the field. In this talk, we will discuss some of our latest efforts to understand catalyst selectivity for CO_2 electro-reduction, including in-depth studies of catalyst surface structure. We then leverage that knowledge to develop catalyst materials and interfaces with improved performance.

491315 Ag-Sn Bimetallic Catalyst with a Core-Shell Structure for CO_2 Reduction

Abstract:

Converting greenhouse gas carbon dioxide (CO_2) to value-added chemicals is an appealing approach to tackle CO_2 emission challenges. The chemical transformation of CO_2 requires suitable catalysts that can lower the activation energy barrier to minimize the energy penalty associated with the CO_2 reduction reaction. First-row transition metals are potential candidates as catalysts for electrochemical CO_2 reduction; however, their high oxygen affinity makes them easy to be oxidized, which could strongly affect the catalytic properties of metal-based catalysts.[1] Recently, we proposed a strategy to synthesize ultra-thin electrocatalysts using a core-shell nanostructure that contains a bimetallic core responsible for high electronic conductivity and an ultra-thin partially oxidized shell for catalytic CO_2 conversion.

In this presentation, we will show a model core-shell electrocatalyst that has a silver-tin bimetallic core and an ultra-thin partially oxidized tin oxide shell (denoted as AgSn/SnO_x) for CO_2 reduction to formate with high Faradaic efficiencies and high current densities.[2] A structure-activity correlation between the AgSn/SnO_x core-shell structure and CO_2 electrocatalytic activity was established by studying a series of Ag-Sn catalysts with different Ag/Sn compositions. More importantly, a volcano-like correlation between the bulk composition and electrocatalytic performance was observed, in which the optimal thickness of the partially oxidized SnO_x shell was ~1.7 nm.

References:

1. Lu, Q., & Jiao, F.* Electrochemical CO_2 reduction: electrocatalyst, reaction mechanism, and process engineering. *Nano Energy* 29, 439-456 (2016).

2. Luc, W., Collins, C., Wang, S. W., Xin, H. L., He, K., Kang, Y. J. & Jiao, F.* Ag-Sn bimetallic catalyst with a core-shell structure for CO_2 reduction. *Journal of the American Chemical Society* 139, 1885-1893 (2017).

501199 Electrochemical CO_2 Reduction over Cuag Bimetallic Electrodes and Well-Defined Surface Alloys with Enhanced Oxygenate Selectivity

Abstract:

Atmospheric CO_2 is a potential source of renewable carbon for the production of fuels and chemicals. However, for this process to be sustainable the hydrogen required must be derived from water and the necessary energy must be supplied by the sun. One approach to this goal is the utilization of renewable electricity to drive the electrochemical CO_2 reduction reaction (CO_2RR). Previous research has shown that the overall efficiency of CO_2RR and the distribution of generated products depend primarily on the electrocatalyst used as the cathode. Cu is the only monometallic electrocatalyst capable of reducing CO_2 into potential fuels or multi-carbon chemicals with a total Faradaic efficiency (FE) in excess of 1%. The principal products formed over polycrystalline Cu are H_2, CH_4, C_2H_4, and C_2H_5OH. However, at potentials cathodic of -1 V vs RHE excessive H_2 evolution leads almost exclusively to the methanation of CO_2. Since multi-carbon products are more valuable precursors to fuels and chemicals than CH_4, recent studies have focused on identifying factors governing the multi-carbon product selectivity. These studies have revealed that the multi-carbon product selectivity is sensitive to the surface morphology, electrolyte composition, and the applied potential. However, these studies have not managed to substantially reduce the fraction of current lost to the parasitic evolution of H_2.

Herein we report our investigations of electrochemical CO_2 reduction over CuAg bimetallic electrodes and well-defined surface alloys, which we have found to exhibit an enhanced activity for the evolution of products derived from CO relative to H_2 compared to pure Cu. Furthermore, we have found that these electrocatalysts display an unusually high selectivity for the formation of carbonyl-containing compounds, such as acetaldehyde. While this bimetallic system was investigated on the hypothesis that CO spillover from phase-segregated domains of Ag to Cu would increase the rate of C-C coupling, further characterization by photoelectron spectroscopy and electrochemical cycling led to a very different conclusion. The revised hypothesis was confirmed by synthesizing well-defined Cu+Ag surface alloys, which reproduced the activity and selectivity trends observed over the bimetallic electrodes. The results are explained in terms of an observable shift in the electronic properties of Cu and the resulting impact that these modifications have on the adsorption energies of key reaction intermediates.

493375 A Carbon Nanotube Supported Gold Catalyst for the Electroreduction of Carbon Dioxide (CO_2)

Abstract:

The electrochemical reduction of carbon dioxide (CO_2) to value added chemicals (such as formic acid, carbon monoxide (CO), ethylene, and ethanol) has been proposed as a potential strategy for the utilization of excess anthropogenic CO_2 emissions.[1, 2] In particular, the electroreduction of CO_2 to CO seems interesting, as CO (either alone or in combination with H_2 *i.e.,* syngas) can be used as a platform chemical to produce a variety of hydrocarbons. However, even after a few decades of research, electrochemical systems that can produce CO *via* the electroreduction of CO_2 at industrially relevant rates (partial current density for CO >150 mA cm^{-2}) and low overall cell overpotentials (absolute value <1V),[3] remain difficult to develop.

In this presentation, we will report a supported gold (Au) catalyst that exhibits remarkably high activity (partial current density for CO ~158 mA cm^{-2}) at low cell overpotentials (~-0.94 V), for the electroreduction of CO_2 to CO. The catalyst was synthesized *via* an *in situ* chemical reduction of the Au precursor (chloroauric acid) on a polybenzimidazole wrapped carbon nanotube support. The electrochemical characterization was performed in a flow electrolyzer to avoid mass transfer limitations associated with the low solubility of CO_2 in aqueous solutions.[4] Furthermore, we will also discuss the effect of electrolyte composition (concentration and pH) with respect to increasing the activity and lowering the overpotential required for the electroreduction of CO_2 to CO on the supported Au catalyst. The mechanistic implications of such a behavior will also be discussed.

References:

[1] A.M. Appel, J.E. Bercaw, A.B. Bocarsly, H. Dobbek, D.L. DuBois, M. Dupuis, J.G. Ferry, E. Fujita, R. Hille, P.J.A. Kenis, C.A. Kerfeld, R.H. Morris, C.H.F. Peden, A.R. Portis, S.W. Ragsdale, T.B. Rauchfuss, J.N.H. Reek, L.C. Seefeldt, R.K. Thauer, G.L. Waldrop, *Chem. Rev.* 2013, 113, 6621-6658.

[2] H.R.M. Jhong, S. Ma, P.J.A. Kenis, *Curr. Opin. Chem. Eng.* 2013, 2, 191-199.

[3] S. Verma, B. Kim, H.R.M. Jhong, S. Ma, P.J.A. Kenis, *ChemSusChem* 2016, 9, 1972-1979.

[4] D.T. Whipple, E.C. Finke, P.J.A. Kenis, *Electrochem. Solid-State Lett.* 2010, 13, B109-B111.

487255 Interplay of Mass Transfer and Local pH Effects in CO_2 Reduction Electrocatalysis on Cu Nanowires

Abstract:

Besides the catalyst structures, the electrocatalysis for CO_2 reduction is also dependent on the electrolyte and CO_2 partial pressure, which can affect the catalytic activity and selectivity via mass transfer and/or local pH effects. While protons are readily available in aqueous solutions, the supply of reactants to the electrode surface is limited by the concentration and diffusivity of dissolved CO_2 molecules, which places a hurdle on the current density and efficiency of CO_2 reduction. The reduction of CO_2, as well as the evolution of H_2, produces hydroxide anions ($OH^{â€“}$) and causes rise of the local pH on the electrode surface.

Here we report on the study of mass transfer and local pH effects on CO_2 reduction by using Cu nanowires as the electrocatalysts. Highly dense Cu nanowires were synthesized by first thermal oxidation of Cu mesh in air, and then reduction by applying a cathodic electrochemical potential or thermal annealing in a reducing atmosphere. Our previous report has shown that the electrochemically reduced (ECR) Cu nanowires are highly active for CO_2 reduction, with favorable production of CO and formate at potentials more positive than -0.5 V (versus reversible hydrogen electrode) and C_2 species (ethanol, ethylene, and ethane) at more negative potentials. We have further performed more systematic studies in the high-overpotential region, i.e., potentials more negative than -0.5 V, where the mass transport of CO_2 and high local pH near the electrode surface are believed to play a significant role in determining the electrocatalytic performance. Simulations have been developed to understand the interplay between the mass transfer and local pH effects and elucidate their roles in the structure-performance relationships of the high-surface-area nanowire catalysts. Based on these results, discussions are provided for optimizing the electrochemical conditions for CO_2 reduction to produce valuable C_2 products.

497824 Measurement of Intrinsic Activity of Electrocatalytic Reduction of CO_2 over Cu

Abstract:

Electrocatalytic reduction of CO_2 offers a sustainable route to mitigate global warming and store intermittent solar and wind energy. Various electrocatalysts have been studied over the past three decades for the conversion of CO_2 to oxygenates such as CO, HCOOH, $C_2H_2O_4$, CH_3OH, and C_2H_5OH; and hydrocarbons such as CH_4, C_2H_4, and C_2H_6. Among these electrocatalysts, Cu is the only catalyst that can produce hydrocarbons. Several electrochemical cells have been proposed in the literature to measure activity and selectivity of the CO_2 reduction reaction. However, all of these measurements were affected by the poor mass transfer of CO_2 to the cathode at potentials more negative than -1 V vs RHE. The limiting factor in the electrochemical reduction of CO_2 is the depletion of dissolved CO_2 coupled with the increase in pH near the cathode. Here we propose a compact design of an electrochemical cell with the cathode being a Rotating Disk Electrode (RDE). The RDE sparged with gaseous CO_2 minimizes the cathode polarization and hence enable the measurement of intrinsic activity. The effect pH and applied potential on the intrinsic activity and selectivity of electrocatalytic CO_2 reduction reaction over Cu will be presented in this talk.

495014 Electrocatalytic Carbon Fixation on Molecular-Functionalized Surfaces

Abstract:

Global warming and related environmental issues have been exacerbated by the accelerated consumption of fossil fuels as energy resources. Utilizing renewable energy input to transform the abundant chemical feedstock carbon dioxide (CO_2) back into fuels and other value-added chemicals and materials can complete a potential carbon-neutral cycle to help the sustainability of our planet. To this end, we are developing hybrid homogeneous-heterogeneous catalysts for electrochemical carbon fixation based on the combination of inorganic molecules and surfaces. Our latest work on the development of systems capable of CO_2 and CO reduction to C-C coupled products will be discussed in this presentation.

494963 Electrochemical Reduction of CO_2 on Metal Doped Graphene

Abstract:

Single atom transition metals embedded at single vacancies of graphene provide a unique paradigm for catalytic reactions.[1,2] We present a density functional theory study of such systems for the electrochemical reduction of CO_2. Using an explicit description of the solvent at the electrochemical interface, the activation energies for key proton-electron transfer steps were determined. On transition metals, the first proton transfer ($*CO + H^+ + e^- \rightarrow *CHO$) has been shown to be the critical rate determining step in C_1 product formation. On these single atom sites, the corresponding barrier scales more favorably with the CO binding energy than for 211 and 111 transition metal surfaces, in the direction of improved activity.[3] Hydrogen evolution intermediates, on the other hand, are found to less stable than those on transition metals, which suggests a lower hydrogen evolution activity. Using a simple microkinetic model based on the calculated electrochemical barriers, we present a rate volcano for the production of methane from CO. All the considered transition metal dopants (Cr, Mo, W, Mn, Fe, Ru, Os, Co, Rh, Ir, Ni, Pd, Pt, and Cu) form stable doped-graphene structures, and a subset of these dopants (Ru, Fe, Ni, Pd, Pt, Rh, Co, and Ni) are not expected to be poisoned by oxide species. The presence of nitrogen near the active site was found to affect intermediate and transition state energies, but variations in these energetics followed the same scaling determined for these materials without nitrogen. We identify promising candidates with high activity, stability, and selectivity for the reduction of CO_2. This work highlights the potential of these systems as improved electrocatalysts over pure transition metals for CO_2 reduction.

1. Krasheninnikov, A.V; Lehtinen, P.O.; Foster, A.S.; Pyykkö P.; Nieminen, R.M. *Phys. Rev. Lett.* **102**, 126807 (2009).
2. Stolbov, S.; Ortigoza, M.A. *J. Chem. Phys.* **142**, 154703 (2015).
3. Liu, X.; Xiao, J.; Peng, H.; Hong, X.; Chan, K.; Nørskov, J.K. *Nat. Comm.* (2017).

484935 Climbing Elementary Steps to the Pinnacle: Surface Science to Heterogeneous Catalysis

Abstract:

For decades our goal has been to make studies of reactivity on well-defined single crystal surfaces relevant to the development of heterogeneous catalysts in much the same spirit as gas phase kinetics and mechanism are important to understanding atmospheric chemistry. As long ago as the mid 1970â€™s significant progress was made in the understanding of partial oxidation of methanol over silver catalysts based on model studies resulting from the graduate work of Prof. Wachs, which were supported years later by DFT-based microkinetic modeling. Recently, based on mechanistic studies in ultrahigh vacuum, metallic gold has emerged as a promising catalyst for selective oxygen-assisted coupling reactions of alcohols and amines to yield industrially important classes of molecules, including esters and amides. The fundamental mechanistic principles determined from the model systems stem directly from the body of work on silver single crystals, but extend well beyond. Consequently, these principles have been used to predict new reactive pathways for other more complex molecular transformations under practical catalytic conditions. Studies ranging from those on single crystal gold to continuum flow and pulse reactors will be discussed. These fundamental reaction paradigms will be extended to the understanding of oxygen-assisted coupling mechanisms on metallic alloy gold catalysts of varying nature across a broad range of studies, including gas phase and liquid phase systems.

502686 Catalyst Synthesis By Atomic Layer Deposition

Abstract:

Atomic Layer Deposition (ALD) has enormous potential for the synthesis of advanced heterogeneous catalysts with control of composition and structure at the atomic scale. ALD can achieve highly uniform catalytically active metal and oxide phases with (sub-) nanometer dimensions. This lecture will provide examples from our laboratory of ALD used to synthesize oxide supports, catalytic oxide overlayers, and metal nanoparticles. These materials have been characterized by SEM, XRF, ICP, UV-Vis absorption spectroscopy, Raman spectroscopy and evaluated for catalysis of oxidative and non-oxidative alkane dehydrogenation, and combustion. We have focused on the synthesis of supported metal nanoparticles and developed a procedure in which the metal and support materials are grown sequentially in each ALD cycle. This method makes possible the synthesis of exceptionally small particles, ca. 0.5 nm. Using additional ALD support layers at the conclusion of the growth, a process we call overcoating, the metal particles can be stabilized against sintering and leaching while still remaining active under harsh conditions in both gas and liquid phase reactions. Through proper annealing procedures the overcoating oxide develops porosity, thus ALD becomes a method for introducing and controlling pore structures.

486110 Single-Site or Not Single-Site?

Abstract:

The rational design and development of catalysts require structure â€“ reactivity relationship approach, hence the need for strategies to obtain well-defined surface sites and their detailed characterization. Here, we discuss the method to control and understand the chemistry at the surface of materials towards the development of well-defined â€“ so-called single-site â€“ heterogeneous catalysts and show how this approach can bring about information about industrial catalysts. We will discuss in particular how the support affects the structure of the surface species.

490474 Promotion of Platinum for Alkane Dehydrogenation: Intermetallic Alloys with Two Catalytically Active Metals

Abstract:

Metal nano-particle alkane dehydrogenation catalysts utilize Pt and are generally promoted by Sn, Zn, Ga or In. While Pt bimetallic nano-particles are highly selective, their exact structure is not generally known, especially in very small nano-particles. Recently, it was shown that intermetallic alloys were are present on PtIn dehydrogenation catalysts and other bimetallic compositions. The increased olefin selectivity was suggested to be due to isolation of the active Pt atoms by inactive promoter atoms. Changes in the turnover rate (TOR) and activation energy also suggested electronic effects.

Several intermetallic alloy structures can also be formed with Pt and a second catalytically active atom. In this study we show that two catalytic metals can be selective for alkane dehydrogenation. The intermetallic structure is dependent on the metal loading. The catalytic performance suggests that the Pt atoms dominate the alkane conversion. At high promoter loading, excess monometallic promoter forms, which has low selectivity and deactivates rapidly. The catalysts have been characterized by in situ synchrotron X-ray absorption and X-ray diffraction to identify the nano-structures formed with different compositions.

492388 Challenges with Model Catalysts

Abstract:

Anchoring of molecular catalysts to solid supports has long offered the prospect of creating heterogeneous catalysts with uniform sites. We have explored over many years the physical, chemical and electronic properties of polyoxometalates â€“ a class of discrete molecular metal oxide catalysts â€“ that are well known for acid and oxidation catalysis. Studies in which these species were dispersed on surfaces of graphite, graphene, nitrogen-containing carbons, and conventional catalyst supports such as silica, have generally shown that they behave as discrete molecules with respect to electronic and chemical properties. We have also shown that spectroscopic techniques can provide useful correlation and prediction of catalyst redox properties. However, our recent studies of catalytic reactions suggest that there are cooperative effects in surface arrays of these molecules, leading to turnover rates that depend on surface concentrations of active sites. These challenge notions about single site catalysts and about appropriate ways to describe their activity.

509819 Nature of Active Selective Oxidation Catalysts: an Operando Approach

Abstract:

Operando spectroscopy has been applied to achieve detailed characterization of catalytic nanoparticles under reaction conditions. This powerful, insightful approach for deriving structure-function relationships will be demonstrated with a number of examples of selective oxidation catalysis.

500170 Experimental and Modeling Studies of Cycle Frequency, Reductant Type and Non-Isothermal Effect on the Performance of a Lean NO_x Trap

Abstract:

Experimental and Modeling Studies of Cycle Frequency, Reductant Type and Non-isothermal Effect on the Performance of a Lean NOx Trap

Allen Wei-Lun Ting, Michael P. Harold and Vemuri Balakotaiah

Department of Chemical & Biomolecular Engineering, University of Houston, Houston, TX 77204, USA

Abstract

The ¡§Di-Air¡¨ system proposed by Toyota researchers shows an overall increase in NOx conversion when using propene as the reductant under fast cycling [1]. In order to clarify this mechanism, a combined experimental and modeling study of fast cycling NOx storage and reduction for emissions control of lean burn gasoline and diesel vehicles is conducted to provide mechanistic insight. NOx generation pathways are analyzed by using H_2 and C_3H_6 as reductants, under different cycle frequencies, and under near-isothermal and non-isothermal conditions. LNT (lean NOx trap) catalysts containing $Pt/BaO/CeO_2/Al_2O_3$ are used in the monolithic reactor experiments. A 1-D two phase model that includes washcoat diffusional effects is used to simulate the cyclic experimental results to provide more details such as the spatio-temporal concentration, site saturation, and temperature profiles.

The mechanism of Di-Air system has been receiving great attention for its NOx conversion enhancement at both low and high feed temperature only by increasing the cycle frequency. The HC-intermediate pathway from which short-lived HC intermediates generated is thought to play important role in this novel system. The essence of this pathway is as follows: The intermediates form from reaction between NOx and propylene and are able to survive through the rich period even at high temperature and reduce NOx in the early lean period. Therefore, the goal of this work is to investigate how this HC-intermediate pathway enhances the NOx conversion.

By feeding H_2 or C_3H_6 as the reductant agents by same reductivity in stoichiometry, the NOx regeneration pathway and the HC NOx regeneration pathway can be compared. Due to the different molecular characteristics of H_2 and C_3H_6, the thermal effect under a non-isothermal environment (aerobic rich feed condition) is studied. The much larger molecular diffusivity of H_2 compared to C_3H_6 shortens the reaction zone. This is observed by measuring the temperature in the front quarter and third quarter of the monolith reactor. Fig. 1a shows a sharper temperature rise and larger axial temperature gradient when using H_2 than C_3H_6 under slow cycling. A 60 ^oC higher temperature rise (250~280 ^oC for H_2) is measured in the first quarter of the monolith under slow cycling when using H_2. This reduces the NOx storage capacity during the early lean period and results in a 10 to 15 % lower NOx conversion, as shown in Fig. 2.

Fast cycling when using H_2 as the reductants has been thoroughly studied in our previous work, exhibiting a better usage of storage sites and a near-steady-state axial temperature profile comparing to slow cycling. However, under fast cycling these is no difference when using H_2 and C_3H_6 in axial temperature profiles (Fig. 1b), so the previously mentioned thermal effect doesn¡¦t exist under fast cycling. Only a 5 ¡V 10 % NOx conversion enhancement can be found when using C_3H_6. This may be contributed from the HC-intermediate mechanism.

In addition to the thermal effect, steam reforming of propylene may also play an important role in NOx regeneration pathways because it converts C_3H_6 to CO and H_2 through H_2O, and oxidation reactions in such non-isothermal environment produces a lot of water.

Therefore, experiments under near-isothermal condition (anaerobic rich feed) with no H_2O in feed were conducted. Fig. 3 shows that the NOx conversions are very similar under this near-isothermal condition at temperatures higher than the ignition temperature. Under fast cycling the enhancement of NOx conversion using C_3H_6 is less than 5 %.

This concludes that the NOx conversion enhancement of Di-air system mainly results from the thermal effect, and HC-intermediate mechanism is relatively minor.

Reference:

[1] Y. Bisaiji, K. Yoshida, M. Inoue, N. Takagi, T. Fukuma, Development of Di-Air - A New Diesel deNOx System by Adsorbed Intermediate Reductants, SAE Int. J. SAE Int. J. Fuels Lubr. 5 (2012) 380¡V388.

Fig. 1 The temperature measured at first quarter (solid line) and third quarter (dashed line) of the catalyst under 305 ^oC feed temperature for H_2 (red) and C_3H_6 case (black) under slow cycling (a, 60/10 s) and fast cycling (b, 6/1 s). Feed condition: Lean: 5 % O_2, Rich: 2.5% O_2, 9 % H_2/ 1 % C_3H_6, Both: 5% CO_2.

Fig. 2 Cycle-averaged NOx conversion with different cycle-averaged catalyst temperature, cycling frequency using H_2 and C_3H_6 as the reductant under non-isothermal condition. Feed condition: Both: 5 % CO_2, 500 ppm NO. Lean: 5 % O_2, Rich: 9 % H_2/ 1 % C_3H_6, 2.5 % O_2.

Fig. 3 Cycle-averaged NOx conversion with different cycle-averaged catalyst temperature, cycling frequency using H_2 and C_3H_6 as the reductant under near-isothermal condition. Feed condition: Both: 5 % CO_2, 500 ppm NO. Lean: 5 % O_2, Rich: 4 % H_2/ 4445 ppm C_3H_6.

502215 Fast Optimal Control of Exothermic Packed-Bed Reactors Via Reduced Order Models

Abstract:

Utilizing volatile renewable energy sources (e.g., solar, wind) for chemical production systems requires a deep understanding of their dynamic operation modes.
Taking the example of a methanation reactor in the context of power-to-gas applications, a dynamic optimization approach is used to identify cooling trajectories for time optimal reactor start-up which simultaneously inhibits distinct hot spot formation. Therefore, we developed a dynamic, two-dimensional model of a packed-bed tube reactor for carbon dioxide methanation which is based on the reaction scheme of the underlying exothermic Sabatier reaction mechanism [1]. However, dealing with large-scale nonlinear dynamical process models (derived from PDEs) often leads to many computational diculties. Facing this issue, snapshot-based model order reduction techniques, such as proper orthogonal decomposition together with the discrete empirical interpolation method (POD-DEIM) generate considerably less complex models, featuring a lower number of states and, furthermore, guarantee an acceptable model error [2].
We illustrate the applicability of POD-DEIM to the methanation rector [3] and show that for dynamic optimization the reduced order model outperforms the full order model in terms of CPU time, memory cost, and even feasibility. Thus, these surrogate models show a vast potential for future PDE based online control applications (e.g., NMPC).

References

[1] Bremer, J., Ratze, K.H., and Sundmacher, K. CO2 Methanation: Optimal Start-Up Control of a Fixed-Bed Reactor for Power-To-Gas Applications. AIChE Journal (2016).
[2] Benner, P., Gugercin, S., and Willcox, K., A Survey of Projection-Based Model Reduction Methods for Parametric Systems. SIAM Review (2015).
[3] Bremer, J., Goyal, P., Feng, L., Benner, P., and Sundmacher, K., POD-DEIM for Ecient Reduction of a Dynamic 2D Catalytic Reactor Model. Computers & Chemical Engineering (2016).

489622 From Particle Resolved CFD to the Transient Modeling of Dynamic Systems: Dispersion in Fixed-Bed Reactors

Abstract:

The modeling of dynamic processes in fixed-bed reactors is of huge interest. Either the overall process can show a transient behavior (e.g. adsorption, desorption or Chemical-Looping) or local and temporal hot spots can be present leading to process safety issues during the start-up phase of otherwise steady-state processes [1]. Particle resolved CFD simulations have been used extensively in the last years (see [2], [3] and [4]) to accurately predict the steady-state conditions of fixed-bed reactors. However, this approach is characterized by a wide range of time scales and a large numerical demand. Hence, it is uncertain if it can be efficiently used for the simulation of the dynamic process behavior in fixed-bed reactors.

On the other hand, the accuracy of simplified models relies on the knowledge of effective parameters like the effective thermal conductivity, the heat transfer coefficient and the dispersion coefficient. Most often these are determined using correlations. In most cases, their application is limited to simple particle shapes and reactors with a large tube-to-particle diameter ratio. To overcome this limitation, a workflow based on particle resolved CFD and particle tracking is proposed to determine axial dispersion coefficients in fixed-bed reactors that can later be used in simplified models to describe the process dynamics.

The effect of the bed morphology, Reynolds number, axial position and tube-to-particle diameter ratio (N) is thoroughly investigated over a wide range of particle shapes. Furthermore, the limitation of the dispersion model for reactors with a very low N is discussed.

References
[1] Adams et al. (2009) *Int. J. Hydrogen Energy,* 34, 8877-8891
[2] Wehinger et al. (2015) *Chem. Eng. Sci., 122*, 197-209
[3] Wehinger et al. (2016) *AIChE J., 62*(12), 4436-4452
[4] Wehinger et al. (2017) *Ind. Eng. Chem. Res., 56*(1), 87-99

494156 N-Butane Partial Oxidation in a Fixed Bed: A Local Selectivity Study By Detailed Numerical Simulations

Abstract:

N-butane partial oxidation in a fixed bed:

A local selectivity study by detailed numerical simulations

Behnam Partopour, Anthony G. Dixon

Department of Chemical Engineering, Worcester Polytechnic Institute,

Worcester, MA, USA, 01609

Maleic anhydride (MA) is one of the most important chemical intermediates which is mainly produced by partial oxidation of n-butane over VPO catalyst. The reaction is fast and highly exothermic and therefore, a highly diluted inlet flow (n-butane < 2%) is used to avoid the reaction runaway limit. Likewise, unreacted n-butane is not recycled. The fixed bed process is carried out in multitubular reactors with narrow tubes (< 6) cooled in a molten salt bath. All these factors lead to a non-optimal process with an overall yield less than 65%.[1]

Recent studies on MA production have been mostly focused on the catalyst active phase nature while there are few reactor scale investigations. Therefore, we have conducted a multi-scale CFD simulation of a packed bed of 807 spherical particles to investigate the interactions of the fixed bed configuration with the reaction, transport and the flow field. In this study a semi-empirical surface kinetics[2] is coupled with the transport and momentum equations in both the fluid and solid phases. It is shown that the temperature increases very rapidly along the tube while due to the cooled tube wall a strong radial temperature gradient forms in the bed. The particles are mainly isothermal and the temperature gradient has the most significant impact on species and reaction profiles along the bed. However, it could be observed that compared to the species mass fractions, the weakly adsorbed oxygen surface coverage is highly affected by the cooled tube wall, and a strong gradient is formed inside each particle. Furthermore, the local selectivity patterns in the bed follow similar patterns to the weakly adsorbed oxygen coverage. These trends are shown in Figure 1.

A statistical approach shows that the correlation is not an injective function and there are more factors contributing to the local selectivity. A similar study for the surface oxygen shows that reduced catalyst (higher surface oxygen coverage) contributes to a lower local MA selectivity as is extensively discussed in the literature, but again that is not the only contributor. Therefore, it could be concluded that while the 3D detailed numerical simulation of the bed shows that all local reaction conditions and particle positions are affecting the local MA selectivity in the bed, temperature and weakly adsorbed oxygen are the dominant factors in shaping the local selectivity patterns.

Figure 1. Contours are shown on a mid-tube plane along the fixed bed

References

1- Trifirò, F., Grasselli, R.K. How the yield of maleic anhydride in n-butane oxidation, using VPO catalysts, was improved over the years. Top. Catal. 2014; 57 (14): 1188–1195.

2- Gascon, J., Valenciano, R., Tellez, C., Herguido, J., Menendez, M. A generalized kinetic model for the partial oxidation of n-butane to maleic anhydride under aerobic and anaerobic conditions. Chem Eng Sci. 2006; 61:6385-6394

495315 Bifurcation Analysis for Allothermic High-Temperature Pyrolysis of Methane in a Moving-Bed Reactor

Abstract:

Multiple steady states are known for numerous reaction engineering examples involving exthermic reactions e.g. in moving bed reactors or reverse-flow reactors. This paper reports studies on the bifurcation analysis and the generation of regions of multiplicity in the operation of moving-bed reactors. The studies were first carried out for a generalized model, which allows the investigation of the effect of various scenarios, such as changes in heat capacity or the number of moles with reaction, as well as the previously unstudied equilibrium and allothermic reactions. Based on these the insights gained from this extension were then applied to an actual reaction, namely the high temperature pyrolysis of methane.

The study of the high temperature pyrolysis of methane is motivated by the search for alternatives to produce hydrogen with low to zero carbon dioxide emissions. Among them, the decarbonization of the fossil energy, especially the high temperature methane pyrolysis, constitutes a promising drop-instrategy. To assess its potential, a research program has been set up to develop a novel two-step process for first generating hydrogen and carbon and then using the former to produce synthesis gas utilizing carbon dioxide. To carry out the pyrolysis of ethane, the moving-bed reactor represents a promising approach, since it has been successfully operated at high temperatures in the metallurgical industry and elsewhere.

The standard models for a moving-bed reactor with a heterogeneous reaction can be easily derived from differential balances under the usual assumptions of negligible axial and radial conduction and diffusion, plug flow in both the solid and gas phases, and temperature-independent properties. The steady-state dimensionless model equations describing a mass and two energy balances for the gas an solid phase. In this work, a generalized dimensionless model for a moving-bed reactor was developed and the influence of several parameters was investigated. The parameter found to have the greatest influence on the multiplicity behavior was the Stanton number. In addition, cases for systems exhibiting changes in moles and heat capacities were studied, showing a larger influence than would have been predicted a priori, mainly because they show the highest influence on the Stanton number.

Furthermore, the insights gained from the generalized model were applied to the allothermal operation of the high-temperature pyrolysis of methane. Here, the reverse reaction at higher pressures was identified as possible source of multiplicity due to its exothermicity. In addition, for a nonlinear heat input it could be shown that multiple steady states can be found for particular cases. The interaction of these two effects was demonstrated in a heat-integrated reactor.

In general, although the case studied was not ideal, the results obtained could be used to show the existence of multiple steady states in other allothermal systems, where the coupling of the reaction and the nonlinear heat input occur in a narrower temperature range as can be seen from the shifted temperature curves.

Finally, the bifurcation methods proved to be a valuable tool when performing parameter studies, and although, depending on the complexity of the system, the simulation can demand considerable computational effort, a more refined coding or the introduction of the Jacobian matrix would dramatically improve the performance.

494854 Catalytic Membrane Reactor for CO_2 Hydrogenation Using H_2-Containing Renewable Streams: Model-Based Feasibility Analysis

Abstract:

Catalytic membrane reactor for CO_2 hydrogenation using H_2-containing renewable streams: Model-based feasibility analysis

Robert Currie and David Simakov

Department of Chemical Engineering, University of Waterloo, Waterloo, ON N2L 3G1, Canada

Converting CO_2 into synthetic fuels and platform chemicals is an attractive pathway to decrease CO_2 emissions and to reduce our dependence on fossil fuels. Thermocatalytic hydrogenation provides advantages of fast reaction rates and high conversion efficiencies (as compared to electro-catalytic and photo-catalytic routes), thus allowing for compact, high-throughput operation. The most fundamental disadvantage is the requirement to supply a pure H_2 stream (water is a source of H_2 in the electro- and photochemical CO_2 reduction). Natural gas reforming is obviously not an option, as this approach will result in the overall positive carbon footprint. Water electrolysis using renewable or surplus electricity is a possibility, but this technology requires high capital investments and operation costs. This study assesses the feasibility of using H_2-containing renewable streams, such as biomass gasification product gases, as a source of H_2 for the thermocatalytic CO_2 conversion to renewable natural gas (Sabatier reaction, $CO_2 + 4H_2 = CH_4 + 2H_2O$).

It is suggested to use a H_2-selective membrane to extract H_2 from a H_2-containing stream *in situ*. The suggested Catalytic Membrane Reactor configuration is based on the distributed H_2 supply via a H_2-selecive membrane along the axial reactor dimension. The heat generated in the highly exothermic Sabatier reaction is simultaneously removed using molten salt as a heat transfer fluid. A dynamic mathematical model was formulated and analyzed by numerical simulations. The model accounts for the heat exchange between the reaction compartment and the cooling tube, for the mass exchange through the H_2-selective membrane, as well as for catalyst deactivation. The reactor fed with a CO_2-rich waste stream (biogas or landfill gas) without upstream CO_2 separation was analyzed. Effects of feed temperature, pressure, and space velocity on the reactor spatiotemporal behaviour were investigated. Catalyst deactivation was found to be significant but not severe, owing to the distributed H_2 supply that leads to more uniform temperature distribution and reduced rate of coking. The simulation results show that, with a proper selection of operating parameters, it is possible to achieve CO_2 conversions over 80% after 1,000 h on stream, producing renewable natural gas from a waste CO_2-rich stream and a H_2-containing biomass gasification product gas.

497356 Experimental and Modeling Studies of Residence Time Distribution in Partially Filled Laminar Flow Reactors

Abstract:

The concept of Residence Time Distribution (RTD) is useful in understanding the amount of time spent by fluid elements in a reactor i.e., the degree of macro-mixing. Residence time has a significant effect on the reaction conversion and hence is an important parameter to be considered while designing reactors. The RTD for a fully filled Laminar Flow Reactor (LFR) is available in the literature. If the flowrate of the reactant in the LFR is not sufficiently high, only a part of the reactor is occupied by the reactant and the rest is filled by air. The RTD for partially filled tubular reactors with different liquid hold up is not reported so far to the best knowledge of the authors. In this work, we study the residence time distribution of partially filled circular millichannel for different liquid hold up both experimentally and theoretically. Towards this, a pulse of NaCl tracer was introduced at the reactor inlet and the exit tracer concentration was measured using a conductivity meter as a function of time. This is carried out for different liquid holdup in the reactor. The residence time distribution is then analyzed in terms of E and F curves for different phase hold up.

The major challenge in modeling partially filled LFR is that neither the Cartesian nor the cylindrical coordinate system can describe the gas-liquid interface and channel wall simultaneously. To overcome this challenge, we employ a special coordinate system known as the bipolar cylindrical coordinate system for the study. The governing equation for fluid flow was derived in the bipolar cylindrical coordinate system and solved analytically using the method of separation of variables [3]. The velocity profiles were analyzed for different liquid hold up. Then, the diffusion free species transport equation considering a unit step tracer injection is solved numerically using two approaches: a) the finite difference method using a first order upwind scheme for the convection term and b) the Lax Wendroff method. The results from the two methods were compared with each other and validated with experiments. Finally, we study how RTD for different liquid holdup influences the conversion for a first order reaction.

489964 Continuous Synthesis of Ethyl Esters from Free Fatty Acids over Metal Oxides in Sub/Supercritical Ethanol

Abstract:

2017 AIChE Annual Meeting

Oct 29th to Nov 3rd, Minneapolis, MN

Continuous synthesis of ethyl esters from free fatty acids over metal oxides in sub/supercritical ethanol

Jiuxu Liu, Yue Nan and Lawrence L. Tavlarides

Department of Biomedical and Chemical Engineering, Syracuse University,

Email: lltavlar@syr.edu (Tavlarides); jliu23@syr.edu(Liu)

Abstract

In our previous study, we proposed a method for biodiesel synthesis by running the transesterification reactions under sub/supercritical conditions with trace amounts of homogeneous catalysts in order to achieve a high product yield and reduce the dosage of catalysts compared to the conventional technology. Another way to avoid using large amount of homogeneous catalysts (which bring burdens in product purification process) is switching to heterogeneous catalysts, acidic or basic. The common basic solid catalysts are hydrotalcites, dolomites, loaded and supported alkaline and alkaline metals and corresponding oxides, and mixed basic metal oxides. The basic solid catalysts reported in the literature face leaching problems. The active compounds can be detected in the product with high concentration, and the catalysts have to be regenerated after a few runs. The solid acid catalysts seem to be more promising compared to basic ones by considering the fact that they are insensitive to free fatty acid content, and esterification and transesterification occurs simultaneously. Commonly used acidic solids in biodiesel synthesis include acidic metal oxides, sulfonic ion-exchange resin, sulfated/ tungstated metal oxides, zeolites, heteropolyacids, amongst others. Similar with basic solids, the reported acidic catalysts deactivate along with reaction due to the leaching problem.

A literature review indicates that several compounds were found to be stable even at alcohol temperature higher than 180 °C including zirconia, titania, silica, and alumina. These simple metal oxides have been commonly used as catalyst supports due to their thermal and mechanical stability. However they did not show any catalytic ability under conventional conditions (T=70 °C, ambient pressure) due to their weak acid/base property.

In this study, we are investigating the possibility and mechanism of continuously synthesizing ethyl-ester biodiesel under sub and supercritical alcohol conditions from fatty acids, which largely exist in cheap feedstocks such as waste cooking oil, catalyzed by simple and stable metal oxides catalysts including ZrO_2, TiO_2 and Al_2O_3. The objectives include: a) kinetics of ethyl-ester biodiesel synthesis with oleic acid at SCTE conditions without the catalysts, b) effect of addition of the solid catalysts on reaction yields, and c) kinetics of the biodiesel synthesis over the selected metal oxides.

At this writing the first aim has been completed, which is the kinetic study of the non-catalytic esterification reaction of oleic acid with sub and supercritical ethanol. The reactions were conducted at temperatures from 175 to 325 °C, residence times from 1 to 8 min, and ethanol-to oleic acid molar ratio of 18:1. The pressure was kept as 200 bar. Under subcritical conditions, biodiesel yield slightly increased from 8% to 33% with temperature changing from 175 to 225 °C at 8 min residence time. Under supercritical conditions, a yield of 60% was reached at 275 °C and 8 min, and increased further to 80% at 325 °C and 8 min. A one-step reversible second-order reaction model was employed to best fit the data, and the activation energy for forward and reverse reaction step was calculated as 55 and 32 kJ/mol, respectively. A negative enthalpy change as -23 kJ/mol indicates that the reaction is exothermic. These kinetic parameters will be used when conducting the kinetic analysis of reaction with presence of the solid catalysts.

Experiments to complete the remaining objectives are in progress and will be reported.

501066 Denitrogenation of Pyridine with Acetic Acid As Hydrogen Source Under Catalytic Hydrothermal Gasification Environment

Abstract:

Biofuel production from microalgae via hydrothermal liquefaction (HTL) produces wastewater with high organic (COD: 40 ~ 100 g/L) and nitrogen content (TN: 15g/L). The majority of nitrogen from the biomass remains in the post-HTL wastewater (PHWW). The most commonly identified chemical compounds in the PHWW are volatile fatty acids (VFAs) and cyclic amines such as pyridines. Pyridines, due to the aromaticity, are chemically stable, and they are also identified as cytotoxicity-inducing chemicals that inhibit the growth of microalgae even in 1% PHWW dilution, when the PHWW is reused as a nutrient source. Hydrodenitrogenation is an effective way to convert the pyridines into inorganic form as ammonia, or organic form as amides, which are better forms of nitrogen for biomass growth, as well as to mitigate the toxicity. Thus it is a beneficial to investigate an effective process of hydrodenitrogenation for recycling the HTL wastewater in large quantity without much dilution needed.

The first step of hydrodenitrogenation of pyridine is hydrogenation of pyridines to its piperidine form, followed by ring-opening and then C-C bond cleavage for ammonia/amine and small hydrocarbon production. Conventional hydrodenitrogenation requires hydrogen gas at high pressure to realize this series of conversions. Due to the large quantity of VFAs in the PHWW, and that VFAs have been extensively studied as model compounds for high hydrogen production in catalytic gasification conditions, it is possible to use VFAs as hydrogen precursor in the hydrodenitrogenation process without apply hydrogen gas.

In this study, acetic acid and pyridine are chosen as model compounds to evaluate the feasibility of using VFA as hydrogen source for hydrodenitrogenation of pyridine. Denitrogenation/gasification will be carried out under 500C reaction temperature, 60 min reaction time, which is previously discovered optimal operation condition for PHWW gasification. Denitrogenation efficiency and gasification yield under different pyridine to acetic acid ratio will be evaluated. Possible reaction pathway, and reaction kinetics will be discussed.

497002 The Non-Catalytic Decomposition of Lignin into Chemical and Fuel Intermediates

Abstract:

The goal of this presentation . . .

In this presentation we will present the results of a study performed to explore the non-catalytic decomposition of lignin into chemical intermediates.

Motivation . . .

Lignin is one of the most abundant biopolymers. On average, 24 wt% of a plantâ€™s structure is composed of lignin. Lignin is produced as a low value by-product from wood pulping processes as well as from ligno-cellulosic biofuel facilities, such as those producing ethanol. Most produced lignin is burnt to recover its energy content. The quest to decompose lignin into fragments that could serve as the building blocks for useful fuel and chemical components has been an active area of research for many years. These efforts have primarily focused on catalytic pyrolytic or biochemical decomposition strategies. However, a commercially feasible approach has been elusive.

Presentation Contents â€“ a Novel Approach to Lignin Decomposition

This presentation will describe a novel, high temperature, high pressure continuous reaction scheme designed and built to study lignin decomposition. A dilute slurry of lignin entrained in a basic water solution was pressurized to near super-critical conditions and heated to its decomposition temperature in a continuous tubular reactor. The reactor outlet stream was cooled against room temperature water and the four resulting phases (gas, polar liquid, non-polar liquid, and solid) were collected and analyzed with a variety of methods. When these initial results are compared to recently published catalytic decomposition results, this novel approach appears to be attractive. A combination of low residence time with appropriate reaction temperature and pressure result in the cleavage of ether bonds and the production of phenolic hydroxyl groups and various organic acids (e.g. vanillic and homovanillic acid) while minimizing the formation of char.

490374 Hydrothermal Liquefaction of Food Waste and Remediation of Aqueous Byproducts

Abstract:

Thermochemical processing via hydrothermal liquefaction (HTL) is capable of processing a broad range of feedstocks, has a favorable fossil energy ratio, and produces an energy-dense liquid bio-oil product and hydrochar co-product with emerging markets. However, there is a well-known obstacle facing current HTL processes that has hampered the adoption of this technology, namely the HTL process produces a considerable amount of aqueous organic byproducts. As such, the carbon yield to bio-oil is reduced and the organic content of the aqueous waste stream exceeds the environmental standards for discharging the water.

The project team is developing HTL technologies to produce bio-oil from commercial and institutional food waste. Using a two-pronged approach, we have demonstrated an improved HTL oil yield while also reducing aqueous contaminants. First, we have successfully identified a stable and active heterogeneous HTL catalyst capable of improving bio-oil yields (dry mass basis) from 19% to 28% compared to traditional homogeneous base catalysts. IR spectra, GC/MS chromatograms, and model compound experiments revealed that the heterogeneous catalyst increased oil molecular weight, increased ketone and ether content, and decreased alcohol content, thereby driving products out of the aqueous phase and into the organic oil phase. Second, we have tested two active hydrothermal gasification (HTG) catalysts capable of converting 93%â€“98% of aqueous organics to permanent gases composed primarily of hydrogen, methane, and carbon dioxide. In addition to reducing aqueous contaminants and simplifying waste water disposal, the hydrogen-rich gas produced from HTG can be recycled to the HTL reactor to improve oil yield and reduce the required residence time.

488218 Hydrothermal Liquefaction of Wastewater Treatment Microalgae in a Pilot-Scale Continuous Flow Reactor

Abstract:

Hydrothermal liquefaction (HTL) is considered as one of most promising techniques for valorization of whole, wet algae biomass. HTL uses subcritical water (270-350 ºC and 8-18 MPa) to convert the organic biomass constituents into an energy-rich bio-crude oil that can be upgraded into liquid fuels. To date, many studies have used batch reactor; shifting to continuous reactors is necessary for progressing to commercial-scale processes due to the higher productivity and lower operating costs of steady-state, continuous systems.

This study reports the modification and preliminary testing of a pilot-scale continuous flow reactor (CFR) for hydrothermal liquefaction of wastewater microalgae at low solids loadings. The goal of this reactor is to perform continuous (>30 liters or >8 hours) HTL on algae slurries with solid algal contents up to 10 wt.%, and to produce char-free bio-crude oils using a dual high-pressure cylinder filter system with blow-down pots to remove solids. CFR systems must address several challenges: smooth flow of homogenous biomass slurry through unit operations, clogging, solid-liquid-gas separation/filtration, multiple unit operations, and process safety and control. Here we report on the progress of the reactor development relative to bio-crude oil production, reaction residence time, and operational stability. In addition to reactor development, HTL product characterization was also conducted using Fourier transform ion cyclotron resonance mass spectroscopy (FT-ICR MS) and fatty acid methyl ester (FAME) analysis by gas chromatography mass spectroscopy (GC/MS for bio-crude oil quality, and oxy-combustion calorimetry and elemental analysis (CHNS) for mass and energy balances.

491786 Experimental Study of Hydrocarbon Trapping over Passive NO_x Adsorber

Abstract:

Experimental Study of Hydrocarbon Trapping over Passive NOx Adsorbers

Sam Malamis, Kyle Karinshak, Michael P. Harold*

Dept. of Chemical & Biomolecular Engineering, University of Houston, Houston, Texas

* corresponding author e-mail: mpharold@central.uh.edu

Introduction

Automotive catalysts including TWC, LNT, and SCR catalysts are all highly effective in reducing NO_x and HC emissions above 200°C [1]. Most engines, however, require a warm-up period during which the catalyst undergoes a “cold start” environment. During this period of up to 200s, large amounts of pollutants such as NOx, N_2O, CO and Hydrocarbons slip through the catalyst into the atmosphere [2]. To mitigate this occurrence, it is beneficial to develop a passive NOx adsorber, or PNA, that can store NOx and hydrocarbons at these low temperatures, and then to release them at higher temperatures to improve the effectiveness of the aftertreatment system [3].

Few comprehensive studies have been performed on this topic, and only recently has there been a movement toward understanding the chemistry behind these low temperature traps. To date, a wide variety of materials has been explored, including palladium on ceria-zirconia (Pd-CZO) [4], CeO_2-M_2O_3 with supported Pt or Pd [5], and Palladium on various zeolites including BEA, MFI, CHA [3]. While it has been reported that Pd-CZO is a potentially effective NO_x trap at low temperatures, it also strongly absorbs SO_2, which inhibits NO_x storage [3,4]. This study focuses on the use of zeolites as PNA materials in order to trap both NOx and HC during the cold start period, with the potential to act as a partial oxidation catalyst upstream of the SCR or LNT. The study mainly explores the effect of temperature and various feed gases on NOx and HC trapping.

Materials and methods

Experiments were performed using 2wt% Pd-BEA zeolite supported on a cordierite monolith. A feed gas was flowed over the catalyst in a vertical quartz-tube reactor, and the effluent species concentrations were measured using FT-IR. The catalyst was first pre-treated in 5% O_2 (bal. Ar) for 30 min at 550°C. To determine the amount of NOx stored on the catalyst during an experiment, a feed gas of 2% O_2 and 400ppm NO was flowed over the catalyst for 5 min at a fixed feed temperature, either 50°C, 80°C or 150°C. After 5 min., the NO was turned off and the reactor was maintained at the adsorption temperature until the concentration of NO dropped to less than 2ppm. To measure the species desorbing after adsorbing on the catalyst surface, a temperature programmed desorption (TPD) was performed at 20°C/min from the adsorption temperature to 550°C. The surface was then pretreated to re-oxidize the Pd to PdO to ensure proper behavior during subsequent experiments [6]. To measure the effect of hydrocarbons on catalyst performance, experiments similar to those described above were performed, using a feed with either 800ppm C_3H_6 and 2% O_2 or a mixture of 400ppm NO, 800ppm C_3H_6 and 2% O_2.

Results

It was found that using a feed of only NO and O_2, Pd-BEA stored between 0.4 and 1.0 mol NOx/gcat*10^4, but released it too early during the TPD (i.e. at a temperature <200°C) (Figure 1). It was also found that as the adsorption temperature increased, the amount of NOx stored and desorbed decreased (Figure 2). Experiments using only propylene in the feed indicate a very strong propensity for the catalyst to trap hydrocarbons at low temperatures (1.5-6.0 mol C_3H_6/gcat*10^4), more so than in trapping NOx (Figures 3). Adding C_3H_6 to the NO feed revealed a competitive adsorption effect of both NOx and C_3H_6, as the stored quantities of these species was significantly lower than the pure feeds, as shown in Figures 2 and 3. While the total amount of stored NOx decreased with the addition of C_3H_6, our studies show that the majority of NOx was released above 200°C, which is a promising result (Figure 1).

Figure 1. Percentage of NOx that desorbs at a catalyst temperature higher than 200°C for feeds with only NO and with added C_3H_6 on 2wt% Pd-BEA. Values are reported for each adsorption temperature. Note that the addition of propylene in the feed significantly delays NOx desorption. Feed: 400ppm NO with or without 800ppm C_3H_6, 2% O_2, bal. Ar.

Figure 2. NOx adsorption and desorption quantities at various adsorption temperatures and for feeds with or without C_3H_6. Values are reported as moles of NOx/g cat*10^4. Note that as the adsorption temperature increases, adsorption/desorption of NOx decreases. Feed: 400ppm NO with or without 800ppm C_3H_6, 2% O_2, bal. Ar.

Figure 3. C_3H_6 adsorption and desorption quantities at various adsorption temperatures and for feeds with or without NO. Adsorption values are reported as moles of C_3H_6/g cat*10^4, while desorption values are reported as moles of C/g cat*10^4. Note that as the adsorption temperature increases, adsorption/desorption of C_3H_6 decreases. Feed: 800ppm C_3H_6 with or without 400ppm NO, 2% O_2, bal. Ar.

References

1. Twigg M: **Progress and future challenges in controlling automotive exhaust gas emissions.** *Applied Catalyst B* 2007, **70**:2-15.

2. Wilenmann M, Favez J, Alvarez R: **Cold-start emissions of modern passenger cars at different low ambient temperatures and their evolution over vehicle legislation categories.** *Atmospheric environment* 2009, **43**:241-2429.

3. Chen HY *et al.*: **Low temperature NO storage of zeolite supported Pd or low temperature diesel engine emission control.** *Catal. Lett.* 2016.

4. Theis JR, Lambert C: **The effects of CO, C_2H_4, and H_2O on the NO_x storage performance of low temperature NO_x adsorbers for diesel applications.** *SAE International* 2017, **10.**

5. Jones S *et al.:* **CeO_2**
2016 **CLEERS Issue.**

498784 CuO/Co_3O_4: A Highly Active and Selective Low Temperature NO Decomposition Catalyst

Abstract:

Abstract:

The direct decomposition of NO is thermodynamically favorable within the temperatures encountered by an automotive catalytic converter; however, activity remains too low for practical application [1]. Platinum group metals (PGM), such as Pd, are active for NO decomposition at temperatures above 600 °C, and attempts to improve their activity below 600 °C have been unsuccessful due oxidation of the metal and subsequent deactivation [2]. Regarding non-PGM catalysts, Cu-ZSM5 is perhaps the most widely known material displaying low temperature (≤ 500 °C) direct NO decomposition activity, however it has propensity to produce an undesired side product: the greenhouse gas N_2O [3]. The spinel oxide Co_3O_4 does not exhibit direct NOx activity as high as Cu-ZSM5, but even at 350 °C it is able to decompose 50% of N_2O to N_2 [4,5]. To address the performance gap below 500 °C, we therefore propose combining the benefits of both CuOx and Co_3O_4 into a single NO decomposition catalyst. Supported copper oxide on cobalt oxide CuOx/Co_3O_4 was synthesized by incipient wetness impregnation. We hypothesized that this material can utilize the N_2O decomposition capabilities of Co_3O_4 to reduce the N_2O generated by NO decomposition over the CuOx species.

The CuOx/Co_3O_4 were synthesized with Cu surface densities ranging from 0.7 to 10.3 Cu/nm^2 and characterized by XRD, XPS, and TEM. The Cu surface density corresponding to the monolayer loading for CuOx/Co_3O_4 was found to be 6.7 Cu/nm^2, therefore, the catalysts set included sub-monolayer, monolayer, or 3D-microcrystalline CuOx species. Thus, this catalyst set formed a strong basis for identifying a direct NOx decomposition structure-activity relationship. The catalysts were evaluated for direct NOx decomposition activity using a packed bed reactor. The maximum activity for NO decomposition by CuOx/Co_3O_4 was found at the CuOx surface density corresponding to monolayer. There is a subsequent decrease in the activity at higher surface densities due to the presence of CuOx microcrystals. When compared to Cu-ZSM5, the specific activity from 300-400 °C is similar, however, the surface area normalized (areal) activity exceeds that of Cu-ZSM5 as the CuOx/Co_3O_4 has a specific surface area that is an order of magnitude lower. Additionally, at or below 400 °C, the N_2 selectivity is higher over CuOx/Co_3O_4 compared to Cu-ZSM5, as a result of decreased N_2O formation. The CuOx/Co_3O_4 catalysts maintain relatively good activity in the presence of O_2 and CO_2. Nitric oxide adsorption/desorption reveals an optimized adsorption strength over CuOx/Co_3O_4. Additional mechanistic insight from *in situ* FTIR will also be presented.

References:

1. H. Falsig, T. Bligaard, C.H. Christensen, J. K. Norskov, *Pure Appl. Chem.*, Vol. 79, No. 11, pp. 1895–1903, 2007.
2. G.K. Reddy, C. Ling, T. Peck, H. Jia, *RSC Adv.*, 2017, 7, 19645
3. B. Moden, P.D. Costa, D. K. Lee, E. Iglesia, *J. Phys. Chem. B* 2002**,** *106,* 9633-9641
4. P.W. Park, J.K. Kil, H.H. Kung, M. C. Kung, *Catal. Today*, Volume 42, Issues 1-2, June 1998, 51-60.
5. L. Xue, H. He, C. Liu, C. Zhang, and B. Zhang. *Environ. Sci. Technol.*, **2009**, 43 (3), 890-895.

488866 Effect of Different Structure Directing Agents on Deactivation of Cu/SAPO-34 during Low-TemperatureÂ NH_3-SCR

Abstract:

Cu/SAPO-34 and Cu/SSZ-13 having CHA structure have attracted significant attention because of their high activity and N_2 selectivity during SCR reaction as well as superior hydro carbon resistance. The Cu/SAPO-34 has shown better hydrothermal durability than Cu/SSZ-13 at high temperature [1]. However, we have earlier observed that Cu/SAPO-34 prepared by the morpholine route deteriorated under water exposure at low temperature, where the activity decreased from 87% conversion to 6% conversion after 9 h of low temperature experiment [2]. Nevertheless, there is still a lack of understanding of the deactivation mechanism of Cu/SAPO-34 during low temperature NH_3-SCR reaction.

In this study, Cu/SAPO-34 using three different templates, i.e., morpholine(MO), trimethylamine (TEA), and tetraethylammonium hydroxide (TEAOH), as structure directing agent were prepared by incipient wetness impregnation (IWI) method. Characterizations of those catalysts suggest that they have different physical/chemical properties as well as various species of Cu. For example, analysis of H_2-TPR profiles of Cu/SAPO-34 catalysts proposed that they contain different amount of Cu species depending on the structure directing agents. Those catalysts were systematically investigated for low-temperature NH_3-SCR system under water exposure in order to find the correlation of their chemical/physical structures and deactivation. It was observed that the activity of Cu/SAPO-34 declined when they were exposed to water at low temperature as observed in the previous study [2]. Interestingly, deactivation of Cu/SAPO-34 catalysts behave differently depending on the choice of template. Additionally we found that activity of Cu/SAPO-34 using MO as template after water exposure at low temperature declined regardless of a variety of regeneration methods including a dry atmosphere at 600 °C. To conclude, the choice of template over Cu/SAPO-34 catalysts significantly influence their activities and deactivation during low temperature NH_3-SCR.

[1] D.W. Fickel, E. Dâ€™Addio, J.A. Lauterbach, R.F. Lobo, Appl. Catal. B: Environ. 102 (2011) 441-448

[2] K. Leistner, L. Olsson, Appl. Catal. B: Environ. 165 (2015) 192-199

497426 A Perovskite-like Catalyst for the Simultaneous Removal of Soot and NO: Effect of the Synthesis Method

Abstract:

Due to the environmental problems associated with the use of diesel engines, and the strengthening of the regulations concerning pollutant emissions in the transport field. The study of the catalytic simultaneous removal of soot and NO at relatively low temperatures (within the range of diesel engine exhaust: 150-450°C) has taken a paramount importance in pollution control area.

Mixed oxides coming from perovskite-like materials, seems to be a promising catalyst for such purpose. In this way, the present work tested the $La_{0.7}Ag_{0.3}MnO_3$ perovskite-like material, prepared for different synthetic routes, for the simultaneous removal of soot and NO. Three different synthetic routes are compared: microwave-assisted hydrothermal (HMW), microwave-assisted (MW) and solid state (SS).

The catalytic activity of the prepared catalysts was tested in a temperature-controlled furnace inside which a soot-catalyst-sand fixed bed was placed. The mixture of carbonaceous material and catalysts was carried out with spatula (i.e. loose contact), keeping always a 1:9 weight ratio. The soot model used in this study consisted of a carbon black, FW200, from Orion Engineering. The soot-catalyst-sand mixture was exposed at a constant heating rate of 10°C/min up to 600°C, under a 2200 ppm NO/ 10% vol.O2/ He atmosphere. The evolution of CO_2, CO, NO, NOx, N_2,O_2 gases with temperature were followed by mass spectrometry and a KIGAZ 310 gas analyzer. The catalyst materials were also characterized by SEM, TEM, XRD, H_2-TPR and O_2-TPD.

The results show a good catalytic behavior of the materials tested for the oxidation of soot. Figure 1 and Table 1 present the temperature at which soot ignition begins (T_{ig}), and the temperatures at which the conversion of soot to CO_2 was 50%, 90% and 100%, named as T50, T90 and T100, respectively. When it is compared the catalytic activity results with the non-catalytic one; it was found an ignition temperature shift of 278°C when the HMW-synthesized catalyst was used, and a shift of 241°C for the MW-synthesized catalyst. Also, the maximum soot oxidation peak, (T50), for all the catalytic systems evaluated in this study was shifted to lower temperatures according to the following decreasing order: non-catalytic > SS > HMW > MW. Another important result regarding to catalytic soot oxidation, it is that no CO emission was found in the gas stream, indicating 100% selectivity toward CO_2. In contrast, for the non-catalytic one, the selectivity CO_2/CO ratio was near one.

Figure 1. CO_2 Yield as a function of soot oxidation temperature over $La_{0.7}Ag_{0.3}MnO_3$ perovskite-like material synthesized by different methodologies (MW, HMW, SS).

Table 1. Ignition temperature (Tig); temperatures at which the soot conversion was 50%, 90% and 100%, respectively; and CO_2 selectivity for catalytic and the non-catalytic test.

Test	Tig (°C)	T50 (°C)	T90 (°C)	T100 (°C)	S_{CO2} (%)
$NO+O_2$	428	634	676	728	N/A
HMW	150	409	465	577	100
MW	187	367	487	665	100
SS	170	439	502	679	100

Regarding NO removal, Figure 2 shows the NO evolution with temperature, for the experiments carried out with $La_{0.7}Ag_{0.3}MnO_3$ synthesized by methodologies. Figure 2 shows that during soot oxidation, the catalysts tested presented some NO adsorption/desorption events followed by a NO reduction toward N_2 around the soot oxidation peak. The catalyst synthesized through MW route shows a prominent desorption event between 210°C and 350°C compared with the other two materials. Nevertheless, its adsorption or reduction capacity was not as good as the one obtained from the HMW-synthesized catalyst.

Figure 2. NO evolution as a function of temperature for each of the synthesized catalysts tested.

Figure 3 shows the CO_2 and NO/NOx profiles as function of temperature when the HMW-synthesized catalyst was used. From figure, it is possible to see that this catalyst was able to remove 45% of the initial NO concentration as N2. Also, it is possible to see that the temperature for the NO/NOx peak reduction was found around 400 °C matching the temperature at which the evolution of CO_2 was maximum. This result shows the ability for the simultaneous removal of soot and NO of the $La_{0.7}Ag_{0.3}MnO_3$ perovskite-like material synthesized by the HMW method.

Figure 3. CO_2, NO/NOx and N_2 profiles as a function of temperature when the HMW-synthesized solid was used as catalyst.

The results presented above demonstrate that the selection of the synthetic route has a significant effect on the catalytic performance of a material. Also, it was evidenced that the perovskite-like material ($La_{0.7}Ag_{0.3}MnO_3$) synthesized by microwave-assisted hydrothermal (HMW) method has an excellent catalytic activity for the simultaneous removal of soot and NO by shifting the soot oxidation peak toward lower temperature and removing up to 45% of the initial NO as N2.

497655 Fast Cycling to Achieve High NO_x Conversion in Lean Exhaust: Role of Ceria

Abstract:

Fast Cycling to Achieve High NOx Conversion in Lean Exhaust: Role of Ceria

Zhiyu Zhou, Michael P. Harold*, Dan Luss**

Department of Chemical and Biomolecular Engineering, University of Houston, Houston, TX 77204

* *mharold@uh.edu,* ** *dluss@central.uh.edu*

The abatement of NOx ($NO+NO_2$) remains a challenge under the continuously more stringent emission standards. Commercialized deNOx systems, NOx storage and reduction (NSR) and selective catalytic reduction (SCR), function well from ~250 °C to ~400 °C, but have unsatisfactory performance outside that temperature range. Toyota researchers recently invented a new deNOx system, Di-Air (diesel NOx aftertreatment by adsorbed intermediate reductants), which involves fast injection of fuel into the exhaust feed to the NSR converter [1]. Previous studies show that ceria increases the deNOx performance of the Di-Air system in both low [2] and high [3] temperature regions. Data suggest that NOx conversion occurs via the reaction of an intermediate produced through the reaction with O_2 and NO [1, 2]. However, a recent study [3] suggests that catalysts containing ceria may involve an alternative pathway of NOx abatement via NO decomposition at high temperatures. We studied in a bench-scale flow reactor system the NO decomposition and reduction on ceria under different frequencies with CO as sole reductant.

In our study, a lean to rich feed time ratio 6 to 1 of was controlled to keep the same fuel penalty (fuel to oxidant ratio). The lean/rich cycling conditions included lean/rich times (in seconds) of 90/15, 60/10, 30/5 and 6/1. Different reaction conditions were applied to study NO conversion behavior under favorable and adverse conditions. 500 ppm NO was added to both rich phase and lean phase. 2.5% CO was only added to rich phase and O_2 with varied concentration was only added to lean phase. In some cases, CO_2 and H_2O were introduced continuously along with NO, O_2 and CO to simulate lean exhaust.

When CO is used as the sole reductant and no H_2O is introduced in the feed, NO is mainly converted to N_2 with only trace amounts of N_2O and NO_2 detected. Figure 1 shows the cycle-averaged NO conversion with four cycle times from 450 °C to 625 °C in two feed conditions. In the first case, when 1400 ppm O_2 is introduced in lean feed, Figure 1(a) shows that increase of lean/rich switching frequency enhances NO conversion. The optimized lean/rich switching frequency is always 6/1 above 500 °C. In the second case, when 2100ppm O_2 is introduced in the lean feed, Figure 1(b) shows that a different optimal lean/rich switching frequency exists for different temperatures. The optimal lean/rich switching frequency is 30/5 at intermediate temperatures (500 °C and 550 °C) and 6/1 at high temperatures (600 °C and 625 °C) respectively. Figure 1 shows that the optimal lean/rich switching frequency changes with reaction conditions (e.g., temperature and feed condition).

Figure 1. (a) NO conversion with different lean/rich switching frequency; (a) rich: 2.5%CO, 500ppm NO; lean: 500ppm NO, 1400ppm NO; (b) rich: 2.5%CO, 500ppm NO, lean: 500ppm NO, 2100ppm O_2

Figure 2 shows the cycle-averaged NO conversion with the same four cycle times at 600 °C with fixed ratio of lean and rich feeds. The short cycle limit represents NO conversion obtained with total mixed lean and rich feeds. The long cycle limit represents weight averaged NO conversion in lean phase and rich phase. The optimal lean/rich switching frequency of 30/5 exceeds those of both short and long cycle time limit. This shows that the lean/rich cycling operation is beneficial to NO conversion.

Figure 2. NO conversion at 600C, with a fixed switching frequency as 90s lean/15s rich; rich: 2.5%CO, 500ppm NO, lean: 500ppm NO, 3600ppm O_2

Figure 3 shows that NO conversion changes when CO_2 and H_2O are fed. By comparing the blue line (reference) with either red line (with H_2O only) or purple line (with CO_2 only), it is obvious that both H_2O and CO_2 inhibit NO decomposition on ceria but H_2O has a smaller detrimental effect than CO_2. The green line (with both CO_2 and H_2O) lies between the red line (with H_2O only) and the purple line (with CO_2 only), which indicates that H_2O can mitigate the detrimental effect of CO_2.

Figure 3. NO conversion at 550C, with a fixed switching frequency as 90s lean/15s rich; (rich: 2.5%CO, 500ppm NO, lean: 500ppm NO, varied O_2 concentration, CO_2 and H_2O in both lean and rich if introduced)

This study on NO decomposition on ceria during lean and rich cycling helps explain the beneficial function of ceria on NOx reduction at high temperatures, which also provide guidance for optimization of catalyst formulation and operation strategies.

Reference

1. Y. Bisaiji et al., *SAE Int. J. Fuels Lubr*. 5 (2012) 380-388
2. Y. Zheng et al., *Catal. Today*, 276 (2016), 192-201.
3. Y. Wang et al., *Top. Catal.*, 59 (2016), 854-860

495352 NO Reduction By CO over CeO_2 Supported Co_xO_y Catalysts

Abstract:

Nitrogen oxides (NO_x) are a main contributorâ€"in addition to hydrocarbons (HC), carbon monoxide (CO), sulfur oxides (SO_x), and particulate matter (PM)â€"to global air pollution [1]. Significant investigations have been performed for catalytic NO_x reduction resulting in technologies such as NH_3-SCR and NO_x storage and reduction [2]. In the last several decades, platinum group metals (PGMs) and zeolites have been extensively applied to NO_x abatement [3]. Although these catalysts exhibit a high level of catalytic activity, NO_x decomposition over PGMs and zeolites has some practical issues such as high cost and high light-off temperature [4]. To resolve these issues, transition metal oxide catalysts have been investigated for low temperature NO_x decomposition reaction by CO with varied O_2 concentrations, which is a primary reaction in three-way catalysis (TWC).

In this study, the series of CeO_2 supported Co_xO_y catalysts were prepared by incipient wetness impregnation method with different metal oxide loadings (0.5 wt%-30 wt%) for NO reduction by CO. Catalytic activity, catalyst physicochemical property, and molecular/electronic structure were investigated by FTIR, BET, Raman spectroscopy, X-ray diffraction (XRD), and X-ray absorption spectroscopy (XAS). No diffraction peaks due to crystalline CoO or Co_3O_4 were observed in XRD patterns under to 10 wt%, while nanocrystaline Co_3O_4 peaks at ~610cm^{-1} was observed in the Raman spectra with 6wt% sample. Both Raman and XRD spectroscopy results provided that 5~6wt% Co_xO_y /CeO_2 contains monolayer coverage. It was observed that both Co_xO_y coverage and oxidation state of Co are closely related to the NO reduction by CO. Monolayer and above monolayer coverage catalysts showed 80~90% CO and NO conversion at 300~400°C with high N_2 and CO_2 selectivity (>90%). Based on the time on stream results, it was also observed that the crystalline Co_3O_4 nanoparticles above monolayer coverage improve the stability of catalyst.

ACKNOWLEDGMENT: We gratefully acknowledge the financial support for this study from the Department of Materials Science & Chemical Engineering at Stony Brook University through start-up research funding.

[1]. K. Skalska, et al. Sci. Total Env., 408 (2010) 3976-3989, [2]. J.H. Kwak, et al., J. Catal., 287 (2012) 203-209, [3]. S. Roy, et al., Applied Energy, 86 (2009) 2283-2297, [4]. J. Liu, et al., Applied Catalysis B: Env., 84 (2008) 185-195

498251 Nb_2O_5 Promoted Base Metal Oxide Catalysts Used for Air Pollution Abatement in Stationary Plants

Abstract:

Catalytic oxidation is a very efficient way for the abatement of low concentrations of Volatile Organic Compounds (VOCs). The most active catalysts used for oxidation are supported noble metals. However the main drawback with noble metals is their high price. Our goal is to consider transition metal oxides as possible replacements for expensive precious metals for low-temperature stationary abatement applications. Current research in our group has found that niobium oxide (Nb_2O_5) is worthy of additional studies as a promoter in altering the oxidation states of base metal oxides. Consequently, bimetallic catalysts supported on Nb_2O_5 were synthesized to achieve the development of stable oxidation. Both Nb_2O_5 and Al_2O_3 carriers were pre-calcined at 800ËšC for 2hr. Then, the calcination temperature for metal-supported catalysts was experimentally determined using temperature programmed decomposition in air using thermogravimetric analysis (TGA). The C_3H_8 oxidation reactions were carried out in a fixed-bed quartz flow reactor at atmospheric pressure. The reaction conditions were as follows: C_3H_8: 0.1 vol. %, O_2: 18.9 vol. % and N_2 balance. Under ambient conditions, the gas hourly space velocity (GHSV) was 65 000 h^{-1} and catalysis volume is 0.1 ml. Effluent gases of were continuously analyzed by micro GC. Isothermal aging tests were conducted to assess the stability of the catalysts at 400 ËšC in the air. Results suggest a strong interaction between niobium oxide and the active components.

499320 N Doped TiO_2 nanoparticles Synthesized By One Step Liquid Flame Aerosol Method (LFSP): Understanding the Effect of Synthesis Parameters and Photocatalytic Degradation of VOCs

Abstract:

Environmental photocatalysis Nitrogen doped titanium dioxide is attracting a continuously increasing. Conventional wet synthesis methods need multiple processing steps, high energy input and long processing times, In contrast, flame synthesis, especially liquid flame spray pyrolysis (LFSP), is a versatile one-step process for synthesizing a variety of anion doped nanoparticles. Nitrogen-doped TiO
synthesis. A combination of various physico-chemical characterizations were performed. The photocatalytic activity and stability of the N-doped TiO_2 samples were evaluated through using the decomposition of phenol as model reaction under visible light irradiation. Characterization results show that the nitrogen dopants have a significant effect on the crystallite size and optical absorption of TiO_2. It is found that the N-doped TiO_2 catalysts have enhanced absorption in the visible light region, and exhibit higher activity for photocatalytic degradation of VOCs. We have produced The UV-vis spectroscopy results of N incorporated TiO_2 showed enhancement of light absorption in the visible range by selected composite materials in the UV range (250-400 nm). We have found the Ti/N ratio of 11% for the materials synthesized from our XPS data. To the best of our knowledge, this is the first report which demonstrates single-step, direct flame synthesis of N-doped TiO_2 and their application as a photocatalyst better performance. The method is extremely simple, low cost, energy efficient, very effective, and can be easily scaled up for large scale production. These interesting results will be discussed in the presentation.

490379 CO_2 Reduction over Metal Cluster/TiO_2 photocatalysts

Abstract:

We previously showed how highly dispersed clusters of Cu on TiO_2 could be synthesized using a redox reaction between Cu^{2+} and Sn^{2+}[1]. Such clusters were active for CO_2 reduction. Several other experiments have also shown CO_2 conversion activity using metal clusters[2,3]. The ability to synthesize supported metal atoms and clusters to initiate CO_2 reduction could lead to cheaper, efficient single-site catalysts. In the current work we modeled using density functional theory (DFT) several metal clusters on TiO_2 in order to understand and predict catalytic activity of these materials.

A key activation step in CO_2 reduction is the formation of bent CO_2 anion species. We initially modeled small Cu_x (x=1-4) clusters on a TiO_2 anatase surface. Our results do indeed show that Cu is able to activate CO_2 into a bent configuration that can be further reduced. Charge analysis indicates that CO_2 becomes negatively charged in a bent configuration, but not in a linear adsorption mode. We analyzed charges on Cu to assign its oxidation state, as well as calculating adsorbed CO vibrational modes, a common experimental method to assign oxidation state of supported metals. Based on these results we extended our studies to other supported metal clusters, such as Pt, Pd, and Ni. By systematically modeling a large number of supported metals we are able to identify how and which metal clusters may activate $CO\hat{A}_2$. Such knowledge is crucial towards refining and designing better catalysts for CO_2 conversion.

(1) Liu, C.; Iyemperumal, S. K.; Deskins, N. A.; Li, G. *J. Photonics Energy* **2016**, *7*, 12004.

(2) Vajda, S.; White, M. G. *ACS Catal.* **2015**, *5*, 7152â€"7176.

(3) Liu, C.; Yang, B.; Tyo, E.; Seifert, S.; DeBartolo, J.; von Issendorff, B.; Zapol, P.; Vajda, S.; Curtiss, L. A. *J. Am. Chem. Soc.* **2015**, *137*, 8676â€"8679.

495756 Unsupported and Supported Au_1-O_x-$(OH)_y$-Na_z Clusters As Stable, Single-Site Gold Catalysts

Abstract:

Mononuclear metal complexes enable the maximum utilization of a metal as a catalyst. Inorganometallic clusters with a single gold cation coordinated with â€“O linkages to a shell of Na cations are shown here to be stable in aqueous solutions, and were used in dried powder (unsupported) form, and on a support, such as titania, for several catalytic reactions. A new facile, one-pot green route was used to synthesize the single-gold atom clusters, using aqueous solutions of $Au(OH)_3$ and NaOH. Incipient wetness impregnation was used to prepare supported catalysts. The single-site gold species catalyze a number of reactions, such as the water-gas shift reaction, the steam reforming of methanol, and the selective dehydrogenation and coupling of methanol to methyl formate (MF) and hydrogen products. Stability up to 250-300 ^{o}C reaction temperatures is demonstrated in cyclic tests over the course of many hours. The preparation method allows for up to 1.0 wt% Au loading as single â€“site isolated species on TiO_2, as shown by STEM and EXAFS. The catalyst is 100% selective for the dehydrogenation of methanol to MF and H_2 below 170 ^{o}C. As a counterpart, 4.8 wt% Au/TiO_2 made by conventional deposition/precipitation methods shows less than 1/10 of 1.0wt% Au_1-O_x-$(OH)_y$-Na_z/TiO_2. However, the apparent activation energy of the reaction is similar on the two catalysts, around 95 kJ/mol, which is close to that of methanol steam reforming on Au catalysts on various supports. Both supported and unsupported Au_1-O_x-$(OH)_y$-Na_z are active for methanol coupling, which is demonstrated by *in-situ* methanol DRIFTS. XPS and XANES analyses show that the gold state in the used catalyst is in the Au(I) state for both supported and unsupported Au_1-O_x-$(OH)_y$-Na_z.

Acknowledgments: The financial support by the DOE/BES under Grant # DE-FG02-05ER15730 is gratefully acknowledged.

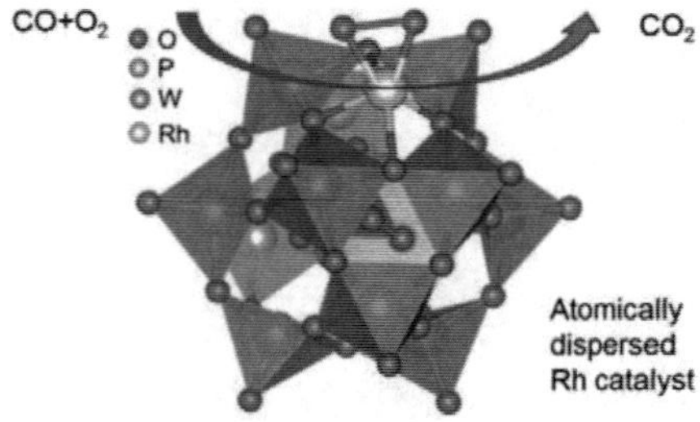

Atomically Dispersed Rhodium on Self-Assembled Phosphotungstic acid: Structural Features and Catalytic CO Oxidation Properties

Bin Zhang, Hiroyuki Asakura, Max J. Hülsey, Ning Yan*

30/10/2017

*Depart of Chemical and Biomolecular Engineering, NUS

LAB of GREEN CATALYSIS

N. Yan et al. *Ind. Eng. Chem. Res.* **2017**, *56*, 3578-3587.

1

1.1 Introduction to heteropoly acid (HPA)

2

Keggin structure, $XM_{12}O_{40}^{n-}$

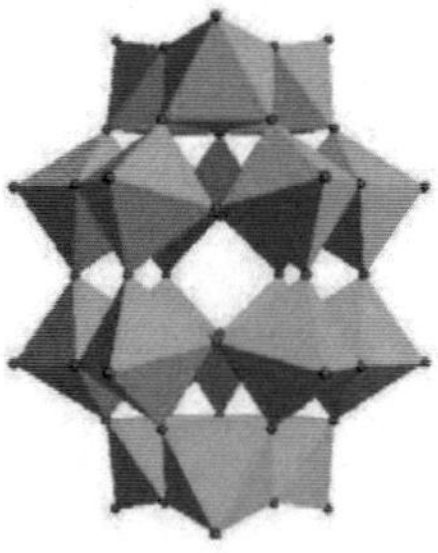

Dawson structure, $X_2M_{18}O_{62}^{n-}$

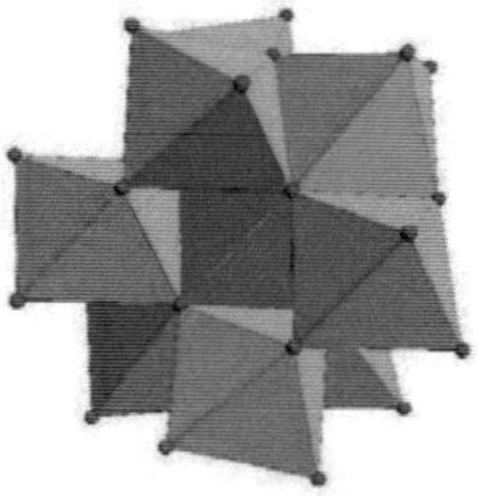

Allman–Waugh structure, $XM_9O_{32}^{n-}$

, etc.

Catalytic applications:

- Homogenous & heterogenous acidic catalyst;
- Homogenous oxidising catalyst.

Advantages:

- High affinity for polar media;
- Strong acidity and oxidising ability;
- Definite structure & various sites;

…

Limitations:

Separation issue, low surface are, etc.

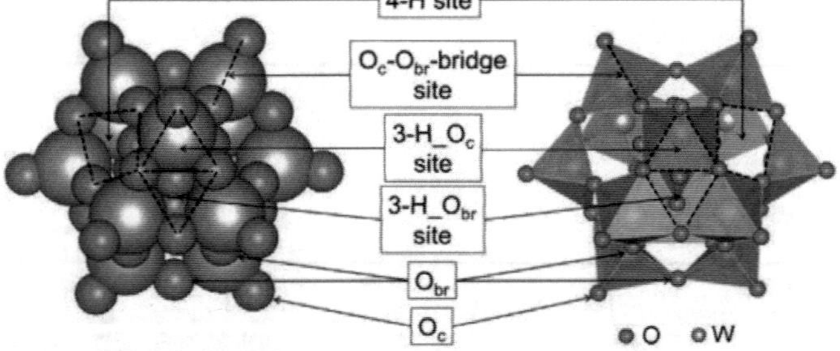

Phosphotungstic acid ($H_3PW_{12}O_{40}$, PTA)

LAB of GREEN CATALYSIS

1.2 Studies on HPA based atomically dispersed catalysts 3

❑ Organometallic complex-HPA/support to metal atom-HPA/support

Activating the catalyst be either first use in the reactor or by a reduction step such as a prehydrogenation; …

U.S. Patent No. 6, 005, 148.

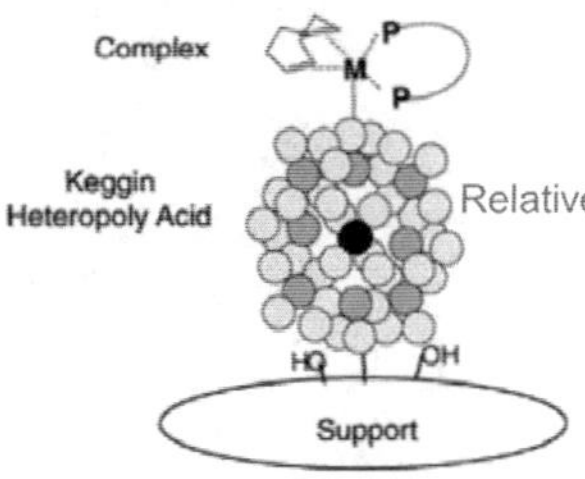

Relatively complicated system

R. L. Augustine et al. *J. Mol. Catal. A: Chem.* **2004**, *216*, 189-197.

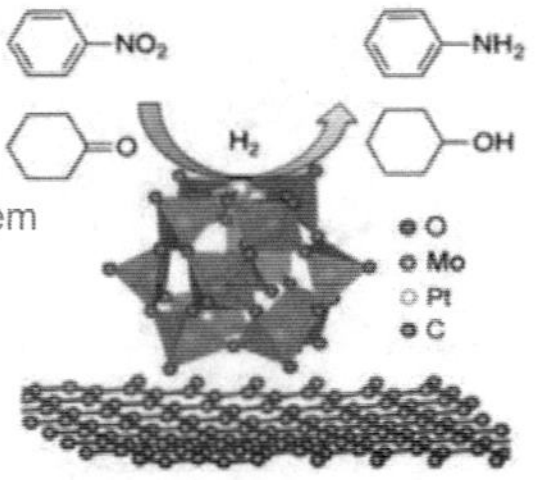

N. Yan et al. *Angew. Chem. Int. Ed.* **2016**, *55*, 8319-8323.

❑ Isolated metal species over HPA salt

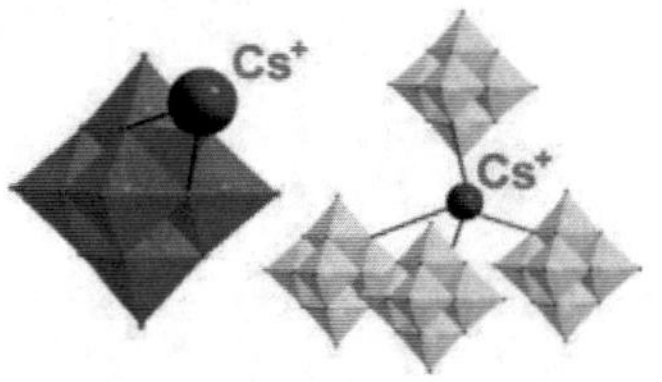

M. Nyman et al. *Phys. Chem. Chem. Phys.* **2017**, *19*, 8715-8725.

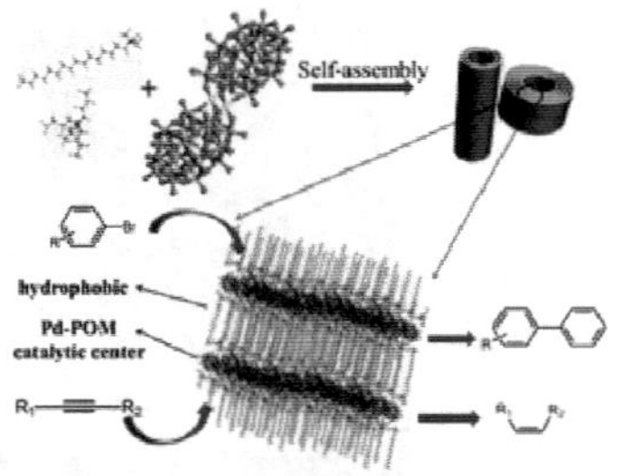

X. Wang et al. *Chem. Sci.* **2016**, *7*, 1011-1015.

LAB of REEN ATALYSIS

2.1 Preparation of our catalyst 4

Phosphotungstic acid (PTA): $H_3PW_{12}O_{40}$

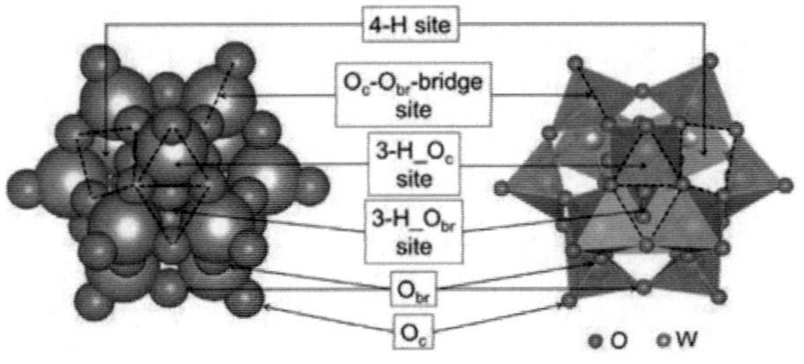

MS signals in ESI-MS spectrum can no longer be observed, suggesting the new species is a neutral Rh1PTA1 complex.

Using the self-assembly method, atomically dispersed Rh catalyst was produced with a Rh loading of 0.9 wt% (0.9Rh/NPTA). Control samples: NPTA and 0.2Rh/NPTA.

ESI-MS spectra

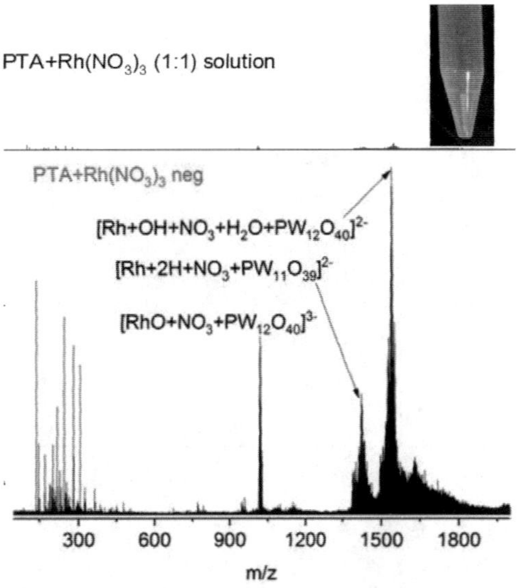

LAB of REEN ATALYSIS

2.2 Catalyst characterization

5

0.9Rh/NPTA

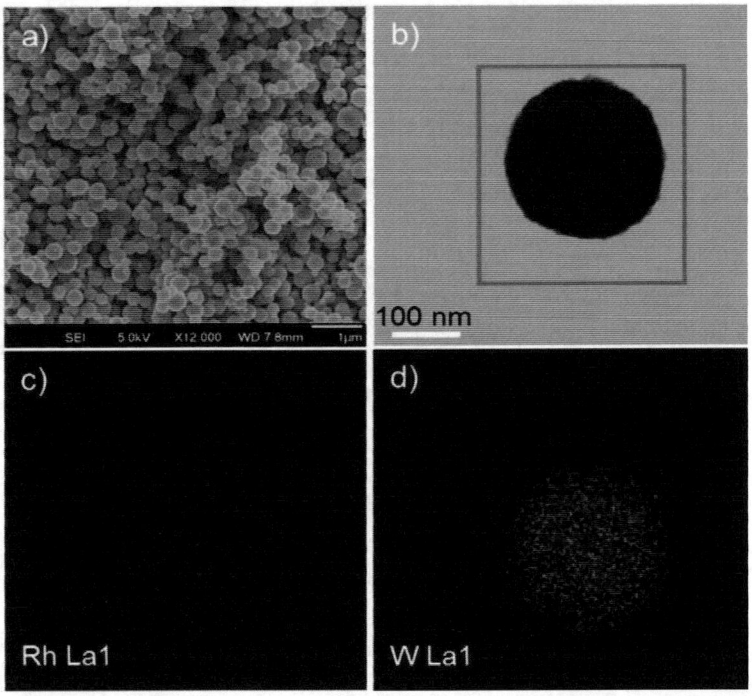

LAB of REEN ATALYSIS

2.2 Catalyst characterization

6

BET

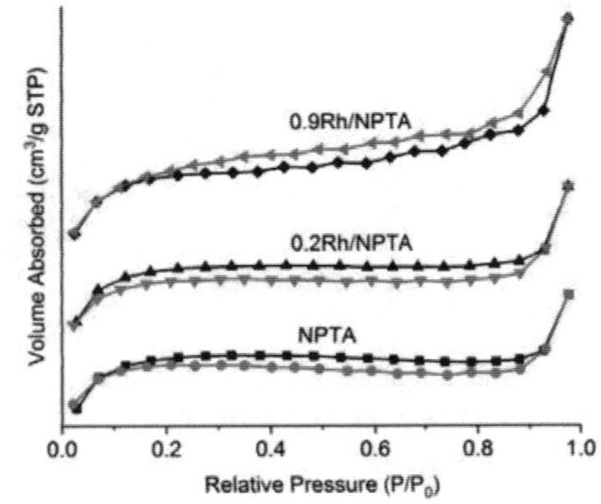

Sample	BET surface area (m^2/g)	Pore volume (cm^3/g)	Average pore size (nm)
NPTA	102	0.061	2.4
0.2Rh/NPTA	98	0.071	2.7
0.9Rh/NPTA	92	0.077	3.4

ATR-IR & XRD

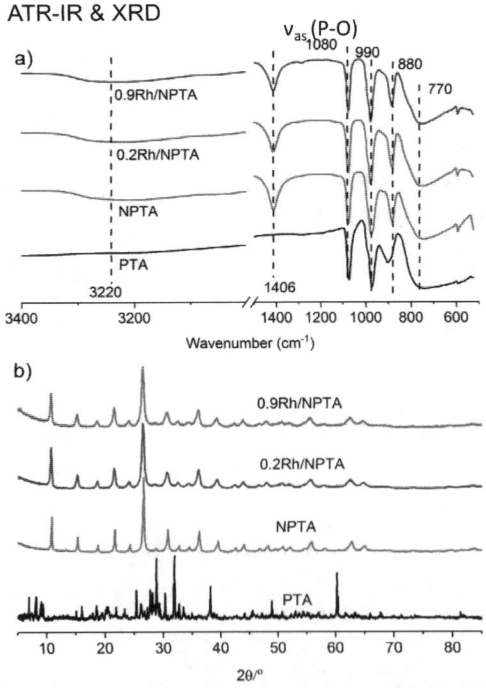

LAB of REEN ATALYSIS

2.2 Catalyst characterization

7

EXAFS

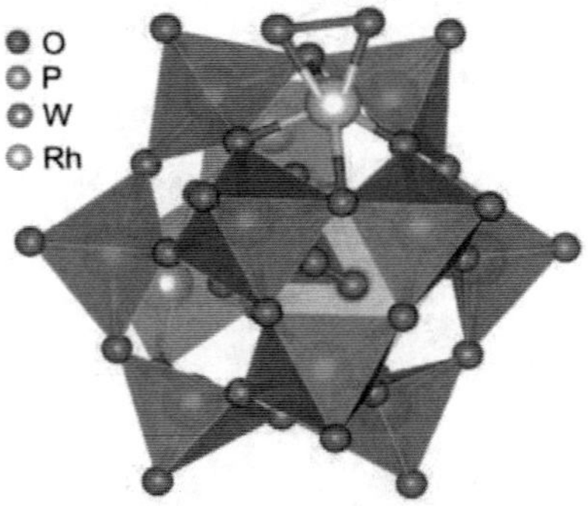

Single Pt atom on PTA

Adsorption site	E_{ad} (eV)
(d) 4-H	-4.96
(e) O_c-O_{br} bridge site)	-3.12
(f) 3-H_O (move to O_{br}-O_{br} bridge site)	-3.24

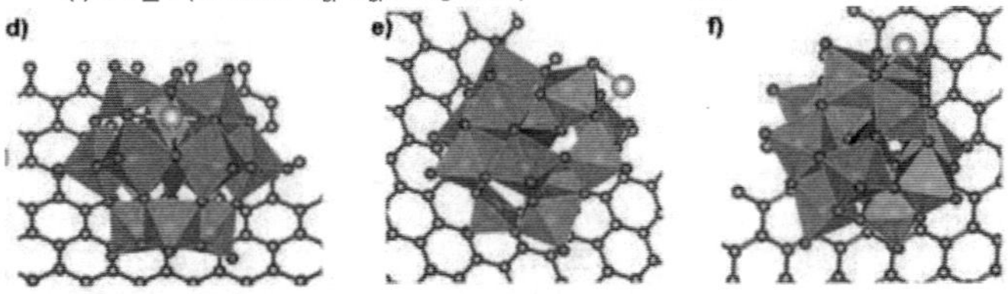

Sample	Shell	C.N.	R (Å)	σ^2 (Å^2)	E_0 (eV)
0.9Rh/NPTA	Rh–Rh	-	-	-	-
	Rh–O	5.6±0.8	2.04±0.01	0.0009±0.0009	-0.3±2.0
0.2Rh/NPTA	Rh–Rh	-	-	-	-
	Rh–O	6.7±1.9	2.02±0.02	0.0027±0.0021	2.5±4.1
Rh foil	Rh–Rh	12	2.69±0.00	0.0036±0.0006	-5.6±1.2

C.N., σ^2, R, and ΔE_0 are the coordination number, interatomic distance, Debye−Waller factor, and shift in the edge energy.

LAB of REEN ATALYSIS

2.2 Catalyst characterization

8

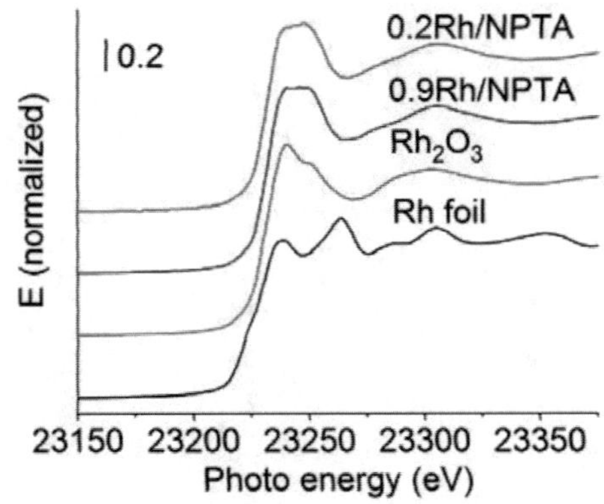

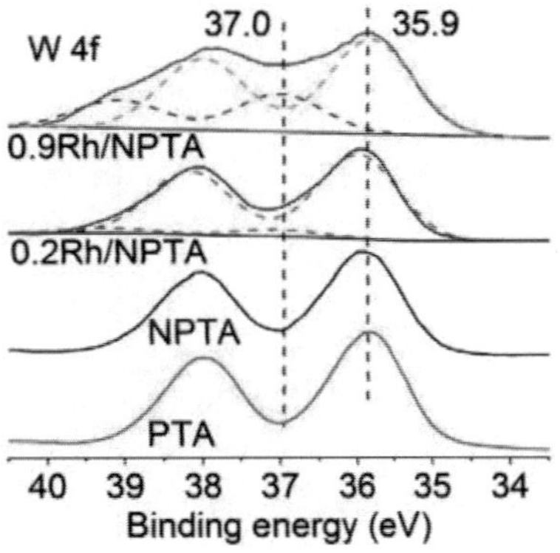

The intensities of white-line at 23240 eV are as follows: Rh foil (1.00), Rh_2O_3 (1.22), 0.9Rh/NPTA (1.18), 0.2Rh/NPTA (1.18).

The ratio between the area of the new peaks and the total area is 1/3.5 for 0.9Rh/NPTA (ICP-OES shows Rh to PTA is 1:4 (molar).

LAB of REEN ATALYSIS

2.3.1 Catalytic performance for CO oxidation reaction

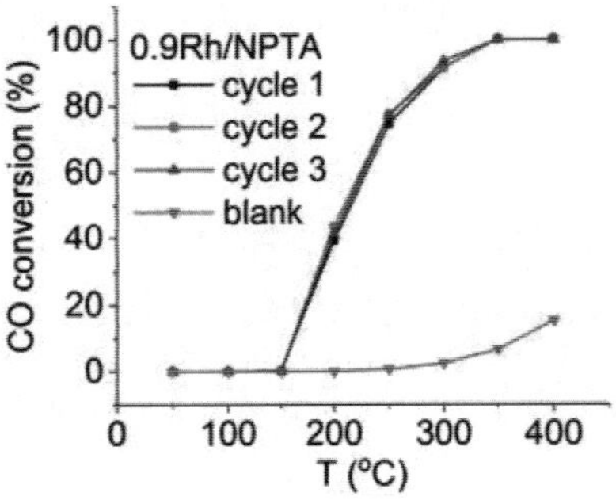

DRIFT spectrum after reaction

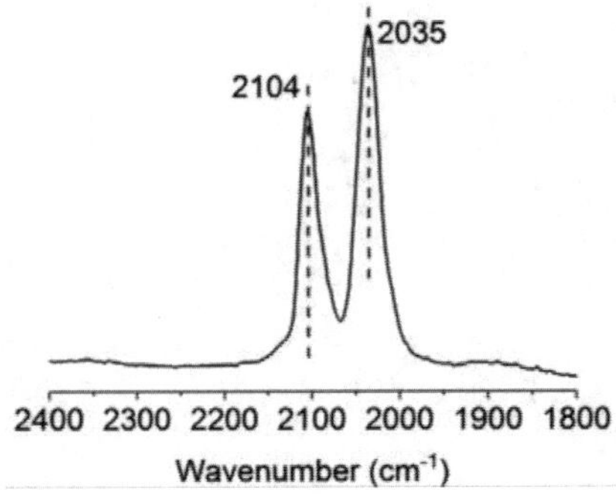

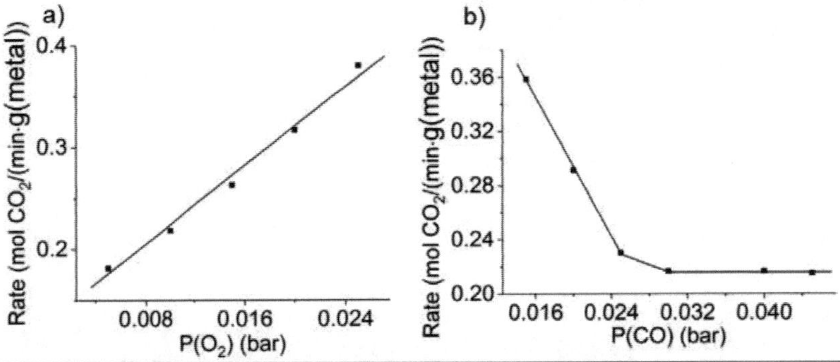

LAB of REEN ATALYSIS

Reaction conditions: 23 mg catalyst, total flow rate: 80 mL/min; gas mixtures: 2.5% CO, 2.5% O_2 and 95% Ar. GHSV = 24000 h^{-1}

2.3.1 Catalytic performance for CO oxidation reaction

CO titration experiments

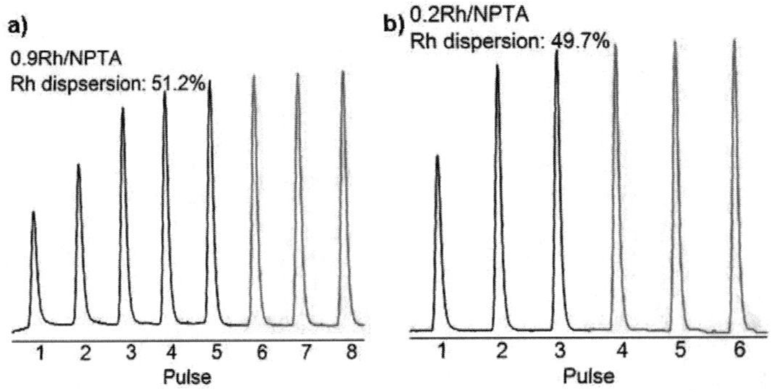

Activation energy derivation

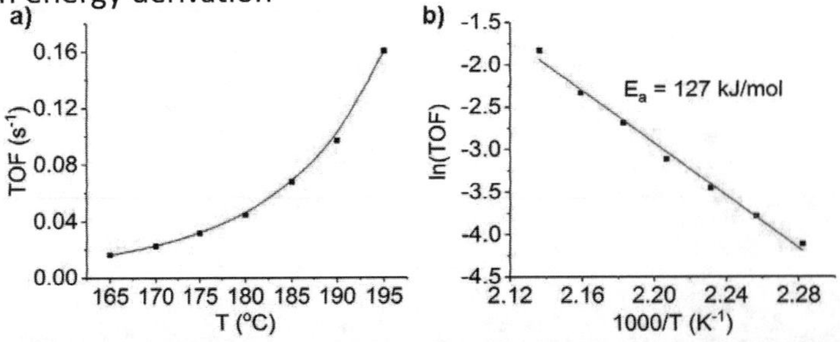

LAB of REEN ATALYSIS

2.3.2 Catalytic pathway studies

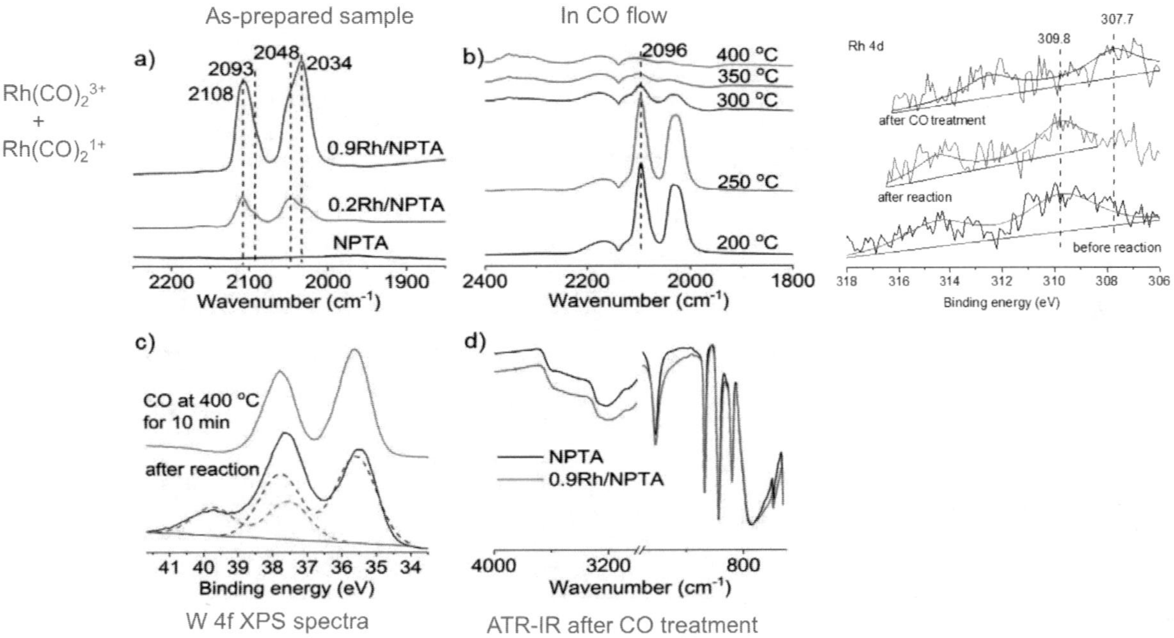

It is speculated that the catalytic cycle starts with partial reduction of Rh^{3+}, achieved via removal of an oxygen atom on PTA by CO to form CO_2.

LAB of REEN ATALYSIS

2.3.3 Catalytic pathway studies

In the presence of $CO + O_2$ (1 : 1)

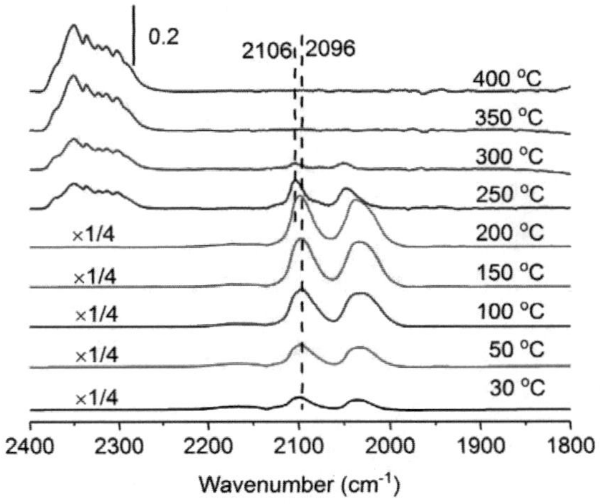

No signal of CO_2 was observed below 200 °C, suggestion the formation of the $Rh(CO)_2$ species is due to the adsorption of CO and O_2 on Rh atoms instead of chemical reactions

In the presence of $CO + O_2$ (1 : 4)

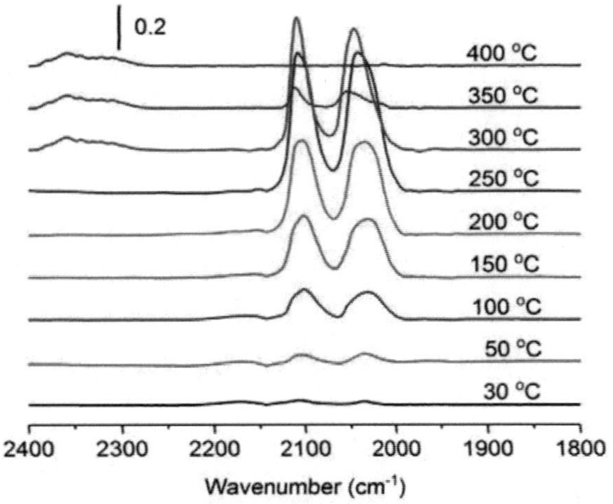

Proving Rh preferentially binds to O_2 at lower temperature while binds to CO stronger at elevated temperatures

LAB of REEN ATALYSIS

2.3.3 Proposed mechanism for CO oxidation over 0.9Rh/NPTA 13

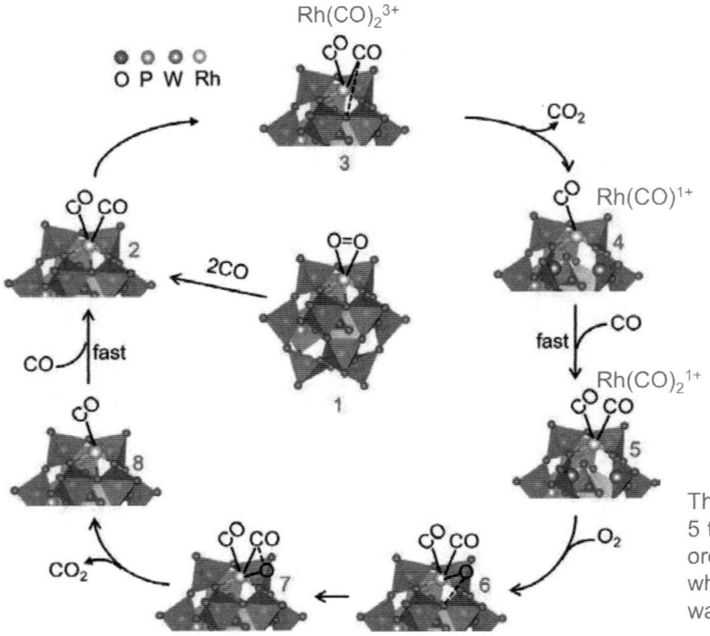

The rate-determining step is proposed to be 5 to 6, as suggested by the positive reaction order of oxygen, and the DRIFT analysis where oxidation of Rh^{1+} back to Rh^{3+} by O_2 was only observed above 200 °C.

LAB of REEN ATALYSIS

2.3.3 Proposed mechanism for CO oxidation over 0.9Rh/NPTA 14

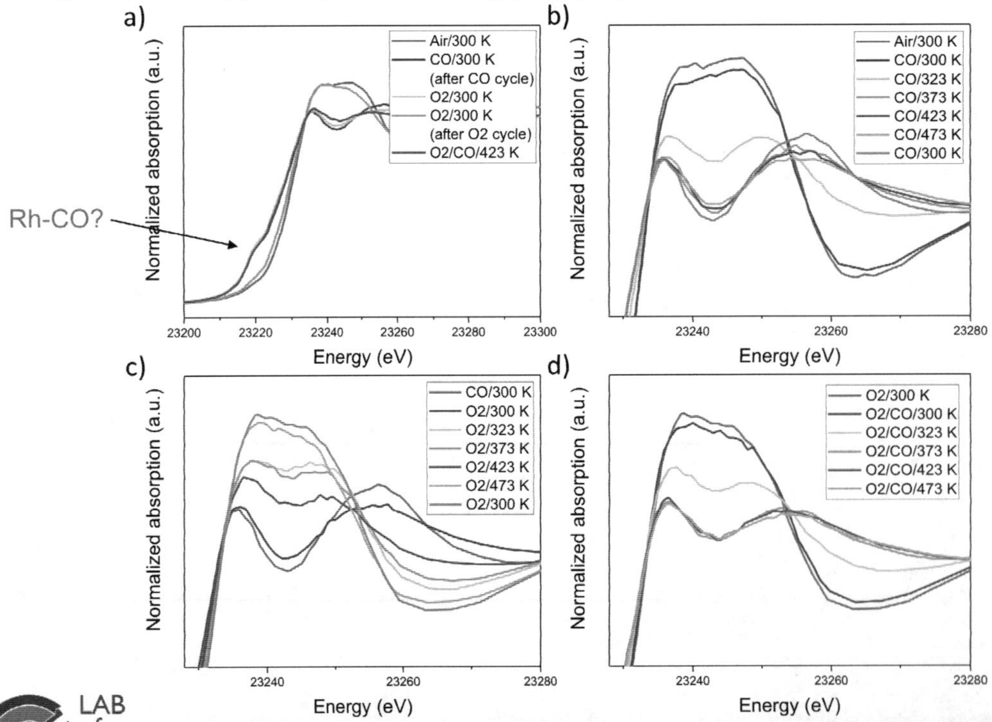

LAB of REEN ATALYSIS

3. Summary

- We have shown that high loading, atomically dispersed Rh catalyst could be achieved by stabilizing Rh on PTA through pure Rh-O interaction with a coordination number of 6 .

- Unlike Rh_2O_3 that is inactive in CO oxidation, 0.9Rh/NPTA exhibits reasonable activity and high thermal stability in CO oxidation at temperatures higher between 150 °C and 400 °C. Rh remains atomically dispersed during the entire cycle, forming dicarbony species with alternating +3 and +1 valent state. In-situ FITR and XAS analyses provided solid evidence.

- PTA does not only act as a support to stabilize single-atom Rh species but also plays a critical role in transferring oxygen to form CO_2 via repeated deoxygenation-oxidation steps. The dynamic geometric and electronic characteristics of the Rh/NPTA catalyst make it worth evaluating in several other reactions.

LAB of REEN ATALYSIS

Acknowledgement

- I would like to thank Dr. Ning Yan for his constructive and patient guidance.

- I am thankful to Dr. Hiroyuki Asakura in ESICB, Kyoto University for XAS measurements and data analyses.

- I express my sincere gratitude to my team members for the cooperation and help.

Thank you for your attention!
Q & A please

LAB of REEN ATALYSIS

501255 High CO Oxidation Activity on Pt Single Atoms and Clusters Supported on $MgAl_2O_4$

Abstract:

The catalytic oxidation of CO on Pt catalysts has practical importance in automotive catalysis and also serves as a probe reaction for fundamental understanding of heterogeneous catalysis. In this study, Pt single-atoms and clusters supported on $MgAl_2O_4$ were prepared by strong electrostatic adsorption (SEA) at constant pH. The synthesis conditions and pretreatment conditions for achieving either single atoms or clusters will be presented. The catalysts were quantitatively characterized by in-situ diffuse reflectance infrared Fourier transform spectroscopy (DRIFTS) with adsorbed-CO as probe molecule to provide information on adsorption strength and geometry. At present, supported Pt nanoparticles have been well studied for CO oxidation, however the mechanism on small clusters and single atoms is still under debate. During CO oxidation, the surface of Pt nanoparticles is poisoned by a high coverage of CO and the reaction shows reaction orders of -1 in CO and 1 in O_2 (rate = $k[O_2]^1[CO]^{-1}$). Here, we report that Pt single atoms and clusters present a very different kinetic behavior from nanoparticles under the same reaction condition. On single atoms, the reaction order is 0.8 in CO but 0 in O_2 (rate = $k[O_2]^0[CO]^{0.8}$). More surprisingly, Pt clusters show positive reaction orders in both CO and O_2 which we hypothesize a result of non-competitive adsorption of CO and O_2 on adjacent sites. Additionally, Pt single atoms show a much higher activity (~25x) and lower apparent activation energy (40 (kJ/mol)) than nanoparticles (80 (kJ/mol)). Using in-situ DRIFTS, we will compare the most stable reaction intermediates on Pt single atoms and clusters and discuss the possible reaction mechanisms leading to the observed reaction orders. The high activity and lack of poisoning by CO make Pt single atoms and clusters ideal candidates for designing low temperature oxidation catalysts.

Ni/Cerium-Titanate Oxide Catalysts for Dry Reforming of Methane

Sachin U. Nandanwar[a,b], Yunkai Zou[a], Joseph H. Holles[b], Jing Zhou[a], Samantha Hulett[c], Michael Cuddy[c]

Department of Chemistry[a] & Chemical Engineering[b]
University of Wyoming, Laramie, WY 82071
Department of Chemistry[c], Northwest College,
Powell, WY 82465

UNIVERSITY OF WYOMING

Introduction

What is the importance of DRM?

- Production of syngas ($CO + H_2$) from the two greenhouse gases (CH_4 and CO_2) over heterogeneous catalysis
- Syngas converted into liquid fuels, ammonia/urea production, synthetic natural gas, and gaseous fuel.

(Wood et al. J. Natural Gas Sci. Eng. 9, **2012**, 196-208)

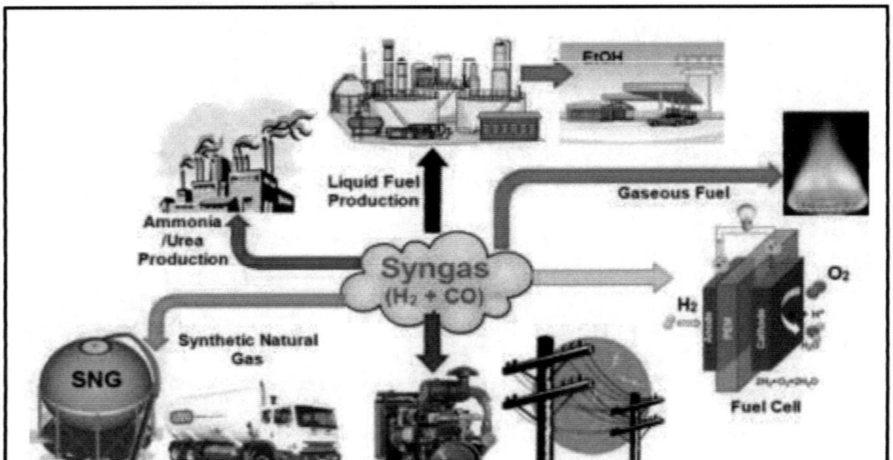

2

Reactions:

Overall DRM reaction:

$$CH_4 + CO_2 \underset{\text{Catalyst}}{\overset{\blacktriangle}{\rightleftharpoons}} 2CO + 2H_2 \quad (\Delta H_{298K} = +247\ kJ\ mol^{-1};\ \Delta G^\circ = 61770 - 67.32T)$$

Other possible reactions

Methane decomposition

$$CH_4 \rightleftharpoons C_{(S)} + 2H_2 \qquad (\Delta H_{298K} = +75\ kJ\ mol^{-1};\ \Delta G^\circ = 2190 - 26.45T)$$

CO disproportionation

$$2CO \rightleftharpoons C_{(S)} + CO_2 \qquad (\Delta H_{298K} = -171\ kJ\ mol^{-1};\ \Delta G^\circ = -39\ 810 + 40.87T)$$

Reserve water gas shift reaction

$$CO_2 + H_2 \rightleftharpoons CO + H_2O \qquad (\Delta H_{298K} = +41\ kJ\ mol^{-1})$$

Steam reforming of CH_4

$$CH_4 + H_2O \rightleftharpoons 3H_2 + CO \qquad (\Delta H_{298K} = +206\ kJ\ mol^{-1})$$

(Pakhare and Spivey Chem. Soc. Rev. 43, **2014**, 7813-7837)

Previous Catalysts Investigated:

$Ni/\gamma\text{-}Al_2O_3$ (Kim et al. Catal. Sci. Technol. 6, **2016**, 2060–2064)

Ni/TiO_2 (Shinde and Madras RSC Adv. 4, **2014**, 4817-4826)

Ni/CeO_2 (Liu et al. Angew. Chem. Int. Ed. 55, **2016**, 7455-9)

$Ni/CeO_2\text{-}ZrO_2$ (Kambolis et al. Appl. Catal. A 377, **2010**, 16-26)

$Pt/CeO_2\text{-}ZrO_2$ (Damyanova et al. Appl. Catal. B 89, **2009**, 149-159)

Ni/ceria–zirconia (Wolfbeisser et al. Catal. Today, 277, **2016**, 234-245)

Existing Catalyst Shortcomings:

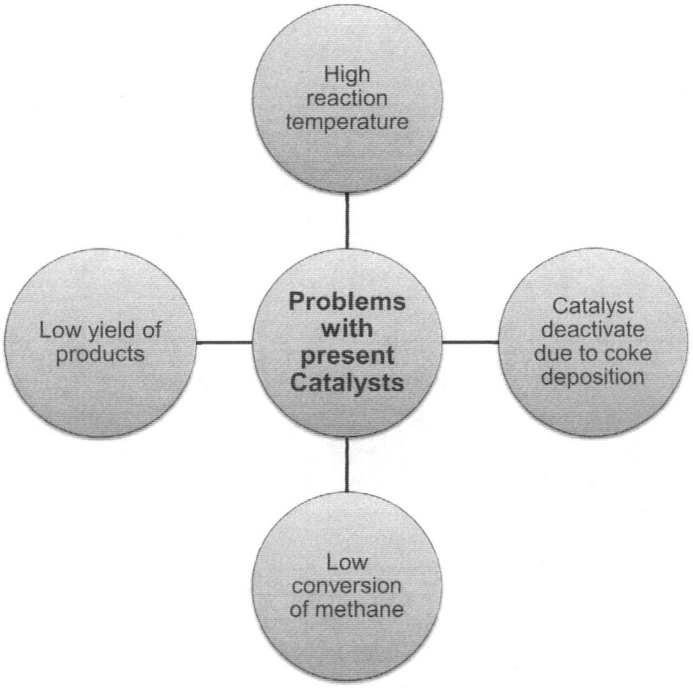

5

Research Objectives:

1
- Design and develop effective metal supported oxide catalysts

2
- CH_4 and CO_2 activation at low temperature
- High product yields

3
- Developed catalyst should have a long term stability

6

Schematic of Bifunctional Mechanism:

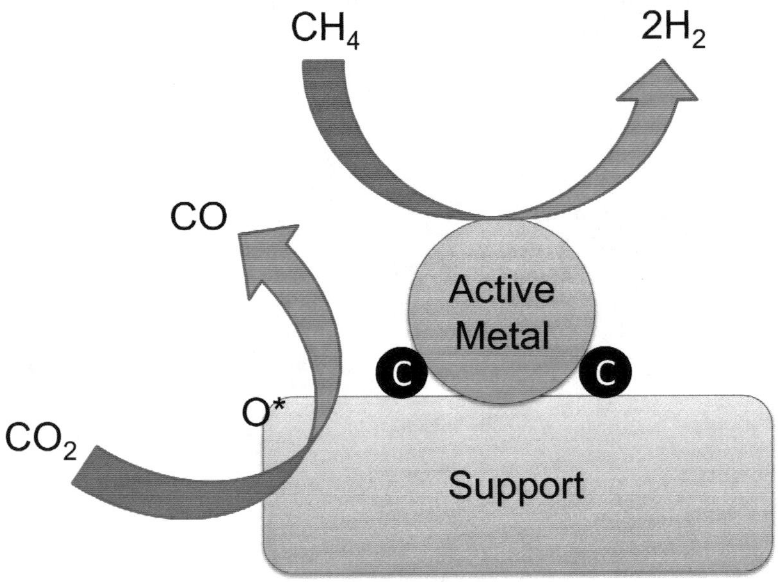

CO_2 activation occurs on the oxide support and/or at the metal-oxide interfaces to form CO+O

CH_4 activation on metal sites

Catalyst deactivates due to deposition of carbon

7

Selection of Active Metal

- ❑ Noble metals: Pt, Rh and Ru are highly active for DRM and more resistance to coke formation than other transition metals
- ❑ But, they are expensive

Ni-based catalysts (Bardford et al. Appl. Catal. A 142, **1996**, 73-96)

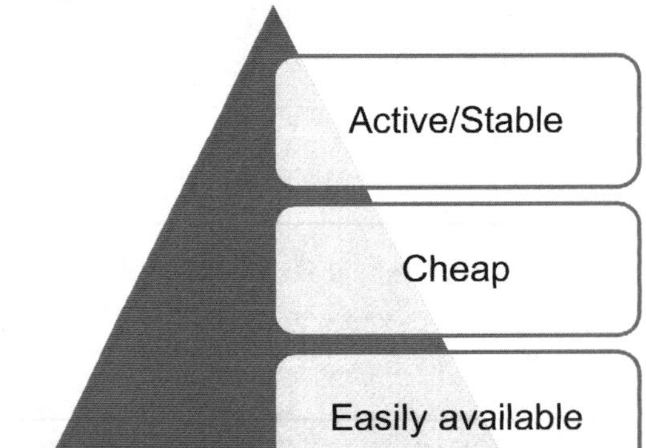

8

Multifunctional Ni/$Ce_{1-x}Ti_xO_y$ Catalysts:

Ni as an active metal for CH_4 activation

Ti-doped ceria as catalytic supports:

Oxygen buffer and good redox properties of ceria

Readily transfers between Ce^{3+} and Ce^{4+}

$$CeO_2 \leftrightarrow CeO_{1.5} + Ov \text{ (oxygen vacancy)}$$

Disperses Ni as small particles and prevents the coke formation

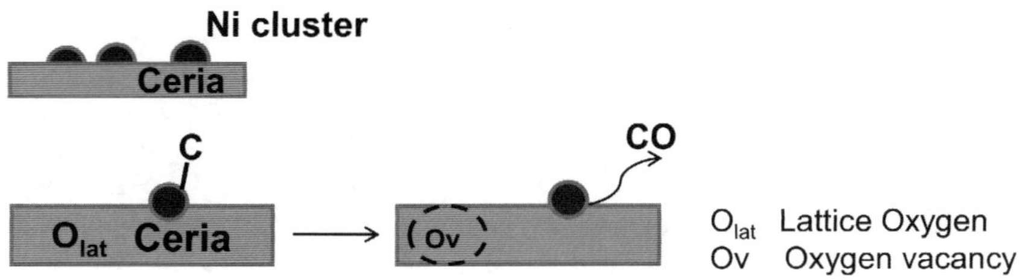

Adding Ti into ceria can enhance its stability and oxygen storage capacities at high temperatures

Research Approach:

Catalyst Synthesis ⟷ **Characterization** ⟷ **Stability Test**

Sol-gel methods

- ✓ $Ce_{1-x}Ti_xO_y$ Support
  varying Ce/Ti ratios
  CeO_2
  $Ce_{0.9}Ti_{0.1}O_y$
  $Ce_{0.8}Ti_{0.2}O_y$
  $Ce_{0.7}Ti_{0.3}O_y$
  $Ce_{0.5}Ti_{0.5}O_y$
- ✓ Ni/$Ce_{1-x}Ti_xO_y$
  0.5-10 wt.% loading of Ni

structure
morphology
composition
surface area
reactant conversion
product yield
...

XRD of $Ni/Ce_{1-x}Ti_xO_y$ Supports:

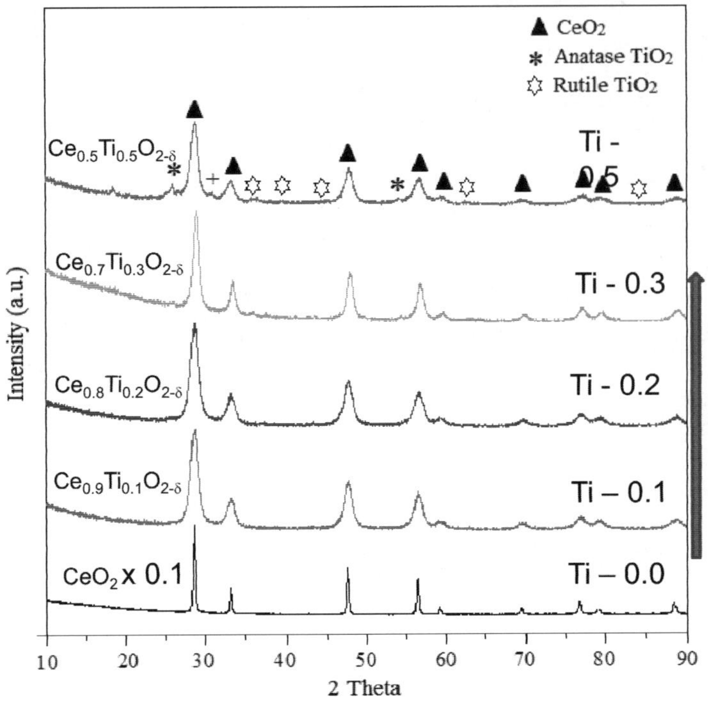

- ❑ CeO_2 peaks at 28.5°, 33.1°, 47.5°, 56.3°, 59.1°, 69.4°, 76.7°, and 79.1° could be attributed to (111), (200), (220), (311), (222), (400), (331), and (420) of the face-centered cubic (fcc) structure.

(Du et al. J. Phys. Chem. C, 116, **2012**, 10009-10016)

- ❑ No Ti peak observed up to 0.3.
- ❑ Small Ti peak observed at 25.4° for Anatase TiO_2.
- ❑ Rutile TiO_2 also observed in the spectra.

11

XRD of Various Ni Loadings on $Ce_{0.7}Ti_{0.3}O_{2-\delta}$:

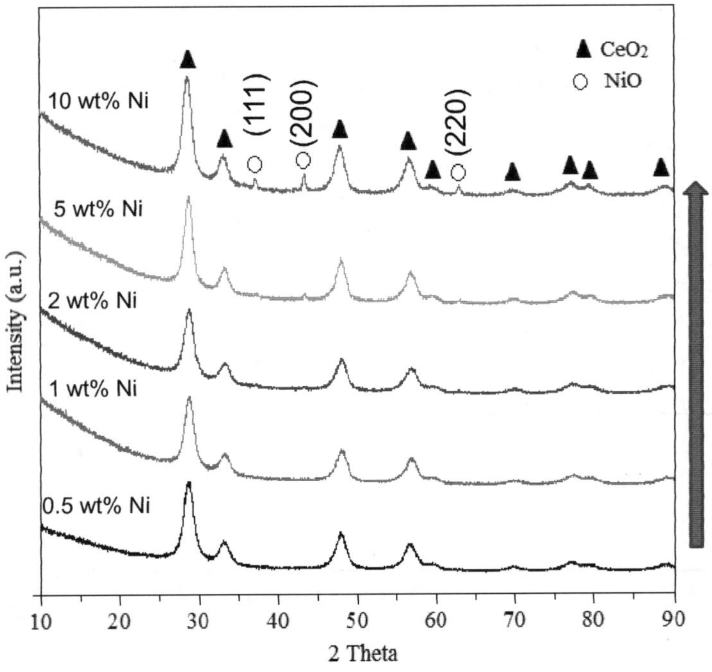

- ❑ Same CeO_2 peaks were observed.
- ❑ No Ni peaks were observed at 0.5 & 2 wt.% Ni loading, indicating good dispersion of Ni nanoparticles on support.
- ❑ At 5 & 10 wt% Ni loading: NiO peak was observed at 2θ = 37.2°, 44.3° and 62.9° for (111), (200) and (220) cubic structure.

(Nadeem et al. Ceram. Int. 40, **2014**,13819–13825)

12

BET Surface Area of Supports and Catalysts:

Entry	Catalyst	BET surface area ($m^2\ g^{-1}$)
1	CeO_2	12
2	$Ce_{0.9}Ti_{0.1}O_{2-\delta}$	13
3	$Ce_{0.8}Ti_{0.2}O_{2-\delta}$	17
4	*$Ce_{0.7}Ti_{0.3}O_{2-\delta}$*	*35*
5	$Ce_{0.5}Ti_{0.5}O_{2-\delta}$	23
6	2 wt.% Ni/CeO_2	23
7	2 wt.% $Ni/Ce_{0.9}Ti_{0.1}O_{2-\delta}$	17
8	2 wt.% $Ni/Ce_{0.8}Ti_{0.2}O_{2-\delta}$	18
9	2 wt.% $Ni/Ce_{0.7}Ti_{0.3}O_{2-\delta}$	26
10	2 wt.% $Ni/Ce_{0.5}Ti_{0.5}O_{2-\delta}$	15

2 wt.% $Ni/Ce_{0.7}Ti_{0.3}O_{2-\delta}$ surface area = 26 $m^2\ g^{-1}$.

13

Experimental Setup:

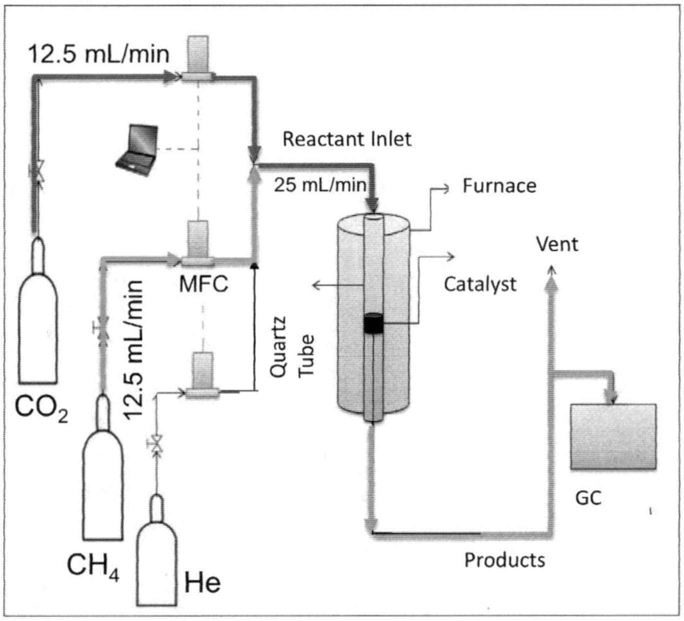

Fresh 200 mg catalyst used in the each experiment.

$$CH_4\ \text{Conversion}\ (\%) = \frac{\text{moles of}\ CH_4\ \text{converted}}{\text{moles of}\ CH_4\ \text{input}} \times 100$$

$$CO_2\ \text{Conversion}\ (\%) = \frac{\text{moles of}\ CO_2\ \text{converted}}{\text{moles of}\ CO_2\ \text{input}} \times 100$$

$$H_2\ \text{Selectivity}\ (\%) = \frac{\text{moles of}\ H_2\ \text{produced}}{2\ \text{moles of}\ CH_4\ \text{reacted}} \times 100$$

$$H_2\ \text{Yield}\ (\%) = \frac{\text{moles of}\ H_2\ \text{produced}}{2\ \text{moles of}\ CH_4\ \text{input}} \times 100$$

$$CO\ \text{Yield}\ (\%) = \frac{\text{moles of CO produced}}{2(\text{moles of}\ CH_4\ \text{input} + \text{moles of}\ CO_2\ \text{input})} \times 100$$

MFC- Mass flow controller
GC – Gas Chromatography

<u>Apparatus</u>:
Furnace: VST 1200, Carbolite Gero
GC: 7890B Agilent Technology

14

CH_4 Activation on Different Catalysts:

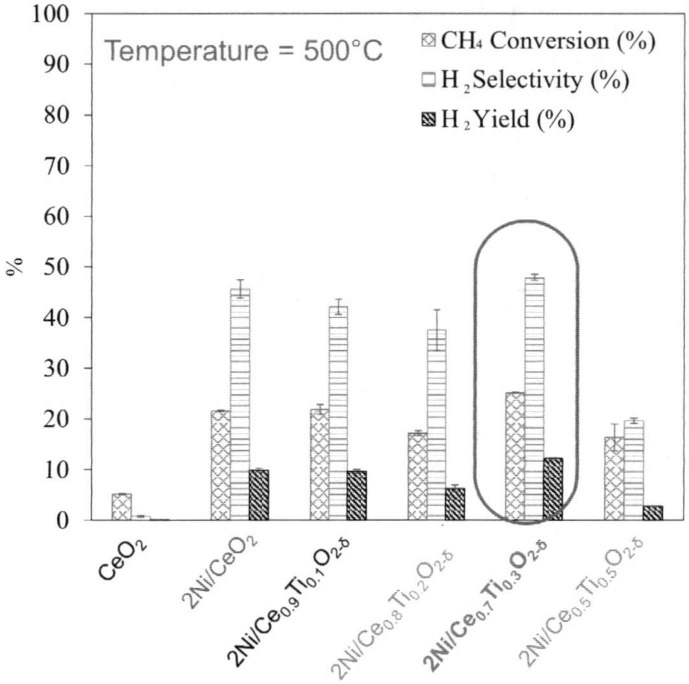

$CH_4 \xrightarrow[500°C]{\text{2 wt\% Ni on Various supports}} C + 2H_2$

The 2 wt. % Ni/$Ce_{0.7}Ti_{0.3}O_{2-\delta}$ observed higher activity compared to other catalysts.

Experimental conditions: Ni loading – 2 wt.%; reaction temperatures – 500°C; amount of catalyst – 200 mg; CH_4 gas flowrate – 12.5 mL min^{-1}.

15

Effect of Reaction Temperatures: DRM

$CH_4 + CO_2 \xrightarrow[\text{2 wt.\% Ni/}Ce_{0.7}Ti_{0.3}O_{2-\delta}]{500 – 800 °C} 2CO + 2H_2$

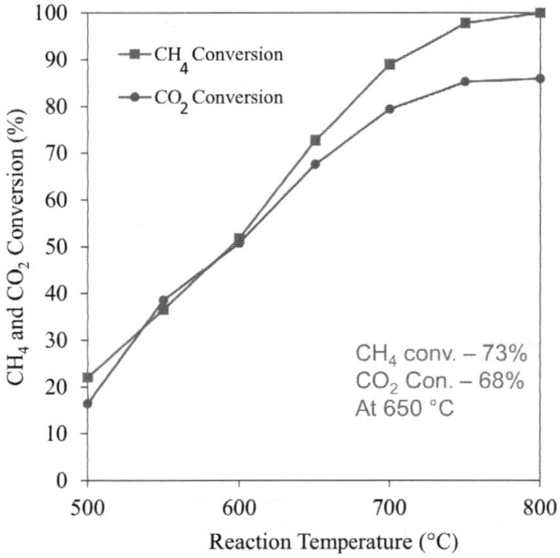

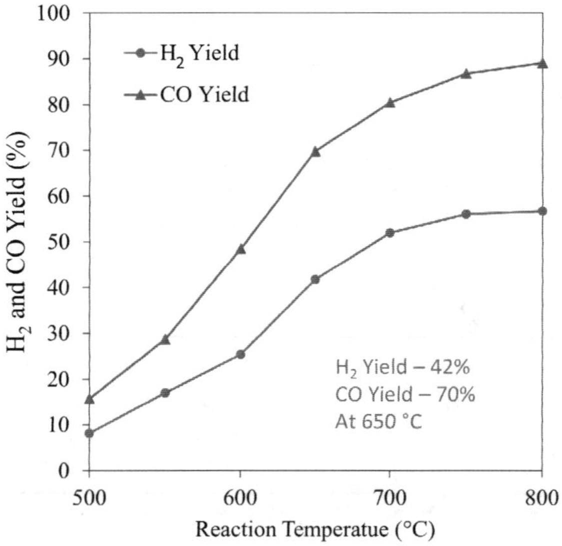

Experimental conditions: Catalyst – 2 wt.% Ni/$Ce_{0.7}Ti_{0.3}O_{2-\delta}$; amount of catalyst – 200 mg; gas flowrate ratio-1:1, CH_4 -12.5 mL min^{-1}, CO_2 – 12.5 mL min^{-1}.

16

Effect of Ni Loading

$$CH_4 + CO_2 \xrightarrow[650°C]{0.5 - 10\ wt.\%\ Ni/Ce_{0.7}Ti_{0.3}O_{2-\delta}} 2CO + 2H_2$$

Entry	Ni Loading (wt.%)	CH_4 Conversion (%)	H_2 yield (%)
1	0.5	40±1	20±1
2	1	62±2	33±1
3	2	73±3	42±1
4	5	71±3	41±1
5	10	73±3	40±1

Experimental conditions: Catalyst – $Ni/Ce_{0.7}Ti_{0.3}O_{2-\delta}$; reaction temperature – 650°C; amount of catalyst – 200 mg; gas flowrate ratio-1:1, CH_4 -12.5 mL min^{-1}, CO_2 – 12.5 mL min^{-1}.

Comparison of DRM Using Various Catalysts:

Catalyst	Temp (°C)	SA (m^2 g^{-1})	Gas flow rate (mL min^{-1})	Reactants conversion (%)		Products Formed (%)			Ref.
				CH_4	CO_2	H_2	CO	H_2/CO	
$Ni/Ce_{1-x}Zr_xO_2$-CP	600	91	25	52	70	63	-	0.6	[Catal. Today, 277, 2016, 234-245][b]
Ni/CeO_2-ZrO_2	700	87	50	38	52	35	47	-	[Apply. Catal. A 377, 2010, 16-26][b]
$Ni/CeO_2-Al_2O_3$-CA	600	135	75	29	40	-	-	0.5	[Apply. Catal. A 500, 2015, 12-22]
Ni/TiO_2	600	-	100	58	59	-	-	0.9	[RSC Adv. 4, 2014, 4817-4726]
Pt/CeO_2-ZrO_2	550	183	100	18	25	10	26	0.5	[Appl. Catal. B, 89 2009, 149-159][b]
Ni/CeO_2-NRs	600	76	-	43	46	71	-	0.9	[Catal. Sci. Technol., 6, 2016, 3594-3605][a]
$Ni/Ce_{0.7}Ti_{0.3}O_{2-\delta}$	**650**	**26**	**25**	**73**	**68**	**42**	**70**	**-**	**This work[b]**

SA - Surface area; a – H_2 and Co Yield; b - H_2 and Co Selectivity; CA – Critic acid method; NRs - Nanorodes
CP- Co-precipitation method

Catalyst Stability:

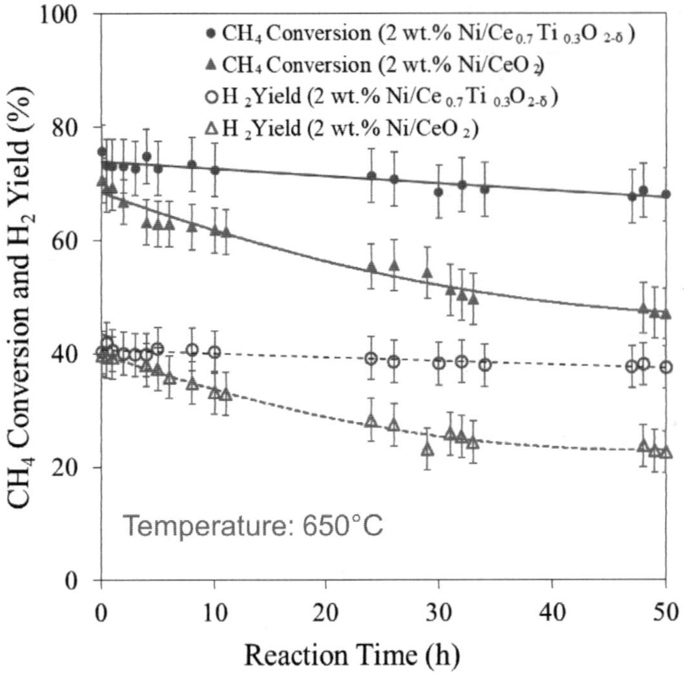

The 2 wt.% Ni/$Ce_{0.7}Ti_{0.3}O_{2-\delta}$ lost 6 percentage point CH_4 conversation activity lost.

This catalyst was active and stable for 50 h.

Experimental conditions: Reaction temperature – 650°C; amount of catalyst – 200 mg; gas flowrate ratio-1:1, CH_4 -12.5 mL min^{-1}, CO_2 – 12.5 mL min^{-1}.

19

Applied Catalysis A: General 377 (2010) 16–26

Contents lists available at ScienceDirect

Applied Catalysis A: General

journal homepage: www.elsevier.com/locate/apcata

ELSEVIER

Ni/CeO_2-ZrO_2 catalysts for the dry reforming of methane

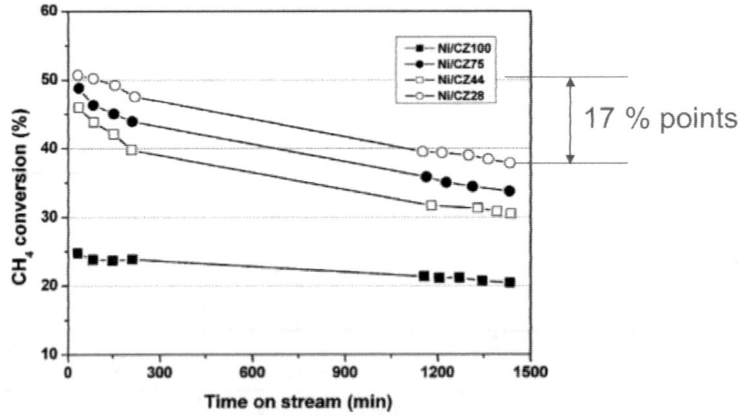

Fig. 9. Results of the stability tests for the NiCZXX catalysts expressed as CH_4% conversion vs time under DRM reaction conditions (Feed: 50%CH_4–50%CO_2, GHSV = 30,000 mL g^{-1} h^{-1}, P = 1 atm, T = 973 K). xx = % molar content of ceria

20

Ce3d XPS Spectra:

Fresh and used 2 wt%. Ni/$Ce_{0.7}Ti_{0.3}O_{2-\delta}$

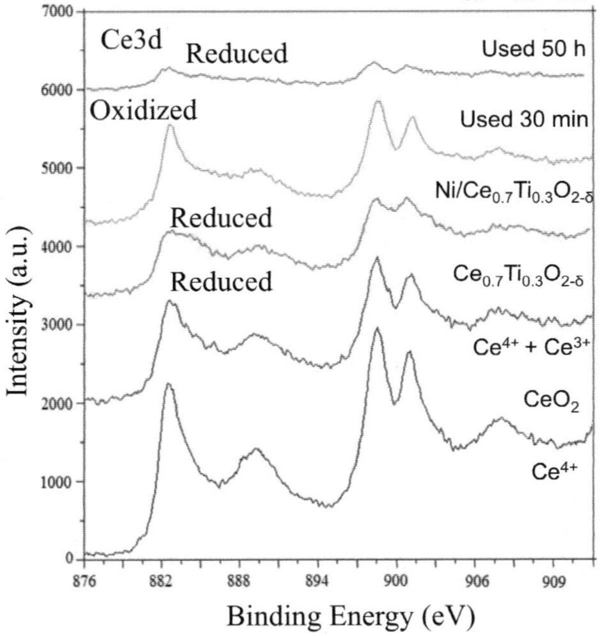

Fresh and used 2 wt.% Ni/CeO_2

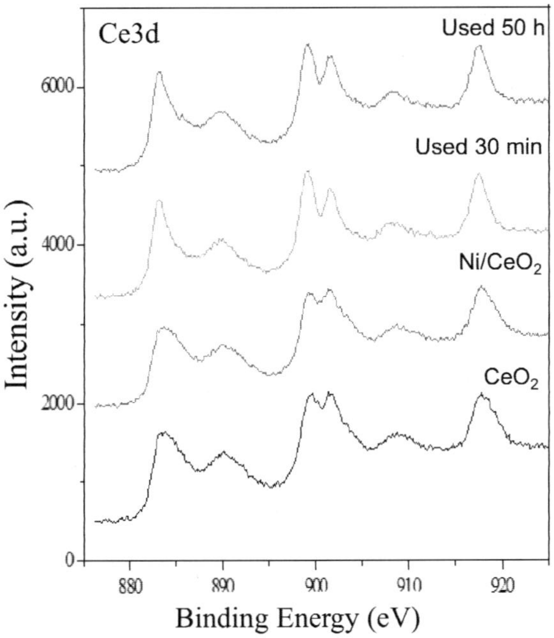

Ce^{4+}: BE 882.1, 889.0, 898.5, 901.1, 907.7 and 916.5 eV.
Ce^{3+}: BE 880.3, 884.7, 898.8 and 903.3

Instrument: Physical Electronics PHI 5800 ESCA

C1s XPS Spectra:

Fresh and used 2 wt%. Ni/$Ce_{0.7}Ti_{0.3}O_{2-\delta}$

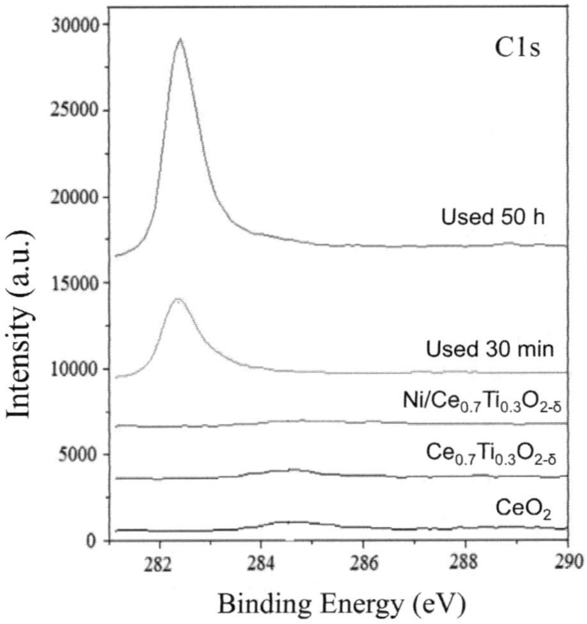

Fresh and used 2 wt.% Ni/CeO_2

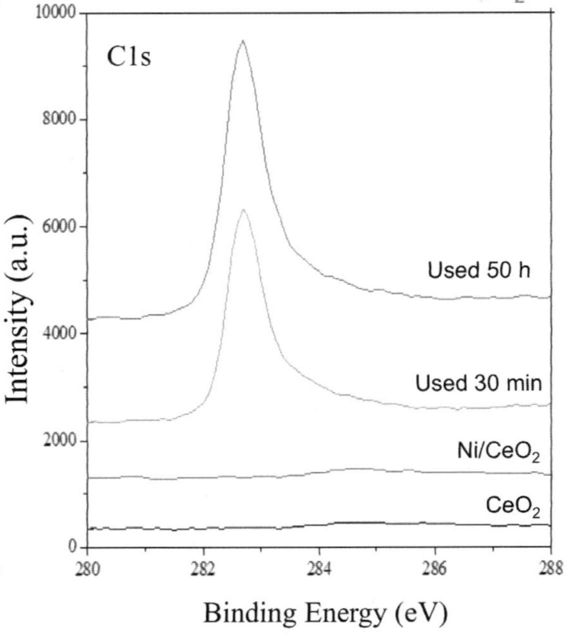

After reaction carbon peak shifted to lower BE due to formation of NiC.

Swaan et al. Catal. Today 21, **1994**, 571-578

XRD of Used 2 wt.% Ni/$Ce_{0.7}Ti_{0.3}O_{2-\delta}$ Catalyst:

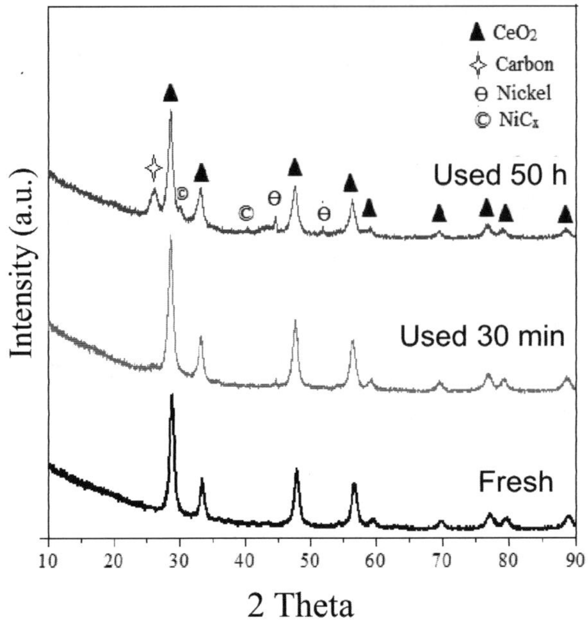

Carbon and Ni peaks were observed after reaction.

23

Conclusions

- We have successfully developed the Ni-supported cerium-titanium oxide catalyst (0.5 – 10 wt% loading).
- The catalyst was analyzed by XRD, SEM, and BET surface area.
- The performance of the catalyst was examined for dry reforming of methane by varying different parameters such as effect of reaction temperatures, Ni loading and stability test.
- The 2 wt. % Ni supported on $Ce_{0.7}Ti_{0.3}O_{2-\delta}$ enhanced the stability and activity of the catalyst.
- Ni/$Ce_{0.7}Ti_{0.3}O_{2-\delta}$ catalyst was stable for 50 h.
- The 2 wt% Ni/$Ce_{0.7}Ti_{0.3}O_{2-\delta}$ was found active catalyst for dry reforming of methane having the **73% CH_4** conversion and **42% H_2** yield.

24

Acknowledgements:

- We would like to thank the "*Wyoming Carbon Engineering Initiative*" for supporting this work.

Carbon Engineering Intiative

25

494784 Tuning the Interfacial Property of Copper-Ceria Catalyst By Indium for Low Temperature CO Oxidation

Abstract:

The development of novel catalysts for low-temperature CO elimination attracts a lot of attention not only in academic research but also in practical application, such as purifying automobile emission and preferential oxidation (PROX) in the proton exchange membrane fuel cell. Supported noble metal catalysts (Au, Pt, Rh, Pd), which show an excellent catalytic performance, have been widely investigated for CO oxidation. However, due to the high cost of the noble metals and remarkable progresses in oxide syntheses, catalytic oxidation of carbon monoxide over base metal oxide catalysts such as Mn_2O_3, Co_3O_4 and CuO has becoming a promising alternative for noble metals.

Among the metal oxide catalysts, CuO-CeO_2 catalyst has been paid much more attention due to their low cost and comparable catalytic activity to the noble metal catalysts in low temperature CO oxidation. It is generally accepted that the high performance for CuO-CeO_2 catalysts is dependent on the synergetic redox property between copper and ceria at the interfacial sites, which is considered as the active sites in CO oxidation reaction.

Up to now, based on the understanding of structure of the active component, two strategies should be employed for the improvement of the catalytic performance of copper-ceria catalysts for CO oxidation. Firstly, the facility for achieving the entities of the Cu^+ species at the Cu-Ce interface, which is favored by an increase in the dispersion and decrease in the particle size. Secondly, the redox properties of the interface sites between partially reduced copper species and the support, this may be the most important factor in determining the catalytic activities when the nature of the ceria-based support is varied.

Therefore, in the present study, the geometry and electronic structure at the interfacial sites were tuned by adding In into the catalyst to enhance the catalytic performance. The structure-performance relationship was thoroughly studied to support fundamental understanding on CO oxidation over CuO-CeO_2 catalyst.

Fig. 1. Temperature-dependent CO oxidation over (¨‹) 5Cu/CeO_2, (¡ö) 0.5In5Cu/CeO_2, (◀) 1In5Cu/CeO_2, (¡ñ) 1.25In5Cu/CeO_2, (¡ø) 2In5Cu/CeO_2, (♦) 3In5Cu/CeO_2 and (insert) T_{100} of various In-Cu/CeO_2 catalysts.

Fig. 2. Binding energy of In $3d_{5/2}$ and Cu $2p_{3/2}$, I_{sat}/I_{pp}, Cu^+(%) and Ce^{3+}(%) as a function of In content in the Cu-In/CeO_2 catalysts.

In this work, In-promoted Cu/CeO_2 catalyst exhibits remarkable catalytic performance for low temperature CO oxidation. A volcano-shape curve is observed for reaction rate vs. In content, in which the 1.25In5Cu/CeO_2 catalyst possesses the lowest temperature of complete conversion (T_{100} = 100 °C). The highest reaction rate (5.93¡Á10^{-6} mol $g_{cat}^{-1}s^{-1}$) over 1.25In5Cu/CeO_2 at 80 °C is obtained. HAADF-STEM images and in situ DRIFTS have proved that an addition of In into the catalysts leads to a decrease in particle size of copper. Copper atoms are diluted by indium atoms indicating that more low-coordination copper sites are exposed and more metal-support interfaces are created, which are considered to be the active sites for this reaction. The adsorption strength of CO on copper is weakened after doping indium, which is assumed to be more favorable for reacting with the adjacent lattice oxygen, preventing the accumulation of surface carbonate-containing products and enhancing the stability and the resistance for CO_2. The electronic structure of the surface copper atoms is changed by accepting electron through the redox equilibrium between copper and indium, which could facilitate the stabilization of Cu^+ sites at the interface. Tuning the interfacial property of Cu/CeO_2 catalyst by indium sheds deep insight for the rational design and controllable synthesis of copper-ceria based catalysts with high catalytic performance.

486017 Renewable Isoprene By Sequential Hydrogenation of Itaconic Acid and Dehydra-Decyclization of 3-Methyl-Tetrahydrofuran

Abstract:

The catalytic conversion of biomass-derived feedstocks to value added chemicals is an important challenge to alleviate the dependence on petroleum-based resources. To accomplish this, the inherently high oxygen content of biomass compounds, such as that of lignocellulosic biomass, requires significant reduction via hydrodeoxygenation strategies. The unsaturated carboxylic acid itaconic acid (IA) can be produced from biomass via fermentation pathways, for example. A pathway of interest is the conversion of IA to isoprene, facilitating the renewable production of an industrially relevant diolefin. IA is successively hydrogenated to yield 3-methyl tetrahydrofuran (3-MTHF), in a one-pot cascade reaction (Figure 1A), where a Pd-Re/SiO_2 bimetallic catalyst results in the highest 3-MTHF yield (80%, Figure 1B). The 3-MTHF is then converted to isoprene, and other pentadienes, through an acid catalyzed vapor-phase dehydra-decyclization. Multiple solid acid catalysts, including aluminosilicates, metal oxides and phosporous modified zeolites, were screened for the dehydra-decyclization step. All silicon phosphorous containing zeolites were found to be the most selective (70% isoprene and 20% pentadienes), where the major side reaction involved a retro-prins condensation of 3-MTHF to butane and formaldehyde. Through kinetic studies, an investigation into the effect of Brønsted acid strength, pore size and operating conditions on the selectivity to isoprene are discussed.

Figure 1. A. Itaconic acid, produced by glucose fermentation, is converted by one-pot hydrogenation to 3-methyl-tetrahydrofuran (3-MTHF). Gas-phase dehydra-decyclization of 3-MTHF with solid acid catalysts yields isoprene. **B.** Three conditions of experimental conversion of IA (5 wt% in H_2O) to 3-MTHF (red) with supported metal catalysts and hydrogen pressure after 24 hours. Products include methyl-butanol (¦, MBO), resulting from the over-hydrogenation of 3-MTHF.

[1] Abdelrahman, O. A.; Park, D. S.; Vinter, K. P.; Spanjers, C. S.; Ren, L.; Cho, H. J.; Zhang, K.; Fan, W.; Tsapatsis, M.; Dauenhauer, P. J. *ACS Catal.* **2017**, *7*, 1428-1431.

498237 Using Microkinetic Analysis to Predict Product Selectivity during Propionic Acid Hydrodeoxygenation over Supported Pt and Ru Catalysis

Abstract:

Recent industrial expansion in the biotechnology sector has raised interest in using succinic acid as an oxygenated platform for commodity chemicals. Most notably, succinic acid serves as a biogenic alternative to maleic anhydride, which is the typical industrial precursor to gamma-butyrolactone, butanediol, and tetrahydrofuran. Each of these can alternatively be prepared by partial hydrodeoxygenation (HDO) of succinic acid; however, each target is susceptible to deep hydrogenation/hydrogenolysis, and selectivity control is nontrivial. The central challenge in partial hydrogenation of succinic acid is that carboxylic acid reduction is difficult relative to the array of secondary and tertiary reactions that occur under similar conditions. These include aldehyde decarbonylation, alcohol hydrogenolysis, alkane hydrogenolysis, and methanation, which broadly result in the formation of low-value carbon oxides and light alkanes.

Although noble metals, such as Pt and Ru, show good activity in hydrodeoxygenation, they generally also activate secondary and tertiary reaction pathways, resulting in the formation of undesired fragmentation and deep hydrogenation products. A popular method for tuning their selectivity is to pair these noble metals with oxophilic "promoter" metals, such as Sn. Unfortunately, our present understanding of how secondary metals impact reaction kinetics to modulate selectivity is murky, and the design of such materials remains largely alchemical. A more rational framework for material synthesis requires an elementary understanding of reaction kinetics governing both carboxylic acid HDO and its myriad competing pathways. Furthermore, we must understand how the addition of promoter metals will perturb this free energy landscape to enhance (or detract from) monometallic selectivity. The goal of the present study is to establish kinetic expectations for carboxylic acid HDO over monometallic Pt and Ru, which provides the necessary foundation for subsequent discussion of how the addition of secondary metals impacts observed selectivity.

As a model system, we focus on propionic acid HDO and attempt to reconcile mechanistic descriptions of HDO chemistry with experimentally observed product formation rates through microkinetic analysis. Because the carboxylic acid HDO occurs alongside multiple side reactions—all of which are generally expected to occur under typical reaction conditions—it is exceedingly difficult to infer rates of primary reactions that consume carboxylic acids. In general, we are only able to directly quantify reaction rates for terminal pathways—such as methanation and alkane hydrogenolysis. Accordingly, we first consider the elementary kinetics of these terminal pathways and subsequently apply these insights as we build toward a quantitative description of carboxylic acid HDO.

490362 Renewable Adipic Acid Production Via Metal-Free Cleavage of C-O Bonds in the Presence of Molecular H_2 in Organic Acid Solvents

Abstract:

A central challenge in biomass valorization is the selective removal of oxygen-containing functional groups from platform chemicals, e.g., 5-hydroxymethylfurfural, to valuable chemicals without overhydrogenating the substrate to undesired products. Supported metal or metal oxide catalysts have shown promise in removing C-O bonds; however, many catalytically active metals/metal oxides are economically costly and can lead to major selectivity challenges, such as overhydrogenation or C-C cracking. Recently, we have developed a system for the selective removal of C-O single bonds, through employment of homogeneous or solid acids with molecular H_2 in organic acid solvents. In this system, cleavage of C-OH and ether bonds occurs readily without reducing higher order oxygenated functional groups such as carboxylic acids. This system is particularly relevant in forming adipic acid (AA), a key polymer precursor for producing Nylon 66, from biomass-derived tetrahydrofuran-2,5-dicarboxylic acid (THFDCA), where we have achieved an AA yield of 89% in the absence of any metal catalyst. In this work, we demonstrate key mechanistic insights in THFDCA hydrogenolysis to AA via computational input, kinetic measurements, and reactivity data, where we shed light onto the interplay between acids, iodides, molecular H_2, and the organic acid solvent.

490678 Influence of Sn Promoter on Pd and Pt Catalysts for Conversion of Heptanoic Acid and Propane

Abstract:

Silica-supported Pd catalysts can be used to selectively remove oxygen from biomass-derived carboxylic acids through decarbonylation. This work aims to improve Pd catalyst stability and selectivity towards desirable products by varying metal particle size as well as by introducing Sn to the catalyst. Catalysts were prepared by ion exchange or incipient wetness impregnation of metal precursors on Davisil 636 silica gel supports, followed by calcination in air and reduction in dihydrogen. Monometallic Pd and Sn catalysts, in addition to chemically-mixed and physically-mixed PdSn catalysts, were tested for activity in heptanoic acid conversion. Two-stage reactor beds were used to probe the effect of Sn proximity on Pd. X-ray diffraction, hydrogen chemisorption, TEM, and XANES-TPR were used to characterize the catalysts. The stability of the monometallic Pd catalyst during decarbonylation of heptanoic acid was influenced by the metal particle size, with smaller particles deactivating slower than larger particles. Whereas catalyst deactivation was reduced by adding Sn to Pd by impregnation, improved stability was also observed when Sn/SiO_2 was physically mixed with Pd/SiO_2. The physically-mixed PdSn initially produced products characteristic of monometallic Pd, however, the two bimetallic catalysts showed similar product selectivities after 10 h on stream. The PdSn catalysts also expanded the reaction network to include dehydration and decarboxylative ketonization, with the primary product shifting from hexene to heptanal. STEM-EDS of physically-mixed PdSn catalyst indicated that Sn migrated to the silica support particles containing Pd. In situ XANES analysis during butyric acid deoxygenation indicated that SnO_x is the primary Sn phase associated with Pd in both the physically-mixed and chemically-mixed catalysts. Ongoing work is exploring Sn promotion of Pt for propane dehydrogenation. The high temperatures and reducing conditions of the alkane dehydrogenation reaction are substantially different from the carboxylic acid reactions and likely result in a more reduced state of the Sn promoter. Bimetallic PtSn catalysts showed improved stability compared to monometallic Pt catalysts which deactivated rapidly with time on stream. X-ray diffraction indicated the presence of Pt_xSn_y in the bimetallic catalysts, suggesting that the alloy phase plays a critical role in improving catalyst stability.

491400 Selectivity Control during the One-Pot Conversion of Aliphatic Carboxylic Acids to Linear Olefins through Tandem Hydrogenation/Dehydration

Abstract:

Olefins are important building blocks for the synthesis of higher value chemicals in the petrochemical industry and they are typically produced from fossil fuels. As an alternative to fossil fuel-based chemicals, biomass-derived carboxylic acids could serve as a viable replacement feed for olefin production. However, the production of olefins from carboxylic acids in one step remains a challenge due to the stability of carboxylic acids and the tendency of olefins to over-hydrogenate during hydrogenation, leading to alkanes.

In the present study, we have studied the one-pot catalytic conversion of linear aliphatic carboxylic acids to linear olefins via tandem hydrogenation/dehydration reactions. Hexanoic acid was converted to a mixture of hexenes over Cu nanoparticles supported on commercial silica-alumina (Siral 70) in a continuous flow reactor. The combination of the hydrogenation properties of earth-abundant Cu and the acid sites on silica-alumina allows hydrogenation and dehydration to occur simultaneously, a process that normally calls for different reaction conditions. At 483 K and 5 bar H_2 gauge pressure, we obtained 98.5% conversion of hexanoic acid, with a molar product distribution of 88.9% hexenes and 11.1% hexane. Remarkably, we observed a brusque selectivity switch to 99.9% hexane at 100% conversion. We hypothesize that the presence of a small amount of hexanoic acid on the surface of the catalyst prevents overhydrogenation of the hexenes. Current efforts are targeted at reaction kinetic studies as well as surface studies using FT-IR spectroscopy to ascertain the nature of this selectivity switch. Process runs using carboxylic acids produced from biomass fermentation will also be carried out to ensure process compatibility.

Catalyst for Conversion of Methyl Ethyl Ketone to Butenes

Zahraa Alauda[a,b], Hayder Alatabi[a,b], Quanxing Zheng[a], Xu Li[a], Keith Hohn[a]

a: Department of Chemical Engineering, Kansas State University, 1005 Durland Hall, Manhattan, KS 66506, United States

b: Department of Chemical Engineering , The University of Technology, Baghdad, Iraq

Abstract

The aim of this work is to figure out a suitable catalyst to convert methyl ethyl ketone (MEK), to butene directly in one step. The catalytic gas phase reaction of MEK was investigated in a fixed bed reactor over Cu-alumina, Cu-zeolite Y sodium (Cu-ZYNa) and Cu-zeolite Y hydrogen (Cu-ZYH) as bifunctional heterogeneous catalysts, which prodvide both hydrogenation and dehydration functionalities. In the reaction, MEK is hydrogenated to 2-butanol which is further dehydrated to produce butene. Experimental results showed that selectivity of butene was the highest over Cu-ZYNa, and it was improved by finding the optimum reaction temperature and the percentage of copper loaded on ZYNa. The highest selectivity of butene (97.9%) was obtained at 270 °C and 20 wt% Cu-ZYNa. Over Cu-alumina, the selectivity of butenes was less than Cu-ZYNa since subsequent hydrogenation of butene occurred to produce butane. Also, it was observed that with increasing the H_2/MEK molar ratio, butane selectivity increased. However, when this ratio was decreased, hydrogenation of butene was reduced but dimerization to C_8 alkenes begane to be favored. The main products over 20% Cu-alumina were butene and butane, and the maximum selectivity of butene (87%) was achieved at an H_2/MEK molar ratio of five. The lowest selectivity of butene was obtained using Cu-ZYH, reaching ~40%. All catalysts were characterized by (TPD-NH_3), (TPD-CO_2), TPR and XPS to probe catalyst acidity, basicity and the reducibility of Cu loaded on supports. It was found that the amount of acidity in Cu-ZYH is much than in Cu-ZYNa. This could have caused the selectivity of butene to decrease as a result of the motivation of dimerization, oligomerization and cracking reactions.

487709 Design of Advanced Nanomaterials for Electrocatalysis

Abstract:

Hydrogen is a promising energy carrier for the replacement of fossil fuels. The hydrogen cycle in the sustainable energy system involves four important reactions, namely, oxygen reduction reaction (ORR), hydrogen oxidation reaction (HOR), oxygen evolution reaction (OER) and hydrogen evolution reaction (HER). Our research emphasizes the catalysts for these key reactions. Here, we introduce a "model surface - nano verification - catalyst design" research system for discovery and production of advanced nanomaterials for sustainable energy and fuel.

491078 Layered Hybrid Structure of Cobalt/N-Doped Carbon Derived from Metal-Organic Frameworks for Electrocatalytic Hydrogen Evolution

Abstract:

This study reports in situ synthesis of Co/N-Carbon hybrid structures via the pyrolysis of layered cobalt based metal–organic frameworks. As excellent hydrogen evolution reaction (HER) catalysts, the 2D hybrid layered structure composed by ultrafine metallic cobalt nanoparticles and N-carbon, showed a small overpotential of 103 mV (vs RHE) at 10mA/cm2, high cycle durability and considerable long-term stability for HER. The superior electrochemical performances of Co/N-carbon for HER can be attributed to such 2D configuration which includes the low H atom binding energy of metallic cobalt, the high conductivity of carbon and the stability of carbon-encapsulated Co nanoparticles. Therefore, we have worked on the synthesis and characterization of the 2D hybrid structure of Co/N-Carbon derived from metal organic frameworks. The as-prepared 2-D sheet structures retained the well-defined morphology of the initial Co-based MOFs, and cobalt nanoparticles (ca. 7.2 nm) were well embedded on the N-carbon. Due to the integration of the superior features of ultrafine cobalt nanoparticles and N-carbon, the resultant Co/N-Carbon showed enhanced electrocatalytic activity and durability when compared to the commercial Pt/C electrocatalyst at the alkaline media. More importantly, Co/N-carbon's electrocatalytic activities in terms of overpotential at 10mA/cm2 for hydrogen evolution have been more advantageous over other nanocarbon-based electrocatalysts.

Acknowledgements: Financial support from Interdisciplinary Graduate School, Nanyang Technological University Singapore is gratefully acknowledged.

References

[1] Walter, M. G.; Warren, E. L.; McKone, J. R.; Boettcher, S. W.; Mi, Q.; Santori, E. A.; Lewis, N. S. Chem. Rev 2010, 110 (11), 6446-6473.

[2] Hou, Y.; Wen, Z.; Cui, S.; Ci, S.; Mao, S.; Chen, J. Adv. Funct. Mater 2015, 25 (6), 872-882.

492501 Comparison of Hydroxide-Mediated and Direct Mechanisms for Alkaline Hydrogen Electrocatalysis

Abstract:

It has long been recognized that the reaction rates of the hydrogen oxidation and hydrogen evolution reactions (HOR and HER) are slower in basic than acidic electrolytes, even though the surface intermediate of adsorbed hydrogen is independent of solution pH. Understanding the root of this observation is critical to designing catalysts for a multitude of electrochemical reactions with relevance to energy conversion and storage. In this work, we undertake a fundamental investigation to determine if adsorbed hydroxide plays an active or spectator role in HER/HOR kinetics.

One proposed explanation for the pH dependence of HER/HOR kinetics, suggested by the research groups of both Gasteiger and Yan, is that in base, hydroxide ions stabilize the Pt-H bond for stronger binding and slower catalysis [1,2]. Central to this hypothesis is an experimentally measured shift with pH in the peak potential for hydrogen underpotential deposition (H-UPD) on the (100) and (110) steps on the Pt surface. However, the Janik and Koper groups have found that the shift in H-UPD potential is caused not by changes to hydrogen adsorption, but instead by competitive adsorption of hydroxide in the presence of alkalai cations [3,4]. Markovic et al have proposed that such hydroxide adsorption is actually necessary to facilitate water recombination/dissociation in base, and that oxophilic sites such as Ni(OH)2 generate a bifunctional mechanism for efficient alkaline hydrogen electrocatalysis [5].

In this work, we investigate specifically the hypothesis that adsorbed hydroxide is an active participant in the alkaline HER and HOR by combining electroanalytical techniques with microkinetic modeling. We develop expressions for time-dependent current and voltage as a function of hydrogen and hydroxide binding strength for the indirect (hydroxide-mediated) and direct (hydroxide-as-spectator) Volmer steps in base. These expressions are compared to experimental cyclic voltammograms of H-UPD on polycrystalline platinum. The free energy of hydroxide adsorption is controlled experimentally by the electrolyte cation via changes in cation solvation strength [6]. Varying the sweep rate results in greater peak separation that can be used to extract rate constants for the adsorption reaction via traditional electroanalysis [7].

Our computations show that at both slow and fast scan rates, either the direct or indirect mechanism can describe the peak potential splitting and peak current. However, hydroxide binding strength affects the mechanisms differently. In the indirect mechanism, stronger hydroxide binding reduces overpotential and, therefore, peak splitting. In the direct mechanism, stronger hydroxide binding decreases available surface sites, leading to lower exchange current densities and greater overpotential. Comparison with experiment shows that on the (110) and (100) facets of polycrystalline platinum, moving from 0.1M KOH to 0.1M LiOH decreases the peak potential location, but increases peak potential splitting. This trend shows that stronger hydroxide binding has detrimental effects on H-UPD kinetics, and that the direct mechanism is therefore more likely to dominate observed behavior.

Our results strongly suggest that adsorbed hydroxide serves as a competitive spectator in the alkaline Volmer step, and that the bifunctional HOR/HER mechanism plays only a minor role at best. This study contributes to resolving a long-standing paradox in electrocatalysis and surface science by determining that oxophilicity is not an accurate descriptor for alkaline hydrogen electrocatalysts. Other parameters, such as water orientation and non-covalent interactions, must play a greater role in overall activity. Future work will discuss these and other factors in the relation between H-UPD kinetics and HER/HOR catalyst activity .

(1) Durst et al. Energy Environ. Sci. 2014, 2255.

(2) Sheng et al. Nat. Commun. 2015, 5848.

(3) McCrum et. al. J. Phys. Chem. C, 2015, 457.

(4) Van Der Niet et al. Catal. Today 2013, 105.

(5) Strmcnik et al, Nat. Chem. 2013, 1.

(6) Danilovic, N. et al. Electrocatalysis 3, 2012, 221.

(7) Angerstein-Kozlowska et al. J. Electroanal. Chem. Interfacial Electrochem. 1979, 1.

497033 Alkaline Hydrogen Evolution from Ni–Mo Intermetallics with High Mo Content

Abstract:

The transition from acidic to alkaline conditions promises considerable cost savings for water electrolysis systems based on polymer ion-exchange membrane electrolytes. Binary transition metal alloys containing Ni and Mo have been known for decades to exhibit excellent performance as alkaline water electrolyzer cathodes,[1] but the specific reason that these alloys are more active toward the hydrogen evolution reaction (HER) than either Ni or Mo alone remains the subject of some considerable debate.[2] Moreover, previously reported mass-specific activities for Ni–Mo alloys are still at least 100-fold lower than that of Pt, which makes them difficult to deploy in a practical electrolysis system that operates at suitably high energy efficiency and power density.

Prior work on Ni–Mo electrocatalysts has suggested that the alloy phase exhibits monotonically increasing HER activity as the bulk Mo mole fraction is increased.[3] However, the accessible composition range for binary Ni–Mo alloys is constrained by the solid solubility of Mo in the Ni fcc lattice, which is only ~25 mol% at 1000 °C. Surface-science experiments on bulk $Ni_{0.8}Mo_{0.2}$ alloys have indicated that the topmost atomic layer becomes considerably enriched in Mo, up to approximately equimolar Mo and Ni.[4] Thus, the question remains as to how much Mo is present in the most active Ni–Mo binary HER electrocatalyst surface, let alone what is the mechanism for enhanced activity.

In this presentation, I will discuss recent and ongoing work aimed at understanding the influence of bulk composition on the activity of nanostructured Ni–Mo alkaline HER catalysts.[5] We synthesized unsupported powders of Ni, $Ni_{0.92}Mo_{0.08}$ alloy, and the intermetallic compound Mo_7Ni_7. Mass-specific activity was found to be highest for the 8 mol% Mo alloy, but these data were clearly convoluted by catalyst morphologies, as the 50 mol% Mo intermetallic exhibited very low surface roughness. Attempts to account for electrochemically active surface area using a straightforward method based on electrochemical impedance spectroscopy (EIS) yielded unexpected results that conflicted with morphological analysis from electron microscopy. We have therefore postulated that Mo undergoes a pseudocapacitive redox transition near the onset potential for the HER, leading to erroneous results from EIS measurements. These results further implicate a plausible bifunctional catalytic mechanism that explicitly involves Mo redox transitions.

(1) Brown, D. E.; Mahmood, M. N.; Man, M. C. M.; Turner, A. K. *Electrochimica Acta* **1984**, *29* (11), 1551–1556.

(2) McKone, J. R.; Marinescu, S. C.; Brunschwig, B. S.; Winkler, J. R.; Gray, H. B. *Chem. Sci.* **2014**, *5* (3), 865–878.

(3) McKone, J. R.; Sadtler, B. F.; Werlang, C. A.; Lewis, N. S.; Gray, H. B. *ACS Catal.* **2013**, *3* (2), 166–

169.

(4) Soriaga, M. P.; Baricuatro, J. H.; Cummins, K. D.; Kim, Y.-G.; Saadi, F. H.; Sun, G.; McCrory, C. C.; McKone, J. R.; Velazquez, J. M.; Ferrer, I. M.; I, C., Azhar. *Surf. Sci.* **2015**, *631*, 285–294.

(5) Csernica, P. M.; McKone, J. R.; Mulzer, C. R.; Dichtel, W. R.; Abruña, H. D.; DiSalvo, F. J. *ACS Catal.* **2017**, 3375–3383.

485062 Increased Photocatalytic Activity of TiO_2 Nanoparticles with Defects for Sustainable Hydrogen Production

Abstract:

Hydrogen gas is touted as a clean-burning energy carrier that can be used in fuel cells for stationary and transportation power at higher efficiency than conventional internal combustion engines. The primary source of H_2today is from nonrenewable natural gas-derived methane steam reforming (MSR). MSR suffers from being a highly endothermic and endergonic reaction (though less so than water splitting), requiring significant energy input as heat generated from burning half of the feed methane, limiting atom efficiency and emitting greenhouse gases. Photocatalytic MSR circumvents the high heat requirement and lowers the reaction temperature, thereby avoiding coke buildup problems as well.

Titanium dioxide (TiO_2) is of particular interest in photocatalysis due to its high photoactivity under ultraviolet illumination and its stability under catalytic reaction conditions. Together with supported metal nanoparticles, TiO_2has been used to catalyze a number of important chemical and photochemical reactions.

We have synthesized defect-rich anatase TiO_2 that shows unique selectivity for hydrogen formation from methane. We have also demonstrated that ultra-violet light illumination boosts the activity and selectivity of the catalyst for hydrogen formation. We have demonstrated hydrogen yield of 27% under steady state conditions using a nickel cocatalyst. Through a combination of experimental and computational techniques, we investigate the mechanisms of defect formation and how these defects activate methane and selectively catalyze hydrogen formation.

486781 Transition Metal Nitride Core-Noble Metal Shell Nanoparticles As Highly CO Tolerant Catalysts

Abstract:

Introduction

Core-shell nanoparticles (NPs) have become increasingly attractive as a catalytic design platform due to their unique ability to simultaneously enhance the activity, increase the durability, and reduce the loading of noble metal (NM) catalysts. However, these materials have been limited by the lack of independent control over the core and shell compositions, shell thickness, and particle size [1]. To address these issues, our group recently developed a novel method for the synthesis of transition metal carbide (TMC) core-NM shell (denoted NM/TMC) NPs with precise control over the core-shell architecture [2]. Group 4-6 TMCs are earth-abundant compounds with high sinter-resistance, electrical conductivity, and chemical stability, making them ideal candidates for core materials [3].

Transition metal nitrides (TMNs) share all of these favorable material properties of TMCs but have rarely been investigated in core-shell materials. Here, we present the ability to synthesize NM/TMN core-shell NPs with varying core and shell compositions and controllable shell thickness via nitridation of the corresponding NM/TMC. The impacts of titanium tungsten nitride (TiWN) and titanium tungsten carbide (TiWC) cores on the electronic properties of a Pt shell are compared, and the effect of these differences on catalytic properties was investigated by measuring the activity for the hydrogen oxidation reaction (HOR) under the presence of CO to assess the CO tolerance. The results illustrate how the use of nitride versus carbide cores offers an effective way to tune the reactivity of NM catalysts.

Materials and Methods

NM/TMC NPs were nitrided under NH_3 at 800°C for 3 h to form NM/TMN NPs. Synthetic details for NM/TMC are provided elsewhere [2]. All materials were supported on carbon. Electrochemical measurements were performed at 30°C on a rotating disk electrode with a Ag/AgCl reference and platinum counter electrode. All potentials are reported versus a reversible hydrogen electrode (RHE). For comparison, catalyst loadings were targeted to have roughly equal Pt surface areas. HOR scans were performed in H_2-sat 0.1 M $HClO_4$ at 2 mV/s and 1600 rpm with iR compensation. To test CO tolerance, the working electrode was first held at 0.025 V until the electrolyte equilibrated with 1000 ppm CO before starting the HOR scan.

Results and Discussion

Pt/TiWC NPs with varying Pt coverage were synthesized using the method previously reported by our group and transformed into Pt/TiWN NPs via nitridation at 800°C. The high sinter-resistance and strong interfacial bonds of Pt/TiWC prevent particle agglomeration and preserve the core-shell structure while the core is converted into a nitride, which is confirmed by XRD, STEM-EDX, TEM, XPS, and CHNS analysis (Figure 1A).

XPS spectra of the core-shell materials reveal shifts in the core electron binding energies, called core level shifts (CLSs), for both the W 4f and Pt 4f signals. These CLSs indicate a significant change in the Pt electronic structure, as shifts in the core level energies of metal overlayers are reflected in similar shifts in the valence d-band center, which is known to correlate strongly with adsorbate binding energies and thus the reactivity [4-5]. The XPS results suggest downward shifts in the Pt d-band center for the core-shell materials compared to a pure Pt surface and an even lower d-band center and weaker binding for Pt/TiWN compared to Pt/TiWC.

To probe the difference in adsorbate binding and reactivity among our core-shell and commercial Pt (denoted Pt_{comm}) catalysts, we used CO as a model adsorbate and measured the HOR activity in the presence of CO (Figure 1B). Strong CO binding is a common problem for Pt catalysts, poisoning the surface and impeding reactivity. In the presence of 1000 ppm CO, the HOR activity of Pt_{comm} is completely suppressed by CO adsorption and is not recovered until above 0.5 V when CO can be oxidized off the surface. In contrast, both core-shell materials retain HOR activity at potentials as low as 0.1 V, thus demonstrating dramatically improved CO tolerance and reduced CO bond strength. Furthermore, 2-3 monolayer (ML) Pt/TiWN displays even greater CO tolerance than 2-3 ML Pt/TiWC, indicative of weaker CO binding as predicted from the XPS CLSs.

Figure 1. A) STEM-EDX map of 2-3 ML Pt/TiWN. B) HOR polarization curves in H_2-sat 0.1 M $HClO_4$ at 1600 rpm with (solid line) and without (dotted line) 1000 ppm CO.

Significance

In addition to Pt/TiWC, the synthetic method can be applied to core-shell NPs of varying NM/TMC compositions, demonstrating a promising way of synthesizing new core-shell architectures with tunable electronic properties that can serve as next generation catalysts.

References

1. Yang, H. *Angew. Chem. Int. Ed.* 50, 2674-2676 (2011).

2. Hunt, S.T., Milina, M., Alba-Rubio, A.C., Hendon, C.H., Dumesic, J.A., and Román-Leshkov, Y. *Science* 352, 974-978 (2016).

3. Zhong, Y., Xia, X., Shi, F., Zhan, J., Tu, J., and Fan, H.J. *Adv. Sci.* 3, 1500286 (2016).

4. Weinert, M., and Watson, R.E. *Phys. Rev. B* 51, 17168-17180 (1995).

5. Kitchin, J.R., Nørskov, J.K., Barteau, M.A., and Chen, J.G. *J. Chem. Phys.* 120, 10240-10246 (2004).

502612 Understanding the Electrochemical Behavior of a Thin, Flexible Micro Fuel Cell

Abstract:

Portable fuel cell systems in the power range of few milliwatts to a few hundred watts have become the focus of vigorous research and development activities. Fuel cells promise to provide more reliable, longer portable power than batteries. Micro fuel cell (mFC) technology can take advantage of fuels with energy densities of an order of magnitude higher than the energy stored in batteries. The air breathing cathode reduces the size and weight of the cathodic plate, and simplifies system requirements since oxygen inlet is fully passive. On the other hand, working conditions for the air breathing cell are more demanding due to the absence of an air flow behind the cathode gas diffusion layer to assist the removal of water.

The importance of micro-structures on polymers for micro fuel cell fabrication is enormous, particularly when considered as a low-cost alternative to the silicon- or glass-based MEMS technologies, for disposable small electronic gadgets. Many polymer-based microfabrication techniques via microinjection molding, casting, and micro-hot embossing have been developed.

Hot embossing provides advantages such as a relatively low-cost for embossing tools, simple operation and higher accuracy in the replication of small features. The embossing master stamp can be a silicon wafer, glass, electroplated nickel mold or other stamp with microfeatures. The hot embossing process introduces less residual stress in the polymer because the polymer stretches for a very short distance from the substrate into microstructure during hot embossing. In addition, the temperature variation range for the polymer is smaller than that required in injection molding, thus can reduce shrinkage during cooling and the friction forces acting on the microfeatures during de-molding. However, still micro-hot embossing is facing challenge in terms of process feasibility, since it is difficult to make the polymer to fill completely into microfeatured geometry of high aspect ratio and it is also delicate to separate the embossed structures from the mold without breakage.

An air-breathing micro fuel cell with direct hydrogen flow through porous anode electrode is realized and reported in this study. The performance of the microdevice with embedded flow channels and electrodes is characterized in ambient conditions. To achieve this goal, we implement a SU-8 microchannel stamp to transfer a pattern into a Nafion 1110 membrane by hot embossing. Nafion 212 thermally seals the whole device as a blanket top layer. The fabrication process of microchannels and micro fuel cell construction is evaluated with optical and scanning electron microscopy (SEM) step by step. Variations of hydrogen feed rate on performance were investigated. The shared-anode characteristic design of the double-sided cell is further studied by separate performance and electrochemical impedance spectroscopy (EIS) measurements taken from each side and compared with data for the complete cell. The maximum power density per superficial (footprint) area is 68.4 mWcm-2 for the shared anode stack. In the present micro fuel cell architecture, the external package is constructed of Nafion polymer, which offers a thin, light and flexible energy source with low cost of manufacturing. The device performance offers a high volumetric and

gravimetric energy density for portable applications.

503670 Digital Alchemy for Assembly Engineeringc

Abstract:

Back in the day, Stuart Churchill was a pioneer in the use of computers for modeling chemical and thermal processes. Since that time, the role of computing in chemical engineering has become indispensable. In this talk, we present new computational methods we call "digital alchemy" that allow for predictive modeling of thermodynamic assembly, and permit the design of building blocks optimized for assembly into target structures with desired properties. In this approach, standard thermodynamic ensembles are generalized to include alchemical variables and their thermodynamic conjugates, allowing for rapid, real-time computational investigation of large, complex design spaces.

503671 Improved Algebraic, Numerical, and Graphical Representations in Fluid Mechanics

Abstract:

The concept of equivalents is applied to fully developed flows through straight channels, and two applications are introduced: the equivalents that produce equal rates of flow through channels of different cross-section, and the diminishment that occurs between the regimes of laminar and turbulent flow. The validity of the eddy diffusivity and the invalidity of the mixing-length as measures of the shear stress due to turbulence have perhaps never before been demonstrated so clearly. The possible choices of primary variables and the consequences are also examined. The objective of these explorations is an improved understanding of the basic fluid mechanics by teachers, students, and industrial practitioners.

503672 The Scaling of Turbulence Near the Wall and the Churchill Turbulent Flux Correlation: Insights with Lagrangian Simulations

Abstract:

The evaluation of mean velocity and mean temperature in wall-bounded turbulent flows has been based on the "classical" theory of turbulence and the law of the wall. The viscous wall parameters, namely the friction velocity, the friction temperature, and the kinematic viscosity of the fluid have been used to scale not only the mean values of velocity, but also other mean turbulent quantities close to the wall, including the Reynolds stresses and the turbulent velocity and temperature fluctuations. Stuart Churchill, as he questioned the validity of the Reynolds analogy between momentum and heat or mass transfer in his works [1], he realized that extending the scaling of momentum to the scaling of scalar quantities in turbulence was not a sound approach. He proposed a scaling of turbulence that depends on the turbulent momentum flux and the turbulent heat flux. In his AIChE Institute Lecture (in Miami, 1998), Churchill presented a summary of his work on this topic, and closed form solutions for the mean velocity and temperature [2]. The turbulent Prandtl number, the most significant parameter for the modeling of turbulent transport not only in theory but also in modern CFD commercial software packages, was also interpreted physically based on these ideas [3]. We have used direct numerical simulation results and results from Lagrangian simulations [4] for the transport of passive scalars in turbulent channel flows to explore Churchill's ideas on turbulent flow scaling and to develop an updated theory for the scaling of heat transport in turbulent flows [5-7]. His passion to remove empiricisms and approximations from the development of predictive correlations in turbulent transport has been an inspiration. In this talk, we will revisit the Churchill scaling in view of our simulation results, and will discuss the counterintuitive finding that wall turbulence can be used to separate rather than mix substances under special circumstances [8].

1. Churchill, S. W., 1996, "A Critique of Predictive and Correlative Models for Turbulent Flow and Convection," Ind. Eng. Chem. Res., 35, pp. 3122–3140.
2. Churchill, S.W., 2000, "Progress in the Thermal Sciences: AICHE Institute Lecture," AIChE Journal, 46(9), pp. 1704-1722.
3. Churchill, S. W., 2002, "A reinterpretation of the Turbulent Prandtl Number," Ind. Eng. Chem. Res., 41, pp. 6393–6401.
4. Papavassiliou, D.V, 2002, "Turbulent transport from continuous sources at the wall of a channel," Int. J. Heat Mass Transfer, 45(17), pp. 3571-3583.
5. Le, P.M., and Papavassiliou, D.V., 2008, "On the scaling of heat transfer using thermal flux gradients for fully developed turbulent channel and Couette flows," Int. Commun. Heat Mass Transfer, 35(4), pp. 404-412.
6. Srinivasan, C., and Papavassiliou, D. V., 2011, "Prediction of Turbulent Prandtl Number in Wall Flows With Lagrangian Simulations," Ind. Eng. Chem. Res., 50(15), pp. 8881–8891.
7. Srinivasan, C., and Papavassiliou, D.V., 2013, "Heat transfer scaling for wall bounded turbulent flows," Applied Mechanics Review 65(3), Art. 031002.
8. Nguyen, Q., Srinivasan, C., and Papavassiliou, D.V., 2015, "Flow induced separation in wall turbulence", Phys Rev E, 91, Art. 033019.

503674 Flow Boiling Using a Piranha Pin Fin Heat Sink

Abstract:

Flow Boiling using a Piranha Pin Fin Heat Sink

Corey Woodcock, Xiangfei Yu, Yoav Peles[1] and Joel Plawsky

Rensselaer Polytechnic Institute; [1]University of Central Florida

The *Piranha Pin Fin* (*PPF*) microdevice is a heat exchanger engineered to dissipate extreme thermal loads using either dielectric fluids such as HFE7000 or water. Fabrication of the device is CMOS compatible and the first generations of the device have been fashioned from silicon substrates. The PPF operates on the principle that heat transfer would be more efficient if heated fluids could be removed nearly as fast as they are created. This process would keep the driving force as high as possible. Thus the heat exchanger is built as two layers; a primary substrate that dissipates the bulk of the heat and shunts the heated fluids to a booster section that continues to heat the fluid and extends the total surface area of the device. Three generations of devices were fabricated. The first two generations were single-layer systems designed to help understand the operation of the primary layer and optimize the shape and location of the pin fins. The third generation was the first, two-layer system to be fabricated and tested.

Figure 1 Representative PPF array

Correlations for the heat transfer coefficient, h_{tp}, based on the Boiling Number and the Weber Number in the PPF design was developed and validated with Gen II. The discrepancy of heat transfer coefficient between predicted and measured values was within 20%. The ratio of two-phase to single-phase pressure drop was correlated with the Boiling Number (Bo) and the Jakob Number (Ja). The discrepancy of the two-phase multiplier between prediction and experimental measurements were within 12%.

The third generation *PPF* heat sinks were the first series to provide both a primary substrate populated with *PPF*s, as well as a booster substrate populated with solid micro pin fins. The substrates are micro-aligned and fusion-bonded to provide a full three dimensional (3D) architecture with bulk silicon properties. Experimental results from the Gen III *PPF* microdevices provide a proof of concept of the true *PPF* heat sink concept with fluid venting into multiple substrates. At low flowrates (G = 670 [kg/s/m^2]) the Gen III *PPF* heat sink has been shown to dissipate heat loads approaching 1.1 [kW/cm^2] at an exit quality of 95%, while at higher flowrates (G = 3,300 [kg/s/m^2]) the thermal limit of the heat sink remains unknown due to exceeding the available power supply at >1.4 [kW/cm^2].

503673 Transport Problems in the Spirit of Stuart Churchill for Teaching and Research at the University of Michigan

Abstract:

Stuart Churchillâ€™s solutions to important heat transport problems have greatly influence both pedagogy and research. The correlation of Churchill and Chu for the natural convection heat transfer to or from a vertical plate, over an astonishing range of Rayleigh numbers is based partly on a similarity solution of a type common in heat, mass, and momentum transfer for problems in chemical engineering. One such problem is asphaltene or wax transport and deposition onto the wall of an oil pipeline, which can be solved using similar methods, as will be discussed. In addition, such problems, and the wealth of their applications make for excellent motivations and examples in the teaching of graduate or undergraduate transport, as will be discussed in the talk.

491908 Understanding the Mechanism and Kinetics of Reactions Catalyzed By Metal Oxides

Abstract:

Metal oxides are used extensively as catalysts to promote a wide variety of chemical reactions, e.g., partial and total oxidation, thermal and oxidative dehydrogenation, aldol condensation, and amination. What make oxides so versatile is the spectrum of active sites contained at their surface, which include Brønsted and Lewis acidic and basic centers. This talk will illustrate how a fundamental understanding of oxide catalyzed reactions can be achieved using a combination of experimental and methods for reactions occurring on bulk and highly dispersed oxides. These illustrations will encompass alkene oxidation ammoxidation, aldol condensation of ketones, alcohol reaction with ammonia to form primary amines, and the oxidation of water.

502394 Controlling Strain and Ligand Effects in Nanoparticle Catalysis By Alloying: Novel Alloys for Electrochemical Oxygen Reduction Reaction

Abstract:

The oxygen reduction reaction (ORR) is the major source of overpotential loss in low-temperature fuel cells. Expensive, Pt-based materials have been found to be the most effective catalysts, but exploration of alternatives has been hampered by stability constraints at the typical operating conditions of low pH and high potential.

I will discuss our studies of elementary mechanism of ORR on various metal electrodes using kinetic and micro-kinetic analysis of reaction pathways and quantum chemical calculations. These studies allowed us to identify the elementary steps and molecular descriptors that govern the rate of ORR. Using these performance descriptors we have been able to identify families of Pt and Ag-based alloys that exhibit superior ORR performance is acid and base respectively.

We have synthesized these alloys to demonstrate the superior ORR activity with rotating disk electrode experiments. We have also performed thorough structural characterization of the bulk and surface properties with a combination of cyclic voltammetry, x-ray diffraction, and electron microscopy with spatially resolved energy-dispersive x-ray spectroscopy and electron energy loss spectroscopy.

489476 A Synopsis of the Production of Propylene Via Oxidative Dehydrogenation of Propane: How Far from Its Commercial Implementation?

Abstract:

Currently, propylene is the second most important olefin and its demand is growing faster than ethylene. Propylene is obtained in well-established industrial processes such as steam cracking of naphtha, and to a smaller extend, via conventional dehydrogenation of propane (PDH). The recent exploitation of methane- and ethane-rich shale gas in North America has caused a shift in the crackers feedstock to ethane instead of naphtha, resulting in the building of more ethane crackers which produce mainly ethylene and almost no propylene. As a consequence, on-purpose PDH technology is being employed increasingly to make propylene. However, PDH suffers inherent disadvantages (demands high reaction temperatures and below atmosphere pressure, expensive catalysts, catalyst regeneration, high capital cost) that make it economically viable only in cases where low-cost liquid-petroleum-gases (LPGs) are available. Despite of other on-purpose technologies (metathesis, methanol to olefins) gaining territory in the production and market of propylene, the oxidative dehydrogenation of propane (PODH) has been historically considered an attractive alternative route to produce propylene and bridge the production-and-demand gap. Nevertheless, PODH has still not emerged from the lab-scale to the industry due to the low propylene yields. In this presentation, I compile and discuss some of the most remarkable advances that have been achieved during last decades in PODH with an emphasis on the catalysts structure-activity/selectivity relationship as well as the reaction mechanism.

501433 Insights into the Molecular Structure of Sulfated Mixed Metal Oxide Catalysts Via Vibrational Spectroscopy

Abstract:

Environmental concerns regarding the use of highly concentrated liquid acids, especially in the selective transformation of hydrocarbons, provide a strong incentive towards the design and development of solid acid catalytic materials as environmental friendly alternatives. Solid â€œsuperâ€ acid catalysts could be considered the acidic materials that are stronger than the acid strength of 100% H_2SO_4. Those catalytic materials are extensively used in oil and petrochemical industries in order to replace highly toxic and non-reusable liquid acids towards economically and environmentally friendly processes. Although aluminosilicates have received considerable attention as solid acid catalysts, many other catalytic materials have been reported to show enhanced acid strength such as sulfated metal and mixed metal oxides (SO_4^{2-}/ZrO_2, SO_4^{2-}/TiO_2, SO_4^{2-}/TiO_2-SiO_2, SO_4^{2-}/ZrO_2-SiO_2).

There is limited understanding regarding the molecular structure of sulfate species developed on zirconia and tin oxides. To that extent, this talk will be focused mainly on Tin-Zirconium sulfated mixed metal oxides. A variety of catalysts have been prepared using co-precipitation/wet impregnation technique spanning different a) Sn:Zr molar ratios as well as b) sulfuric acid concentrations. Utilizing Raman and FTIR spectroscopies under hydrated and/or dehydrated conditions, the sulfate species developed depending on the composition of the mixed metal oxide are studied. The relative distribution of bidentate, tridentate and polymeric sulfate species is a strong function of the Sn:Zr molar ratio but not of the initial H_2SO_4 concentration. We also observe a strong dependence between the calcination temperature and the surface sulfate species are deposited on the mixed metal oxide. This in turn underscores that the sequence steps on which the sulfation applied controls not only the crystalline structure of the mixed metal oxide but also the relative distribution of acid functionalities. The catalytic evaluation of the aforementioned materials on the alkylation of phenol will be also discussed with respect to the intrinsic reactivity of the surface sulfate species.

486466 Supported Metal Oxides As Model Oxide Catalysts

Abstract:

Supported Metal Oxides as Model Oxide Catalysts

Israel E. Wachs

*Operando*Molecular Spectroscopy & Catalysis Laboratory

Department of Chemical & Biomolecular Engineering

Lehigh University, Bethlehem, PA 18015

Supported metal oxide catalysts consist of an active metal oxide surface phase dispersed on a high surface area oxide support. The molecular nature of the active surface metal oxide phase can be varied from isolated sites to oligomeric sites to nanoparticles by controlling the metal oxide surface coverage and the synthesis method. Unlike supported metal catalysts, supported metal oxide catalysts are model oxide systems that allow for determining the (1) influence of metal oxide domain size on the specific reactivity (TOF), (2) number of catalytic active sites involved in the kinetic rate-determining-step, (3) influence of oxide ligands (isolated, oligomeric or nanoparticles) on the specific reactivity, and (4) influence of adjacent secondary surface metal oxides on the molecular structure and specific reactivity of the primary surface metal oxide. The surface chemistry and kinetics of supported metal oxide catalysts will be presented to highlight the fundamental insights provided by these model oxide catalysts.

509054 Solve this! Fundamental Approach to Problem Solving in Industrial Processes I (Invited Talks)

Abstract:

Solving complex plant problems is an everyday exercise that affects all projects and budgets, and it is very rarely introduced to students at any level. In fact, by the time young engineers are few years into working in industry, most have lost the engineering skills that they developed in school. We hope to encourage future engineers to retain those skills and learn the importance of plant problem-solving.

492840 The Role of Oxygen Vacancies of CeO_2 Supported Catalyst on Catalytic Performance of the CO_2 to CO

Abstract:

The performance of a catalyst depends on many factors. It is usually hard to discuss these effects separately due to their interwoven relationship. Here, we focus on the catalytic activity of nanosized Cu supported on cerium oxide by the space-confined method and mesoporous silica, whereby Cu nanocatalysts are deliberately controlled under the same loading, particle size and coordination number to highlight the interactions caused by different supports. It is interesting to note that the redox temperature of cerium oxide is dramatically decreased from >600°C to <300°C by CuCeOx solid solution so that its conversion is 18% with 100% selectivity approaching the equilibrium conversion 23% of the CO_2 to CO at 1 bar and 300°C the by flow reactor.

Oxygen vacancies generated by the reducible cerium oxide were found to be able to contribute more carbonate with bidentate type absorption. The space confined method was capable of generating high Cu dispersion (~50%) from nano-size CuCeOx solid solution. The Cu maintains the CeO_2 support to keep the oxygen vacancies recycling by hydrogen spill over behavior at 300°C when the reaction of CO_2 to CO oxidized the catalyst. However, similar metal-support was not seen in the mesopeorous silica system.

We have demonstrated that the active sites locate in the proximity between support and reduced nano-sized metal by synchrotron analysis, TEM mapping, in-situ FTIR and catalytic reaction. Furthermore, the work presented illustrates the significant role of oxygen vacancies in the metal-support interaction and how the potential of oxide supported nanocatalysts can be further explored and tuned by for the space confined method and thermal treatment for CO_2 to CO.

Advanced Ni-CeO_2 Nanocatalysts for Chemical CO_2 Recycling

*T. Smith, T. Stroud, E. Le Saché, H. Arellano-Garcia, T.R. Reina**

Department of Chemical and Process Engineering, University of Surrey, Guildford, GU2 7XH, UK

*Corresponding author: t.ramirezreina@surrey.ac.uk

Introduction

Carbon dioxide is one of the major issues facing the world currently, the adverse effects on the environment that CO_2 emissions cause are well documented. Major steps are being taken by industry to mitigate these effects, with a proportion of the CO_2 produced being captured each year. This scheme to reduce carbon emissions into the atmosphere is part of a worldwide initiative, whereby the carbon captured can be stored, but there is still a lot of progress to make within this field before this is a truly viable option. There is however, another exciting alternative which is the chemical upgrading of CO_2 to obtain fuels and chemicals [1] [2]. This way, the depletion of CO_2 emissions is accompanied by an extra motivation in terms of the generation of added-value products.

Among the different alternatives for CO_2 utilisation, dry reforming of methane is a promising route leading to the production of syngas (H_2 and CO) [3]. This is a highly useful and valuable intermediary, which can be used in processes like the Fischer-Tropsch Synthesis to create long chain hydrocarbons, such as diesel [4].

The issue that arises in the dry reforming process, is the deactivation of the catalysts used in the reaction of the feed chemicals. This deactivation is due to sintering of the active phase and/or carbon deposition [3] [5]. In order to overcome these problems noble metal based catalysts have been employed but their cost make them unsuitable for a real application.

The aim of this work is to develop a highly active dry reforming catalyst based on Ni. Considering the disadvantages of Ni in terms of deactivation, we have developed multicomponent materials based on monometallic and bimetallic Ni/CeO_2/Al_2O_3 catalysts with excellent features for reforming reactions. In this study, all the catalysts contain 10 %wt. Ni and 20 %wt. CeO_2. As for the bimetallic formulations, Pt and Sn have been selected as promoters with a nominal loading of 0.3 and 0.8 % wt. respectively.

Experimental

The catalysts were prepared by wet sequential impregnation using alumina (Sasol) and the relevant precursors (Sigma Aldrich) employing a Rotovap, with a session of drying and calcining between each step of the impregnation. The drying was performed at 80°C and was left overnight, the calcination was performed at 700°C for 4 h.

The dry reforming of methane was carried out at atmospheric pressure in a quartz tube reactor. 0.1 g of catalyst was used per reaction supported upon a reactor bed of quartz wool. The catalyst was reduced in situ at 800°C for 1 h, in a 100 mL/min flow containing 20% H_2 and 80% N_2. The reaction was then conducted at 700°C for over 20 h, in a 100 ml/min flow containing 12.5% CH_4, 12.5% CO_2 and 75% N_2 (WHSV = 60,000 mL g^{-1} h^{-1}). The gas products of the reaction where analysed using an ABB AO2208 gas analyser. The spent catalysts were recovered and characterized.

All the fresh catalyst samples have been characterized fully, by XRD, H_2-TPR and N_2 adsorption. The spent samples have also been characterized fully, post reaction, by XRD, Raman and SEM.

Results and Discussion

The textural properties of the sample catalysts are shown below in **Table 1**. What can be observed from the data is that the specific surface area as well as the pore volume of the catalyst is reduced by the addition of ceria. However, the addition of Sn and Pt does not affect the textural properties of the catalysts. The decrease in the surface area can be attributed to ceria species covering the surface of Al_2O_3, this is corroborated by the decrease in pore volume.

***Table 1:** Textural properties of the catalyst samples.*

Catalyst	BET (m^2 g^{-1})	Pore Volume (cm^3 g^{-1})
Ni/Al	164	0.42
Ni/CeAl	116	0.28
SnNi/CeAl	120	0.29
PtNi/CeAl	118	0.28

The XRD data shown below in **Figure 1** provides insight into the chemical structure of the prepared catalyst samples. The absence of any crystalline peak attributed to Ni species indicates a good metallic dispersion and small particle sizes. However, the presence of $NiAl_2O_4$ cannot be fully discarded since the spinel peaks overlap with those of the bare Al_2O_3. In addition, the presence of ceria fluorite type structures is evidenced in the XRD patterns. As for the bimetallic samples, the XRD depicted for PtNi/CeAl is comparable to that of the Ni/CeAl base catalyst except for the small PtO peak at 2Θ = 40°, while the SnNi/CeAl catalyst shows a very similar XRD pattern to Ni/CeAl.

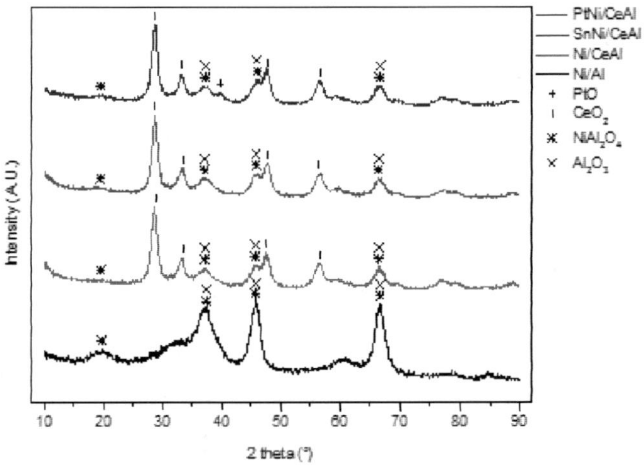

***Figure 1:** XRD patterns of the catalysts.*

The redox properties of the catalysts are shown in the H_2-TPR profiles displayed in **Figure 2**. All the profiles present, contain a peak at high temperatures (650-850°C), this is due to the bulk reduction of the $NiAl_2O_4$ spinels, which due to their structure require a high temperature to be reduced. The later agrees with the XRD data. The three catalysts containing ceria exhibit a shoulder at 850°C as well as much larger H_2 consumption peak than the base Ni/Al catalyst. This is attributable to the bulk ceria reduction occurring in the temperature range 700-900°C. Also, a reduction process at 250°C on the PtNi/CeAl profile was identified and attributed to the reduction of PtO.

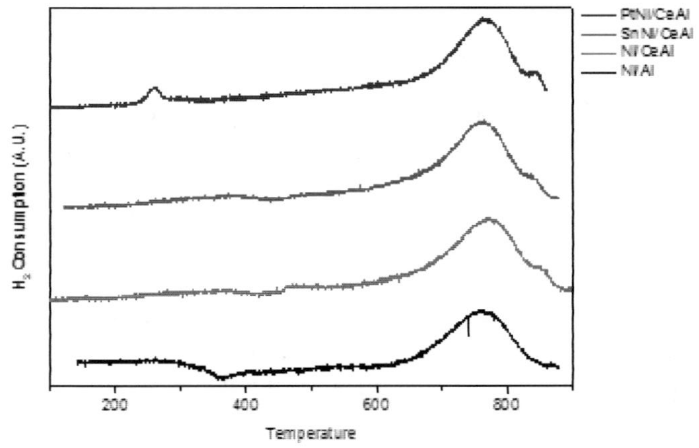

Figure 2: H_2-TPR profiles of the catalysts.

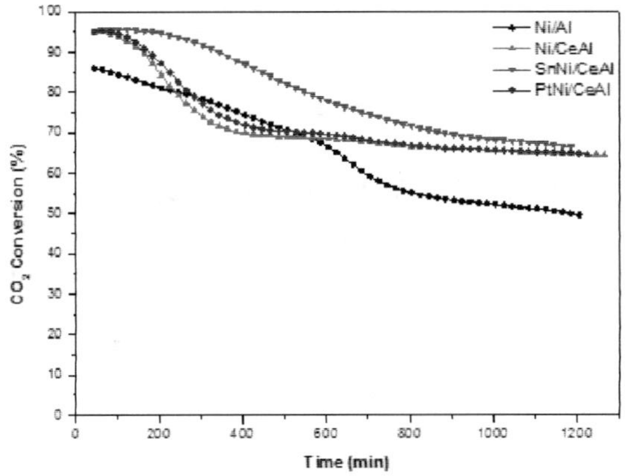

Figure 4: CO_2 conversions of the catalysts.

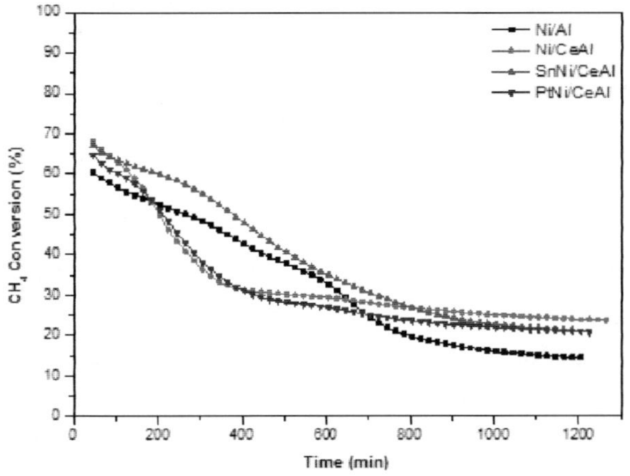

Figure 3: CH_4 conversions of the catalysts.

Figure 3 and **Figure 4** below show the CH_4 and CO_2 conversions for all the catalyst after 20+ h on the reaction stream. It can be seen that the addition of ceria helps to significantly stabilise the catalyst due to its high oxygen storage capacity (OSC), in comparison the Ni/Al conversions are observed to drop especially after 10 hours. This is attributed to the formation of hard carbon on the surface of the catalyst blocking the accessibility of the molecules to the active sites. The effect the promoters have on the conversions differs, the Sn promoted catalyst has higher conversions to begin with before stabilising at similar conversions to the other ceria based catalyst. This has been attributed to the smaller particle size and higher particle dispersion the Sn promotion causes. By contrast the addition of Pt as a promoter hardly affects the catalytic performance of the Ni/Ce/Al sample. The Sn promoted catalyst also had the highest H_2/CO ratio to begin with, of 0.97, before stabilising at a similar value to the other catalysts of around 0.75.

The characterisation performed post reaction on the spent catalysts (not shown for sake of briefness) concur with the conversion hypothesis. Large concentration of hard carbon were deposited on the Ni/Al while smaller concentrations and softer carbon haven been found on the ceria and ceria-tin promoted catalysts.

Conclusions

This work evidences that multicomponent Ni based catalysts can be efficiently utilised for the chemical CO_2 recycling via dry reforming. High levels of CO_2 conversion and good H_2/CO ratios have been achieved in general. The bimetallic Ni-Sn system supported on CeO_2-Al_2O_3 was the most promising material among the studied catalysts. The excellent behaviour of this catalyst is due to its improved redox properties and the Ni-Sn synergy which prevents active phase sintering and carbon poisoning.

Further studies are currently being undertaken by our group to discern the effect of the operating conditions as well as the long-term stability of these advanced materials, in order to explore their potential application in real CO_2 utilisation processes.

References

1. Pires JCM, Martins FG, Alvim-Ferraz MCM, Simoes M. Recent Developements on Carbon Capture and Storage: An Overview. Chemical Engineering Research and Design. 2011; 89(9): p. 1446-1460.

2. Leung DYC, Caramanna G, Maroto-Vler MM. An Overview of Current Status of Carbon Dioxide Capture and Storage Technologies. Renewable and Sustainable Energy Reviews. 2014; 39: p. 426-443.

3. Pakhare D, Spivey J. A Review of Dry (CO2) Reforming of Methane over Noble Metal Catalysts. Chemical Society Reviews. 2014; 43: p. 7813-7837.

4. de Araujo GC, de Lima SM, Assaf JM, Pena MA, Fierro JLG, Rangel MdC. Catalystic Evaluation of Perovskite-Type Oxide LaNi1-xRuxO3 in Methane Dry Reforming. Catalysis Today. 2008; 133: p. 129-135.

5. Charisiou ND, Siakavelas G, Papageridis KN, Baklavaridis A, Tzounis L, Avraam DG, et al. Syngas Production via the Biogas Dry Reforming Reaction Over Nickel Supported on Modified with CeO2 and/or La2O3 Alumina Catalysts. Journal of Natural Gas Science and Engineering. 2016; 31: p. 164-183.

497365 One-Step Synthesis of Transition Metal Doped Titanium Dioxide Catalysts for Oxygen Removal

Abstract:

Carbon dioxide (CO_2) capture and sequestration have been considered as promising techniques to alleviate the emission of greenhouse gas. However, it is difficult to cost effectively deploy them in conventional coal-fired combustion plants due to the diluted CO_2 concentration in the flue gas, requiring additional methodologies for concentrating the stream. An oxy-fuel combustion technique which uses pure oxygen (O_2) to obtain a concentrated CO_2 stream has been developed to solve this problem, and has obtained a sequestration-ready gas stream. However, this gas stream contains O_2 as well as CO_2, which can cause damage to the gas infrastructure through corrosion. To overcome this shortcoming, catalytic O_2 removal system with hydrocarbons has been recently applied. In this study, we develop 1) a one-step synthesis of transition metal doped catalysts and 2) an efficient catalytic O_2 removal system with methane (CH_4). To achieve this, we first synthesize different transition metal (Cu, Fe, Ir, Ni, Mn, Rh, Ru) doped titanium oxide (TiO_2) catalysts in a flame aerosol reactor. Compared to liquid phase synthesis methods such as wet impregnation and sol-gel, flame synthesis is a gas phase technique that can produce homogeneous nanoparticles and can be scaled up for high throughput production. The different catalysts are characterized by BET, X-ray diffraction, scanning electron microscopy and transmission electron microscopy to relate nanoparticle characteristics to performance. Second, we test the synthesized catalysts for O_2 removal with CH_4 to optimize the catalyst. Third, we optimize the synthesis conditions (dopant concentration (0-5 wt.%), fuel flow rate (0.8-1.8 LPM)) to get the highest O_2 removal efficiency.

494289 Carboxylation of Propylene Oxide to Propylene Carbonate

Abstract:

Polycarbonates, a class thermoplastic polymer, have wide ranging applications in automotive & transportation, construction, packaging, consumer goods, optical media, electrical & electronics and medical devices. Conventionally, manufacturing of polycarbonates involves phosgenation of bis-phenol-A, which has drawbacks due to toxic phosgene and corrosive products like HCl in stoichiometric amounts. Alternative non-phosgene route for polycarbonates involves carboxylation of propylene oxide as a key step. This is also an example of CO_2 utilization for value added chemicals. Cyclic carbonates are also widely used as aprotic solvents, anti-foaming agents, antifreeze, plasticizers, and monomer for various commodity polymers. Several studies reported carboxylation of epoxides employing either homogeneous or heterogeneous catalysts. Reported heterogeneous catalysts include metal oxides, zeolites, polymer supported quaternary onium salts, ion exchange resins, and polymer supported ionic liquids. However, the main challenges in this reaction are low catalyst activity, and high CO_2 pressure. In this study, we systematically investigated the influence of solvents on ion exchange resin catalyzed carboxylation of propylene oxide with the aim of enhancing the activity and selectivity. Intrinsic kinetics of carboxylation of propylene oxide has also been studied in a batch slurry reactor to understand the effect of reaction conditions and develop a suitable rate equation.

The experiments for solvent effects and kinetic studies were carried out in a stirred pressure reactor with 100 cm^3 capacity with provisions for control of agitation speed, temperature and sampling of liquids. In these experiments, the CO_2 pressure in the reactor was kept constant by continuous supply through a CO_2 reservoir using a constant pressure regulator such that the temporal reaction progress was followed from the pressure decrease in the reservoir. At the end of each experiment, liquid products were analyzed for PO and PC to assess the material balance. The effect of solvents and the order of activity is: Methanol > Ethanol > Isopropyl alcohol > Dimethyl formamide > N-methyl-2-pyrrolidone > Propylene carbonate > Dimethyl carbonate and it was found that using methanol as a solvent rate of carboxylation were enhanced by a factor of 3. The polar protic solvents (such as linear alcohols) are strong nucleophiles, strong hydrogen-bond donors and acceptors and interact strongly with the electron-deficient electrophiles, epoxides and CO_2. In contrast, polar aprotic solvents are weak nucleophiles compared to the polar protic solvents resulting in low activity. These conclusions, based on CO_2 solubility measurements in methanol and PC solvents along with kinetic studies of PO carboxylation, will be presented and discussed.

Improving the Performance of Sn Promoted Ni/Al_2O_3 Catalysts for the Dry Reforming of Methane

*T. Stroud, T. Smith, E. Le Saché, H. Arellano-Garcia, T.R. Reina**

Department of Chemical and Process Engineering, Faculty of Engineering and Physical Sciences, University of Surrey, Guildford, UK

Corresponding author: t.ramirezreina@surrey.ac.uk

Introduction

Prevention of CO_2 reaching the upper atmosphere is a topic of great interest in the world today. Co-currently, green energy production and chemical feedstock sources are being highly researched, where combining these issues represents the most efficient solution. Dry reforming of methane (DRM) represents one such solution:

$$CH_4 + CO_2 \rightleftharpoons 2CO + 2H_2$$

$$\Delta H_{298K} = +247 \text{ kJ/mol} \qquad \Delta G_{298K} = +170 \text{ kJ/mol}$$

The products of this reaction (known colloquially as syngas) are a highly useful precursor for a number of chemical processes, such as the production of methanol and synthesis of liquid fuels via Fischer-Tropsch (FTS).

Given the endothermic nature of the reaction, high reaction temperatures are required to achieve favourable conversions of the reactants to syngas. Unfortunately, elevated temperatures also promote the formation of solid carbon via a number of further side reactions [1]. This carbon then deposits on the surface of the catalyst which (alongside metal sintering) causes the catalyst to deactivate.

Hence an effective catalyst must resist sintering and be resilient towards the formation of carbon upon the surface; whilst also being relatively inexpensive, such that it can be realistically employed within the wider industry. Nickel has previously been identified as a viable active metal to use, whilst an alumina (Al_2O_3) support helps disperse the Ni over a wider surface area, improving the activity over that of just Ni alone [1].

However, this still suffers from severe carbon deactivation and hence there remains potential for improvement. The aim of this work is to develop advanced catalysts for chemical CO_2 recycling via DRM using tin and ceria as promoters of Ni/Al_2O_3

Experimental

Catalysts were prepared by the method of sequential impregnation, where a base of 10 wt.% Ni was first prepared by adding the appropriate amount of $Ni(NO_3)•6H_2O$ (Sigma-Aldrich) to an alumina (SASOL) base, using acetone (Sigma-Aldrich) as the solvent. This was then thoroughly mixed for 30 minutes, before evaporating the acetone off with the aid of a vacuum pump. The sample was then further air dried in an oven overnight at 80 °C, after which calcination was performed at 700 °C for 4 hours. The resultant Ni/Al catalyst was employed as a reference material in this paper. Following the same preparation method, 3 promoted catalysts were prepared: Sn 0.02 mol.% Ni/Al, Sn 0.04 mol.% Ni/Al, Sn 0.02 mol.% Ni/Ce-Al. The tin was added using a $SnCl_2$ (Sigma-Aldrich) precursor and the ceria by $Ce(NO_3)_2•H_2O$ (Sigma-Aldrich) aiming for a nominal CeO_2 loading of 20 wt.%.

Experiments were run in a quartz tube reactor at atmospheric pressure, utilising quartz wool as a reactor bed. An electric fired oven was used to ramp up (15 °C/min) and maintain the temperatures used for reduction (800 °C) and reaction (700 °C). Reduction was performed in a total flow of 100 ml/min for 1 hour in an atmosphere consisting 20 % H_2 (BOC) and 80 % N_2 (BOC).

Each reaction was run for over 20 hours, using 0.1 g of catalyst and a total inlet flow of 100 ml/min, corresponding to a weight hourly space velocity (WHSV) of 60, 000 ml/(g-cat h). Of the inlet flow, 12.5 % was CO_2 (BOC), 12.5 % CH_4 (BOC) and the remainder being that of N_2.

All fresh catalysts were characterised by means of XRD, H_2-TPR and N_2 physisorption (BET) analysis. Selected spent catalysts also underwent SEM and Raman.

Results and Discussion

H_2-TPR analysis (not shown here) confirms that a reduction temperature of 800 °C is sufficient to reduce the metal oxides to their elemental forms, whilst the XRD data (**Figure 1**) shows the Ni particles have been well dispersed. The increase of Sn loading as a promoter appears to have little effect on the crystalline structure of the samples, whereas cerium shows typical diffraction peaks of CeO_2 fluorite phases.

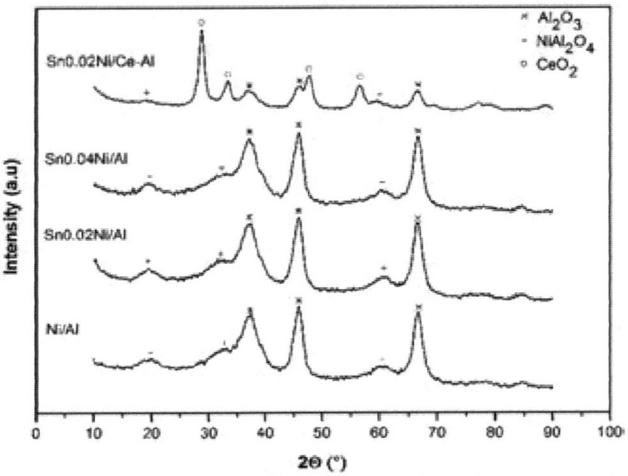

Figure 1 XRD profiles of the sample catalysts

N_2 physisorption results (**Table 1**) show the addition of Sn does not affect the textural properties of the reference Ni/Al material. However CeO_2 reduces the surface area and volume of the pores, which is down to the CeO_2 nanoparticles covering the surface of the primary support (Al_2O_3).

Table 1 Results of N_2 physisorption analysis on fresh catalysts

Sample	S_{Bet} (m^2/g)	V_{Pore} (cm^3/g)	D_{Pore} (nm)
Ni/Al	164	0.42	10.1
Sn0.02 Ni/Al	173	0.42	9.8
Sn0.04 Ni/Al	172	0.42	9.8
Sn0.02 Ni/Ce-Al	120	0.29	9.6

In terms of catalytic performance, both methane and CO_2 conversions (**Figure 2** and **Figure 3** respectively) are improved by the addition of Sn, showing less deactivation with time when compared with the un-promoted Ni/Al catalyst. Initially, both the un-promoted and 0.04 Sn sample appear to perform better than the 0.02 Sn catalyst, but this quickly changes around the 500 minute mark for CH_4 conversion and 450 minute mark for CO_2. Increasing the quantity of Sn from 0.02 mol.% to 0.04 mol.% exhibits a decrease in performance with time. This is theorised to be down to the extra Sn blocking the active sites [2], therefore reducing performance; hence a ratio of 0.02 mol.% of Sn was selected as the optimum to be further improved upon.

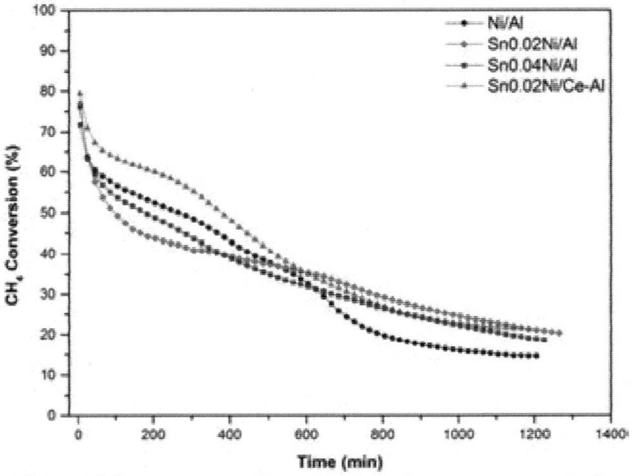

Figure 2 Methane conversions for the catalysts over a period of approximately 20 hours

Ceria is an excellent redox promoter due to its high oxygen storage capacity [3]. Therefore by adding CeO_2, the amount of surface oxygen available increases [1]. This helps with the removal of carbon species from the catalyst surface, increasing the lifespan and improving activity. This has been observed in the case of the multicomponent catalyst (Ni-Sn/Ce-Al), with conversions being greater for longer periods of time in the case of methane. An impressively high peak of 95 % CO_2 conversion can be seen, with a seemingly stable 66 % conversion after approximately 20 hours of on-stream operation. Lower conversions of methane (around 20 % after 20 hours) in comparison to CO_2 are to be expected, due to decomposition of CH_4 being the rate determining step and also the primary depositor of carbon [4].

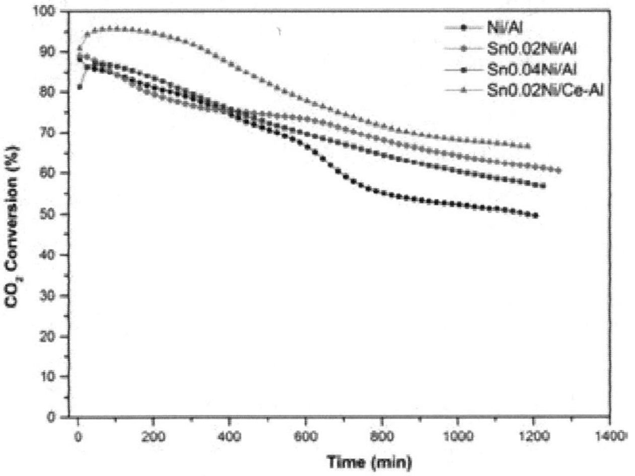

Figure 3 Carbon dioxide conversions for the catalysts over a period of approximately 20 hours

For the H_2/CO ratio (**Figure 4**), the Ce doped sample appears the best candidate early on, peaking at a ratio of 0.97 after *ca.* one hour of operation. However this eventually decreases down to the same level (a ratio of around 0.75) of the un-promoted and 0.04 Sn samples after 20 hours. The Sn0.02 Ni/Al catalyst shows a slightly more favourable ratio of 0.81 after 20 hours, showing the addition of tin is also effective at improving the H_2/CO ratio; with the optimum ratio again being 0.02 mol.%.

With the addition of ceria, a slight drop in performance is observed when compared with the 0.02Sn promoted catalyst. At the end of the experiment, a higher conversion of CO_2 is present for Ce, with a similar methane conversion. This therefore results in a comparatively greater abundance of CO being present in the product stream (as the H_2 is only formed by methane decomposition), giving a lower ratio. The observed trend could be due to the ability of ceria to catalyse the reverse water shift reaction, RWGS ($CO_2 + H_2 \rightleftharpoons H_2O + CO$) which explains greater CO_2 conversion and the smaller H_2/CO ratio for the ceria promoted sample.

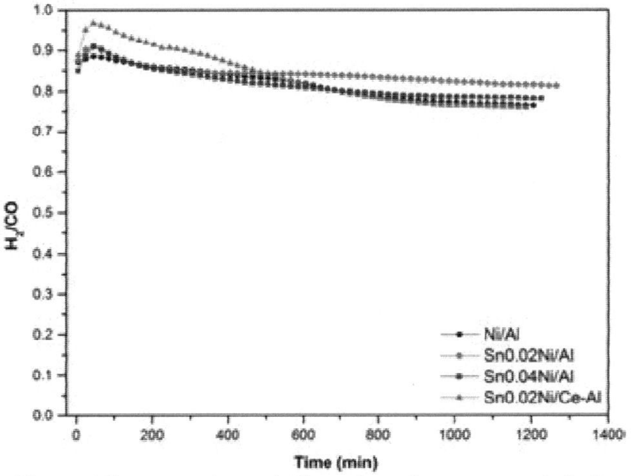

Figure 4 Syngas product ratio for each catalyst over a period of approximately 20 hours

Conclusions

The addition of tin as a promoter to a nickel alumina catalyst shows a marked improvement in conversions of both carbon dioxide and methane to syngas. However, adding too large amounts of tin can have a detrimental effect on catalytic performance by restricting access of reactants to the active nickel sites. It has been found that 0.02 mol.% is an optimum tin loading.

Addition of ceria to the alumina support increases resistance to carbon formation on the surface, exhibiting higher conversions for a longer period of time. In parallel, ceria aids the RWGS reaction boosting the CO_2 conversion. When included together, tin and ceria work synergistically, resulting in a very promising catalyst for dry reforming reactions.

References

1. Charisiou N.D., Baklavaridis A., Papadakis V.G., Goula M.A. *Syn Gas Production via the Biogas Reforming Reaction Over Ni/MgO–Al2O3 and Ni/CaO-Al2O3 Catalysts.* Waste and Biomass Valorization. 2016; 7: p. 725-736.

2. Hou Z., Yokota O., Tanaka T., Yashima T. *Surface Properties of a Coke-Free Sn Doped Nickel Catalyst for the CO2 Reforming of Methane.* Applied Surface Science. 2004; 233: p. 58-68.

3. Reina T.R., Ivanova S., Centeno M.A., Odriolzola J.A. *Boosting the Activity of a Au/CeO2/Al2O3 Catalyst for the WGS Reaction.* Catalysis Today. 2015; 253: p. 149-154.

4. Wei J., Iglesia E. *Isotopic and Kinetic Assessment of the Mechanism of Reactions of CH4 With CO2 or H2O to Form Synthesis Gas and Carbon on Nickel Catalysts.* Journal of Catalysis. 2004; 224(2): p. 370-383.

498418 The Effects of Aging on Steady State Ammonia Oxidation over Pt/Al_2O_3

Abstract:

Three-way catalyst (TWC) light-off (LO) performance is critical in cost-effectively meeting emission standards. Improvements in catalyst LO should include catalyst modifications that minimize inhibition effects of exhaust species such as CO and hydrocarbons (HCs), lower precious group metal (PGM) loading, or use a lower cost PGM such as Pd. Alumina (Al_2O_3) and zirconia (ZrO_2) supports provide high surface area and stability. Furthermore, ceria (CeO_2) provides the key benefit of oxygen storage, compensating for deviations from stoichiometric TWC operating conditions by supplying oxygen for CO and HC oxidation under rich transient conditions. Inclusion of ceria in PGM catalysts enhances CO and HC conversion and minimizes self-inhibition; this improved activity is mainly attributed to ceria containing more efficient oxygen activation sites at the PGM-support interface relative to other support materials such as Al_2O_3 [1-4].

Monolith catalysts having washcoats with 1 wt% Pd, 3 g/in^3 loading, 400 cpsi and formulations Pd/Al_2O_3, Pd/CZO, and $Pd/CZO/Al_2O_3$ (17 wt% CZO, balance Al_2O_3) were synthesized and compared. Catalyst performance experiments were conducted to study the role of Pd and support compositions and loadings on inhibition during oxidation of CO, acetylene (C_2H_2), and propylene (C_3H_6) under near-stoichiometric conditions ($\lambda = 1.01$). Transient and steady-state bench scale reactor studies with simulated exhaust gas mixtures were conducted to extract oxidation kinetics, compare LO behavior, and develop a predictive model for understanding and optimizing catalyst performance.

Catalyst activities were compared via temperature-programmed oxidation experiments. Figure 1 compares CO LO curves using feed concentrations of 0.5% CO and 1% CO. The Pd/CZO catalyst was tested when "fresh" (initial tests after synthesis) and 10+ months later ("deactivated"). Using all three catalysts, LO temperatures increase with feed CO concentration, showing that CO oxidation is self-inhibiting. With increasing CZO content, LO temperatures decrease, demonstrating the promotional effects of using ceria.

Figure 2 shows C_2H_2 LO (feed concentrations of 375 ppm C_2H_2 and 750 ppm C_2H_2) and C_3H_6 LO (feed concentrations of 250 ppm C_3H_6 and 500 ppm C_3H_6) using the deactivated Pd/CZO catalyst. Like CO, C_2H_2 and C_3H_6 are self-inhibiting. For the same concentrations of carbon in the feed mixture (750 ppm C_1 or 1500 ppm C_1), C_2H_2 oxidation occurs at higher temperatures than C_3H_6 oxidation.

Exhaust hydrocarbons propylene and acetylene reveals some interesting differences during co-oxidation with CO. The individual LO features of 1% CO and 750 ppm C_2H_2 in Figures 1 and 2 are compared to their co-oxidation (1% CO + 750 ppm C_2H_2) in Figure 3. CO lights off at higher temperatures in the mixture, i.e. C_2H_2 inhibits CO oxidation. During co-oxidation, C_2H_2 oxidation is inhibited by the presence of CO below

~15% conversion, but is otherwise improved compared to oxidation of C_2H_2 alone. Whereas 1% CO lights off before 750 ppm of C_2H_2 during individual oxidation, the reverse is true during co-oxidation. This is in contrast to earlier findings of CO + C_3H_6 trends in which was found that CO inhibits C_3H_6; i.e. propylene conversion follows CO light off (see below).It is instructive to compare the behavior of the two hydrocarbons at the same C1 concentration. Though the same feed concentration of 1% CO + 1500 ppm C_1 (from HCs) is used, C_3H_6 and C_2H_2 co-oxidation with CO exhibits different trends in terms of LO temperatures and order of species oxidation. As shown in Figure 4, using the deactivated Pd/CZO catalyst, 1% CO lights off before 500 ppm C_3H_6 both during individual oxidation and mixture LO. During co-oxidation, 500 ppm C_3H_6 is 50% converted at ~ 246 °C, compared to ~ 266 °C for 750 ppm C_2H_2.

Differential kinetics studies were conducted to quantify reaction orders and activation energies. The reaction order with respect to CO was found to be approximately −1 using Pd/Al_2O_3 and $Pd/CZO/Al_2O_3$ and 0 to slightly negative using Pd/CZO, consistent with literature values [5-7]. The activation energies for CO oxidation were found to be 134-213 kJ/mol using Pd/Al_2O_3 and $Pd/CZO/Al_2O_3$, and 25-70 kJ/mol using Pd/CZO. Using all three catalysts, the reaction order with respect to C_3H_6 was found to be approximately −0.89 to −0.75 and the activation energy in the range of 72-105 kJ/mol, approximately half of that observed for CO oxidation and similar to literature values [5]. Using the Pd/CZO catalyst, the reaction order with respect to C_2H_2 was found to be approximately −0.52 to −0.96 and the activation energy of C_2H_2 oxidation approximately 73 kJ/mol.

Steady-state and transient CO, C_2H_2, and C_3H_6 oxidation experiments using fresh Pd/Al_2O_3, Pd/CZO, and $Pd/CZO/Al_2O_3$ monolith catalysts were conducted to compare LO behavior and kinetic parameters (i.e. reaction orders and activation energies). CO LO temperatures decreased with increasing proportion of ceria used in the TWC, and C_2H_2 and C_3H_6 LO studies using Pd/CZO demonstrated self- and mutual-inhibition during species oxidation. Kinetics studies demonstrated further the self-inhibiting oxidation behavior.

Ongoing work involves the use of a mechanistic-based kinetic model incorporated into a low-dimensional monolith model [8-9]. The global reactor models are used with CO, C_2H_2, and C_3H_6 oxidation on Pd with steps involving ceria. The 1+1 dimensional model formulation comprises coupled species balances with transverse-average equations having external and internal mass transfer coefficients to verify experimental LO behavior during individual and co-oxidation using catalysts and feed mixtures of different formulations. Finally, experiments involving C_2H_2 oxidation on Pd/Al_2O_3 and $Pd/CZO/Al_2O_3$ catalysts and other hydrocarbons including toluene are being conducted to investigate inhibition during LO, extract kinetic parameters for oxidation reactions, and provide a basis for validating modeling simulations.

Figure 1: CO conversion (%) versus monolith temperature using Pd/Al_2O_3, Pd/CZO/Al_2O_3, and Pd/CZO catalysts for feed concentrations of 0.5% and 1% CO.

Figure 2: C_2H_2, C_3H_6 conversion (%) versus monolith temperature using deactivated Pd/CZO for feed concentrations of 375 ppm C_2H_2, 750 ppm C_2H_2, 250 ppm C_3H_6, and 500 ppm C_3H_6.

Figure 3: CO, C_2H_2 conversion (%) versus monolith temperature using deactivated Pd/CZO for feed concentrations of 750 ppm C_2H_2 and/or 1% CO.

Figure 4: CO, C_3H_6 conversion (%) versus monolith temperature using deactivated Pd/CZO for feed concentrations of 500 ppm C_2H_2 and/or 1% CO.

References

[1] Fernandez-Garcia, M.; Martinez-Arias, A.; Iglesias-Juez, A.; Hungria, A.B.; Anderson, J.A.; Conesa, J.C.; and Soria, J. *App. Cat. B* 2001, 3**1**, 39-50.

[2] Martinez-Arias, A., Fernandez-Garcia, M.; Iglesias-Juez, A.; Hungria, A.B.; Anderson, J.A.; Conesa, J.C.; and Soria, J.. *App. Cat. B* 2001, **31**, 51-60.

[3] Sharma, S.; Hegde, M.S.; Das, R.N.; and Pandey, M. *App. Cat. A* 2008, **337**, 130-137.

[4] Harmsen, J.M.A.; Hoebink, H.B.J.; and Shouten, J.C. *Chem. Eng. Sci.* 2001, **56**, 2019-2035.

[5] Yao, Y-.F. Y. *J. Catal.*, 1984, **87**, 152-162.

[6] Cant, N.W.; Hicks, P.C.; and Lennon, B.S. *J. Catal.* 1978, **54**, 372-383.

[7] Rainer, D.R.; Koranne, M.; Vesecky, M.; and Goodman, D.W. *J. Phys. Chem. B* 1997, **101**, 10769-10774.

[8] Joshi, S.; Harold, M.P.; Balakotaiah, V. *AIChE J.,* 2009, **55,** 1771-1783.

[9] Raj, R.; Harold, M.P.; Balakotaiah, V. *Chem. Eng. J.*, 2015, **281**, 322-333.

494315 Two-Step Selective Bromination of Thiophenic Compounds for Gasoline Desulfurization Under Mild Conditions

Abstract:

The removal of trace amount of refractory thiophenic compounds from fuels has raised intense concern globally. The development of less energy-intensified desulfurization process is under great demand. In this work, we proposed a new approach of two-step selective bromination of thiophenic compounds for gasoline desulfurization under mild conditions. The Ag modified MoO_3 nanorods (AMO) were synthesized and characterized by XRD, HRTEM and N_2 adsorption, and evaluated in selective bromination of thiophenic compounds. Results showed that (a) thiophenic compounds in model gasoline can be selectively brominated over AMO catalyst with the addition of Br_2, and their bromination production can be removed selectively by followed-up extraction. (b) At the optimized Ag loading of 10%, the apparent rate constants for T and BT reached as high as $95.8h^{-1}$ and $142.7h^{-1}$, respectively under ambient conditions, much higher than those cases in oxidative desulfurization, adsorptive desulfurization, etc. Product analyses showed that the portion of dibromo-thiophenes increased dramatically with increasing loading of Ag, due likely to the formation of $MoAg_2O_4$. (c) The bromination selectivity of thiophenic compounds reached four orders of magnitude than those of aromatic compounds in fuel. Meanwhile, the competition from olefins can be strategically suppressed by HBr pretreatment. (d) With the integration of the 2nd-step of extraction, desulfurization capacity of thiophenes reached as high as 30 mg-S/g-cat, while the integrated system showed excellent bromination desulfurization capacity of thiophene at ~260 mg-S/g-cat under high fuel-to-catalyst mass ratio of 5000. The adsorption of various bromides over $MoAg_2O_4$ was simulated to elucidate the desulfurization mechanism. The fast kinetics, high desulfurization capacity and selectivity of the proposed two-step approach under mild conditions make it promising for gasoline desulfurization.

Keywords: desulfurization, gasoline, selectivity, Ag modified MoO_3

494394 Improvement of Low Temperature Activity for CO Oxidation on Ceria Catalyst By Transition Metal Substitution

Abstract:

Recently, reduction of exhaust gas has been an important issue with increasing the sales of automobiles. CO reduction in various exhaust gases is particularly critical because it damages people and environment. A ceria-based catalyst that can reduce automotive emissions instead of three-way catalysts have been developed. When the temperature is lowered, however, the efficiency is rapidly decreased. In order to improve the activity of ceria catalyst, we synthesized transition metal (TM)-doped ceria. In this study, we examined four transition metal (TM=Cu, Co, Mn, and Fe) doped and their co-doped ceria. TM-doped ceria ($Ce_{0.95}TM_{0.05}O_{2\text{-î'}}$) were synthesized by EDTA-citrate complexing method. By means of this method, we obtained highly pure and homogeneous nano-powder. The nanoparticles were then characterized by X-ray diffraction (XRD), scanning electron microscopy (SEM), and energy dispersive spectroscopy (SEM-EDS). To clarify the positive effects of doping, the surface area was evaluated by Brunauer-Emmett-Teller (BET) measurement. It turns out that the TM-doped ceria powders had higher specific surface area than that of undoped ceria, and correspondingly, the larger specific surface area enhanced the catalytic activity. Our density functional theory results demonstrated that this is attributed to the different interactions between dopant and support. Our results will be helpful for effectively exploring the TM dopant candidates of ceria and determining their optimal concentration for the enhanced catalytic activity.

487852 Characterization of the Active Site and Mechanism for 1,6-Hexanediol Production from Tetrahydrofuran-Dimethanol over Pt-Based Catalysts

Abstract:

Characterization of the Active Site and Mechanism for 1,6-Hexanediol Production from Tetrahydrofuran-Dimethanol over Pt-Based Catalysts

Terminal diols such as 1,4-butanediol (1,4-BDO) and 1,6-hexanediol (1,6-HDO) are important building-block chemicals and find application in the synthesis of specialty chemicals and a variety of polymers, primarily in polyesters and polyurethanes, and polyamides such as nylon-6,6.[1,2] Currently, 1,6-HDO is produced industrially from cylohexanone/cyclohexanol (KA oil) by oxidation with nitric acid to form adipic acid, followed by hydrogenation of the dimethoxy ester to yield 1,6-HDO. The attractiveness of the overall process is tempered by low conversions (4-8% for the oxidation of KA oil), difficult separations, the use of non-renewable fossil feedstocks, and emission of greenhouse gases, namely N_2O.[3] Various attempts to synthesize 1,6-HDO, as well as other alpha,omega-diols, from biomass resources have been reported. Using supported catalysts that contain precious metals with a reducible metal oxide such as Rh-ReO_x,[4-9] and Ir-ReO_x,[10,11] tetrahydropyran-2-methanol (THP2M), a biomass-derived feedstock,[12] can be converted into 1,6-HDO with high selectivity. However, the high cost of the catalyst, in combination with a low productivity, limits the industrial applicability of those systems.[13] More recently, it has been shown that 1,6-HDO can be produced from THP2M without the use of precious metals. However, 1,6-HDO selectivity only reached ~40%.[14] Therefore, this route is not likely to be industrially applicable soon either.

In an effort to make the route to 1,6-HDO from biomass more industrially viable, our group has developed a complete reaction network producing 1,6-HDO from woody biomass.[15] The early steps in this process convert biomass to tetrahydrofuran-dimethanol (THFDM) with high selectivity, and will likely be discussed elsewhere at this conference. In this study, we focus on the final steps to 1,6-HDO from THFDM, shown in Scheme 1.

Scheme 1. Reaction sequence to 1,6-hexanediol (1,6-HDO) from tetrahydrofuran-dimethanol (THFDM)

We have developed a Pt-based catalyst that performs each step of the above reaction sequence with a total 1,6-HDO selectivity >90%. In comparison to the state-of-the-art, a previously described Rh-ReO_x/C catalyst is not active for this reaction. Moreover, while Rh-ReO_x/C is active for 1,6-HDO production from THP2M with selectivity >90%, the reaction rate is an order of magnitude lower than the total production rate of 1,6-HDO from THFDM over our Pt-based catalysts. Our catalyst, therefore, produces 1,6-HDO from a further upstream product, and at a much higher rate than the state-of-the-art Rh-ReO_x/C system.

In situ characterization of this catalyst has revealed important information with regards to the active site and mechanism for 1,6-HDO production over our Pt-based catalysts. Our primary findings are that the oxidation state and acidity of the oxophilic additive (Mo, V, W, Re, for example) are crucial for this reaction. All catalysts are reduced with H_2 prior to reaction, so the same conditions were used for characterization. We studied these catalysts with pyridine FTIR spectroscopy to obtain the Bronsted/Lewis acid ratio in each of our Pt-based catalysts, and compare that to reactivity data we had already compiled for the reaction of interest. As we observed that Bronsted acidity is correlated to the turnover frequency. Moreover, we found that this ratio increases in all catalysts during pre-reduction, a necessary step to achieve any reactivity. This data seems to indicate that the Bronsted acid site necessary for this reaction to take place, which are also confirmed by XPS and Raman characterizations. Furthermore, we are able to alter and characterize the distance between Pt nanoparticles and the oxophilic additive by changing the catalyst synthesis technique. Based on these studies we proposed a different mechanism to those which have been proposed previously, which will be discussed in more detail during the conference.

1 P. Werle, M. Morawietz, S. Lundmark, K. Sšrensen, E. Karivinen and J. Lehtonen, Alcohols, Polyhydric, in *UllmannÕs Encyclopedia of Industrial Chemistry*, 2012.

2 W. Tianfu, M. S. Ide, M. R. Nolan, R. J. Davis and B. H. Shanks, *Energy Environ. Focus*, 2016, **5**, 13Ð17.

3 S. Van de Vyver and Y. Roman-Leshkov, *Catal. Sci. Technol.*, 2013, **3**, 1465Ð1479.

4 M. Chia, Y. J. Pag‡n-Torres, D. Hibbitts, Q. Tan, H. N. Pham, A. K. Datye, M. Neurock, R. J. Davis and J. A. Dumesic, *J. Am. Chem. Soc.*, 2011, **133**, 12675Ð12689.

5 M. Chia, B. J. OÕNeill, R. Alamillo, P. J. Dietrich, F. H. Ribeiro, J. T. Miller and J. A. Dumesic, *J. Catal.*, 2013, **308**, 226Ð236.

6 P. U. Karanjkar, S. P. Burt, X. Chen, K. J. Barnett, M. R. Ball, M. D. Kumbhalkar, J. B. Miller, I. Hermans, J. A. Dumesic and G. W. Huber, *Catal. Sci. Technol.*, 2016, **6**, 7841Ð7851.

7 Y. Kojima, S. Kotani, M. Sano, T. Suzuki and T. Miyake, *J. Jpn. Pet. Inst.*, 2013, **56**, 133Ð141.

8 K. Chen, S. Koso, T. Kubota, Y. Nakagawa and K. Tomishige, *ChemCatChem*, 2010, **2**, 547Ð555.

9 Y. Nakagawa, M. Tamura and K. Tomishige, *ACS Catal.*, 2013, **3**, 2655Ð2668.

10 S. Koso, I. Furikado, A. Shimao, T. Miyazawa, K. Kunimori and K. Tomishige, *Chem. Commun.*, 2009, **15**, 2035Ð2037.

11 Y. Nakagawa, Y. Shinmi, S. Koso and K. Tomishige, *J. Catal.*, 2010, **272**, 191Ð194.

12 B. Xiao, M. Zheng, X. Li, J. Pang, R. Sun, H. Wang, X. Pang, A. Wang, X. Wang and T. Zhang, *Green Chem.*, 2016, **18**, 2175Ð2184.

13 Z. J. Brentzel, K. J. Barnett, K. Huang, C. T. Maravelias, J. A. Dumesic, and G. W. Huber, *ChemSusChem.*, 2017, **10**, 1-6.

14 S. P. Burt, K. J. Barnett, D. J. McClelland, P. Wolf, J. A. Dumesic, G. W. Huber, and I. Hermans, *Green Chem.*, 2017, **19**, 1390-1398.

15 J. He, K. Huang, K. J. Barnett, S. Krishna, D. M. Alonso, Z. Brentzal, S. P. Burt, T. W. Walker, W. Banholzer, C. T. Maravelias, I. Hermans, J. A. Dumesic, and G. W. Huber, *Faraday Discussions*, 2017.

492877 Mechanism and Kinetics of 1-Dodecanol Etherification and Dehydration over Tungstated Zirconia

Abstract:

As atmospheric carbon dioxide levels continue to rise as a result of burning fossil fuels, there is a growing interest in developing renewable fuels derived from biomass to meet the worldâ€™s increasing energy demands. Ethers have attracted recent interest as diesel additives and specialty chemicals due to their high cetane numbers and excellent lubricant properties, and can be produced via direct etherification of biomass-derived alcohols in the liquid phase. The competing reaction for alcohol dehydration over an acid catalyst is unimolecular dehydration to form alkenes, which is thermodynamically favored above approximately 350 K. In order to optimize the direct etherification of long chain alcohols in the liquid phase, it is necessary to develop an understanding of the mechanism and kinetics of etherification and dehydration reactions. In this study, tungstated zirconia was identified as a selective solid acid catalyst for the liquid phase etherification of 1-dodecanol. Through kinetic modeling and mechanistic probing, this study suggests that a cooperative effect between Brønsted and Lewis acid sites on tungstated zirconia enhances the selectivity to ether by increasing the surface concentration of adsorbed alcohol molecules, promoting bi-molecular etherification over unimolecular dehydration. Kinetic isotope effects for linear alcohol dehydration were measured in order to elucidate the rate limiting steps in the mechanism. Effects of alcohol concentration and product inhibition were measured and fit to kinetic models consistent with the proposed mechanisms. In addition, acid site characterization and selective poisoning experiments were used to probe the role of the acid sites and support the proposed mechanism.

WISCONSIN
UNIVERSITY OF WISCONSIN–MADISON

A Three-Step Catalytic Pathway for the Scalable Production of 1,5-Pentanediol from Biomass-Derived Tetrahydrofurfuryl Alcohol

Kevin J. Barnett[1], Zachary J. Brentzel[1], Kefeng Huang[1], Ling Li[1,2], Guozhu Liu[2], Christos T. Maravelias[1], James A. Dumesic[1] and George W. Huber[1]

(1)Department of Chemical and Biological Engineering, University of Wisconsin-Madison, Madison, WI,
(2)School of Chemical Engineering and Technology, Tianjin University, Tianjin, China

AIChE National Conference: October 30, 2017

C4-C6 α,ω-Diols Production and Market

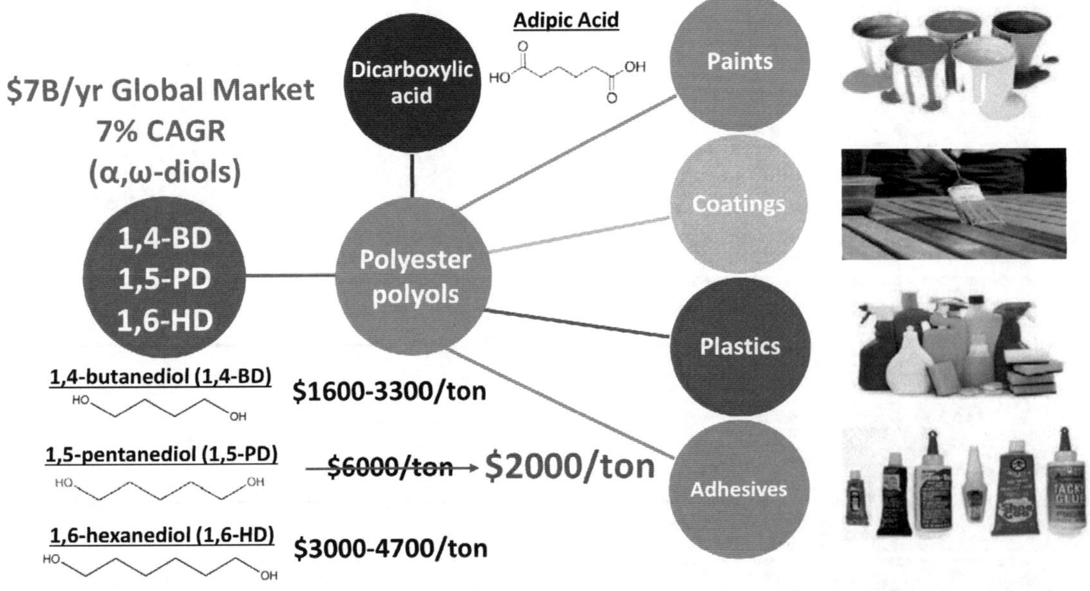

2

1,5-PD Production from Biomass

- Direct Hydrogenolysis Approach:
  Hydrogenolysis over bimetallic, noble-metal catalysts (i.e. RhRe)[1,2]
  - Uneconomical due to high catalyst costs

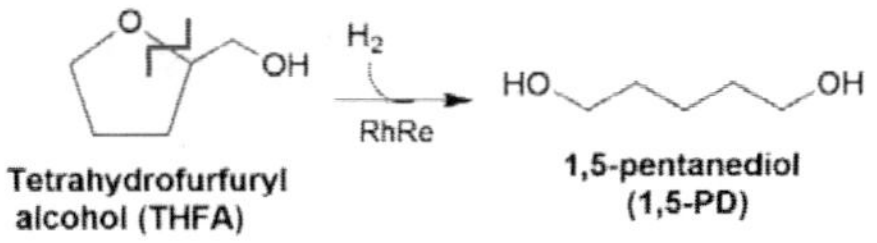

[1] M. Chia et al. *J. Am. Chem. Soc.* **2011**, *133*, 12675–89.
[2] Y. Nakagawa et al. *Catal. Today* **2012**, *195*, 136–143.

- New Approach: 3-step **D**ehydration-**H**ydration-**H**ydrogenation (**DHH**) pathway

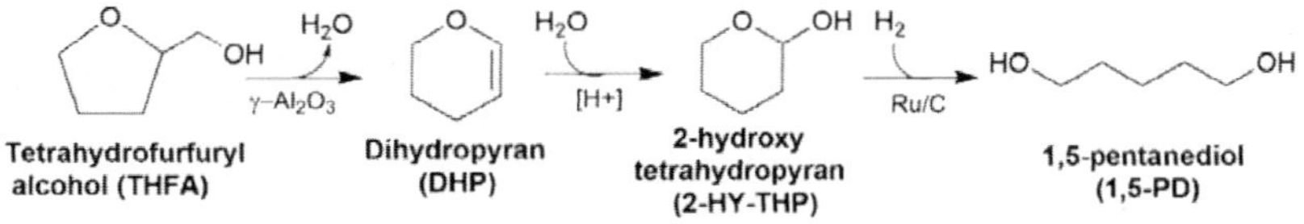

- Previously attempted by Geller et. al in 1946[3]

[3] *L. E. Schniepp, H. H. Geller, J. Am. Chem. Soc.* ***1946****, 68, 1646–1648.*

3

DHH pathway for 1,5-PD production

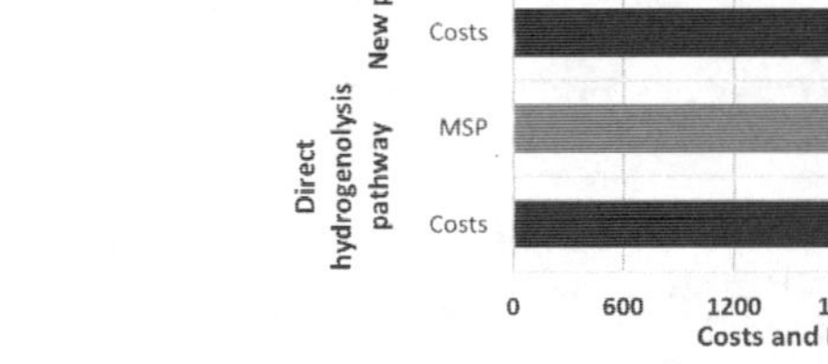

- Brentzel et. al. Chemicals from Biomass: Combining Ring-Opening Tautomerization and Hydrogenation Reactions to Produce 1,5-Pentanediol from Furfural. *ChemSusChem* **2017**.*10,* 1351-1355.
- Huang et al. Conversion of Furfural to 1,5-Pentanediol: Process Synthesis and Analysis. *ACS Sustainable Chemistry and Engineering.* **2017**. *5* (6), 4699-4706.
- Huang et al. Improving Economics of Lignocellulosic Biofuels: An Integrated Strategy for Coproducing 1,5-Pentanediol and Ethanol. **2017**. *In revision.*

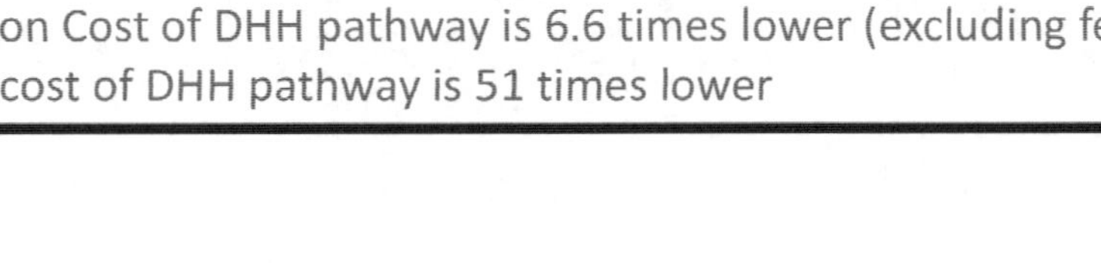

- Production Cost of DHH pathway is 6.6 times lower (excluding feedstock)
- Catalyst cost of DHH pathway is 51 times lower

4

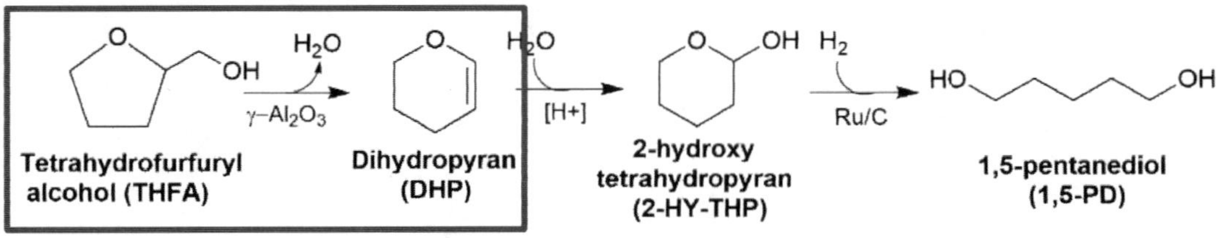

THFA DEHYDRATION

5

THFA Dehydration to Dihydropyran

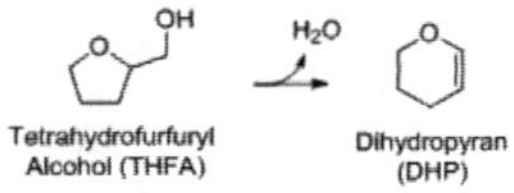

- THFA dehydrated over γ-Al_2O_3 in vapor phase
- Catalyst regenerated with 500˚C calcination step

Regeneration recovers surface area lost to coking

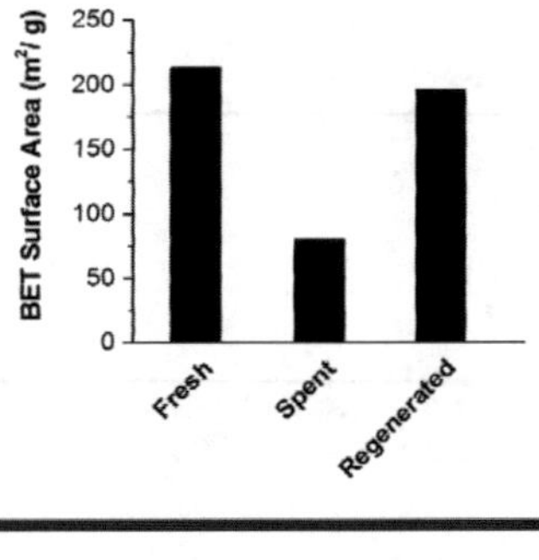

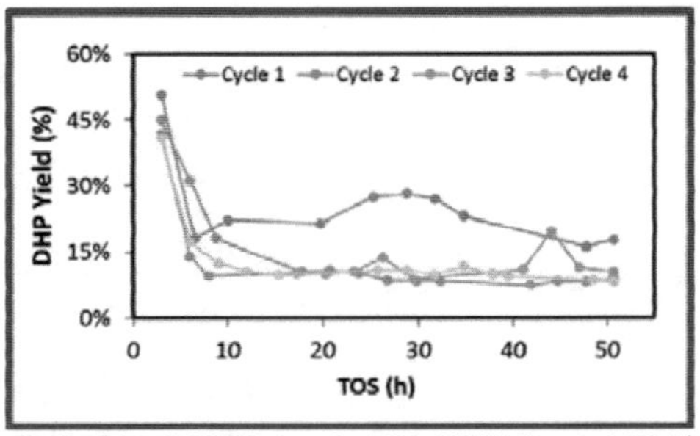

Reaction Conditions: T=375˚C, P=1atm, Feed=100wt% THFA, Flowrate=0.028mL/min

- ***Catalyst nearly completely regenerable***
  - Only 2.5% loss in activity from Cycle 3 to Cycle 4
- ***90% DHP Yields achieved at high conversion***

6

DHH Pathway Advantages

- **Step 1:** THFA Dehydration
  - Low-cost γ-Al_2O_3 catalyst
  - Catalyst regenerable

Reactant Concentration	Product Yield
100%	90%

7

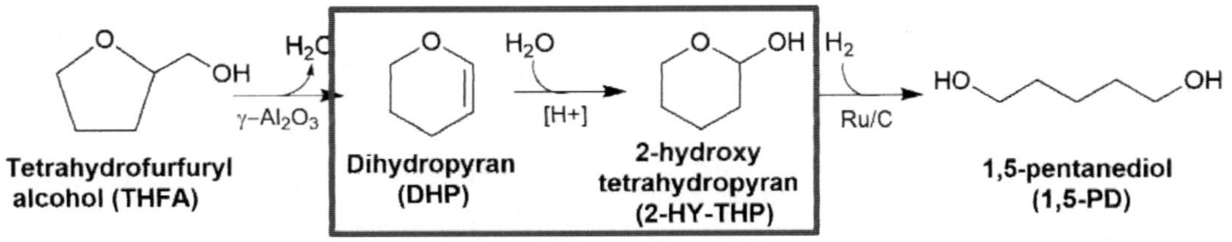

DIHYDROPYRAN HYDRATION

8

DHP Hydrates to 1,5-PD precursors in >99% Yields

- DHP hydrated in water with <u>no catalyst added</u>

Temperature (°C)	Reaction Time (h)	2-HY-THP Yield	5-tetrahydropyranyloxy-pentanal Yield	2-tetrahydropyranyl ether Yield	Overall Yield
70	4	84.5%	5.3%	1.9%	91.7%
100	4	92.0%	6.6%	1.2%	99.8%
130	4	89.5%	6.4%	1.0%	96.9%
160	4	65.4%	4.2%	1.3%	70.9%

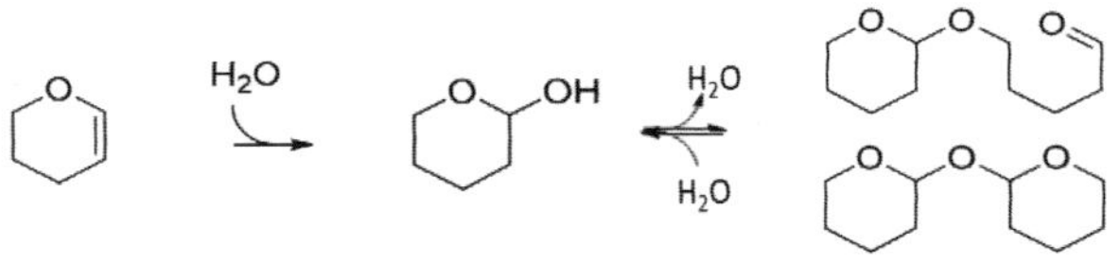

<u>2-phase Batch Reaction</u>

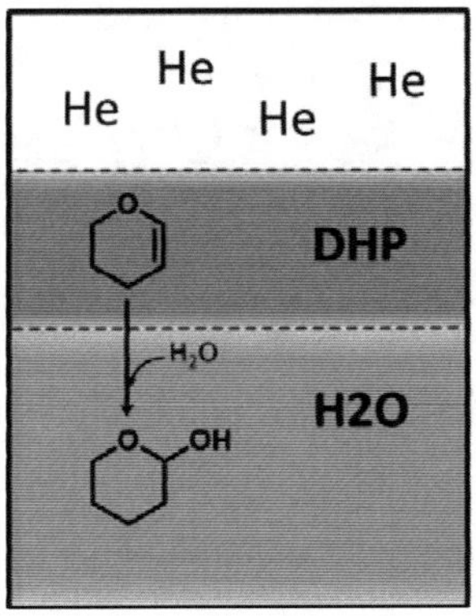

Reaction Conditions: P=250 psi He, Feed=20wt% DHP/H_2O

9

DHP Hydration is Autocatalytic

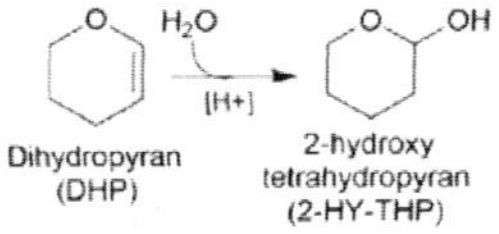

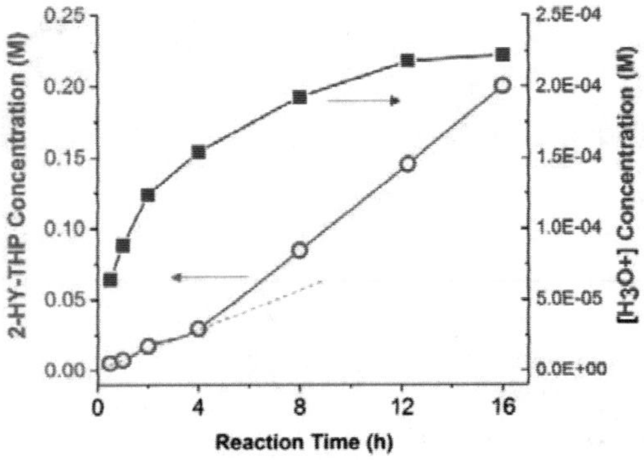

Reaction Conditions: T=25°C, P=500 psi Ar, Feed=20wt% DHP/H_2O

- Performed DHP hydration at 25°C to study initiation period of reaction
- Reaction rate doubles after 4h due to increase in solution acidity
- Small amount of acidic species (likely 5-hydroxyvaleric acid) form
- Initiation period very short! Occurs at <1% conversions

10

Acidic solid coke autocatalyzes DHP hydration in flow reactor

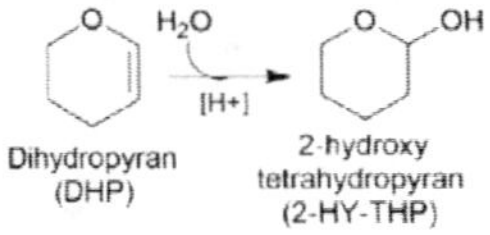

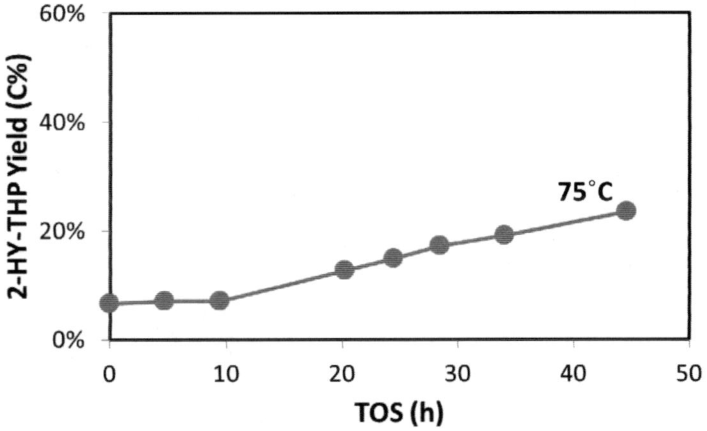

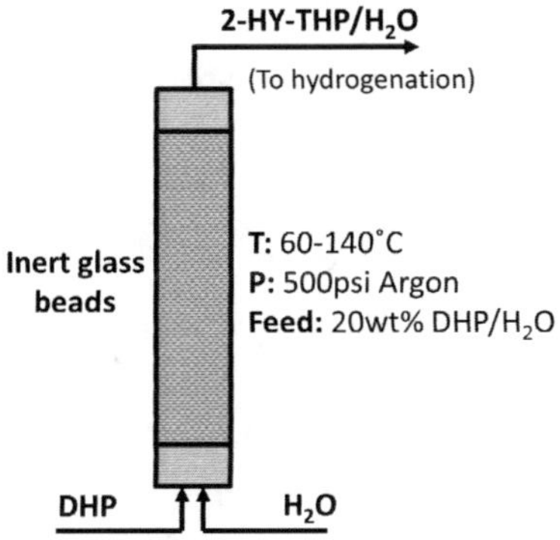

- DHP hydration over inert glass beads performed in continuous flow reactor

11

Acidic solid coke autocatalyzes DHP hydration in flow reactor

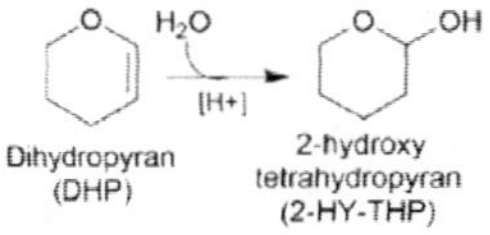

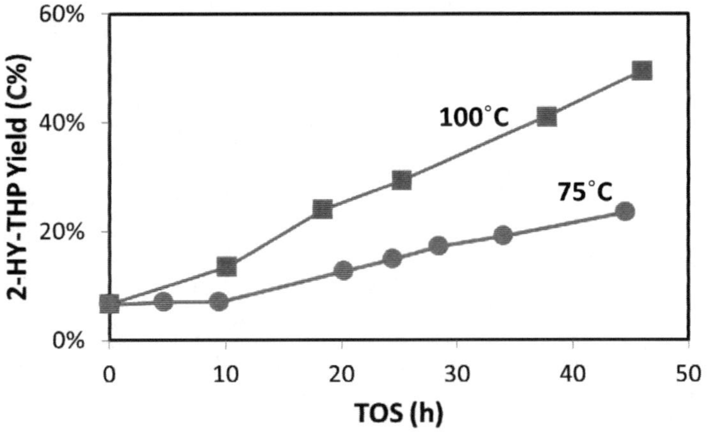

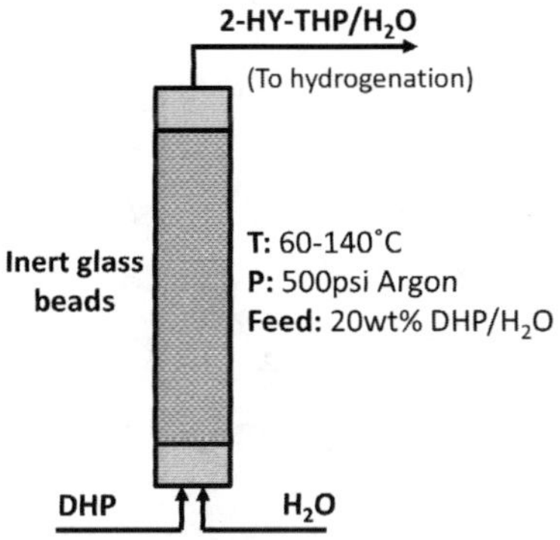

- Increased acidic coke formation at higher temperature increases hydration rate

12

Acidic solid coke autocatalyzes DHP hydration in flow reactor

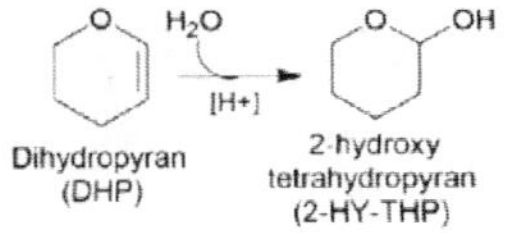

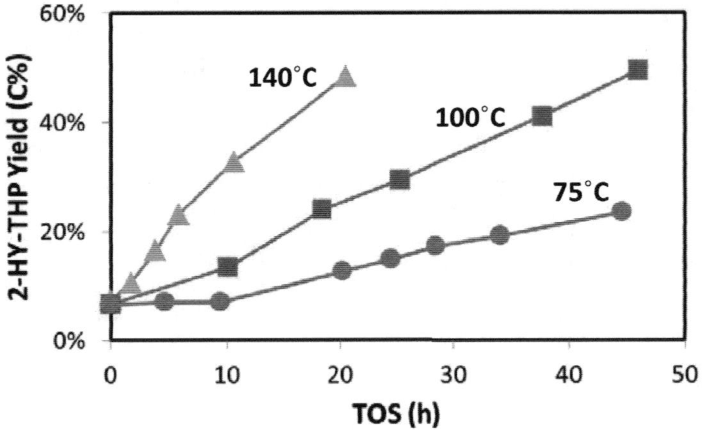

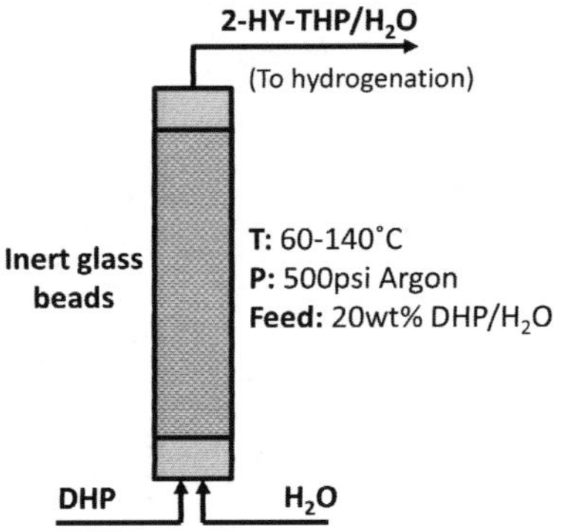

- Increased acidic coke formation at higher temperature increases hydration rate 13

Acidic solid coke autocatalyzes DHP hydration in flow reactor

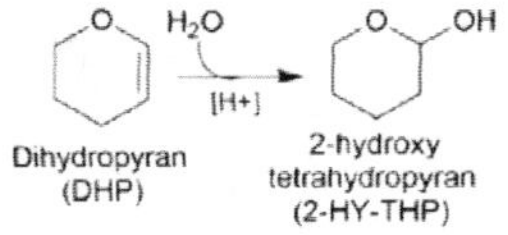

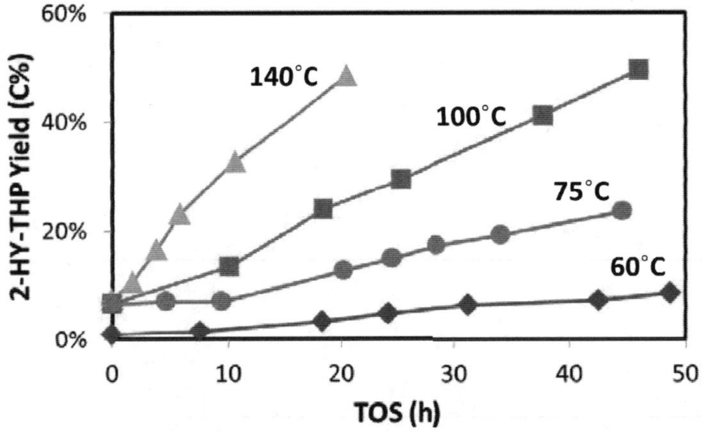

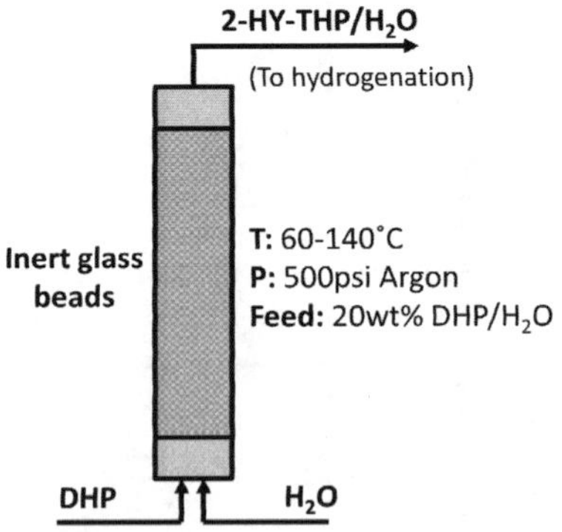

- Increased acidic coke formation at higher temperature increases hydration rate 14

Solid-acid Catalyzed DHP Hydration

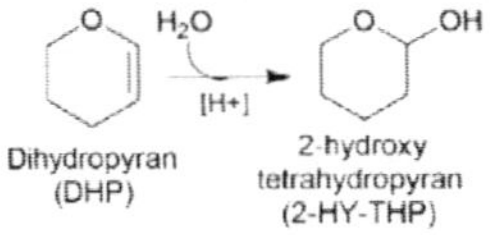

- **Goal:** Improve hydration rates by using solid acid catalysts

- Solid acid catalysts screened for DHP hydration in batch reactors:

Catalyst	Reaction Rate (μmol/gcat-min)	Bronsted site TOF (s-1)
No Catalyst	0	-
γ-Al2O3	4	0.00038
HZSM5	2822	0.115
Amberlyst-70	26017	0.152

Reaction Conditions: T=50°C, P=500psi Ar, Feed=0.5wt% DHP/H2O, Reaction Time=1h

← **No activity contribution from autocatalytic hydration**

- Solid acids show high activity at condition where autocatalytic hydration does not occur
- Hydration rates correlate with Brønsted site density

15

Why does DHP so readily hydrate?

- DHP hydration rates >> cyclohexene hydration rates
- **Hypothesis:** due to formation of stable oxocarbenium transition state

Cyclohexene Mechanism

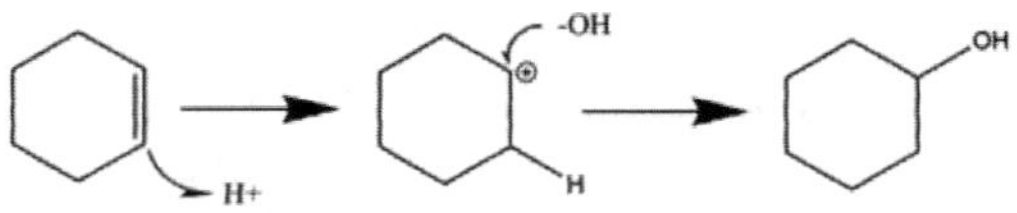

2° Carbocation

Dihydropyran Mechanism

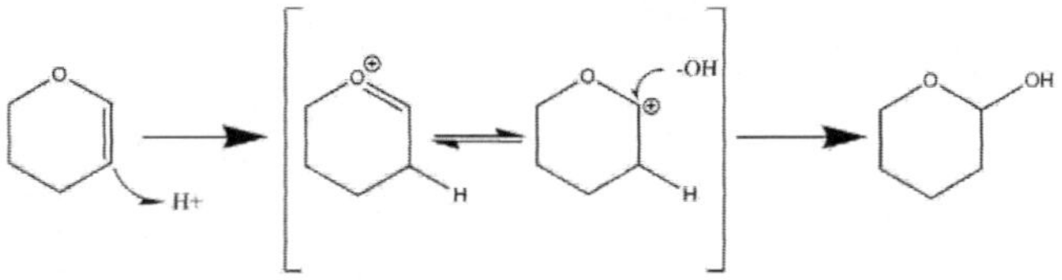

Oxocarbenium

16

Why does DHP so readily hydrate?

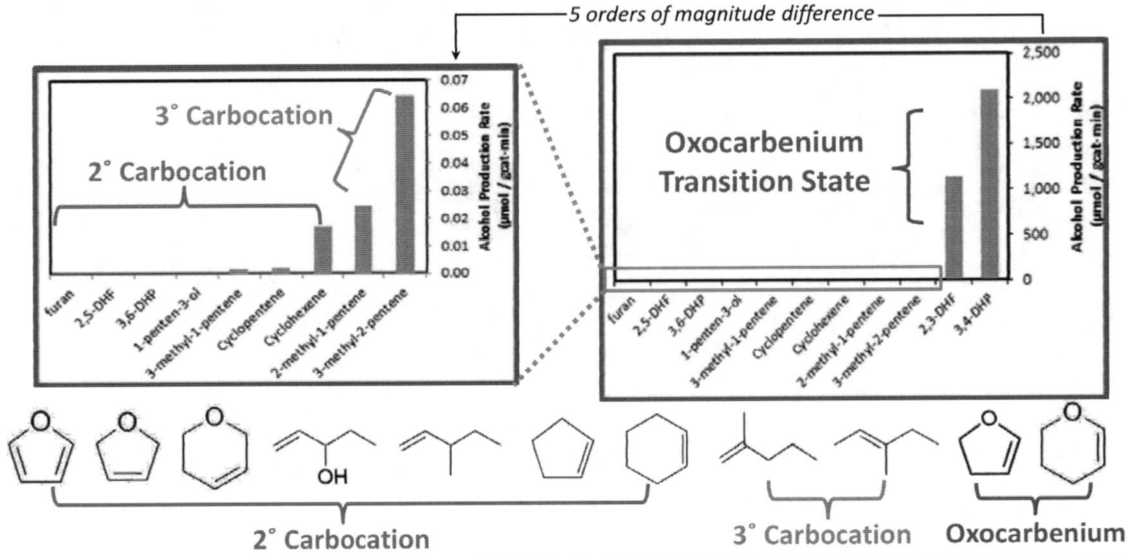

17

DHH Pathway Advantages

	Reactant Concentration	Product Yield
Step 1: THFA Dehydration	100%	90%
Step 2: DHP Hydration	50%	98%

- **Step 1:** THFA Dehydration
  - Low-cost γ-Al_2O_3 catalyst
  - Catalyst regenerable
- **Step 2:** DHP Hydration
  - DHP hydrates with no catalyst
  - Rate increased by *in situ* formation of acids
  - Low-cost solid acid catalysts
  - DHP readily hydrates due to stable transition state

18

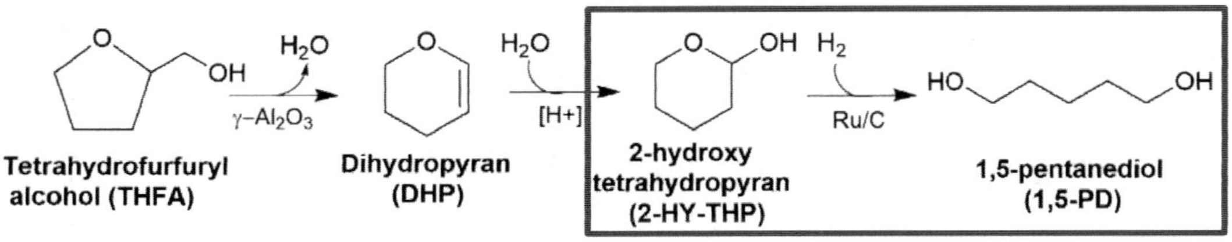

2-HY-THP HYDROGENATION

2-HY-THP and Dimers convert to 1,5-pentanediol at ~100% Yields

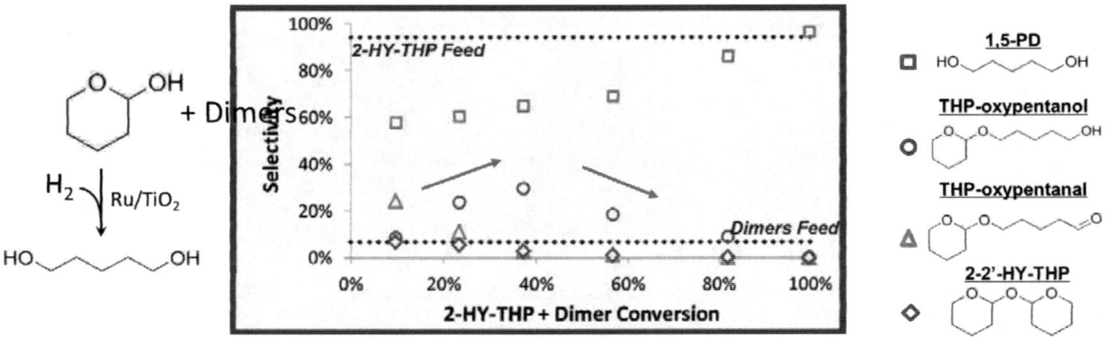

Reaction Conditions: Feed=20wt% 2-HY-THP/H_2O, Catalyst=1% Ru/TiO2, T=120˚C, P=650psi H_2

- Dimers increase at low conversion due to shift in equilibrium
- 97% 1,5-PD yields achieved at complete conversion

Ring-opening Tautomerization of 2-HY-THP into 5-hydroxyvaleraldehyde: NMR Studies

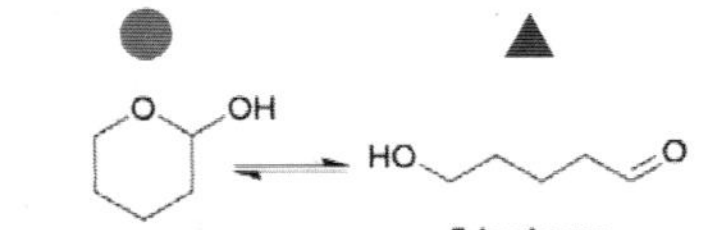

2-hydroxy tetrahydropyran (2-HY-THP)

5-hydroxy valeraldehyde (5HVal)

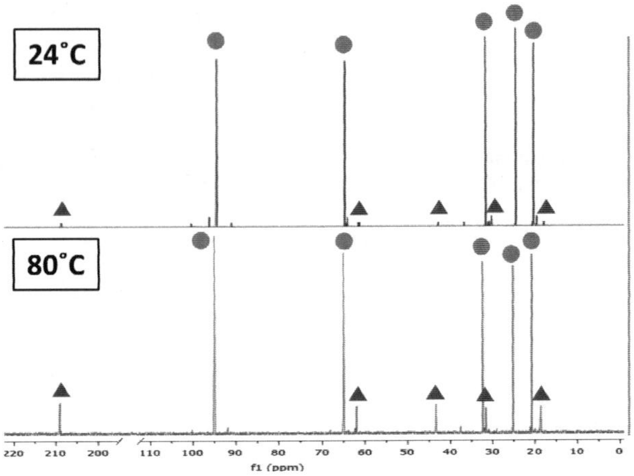

- Variable temperature quantitative ^{13}C NMR studies confirm the existence of 5-hydroxyvaleraldehyde (5HVal)
- Higher concentrations of 5HVal are observed at elevated temperatures
  - Tautomerization reaction is endothermic

21

Ring-opening Tautomerization of 2-HY-THP into 5-hydroxyvaleraldehyde: NMR Studies

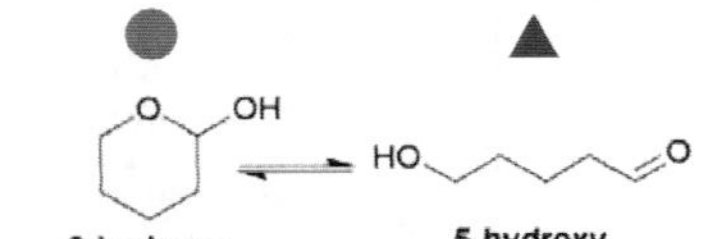

2-hydroxy tetrahydropyran (2-HY-THP)

5-hydroxy valeraldehyde (5HVal)

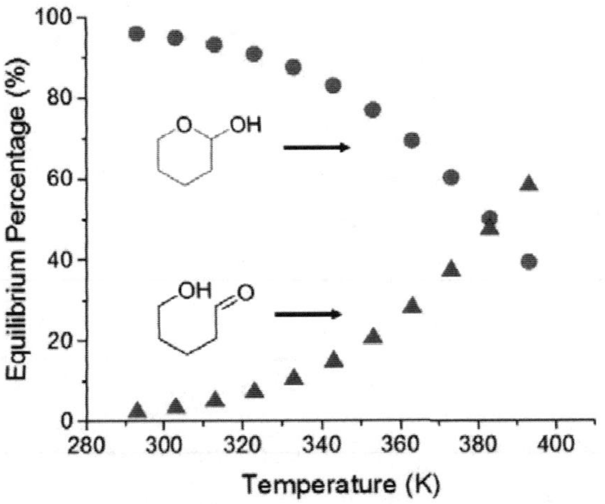

- Variable temperature quantitative ^{13}C NMR studies confirm the existence of 5-hydroxyvaleraldehyde (5HVal)
- Higher concentrations of 5HVal are observed at elevated temperatures
  - Tautomerization reaction is endothermic

22

Tautomerization vs. Hydrogenolysis

- <u>Direct Hydrogenolysis Approach</u>:
  Hydrogenolysis over bimetallic, noble-metal catalysts (i.e. RhRe)

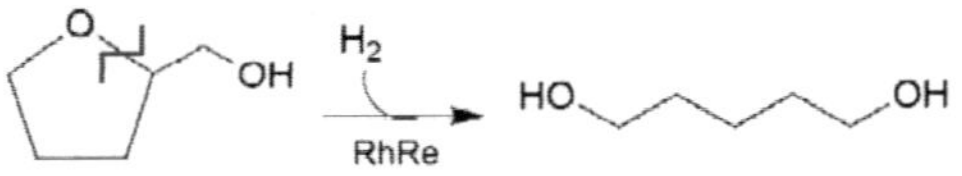

- <u>New Approach</u>: Ring-opening tautomerization and hydrogenation

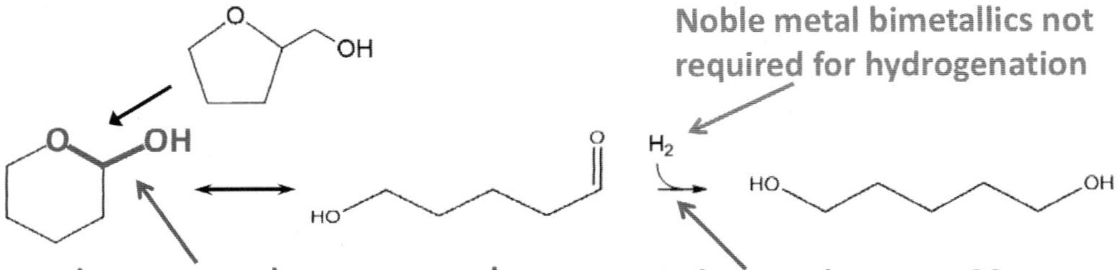

Hemiacetal groups can homogeneously tautomerize to their ring-opened form

Hydrogenation rate >80x faster than hydrogenolysis

23

DHH chemistry applicable to C4 and C6 streams

- Gamma-butyrolactone to 1,4-butanediol

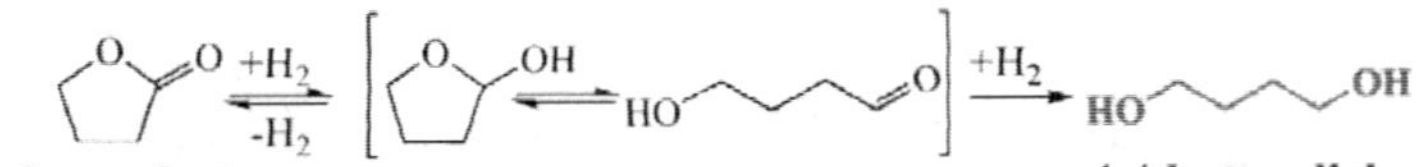

γ-butyrolactone (GBL) **2-HTHF** **4-HB** **1,4-butanediol (BDO)**

Z. Huang et al. <u>Hydrogenation of γ-butryolactone to 1,4-butanediol over bimetallic $CuCo/TiO_2$ catalysts</u>. *Submitted.*

- Tetrahydropyran-2-methanol to 1,6-hexanediol

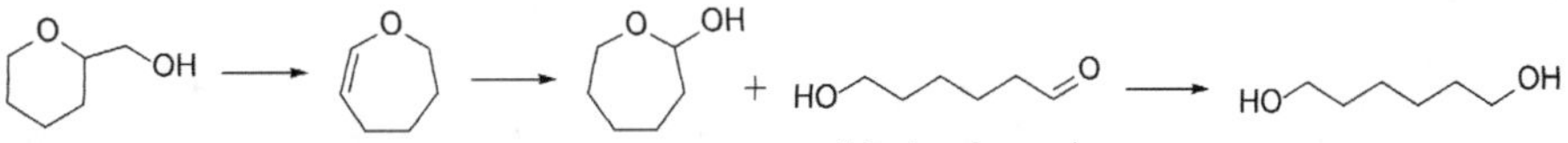

THP2M | 2,3,4,5-tetrahydrooxepine (THO) | 2-oxepanol (OXL) | 6-hydroxyhexanal (6HDHX) | 1,6-hexanediol (1,6-HDO)

S. P. Burt et al. <u>Production of 1,6-hexanediol from tetrahydropyran-2-methanol by dehydration-hydration and hydrogenation</u>. *Green Chemistry.* ***2017**, 19, 1390.*

24

DHH Pathway Advantages

- **Step 1:** THFA Dehydration
  - Low-cost γ-Al_2O_3 catalyst
  - Catalyst regenerable
- **Step 2:** DHP Hydration: Autocatalytic
  - DHP hydrates with no catalyst
  - Rate increased by *in situ* formation of acids
  - Low-cost solid acid catalysts
  - DHP readily hydrates due to stable transition state
- **Step 3:** 2-HY-THP Hydrogenation
  - Tautomerization of 2-HY-THP to aldehyde
  - Lower-cost, monometallic catalysts

	Reactant Concentration	Product Yield
Step 1	100%	90%
Step 2	50%	98%
Step 3	50%	99%

87% Overall Yield

25

Conclusions

- DHH pathway: 51x lower catalyst cost than direct hydrogenolysis
- High DHP hydration rates over solid acids or with no added catalyst
- Ring-opening tautomerization allows for high hydrogenation rates over lower-cost monometallic catalysts
- 1,5-PD minimum selling price <$2000/ton from furfural
  - Current 1,5-PD price = $6000
  - Historic 1,6-HD price = $3000-4700
- In process of commercializing 1,5-PD process

Pyran

26

Acknowledgements

Advisor: Professor George Huber

Collaborators: Zachary Brentzel, Kefeng Huang, Professor James Dumesic, and Professor Christos Maravelias

Huber Group Members

Funding: Department of Energy - Bioenergy Technologies Office (DOE BETO)

495407 Continuous Condensed Phase Conversion of Ethanol to Higher Alcohols over Bimetallic Catalysts

Abstract:

The production of bio-derived fuels and additives is one of the major avenues for reducing consumption of fossil fuels. Ethanol plays a central role in the bio-fuel arena as a gasoline substitute and oxygenate additive. It can also be used in the production of n-butanol, which has advantages over ethanol in terms of its applications in fuels. Such benefits include higher miscibility with gasoline, more resistance to water contamination, and chemical properties such as energy density, stoichiometric air-fuel ratio, heat of vaporization, research octane number, and motor octane number that are closer to gasoline [1][2]. The general pathway for the production of bio-butanol from bio-ethanol is known as the Guerbet reaction. There have been several studies in recent years for conducting Guerbet reaction using numerous catalysts and reaction conditions [3]. Most of these experiments were done in batch systems or gas phase continuous packed bed reactors. In this study, the previous results obtained from our research group in a batch reactor [4][5] have been extended to a condensed-phase continuous reactor using similar catalysts (8% Ni/γ-Al_2O_3 modified with La_2O_3) and varying reaction parameters such as temperature, WHSV, and feed composition. Also, the catalyst has been modified using bi-metallic (Ni-Cu, Ni-Pt, Ni-Pd, and Ni-Co), and tri-metallic catalysts. Results show that adding Cu to Ni catalyst improves reaction selectivity to higher alcohols because of reduced byproduct gas formation. i. However, a decreases in catalyst activity is observed because Cu has a lower activity for alcohol dehydrogenation that is the initial step in the Guerbet reaction. Ethanol conversion of 40% and higher alcohol selectivity of 80% were observed at optimal condition for the Ni-Cu bimetallic catalysts.

References

[1] R. L. Wingad, P. J. Gates, S. T. G. Street, and D. F. Wass, "Catalytic Conversion of Ethanol to n-Butanol Using Ruthenium P-N Ligand Complexes," *ACS Catal.*, vol. 5, no. 10, pp. 5822–5826, 2015.

[2] H. S. Ghaziaskar and C. (Charles) Xu, "One-step continuous process for the production of 1-butanol and 1-hexanol by catalytic conversion of bio-ethanol at its sub-/supercritical state," *RSC Adv.*, vol. 3, no. 13, pp. 4271–4280, 2013.

[3] D. Gabriëls, W. Y. Hernández, B. Sels, P. Van Der Voort, and A. Verberckmoes, "Review of catalytic systems and thermodynamics for the Guerbet condensation reaction and challenges for biomass valorization," *Catal. Sci. Technol.*, vol. 5, pp. 3876–3902, 2015.

[4] T. L. Jordison, L. Peereboom, and D. J. Miller, "Impact of Water on Condensed Phase Ethanol Guerbet Reactions," *Ind. Eng. Chem. Res.*, vol. 55, no. 23, pp. 6579–6585, 2016.

[5] T. L. Jordison, C. T. Lira, and D. J. Miller, "Condensed-Phase Ethanol Conversion to Higher Alcohols," *Ind. Eng. Chem. Res.*, vol. 54, no. 44, pp. 10991–11000, 2015.

501011 The Effects of Metal-Acid Site Proximity on Bifunctional Isomerization of Alkanes and Deoxygenation of Alcohols

Abstract:

Hydroisomerization of alkanes occurs through kinetic cascades mediated by alkene intermediates that require a (de)hydrogenation metal function within diffusion distances of the kinetically-relevant acid function. Hydrodeoxygenation of alkanols similarly requires a hydrogenating metal function proximal to kinetically relevant acid sites to scavenge dehydrated product alkenes and prevent catalyst deactivation with no influence on rates. In practice, the location and distribution of metal clusters with respect to acid sites—and thus their proximity—is seldom well-characterized because metal precursors are deposited onto bound solid acids that also contain inert metal-oxide binders during synthesis protocols. The accepted intimacy criterion for bifunctional isomerization [1] requires that the metal and acid functions reside at distances (~100 um for mesoporous solid acids) that prevent concentration gradients of alkenes—derived from reactant alkanes in isomerization and formed from oxygenates via dehydration for hydrodeoxygenation—within acid domains. Indeed, for zeolite crystals with an extracrystalline metal function present in the form of Pt/SiO_2, distances of ~150 nm are sufficient to maintain constant alkene reactant concentration within acid domains, as shown by isomerization rates that are unaffected by the size or intracrystalline site density of zeolite crystals. We find, however, that as metal-acid distances approach the nanometer scale via introduction of the metal function as intracrystalline clusters using exchange methods, isomerization rates are markedly higher. These remarkable yet previously unexplained effects of site proximity and their mechanistic underpinnings are addressed in this study. Turnover rates (per H^+) (i) for n-heptane isomerization on Pt-exchanged MFI and on physical mixtures of Pt/SiO_2 with MFI (of varying site density ($[H^+]$) and crystal size) and (ii) for 4-methylcyclohexanol, 2-butanol, and 2-propanol hydrodeoxygenation on physical mixtures of Pt/SiO_2 with FAU and BEA and on physical mixtures of Pt/SiO_2 with Pt-exchanged FAU and Pt-exchanged BEA. We also report turnover rates for probe reactions selective to each of the two functions to assess their monofunctional reactivities as the distances separating them decrease. Acid sites are characterized by turnover rates for methanol dehydration [2], a reaction that sensitively probes acid reactivity and any consequences of changes in acid strength or confining environment as a result of proximity; the metal function is assessed using CO oxidation rates [3], which are able to report on changes in electronic properties of Pt clusters as a result of their proximity to zeolitic protons.

n-Heptane isomerization turnover rates (548 K, 60-101 kPa H_2, 5 kPa n-heptane: ~70 molec. $(H^+ ks)^{-1}$) on MFI (of different crystallite size and site density; mixed with Pt/SiO_2) are invariant with the Thiele modulus, the dimensionless quantity that rigorously describes reactant concentration gradients, until large values of the Thiele parameter (corresponding to metal-acid separation of ~250 nm) at which point rates decrease by about twofold. Thus, smaller metal-acid distances lead to negligible intracrystalline heptene concentrations, the required intimacy criterion [1] has been satisfied for these bifunctional catalysts with metal function present only as extracrystalline Pt/SiO_2. Alkanol dehydration turnover rates (393 K, 2 MPa H_2, 2.5 kPa 4-methylcyclohexanol: ~25 molec. $(H^+ ks)^{-1}$) were measured on BEA and FAU (0.60 and 0.74 nm apertures, respectively, compared to 0.47 nm in MFI [4]). Dehydration turnover rates are much lower than those for n-heptane isomerization, which, taken together with the larger molecular diffusivities afforded by large pore frameworks, precludes the presence of concentration gradients of methylcyclohexene products. Concentrations of products measured are much lower than those predicated by equilibrated reactant dehydration, indicating that further decreases in metal-acid site proximity should be inconsequential. In contrast to these expectations, we measure unexpectedly significant increases (by factors of 2-9) in reaction rate constants (Figure 1a for 2-butanol and 4-methylcyclohexanol on BEA, BEA-Pt/SiO_2, and PtBEA-Pt/SiO_2; Figure 1b for n-heptane isomerization over MFI-Pt/SiO_2 and PtMFI-Pt/SiO_2 for MFI of different site density) for both isomerization and hydrodeoxygenation when Pt clusters and H^+ reside within the intracrystalline framework at distances of ~5-20 nm—as estimated assuming uniform distribution of each of the two functions during synthesis procedures.

Figure 1. **(a)** Measured dehydration turnover rate constants (per H^+) for 2-propanol, 2-butanol, and 4-methylcyclohexanol on BEA, BEA-Pt/SiO_2, and PtBEA-Pt/SiO_2 (393 K; 0.5-2.5 MPa H_2; 0.1-10 kPa alkanol). Inset equations are rate expressions. *2-Propanol dehydration is independent of P(H_2) for PtBEA-Pt/SiO_2. **(b)** Measured n-heptane isomerization turnover rate constants (per H+) on MFI-Pt/SiO_2 and PtMFI-Pt/SiO_2 for MFI of varying proton density (548 K; 60-101 kPa H_2; 0.1-25 kPa n-heptane).

The nanoscale proximity between metal and acid function could potentially result in cooperative behavior changing the inherent monofunctional reactivity. Alkene skeletal isomerization and alkanol dehydration steps limit the respective rates of alkane isomerization and alkanol hydrodeoxygenation. Any potential changes in acid reactivity, caused by modification of acid strength and local environment for acid sites as a result of proximate Pt clusters, were assessed using methanol dehydration turnover rates (per H^+; 433 K) were measured on both MFI and on PtMFI (Figure 2a). Turnover rates and rate constants (inset in Fig. 2a) are the same, consistent with unperturbed reactivity and properties of the acid sites. CO oxidation turnover rates and rate constants (per Pt_s; 443 K) on Pt/SiO_2 and Pt-MFI are also the same (Figure 2b and inset), indicating that the reactivity and properties of Pt clusters are unaffected by their confinement or by their proximity to acid sites within intracrystalline regions. Thus, we conclude that the monofunctional reactivity of the individual metal and acid functions are unaffected by their close proximity.

Figure 2. **(a)** Measured methanol dehydration turnover rate (per H^+) on MFI and PtMFI (433 K). Dashed curve represents average regressed fit to the inset rate expression; regressed values of first- and zero-order rate constants inset. **(b)** Parity plot for CO oxidation turnover rates (per Pt_s) on Pt/SiO_2 and PtMFI (443 K). Rate expression and values of effective rate constants inset. Dashed line represents parity.

These significant rate enhancements for isomerization and dehydration reactions must occur via highly reactive intermediate species. We surmise that these short-lived intermediates are present only at low concentrations and thus require metal clusters within nanometer distance from acid sites. Rate enhancements for 2-butanol and 4-methylcyclohexanol but not 2-propanol dehydration (Figure 1a) shed light on the identity of such reactive intermediates. Additionally, alkanol dehydration turnover rates are independent of $P(H_2)$ on BEA and on BEA-Pt/SiO_2; however, turnover rates for 2-butanol and 4-methylcyclohexanol become inversely dependent on $P(H_2)$ when Pt is present in nanometer proximity from acid sites (Fig. 1a inset). These kinetic data, together with insights from density functional theory, implicate unsaturated alkenone/alkenol analogs of reactants as plausible intermediates for dehydration reactions. Such a pathway is unavailable for 2-propanol because the analogous alkenol cannot undergo dehydration and consequently does not impact its measured rates. The identity of highly reactive intermediates mediating isomerization remains under investigation. Unlike for alkanol dehydration, isomerization rates show no change in $P(H_2)$ dependence with encapsulation of Pt clusters, thus excluding diene-derived intermediates. We continue to apply density functional theory treatments to assess the plausibility of species as alkenederived intermediates for isomerization.

We have addressed previously uninterpreted enhancements in reaction rates as they arise from nanoscale metal-acid site proximity. Such insights can be applied to optimize industrial catalysts for these and other analogous reactions.

References

[1] P.B. Weisz, Adv. Catal. 13 (1962) 137–190. doi:10.1016/S0360-0564(08)60287-4.

[2] A.J. Jones, R.T. Carr, S.I. Zones, E. Iglesia, J. Catal. 312 (2014) 58–68. doi:10.1016/j.jcat.2014.01.007.

[3] A.D. Allian, et al., J. Am. Chem. Soc. 133 (2011) 4498–4517. doi:10.1021/ja110073u.

[4] C. Baerlocher, L. McCusker, Database of Zeolite Structures. http://www.iza-structure.org/databases/.

The authors thank Zhichen (Jane) Shi for measurement of diffusion timescales of zeolite samples. Additionally, the authors acknowledge financial support from Chevron Energy Technology Company and computational resources from the National Science Foundation's Extreme Science and Engineering Discovery Environment (ACI-1053575).

501210 The Influence of Solvent on Acid Catalyzed Dehydration of Model Polyols

Abstract:

The catalytic conversion of biomass and biomass-derived intermediates to energy-dense fuels and chemicals is often carried out in solvents or complex liquid mixtures. The reaction media can play an important role in influencing the catalytic molecular transformations. Herein we examine the catalytic dehydration of different model alcohols in solvent mixtures comprised of water and a polar organic solvent such as DMSO and GVL. Experiments along with theoretical simulations show that the acid-catalyzed dehydration reactions can be significantly promoted without product degradation by the appropriate choice of the solvent system. The net effect yields a significant improvement in both selectivity and reactivity over the reactions in these specific aqueous based systems. Molecular dynamics simulations were carried out to establish the faithful models of the co-solvent systems. These resulting structures provided input for more rigorous ab initio molecular dynamics simulations to follow the free energies changes for the detailed reaction paths involving proton addition, water elimination and proton abstraction to form the corresponding olefin products. The results show that the extent of solvation for different reactant alcohols can be rather different. There are non-uniform changes that occur with increasing the composition of the co-solvent or changing the alcohol reactant. In some cases there is a significantly lowering of activation barriers for the different alcohols while in others the barrier increases. These solvent effects were subsequently utilized to help promote the production of hydroxymethylfurfural (HMF) from fructose, while limiting the subsequent degradation of HMF to levulinic acid. The results provide important insights that may help to provide guidance into solvent selection for the selective production of targeted chemicals, along with the development of new acidic catalytic systems.

496817 Generalizing the Design Principles of Pt-Based Alloy Catalysts with Improved ORR Performance and Durability

Abstract:

Since the discovery of Pt skin type Pt_3Ni catalysts with superior performance towards oxygen reduction reaction,[1] there has been a blossoming of Pt-based alloy catalysts for ORR and various other applications.[2,3] In last couple years, the focus has slowly shifted from catalysts with monolayer skins to those with thicker Pt overlayers for the sake of improving the stability.[4-6] But the decrease in the activity has been observed with increasing overlayer thickness. Thus, how to increase the stability of catalysts, with no harm on the activity, or even simultaneously increase activity, have been the focus of the current catalyst development, but few systematic design principles have been established to guide these efforts. It is worsened by a lacking of atomic-scale understanding of water/electrode interface and consequently the bonding nature of ORR intermediates. For example, it is still in the debate regarding whether hydrated OH or non-hydrated OH is the plausible ORR intermediate.

In the present talk, we will start by determining the atomic-scale structure of oxygenated species at water/electrode interfaces under electrochemical ORR conditions, by combining a careful calibration of Density Functional Theory (DFT)-determined energetics, the liquid water/Pt interfaces generated from *ab-initio* molecular dynamics (AIMD) simulations and a detailed simulation of X-ray Photoelectron Spectroscopy (XPS) signatures.[7] In the second part, we will investigate the structure and the stability of Pt overlayers with various thickness. We will further analyze their activity based on a scheme established through above calibration. Then, general design principles will be proposed for searching for ORR catalysts with improved stability and activity. We will close by comparing the predicted stability and activity of various catalysts with those available in the literature and from our collaborators.

[1] V.R. Stamenkovic, B. Fowler, B.S. Mun, G.F. Wang, P.N. Ross, C.A. Lucas, N.M. Markovic, Science 315 (2007) 493.

[2] J. Greeley, I.E.L. Stephens, A.S. Bondarenko, T.P. Johansson, H.A. Hansen, T.F. Jaramillo, J. Rossmeisl, I. Chorkendorff, J.K. Nørskov, Nat. Chem. 1 (2009) 552.

[3] P. Strasser, S. Koh, T. Anniyev, J. Greeley, K. More, C. Yu, Z. Liu, S. Kaya, D. Nordlund, H. Ogasawara, M.F. Toney, A. Nilsson, Nat Chem 2 (2010) 454.

[4] M. Escudero-Escribano, P. Malacrida, M.H. Hansen, U.G. Vej-Hansen, A. Velázquez-Palenzuela, V. Tripkovic, J. Schiøtz, J. Rossmeisl, I.E.L. Stephens, I. Chorkendorff, Science 352 (2016) 73.

[5] N. Todoroki, H. Watanabe, T. Kondo, S. Kaneko, T. Wadayama, Electrochim. Acta 222 (2016) 1616.

[6] M. Asano, R. Kawamura, R. Sasakawa, N. Todoroki, T. Wadayama, ACS Catal. 6 (2016) 5285.

[7] Z. Zeng, J. Greeley, Nano Energy 29 (2016) 369.

490188 Computational Insights into Thermal and Photoresponse of Pd/C and Pd/N Co-Doped TiO_2Â using DFT

Abstract:

TiO_2 has attracted increasing attention during the past few decades as a promising catalytic and photocatalytic material due to its unique physicochemical properties. Specifically, in heterogeneous catalysis, it has been used as a catalytic material for treatment of environmental pollutants (greenhouse gases), photocatalytic reduction of CO_2 with water vapour and photodegradation of dyes. In recent years, doping of transition metal and nonmetal ions in TiO_2 has been reported to enhance the catalytic and photocatalytic performance of TiO_2. The main objective of this work is to study the effect of co-doping on the structure-property relationships of anatase TiO_2 using density functional theory (DFT+U) approach. Palladium ion and carbon (C)/Nitrogen (N) were co-doped in TiO_2 bulk and surface (001) and (100) planes in order to investigate the thermal and photoresponse of TiO_2. Differential bond length formations and associated net oxygen activations (NOA) were determined for the bulk and surface exposed Pd/C and Pd/N doped TiO_2. Effects of anion vacancies on the bond length distributions and net oxygen activations were found for the co-doped structures. NOA was found to be higher for Pd/N co-doped catalysts with 18.8 % and 28% in case of bulk and surface exposed codoped TiO_2 when compared to 17 % and 25 % in case of bulk and surface exposed Pd/C co-doped TiO_2. Band structure analyses were performed in order to examine the photoresponse of co-doped TiO_2. Band gaps obtained from the band structure analyses were in a range of 1.9-3 eV for Pd/C-doped TiO_2 whereas 2.7-3.1 eV for Pd/N-doped TiO_2, respectively, thereby establishing the importance of codoping for thermal and photocatalytic behaviour of TiO_2-based materials.

491746 Transition Metal-Doped C_2N as Active Catalysts for the Oxygen Reduction Reaction

Abstract:

Meeting ever-increasing worldwide energy demands in a sustainable way is one of the most pressing challenges of our times. The oxygen reduction reaction (ORR) plays a critical role in pollution-free energy conversion technologies, such as hydrogen fuel cells and metal-air batteries. Platinum (Pt) is one of the best known catalysts for ORR, but the high cost of this precious metal and the sluggish ORR kinetics inhibit the widespread growth and implementation of fuel cell technologies. [1,2]

Two-dimensional (2D) transition metal-doped carbon nitride catalysts have emerged as abundant and low-cost alternatives to Pt for ORR. [3] A more recent member of this family, C_2N, [4] exhibits several promising features for ORR, including a conductive framework and the potential to create and tune active sites through controlled transition metal doping. [5]

In this work, we use density functional theory (DFT) to investigate the ORR mechanism for C_2N catalysts doped with various transition metals. We find that the choice of metal dopant significantly affects the coverages of the ORR intermediates (*OOH, *O, *OH) on this material. These coverages directly influence the predicted ORR mechanism and overpotentials, which determine catalytic activity. By including explicit solvent layers in our predictions, we also investigate solvation effects on the binding energies. By studying the ORR mechanism and activity of various metal-doped C_2N catalysts, we identify critical material properties that influence the performance of single site 2-D catalysts for electrochemical reactions.

References:

[1] Xia, W., Mahmood, A., Liang, Z., Zou, R., & Guo, S. (2016). Earthâ€Abundant Nanomaterials for Oxygen Reduction.Â *Angewandte Chemie International Edition*,Â *55*(8), 2650-2676.

[2] United States Department of Energy. (2016). Comparison of Fuel Cell Technologies. https://energy.gov/eere/fuelcells/comparison-fuel-cell-technologies.

[3] Zheng, Y. et al. (2017). Molecule-Level g-C_3N_4 Coordinated Transition Metals as a New Class of Electrocatalysts for Oxygen Electrode Reactions. *Journal of the American Chemical Society*, 139(9), 3336-3339.

[4] Mahmood, J., et. al. (2015). Nitrogenated holey two-dimensional structures. *Nature communications*, *6*, 6486.

[5] Li, X., Zhong, W., Cui, P., Li, J., & Jiang, J. (2016). Design of Efficient Catalysts with Double Transition Metal Atoms on C_2N Layer. *The Journal of Physical Chemistry Letters*, *7*(9), 1750-1755.

496687 Electrochemical Synthesis of Hydrogen Peroxide Via Water Electrolysis: Challenges and Opportunities

Abstract:

Electrochemical production of hydrogen peroxide (H_2O_2) from water oxidation could provide a way to produce a chemically valuable product from an earth abundant material. Today there are only a few experimental studies that produce measurable amount of hydrogen peroxide through electrochemical oxidation of water.[1,2] A major challenge is that the electrocatalyst must suppress the thermodynamically favored four-electron pathway (O_2 evolution) and direct selectivity towards two-electron path (H_2O_2 evolution).[3] We have recently developed a complete understanding of water oxidation incorporating the formation of hydroxyl radical, a one-electron water oxidation reaction.[4] Based on thermodynamic analysis, we demonstrated leads to OH radical, H_2O_2 and O_2 evolution and rationalized product selectivity in several different experimental reports.

In this work, by combining the above knowledge and density functional theory (DFT) calculations we search for other materials to identify suitable electrocatalyst with high selectivity towards two-electron path (H_2O_2 evolution).

References

(1) Fuku, K.; Sayama, K.; Fujishima, A.; Honda, K.; Maeda, K. *Chem. Commun.* **2016**, *52* (31), 5406–5409.

(2) Izgorodin, A.; Izgorodina, E.; MacFarlane, D. R. *Energy Environ. Sci.* **2012**, *5* (11), 9496.

(3) Viswanathan, V.; Hansen, H. A.; Nørskov, J. K. *J. Phys. Chem. Lett.* **2015**, *6* (21), 4224–4228.

(4) Siahrostami, S.; Li, G.-L.; Viswanathan, V.; Nørskov, J. K. *J. Phys. Chem. Lett.* **2017**, 1157–1160.

497976 Modeling Electrochemical Reactions:Â DFT-Based Models Including Explicit Solvation, Electrolyte, and Electrochemical Potential

Abstract:

Modeling Electrochemical Reactions: DFT-Based Models Including

Explicit Solvation, Electrolyte, and Electrochemical Potential

Jason D. Goodpaster*

Department of Chemistry, University of Minnesota Twin Cities, Minneapolis, Minnesota 55455, USA

**Corresponding author: jgoodpas@umn.edu*

I report an electrochemical model based on periodic Kohn-Sham Density Functional Theory (DFT) that includes the effects of the electrochemical potential, solvent, and electrolyte. In this model, the electrode surface and all adsorbates are treated using DFT. In the region close to the surface, the solvent is treated explicitly, whereas in the region far from the surface the solvent is treated using a continuum dielectric. The electrolyte ions are implicitly represented by a linearized Poisson-Boltzmann model.[1,2] The electrode potential was calculated from the Fermi energy. The Fermi energy can be varied by changing the number of electrons in the simulation cell; therefore, the electrochemical potential was set by calculating the number of electrons the simulation cell required for a specific Fermi energy.[3] Using this model, free energy profiles for a reaction can be calculated as a function of the applied potential. The free energy of activation for each elementary step can then used to determine the rate coefficient for that step and combined with microkinetic models to determine the rate of product formation.

I applied this model to investigate the mechanistic pathways by which CO_2 on copper and silver surfaces are electrochemically reduced.[3,4] I compare this model to vacuum DFT, implicit solvation DFT, explicit solvation DFT, and electrochemical potential DFT models. Our results demonstrated that including electrochemical potential is critical to accurate calculations of free energy barriers and kinetics for electrochemical reactions. Additionally, we show that for some reactions explicit solvation plays an important role in solvating adsorbed intermediates. Finally, I discuss progress in using these free energies for microkinetic modeling.

References:

1. K. Letchworth-Weaver and T. A. Arias Phys. Rev. B 86, 075140 (2012).
2. K. Mathew, R. Sundararaman, K. Letchworth-Weaver, T. A. Arias, and R. G. Hennig. J. Chem. Phys. 140, 084106 (2014).
3. J. D. Goodpaster, Alexis T. Bell, and M. Head-Gordon J. Phys. Chem. Lett. 7, 1471 (2016).

4. J. D. Goodpaster, Alexis T. Bell, and M. Head-Gordon *in prep.*

491234 Efficiently Determining Solvent Effects at Metal-Water Interfaces

Abstract:

Accurately modeling interfaces between liquid and solid phase is one important frontier in computational catalysis. Explicitly modeling these interfaces can be challenging due to the large number of molecules needed, as well as the large number of possible configurations. Implicit or continuum solvent models can be much quicker, but have only recently been included in plane wave density functional theory (DFT) codes. In contrast, implicit solvation methods are much more mature in molecular DFT codes. How implicit solvation methods perform in modeling periodic catalytic systems is still unclear. In this work we systematically evaluated how interfaces between water can metals can be treated using implicit solvation methods.

In total we considered 41 different adsorbates and used three different solvation methods implemented in the NWChem, VASP, and JDFTx codes. Agreement between the three programs was strong. We classified the adsorbates into five different groups, and found that trends within each group were consistent. For instance, adsorbed atomic or weakly polar species showed almost negligible solvation effects, while adsorbed aromatic compounds were strongly stabilized by the presence of water solvent (up to 0.44 eV). We also modeled four prototypical reactions in vacuum and water: oxygen reduction reaction, formic acid oxidation, C-C cleavage, and water-gas shift. Reaction energies changed up to 0.23 eV when solvent was present. We analyzed the contributing energy terms to the solvation energy of adsorbed species, and found that relevant desciptors (e.g. Bader charge of adsorbates, dipole moments) are correlated to these solvation energy terms. We furthermore created several artificial neural networks, where we evaluated potential descriptors for predicting solvation energies of adsorbates. Our best neural network indicated that Bader charge of adsorbed species, solvation energies of gas species, gas dipole moments, and molecular surface area of adsorbates are all simple descriptors that can be used to predict solvation energies of adsorbates. Our work provides guidelines on when solvation effects may be important for surface chemistry, and also provides valuable insights on modeling such effects in an efficient manner using implicit solvation.

498550 Machine-Learning Model Development for Electrocatalyst Discovery

Abstract:

Machine-Learning Model Development for Electrocatalyst Discovery

Zheng Li, and Hongliang Xin*

*hxin@vt.edu

Session: Catalysis and Reaction Engineering Division- Electrocatalysis and Photoelectrocatalysis

Transition metal alloys, metal oxides and molecular complexes have shown great promise for catalyzing many chemical reactions. For a given type of catalyst, its surface reactivity can be tailored by varying geometric characteristics, metal ligands, organic functional groups and extrinsic factors (e.g., solvation, support) in the vicinity of active sites [1]. With hierarchical complexities in catalyst design, a priori estimation of chemical reactivity of surface atoms is attractive. However, it is very time-consuming and costly, if at all possible, to search for highly optimized catalyst structures by the empirical testing and/or quantum-chemical calculations. Therefore, we tackle this problem by developing the predictive, data-driven machine-learning-augmented models (e.g., artificial neural network, kernel ridge regression) that can unravel the complex, non-linear interactions between the electronic structure properties with reactivity descriptors, thus enabling fast and accurate predictions of surface reactivity of a broad range of material space without resorting to the expensive quantum-chemical calculations.

Since the efficiency and accuracy of machine-learning models are strongly dependent on the input features, we develop various types of data representation of catalyst surfaces. For metal alloys and metal oxides, we engineer numerical representation of surface metal atoms by using geometry and electronic structure based primary features [2-4] including the local electronegativity and the d-band features that are dependent on the surroundings of an adsorption site, together with the intrinsic properties of active metal atoms including the electronegativity, ionic potential, electron affinity, etc. For the description of molecular complexes, we apply the Coulomb matrix [5], which is a simple matrix form crystal structure representation based on the molecular complex structure coordination and atoms physical properties. Another type of representation method for molecular complexes is based on the similarity indexing of molecular complexes with the associating functional groups and structure components. The machine-learning-augmented models exhibit outstanding prediction power in screening for the bimetallic, metal oxide and molecular complexes surfaces with desired adsorption properties. Compared with the traditional high-throughput computational and experimental trial-and-error approach, the machine-learning models have great potential in accelerating the discovery of new catalytic materials with desired properties. Sensitivity analysis of the machine-learning models can shed light on the nature of chemical bonding and its governing factors, providing chemical insights into catalytic processes and fundamental knowledge base of molecule-catalyst interactions.

[1] Vojvodic, Aleksandra, and Jens K. Nørskov. 2015. "New Design Paradigm for Heterogeneous Catalysts." *National Science Review* 2 (2): 140–49.

[2] Z. Li, X. Ma, and H. Xin, Feature engineering of machine-learning chemisorption models for catalyst design, Catalysis Today. 280, Part 2 (2017) 232–238.

[3] X. Ma, Z. Li, L.E.K. Achenie, and H. Xin, Machine-Learning-Augmented Chemisorption Model for CO2 Electroreduction Catalyst Screening, J. Phys. Chem. Lett. 6 (2015) 3528–3533.

[4] Z. Li, S. Wang, W. Chin, and H. Xin, Rapid Screening of Bifunctional Alloy Catalysts Enabled by Machine Learning, Submitted to Journal of Materials Chemistry A. (2017).

[5] M. Rupp, A. Tkatchenko, K.-R. Müller, and O.A. von Lilienfeld, Fast and Accurate Modeling of Molecular Atomization Energies with Machine Learning, Phys. Rev. Lett. 108 (2012) 58301.

500991 Revealing Catalytic Active Units of Ru for NH_3 Decomposition By *Operando* XPS

Abstract:

Revealing Catalytic Active Units of Ru for NH_3 Decomposition by *operando* XPS

Weiqing Zheng,[1] Jian Zhang,[2] Dang Sheng Su,[3] Robert Schlögl[4]

1. Catalysis Center for Energy Innovation, University of Delaware, 221 Academy St. Newark, DE 19716

2. Institute of New Energy Technology, Ningbo Institute of Materials Technology and Engineering, Chinese Academy of Sciences, 1219 Zhongguan West Road, Ningbo 315201, China

3. Shenyang National Laboratory for Material Science, Institute of Metal Research, Chinese Academy of Sciences, 72 Wenhua Road, Shenyang 110016, China

4. Department of Inorganic Chemistry, Fritz Haber Institute of the Max Planck Society, 4-6 Faradayweg, Berlin 14195, Germany

Ruthenium-based catalysts exhibit the most promising catalytic properties both in ammonia synthesis and decomposition. There is consensus in the literature that the B_5 sites on Ru particle strongly influence the activity of ammonia decomposition. Yet, the nature of the active state in the Ru system is still heavily debated, since the appearance of the B_5 sites does not solely depend on particle size, but also on particle shape. Herein we probe the metal-carbon support interaction focusing on the morphology (size and shape) of Ru nanoparticles by atomic resolution transmission electron microscopy (TEM) and high pressure X-ray photoelectron spectroscopy (*in situ* XPS).

We found the morphology of Ru NPs strongly depends on the local geometry of carbon. We have deposited Ru nanoparticles on different pretreated graphite surfaces and further investigated by atomic resolution TEM imaging carried out on a Cs-corrected TEM. The Ru atomic clusters are immobilized on the dispersed oxidized vacancies on carbon surface, and tend to sit on locally disordering sites on the graphene layers. These atomic Ru clusters have a much higher density of high-energy sites (defects), leading to a strong Ru-N interaction and a shortage of sites for adsorbed H atoms, which will result in a lower turnover frequency of ammonia decomposition copared to the well-order Ru nanoparticles. The core-level photoelectron N1s spectra obtained at 400 °C in 0.5 mbar NH_3 show a mixture of N states. The peak at about 400.2 eV has been assigned to a molecular adsorbed N_2,[1] which only appears on the disordered Ru nanoparticles.

The combinative characterizations can lead to a direct conclusion that the performance difference of carbon supported

Ru catalysts for ammonia decomposition is dependent on the local graphitic order of carbon supports. The highly graphitic carbon support can promote the formation of well-order Ru nanoparticles with large terraces for H_2 desorption and steps as active sites for NH_3 dissociation.

References

[1] R. Schlögl, *Angew. Chemie*, **2003**, *42*, 2004-2008

500739 ÂQuantitative and Atomic-Scale View of CO-Induced Pt Nanoparticle Surface Reconstruction at Saturation Coverage and Implications for CO Oxidation Structure Sensitivity

Abstract:

Atomic-scale insights into how different geometric configurations of surface atoms on supported metal nanoparticles catalyze chemical reactions are critical for the optimization of chemical conversion processes. It is well-known that the adsorption of reactive species can cause reconstruction of metal surfaces. Thus, characterizing metallic surface structures under reaction conditions at the atomic scale is critical for understanding the structure sensitivity of various chemical reactions. Elucidation of surface structures of high surface area oxide supported metal nanoparticles has been limited by less than atomic resolution typically achieved by environmental transmission electron microscopy (TEM) when operated under realistic conditions. In addition, a lack of agreement between theoretical structure sensitivity and correlated experimental measurements providing quantitative information about the distribution of exposed surface atoms under relevant reaction conditions, as has been the case for CO oxidation on Pt nanoparticles, precludes the correct determination of the active site under reaction conditions.

We overcome these limitations by correlating density functional theory predictions of adsorbate-induced surface reconstruction visually with atom-resolved imaging by *in-situ* TEM and quantitatively with sample-averaged measurements of surface atom configurations by *in-situ* quantitative infrared (IR) spectroscopy, all at identical saturation adsorbate coverage. This has recently been demonstrated for platinum (Pt) nanoparticle surface reconstruction induced by CO adsorption at saturation coverage and elevated (>400 K) temperature, which is relevant for the CO oxidation reaction under cold-start conditions in the catalytic converter.[1] Through our correlated approach, it was observed that the truncated octahedron shape adopted by bare Pt nanoparticles undergoes a reversible, facet selective reconstruction due to saturation CO coverage, where {100} facets roughen into vicinal stepped high Miller index facets, while {111} facets remain intact. In addition, we apply the same *in-situ* quantitative IR technique during CO oxidation on various sizes of Pt nanoparticles to measure the concentration of CO bound to under-coordinated (UC) and well-coordinated (WC) Pt sites and relate the turnover frequency of CO oxidation to the concentration of UC and WC Pt sites.[2] We have shown that CO oxidation on Pt nanoparticle surfaces is structure sensitive, where WC Pt atoms are the active site, but that this effect is masked by the observed adsorbate-induced reconstructions observed under reaction conditions. Our results provide a complete description of the observed trends in the structure sensitivity of CO oxidation on Pt nanoparticles of various sizes, in agreement with experimental observations and theoretical predictions.

References:

(1) Avanesian, T.; Dai, S.; Kale, M. J.; Graham, G. W.; Pan, X.; Christopher, P. *J. Am. Chem. Soc.* **2017**, *139*, 4551.

(2) Kale, M. J.; Christopher, P. *ACS Catal.* **2016**, *6*, 5599.

500404 *Operando* Electrochemical Grazing Incidence X-Ray Absorption and Diffraction for CO_2 Reduction on Aupd, Pd and Au Electrodes

Abstract:

While the electrochemical reduction of carbon dioxide (CO_2) to viable fuels and chemicals remains an appealing approach to partner with renewable, carbon-neutral energy sources, there remains a lack of fundamental understanding of how the surface of the catalyst changes during electrochemical experiments. Recent reports in the literature indicate that the surface of the catalyst can change drastically under reducing conditions, which suggests that the active catalyst during electrochemical CO_2 reduction conditions may be very different than the catalyst characterized *ex situ*. Consequently, it is critical to accurately probe the surface of the catalyst in these relevant *operando* conditions, as it would allow one to observe structural, electronic and orientation changes in the catalyst and to acquire key insights into the CO_2 reduction reaction. Our work reports the design and development of such an experiment at the Stanford Synchrotron Radiation Lightsource (SSRL) using electrochemical grazing incidence x-ray absorption spectroscopy (GIXAS) and electrochemical grazing incidence x-ray diffraction (GIXRD). Using these two techniques, we report changes in the surface electronic structure and lattice orientation of thin film $Au_{55}Pd_{45}$, Pd and Au catalysts during electrochemical CO_2 reduction.

500062 Chemical Transient Kinetics in Studies of the CO Hydrogenation Mechanism over Co-Based Catalysts

Abstract:

Despite extensive studies on the mechanism of the catalytic hydrogenation of CO a generally accepted reaction mechanism responsible for the formation of chain-lengthened paraffins, olefins and oxygenates is still not available. C-C coupling and CO-insertion remain the two main mechanisms considered for the reaction. The distinguishing feature of these two mechanisms is whether cleavage of the C-O bond in the CO molecule occurs before (leading to C-C coupling) or after incorporation of the C_1 monomer (CO-insertion) into the growing hydrocarbon chain. The present contribution is aimed at scrutinizing which of the mechanistic suggestions are compatible with the experimental results obtained by quantitative Chemical Transient Kinetics (CTK) [1]. In particular, we will compare results obtained using model catalysts such as Co/MgO with those obtained for bimetallic CoMn-based ones which recently showed promising catalytic performance in the high-pressure synthesis to tunable product classes, including aldehydes and alcohols [2].

CTK experiments start with a catalyst in dynamic adsorption-desorption equilibrium with hydrogen. A "build-up" transient is then initiated by switching to CO/H_2 mixtures. Similarly, switching back to H_2-only conditions, after establishing steady-state reaction conditions, allows monitoring how the FT catalytically active phase is being scavenged. Build-up and back-transient measurements are usually performed at atmospheric pressure conditions in the presence of reference noble gas to allow a quantitative evaluation in terms of molecular fluxes. All catalysts were prepared using the oxalate method of co-precipitation [1, 2]. Oxalates provide a metal organic framework (MOF) which may be thermally decomposed into nanosized particles with core-shell structure [1].

Our studies with Co/MgO as well as pure Co and Ni model catalysts allowed the atomic amounts of carbon, oxygen and hydrogen during the "build-up" to be determined. This way we showed that the monolayer limit was surpassed even before reaching the steady state. Thus, the catalyst surface did not provide metallic sites under steady state conditions. Such surface atom counting turned out to be difficult for CoMn catalysts. According to electron microscopy studies [2], major amounts of Mn in these catalysts were present as Mn_5O_8 which may undergo partial reduction under overall reducing conditions of the FT process.

Furthermore, for all of the investigated catalyst systems, chain-lengthened hydrocarbons appeared in sequence with considerable delay times (about 20 s) when switching from dynamic hydrogen adsorption-desorption to reactive FT conditions. Besides alkanes, CoMn catalysts also produced olefins under atmospheric pressures. A quantitative evaluation of the delay times during build-up clearly showed olefins form later as compared to alkanes. We therefore conclude that olefins cannot be the precursors of alkane formation. In any case, gaseous CO was necessary for chain lengthening to occur. This behavior is in agreement with a CO insertion mechanism. The provided interpretation was supported by the finding that during back-transients from FT to non-reactive conditions CO disappeared with the same time constant as

Ar reference gas. Strikingly, plotting the Anderson-Schulz-Flory" (ASF) chain lengthening probability α for hydrocarbons clearly showed proportionality to the CO pressure but non-monotonous behavior as a function of accumulating carbon (or CH_x) as would be expected if CH_x insertion or C-C coupling were occurring.

The largely delayed appearance of gaseous CO (time-correlated with chain lengthening) after switching molecular fluxes from H_2 to $CO+H_2$ is considered to be due to the irreversible chemisorption of CO. Since no oxygen containing species (CO_2 and H_2O) appeared either, at these short reaction times, we consider the formation of surface hydroxyl groups as target of a CO insertion is key to the formation of the C1 "most abundant surface intermediate". Combing transient measurements with DRIFTS (Diffuse Reflection Infra-red Fourier Transformation Spectroscopy) clearly revealed the formation of formate-type species.

As to the back-transients, alkanes are observed to pass through a maximum intensity when switching from $CO+H_2$ to H_2. On the other hand, olefins do not show this behavior and fade away exponentially immediately after switching gases, i.e. their time constant in a first-order desorption process is shorter than that for alkanes. The consequences of this observation with regard to the reaction mechanism will be discussed.

In summary, we provide clear evidence for the occurrence of a CO insertion mechanism leading to chain-lengthened hydrocarbons in the Fischer Tropsch reaction. The surfaces of Co, Ni, Co/MgO and Co/Mn_5O_8 catalysts do not provide metallic sites during the synthesis. Chain lengthening occurs time-correlated with CO gas. The ASF chain lengthening probability is directly proportional to the CO transient pressures.

References

1. Schweicher, J., A. Bundhoo, and N. Kruse., J. Am. Chem. Soc. 134 (2012) 16135–16138.

2. Xiang, Y., Kruse, N., Nature Commun. 7, 13058 (2016).

501427 *In situ* FTIR Studies of Ethanol Conversion on Zr-KIT-5 and Zr-KIT-6 Catalysts

Abstract:

Zr-KIT-5 and Zr-KIT-6 have been found to be active for acid catalyzed reactions such as Prins reaction, Meerweinâ €"Ponndorfâ€"Verley (MPV) reduction, Hantzsch reaction, alkylation, alcohol dehydration, and lignin depolimerization. In this latter reaction, Zr-KIT-5 has been particularly found to be more active than Zr-KIT-6 (at the same level of Zr content) and commercial H-ZSM-5 (Si/Al=11.5) catalysts. Here, we have used model ethanol conversion on Zr-KIT-5, Zr-KIT-6, and H-ZSM-5 catalysts to assess their acid properties. In situ FTIR-MS temperature programmed desorption (TPD) of acid site probes (e.g., pyridine) and operando FTIR-GC during ethanol conversion at intermediate temperatures (300-450 Â°C) showed that Zr-KIT-6 and H-ZSM-5 catalysts are primarily composed of Lewis and BrÃ¸nsted acid sites, respectively. On both catalysts, ethylene, the unimolecular dehydration product, was found to be the main reaction product; however, diethyl ether/acetaldehyde and diethyl ether/C4 olefins by-products formed as a result of alcohol bimolecular dehydration, alcohol dehydrogenation, and olefin dimerization on Zr-KIT-6 and H-ZSM-5, respectively. At the same Zr content, operando FTIR-GC of Zr-KIT-5 showed that this catalyst contains an acid site density significantly larger than that of Zr-KIT-6 and consequently resulted in a significantly higher catalytic activity. In situ pyridine-TPD and FTIR results showed apparent similar Lewis and Brønsted acid sites distribution, however, product selectivities with significant C4 olefin production resembled those of the H-ZSM-5 catalyst. These results point out to a synergy between Lewis and strong Brønsted acid sites in Zr-KIT-5. In situ FTIR suggested, however, that only a small fraction of these acid sites are responsible for ethanol conversion. Possible ethanol conversion mechanisms will be discussed.

488646 Mechanistic and Spectroscopic Evidence for Reactive Intermediate Structures during C-O Bond Rupture in Small Oxygenates on Metal Phosphide Clusters

Abstract:

Selective hydrogenolysis of C-O bonds in small oxygenates is a critical step for the conversion of biomass-pyrolysis oils into high value platform chemicals, such as α,ω-diols from furanic and pyranic species.[1] Metal phosphides (MPs) are promising catalysts for C-O bond rupture as they are inexpensive, chemically and thermally stable, and selectively cleave C-X bonds (where X=S, N, or O) over C-C bonds.[2,3]

Here, we combine kinetic analysis of reaction networks, modulation excitation *in situ* infrared spectroscopy (MES), and density functional theory (DFT) calculations to describe the mechanism and reactive intermediates for C-O bond rupture over MPs. Rate measurements of C-O bond rupture within 2-methyltetrahydrofuran (2-MTHF; 1-50 kPa 2-MTHF; 0.1-6 MPa H_2; 543 K) and DFT calculations show adsorption and dehydrogenation steps of 2-MTHF are quasi-equilibrated and precede kinetically relevant C-O bond rupture.[4] The reactive species have lost 1-2 H-atoms from the fully hydrogenated reactant and are coordinated to Ni ensemble active sites within Ni-rich terminating planes of Ni_xP surfaces (where x = 2-2.4). Ratios of sterically-hindered C-O (^{3}C-O) bond rupture rates to those for unhindered C-O (^{2}C-O) bonds increase with $[H_2]^{1/2}$, which suggests H-content of the reactive intermediates for ^{3}C-O and ^{2}C-O rupture differ by one H-atom. Selectivities for ^{3}C-O bonds are higher than for ^{2}C-O bonds over Ni_xP and much greater than the ^{3}C-O bond selectivities on Ni. Measured and DFT-predicted apparent activation enthalpies ($\Delta H^‡$, 473-583 K) for C-O bond rupture indicate that the phosphorus content decreases the $\Delta H^‡$ for ^{3}C-O bond rupture relative to that of ^{2}C-O bond rupture (i.e., $\Delta H^‡_{2CO}$ - $\Delta H^‡_{3CO}$ increases with phosphorus content). Selectivity differences between specific C-O bonds within 2-MTHF reflect differences in the H-content of reactive intermediates, activation enthalpy barriers, and phosphorus content of Ni and Ni_xP clusters.

Direct experimental evidence for the structure of reactive intermediates would provide clear connections between the relationship between C-O bond rupture selectivity, surface coordination, and enthalpic barriers over Ni_xP clusters. MES experiments coupled with phase sensitive detection reveal spectral features of reactive intermediates by suppressing those of inactive surface species.[5] Pure-component spectra and surface coverages (extracted by multivariate curve resolution-alternating least squares algorithm[6]) indicate the presence of distinct reactive surface intermediates during C-O bond rupture of 2-MTHF on Ni_xP, whose composition and surface coordination are consistent with those implied by kinetic measurements and predicted by DFT calculations. Modulations in H_2 pressure that are slower than 2 ks^{-1} (i.e., the measured turnover rate for C-O bond rupture in 2-MTHF) allow all intermediates to form and oscillate at the applied frequency; however, experiments performed with modulation frequencies much faster than 2 ks^{-1} (e.g., 2.9 ks^{-1} – 0.13 s^{-1}) suppress the creation of reactive intermediates whose barriers for formation correspond to characteristic timescales of formation that are greater than the period of the modulation. Such experiments show that the compositions of the most abundant reactive intermediate (MARI) on Ni and Ni_xP clusters during C-O bond rupture of 2-MTHF are identical; however, the MARI changes from $Ni_3(\mu^3\text{-}C_5H_{10}O)$ to $Ni_3(\eta^5\text{-}C_5H_{10}O)$ with increasing phosphorus content. The change in binding configuration with phosphorus content is likely the fundamental cause for increased selectivity towards ^{3}C-O bond rupture. The combination of these kinetic, spectroscopic, and theoretical techniques provides insight into underlying causes of selectivity across Ni and Ni_xP clusters.

References

(1) Chia, M.; Pagan-Torres, Y. J.; Hibbitts, D.; Tan, Q. H.; Pham, H. N.; Datye, A. K.; Neurock, M.; Davis, R. J.; Dumesic, J. A. *J. Am. Chem. Soc.* **2011**, 133, 12675.

(2) Oyama, S. T.; Wang, X.; Lee, Y. K.; Chun, W. J. *J. Catal.* **2004**, 221, 263.

(3) Rensel, D. J.; Rouvimov, S.; Gin, M. E.; Hicks, J. C. *J. Catal.* **2013**, 305, 256.

(4) Witzke, M. E., Almithn, A.; Coonrod, C. L., Hibbitts, D. D.; Flaherty, D. W. *J. Catal.* to be submitted.

(5) Urakawa, A.; Burgi, T.; Baiker, A. *Chem. Eng. Sci.* **2008**, 63, 4902.

(6) Jaumot, J.; de Juan, A.; Tauler, R. *Chemometr. Intell. Lab.* **2015**, 140, 1.

Insight into the structure evolution of iron-based catalysts for hydrogenation of CO/CO_2 using *Operando* techniques

*Yulong Zhang[a], Yang Sun[a], Zhengpai Zhang[a], Pengfei Tian[a], Jing Xu[a], Hui-Ping Li[b] and Yi-Fan Han[a,b]**

[a] *State Key Laboratory of Chemical Engineering, East China University of Science and Technology, Shanghai 200237, China*

[b] *Research Center of Heterogeneous Catalysis and Engineering Sciences, School of Chemical Engineering and Energy, Zhengzhou University, Zhengzhou 450001, China*

** Corresponding author phone: (+86) 21-64251928; fax: (86) 21-64251928;*

E-mail: yifanhan@ecust.edu.cn

Abstract

Operando techniques have opened up new vistas for visualization of the active phases/sites evolution during reactions, especially at high pressures and high temperatures. The rational design and controllable synthesis should still be the ultimate goal for chemists and chemical engineers. The greatest challenge is how to establish the relationship between structure and performance of industrial catalysts. Recently, with the combination of operando Raman spectroscopy (ORS) and some other techniques, such as in situ XRD, Temperature-Programmed techniques and in situ DRIFTS, we studied the complex reaction systems like F-T synthesis and CO_2 hydrogenation under harsh conditions For the FTS reaction, during activation process in different atmospheres, the phase transition of α-Fe_2O_3--Fe_3O_4--Fe_5C_2 was observed in both CO and syngas (H_2/CO) atmosphere. Moreover, various iron oxide/carbides aiming at different products have been also investigated by operando Raman spectroscopy during the CO_2 hydrogenation reaction. The preliminary relationship of catalyst structure and performance of iron catalysts under harsh working conditions was established. This work directs us to explore the reaction mechanism and design the next generation industrial catalysts.

Keywords: FTS; CO_2 hydrogenation; iron catalysts; operando techniques; structure and performance

Introduction

Catalysis is a fascinating science for people highly involved. What's more, more than 85 % of the fuels and chemicals are produced from catalytic processes [1]. Up to now, the catalyst R & D is mainly based on trial-and-error works. As Schlögl declared "In-situ investigations with complementary methods are not a „luxury" but rather essential parts of the formulating a hypothesis on the function of materials" [2], it is necessary to use the operando techniques to enhance the catalyst development effective. In all existing in situ techniques, operando experiment is a straightforward way to determine the dynamic structure of catalysts and can offer reliable information for the rational design of catalysts by characterizing the catalytic under real reaction conditions.

For instance, high temperature usually leads to the sintering and agglomeration of individual catalyst particles and the reduction on the surface adsorption sites, then further deteriorates the catalytic performance [3， 4]. Hence, for the study on dynamic structures of industrial catalysts, the development of operando techniques operating under harsh conditions is of paramount importance.

The operando Raman spectroscopy is naturally born for this aim not only because it is insensitive to CO_2 and H_2O, but also for those species have hampered most of spectroscopy applied in operando measurements [5].

In this paper, after long-term continuous efforts, operando Raman spectroscopy has been applied to monitor the structural evolution of iron-based catalysts in the FTS and CO_2 hydrogenation reaction under high temperature (260 ℃ ~ 350 ℃) and high pressure (2.5 ~ 3.0 MPa) which is typical harsh conditions in the process of industrial reactions.

Experiment Sections

Catalysts: High purity of α-Fe_2O_3& γ-Fe_2O_3 (Aladdin, 99.5 %) and Fe_3O_4 (Aladdin, 99.5 %) were used as reference samples. Also, a high purity of α-Fe_2O_3 & γ-Fe_2O_3 (Aladdin, 99.5 %) were used as the catalyst precursors for the F-T synthesis and CO_2 hydrogenation reaction.

Iron carbides: Fe_5C_2 and Fe_3C were synthesized on-line by exposing the α-Fe_2O_3 (Aladdin, 99.5 %, 30 nm) to a pure CO flow (99.995 %, 25 mL/min) and heating up to 350 ℃ and 450 ℃ at the temperature rate of 2 ℃/min and keeping for 2 h, respectively.

ORS activation and reaction: The laser Raman spectra were recorded with a home-build operando setup using a confocal Raman spectroscopy (Lab RAM HR, Horiba J.Y.) equipped with a high-grade Leica microscope (long working distance objective 50x).

XRD patterns were recorded using Riguku D/max 2550 diffractometer with accelerated voltage 40 kV and detector current 100 mA. Cu-Kα (λ = 51.540589 Å) radiation was used for a continuous scanning with the step-size of 0.01° over a 2 theta range 10 - 80° with a scan speed of 4° /min.

Results and discussion

Using the operando Raman spectroscopy (ORS), we have successfully determined active phases for the iron-based FTS reaction under industrial conditions. The structure of α-Fe_2O_3 catalyst was observed to experience extensive restructuring processes under activation and reaction processes (Fig.1). α-Fe_2O_3 was transformed into Fe_3O_4 slowly in a H_2 flow and the main phase during reaction is the mixture of Fe_3O_4 and Fe. CO- and H_2/CO-activated catalysts showed a transformation of α-Fe_2O_3→Fe_3O_4→Fe_5C_2; Meanwhile Fe_3O_4 is the key transition state during reaction. The catalytic activity followed an order of CO-activated > H_2/CO-activated ≈ H_2-activated (Fig.2). The relationships of structure and performance indicate: (1) iron carbides should be responsible for its lower FTS activity; (2) Fe_3O_4 plays an important role for FTS; (3) carbonaceous species involved positively or negatively affect the performance of iron-based FTS catalysts.

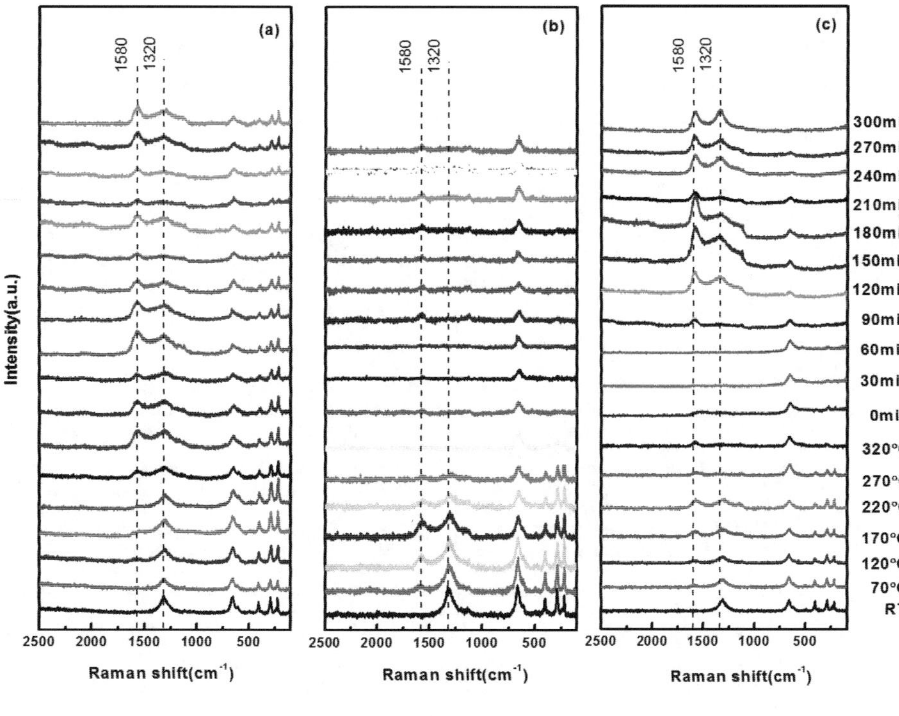

Fig.1 In situ Raman spectra of α-Fe_2O_3 during activation processes in (a) 10 % H_2/Ar; (b) 10 % CO/Ar; (c) 5 % CO/5 % H_2/Ar. The spectra at 350 ℃ were recorded in the time interval of 30 min with the flow of 8 mL/min^{-1}, corresponding to a gas hourly space velocity (GHSV) of ≈ 30000 h^{-1}.

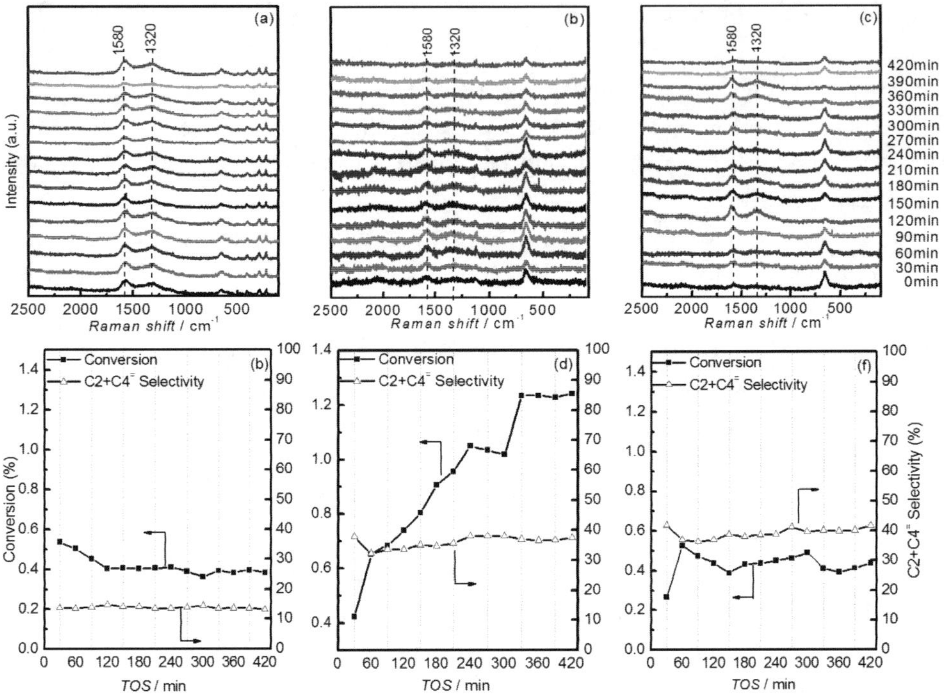

Fig.2 Operando Raman spectra of α-Fe_2O_3 (260 ℃ , 3.0 MPa) were recorded in the time interval of 30 min for 7 h with the flow of 3.2 ml/min, corresponding to a gas hourly space velocity (GHSV) of 12000 h^{-1}; (a, b) 10 % H_2/Ar; (c, d) 10 % CO/Ar; (e, f) 5 % CO/5 % H_2/Ar. (a, c, e) represent the structural evolution of the catalyst and (b, d, f) represent the corresponding catalytic performance.

Moreover, with the combination of operando Raman spectroscopy (ORS) and some other techniques, such as in situ XRD, Temperature-Programmed techniques and in situ DRIFTS, we also studied the complex reaction systems under harsh conditions---CO_2 hydrogenation. Various iron oxide/carbides aiming at different products have been investigated by *operando* Raman spectroscopy during the CO_2 hydrogenation reaction (Fig. 3).

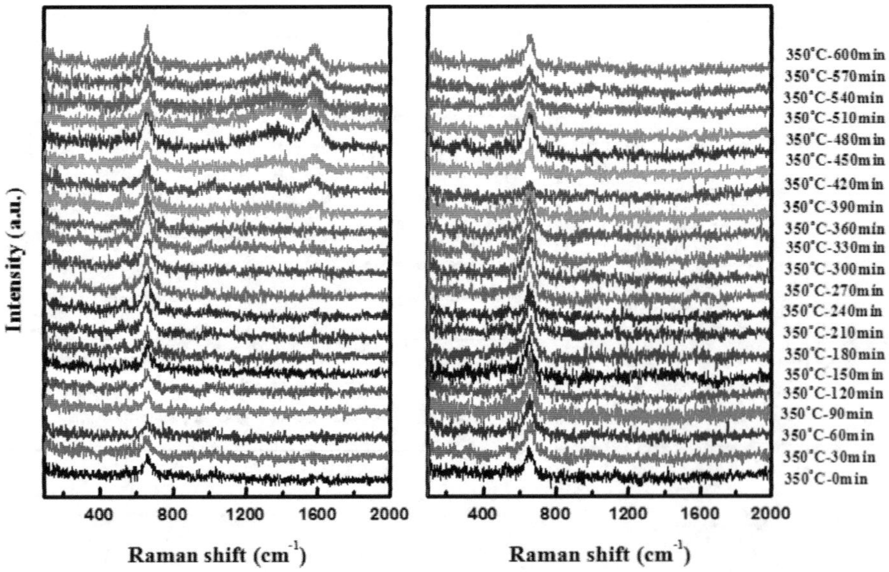

Fig.3 Operando Raman spectra of (a) 10 % H_2/Ar reduced γ-Fe_2O_3 (b) 10 % H_2/Ar reduced α-Fe_2O_3 during the reaction processes (5 % CO_2/15 % H_2/Ar, 350 ℃ , 25 bar) were recorded in the time interval of 30 min for 10 h with the flow of 3 ml/min, corresponding to a GHSV of 6000 ml/(g·h).

Also, in situ XRD was applied for CO_2 hydrogenation reaction under 1 bar and 350 ℃. Fig.4 shows the formation of iron carbides within 5 hours, and no iron oxide was observed during all the reaction process, which ensure the inference that iron carbides could be directly transformed from metallic iron. The preliminary relationship of catalyst structure and performance of iron catalysts under harsh working conditions was established (scheme 1).

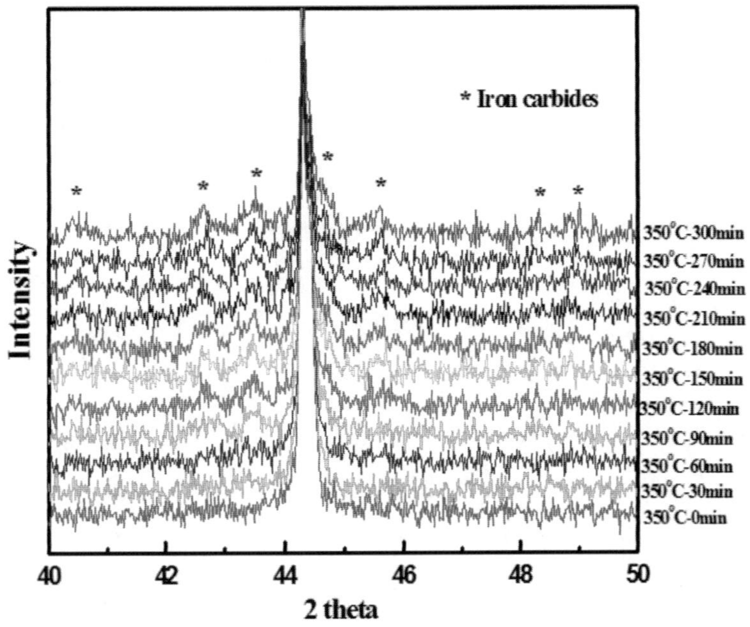

Fig.4 In situ XRD patterns for metallic iron during the 5 % CO_2/15 % H_2/Ar (1 bar, 350 ℃) reaction

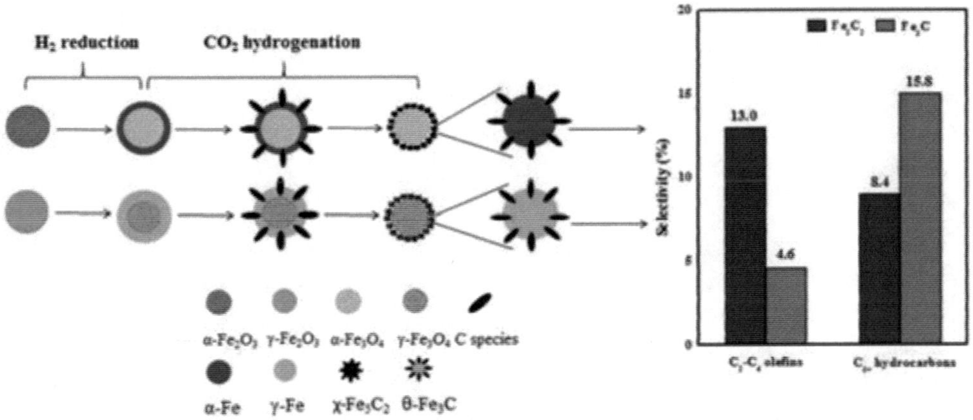

Scheme 1. Structure evolution of α-Fe_2O_3 & γ-Fe_2O_3 during both activation and reaction processes.

Conclusions

The relationship of structure and performance of iron-based catalysts under harsh reaction conditions (F-T synthesis and CO_2 hydrogenation reaction) have been built by using operando techniques. For the FTS reaction, during activation process in different atmospheres (10 % H_2/Ar, 10 % CO/Ar and 5 % H_2/5 % CO/Ar), the phase transition of α-Fe_2O_3--Fe_3O_4--Fe_5C_2 was observed in both CO and syngas atmospheres. Various iron oxide/carbides aiming at different products have been also investigated by operando Raman spectroscopy.

The application of operando techniques will greatly enhance the establishment of the relationship of structure and performance of a catalyst, which is of paramount importance to understand the origin of active sites and reaction pathways.

References

1 Olah GA: Beyond oil and gas: The methanol economy. Angewandte Chemie International Edition (2005) 44(18):2636-2639.

2 Schlögl R: Heterogeneous catalysis. Angewandte Chemie (2015) 127(11):3531-3589.

3 Zhang, Y.; Fu, D.; Xu, X.; Sheng, Y.; Xu, J.; Han, Y.F.* Current opinion in chemical engineering, 2016, 12, 1-7.

4 Torres Galvis, H. M.; Johannes, H. B.; Chaitanya, B. K.; Ruitenbeek, M.; Iulian Dugulan, A.; de Jong, K. P.* Science,2012, 335 (6070), 835.

5 Fu,D.; Dai,W.; Xu,X.; Mao,W.; Su,J.; Zhang,Z.; Shi,B.; Smith,J.; Li,P.; Xu,J.; Han,Y.F.* ChemCatChem, 2015, 7(5), 752.

495682 Modeling of Stationary Phase in Microalgae Growth Using the Population Balance Equation

Abstract:

Microalgae are being widely studied because of their versatility to some of the main needs of humanity such as energy, food and fine chemicals [1]. These kinds of cells can be found in every corner of the globe and they are considered among the fastest growing plants on Earth. Sometimes, they are unwanted and, for that reason, defining what happens in the stationary phase could play an important role in a better understanding of how these species survive in very adverse conditions.

Microalgae growth can be classified into different phases, commonly named lag, exponential, linear, stationary and decaying phases [2]. To maximize the production, it is necessary to define which are the best living conditions. For that purpose, experiments can be performed and the best environmental conditions (temperature, nutrients and light intensity to mention some) can be defined. However, experiments are quite time consuming and, often, require expensive instruments. Therefore, it is useful to develop models able to predict the growth of microalgae. This can be particularly useful for a further optimization of the production process.

Many models have been utilized to predict the growth, but most of them canâ€™t give much information about the life cycle of microalgae [2]. Population balance has been utilized on numerous fields of engineering and science [3]. Recently the population balance equation (including the breakage equation) has been applied to cellular growth and microalgae growth in particular [4]-[6]. Especially, having more information about the growth, the division rate as well as the size distribution can be quite advantageous when the objective is to optimize the whole system. For example, knowing the size of the cells can enhance the harvesting technology, which is among the most energy consuming steps before obtaining any final product from microalgae [7]. Therefore, it can be useful to know the behavior of the population during the stationary phase, in which the highest concentration of cells is reached. At this point the growth rate of the whole population is close to zero, considering also that part of the population would die. For that reason, a better understanding of the growth of living cells as well as the content of dead cells may give some valuable information related to the life cycle of microalgae. Differently from the exponential and linear phases, where the number of dead cells can be neglected, in the stationary phase that number plays an important role. In particular, maintaining the stationary phase for a prolonged period can give indication about this phase and explain why the decaying period does not occur immediately after the linear phase.

Numerous methods can be used to solve the population balance equation (PBE) [3], [8]. In our case, to solve the PBE, which is a integro-partial differential equation, a discretized method has been applied. The model considers dead cells and living population and it can predict the total population (alive and not) as well as the size distribution dynamically. Other than that, it can provide additional parameters such as the growth rate, the division rate at different cellular sizes and the number of daughter cells. The model has been validated experimentally using some batch photobioreactors, in which a unicellular species of microalgae (*Chlorococcum*) was grown. The temperature was maintained constant using a temperature controller, while nutrients were introduced only in the beginning of the experiment. Cell concentration and size distribution was measured daily; the experimental data were then utilized for parameter estimation applying the population balance equation. The results are in agreement with experiments and additional investigation will be performed in the future.

References

[1] E. M. Grima *et al*, "Recovery of microalgal biomass and metabolites: process options and economics," *Biotechnol. Adv.*, vol. 20, pp. 491-515, 2003.

[2] E. Lee, M. Jalalizadeh and Q. Zhang, "Growth kinetic models for microalgae cultivation: a review," *Algal Research*, vol. 12, pp. 497-512, 2015.

[3] D. Ramkrishna, *Population Balances: Theory and Applications to Particulate Systems in Engineering.* Academic press, 2000.

[4] M. A. Hjortsø, *Population Balances in Biomedical Engineering.* McGraw-Hill, 2006.

[5] A. Bertucco *et al*, "Population balance modeling of a microalgal culture in photobioreactors: Comparison between experiments and simulations," *AIChE J.*, vol. 61, pp. 2702-2710, 2015.

[6] E. Pahija *et al*, "Microalgae growth determination using modified breakage equation model," *Computer Aided Chemical Engineering*, vol. 37, pp. 389-394, 2015.

[7] A. Concas, M. Pisu and G. Cao, "A novel mathematical model to simulate the size-structured growth of microalgae strains dividing by multiple fission," *Chem. Eng. J.*, vol. 287, pp. 252-268, 2016.

[8] R. J. LeVeque, *Finite Volume Methods for Hyperbolic Problems.* Cambridge university press, 2002.

499688 Avoiding Fluidization in 3-Phase Packed Bed Reactors

Abstract:

In packed bed reactors, care must be taken to operate the bed well below the fluidization velocity in order to avoid motion of the particles and subsequent deterioration of the catalyst. For 2-phase reactors (gas-solid or liquid-solid), the physics are well understood. However, in co-current upflow reactors where a gas-liquid mixture is passed through a bed of packed solids, the bed becomes partially or completely fluidized at conditions well below the minimum fluidization velocity of either the gas or liquid phase alone.

In this study, a series of experiments varying the particle size, density, viscosity, and relative gas-liquid flowrates are varied to characterize the minimum fluidization conditions for the system. Computational Fluid Dynamics simulations in OpenFOAM are used to characterize the forces acting on the particles to elucidate the relevant physics.

501761 Residence Time Distribution Studies in a Laboratory Scale Trickle-Bed Reactor for Liquid Phase Hydrogenation of Acetylene

Abstract:

Residence Time Distribution Studies in a Laboratory Scale Trickle-Bed Reactor for Liquid Phase Hydrogenation of Acetylene

Humayun Shariff, Premkumar Kamalanathan, and Muthanna H. Al-Dahhan

Multi-phase Reactors Engineering and Applications Laboratory(mReal), Department of Chemical and Biochemical Engineering, Missouri University of Science and Technology, Rolla, MO, USA

Liquid phase hydrogenation of acetylene for ethylene production has recently gained interest for its safe operation and better catalyst life in comparison with gas phase process [1]. A selective solvent is used as the liquid medium which not only selectively absorbs acetylene from a gas mixture but also reduces the formation of green oil due to oligomerization in the reaction. The gas phase hydrogenation of acetylene itself is a highly exothermic reaction leading to thermal runaway [2]. In industries, trickle bed reactors are widely used for gas-solid-liquid contacting due to the inherent advantages of high pressure operations and high catalyst to liquid ratio [3]. Naturally, trickle bed reactors are the choice for the liquid phase hydrogenation of acetylene. Due to the complex flow behavior in TBRs, in addition to the kinetics, factors such as wettability of the liquid, mixing, holdup (gas and liquid) affect the conversion and selectivity. As part of our continued effort to understand and develop the scale models for liquid phase hydrogenation, in this work, global mixing is quantified using Residence Time Distribution (RTD) studies. Thereby, the performance of the reaction in TBR. RTD experiments were investigated in a stainless-steel lab scale TBR (1 inch I.D. and 60 cm long) with 2mm spherical particles of 30 cm catalyst bed length. RTD studies were conducted at the similar bed conditions as the reaction studies of liquid phase hydrogenation of acetylene for ethylene production. The experiments were conducted for liquid velocities of $0.10 \times 10^{-3} - 3\times10^{-3}$ m/s and for gas superficial velocities of 0.02 – 0.45 m/s. RTD were determined using conductivity measurements from the pulse-input liquid-tracer injection. Mean residence time and standard deviation were determined from the C - curve. Axial dispersion model (ADM) was used to obtain the dispersion coefficient by regressive fitting. D_{ax} was compared with the available correlations from literature to predict the reactor performance for a developed reactor scale model. The influence of the liquid and gas velocities on the model parameters is also studied.

Keywords: Trickle Bed Reactor, Liquid phase hydrogenation, Acetylene, Residence time distribution, Axial Dispersion

References

[1] Edvinsson, R. K., Holmgren, A. M., & Irandoust, S. (1995) *Ind. & Engg. Chemistry Research*, 34(1), 94-100

[2] Borodziński, A., & Bond, G. C. (2006) *Catalysis Reviews*, *48*(02), 91-144

[3] Al-Dahhan, M.H., et al. (1997) *Ind. & Engg. Chemistry Research 36*(8), 3292-3314

494676 Numerical Simulation of the Bubble Column with a CFD-PBM Coupled Model: Importance of the Drag Force of Bubble Swarms

Abstract:

Introduction

The drag coefficient of bubble swarms is different from that of single bubbles for the existence of the bubble interaction [1-3]. Some researchers [3-4] studied the rising of bubbles in bubbly flow and found that the velocity of small bubbles decreased in the bubble swarms due to hindering effect. Meanwhile, some other researchers [5-6] found the rising velocity of large bubbles increased due to wake accelerating effect. To account for these effects, some correlations were proposed for the relatively velocity or drag coefficient of bubble swarms. Actually, both the hindering effect of small bubbles and wake accelerating effect of large bubbles exist in bubble swarms [1-2], and the drag coefficient should be related to both the bubble diameter and local gas holdup. However, the applicable condition range of the existing drag models and correlations are still limited. The present work aimed to propose a drag model applicable in a wider range of conditions by correlating with the bubble size distribution and local gas holdup and including both the hindering effect and wake accelerating effect. The performance of this drag model was investigated in the numerical simulations of a bubble column with a CFD-PBM (population balance model) coupled model.

Model Development

The numerical simulations were carried out with the CFD-PBM coupled model, which was similar to that used in our previous work [7], except that a new drag model was employed, as shown in Figure 1. The average drag coefficient of single bubbles was first calculated based on the local gas holdup and bubble size distribution. Then the lumped quantity of bubble swarms effect was modified with a drag correction factor C_D/C_{D0}, which based on the bubble size distribution and local gas holdup and included both hindering effect of small bubbles and wake accelerating effect of large bubbles in bubble swarms.

Figure 1. Schematic of the CFD-PBM coupled model with drag force of bubble swarms.

Results and Discussion

The calculated drag correction factor was plotted as a function of the local gas holdup for different superficial gas velocities, as shown in Figure 2(a). With increasing local gas holdup, C_D/C_{D0} increased linearly in the low gas holdup range, then increased to its maximum (1.5 at 0.06 m/s), and significantly decreased in the high gas holdup range. This complex variation was attributed to the change of the governing factor from the hindering effect of small bubbles to the wake accelerating effect of large bubbles. Although the variation trend for the drag correction factor was similar at different superficial gas velocities, the value of C_D/C_{D0} depended on the gas velocity, especially at low local gas holdup. The reason was that the drag correction factor depended not only the local gas holdup, but also the bubble size distribution that can be significantly different even at a certain local gas holdup.

To provide a concise diagram, we chose the simulated data of U_g = 0.06 m/s to compare with the experimental results from literatures [2-3]. Note that the data of Roghair et al. [3] were from numerical experiments for d_b = 4 mm. The complex variation trend of C_D/C_{D0} was similar, as shown in Figure 2(b). The slight deviation was attributed to the different bubble size distributions.

Figure 2. (a) Simulated drag correction factors at different superficial gas velocities; (b) Comparison of simulated and experimental drag correction factors.

Conclusions

Both the hindering effect and wake accelerating effect exist for bubble swarms in a bubble column. With increasing local gas holdup, the drag correction factor increased and then decreased, which was attributed to the change in the governing factor from the hindering effect of small bubbles to the wake accelerating effect of large bubbles. The drag model based on the bubble size distribution and local gas holdup was applicable to a wider range of conditions.

Acknowledgement

The authors thank the financial supports by the National Natural Science Foundation of China (No. 21476122).

References

[1] Ishii, M., Zuber, N. AIChE J. 1979, 25: 843-855.

[2] Simonnet, M., Gentric, C., Olmos, E., Midoux, N. Chem. Eng. Sci. 2007, 62: 858-866.

[3] Roghair, I., Annaland, M.V.S., Kuipers, H.J.A.M. AIChE J. 2013, 59: 1791-1800.

[4] Lane, C.D., Parisien, V., Macchi, A., Donaldson, A.A. Chem. Eng. Sci. 2016. 152: 381-391.

[5] Thakre, S.S., Joshi, J.B. Chem. Eng. Sci. 1999. 54: 5055-5060.

[6] Krishna, R., Urseanu, M.I., Baten, J.M.V., Ellenberger, J. Chem. Eng. Sci. 1999. 54: 171-183.

[7] Wang, T., Wang, J., Jin, Y. AIChE J. 2006. 52: 125-140.

491776 Low-Order Modeling of the Vapor Phase Upgrading of Fast Pyrolysis Bio-Oil in a Bubbling Fluidized Bed Reactor

Abstract:

Recently, the conversion of raw biomass into valorized fuels and chemicals has received a tremendous amount of attention, with a number of approaches being considered. One approach is fast pyrolysis of raw biomass followed by catalytic vapor phase upgrading of the raw bio-oil. A bench scale system for evaluating this strategy and guiding future scale-up efforts has been built at the National Renewable Energy Laboratory (NREL).[1, 2] Although initial results are promising, the complex coupling between the reactor hydrodynamics and the kinetics of pyrolysis and upgrading inhibits efforts to first identify optimal operating conditions and subsequently transfer this knowledge to larger-scale reactors. A predictive low-order model of these reactors would facilitate a search for more effective operating conditions and provide insights into some potential challenges in scaling-up the system.

We have developed a model of the catalytic vapor phase upgrader at NREL. This model employs a network of ideal stirred tank reactors to represent the different zones (bubbling bed, freeboard above the bed, solids inlet/outlet in the middle of the freeboard, and freeboard above the solids inlet) in the real reactor. Aside from a few adjustable parameters (the fraction of solids in the freeboard regions and the solids inlet/outlet region), all necessary parameters are specified with correlations drawn from the fluidized bed literature. We also incorporate an existing kinetic scheme for upgrading bio-oil on ZSM 5.[3, 4]

We employ uncertainty quantification to estimate the reliability of our predictions and global sensitivity analysis to determine the parameters with the most influence on reactor operation. Based on these results, we offer suggestions to improve reactor performance and highlight areas requiring more fundamental research such as the need for improved kinetic schemes and better fluidization correlations.

References

1. Iisa, K., et al., Energy & Fuels 30, 2144 (2016).
2. Yung, M.M., et al., Energy & Fuels 30, 9471 (2016).
3. Adjaye, J.D. and N.N. Bakhshi, Fuel Processing Technology 45, 161 (1995).
4. Adjaye, J.D. and N.N. Bakhshi, Fuel Processing Technology 45, 185 (1995).

500523 A Sustainable Waste Management Alternative: Catalytic Gasification

Abstract:

This research focuses on advancing the current knowledge of a catalytic gasification process as a potential in-situ resource utilization and waste management system. This research has significance in a variety of engineering applications, but is of particular relevance towards municipal waste management and as an in-situ resource utilization alternative for advancing space exploration.

In this technology, typically referred to as Trash to Supply Gas (TtSG), the catalytic gasification reaction mechanism evolves through four reactions. The first two reactions are liquid-phase oxidation reactions of long-chain polymers (or substrates). The liquid phase oxidation reactions produce carbon monoxide, carbon dioxide, and water. The oxidation reactions are complemented by two gas-phase reactions: the Water Gas Shift (WGS) and the Sabatier (or methanation) reaction. These two reactions are the main stages in the pathway of producing hydrogen and methane in this technology.

The focus of this research is elucidating the reaction kinetics for the liquid phase oxidation reactions. A detailed analysis of the gas-phase kinetics where the Sabatier and Water-Gas Shift reactions complement each other to complete the Wet Thermal Catalytic Oxidation of long-chain polymers to syngas is presented. Current efforts including the connection of transport phenomena with gas phase kinetics for the formulation of an overall model are addressed in this paper.

Results obtained in a research grade laboratory reactor with two model substrates: cellulose and polyethylene, are presented. Cellulose and polyethylene were chosen as model substrates because they exhibit marked differences under the current reaction conditions being studied. Cellulose remains solid during catalytic gasification experiments, while polyethylene melts prior to reaction conditions. The reaction time is varied between 2 and 24 hours. Gas products are collected and analyzed with Gas Chromatography. In addition, characterization of the liquid and solid residuals includes Differential Scanning Calorimetry, Thermo-Gravimetric Analysis, and Scanning Electron Microscopy. The results clearly demonstrate the potential of catalytic gasification as a sustainable waste management alternative.

488565 Design andÂ Optimization of Multiphase FCC Regenerator Hydrodynamics Coupled with Reaction Kinetics

Abstract:

Fluidized catalytic cracking(FCC), profoundly referred as â€œgasoline machineâ€. Transforms relatively dense hydrocarbons to monetary worth gasoline with a elevated octane rating & other dominant petrochemical feed stock. FCC regenerator is utilized in the retrieval of solid FCC catalyst reactivity, & furthermore enhances significantly catalyst temperature to impart thermal energy to endothermic catalytic cracking reactions in FCC riser. It is of paramount importance in determining FCC process economics. In FCC regenerator, there coexists multiphase flow regimes, a heavier dense typically employed in the fast fluidization regime characterized by high circulation rates prompt to the non-uniform dispersal of particle velocity and solid holdup, a intermediate regime characterized by cluster formation of the catalyst which is the phenomena of catalyst segregation and backmixing, & an upper dilute transport regime, scattered flow region having reactor cyclone systems, which handles gas-solid separation. The challenge is to maximize catalyst regeneration by thorough dispatch of spent catalyst and air in FCC regenerator & mitigate thermal/mechanical degradation of catalyst. To distinguish the hydrodynamic effects on selectivity and conversion, including attaining perception into the performance of FCC regenerator. The laws of thermodynamics, phase & chemical reaction equilibrium(coke burn-off from the FCC catalyst particles in regenerator both homogenous & hetrogenous reactions) will be coupled with the hydrodynamics. These three-dimensional multiphase numerical flow simulations assimilating chemistry modeling and multi-species transport along with
incorporating radiative heat transfer will help in dictating operating framework of FCC regenerator. FCC regenerator design, optimization & scale-up is still predominantly empirical, owing to slender comprehension of intrinsic multiphase flow in regenerator. Impact of dominant closure models such as viscous/turbulence stress models, boundary conditions, drag models, and chemical reaction kinetic models on forecasting conversion needs to be sternly scrutinized. The multiphase Euler - Euler approach, with kinetic theory of granular flow is employed for CFD study of FCC regenerator. It includes drag model & restitution coefficient for the gas - solid interaction drag force & to define the solid - solid collision forces respectively. A chemical reaction model is utilized which includes coke combustion kinetic equations with apt kinetic constants to simulate the gas phase & interphase reactions in the regenerator. FCC regenerator multiphase simulations predict the phase pressure drop, effect of gas velocity on residence time distribution. Turbulence characteristics namely, ensemble averaged velocity statistics, dissipation of turbulent momentum by fluid velocity fluctuations which helps in analyzing heat and mass transfer properties in turbulent suspensions & turbulence amplification/attenuation due to fluid-solid interaction. It further explores the combustion characteristics such as gas species i.e, mass fraction, the particle temperature, solid-holdup. The simulation results also provide detailed quantitative mapping of the regenerator behavior including afterburning, & predict clustering behavior of the FCC catalysts which results in temperature hotspots accompanying
with higher vapor compositions, which causes irreversible attenuation of FCC catalyst reactivity.

502458 Mechanism and Kinetics of Direct Acylation of Furanic and Phenolic Species with Carboxylic Acids

Abstract:

The direct coupling of carboxylic acids with biomass derived species is important for the production of renewable fuels and specialty chemicals.[1] For fuels production, this reaction has the potential to significantly increase yields to liquid products from biomass by directly coupling light acids into the gasoline/diesel fuel range.[2] This reaction also has also gained attention recently for its potential to form specialty commodity chemicals such as renewable surfactants. Here we discuss the mechanism of direct activation of carboxylic acids with renewable furanic and phenolic species over zeolites.[3] Through a combination of detailed vapor phase reaction kinetics, isotope labeling experiments, temperature programmed experiments, and DFT calculations we reveal conditions necessary for selective dehydration of acids to form surface acyl groups over H-ZSM-5 and H-Beta zeolites.[4] In the presence of only carboxylic acids, the rate limiting step is the carbon-carbon coupling step, with dehydration to form surface acyl species requiring a lower barrier. We then evaluate the nature of the kinetically relevant transition state required for coupling surface acyl species a variety of substrates. The barrier for acylation is contrasted with coupling with a second carboxylic acid. For several renewable furanic and phenolic species, the barrier for the C-C coupling step is reduced to a value comparable to or lower than that required for dehydration to form surface acyls, making dehydration the rate determining step. This significantly shifts the dependence of the acylation reaction rate with respect to water approaching an order of -1. By comparing a variety of furanic, phenolic, and aromatic substrates, a general correlation is developed between the electronic structure of the substrate and the net barrier for C-C coupling, reveling the shift in degree of rate control for the dehydration vs. C-C coupling steps for the various substrates.

[1] A. Gumidyala, B. Wang, S. Crossley, *Science Advances* **2016**, *2*, e1601072.

[2] J. A. Herron, T. Vann, N. Duong, D. E. Resasco, S. Crossley, L. L. Lobban, C. T. Maravelias, *Energy Technology* **2017**, *5*, 130-150.

[3] D. S. Park, K. E. Joseph, M. Koehle, C. Krumm, L. Ren, J. N. Damen, M. H. Shete, H. S. Lee, X. Zuo, B. Lee, W. Fan, D. G. Vlachos, R. F. Lobo, M. Tsapatsis, P. J. Dauenhauer, *ACS Central Science* **2016**, *2*, 820-824.

[4] A. Gumidyala, T. Sooknoi, S. Crossley, *Journal of Catalysis* **2016**, *340*, 76-84.

486993 Approaches and Software Tools for the Development of Molecular-Level Kinetic Mega Models

Abstract:

Many energy and petrochemical processes involve the reactions of large numbers of species. Modeling these systems at the molecular-level can therefore involve model sizes reaching the mega level in terms of reactions and/or species. Model development at this scale becomes time consuming in each of the building, solving, optimization and editing phases. Advances in computer science have allowed for the development of new modeling frameworks that can be tailored to decrease the time spent and dynamically scale to variations in reaction system sizes that reach up to and past the mega level, all while retaining chemical engineering fundamentals. The Dynamic Model Builder (DMB) is a C++ based modeling framework developed to account for these needs. DMB functions independently of program compilation, allowing for easy use and application portability.

To illustrate the robustness of this modeling framework a lignin structure was developed for modeling pyrolysis between the temperatures of 300-500 °C. To bypass the combinatorial issues created by the large number of reactive sites on the lignin structure a network merging approach was developed and utilized. In the merging approach, reaction chemistries were split into two reaction networks that were later merged into one. For lignin pyrolysis, the Î±-O-4 and Î²-O-4 linkage cleavage reaction network (RN1) and the vinyl degradation reaction network (RN2) were merged. The products from RN1 were used as the input to RN2. It was found that RN2 is the root cause of a combinatorial explosion, so its growth was limited by reactant carbon number. Multiple tractable reaction networks were produced based on increasing the carbon range. Greater carbon ranges led to mega level growth of the reaction network. Therefore, DMB was used to model the reaction networks produced from the merging approach. In DMB, linear free energy relationships were used to decrease model kinetic parameters. Optimization of kinetic parameters was achieved from literature experimental data and produce a model output that shows good agreement with experimental values.

495528 First Principles Insights into the Selective Conversion of Glycerol to 1,3-Propanediol: The Synergistic Effect of Metal and Metal Oxide

Abstract:

Glycerol, a by-product of the biodiesel production process, is a cheap and green feedstock to produce the high value 1,3-Propanediol (1,3 PDO), a monomer in the production of Polytrimethylene terephthalate (PTT) which has superior properties compared to the polyethylene and polybutylene counterparts. Mechanistic insights into the liquid phase hydrogenolysis of glycerol for the formation of 1,3 PDO on MO_x (metal oxide like ReO_x or WO_x) modified M1 (active metals like Ir/Ru/Pt) are critical to improve its yield, which is currently limited to around 40%. ReO_x modified Ir catalyst is reported to be the best among the investigated catalyst systems for this reaction and hence, unravelling the specific roles of the metal oxide and the active metal will also help in identifying cheaper and more efficient MO_x/M1 combinations for the reaction. Density functional theory (DFT) based calculations are performed to compute reaction pathways and activation barriers associated with the key steps in glycerol hydrogenolysis on clean Ir and ReO_x modified Ir catalysts as model catalyst systems. Our calculations suggest that the low conversion of glycerol, as observed on Ir alone, is due to the extremely high activation barrier associated with the OH removal of glycerol. The selectivity towards 1,2-PDO on Ir alone could be due to the lower activation barrier (lower by ~37 kJ mol^{-1}) for removal of the primary OH of glycerol than that of the secondary OH. On ReO_x/Ir catalyst, the partially reduced ReO_x clusters dispersed on metallic Ir easily abstracts the hydroxyl hydrogen of glycerol and facilitates its attachment to the cluster as an alkoxide. The preferential attachment of glycerol through its primary OH groups in the early stages of the reaction allows the direct hydro-deoxygenation of the un-attached secondary OH of glycerol on the Ir catalyst, converting it to 1,3-PDO. Progressive reduction of the ReO_x clusters in course of the hydrogenolysis reaction lowers the preference of attachment of glycerol through its primary hydroxyls and this explains the drop in 1,3-PDO selectivity with the increase in conversion of glycerol. Since the MO_x cluster is attached to the active metal M1 by direct metal-metal interactions, stronger interactions facilitate the formation of finely dispersed MO_x clusters which are crucial for the direct hydrogenolysis mechanism. Thus, metal-metal interaction strength is suggested to be a suitable guide for the choice of metal oxide to be dispersed on the active metal in the mixed catalyst system.

492831 Analyzing Reaction Networks and Pathway Kinetics Via Metadynamics Simulations

Abstract:

Investigating reacting systems can be challenging due to the large number of species and pathways involved in these networks. A common obstacle to computational studies of reacting systems is forming a comprehensive scheme of all the relevant species and transition states. Molecular dynamics (MD) offers the ability to characterize the thermodynamics and kinetics of the different states and pathways involved in a particular system at high resolution. However, the computational cost associated with MD simulations often limits its application in this area. While a variety enhanced sampling methods, such as metadynamics (MetaD),[1] have been developed to reduce the computational cost of simulations, these methods typically require identification of a reaction coordinate of relevant degrees of freedom for transitions. In this poster, we highlight developments to the MetaD family of methods that reduce the level of chemical insight required and expedite sampling of transition paths, making this method well-suited for studying complex reaction systems.

We present a novel workflow of exploring a large, complex reaction network by using the SPRINT coordinates[2] of all the atoms in the system as collective variables (CVs), and biasing them following the Parallel Bias MetaD method. We apply this framework to study the decomposition of g-ketohydroperoxide (the Korcek reaction)[3] described by the PM6 Hamiltonian. Using an ensemble of simulations we are able to reconstruct a reaction network of the participating species and transition pathways that encompass steps of the documented Korcek reaction mechanism,[3] as well as other fundamental chemistries.

Additionally, we discuss recent advances in sampling transition events within the infrequent MetaD framework, a variant of MetaD used to calculate transition times of rare events from biased simulations, to aid in its application to reacting systems.[4] Because transition events are stochastic, many events must be sampled to get a converged estimate of the rate of transition.[5] For systems with multiple competing pathways, adequately sampling all of the relevant pathways can be challenging and lead to an excessive number of simulations to be carried out. We illustrate a method that isolates individual pathways of such systems to reduce the number of simulations needed, without corrupting the recovered transition rate, and apply it to two model systems.[6] Furthermore, we also discuss how the use of two generic CVs, SPRINT coordinates and collective-variable driven hyperdynamics,[7] can be suitable CVs for infrequent MetaD, despite requiring a relatively low level of knowledge to define.[8] We individually bias these generic CVs, and compare them to typical CV (bond distances), to recover the reaction rates of an SN2 reaction and a Diels Alder reaction. For both systems, we show that for biasing each CV leads to the same rate of reaction and energy barrier.

The poster concludes with a perspective on how these methods can be applied together to discover and then characterize the kinetics of pathways in large, complex reaction networks.

References:

(1) Valsson, O.; Parrinello, M. Variational Approach to Enhanced Sampling and Free Energy Calculations. *Phys. Rev. Lett.* **2014**, *113* (9), 1–5.

(2) Pietrucci, F.; Andreoni, W. Graph Theory Meets Ab Initio Molecular Dynamics: Atomic Structures and Transformations at the Nanoscale. *Phys. Rev. Lett.* **2011**, *107* (8), 85504.

(3) Jalan, A.; Alecu, I. M.; Aguilera-Iparraguirre, J.; Merchant, S. S.; Yang, K. R.; Merhcant, S. S.; Truhlar, D. G.; Green, W. H. New Pathways for Formation of Acids and Carbonyl Products in Low Temperature Oxidation. *J. Am. Chem. Soc.* **2013**, *135*, 11100–11114.

(4) Tiwary, P.; Parrinello, M. From Metadynamics to Dynamics. *Phys. Rev. Lett.* **2013**, *111* (23), 230602.

(5) Salvalaglio, M.; Tiwary, P.; Parrinello, M. Assessing the Reliability of the Dynamics Reconstructed from Metadynamics. *J. Chem. Theory Comput.* **2014**, *10* (4), 1420–1425.

(6) Fu, C. D.; L. Oliveira, L. F.; Pfaendtner, J. Determining Energy Barriers and Selectivities of a Multi-Pathway System with Infrequent Metadynamics. *J. Chem. Phys.* **2017**, *146* (1), 14108.

(7) Bal, K. M.; Neyts, E. C. Merging Metadynamics into Hyperdynamics: Accelerated Molecular Simulations Reaching Time Scales from Microseconds to Seconds. *J. Chem. Theory Comput.* **2015**, *11* (10), 4545–4554.

(8) Fu, C. D.; Oliveira, L. F. L.; Pfaendtner, J. Assessing Generic Collective Variables for Determining Reaction Rates in Metadynamics Simulations. *J. Chem. Theory Comput.* **2017**, acs.jctc.7b00038.

Reaction Network and Mechanistic Assessment of Acrolein Oxidation on an Industrial Oxide Catalyst

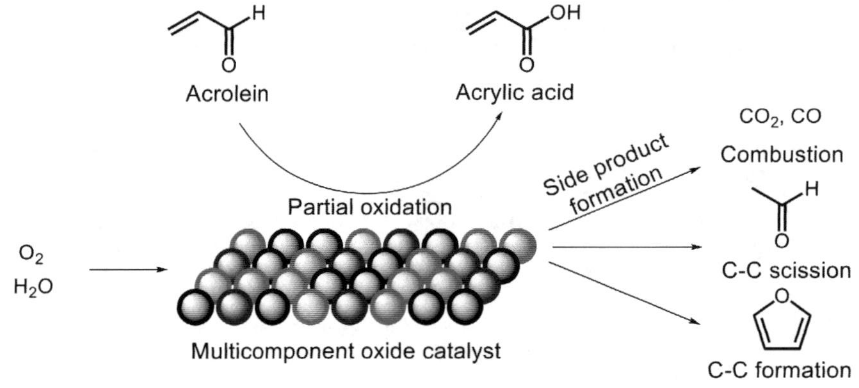

AIChE Annual Meeting 2017, Minneapolis, MN
Reaction Path Analysis II

UNIVERSITY OF MINNESOTA
Driven to Discover

Jacob Miller and Aditya Bhan
30 October 2017

Industrial acrylic acid uses and manufacturing process

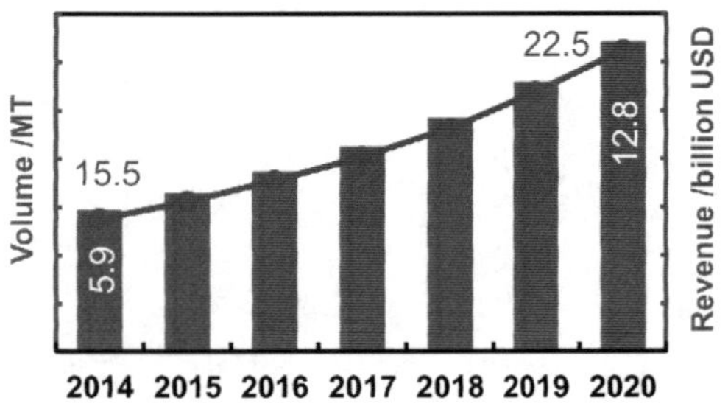

Zion Research Analysis, 2016
Tsuneki et al. (2016) *US 9422377 B2*

2

Industrial acrylic acid uses and manufacturing process

World acrylic acid production

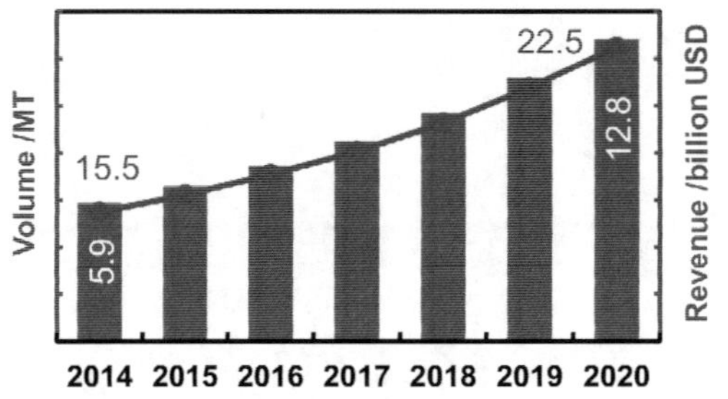

- Superabsorbent polymers
- Esterification-surface coatings

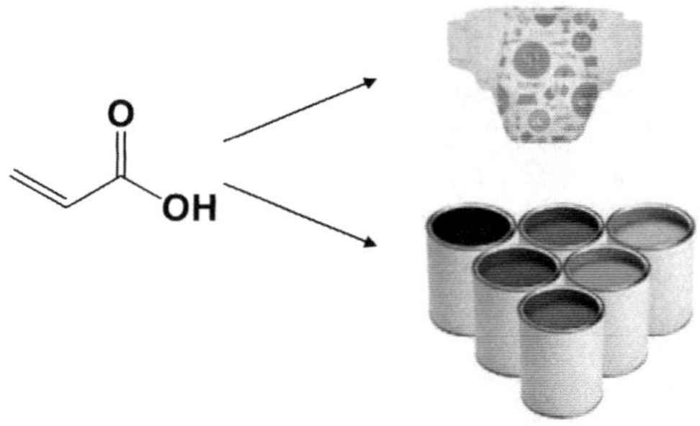

Zion Research Analysis, 2016
Tsuneki et al. (2016) *US 9422377 B2*

Industrial acrylic acid uses and manufacturing process

World acrylic acid production

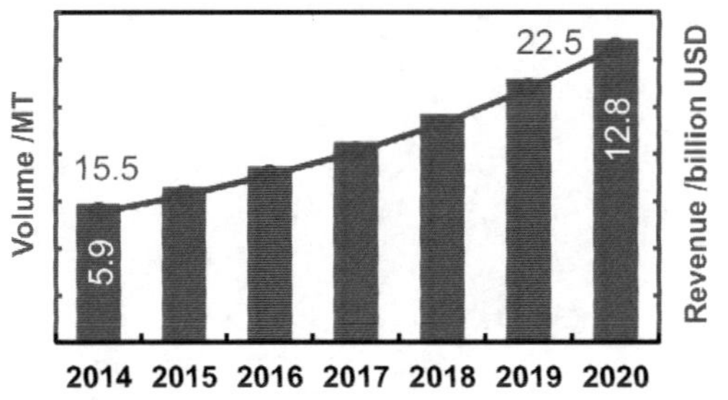

- Superabsorbent polymers
- Esterification-surface coatings

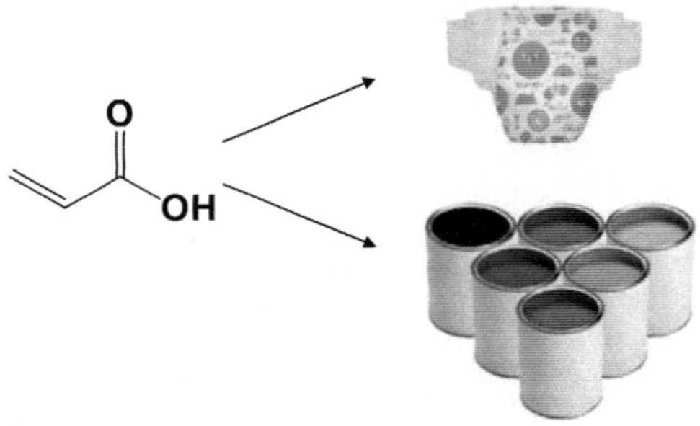

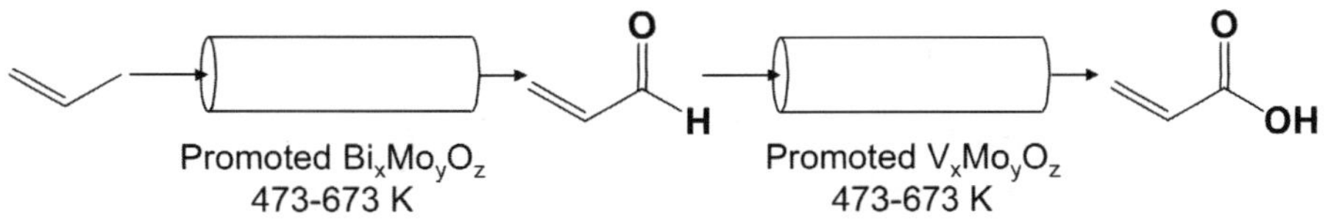

Zion Research Analysis, 2016
Tsuneki et al. (2016) *US 9422377 B2*

Industrial acrylic acid uses and manufacturing process

World acrylic acid production

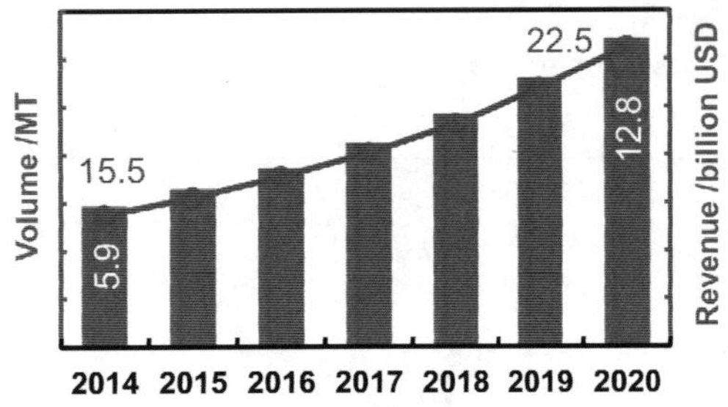

- Superabsorbent polymers
- Esterification-surface coatings

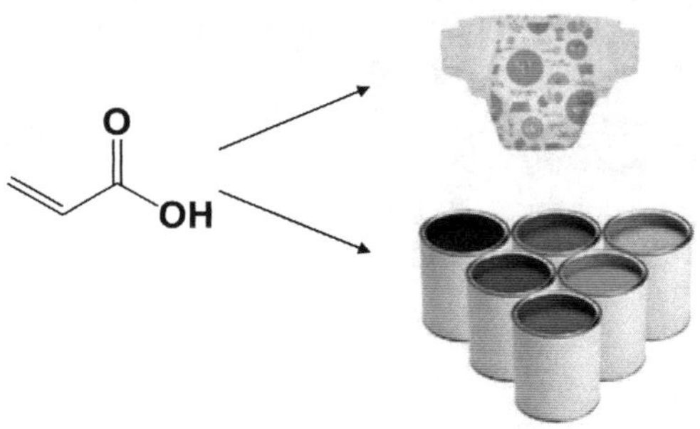

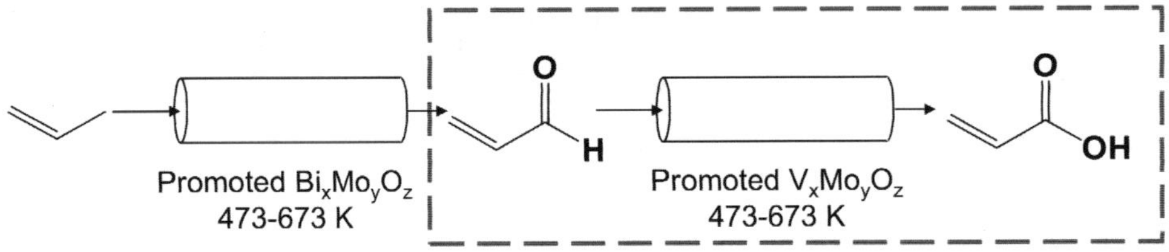

Focus: Acrolein oxidation

Zion Research Analysis, 2016
Tsuneki et al. (2016) *US 9422377 B2*

Acrolein oxidation on industrial catalyst

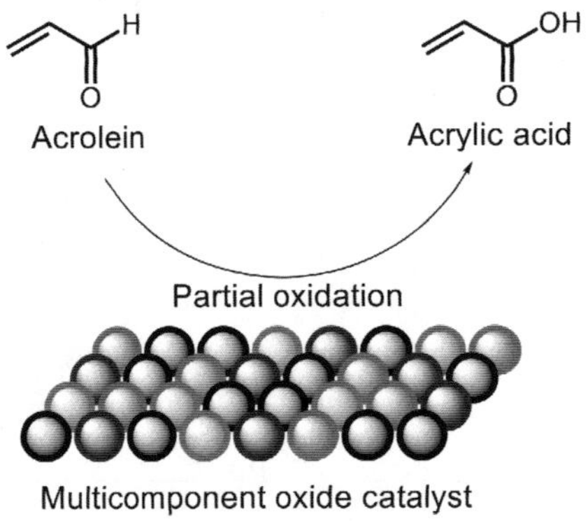

$S_{Acrylic\ acid} > 98\%$

- Multiphase, reducible oxide
- ~10 m^2/g
- $V_xMo_yO_z$; Typical promoters: Cu, W, Sb, Fe, Nb, Bi

Tsuneki et al. (2016) *US 9422377 B2*
Devaux et al. (2013) *US 2013/0324758 A1*
Charendorff et al. (2013) *US 2013/0281737 A1*
Towler and Sinnott, *Chemical Engineering Design - Principles, Practice and Economics of Plant and Process Design*

Acrolein oxidation on industrial catalyst

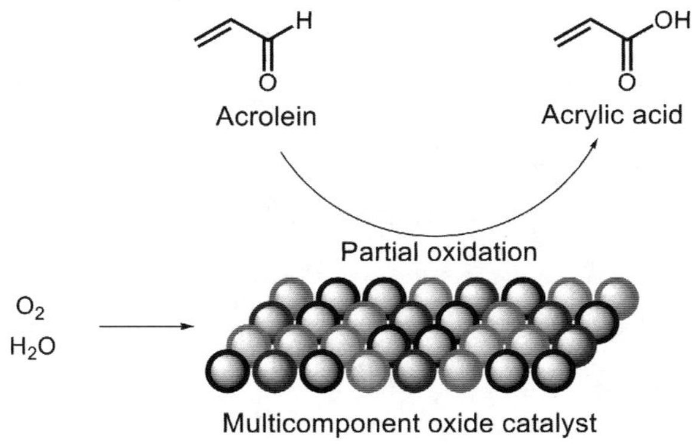

$S_{Acrylic\ acid} > 98\%$

- Multiphase, reducible oxide
- ~10 m^2/g
- $V_xMo_yO_z$; Typical promoters: Cu, W, Sb, Fe, Nb, Bi

Tsuneki et al. (2016) *US 9422377 B2*
Devaux et al. (2013) *US 2013/0324758 A1*
Charendorff et al. (2013) *US 2013/0281737 A1*
Towler and Sinnott, *Chemical Engineering Design - Principles, Practice and Economics of Plant and Process Design* 7

Acrolein oxidation on industrial catalyst

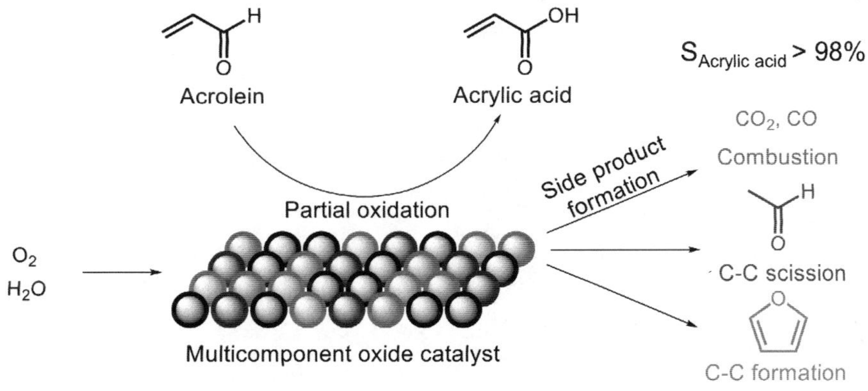

$S_{Acrylic\ acid} > 98\%$

$S_{Side\ products} < 2\%$

- Multiphase, reducible oxide
- ~10 m^2/g
- $V_xMo_yO_z$; Typical promoters: Cu, W, Sb, Fe, Nb, Bi

Tsuneki et al. (2016) *US 9422377 B2*
Devaux et al. (2013) *US 2013/0324758 A1*
Charendorff et al. (2013) *US 2013/0281737 A1*
Towler and Sinnott, *Chemical Engineering Design - Principles, Practice and Economics of Plant and Process Design* 8

Acrolein oxidation on industrial catalyst

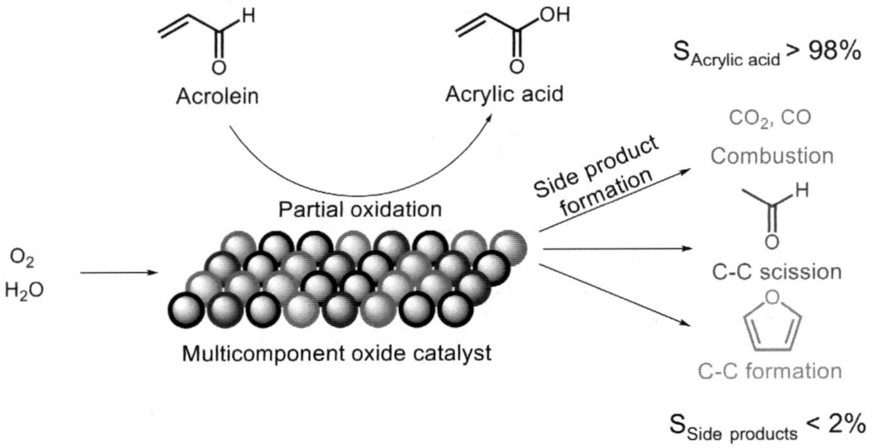

Goal: Identify mechanistic pathways to side product evolution

Tsuneki et al. (2016) *US 9422377 B2*
Devaux et al. (2013) *US 2013/0324758 A1*
Charendorff et al. (2013) *US 2013/0281737 A1*
Towler and Sinnott, *Chemical Engineering Design - Principles, Practice and Economics of Plant and Process Design*

Acrolein oxidation on industrial catalyst

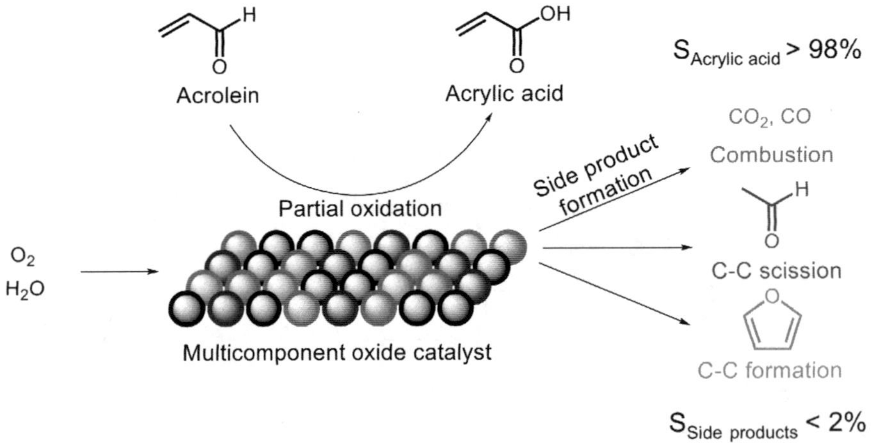

Goal: Identify mechanistic pathways to side product evolution

Obstacle: Exothermic reaction ($\Delta H°= -266$ kJ/mol)

Tsuneki et al. (2016) *US 9422377 B2*
Devaux et al. (2013) *US 2013/0324758 A1*
Charendorff et al. (2013) *US 2013/0281737 A1*
Towler and Sinnott, *Chemical Engineering Design - Principles, Practice and Economics of Plant and Process Design*

Gradientless recirculating batch reactor unit

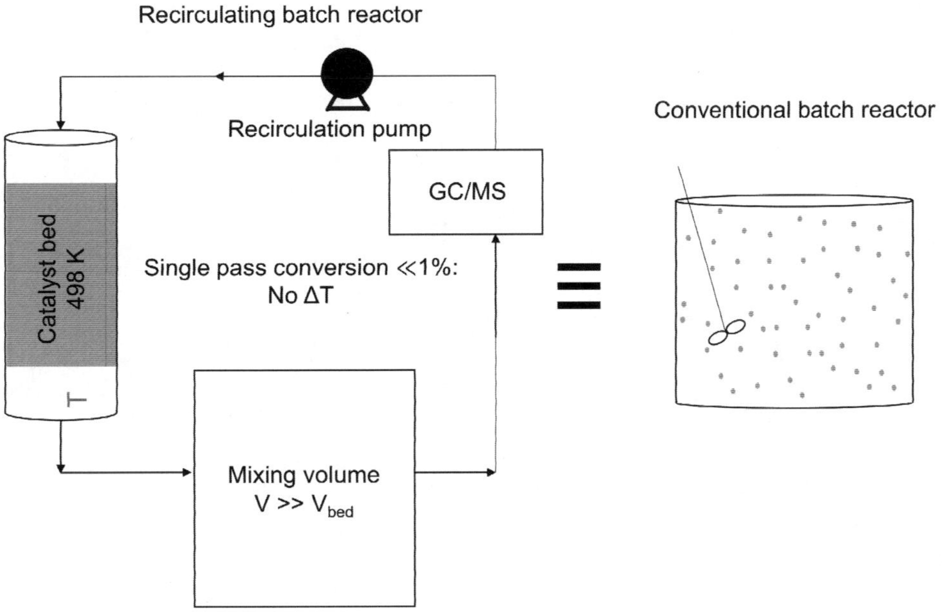

11

Acrolein oxidation product selectivities

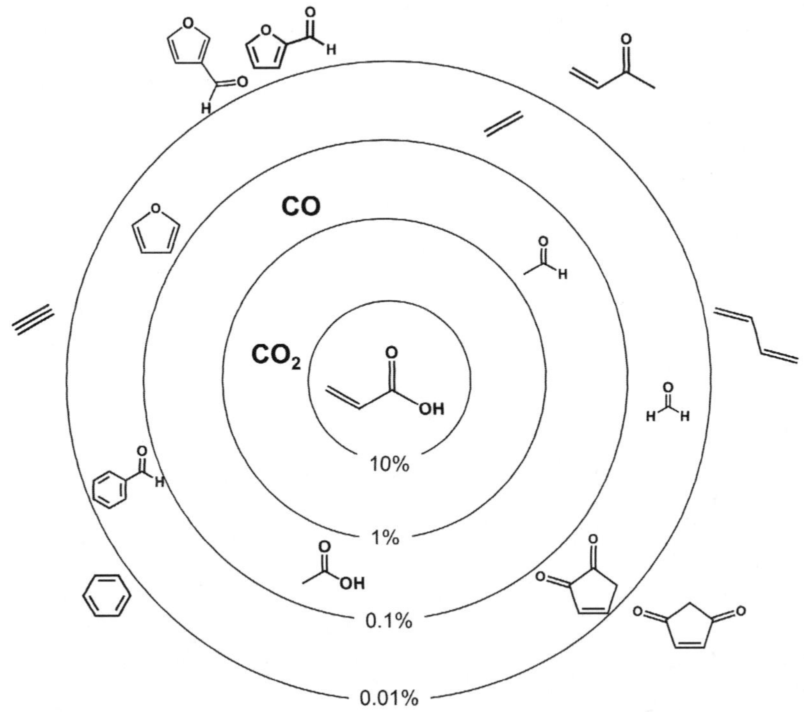

4.2 mg mixed-metal oxide catalyst, T=498 K, 2.0 kPa acrolein, 3.7 kPa O_2, 4.5 kPa water, $X_{Acrolein}$=0.2

12

Acrolein oxidation product selectivities

- Combustion (C_1)

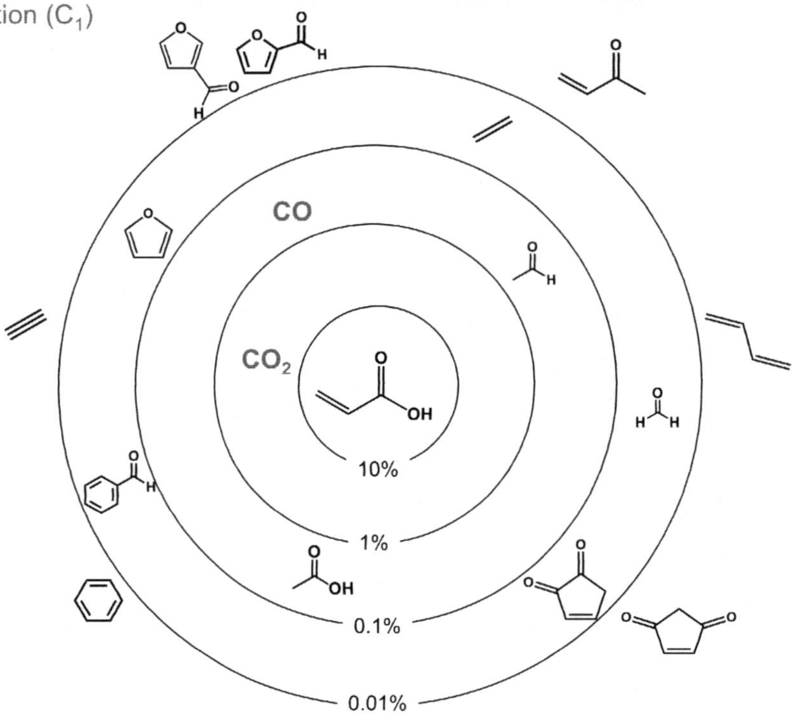

4.2 mg mixed-metal oxide catalyst, T=498 K, 2.0 kPa acrolein, 3.7 kPa O_2, 4.5 kPa water, $X_{Acrolein}$=0.2

13

Acrolein oxidation product selectivities

- Combustion (C_1)
- C-C scission ($C_{1\text{-}2}$)

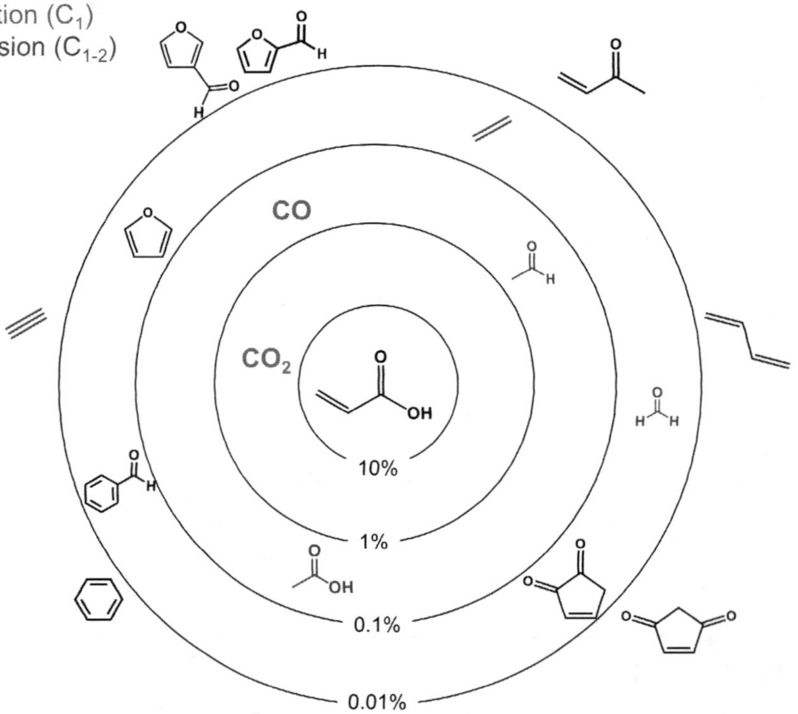

4.2 mg mixed-metal oxide catalyst, T=498 K, 2.0 kPa acrolein, 3.7 kPa O_2, 4.5 kPa water, $X_{Acrolein}$=0.2

14

Acrolein oxidation product selectivities

- Combustion (C_1)
- C-C scission ($C_{1\text{-}2}$)
- C-C formation (C_{4+})

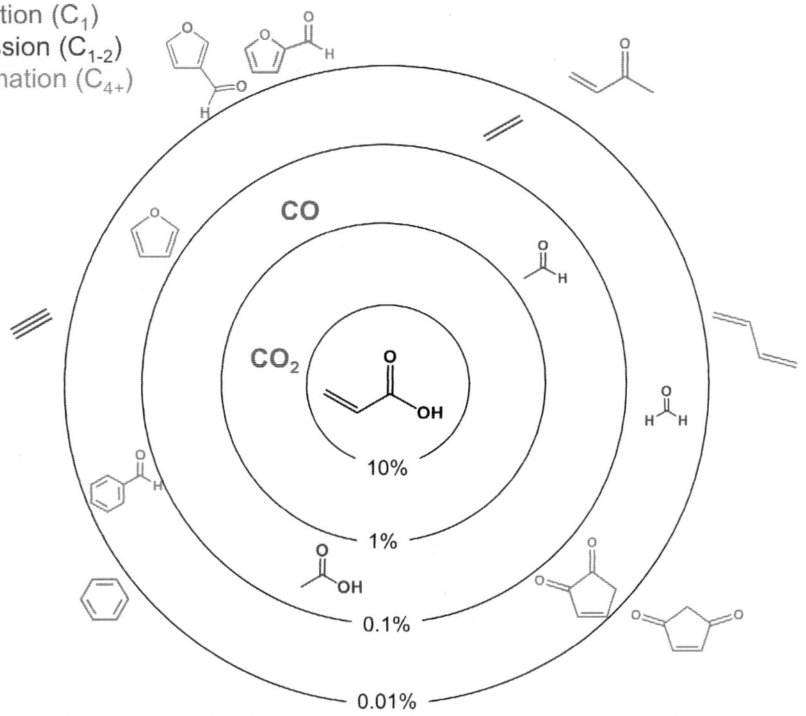

4.2 mg mixed-metal oxide catalyst, T=498 K, 2.0 kPa acrolein, 3.7 kPa O_2, 4.5 kPa water, $X_{Acrolein}$=0.2

15

O_2 and H_2O co-feeds reveal role of O_L and OH surface species

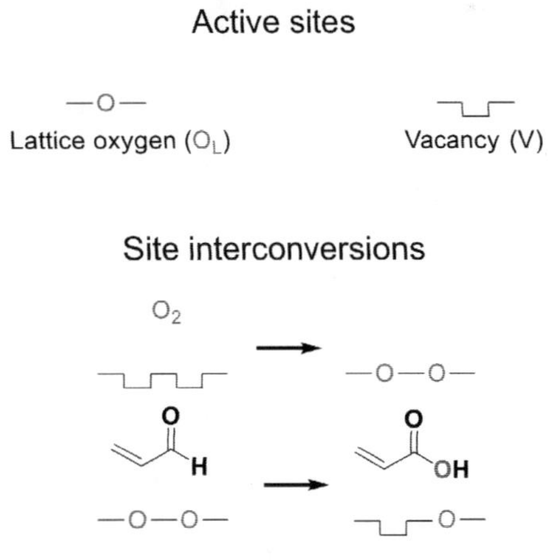

Mars and van Krevelen. *Chem. Eng. Sci.* 1954, *3*: 41–59
Drochner et al. *Top. Catal.* 2016, *59*. 1518-1532
Petzold et al. *ChemCatChem.* 2014, *6*: 2053-2058

16

O_2 and H_2O pressure variations reveal role of O_L and OH surface species

Active sites

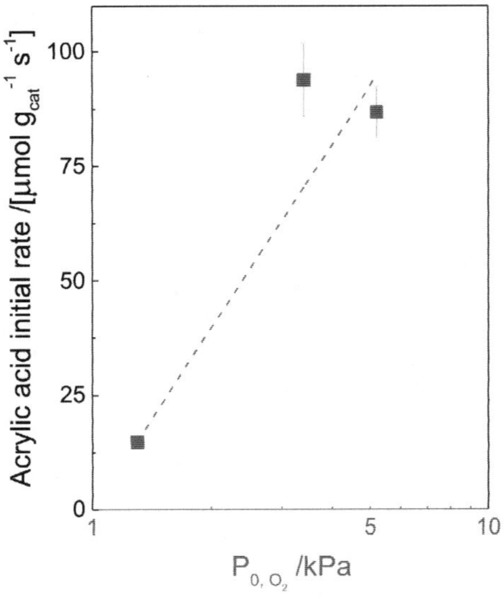

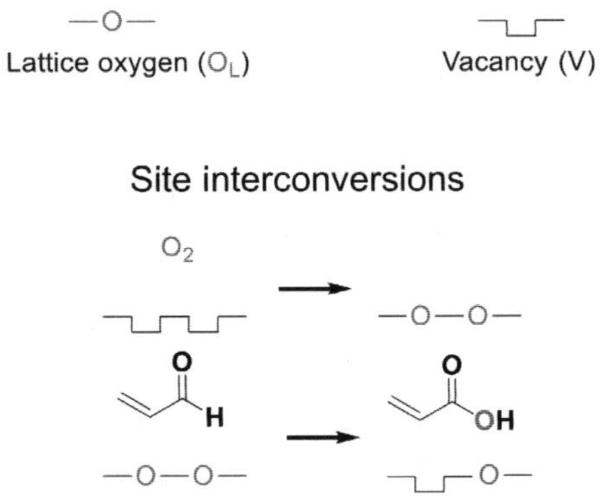

4.2 mg mixed-metal oxide catalyst, T=498 K; $P_{Acrolein}$=1.4 kPa, P_{H2O}=3.9 kPa

Mars and van Krevelen. *Chem. Eng. Sci.* 1954, *3*: 41–59
Drochner et al. *Top. Catal.* 2016, *59*. 1518-1532
Petzold et al. *ChemCatChem.* 2014, *6*: 2053-2058

17

O_2 and H_2O pressure variations reveal role of O_L and OH surface species

Active sites

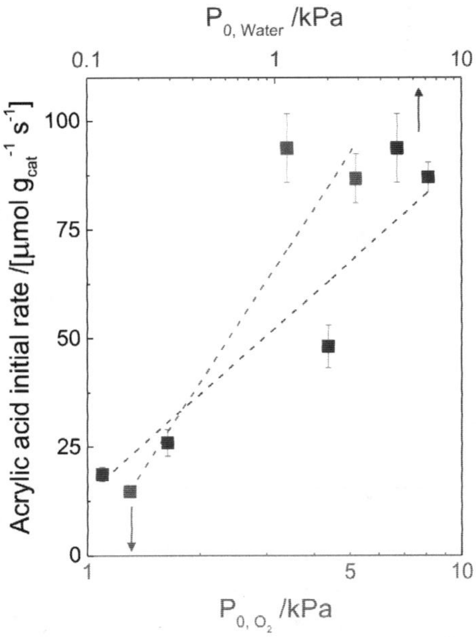

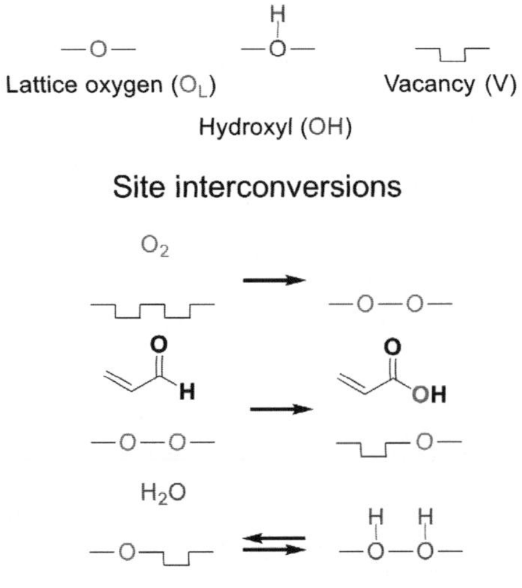

4.2 mg mixed-metal oxide catalyst, T=498 K; $P_{Acrolein}$=1.4 kPa, P_{H2O}=3.9 kPa
4.2 mg mixed-metal oxide catalyst, T=498 K; P_{ACR}=1.5 kPa, P_{O2}=3.3 kPa

Mars and van Krevelen. *Chem. Eng. Sci.* 1954, *3*: 41–59
Drochner et al. *Top. Catal.* 2016, *59*. 1518-1532
Petzold et al. *ChemCatChem.* 2014, *6*: 2053-2058

18

Surface reaction mechanisms

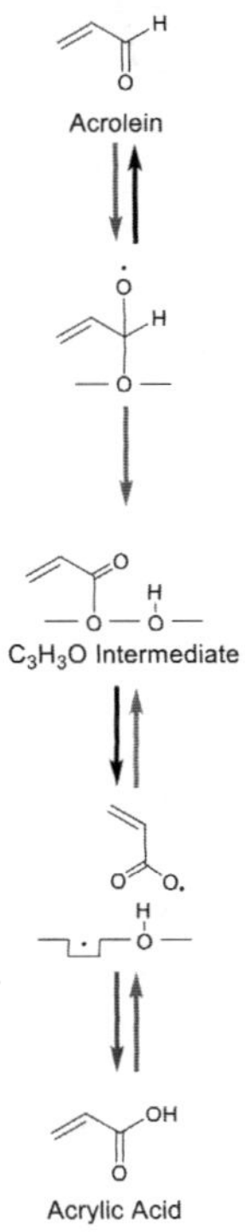

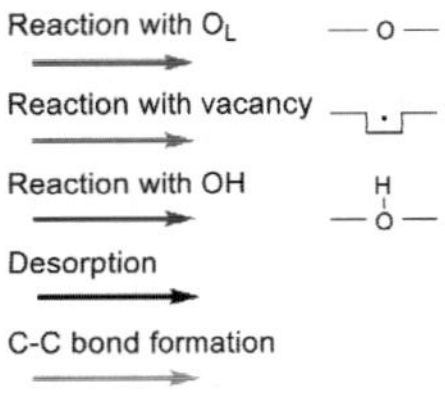

19

Trends in product selectivities guide hypotheses of pathways

Batch reactor feed:

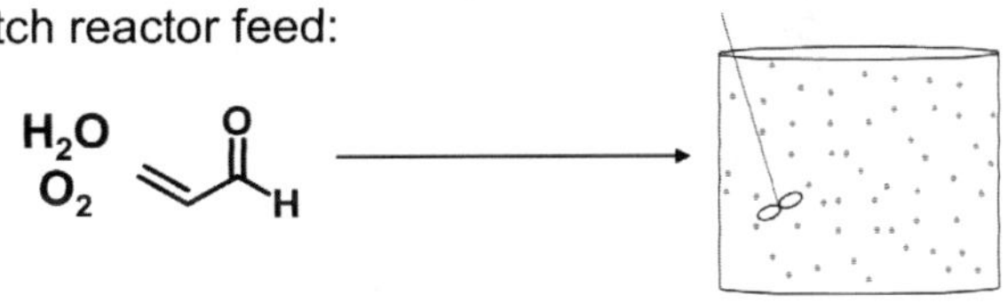

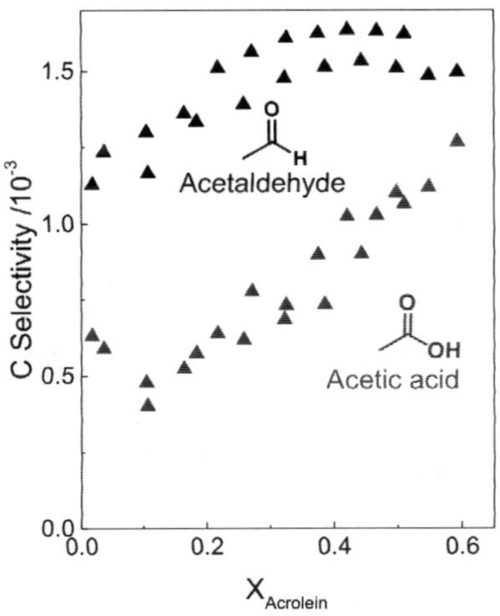

4.2 mg mixed-metal oxide catalyst, T=498 K, 1.7-1.9 kPa acrolein, 3.2-3.7 kPa O_2, 3.7-4.6 kPa water

20

Trends in product selectivities guide hypotheses of pathways

Batch reactor feed:

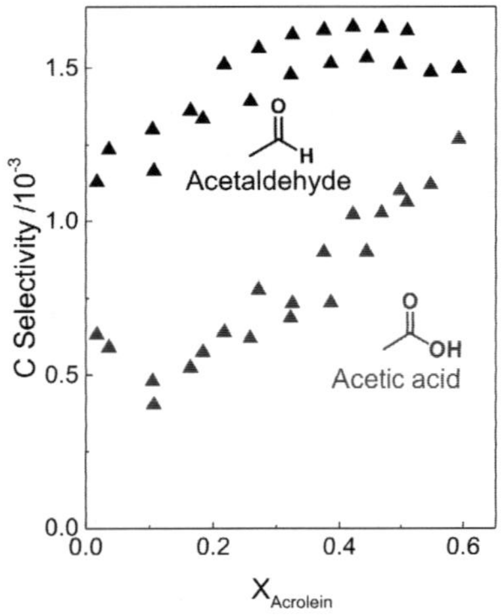

Hypothesize pathway

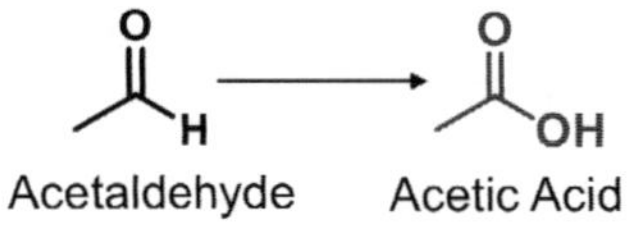

4.2 mg mixed-metal oxide catalyst, T=498 K, 1.7-1.9 kPa acrolein, 3.2-3.7 kPa O_2, 3.7-4.6 kPa water

21

Trends in product selectivities guide hypotheses of pathways

Batch reactor feed+co-feed:

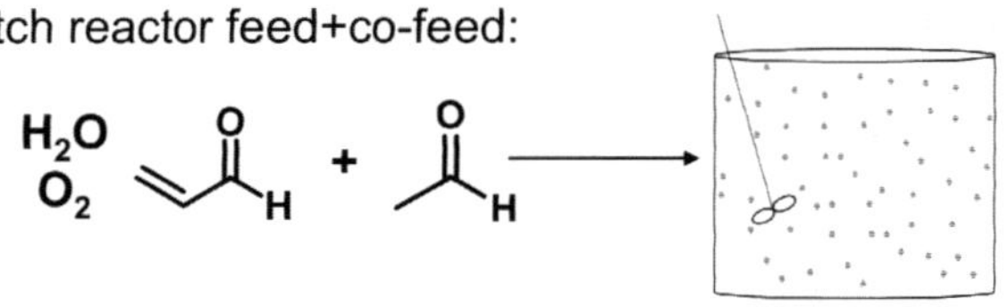

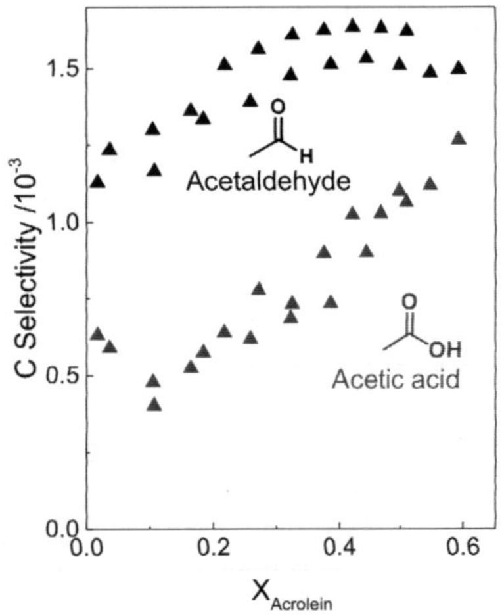

Hypothesize pathway

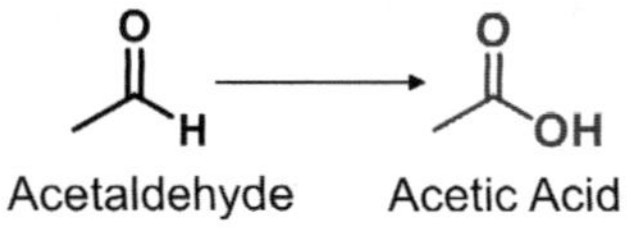

4.2 mg mixed-metal oxide catalyst, T=498 K, 1.7-1.9 kPa acrolein, 3.2-3.7 kPa O_2, 3.7-4.6 kPa water

22

Trends in product selectivities guide hypotheses of pathways

Batch reactor feed+co-feed:

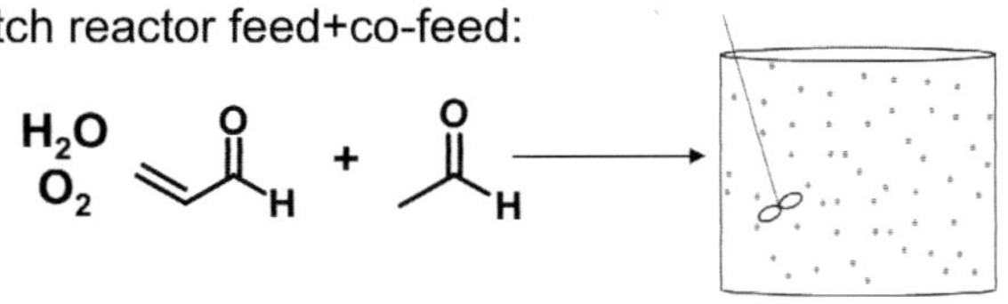

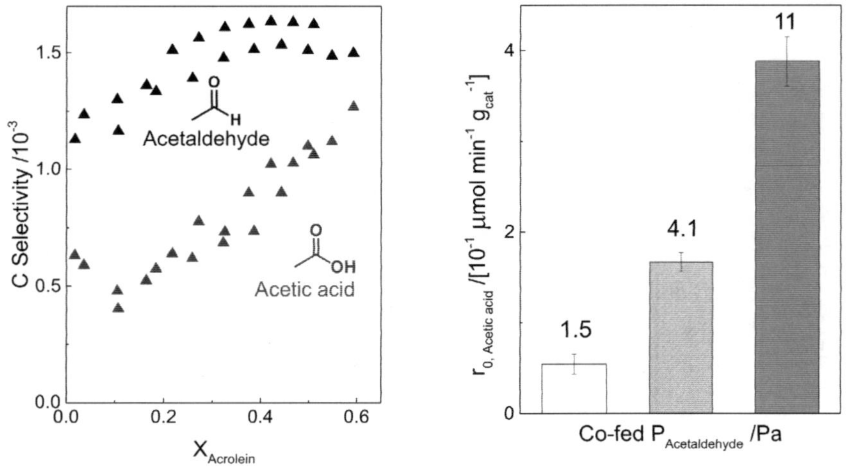

4.2 mg mixed-metal oxide catalyst, T=498 K, 1.7-1.9 kPa acrolein, 3.2-3.7 kPa O_2, 3.7-4.6 kPa water 23

Trends in product selectivities guide hypotheses of pathways

Batch reactor feed+co-feed:

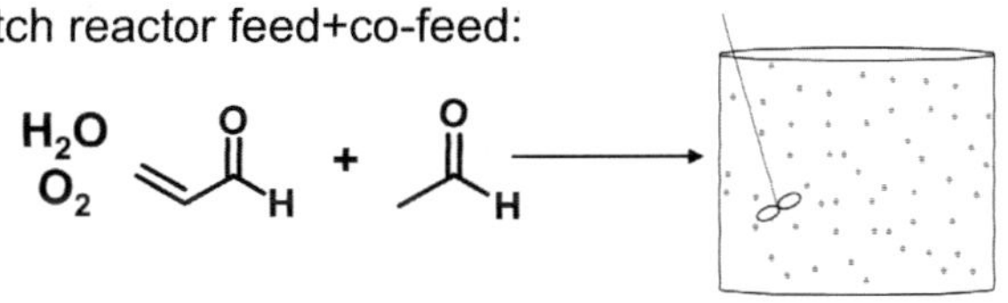

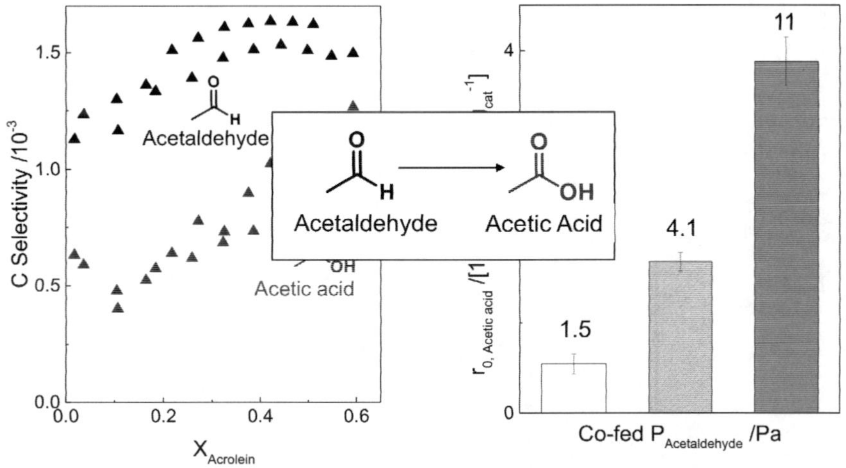

4.2 mg mixed-metal oxide catalyst, T=498 K, 1.7-1.9 kPa acrolein, 3.2-3.7 kPa O_2, 3.7-4.6 kPa water 24

Surface reaction mechanisms

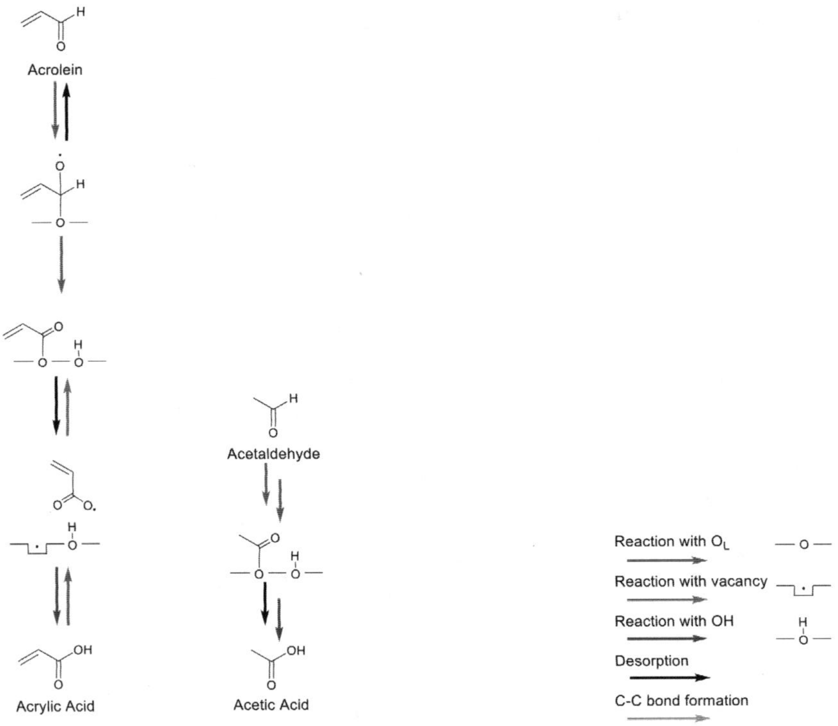

25

Acrylic acid co-feed experiment reveals participation in many pathways

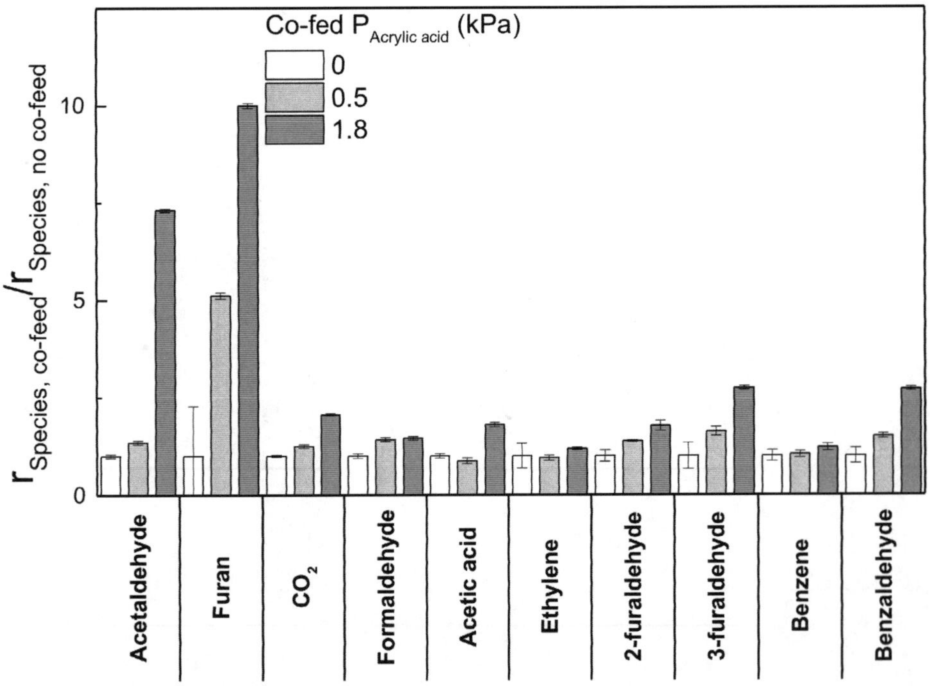

2.4 mg mixed-metal oxide catalyst, T=498 K, 1.7-1.9 kPa acrolein, 3.2-3.7 kPa O_2, 3.7-5.1 kPa water

26

Acrylic acid co-feed experiment reveals participation in many pathways

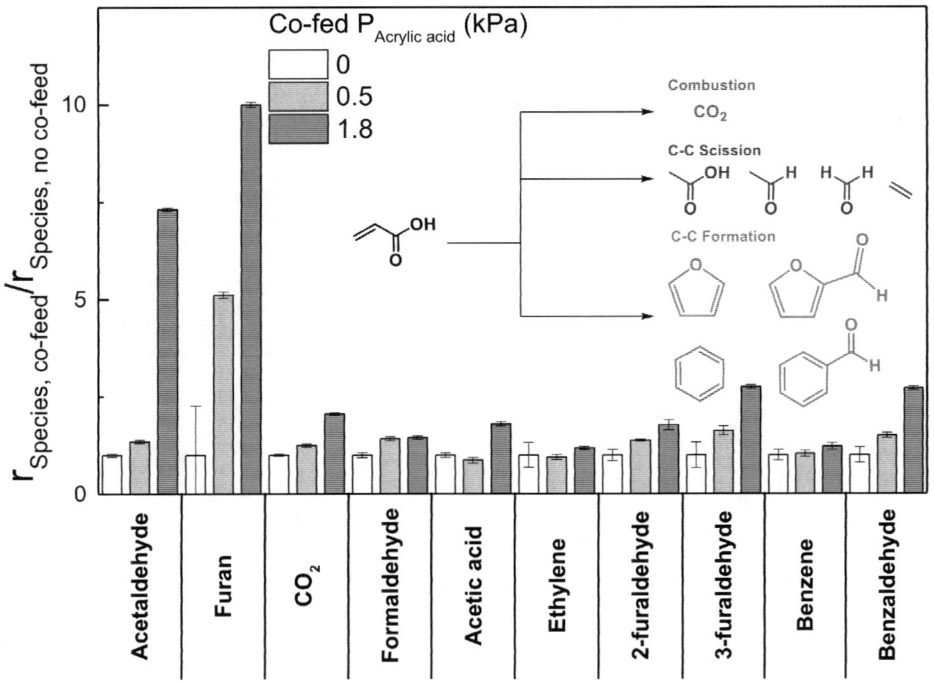

2.4 mg mixed-metal oxide catalyst, T=498 K, 1.7-1.9 kPa acrolein, 3.2-3.7 kPa O_2, 3.7-5.1 kPa water 27

Surface reaction mechanisms

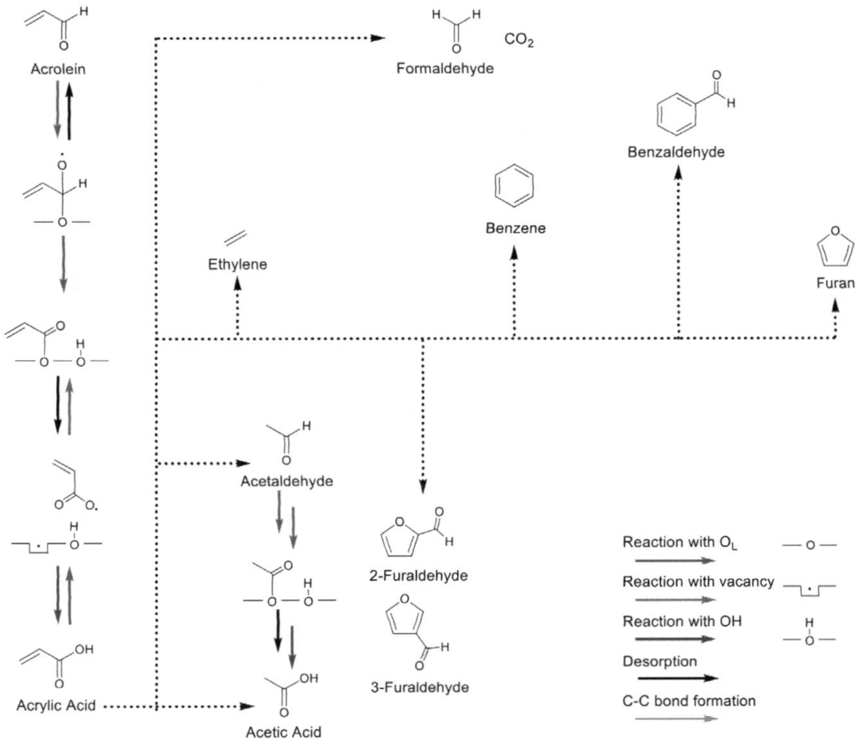

28

Isotopic labeling co-feeds point to existence of C_2 surface species

Ex: Furan

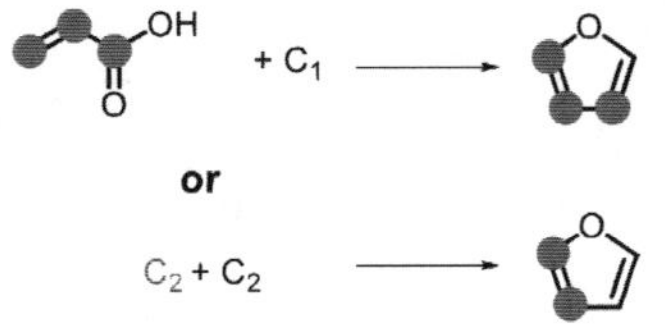

● = From acrylic acid

29

Isotopic labeling co-feeds point to existence of C_2 surface species

Ex: Furan

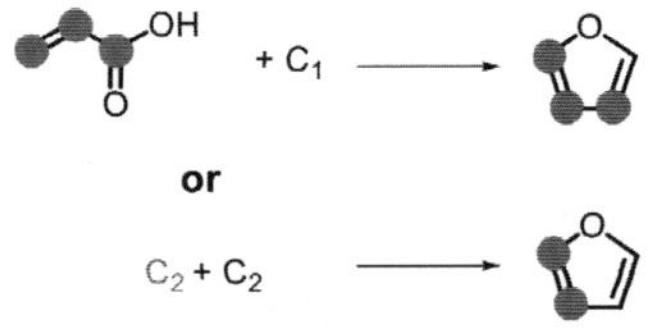

● = ^{13}C label

Isotopic Co-feed:

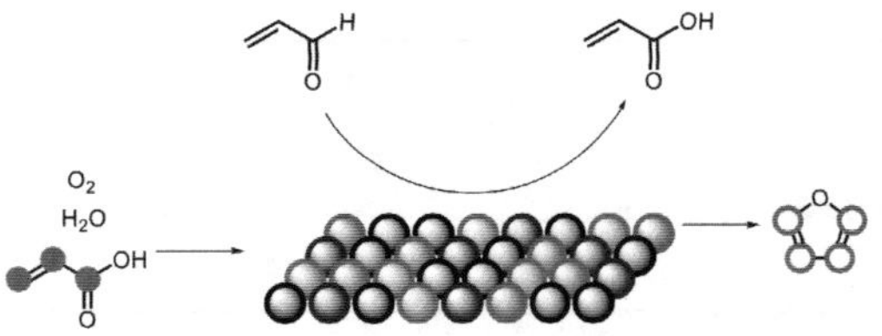

Price and Iglesia. *Ind. Eng. Chem. Res.* 1989, *28*: 839-844

30

Isotopic labeling co-feeds point to existence of C_2 surface species

Ex: Furan

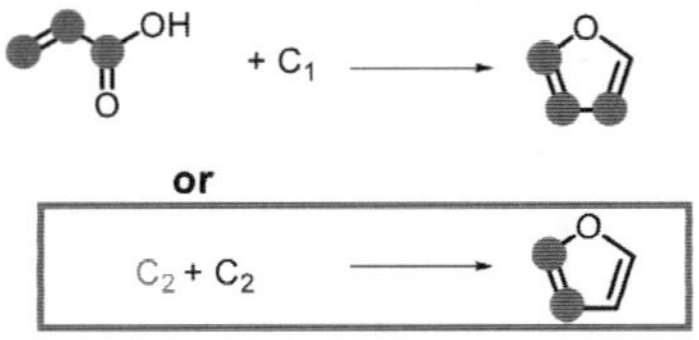

● = ^{13}C label

Acrylic acid→C_2 surface species

Isotopic Co-feed:

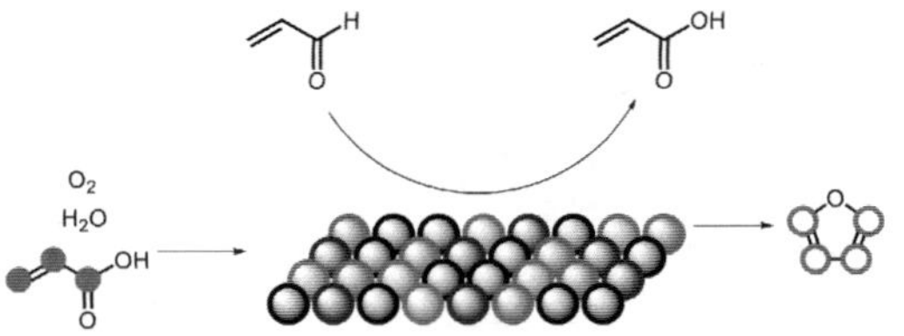

Isotopic labeling of furan

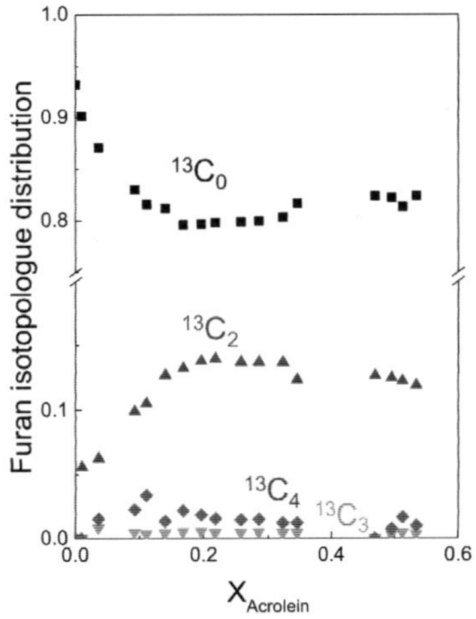

Note: Corrected for naturally occurring $^{13}C_1$

Initial conditions: $P_{Acrolein}$=2.2 kPa, $P_{Acrylic\ acid}$=0.1 kPa, P_{O2}=3.3 kPa, P_{H2O}=5.1 kPa; 4.1 mg mixed-metal oxide catalyst, 498K 31

Isotopic labeling co-feeds point to existence of C_2 surface species

Ex: Furan

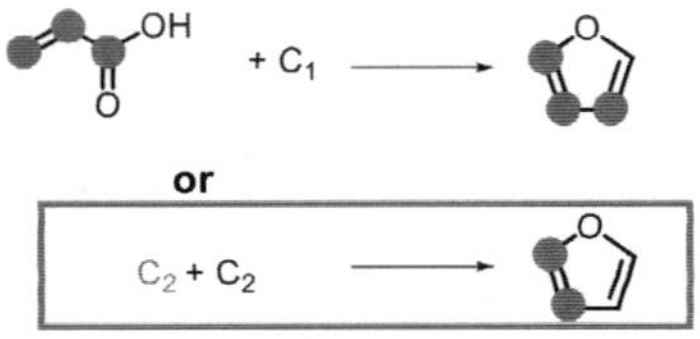

● = ^{13}C label

Acrylic acid→C_2 surface species

Isotopic Co-feed:

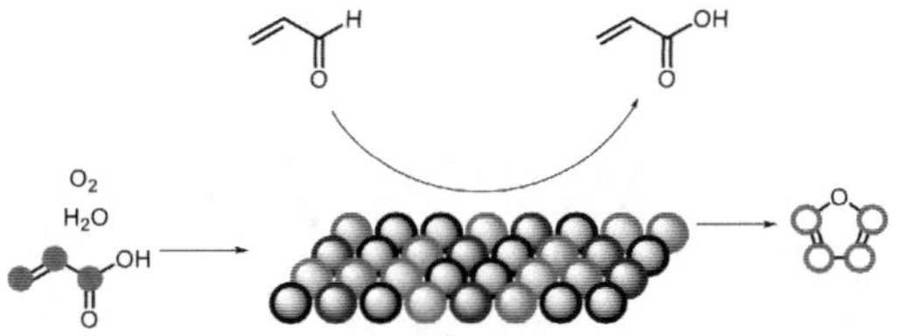

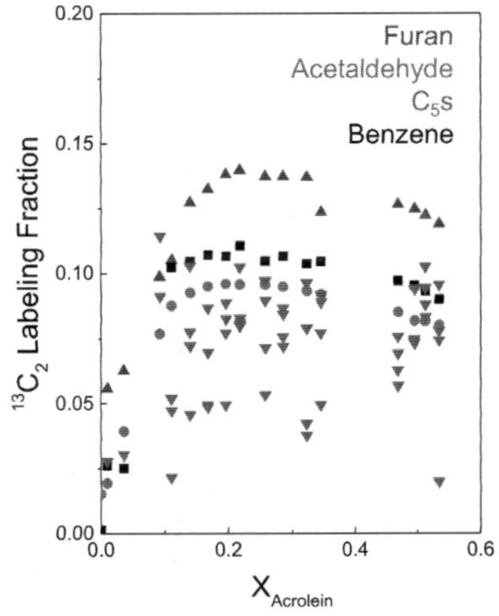

Initial conditions: $P_{Acrolein}$=2.2 kPa, $P_{Acrylic\ acid}$=0.1 kPa, P_{O2}=3.3 kPa, P_{H2O}=5.1 kPa; 4.1 mg mixed-metal oxide catalyst, 498K 32

Surface reaction mechanisms

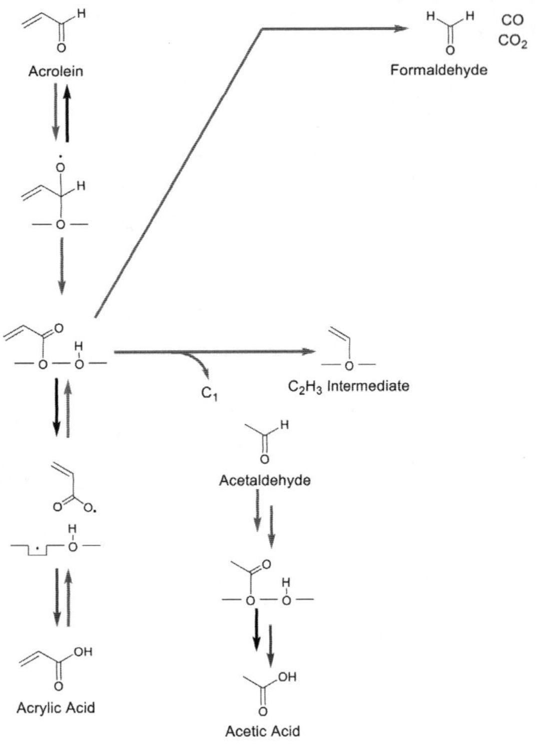

Bui et al. *ACS Catal.* 2016, *6* (10), 6567-6580
Bui and Bhan *Applied Cat. A: Gen.* 2017, *546:* 87-95

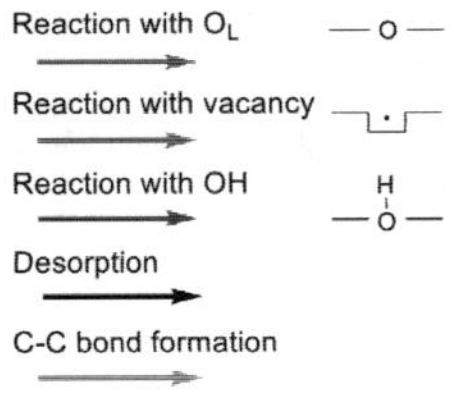

33

Surface reaction mechanisms

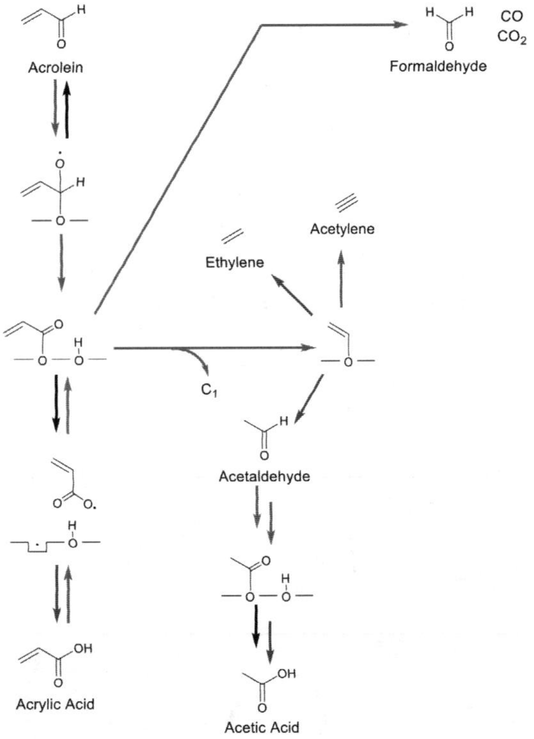

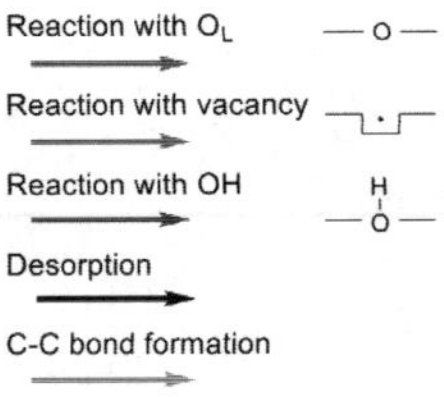

34

Surface reaction mechanisms

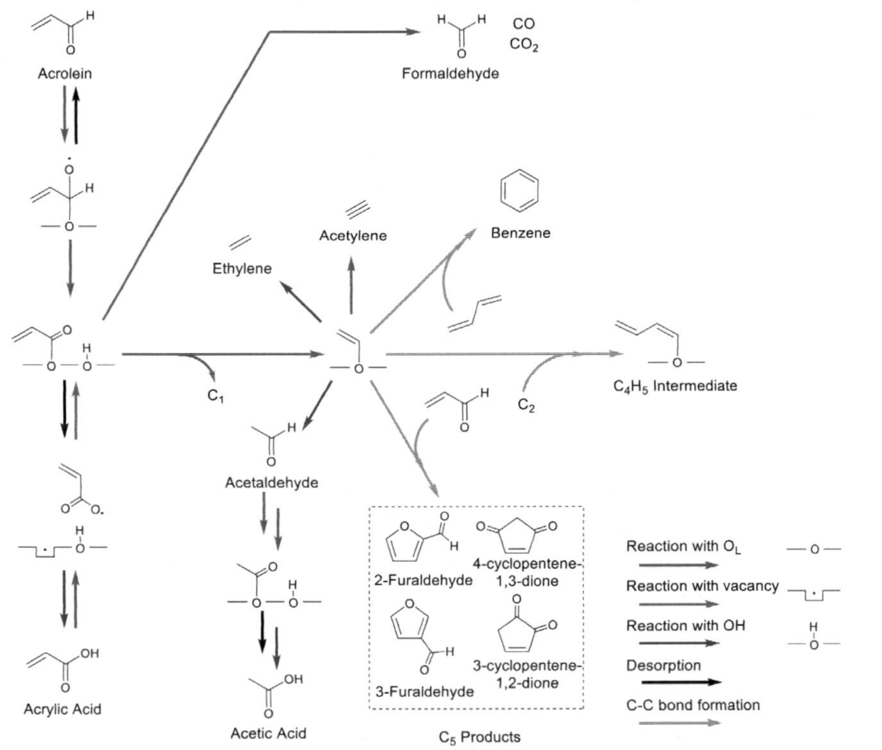

35

Surface reaction mechanisms

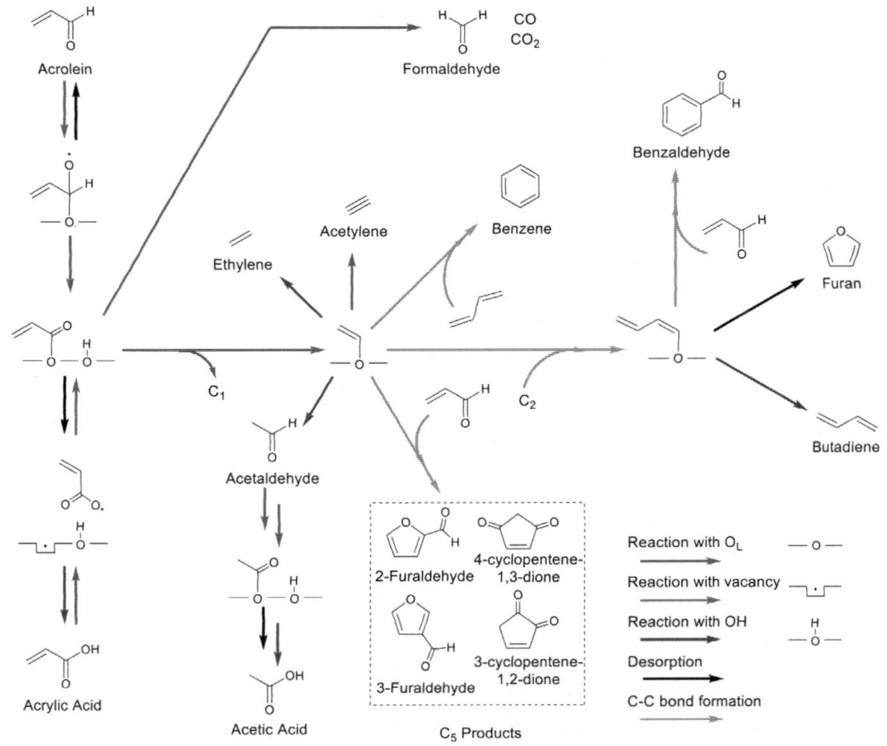

36

Kinetic model development

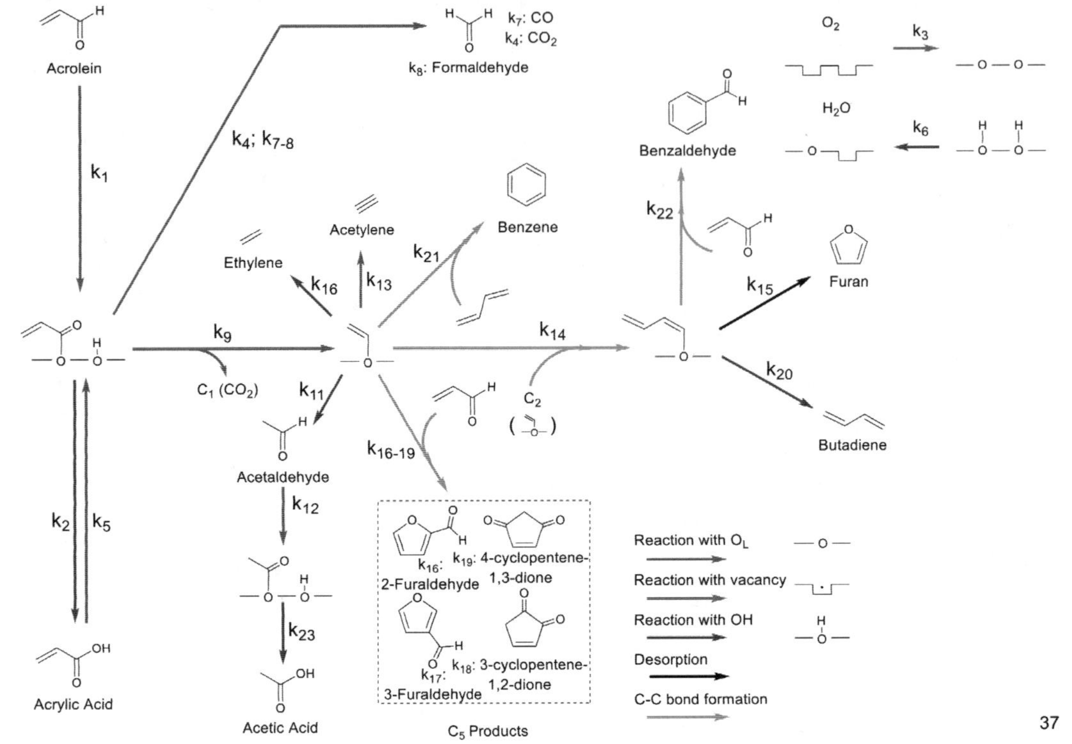

37

Kinetic model development

Gas-phase species balance equations…

$$\frac{1}{m_{cat}}\frac{d}{dt}n_{Acrolein} = -k_1P_{Acrolein}[O_L]^2-(k_{16}+k_{17}+k_{18}+k_{19})P_{Acrolein}[C_2H_3*]-k_{22}P_{Acrolein}[C_4H_5*]$$

$$\frac{1}{m_{cat}}\frac{d}{dt}n_{AcrylicAcid} = k_2[C_3H_3O*][OH] - k_5P_{AcrylicAcid}[OL]$$

$$\frac{1}{m_{cat}}\frac{d}{dt}n_{O_2} = -k_3P_{O_2}[V]$$

$$\frac{1}{m_{cat}}\frac{d}{dt}n_{CO_2} = 3k_4[C_3H_3O*][O_L] + 2k_8[C_3H_3O*][O_L] + k_9[C_3H_3O*][O_L]$$

…and surface species balance equations

$$\begin{aligned} 0 = \tfrac{1}{m_{cat}}\tfrac{d}{dt}[O_L] &= -2k_1P_{Acrolein}[O_L]^2 + k_2[C_3H_3O*][OH] + k_3P_{O_2}[V] \\ &-6k_4[C_3H_3O*][O_L] - k_5P_{AcrylicAcid}[O_L] + k_6[OH]^2 - 3k_7[C_3H_3O*][O_L] - 4k_8[C_3H_3O*][O_L] \\ &-k_9[C_3H_3O*][O_L] + 2k_{10}[C_2H_3*][OH] + k_{11}[C_2H_3*][OH] - 2k_{12}P_{CH_3CHO}[O_L]^2 - k_{15}[C_4H_5*][O_L] \\ &- 3(k_{16} + k17 + k_{18} + k_{19})P_{Acrolein}[C_2H_3*] + 2k_{20}[C_4H_5*][OH] \\ &- 2k_{22}P_{Acrolein}[C_4H_5*] + k_{23}[C_2H_3O*][OH] \end{aligned}$$

38

Kinetic model development

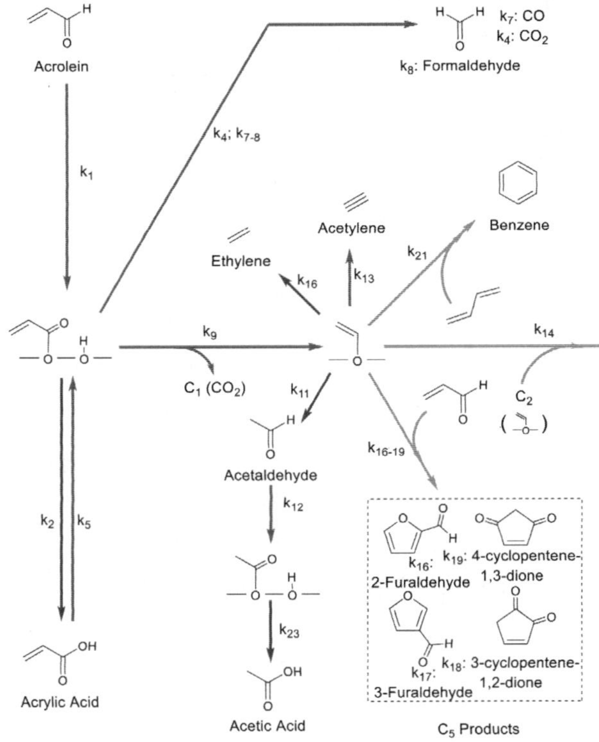

Model specifications:

- 19 gas phase species, 7 surface intermediates
- 11 batch runs; >3000 total data points
- 23 parameters

Modeling results

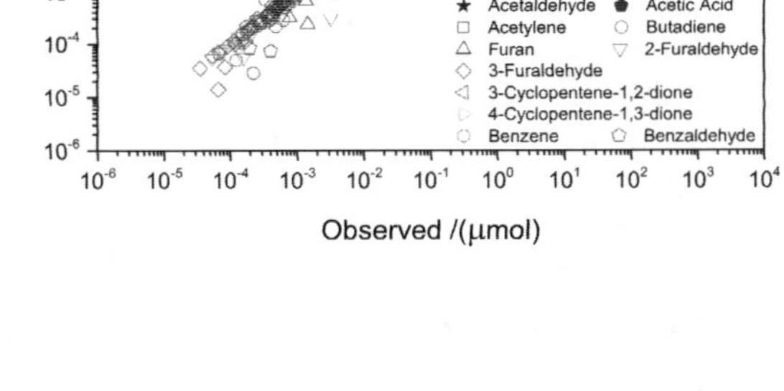

Modeling results

Surface species coverage

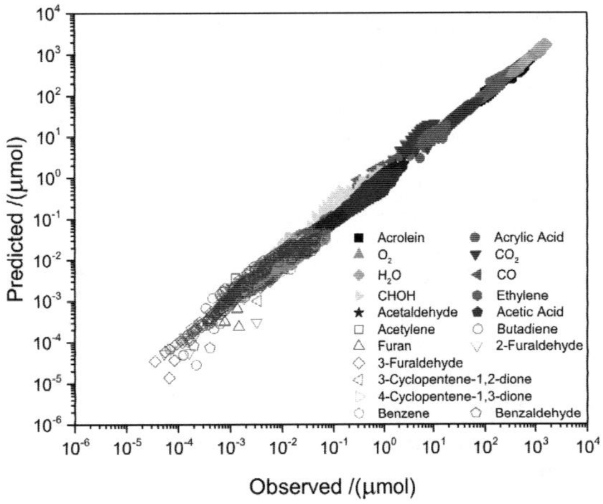

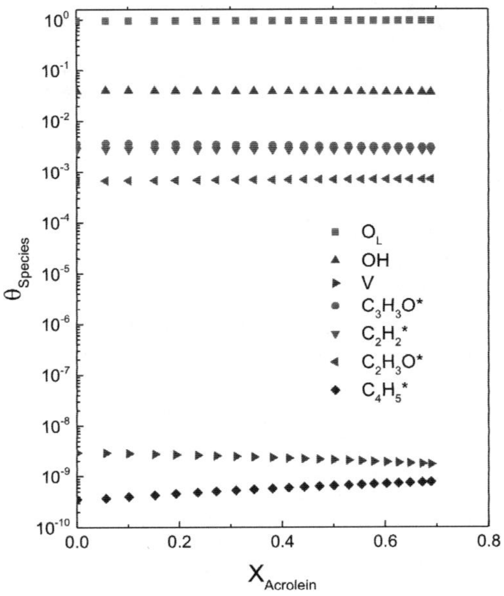

4.3 mg mixed-metal oxide catalyst, T=498 K; $P_{Acrolein}$=2.2 kPa, P_{H2O}=4.2 kPa, P_{O2}=3.4 kPa

41

Conclusion

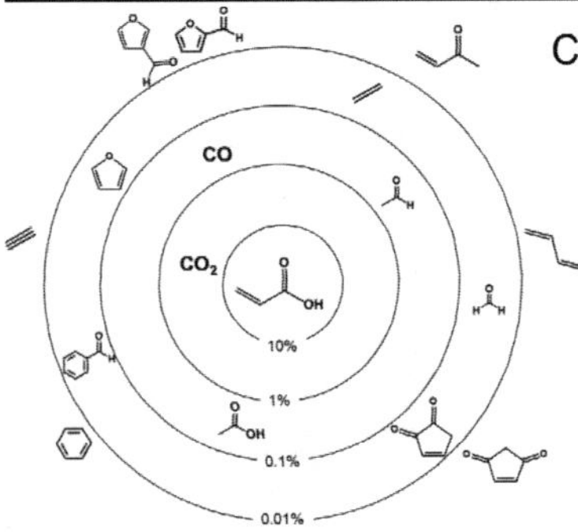

Complex chemistry

42

Conclusion

Complex chemistry

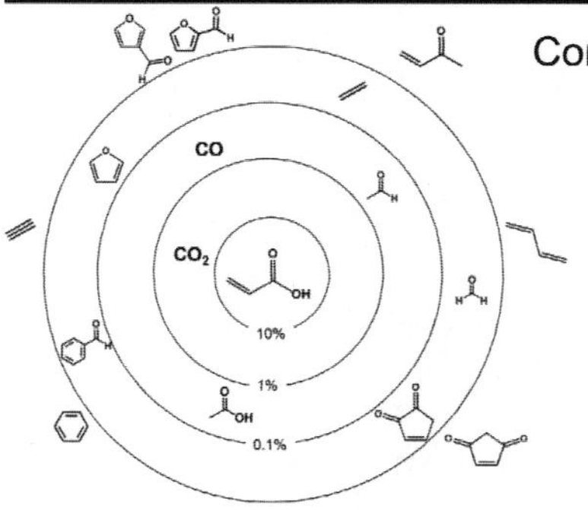

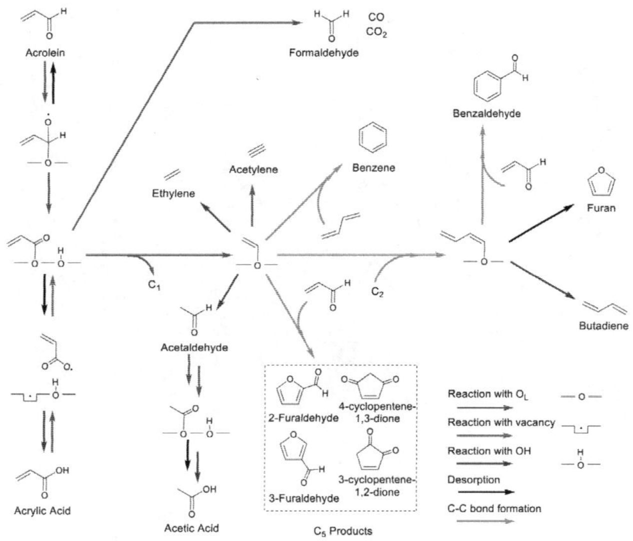

Mechanistic scheme

Conclusion

Complex chemistry

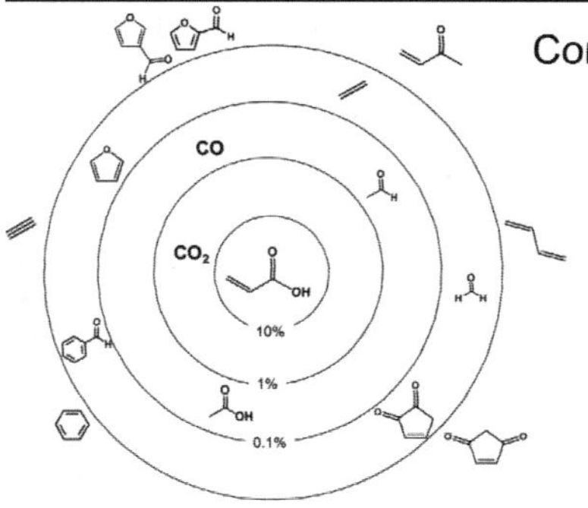

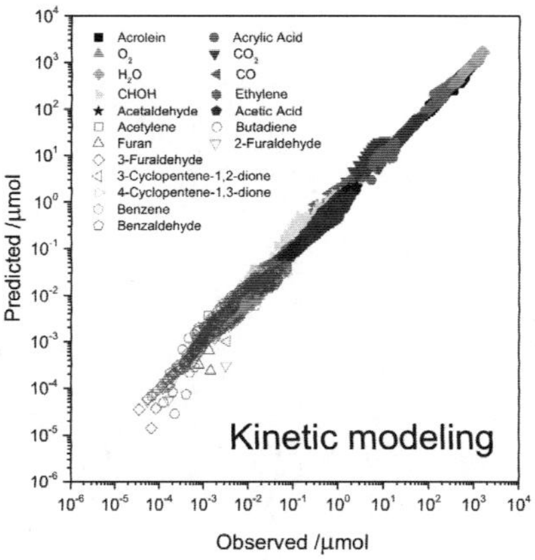

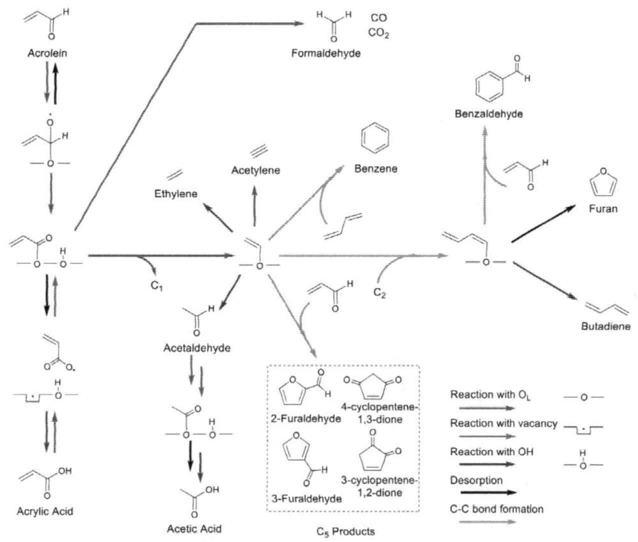

Mechanistic scheme

Acknowledgements

UNIVERSITY OF MINNESOTA
Driven to Discover℠

DOW ®

® Trademark of the Dow Chemical Company

45

Fits of data: C_3s, O_2, C_1s

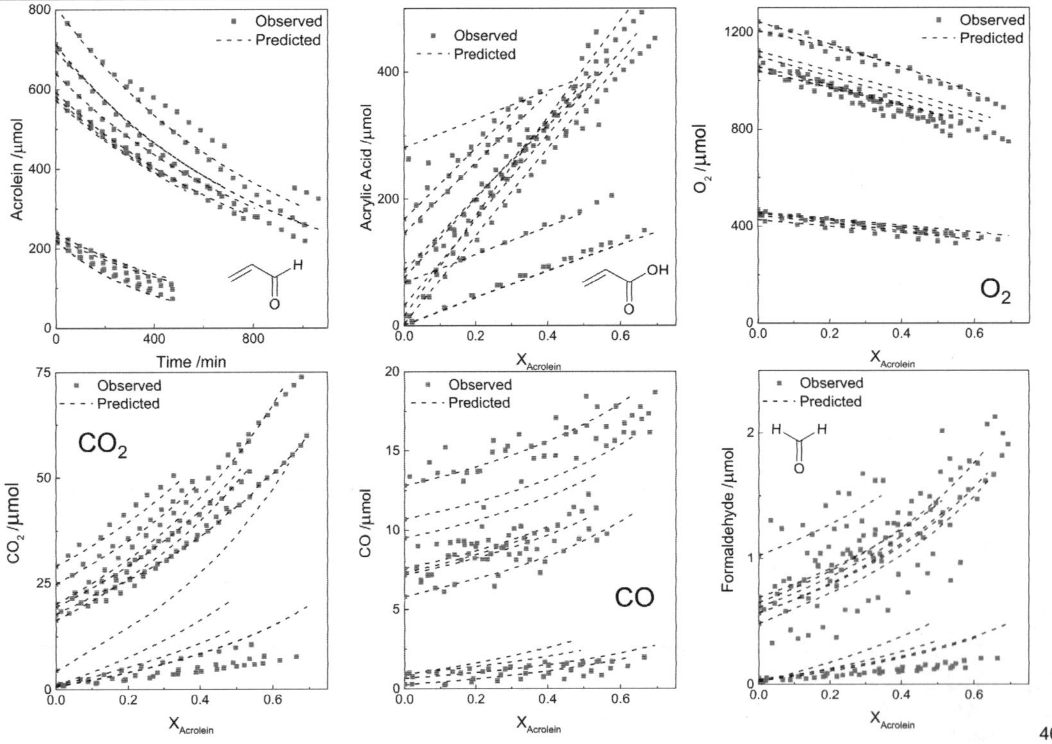

46

Fits of data: C_2s, C_4s

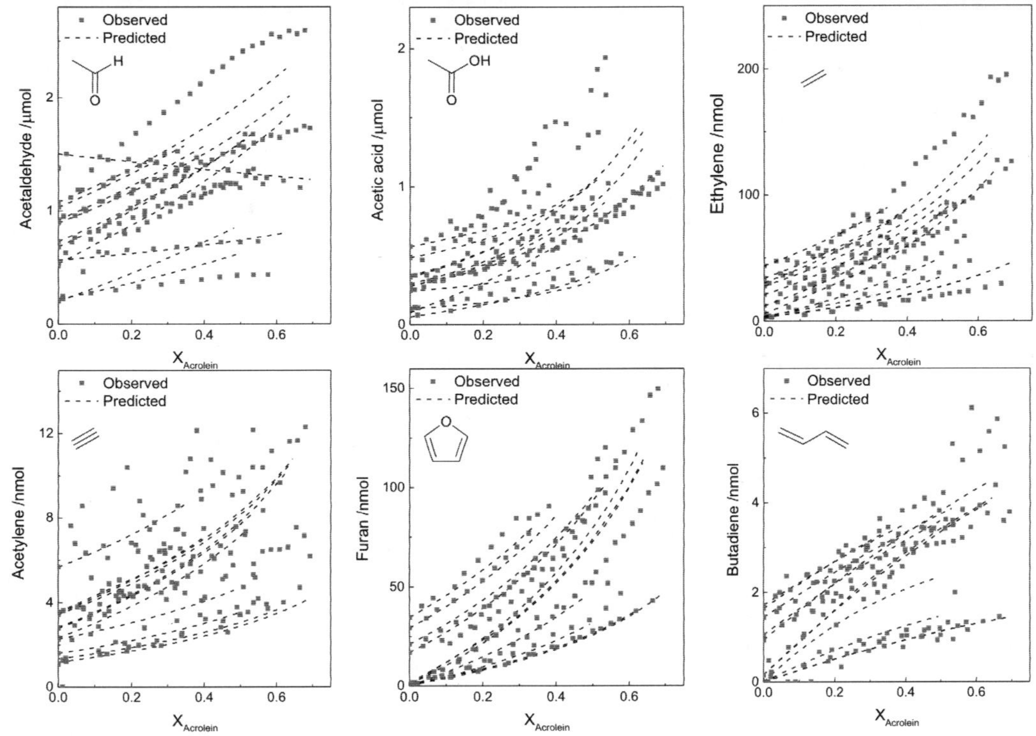

47

Fits of data: C_5-C_7s

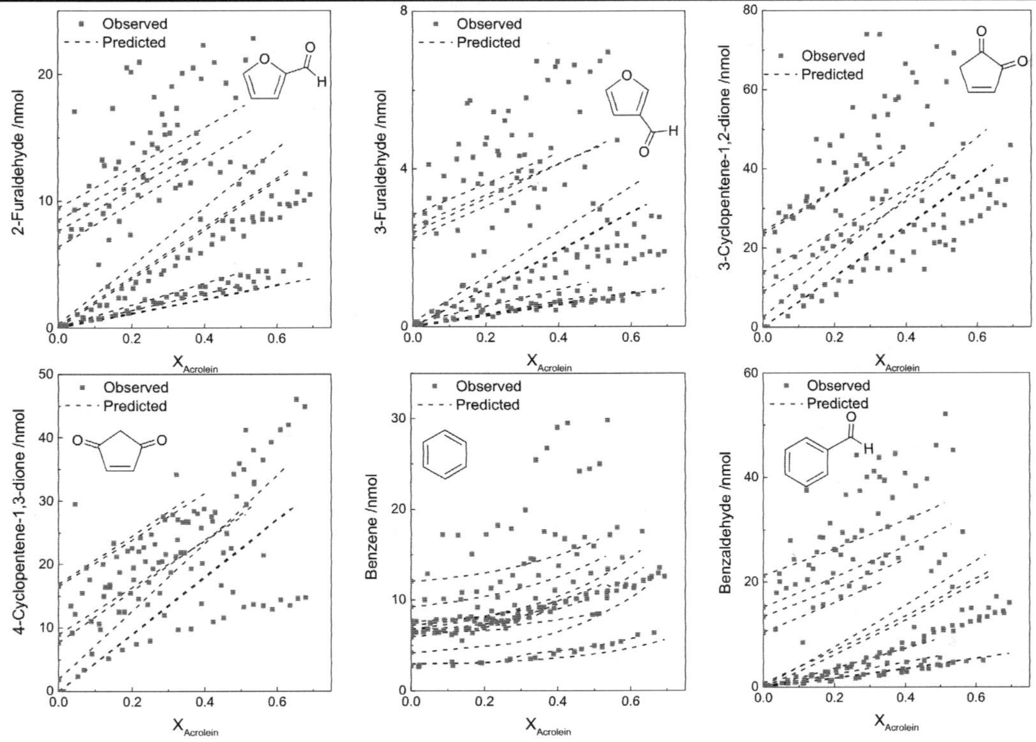

48

Model balance equations: C_1-C_3

$$\frac{1}{m_{cat}}\frac{d}{dt}n_{Acrolein} = -k_1 P_{Acrolein}[O_L]^2 - (k_{16}+k_{17}+k_{18}+k_{19})P_{Acrolein}[C_2H_3*] - k_{22}P_{Acrolein}[C_4H_5*]$$

$$\frac{1}{m_{cat}}\frac{d}{dt}n_{AcrylicAcid} = k_2[C_3H_3O*][OH] - k_5 P_{AcrylicAcid}[OL]$$

$$\frac{1}{m_{cat}}\frac{d}{dt}n_{O_2} = -k_3 P_{O_2}[V]$$

$$\frac{1}{m_{cat}}\frac{d}{dt}n_{CO_2} = 3k_4[C_3H_3O*][O_L] + 2k_8[C_3H_3O*][O_L] + k_9[C_3H_3O*][O_L]$$

$$\frac{1}{m_{cat}}\frac{d}{dt}n_{H_2O} = k_4[C_3H_3O*][O_L] + k_6[OH]^2 + k_7[C_3H_3O*][O_L]$$

$$\frac{1}{m_{cat}}\frac{d}{dt}n_{CO} = 3k_7[C_3H_3O*][O_L]$$

$$\frac{1}{m_{cat}}\frac{d}{dt}n_{CHOH} = k_8[C_3H_3O*][O_L]$$

$$\frac{1}{m_{cat}}\frac{d}{dt}n_{C_2H_4} = k_{10}[C_2H_3*][OH]$$

$$\frac{1}{m_{cat}}\frac{d}{dt}n_{CH_3CHO} = k_{11}[C_2H_3*][OH] - k_{12}P_{CH_3CHO}[O_L]^2$$

$$\frac{1}{m_{cat}}\frac{d}{dt}n_{CH_3COOH} = k_{23}[C_2H_3O*][OH]$$

$$\frac{1}{m_{cat}}\frac{d}{dt}n_{C_2H_2} = k_{13}[C_2H_2*][O_L]$$

49

Model balance equations: C_4-C_7

$$\frac{1}{m_{cat}}\frac{d}{dt}n_{C_4H_6} = k_{20}[C_4H_5*][OH] - k_{21}P_{C_4H_6}[C_2H_3*]$$

$$\frac{1}{m_{cat}}\frac{d}{dt}n_{C_4H_4O} = k_{15}[C_4H_5*][O_L]$$

$$\frac{1}{m_{cat}}\frac{d}{dt}n_{2-Furaldehyde} = k_{16}P_{Acrolein}[C_2H_3*]$$

$$\frac{1}{m_{cat}}\frac{d}{dt}n_{3-Furaldehyde} = k_{17}P_{Acrolein}[C_2H_3*]$$

$$\frac{1}{m_{cat}}\frac{d}{dt}n_{3-Cyclopentene-1,2-dione} = k_{18}P_{Acrolein}[C_2H_3*]$$

$$\frac{1}{m_{cat}}\frac{d}{dt}n_{4-Cyclopentene-1,3-dione} = k_{19}P_{Acrolein}[C_2H_3*]$$

$$\frac{1}{m_{cat}}\frac{d}{dt}n_{C_6H_6} = k_{21}P_{C_4H_6}[C_2H_3*]$$

$$\frac{1}{m_{cat}}\frac{d}{dt}n_{C_6H_5CHO} = k_{22}P_{Acrolein}[C_4H_5*]$$

50

Model balance equations: O_L, OH, V

$$0=\frac{1}{m_{cat}}\frac{d}{dt}[O_L]=-2k_1P_{Acrolein}[O_L]^2+k_2[C_3H_3O*][OH]+k_3P_{O_2}[V]-6k_4[C_3H_3O*][O_L]$$
$$-k_5P_{AcrylicAcid}[O_L]+k_6[OH]^2-3k_7[C_3H_3O*][O_L]-4k_8[C_3H_3O*][O_L]-k_9[C_3H_3O*][O_L]$$
$$+2k_{10}[C_2H_3*][OH]+k_{11}[C_2H_3*][OH]-2k_{12}P_{CH_3CHO}[O_L]^2-k_{15}[C_4H_5*][O_L]$$
$$-3(k_{16}+k17+k_{18}+k_{19})P_{Acrolein}[C_2H_3*]+2k_{20}[C_4H_5*][OH]-2k_{22}P_{Acrolein}[C_4H_5*]+k_{23}[C_2H_3O*][OH]$$

$$0=\frac{1}{m_{cat}}\frac{d}{dt}[OH]=k_1P_{Acrolein}[O_L]^2-k_2[C_3H_3O*][OH]+k_4[C_3H_3O*][O_L]+k_5P_{AcrylicAcid}[O_L]$$
$$-2k_6[OH]^2+k_7[C_3H_3O*][O_L]+k_8[C_3H_3O*][O_L]-k_{10}[C_2H_3*][OH]-k_{11}[C_2H_3*][OH]+$$
$$k_{12}P_{CH_3CHO}[O_L]^2+k_{13}[C_2H_2*][O_L]+k_{14}[C_2H_3*]^2+k_{15}[C_4H_5*][O_L]+$$
$$3(k_{16}+k_{17}+k_{18}+k_{19})P_{Acrolein}[C_2H_3*]-k_{20}[C_4H_5*][OH]+k_{21}P_{C_4H_6}[C_2H_3*]$$
$$+3k_{22}P_{Acrolein}[C_4H_5*]-k_{23}[C_2H_3O*][OH]$$

$$0=\frac{1}{m_{cat}}\frac{d}{dt}[V]=k_2[C_3H_3O*][OH]-2k_3P_{O_2}[V]+6k_4[C_3H_3O*][O_L]-k_5P_{AcrylicAcid}[O_L]$$
$$+k_6[OH]^2+3k_7[C_3H_3O*][O_L]+4k_8[C_3H_3O*][O_L]+k_9[C_3H_3O*][O_L]+k_{11}[C_2H_3*][OH]$$
$$+k_{12}P_{CH_3CHO}[O_L]^2+k_{15}[C_4H_5*][O_L]+(k_{16}+k_{17}+k_{18}+k_{19})P_{Acrolein}[C_2H_3*]+k_{23}[C_2H_3O*][OH]$$

51

Model balance equations: C-containing surface species; Parameter values

$$0=\frac{1}{m_{cat}}\frac{d}{dt}[C_3H_3O*]=k_1P_{Acrolein}[O_L]^2-k_2[C_3H_3O*][OH]-k_4[C_3H_3O*][O_L]+k_5P_{AcrylicAcid}[O_L]$$
$$-k_7[C_3H_3O*][O_L]-k_8[C_3H_3O*][O_L]-k_9[C_3H_3O*][O_L]$$

$$0=\frac{1}{m_{cat}}\frac{d}{dt}[C_2H_3*]=k_9[C_3H_3O*][O_L]-k_{10}[C_2H_3*][OH]-k_{11}[C_2H_3*][OH]-k_{13}[C_2H_2*][O_L]-$$
$$k_{14}[C_2H_3*]^2-(k_{16}+k_{17}+k_{18}+k_{19})P_{Acrolein}[C_2H_3*]-k_{21}P_{C_4H_6}[C_2H_3*]$$

$$0=\frac{1}{m_{cat}}\frac{d}{dt}[C_4H_5*]=k_{14}[C_2H_3*]^2-k_{15}[C_4H_5*][O_L]-k_{20}[C_4H_5*][OH]-k_{22}P_{Acrolein}[C_4H_5*]$$

$$0=\frac{1}{m_{cat}}\frac{d}{dt}[C_2H_3O*]=k_{12}P_{CH_3CHO}[O_L]^2-k_{23}[C_2H_3O*][OH]$$

Parameter values

k_1= 87.4 µmol kPa^{-1} g_{cat}^{-1} min^{-1}	k_9= 184 µmol g_{cat}^{-1} min^{-1}	k_{17}= 0.192 µmol kPa^{-1} g_{cat}^{-1} min^{-1}
k_2= 1.14*10^7 µmol g_{cat}^{-1} min^{-1}	k_{10}= 230 µmol g_{cat}^{-1} min^{-1}	k_{18}= 2.54 µmol kPa^{-1} g_{cat}^{-1} min^{-1}
k_3= 1.00*10^{10} µmol kPa^{-1} g_{cat}^{-1} min^{-1}	k_{11}= 4490 µmol g_{cat}^{-1} min^{-1}	k_{19}= 1.80 µmol kPa^{-1} g_{cat}^{-1} min^{-1}
k_4= 991 µmol g_{cat}^{-1} min^{-1}	k_{12}= 57.5 µmol g_{cat}^{-1} min^{-1}	k_{20}= 63.6 µmol g_{cat}^{-1} min^{-1}
k_5= 115.5 µmol kPa^{-1} g_{cat}^{-1} min^{-1}	k_{13}= 0.604 µmol kPa^{-1} g_{cat}^{-1} min^{-1}	k_{21}= 7.28*10^4 µmol kPa^{-1} g_{cat}^{-1} min^{-1}
k_6= 2750 µmol g_{cat}^{-1} min^{-1}	k_{14}= 4100 µmol g_{cat}^{-1} min^{-1}	k_{22}= 3.75 µmol kPa^{-1} g_{cat}^{-1} min^{-1}
k_7= 122 µmol g_{cat}^{-1} min^{-1}	k_{15}= 26.5 µmol g_{cat}^{-1} min^{-1}	k_{23}= 1.00*10^{10} µmol g_{cat}^{-1} min^{-1}
k_8= 78.4 µmol g_{cat}^{-1} min^{-1}	k_{16}= 0.771 µmol kPa^{-1} g_{cat}^{-1} min^{-1}	

52

Model balance equations: C_1-C_3

$$\frac{1}{m_{cat}}\frac{d}{dt}n_{Acrolein} = -k_1 P_{Acrolein}[O_L]^2 - (k_{16}+k_{17}+k_{18}+k_{19})P_{Acrolein}[C_2H_3*] - k_{22}P_{Acrolein}[C_4H_5*]$$

$$\frac{1}{m_{cat}}\frac{d}{dt}n_{AcrylicAcid} = k_2[C_3H_3O*][OH] - k_5 P_{AcrylicAcid}[OL]$$

$$\frac{1}{m_{cat}}\frac{d}{dt}n_{O_2} = -k_3 P_{O_2}[V]$$

$$\frac{1}{m_{cat}}\frac{d}{dt}n_{CO_2} = 3k_4[C_3H_3O*][O_L] + 2k_8[C_3H_3O*][O_L] + k_9[C_3H_3O*][O_L]$$

$$\frac{1}{m_{cat}}\frac{d}{dt}n_{H_2O} = k_4[C_3H_3O*][O_L] + k_6[OH]^2 + k_7[C_3H_3O*][O_L]$$

$$\frac{1}{m_{cat}}\frac{d}{dt}n_{CO} = 3k_7[C_3H_3O*][O_L]$$

$$\frac{1}{m_{cat}}\frac{d}{dt}n_{CHOH} = k_8[C_3H_3O*][O_L]$$

$$\frac{1}{m_{cat}}\frac{d}{dt}n_{C_2H_4} = k_{10}[C_2H_3*][OH]$$

$$\frac{1}{m_{cat}}\frac{d}{dt}n_{CH_3CHO} = k_{11}[C_2H_3*][OH] - k_{12}P_{CH_3CHO}[O_L]^2$$

$$\frac{1}{m_{cat}}\frac{d}{dt}n_{CH_3COOH} = k_{23}[C_2H_3O*][OH]$$

$$\frac{1}{m_{cat}}\frac{d}{dt}n_{C_2H_2} = k_{13}[C_2H_2*][O_L]$$

53

Model balance equations: C_4-C_7

$$\frac{1}{m_{cat}}\frac{d}{dt}n_{C_4H_6} = k_{20}[C_4H_5*][OH] - k_{21}P_{C_4H_6}[C_2H_3*]$$

$$\frac{1}{m_{cat}}\frac{d}{dt}n_{C_4H_4O} = k_{15}[C_4H_5*][O_L]$$

$$\frac{1}{m_{cat}}\frac{d}{dt}n_{2-Furaldehyde} = k_{16}P_{Acrolein}[C_2H_3*]$$

$$\frac{1}{m_{cat}}\frac{d}{dt}n_{3-Furaldehyde} = k_{17}P_{Acrolein}[C_2H_3*]$$

$$\frac{1}{m_{cat}}\frac{d}{dt}n_{3-Cyclopentene-1,2-dione} = k_{18}P_{Acrolein}[C_2H_3*]$$

$$\frac{1}{m_{cat}}\frac{d}{dt}n_{4-Cyclopentene-1,3-dione} = k_{19}P_{Acrolein}[C_2H_3*]$$

$$\frac{1}{m_{cat}}\frac{d}{dt}n_{C_6H_6} = k_{21}P_{C_4H_6}[C_2H_3*]$$

$$\frac{1}{m_{cat}}\frac{d}{dt}n_{C_6H_5CHO} = k_{22}P_{Acrolein}[C_4H_5*]$$

54

Model balance equations: O_L, OH, V

$$0 = \frac{1}{m_{cat}}\frac{d}{dt}[O_L] = -2k_1 P_{Acrolein}[O_L]^2 + k_2[C_3H_3O*][OH] + k_3 P_{O_2}[V] - 6k_4[C_3H_3O*][O_L]$$
$$-k_5 P_{AcrylicAcid}[O_L] + k_6[OH]^2 - 3k_7[C_3H_3O*][O_L] - 4k_8[C_3H_3O*][O_L] - k_9[C_3H_3O*][O_L]$$
$$+2k_{10}[C_2H_3*][OH] + k_{11}[C_2H_3*][OH] - 2k_{12}P_{CH_3CHO}[O_L]^2 - k_{15}[C_4H_5*][O_L]$$
$$-3(k_{16}+k17+k_{18}+k_{19})P_{Acrolein}[C_2H_3*] + 2k_{20}[C_4H_5*][OH] - 2k_{22}P_{Acrolein}[C_4H_5*] + k_{23}[C_2H_3O*][OH]$$

$$0 = \frac{1}{m_{cat}}\frac{d}{dt}[OH] = k_1 P_{Acrolein}[O_L]^2 - k_2[C_3H_3O*][OH] + k_4[C_3H_3O*][O_L] + k_5 P_{AcrylicAcid}[O_L]$$
$$-2k_6[OH]^2 + k_7[C_3H_3O*][O_L] + k_8[C_3H_3O*][O_L] - k_{10}[C_2H_3*][OH] - k_{11}[C_2H_3*][OH] +$$
$$k_{12}P_{CH_3CHO}[O_L]^2 + k_{13}[C_2H_2*][O_L] + k_{14}[C_2H_3*]^2 + k_{15}[C_4H_5*][O_L] +$$
$$3(k_{16}+k_{17}+k_{18}+k_{19})P_{Acrolein}[C_2H_3*] - k_{20}[C_4H_5*][OH] + k_{21}P_{C_4H_6}[C_2H_3*]$$
$$+ 3k_{22}P_{Acrolein}[C_4H_5*] - k_{23}[C_2H_3O*][OH]$$

$$0 = \frac{1}{m_{cat}}\frac{d}{dt}[V] = k_2[C_3H_3O*][OH] - 2k_3 P_{O_2}[V] + 6k_4[C_3H_3O*][O_L] - k_5 P_{AcrylicAcid}[O_L]$$
$$+k_6[OH]^2 + 3k_7[C_3H_3O*][O_L] + 4k_8[C_3H_3O*][O_L] + k_9[C_3H_3O*][O_L] + k_{11}[C_2H_3*][OH]$$
$$+k_{12}P_{CH_3CHO}[O_L]^2 + k_{15}[C_4H_5*][O_L] + (k_{16}+k_{17}+k_{18}+k_{19})P_{Acrolein}[C_2H_3*] + k_{23}[C_2H_3O*][OH]$$

Model balance equations: C-containing surface species; Parameter values

$$0 = \frac{1}{m_{cat}}\frac{d}{dt}[C_3H_3O*] = k_1 P_{Acrolein}[O_L]^2 - k_2[C_3H_3O*][OH] - k_4[C_3H_3O*][O_L] + k_5 P_{AcrylicAcid}[O_L]$$
$$- k_7[C_3H_3O*][O_L] - k_8[C_3H_3O*][O_L] - k_9[C_3H_3O*][O_L]$$

$$0 = \frac{1}{m_{cat}}\frac{d}{dt}[C_2H_3*] = k_9[C_3H_3O*][O_L] - k_{10}[C_2H_3*][OH] - k_{11}[C_2H_3*][OH] - k_{13}[C_2H_2*][O_L] -$$
$$k_{14}[C_2H_3*]^2 - (k_{16}+k_{17}+k_{18}+k_{19})P_{Acrolein}[C_2H_3*] - k_{21}P_{C_4H_6}[C_2H_3*]$$

$$0 = \frac{1}{m_{cat}}\frac{d}{dt}[C_4H_5*] = k_{14}[C_2H_3*]^2 - k_{15}[C_4H_5*][O_L] - k_{20}[C_4H_5*][OH] - k_{22}P_{Acrolein}[C_4H_5*]$$

$$0 = \frac{1}{m_{cat}}\frac{d}{dt}[C_2H_3O*] = k_{12}P_{CH_3CHO}[O_L]^2 - k_{23}[C_2H_3O*][OH]$$

Parameter values

k_1= 87.4 µmol kPa^{-1} g_{cat}^{-1} min^{-1}	k_9= 184 µmol g_{cat}^{-1} min^{-1}	k_{17}= 0.192 µmol kPa^{-1} g_{cat}^{-1} min^{-1}
k_2= 1.14*10^7 µmol g_{cat}^{-1} min^{-1}	k_{10}= 230 µmol g_{cat}^{-1} min^{-1}	k_{18}= 2.54 µmol kPa^{-1} g_{cat}^{-1} min^{-1}
k_3= 1.00*10^{10} µmol kPa^{-1} g_{cat}^{-1} min^{-1}	k_{11}= 4490 µmol g_{cat}^{-1} min^{-1}	k_{19}= 1.80 µmol kPa^{-1} g_{cat}^{-1} min^{-1}
k_4= 991 µmol g_{cat}^{-1} min^{-1}	k_{12}= 57.5 µmol g_{cat}^{-1} min^{-1}	k_{20}= 63.6 µmol g_{cat}^{-1} min^{-1}
k_5= 115.5 µmol kPa^{-1} g_{cat}^{-1} min^{-1}	k_{13}= 0.604 µmol kPa^{-1} g_{cat}^{-1} min^{-1}	k_{21}= 7.28*10^4 µmol kPa^{-1} g_{cat}^{-1} min^{-1}
k_6= 2750 µmol g_{cat}^{-1} min^{-1}	k_{14}= 4100 µmol g_{cat}^{-1} min^{-1}	k_{22}= 3.75 µmol kPa^{-1} g_{cat}^{-1} min^{-1}
k_7= 122 µmol g_{cat}^{-1} min^{-1}	k_{15}= 26.5 µmol g_{cat}^{-1} min^{-1}	k_{23}= 1.00*10^{10} µmol g_{cat}^{-1} min^{-1}
k_8= 78.4 µmol g_{cat}^{-1} min^{-1}	k_{16}= 0.771 µmol kPa^{-1} g_{cat}^{-1} min^{-1}	

Parameter estimability analysis

Methodology:

- Generate 4186 x 23 matrix of derivatives: $\frac{\partial n_{i,t}}{\partial k_j}$ via model forward simulation
- Select column with highest norm, orthogonalize by this column, evaluate residual
- Re-evaluate norms of residual matrix, orthogonalize matrix, continue iterations until:
  - Norms of all columns are 0
  - Norms of all columns below prescribed cutoff value
- If all parameters are eliminated before norms of all columns 0 or below cutoff, then all parameters are estimable

Bui et al. *ACS Catal.* 2016, *6* (10), 6567-6580
Eghtesadi and McAuley *Ind. Eng. Chem. Res.* 2014, *53*, 6033-6046

Parameter estimability analysis

Columns: Norms of each column in matrix of derivatives for each iteration

k_1	19.66	19.65	**16.47**	0	0	0	0	0	0	0	0	0	0	0	0	0	0	0	0	0	0	0	0	0
k_2	21.70	21.69	12.62	10.69	10.49	10.42	**10.21**	0	0	0	0	0	0	0	0	0	0	0	0	0	0	0	0	0
k_3	12.55	12.55	12.55	10.96	10.95	**10.91**	0	0	0	0	0	0	0	0	0	0	0	0	0	0	0	0	0	0
k_4	23.15	23.15	8.67	8.56	8.32	8.31	8.26	8.24	8.08	**8.08**	0	0	0	0	0	0	0	0	0	0	0	0	0	0
k_5	9.82	9.81	6.91	6.50	6.40	6.40	6.39	3.84	3.81	3.80	3.80	3.80	3.79	3.74	3.74	3.74	3.74	3.73	**3.73**	0	0	0	0	0
k_6	27.76	27.76	9.03	8.97	8.54	8.54	8.52	8.33	7.90	7.88	6.23	5.96	5.59	5.56	5.16	5.16	**5.16**	0	0	0	0	0	0	0
k_7	5.01	5.01	4.24	4.22	4.21	4.21	4.21	4.17	4.17	4.17	4.16	4.16	4.16	4.16	4.15	4.11	4.11	**4.09**	0	0	0	0	0	0
k_8	7.03	7.03	6.81	6.69	6.69	6.69	6.66	6.36	6.35	6.35	6.20	6.20	6.20	6.20	**6.20**	0	0	0	0	0	0	0	0	0
k_9	44.14	**44.14**	0	0	0	0	0	0	0	0	0	0	0	0	0	0	0	0	0	0	0	0	0	0
k_{10}	8.33	8.33	8.32	8.26	8.26	7.93	7.89	7.85	7.85	7.80	**7.60**	0	0	0	0	0	0	0	0	0	0	0	0	0
k_{11}	34.81	34.81	9.38	9.37	7.45	7.22	7.20	7.17	7.17	7.14	6.93	6.76	6.73	**6.73**	0	0	0	0	0	0	0	0	0	0
k_{12}	12.02	11.95	11.91	11.88	6.12	6.11	6.11	6.11	6.11	6.11	6.10	6.10	6.10	6.10	6.04	**6.04**	0	0	0	0	0	0	0	0
k_{13}	5.94	5.94	5.94	5.94	5.94	5.94	5.94	5.94	5.94	5.94	5.93	5.93	5.93	5.93	5.91	5.91	1.43	1.43	1.43	**1.41**	0	0	0	0
k_{14}	17.97	17.97	12.38	12.21	**12.05**	0	0	0	0	0	0	0	0	0	0	0	0	0	0	0	0	0	0	0
k_{15}	15.03	15.03	11.60	11.55	11.51	9.23	9.23	9.16	7.01	7.01	7.00	7.00	**6.93**	0	0	0	0	0	0	0	0	0	0	0
k_{16}	11.52	11.52	10.47	10.44	10.43	9.50	9.50	**9.30**	0	0	0	0	0	0	0	0	0	0	0	0	0	0	0	0
k_{17}	9.76	9.75	9.59	9.52	9.52	9.12	9.11	9.10	9.10	0.48	0.48	0.48	0.47	0.44	0.44	0.44	0.44	0.44	0.44	0.43	0.43	**0.41**	0	0
k_{18}	**565**	0	0	0	0	0	0	0	0	0	0	0	0	0	0	0	0	0	0	0	0	0	0	0
k_{19}	9.77	9.77	9.63	9.57	9.56	9.15	9.15	9.14	**9.14**	0	0	0	0	0	0	0	0	0	0	0	0	0	0	0
k_{20}	10.51	10.51	10.21	10.03	10.02	9.47	9.43	9.23	7.25	7.23	7.02	7.00	2.40	1.05	1.05	1.04	1.04	1.04	1.04	1.00	**1.00**	0	0	0
k_{21}	10.46	10.46	10.04	9.84	9.84	9.18	9.15	8.96	7.15	7.14	6.93	6.92	1.77	0.96	0.96	0.96	0.96	0.96	0.96	0.92	0.91	0.34	**0.34**	0
k_{22}	8.76	8.76	8.72	8.63	8.63	8.63	8.60	8.59	7.21	7.19	7.02	**7.01**	0	0	0	0	0	0	0	0	0	0	0	0
k_{23}	12.56	12.52	12.44	**12.40**	0	0	0	0	0	0	0	0	0	0	0	0	0	0	0	0	0	0	0	0

Orange highlight= Parameter eliminated this iteration. Conclusion: All ks are estimable.

Bui et al. *ACS Catal.* 2016, *6* (10), 6567-6580
Eghtesadi and McAuley *Ind. Eng. Chem. Res.* 2014, *53*, 6033-6046

Isotopic labeling co-feeds point to existence of C_2 surface species

Ex: Furan

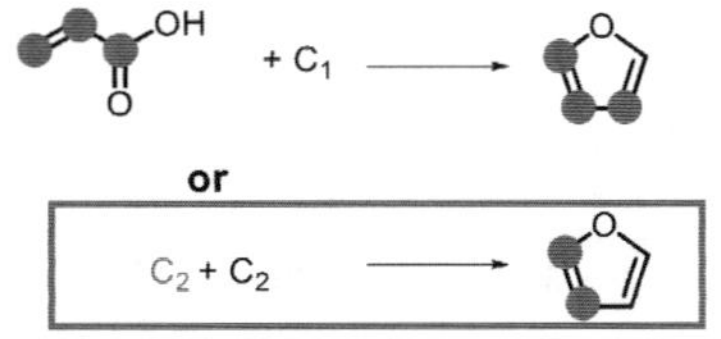

● = ^{13}C label

Acrylic acid→C_2 surface species

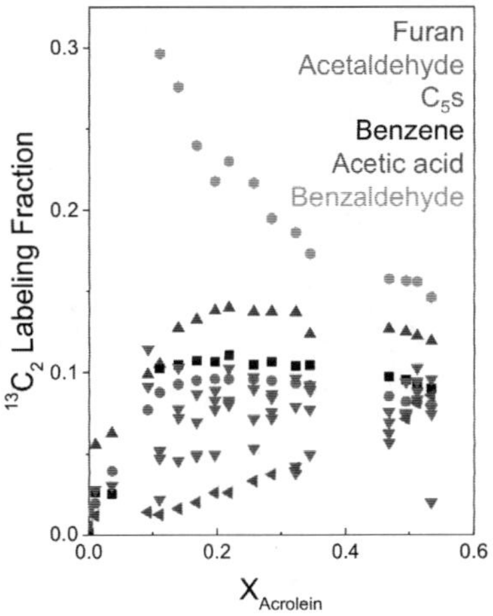

Isotopic Co-feed:

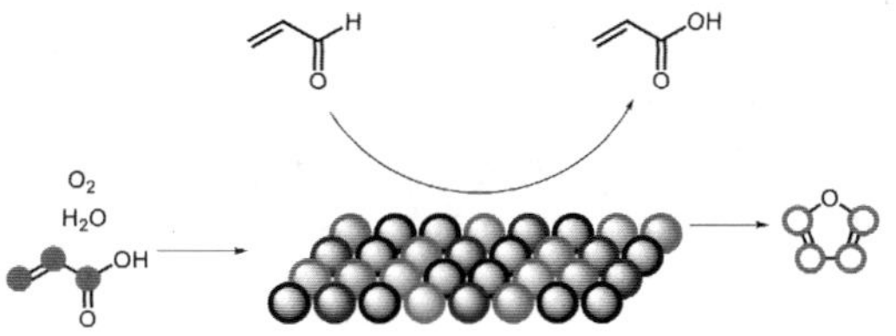

Initial conditions: $P_{Acrolein}$=2.2 kPa, $P_{Acrylic\ acid}$=0.1 kPa, P_{O2}=3.3 kPa, P_{H2O}=5.1 kPa; 4.1 mg mixed-metal oxide catalyst, 498K 59

Selectivity analysis and co-feeds

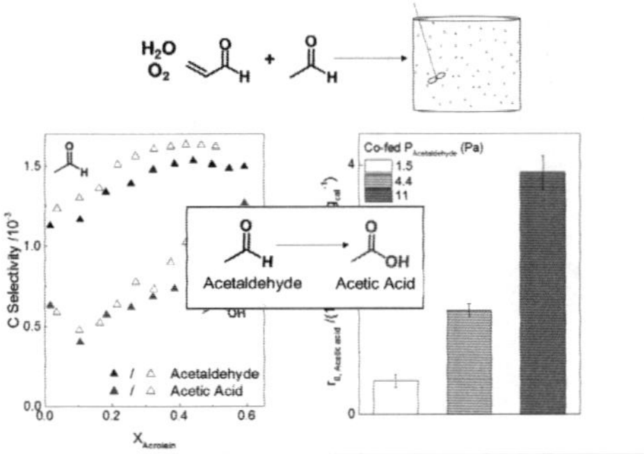

Active sites

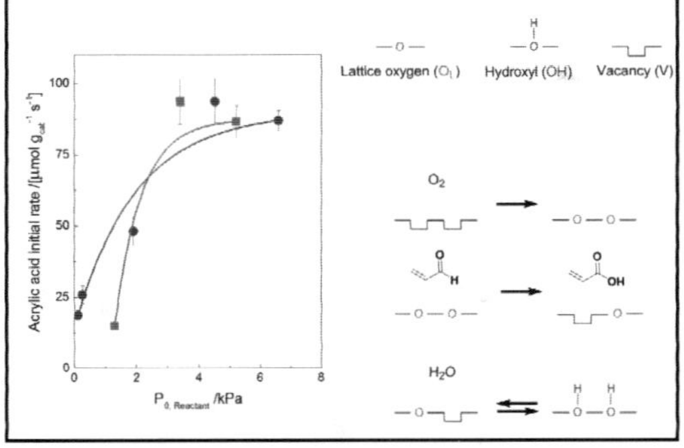

Isotopic labeling

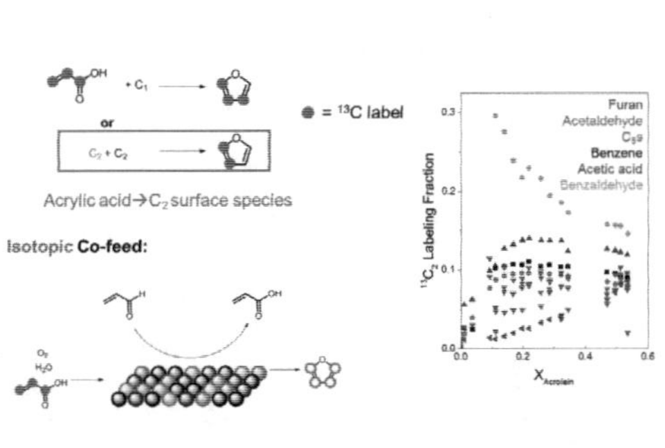

Kinetic modeling

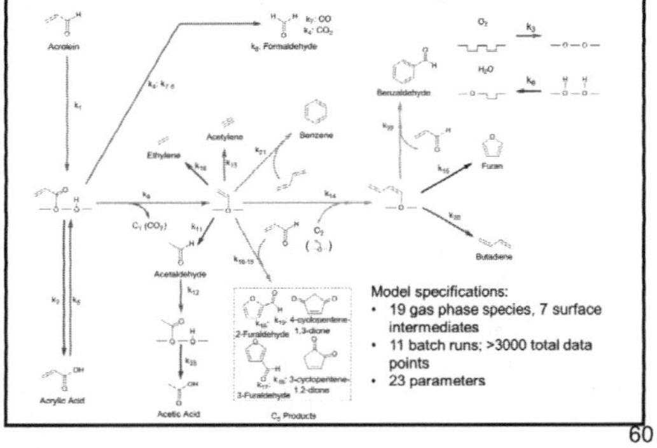

60

491938 The Mechanism of Isobutylene Polymerization: New Insight into Proton-Catalyzed Polymerizations

Abstract:

Low molecular weight polyisobutylenes (PIBs) are crucial additives for lubricants, adhesives, and fuels. Industrial processes for PIB polymerization have existed for over 50 years, but very little is known about the atomic-level details of the reaction mechanism. â€œConventional PIBâ€, which contains internal double bonds (trisusbtituted or tetrasubstituted), is routinely produced by cationic polymerization co-initiated by $AlCl_3$ and adventitious water. Although many different PIB catalysts have been reported in the literature, there is virtually no atomic scale understanding of these reaction mechanisms and therefore limited fundamental insight on how to model and control the reaction process. In this presentation we report atomic scale reaction mechanisms for PIB polymerization with $AlCl_3$ discovered from quantum chemistry modeling. Contrary to textbook mechanisms for this process, which assumes initiation by a single $AlCl_3$ and a single water molecule, we found that the conventional cationic polymerization involves at least two $AlCl_3$ molecules and a single water molecule. We have explored potential initiation, propagation, and termination steps for PIB polymerization and have mapped out potential reaction pathways and barriers.

500850 Unifying Mechanistic Analysis of the Primary Factors Controlling Selectivity in Fructose Dehydration By Acid Catalysts

Abstract:

US interest in biomass as an alternative energy source continues to increase in demand with growing political, economic and environment concerns associated with the consumption of fossil fuels. One potential route for converting cellulosic biomass to fuels is through the dehydration of fructose to produce 5-hydroxymethylfurfural (HMF), followed by the catalytic conversion of HMF to renewable fuels, such as dimethylfuran (DMF). Economically viable production of HMF from fructose requires high selectivities and yields. However, mechanistic aspects of fructose conversion to HMF are still unclear, making the design of catalysts that enable high yield HMF production difficult. Various reports have proposed different factors that control selectivity including the acid type, fructose tautomer distribution, and solvent type. In order to develop detailed insights into the factors that control HMF yields, we executed extensive kinetic analysis using a variety of solvent systems, reaction conditions, acid types and acid concentrations in a high-throughput reactor to determine the mechanism and selectivity controlling factors in fructose dehydration to HMF by homogeneous acid catalysts. Detailed NMR analysis of the relative distribution of fructose tautomers as a function of solvent type and temperature were also executed. Through these studies we strongly support the hypothesis that selectivity in fructose conversion to HMF is controlled by the equilibrium between the tautomeric forms of fructose in solution, by directly correlating HMF selectivity with the distribution of furanose and open-chain forms of fructose. We will describe the implication of these mechanistic insights in the context of the design of heterogeneous systems that allow for operation of this reaction with high selectivity in low boiling point solvents.

496646 Reaction Coupling of Nitrobenzene Hydrogenation with Ethylbenzene Dehydrogenation: Reaction Pathways and Kinetics

Abstract:

The production of aniline from nitrobenzene hydrogenation and styrene from ethylbenzene dehydrogenation are two important reactions in the petrochemical industry. Nitrobenzene hydrogenation is highly exothermic, which does not require significant energy. However, its main cost comes from the required external hydrogen, primarily produced from steam reforming of methane. On the other hand, ethylbenzene dehydrogenation is highly endothermic and thermodynamically unfavorable with limited equilibrium conversion. Coupling these two reactions has the potential benefits of removing the need for hydrogen in nitrobenzene hydrogenation and increasing the yields and/or reducing the reaction temperature of ethylbenzene dehydrogenation. In this work, the coupling of nitrobenzene hydrogenation with ethylbenzene dehydrogenation has been performed over MoO_3 catalysts, which has shown high deoxygenation and hydrogen adsorption activities in the literature. A hypothesized reaction mechanism is tested by our experimental findings. Our results show that the active sites on the MoO_3 catalyst capture oxygen atoms from nitrobenzene first to produce a phenylhydroxylamine intermediate. The surface adsorbed hydrogen atoms, produced from hydrogen abstraction of ethylbenzene, then react with phenylhydroxylamine to form aniline. Due to this hydrogen transfer process, the equilibrium of ethylbenzene dehydrogenation is shifted such that the conversion is increased. In addition, the hydrogenolysis reaction leading to side products (benzene and toluene) is reduced, improving the selectivity of styrene.

494900 Introducing Catalytic Diversity into Single-Site Zeolites of Fixed Composition Via Synthetic Control of Active Site Proximity

Abstract:

Introducing Catalytic Diversity into Single-Site Zeolites of Fixed Composition via Synthetic Control of Active Site Proximity

John R. Di Iorio, Claire T. Nimlos, Rajamani Gounder*

Charles D. Davidson School of Chemical Engineering, Purdue University, 480 Stadium Mall Drive, West Lafayette, IN 47907, USA

Distinct Brønsted acid site (H^+) ensembles within a given zeolite arise from differences in the arrangement (Al-O(-Si -O)$_x$-Al) of framework Al atoms between isolated (x = 3) or paired configurations (x = 1, 2), the latter defined functionally by their ability to exchange extraframework divalent cations. The effects of proton proximity on hydrocarbon catalysis have been previously studied in MFI zeolites,[1-2] but the concomitant changes in Al distribution among different crystallographically distinct tetrahedral sites (T-sites) and void environments (i.e., straight and sinusoidal channels, and their intersections) have precluded unambiguous kinetic assessments of proton proximity in MFI. Here, we use chabazite (CHA) zeolites, a high-symmetry framework with only 1 unique lattice T-site, to develop a predictive relationship that describes how synthetic methods can be developed to prepare zeolites, of fixed topology and elemental composition, with systematically different framework Al arrangements. Such arrangements, in turn, generate different proton active site ensembles that markedly influence turnover rates for the Brønsted acid-catalyzed dehydration of methanol to dimethyl ether, a known probe reaction of solid acids.[3] Specifically, we extend previously reported synthetic procedures[4] to crystallize CHA zeolites at similar composition (Si/Al = 15), but with fractions of paired Al sites that systematically varied from the limit of site isolation (0% paired Al) to nearly half of the sites (44% paired Al) in paired arrangements. This series of CHA zeolites is used to demonstrate that first and zero-order rate constants (415 K, per total H^+) of methanol dehydration are one order of magnitude larger on paired than on isolated protons. *In situ* IR spectra measured during steady-state dehydration catalysis are used to observe the formation of surface methoxy species at paired Al sites in CHA, which are not observed on isolated protons in CHA or MFI, providing evidence that alternate dehydration pathways are accessible on paired protons in CHA zeolites. These results represent a step toward the synthetic control of active site atomic arrangement and establish a predictive synthesis-structure-function relation for Brønsted acidic CHA zeolites, demonstrating that catalytic diversity can arise solely from differences in the atomic arrangement of active sites, even among single T-site zeolites of fixed composition.

References:

1. Janda, A.; Bell, A. T. *J. Am. Chem. Soc.* **2013**, *135*, 19193-19207

2. Bernauer, M.; Tabor, E.; Pashkova, V.; Kaucký, D.; Sobalík, Z.; Wichterlová, B.; Dedecek, J. *J. Catal.* **2016**, *344*, 157-172.

3. Carr, R. T.; Neurock, M.; Iglesia, E. *J. Catal.* **2011**, *278*, 78-93.

4. Di Iorio, J. R.; Gounder, R. *Chem. Mater.* **2016**, *28*, 2236-2247

493983 Probing the Site-Pairing Preference of Framework Aluminum in ZSM-5 with $Ga(CH_3)_3$

Abstract:

Probing the Site-Pairing Preference of Framework Aluminum in ZSM-5 with $Ga(CH_3)_3$

Kyle Groden[a], Zachary Jones[b], Renqin Zhang[a], Susannah Scott[b], Jean-Sabin McEwen[a]

[a] Washington State University, WA, 99164;

[b] University of California, Santa Barbara, CA, 93106;

Figure 1. Comparison of the computed and the simulated XANES for methylene bridged Ga and how the Al distribution influences the Ga-Ga distance within a zeolite framework. The computed Ga-Ga distance is 2.96 Å, in strong agreement with experimental findings.

ZSM-5 is a zeolite catalyst marked with great versatility in catalytic application as a result of its flexibility in cation incorporation. Due to this property, the structure of ZSM-5 has been under question for its ability to accept divalent cations, implying that the substituted aluminum sites exist in paired configurations at specific locations within the zeolite framework. Trimethylgallium was used to probe these sites, as the dimethyl gallium fragments that exist in these locations following impregnation were thought to bind to nearby fragments via terminal methyl groups. Experimental extended x-ray absorption fine structure (EXAFS) spectroscopy data supports this "bridging" configuration due to the presence of potential gallium-gallium scattering signals. Zeolite cluster models were created containing potential sites for aluminum pairings with dimethyl gallium fragments, with optimized geometrical conformations obtained from first principles calculations based on density functional theory (DFT) implemented in the VASP program.[1,2] As methyl-bridged structures were unfavorable, additional cluster and periodic models were constructed to explore the possibility of carbon-hydrogen bond cleavage due to the reactivity of the methyl groups bound to the gallium centers. This reaction results in a methylene (CH_2) group joining the gallium centers in contrast with the methyl bridges previously hypothesized. X-ray absorption near edge spectra (XANES) were computed using CASTEP[3], another DFT program, for many of these configurations examined, which allows for a direct comparison with the synchrotron-based experimental results. As can be seen from Figure 1, the agreement between the gallium-gallium bond distances and spectral results from our calculations and those measured experimentally imply that paired aluminum sites are indeed present within the ZSM-5 architecture.

References:

[1] Kresse, G.; Furthmüller, J.*Phys. Rev. B* **1996**, *54* (16), 11169.

[2] Kresse, G.; Hafner, J. *Phys. Rev. B* **1993**, *47* (1), 558.

[3] Clark, S. J.; Segall, M. D.; Pickard, C. J.; Hasnip, P. J.; Probert, M. I. J.; Refson, K.; Payne, M. C. *Zeitschrift fur Krist.* **2005**, *220* (5-6), 567–570.

496765 Quantitative Analysis of Catalysis in Hierarchical Zeolites and Surface Topology Effect Based on MFI and MWW Frameworks

Abstract:

Recent advances in hierarchical zeolites have demonstrated the possibility of synthesizing zeolites with microporous domains down to single-unit-cell. The improved catalytic performances of such zeolites were mostly associated with the reduced diffusion limitation in microporous regions, and the better access to active sites on external surface. Here, liquid-phase alkylation, self-etherification and de-etherification reactions are performed at mild conditions using two 10MR acid zeolites, H-self-pillared-pentasil zeolite (H-SPP) and H-MCM-22. A kinetic study with quantitative analysis of above reactions shows the dominant contribution to catalytic performance from external active sites in comparison to their 3D counterparts, and therefore the increasing importance of surface topology effect in such hierarchical zeolites.

496772 Multi-Faceted Approach to Optimize ZSM-11 Catalysts for Methanol-to-Hydrocarbon Reactions

Abstract:

ZSM-11 (MEL) is an emerging catalyst for a wide variety of applications. The MEL framework is very similar to ZSM-5 (MFI), which is one of the most widely used zeolite catalysts in industry. Recent studies have shown that ZSM-11 has better catalytic performance in several different reactions compared to ZSM-5 [1-3], possibly due to improved intra-crystalline diffusion within the less-tortuous straight channels of the MEL topology; however, there are relatively few systematic studies of ZSM-11 synthesis, specifically with respect to tailored physicochemical properties such as size, shape, and Si/Al ratio (SAR). Here, we will present a systematic study of ZSM-11 catalyst preparation and testing for C_1 upgrading reactions. Through the judicious selection of synthesis parameters, we successfully developed a facile protocol to generate a library of ZSM-11 catalysts with tunable physicochemical properties (e.g., size, shape, and SAR). These methods for tailoring zeolite crystallization will be discussed along with catalytic studies of methanol-to-hydrocarbon (MTH) reactions. Collectively, our findings are part of a broader initiative to establish quantitative structure-performance relationships as a platform for optimizing zeolite catalysts.

[1] Kustova, M. Y.; Rasmussen, S. B.; Kustov, A. L.; Christensen, C. H., *Applied Catalysis B-Environmental* **2006**, *67* (1-2), 60-67.

[2] Zhang, L.; Liu, H.; Li, X.; Xie, S.; Wang, Y.; Xin, W.; Liu, S.; Xu, L., *Fuel Processing Technology* **2010**, *91* (5), 449-455.

[3] Bleken, F.; Skistad, W.; Barbera, K.; Kustova, M.; Bordiga, S.; Beato, P.; Lillerud, K. P.; Svelle, S.; Olsbye, U., *Physical Chemistry Chemical Physics* **2011**, *13* (7), 2539-2549.

500785 Atomic-Level Insights into Molecular Adsorption and Reaction Properties at Distinct Heteroatom Sites in Aluminosilicate Zeolite Catalysts

Abstract:

Heteroatom-containing zeolites are of technological interest because they exhibit high surface areas, highly uniform sub-nanometer pore dimensions, and high catalytic activities for diverse applications, including hydrocarbon conversion (e.g., fluidized catalytic cracking, FCC, and methanol-to-olefin, MTO, processes). Heteroatoms, such as B or Al, in zeolite frameworks are associated with molecular adsorption and reaction sites, and therefore the types and distributions of heteroatom environments strongly influence the macroscopic reaction properties of zeolite catalysts. However, measurements of local heteroatom environments, their distributions in zeolite frameworks, and their interactions with adsorbed molecular guest species, such as structure-directing organic molecules, reactants, or catalytic reaction products, have been extremely challenging. This has been due, in part, to non-stoichiometric zeolite framework compositions and distributions of both framework and extra-framework species that lack long range order, which are often exacerbated by post-synthesis treatments, such as calcination, steaming, or exposure to reaction conditions. Nevertheless, detailed atomic-level insights on heteroatom environments and their interactions with adsorbed molecular guest species are provided by solid-state nuclear magnetic resonance (NMR) spectroscopy, correlated with complementary transmission electron microscopy, X-ray diffraction, and catalytic reaction analyses. Notably, two-dimensional dynamic-nuclear-polarization (DNP)-enhanced NMR techniques provide substantially increased NMR signal sensitivity (ca. 100x) and resolution, which enables detection and correlation of specific heteroatom environments with the adsorbed molecules with which they interact. Such measurements have, until now, been infeasible by scattering or other spectroscopic techniques. Results will be presented for methanol, toluene, and carbonaceous deactivation products adsorbed on aluminosilicate zeolites Y and SSZ-13, which are associated with technologically important cracking or methanol-to-olefin reactions. These analyses identify heteroatom sites in the framework and at exterior surfaces that have distinct adsorption and reaction behaviors, providing useful new insights for the design of highly effective and stable catalysts.

500318 Synthesis and SCR Testing of SSZ-39

Abstract:

Synthesis and SCR Testing of SSZ-39

Ross Ransom[1], Jonathan Coote[1], Roger Moulton[2], Feng Gao[2], and Daniel F. Shantz[1]*

[1]Tulane University, New Orleans, Louisiana 70118 (USA)

[2]SACHEM, Inc., Austin, Texas 78704 (USA)

SSZ-39, the aluminosilicate anologue of the AEI ($AlPO_4$-18) topology, can be prepared in a straightforward manner using faujasite as an aluminum source[1]. This material shows great promise as a catalyst for deNOx catalysis,[2,3] and thus a practical, simple, and scalable synthesis is something that would be highly desirable. A variety of organic SDAs have been shown to afford SSZ-39 at various levels of efficacy, including N,N-dimethyl-3,5-dimethylpiperidinium, the organic SDA employed here. Prior work has shown that not only does this SDA lead to SSZ-39 formation, but has suggested that the isomer identity has a significant effect on growth kinetics. [4]

This talk will describe how the synthesis conditions used impact the growth kinetics and composition of the final material obtained from synthesis. There are four main findings from our synthesis study.[5]Â The first is that using a faujasite with a Si/Al = 2.7 it is possible to form phase pure SSZ-39 across a wide gel Si/Al ratio (15-90), something not previously reported. Second, the yield from these syntheses changes consistently and clearly shows that the aluminum is the limiting reagent, and in all syntheses the yield based on aluminum is over 90%. Third, the trans isomer effect reported previously in the literature is confirmed and quantified across a wide range of cis/trans ratios in this work. Fourth, at an Si/Al = 45 one observes a clear increase in the Si/Al of the material obtained (from 8.6 â€“ 10.5) as one increases the trans isomer content of SDA.

The talk will then describe our initial results in evaluating this material for SCR, showing how the SSZ-39 properties such as Si/Al, which can be controlled based on our synthesis studies, impact the catalytic properties of the SSZ-39 samples. These results will be put into the context of the voluminous literature around SSZ-13.

References

1. Maruo, T., Yamanaka, N., Tsunoki, N., Sadakane, M., Sano, T. . *ChemLett* **2014**, *43*, 302.
2. Albarracin-Caballero, J. D., Khurana, I., Di Iorio, J. R., Shih, A. J., Schmidt, J. E., Dusselier, M., Davis, M. E., Yezerets, A., Miller, J. T., Ribeiro, F. H., Gounder, R. *Reaction Chemistry and Engineering* **2017**, *2*, 168.
3. Moliner, M., Franch, C., Palomares, E., Grill, M., Corma, A. *ChemComm* **2012**, 8264.
4. Dusselier, M., Schmidt, J.E., Moulton, R., Haymore, B., Hellums, M., Davis, M.E. *Chem. of Materials* 27, 7

   (2015).
5. Ransom, R., Coote, J., Moulton, R., Gao, F., Shantz, D. F. *Ind. Eng. Chem. Res.* ASAP Article.

500518 Determining Siting Preference of Exchanged Fe Ions in Fe-SSZ-13 Zeolite through Density Functional Theory andÂ Ab Initio Molecular Dynamics

Abstract:

Fe-exchanged zeolites are known to be catalytically active for various reactions, for example, N2O decomposition, paritial methane oxidation, selective catalytic reduction of NOx, etc. In spite of great efforts spent on understanding catalytic mechanisms of these reactions over Fe-zeolites, identification of available Fe species and their siting preference, even under ex situ conditions, is still under debate. Several factors complicate the site identification, including: (1) potential coexistence of Fe with different nuclearities, ranging from monomeric cations to oligomeric Fe complexes and Fe oxides, (2) coexistence of Fe with oxidation states (Fe2+ and Fe3+) and (3) uncertainties in interpretations of experimental characterization results. The available literature [1] suggests that in low-loaded SSZ-13 samples, each Fe cation charge-compensate single framework Al, and that these cations are present primarily as isolated and dimeric species.

Here we present a theoretical approach to determine Fe siting preference in Fe-SSZ-13 zeolite using density functional theory (DFT) and ab Initio molecular dynamics (AIMD). Relative stabilities of exchanged Fe monomers in isolation and dimers ligated by arbitrary combinations of O and OH ligands at isolated Al and different paired Al sites are determined by construction of reactions that interconvert these species. AIMD and time-dependent DFT are combined to compute ensemble-averaged UV-visible spectra of different Fe species. We show that both Fe2+ and Fe3+ have strong tendency to form hydroxo-bridged dimers at paired Al sites, and OH-ligated Fe3+ monomer and dimers have qualitatively similar computed UV-visible spectra, suggesting previous UV-vis based assignments of isolated Fe3+ cation preference may need to be revisited. Through first-priciples thermodynamic analysis, we identify potential active sites by determining abundant Fe species at different Al sites (either isolated or paired) under conditions of catalyst activation for partial methane oxidation and other reactions. These findings help in better understanding of relevant active sites as well as redox mechanism in Fe-SSZ-13 catalysis.

Reference

[1] Gao, F., Zheng, Y., Kukkadapu, R.K., Wang, Y., Walter, E. D., Schwenzer, B., Szanyi, J. and Peden, C. H. G. ACS. Catal. 6, 5 (2016)

485484 Fabrication of Bifunctional Acid-Base Catalyst By Organic Pillared MFI Nanosheets

Abstract:

Organic pillared MFI zeolite has been synthesized by detemplation of organic templates and intercalation of arylic silsesquioxane (1,4-bis(triethoxysilyl)benzene, BTEB) molecules between multilamellar MFI nanosheets. The acid extraction and UV light radiation were sequentially employed for removal of organic templates located not only between MFI layers but also in the zeolite micropores. The removal of organic templates by non-thermal calcination method prevents the condensation of external silanol groups of zeolitic nanosheets stacked next to each other, which allows intercalation of BTEB molecules between the zeolitic nanosheets and their successive reaction with the silanol groups to form the organic pillared zeolite structure. A post-treatment amination of the resultant zeolite sample introduced amino groups in the organic BTEB pillars. The acid sites from the zeolite framework aluminium and basic sites from the amino groups in BTEB pillars endow organic pillared MFI bifunctionality for cascade catalytic reactions. The synthesis of organic pillared MFI is a new addition to the structural modifications of lamellar zeolite with more open structures for processing bulky molecules. The resultant inorganic-organic hybrid zeolite structure creates new opportunities for potential applications of lamellar zeolites with tailored properties.

490696 Production of Furandicarboxylic Acid (FDCA) Using Biomass Derived Î³-Valerolactone

Abstract:

The demand for bio-based polymers is increasing globally due to their lower carbon footprint and independence from fossil fuels. 2,5-Furandicarboxylic acid (FDCA) is a potential bio-based alternative to terephthalic acid for the production of bio-based polymer polyethylene furanoate (PEF). However, the economical production of FDCA from carbohydrates must address technological challenges, such as instability of hydroxymethylfurfural (HMF), low solubility of FDCA, and requirement of stoichiometric amounts of base for the oxidation of HMF to FDCA.

We will report a process for production of FDCA from fructose which utilizes a solvent system comprising of Î³-valerolactone (GVL) and water to address the aforementioned challenges. In our process, a high concentration of fructose (15 wt%) is dehydrated to HMF at high yield (>70%), and HMF is then oxidized to FDCA over a heterogeneous catalyst (Pt/C) without intermediate separation. We demonstrate that the GVL/H_2O solvent system has high FDCA solubility which not only leads to high FDCA yield but also eliminates the need of a homogeneous base during HMF oxidation. We further show that FDCA can be separated from the GVL/H_2O solvent system by crystallization, and a highly pure (>99% purity) FDCA product is obtained. Furthermore, we demonstrate that FDCA is an effective catalyst for fructose dehydration in GVL/H_2O, which obviates the need of a mineral acid catalyst, thereby improving both the economic and the environmental impact of the process. We estimate through preliminary techno-economic modeling that the overall process is cost-competitive to state of art FDCA production processes.

492378 Spectroscopic and Modeling Study of Aluminum Active Species in Glucose Isomerization

Abstract:

The need to reduce greenhouse gas emissions and our dependence on fossil fuels has prompted considerable research in the production of fuels and chemicals from lignocellulosic biomass.[1] Among the top biomass-derived chemicals, 5-hydroxymethyl furfural (HMF) has gained great interest as a platform chemical.[2] HMF synthesis includes the hydrolysis of cellulosic biomass to glucose, glucose isomerization to fructose, and fructose dehydration to HMF.[1] Currently, HMF production is not commercial because industrial glucose isomerization uses enzymes, which are costly and incompatible with the Brønsted acidity necessary for fructose dehydration.[3] To overcome this challenge, homogeneous metal chlorides (e.g., $CrCl_3$, $AlCl_3$) have demonstrated effectiveness in direct glucose conversion to HMF in a single pot.[4] This rather unexpected result was rationalized by tandem catalysis where the metal species catalyzes glucose isomerization via Lewis acidity and hydrolyzes to produce Brønsted acid sites (H_3O^+) necessary for dehydration.[4] Direct, in situ experimental evidence in support of multiple catalytic species has though been lacking. This in turn hinders our ability to develop molecular descriptors and ultimately to predict better homogeneous and heterogeneous catalysts. In this work, we investigate the effect of combining Lewis ($AlCl_3$) and Brønsted (HCl) acids in glucose isomerization. For the first time, we demonstrate the interplay between the acids and from a practical position, optimize the concentration of the Lewis and Brønsted acids to maximize the formation of fructose. Simultaneously, the aluminum speciation in water is explored using in situ ^{27}Al nuclear magnetic resonance (NMR), dynamic light scattering (DLS), and inductively coupled plasma-mass spectrometry (ICP-MS), and the glucose-to-fructose isomerization is correlated with the various Al species. An $AlCl_3$ speciation model combined with kinetic results indicates that the hydrolyzed Al(III) complex $[Al(H_2O)_4(OH)_2]^{1+}$ is the active species in glucose isomerization and acts as a Lewis acid-Brønsted base bifunctional site.

References:

[1] van Putten, R. J., et al., *Chem Rev* **2013,** *113*(3), 1499-1597.

[2] Corma, A., et al. *Chem Rev* **2007,** *107*(6), 2411-2502.

[3] Nguyen, H., Vlachos, D., et al., *Acs Catal* **2016,** *6*(3), 1497-1504.

[4] Choudhary, V., Vlachos, D., et al., *J Am Chem Soc* **2013,** *135* (10), 3997-4006.

495639 Hydrodeoxygenation of Furfural over Multifunctional Catalysts

Abstract:

Hydrodeoxygenation (HDO) of biomass-derived platform molecules, such as furfural and hydroxymethylfurfural (HMF) is required for their utilization as fuels or chemical precursors. In our previous work [1,2], we have shown the utility of trifunctional Ru/RuO_x catalysts for the HDO of furfural and HMF. Based on theoretical and experimental studies, we have shown that these reactions require the presence of a metal and an oxide and proceed through a Lewis-acid-mediated transfer hydrogenation from isopropanol to furfural over the RuO_2, followed by the deoxygenation of furfuryl alcohol over O vacancies on the RuO_2 surface. These vacancies are generated in the reaction of surface oxygen with hydrogen produced in situ from the dehydrogenation of isopropanol over the metal sites [3].

However, the 2-methylfuran (2MF) product may further react over Ru, giving undesirable 2-methyltetrahydrofuran (MTHF) and 2-pentanol, and consuming valuable hydrogen [4]. Moreover, RuO_2 is reduced to metallic Ru under reaction conditions, resulting in the suppression of the HDO reaction. For these reasons, we have developed an alternative catalyst for the HDO of furfural to 2MF. To do this, we correlate the intrinsic HDO rates and the catalyst stability over a series of oxides with bulk and surface DFT descriptors and we choose an active and stable oxide. Extensive catalyst characterization, including microscopy, XPS, XRD, and XAS, is conducted. In a parallel effort, we correlate the selectivity of monometallic and bimetallic catalysts with the descriptors using correlations for thermochemistry and kinetics, such as group additivity and Polanyi relations. These studies demonstrate volcano like behavior and reveal best materials for HDO of biomass molecules.

References:

1. Panagiotopoulou, Vlachos. Appl. Catal A: General, 480 (2014) 17-24

2. Jae, et al. ChemCatChem, 6 (2014) 848-856

3. Mironenko and Vlachos. J. Am. Chem. Soc. (2016) 2016, 138, 8104â€"8113

4. Gilkey, et al. (2016) ChemSusChem, 9, 3113-3121

498198 Functionalization of 5-Hydroxymethylfurfural By Selective Etherification

Abstract:

There is a desire to use biomass to replace oil as the feedstock for modern chemicals and products. In this regard, polymers represent an attractive class of potential biobased materials. In this study, we seek to create tunable polymers using 5-hydroxymethylfurfural (HMF) as a renewable platform chemical. The properties of the final polymer can be tuned by modifying the side-chain of the monomer, which is added by the etherification of HMF with an alcohol. The R-group of the alcohol is then carried through the subsequent reactions to become the side-chain of the monomer. Here the etherification reaction was investigated first, by evaluating several catalysts with a variety of active sites and pore structures for their activity in this reaction and second by evaluating the interplay of R-group identity and active site structure on the etherification kinetics.

Several catalysts were evaluated including, H-FAU, H-MFI, H-BEA, H-MOR, amorphous SiO_2-Al_2O_3, Amberlyst-15, hydrotalcite, tungsten oxide, and Î³-Al_2O_3. It was found that catalysts with BrÃ¸nsted acid sites (*i.e.*, H-FAU, H-MFI, H-MOR, SiO_2-Al_2O_3, and H-BEA Zeolites, and Amberlyst-15) showed the highest selectivities to the ether product at 70-80% HMF conversion. These same catalysts also showed high production rates, most notably Amberlyst-15 and H-BEA. Catalysts possessing either basic sites (e.g., hydrotalcite) or Lewis acid sites (e.g., WO_3 and Î³-Al_2O_3) were an order of magnitude less active and less selective than those possessing BrÃ¸nsted sites. The ether production rates using H-BEA and Amberlyst-15 are very similar; however, H-BEA is ~40% more selective than Amberlyst-15. The improved selectivity observed for H-BEA-Zeolite, compared with Amberlyst-15, is hypothesized to be due to the ability of the zeolite to stabilize reaction intermediates. To further understand the capability of H-BEA several different SiO_2:Al_2O_3 ratios were evaluated for the etherification, revealing that the turnover frequency for HMF etherification is approximately constant and suggesting the absence of strong transport limitations with this zeolite morphology.

In the second portion of this study ethanol, butanol, phenol, and cyclohexanol were used to evaluate the production of a variety of ether products. Increasing the size of the alcohol (*e.g.*, cyclohexanol or phenol vs. ethanol or butanol) was found to have little effect on the ether production rate, suggesting that this rate is governed by the interaction between HMF and the catalyst. Based on these results, we hypothesize that the structure of H-BEA zeolite leads to improved activity for etherification, possibly due to the size compatibility between HMF and the pores of H-BEA. We suggest that Brønsted acidity is needed for etherification, and catalyst morphology may be highly important for obtaining high selectivity.

498584 Inhibition of Xylene Isomerization in the Production of Renewable Aromatic Chemicals from Biomass-Derived Furans

Abstract:

High selectivities (>90%) have been achieved with an H-BEA catalyst in the Diels-Alder cycloaddition, and subsequent dehydration, of 2,5-dimethylfuran (DMF) and ethylene to produce *p*-xylene.[1-4] This success is unexpected due to the ability of xylene to isomerize in the presence of zeolite acid catalysts at similar conditions. In this work, three possible mechanisms describing the conversion of 2,5-dimethylfuran to *p*-xylene were evaluated, where the dominant mechanism was indicated as the homogeneous, uncatalyzed cycloaddition of DMF and ethylene, followed by an acid catalyzed dehydration (Figure 1a).[5] The inhibition of xylene isomerization was attributed to large active site coverage of 2,5-hexanedione, the ring opened product of DMF. This is supported by the similar adsorption enthalpies of 2,5-hexanedione and the cycloadduct versus the less favorable adsorption enthalpy of *p*-xylene (Figure 1b). Though all reaction components were independently examined, only the presence of 2,5-dimethylfuran and 2,5-hexanedione, even at 1/50th the concentration of xylene, inhibited transalkylation and methyl shift reactions of *p*-xylene on H-Y (Si/Al 2.6) and H-BEA (Si/Al 12.5) zeolites. A combination of experimental techniques (thermogravimetric analysis, TGA, diffuse reflectance infrared Fourier transform spectroscopy, DRIFTS, and ^{27}Al NMR) and ONIOM computational method allowed for the investigation of adsorption energetics of relevant species on zeolite acid active sites.

Figure 1. Proposed mechanisms of reaction of 2,5-dimethylfuran and ethylene to *p*-xylene and adsorption enthalpies of compounds in reaction mixture. A) Proposed catalytic cycles converting dimethylfuran to *p*-xylene. Brønsted acid sites within H-BEA and H-Y zeolite catalyze conversion of DMF and ethylene to *p*-xylene and water. B) Heat of adsorption versus proton affinity in H-Y zeolite. The heat of adsorption correlates linearly with proton affinity above 200 kcal mol^{-1}. Deviation from the linear proton affinity prediction exists below 200 kcal mol^{-1}.

[1] Williams, C. L.; Chang, C.-C.; Do, P.; Nikbin, N.; Caratzoulas, S.; Vlachos, D. G.; Lobo, R. F.; Fan, W.; Dauenhauer, P. J. *ACS Catal.*, 2012, 2, 935-939.

[2] Chang, C.-C.; Green, S. K.; Williams, C. L.; Dauenhauer, P. J.; Fan, W. *Green Chem*. 2014, 16, 585-588.

[3] Patet, R. E.; Nikbin, N.; Williams, C. L.; Green, S. K.; Chang, C.-C.; Fan, W.; Caratzoulas, S.; Dauenhauer, P. J.; Vlachos, D. G. *ACS Catal.*, 2015, 5, 2367-2375.

[4] Williams, C. L.; Vinter, K. P.; Chang, C.-C.; Xiong, R.; Green, S. K.; Sandler, S. I.; Vlachos, D. G.; Fan, W.; Dauenhauer, P. J. *Catal. Sci. Tech.*, *6*, 178-187.

[5] Williams, C. L.; Vinter, K. P.; Patet, R. E.; Chang, C.-C.; Nikbin, N.; Feng, S.; Wiatrowski, M. R.; Caratzoulas, S.; Fan, W.; Vlachos, D. G.; Dauenhauer, P. J. *ACS Catal.*, 2016, 6, 2076-2088.

501076 Renewable Butadiene Production from Tetrahydrofuran over HZSM-5

Abstract:

Butadiene is an indispensable monomer to the production of synthetic rubber.[1] It is also used in other important copolymers with unique characteristics, such as styrene-butadiene rubber (SBR) and the impact-resistant ABS (acrylonitrile-butadiene-styrene) plastic. In the classic petrochemical approach butadience is recovered after separation of the C4 cut from naphtha crackers, where ethylene is the main product. However, the recent shale gas revolution increases reliance on non-conventional, shale-derived feeds for ethylene production, which leads to a shortage in C3/C4 olefins including butadiene.[2] Therefore, new, on-purpose butadiene production routes that are not coupled to ethylene production need to be developed.

In this work, we present a bio-based route to butadiene from tetrahydrofuran (THF), which can be produced by decarbonylation and reduction of biomass-derived furfural. The Brønsted-acidic zeolite HZSM-5 has shown good selectivity for THF dehydration to butadiene at low conversions, but the selectivity decreases at high conversions, due to byproducts such as formaldehyde and propene.[3] In this paper, we present different decomposition pathways of THF on HZSM-5 using density functional theory (DFT) calculations, and elucidate reaction mechanisms for the formation of butadiene, formaldehyde and propene. . In addition, we develop a microkinetic model in order to identify key surface intermediates and the rate-limiting step and compare with experimentally obtained apparent activation energies and kinetics.

References:

1. Wagemann, K., *Chemie Ingenieur Technik* **2014,** *86* (12), 2115-2134.

2. Bruijnincx, P. C. A.; Weckhuysen, B. M., *Angewandte Chemie-International Edition* **2013,** *52* (46), 11980-11987.

3. Abdelrahman, O. A.; Park, D. S.; Vinter, K. P.; Spanjers, C. S.; Ren, L.; Cho, H. J.; Vlachos, D. G.; Fan, W.; Tsapatsis, M.; J., D. P., *ACS Sustainable Chemistry and Engineering* **2017**, DOI: 10.1021/acssuschemeng.7b00745.

502462 Reaction Mechanisms Responsible for the Selective Vapor Phase Rearrangement of Furanics

Abstract:

The conversion of the abundant but highly reactive furfural to more stable fuel precursors is an important challenge in the conversion of biomass to higher value products. Cyclopentanone is a stable building block that can serve both as a commodity chemical and as a building block for diesel and jet fuels. The ring rearrangement of furfural to cyclopentenone in the aqueous phase in the presence of an acid catalyst is a well-studied reaction, commonly referred to as Piancatelli rearrangement, that is believed to require the presence of a condensed aqueous phase to proceed. Here we report alternative mechanisms for the conversion of furfural to cyclopentenone and cyclopentanone in a single vapor phase upgrading step using multifunctional catalysts. Direct vapor phase upgrading enables catalytic strategies that can be coupled with pyrolysis and torrefaction steps that donâ€™t require a condensed phase to proceed, where undesirable humin formation may occur. We illustrate how selective sites present at the interface of Ru nanoparticles on a TiO_2 support can facilitate this rearrangement in the vapor phase at temperatures as high as 400°C in a flow reactor, both with model compounds and real biomass vapors. We further show how water molecules can interact with these interfacial sites to modify the chemistry, which we propose is due to a different mechanism than the established Piancatelli rearrangement. We show that acidic coadsorbates that promote this reaction in the condensed aqueous phase, such as acetic acid, inhibit the vapor phase rearrangement by competing for active sites. This behavior is expressed in the context of a kinetic model, with a mechanism proposed involving binding to the metal as well as interaction with the interfacial site. The requirement of an interfacial site for this reaction is contrasted over supports with varying acidity, such as TiO_2 vs. Al_2O_3. We compare this chemistry over Ru with that over Pd and Ni to generate an overall correlation with ring rearrangement rate with the metal-carbon bond strength.

Interfacial sites and oxide overlayers manipulate selectivity for selective C-O cleavage and structure sensitive decarbonylation reactions as well. By systematically controlling the degree of oxidation while maintaining nanoparticle size, we illustrate the importance of partially oxidized sites on the Ni surface, and how these sites can influence the rate of C-O cleavage and decarbonylation reactions.

498959 Implications of High Accuracy Thermochemical Kinetics for H + CH_3 (+M) â‡„ CH_4 (+M) on Combustion Models

Abstract:

CH_3, the simplest alkyl radical, is an extremely stable intermediate that persists in large concentrations even in high temperature chemically reacting systems such as flames. Radical-radical recombination with H-atoms (R1: H + CH_3 â‡„ CH_4) is the dominant process for CH_3 removal in high temperature combustion and flame processes. As such, the net result of this reaction is to terminate reactive H-atoms and CH_3 radicals in flames. Numerous studeies have been conducted to directly measure k_1, however, k_1 has only been measured at low temperatures (< 700 K) while all data at high temperatures (> 1300 K) relevant to the pre-flame regime have been measured in the reverse direction. Hence, reconciling k_1 and k_{-1} requires accurate equilibrium constants spanning an extended range of temperatures. In the present work, we include all relevant anharmonic corrections to calculate an accurate Active Thermochemical Tables (ATcT) partition function for CH_3. The resulting nonrigid-rotator-anharmonic-oscillator partition function is used to determine thermochemical parameters (heat capacity, entropy, and enthalpy increment values) for CH_3 and K_{eq} values for reaction R1 over a wide temperature range (200-6000 K). With the newly evaluated equilibrium constant, literature experiments and theory (based on two-dimensional master equation calculations with a first principles energy and angular momentum transfer kernel) are used to obtain an accurate representation of the kinetics for the title reaction. In particular, the present work provides fits to $k_1(T,P)$ for N_2, Ar, and He as colliders for direct use in combustion modeling. The results of the new rate constant evaluation for R1 has a dramatic impact on combustion modeling. In particular, we assess the impact of incorporating anharmonic thermochemistry for CH_3 and the updated fits for k_1 (and k_{-1}) in current widely used literature models for simulations of CH_4/air and n-heptane/air laminar flame speeds and find that flame speed predictions can be affected by as much as 50%.

This work was supported by the Office of Basic Sciences, Division of Chemical Sciences, Geosciences, and Biosciences, U. S. Department of Energy, under Contract No. DE-AC02-06CH11357.

496829 Estimation of Thermodynamic Properties of Polycyclic Molecules By a Linear Regression Model

Abstract:

Though recent advances in *ab initio* methods have opened the door toward accurate calculations of molecular thermochemistry, large-scale theoretical studies are still often limited, at least initially, to the use of empirical methods to rapidly screen out unimportant species, so only the important species are the subject of CPU-time intensive quantum chemistry calculations. Among the empirical methods developed in the past decades, the Benson group additivity scheme is one of the quickest and most convenient methods to determine thermodynamic properties of molecules without needing the 3D molecular structures. It has achieved great success in accurate prediction of thermochemistry of small molecules and has been widely adopted in modeling software for on the fly prediction of thermodynamic parameters. However, because this additivity scheme simply uses the properties of individual chemical groups independently to calculate the entire property for a molecule, the contribution of the overall molecular structure to the property is usually not taken into account. This problem manifests itself for strained structures and can cause significant errors for polycyclic molecules with fused rings. Therefore, the application of the additivity method is often restricted to simple chemical systems without the presence of polycyclic species.

To address this issue, we have developed an improved model to predict the thermochemistry of polycyclic species.Â A regularized linear model was chosen to resemble the simplicity and interpretability of the additivity method. However, instead of following the definitions of the chemical groups of Bensonâ€™s scheme, we generated a comprehensive list of identifiers containing local (atoms, bonds, and angles) and/or nonlocal (rings) structural information and used these identifiers to compose feature vectors for molecules.Â The identifiers carrying nonessential information were eliminated by L1 regularization during model training so that those remained in the final model represent an optimum set of chemical units for the calculation of the thermodynamic property of interest with the contribution of the cyclic structures. These chemical units are human interpretable and conceptually equivalent to the chemical groups defined in Bensonâ€™s scheme but are selected by the model objectively without human intervention.

For a training set of 25,716 cyclic and polycyclic organic molecules made up of C, H, and O atoms, 408 identifiers were found to be needed for the calculation of formation enthalpies. The transferability of the trained model was validated on an independent test set of 2,858 molecules with a mean absolute error of 1.88 kcal/mol.

490830 Hydrogen Binding Versus Dissociation during Soot Formation: Insights from Reactive Molecular Dynamics

Abstract:

It has been proposed that two reactions are prominent in the earliest stages of soot nanoparticle formation; hydrogen abstraction via collision between a growing PAH (polyaromatic hydrocarbon) cluster and a hydrogen atom (yielding H_2 and a dehydrogenated cluster) and carbon addition via the binding of an alkane/alkene/alkyne chain to the cluster. We examine the first part of this process via reactive molecular dynamics using REAXFF C-H potentials. In simulations, cluster structures determined using the previously developed Atomistic Model for Particle Inception (AMPI) code of Violi and coworkers are equilibrated (at a predefined temperature) and a hydrogen atom with a specific velocity and impact parameter is driven towards the cluster. Depending upon the velocity, impact parameter, and collision orientation, we find that pure hydrogen abstraction is rare and instead at low velocities hydrogen binds to clusters. At increasing velocities, hydrogens rebound from the cluster surface and can drive hydrogen facilitated collision induced dissociation (CID) of the cluster. CID events yield smaller alkane/aklene/alkyne chains, which may serve to bind other clusters. Integration over all impact parameters and velocities for collisions (in accordance with the Maxwell Boltzmann distribution) yields the hydrogen binding rate coefficient and hydrogen induced dissociation rate coefficient for a given cluster structure. We discuss how these rates vary with temperature and cluster size.

499595 Effect of Composition Variation of Syngas Mixtures on NO_x Formation at High Pressure

Abstract:

Increasing interest in syngas combustion in gas turbines has led to highlighting the environmental feasibility, in particular the control of nitrogen oxide (NO_x) emissions. Trace amounts of NO_x species may be present during combustion and lean premixed (LPM) combustion method is one of the effective NO_x control technologies for gas turbines. Despite the proven feasibility of syngas, these fuels face variation in their composition due to different feedstock and production methods. Syngas can contain small amounts of hydrocarbons, such as methane, ethane and ethylene, and the presence of hydrocarbon species in the fuel can affect the NO_x emissions in the post-flame region. A detailed investigation of high-pressure syngas combustion is required to provide both practical information regarding NO_x emissions from gas turbine systems and fundamental understanding of reaction pathways to verify and develop chemical kinetic models.

In this study, a high-pressure burner facility capable of operating at pressures up to 15 bar, was designed and fabricated to simulate syngas combustion in a gas turbine. Syngas consisting mainly of H_2 and CO was blended with components such as nitric oxide (NO), nitrogen dioxide (NO_2) and trace amount of hydrocarbons prior to combustion at high pressure. Hot combustion products were supplied to the high pressure vessel by a flat flame burner operating with lean, premixed syngas-air mixtures with various H_2/CO ratios ranging between 0.25-1.0 and equivalence ratios between 0.5-1.0. Detailed NO_x speciation measurements in the post-flame region were conducted for different fuel compositions at various pressures up to 10 bar using Fourier transform infrared (FTIR) spectrometer. The experimental data generated from this project can be used to validate/develop kinetic models to predict NO_x emissions.

Numerical modelling of industrial burners for reduction of NO_x emissions using flamelet methods in combination with a newly developed postprocessor for fast and accurate emission prediction

DDI Werner Pollhammer, K1-MET GmbH, Linz, Austria
DI Dr. Christoph Spijker, Montanuniversitaet, Leoben, Austria
DI Michael Koller, Ebner Industrieofenbau GmbH, Leonding, Austria
Univ.-Prof. DI Dr. Harald Raupenstrauch, Montanuniversitaet, Leoben, Austria

Abstract

Current trends of research on combustion systems show increased interests on prediction of pollution. To meet the demands of modern combustion systems such as industrial burners detailed studies of their behaviour is needed. One of these pollutants which are commonly of interest is NO_x. The aim of the work is to develop a numerical toolbox that is able to predict NO_x emissions of industrial burners by numerical simulation using Computational Fluid Dynamics (CFD).

A set of numerical equations is solved for flow, temperature, pressure and species. Using CFD it is possible to model a combustion system using different approaches like flamelet methods or detailed chemical reaction systems. To predict the emission of NO_x from a flame a detailed chemical reaction system is necessary such as the GRI 3.0. mechanism [1]. The problem of using this mechanism is that it can be applied only for small geometries to avoid long calculation time or processing costs. Using a detailed chemical reaction system for every species a few chemical reactions have to be solved. Using the GRI 3.0. mechanism 325 reactions for 53 species have to be solved [2]. Also with modern computers it is not affordable to model industrial burners using this approach which would calculate NO_x emissions well.

The second main possibility for simulation of burners is to use flamelet methods [3][4]. Choosing this approach the combustion system is pre-calculated and reaction data will be stored in so called lookup tables. During simulation this reaction data will be accessed by reading them from the lookup tables. This approach allows fast calculation of temperature and flow field of gas burners as well as fast chemical reactions like formation of CO_2, H_2O or CO. This is the standard procedure for simulation of industrial burners for analysis of flame shape and temperature fields. The problem is that the described approach is not applicable for slow chemical reactions such as formation of NO_x. Therefor a detailed chemical reaction system is necessary. The problem is that on the one hand there is the detailed chemical reaction system which would predict NO_x emissions very well but takes too long for calculation. On the other hand there is the flamelet method which is fast but delivers only for flow fields, temperature fields as well as some species fields right results but models the formation of NO_x not very well. The idea now is to combine the benefits of both approaches and to postprocess the results out of the flamelet calculation using a detailed chemical reaction system. This means temperature and flow field are calculated by the fast flamelet calculation. After convergence of this calculation the solution for the flow field and temperature field as well as for some species fields are known. Subsequently for slow chemical reactions such as NO_x production, which cannot be predicted by the flamelet method, the postprocessor calculates the reactions using the already existing temperature and flow field from the flamelet calculation. This combination enables a much faster calculation of NO_x formation in the flame and combines the speed of flamelet method with the accuracy of using detailed chemical reaction systems with reasonable calculation costs [5].

Model description

For a more precise description of radiation effects it was necessary to implement the Discrete Ordinate Model (DOM) into the flamelet solver of OpenFOAM [6]. The standard radiation model is only valid for free stream flames with a constant surrounding wall temperature. The radiation model was used in combination with a Weight Sum of Gray Gases Model (WSGGM) which calculates the optical density of the gas by applying a mixture of temperature dependent absorption coefficients for calculation of the radiation beams [7].

For flamelet calculation a steady laminar flamelet for non-adiabatic combustion was used out of literature [3][4]. By defining a range for enthalpy defects in flamelet generation heat transfer in the system and out of the system can be taken into account. Flamelets were generated for different scalar dissipation rates which takes mixing time scales into account and for different enthalpy defects.

For modelling of turbulent effects the k-ε-model was used with a slight change according to literature data. For a more precise prediction of flame acceleration the constant $C_{2\varepsilon}$ was modified to 1.8. [8]. Out of data from Sandia Flame D boundary conditions for turbulence were generated for turbulent kinetic energy k and dissipation of it named ε [1].

The NO_x postprocessor uses a detailed chemical reaction approach the so called partially stirred reactor model implemented in OpenFOAM for calculation of chemical reaction [9]. It takes chemical reaction kinetics, for instance out of GRI 3.0., as well as turbulent mixing effects into account [2]. The solver uses a conservation equation for each species and chemical reaction is implemented as a source term in every species equation. This means for chemical reaction in every mesh cell the whole reaction mechanism consisting of hundreds of Arrhenius equations has to be solved.

To reduce the consumption of calculation power a special strategy of solving chemistry was implemented. As a matter of fact that the system seeks for a steady state solution that means after a certain time species composition will not chance any more in the cells. While post processing there is no change of temperature and turbulence because these fields are calculated by the previous flamelet calculation and they are constant during post processing. The reaction rates of chemistry then only depend on species concentration. If steady state is nearly reached in a cell and the species concentration varies just a bit there is no big difference between reaction rates of the previous and actual time step. An algorithm which analyses these changes was implemented. Using this algorithm the chemical reaction source terms were tabled from time step to time step. If the change in species is lower than 2 percent of the relative concentration of every species in a cell then the source terms of the previous time step is used. Using this algorithm about 95 percent of chemistry can be accessed by tabling and only 5 percent of the cells have to be calculated. This leads to huge saving of computational power and to a big speed up in simulation.

To avoid unphysical states there is also a check if one of the species concentrations goes under zero which is physically incorrect. If this occurs then chemistry will be calculated in the next time step. To avoid pseudo steady state solutions every 200 time steps the whole domain will be calculated.

The model validation was done by means of a comparison with measurements on the Sandia Flame D [1]. Therefor a two dimensional mesh consisting of 200 000 hexahedra was generated using blockMesh and analysed in terms of mesh independency. The benefit of validation on Sandia Flame D is that temperature, velocity, turbulence and species decomposition is also known at certain points in the flame. This means it is feasible to analyse the whole flame shape and its position dependent characteristics and not only the properties at the outlet. For reliable results it is important to be able to predict the temperature maximum very well. Comparison of species decomposition shows very good agreement in terms of oxidizer consumption and CO_2 formation [1]. Using the newly developed postprocessor a very accurate description of NO_x formation is possible which fits well to literature data [1].

Model validation on Sandia Flame D and results

First the flamelet model was used to describe the temperature, velocity and turbulence field of the flame. The following Figure 1 shows the solution of the Sandia Flame D case and its temperature field.

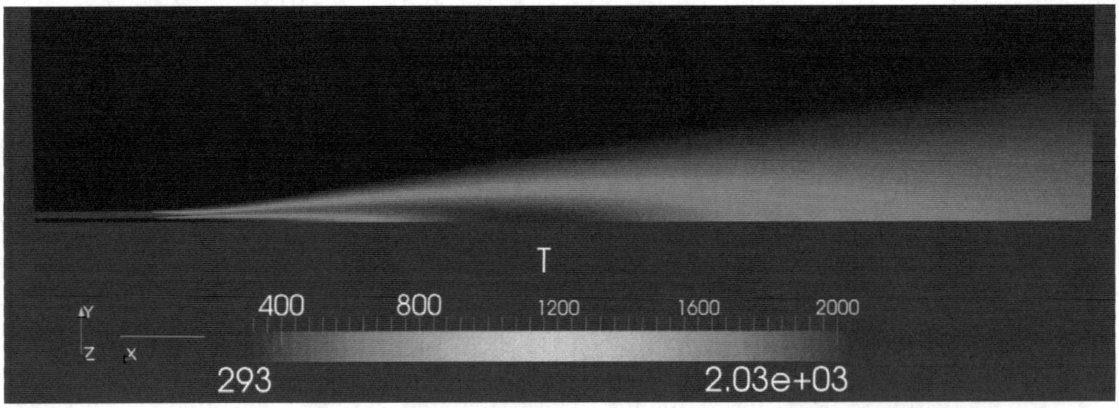

Figure 1: Temperature distribution in Sandia Flame D calculated using flamelet model

This simulation results were then compared to literature data [1] which is displayed in Figure 2. As it can be seen is that the temperature along the rotational axis fits very good to measurements. This is very important for an accurate prediction of nitrogen oxide formation.

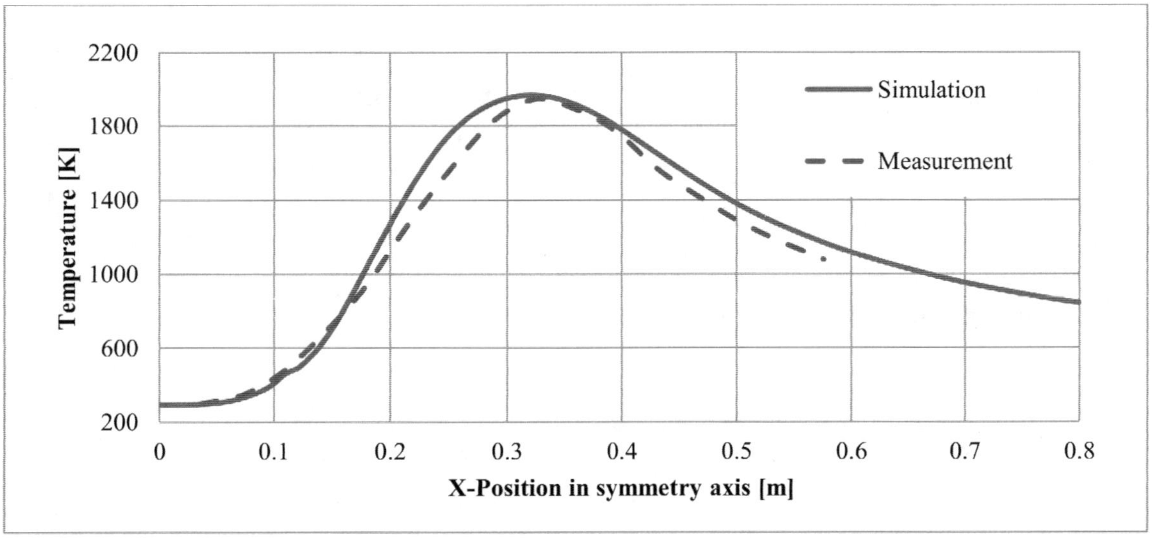

Figure 2: Plot of flame temperature comparing simulation and measurement data from Sandia Flame D [1]

Using this data for temperature, velocity, turbulence and species which can be predicted by flamelet the post processing was performed to calculate species like NO, N_2O, NO_2 which cannot be predicted accurate enough by flamelet. For this species a detailed chemical reaction mechanism was used. In this case the simulations were performed using GRI 3.0. [2]. The following Figure 3 shows the distribution of NO in the Sandia Flame D with three sections through the flame. At these positions a comparison with measurement data was made which is displayed in Figure 4.

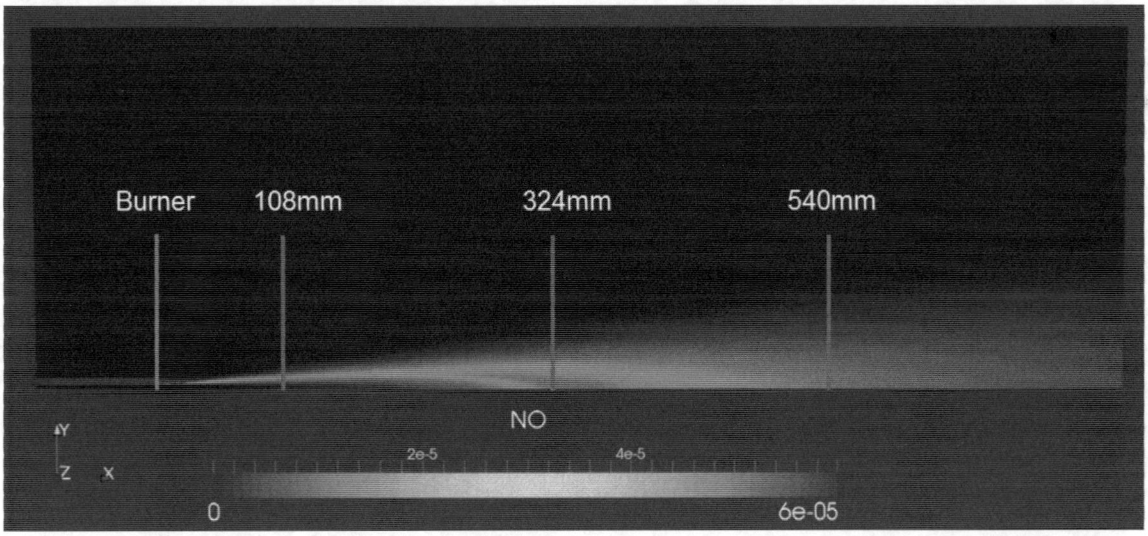

Figure 3: NO field of Sandia Flame D after convergence of postprocessor

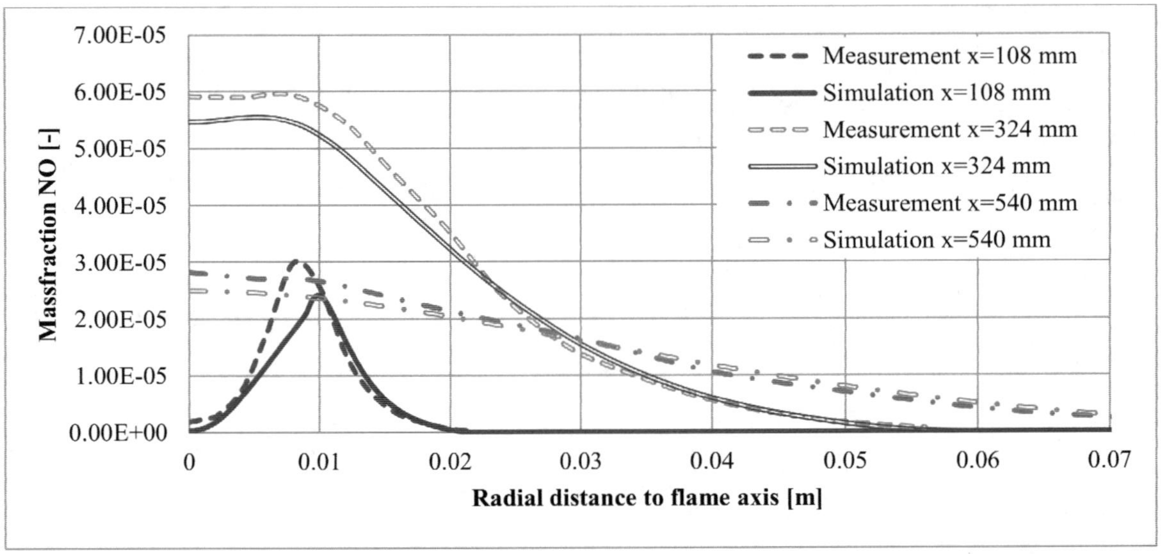

Figure 4: Comparison between simulation and measurement at three cuts through Sandia Flame D as shown in Figure 3 [1]

The result of the first validation is that the generated postprocessor is able to calculate nitrogen oxide formation in Sandia Flame D very well. It predicts slightly lower concentration but the accordance is good in the whole flame. So the developed model can be used for detailed investigations on nitrogen oxide formation in industrial burners.

Model validation on test chamber and results

A second model validation was done by means of an industrial test where a burner was operated in a test chamber. Measurements of wall temperatures by thermocouples and a detailed exhaust gas analysis were performed. Out of the measurements of 20 thermocouples at different positions on the furnace wall a temperature profile was generated to implement it as boundary condition for the flamelet calculation. This lab test was modelled using the developed simulation toolbox. A mesh of the whole test geometry was generated using cfMesh [10]. Out of mass flow and temperature measurements boundary conditions for simulation were calculated. The first validation for the model was a comparison between exhaust gas temperature in the lab test and simulation. These data were then used to start the NO_x post processing. Therefor the developed transient postprocessor was used to calculate the formation of NO_x and various other species which are not predicted well by the flamelet method. At that point it is only necessary to calculate the chemical reaction source terms without solving flow and temperature fields which are now constant in this calculation and already known from the flamelet calculation. This way the simulation is much faster and calculation of NO_x formation now is possible in affordable time using the detailed chemical reaction mechanism with the benefit of a much more accurate description of NO_x formation. For the whole geometry a mesh independency study was performed on the one hand for the flamelet model and on the other hand for the postprocessing. The mesh was generated with a cell size which delivers mesh independent results. For the test chamber with a height of 2.3, a width of 1.7 and a depth 2.5 meters the mesh consists of 5.26 million polyhedral cells.

Figure 5 shows the result of the post processing of the test chamber simulation.

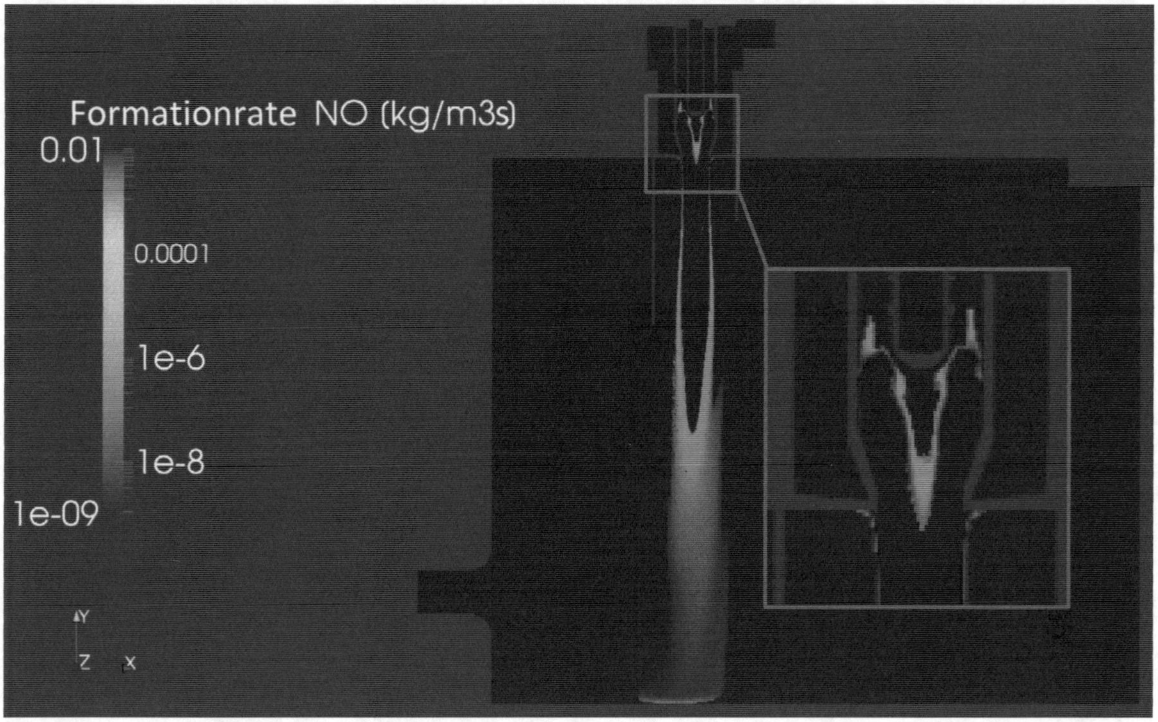

Figure 5: Areas of NO formation in test chamber, calculated using developed postprocessor

This test was performed twice. One test was made with air at ambient temperature and one with preheated air with 199°C.

Table 1: Comparison between simulation results and measurements in test chamber

Formation of NO_x	**Simulation [ppm]**	**Measurement [ppm]**
Cold burner air with 28°C	63	65
Preheated burner air with 199°C	78	92

As it can be seen in Table 1 there is also a good accordance for the postprocessor with measurements in the big test chamber. The case with air at 28°C shows the best accordance. The second lab test using preheated air shows the right trend by increasing NO_x formation in terms of the cold air case but shows lower NO_x emissions like the measurement at the test chamber. This effect of lower NO_x concentrations can also be seen in Figure 4 that the model predicts lower nitrogen oxide emissions than the measurements. Important is that the model predicts relative improvements and trends. So now the model was used to perform a geometry study and to test different improvements of the burner geometry by changing the geometry and using the postprocessor for analyzation.

Outlook

Now the model is used to analyse the formation zones of nitrogen oxides in the burner. The next steps are further improvements of the code efficiency and to find more strategies to reduce the calculation time. After some improvement steps of simulation a prototype of the new developed burner will be build a new chamber test will be performed to check these improvements also in a real test.

Acknowledgements

The authors gratefully acknowledge the funding support of K1-MET GmbH, metallurgical competence center. The research program of the competence center K1-MET is supported by COMET (Competence Center for Excellent Technologies), the Austrian program for competence centers. COMET is funded by the Federal Ministry for Transport, Innovation and Technology, the Federal Ministry for Science, Research and Economy, the province of Upper Austria, Tyrol, and Styria and the Styrian Business Promotion Agency.

References

[1] Barlow R, Frank J. *Piloted CH4/Air Flames C, D, E, and F – Release 2.1.* Available at: http://www.sandia.gov/TNF/DataArch/FlameD/SandiaPilotDoc21.pdf; 2007 [accessed 07.04.2017].

[2] Smith GP, Golden DM, Frenklach M, Moriarty N, Eiteneer B, Goldenberg M, Bowman T, Hanson RK, Song S, Gardiner WC, Lissianski, Jr., Vitali V., Qin Z. *GRI Mech 3.0.* Available at: http://www.me.berkeley.edu/gri_mech/; 1999 [accessed 11.04.2017].

[3] Cuoci A, Frassoldati A, Faravelli T, Ranzi E. Formation of soot and nitrogen oxides in unsteady counterflow diffusion flames. *Combustion and Flame* 2009;156:2010–22, doi:10.1016/j.combustflame.2009.06.023.

[4] Cuoci A, Frassoldati A, Faravelli T, Ranzi E. Kinetic Modeling of Soot Formation in Turbulent Nonpremixed Flames. *Environmental Engineering Science* 2008;25:1407–22, doi:10.1089/ees.2007.0193.

[5] Spijker C. *Unsteady Laminar Flamelet Modellierung zur Beschreibung von Mündungsmischbrennern*. Masterthesis, Montanuniversitaet. Leoben; 2010.

[6] Thynell ST. Discrete-ordinates method in radiative heat transfer. *International Journal of Engineering Science* 1998;36:1651–75, doi:10.1016/S0020-7225(98)00052-4.

[7] Barlow RS, Karpetis AN, Frank JH, Chen J-Y. Scalar profiles and NO formation in laminar opposed-flow partially premixed methane/air flames. *Combustion and Flame* 2001;127:2102–18, doi:10.1016/S0010-2180(01)00313-3.

[8] Pfeiler C, Spijker CJ, Raupenstrauch H. CFD als Werkzeug in der Industrieofentechnik. *Berg Huettenmaenn Monatsh* 2011;156:347–52, doi:10.1007/s00501-011-0018-z.

[9] OpenFOAM. *OpenFOAM 2.4.0.* Available at: https://openfoam.org/; 2017.

[10] cfMesh. *Professional Meshing Solutions - cfmesh.com.* Available at: http://cfmesh.com/; 2017 [accessed 11.04.2017].

491775 A CFD Model-Based Optimization of a Low NO_X Burner in Integrated Process System

Abstract:

Although process burners are frequently used in many industries such as refeinring, petrochemical and chemical processes, there is a serious concern about emission control of these equipment. Recently, the deveopment of new numerical approaches such as computational fluid dynamics (CFD) has led to significant improvements in emission control of industrial processes. The proper analysis and testing leads to more efficient combustion, less physical testing and lower cost for full scale equipment burners.

In this study, we used CFD simulations to design, develop and optimize a low NOx burner which produces heat for methane steam reforming. The CFD method helped to design a burner with angled air holes arranged in a swirl pattern. This optimized design increased the mixing near the burner plate that led to lower peak flame temperature and therefore lower NOx emissions. The designed burner was built and experimentally tested and showed a major NOx reduction in steam reforming hydrogen generation process. In addition, reformer performance improved significantly due to the better temperature distribution by the new CFD optimized burner.

496078 Mapping the Stable Region of Operation of Homogenously Charged Compression Ignition (HCCI) Engine for Methane and DMF Fuel Blend

Abstract:

Abstract:

The emission legislations are transmuting to be rigid in accordance to reduce the emissions from the IC engines. Research has been focused to ameliorate both Spark Ignition and Compression Ignition engine in order to make them more efficient. Over the previous decade, major research has been focused on Homogenous Charge Compression Ignition (HCCI) engines. This engine has the potential to consummate the eminent feature of Spark Ignition engine, with no emission of NOx and Soot leading to cleaner combustion, the best of CI engine with higher efficiency and low fuel combustion along with lower carbon dioxide emissions[1].

The focus of this work is to map the stable regions of operation of HCCI engine using Methane and Dimethyl Furan blend as fuel. DMF is comparable to gasoline since it has an energy density of 40% higher than ethanol. Also, Burning DMF requires 33% less air compared to the same amount of gasoline since its stoichiometric air to fuel ratio is 10.72. The Simulations of HCCI engine were carried out in CHEMKIN, where the model used was a closed transient CSTR[2]. The main objective of the study was to depict the stable regions of operation of HCCI engine with and without Exhaust gas recirculation. EGR refers to a part of the exhaust stream recycled back to the engine inlet along with the fuel mixture[3].

Parameters such as EGR Temperature, EGR percentage and equivalence ratio were varied to study the effect on the sustainable emissions at the exhaust for CO, CO_2 and NOx. Detailed Kinetics of DMF obtained from NCBI was used. Initial understanding of the working of HCCI engine was done using Methane as source fuel, keeping EGR temperature and HCCI engine temperatures as a constant at 600K and 447K and the equivalence ratio were varied from 0.5 to 0.7 and the EGR percentage were varied from 0 to 30%. The in-cylinder temperature, pressure and species profile were obtained as a function of Crank Displacement angle (CAD).

The Figure (1) shows the effect of equivalence ratio on the in-cylinder temperature, pressure and species profile for a engine temperature of 447K and EGR temperature of 600K for 30% EGR. Increase in equivalence ratio states the presence of higher amount of fuel where as smaller ratios represent deficit of fuel in the mixture. It could be observed that Peak temperature and pressure increases with increase in equivalence ratio. The mole fraction of CO increase with increase in equivalence ratio, this is because of the lack of oxygen available for complete combustion. In case of Thermal NO production rate, it increases with increase in equivalence ratio due to increase in temperature which leads to increased production of NO at higher temperatures. It was observed from the initial study using Methane as a fuel that, for Lesser exhaust emissions, the fuel must be lean with a higher percentage of EGR.

Fig 1: Variation of In-cylinder temperature, pressure, mole fraction of CO, CO_2 and Thermal NO production rate with Equivalence ratio

The study was further extended to DMF as an additive to Methane and the parameters such as EGR percentage, Equivalence ratio and EGR temperature were varied to find the stable region of operation of HCCI engine for Methane and DMF blend. The effect of Equivalence ratio on the in-cylinder temperature, pressure and species properties for 0% EGR for an engine temperature of 447K and EGR temperature of 600K were studied initially. The figure (2) shows the effect of Equivalence ratio on Peak CO_2 mole fraction in the cylinder for different percentage of Methane and DMF blend. It could be observed that the peak mole fraction of CO_2 increases with increase in Equivalence ratio and also increases with increase in blend percentage. The latter is due to the increase in the carbon availability in the cylinder with increase in fuel.

Fig 2: The Variation of Peak Mole Fraction of CO_2 with percentage of the DMF blend for different Equivalence ratios

As a next step, it is intended to study the effect of EGR percentage and EGR mixing temperature on Methane and DMF blend as shown in the Table 1 and analyse the Peak in-cylinder temperature, pressure and species properties for the different parameters combinations and map the stable region of operation of HCCI engine for Methane and DMF fuel blend.

Table 1: Parameters to be estimated for the Blend

Blend	**Parameters**
Methane & DMF blend	EGR Percentage – 0%, 10%, 20% & 30% EGR temp – 400K, 500K & 600K Equivalence ratio – 0.5, 0.6 & 0.7 Blend Ratio – 0%, 50% & 90%

Keywords:

Exhaust Gas Recirculation, DMF, Crank Rotation angle.

References:

1. Harrison R. M. ed. (1996) Pollution, Causes, Effects and Control. 3rd Ed. Cambridge, The Royal Society of Chemistry.

2. Xu H., Liu, M., Gharahbaghi, S., Richardson, R., Wyszynski, M., Megaritis, T., "Modelling of HCCI Engines: Comparison of Single-zone, Multi-zone and Test Data", SAE 2005-01-2123

3. Heywood JB, Internal Combustion Engine Fundamentals, Mc Graw Hill; 1988

500365 Rational Design and Synthesis of Surfactant-Free Pt-Based Ternary Nanoalloys As Superior Oxygen Reduction Reaction Electrocatalysts

Abstract:

Efficient and low cost electrocatalysts are indispensable for electrochemical oxygen reduction reactions (ORR) during proton exchange membrane fuel cell (PEMFC) operations. Yet, the challenge for solution phase synthesis of nanocatalysts with low precious metal loading, high durability and catalytic activities remains in tailoring their architectures and compositions without using unwanted chemicals and surfactants/ligands that are known to impede their active catalytic sites. We present to synthesize ternary nanoalloys (NAs) of Pt with transition metals (Co, Cu, Ni, Ti, Ru, Mn) as ORR electrocatalysts using our recently developed laser ablation synthesis in solution-galvanic replacement reaction (LASiS-GRR) technique as a facile and surfactant-free nanomanufacturing route. The specific choice of the elemental compositions is driven by the respective target metal/metal salt redox potential gaps as well as the target metal and/or, metal oxide solubility in desired acids. The high-energy thermodynamics of the LASiS-GRR process enables control of the sizes, elemental compositions and distributions of the ternary NAs through systematic tuning of initial metal salt concentration, pH, laser fluence and ablation time. Specifically, the PtCuCo NAs synthesized with an elemental composition of 72:12:16 (Pt:Co:Cu) indicates the best ORR catalytic activity. The NAs largely possess shell-core structure with the shell comprising mostly of Pt and minor amount Cu, along with a uniformly alloyed PtCuCo core. Mass and specific activities for ORR performance of the NAs indicate a 4 and 6.5-fold improvements respectively over the corresponding activities of commercial Pt/C. We attribute the enhanced activity to 1) our surfactant/ligand-free synthesis technique that prevents catalytic site degradations; and 2) minor alloying of the second transition metal Cu that shifts back the Pt d-band center to an optimal position between that of Pt and the PtCo binary NAs, thereby tuning their binding affinities for both oxygen and oxygenated species. Finally, this work establishes the versatility of the LASiS-GRR technique through the synthesis of other ternary NAs (PtRuNi, PtCoMn, PtNiTi) that also exhibit reasonably good ORR activities.

496706 Improved Activity and Durability of Ionic Liquid Composite Nanoporous Nanoparticle Electrocatalysts for Oxygen Reduction Reaction

Abstract:

The development of efficient and stable catalysts for the cathodic oxygen reduction reaction (ORR) is one of the greatest challenges for the exploitation of polymer electrolyte membrane fuel cells (PEMFCs). Dealloying has shown increasing utility in the field of electrocatalysis as a tool for the synthesis and development of nanoporous materials possessing high surface-to-volume ratios with controlled morphology and compositional gradient (core-shell structure) [1]. After electrochemical dealloying, the open, bicontinuous, three-dimensional nanoporous nanoparticle electrocatalysts exhibit dramatically enhanced electrocatalytic properties [2,3]. We have taken advantage of the free volume within these nanoporous nanoparticles and created a composite catalyst architecture through the incorporation of a hydrophobic, protic ionic liquid (IL), specifically [MTBD][beti]. Modification of the interface through the formation of IL thin film on the surface of an electrocatalyst yields enhanced catalytic rates by chemically biasing reactants to the surface and changing the degree of solvation/stabilization of adsorbed intermediates.

In the development of efficient electrocatalysts for ORR, durability is too often ignored in the pursuit of higher activities. Here we studied the manner by which ILs impact the stability of 3-dimensional, nanoporous nanoparticle electrocatalysts. In addition to the standard mechanisms of catalyst degradation including Pt dissolution/Ostwald ripening and coalescence/aggregation, nanoporous materials can lose electrochemically active surface area (ECSA) through surface smoothening driven coarsening. We use a combination of in-situ and ex-situ experimental techniques, under accelerated degradation conditions, to develop insight into the structural and compositional evolution of nanoporous PtNi nanoparticles (np-NiPt) formed through the dealloying of $Pt_{20}Ni_{80}$precursor nanoparticles, and demonstrate that facilitated diffusion of stable surface atoms under relevant electrochemical conditions will drive evolution of the particle morphology from a porous to a nonporous structure. With a better understanding of the interplay between nanoporous structure coarsening and transition metal loss, we have developed strategies to mitigate coarsening through incorporating ILs at the metal/electrolyte interface to directly mitigate electrochemically enhanced surface diffusion, which leads to ECSA loss. The hydrophobicity of these ILs is known to change the interaction of water with the surface, positively shifting the onset potential of Pt oxidation [3]. The interfacial IL acts to limit the charge dependent formation of a surface metal/electrolyte anion complex which is responsible for the potential dependent enhancement in surface diffusion. The oxygen reduction reaction activity, ECSA and residual transition metal content of these complex composite nanoporous electrocatalysts was tracked as a function of cycle number during accelerated degradation testing providing insight into the mechanism of catalyst degradation and the effect of ILs on catalyst stability.

References

[1] Snyder, J.; McCue, I.; Livi, K. & Erlebacher, J., *J. Am. Chem. Soc.* **134,**8633-8645 (2012).

[2] Snyder, J.; Fujita, T.; Chen, M. W. & Erlebacher, J., *Nature Mater.* **9,**904-907 (2010).

[3] Snyder, J.; Livi, K. & Erlebacher, J., *Adv. Funct. Mater.* **23,** 5494-5501 (2013).

489132 Particle Atomic Layer Deposition for Stabilization of Oxygen Reduction Catalysts

Abstract:

As a highly efficient, portable, non-greenhouse gas emitting source of electrical power, the proton exchange membrane (PEM) fuel cell holds great potential as a replacement for the internal combustion engine. One of the most significant development challenges in the commercialization of the PEM fuel cell is the long-term durability of the oxygen reduction reaction (ORR) catalyst. Cathode potential cyclingâ€"resulting from the variable voltage loads imposed during vehicular operationâ€"is known to promote agglomeration and growth of the Pt nanoparticle catalyst, which in turn degrades the power output of the fuel cell. In this work, atomic layer deposition (ALD) was used to deposit protective TiO_2 films onto both commercial and ALD-synthesized Pt/C catalyst in order to increase its electrochemical durability. Deposition half-reactions during fluidized bed particle ALD were observed via in-situ mass spectrometry, and chemisorption analysis was used to quantify changes in Pt surface area following TiO_2 deposition. Rotating disk electrode analysis indicated that electrochemical surface area (ECSA) and activity were slightly depressed by the addition of TiO_2 ALD layers. Activity retention during electrochemical durability testing was modestly improved, with TiO_2-coated catalysts retaining upwards of 60% of their initial activity. Examination of catalysts before and after testing also showed that TiO_2 ALD barriers were effective at preventing Pt agglomeration under fuel cell operating conditions. Further testing of the catalysts in membrane electrode assemblies was used to quantify effects of TiO_2 ALD on the hydrophobicity of the catalysts, which has been previously demonstrated as an important factor in fuel cell device operation with ALD-fabricated catalysts.

500977 A Mechanistic Insight into the Electrocatalytic Behavior of Mixed Oxide Cathode Catalysts for Li-O_2 Batteries

Abstract:

The recent advancements in renewable energy generation will require investments in the development of energy storage technologies. Generally, the generated renewable energy can be stored in multiple forms, such as chemically (i.e., water splitting to form hydrogen fuel), mechanically (i.e., flywheel that converts electrical to rotational energy), or electrochemically (i.e., batteries that can store electrical energy). Among the many different electrochemical energy storage technologies, lithium-oxygen (Li-O_2) batteries exhibit the highest theoretical energy density (11,680 Wh/kg) [1]. Unlike the current state-of-the-art Li-ion batteries, where the technology has been explored for many years and reached a ceiling performance, Li-O_2 batteries are still far from reaching the theoretical energy density, thus are still under wide investigation [2]. In order to improve the performance of these systems, a fundamental understanding of the reaction mechanism involved in both the oxygen reduction reaction (ORR) and oxygen evolution reaction (OER) during discharge and charge respectively, at the cathode of these systems is necessary. It is in part due to the reported discharge product, Li_2O_2, being a high bandgap insulator, thereby making the electron transfer during OER extremely challenging [1]. We have shown previously that the performance of the Li-O_2 battery cathodes can be improved upon addition of non-precious metal oxide catalysts, known as Ruddlesden-Popper (R-P) oxides [3]. To optimize the performance of these catalytic materials, a fundamental understanding of the mechanism that leads to this improvement is required. In this contribution, we combine electrochemical and spectroscopic techniques to unravel the complex chemistry involved in Li-O_2 cathodes containing R-P oxide electrocatalysts. The effect of the nature of the discharged product species during ORR and its effect on lowering the overpotential losses during OER will be discussed. We finalize by devising ways of enhancing the performance of R-P oxides for oxygen reduction and evolution in Li-O_2 batteries.

References

1. Girishkumar, G., et al., *Lithium - Air Battery: Promise and Challenges.* Journal of Physical Chemistry Letters, 2010. **1**(14): p. 2193-2203.

2. Armand, M. and J.M. Tarascon, *Building better batteries.* Nature, 2008. **451**(7179): p. 652-657.

3. Nacy, A., X.F. Ma, and E. Nikolla, *Nanostructured Nickelate Oxides as Efficient and Stable Cathode Electrocatalysts for Li-O-2 Batteries.* Topics in Catalysis, 2015. **58**(7-9): p. 513-521.

494499 Characterization of Feni Alloy Nanoparticles for Oer As a Function of Iron-Nickel Composition

Abstract:

Iron-doped nickel hydroxide catalysts are now recognized as the leading catalyst composition for alkaline electrochemical water splitting, and in particular, the anodic reaction of oxygen evolution [1]. The iron-nickel bimetallic composition of these hydroxide catalysts has been demonstrated to be critical for optimal catalyst performance [2]. However, few studies thus far have investigated if the same trend in catalyst activity is observed when catalyst morphology is scaled from bulk films to the nanoscale [3], where the overall field of iron-nickel hydroxide/oxide catalysts is in its infancy in terms of developing, synthesizing, and fully characterizing these nanoparticulate bimetallic hydroxides. Further, preliminary stability studies performed on iron-nickel [4] and iron-cobalt [5] thin films suggests that an evaluation of the stability of the bimetallic catalysts as a function of composition will be critical to identifying the best-performing catalyst compositions to pursue. In particular, Burke et al. report that while iron atomic compositions of less than 54 % resulted in high catalytic activities, the activities of these iron-cobalt bimetallic catalysts decreased by up to 62 % over two hours [5]. In the iron-nickel catalyst field, little work has evaluated how catalyst stability is affected by bimetallic composition in either thin film or nanocatalyst morphologies. Thus far, most studies have primarily focused on how the iron-nickel bimetallic composition affects catalyst performance from the perspective of lowering the overpotential, thereby lowering the kinetic limitation to the oxygen evolution reaction [2, 6].

In this talk, our ongoing work to develop and characterize a suite of bimetallic iron-nickel alloy hydroxide nanoparticle catalysts for the oxygen evolution reaction will be discussed. In particular, results will be presented for three iron-nickel compositions (1:5, 1:1, and 5:1 mol:mol Fe:Ni) and trends in both overpotential and stability will be discussed. Nanoparticle characterization via high resolution transmission electron microscopy, elemental analysis, and synchrotron-based x-ray absorption spectroscopy will be presented and discussed as a function of bimetallic composition. The role of nanoparticle structure in controlling activity and stability will be discussed, and electrochemical performance data will be used to demonstrate that both activity (i.e., overpotential) and stability (i.e., degradation rate as mV/hr) are critical to evaluate as we develop these catalysts for water splitting.

References

[1] D. Friebel, M.W. Louie, M. Bajdich, K.E. Sanwald, Y. Cai, A.M. Wise, M.-J. Cheng, D. Sokaras, T.-C. Weng, R. Alonso, R.C. Davis, J.R. Bargar, J.K. Norskov, A. Nilsson, A.T. Bell, Identification of highly active Fe sites in (Ni,Fe)OOH for electrocatalytic water splitting, J. Am. Chem. Soc., 137 (2015) 1305â€"1313.

[2] L. Trotochaud, S.L. Young, J.K. Ranney, S.W. Boettcher, Nickel-iron oxyhydroxide oxygen-evolution electrocatalysts: The role of intentional and incidental iron incorporation, J. Am. Chem. Soc., 136 (2014) 6744-6753.

[3] J.A. Bau, E.J. Luber, J.M. Buriak, Oxygen Evolution Catalyzed by Nickelâ€"Iron Oxide Nanocrystals with a

Nonequilibrium Phase, ACS Applied Materials & Interfaces, 7 (2015) 19755-19763.

[4] M. Gong, Y. Li, H. Wang, Y. Liang, J.Z. Wu, J. Zhou, J. Wang, T. Regier, F. Wei, H. Dai, An Advanced Niâ€"Fe Layered Double Hydroxide Electrocatalyst for Water Oxidation, Journal of the American Chemical Society, 135 (2013) 8452-8455.

[5] M.S. Burke, M.G. Kast, L. Trotochaud, A.M. Smith, S.W. Boettcher, Cobalt-iron (oxy)hydroxide oxygen evolution electrocatalysts: The role of structure and composition on activity, stability, and mechanism, J. Am. Chem. Soc., 137 (2015) 3638-3648.

[6] M.W. Louie, A.T. Bell, An investigation of thin-film Ni-Fe oxide catalysts for the electrochemical evolution of oxygen, J. Am. Chem. Soc., 135 (2013) 12329-12337.

495038 Active Sites in Nitrogen-Doped Carbon Nanostructures for Oxygen Reduction and Oxygen Evolution Reactions

Abstract:

Regenerative polymer electrolyte membrane (PEM) fuel cell technology has a promising future for sustainable power generation. In regenerative PEM fuel cells, oxygen reduction (ORR) and oxygen evolution (OER) are two important reactions that take place in the fuel cell mode and electrolyzer mode, respectively. State-of-the-art catalysts used for these reactions – Pt for ORR and Ir (or Ru) for OER – are expensive and found in limited reserves. Also, both these catalysts are active only for one of the two reactions, thus necessitating the development of cheaper materials that can serve as efficient bifunctional catalysts for both ORR and OER in regenerative PEM fuel cells. We have reported nitrogen doped carbon nanostructures, denoted as 'CN_x' [1], to be active, selective and stable catalysts for ORR in acidic medium. Our current focus is to further evaluate the applicability of CN_x as bifunctional ORR and OER catalysts, and also understand the nature of active sites in these materials.

Electrocatalytic activity measurements using rotating ring disk electrode (RRDE) show significant ORR and OER activity of CN_x compared to state-of-the-art Pt and Ir-based catalysts. Analysis of bifunctional electrocatalytic activity demonstrates much better bifunctional characteristics of CN_x compared to Ir/C and Pt/C [2].

Use of poisoning probes is a useful methodology to systematically examine the nature of active sites in catalytic materials. However, due to the absence of a metal centered active site, CN_x materials are not susceptible to poisoning by CO, H_2S or CN^- [3, 4]. We have recently identified phosphate anions ($H_2PO_4^-$) as probe molecules for poisoning CN_x catalysts in an acidic ORR environment [5]. In-situ RRDE measurements show a linear decrease in ORR activity with increasing $H_2PO_4^-$ anion concentration. Transmission IR spectroscopy and Raman spectroscopy reveal the presence of $H_2PO_4^-$ species on H_3PO_4–soaked CN_x. A linear decrease in ORR kinetic current with a decrease in pyridinic nitrogen content obtained using X-ray photoelectron spectroscopy (XPS) is observed. This poisoning phenomenon is consistent with two possible active site models: (i) pyridinic N as the active site which is rendered inactive by protonation [6, 7], and (ii) carbon sites adjacent to pyridinic N as the active site (i.e. pyridinic N as a marker for the active site) wherein poisoning is caused by a site-blocking effect due to adsorption of $H_2PO_4^-$ species on carbon. OER activity has also been found to increase with increasing pyridinic nitrogen site density in CN_x samples.

References

[1] P.H. Matter, U.S. Ozkan, Non-metal catalysts for dioxygen reduction in an acidic electrolyte, Catal. Lett., 109 (2006) 115-123.

[2] K. Mamtani, D. Jain, D. Dogu, A. C. Co, U. S. Ozkan, Insights into Oxygen Reduction Reaction (ORR) and Oxygen Evolution Reaction (OER) Active Sites for Nitrogen-doped Carbon Nanostructures (CNx) in Acidic Media, Applied Catalysis B: Environmental, Under Review, (2017).

[3] D. von Deak, D. Singh, J.C. King, U.S. Ozkan, Use of carbon monoxide and cyanide to probe the active sites on nitrogen-doped carbon catalysts for oxygen reduction, Appl. Catal. B-Environ., 113-114 (2012) 126-133.

[4] D. von Deak, D. Singh, E.J. Biddinger, J.C. King, B. Bayram, J.T. Miller, U.S. Ozkan, Investigation of sulfur poisoning of CNx oxygen reduction catalysts for PEM fuel cells, J. Catal., 285 (2012) 145-151.

[5] K. Mamtani, D. Jain, D. Zemlyanov, G. Celik, J. Luthman, G. Renkes, A.C. Co, U.S. Ozkan, Probing the Oxygen Reduction Reaction Active Sites over Nitrogen-Doped Carbon Nanostructures (CN x) in Acidic Media Using Phosphate Anion, ACS Catalysis, 6 (2016) 7249-7259.

[6] G. Liu, X. Li, P. Ganesan, B.N. Popov, Studies of oxygen reduction reaction active sites and stability of nitrogen-modified carbon composite catalysts for PEM fuel cells, Electrochim. Acta, 55 (2010) 2853-2858.

[7] X. Li, G. Liu, B.N. Popov, Activity and stability of non-precious metal catalysts for oxygen reduction in acid and alkaline electrolytes, J. Power Sources, 195 (2010) 6373-6378.

497178 Atomic Iron-Dispersed Electrocatalysts Derived from Metal-Organic Framework for Oxygen Reduction

Abstract:

To significantly reduce the cost of proton exchange membrane (PEM) fuel cells, current Pt must be replaced by platinum-metal-group (PGM)-free catalysts for the oxygen reduction reaction (ORR) in acid. We report here a new class of high-performance atomic iron-rich catalyst via chemical doping of iron ions into zinc-zeolitic imidazolate framework (ZIF), a type of metal-organic framework (MOF). The novel synthetic chemistry enables accurate size control of Fe-doped ZIF catalyst particles with a wide range from 20 to 1000 nm without changing chemical property. This provides a great opportunity to increase the density of active sites. An optimal size at 50 nm exhibited the highest activity; further reducing size to 20 nm results in particle agglomeration with decreased active sites. In addition to the size effect, we carefully explored the thermal activation mechanism to provide insight into the origin of active site formation. A temperature of 800oC was identified as the critical point to generate active sties with new bond formation associated with nitrogen doping, carbon, and iron. 1100 oC was found for the best activity with maximum density of active sites along with optimal porosity, graphitic nitrogen doping, and carbon structures. We also found the control of optimal iron doping is crucial for performance enhancement without the formation of inactive iron cluster. Overall, the best performing catalyst is enriched in atomic iron sites embedded into carbon structures, achieving a new performance milestone for the ORR including a half-wave potential of 0.84 V vs RHE in 0.5 M H2SO4 and only 20 mV loss after 10,000 cycles in O2 saturated electrolyte, which has been approaching to Pt in the challenging acidic media. Thus, the chemically Fe doped MOF catalysts are able to further narrow down the performance gap with Pt catalysts in more challenging acidic media, holding great promise to advance PGM-free catalyst as a replacement of Pt for PEM fuel cells.

494942 Multiconfiguration Pair-Density Functional Theory for Computational Catalysis

Abstract:

Multiconfiguration pair-density functional theory[1],[2] is a generalization of Kohnâ˘'Sham density functional theory in that the electron kinetic energy and classical electrostatic energy are calculated from a reference wave function, with the rest of the energy obtained from a density functional.Â By combining the advantages of wave function theory and density functional theory it provides a better treatment of strongly correlated systems. Some recent applications of the new method in computational catalysis will be presented.

[1] L. Gagliardi, D. G. Truhlar, G. Li Manni, R. K. Carlson, C. E. Hoyer, and J. Lucas Bao, Multiconfiguration Pair-Density Functional Theory: A New Way To Treat Strongly Correlated Systems, *Acc. Chem. Res.*, 50, pp 66â€"73 (2017)

[2] A. M. Sand, D. G. Truhlar, and L. Gagliardi, Efficient algorithm for multiconfiguration pair-density functional theory with application to the heterolytic dissociation energy of ferrocene, *J. Chem. Phys.*, 146, 034101 (2017)

493670 Quantum Mechanical Description of Excited-State Heterogeneous Catalysis Via Embedded Correlated Wavefunction Methods

Abstract:

Nanoplasmonics is recognized to be potentially transformative for the field of heterogeneous catalysis. Localized surface plasmon resonances (LSPRs) can greatly influence chemical processes occurring on the metal catalystsâ€™ surfaces, however the mechanism of enhancement is yet to be understood. First-principles quantum mechanics modeling is a widely used tool to understand heterogeneous catalysis, however, â€œgarden-varietyâ€ density functional theory (DFT) is not suitable for the accurate description of such excited-state phenomena. Additionally DFT is known to have poor predictive ability when it comes to charge-transfer and bond dissociation processes. We thus employ embedded correlated wavefunction methods[1] (*e.g.*, multi-configurational second order perturbation theory embedded in a density-functional-derived embedding potential) to explore the feasibility of light-driven catalysis on metallic nanoparticles. We evaluate if local electronic excitations (with energies at or below the LSPR frequency) centered at the surface reaction sites, can enable overcoming energy barriers for reactions that are otherwise kinetically forbidden to proceed in the ground state. N_2 dissociation on Fe- and Mo-doped Au(111) surfaces and CH_4activation on Ru-doped Cu(111) will be presented as examples.

[1] F. Libisch, C. Huang, and E. A. Carter, *Acc. Chem. Res.*, **47**, p. 2768 (2014).

500172 Incorporation of Linear Scaling Relations into Automatic Mechanism Generation for Heterogeneous Catalysis

Abstract:

To predict the selectivity and reactivity of novel catalysts at industrially relevant conditions requires a detailed microkinetic mechanism, comprising many elementary reactions. Recent advances such as such as the work of Ulissi et al [5] use a combination of scaling relations, machine learning, and DFT calculations, to gradually refine a microkinetic model until the rate limiting steps have been calculated with sufficient accuracy to be confident that they are correctly identified. However, such a system requires as input a comprehensive kinetic model containing all the possible pathways. Our recently developed Reaction Mechanism Generator for Heterogeneous Catalysis (RMG-Cat) [3], built upon the open-source RMG software primarily used for gas-phase pyrolysis and combustion [1,2], can provide such mechanisms ab inito: the user supplies just the initial conditions (eg. reactant composition, temperature, pressure) and the software predicts all the possible reactions, estimates the thermochemical and kinetic parameters, solves the governing equations, and decides which reaction pathways to include and explore further. RMG-Cat makes its decisions regarding which pathways to explore and which to ignore, using the estimated parameters, so it is important that the estimates are reasonable, even if the important parameters will be refined with more accurate calculations later in the model development process.

Linear Scaling Relations (LSRs) can provide reasonable estimates of adsorption energies in a very computationally efficient manner [4-6]. We have now implemented linear scaling relationships for the estimation of adsorption energies in the RMG-Cat software. Our database of parameters is organized in a hierarchical tree structure, enabling detailed functional group descriptions to be used when data are available and more general descriptions to be used when necessary. We include parameters to describe many adsorbates on a range of metal surfaces, and a framework to re-train the parameters whenever new data are available.

[1] Gao, C.W., Allen, J.W., Green, W.H., West, R.H. "Reaction Mechanism Generator: Automatic construction of chemical kinetic mechanisms". Computer Physics Communications, 203, 212-225, (2016) http://doi.org/10.1016/j.cpc.2016.02.013

[2] RMG - Reaction Mechanism Generator, open-source software, RMG-Py version XXX http://reactionmechanismgenerator.github.io

[3] Goldsmith, C. F., West, R. H. "Automatic Generation of Microkinetic Mechanisms for Heterogeneous Catalysis" J. Phys. Chem. C. http://doi.org/10.1021/acs.jpcc.7b02133

[4] Medford, A. J., Lausche, A. C., Abild-Pedersen, F., Temel, B., Schjødt, N. C., Nørskov, J. K., Studt, F. "Activity and Selectivity Trends in Synthesis Gas Conversion to Higher Alcohols" Topics in Catalysis (2013) 57, 135-142

[5] Ulissi, Z. W., Medford, A. J., Bligaard, T., Nørskov, J. K. "To address surface reaction network complexity using scaling relations machine learning and DFT calculations" Nature Comm. (2017) 8, 14621-14627

[6] Hummelshøj, J. S., Abild-Pedersen, F., Studt, F., Bligaard, T., Nørskov, J. K. "CatApp: A Web Application for Surface Chemistry and Heterogeneous Catalysis" Angewandte Chemie International Edition (2011) 51, 272-274

493221 DFT+U-Inspired Functional for Improved Modeling of Molecules and Solids

Abstract:

Accurate description of chemical bond energetics is a pivotal part of computational chemistry and catalysis. It ultimately determines how successful quantum mechanical models are and how reliable their predictions. Density functional theory (DFT) together with the generalized gradient approximation (GGA) is inarguably the method of choice for modeling surface phenomena in heterogeneous catalysis due to its optimal accuracy-to-cost ratio. Despite the success of GGA functionals (PBE, RPBE, PW91) in describing chemistry on metal surfaces, the GGA-inherent self-interaction error poses challenges for accurate treatment of processes on semiconductor surfaces, such as oxides.

Herein, we reveal the inadequacy of the PBE functional for describing experimental trends on metal oxide surfaces (e.g., RuO_2, TiO_2) for the vacancy-catalyzed[1] furfuryl alcohol hydrodeoxygenation, using first principles microkinetic modeling. We attribute the error to nonsystematic deviations of metal-oxygen bond energetics from experimental values. By utilizing adiabatic connection and the nth order perturbation theory, we arrive at an orbital-occupancy-dependent correction to the PBE functional that holds promise in mitigating the DFT accuracy problem. Corrections are formulated in the spirit of the DFT+U method. The new PBE-based parameter-free functional demonstrates clear improvement over pure PBE in describing dissociation energetics of molecules, NiO local magnetic moment and density of states, and cohesive energies of metal oxides, at minimal extra computational cost.

References:

1. Mironenko, A.V., Vlachos, D.G. "Conjugation-Driven "Reverse Mars–van Krevelen"-Type Radical Mechanism for Low-Temperature C–O Bond Activation." *Journal of the American Chemical Society* 138.26 (2016): 8104-8113.

497250 Development of "Surrogate" Hybrid Functionals Based on Electron Density Convolutions

Abstract:

In this work we explore the potential of a new data-driven approach to the design of exchange-correlation (XC) functionals. The approach, inspired by "surrogate" functions from optimization and convolutional filters in computer vision, utilizes convolutions of the electron density in order to form a feature space to represent local electronic environments. These features are orbital free, and provide a systematic route to including information at various length scales. Dimensional reduction and regression models are utilized to quantitatively connect these orbital-free descriptors to the exchange-correlation energies and potentials extracted from hybrid functionals. In this talk the approach will be introduced and demonstrated for the construction of a PBE-B3LYP surrogate functional that is tested for small hydrocarbon molecules. The results suggest that this is a promising new route for designing exchange-correlation functionals and embedding higher level physics into density functional theory.

495033 Accurate Correction of DFT Delocalization Error in Transition Metal Catalysis

Abstract:

Semi-local density functional theory (DFT) suffers from delocalization error that prevents the accurate treatment of systems such as transition metal oxides in the solid state and transition metal complexes. Rigorously, delocalization error is defined as deviation from piecewise linearity of the energy with respect to fractional electron removal or addition. We will describe the extent to which popular approximate treatments for alleviating self-interaction error, such as global and range-separated hybrids as well as DFT+U, alleviate delocalization error and recover the derivative discontinuity. We recently showed for the first time that DFT+U eliminates delocalization error and recovers the derivative discontinuity, but not at the value of U typically employed in most simulations. Even when the former delocalization error is eliminated through DFT+U or exchange mixing in functionals, we have recently identified residual density delocalization error, when compared to accurate correlated wavefunction theory references. We will present recent extensions to this work on representative test cases in both catalytic transition metal complexes as well as in representative bulk and slab models of transition metal oxides. Although we identified DFT+U and Hartree-Fock exchange to equivalently alter properties of the density and energetics in molecules, we will present new, periodic-table-dependent, divergent behavior in the solid state. We describe the interplay between in models and quantities relevant for heterogeneous catalysis, including vacancy formation energies, adsorption energies, surface formation energies, and properties of the electron density. We identify where these functional tuning strategies provide improvements in properties consistently between DFT+U and hybrid exchange tuning and with respect to experiment. Finally, time permitting, we present new functional development strategies we have advanced that address both delocalization error due to fractional charge and fractional spin error. Prior methodological developments, including hybrid exchange and DFT+U, have eliminated delocalization error at the cost of increasing fractional spin error, which is correlated to increased errors in transition states in catalysis.

492779 The Effects of Oxidation and Transition Metal-Doping on the Structure and Properties of Pt-Ni Nanoparticles

Abstract:

We present the use of *ab-initio* calculations and the cluster expansion method to build a direct bridge between atomic-scale structures and catalytic properties for Pt-Ni nanoparticles for the oxygen reduction reaction (ORR). Although the practical use of Pt-Ni catalysts is limited by Ni dissolution in operating conditions, it has been shown that it is possible to stabilize octahedral Ptâ€"Ni nanoparticles by doping them with a small amount of Mo (around 1.6% mole fraction). Using a cluster expansion built on nanoparticles, we show that the combination of the stabilization of sites with low coordination (e.g. edge and vertex sites), and/or the reduction of equilibrium surface Ni composition due to Mo doping greatly stabilizes the shape and composition of Pt-Ni nanoparticles. Our calculations also reveal insights into the shape evolution of Ptâ€"Ni nanoparticles: the preferential oxidation of edges can make (111) face sites more vulnerable to dissolution than edge sites, which may contribute to the observed formation of Pt-Ni nanoframes and nanoparticles with concave surfaces. To facilitate the rational design of Pt-Ni based nanoparticles with optimized structures for the ORR, we have extended our approach to model the oxygen binding energy on the surface of nanoparticles with experimentally observed sizes (~5nm), enabling the prediction of ORR current as a function of particle structure.

489522 The Influence of Hubbard U Parameter in Simulating Adsorption and Reactivity on CuO Surface(s): A Combined Theoretical and Experimental Study

Abstract:

Transition metal oxides are an important class of catalytic materials widely used in the chemicals manufacturing and processing industry owing to their low cost, high surface area, low toxicity and easily tuneable surface and structural properties. For these strongly correlated transition metal oxides, standard approximations in the density functional theory (DFT) exchange-correlation functionals fail to describe the electron localization accurately, due to the intrinsic errors arising from electron self-interactions. DFT+U method is a widely used extension of DFT, where the Hubbard U term is an onsite potential which puts a penalty on electron delocalization, successfully describing these systems at only slightly higher computational cost than standard DFT methods. The U-value is usually chosen based on its accuracy in reproducing bulk properties like lattice parameters and band structure. However, chemical reactions on transition metal oxide surfaces involve complex surface-adsorbate interactions and using the bulk properties based U-values in a locally changing surface environment may not describe reaction energetics correctly. Hence, in the current benchmarking work, using CuO as a model transition metal oxide, we perform DFT+U calculations to investigate the dissociative chemisorption of H_2 on it. it is observed that the U-value changes the calculated adsorption energies by over 100 kJ mol^{-1}. Chemisorption experiments are performed at 30-50 °C to ensure that adsorption products do not leave the system. The (dissociative) adsorption configurations along the reaction coordinate are determined by performing in-situ infrared analysis at adsorption temperatures. For the experimentally validated equilibrium configuration, adsorption energies are comupted using DFT+U for progressive U values and compared with experimental adsorption energies. We reveal that the commonly used U-value of 7 eV (fitted against CuO bulk properties) overestimates the adsorption energy by 20-40 kJ mol^{-1}. The U-value between 4.5-5.5 eV correctly predicts the adsorption of H_2 on CuO. The DFT+U benchmarking procedure elucidated in this paper, encapsulates surface-adsorbate interactions, surface reactivity and dynamic surface reaction environment and thus, provides an appropriate U-value to be used to model reactions on metal oxide surfaces.

485503 Application of Structured Porous Reactors for Heat Sensitive Liquid-Liquid Reactions

Abstract:

For biphasic transformations, packed beds represent a versatile reactor system to improve mixing and mass transfer. However, the main drawbacks limiting their use, especially for scale-up applications, is their large associated pressure drop, flow mal-distribution resulting in non-uniform contact time, and attrition of the small particles. An interesting alternative to overcome these drawbacks is the use of porous structures based on open cell metal foams, which combine the improved transport processes of the micro-scale with the throughput of milli-scale reactors [1].

In terms of multiphase flow, open cell metal foams were primarily investigated for gas-liquid systems to study hydrodynamics and mass transfer in both co-current and counter-current operation in solid foam packings [2]. In our previous work, we have investigated liquid-liquid flow hydrodynamics and mass transfer in various structured and well-defined porous media similar to open cell foams [3]. We have found that depending on the fluid properties, different design parameters of the porous structures play a crucial role in determining the overall mass transfer performance. In general, it is observed that the porous reactors enhance slug breakup, resulting in lower mean slug lengths for both phases and an associated enhancement in surface renewal velocities. The designed porous milli-scale reactors provide enhanced mass transfer performance, with a two orders of magnitude reduced energy dissipation compared to conventional milli-scale packed bed reactors.

We extend this study to investigate the heat transfer performance of these structured porous reactors. In single phase flow, we use planar laser-induced fluorescence (PLIF) and apply a two-color ratiometric technique which allows the in-situ determination of the fluid temperature field [4]. In addition, the effect of different design parameters on two-phase flow heat transfer is analyzed globally.

Based on the results of the heat and mass transfer study, we have applied these novel designed porous reactors for the amination reaction of aryl halides in the presence of a phase transfer catalyst (PTC). The palladium catalysed amination reaction of aryl halides is one of the important transformations for pharmaceutical industry, and the biphasic amination with use of a PTC provides a solution to the problem of solid formation in continuous flow reactors [5]. The effect of mixing on the yield of the C-N cross coupling reaction is studied in detail at optimum reaction conditions. Finally, the performance of the designed porous reactors is assessed by comparing with an equivalent void volume milli-scale packed bed reactor.

References:

1. Hutter, C., et al., Axial dispersion in metal foams and streamwise-periodic porous media. Chemical Engineering Science, 2011. 66(6): p. 1132-1141.

2. Stemmet, C.P., et al., Hydrodynamics of gasâ€"liquid counter-current flow in solid foam packings. Chemical Engineering Science, 2005. 60(22): p. 6422-6429.

3. Potdar, A., et al., Designed porous milli-scale reactors with enhanced interfacial mass transfer in two-phase flows. React. Chem. Eng., 2017.

4. Kuhn, S. and Rudolf von Rohr, Ph., Experimental study of heat flux in mixed convective flow over solid waves. Experiments in Fluids, 2008. 44(6): p. 973-984.

5. Naber, J.R. and Buchwald, S.L., Packed-Bed Reactors for Continuous-Flow C-N Cross-Coupling. Angewandte Chemie International Edition, 2010. 49(49): p. 9469-9474.

493775 Heterogeneous Interfaces Confined in Microreactors and Characterized By *in situ* Spectroscopic Techniques

Abstract:

Engineering novel tools for the discovery of science and translation of the new knowledge from the laboratory to application are societal challenges. In this work, we help address these challenges by applying microreaction engineering principles with *in situ* Raman and UV-vis spectroscopy and microscopy. Engineering online analytics with continuous-flow microsystems has the potential to reduce the amount of chemical waste generated, minimize the building space and energy requirements, expedite information, yield more accurate predictive mathematical models, and enable safer handling of hazardous compounds during scientific discovery, development, and manufacture. This so called â€œprocess intensificationâ€ has merit to revolutionize the way we currently discover, develop, and manufacture useful materials and compounds.

This presentation will summarize three examples of why microflow reactors with *in situ* analytics can advance society. Confined gas-liquid and liquid-liquid interfaces behave differently than unconfined ones created in traditional chemical manufacturing processes. Using *in situ* Raman spectroscopy, the role of tangential fluid motion at static confined methane-water and aqueous-nonpolar interfaces will be reported. Remarkably, the interfaces are made up of layers that influence molecular transport. In our second example, we have begun to examine the influence of the Al_2O_3:SiO_2 ratio (an important characteristic of hydrocarbon reservoir mineralogy) on asphaltenes damage in quartz packed-bed microreactors using *in situ* Raman spectroscopy, UV-vis spectroscopy, and pressure sensors. Hydrocarbon reservoirs are complex in their mineralogy and chemistry. The outcomes of stimulation chemistry on asphaltenes-damaged reservoirs are not yet well understood, due to a lack of available information in the field and limited laboratory techniques for the discovery of asphaltenes-reservoir interactions. Microfluidics offer an effective platform for rapid, *in situ* characterizations at experimental conditions that simulate sandstone reservoirs. In our third example, cryogenic flash crystallizations are often difficult to control, and therefore understanding mixed-heat-transfer-limited and crystallization-limited kinetics can be an arduous endeavour in the laboratory. By *in situ* Raman spectroscopy and microscopy, we have discovered the transition from heat-transfer-limited to crystallization-limited kinetics of methane (sI) hydrates. Our findings could help understand strategies for natural gas storage, flow assurance in hydrocarbon and natural gas production, and the continuous manufacture of nanomaterials. Finally, a few emerging trends in catalysis and reaction engineering will be highlighted.

495133 Reaction Mechanism and Microkinetics of the Cobalt Catalyzed Fischer–Tropsch Synthesis

Abstract:

Reaction Mechanism and Microkinetics of the Cobalt Catalyzed Fischer–Tropsch Synthesis

G. T. Kasun Kalhara Gunasooriya,[†] Mark Saeys[†]

[†]Laboratory for Chemical Technology, Ghent University, Ghent, Belgium

Fischer-Tropsch synthesis (FTS) transforms synthesis gas, a mixture of CO and H_2, to long-chain hydrocarbons and water. Supported cobalt catalysts are often preferred due to their high activity, selectivity towards long-chain hydrocarbons, and low CO_2selectivity. The detailed reaction mechanism, a complex combination of C-C bond formation, C-O scission and hydrogenation steps, remains intensely debated.[1, 2] FTS turnover frequency (TOF) is independent of the particle size above 10 nm,[3] and therefore the kinetically relevant steps occur on terraces. However under FTS conditions, cobalt catalyst surfaces undergo a massive reconstruction as observed by STM.[4] The synergistic adsorption of carbon and CO, with the high CO coverage make it thermodynamically favorable to break up Co terraces and create step sites and nano-islands.[5,6] To gain insights into the rate-limiting reactions, the most abundant surface species and the factors governing selectivity, a dual-site microkinetic model accounting for catalyst structure, coverage and reaction conditions is required.

A dual-site microkinetic model based on first principle calculations was constructed considering both the terrace sites and the edges of the self-assembled nano-islands. This model incorporates adsorption of species on both the terrace sites and the edges of the self-assembled nano-islands, and diffusion between those sites. C-O activation is a crucial step in FTS and several pathways have been proposed. Direct CO dissociation is a difficult step on Co terraces and barriers greater than 220 kJ/mol are consistently reported. In addition to direct CO dissociation, CO can be activated by insertion into growing hydrocarbon chains (CO insertion mechanism [2]), or through hydrogenation (hydrogen-assisted CO activation). Hydrogenation at the O-atom in CO* and RCO* is a difficult step, while hydrogenation at the C-atom leads to oxygenates on Co terraces. In the hydroxyl-assisted route, CHOH is formed via transfer of a proton from a surface OH group to the O-atom of CO. CH-OH then undergoes a rate-limiting C-O scission step to regenerate the OH group and a CH species.[7] Moreover, the thermodynamically favorable self-assembly of nano-islands introduces edge sites on the cobalt terraces. Evaluation of various possible pathways show that a sequence involving the creation of a free site through C hydrogenation, CH migration to terraces, and direct CO dissociation is a kinetically viable step at the edges of the nano-islands. Microkinetic modeling reveals that coupled H- and OH-assisted C-O activation is preferred over direct CO dissociation and H-assisted CO activation on terraces. Moreover, both coupled H- and OH-assisted C-O activation and direct CO dissociation at the edges of the nano-islands form the CH species that initiate chain growth. However, kinetic parameters calculated at low CO coverage leads to high methane selectivity. At high CO coverage, limited

stabilities of CH groups further reduce their coverage. Detailed microkinetic model again reflects the complexity of FTS mechanism providing opportunities to control activity and selectivity.

References

[1] van Santen, R.A., Markvoort, A.J., Filot, I.A.W., Ghouri, M.M., Hensen, E.J.M. *Phys. Chem. Chem. Phys.*, **2013**, 15, 17038

[2] Zhuo, M., Tan, K.F., Borgna, A., Saeys, M., *J. Phys. Chem. C* **2009,** 113, 8357

[3] den Breejen, J. P., Radstake, P. B., Bezemer, G. L., Bitter, J. H., Frøseth, V., Holmen, A., Jong, K. P. d. *J. Am. Chem. Soc.* **2009**, 131, 7197

[4] Wilson, J.; De Groot, C. *J. Phys. Chem.* **1995**, 99, 7860

[5] Banerjee, A.; van Bavel, A. P.; Kuipers, H. P. C. E.; Saeys, M. J. *Phys. Chem. Lett.* **2016**, 7, 1996–2001

[6] Gunasooriya, G.T.K.K.; van Bavel, A.P.; Kuipers, H.P.C.E.; Saeys, M. *Surf. Sci.* **2015**, 642, L6

[7] Gunasooriya, G.T.K.K., van Bavel, A.P., Kuipers, H.P.C.E, Saeys, M., *ACS Catal.* **2016**, 6, 3660

489518 Interplay of Reaction and Transport within Biomass Particle during Fast Pyrolysis: Development of Reaction and Transport Models and Their Non-Dimensionalization

Abstract:

Pyrolysis of biomass and waste converts it to bio-oil, which can be further upgraded to fuels and chemicals.[1] Decomposition of the biomass/waste feed particle, within the pyrolysis reactor, governs the product distribution, and hence the yield and the composition of bio-oil. Complex and often unknown decomposition chemistry, high temperatures and short residence times of the particle in the reactor and an interplay between kinetics and heat and mass transport effects within the particle make it extremely difficult to understand the particle level decomposition of biomass. In the present work, we present a particle level model comprising of energy and reactant and products species mass conservation equations, as heat and mass transport occurs simultaneously along with pyrolysis reactions, considering the entire particle as a control volume. Conservation equations consist of generation, convective and diffusive transport and accumulation terms. Unlike using lumped models for the pyrolysis reactions[2-5], component specific kinetic models which are fitted using transport limitation free experimental data, are incorporated into the mass conservation equations. Changes in particle porosity and thermal conductivity and shrinkage of the particle during the pyrolysis process are also taken into account. Further, the model equations and boundary conditions are non-dimensionalized to group difficult to measure variables/properties/parameters into non-dimensional numbers like Biot, Pyrolysis, Peclet, Damkohler and Sherwood numbers that represent characteristic reaction and transport timescales. The non-dimensionalization would provide an independent platform to compare various pyrolysis experiments performed under different operating conditions.

(1) Aguado, R.; Olazar, M.; Gaisán, B.; Prieto, R.; Bilbao, J. *Industrial & Engineering Chemistry Research* **2002**, *41*, 4559.

(2) Younan, Y.; van Goethem, M. W. M.; Stefanidis, G. D. *Computers & Chemical Engineering* **2016**, *86*, 148.

(3) Corbetta, M.; Frassoldati, A.; Bennadji, H.; Smith, K.; Serapiglia, M. J.; Gauthier, G.; Melkior, T.; Ranzi, E.; Fisher, E. M. *Energy & Fuels* **2014**, *28*, 3884.

(4) Haseli, Y.; van Oijen, J. A.; de Goey, L. P. H. *Journal of Analytical and Applied Pyrolysis* **2011**, *90*, 140.

(5) Ciesielski, P. N.; Crowley, M. F.; Nimlos, M. R.; Sanders, A. W.; Wiggins, G. M.; Robichaud, D.; Donohoe, B. S.; Foust, T. D. *Energy & Fuels* **2015**, *29*, 242.

495032
Kinetics of Glycerol Conversion to Hydrocarbon Fuels over Pd/H-ZSM-5 Catalysts

Abstract:

As the primary byproduct of biodiesel production, glycerol conversion to alternative fuels and chemicals is important to enhance process economics. In our recent work, bifunctional metal supported H-ZSM-5 catalysts were used to convert glycerol to aromatic hydrocarbon fuels. An optimal ~60% yield of hydrocarbon fuels was obtained over Pd/H-ZSM-5 catalyst. In the present work, based on experimental data over this catalyst, a lumped reaction network and kinetic model are developed. The rate constants, reaction orders and activation energies are obtained for each reaction step over the temperature range 300-450 °C. The predicted values from the kinetic models match well with experimental data for glycerol conversion up to ~90%.

491204 Reaction of C_2H_4 Under Lower Temperature Fischer-Tropsch Conditions on a TiO_2 Supported Cobalt Catalyst with Co-Feeding of H_2 and Syngas

Abstract:

Reaction of C_2H_4 under Lower Temperature Fischer-Tropsch Conditions on a TiO_2 Supported Cobalt Catalyst with Co-feeding of H_2 and Syngas

Xiaojun Lu*, Diane Hildebrandt, Yusheng Zhang, and Xinying Liu

Material and Process Synthesis; College of Science, Engineering, and Technology, University of South Africa

Co-feeding of C_2H_4 to the Fischer-Tropsch reaction system has been studied some years back to investigate the effect of secondary reactions of the olefins in the Fischer-Tropsch reaction. Some important phenomena were observed and the conclusions were not consistent due to the complexity of the Fischer-Tropsch reaction. Researchers often found that it is very difficult to investigate certain aspects independently in such complex reaction system. This work tries to investigate one aspect at a time in the complex system. Three groups of experiments were designed. The first group of experiments was designed to investigate the reaction of C_2H_4 only on the supported Co catalyst. The purpose was to investigate the possible secondary reactions of C_2H_4 without other reactants. The second experimental design was to add H_2 into the reaction system when keeping the feed of C_2H_4. This was to investigate the reactions of C_2H_4 with the presence of H_2. The C_2H_4 to H_2 ratio was adjusted in a wide range so that excessive amount of both of these two reactants could be achieved. The third experimental design was to co-feed synthesis gas ($H_2/CO/N_2$) when keeping the feed of C_2H_4. The amount of syngas fed into the reaction system was controlled to ensure that the main feed was still C_2H_4. This is different from the feed ratio chose by other researcher in the past as the amounts of C_2H_4 co-fed by them was always far less than the syngas feed.

The results we have obtained in these experiments are of big interests. C_2H_4 was found not to react under the conditions applied when no other feed was introduced into the reactor; while it reacts quickly when H_2 was co-fed. The reaction temperature, when H_2 was co-fed, could be as low as 100°C and only products of C_1 to C_5 were found. These results suggest that C_2H_4 can undertake several reaction pathways to form different products. Fischer-Tropsch synthesis reaction was found to occur, with the presence of C_2H_4, when the operating temperature was at 140°C, which is much lower than the temperature (higher than 180°C with Co as the catalyst) required for normal Fischer-Tropsch synthesis. The reaction temperature required for reactions to happen changed to around 190°C when

the feed of C_2H_4 was stopped. The presence of C_2H_4 can promote the Fischer-Tropsch synthesis by lowering the reaction temperature. This has not been reported in any existing literature. The products distribution is quite different when both C_2H_4 and syngas were fed and this leads to the understanding of the reaction mechanisms for the Fischer-Tropsch Synthesis.

Keywords: Fischer-Tropsch Synthesis, C_2H_4 reaction, Reaction Mechanism

497184 High Throughput Study of Catalysis on Pd-Based Alloys

Abstract:

Alloys of transition metals often makes better catalysts than their pure components. It is challenging to understand the reasons behind this fact and to design new multicomponent catalysts with desirable and predictable properties due to the need to perform many catalyst preparation, characterization and reactivity measurements across composition space. This search can be accelerated using Composition Spread Alloy Films (CSAFs) which are thin multicomponent films that have composition gradients parallel to their surfaces, $A_xB_yC_{1-x-y}$ with x = 0 â†' 1 and y = 0 â†' 1-x. Study of alloy catalysis across composition space using a CSAF requires a multichannel reactor system that can be used to run steady state catalytic reactions at many different positions or alloy compositions on the CSAF. For this purpose we have developed a 100 channel microreactor array that can sample product distributions from 100 different alloy catalysts of about 10 minutes. $Cu_xAu_yPd_{1-x-y}$ CSAFs spanning all of binary and ternary composition space have been prepared using a rotating shadow mask CSAF deposition tool which is designed and developed in our group. CSAF surface composition and valence electron energy distributions are measured using X-ray Photoemission Spectroscopy (XPS).

High-throughput techniques have been used to characterize the activity of ethylene hydrogenation reaction over ternary alloy catalysts. Ethylene hydrogenation reaction has been chosen for this work since it is a mechanistically simple reaction and it has been investigated on binary alloy catalysts before. It is a simple reaction producing only ethane, but it has also been referred as deceptively simple reaction, which in fact possesses an extremely rich and complicated chemistry. Horiuti-Polanyi mechanism is widely accepted for this reaction. In this mechanism, ethylene adsorbs onto catalyst surface molecularly and H_2 adsorbs dissociatively. Adsorbed H atoms combine with adsorbed ethylene to form ethyl groups on the surface. As the last step, another H atom is added to the ethyl group, ethane is formed and immediately desorbed from the surface.

In order to study this reaction, as a first step, H_2 dissociation properties were investigated over ternary CSAFs via characterizing H_2-D_2 exchange activity versus composition at various temperatures, partial pressures and flow rates. The relationship between alloy catalyst activity and electronic structure has been investigated experimentally across a broad, continuous span of $Cu_xAu_yPd_{1-x-y}$ composition space. The CSAF was used as a catalyst library with a multichannel microreactor to measure H_2-D_2 exchange kinetics at 100 discrete compositions on the CSAFs over a temperature range of 333 â€“ 597 K at atmospheric pressure. H_2 conversion was chosen to be the indicator of activity. It was found that H_2-D_2 exchange activity varies across the CSAF and it tends to increase with increasing Pd content. A microkinetic model that has been validated using a number of single component Cu-Pd catalysts in a fixed bed reactor was used to estimate the energy barriers to dissociative adsorption ( ) and associative desorption ( ) of H_2 as functions of alloy composition, x and y. On the $Cu_xAu_yPd_{1-x-y}$ CSAF, increasing Pd content from 0 to 1 was found to decrease adsorption barrier from 0.44 to 0.12 eV. Increasing Pd content from 0.25 to 1 was found to increase

desorption barrier from 0.4 to 0.74 eV which suggests H_2-D_2 exchange reaction is limited by H_2 desorption step within this Pd content. Spatially resolved X-ray photoelectron spectra were obtained from the CSAFs and used to estimate the energy of the valence-band center as a function of alloy composition. The v-band center shifted monotonically from -3.4 to 5.6 eV across the $Cu_xAu_yPd_{1-x-y}$ CSAF. The barrier to dissociative adsorption of H_2 was found to decrease as the v-band energy increases. This data proves that we have successfully developed a high-throughput experimental strategy for understanding and designing ternary alloy catalysts. And it provides the first experimental correlation of elementary reaction barriers with valence band energy across a continuous span of alloy composition space.

The same experimental strategy was then used to study ethylene, ethyl and ethane adsorption and desorption characteristics. The preliminary results showed that ethylene hydrogenation activity can be monitored using our high-throughput methodology and it was found to be low at Cu- and Au-rich region of $Cu_xAu_yPd_{1-x-y}$ CSAF and high across Pd-rich region of the ternary CSAF. The kinetics of ethylene hydrogenation will be measured across $Cu_xAu_yPd_{1-x-y}$ CSAF so that the kinetic analysis can be performed across a wide composition range. The obtained rate law will provide insight into the reaction mechanism across alloy compositions.

504321 Methanol to Propylene Process: An Integration of Catalyst Development and Reactor Engineering

Abstract:

Methanol, which can be produced from coal or natural gas, is one of the most important C1 chemicals. The utilization of methanol to make light olefins including ethylene and propylene has been recently commercialized. However, in many applications, it is desired to maximize the yield of propylene, and thus methanol to propylene (MTP) process becomes increasingly interesting. It is normally hard to achieve high selectivity to propylene in methanol conversion. In this work, we have developed a multi-functional catalyst, over which three reactions, i.e. methanol conversion, ethylene alkylation, and C4+ cracking, can be catalyzed. By integrating these three reactions, we attempted to develop a MTP process with the selectivity to propylene higher than 75%. Evaluation of this catalyst has been carried out in laboratory-scale micro-fluidized bed reactor. It shows that the selectivity to propylene of 50% has been achieved in the methanol conversion reaction. We also studied the ethylene alkylation reaction and C4+ cracking reaction, and identified the optimal operation conditions. Based on the results obtained in the micro-fluidized bed reactor, we proposed an integrated process to maximize the selectivity of propylene. In achieving this goal, a turbulent fluidized bed reactor, in which the optimal operation conditions suitable for both methanol conversion and ethylene alkylation have been considered, has been designed and constructed in our pilot plant. The reactor can be operated with 300t/a methanol feed rate. Pilot plant experiments have been conducted for validating the process concept and scaling up the reactor. It shows the selectivity to propylene of 75%, in addition to a selectivity of ethylene of 10%, can be achieved in the pilot plant. This MTP process can be seen as a successful example of the integration of catalyst development and reactor engineering.

504320 Direct Production of Value-Added Chemicals Via Fischer-Tropsch Synthesis

Abstract:

Fischer-Tropsch synthesis (FTS) is a heterogeneous catalytic process for sustainable production of fuel and value-added chemicals via syngas derived from coal, biomass and natural gas. It is considered as an attractive alternative non-petroleum-based production route. FTS is a very complicated catalytic reaction process with co-production of paraffin, olefin, alcohol and other products. The direct synthesis of high value-added olefins and higher alcohols from syngas has drawn particular attention due to due to its process simplicity, low energy consumption and clean utilization of carbon resource. For the purpose of high selectivity to specific olefins or higher alcohols, it is very desirable to regulate the reaction network for certain reactions to take place preferentially. Obviously, it is necessary to control CO activation and then both carbon chain growth and termination for high selectivity to target product by development new catalyst systems and optimization of reaction conditions.

Reaction network during syngas conversion was firstly investigated by probe molecules techniques [1]. Alcohol, aldehyde, olefin as probe molecules were in-situ introduced into the reaction system. Several parallel reactions and second reactions were clarified. The existence of surface acyl species indicated that the insertion of CO rather than hydroxyl was the underlying mechanism for higher alcohols synthesis over modified FT catalysts along with the support from surface simulation. In addition, the dual active sites for higher alcohols formation was further investigated [2]. Several kinds of dual active sites (Cu/FeC_x, CuCo, and Co/Co_2C) were controllably prepared for higher alcohol synthesis. Both Cu/FeC_x and Co/Co_2C could afford high selectivity for long chain alcohols, while only short chain alcohols were produced for CuCo. The difference of the detailed surface structure and the corresponding catalytic performance was discussed in detail.

Fischer-Tropsch to olefins (FTO) by using new active phase was also investigated. Co_2C nanoprisms with exposed facet of {101} and {020} were prepared and tested for syngas conversion [3]. It is widely accepted the formation of Co_2C is considered to be one of the main reasons for deactivation of the Co-based FT reaction because it possesses very low activity for CO hydrogenation, with methane and CO_2 as the main products. However, due to the strong facet effect of Co_2C, Co_2C nanoprisms afforded very high selectivity for the production of lower olefins (60.8 C%), while little methane (about 5.0 C%) was generated with the ratio of olefin/paraffin amongst the C_{2-4} products being as high 30. The product distribution deviates markedly from the classical ASF distribution with the highest selectivity to propylene. The effect of several promoters and the model for modified ASF distribution were discussed [4]. In addition, the nanoeffects of Co_2C were studied in details to reveal the structure-performance relationship.

References

[1] *Catalysis Science & Technology*ï¼Œ2015, 5, 4224

[2] *Catalysis Science & Technology*ï¼Œ2013, 3, 1591

[3] *Nature*, 2016, 538, 84..

[4] *ACS Catalysis*, 2017, 7, 3622.

504319 An Investigation into the Effect of Li and Mn Promotions on the Activity and Selecivities to Olefins and Alcohols of Co@Co_2C/Activated Carbon (AC) for Fischer-Tropsch Reaction

Abstract:

Abstract:

Utilization of non-precious transition metals for high alcohols and olefins synthesis is of a great importance in heterogeneous catalysis. We synthesized successfully Li and Mn promoted AC supported cobalt metal-carbide (Co@Co_2C/AC) catalysts, which present remarkable activity of more than 200g/kg-cat.h and selectivities of ca. 65 C% both for alpha-alcohols and alpha-olefins via the Fischer-Tropsch reaction at the mild conditions. The formation of the stable cobalt carbide and the Co@Co_2C interface are found essential for the observed reactivity. The results of characterization from In-situ EXAFS, in-situ XRD, HRTEM, XSP, and chemo-adsorption of H_2 and CO experiments showed that Mn promotes the formation of Co_2C species during Fischer-Tropsch reaction and Li increased the CO uptake, while both promoters have little impact on the amount of H_2 uptakes. Density functional theory calculations show that Co_2C is highly efficient for CO non-dissociative adsorption, behaving as noble-metal-like, whereas the Co metal is highly active for CO dissociative adsorption and the subsequent carbon-chain growth. The interface between the cobalt metal and its carbide phase as well as the dual sites available at the interface for facile CO insertion to C_nH_m-groups formed on the Co sites. The remarkably high alpha-alcohol and alpha-olefins selectivities from the Li and Mn promotion effects, namely the formation of more interfacial active sites between Co and Co_2C and enhancement of the ratio for chemo-adsorption CO to H_2 uptake on the surface of Co@Co_2C/AC catalysts, paves a new way for the design of the new catalysts for synthesizing various high value-products in syngas applications.

504154 Direct Production of Gasoline-Range Hydrocarbons from Carbon Dioxide over Iron-Based Multifunctional Catalysts

Abstract:

Converting CO_2 from a detrimental greenhouse gas into value-added liquid fuels not only contributes to mitigating CO_2 emissions, but also reduces dependence on petrochemicals. Unfortunately, the activation of CO_2 and its hydrogenation to hydrocarbons or other alcohols are challenging tasks because CO_2 is a fully oxidized, thermodynamically stable and chemically inert molecule. Most research to date, not surprisingly, is focusing on selective hydrogenation of CO_2 to short-chain products, while few studies on long-chain hydrocarbons, such as gasoline-range (C_5–C_{11}) hydrocarbons. The key to this process is to search for a high efficient catalyst.

Herein, we have succeeded in preparing high efficient, stable, and multifunctional Na–Fe_3O_4/Zeolite catalysts for the direct production of gasoline-range (C_5–C_{11}) hydrocarbons from CO_2 hydrogenation. This catalyst displayed record selectivity towards C_5–C_{11} hydrocarbons (78%) as well as low CH_4 and CO selectivity under industrial relevant conditions, and gasoline fraction are mainly isoparaffins and aromatics thus favouring the octane number. Moreover, the composition of C_5–C_{11} can be tuned by the choice of zeolite type and the integration manner of multifunctional catalyst. When Na–Fe_3O_4 nanocatalyst was matched with HZSM-5 zeolite and integrated by granule-mixing manner, up to 63% of aromatics in gasoline fraction were produced. While Na–Fe_3O_4 was matched with HMCM-22 zeolite, up to 60% of isoparaffins in gasoline fraction were obtained under the dual-bed configuration. Not only that, the multifunctional catalyst exhibited a remarkable stability for 1,000 h on stream, which definitely has the potential to be a promising industrial catalyst for CO_2 utilization to liquid fuels.

In-depth characterizations indicate that, during CO_2 hydrogenation reaction, this catalyst enables RWGS over Fe_3O_4 sites, olefin synthesis over Fe_5C_2 sites, and oligomerization/aromatization/isomerization over zeolite acid sites. The concerted action of the active sites calls for precise control of their structures and proximity. This study paves a new path for the synthesis of liquid fuels by utilizing CO_2 and H_2. Furthermore, it provides an important approach for dealing with the intermittency of renewable sources (sun, wind and so on) by storing energy in liquid fuels.

References

1. G. A. Olah, A. Goeppert and G. S. Prakash, *J. Org. Chem.*, 2008, **74**, 487-498.

2. R. W. Dorner, D. R. Hardy, et al., *Energy Environ. Sci.*, 2010, **3**, 884-890.

3. W. Wang, S. Wang, X. Ma and J. Gong, *Chem. Soc. Rev.*, 2011, **40**, 3703-3727.

4. F. Jiao, J. Li, X. Pan, et al., *Science*, 2016, **351**, 1065-1068.

5. P. Zhai, C. Xu, R. Gao, et al., *Angew. Chem. Int. Ed.*, 2016, **55**, 9902-9907.

6. J. Wei, Q. Ge, R.Yao, et al., *Nat. Commun.*, 2017, **8**, 15174.

Innovative Syngas Production Catalyst for Utilization of CO_2 (CT-CO_2AR™)

Takenori Kanda, Osamu Hirohata, Tomoyuki Mikuriya and Fuyuki Yagi
Chiyoda Corporation, Kanagawa, Japan

1. Introduction

Production of synthesis gas from natural gas and conversion of synthesis gas to fuels and chemicals have become increasingly attractive from the viewpoint of energy resource diversification. The technology of synthesis gas production by the reforming of natural gas with CO_2 (hereafter referred to as "CO_2-reforming") is an effective solution for the issues of efficient energy resource utilization and environmental preservation.

Chiyoda has developed and commercialized a new noble metal based reforming catalyst named CT-CO_2AR™ which has a higher resistance to carbon formation and produces synthesis gas more effectively compared with commercial reforming catalyst.

In this paper, we will elaborate on the characteristics and the competitiveness of the CT-CO_2AR™ catalyst and the results of the operation of the commercial plant.

2. Characteristics of CT-CO_2AR™ catalyst

(1) The characteristics of CO_2/steam reforming

Synthesis gas is used for hydrogen production and as the feed stock of chemicals, and the H_2/CO ratio of synthesis gas depends on the purpose.

For example, hydrogen production requires an H_2/CO ratio of more than 3.0. For ethylene glycol production, the required H_2/CO ratio is 2.0. For chemical production through oxo synthesis, the required H_2/CO ratio is 1.0. For the feedstock of acetic acid or MMA requires only CO or as low as possible H_2/CO ratio.

When hydrocarbons such as natural gas or naphtha and steam are used as reformer feedstocks, the steam reforming reaction shown in Eq. 1 is normal and the H_2/CO ratio of produced synthesis gas becomes more than 3.0.

$$CH_4 + H_2O \Leftrightarrow 3\,H_2 + CO \qquad (1)$$

However, when CO_2 and steam are used as reforming agents, the CO_2 reforming reaction shown in Eq. 2 occurs.

$$CH_4 + CO_2 \Leftrightarrow 2\,H_2 + 2\,CO \qquad (2)$$

Therefore, by changing the ratio of feed CO_2 and steam, various H_2/CO ratios of synthesis gas can be produced.

However, during the reforming reactions, the carbon formation reaction shown in Eq. 3 occurs on the catalyst surface. Deposited carbon causes the deactivation of the catalyst and the occlusion of the catalyst tube[1)] leading to an unstable operation.

$$CH_4 \Leftrightarrow C\,(s) + 2\,H_2 \qquad (3)$$

To prevent carbon formation for a stable operation, commercial Ni based reforming catalysts require steam and CO_2 greater than the stoichiometric ratio.

Fig. 1 shows the relation between the operation condition and risk of carbon formation[2)]. The bold line in the center represents carbon limit line at 600 ℃. In the black area to the left of this line, the process gas has enough high thermodynamic potential for carbon formation at 600 ℃. Operation around this line cannot be carried out with commercial Ni based reforming catalysts. They are usually operated within the excess steam carbon ratio (S/C) zone (blue area).

On the other hand, our new commercialized noble metal based reforming catalyst (CT-CO_2AR™) enables a stable operation under a lower S/C condition (red area) because it has a higher resistance to carbon formation.

In next section, the advantages of a lower S/C operation will be shown.

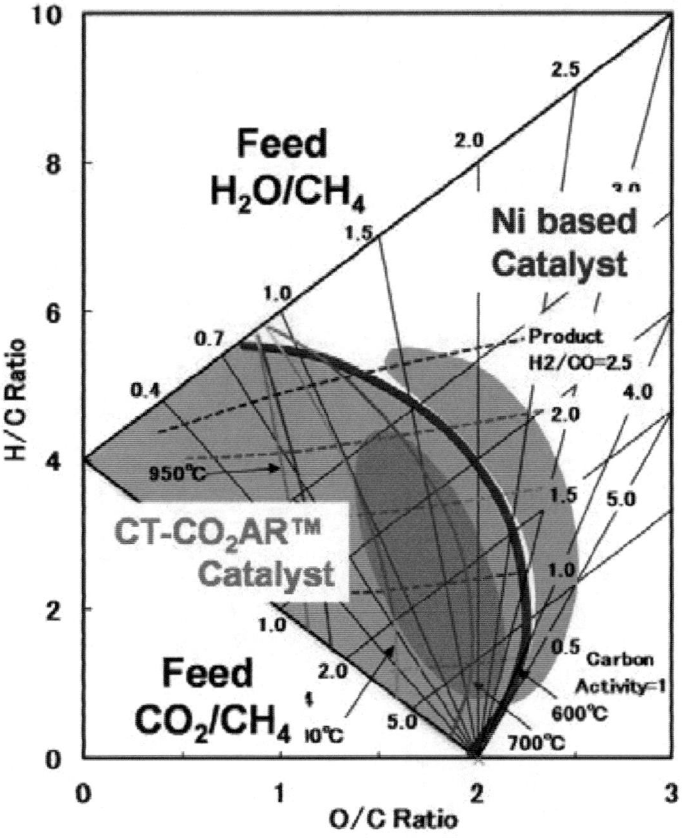

Fig. 1. Carbon Limit Diagram

(2) The advantages of a lower S/C ratio operation

In the case of the synthesis gas production process shown in Fig. 2, the advantages of a lower S/C operation are as follows:

1. Since the amount of steam and CO_2 is reduced, the heat duty to heat steam and CO_2 can be reduced. Therefore, the fuel consumption of the reformer can be reduced.
2. Fig. 3 shows the comparison of synthesis gas concentration at the reformer outlet between a high S/C (=2.2) operation and a low S/C (=1.0) operation in the case of $H_2/CO=1.0$
   In the case of a low S/C operation, the amount of CO_2 can be reduced and as a result, the system volume of CO_2 removal is downsized.
   In addition, because of lower CO_2/C, the capacity of the CO_2 compressor is downsized.
   Therefore, both the CAPEX and the OPEX of CO_2 removal and CO_2 compressor are reduced.
3. Fig. 3 shows that a lower S/C operation can reduce the amount of H_2O at the reformer outlet and it is possible to reduce the amount of process condensate.
   In the case of a low S/C operation, the total amount of produced steam is less than that of a high S/C operation. However, the amount of steam uses as the reforming agent in a low S/C operation is significantly less than that in a high S/C operation.
   As a result, a low S/C operation can increase the amount of export steam.

From these reasons, the CT-CO_2AR™ catalyst is preferable to a commercial Ni based catalyst.

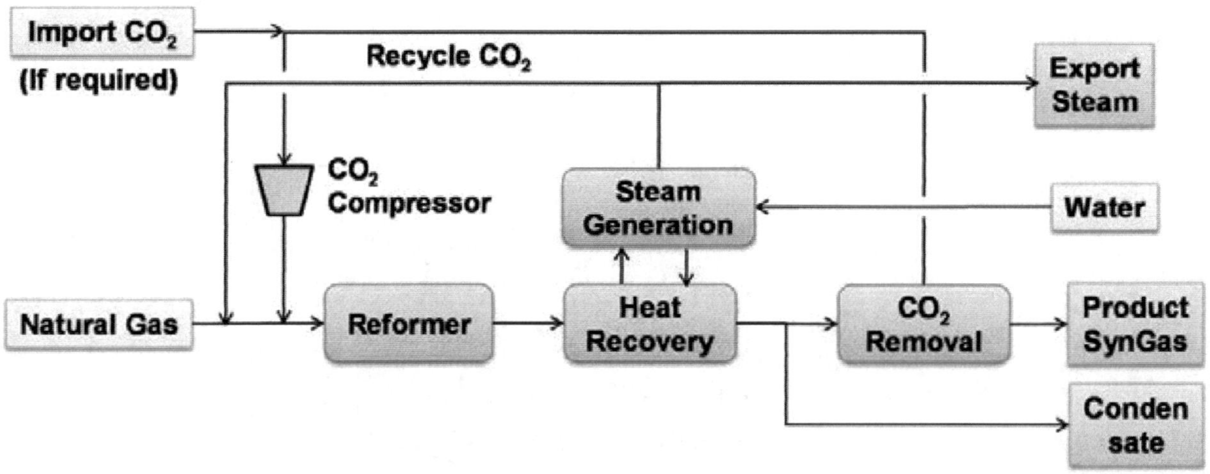

Fig. 2. Block Flow Diagram of Syngas production

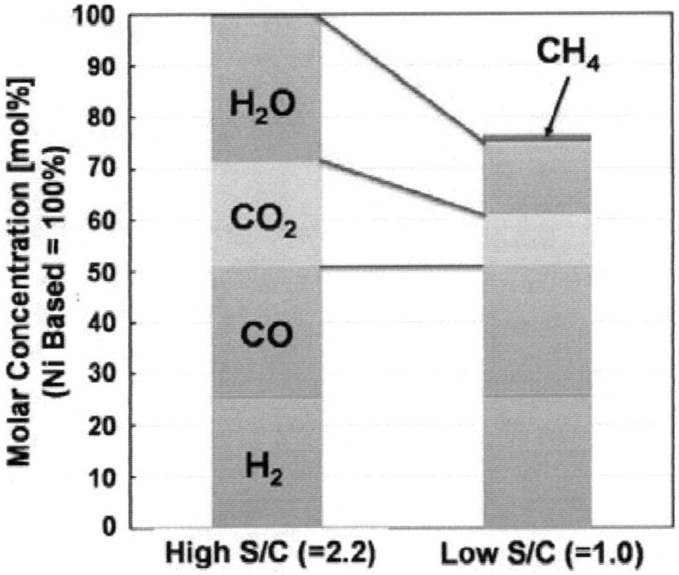

Fig. 3. Typical Example of synthesis gas composition in $H_2/CO=1.0$

(3) The characteristics of the CT-CO_2AR™ catalyst

Fig. 4 shows the result of EPMA (Electron Probe Micro Analyzer) of the CT-CO_2AR™ catalyst. Generally, the surface reaction on the reforming catalyst is the rate-determining step. Therefore, to reduce the amount of active metal, the active metal is supported on only the outer surface of the carrier like egg-shell.

Fig. 5 shows the accelerated experiment result. The CT-CO_2AR™ catalyst keeps its activity due to its high resistance to carbon formation.

We built the reaction kinetics model shown in Fig. 6 to estimate the behavior of the CT-CO_2AR™ catalyst.

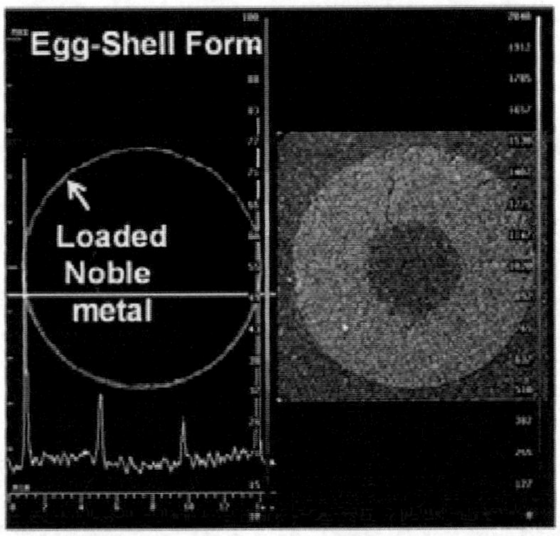

Fig. 4. EPMA result of CT-CO_2AR™ Catalyst

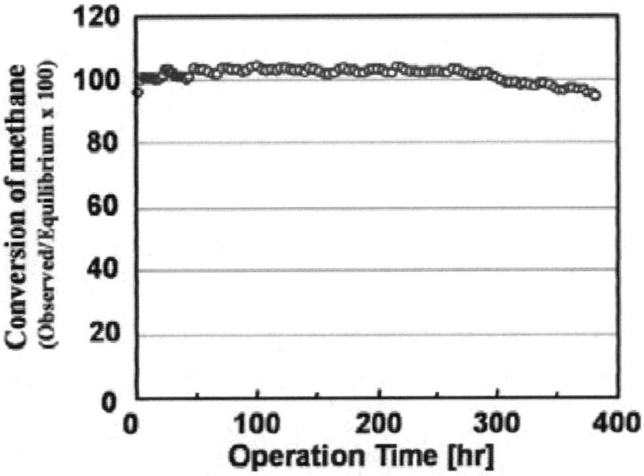

Fig. 5. Result of accelerated experiment

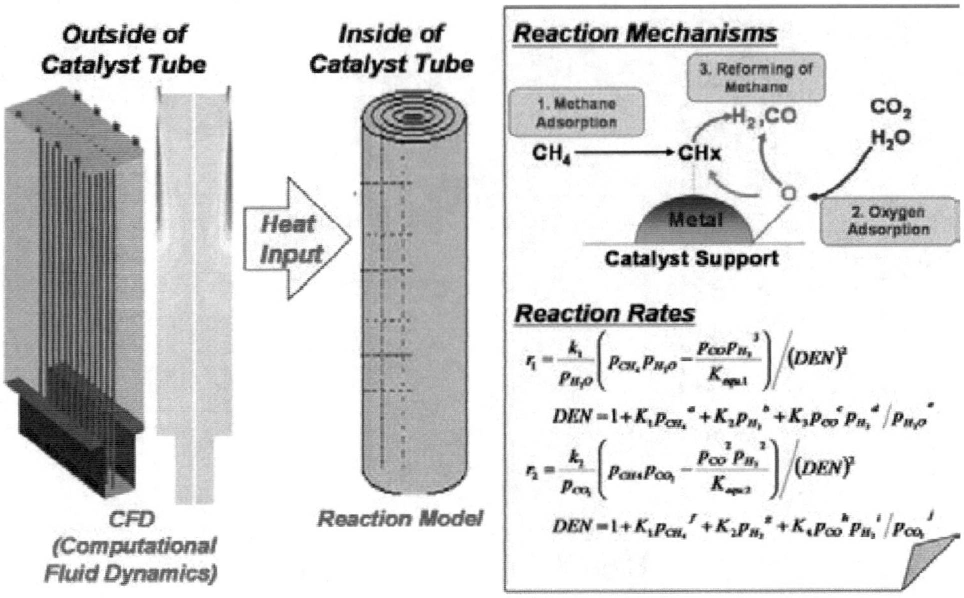

Fig. 6. Reaction model of CT-CO_2AR™ catalyst

3. Advantages of CT-CO_2AR™ catalyst

(1) Study Basis

Regarding the competitiveness of the CT-CO_2AR™ catalyst compared to a commercial Ni based reforming catalyst, we have compared the CT-CO_2AR™ catalyst and Ni based catalysts from the following three points of view based on the block flow shown in Fig. 2.

1. Total energy input (feed + fuel hydrocarbon + utility consumption)
2. Manufacturing costs of synthesis gas
3. CO_2 Emission per final products

[CO_2 Emission] = [CO_2 in Flue and Utility] - [Import CO_2] + [CO_2 in downstream process to produce the final products]

In this study we considered cases with H_2/CO ratios of 1.0 and 0.5. The conditions and estimated final product of each case are listed in Table 1.

(2) Study Result

From Fig.7-9 show the results of the process study. In the case of H_2/CO=1.0, compared to Ni based catalysts, the CT-CO_2AR™ catalyst can cut total energy input by 11% (Fig. 7) manufacturing cost by 12% (Fig. 8).Regarding CO_2 emission, the CT-CO_2AR™ catalyst can achieve a reduction of 22% compared with a Ni based catalyst (Fig. 9). In the case of H_2/CO=0.5, CT-CO_2AR™ catalyst is more competitive. Approximately 30% of both total energy input and manufacturing cost can be reduced and 76% of the amount of CO_2 emission can be reduced. From this study, the CT-CO_2AR™ catalyst is competitive compared with Ni based catalysts both in terms of CAPEX and OPEX.

Table 1. Process study conditions

H_2/CO Ratio	Catalyst	Feed S/C	Amount of (H_2+CO)	Final Products	Capacity
[-]	[-]	[-]	[Nm3/h]	[-]	[ton/hr]
CO (0.5)	Ni Based	3.3	37,500 (25,000 as CO)	Acetic Acid	500,000
	CT-CO_2AR™	0.4			
1.0	Ni Based	2.2	25,000	n-butanol (oxo synthesis)	270,000
	CT-CO_2AR™	1.0			

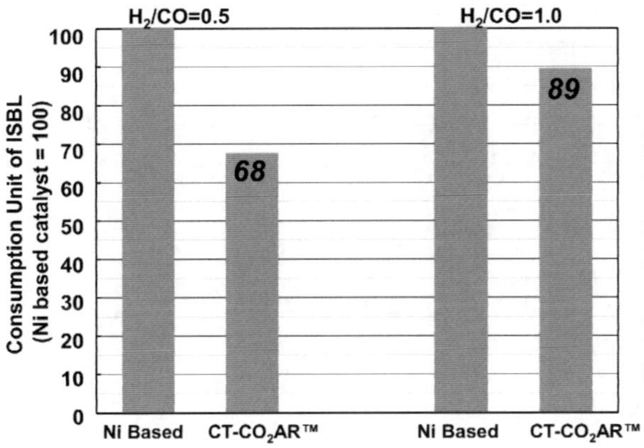

Fig. 7. Total Energy Input

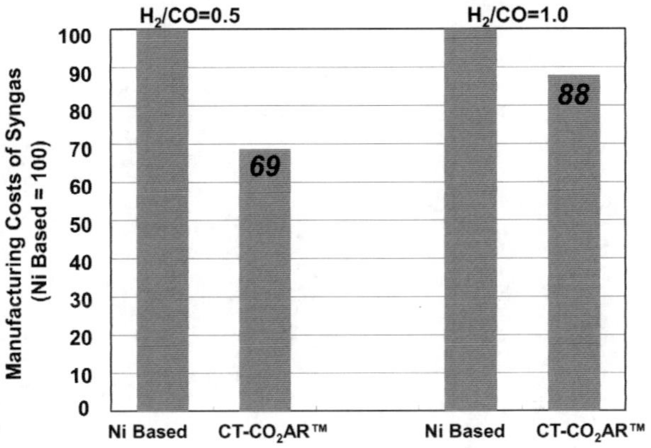

Fig. 8. Manufacturing Costs of Syngas

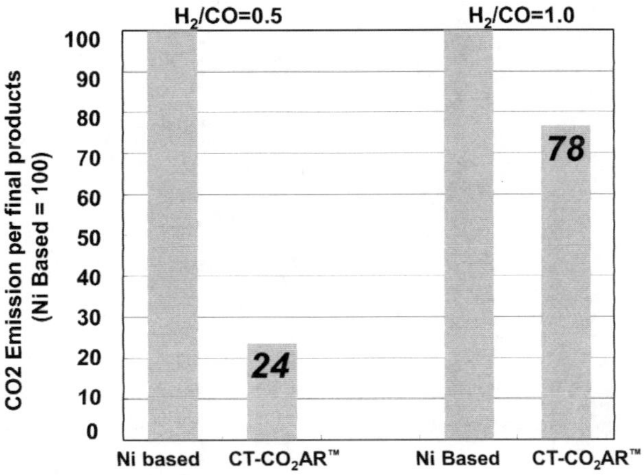

Fig. 9. CO_2 Emission per final products

4. Application to Commercial Plant

The CT-CO_2AR™ catalyst has been operated in a commercial plant in Japan since 2014. The information of the commercial plant is listed in Table 2. In this project, the reforming catalyst has been changed to the CT-CO_2AR™ catalyst and the S/C ratio at the inlet of the prereformer has reduced from 2.2 to 1.4 without changing the amount and specification of the produced synthesis gas.
Regarding the comparison between analysis result using the reaction model and the operation result, a pressure drop through the reformer tube is shown in Fig. 10 and the methane concentration of the synthesis gas is shown in Fig.11.
Since a significant increase in the pressure drop and methane concentration was not observed, a stable operation was achieved during the whole operation time.
In addition, our reaction model can accurately predict the performance of the CT-CO_2AR™ catalyst.

Table 2. Information about the commercial plant

Client's Name :	KH Neochem Co., Ltd
Plant Location :	Yokkaichi, Japan
Product :	Synthesis gas H_2/CO ratio=1.0
Product capacity :	5,000 Nm^3/hr (as H_2+CO)
Project Scope :	Catalyst supply to the existing plant
Feed Hydrocarbon :	Natural gas or Naphtha

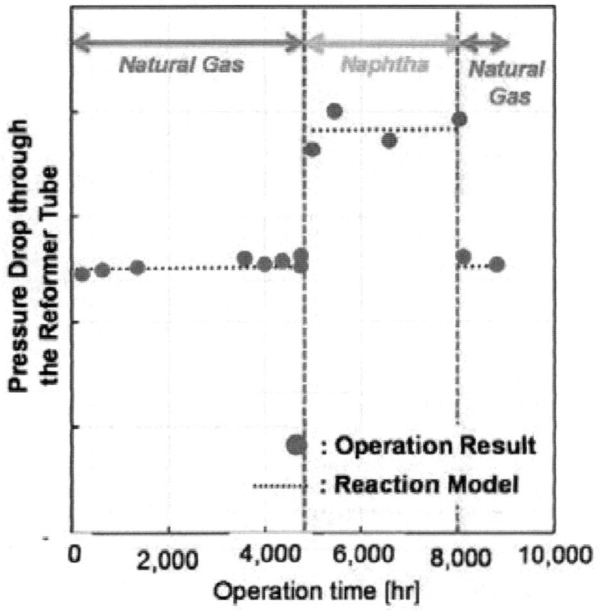

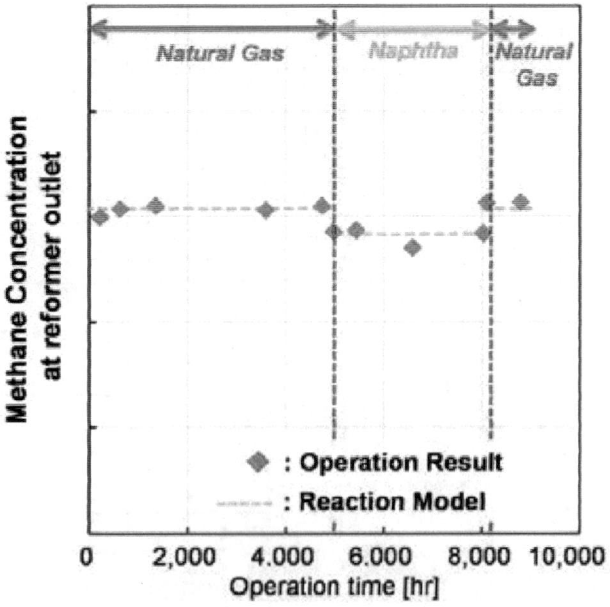

Fig. 10. Pressure difference of a reformer tube Fig. 11. Methane concentration at reformer outlet

Fig. 12 shows the comparison of the reformer duty and the amount of input natural gas between before and after replacing the CT-CO_2AR™ catalyst. By replacing the CT-CO_2AR™ catalyst, 10% of the natural gas input is reduced and this result matches the result of process study shown in section 3 (Advantages of the CT-CO_2AR™ catalyst).

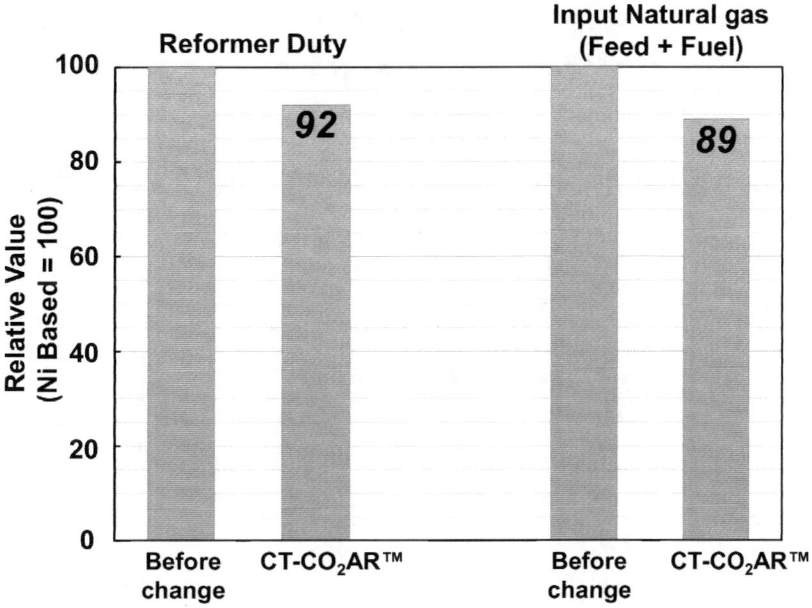

Fig. 12. Comparison result of commercial plant operation

5. Conclusion

The CT-CO_2AR™ catalyst enables a lower S/C condition operation because of its higher resistance to carbon formation.

Compared with a commercial catalyst, the CT-CO_2AR™ catalyst demonstrates high competitiveness in both manufacturing cost and CO_2 emission in the case of H_2/CO ratio = 0.5 and 1.0, especially, 0.5. The CT-CO_2AR™ catalyst was installed in a commercial plant in 2014 and has been operating stably. 10% reduction of natural gas input was achieved in the commercial operation.

6. References

1) J.R. Rostrup-Nielsen, J. H. B. Hansen, *J. Catal.*, **144**, 38 (1993)

2) N.R.Udengaard, J.-H. Bak Hansen and D.C.Hanson, Oil & Gas J., Mar.9(1992)

Solar Thermochemical Splitting of H_2O and CO_2 using Nonvolatile Metal Oxides

(Principal Author)
Rahul Bhosale
Department of Chemical Engineering
College of Engineering
Qatar University, Doha, Qatar
rahul.bhosale@qu.edu.qa

(Other Authors)
Gorakshnath Takalkar
Department of Chemical Engineering
College of Engineering
Qatar University, Doha, Qatar

Anand Kumar
Department of Chemical Engineering
College of Engineering
Qatar University, Doha, Qatar

Fares A. AlMomani
Department of Chemical Engineering
College of Engineering
Qatar University, Doha, Qatar

Prepared for Presentation at
American Institute of Chemical Engineers
2017 Fall Meeting
Minneapolis, MN, USA
October 29 – November 3, 2017

Solar Thermochemical Splitting of H_2O and CO_2 using Nonvolatile Metal Oxides

Rahul R. Bhosale, Anand Kumar, Gorakshnath Takalkar, Fares A. AlMomani

Department of Chemical Engineering
College of Engineering
Qatar University, Doha, Qatar

Keywords: Thermochemical Reactions, Metal Oxides, Solar Fuels, H_2O/CO_2 Splitting

Abstract

In this study, a Co ferrite/ZrO_2 ceramic nanomixture was synthesized by the propylene-oxide-assisted sol–gel method. Cobalt, iron, and zirconium nitrate salts were used as the metal precursors, and ethanol and propylene oxide were used as the solvent and gelation agent, respectively. The prepared gels were aged, dried, and annealed at 1000°C for different dwell times. The annealed Co ferrite/ZrO_2 powder was analyzed by powder X-ray diffraction (PXRD), Brunauer–Emmett–Teller surface area analysis (BET), and energy-dispersive spectroscopy (EDS),. The characterization results are reported in detail.

1. Introduction

Metal-oxide-based solar thermochemical water and carbon dioxide splitting is considered a promising technology to satisfy future energy demands [1]. Zinc oxide [2-5], iron oxide [5-6], tin oxide [7-8], doped iron oxides (ferrites) [9-14], ceria-based redox materials [15-19], and perovskites [20-24] are used for splitting. Among these oxides, ferrites have been used extensively for solar thermochemical water splitting; however, their use is extremely limited for CO_2 splitting.

However, the major issue limiting the use of ferrites in solar thermochemical cycles is the increased size of the ferrite grains obtained as a result of high-temperature processing. In this study, a Co ferrite/ZrO_2 ceramic mixture was synthesized by the propylene-oxide-assisted sol–gel method [25-27], exhibiting mixing of Co ferrite and ZrO_2 with a uniform distribution. The effect of annealing time on the phase and chemical composition, crystallite size, specific surface area (SSA), pore volume, and morphology of the Co ferrite/ZrO_2 ceramic nanomixture was investigated, and the results were reported in detail.

2. Experimental

2.1 Sol–Gel Synthesis

Metal precursors of cobalt, iron, and zirconium were first mixed in a specific weight ratio and dissolved in ethanol using a sonication bath. Once a visibly clear solution was obtained, a predetermined amount of propylene oxide was slowly added, affording a gel. The gel was aged, dried, and annealed at 1000°C (at various annealing times) in air. Fig. 1 shows the flow chart for the synthesis of the Co ferrite/ZrO_2 ceramic nanomixture.

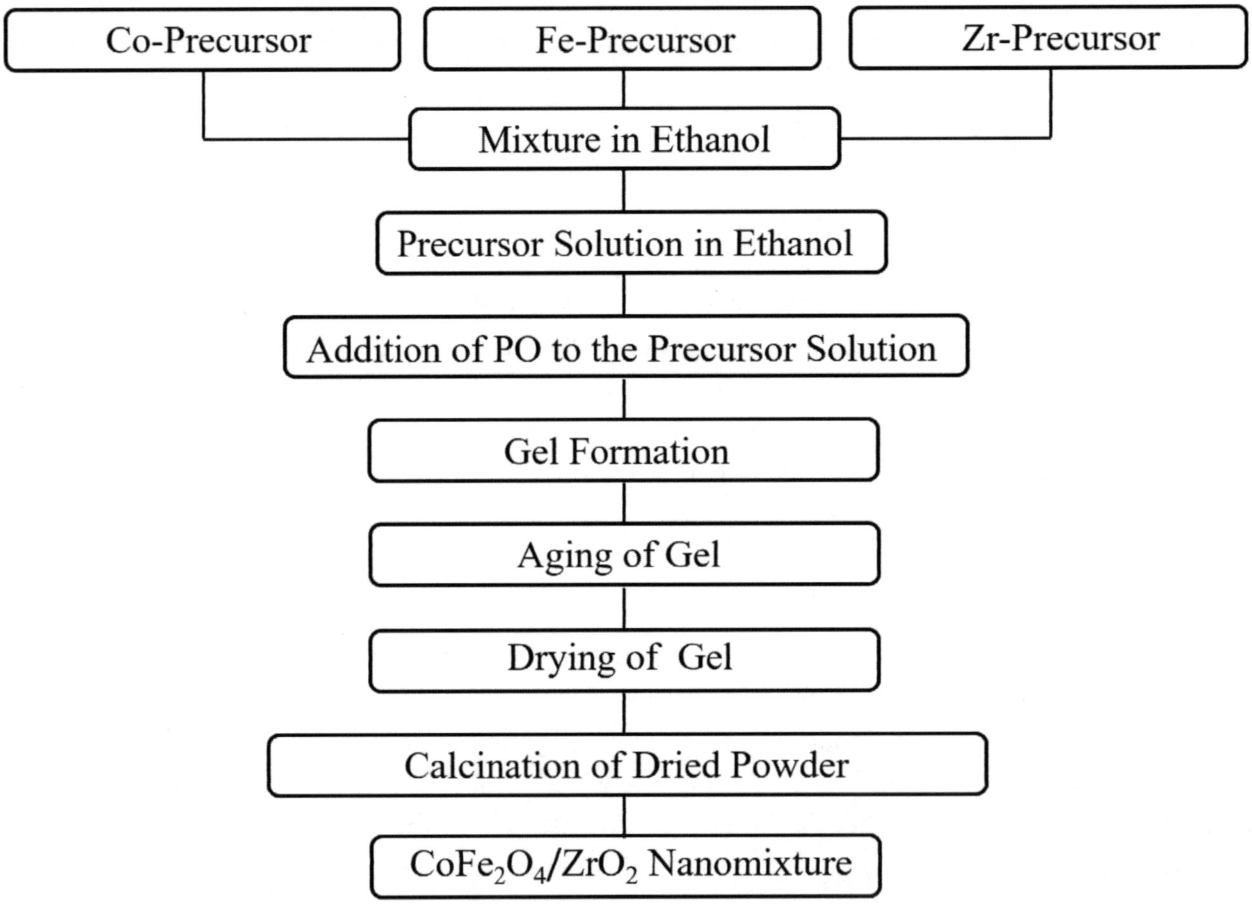

Fig. 1. Flow chart for the synthesis of the Co ferrite/ZrO_2 ceramic nanomixture by the sol–gel method.

2.2 Characterization

To investigate the effects of annealing on the physiochemical properties of the Co ferrite/ZrO_2 ceramic nanomixture, the Co ferrite/ZrO_2 powder obtained after annealing was characterized by powder X-ray diffraction (PXRD), Brunauer–Emmett–Teller (BET) surface area analysis, scanning electron microscopy (SEM) with a field-emission gun, X-ray energy-dispersive spectroscopy (EDS), and high-resolution transmission electron microscopy (HR-TEM).

3. Results & Discussion

The effect of annealing time on the physicochemical properties of the sol–gel-derived Co ferrite/ZrO_2 ceramic nanomixture was investigated. For this purpose, the Co ferrite/ZrO_2 gels (prepared using 20 mL of propylene oxide, aged for 24 h, and dried at 100°C) were annealed at 1000°C for 1–5 h. The calcined powders thus obtained were characterized by PXRD and EDS to determine the phase and chemical composition. The results shown Fig. 2 clearly indicated that the phase and chemical composition of the sol–gel-derived Co ferrite/ZrO_2 ceramic nanomixture is unaffected by the increase in the annealing time.

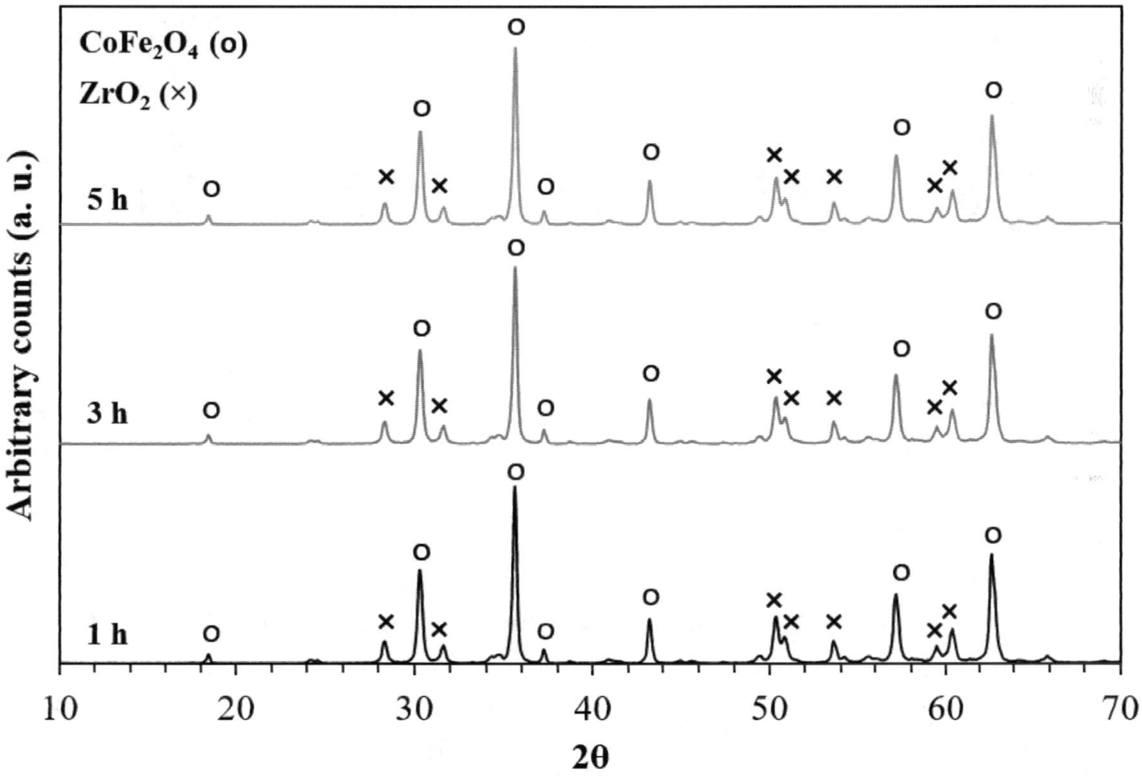

Fig. 2 – Effect of the annealing time on the phase composition of sol–gel-derived Co ferrite/ZrO_2 ceramic nanomixture.

The effect of the annealing time on the crystallite size of the sol–gel-derived Co ferrite/ZrO_2 ceramic nanomixture was also determined. Fig. 3 shows the results obtained. With the increase in the annealing time from 1 to 3 h, the crystallite size increases by 14.86 nm. With the further increase in the annealing time from 3 to 5 h, the crystallite size further increases by 22.83 nm. In summary, with the increase in the annealing time from 1 to 5 h, the size of the Co ferrite/ZrO_2 crystallites increases from 53.66 to 91.35 nm.

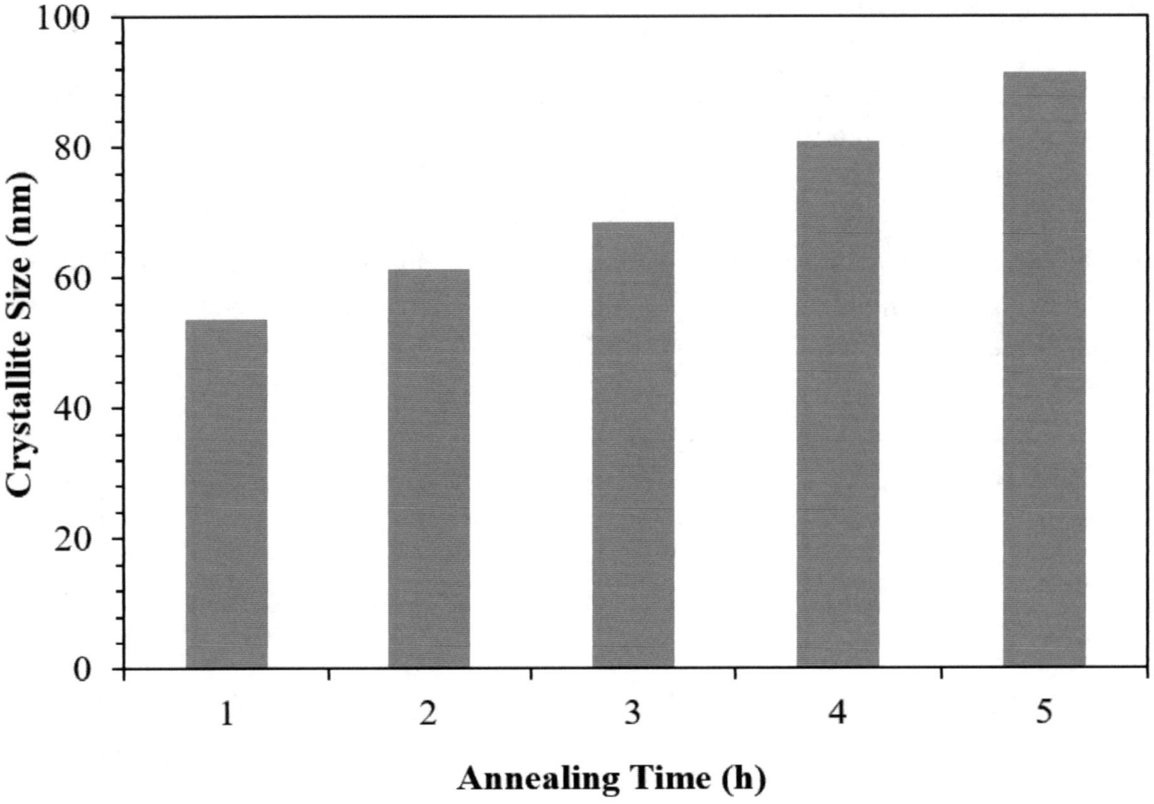

Fig. 3 – Effect of the annealing time on the crystallite size of the sol–gel-derived Co ferrite/ZrO_2 ceramic nanomixture.

In addition to the composition and crystallite size, the SSA and pore volume are also crucial parameters to be investigated. The results obtained by BET analysis indicated that with the increase in the annealing time from 1 to 3 h, the SSA decreases by 14.33%. Similarly, with the increase in the annealing time from 3 to 5 h, the SSA decreases by 13.57%. With the increase in the annealing time from 1 to 5 h, the pore volume decreases from 1.52×10^{-3} cm^3/g to 1.03×10^{-3} cm^3/g.

4. Conclusion

In this paper, a Co ferrite/ZrO_2 ceramic nanomixture, with a pure phase composition, high SSA, pore volume, nanoparticulate morphology, and a uniform distribution of Co ferrite and ZrO_2 in the mixture, was synthesized by the propylene-oxide-assisted sol–gel method. The phase and chemical composition of the Co ferrite/ZrO_2 nanomixture was not affected under all of the annealing conditions. With the increase in the annealing time from 1 to 5 h at an annealing temperature of 1000°C, the crystallite size increased by 37.69 nm. with the increase in the annealing time from 1 to 5 h, the SSA and pore volume decreased by 2.59 m^2/g and 0.484×10^{-3} cm^3/g, respectively.

5. Acknowledgments

This publication was made possible by the NPRP grant (NPRP8-370-2–154) from the Qatar National Research Fund (a member of Qatar Foundation). The statements made herein are solely the responsibility of author(s).

References

[1] J.R. Scheffe, A. Steinfeld, Oxygen exchange materials for solar thermochemical splitting of H_2O and CO_2: a review, Materials Today 17 (2014) 341-348.

[2] P.G. Loutzenhiser, A. Steinfeld, Solar syngas production from CO_2 and H_2O in a two-step thermochemical cycle via Zn/ZnO redox reactions: thermodynamic cycle analysis, International Journal of Hydrogen Energy 36 (2011) 12141-12147.

[3] A. Stamatiou, P. Loutzenhiser, A. Steinfeld, Solar syngas production from H_2O and CO_2 via two-step thermochemical cycles based on Zn/ZnO and FeO/Fe_3O_4 redox reactions: Kinetic analysis, Energy & Fuels 24 (2010) 2716-2722.

[4] R. R. Bhosale, A. Kumar, F. AlMomani, R. B. Gupta, Solar thermochemical $ZnO/ZnSO_4$ water splitting cycle for hydrogen production, International Journal of Hydrogen Energy (2017) in-press.

[5] P. Loutzenhiser, M. Gálvez, I. Hischier, A. Stamatiou, A. Frei, A. Steinfeld, CO_2 splitting via two-step solar thermochemical cycles with Zn/ZnO and FeO/Fe_3O_4 redox reactions II: Kinetic analysis, Energy & Fuels 23 (2009) 2832-2839.

[6] R. R. Bhosale, A. Kumar, van den Broeke, Leo JP, S. Gharbia, D. Dardor, M. Jilani, J. Folady, M.S. Al-Fakih, M.A. Tarsad, Solar hydrogen production via thermochemical iron oxide–iron sulfate water splitting cycle, International Journal of Hydrogen Energy 40 (2015) 1639-1650.

[7] P. Charvin, S. Abanades, F. Lemont, G. Flamant, Experimental study of $SnO_2/SnO/Sn$ thermochemical systems for solar production of hydrogen, AIChE Journal 54 (2008) 2759-2767.

[8] R. R. Bhosale, A. Kumar, P. Sutar, Thermodynamic analysis of solar driven SnO_2/SnO based thermochemical water splitting cycle, Energy Conversion & Management 135 (2017) 226-235.

[9] R. R. Bhosale, A. Kumar, F. AlMomani, U. Ghosh, P. Sutar, G. Takalkar, A. Ashok, I. Alxneit, Effectiveness of Ni incorporation in iron oxide crystal structure towards thermochemical CO_2 splitting reaction, Ceramics International 43 (2017) 5150-5155.

[10] C.C. Agrafiotis, C. Pagkoura, A. Zygogianni, G. Karagiannakis, M. Kostoglou, A.G. Konstandopoulos, Hydrogen production via solar-aided water splitting thermochemical cycles: Combustion synthesis and preliminary evaluation of spinel redox-pair materials, International Journal Hydrogen Energy 37 (2012) 8964-8980.

[11] R. R. Bhosale. A. Kumar, F. AlMomani, I. Alxneit, Propylene oxide assisted sol–gel synthesis of zinc ferrite nanoparticles for solar fuel production, Ceramics International 42 (2016) 2431-2438.

[12] M. Roeb, N. Gathmann, M. Neises, C. Sattler, R. Pitz-Paal, Thermodynamic analysis of two-step solar water splitting with mixed iron oxides, International Journal of Energy Research 33 (2009) 893-902.

[13] R.R. Bhosale, R.V. Shende, J.A. Puszynski, Thermochemical water-splitting for H_2 generation using sol-gel derived Mn-ferrite in a packed bed reactor, International Journal of Hydrogen Energy 37 (2012) 2924-2934.

[14] J.R. Scheffe, J. Li, A.W. Weimer, A spinel ferrite/hercynite water-splitting redox cycle, International Journal of Hydrogen Energy 35 (2010) 3333-3340.

[15] R.R. Bhosale, A. Kumar, F. AlMomani, I. Alxneit, Sol–gel derived CeO_2–Fe_2O_3 nanoparticles: Synthesis, characterization and solar thermochemical application, Ceramics International 42 (2016) 6728-6737.

[16] R.R. Bhosale, A. Kumar, F. AlMomani, U. Ghosh, S. Al-Muhtaseb, R. Gupta, I. Alxneit, Assessment of $Ce_xZr_yHf_zO_2$ based oxides as potential solar thermochemical CO_2 splitting materials, Ceramics International 42 (2016) 9354-9362.

[17] J.R. Scheffe, A. Steinfeld, Thermodynamic analysis of cerium-based oxides for solar thermochemical fuel production, Energy & Fuels 26 (2012) 1928-1936.

[18] W.C. Chueh, C. Falter, M. Abbott, D. Scipio, P. Furler, S.M. Haile, A. Steinfeld, High-flux solar-driven thermochemical dissociation of CO_2 and H_2O using nonstoichiometric ceria, Science 330 (2010) 1797-1801.

[19] P. Furler, J.R. Scheffe, A. Steinfeld, Syngas production by simultaneous splitting of H_2O and CO_2 via ceria redox reactions in a high-temperature solar reactor, Energy & Environmental Science 5 (2012) 6098-6103.

[20] J.R. Scheffe, R. Jacot, G.R. Patzke, A. Steinfeld, Synthesis, Characterization, and Thermochemical Redox Performance of Hf^{+4} , Zr^{+4} , and Sc^{+3} Doped Ceria for Splitting CO_2, The Journal of Physical Chemistry C 117 (2013) 24104-24114.

[21] J.R. Scheffe, D. Weibel, A. Steinfeld, Lanthanum–strontium–manganese perovskites as redox materials for solar thermochemical splitting of H_2O and CO_2, Energy & Fuels 27 (2013) 4250-4257.

[22] A.H. McDaniel, E.C. Miller, D. Arifin, A. Ambrosini, E.N. Coker, R. O'Hayre, W.C. Chueh, J. Tong, Sr-and Mn-doped $LaAlO_{3-\delta}$ for solar thermochemical H_2 and CO production, Energy & Environmental Science 6 (2013) 2424-2428.

[23] M. Gálvez, R. Jacot, J. Scheffe, T. Cooper, G. Patzke, A. Steinfeld, Physico-chemical changes in Ca, Sr and Al-doped La–Mn–O perovskites upon thermochemical splitting of CO_2 via redox cycling, Physical Chemistry Chemical Physics 17 (2015) 6629-6634.

[24] R. R. Bhosale, A. Kumar, A. Ashok, P. Sutar, G. Takalkar, M. Khraisheh, F. AlMomani, U. Ghosh, La-Based Perovskites as Oxygen-Exchange Redox Materials for Solar Syngas Production, MRS Advances (2017) in-press.

[25] R. R. Bhosale, R. Khadka, J. Puszynski, R. V. Shende, H_2 generation from two-step thermochemical water-splitting reaction using sol-gel derived $Sn_xFe_yO_z$, Journal of Renewable & Sustainable Energy 3 (2011) 063104-1 – 063104-12.

[26] R. Bhosale, R. Shende, J. Puszynski, H_2 generation from thermochemical water splitting using sol-gel derived Ni-ferrite, Journal of Energy and Power Engineering 4 (2010) 27-38.

[27] R.R. Bhosale, R.V. Shende, J.A. Puszynski, H_2 Generation From Thermochemical Water-Splitting Using Sol-Gel Synthesized Zn/Sn/Mn-doped Ni-Ferrite, International Review of Chemical Engineering 2 (2010) 852-862.

500493 Critical Issues in the Development of Commercial Natural Gas to Industrial Chemicals Bioprocesses

Abstract:

The low cost and abundant supply of natural gas make it America's greatest untapped energy resource, leading to the development of technologies for natural gas conversion to compounds of greater value. Intrexon has developed the first natural gas-to-liquids bioconversion platform powered by methantrophic fermentation, which applies synthetic biology to program bacteria and produce higher value materials. Our innovative genetic toolbox allows us to engineer the biology of the methanotroph to more efficiently convert natural gas into a wide range of valuable chemicals and fuels. These applications include isobutanol, farnesene, 1,4-butanediol, 2,3-butanediol, isoprene and isobutyraldehyde – the building blocks for industrial and consumer products like fuels, synthetic rubber, acrylics, resins and spandex.

Using a methanotrophic bioprocess to produce fuels and chemicals has several advantages. Natural gas bioconversion offers scaling opportunities beyond traditional gas-to-liquid (GTL) facilities, as well as biomass feedstock facilities. The theoretical yields of methanotroph/methane exceed that of yeast/sugar approaches by a wide margin. Biocatalytic processes that achieve high yields under milder processing conditions allow for reduced capital and operation expenditures, offering greater economic viability. To successfully bring this technology to commercial scale, consideration must be given not only to optimizing the bacterial metabolic pathway, but to key issues in the fermentor design, downstream processing, scale-up and overall plant economics.

At any scale, aerobic natural gas fermentation must overcome the challenge of gas-to-liquid transfer of low solubility gases to increase bioavailability of gas substrates to the bacteria. A low gas-to-liquid mass transfer rate reduces the bacterial productivity and process yield, which in turn increases fermentor size, capital cost, operating costs and may add additional equipment necessary to recycle gases. During commercial scale-up of gas fermentation processes, these factors can become cost prohibitive. Standard stirred tank reactors become unfeasible for use at larger scales, as the agitator power input and size become excessive. There are many fermentor designs currently available to increase the overall mass transfer coefficient of these gases. However, capital and operating expenditures of the reactor are critical metrics for evaluating commercialization feasibility. Unique fermentor configurations can greatly increase capital cost by raising the cost of fabrication and decreasing the ability to receive competitive bids. These unique fermentors also may not create a homogenous, well-mixed broth – which increases the cost of controls and instrumentation. The economic feasibility of gas transfer designs must be balanced with the biologic effects of reactor forces on the bacteria. Methods of increasing mass transfer often include increasing turbulence in the reactor, which can create detrimental effects on the bacteria through high shear forces, high pressure and vacuum areas, and dissolved gas substrate inhibition. Careful equipment selection and pilot trials are necessary to determine the various effects of gas transfer methods on mass transfer rates and cell viability.

Downstream process selection for commercialization is a balance of biologic limitations, technical feasibility, capital cost, operating cost and the risk of cutting-edge over well-established techniques. The developing bioprocess must be iterated upon the limitations of strain engineering and practicality of downstream processing techniques. Fermentation broth concentrations may be low to product inhibition on the cell, but this increases the size and energy requirement of downstream equipment. Cutting-edge technologies can increase the efficiency of low titer separation and reduce energy inputs, which can be capital intensive and unproven at commercial scales. Genetic engineering tools can increase the tolerance of the bacteria to products. However, the directed or natural evolution process requires time.

A minimum broth titer must be determined through techno-economic modeling of processes at each stage of process development and refinement. Bioprocesses also create metabolic byproducts which can add unit operations, and therefore cost, to the downstream purification process. Some byproducts may originate from the genetic pathway or from commercial feedstocks, as they are often lower in purity and have higher variability than those at the research scale. Additional downstream steps must be evaluated to determine if they are technically and economically feasible to meet final product specifications for sale, or if the byproduct genetic pathways can be manipulated to terminate production and not add cost to the overall process.

Furthermore, certain tools that aid in genetic engineering and strain development at the research and development scale must be eliminated or retooled to create commercially viable strains. During development, production pathways are often introduced to the bacteria through plasmids which carry antibiotic resistance. To avoid the material cost and downstream separation of the antibiotic at commercial scale, the genetic material must be integrated into the chromosome. This genomic integration may affect the strain's capability to produce product at previously observed yields, titers and rates. Another common development tool is that production pathways may be turned on or off using an externally applied chemical inducer. At scale, these chemical inducers can add material cost as well as affect downstream processes and product purity. Induction systems may be completely removed for commercial scale or based off an easily controllable environmental switch.

Favorable plant economics are necessary to move forward with commercialization of a technology. Scale is carefully selected based on a comparison of the production rate, fermentation productivity and yield targets to revenue, capital and operating costs. The plant location is an important selection factor, as costs are affected by utility rates in the area and/or co-location of the plant with shared infrastructure.

Traditional engineering workflows, specifically front-end loading strategy, were used to shape the commercial scale process design of Intrexon's natural gas to chemical plants. The biologic, technical and economic commercial feasibilities of the end-to-end natural gas to isobutanol process were critical to Intrexon's plant design. Laboratory scale learnings were applied to select technology that was commercially feasible at demonstration scale. This demonstration scale design was scaled down to pilot scale, where the operation will yield learnings into refinement of at-scale design, equipment selection, empirical data on heating and cooling, and controls strategies and feedback to further iterations on genetics.

495124 Direct, Single-Pass Thermocatalytic Upgrading of Biogas and Landfill Gas into Renewable Natural Gas over the Ultra-Low Loading Ru/Î³-Al_2O_3 Catalyst

Abstract:

Direct, single-pass thermocatalytic upgrading of biogas and landfill gas into renewable natural gas over the ultra-low loading Ru/g-Al_2O_3 catalyst

Yichen Zhuang and David Simakov

Department of Chemical Engineering, University of Waterloo, Waterloo, ON N2L 3G1, Canada

Converting CO_2 into synthetic renewable fuels is an attractive pathway to decrease CO_2 emissions and to reduce our dependence on fossil fuels. Thermocatalytic hydrogenation provides advantages of fast reaction rates and high conversion efficiencies (as compared to electro-catalytic and photo-catalytic routes), thus allowing for compact, high-throughput operation. One option is to convert CO_2 into synthetic methane, so-called renewable natural gas (RNG) via the Sabatier reaction ($CO_2 + 4H_2 = CH_4 + 2H_2O$). Water electrolysis using renewable or surplus electricity can be used to provide a non-fossil source of H_2 required for the reaction. There are multiple sources of concentrated CO_2-containing streams including biogas and landfill gas (up to 50 mol% CO_2). It is highly desirable to develop a technology that can convert such CO_2-rich streams directly, without costly upstream CO_2 separation.

One of the most fundamental challenges is the development of a catalyst, which is highly active, selective to CH_4 formation, stable against coking, and low-cost at the same time. The selectivity becomes a major challenge when a mixed feed (biogas or landfill gas contain up to 50 mol% CH_4 in addition to CO_2) is used instead of purified CO_2 feed. In this study, a highly selective Ru-based catalyst with ultra-low Ru loadings (less than 0.5 wt%) is reported. A series of catalysts with Ru loadings ranging from 0.05-0.5 wt% were synthesized and evaluated for their catalytic activity in selective conversion of CO_2 in the mixed CH_4/CO_2 feed. CO_2 conversions up to 80% were achieved with nearly complete selectivity to CH_4 generation (left plot in the figure below). Importantly, the catalysts showed good stability, without any notable decline in performance over 60 hours on stream at very high space velocity of GHSV = 90,000 1/h (right plot in the figure below). A variety of analytical techniques were employed in order to investigate the mechanism of the superior selectivity and activity, including electron microscopy, thermal gravimetric analysis, temperature programmed reaction, chemisorption, and X-ray photon spectroscopy. Our results point out at the strong structure-activity relationship and the importance of the active phase interaction with the support.

492217 Energy Quality Factor and Exergy Destruction Processes Analysis for a Proposed Polygeneration System

Abstract:

A new polygeneration system was established through combining coal pyrolysis with semicoke combustion, semicoke gasification and a steam Rankine cycle, and then the calculation methodology was developed to calculate the system. The energy and exergy efficiencies of the system are 65.9 % and 75.1 %, respectively; and the standard energy quality factor Î±Î¸ values of the involved 17 components and the energy quality factor Î± of each system stream were obtained. By comparing the differences among the standard energy quality factor Î±Î¸ values of the related 17 components, the features of the energy grade of the components at the standard conditions were elucidated. Then the effect of streamÂ¢s composition and temperature on the actual energy quality factor was discussed. Combining the concept of energy quality factor, the processes of exergy destruction in system blocks were visually expressed via a kind of diagram; and the reasons of exergy destruction were tried to be explained from the perspective of energy quality factor.

491941 Mechanistic Implications of Low-Pressure Feeds for Methanol-to-Olefins Conversion on MFI

Abstract:

The production of methanol and its subsequent conversion on zeolites to produce important petrochemical precursors, such as ethylene and propylene, provides a route for the commercial-scale valorization of unconventional carbon sources such as coal, biomass, and natural gas. The complex network of reactions governing this chemistry can be summarized by a dual catalytic cycle description where olefins-based chemistries of methylation and ß-scission are interconnected with the aromatics-based chemistries of methylation and dealkylation via hydrogen transfer and cyclization. Transient isotopic labeling studies implicate ethylene production predominantly from the aromatics-based cycle, thereby, suggesting that process or material parameters that result in the selective propagation of the olefins-based cycle would in turn result in high propylene-to-ethylene ratios. We implicate inlet methanol partial pressure as the crucial process parameter for preferential propagation of the olefins-based cycle over its aromatics counterpart, regardless of the reaction temperature, on H-ZSM-5. We report

(i) ~10-fold increase in the propylene-to-ethylene ratio (~2.7 – 27) as the inlet methanol pressure was decreased from 52 to 0.6 kPa; carbon selectivities towards propylene increased from ~29 to 41%C when compared at iso-conversion (~30 %C) levels. This effect persisted regardless of the reaction temperature (623 – 773 K). These results clearly suggest that low inlet pressures of methanol can effectively decouple the olefins- and aromatics-based cycles;

(ii) The previously unrecognized role of formaldehyde, obtained from persistent methanol dehydrogenation events, in initiating the aromatics-based cycle. A small co-feed of formaldehyde (~24 Pa) with 0.6 kPa of methanol resulted in ~15-fold increase in ethylene selectivity (from ~1.5 to 22 %C) while the propylene-to-ethylene ratio dropped from ~27 (in the absence of formaldehyde) to ~1 (in the presence of formaldehyde), thereby, advocating suppressed formaldehyde production at low inlet methanol pressures as the critical factor in selective propagation of the olefinic cycle.

These mechanistic studies provide guidance for process conditions to minimize the formation of undesirable products during methanol-to-hydrocarbons conversion on zeolites.

498789 Catalysis on Microporous Solid Acids: Mechanism and Catalyst Descriptors for the Coupling of Alkenes and Alkanones

Abstract:

The conversions of light alkenes and oxygenates on solid Brønsted acids are mediated by ion-pair transition states (TS) [1,2]. The stability (with respect to their relevant kinetic precursors) of these ion-pairs, consisting of organic cations and inorganic anions formed upon proton transfer, determine turnover rates for these reactions and also those for subsequent conversions of the primary products formed, such as isomerization, β-scission, hydride transfer, cyclization and the formation of unsaturated species that can render protons unreactive. The inorganic catalyst host affects TS stability by the ability of its anion to stabilize negative charge (acid strength) and of its voids to establish van der Waals contacts with the organic guest; such contacts depend sensitively on the matching of the size and shape of host and guest. The stability of the cationic moiety at the transition state also depends on its ability to accept the proton (proton affinity) and then to interact electrostatically with the localized negative charge at the framework and via dispersion forces through van der Waals contacts with the surrounding framework. These confinement and electrostatic effects are probed here for types of reactions mediated by TS structures that differ in size, shape, or charge from each other and from their relevant precursors on solids with diverse acid strength and confining environments. Density functional theory (DFT) methods that include pertinent dispersive interactions are used to provide insights that are inaccessible from experiment alone, such as the nature and magnitude of relevant energetic descriptors in the form of van der Waals confinement energies, proton affinities of organic molecules, and deprotonation energies.

In this work, we report turnover rates (per proton) for C_2-C_4 alkene coupling (503 K, 10- 500 kPa alkene) and for aldol condensation (473 K, 0.1-10 kPa acetone) reactions on acids with different acid strength and confining environments (H-Al-TON, H-X-MFI (X=Al, Ga, Fe, and B isomorphously substituted samples), H-Al-BEA, H-Al-FAU, H-Al-MCM-41, and supported Keggin polyoxometalate clusters ($H_3PW_{12}O_{40}/SiO_2$)). For both reactions, turnover rates are proportional to the pressure of the respective reactants on all solid acids and the first-order rate constants reflect an energy barrier between bimolecular C-C bond-forming TS and protons that are saturated with less-charged precursors, consistent with infrared spectra and DFT-simulations; these precursors include covalently-bound alkoxides during alkene coupling reactions and H-bonded acetone during condensation reactions.

The measured rate constants are sensitive to changes in acid strength, in the case of unconfined acid sites in polyoxometalates and silica-alumina and of similarly-confined sites within MFI frameworks with Al, Ga, Fe, or B heteroatoms; here, acid strength is described in terms of deprotonation energy, a theoretically accessible value. In both cases, rate constants decreased exponentially with increasing deprotonation energy, consistent with kinetically-relevant energy barriers between charged, ion-pair TS and less charged, precursors.

Measured condensation and alkene coupling rate constants on microporous and mesoporous aluminosilicates, which contain acid sites of similar strength, reach maximum values as the framework voids and the TS structures become of similar size. The consequences of such "good fits" for reactivity are quantitatively described first based on spherical geometry by comparing the diameter of a sphere with volume equivalent to the van der Waals volume of the TS to a characteristic void size defined by the largest included diameter within the pore network. We find, however, that to describe these interactions both the size *and the shape* of the zeolite void and the TS must be characterized. The incomplete nature of such geometrical descriptors is resolved by using van der Waals interaction energies (E_{vdw}) obtained by ensemble-averaging Lennard-Jones interactions between each framework O-atom and each atom in the organic moiety (TS or precursor) at each crystallographically distinct T-site for each of the frameworks examined in this study.

The alkene coupling TS structures for alkenes of different chain length vary in size, shape, and proton affinity, thus allowing us to dissect effects of confinement and intrinsic TS stability. Their different size also show that oxide frameworks distort locally upon containment of TS structures and alkoxides as their size approaches that of the confining voids. These distortions are quantified on a geometric basis as the mean O-atom displacement and, energetically as E_{dist}; both metrics indicate that the distortions enhance van der Waals contacts for the host-guest structures. These distortions, however, bring forth energetic penalties that attenuate and ultimately negate the benefits of more effective van der Waals contacts, consistent with the repulsive component of E_{vdw} that dominates for the large isobutene coupling TS in the 10-membered ring channel of TON.

Here, we provide a framework via thermochemical cycles to understand the effects of these catalyst and mechanistic descriptors (DPE, E_{vdw}, E_{dist}, proton affinity) on the relevant energy barriers. The similar conclusions drawn for both alkene coupling and alkanone condensation reflect the resemblance of chemistry, i.e. C-C bond formation via bimolecular transition states, but, moreover, a commonality of the analysis using energetic descriptors for catalysis on microporous acids. This universality begets a tool to predict reactivity of catalysts and reactants not studied here.

[1] Sarazen, M. L., Doskocil, E. Iglesia, E., J. Catal. 344, 553 (2016)

[2] Herrmann, S., Iglesia, E., J. Catal. 346, 134 (2017)

Financial support from BP through the ICC Program and from the National Science Foundation Graduate Research Fellowship Programs. Computational resources from National Science Foundation's Extreme Science and Engineering Discovery Environment (XSEDE; ACI- 1053575) are gratefully acknowledged.

486644 Structure-Property Relationships for Unidimensional, Large and Extra-Large Pore Zeolites Using Alkane Hydrocracking and Hydroisomerization As Probe Reactions

Abstract:

Structure-Property Relationships for Unidimensional, Large and Extra-Large Pore Zeolites Using Alkane Hydrocracking and Hydroisomerization as Probe Reactions

Viktor J. Cybulskis[1], Stacey I. Zones[2], Tracy M. Davis[2], Cong-Yan Chen[2], Michael W. Deem[3] and Mark E. Davis[1]

[1]Chemical Engineering, California Institute of Technology, Pasadena, CA 91125, USA

[2]Chevron Energy Technology Company, Richmond, CA 94802, USA

[3]Bioengineering, Rice University, Houston, TX 77005 USA

High-silica zeolites (Si/Al > 30) are versatile catalytic materials in petroleum refining processes due to their thermal stability, shape selectivity, and chemical functionality. However, of the 232 known [1] and more than 330,000 predicted [2] zeolite structures, the global refining market is dominated by five main frameworks: FAU, MOR, MFI, *BEA and FER [3]. While these zeolites often deliver satisfactory commercial performance, the development of materials with improved function or entirely new structures for future applications requires a deeper understanding of how specific properties, such as composition, framework topology, and acid site function, can influence the product distribution for a given reaction.

For the present study, large and extra-large pore zeolites were prepared from the corresponding structure-directing agents (SDAs) shown in Table 1 to investigate the effects of the organo-cation guest molecules and synthesis conditions on zeolite crystallinity, microporosity, and heteroatom incorporation. Unidimensional (1D) structures were chosen so that the zeolite pore opening could be used to adequately describe the confining size. Hydrocracking and hydroisomerization of *n*-hexane (nC_6) and *n*-decane (nC_{10}) were used as probe reactions on these Pt-exchanged, 12- and 14-membered ring (MR) zeolites to examine the effect of pore structure on shape selectivity by comparing the distributions and yields of branched C_6 and C_{10} isomers. Elucidation of these structure-property relations support *de novo* computational design efforts to predict chemically-synthesizable SDAs and produce a novel, 1D zeolite framework with tailored catalytic properties.

Table 1. 1D, 12- and 14-MR zeolites synthesized from the corresponding SDAs

References:

[1] C. Baerlocher and L.B. McCusker, Database of Zeolite Structures, http://www.iza-structure.org/databases/. Accessed March 27, 2017 (2017).

[2] R. Pophale, P.A. Cheeseman and M.W. Deem, *Phys. Chem. Chem. Phys.* 13, 12407 (2011).

[3] W. Vermeiren and J.P. Gilson, *Top. Catal.* 52, 1131 (2009).

501594 Modeling Complex Reactions in Zeolites: Effects of Acid Site Location, Framework, and Reagent Structure on Methanol-to-Hydrocarbon Reactions

Abstract:

Acid-catalyzed reactions within proton-form zeolites are mediated by cationic transition states which (along with their precursors) are stabilized by electrostatic interactions with the conjugate base and are 'solvated' by the zeolite framework through van der Waals interactions. The diversity of zeolite frameworks has sustained decades of research into determining and synthesizing optimal zeolite materials for many reactions, including the conversion of methanol into hydrocarbons (MTH). The size and topology of zeolite pores can shift selectivity during MTH from light olefins (in TON zeolites) to gasoline-range hydrocarbons (in MFI zeolites), but zeolites relevant to industrial practice for MTH processes rapidly deactivate. Acid site locations within zeolites are often poorly controlled/understood, leading to difficulties in modeling zeolite-catalyzed reactions with theoretical tools, because of the configurational complexity of such reactions. Here, we investigate reactions within methanol-to-hydrocarbons reactions and show how they are affected by the local zeolite environment (by altering acid site location within a framework or by changing the framework entirely). Ultimately, the methods outlined here will enable rigorous screening of zeolite materials to identify synthesis-targets for controlling heteroatom location and new framework synthesis.

Acid site location directly impacts molecular adsorption energies, reaction energies, and activation barriers. Here, we examine the effects of acid site location using novel methods that rigorously examine molecular adsorption and reaction pathways at all acid sites within zeolite frameworks. Methanol adsorption free energies, calculated using dispersion-corrected energies within a periodic model of MFI, for example, vary from -126 kJ mol^{-1} to -16 kJ mol^{-1} at the 48 Al–O site pairs which define acid sites within MFI. Intrinsic free energy barriers for methanol dehydration to form surface methoxy species (CH_3-Z) similarly vary from 80 kJ mol^{-1} to 150 kJ mol^{-1}. Varying acid site location within MFI has the same impact on intrinsic rates as changing the entire zeolite framework, indicating the importance of NMR and synthesis techniques which attempt to locate and control the location of heteroatoms within zeolite frameworks. In this work, we will present methanol and ethanol dehydration pathways within all acid sites of TON, CHA, and MFI and use these data to demonstrate that rates, selectivities, and even mechanisms can be affected by the local zeolite environment by changing the framework or acid site location.

Methanol-to-hydrocarbon reactions are also impacted by the local zeolite environment, by certain environments permitting the formation of aromatic species which constitute a "hydrocarbon pool" of species which co-catalyze further reaction cycles—but can also grow into coke precursors, leading to catalyst deactivation. This "hydrocarbon pool" consists of methylated benzenes (from benzene to hexamethylbenzene, 13 species) and each of these species can 1) by methylated by methanol, 2) undergo intramolecular C–C bond rearrangements to form alkyl substituents, and 3) dealkylate to form light alkene products. The total number of elementary steps involved is enlarged by the large set of possible co-catalysts. Here, we use structural refinement methods to examine reaction energetics all co-catalysts for each elementary step. The free energy barrier for methylating an aromatic ring, for example, increases from 80 kJ mol^{-1} to form toluene to 140 kJ mol^{-1} to form hexamethylbenzene, indicating that methylation becomes more difficult as reagents become more sterically-crowded by methyl-substituents. These effects of co-catalyst structure, along with kinetic Monte Carlo methods, allow us to report not only the mechanism of light alkene formation during MTH, but also the hydrocarbon pool species predominantly responsible for light alkene production.

These two related topics focus on computational methods and strategies for examining reactions solvated by diverse zeolite environments (within one and across multiple frameworks) and for examining the effects of reagent structure through systematically varying –H and alkyl substituents in hydrocarbon pool species which facilitate methanol-to-hydrocarbon reactions. These approaches lay the groundwork for examining the complete MTH pathways in all known and predicted zeolite structures to ultimately use computation to direct zeolite synthesis efforts by identifying promising novel materials in which the hydrocarbon pool can be formed without concomitant coke buildup and catalyst deactivation.

497790 Simple Characterization of Solid Acid Catalysts By Reactive Gas Chromatography

Abstract:

One of the more common methods for characterizing solid acid catalysts is the temperature programmed desorption (TPD) of alkylamines, which selectively decompose to an alkene and ammonia over Brønsted acid sites. TPD is traditionally carried out either in a flow reactor or microbalance, coupled with a mass spectrometer for identification and/or quantitative purposes. Reactive gas chromatography is a new technique to measure Brønsted acid site densities of solid acid catalysts, which condenses the typical experimental setup required for a TPD into a single, fully automated gas chromatograph (GC). Reactions are carried out in a microcatalytic reactor, housed within a temperature controlled GC inlet liner, products are then separated in a GC column and quantified by a flame ionization detector (Figure 1a). This eliminates the need for more complex instrumentation such as on-line mass spectrometers or a flow reactor apparatus.

An investigation of the RGC method was carried out by characterizing ZSM-5, BEA and SPP zeolites of varying Si/Al ratios, utilizing alkylamine decomposition to selectively count Brønsted acid sites. A comparison between RGC measurements and conventional TPD methods show excellent agreement (Figure 1b), confirmed by comparison with pyridine in-situ titrations under Brønsted catalyzed reaction conditions. Reactive gas chromatography measurements performed using n-propylamine, isopropylamine and tert-butylamine were found to yield identical Brønsted acid site densities. Given the relatively low detection limits of a flame ionization detector, the method of reactive gas chromatography is found to be highly sensitive. Siliceous materials with Brønsted acid site densities of ~1 μmol g_{cat}^{-1} could be reliably measured.

Figure 1. Method of Reactive Gas Chromatography for Solid Acid Titration. A. Solid acid samples loaded into the inlet liner (red rectangle) of a gas chromatograph are titrated with amines, introduced via an automatic liquid sampler (ALS) syringe, adsorbed and reacted on solid acid sites to produce propylene and ammonia. Resultant products are quantified by a flame ionization detection (FID). **B.** Comparison of zeolite measured Brønsted acid site density between reactive gas chromatography and literature values. Catalysts include ZSM-5 (Si/Al 11.5, 40, 140: ●) and BEA (Si/Al 12.5, 19: ¦).

497279 Understanding Effective Diffusion Length Theory in Nanoscale Zeolites

Abstract:

Understanding effective diffusion length theory in nanoscale zeolites

<u>Xiaoduo Qi, Vivek Vattipalli, Paul Dauenhauer and Wei Fan</u>

Department of Chemical Engineering, University of Massachusetts Amherst, 686 N Pleasant Street, Amherst, 01002

Zeolites represent a class of crystalline aluminosilicates with well-defined micropores in the framework. Due to their unique molecular sieving and catalytic properties, zeolites have been extensively utilized as catalyst to facilitate chemical reactions. However, as the micropore size is comparable to the molecular dimension, the catalytic performance of the zeolites is often limited by the slow intra-crystalline molecular transport within the micropore channel when bulky molecules try to access the active site. To overcome such a transport limitation, considerable efforts have been made with the aim of reducing the contribution of molecular transport to the overall rate of reaction.[1] Among all other methods, synthesizing zeolite nanoparticle has been considered as one of the most promising strategy due to simple and well controlled preparation procedure. As the radius of the zeolite crystal is minimized to nanometers, it is expected that the characteristic diffusion length of such material is extremely short, leading to a rapid micropore diffusion process.

However, diffusion studies done by ZLC technique have shown that for zeolite nanoparticles, the effective micropore diffusion length can be much longer than the radius of the crystal due to a secondary diffusion step on the zeolite surface which leads to the occurrence of pore re-entry[2]. Despite the fact that the effective diffusion length phenomena can greatly hinder the mass transport within zeolite nanocrystals, the origin of such negative effect has not been well studied. In this work, four adsorbates (Cyclohexane, Methylcyclohexane, Ethylcycllohexane, and cis-dimethylcyclohexane) with different adsorption energies to MFI zeolite were used as probing molecules to investigate the cause of the effective diffusion length effect. It shows that the significance of such effect strongly depends on the adsorption strength of diffusion molecule to the zeolite surface. More specifically, as the adsorption becomes stronger, it is harder for diffusing molecule to completely desorb from the zeolite surface after it exits the micropores. As a result, this molecule will continue to diffuse along the surface and repeatedly diffuse within the pore, leading to a longer diffusion length. This work provides solid experimental evidence suggesting that the effective diffusion length effect is strongly related to guest/host adsorption. This fundamental understanding of micropore diffusion process within zeolite nanoparticles also enables us to optimize the mass transport by enhancing the adsorptive property of the zeolite.

[1] Chang, C.-C.; Teixeira, A. R.; Li, C.; Dauenhauer, P. J.; Fan, W., Enhanced Molecular Transport in Hierarchical Silicalite-1. Langmuir 2013, 29, (45), 13943-13950.

<>[2] Vattipalli, V.; Qi, X.; Dauenhauer, P. J.; Fan, W., Long Walks in Hierarchical Porous Materials due to Combined Surface and Configurational Diffusion. Chemistry of Materials 2016, 28, (21), 7852-7863.

498345 3D-Printed Zeolitic Scaffolds forÂ Selective Transformation of Light Alcohols to Light Olefins

Abstract:

This study aims at developing novel structured zeolite monoliths with various compositions by using three dimensional (3D) printing techniques to promote the production of light olefins and catalyst stability in selective transformation of light alcohols. 3D-printed ZSM-5 and silicalite-1 monoliths were coated with a thin layer of SAPO-34 zeolite. Characterization results indicated that the 3D-printed monoliths were highly porous and their micropore surface area and pore volume further increased by consequent coating with SAPO-34 zeolite. The catalytic results showed that the thin layer of SAPO-34 on 3D-printed MFI structures impeded the gaseous mas transport but favored the production of light olefins. The total acidity of SAPO-34/3D-printed ZSM-5 structure was higher than both SAPO-34/3D-printed silica and uncoated 3D-printed ZSM-5 monolith which favored the light olefins production. Coating 3D-printed HZSM-5 monolith with SAPO-34 enhanced the ability to produce olefins while circumvented the quick coke formation. These results demonstrated that selectivity and reaction rate in dehydration of light alcohols to light olefins can be modulated by changes in catalyst topology and morphology.

494815 The Role of Surface Barriers As Dominant Transport Mechanism in Hierarchically Structured Zeolites-- Application to the Alkylation of Benzene with Ethylene

Abstract:

The meso-macropore network of a hierarchically structured zeolite catalyst is numerically optimized to maximize the volume-integrated reaction yield in the ethylation of benzene to produce ethylbenzene over zeolite H-ZSM-5. A hierarchical approach is used at multiple length scales to determine the optimal pore network properties. The maximum volume-integrated reaction yield of the hierarchically structured zeolite catalyst containing meso- and macropores is nearly twice the yield of a zeolite pellet containing only macropores, at the same macroporosity. To bridge the gap between modeling and experiments, a series of physical mixtures of ZSM-5 crystals and mesoporous silica, containing different weight fractions of zeolite is synthesized and used in fixed bed reactor experiments to determine the optimal pellet structure to maximize the conversion of ethylene. Comparison with reactor simulations of the zeolite composites shows that the performance of the zeolite composites might be limited by surface barriers at the external surface of the zeolite crystals, rather than by diffusion limitations within the meso-macropore network of the pellets.

References:

Rao, S. M.; Saraci, E.; Glaeser, R.; Coppens, M.-O. Chem. Eng. J. 2017 (http://doi.org/10.1016/j.cej.2017.04.015)

488270 Stabilizing Pd Particles on Nitrogen-Doped Carbon Supports Under Hydrothermal Conditions

Abstract:

Carbon coating has been applied to improve the hydrothermal stability of metal oxide supports for use in aqueous-phase biomass reaction systems. Previously, incorporation of oxygen heteroatoms in the carbon coating were found to provide some enhancement the hydrothermal stability of Pd particles on these carbon-coated materials. To further improve the hydrothermal stability of these Pd particles, nitrogen-doped carbon supports were synthesized. Compared with merely carbon supported Pd catalyst, better dispersion and stability of the particles were achieved with the nitrogen-doped carbon. Unlike with undoped carbon, the Pd particles had limited sintering after extended exposure to hydrothermal reactions conditions and Pd leaching was greatly reduced. Through detailed solid-state ^{13}C and ^{15}N NMR analysis, the carbon surface chemistry was investigated and systematically compared for varying degrees of N incorporation. Previous studies have suggested that heteroatom incorporation could improve stability, but the NMR characterization techniques used in this study allowed for more detailed and quantitative characterization of N species present in the carbon. Better stability was ascribed to nitrogen functional groups including pyridine, pyrrole and secondary amine groups. This work will help to develop a more systematic strategy for improving hydrothermal stability of metal particles by heteroatom doping in carbon supports.

502131 Reaction Mechanism for the Conversion of Î³-Valerolactone (GVL) over a Ru Catalyst

Abstract:

Conversion of biomass-derived gamma-valerolactone (GVL) to valuable chemicals has been studied extensively, and understanding the reaction mechanism is very valuable for improving turnover rates and selectivities. Recently, it was shown that, in an aqueous phase over a Ru/C catalyst, GVL could be converted to 1,4-pentanediol, 2-butanol and 2-pentanol in parallel paths[1]. Here, we report first-principles density functional calculations, through which we show in detail the reaction pathways of GVL conversion on a Ru(0001) surface[2]. We find that GVL undergoes a ring-opening reaction rather easily, and that the rate-limiting step towards the formation of 1,4-pentanediol and 2-pentanol is the hydrogenation step. The high energy barrier for this step is caused by a strong interaction between Ru and the unsaturated acyl intermediate that is formed after opening the ring. Among all the primary products, formation of the 2-butanol has the smallest activation barrier, while the slowest step is the C-C bond cleavage in the decarbonylation step. We further show that the same acyl intermediate after ring opening of GVL can also be formed by dehydrogenation of 1,4-pentanediol with a moderate activation barrier, which suggests both 2-pentanol and 2-butanol can also be produced in secondary steps.

Acknowledgements

The calculations were performed at NERSC and OU Supercomputing Center for Education and Research (OSCER). We appreciate support from the U.S. DOE through a DOE/EPSCOR grant (DESC0004600).

1. Rozenblit, A.; Avoian, A. J.; Tan, Q.; Sooknoi, T.; Resasco, D. E., Reaction mechanism of aqueous-phase conversion of Î³-valerolactone (GVL) over a Ru/C catalyst. *Journal of Energy Chemistry* **2016,** *25* (6), 1008-1014.

2. Bababrik, R. M.; Wang, B.; Resasco, D., Reaction mechanism for the conversion of Î³-valerolactone (GVL) over a Ru catalyst: a first-principles study. *Industrial & Engineering Chemistry Research* **2017,** *56* (12), 3217-3222.

495140 Hydrodeoxygenation of Biomass Derivatives on Metal-Modified Molybdenum Carbides

Abstract:

Hydrodeoxygenation of Biomass Derivatives on Metal-Modified Molybdenum Carbides

Weiming Wan, Jingguang G. Chen

Department of Chemical Engineering, Columbia University, New York, NY, 10027

Biomass has attracted significant attentions as an alternative resource for fuel and chemicals. However, biomass derivatives have a high O/C ratio and low energy density that make them unsuitable for use in the fuel and chemical sectors.[1] Thus, the hydrodeoxygen (HDO) reaction is necessary to upgrade these biomass derivatives to value-added products. For example, furfural and glycerol are two typical biomass derivatives. Through HDO reactions, furfural can be converted to 2-methylfuran, a promising fuel additive, and glycerol can be converted to propylene or propanediol, an important monomer.

Transition metal carbides are novella class of interesting materials that often demonstrate unique catalytic properties. Due to the high oxygen binding energy, which helps to break the C-O bond, many metal carbides have shown activities towards HDO reactions. Mo_2C is a typical transition metal carbide and it shows high HDO activity for oxygenates such as furfural and propanol.[2],[3] However, the high oxygen binding energy makes it hard to remove the surface oxygen to complete the catalytic cycle. Metal modifiers can be used to adjust the oxygen binding energy. The strong oxophilic metal modifiers, such as Co and Fe, have slightly lower oxygen binding energies than Mo_2C, and the modified Mo_2C surface could have a similar HDO selectivity with better stability. In contrast, the weak oxophilic metal modifiers, such as Cu and Pt, have much lower oxygen binding energies than Mo_2C, and the modified Mo_2C surface should have a different reaction pathway and selectivity.

The current work combines density functional theory (DFT) calculations and surface science experiments using temperature programmed desorption (TPD) and high-resolution electron energy loss spectroscopy (HREELS) under ultrahigh vacuum (UHV) conditions. The main objective is to study the effect of metal modifiers on the adsorption geometry and selective HDO of furfural and glycerol on the Mo_2C surface. The Mo_2C surface was prepared by cycles of ethylene treatment on the Mo(110) single crystal, and the metal layer was prepared by physical vapor deposition.

In the furfural HDO reaction, compared with the Mo_2C surface, furfural has a similar adsorption geometry on Co-modified Mo_2C, suggesting the same reaction pathway on both surfaces. The lower oxygen binding energy of the Co/Mo_2C surface improves the catalytic stability in the TPD experiments. On the other hand, Pt/Mo_2C has lower oxygen binding energy than Co/Mo_2C and the furan ring is flat on the Pt/Mo_2C surface. The flat furan ring and the low oxygen binding energy help produce more furan through decarbonylation.

In the glycerol HDO reaction, the Mo_2C surface can break all the C-O bonds to make propylene. Similarly, the Fe/Mo_2C surface can produce propylene from glycerol with a better stability. On the other hand, the Cu/Mo_2C surface can selectively break the C-O bonds. With a high Cu coverage, the Cu/Mo_2C surface can break one C-O bond to form acetol; with a low Cu coverage, the Cu/Mo_2C surface can break two C-O bonds to form ally alcohol.

This work provides guidance for designing metal-modified carbide HDO catalysts and a detailed discussion of the DFT calculations and the surface science experiments will be presented.

References:

[1] K. Xiong, W. Yu, D. G. Vlachos, J. G. Chen. *ChemCatChem* **2015**, *7* (9), 1402.

[2] K. Xiong, W. Lee, A. Bhan, J. G. Chen. *ChemSusChem* **2014**, *7* (8), 2146.

[3] H. Ren, W. Yu, M. Salciccioli, Y. Chen, Y. Huang, K. Xiong, D. G. Vlachos, J. G. Chen. *ChemSusChem* **2013**, *6* (5), 798.

496760 Structure-Activity Relations on Î³-Al_2O_3: From Alcohol Dehydration to Alkane Dehydrogenation

Abstract:

Structure-Activity Relations on Î³-Al_2O_3: From Alcohol Dehydration to Alkane Dehydrogenation

Pavlo Kostetskyy and Giannis Mpourmpakis

Department of Chemical Engineering, University of Pittsburgh, Pittsburgh, PA 15261, USA

Abstract:

Nonoxidative dehydrogenation of alkanes is an important chemical reaction, as it can be used for production of alkenes â€“ building blocks for a range of plastics and chemicals[1-2]. Several metal oxides, including gamma-alumina, have exhibited dehydrogenation activity, showing promise as candidates for the upgrade of alkanes. In this work we use computational chemistry methods to elucidate the surface mechanism of ethane, propane, n-butane and i-butane dehydrogenation pathways on gamma-alumina. Based on our calculations, it was shown that a concerted mechanism is energetically preferred for alkene formation, with a carbenium-ion-like transition state forming on surface acid-base pairs. A dehydrogenation model was developed as a screening tool for oxide catalysts, based on physicochemical properties of the oxide surface and stability of the reacting hydrocarbons. The dehydrogenation model was developed based on methodology previously applied in structure-activity relations (SAR) for alcohol dehydration on metal oxides[3-4]. The Carbenium Ion Stability (CIS), shown to be a descriptor in alcohol dehydration[3-5], was used as a quantitative descriptor in alkane dehydrogenation and was found to correlate with the calculated activation energy barriers for the hydrocarbons in question. Increased hydrocarbon substitution (branching) was found to decrease the calculated reaction barriers, based on the CIS at the transition state. Finally, SARs developed for alcohol dehydration on various metal oxides were found to accurately capture the trends in alkane dehydrogenation, accounting for catalyst acid-base surface properties and CIS of intermediates at the transition state.

References:

1. Coperet, C., *Chemical Reviews* **2010,** *110*, 656-680.

2. Joubert, J.; Delbecq, F.; Sautet, P., *Journal of Catalysis* **2007,** *251*, 507-513.

3. Kostestkyy, P.; Yu, J.; Gorte, R. J.; Mpourmpakis, G., *Catal. Sci. Technol.* **2014,** *4*, 3861-3869.

4. Kostetskyy, P.; Mpourmpakis, G., *Catal. Sci. Technol.* **2015,** *5*, 4547-4555.

5. Roy, S.; Mpourmpakis, G.; Hong, D.-Y.; Vlachos, D. G.; Bhan, A.; Gorte, R. J., *Acs Catalysis* **2012,** *2*, 1846-1853.

501410 Catalytic Non-Aqueous Processing and Hydrogenation of Chitin and Amino-Based CarbohydratesÂ Using Ru Carbene Organometallics Immobilized on Active Hydrotalcites

Abstract:

Although cellulose has garnered significant attention over the past decades in the search for renewable fuel sources that can replace petroleum fuels, chitin can be catalytically converted into similarly valuable products as well. Chitin is the second most abundant carbohydrate and only differs from cellulose by an acetyl amine on the C5 position of cellulose. In comparison to traditional mineral acids, tunable hydrotalcites (HT) promote glycosidic bond cleavage by its basic sites while dehydration of the monomer is catalyzed by its Lewis acid sites. Additionally, immobilization of homogenous catalysts onto HT create highly selective single site heterogeneous catalysts (SSHC-HT). The main challenge in the development of such a process was to pair a stable solvent that would dissolve chitin with an active heterogeneous catalyst. To overcome this challenge, a total of five systems were studied: gamma-valerolactone (GVL), GVL/levulinic acid and formic acid (as both solvent and acid), tetrahydrofuran, ionic liquids, and water. A 3:1 ratio of Mg:Al HT was used as a basis for all catalytic reactions while substitutions of Al with Fe, Cr, Ni, and Co were tested for improved catalytic activity. Hydrogenation of products were carried out using Ru and Pd HT. A wide array of products, as cited in literature under similar reaction conditions, can be generated, including 5-(hydroxymethyl)furfural, levulinic acid, amino furans, and ketones. Products were identified using LCMS, HPLC, NMR, GCMS, and FTIR.

495616 Selective C-C Bond Scission of Biomass-Derived Oxygenates Using Cerium Oxide-Supported Ruthenium Catalyst

Abstract:

Selective C-C Bond Scission of Biomass-Derived Oxygenates Using Cerium Oxide-Supported Ruthenium Catalyst

Tomoo Mizugaki,[1] Kohei Uesugi,[1] Kodai Nitta,1 Zen Maeno,[1] Takato Mitsudome,[1] Koichiro Jitsukawa,[1] and Kiyotomi Kaneda*[1,2]

1. Graduate School of Engineering Science, Department of Materials Engineering Science, Osaka University

2. Research Center for Solar Energy Chemistry, Osaka University

kaneda@cheng.es.osaka-u.ac.jp

Current requirements to reduce carbon dioxide emissions have led to the utilization of renewable resources, such as plant biomass, as carbon-neutral feedstocks for commodity chemicals. To date, much effort has been devoted to the direct C–O bond cleavage of high-oxygen containing biogenic polyols, such as glycerols, to produce valuable chemicals by hydrogenolysis and deoxydehydration. On the other hand, C–C bond cleavage has not yet been widely researched despite its great potential to extend the utility of biomass-derived chemicals. The existing C–C bond cleavage methods often suffer from low selectivity toward the desired chemicals and limited substrate scope. Therefore, the development of highly selective and versatile C–C bond scission catalysts able to work under mild conditions is highly desired to open new routes for industrially important chemicals from a wide range of biomass derivatives.

In this work, we chose levulinic acid as the model substrate, which is an important platform biomass-derivative leading to a variety of useful C5 chemicals such as g-valerolactone (GVL),[1] 1,4-pentanediol (1,4-PeD),[1] and 2-methyltetrahydrofuran (MTHF).[2] However, there are fewer examples of the transformation of LA into C4 chemicals through one-carbon scission reactions. We found that cerium oxide-supported ruthenium nanoparticles (Ru/CeO_2) efficiently promoted the selective one-carbon scission of levulinic acid to give 2-butanol in over 80% yield under 3 MPa of H_2 at 150 ^oC in water. The high generality of this method is demonstrated by its broad substrate scope, where the cleavage of C–C bonds chemospecifically occurs at positions adjacent to the carboxyl, ester, and hydroxymethyl groups of oxygenated compounds (Scheme 1).

References

[1] T. Mizugaki, Y. Nagatsu, K. Togo, Z. Maeno, T. Mitsudome, K. Jitsukawa, K. Kaneda, *Green Chem.* **2015**, *17*, 5136–5139.

[2] T. Mizugaki, K. Togo, Z. Maeno, T. Mitsudome, K. Jitsukawa, K. Kaneda, *ACS Sustainable Chem. Eng.* **2016**, *4*, 682–685.

500440 Large Eddy Simulation of Soot Formation in Oxy-Coal Combustion

Abstract:

Oxy-Coal combustion is a promising clean coal technology. Typical combustion burns coal in air, resulting in a product mixture of water, nitrogen, and carbon dioxide. Separating carbon dioxide from nitrogen is costly. In oxy-fired combustion, nitrogen is first separated from air and a mixture of coal, oxygen, and (optionally) recirculated flue gases is burned. The resultant products are water and carbon dioxide, which are easily separated for subsequent sequestration. Oxy-fired combustion often occurs at high temperatures and under flow and thermal conditions that differ somewhat from combustion in air. Radiative heat transfer is a key process in such combustion as it is a primary source of heat transfer in steam generation applications. Radiation is present from gaseous species, ash, coal char, and soot, with soot being a significant contributor. Unfortunately, soot formation is a complex process. While much research has been done on soot formation in general, relatively little research has been published on soot formation in coal combustion. In oxy-fire systems with high water and carbon dioxide concentrations, and at high temperature, soot consumption via gasification may be significant. (Normally so consumption is dominated by oxidation reactions.)

We present a series of large eddy simulations in an oxy-coal combustor in which sensitivity to the soot formation mechanism is presented. Differences is flow, temperature, and radiative heat transfer are shown. We compare results using a previously published coal soot model, as well as extensions that include optimized gasification and oxidation rates. An advanced physics-based soot formation from coal tar mechanism is presented.

Ash aerosol and ash deposit formation during high temperature oxy-combustion of petroleum coke

Yueming Wang [a], Xiaolong Li [a], Jost Wendt [a]

[a] Department of Chemical Engineering, University of Utah, Utah, USA

Background

What is petroleum coke

- Peroleum coke (petcoke) is a byproduct of petroleum refining process

Advantages

- Petcoke is available in large amounts with low price
- Petcoke has low ash content and high heating value

Challenges

- The volatile content is very low in petcoke making it hard to be burned efficiently
- The sulfur content is very high and can cause corrosion and pollution

2

Research Objectives

- Assess the combustion efficiency of petroleum coke under pulverized coal furnace
- Uncover the effect of flame temperature on combustion performance of petroleum coke
- Study the formation mechanisms of ash aerosol and ash deposition (fouling)
- Develop the relationship between ash deposition rates and concentration of the ash aerosol

THE UNIVERSITY OF UTAH

3

Oxy-Fuel Combustor (OFC)

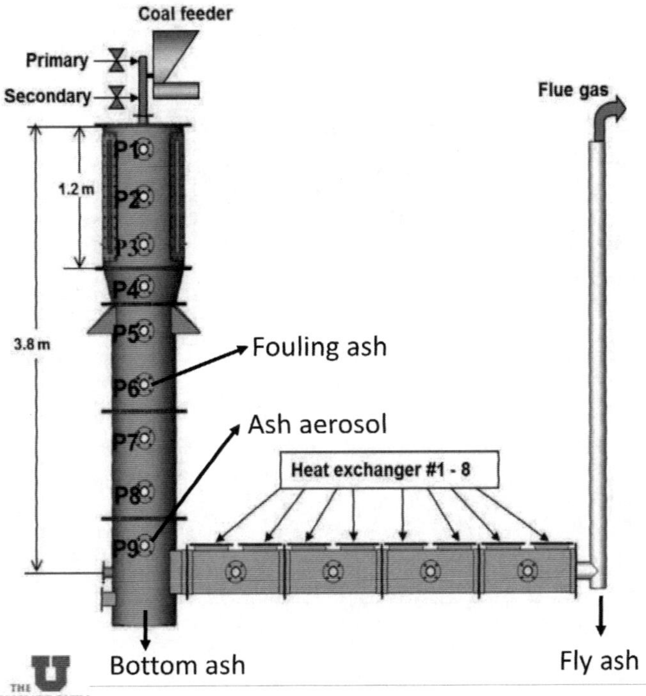

- 100 kW pilot scale
- Self-sustained down-fired
- Systematically controlled
- Temperature profile and particle concentration comparable with full scale units

THE UNIVERSITY OF UTAH

4

Fouling ash sampling system

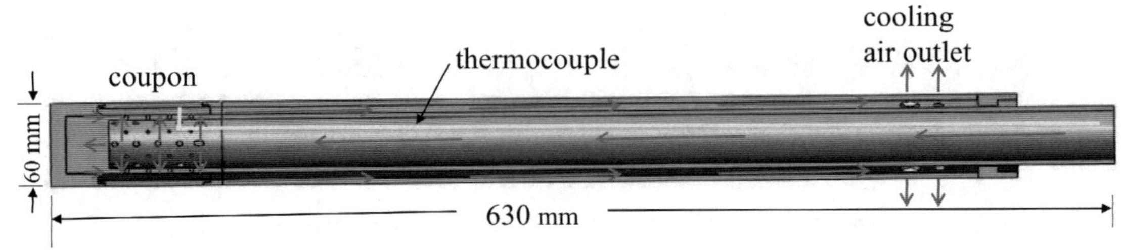

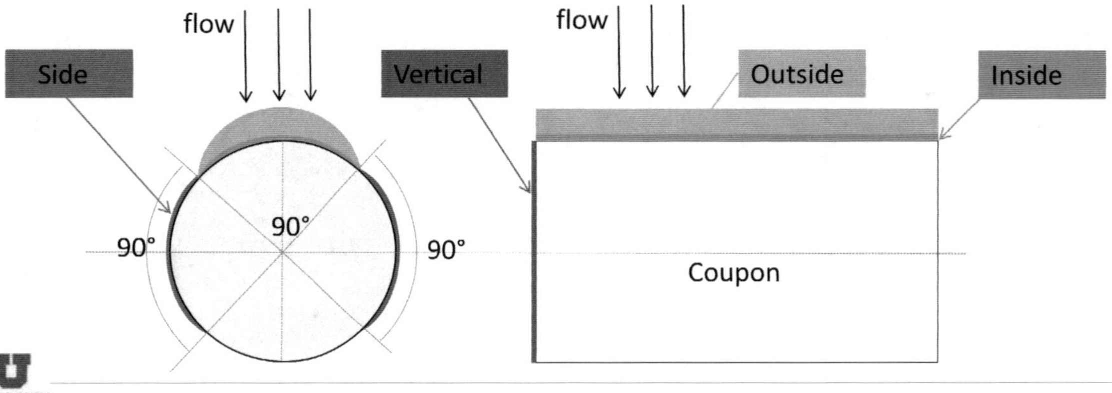

Ash aerosol sampling system

BLPI:
0.0324 - 15.7 μm

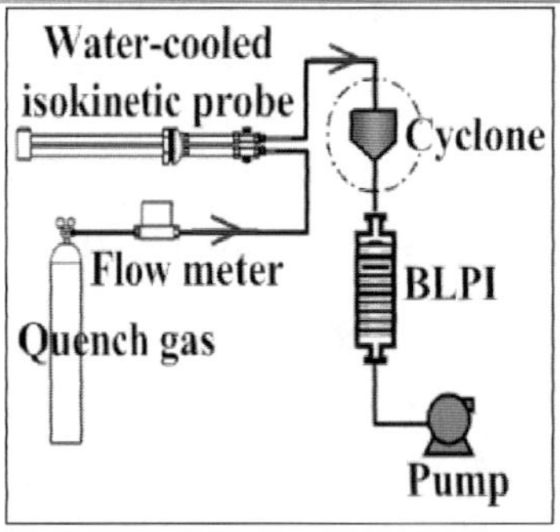

\+

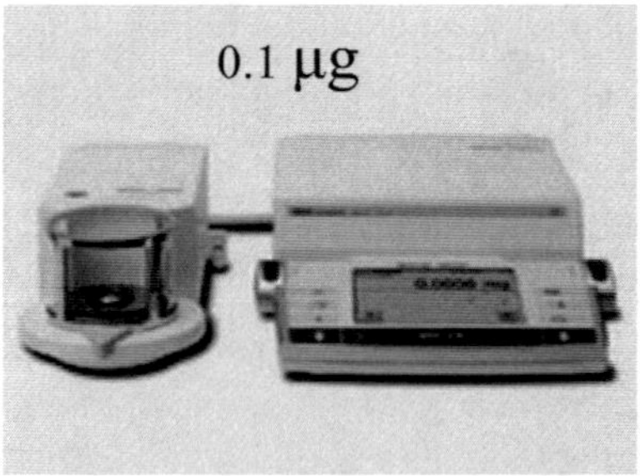

SMPS:
0.0143 - 0.6732 um
APS:
0.532 - 20 um

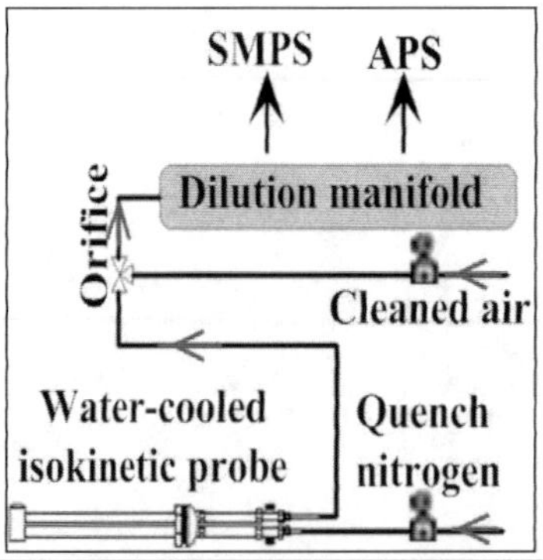

Combustion parameters

Composition Analysis

ASH (%)	C (%)	H (%)	N (%)	S (%)	O (diff) (%)	H_2O (%)	Volatile (%)	Fixed Carbon (%)	HHV (kJ/kg)
2.99	82.51	6.02	1.71	5.65	0.49	0.57	10.18	86.26	34447.6

Al_2O_3 (%)	CaO (%)	Fe_2O_3 (%)	MgO (%)	K_2O (%)	SiO_2 (%)	Na_2O (%)	SO_3 (%)	TiO_2 (%)	NiO (%)	V_2O_5 (%)
19.4	4.22	7.02	0.66	1.17	46.7	0.72	3.77	0.63	1.26	8.24

Combustion conditions

Tests conditions	OXY70	Air
Feeding rate of petroleum coke (kg/h)	1.36	1.36
Feeding rate of natural gas (kg/h)	1.36	1.36
Excess O_2 (vol%_dry)	3	3
Adiabatic flame temperature (K)	3095	2312
Flow rate of flue gas (m^3/h_standard condition)	11.85	37.73

THE UNIVERSITY OF UTAH

7

Temperature profile of flue gas

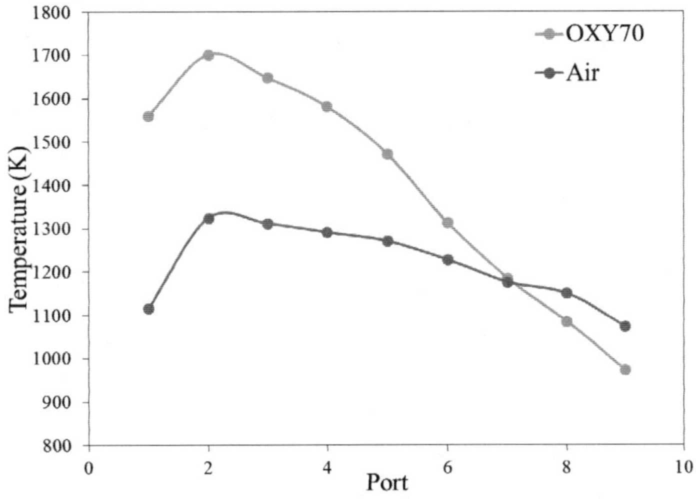

- Peak temperature of OXY70 (1701 K) is much higher than Air (1324 K)
- Peak temperatures for both cases are much lower than adiabatic flame temperatures.

THE UNIVERSITY OF UTAH

8

Aerosol particle size distribution (PSD)

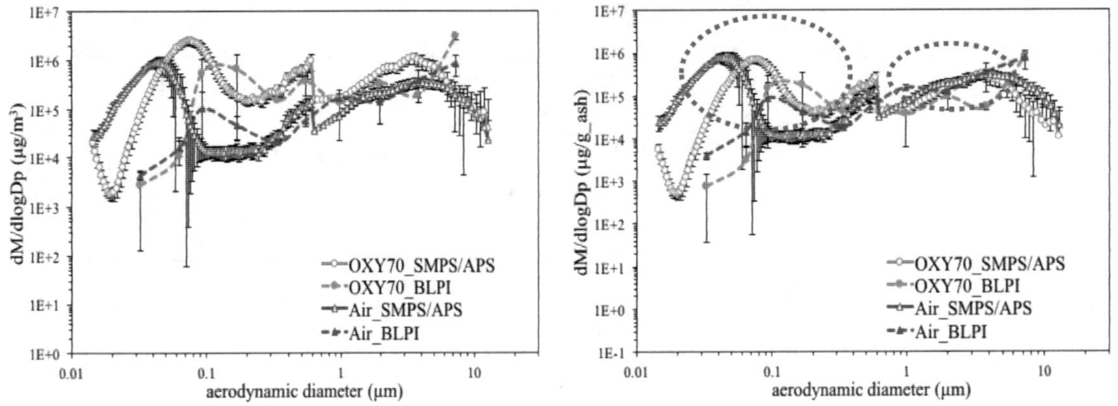

- BLPI data match with SMPS/APS data
- High temperature oxy-firing generates more submicron particles (99.1 mg/g_ash) than air-firing (72.4 mg/g_ash), but supermicron particles are not affected by temperature

THE UNIVERSITY OF UTAH

9

Aerosol size segregated composition

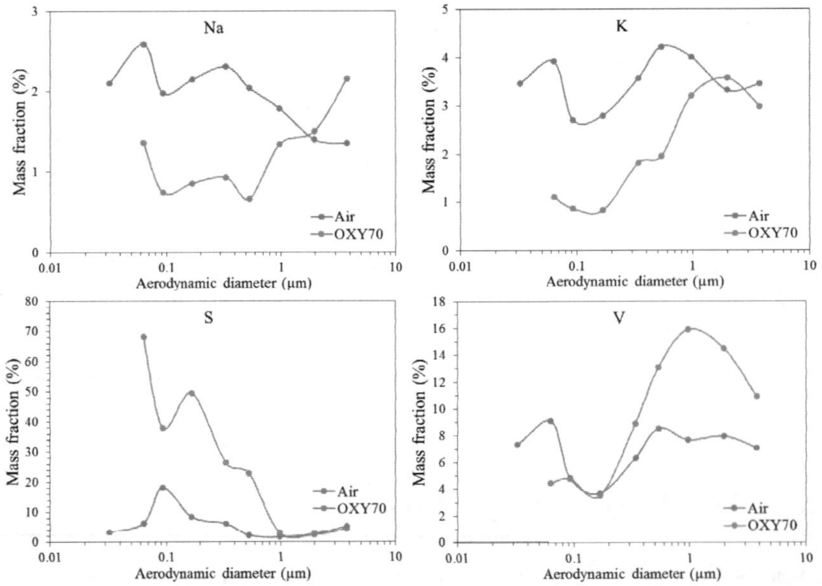

- More alkali metals are scavenged from submicron particles by alumina-silica under high temperature condition

THE UNIVERSITY OF UTAH

10

Composition of fouling ash

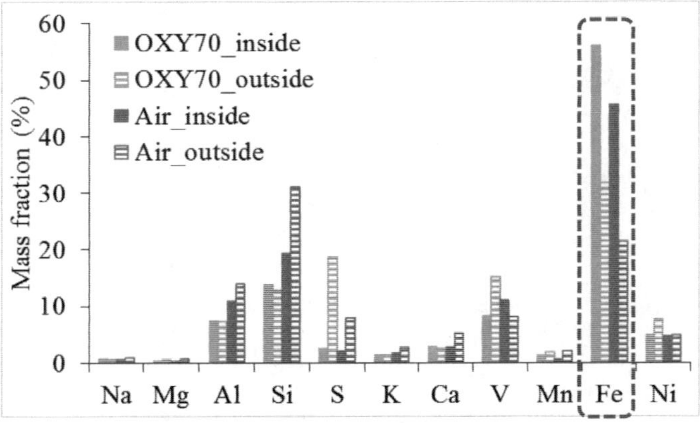

Al_2O_3 (%)	CaO (%)	Fe_2O_3 (%)	MgO (%)	K_2O (%)	SiO_2 (%)	Na_2O (%)	SO_3 (%)	TiO_2 (%)	NiO (%)	V_2O_5 (%)
19.4	4.22	**7.02**	0.66	1.17	46.7	0.72	3.77	0.63	1.26	8.24

- Composition of deposits are dominated by Fe

THE UNIVERSITY OF UTAH

SEM and EDS analysis for deposits

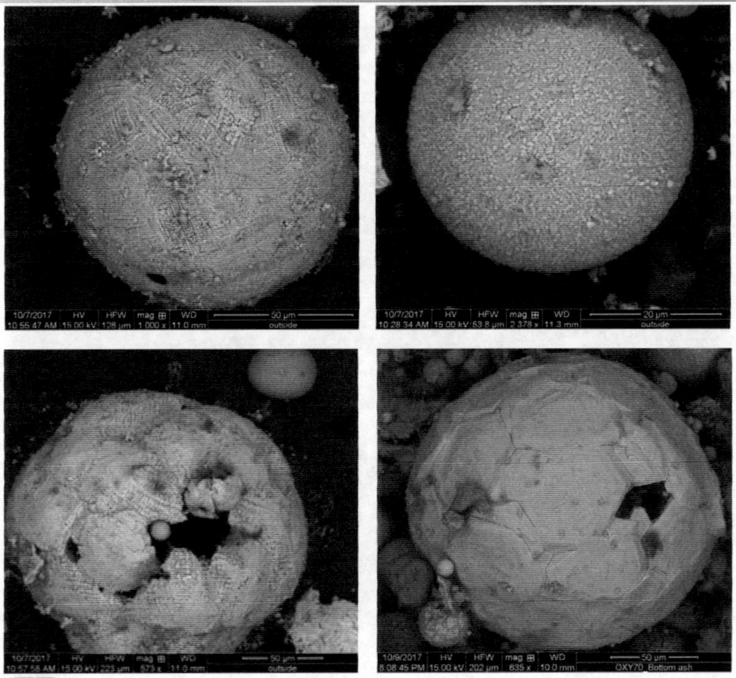

Na	0
Mg	0.17
Al	1.63
Si	2.30
S	1.05
K	0.33
Ca	0.82
V	1.33
Mn	0.61
Fe	**90.01**
Ni	1.75

THE UNIVERSITY OF UTAH

Composition of fouling ash

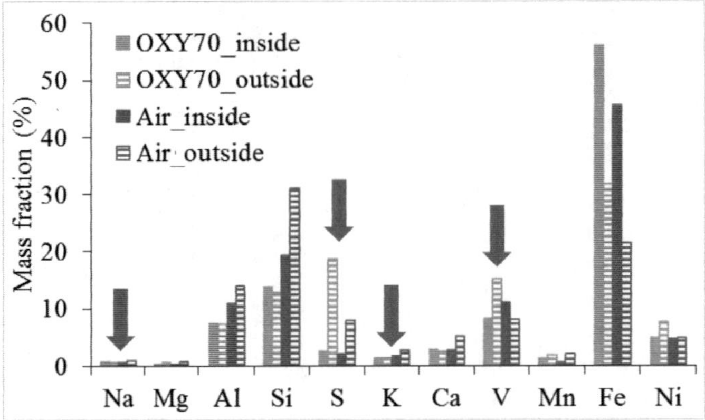

➤ Composition of deposits are dominated by Fe

➤ OXY70 deposits have more S and V, but less Na and K than Air, which are consistent with aerosol composition analysis

THE UNIVERSITY OF UTAH

13

Sulfur and carbon analysis

Condition	OXY70		Air	
Content	S (%)	C (%)	S (%)	C (%)
outside deposit	3.21	79.0	1.13	28.0
fly ash	3.43	11.9	1.55	13.3
bottom ash	3.12	68.8	1.89	36.8

➤ Air firing seems surprisingly more efficient to burn petcoke than OXY70

➤ Sulfur content in OXY70 is higher than Air case

THE UNIVERSITY OF UTAH

14

Correlation of deposition rates with PM1

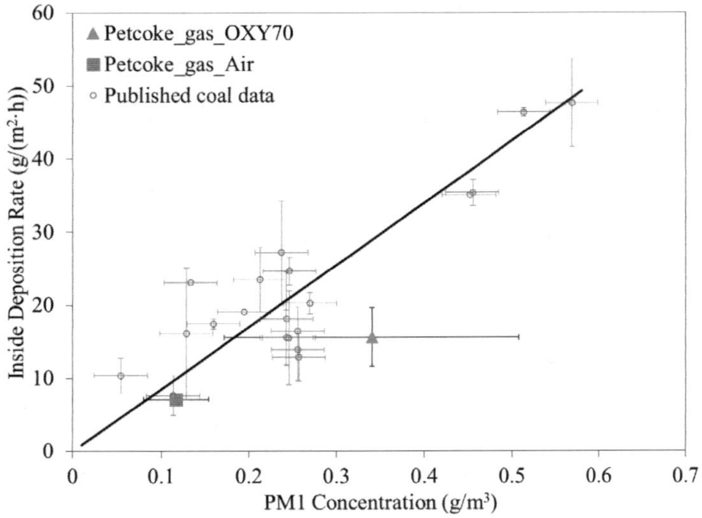

Published data from "Zhan, Z.; Wendt, J. O. L. Energy & Fuels **2017**, 31 (3), 2198–2202."

- Inside deposition rates of petcoke show positive relationship with concentration of PM1, which is consistent with previous findings in coal.

THE UNIVERSITY OF UTAH

15

Conclusions

- More alkali metals are scavenged from ultrafine fly ash by alumina-silica under high temperature condition
- The composition of fouling ash can be predicted by ash aerosol
- Fe is enriched in fouling ash, especially for high temperature condition
- Mixing between oxidant gas and petcoke is more important for carbon burnout than flame temperature
- Inside deposition rates can be predicted by the concentration of PM1

THE UNIVERSITY OF UTAH

16

Acknowledgements

- National Science Foundation under Award 1603249
- Praxiar for contributing the O_2 and CO_2 supply
- Brittany Walker, Dave Wagner, Paulo Perez and Brian van Devener

THE UNIVERSITY OF UTAH

17

Thank you!

THE UNIVERSITY OF UTAH

18

499074 Mercury Oxidation over Cu-SSZ-13 Catalysts under Flue Gas Conditions

Abstract:

Increasing concerns over mercury (Hg) emissions from coal-fired power plants have led the creation of new emission standards such as the recent Mercury and Air Toxic Standards, which requires a 90% reduction of Hg emissions. These new standards have led to the implementation of various novel technologies to control Hg emissions. Selective catalytic reduction (SCR) catalysts have shown promise as bifunctional catalysts for limiting both nitrogen oxide (NO_x) and Hg emissions through simultaneous NO_x reduction and Hg oxidation. The catalytic oxidation of elemental Hg, which is difficult to capture without dedicated Hg capture systems, to Hg^{+2} via HCl enables the capture of Hg^{2+} in the existing wet flue gas desulfurization unit as it is water soluble. Previous studies have investigated the possibility of utilizing current commercial SCR catalysts to fill this role, however these catalysts often have unsatisfactory performance when it comes to reducing Hg pollution to the levels required by current emission standards. To develop the next generation of SCR catalysts, a copper (Cu) exchanged small pore zeolite (Cu-SSZ-13) was investigated for its possible application in coal-fired power plants as a new bifunctional SCR and Hg oxidation catalyst. Cu-SSZ-13 has shown promise as a SCR catalyst for diesel exhaust, but has not been studied in coal-fired power plants.

In this study, the activity of Cu-SSZ-13 was investigated for its ability to oxidize Hg under a variety of flue gas conditions. The Cu-SSZ-13 catalyst was compared to SSZ-13 to understand the effect of Cu in the Hg oxidation mechanism. The Hg oxidation performances of both zeolites were compared to that of a commercial SCR catalyst (V-Ti-W), which served as a benchmark. All catalysts required hydrogen chloride (HCl) to exhibit any Hg oxidation with the zeolites outperforming the commercial catalyst. However, the activity of all catalysts significantly decreased in the presence of either sulfur oxides (SO_x) or SCR reactants (nitrogen oxide (NO)+ammonia (NH_3)) in the flue gas, with SO_x having a greater effect on Cu-SSZ-13 and SCR reactants having a greater one on the commercial catalyst. Additionally, it was determined that Cu was not the active site for Hg oxidation but was necessary for the SCR reaction. In fact, in the presence of SO_2, Cu actually contributed to the deactivation of the catalyst for Hg oxidation, likely due to the formation of sulfates as confirmed by XPS analysis.

502807 Mediated Oxycombustion with Integrated Uncoupled Oxygen Supply Â (MOBIUS)

Abstract:

Combustion systems that circumvent or reduce the energy penalty for carbon dioxide separation are being vigorously investigated. The approach we have pioneered aims at synergistically combining the basic objectives of three such technologies, namely oxy-combustion, chemical looping and liquid anode DCFC, and transforming them into a single lean, thoroughly optimized single unit process - Mediated Oxycombustion with Integrated Uncoupled Oxygen Supply (MOBIUS) - through innovations that overcome the bottlenecks of each of these technologies such as a) controlled combustion in a molten oxide to overcome the issues related to high temperatures (oxycombustion); b) obviating the need of solids transport between reactors (CLC); and c) avoiding operational issues such as electrical resistance, ash separation, impurity tolerance etc (DCFC). The system (simplistic concept in Figure 1) synergistically combines the effective aspects of the above mentioned technologies along with innovations to overcome the hurdles faced by these technologies. In MOBIUS, the solid fuel is combusted in the absence of any nitrogen as in oxy-combustion. However, the carbon is not combusted directly by oxygen but by a metal oxide as in a CLC. Unlike conventional CLC, the MeO_x and Me in the proposed Chemical Looping with Oxygen Uncoupling (CLOU) systems are in the liquid phase to improve the combustion rates and the mobility of the metal and the oxide. The metal is re-oxidized at the walls of the combustor with oxygen that is uncoupled from the reactor wall material. The individual subsystem studies will be presented and the integration of the various subsystems into a single reactor system will be discussed. Results from the subsystem will include the combustion rates, metal oxidation rates, oxygen uptake, transport and release rates under various conditions and composition of the membrane (during the membrane evaluation). The long terms operation of the integrated system will be demonstrated.

497932 Mobility Size and Effective Density of Carbonaceous Aerosols

Abstract:

Mobility Size and Effective Density of Carbonaceous Aerosols

Georgios A. Kelesidis, Eirini Goudeli, Sotiris E. Pratsinis

Particle Technology Laboratory, Institute of Process Engineering,

Department of Mechanical & Process Engineering, ETH Zurich, Switzerland

The mobility size estimation of evolving fractal-like soot aggregates is not trivial, as it depends on their structure as well as the number and size of their constituent primary particles. A scaling law describing the mobility size of agglomerates consisting of monodisperse primary particles in point contact (Sorensen, 2011) overestimates the mobility diameter of mature soot aggregates from diffusion flames for a given number of primary particles (Rissler et al., 2013). Rissler et al. (2013) attributed this difference to the primary particle polydispersity and chemical bonding induced by surface growth (aggregation), neglected in the above scaling law (Sorensen, 2011). Typically population balance models for soot formation estimate the mobility diameter assuming it is equal to the gyration diameter (Yapp et al., 2015).

Here, the Discrete Element Model of nascent soot agglomeration and surface growth by acetylene pyrolysis (Kelesidis et al., 2017) is extended to investigate the dynamics of mature soot formation at longer residence times and different flow regimes than nascent soot in the absence of soot oxidation. The evolution of soot mean gyration, mobility and primary particle diameters as function of the mean number of primary particles per aggregate are elucidated. The common assumption that the mobility is equal to the gyration diameter leads to 40 % underestimation of the nascent soot mobility size and 30 % overestimation of the mature soot aggregate mobility diameter.

Relationships for the soot mobility size and effective density are derived by following the evolution from nascent to mature soot morphology and compared to the correlation of Sorensen (2011) and mass-mobility measurements in quenched diffusion flames (Rissler et al., 2013; Yon et al., 2015). The scaling law for agglomeration in the absence of surface growth (Sorensen, 2011) overestimates the measured mature soot aggregate mobility size by 37 % and underestimates its effective density by 44 %. Accounting, however, for the primary particle polydispersity and chemical bonding or aggregation induced by surface growth reduces the deviation from the mass-mobility measurements down to 7 and 11 % for the mobility size and the effective density, respectively.

References:

Kelesidis, G.A., Goudeli, E., Pratsinis, S.E. (2017) *Proc. Combust. Inst.* **36**, 29-50.

Rissler, J., Messing, M.E., Malik, A.I., Nilsson, P.T., Nordin, E.Z., Bohgard, M., Sanati, M., Pagels, J.H. (2013)

Aerosol Sci. Technol. **47**, 792-805.

Sorensen, C.M. (2011) *Aerosol Sci. Technol.* **45**, 765-779.

Yapp, E.K.Y., Chen, D., Akroyd, J., Mosbach, S., Kraft, M., Camacho, J., Wang, H. (2015) *Combust. Flame* **162**, 2569-2581.

Yon, J., Bescond, A., Ouf, F.X. (2015) *J. Aerosol Sci.* **87** (2015) 28-37.

499699 Deciphering the Chemical and Physical Mechanisms Active in Elevated Temperature Photocatalytic Synthesis of Hydrocarbons and Ammonia

Abstract:

The photocatalytic synthesis of hydrocarbons, oxygenates, and NH_3 from CO_2, N_2, NO/NO_2, and H_2O is a truly sustainable route to fuels and building block chemicals production. Unfortunately, the community still lacks the ability to rationally improve overall rates of reaction and control product distributions due to a lack of reaction mechanism understanding. Endemic issues with organic contamination in experimental studies has further contributed to this lack of progress. Our combined computational/experimental studies of photocatalytic CO_2 reduction by H_2O in organic-free reaction conditions at an appreciable temperature range (gas phase reaction at ~100-400°C+) have shed significant light upon the surface chemical and electronic properties that affect these reaction mechanisms. Moreover, studies have also isolated new H-transfer mechanisms. Both of these developments have the power to significantly aid in the design of these materials. The new H-transfer mechanisms and an understanding of the role of surface chemistry have further enabled control over the selectivity of surface-bound atomic H between hydrogenation and H_2 evolution -- a critical efficiency aspect of the system. Additionally, key carbon-containing reaction intermediates that allow for carbon-carbon coupling have been isolated as well. Catalysts shown to be active at elevated temperatures indicate semiconductors with a high Debye temperature, robust bulk bonding, and elevated surface chemical reactivity may lend additional tunable features for photocatalytic synthesis mechanisms. With the ultimate aim of producing higher hydrocarbons, oxygenates, and ammonia photocatalytically from simple thermodynamic-minimum reactants, these studies have mapped the course forward for materials selection and mechanistic understanding.

502400 A High-Throughput Computational Screening Approach for Solar Fuels Photoelectrocatalysis

Abstract:

In recent years, theoretical overpotential has become increasingly popular as a metric for characterizing the suitability of oxygen evolution electrocatalysts using density functional theory (DFT). In addition, DFT has become an invaluable tool for characterizing photoabsorbing materials in a number of high-throughput screening studies aimed at designing materials for solar fuels production from stability and bandstructure criteria. In this work, we present an approach which integrates calculations of theoretical overpotential with those of band-gap and band-edge positions on mixed metal oxides. Given that the prospect of discovering lone materials fulfilling the criteria of stability, catalytic activity, and photoabsorption efficiency, we also report on how materials with suitable surface chemistry may be more effectively paired with suitable photoabsorbing substrates using interfacial matching. Lastly, we benchmark our approach comparing potential combined metrics with experimentally measured photoelectrocatalytic activity.

499265 Computational Insights into Photo(electro)Chemical Nitrogen Fixation over Titania Catalysts

Abstract:

Photocatalytic nitrogen fixation is a form of artificial photosynthesis capable of enabling distributed production of fixed nitrogen at near-ambient conditions from the nitrogen, water, and photons that are readily available from the environment. The process has been demonstrated over titania catalysts through both reductive and oxidative pathways, although the currently observed nitrogen fixation rates are too low to be practical for fertilizer synthesis and there is little fundamental understanding of the reaction mechanism or chemical phenomena that enable the process. Interestingly, the process occurs under environmental conditions and has been speculated to play a significant role in the global nitrogen cycle owing to the prevalence of $TiO2$ in naturally occurring sands and man-made pigments. Despite the wide-ranging potential impacts of this reaction, first demonstrated nearly 75 years ago, the process has received relatively little attention and has not been studied with modern techniques. This talk will briefly review the historical context and current progress toward photocatalytic nitrogen fixation, followed by a theoretical perspective into the thermodynamic driving forces and likely reaction mechanisms, with an emphasis on the potential for developing photoelectrochemical processes. The computational results will be complemented by recent experiments demonstrating the feasibility of the reaction.

501308 Engineering the Interface to Improve the Efficiencies of Insulator-Protected Semiconductors for Photoelectrochemical Cells

Abstract:

Recent improvements in solar fuels production efficiencies have been enabled by the development of protection schemes for low bandgap semiconductors that would otherwise be unstable in the harshly oxidative environment of an electrochemical cell. One common protection strategy incorporates an insulator layer placed between a metal electrocatalyst and semiconductor. Recent research has shown that the characteristics of the insulator layer have a large effect on the overall efficiency of a photoelectrochemical cell. In this contribution, we will provide experimental evidence that demonstrates how by tuning the properties of the insulator layer we can better control charge transfer across the interface in order to improve efficiencies. We will also present a model to describe the insulator layerâ€™s effect on charge transfer mechanisms. Provided with these modeling insights, we will discuss which experimental systems can benefit from the most from insulator tuning in order toÂ expand the flexibility of material selection for a photoelectrochemical device.

492457 The Role of Surface-Bound Dihydropyridine Analogs in Pyridine-Catalyzed CO_2 Reduction over Semiconductor Photoelectrodes

Abstract:

The (photo)electro-catalytic reduction of CO_2 to useful fuels via energy from sunlight has received significant attention as a promising route for generating carbon-neutral fuels and value-added chemicals. Experimental evidence suggests that pyridine (Py) is an effective co-catalyst during CO_2 reduction over GaP, CdTe, and $CuInS_2$ semiconductor electrodes. Identifying the role of Py during catalytic reduction is essential for optimizing the design of such photocatalytic processes. We propose a general reaction mechanism for the Py-catalyzed reduction of CO_2 over GaP(111), CdTe(111), and $CuInS_2$(112) photoelectrode surfaces. This mechanism proceeds via formation of a surface-bound dihydropyridine (DHP) analog, which is a newly postulated intermediate in the Py-catalyzed mechanism. Using density functional theory, we calculate the standard reduction potential related to the formation of the DHP analog, which demonstrates that it is thermodynamically feasible to form this intermediate on all three investigated electrode surfaces under photo-electrochemical conditions. Hydride transfer barriers from the intermediate to CO_2 demonstrate that the surface-bound DHP analog is as or even more effective at reducing CO_2 to HCOO- as the $DHP_{(aq)}$ molecule in solution. This intermediate is predicted to be both stable and active on many varying electrodes, therefore pointing to a mechanism that can be generalized across a variety of semiconductor surfaces and explains the observed electrode dependence of the photocatalysis.

490917 Surface Plasmon-Assisted Photoelectrochemical CO_2 Reduction on Well-Defined Nanostructured Silver Electrodes

Abstract:

Plasmonic hot-carriers and the strong local electric fields produced by plasmon excitation may open new pathways for CO_2 reduction, resulting in increased product selectivity and lower overpotential. Plasmon decay generates â€œhotâ€ electrons which can be transferred selectively to a surface molecule for direct reduction. Simultaneously, the enhanced electric fields can alter the electronic coupling with surface adsorbed molecules, thereby changing the binding energy of these species and the catalytic properties of the plasmonic metals.[1-3] Plasmon-assisted photoelectrochemical CO_2 reduction was demonstrated when optically excited surface plasmons at the roughened surface of a Ag electrode in CO_2-containing solutions produced strong enhancement of the cathodic current.[4-5]

Here we reveal the mechanism of plasmonic hot-carriers interacting with CO_2 and associated adsorbates on plasmonically-active Ag nanopyramid catalysts. Photocurrent measurements and photoelectrochemical spectroscopy were performed on the plasmonic Ag nanopyramid photocatalysts in electrolyte saturated with CO_2 and Ar. We observed that the photocatalytic reaction is much stronger in the presence of CO_2 and is resonantly enhanced at a specific potential. In Ar-saturated electrolyte, no characteristic feature of photocurrent was observed. These results indicate that the hot-carriers created by the localized surface plasmon resonance effect are reacting specifically with CO_2 molecules and CO_2 reduction intermediates. Further, the photocurrent onset potential was significantly reduced with respect to the dark current. Our findings suggest that the plasmonic hot-carriers behave as an effective co-catalyst with Ag and may play a critical role in discovering a CO_2reduction process that forms readily usable hydrocarbon fuels.

Scanning electron microscopy (SEM) was used to confirm the shape, size, and distribution of the silver nanopyramids. Photocurrent measurements were performed in a custom glass photoelectrochemical cell with a quartz window that allows for front- or back-illumination of the electrode. The two electrode chambers are separated by a membrane and can achieve a gas-tight seal that allows for precise pressure measurements. The cell was coupled to a gas chromatograph (GC) that allows for detection and quantification of gaseous products. *Ex-situ* proton nuclear magnetic resonance spectroscopy (^1H NMR) and high performance liquid chromatography (HPLC) were used to quantify liquid products.

References
[1] A. O. Govorov, et. al., *Nano Today* 9, 85 (2014)
[2] R. Sundararaman, et. al., *Nature Comm.* 5, 5788 (2014)
[3] S. Mukherjee, et. al., *J. Am. Chem. Soc.* 136, 64 (2014)
[4] R. Kostecki, et. al., *J. Chem. Phys. Lett.* 194, 386 (1992)

[5] R. Kostecki, et. al., *J. Appl. Phys.*77, 4701 (1995)

This material is based upon work performed by the Joint Center for Artificial Photosynthesis, a DOE Energy Innovation Hub, supported through the Office of Science of the U.S. Department of Energy under Award Number DE-SC0004993.
The work is supported by the National Science Foundation Graduate Research Fellowship under Grant No. 1106400. Work at the Molecular Foundry was supported by the Office of Science, Office of Basic Energy Sciences, of the U.S. Department of Energy under Contract No. DE-AC02-05CH11231.

490812 Artificial Leaf for Carbon Dioxide Photo-Reduction to Fuels

Abstract:

An artificial leaf that mimics photosynthesis process is a promising approach to directly convert solar energy into energy-rich chemicals[1–4]. Thus far, this concept has been successfully developed for the hydrogen generation process[2–4]. Here, we have developed an artificial leaf for the conversion of carbon dioxide (CO_2) into synthesis gas in an acidic electrolyte using the energy from a sun simulator. A configuration of the artificial leaf was achieved by using cobalt, as an oxygen-evolving catalyst and molybdenum disulfide (MoS_2) catalyst for CO_2 reduction reaction deposited on the illuminating and back-substrate surfaces of a photovoltaic cell, respectively. We have performed detail experimental studies such as linear sweep voltammetry (LSV) and chronoamperometry (CA), in-situ differential electrochemical mass spectroscopy (DEMS), and gas chromatography (GC) followed by some characterization techniques e.g. scanning electron microscopy (SEM), X-ray photoelectron spectroscopy (XPS) and energy dispersive spectroscopy (EDS) to elucidate the performance and stability of our developed artificial leaf. The experimental analyses indicate that our custom made artificial leaf cell can directly convert the solar energy to synthesis gas with a catalytic conversion efficiency of 75%. Our results reveal that the operation of the artificial leaf cell is not restricted by the CO_2 reduction reaction according to the remarkable activity of MoS_2 NFs catalysts[5]. However, it is rather limited by the oxygen evolution reaction (water splitting) at the anode side. This finding could potentially open a new route in the solar-driven CO_2 conversion field where the CO_2 reduction reaction is known to be the bottle neck of this process.

References:

1. Asadi, M. *et al.* Nanostructured transition metal dichalcogenide electrocatalysts for CO_2 reduction in ionic liquid. *Science.* **353,**467–470 (2016).

2. Kanan, M. W. & Nocera, D. G. In Situ Formation of an Water Containing Phosphate and Co^{2+}. *Science.* **321,**1072 (2008).

3. Reece, S. Y. *et al.* Wireless Solar Water Splitting Using Silicon-Based Semiconductors and Earth-Abundant Catalysts. *Science.* **334,**645–648 (2011).

4. Qiu, F. *et al.* Photocatalytic Hydrogen Generation by CdSe/CdS Nanoparticles. *Nano Lett.* **16,**5347–5352 (2016).

5. Abbasi, P. *et al.* Tailoring the Edge Structure of Molybdenum Disulfide toward Electrocatalytic Reduction of Carbon Dioxide. *ACS Nano.* **11**, 453–460 (2017).

500536 Understanding Photoelectrochemistry on Epitaxial Oxides through Surface Electronic Structure

Abstract:

The intermittent nature of renewable energy sources requires a clean, scalable means of converting and storing energy. One earth abundant storage option is water electrolysis, storing energy in the bonds of O_2 and H_2. Photoelectrochemical (PEC) cells based on semiconductor/liquid interfaces can convert sunlight to chemical fuels without external circuitry, such as "splitting" water into O_2 and H_2 upon illumination of a photoabsorber and catalyst. Tuning the location of the band edges of a semiconductor tailors the wavelength of photoabsorption and the charge transfer to adsorbed species. The efficacy of conversion depends in part on the rectifying properties of semiconductor–electrolyte junctions, as the band bending present in the semiconductor space-charge region from Fermi level equilibration drives the separation of electron–hole pairs. However, the relationship between band structure, band bending/flattening, photocatalytic activity, and surface functionality (such as hydroxylation properties and hydrophilicity) is not well understood. Improving our understanding of the semiconductor/electrolyte interface is crucial in the design of efficient photoelectrode materials for harvesting solar energy in a PEC system.

In the particular case of the semiconductor Ge, favorable conduction band alignment with the hydrogen evolution reaction (HER) suggests the potential for efficient solar-to-hydrogen conversion with visible light. However, material instability in aqueous environments and the potential for photocorrosion limits its utility in PEC applications. In order to protect the photoabsorber surface while maintaining favorable alignment of the conduction band, we epitaxially grow $SrTiO_3$ on top of p-Ge. We will present studies of such model oxide photoelectrodes grown by molecular beam epitaxy (MBE) on single crystal substrates that display a known crystallographic orientation, surface area, path for charge transport, and strain. Photoelectrochemical measurements on these heterostructures can establish the intrinsic activity of oxide catalysts in a way that cannot be realized with polydisperse nanoparticle systems. Insight into the band bending between the substrate and oxide overlayer, as well as at the semiconductor surface, can be obtained from X-ray photoelectron spectroscopy (XPS).[1] Measurement of XPS at ambient pressures (AP-XPS) can further elucidate the relationship between adsorbates and surface band bending.[2] This fundamental insight will build understanding necessary for the design of active, earth-abundant photocatalysts that can be integrated into PEC devices for efficient conversion of solar energy into chemical fuels.

References

1. S.A. Chambers, Y. Du, R.B. Comes, S.R. Spurgeon, P.V. Sushko, *Applied Physics Letters,* **110**, 082104 (2017).

2. K.A. Stoerzinger, R. Comes, S.R. Spurgeon, S. Thevuthasan, K. Ihm, E.J. Crumlin, S.A. Chambers. *J. Phys. Chem. Lett.* **8**, 1038 (2017).

499037 Simulating Solvent Effects in Catalytic Systems

Abstract:

Many of the catalytic transformations involved in the conversion of chemicals and biomass-derived intermediates are carried out in solvents that can significantly influence the catalytic chemistry, the kinetics and the underlying mechanisms. Solvents can act to stabilize or destabilize reaction intermediates and transition states and, in some cases, they are intimately involved in co-catalyzing reactions. We will discuss methods to simulate solvent structure and its influence on the free energy for catalytic transformations that occur in the solution phase for mixed solvent systems, on metal surfaces as well as within nanoporous architectures.

497801 Electronic Excitations in Thermal Heterogeneous Catalysis

Abstract:

Most studies of thermal catalysis assume that the system remains in the electronic ground state. However, electronic excitations are easily induced in metal surfaces, due to the lack of a band gap. Here, we show that processes on catalytic surfaces can induce electronic excitations, and we examine the possible effects of these processes on surface chemistry.

We performed nonadiabatic dynamics using real-time, time-dependent density functional theory, propagating the nuclei using Ehrenfest dynamics. We study several trajectories of an N_2 molecule interacting with Ru nanoparticles ranging in size from 13 to 147 atoms. These simulations show that, during the adsorption and dissociation processes, a significant amount of energy can be dissipated into electronic degrees of freedom, comparable to the amount of energy dissociated into ionic vibrations. This suggests that the electronic structure of catalytic surfaces may, transiently, be quite different from the ground state. We characterize the electronic structure and spatial character of the excitations, and we examine how they can affect the surface chemistry of other species reacting on the nanoparticle.

500631 Large-Scale Nonadiabatic Molecular Dynamics Enabled By Machine Learning

Abstract:

Transferring energy from solid surfaces to chemical bonds of adsorbed species is a fundamental process in heterogeneous catalysis. To a great extent, the dynamics of molecule-surface interactions driven by thermal energy, i.e., heat, can be well described within the framework of the Born-Oppenheimer approximation. New reaction channels[1] often open up in response to electronic excitations via surface-mediated energy transfer. In this process, the energetic electrons or holes can scatter into surface species and heat up vibrational modes of adsorbates. This process involves not only the electronic ground state but also the excited state, and it is nonadiabatic with respect to the nuclear motions[2].

In the weak coupling limit, the nonadiabatic energy transfer on metal surfaces can be modeled using only the ground-state potential energy surfaces and the nonadiabacity is taken into account with electronic friction contributions in Langevin dynamics by solving Newtonian equations with a thermal fluctuation term[2], [3]. The electronic friction via partial populating and depopulating of the adsorbate ground state right above the Fermi level is governing the energy transfer. The crucial parameter here is the electronic friction of an excited charge with nuclear motions of surface adsorbates. However, it is extremely time consuming to compute electron-phonon coupling strength for large systems.

In this talk, we will discuss our most recent development of a machine-learning nonadiabatic molecular dynamics approach that uses predicted forces and electronic friction coefficients by 'learning from data'. Machine learning algorithms, such as the artificial neural networks[4] can use past trajectories as training datasets for fast and accurate prediction of forces and electronic friction coefficients, thus allows us to perform statistical analysis of many trajectories. We benchmark the approach for activation of oxygen from hollow to bridge because of strong interests in understanding laser-induced chemistry of CO oxidation on metal surfaces. By comparing the Head-Gordon/Tully electronic friction with the Local Density Friction Approximation (LDFA), we found that the LDFA is mode unselective, thus giving identical coefficients along the lateral and vertical directions to surface and no oxygen activation is observed, while the Head-Gordon/Tully friction can differentiate the electron dragging force in vertical and lateral modes and capture the laser-induced oxygen activation on Ru [2].

[1] M. Bonn *et al.*, "Phonon- Versus Electron-Mediated Desorption and Oxidation of CO on Ru(0001)," *Science*, vol. 285, no. 5430, pp. 1042–1045, Aug. 1999.

[2] M. Head-Gordon and J. C. Tully, "Molecular dynamics with electronic frictions," *J. Chem. Phys.*, vol. 103, no. 23, pp. 10137–10145, Dec. 1995.

[3] M. Brandbyge, P. Hedegård, T. F. Heinz, J. A. Misewich, and D. M. Newns, "Electronically driven adsorbate excitation mechanism in femtosecond-pulse laser desorption," *Phys. Rev. B*, vol. 52, no. 8, p. 6042, 1995.

[4] A. Khorshidi and A. A. Peterson, "Amp: A modular approach to machine learning in atomistic simulations," *Comput. Phys. Commun.*, vol. 207, pp. 310–324, Oct. 2016.

500076 Modeling of Segregation on Au-Pd (111) Surfaces with Monte Carlo Simulations and Neural Network Atomic Potentials

Abstract:

The simulation of segregation in multi-component alloy surfaces is challenging with atomistic approaches because of the need to model a very large number of possible configurations with a high degree of accuracy. Density functional theory (DFT) is too expensive to use directly, and atomistic potentials are often a compromise between accuracy and computational speed. Neural networks (NN) are a promising alternative to traditional atomistic potential. Using the Behler-Parrinello framework, we were able to create a potential capable of predicting highly-accurate energies for any configuration of a AuPd(111) slab. The fully trained NN spanning all configurations and lattice constants of a AuPd binary alloy is trained from only 3,914 DFT calculations, each with only 7-21 atoms. From these relatively small slabs, Monte-Carlo simulations can be performed on 10 x 10 x 15 atom slabs with average errors under 5 meV/atom.

Mean compositions from the Monte-Carlo simulations were then used to construct segregation profiles spanning bulk compositions between 10 and 90% Au, and at temperatures ranging from 700-1000 K. These simulations result in excellent agreement with available experimental low-energy ion scattering spectroscopy data, which is sensitive only to the composition of the top layer of a metal surface. Site distributions were computed and compared to random distributions, indicating the presence of some short-range ordering favoring the formation of Au-Pd surface bonds. These trends in site distribution are also in excellent agreement with available scanning tunneling microscopy data, further validating our model. These profiles can also be fit to the Langmuir-McLean formulation of the Gibbs-isotherm with a model for the enthalpy of segregation. Based on the trends observed in the calculated enthalpy of segregation across all bulk compositions, it is not clear how one would derive the same trend using more traditional techniques, such as course-grained models based on dilute limit segregation energies alone. The techniques derived in this work are easily implemented, even with limited computational recourses. Due to the flexible nature of the NN, results from this work are also useful in application to more complex systems with adsorbates.

488959 Accelerating Electronic Structure Calculations with Machine Learning

Abstract:

Electronic structure calculations have produced mechanistic insights and have led to systematic design principles for catalysis. However, a typical atomistic calculation involves much computational "waste": often, thousands of single-point electronic structure calculations are performed to produce each one or two configurations which are ultimately published. Recently, methods have been developed which can "learn" from the output of electronic structure calculations, allowing for the prediction of the energies and forces of atomic configurations that they have not been exposed to. In this talk, we will discuss new algorithms that can use such machine learning techniques to strategically accelerate and extend electronic structure calculations, through such routines as transition-state searches and hybrid quantum-mechanics/molecular-mechanics approaches. These methods will be introduced in terms of our open-source machine-learning package, "Amp".

499604 A Data-Driven *in silico* Research Paradigm for the Rational Design of Catalyst Systems and the Exploration of Chemical Space

Abstract:

Trial-and-error research approaches are increasingly ill equipped to meeting the complex challenges involved in the discovery and design of next-generation chemistry. Our work recognizes the great opportunities that are arising with the shift towards data-driven *in silico* research and a rational design paradigm. These approaches are poised to mitigate the inefficiencies, shortcomings, and limitations of traditional trial-and-error research. However, the notion to utilize modern data science in the chemistry context is so recent that much of the basic infrastructure has not yet been developed, or is still in its infancy. The existing tools and expertise tend to be in-house, specialized, or otherwise unavailable to the community at large. Data science is thus in practice beyond the scope and reach of most researchers in the field. Our work aims to chart new paths in this area by creating an open, general-purpose software ecosystem designed to overcoming this situation, filling the prevalent infrastructure gap, and thus making data-driven research a viable and widely accessible proposition. Our software ecosystem fuses *in silico* modeling (in particular computational quantum chemistry), high-throughput screening techniques, and Big Data analytics into an integrated research infrastructure. We will demonstrate the utility of this approach in the context of the development of molecular catalysts for solar water splitting.

UCL

Overcoming the Compromise between Accuracy and Efficiency in Modelling Catalytic Kinetics

Miguel Pineda and Michail Stamatakis
Department of Chemical Engineering
University College London

New Developments in Computational Catalysis II session
2017 AIChE Annual Meeting
Tuesday October 31, 2017

Motivation

A popular and successful formalism used to analyse catalytic systems is the so-called chemical kinetic approach (deterministic and stochastic)

Micro-kinetic modelling (MKM)

Efficiency but inaccuracy

$$\frac{d\theta}{dt} = r_+ - r_-$$

Well-mixed
Mean-field

- Not correlations (thermodynamically or kinetically driven)
- Not stochasticity

Kinetic Monte Carlo (KMC) method

Accuracy but inefficiency

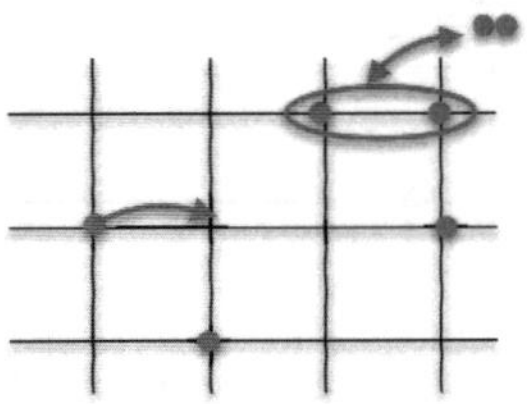

$$\frac{dP_\sigma}{dt} = \sum_\beta [k_{\sigma\beta}P_\beta - k_{\beta\sigma}P_\sigma]$$

- Correlations (thermodynamically or kinetically driven)
- Stochasticity and detailed models of complex catalysis

Developing accurate and efficient methods that take into account the best of MKM and KMC

Spatial Correlations

- Cluster mean-field approximation (CMA) adapted for use in Chemical Kinetics
- Reduced model for NO oxidation and NO_2 reduction on $Pt(111)$

+ a cloud of atoms

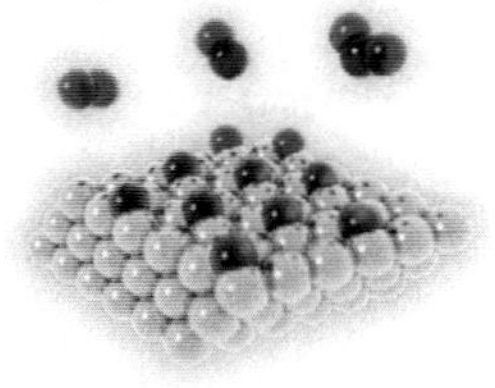

Stochasticity

- Reflected chemical Langevin equations (RCLE)
- Validation in a representative reaction mechanism with a single species

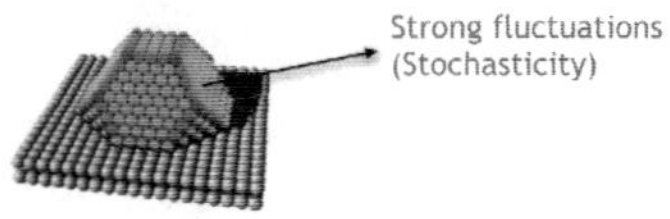

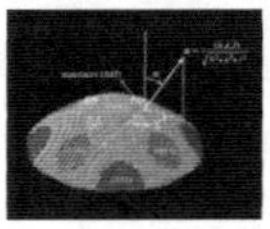

Field emitter tips

McEwen et. al. PNAS., **2009**, *106*, pp 3006

• Spatial correlations

•Cluster mean-field approximation: Lattice gas Hamiltonian formulation (LGHF) or cluster expansion

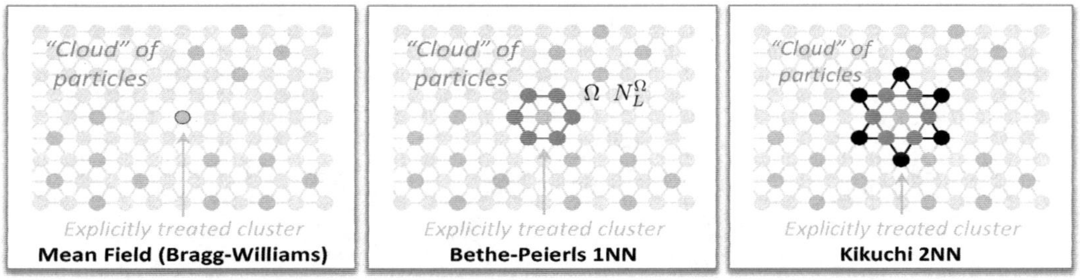

$\sigma_i = 0, 1 \qquad \sigma = \left\{\sigma_1, \sigma_2, \cdots, \sigma_{N_L^\Omega}\right\}$

$$H_C(\sigma) = H_\Omega(\sigma) + \sum_{i=2}^{N_L^\Omega} a_i\sigma_i + \frac{1}{2!}\sum_{i=2}^{N_L^\Omega}\sum_{\substack{j=2\\ j\neq i}}^{N_L^\Omega} b_{ij}\sigma_i\sigma_j + \cdots$$

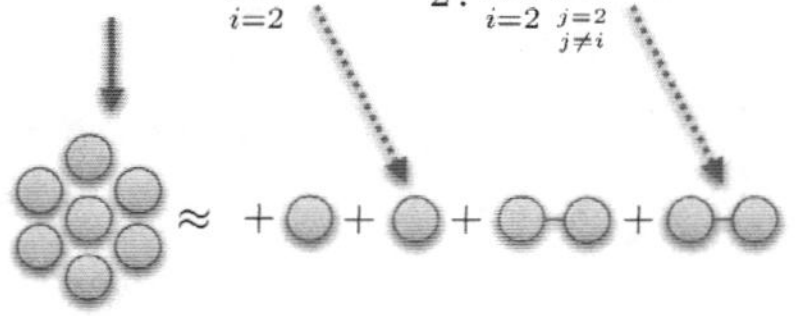

$a_i, b_{i,j} \longrightarrow$ Effective fields determined self-consistency by imposing:

- Equality of coverages of all sites
- Equality of correlation functions for symmetrically equivalent cluster of sites

• Benchmark system

Reduced model for NO oxidation and NO_2 reduction on $Pt(111)$

1) $NO_{(gas)} + O^* \underset{k^{red}}{\overset{k^{oxi}}{\rightleftarrows}} NO_{2(gas)} + *$

2) $O_{2(gas)} + 2* \underset{k^{des}}{\overset{k^{ads}}{\rightleftarrows}} 2O^*$

3) $O^* + * \underset{k^{diff}}{\overset{k^{diff}}{\rightleftarrows}} * + O^*$

- Equilibrated interconversion between NO and NO_2 allows us to assume

$$\mu_{O^*} = \mu_{NO_2} - \mu_{NO}$$

→ Kinetic observations

- Strong repulsive interactions and very fast diffusion of oxygen atoms led to oxygen adlayer quasi-equilibration (superlattice ordering)

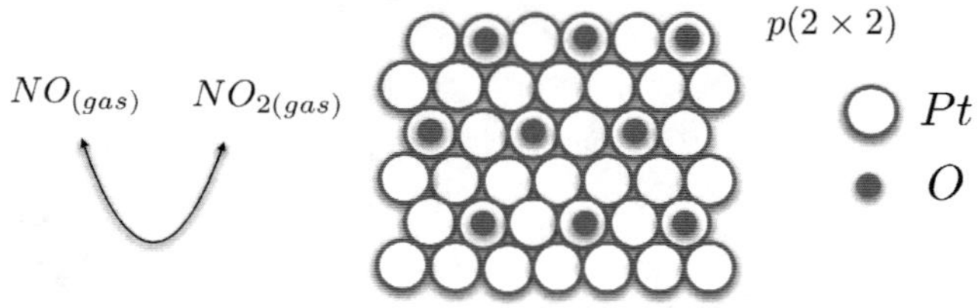

Getman et. al. Physical Review Letters., **2009**, *102*, pp 076101
Nielsen et. al. Journal of Chemical Physics., **2013**, *139*, pp 224706
Wu et. al. J. Cata., **2012**, *286*, pp 88

Rate-limited process → $O_{2(gas)} + 2* \underset{k^{des}}{\overset{k^{ads}}{\rightleftarrows}} 2O^*$

NO oxidation rate (TOF)

$$TOF = R_{O_2,ads} - R_{O_2,des} \qquad \beta \longrightarrow \sigma$$

$$R_{O_2,ads} = \frac{2}{N_L^{\Omega}} \sum_{\sigma\beta} P_\beta k_{\sigma\beta} \qquad R_{O_2,des} = \frac{-2}{N_L^{\Omega}} \sum_{\sigma\beta} P_\beta k_{\sigma\beta}$$

Probability distribution:

$$P_\sigma = \frac{1}{Q} e^{-\frac{(H_C(\sigma) - \mu_{O^*} N_\sigma)}{k_B T}}$$

The reaction rates:

$$k_{\sigma\beta} = k_{\sigma\beta}^{ads} = A e^{-\frac{E_a^{\ddagger}(\beta)}{k_B T}}$$

$$k_{\sigma\beta} = k_{\sigma\beta}^{des} = A' e^{-\frac{E_d^{\ddagger}(\beta)}{k_B T}}$$

BEP relationship adsorption barrier:

$$E_a^{\ddagger}(\beta) = E_a^{\ddagger}(0_a) + \omega(\Delta E_{rxn}(\beta) - \Delta E_{rxn}(0_a))$$

$$\Delta E_{rxn}(\beta) = H_C(\sigma) - H_C(\beta) + \Delta E_{O_2}$$

$$\Delta E_{O_2} = 002eV \qquad \omega = 1$$

Nielsen et. al. Journal of Chemical Physics., **2013**, *139*, pp 224706

1st nearest neighbour O*···O* interactions

$h_1 = -1.200eV$

$J_{ij} = J = 0.300eV$

$h_0 = -0.027eV$

K2NNC1: first nearest neighbour corrections

K2NNC2: second nearest neighbour corrections → "Ghost interactions"

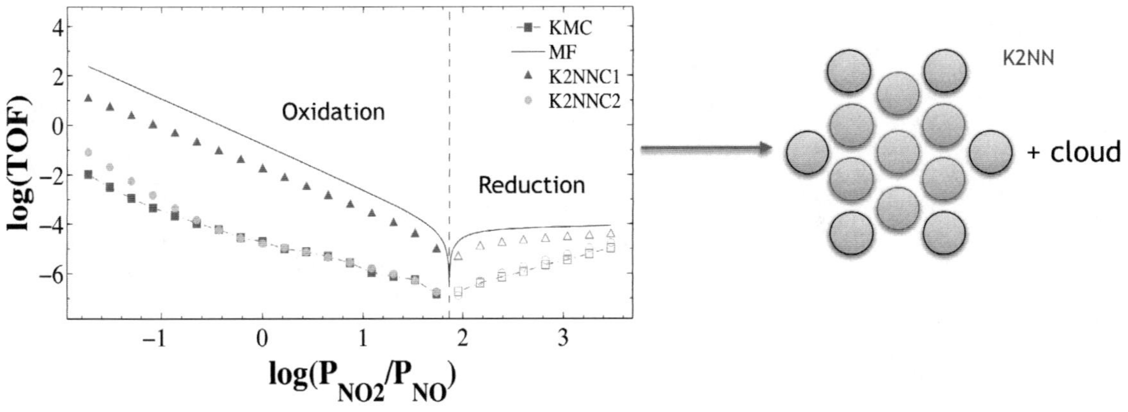

Computational efficiency

Method	Cost in CPU seconds
KMC	2.1×10^7
MF	4.3×10^{-1}
K2NNC1	1.5
K2NNC2	2.6

Tremendous computational gains!
Negligible loss of accuracy.

M. Pineda and M. Stamatakis.
Journal of Chemical Physics., **2017**, *147*, pp 024105

- Stochasticity

Let us neglect repulsive interactions and consider very fast diffusion of adsorbates

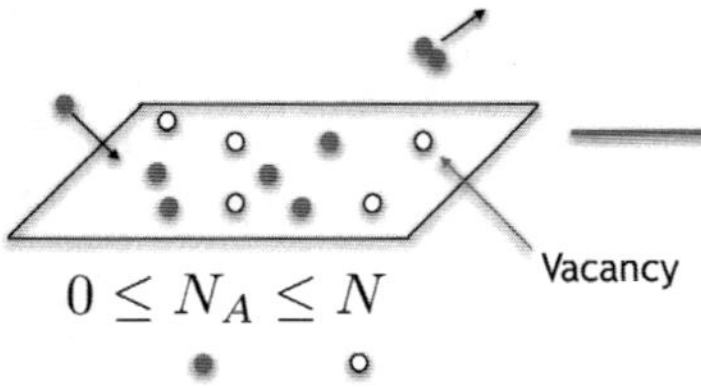

$0 \leq N_A \leq N$

Chemical master equations (CME)

$$P(N_A; t)$$

Discrete-state continuous-time description

Computationally expensive

Trajectories and probability distributions

In the limit of N very larger

$$y = \epsilon N_A \qquad \epsilon = \frac{1}{N}$$

$$0 \leq y \leq 1$$

Chemical Fokker-Plank equation (CFPE)

$$P(y; t)$$

Continuous-state continuous-time description

Computationally efficient

Only probability densities and average values

The chemical Langevin equations (CLE)

$$dy = a(y)dt + \sqrt{\epsilon b(y)}dW$$

Numerical solution: Euler-Maruyama (EM) scheme

$$y_{i+1} = y_i + a(y_i)\Delta t + \sqrt{\epsilon b(y_i)\Delta t}\mathcal{N}(0,1)$$

Gaussian

$$y_i \approx 0 \longrightarrow y_{i+1} < 0$$

$$y_i \approx 1 \longrightarrow y_{i+1} > 1$$

Solutions of the CLE not comparable to solutions of the CFPE

Reflected chemical Langevin equation (RCLE)

$$y = x + k$$

$$x \in (0,1)$$

Minimal process that force the solution to remain in the unit interval

$$dy = a(y)dt + \sqrt{\epsilon b(y)}dW + dk$$

Solution of the CFPE

$$P_{st}(y) = \frac{C}{b(y)} exp\left\{\frac{2}{\epsilon}\int_0^y dx \frac{a(x)}{b(x)}\right\}$$

• Benchmark system

1) $A_{(gas)} + *_s \underset{k^{des}}{\overset{k^{ads}}{\rightleftarrows}} A_s$

2) $A_{(s)} \overset{k^{rxn1}}{\longrightarrow} *_{(s)} + B_{(gas)}$

3) $A_{(s)} + A_{(n)} \overset{k^{rxn2}}{\longrightarrow} *_{(s)} + *_{(n)} + A_{2(gas)}$

Euler-Lepingle (EL) scheme for two-side barrier

$y \in [0, 1]$

- Gaussian variable and a independent exponential variable

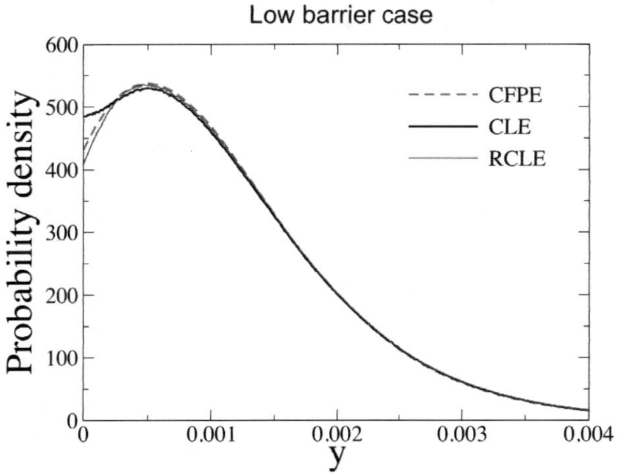

D. Lepingle, Mathematics and Computers in Simulations., **1995**, *38*, pp 119

Comparison of the RCLE with the CME and CFPE

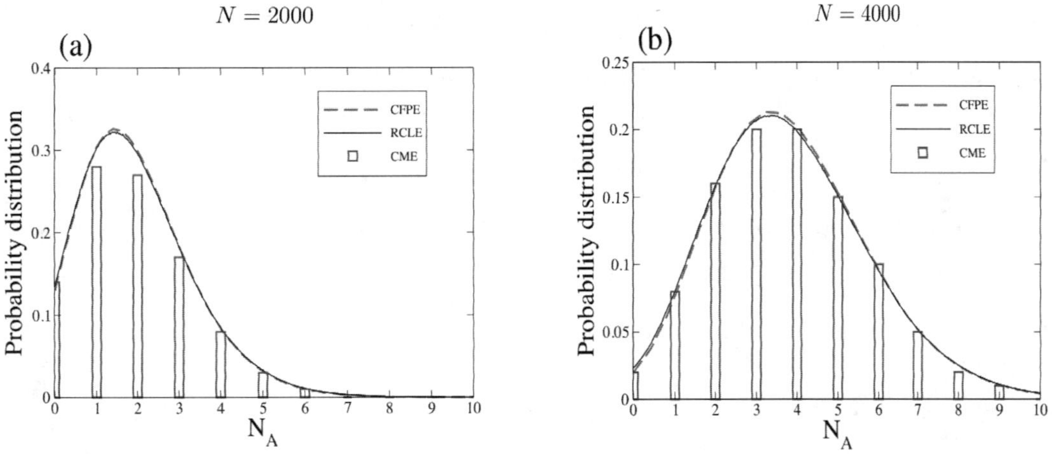

CME solutions are obtained solving the evolution equation for $P(N_A; t)$ at steady state

Computational efficiency

CME solutions are obtained using the stochastic simulation algorithm (SSA) and stopped at t=1

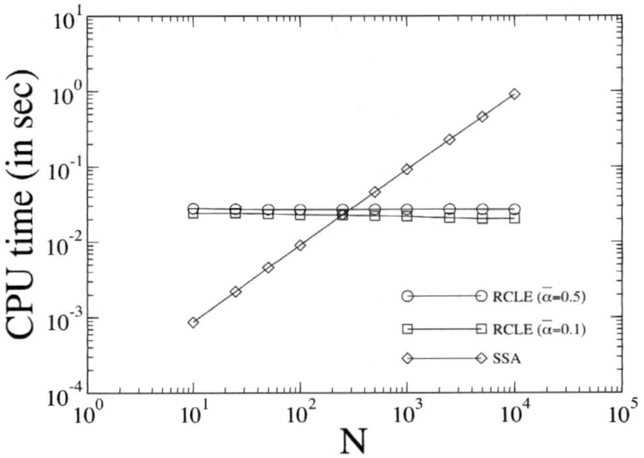

Low barrier case

M. Pineda and M. Stamatakis. Submitted for publication., **2017**

Computational gains!

Reproduce very good the well-mixed CME and CFPE.

Ongoing work

- Coupling the two methods in order to have an efficient and accurate approach involving spatial correlations and stochasticity

Funding

- Leverhulme Trust: RPG-2014-161

Computational resources

- Legion@UCL

14

496191 CO Adsorption on Platinum and Cobalt: Site Preference and Coverage Effects

Abstract:

CO Adsorption on Platinum and Cobalt: Site Preference and Coverage Effects

G. T. Kasun Kalhara Gunasooriya,† Mark Saeys†

†Laboratory for Chemical Technology, Ghent University, Technologiepark 914, 9052 Gent, Belgium

Calculation of correct site preference, adsorption energies and distribution of the adsorbates forming stable phases remain important challenges for DFT.[1] In this presentation we demonstrate that van der Waals functionals[2] provide an accurate description for these challenges on platinum and cobalt catalyst surfaces. Platinum catalysts are widely used in CO oxidation in automotive exhausts and as anode catalysts for direct methanol fuel cells. Cobalt catalysts are the preferred choice for Fischer-Tropsch synthesis (FTS), the conversion of synthesis gas, a mixture of CO and H_2, to long-chain hydrocarbons.

At low CO coverage, CO adsorbs at the top sites on Pt(111) and Co(0001) facets. Low coverage CO vdW-DF adsorption energy agrees well with the experimental data and the site preference is correctly predicted as top site. PBE functional overestimates CO adsorption energy and favors hollow over top site as most stable configuration. Interestingly, low coverage CO BEEF-vdW adsorption energy at the top site is close to the respective vdW-DF adsorption energy. However, the BEEF-vdW indicates bridge site to be the most favorable over hollow and top sites on Pt(111) facet. With increase in CO coverage, CO adsorbs both on top and bridge sites on the Pt(111) and Co(0001) facets. The observation of change in site preferences with increase in CO coverages is remarkable. To elucidate the change in stabilities of top and bridge sites with increase in CO coverage, a Bader charge and Natural Bond Orbitals (NBO) analysis was conducted. Charge analysis indicate that, CO adsorption on top site is more sensitive to surface charge compared to bridge site. Bonding analysis indicate that increase in surface charge, increases Pauli repulsion between metal dz^2 states and the filled CO 5σ orbital. Thus indicating that charge is the essence in determining the preferred adsorption site. This effect was further experimentally observed on TiO_2supported Pt clusters.[3]

CO adsorption is often described by a Langmuir isotherm, sometimes accounting for a gradual decrease in the adsorption energy with coverage. Here, we show that CO adsorption on cobalt does not follow this typical behavior. Instead, adsorption on Co(0001) is dominated by two surface phases. At low pressures, the (√3×√3)R30°–CO structure is the stable phase, and CO forms (√3×√3)R30°–CO islands for coverages below 1/3 ML because of attractive CO–CO interactions. However, a similar attractive interaction was not observed on Pt or with PBE and RPBE functionals. Charge and bonding analysis indicate that, these attractive interactions are due to decrease in charge at the diagonal Co atoms, hence reducing the Pauli repulsion between the partially filled Co dz^2 state and the CO 5σ orbital. Increasing the pressure does not gradually increase the coverage beyond 1/3 ML. Instead, a transition to a high coverage (2√3×2√3)R30°–7CO surface structure occurs at 0.1 mbar at room temperature and at 21 bar at 500 K. First principle phase diagram calculations with the van der Waals functional confirm the stability of these two phases and show a true first-order phase transition between the two phases.[4,5] It is important to note that CO adsorption introduces five new low frequency modes, i.e., vibrations associated with frustrated rotation and translation, with frequencies in the range of 40 to 400 cm^{-1}. Those modes contribute significantly to the adsorption entropy and need to be computed accurately. For comparison, CO adsorption on Pt(111) was also studied, where a similar phase transition is not observed.

References

[1] Schimka, L., Harl, J., Stroppa, A., Grüneis, A., Marsman, M., Mittendorfer, F., Kresse, G. *Nat. Mater.* **2010**, 9, 741

[2] Dion, M.; Rydberg, H.; Schroder, E.; Langreth, D. C.; Lundqvist, B. I. *Phys. Rev. Lett.* **2004**, *92*, 246401

[3] Gunasooriya, G. T. K. K., Seebauer, E. G., Saeys, M. *ACS Catal.*2017, 7, 1966

[4] Beitel, G. A.; Laskov, A.; Oosterbeek, H.; Kuipers, E. W. *J. Phys. Chem.* **1996**, *100*, 12494

[5] Gunasooriya, G.T.K.K.; van Bavel, A.P.; Kuipers, H.P.C.E.; Saeys, M. *Surf. Sci.* **2015**, 642, L6

492482 Revealing Cellulose Pyrolysis Initiation Kinetics Via Novel Milisecond Pulse Reactor System

Abstract:

Biomass pyrolysis is an appealing route for producing fuel and chemicals from renewable resources. Detailed chemical models of pyrolysis chemistry are necessary to optimize the production and composition of bio-oil during pyrolysis. Cellulose, the most abundant biopolymer in lignocellulosic biomass, initially reacts through a short-lived liquid intermediate species which controls the chemistry and determines the composition of bio-oil [1]. In this work, a novel millisecond pulse reactor system known as PHASR (Pulse Heated Analysis of Solid Reactions)[2] is used to study the kinetics of the initiation chemistry of cellulose between 385 ^{0}C and 500 ^{0}C [3]. Using a cellulose surrogate, Î±-cyclodextrin[4], the energetics for the initiation chemistry were characterized by measuring conversion from 20 ms to 20 sec. Cellulose decomposition initiation undergoes two distinct kinetic regimes of intra-chain scission; low-temperature glycosidic bond cleavage (T< 467 ^{0}C) with a low apparent activation energy and high-temperature glycosidic bond cleavage (T>467^{0}C).

References:

1.Dauenhauer, P. J.; Colby, J. L.; Balonek, C. M.; Suszynski, W. J.; Schmidt, L. D., Reactive boiling of cellulose for integrated catalysis through an intermediate liquid. *Green Chemistry* **2009**, 11, (10), 1555-1561.

2.Krumm, C.; Pfaendtner, J.; Dauenhauer, P.J., Millisecond pulsed films unify the mechanisms of cellulose fragmentation. *Chemistry of Materials* **2016**, 28, 3108-3114.

3. Zhu, C.; Krumm, C.; Facas, G. G.; Neurock, M.; Dauenhauer, P.J., Energetics of cellulose glycosidic bond cleavage. *Reaction Chemistry and Engineering* **2017**, 2, 201-214.

4. Mettler, M. S.; Mushrif, S. H.; Paulsen, A. D.; Javadekar, A. D.; Vlachos, D. G.; Dauenhauer, P. J., Revealing pyrolysis chemistry for biofuels production: Conversion of cellulose to furans and small oxygenates. *Energy & Environmental Science* **2012**, 5, (1), 5414-5424.

The Roles of Au clusters and Co-ZSM-5 in the Low Temperature Aerobic Selective Oxidation of Neat Ethylbenzene

Anyang Peng,[a] Linping Qian,[b] Ruiyi Yan,[a,c] Mayfair C. Kung[a]* and Harold H. Kung[a]*

[a]*Chemical and Biological Engineering Department, Northwestern University, Evanston, IL 60208, USA*
[b]*Department of Chemistry and Shanghai Key Laboratory of Molecular Catalysis and Innovative Materials, Fudan University, Shanghai, 200433, PR China*
[c]*Institute of Process Engineering, Chinese Academy of Sciences, Beijing, 100190, PR China*

Introduction & Backgrounds

- Au nanoparticles have unique catalytic properties in hydrocarbon oxidation:
  - propene epoxidation[1],
  - styrene oxidation[2],
  - Ethylbenzene oxidation[3].

- Limited examples of reactions using O_2 as oxidant. We have recently showed cyclooctene (c-$C_8^=$) epoxidation with O_2 under mild reaction conditions.[4]

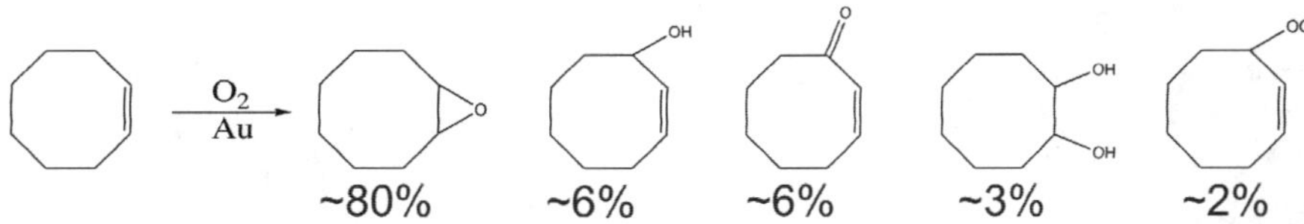

~80% ~6% ~6% ~3% ~2%

[1]Angew. Chem. Int. Ed. 2009, 48, 7862
[2]J. Mol. Catal. A: Chem., 2005, 240, 40.
[3]Applied Catalysis A: General 435–436 (2012) 19.
[4]Nature Communications **8**, 14881 (2017)

Au Behaving in Unusual Catalytic Mode

- *c*-$C_8^=$ Unique features
  - Soluble atomic Au clusters *derived from Au nanoparticles*

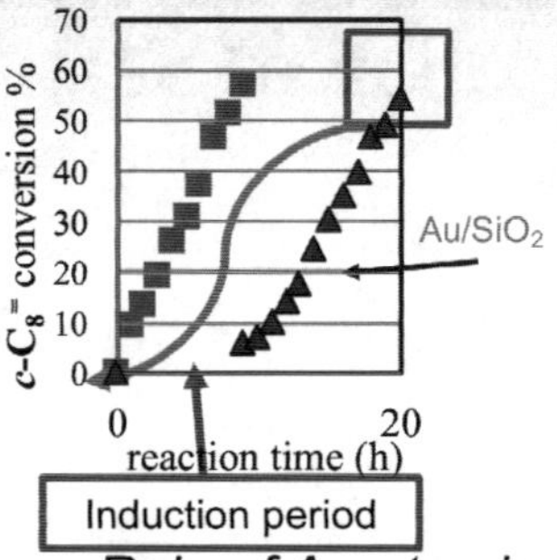

Aberration-corrected electron microscopy images of filtrates

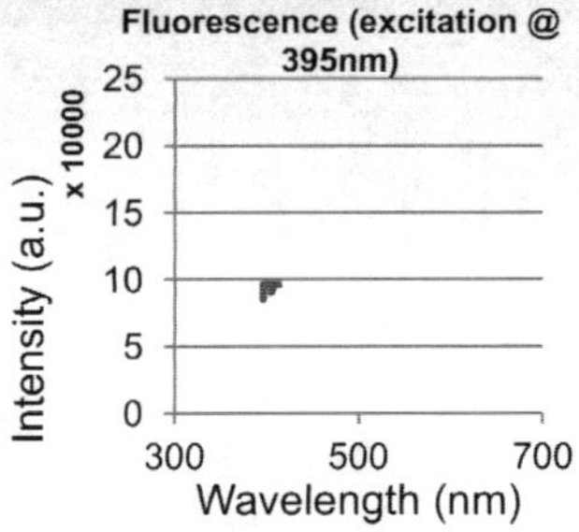

  - *Role of Au atomic clusters: generation of radical initiators.*

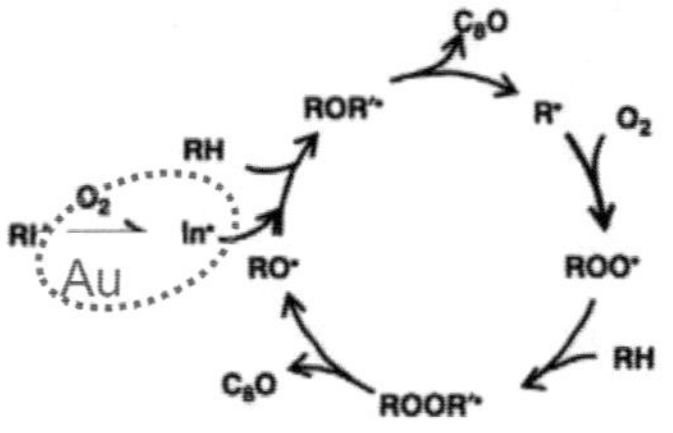

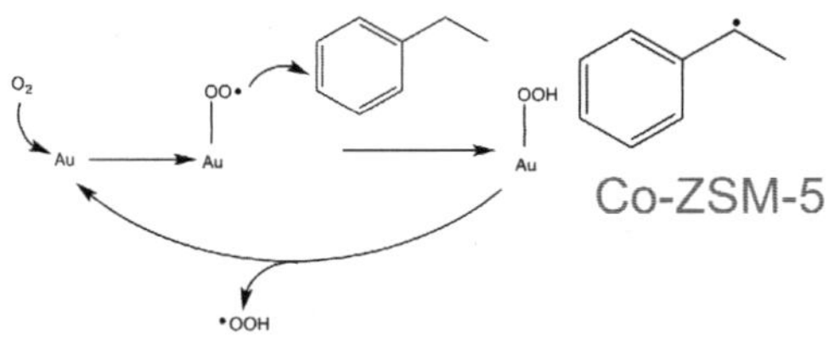

Synergistic Effect in Ethylbenzene oxidation

- Reaction products are identified by GC-MS and NMR.
- Quantification of products is achieved by ^{1}H-NMR with decane as the internal standard.
- After compensating evaporation loss, no carbon imbalance is observed.

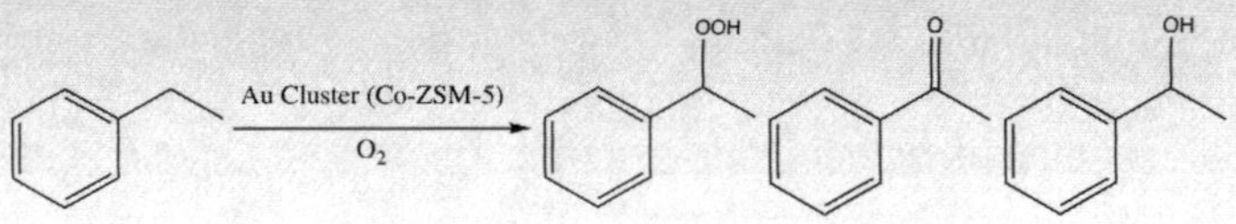

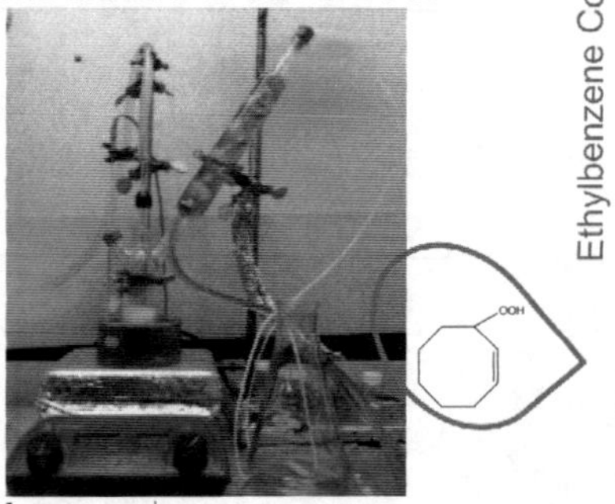

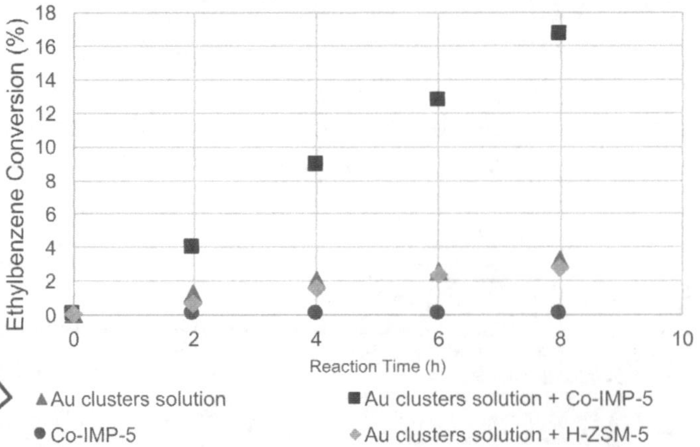

Co-IMP-5: Co-ZSM-5 prepared by impregnation with 5% Co loading

Role of Au in Ethylbenzene oxidation
Differentiate Au from *c*-C_8 Hydroperoxide

Four types of solutions: with and without Au.

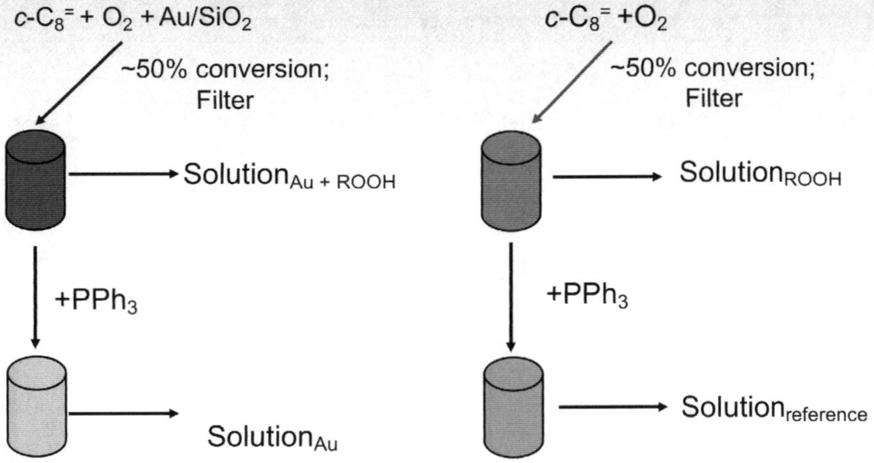

$$PPh_3 + C_8H_{13}OOH = C_8H_{13}OH + PPh_3O$$

After addition of PPh_3, peroxide concentration as determined by peroxide test strip MQuant is below 0.02M (>90% removal).

Role of Au in Ethylbenzene Oxidation
comparison of solutions with and without Au in the presence of triphenylphosphine (PPh3)

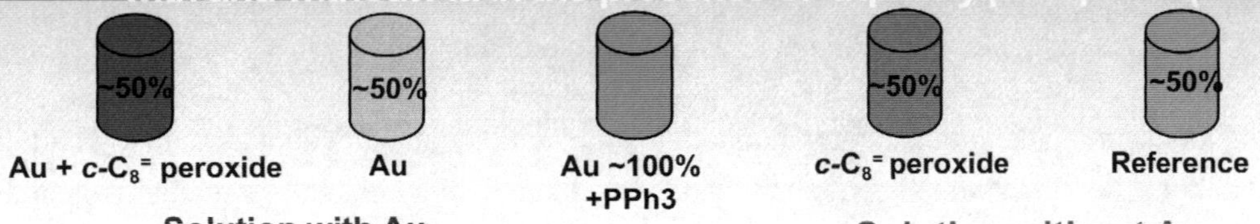

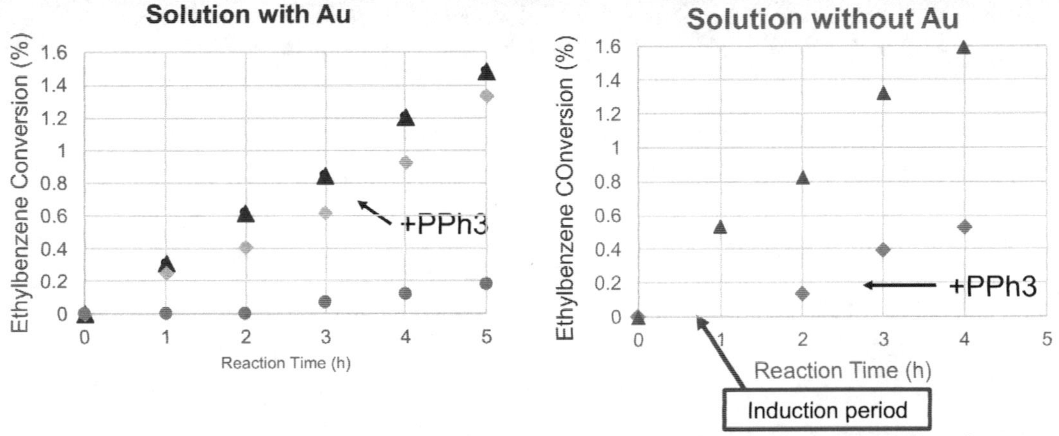

Conclusion: Au is important to initialize the ethylbenzene oxidation by continuously supplying *c*-C_8 hydroperoxide from *c*-$C_8^=$ oxidation!!.

Synergistic Effect

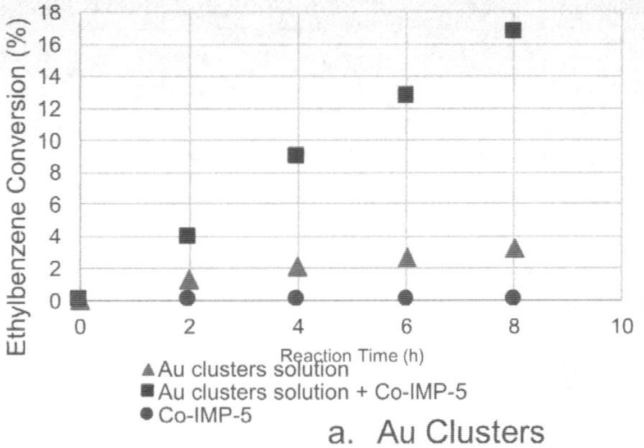

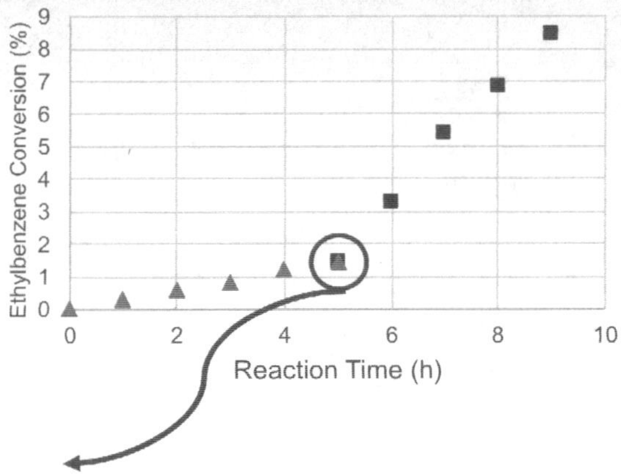

a. Au Clusters
b. 1-phenylethyl hydroperoxide (PEHP)
c. *c*-$C_8^=$ hydroperoxide

Role of Co in the Synergistic Effect

Au is not essential to enhance the rate

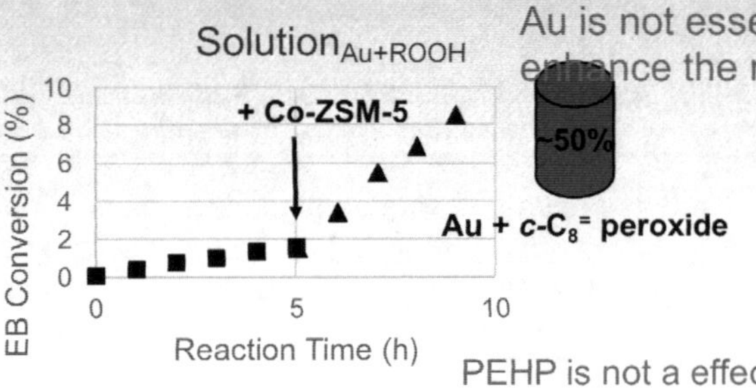

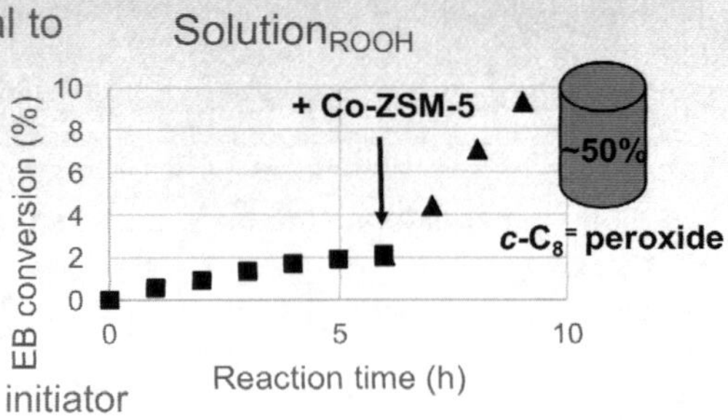

PEHP is not a effective initiator under the reaction condition

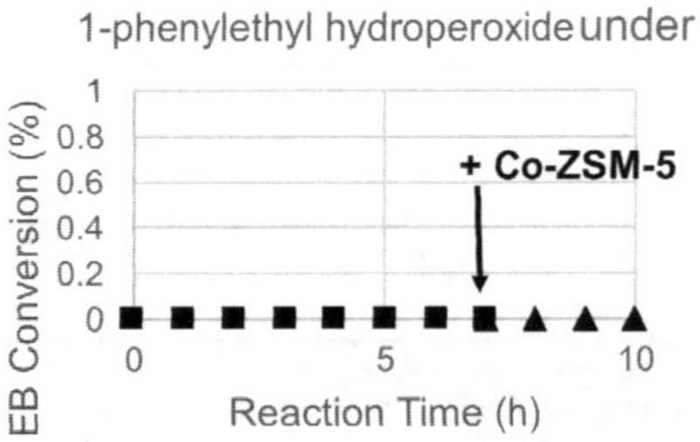

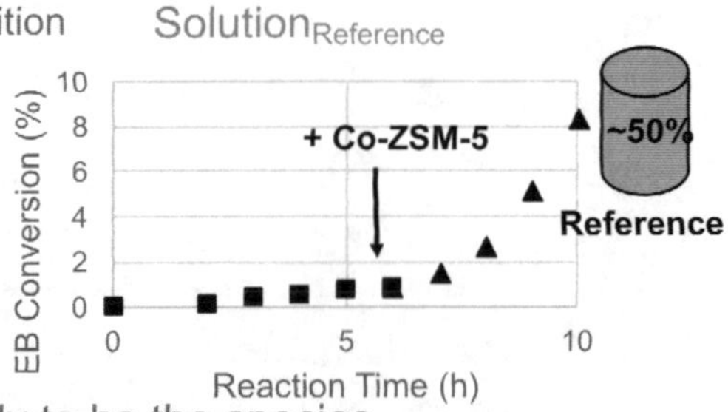

Conclusion: Cyclooctene hydroperoxide is likely to be the species interacting with Co to give the synergistic effect.

Uniqueness of c-C_8 hydroperoxide

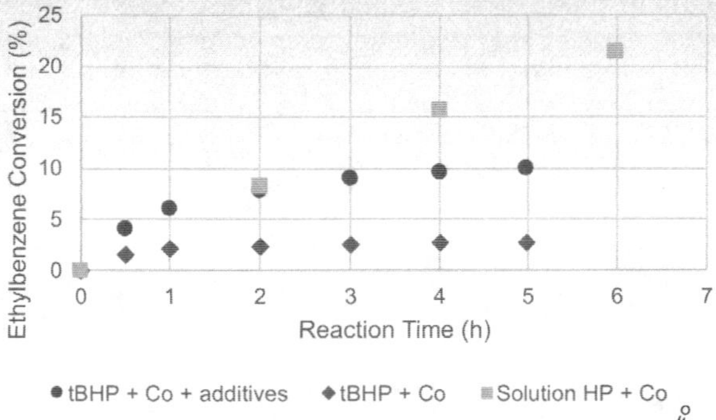

Type of initiator	Amount of initiator	Amount of product
c-C_8 hydroperoxide with other reaction products	**0.6** mmol	**12** mmol
tBHP with additives	5.5 mmol	6 mmol
tBHP	5.5 mmol	2 mmol

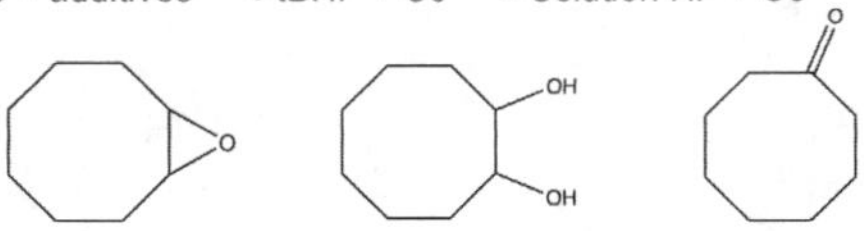

Conclusion: c-C_8 hydroperoxide is not reacting with ethylbenzene in a stoichiometric manner. It's a mediator between Au and Co in the synergistic effect. Other additives generated from c-$C_8^=$ oxidation can increase the efficiency of peroxide initiator.

Summary

- A synergistic effect between Au and Co has been successfully demonstrated in the selective oxidation of Ethylbenzene.
- Solubilized Au nanoclusters are initiators different from traditional free radical initiators.
- Cyclooctene hydroperoxide is most likely to be the mediator between Au and Co. Some oxidation products from c-$C_8^=$ oxidation can increase the efficiency of peroxide initiator.

Acknowledgement

U.S. Department of Energy, Office of Science, Office of Basic Energy Sciences under Award Number DOE DE-FG02-03-ER15457

Prof. Harold Kung
Prof. Mayfair Kung

Dr. Chloe Wu
Dr. Linping Qian
Dr. Ruiyi Yan
Dr. DelRae Becker

Scott Nauert
Rob Brydon

Nature of Co

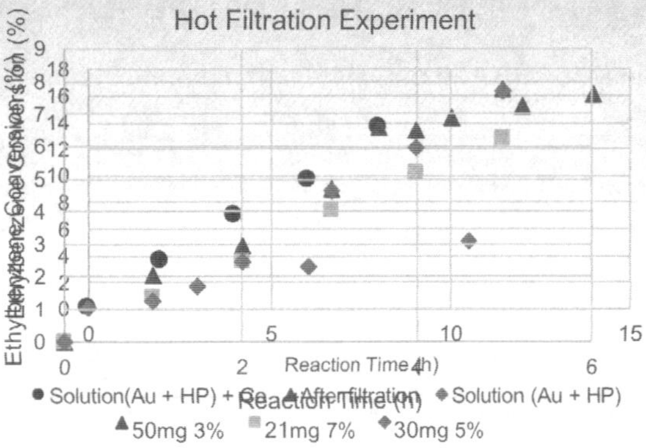

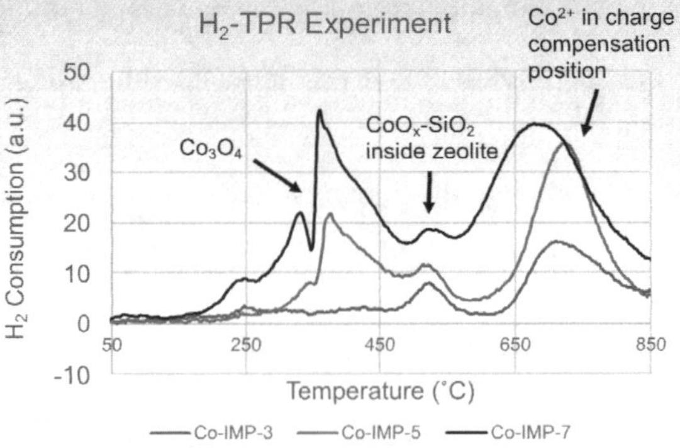

Conclusion: Catalytic relevant Co species are located on the zeolite support. Co^{2+} in charge compensation position and Co_3O_4 are 2 major Co species and they have similar catalytic activities.

494140 Oxidation of Dibenzothiophene in Diesel By *in situ* Generation of Hydrogen Peroxide in a Trickle Bed Electrochemical Reactor

Abstract:

An advanced integrated technology is proposed for oxidative desulfurization of diesel fuel in a trickle bed electrochemical reactor(TBER). In this process a strong and green oxidizing agent hydrogen peroxide was *in situ* generated by oxygen cathodic reduction, which was further employed for oxidation of dibenzothiophene(DBT) in diesel. The oxidized DBT can be easily removed by extraction with polar solvents. The process is environmentally friendly and eliminates the requirement cost for introduction of liquid oxidants in oxidative desulfurization process including hydrogen peroxide, organic hydroperoxide and organic peroxide from outside sources that require transportation, storage expenses, handling and safety precautions. The TBER cathode composed of carbon black and polytetrafluorethylene(C-PTFE) was initially developed to increase hydrogen peroxide generation by new bed design that showed a significant increase in hydrogen peroxide concentration. This performance is attributed to the good distribution of electrolyte and oxygen on the cathode bed and the effective mas transport of oxygen from the gas phase to the electrolyteâ€"cathode interface. Factors for hydrogen peroxide electrogeneration were studied, involving, cell potential, flow mode, electrolyte concentrations flow rates, duration, diesel ratio and temperatures. The produced data indicated the possibility of *in situ* oxidation of DBT in diesel using divided TBER.

495475 Fundamentals of the Removal of Sulfur Compounds from Gaseous Streams Via Reactive Sorption with Copper Oxide

Abstract:

Hydrocarbon streams derived from natural gas and oil processing are often contaminated by sulfur containing compounds, such as hydrogen sulfide and alkyl-mercaptans, which affect the quality and specifications of the product streams, pose a corrosion hazard in transfer pipelines and equipment, and conflict with health and environmental regulations. Several removal techniques have been studied and implemented in industrial purification processes ranging from gas-alkaline liquid absorption and physisorption on molecular sieve materials. Chemical reaction of sulfur molecules with copper oxide is an attractive purification strategy (especially for dilute concentrations) because the reactions are thermodynamically favorable at low temperature and the sulfur is sequestered as solid Cu-sulfur products. This study aims to elucidate the fundamentals and mechanistic details of these gas-solid reactions. Specifically, the results from this study led to semi-empirical correlations between reaction conditions (contaminant concentration, temperature, and presence of other contaminants) and the reaction rates and saturation capacities of the solid phase. Moreover, the effect of structural differences resulting from different preparation methods and starting precursors on the performance of copper oxide adsorbents in reactive sorption experiments was investigated to develop a structure-function relationship that can help in future design and tuning of adsorbent properties. A combination of thermodynamic calculations, fixed bed sorption experiments, and advanced characterization techniques are used to achieve this goal.

Apparent H_2S uptake kinetics and contaminant capacities were determined from contaminant breakthrough profiles collected from fixed beds of sorbents. The breakthrough curves were analyzed using irreversible adsorption models based on linear driving force approximations. This assumption leads to an analytical solution to the contaminant mass balance that consists of two parameters: the saturation concentration of contaminant in solid phase and an apparent sorption rate parameter. The rate parameter can be de-convoluted into contributions from bulk diffusion, pore diffusion, and reaction-diffusion phenomena within the solid. The effect of agglomerate size was probed using a commercially produced Cu-Zn-Al oxide material. As the agglomerate size increased from 600 to 1000 microns, the sorption rate parameter decreased from 1.04×10^{-4} s^{-1} to 6.05×10^{-5} s^{-1}. The value of the sorption parameter for agglomerates larger than 550 microns were only weakly dependent on temperature and pressure and were in close agreement with theoretically calculated values of the contribution from diffusion within the agglomerate macropores. For agglomerate sizes less than 500 microns, the rate parameter is less sensitive to decreasing agglomerate size, and the rate parameter eventually becomes independent of particle size at 196 microns. The results from these agglomerate size studies indicate that bulk and pore diffusion are the dominant resistances to H_2S uptake for agglomerates that are larger than ~500 microns, however, as agglomerate size decreases, bulk and pore diffusion resistances become negligible compared to reaction-diffusion phenomena within the copper oxide and copper sulfide phases.

The effects of varying reaction conditions (temperature, pressure, inlet concentration of H_2S) on the rate parameter were also studied. Varying the total pressure (1-2 atm) while maintaining constant temperature and inlet concentration had negligible effect on the rate parameter while increasing the inlet concentration (930-2200 ppm) of H_2S led to an increase in the parameter from $3.74x10^{-5}$ s^{-1} to $1.1x10^{-4}$ s^{-1}. Semi-empirical relations between concentration and the reaction rate parameter were developed. The rate parameter has an apparent 0.5 order with respect to inlet H_2S concentration. The saturation capacity remained constant at an average of 12.8 wt% (g of H_2S/100 g of sorbent) over a concentration range of 930-2200 ppm. Moreover, increasing temperature (21-150Â°C) increased the saturation capacity from 11 wt% to 42 wt%. The rate parameter exhibits an Arrhenius-type dependency on temperature with a pseudo-activation energy of 18 kJ/mol.

To test the applicability of linear driving force models for the reactive sorption of other sulfur containing compounds, reactive sorption experiments were conducted using methyl and ethyl mercaptan at varying temperatures (21-200Â°C) on the commercially produced Cu-Zn-Al material. Unlike the CuO-H_2S reactions that yielded the same products throughout the experimentsâ€™ duration and with varying conditions, CuO-mercaptan experiments resulted in various products. Examining the concentration profiles of different species identified the dominant reaction/s under given conditions. At room temperature, methyl and ethyl mercaptans were physically adsorbed by the aluminum base of the materials and breakthrough happened fast with no other sulfur containing compounds detected in effluent streams which suggests that, although thermodynamically favored, the rate of formation of dimethyl and diethyl sulfide at room temperature is too low for any reaction to be detected. At 120°C the methyl mercaptan reactive sorption experiment initially produces water and a high molecular weight sulfur containing compound which indicates the reaction: $2CH_3SH$ + 2 CuO à Cu_2O +H_2O + CH_3S-SCH_3. Over time, the water concentration in the effluent decreases with a corresponding increase in CH_3SCH_3 and methyl mercaptan which suggests that the reaction $2CH_3SH$ + CuO à CH_3SCH_3 + CuS is taking place at a low rate. The spent sample of this experiment showed a color change from black to orange which indicates the formation of Cu_2O via $2CH_3SH$ + 2 CuO à Cu_2O +H_2O + CH_3S-SCH_3. Increasing temperature to 200°C yielded CH_3SCH_3 and CH_3SH throughout the reaction duration which suggests that at 200°C the $2CH_3SH$ + 2 CuO à Cu_2O +H_2O + CH_3S-SCH_3reaction pathway is both kinetically and thermodynamically favored.

Fresh and spent nanostructures were analyzed via XRD and SEM to identify composition, crystallite agglomerate size, and crystallite size (using the Sherrer Equation). XRD patterns of fresh CuO nanostructures showed the characteristic peaks of tenorite, the monoclinic structure of CuO, for both nanofibers and nanoparticles regardless of precursor. Furthermore, nanofibers and nanoparticles consisted of CuO crystallites of similar size (16-18 nm). XRD analysis of the commercial materials also indicated monoclinic CuO with ~20 nm crystallites. SEM images revealed that the nanofibers exist as approximately rectangular solid shaped agglomerates of crystallites that are 250-300 nm in size and arranged in a one-dimensional stacking pattern. Nanoparticles and commercial Cu-Zn-Al oxide also consisted of 250-300 nm sized crystallite agglomerates, however, these agglomerates are packed into sphere like structures that are 5-10 microns in size.

To further probe the reaction between H_2S and CuO at the structural/morphological level, H_2S removal using CuO nanofibers and nanoparticles was compared. Nanofibers (230 nm diameter) and nanoparticles exhibited similar maximum capacities (2.4 wt%), however, sorption rate parameters for nanofibers (3.0 x 10^{-3} s^{-1}) were 1.7 times larger than those for nanoparticles (1.7 x 10^{-3} s^{-1}). This faster sorption rate is likely the result of the higher intra-agglomerate void fraction in the nanofibers compared to the nanoparticles. Moreover, the effects of changing nanofiber diameter on sorption rate and saturation capacity was studied. It was found that an increase of diameter (230-600 nm) resulted in a drop in the saturation capacity (2.4-0.3 wt%) and a decrease in the sorption rate parameter from 3.0 x 10^{-3} s^{-1} to 1.4 x 10^{-4} s^{-1}. This decrease can be explained by the decrease in the intra-agglomerate void space which decreases accessibility to the reactive domains. The temperature effects were probed in CuO nanofibers and nanoparticles in similar fashion to the commercial Cu-Zn-Al. Increasing the temperature from 21°C to 200°C increased the saturation capacity for both 600 nm nanofibers (from 0.3 wt% to 38.8 wt%) and CuO nanoparticles (from 2.4 wt% to 40 wt%) very close to the stoichiometric limit of 42 wt% for complete conversion of CuO to CuS. Moreover, the concentration effects on rate parameter and saturation capacity were probed in CuO nanoparticles. In the range of 250-1000 ppm, the increase in inlet concentration increased the rate parameter from $6.95x10^{-5}$ to $1.6x10^{-3}$while for higher concentrations (1000-2000 ppm) the dependency becomes weaker and the rate parameter eventually becomes invariant to concentration.

The results from these studies indicate that equations derived from linear driving force approximations can be used as predictive models for the removal of sulfur compounds via reactive sorption with CuO. Mass transfer is the predominant resistance to contaminant removal for large particles of sorbent material, and thus, reaction rate parameters for linear force driving models can be calculated from theoretical relationships. Reaction-diffusion phenomena within the solid are more complex and require semi-empirical relationships to determine rate parameters for linear driving force models. Studies of CuO from different synthesis methods reveals that the fundamental kinetics and mechanism of these gas-solid reactions are highly sensitive to the nanostructure of CuO.

494437 The Mechanism and Kinetics of Magnesium Oxide Carbothermic Reduction

Abstract:

Compared to the commercially established production routes (FeSi reduction of MgO and electrolysis of $MgCl_2$), the carbothermic reduction of MgO has been considered as an attractive candidate for production of Mg due to its potentially lower energy consumption and CO_2 emissions. In this process MgO-C blends are heated to temperatures that, depending on pressure, range from 1350°C (at 10 mbar) to 1750°C (at 1 bar), which results in production of gaseous Mg and CO. As at lower temperatures the products recombine readily, rapid quench of the product gas is needed to increase the yield of Mg. To identify the key parameters that control the reduction rate, we investigated the prevailing pathway and kinetics of Mg production at 1400°C and total pressure of 10 mbar. Accurate kinetic data has been acquired utilizing a novel experimental method that precluded confounding the product gas composition by the partial product recombination in the quencher. By varying reactant molar ratio, morphology, and physical proximity, we have conclusively elucidated the rate-controlling steps along the course of reaction. Our findings imply that the Mg production rate is initially controlled by the removal of the O_2, originating from thermal dissociation of MgO, via reaction with C. The further progress of reaction is hindered by a gradual sintering of MgO that increases the overall O_2 transport resistance, thereby decreasing the Mg production rate. In this stage the Mg is mainly produced by the MgO reduction with CO yielding CO_2 as intermediate, which is then removed from the reaction site via reaction with C that replenishes the CO to sustain the further MgO reduction.

501444 Enhancing the Rate of Magnesium Oxide Carbothermal Reduction By Catalysis, Mechanical Milling, and Vacuum

Abstract:

Primary magnesium metal production by carbothermal reduction of magnesium oxide can be an efficient alternative over current silicothermic reduction methods. Catalyst addition, mechanical milling, and vacuum reduction were investigated as methods to increase the reactivity of precursor pellets. When using hard burn magnesia (0.6 m^2/g), catalyst addition and mechanical milling of powders prior to pelletization increased pellet reactivity during reduction at 1550°C and 1 kPa. Further increasing vacuum in the reaction chamber to 0.1 kPa resulted in catalyst addition having no effect on the reaction rate. When using soft burn magnesia (147 m^2/g), mechanical milling showed increased reactivity, but catalyst addition hindered the reaction rate. Catalysis was determined to proceed by a gas-solid reaction pathway, with CO_2 as the likely reaction intermediate, and not by the direct solid-solid reaction of carbon with magnesium oxide. For catalyst addition to improve the reactivity of pellets, the catalysis of the gas-solid reaction pathway must outweigh any hindrance to the solid-solid reaction pathway.

507788 UOP Advanced MTO Technology â€“ A New Route for the Production of Light Olefins

Abstract:

The UOP Advanced Methanol-to-Olefins (MTO) Technology combines the UOP/Hydro MTO process for the conversion of methanol to light olefins with the Total Petrochemicals/UOP Olefin Cracking Process (OCP) to convert C_{4+} by-products into additional ethylene and propylene. The overall process provides a cost-advantaged route for the utilization of natural gas, coal or biomass to produce these high volume petrochemicals. Fundamental understanding of catalyst structure-performance relationships and the MTO reaction mechanism have guided the development of this technology. The integration of the MTO and OCP processes provides close to 90% overall carbon selectivity to ethylene and propylene from methanol. Recent progress in the commercialization of this technology will also be reviewed.

507789 On the Reaction Mechanism and the Nature of the Active Site for Standard Selective Catalytic Reduction of NO_x on Cu/SSZ-13 Zeolites

Abstract:

Cu/SSZ-13 zeolites are used commercially in mobile applications for the selective reduction of NO_x with ammonia. Our data suggest that the catalytic cycle is a redox one involving Cu(I) and Cu(II) species. During preparation, the Cu(II) species are the most stable and reside preferentially on six-member rings containing two aluminum sites. The Cu(II) ions first populate 2Al sites before the other remaining unpaired sites with 1Al are populated, Cu(II)OH. These sites could be counted ex situ through vibrational and X-ray absorption spectroscopies (XAS) and chemical titrations. In spite of an apparent chemical difference, the two sites had similar standard SCR reaction turnover rates, apparent activation energies and apparent reaction orders at SCR conditions, even on zeolite frameworks other than SSZ13. O*perando* XAS experiments showed similar Cu(I) â€" Cu(II) fractions at steady state, demonstrating their similarity for SCR redox chemistry. Finally, this study also illustrates the necessity of a collaboration among experts in synthesis, characterization, kinetics and theory for a chance to understand the catalytic cycle.

507790 Mechanisms for C-C Bond Cleavage and formation during Acrolein Production on a Mixed Metal Oxide Catalyst

Abstract:

We report a reaction network detailing the mechanistic origins of 20 C_2 â€“ C_6 byproducts observed during the oxidation of propylene to acrolein at 623 K on a molybdenum-based catalyst promoted with cobalt and nickel used in the commercial production of acrolein. The carbon backbone of propylene is preserved in the sequential oxidation of propylene to allyl alcohol, acrolein, and acrylic acid, as well as propylene oxidation to acetone and propanal via water-mediated pathways. Transient kinetic measurements in conjunction with co-feed experiments of C_2 and C_3 aldehydes and carboxylic acids show that decarbonylation and decarboxylation reactions, reactions of organic compounds with surface-adsorbed oxygen species, and total combustion reactions are the three mechanisms for C-C bond cleavage. C-C bond formation reactions that result in C_4 â€“ C_6 byproducts occur via: (i) the addition reaction of a propylene-derived surface allyl (C_3H_5) species with formaldehyde to form C_4 products and with propylene and/or allyl alcohol to form C_6 products, or (ii) the addition reaction of an acrolein (acrylic acid)-derived surface ethenyl (C_2H_3) intermediate with propylene to form pentadiene and with acrolein to form C_5 cyclic oxygenates.

507791 Conversion of Methane to Methanol and Ethanol in a Single Reactor

Abstract:

Methane is the most abundant resource of organic carbon. However, many methane reserves are in remote locations, where the economy of scale for steam reformers cannot be realized and distances to chemical plants are too large for efficient transportation through pipelines.

To provide an option for converting methane to liquid intermediates, we developed a bi-functional catalyst that combines Lewis acidic NiO clusters with a redox active ceria-zirconia (CZ) support. Methane is activated on Lewis acid sites, and under the right conditions, surface methyl or methoxy groups can couple to higher alkyl/alkoxy chains. In the presence of steam, the surface groups can be hydrolyzed to alcohols. Specifically, methanol and ethanol were detected in our study. The redox active CZ support activates O_2 to enable a net selective oxidation reaction, providing a thermodynamic driving force. Besides alcohols, CO_2 and H_2 are formed as by-products. Importantly, formation of alcohols occurs at 450 °C under steady state conditions with a turnover frequency of at least 50 h^{-1}. This is a clear improvement from previously reported catalysts, which required a high-temperature calcination step for every turnover. The nature and synergy of active sites as well as the mechanism of the reaction will be discussed.

Controlled Catalyst Binding Strength Using Bimetallic Overlayer Structures for Improved Reactivity

Joseph H. Holles

Department of Chemical Engineering
University of Wyoming
Laramie, WY 82071

UNIVERSITY OF WYOMING

D-Band Center

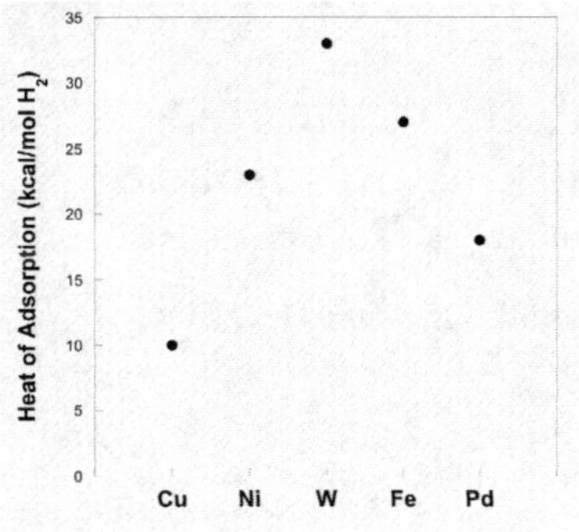

- Heat of adsorption can be measured for the elements.
- What elemental property varies continuously in a way for us to correlate with adsorbate bond strength?

First-principles quantum calculations using computers allows determination of the density of states of the d-electrons (the bonding electrons).

The center of the d-band (arrow) can be representative of the "bonding availability" for that element.

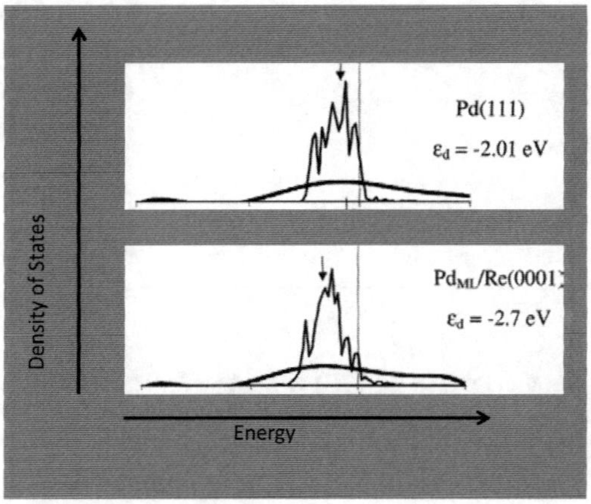

Correlation of Properties

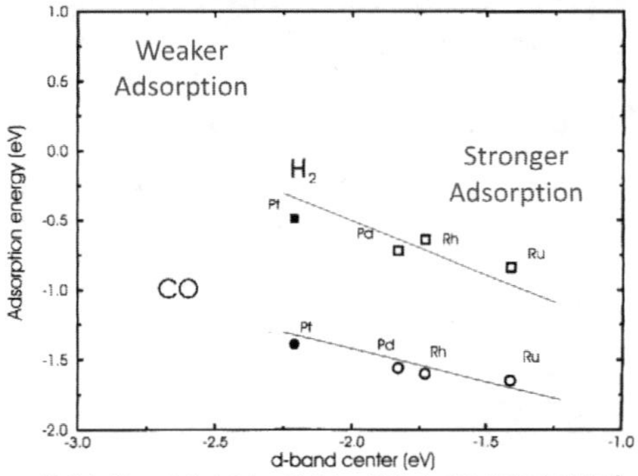

[E. Christoffersen, P. Liu, A. Ruban, H.L. Skriver, J.K. Norskov, J. Catal. 199 (2001) 127.]

Adsorption energy and d-band center are well correlated for the elements.
The line slope is different for each adsorbate.

Unfortunately, there are only so many elements.

Question:
Perhaps Pt is a pretty good catalyst for a reaction of interest.

But:
It is known to bind a reactant too strongly and block sites.

Can we find a way to go beyond the elements?

Addition of Overlayers

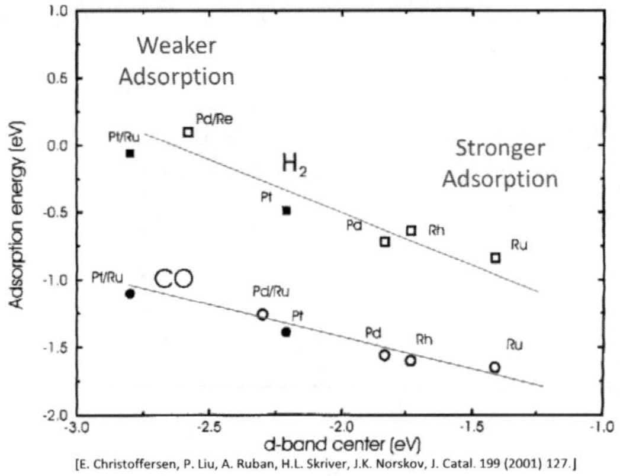

[E. Christoffersen, P. Liu, A. Ruban, H.L. Skriver, J.K. Norskov, J. Catal. 199 (2001) 127.]

Calculated variations in CO and H_2 chemisorption energies as a function of the center of the d-band.

- Allows choice of overlayer material to change bonding characteristics of reactants and products:
  - adsorb more strongly/weakly as desired.

Catalyst Design

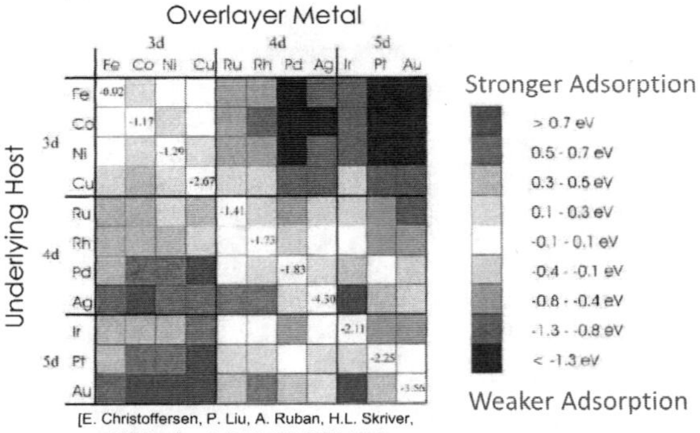

[E. Christoffersen, P. Liu, A. Ruban, H.L. Skriver, J.K. Norskov, J. Catal. 199 (2001) 128.]

- First principle computational techniques have been used to predict d-band shifts and adsorption properties

Goal:
Identify reactions where specific binding issues are known to affect reaction rates.

Solution:
Use overlayer catalysts to address these binding issues and increase reaction rates.

Heat of H_2 Adsorption

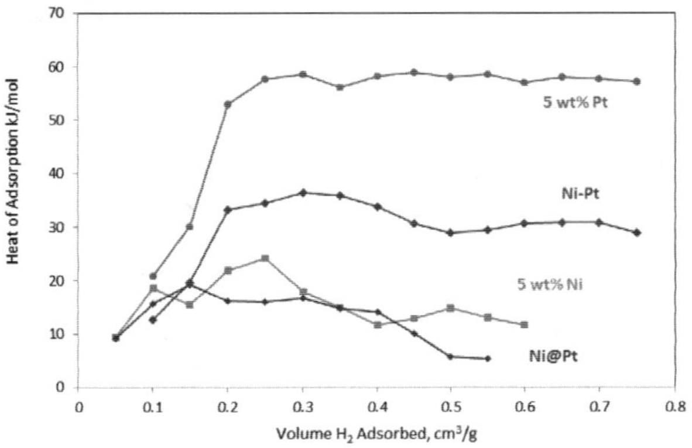

- Characterization tool to examine catalyst binding properties.
- Obtain H_2 adsorption isotherms at multiple temperatures.
- Use Clausius-Clapeyron equation to calculate heat of adsorption as a function of coverage.

Can we change the binding properties of Pt by overlaying it on Ni?

- Ni@Pt catalyst shows deceased heat of adsorption compared to pure Pt, as desired and consistent with computational studies.
- Ni@Pt catalyst displays different behavior than the non-structured Ni-Pt bimetallic.

Ethylene Hydrogenation

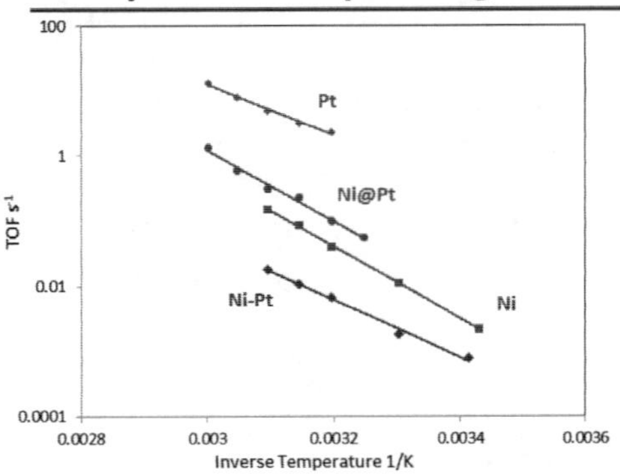

- Overlayer deposition decreases the turnover frequency from Pt, indicating decreased H surface coverage and H adsorption compared to pure Pt.
- Consistent with computational studies predicting overlayer catalysts will decrease d-band center and adsorption strength.

- Characterization tool to examine catalyst binding properties.
- Reaction is known to be a function of H surface coverage. Decreased rate indicates decreased H coverage and decreased H binding strength.

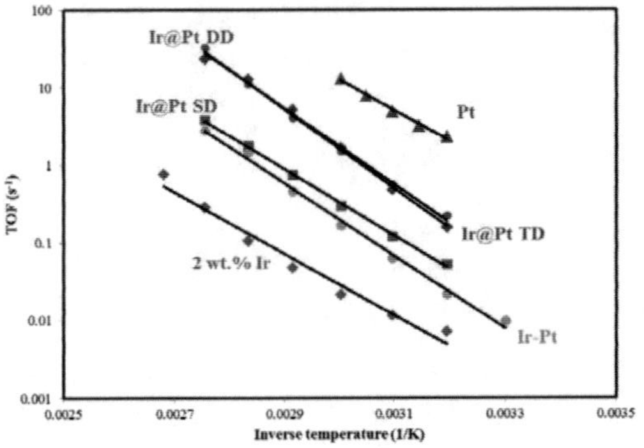

Predict higher coverage of Pt on Ir compared to Pt on Ni.

Aqueous Phase Hydrodeoxygenation (APH)

- Hydrodeoxygenation:
  - Oxygen within biomass significantly reduces the energy density.
  - A possible approach to produce biomass-derived fuels.

- Glycerol ($C_3H_8O_3$):

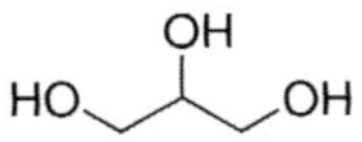

  - Byproduct from the production of biodiesel
  - One of the 12 potential biomass-based platform chemicals recognized by DOE
  - Desired products:
    - Hydrocarbons: fully-deoxygenated

Strong adsorption of H_2 and CO products are known to block sites and limit activity.

Hydrocarbon Production

For Ni@Pt and Co@Pt catalysts

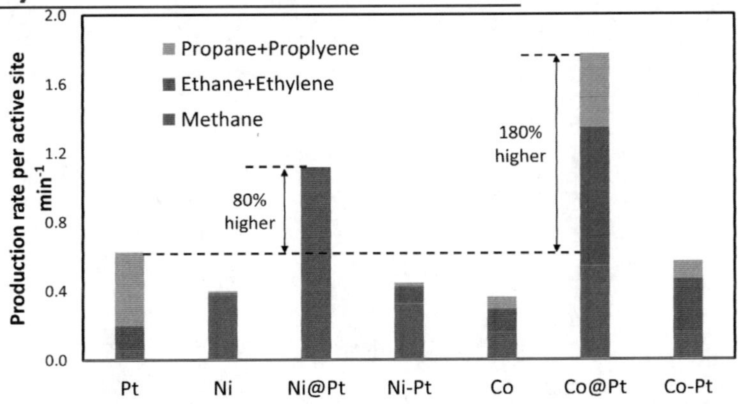

- Both Ni@Pt and Co@Pt behaved distinctly from bimetallic alloys in terms of reactivity and selectivity.
- Co@Pt shows less C-C bond breaking than Ni@Pt.
- Hydrocarbon production for overlayer catalysts is more pronounced.

Glycerol APH vs. Ethylene Hydrogenation

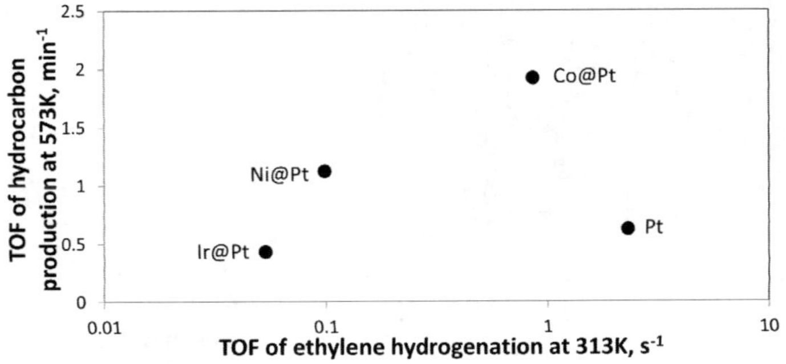

- High ethylene hydrogenation TOF indicates strong hydrogen adsorption strength, which has been suggested to reduce the catalysts reactivity.
- Low ethylene hydrogenation TOF is likely to inhibit the hydrocarbon production rate, as hydrogenation reactions are crucial intermediate steps in glycerol APH.

Hydrocarbon Production

For Ir@Pt catalysts

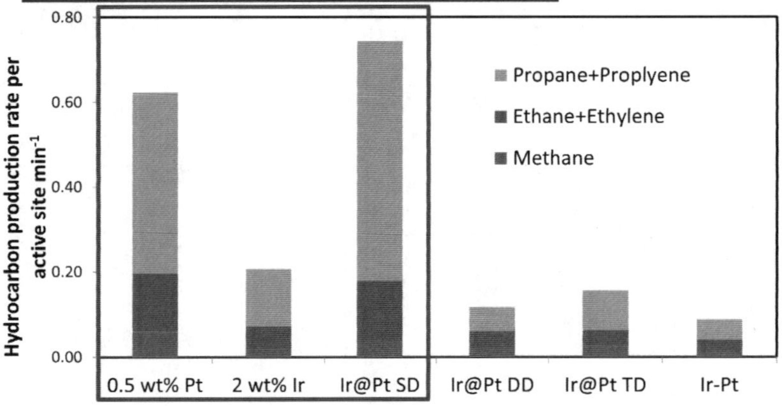

Prediction:
Due to atom sizes being more similar, Pt should more easily cover Ir in an overlayer structure compared to Ni or Co.

Result:

Ni@Pt	26% Coverage
Co@Pt	6%
Ir@Pt	43%

- Ir@Pt SD showed higher TOF for hydrocarbon production and higher C_3 hydrocarbon selectivity compared to pure Pt.
- Double and triple deposition samples showed lower reactivity than the single deposition sample.

HAADF-TEM Imaging for Ir@Pt

Elemental Mapping

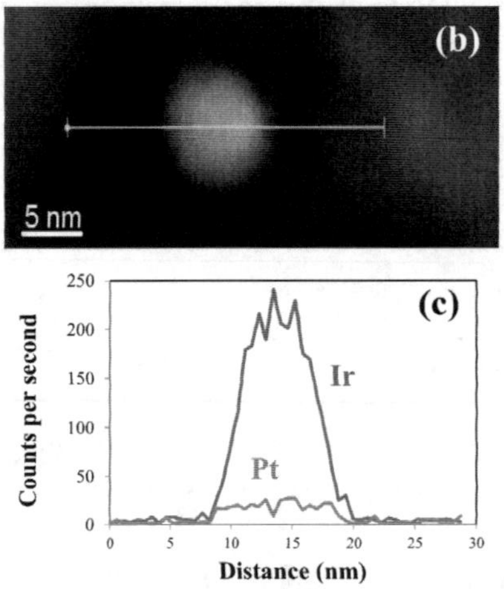

Elemental Profiling

- Elemental mapping shows the Pt well dispersed (i.e., not clumped) on the Ir base particle.
- Elemental profiling shows the Pt concentration consistent across the Ir base particle.

Evidence for Electronic Change

- XANES can be used to investigate electronic structure.
- White line height changes as function of d-band.
    - Decreased height → d-band narrowed.
    - Increased height → d-band broadened.

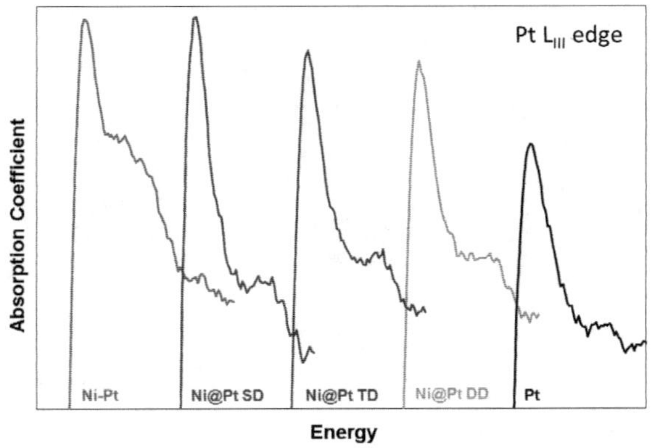

- White line height for Ni@Pt catalysts changes as additional Pt is deposited.
- White line height increased compared to Pt, as expected for overlayer catalysts.
- EXAFS coordination number and interatomic distance for Ni-Pt bimetallic shows Pt in a Ni matrix.
- EXAFS coordination number and interatomic distance for Ni@Pt overlayer catalysts shows Pt consistent with a surface structure.
- Ni@Pt TD shows evidence for Pt agglomeration.

Ni@Pt catalysts are clearly not "standard" bimetallic catalysts.

APR Reforming of Lactose

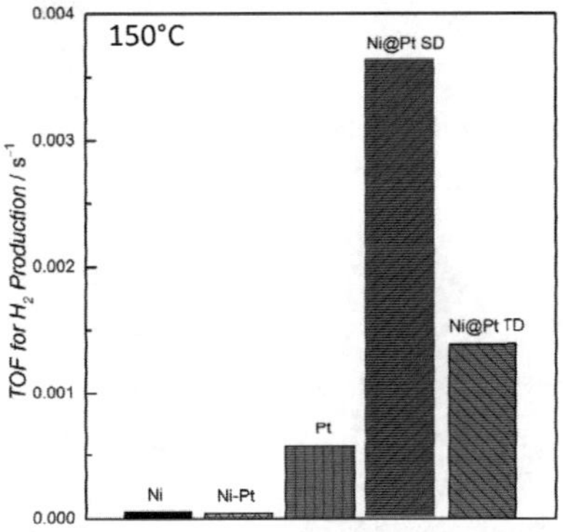

- When normalized by Pt sites the Ni@Pt catalyst shows improved APR activity.
- Ni@Pt TD shows decreased activity compared to Ni@Pt SD, consistent with Pt agglomeration observed by XAS.

- CO and H_2 products are known to adsorb strongly and block sites.
- Decrease Pt binding strength to increase activity?
- Ni@Pt should decrease Pt binding strength.

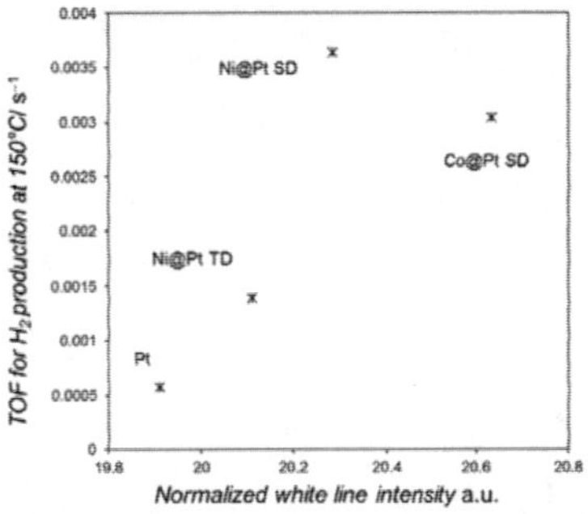

Correlation of H_2 production with white line intensity.

Guaiacol HDO Results

- Guaiacol is known to adsorb strongly and block sites.
- Decrease Pd binding strength to increase activity?
- Ni@Pd should decrease Pd binding strength.

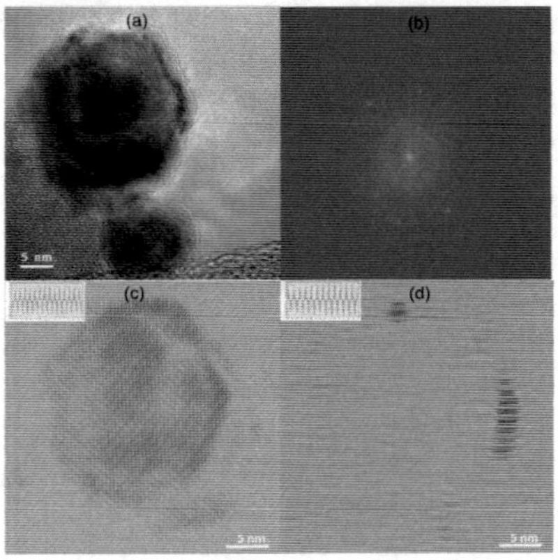

- (c) shows inverse Fourier transform lattice fringes consistent with Ni base metal particle.
- (d) shows lattice fringes around the particle consistent with Pd and show evidence for surface deposition of the Pd.

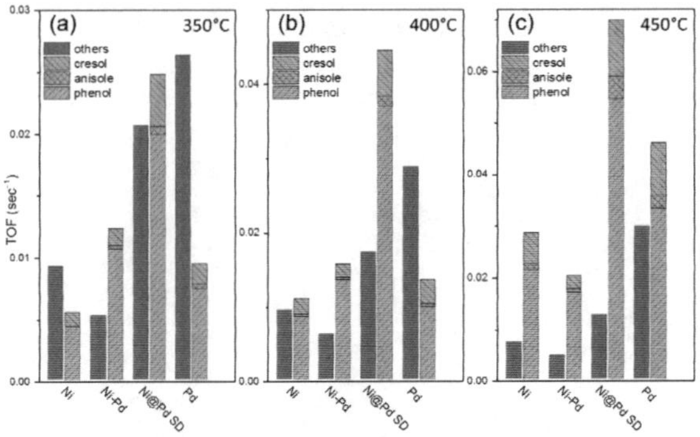

Ni@Pd catalyst shows improved guaiacol HDO activity compared to pure Pd catalyst.

Catalyst Stability

- For Guaiacol HDO.

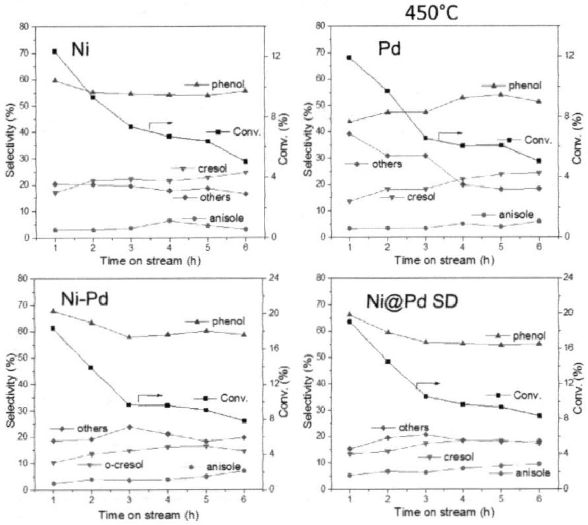

- Initial conversion for the Ni and Pd catalysts is ~12%.
- Conversion for the Ni and Pd catalyst decreases over 6 hours to ~5%.
- Conversion for the Ni@Pd catalyst starts at 19% and ends greater than 8%.

Ni@Pd was able to maintain higher conversion than pure Pd over a longer time.

Change of Direction

All work so far has focused on reducing binding strength. Can we increase binding strength?

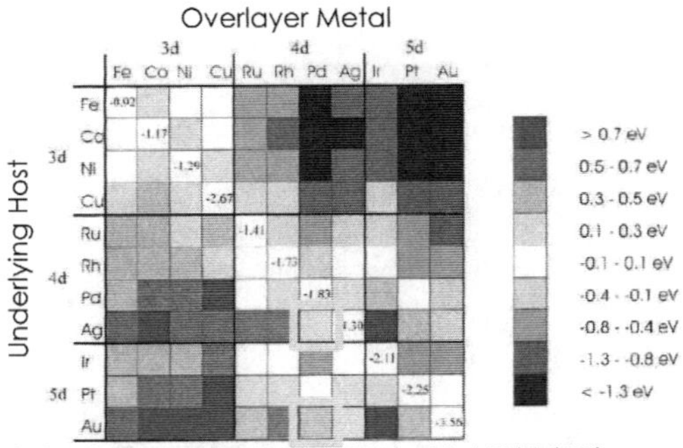

[E. Christoffersen, P. Liu, A. Ruban, H.L. Skriver, J.K. Norskov, J. Catal. 199 (2001) 128.]

- Literature results suggest that Au@Pd or Ag@Pd would increase binding strength.
- We could then expect ethylene hydrogenation activity to increase.

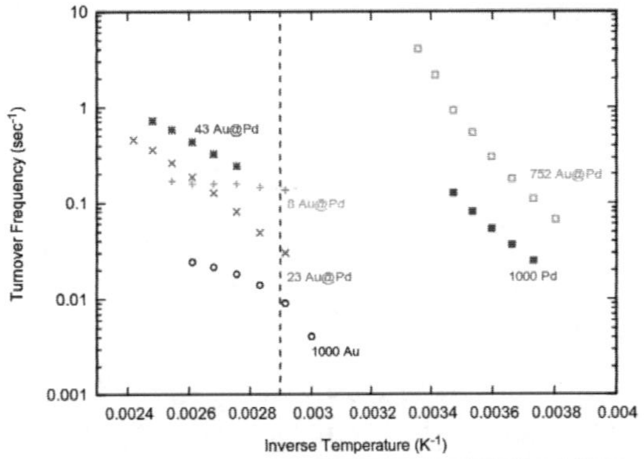

Pd overlayer on Au does indeed increase ethylene hydrogenation activity compared to pure Pd.

Conclusions

- Heat of adsorption can be deceased, as predicted, for specific overlayer catalysts.
- Consistent with decreased adsorption strength, ethylene hydrogenation activity can be decreased.
  - These two techniques serve as descriptors for overlayer catalysts.
- Overlayer catalysts to decrease binding strength resulted in increased reactivity for three systems with known binding issues:
  - Glycerol APH using Ni@Pt catalysts
  - Lactose APR using Ni@Pt catalysts
  - Guaiacol HDO using Ni@Pd catalysts.
- Overlayer catalysts can also increase binding strength Au@Pd compared to pure Pd.

Acknowledgements

Funding provided by National Science Foundation, Chemical, Bioengineering, Environmental and Transport Systems [CBET-0933017]

National Science Foundation
WHERE DISCOVERIES BEGIN

University of Wyoming, Academic Affairs Energy GA

University of Wyoming, School of Energy Resources GA

Doctoral Students:
- Michael Latusek
- Michael Skoglund
- Allen Morris
- Chen Zhang
- Qinghua Lai

Masters Students:
- Patrick Duncan
- Shamil Sabirzyanov
- Cami Andrie Jackson
- Heather Young Olson

Undergraduate Students:
- Weston Hubele
- Samuel van Meter
- Kaitlyn McKim

The roles of Re and Cs promoters for Ag-based, high selectivity ethylene oxide catalysts

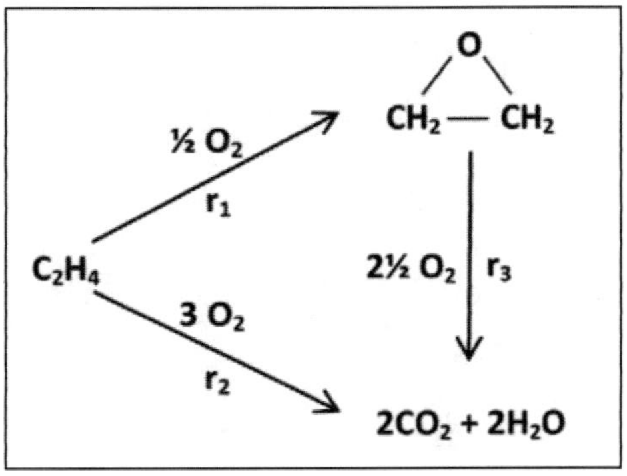

John Monnier, Weijian Diao, Jayson Keels, Christopher DiGiulio, and Melanie Schaal

Department of Chemical Engineering

University of South Carolina

2017 AIChE Annual Meeting

October 29 - November 3, 2017

UNIVERSITY OF SOUTH CAROLINA

US production volume from heterogeneously-catalyzed oxidation of hydrocarbons

(based on 2007 data and updated to ~2014))

- Oxidation represents ~25% of volume of top 50 organic chemicals.
- Allylic olefin oxidation products include acrylonitrile, acrylic acid, methacrylic acid.
- Paraffin oxidation primarily maleic anhydride.
- Aromatic oxidation includes phenol and oxidations at benzylic positions.
- Only <u>one</u> reaction accounted for ~44% of volume—ethylene oxide!
- Another 10-15 Blb/yr to come in line on TEX and LA gulf coast by 2019—cheap C_2H_4.
- Will boost production to 75B lb/yr.

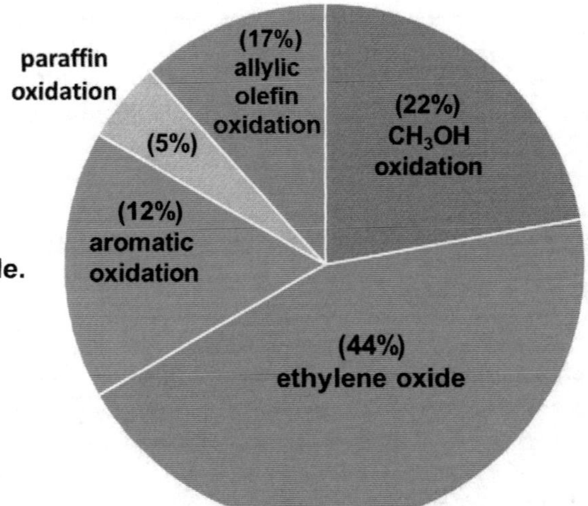

UNIVERSITY OF SOUTH CAROLINA

Very brief history of EO catalysts

Some figures:

- In 2014, 50B lbs/yr of EO and derivatives were produced globally.
- One of the "top 25" volume industrial chemicals (No. 6 in latest ranking).
- 2nd largest volume chemical produced where hetero-atom added to C_xH_y.

Catalyst development history

- Direct epoxidation of C_2H_4 using Ag and O_2 discovered by Lefort in 1931 (S_{EO} ~ 50%).
- Commercialized by Union Carbide in 1937.
- Shell, SD, and Dow/Carbide are currently major catalyst producers.
- Until mid-80's catalysts were "high activity", Cs-promoted, Ag catalysts with S_{EO} < 80%.
- In 1987, Shell introduced first "high selectivity" Re-promoted catalyst.
- First Re patent in 1984. <10 journal articles have been published about role of Re.
- > 200 journal articles for non-Re, Ag catalysts published during the same time period.

UNIVERSITY OF SOUTH CAROLINA

Value of high selectivity EO catalysts

- **For 50B lbs/yr of production**
- **Recent C_2H_4 prices @ \$980/metric ton, or 0.45\$/lb.**

	C_2H_4 amount	C_2H_4 cost	Capital costs for added capacity due to select. loss	Δ CO_2 production
For 78% EO process	41B lbs	\$18.3B	\$1B	+15.7B lbs
For 89% EO process	36B lbs	\$16.0B	--	---
Savings:		2.3B	\$10B for 1st yr	Cost of CO_2 capture and removal

	O_2 amount	Cost at \$36/ton
For 78% process	50.2B lb	\$0.94B
For 89% process	32B lbs	\$0.60B

UNIVERSITY OF SOUTH CAROLINA

Background of ethylene epoxidation reaction

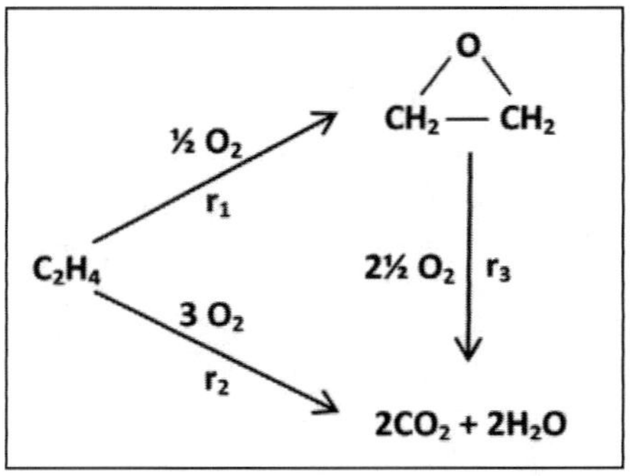

	ΔH (kJ/mol)
• r_1 is addition of O to C_2H_4.	−105
• r_2 is direct combustion of C_2H_4.	−1326
• r_3 is sequential combustion of EO.	−1222

- **Ag, Cs, Re** are major components for current generation EO catalysts.
- Ag is only metal that adsorbs and dissociates O_2 to give O species capable of electrophilic addition, rather than normal nucleophilic O which forms CO_2.
- After many yrs of arguments, atomic O and not molecular O_2, confirmed to be active oxygen for electrophilic addition to C=C.
- Cs effect has been argued to be either site blocking or electronic. Most data support electronic effect.
- We argue that Re effect also electronic.

UNIVERSITY OF SOUTH CAROLINA

Implications of consecutive-parallel pathways

(for old generation of Cs-Ag catalysts)

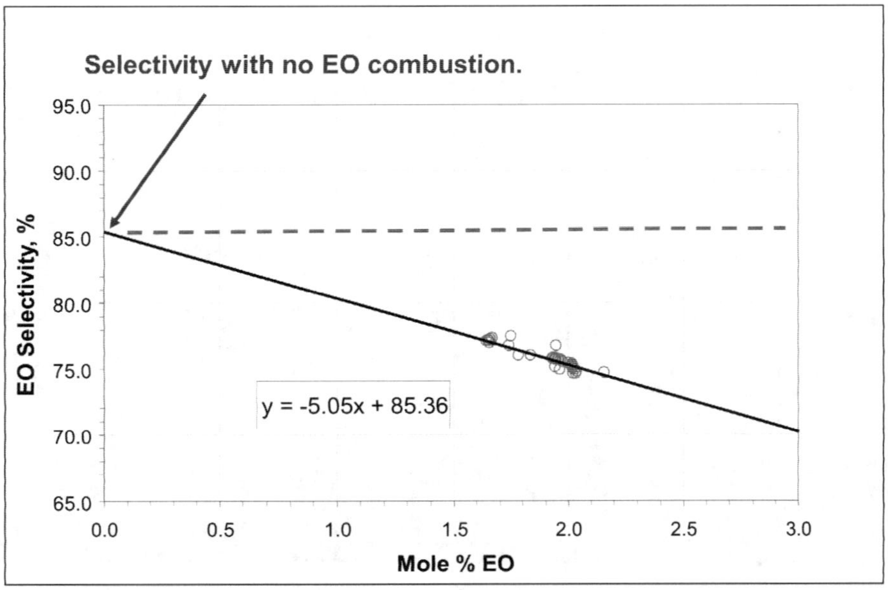

UNIVERSITY OF SOUTH CAROLINA

First disclosure of Re-promoted catalysts

Lauritzen, Shell Chemical, US4766105 (1984).

Salt	Amount	Cat A	Cat B	Cat C	Cat D	Cat E
Ag oxalate	Ag, %	13.6	13.5	13.5	13.5	13.8
CsOH	Cs, ppm	0	230	420	410	0
NH_4ReO_4	Re, ppm	0	0	372	186	186
$(NH_4)_2SO_4$	S, ppm	0	0	0	32	32
Performance						
EO Sel, %		74.6	80	81.9	82.9	61.2
T_{40}, degC		229	242	248	253	232

- **Use of Re requires more Cs.**
- **Re is necessary but not sufficient for optimum performance.**

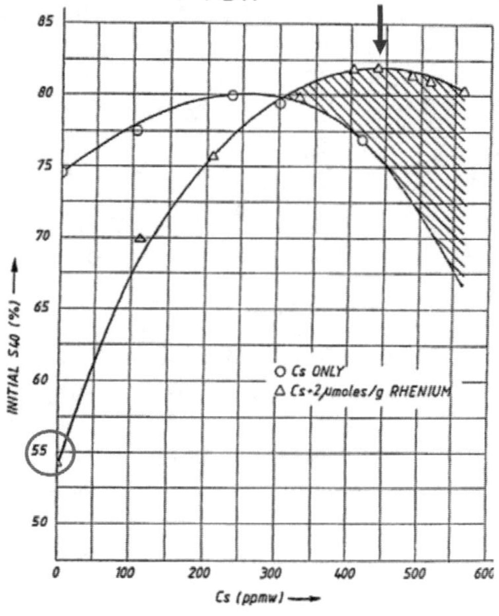

UNIVERSITY OF SOUTH CAROLINA

Catalyst Preparation

- α-Al_2O_3 catalyst carrier (8mm rings, BET surface area = 0.60 m^2g^{-1}, and pore volume = 0.53 cm^3/g) from Norpro-St. Gobain.
- Ag source was ethylenediamine-solubilized **$Ag_2C_2O_4$**.
- **$CsNO_3$**, **NH_4ReO_4** and, optionally, **$(NH_4)_2MoO_4$** and **$(NH_4)_2SO_4$** added to $Ag_2C_2O_4$-containing impregnation solution.
- Rings tumbled dry at 60°C and then calcined in forced air flow (100 L/min) at 260°C for a total of 5 – 7 mins.
- For evaluation, catalyst rings broken and sieved to 20/40 mesh (850 – 425 μm).

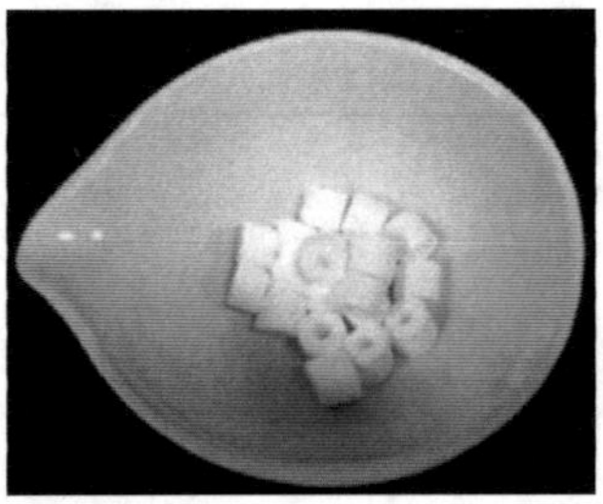

α - Al_2O_3 ring

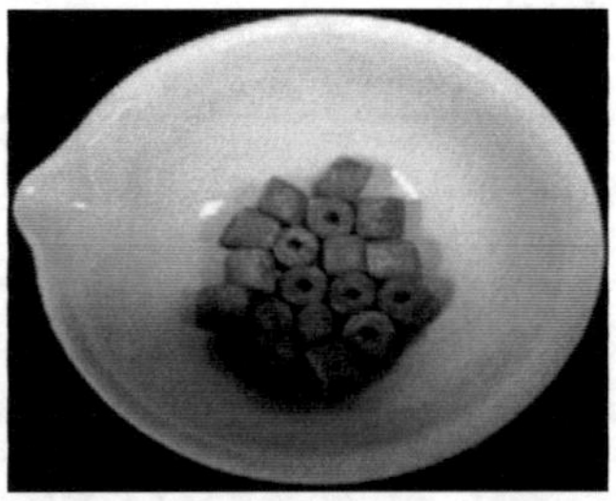

Ag-Cs-Re-Mo/α-Al_2O_3 catalyst

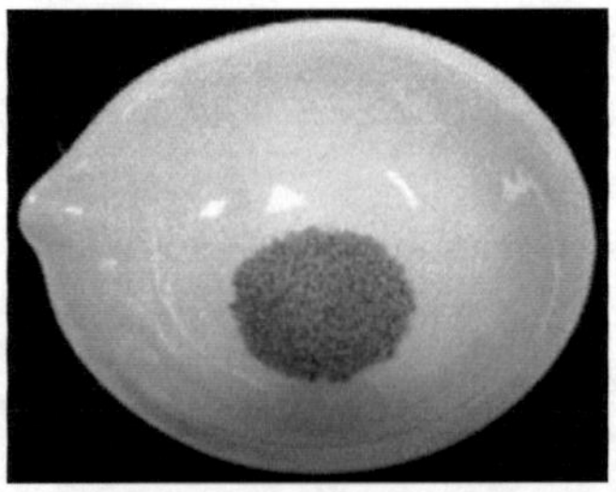

Sieved catalyst for reaction

UNIVERSITY OF SOUTH CAROLINA

Catalyst Evaluation Protocol

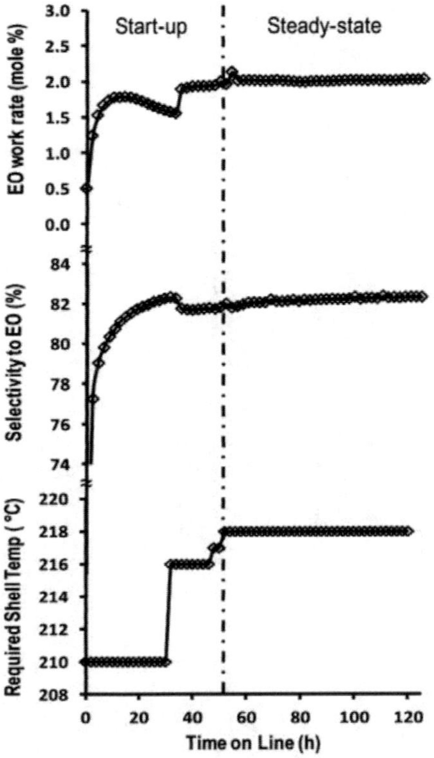

- Catalyst wt: 2.0 g of sieved 20/40 mesh particles.
- <u>Reaction conditions</u>:
  - ¼" OD SS reactor with 1" press-fit Al shell.
  - Feed: 8% O_2, 25% C_2H_4, bal. CH_4, 1 – 5 ppm EtCl.
  - GHSV = 4500 hr^{-1}.
  - P = 250 psig.
- Start-up conditions typically 210°C and 2 ppm EtCl.
- After 30-50 hrs, temp adjusted to achieve 2.0 mole % EO in product stream.
- Catalyst evaluated for period of time to obtain steady-state performance (70 – 400 hrs).
- Reaction times up to 1200 hrs.
- Each reactor sampled every 2 hrs.

UNIVERSITY OF SOUTH CAROLINA

Optimization of Cs, Re and Mo

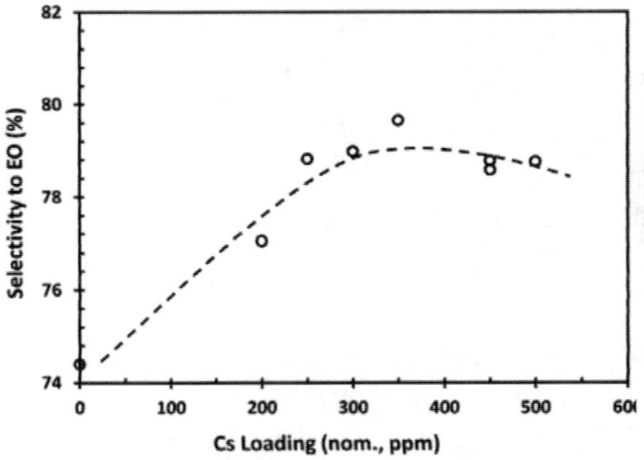

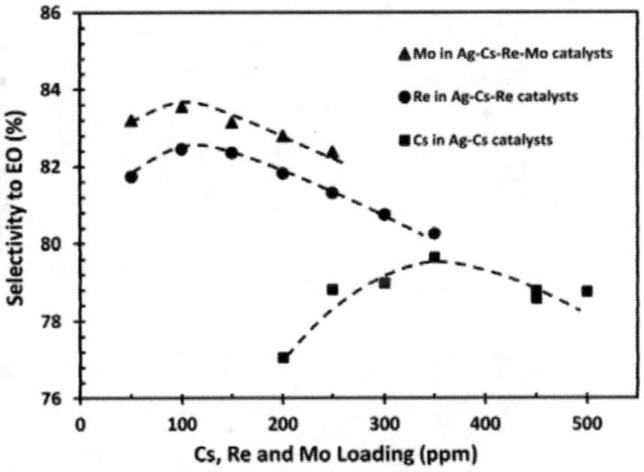

Catalyst Composition	C_2H_4 Conversion (%)	EO work Rate (mole %)	EO Selectivity (%)	Temp. (°C)
12% Ag	10.5	1.94	74.5	224
12% Ag, 350 ppm Cs	10.3	2.00	79.7	224
12% Ag, 350 ppm Cs, 200 ppm Re	10.1	2.04	81.7	214
12% Ag, 350 ppm Cs, 200 ppm Re, 150 ppm Mo	9.8	2.02	83.1	224

UNIVERSITY OF SOUTH CAROLINA

Optimization of Cs, Re and Mo

Cumulative effect of high-valent Re, Mo, S promoters
(different families/series of catalysts)

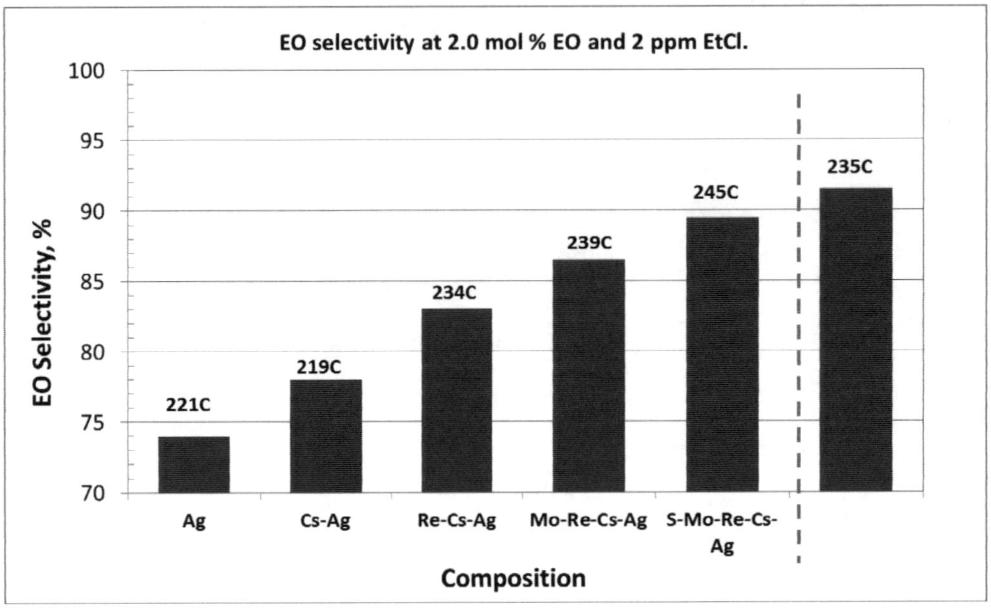

UNIVERSITY OF SOUTH CAROLINA

SEM images for EO catalysts

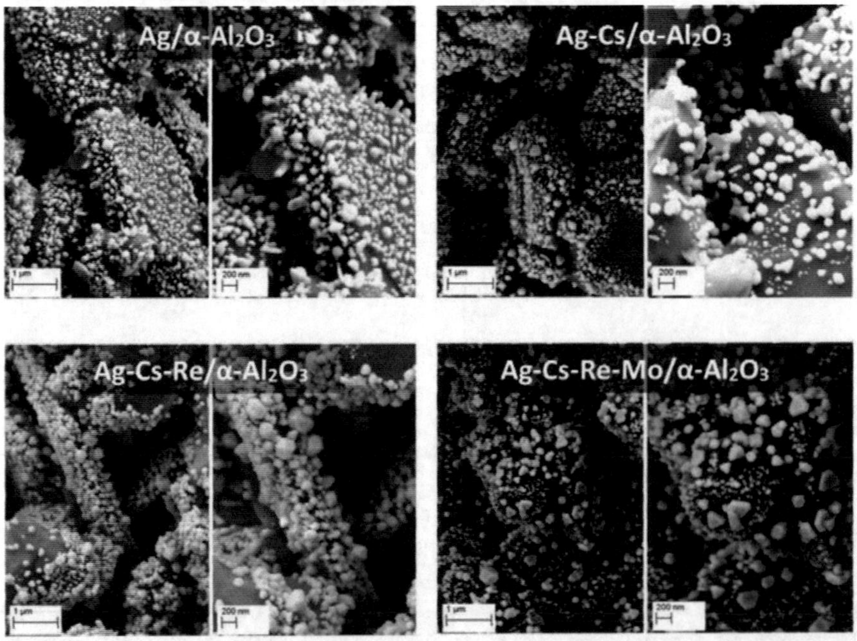

- Wide variety of geometric shapes, including hexagons, octagons, pyramids, triangular platelets, spherical particles.
- No obvious trends in particle sizes or shapes were observed.
- Ain't no nanoparticles here!

UNIVERSITY OF SOUTH CAROLINA

XPS of Ag $3d_{5/2}$ in 12% Ag/α-Al_2O_3 catalyst

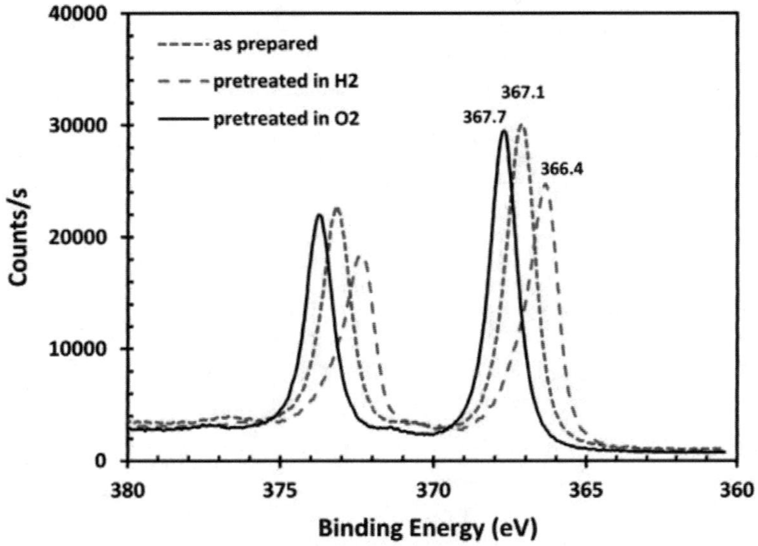

- <u>The $3d_{5/2}$ values:</u>

  After reduction at 280°C for 12 hrs = 366.4 eV **(Ag^o)**

  After calcination at 280°C for 12 hrs = 367.7 eV **(Ag^+)**

  As prepared = 367.1 eV **($Ag^{\delta+}$)**

- **Use as reference values for promoted catalysts.**

UNIVERSITY OF SOUTH CAROLINA

XPS of Ag $3d_{5/2}$ in Re, Cs-Ag/α-Al_2O_3 catalyst at <u>rxn</u> conditions

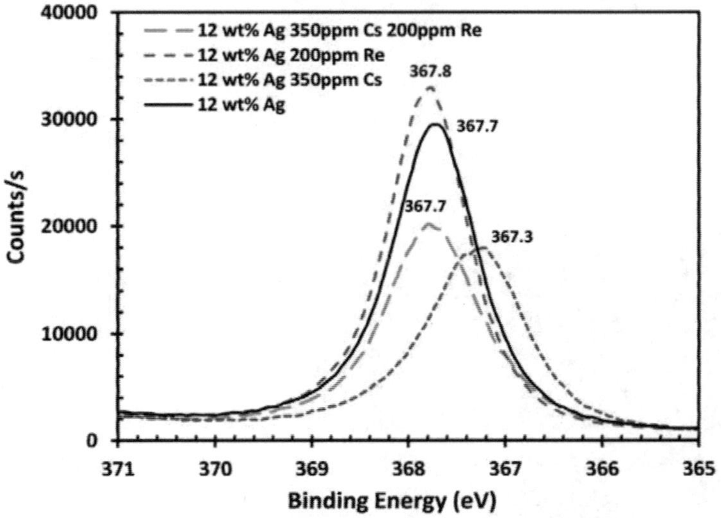

Catalyst	Ag $3d_{5/2}$ (eV)
Ag	367.7
Cs-Ag	367.3
Re-Ag-Cs	367.7
Re-Ag	367.8

- Addition of 350 ppm Cs lowers Ag $3d_{5/2}$ BE from 367.7 eV → 367.3 eV, suggesting e^- transfer from highly polarizable Cs^+ to e^- - deficient $Ag^{\delta+}$.
- Ag $3d_{5/2}$ BE value shifted back to 367.7 eV when 200 ppm Re is added to Cs-Ag, indicating interaction of Re^{n+} cation with $Ag^{\delta+}$ offsets opposing effect of Cs^+.

UNIVERSITY OF SOUTH CAROLINA

XPS of Re $4f_{7/2}$ in Re-promoted Cs-Ag/α-Al_2O_3 catalyst

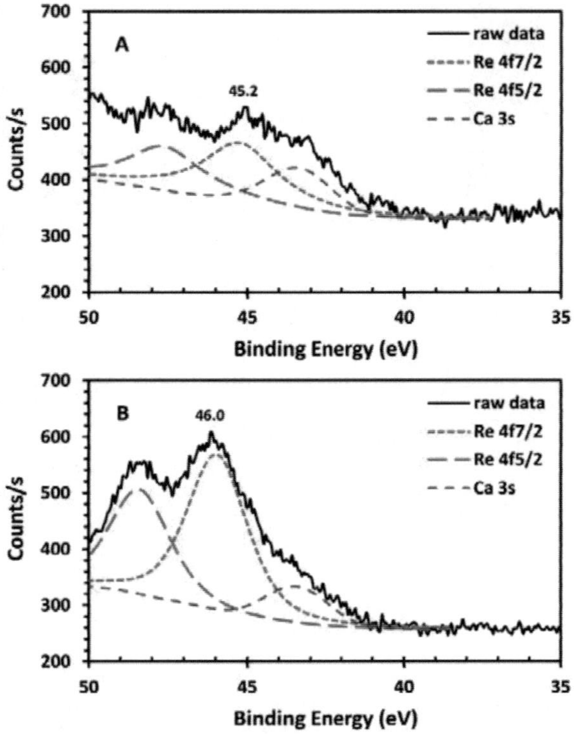

A. As prepared: BE at 45.2 eV.

Both Re^{6+} and Re^{7+} are present.

B. After calcination at 280°C:

BE = 46.0 eV.

Re oxidized to Re7+

UNIVERSITY OF SOUTH CAROLINA

Effect of alkali metal cations on EpB formation

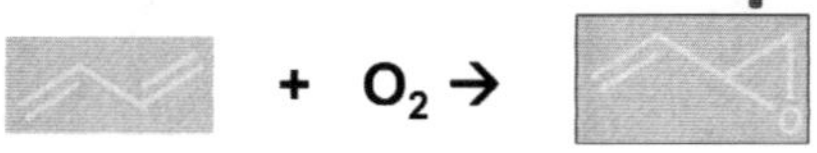

Selectivities at optimum loading

none	45%
CsCl	94.3%
RbCl	93.5%
TlCl	90.0%

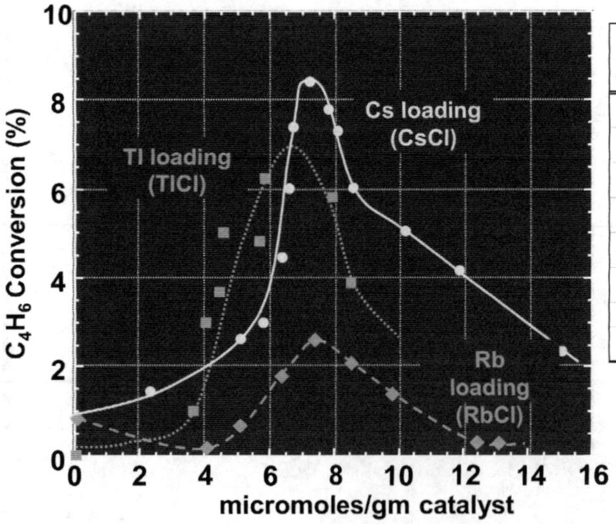

Cation (+1)	Ionic radius (Å)	Pauling polarizability
Lithium	0.78	0.03
Sodium	0.98	0.03
Potassium	1.33	1.1
Rubidium	1.49	1.9
Cesium	1.65	2.9
Thallium	1.40	4.3
Silver	1.26	

Monnier, J. Catal., 226, 401 (2004).

UNIVERSITY OF SOUTH CAROLINA

Physical/electronic promoters of Re and co-promoter oxy anions

Promoter	Oxid. state of central atom	Ionic radius (A)	M-O bond length (A)	Geometry
ReO_4^-	7	0.53	NA	T_d
$ReCl_6^{2-}$	4	0.63	NA	O_h
MoO_4^{2-}	6	0.59	~1.7	T_d
WO_4^{2-}	6	0.60	~1.7	T_d
SO_4^{2-}	6	0.29	1.49	T_d
CrO_4^{2-}	6	0.44	1.60	T_d
ClO_4^-	7	0.27	1.46	T_d
SeO_4^{2-}	6	0.42	NA	T_d

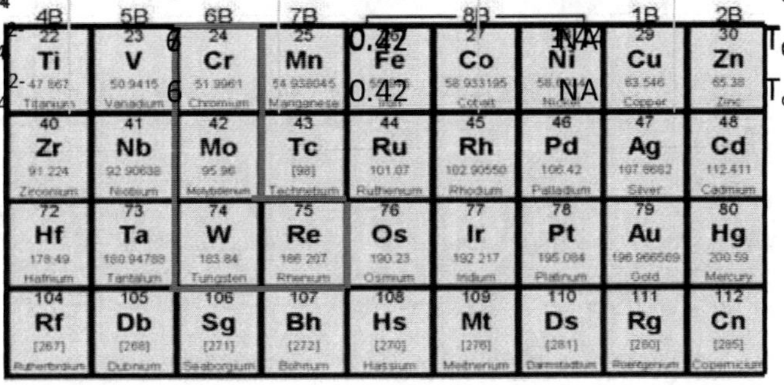

UNIVERSITY OF SOUTH CAROLINA

Cooperative roles of Cs and Re

If Re is responsible for high selectivity, is Cs even needed?

Catalyst Composition	EO conc (mole %)	EO Select (%)	Temp (°C)	C_2H_4 conv (%)	O_2 conv (%)
(1) 12% Ag	1.94	74.5	224	10.4	28.7
(2) 12% Ag, 350 ppm Cs	2.00	79.7	224	10.3	26.7
(3) 12% Ag, 100 ppm Re	1.93	49.9	262	15.9	60.5
(4) 12% Ag, 350 ppm Cs, 100 ppm Re	2.03	82.4	218	9.8	23.5
(3') 12% Ag, 100 ppm Re	1.6	60.7	250	11.4	41.5
(3'') 12% Ag, 100 ppm Re	0.35	79.5	225	1.8	4.4

- Re is necessary, but not sufficient.
- Cs offsets the selectivity diminishing effects of Re-only promoted Ag. EO select increases from 49.9% to 82.4% when Cs is added to Re-Ag catalyst.
- Why? From XPS and composition-performance correlations, Cs and Re must promote different elementary steps of the reaction mechanism.

UNIVERSITY OF SOUTH CAROLINA

Proposed mechanism of high selectivity EO formation

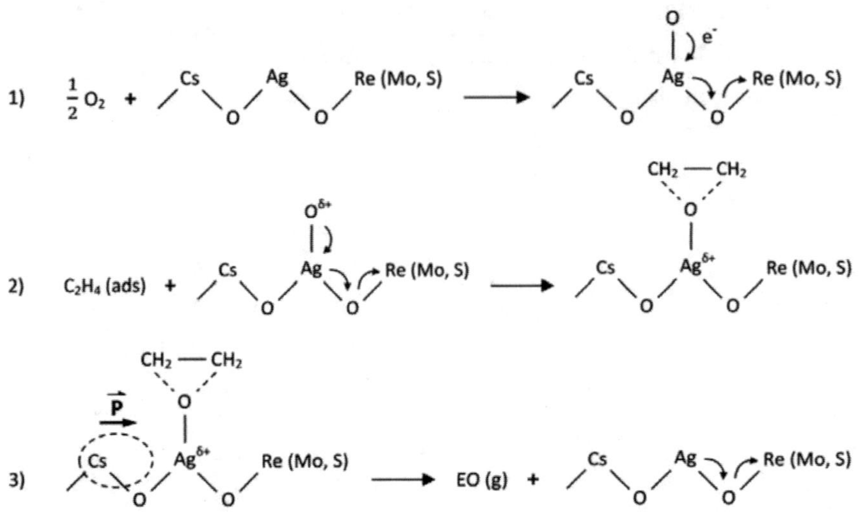

- Step 1: High-valent Re^{7+} draws e^- density from Ag to increase electrophilicity of adsorbed O.
- Step 2: Electrophilic attack of O at the e^- rich C=C bond favored over nucleophilic attack at C-H bond.
- Step 3: Cs assists in desorption of EO due to high polarizability, giving higher EO selectivity.

UNIVERSITY OF SOUTH CAROLINA

More complete reaction network

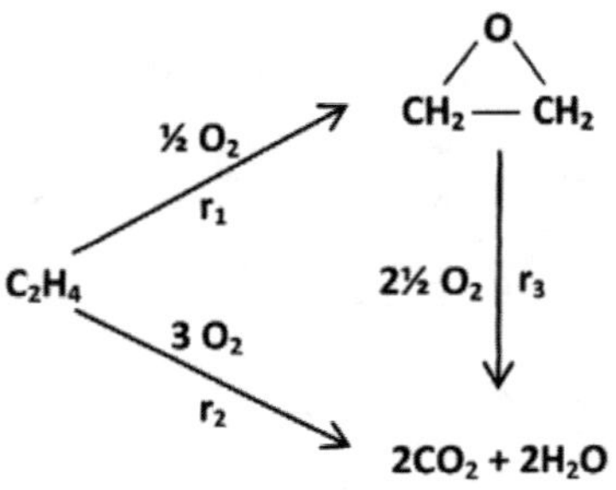

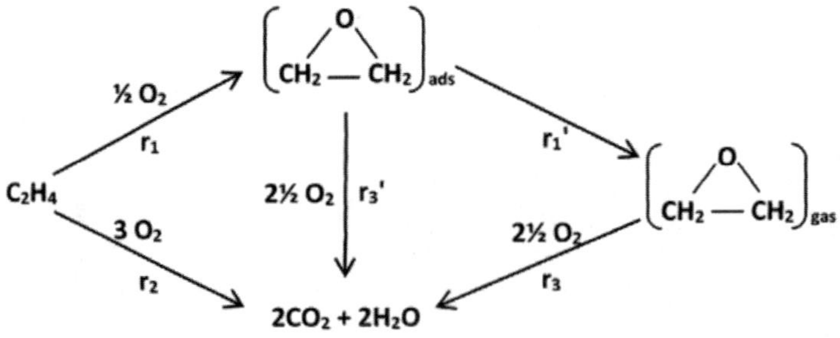

r_1 : electrophilic O attack at C=C.
r_2: nucleophilic O attack at C-H.
r_3: oxidation of gas phase EO (CH_3CHO).

r_1: Re increases electrophilic attack at C=C. (good)
r_2: Re decreases nucleophilic attack at C-H. (good)
r_3': Re increases combustion of adsorbed EO. (bad)
r_1': Cs increases desorption of EO. (good)

$EO/CO_2 = r_1/(r_2 + r_3)$

$EO/CO_2 = (r_1 + r_1')/(r_2 + r_3 + r_3')$

UNIVERSITY OF SOUTH CAROLINA

Illustration of modified reaction network

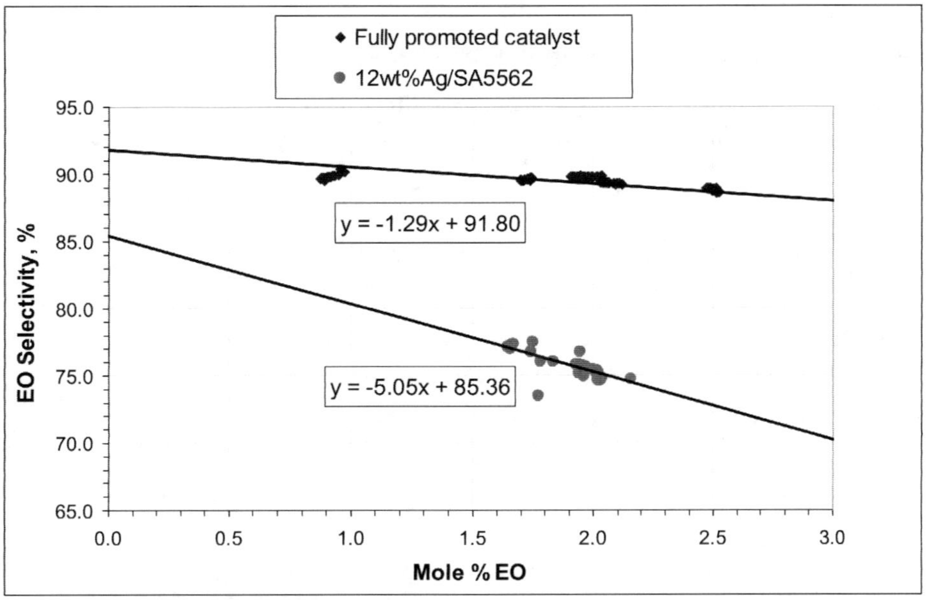

- **Re and co-promoters lower nucleophilic loss from 15 → 8%.**
- **Flatter slope indicates lower combustion of EO.**

UNIVERSITY OF SOUTH CAROLINA

Effect of Re on rate of catalyst activation

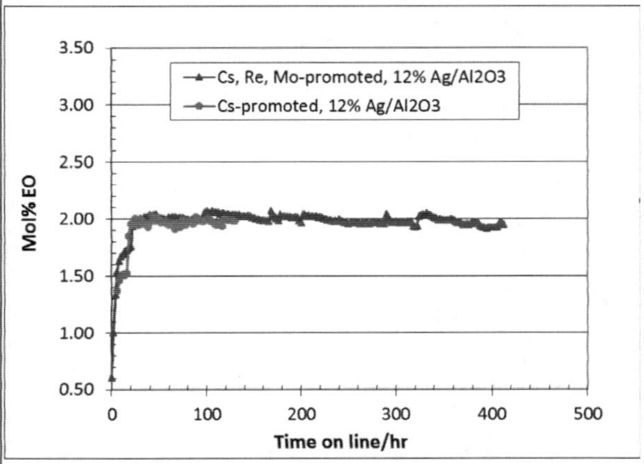

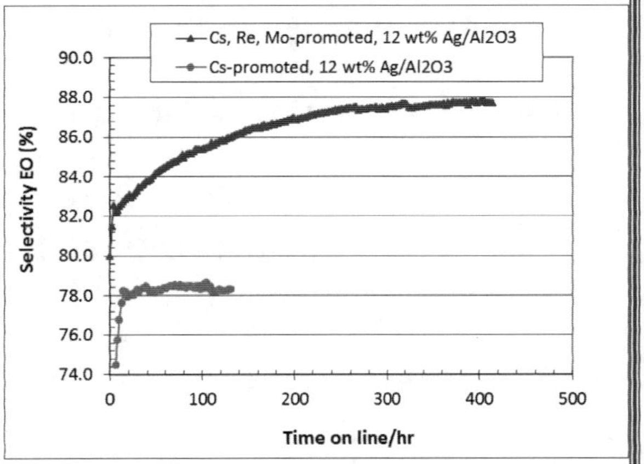

<u>Catalysts</u>:

(1), 350 ppm Cs, 12% Ag/α-Al_2O_3. Temp 199 → 214°C to maintain 2% EO.

(2), 350 ppm Cs, 350 ppm Re, 100 ppm Mo, 12% Ag/ α-Al_2O_3. Temp 234 → 239°C to maintain 2% EO.

Why long activation period for Re-modified catalysts?

- Re effect is realized only after many hrs on-line.
- Very different surface free energies (SFE) for this catalyst. At ~230°C,

Material	σ, (ergs/cm²)
Ag	1302
Al_2O_3	832
MoO_3	97
Re_2O_7	38

- Surface thermodynamics seeks lowest free energy of complete system.
- Lowest surface energy when ReO_x is distributed on Ag (highest SFE).
- Highest EO selectivity when Ag surface is fully promoted.

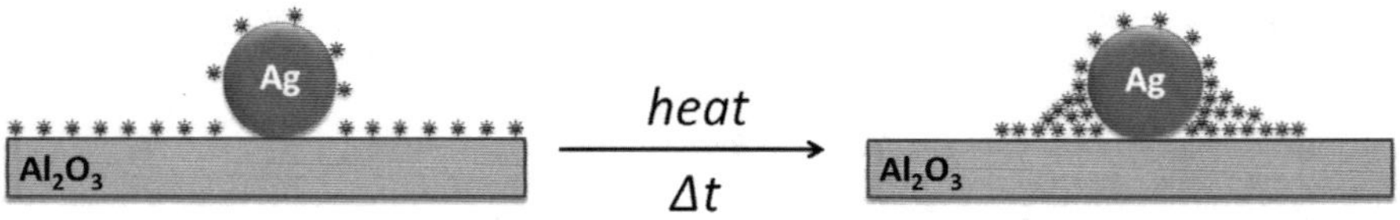

UNIVERSITY OF SOUTH CAROLINA

Conclusions

- Re and Cs promoters in Ag catalysts improve EO selectivity.
- Coherent mechanism explaining roles of Re and Cs has been developed.
- Both roles are electronic in nature.
  --Re increases electrophilicity of O adsorbed on adjacent Ag site.
  --Cs lowers desorption energy of adsorbed EO precursor.
- Re migrates from support to Ag—longer activation period for high select.
- No fundamental reasons why EO selectivity can't be improved further.
  -- Must ensure all Ag surface sites are modified by high-valent oxyanions.
  -- All surface sites also need Cs to lower EO desorption energy.

UNIVERSITY OF SOUTH CAROLINA

Acknowledgements

Former students

Jayson Keels

Melanie Schaal (now at UOP)

Jay Rebelli (now at Evonik)

Travis Gott (post-doc, now at Eastman)

Chris Digiulio (now at UOP)

Weijian Diao (Idaho Nat'l Labs, now Res. Prof at USC)

Funding:

University of South Carolina Aspire II Fund

Industrial support

Thanks again to CRE Division of AIChE

UNIVERSITY OF SOUTH CAROLINA

494281 The Surprising Accuracy of Dispersion-Corrected Ggas in the Prediction of Dissociation Barriers on Transition Metal Surfaces

Abstract:

The development of density functional theory (DFT) for accurate description of molecule-surface interactions, while essential for heterogeneous catalysis and energy conversion, is severely limited by the absence of well-defined, chemically accurate benchmark databases for reactions on surfaces. Therefore, unlike gas phase barriers for which density functionals have been extensively benchmarked, limited information is available on computational accuracies for surface reactions. We have developed a database of ten experimental apparent barrier heights (SBH10) for dissociation of small molecules such as H_2, N_2, and CH_4 on transition metal surfaces. The barriers are carefully chosen from combined molecular beam experiments and quantum dynamics simulations, laser assisted associative desorption (LAAD) experiments, and thermal rate measurements.

We have also benchmarked the performance of a dispersion-corrected GGA, BEEF-vdW, a metaGGA, MS2, and a screened hybrid functional, HSE06. In the gas phase, GGAs systematically underestimate barriers due to self-interaction errors that lead to incorrect charge separation at the transition state. By including exact exchange, hybrid functionals correct for these errors and lead to better barrier estimates. The trends are reversed for reactions occurring on transition metal surfaces. Our benchmarking study shows that the BEEF-vdW functional (mean error < 0.2 eV) outperforms MS2 and HSE06, which underestimate barriers by 0.3 eV, and 0.5 eV, respectively. This is because transition states for dissociative adsorption closely resemble the final chemisorbed states, and errors in barrier heights mirror errors in chemisorption energies. In other words, dissociation transition states are bound to the surface, while gas phase transition states are isolated species. The key driver for functional accuracy for reactions on surfaces, therefore, is the description of surface-adsorbate interactions, and not charge separation or self-interaction correction. A functional that can accurately predict chemisorption energies, like BEEF-vdW, is also reliable for estimating barrier heights on surfaces.

Adsorbate Vibrations on Transition Metal Surfaces: Applications and Theory

Joshua Lansford and D. G. Vlachos

2017 AIChE Annual Meeting, Minneapolis, MN
Computational Catalysis I: Fundamentals

University of Delaware
October 31, 2017

Advances in First-Principles Modeling

- Rapid advances over the past 20 years
  - Computer power and accurate methods
- Insights into mechanisms and kinetics using Density Functional Theory (DFT)
  - Adsorption
  - Spectroscopy (vibrational frequencies)
  - Reaction barriers
  - Rate constants
- **Microkinetic models enable direct comparison to experimental data**
- **Catalyst discovery using microkinetics is impractical**
- **The materials gap (actual catalyst structure, and active site) prohibits predictive modeling**

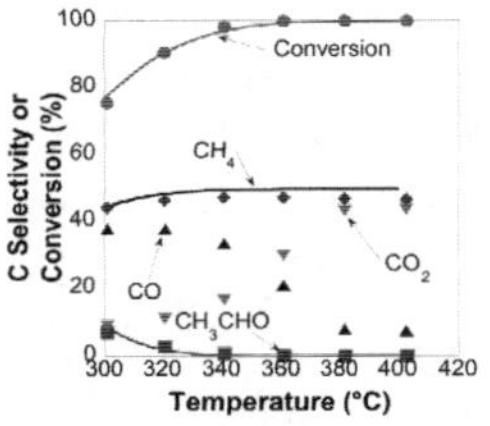

2

Adsorption Energy Linear Scaling Relations

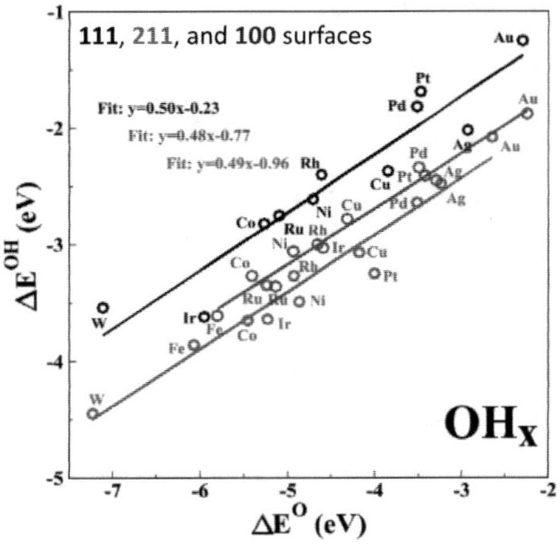

$$\Delta E^{AHx} = m_E \Delta E^A + b_E$$

Linear scaling relations improve our understanding of surface reactions and **enable catalyst screening**[1]

Extension of these relations to entropies is lacking

[1] F. Abild-Pedersen et al., Phys. Rev. Lett. 99, 4 (2007).

Catalysis Center for Energy Innovation
UNIVERSITY OF DELAWARE

Spectroscopy is Very Surface Sensitive Technique

CO on **Pt (A)** and **Pd (B)** electrodes[1]

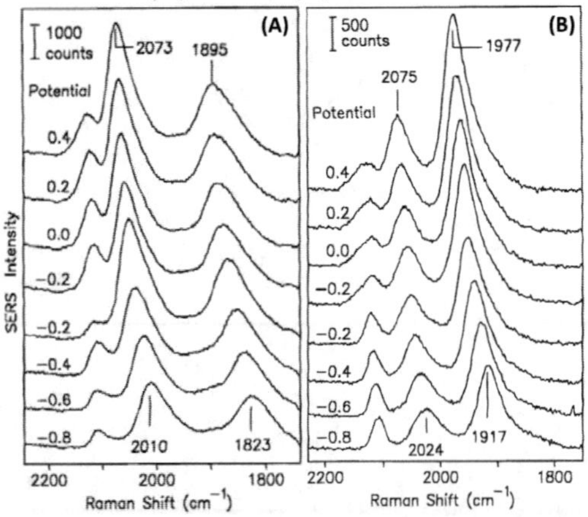

Exp. CO Frequency on Pt(111)[2]	
Site	**Freq. [cm^{-1}]**
atop	2070
bridge	1830
fcc	1760

Frequencies depend on metal and site

[1] Zou et al., J. Phys. Chem. 100 (1996)
[2] Dabo et al., J. Am. Chem. Soc. 129 (2007)

Objectives

- **Exploit scaling relations** from experiments and first-principles calculations
- **Develop** theory for vibrations
- **Assess effect of entropy** on coverages and microkinetics
- **Deduce adsorption sites and ambiguous species**

Experimental Evidence for Frequency Scaling

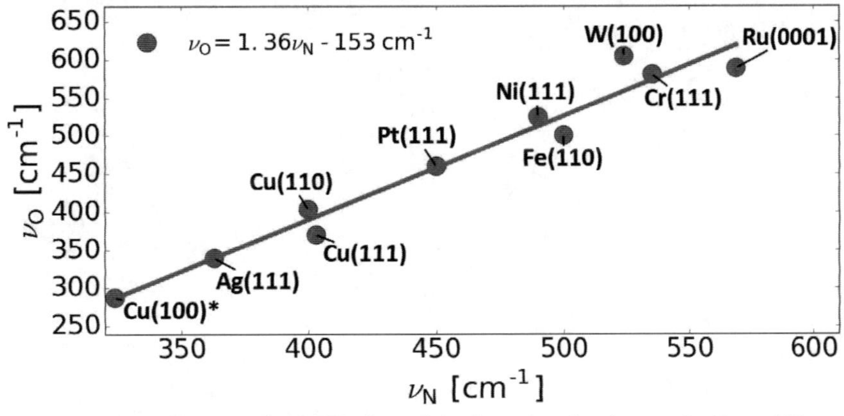

- Experimental HREEL data for chemisorbed atomic O and N

Data from 40+ different HREEL studies suggest frequencies scale across metals

J.L. Lansford, A.V. Mironenko, D.G. Vlachos, Nat. Comm. Scaling Relationships and Theory for Vibrational Frequencies of Adsorbates on Transition Metal Surfaces (Accepted)

Computational Methods

- Periodic Density Functional Theory (DFT) using the Vienna Ab initio Simulation Package (VASP)[1]
- RPBE functional with D3 dispersion corrections[2,3]
- 4-layer slabs with the bottom 2 layers fixed and 16 atoms per layer
- All vacuum space 20 Å
- 400 eV cutoff with 3x3x1 k-point sampling
- Force cutoff 0.02 eV
- Frequencies determined by diagonalizing a mass-weighted Hessian under the harmonic approximation
- **Algorithmic criteria developed for assigning frequencies to normal modes**

[1] G. Kresse and J. Furthmüller, Phys. Rev. B. 54, 11169 (1996).
[2] B. Hammer, Phys. Rev. B. 59, 7413 (1999).
[3] S. Grimme, J. Antony, S. Ehrlich, and H. Krieg, J. Chem. Phys. 132, 154104 (2010).

Catalysis Center for Energy Innovation
UNIVERSITY OF DELAWARE

Scaling of Normal Mode Vibrations from DFT

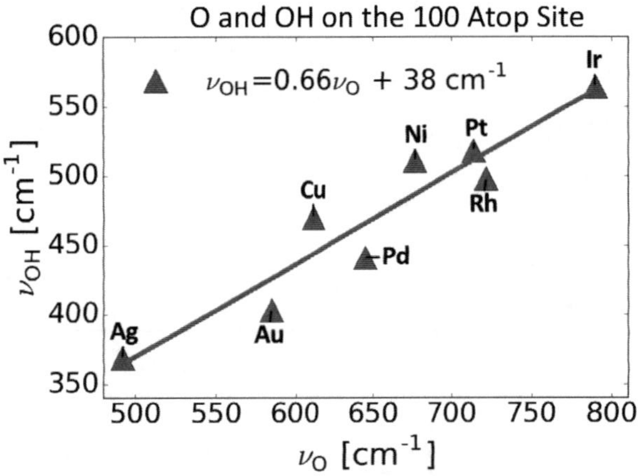

Over 500 DFT calculations reveal vibrational scaling relations (VSRs)

J.L. Lansford, A.V. Mironenko, D.G. Vlachos, Nat. Comm. Scaling Relationships and Theory for Vibrational Frequencies of Adsorbates on Transition Metal Surfaces (Accepted)

Linear Scaling Relationships (LSRs)

$$\Delta E^{AHx} = m_E \Delta E^A + b_E$$

Energy Is Additive for Transition Metals

$$\Delta E = \Delta E_{sp} + \Delta E_d$$

Effective Medium Theory (EMT)[1,2]

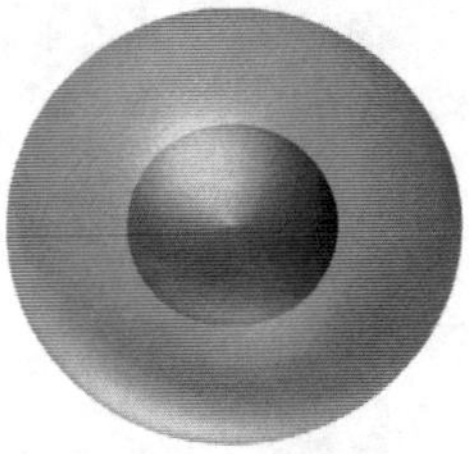

Linear Muffin Tin Orbital Theory (LMTO)[3,4]

[1] J. K. Nørskov and N. D. Lang, Phys. Rev. B. 21, 2131 (1980).
[2] J. K. Norskov, J. Chem. Phys. 90, 7461 (1989).
[3] O. K. Andersen and O. Jepsen, Phys. Rev. Lett. 53, 2571 (1984).
[4] B. Hammer and J. K. Nørskov, Surf. Sci. 343, 211 (1995).

Competitive Nature of sp and d Bands

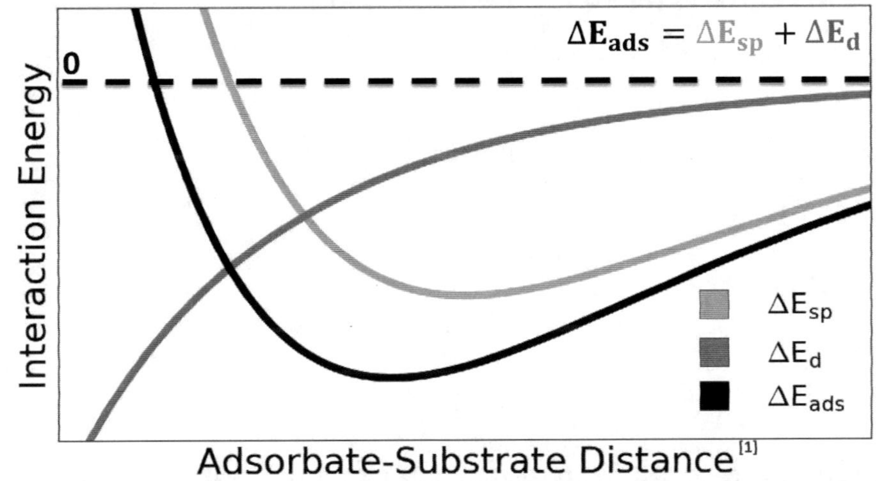

- ΔE_d at equilibrium taken from the Newns-Anderson model[2]; curves produced from EMT and LMTO

[1] J.L. Lansford, A.V. Mironenko, D.G. Vlachos, Nat. Comm. Scaling Relationships and Theory for Vibrational Frequencies of Adsorbates on Transition Metal Surfaces (Accepted)
[2] B. Hammer, Faraday Discuss. 110, 323 (1998).

Extending Adsorption Scaling to Vibrations

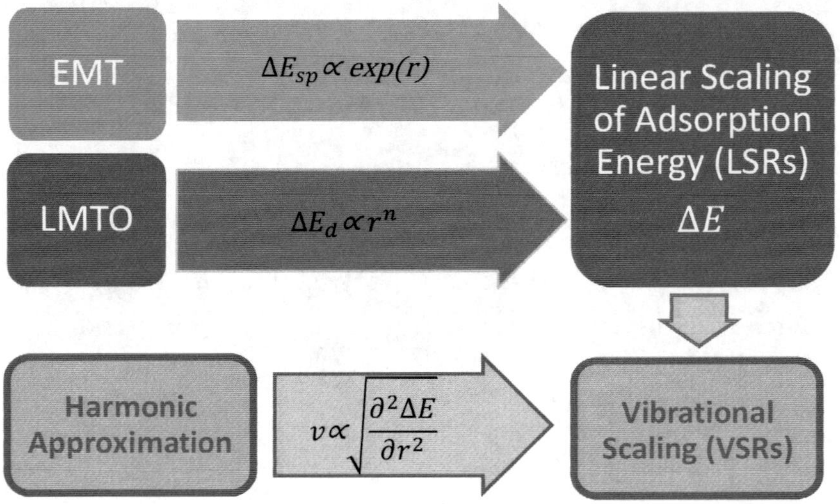

Vibrational Scaling Relation (VSR)

$$(v^{AHx})^2 = m_v (v^A)^2 + b_v$$

Deriving the VSR in terms of the LSR

$$m_v = m_E \frac{\mu^A}{\mu^{AHx}} \left(\frac{R_n^{AHx}}{R_n^A} \right)$$

where

$$R_n = f(bond\ distance, orbital\ character)$$

We can predict m_v from a single DFT calculation!

J.L. Lansford, A.V. Mironenko, D.G. Vlachos, Nat. Comm. Scaling Relationships and Theory for Vibrational Frequencies of Adsorbates on Transition Metal Surfaces (Accepted)

Predicted vs. Actual Slopes of Correlations

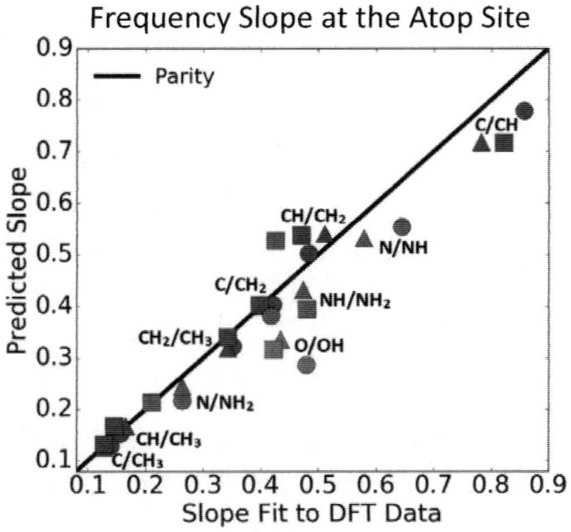

Predicted Slope

$$m_v = m_E \frac{\mu^A}{\mu^{AHx}} \left(\frac{R_n^{AHx}}{R_n^A} \right)$$

Slope Fit to DFT Data

$$(v^{AHx})^2 = m_v (v^A)^2 + b_v$$

Standard Error in Regressed Slope		
Species	LSR	VSR
O/OH	0.13	0.06
NH/NH_2	0.03	0.05
CH_2/CH_3	0.05	0.05

Impact of Vibrational Correlations on Modeling

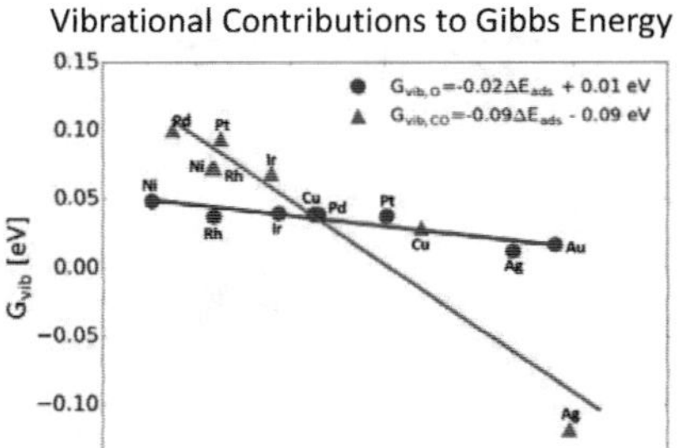

$$\theta_{CO} = \frac{K_{CO} P_{CO}}{1 + K_{CO} P_{CO} + \sqrt{K_O P_{O_2}}}$$

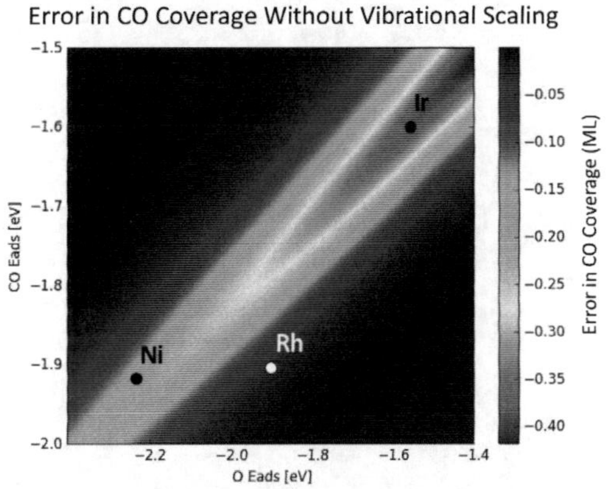

400 K, 1 bar and 100 ppm CO in O_2

Interpreting Complex Spectra on Bimetallics

NH$_X$ on the 100 Atop Site

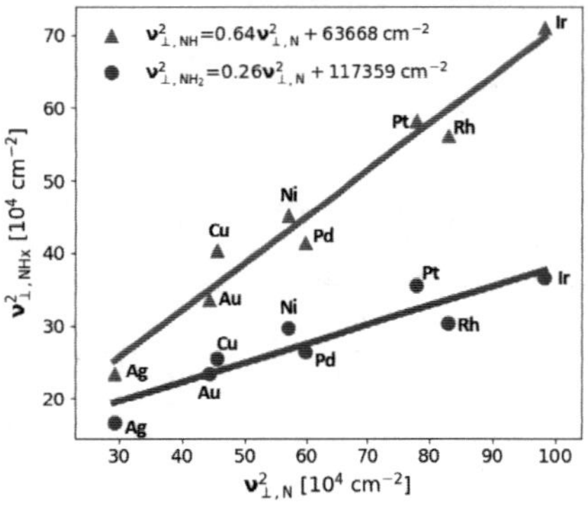

Scaling from DFT[1]

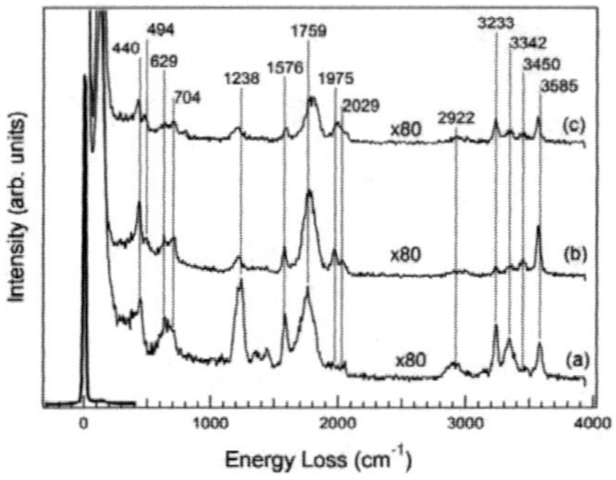

Ammonia decomposition on Ni-Pt-Pt at various temperatures[2]

[1] J.L. Lansford, A.V. Mironenko, D.G. Vlachos, Nat. Comm. Scaling Relationships and Theory for Vibrational Frequencies of Adsorbates on Transition Metal Surfaces (Accepted)
[2] D. A. Hansgen, D. G. Vlachos, and J. G. Chen, Nat Chem 2, 484 (2010).

Identifying the O_2 Adsorption Site and Normal Modes

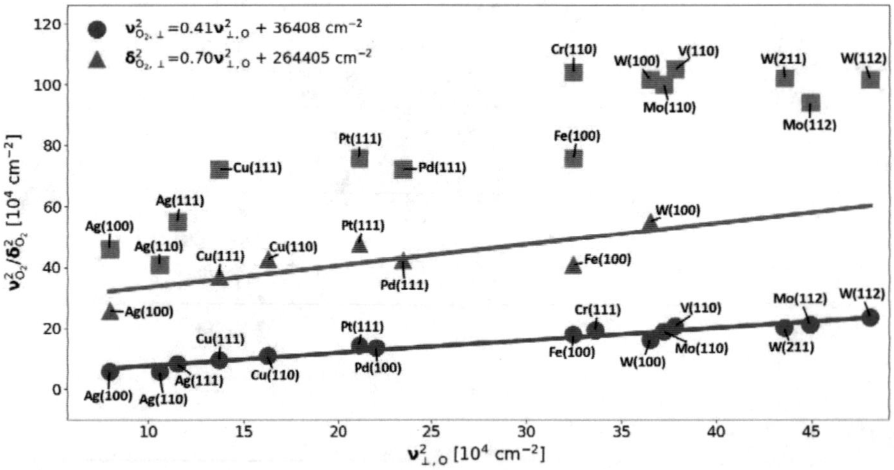

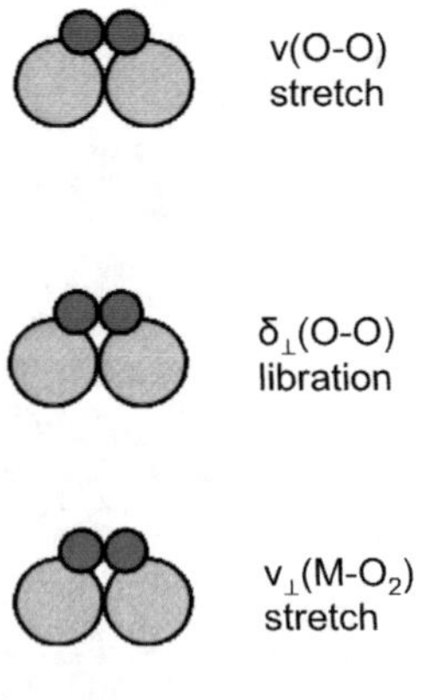

- Experimental HREEL data for O_2 and O

J.L. Lansford, A.V. Mironenko, D.G. Vlachos, Nat. Comm. Scaling Relationships and Theory for Vibrational Frequencies of Adsorbates on Transition Metal Surfaces (Accepted)

Summary

- We develop vibrational scaling relations **(VSRs)** and **corresponding theory**
- We **demonstrate** the **predictive power** of our theory and **impact** of vibrational scaling on thermodynamics
- **Scaling frequencies** can be used to **identify** adsorption site and species

Acknowledgements

Funding from DARPA

Professor Dionisios G. Vlachos for advisement

The Vlachos group

494955 Genetic Algorithm Enhanced By Atomistic Neural Network: Pt Clusters at the H_2 Atmosphere as an Example

Abstract:

Revealing the catalyst structures at the atomic level is essential to rationally improve or design new efficient catalyst. Accordingly, it is a key goal for theoretical catalysis studies. Although theoretical calculations can reveal the catalyst structures in atomic details, which is beyond the capability of most current experimental techniques, there are still some challenges. First, theoretical calculations based on first-principles calculations are very expensive, therefore the simulations are normally limited within several hundred atoms; Secondly, realistic catalyst structures are quite diverse and they are easily affected by reactants, products et al[1]. The catalyst might be associated to several competing low energy structures, and not to a single unique structure, which in turn needs more computations. Finally the exploration of potential structures on a system with a large number of degrees of freedom is computationally complex[2]. All of those factors hamper the understanding of the relationship between reaction activity and catalyst structures.

We proposed two strategies to address above obstacles. Firstly, in order to investigate all the possible catalyst structures, we devise a scheme to enhance the capability of normal genetic algorithm (GA)[3]. Considering that the most stable structure is not necessary the most active catalyst structure, a comprehensive exploration of many low energy catalyst isomers is vital. With this in mind, our enhanced GA is aimed at finding all the possible structures that could have a reasonable population in catalytic temperatures rather than only locating the global minimum. Secondly, since the genetic algorithm will exploit a large number of structure optimizations, it is time-consuming to carry out all optimizations by DFT directly. Therefore, we accelerated the computationally expensive structure optimizations with an atomistic neural network (ANN)[4], which is trained before performing GA optimizations. An iterative scheme is also utilized to efficiently find the optimal training set structures during the ANN training procedures.

With those tools, we studied the structures of platinum clusters covered with hydrogen adsorbates (Pt_nH_x). For every composition, we can find all the possible structures whose energy lies between the E_{gm} (gm=global minimum) and E_{gm}+0.5 eV. Those structures cover all the catalytically relevant structures that may appear during catalytic reactions, therefore we can locate the most important catalytic species from the structure population. The influence of the supports is in consideration. In the end, we can improve our understanding of Pt clusters at more realistic conditions.

All in all, we presented a combined routine, which takes advantages of two cutting-edge techniques: global optimizations and neural-network methods. This method can efficiently find out the low energy structures for a large system without loss of energy accuracy vs DFT. Then the method is utilized for studying a realistic system-Pt cluster at the atmosphere of H_2. Besides the example in this study, this scheme can be easily applied to similar catalytic problems benefiting from the generality of genetic algorithm and atomistic neural network.

1. Mager-Maury, C.; Bonnard, G.; Chizallet, C.; Sautet, P.; Raybaud, P., H -induced reconstruction of supported Pt

2

clusters: metal-support interaction versus surface hydride. *ChemCatChem* **2011,** *3* (1), 200-207.

2. Zhai, H.; Alexandrova, A. N., Ensemble-Average Representation of Pt Clusters in Conditions of Catalysis Accessed through GPU Accelerated Deep Neural Network Fitting Global Optimization. *J Chem Theory Comput* **2016,** *12* (12), 6213-6226.

3. Vilhelmsen, L. B.; Hammer, B., A genetic algorithm for first principles global structure optimization of supported nano structures. *J Chem Phys* **2014,** *141* (4), 044711.

4. Artrith, N.; Hiller, B.; Behler, J., Neural network potentials for metals and oxides - First applications to copper clusters at zinc oxide. *Physica Status Solidi (b)* **2013,** *250* (6), 1191-1203.

500606 Prediction of Chemisorption Energies By Gaussian Processes

Abstract:

Design of heterogenous catalysts is going through a revolution due to the development of quantum chemical computations and descriptor based approaches, which reduce the dimensionality of the design problem based on a detailed physical understanding.[1]

Chemisorption energies of reaction intermediates are most often the descriptors of choice. Meanwhile, a bottleneck is currently caused by the CPU time needed to calculate energetics of these intermediates and transition states from atomic structures. It is therefore feasible to use machine learning methods to predict chemisorption energies, particularly as the amount of data grows.[2]

For catalyst screening purposes, a Gaussian Process (GP) has advantageous properties, the biggest of which is the capability of error estimates using Bayesian statistics.[3] Simple linear models are already sufficient for describing correlation between adsorbate energies, when they bind through the same atom to various surfaces. Simple GP models may therefore be made with the advantage of being highly interpretable, which provides physical insight. Typical numbers of datapoints in calculated chemisorption energy databases has yet to reach the regime of "big data" and GP's may therefore also prove to be sufficiently efficient.

We present our framework for modeling chemisorption energies using a Gaussian Process and we present applications in elucidating trends and designing metal catalyst surfaces for relevant problems in modern thermal catalysis.

1. Jens K. Nørskov et al. "Towards the computational design of solid catalysts". In: Nature chemistry 1.1 (2009), pp. 37–46.

2. Zachary W. Ulissi et al. "To address surface reaction network complexity using scaling relations machine learning and DFT calculations". In: Nature Communications 8 (2017).

3. Carl E. Rasmussen and Christopher K. I. Williams. "Gaussian processes for machine learning. 2006". In: The MIT Press, Cambridge, MA, USA 38 (2006), pp. 715–719.

494916 New Computational Tools for High-Throughput Discovery in Transition Metal Catalysis

Abstract:

Virtual high throughput screening has emerged as a powerful tool for the discovery of new materials. Nevertheless, screening of inorganic complexes is presently limited by the high cost of accurate property evaluation and the large, complex search space. To address these challenges, we have recently developed molSimplify[1], an open-source toolkit for the automated generation of highly accurate inorganic molecular structures and property analysis to accelerate high throughput screening of inorganic complexes. In conjunction with our structure generation toolkit, we have also developed and implemented a robust and automated workflow involving: (1) selection of a test set of ligands maximizing relative molecular diversity[2], (2) prediction of electronic structure outputs with a neural network prior to simulation for pre-screening and (3) screening of multi-million molecule organic libraries according to the discovered design rules to discover experimentally accessible ligand candidates[2]. We demonstrate the broad utility of our integrated approach to inorganic complex discovery across the following applications: design of redox active functionalized ferrocenium complexes for selective ion separation[3], discovery of octahedral Fe(II/III) redox couples with nitrogen ligands[4], and design of single-site Fe(II) catalysts for oxygen activation. We also demonstrate the potential for molSimplify to accelerate catalytic property evaluation by automatically generating high-quality transition state guesses for common reactions in inorganic chemistry.

[1]E. I. Ioannidis, T. Z. H. Gani, and H. J. Kulik "molSimplify: A toolkit for automating discovery in inorganic chemistry" J. Comput. Chem. 37, 2106-2117 (2016).

[2]J. Y. Kim, A. H. Steeves, and H. J. Kulik "Harnessing Organic Ligand Libraries for First-Principles Inorganic Discovery: Indium Phosphide Quantum Dot Precursor Design Strategies" Chem. Mater., just accepted (2017).

[3]T. Z. H. Gani, E. I. Ioannidis, and H. J. Kulik "Computational Discovery of Hydrogen Bond Design Rules for Electrochemical Ion Separation" Chem. Mater., 28 6207-6218 (2016).

[4]J.P. Janet, T. Z. H. Gani, A. H. Steeves, E. I. Ioannidis, and H. J. Kulik "Leveraging Cheminformatics Strategies for Inorganic Discovery: Application to Redox Potential Design" Ind. Eng. Chem. Res., just accepted (2017).

493380 More Accurate Depiction of Adsorption Energy on Transition Metals Using Work Function As One Additional Descriptor

Abstract:

Transition metals (TMs), benefiting from their active interactions with chemical species owing to partial occupancy of the d bands, represent one most important category of catalysts and have been widely used in many reactions. The catalytic properties of TMs in a chemical reaction are primarily determined by adsorption energy (E_{ads}) of the involved species. In this regard, it is of high research significance to get accurate E_{ads} values, however, which is often challenging be measured experimentally. Active research interests have been devoted to determine Eads by establishing an exact relationship with measurable catalyst parameters like the d-band (εd) theory, which was based on the Newns-Anderson model and suggested for use by Hammer and Nørskov. It describes a simple E_{ads}–ε_d linear correlation and has been widely adopted for application, which turns out to work well for certain adsorption systems, for instance H_2 and CO adsorption, but becomes less effective for many other systems, such as O, OH and OOH adsorption. These controversial results suggest the current E_{ads}–$\varepsilon_d$linear correlation could be oversimplified to describe these adsorption systems and εd might not be the only influencing parameter, thus improvements are needed.

Considering the fact that band hybridization and electron transfer could occur simultaneously when a molecule adsorbs to a TM surface, we propound that E_{ads} contains a mixture of covalent and ionic contributions (i.e., $E_{ads}=E_{covalent} + E_{ionic}$). The $E_{covalent}$ term would be describable using the ε_d parameter as it accounts for the band hybridization, whereas the E_{ionic} term is determined by the charge transfer between TM and adsorbing molecules and thus would be related to the work function (W) parameter. Herein by taking both the ionic and covalent contributions into account, we suggest a new description model for E_{ads} by dividing it into $E_{covalent}$ and E_{ionic} components and adding W as one additional descriptor. We study the adsorption of O, OH, and OOH groups on TM surfaces using the new E_{ads}–(ε_d, W) model with the explicit formula as $E_{ads} = f(\varepsilon_d, W) = z_0 + a \times \varepsilon_d + k \times (W - W_0)^2$ and achieve significantly better goodness of fit compared to the current E_{ads}–ε_d model. We further demonstrate the new E_{ads}–(ε_d, W) model can better predict activity property of TMs in oxygen reduction reaction (ORR) benefiting from a more accurate description of the oxygen binding energy (E_O).

496405 Molecular Simulation Study of How the Structure of Liquid Water Affects the Free Energies of Adsorption and Reaction in Aqueous Phase Heterogeneous Catalysis

Abstract:

There are many applications of aqueous phase heterogeneous catalysis, including aqueous phase reforming (APR), Fischer-Tropsch synthesis, electrocatalytic oxygen reduction, etc., and a grand challenge in catalysis research is understanding how water influences the chemistry, mechanism, thermodynamics, and kinetics of reactions occurring on metal catalyst surfaces under liquid H_2O. Recent work by our group and others has shown that H_2O molecules form hydrogen bonds (HB) with reaction intermediates, alter their adsorption energies, co-catalyze certain reactions, affect kinetic barriers and reaction mechanisms, amongst other phenomena. We are interested not only in the roles that liquid H_2O molecules play, but also how these phenomena influence the rates and equilibria of catalytic surface reactions. For example, we previously found that hydrophilic adsorbates form stronger and more stable HB with H_2O molecules than hydrophobic adsorbates, which we postulated would lead to greater degrees of order in the interfacial water structure around hydrophilic adsorbates than for hydrophobic ones. We further postulated that this would give rise to a free energy effect in reactions where hydrophilic adsorbates are converted into less hydrophilic adsorbates or even hydrophobic ones, e.g., in APR reactions where sugar alcohol molecules are sequentially deprotonated down to CO_2 and H_2 (and other analogous reactions). In this work, we use density functional theory (DFT) and methods in molecular dynamics (MD) and statistical mechanics to study these effects in the context of the decompositions of CH_3OH and NH_3 over Pt catalysts. Specifically, we compute the free energies of different surface intermediates along the pathways for these reactions under liquid H_2O using the following methods: potential of mean force (PMF), calculated from the radial distribution function (RDF), which gives the free energy due to configurational fluctuations in the liquid H_2O structure, and free energy perturbation (FEP), thermodynamic integration (TI), and implicit solvation method for surfaces (iSMS), which give the free energies of solvation for the different adsorbates. A key step in the PMF/RDF, FEP, and TI methods is producing an explicit structure of liquid H_2O at the catalyst interface, in order to identify how it influences catalytic thermodynamics. We provide values for these interfacial free energy contributions relative to the free energies of reaction and discuss how they influence the thermodynamics of adsorption to the surface from aqueous phase and reaction on the surface under liquid H_2O. We summarize by discussing how these contributions influence the equilibrium and rate constants in aqueous phase heterogeneously catalyzed reactions.

498622 Durable, Self-Hydrating, Tungsten Carbide-Based Composite Polymer Electrolyte Membrane Fuel Cells

Abstract:

Proton conductivity of the polymer electrolyte membranes in fuel cell (PEMFC) dictates the performance and requires sufficient water management. Here, we introduce a simple scalable method to produce well-dispersed transition metal carbide (TMC) nanoparticles with narrow size distribution (3-5 nm). This protocol can be extended to the synthesis of other early transition metal carbides. Extensive characterizations including HRTEM, STEM tomography, SEM, XPS, XRD and TG are conducted.

WC nanoparticles incorporated into a recast Nafion membrane demonstrate significant improvement of the PEMFC power density compared to the baseline Nafion membrane. Based on both theoretical and experimental results, the WC composite membrane catalyzes the reaction of crossover H_2/O_2 and provides sufficient water management within the membrane, improving the proton conductivity. Tomographic studies using FIB-SEM indicate large pinholes forming around Pt nanoparticles and provide support that the membrane/Pt catalyst facilitates the production of radicals that leach from Nafion. Superior to Pt, incorporated WC nanoparticles in the membrane result in considerable durability of the membrane. DFT calculations provide insights into the mechanisms by which WC improves performance.

492100 Teaching an Old Electrocatalyst to Do New Tricks

Abstract:

Electrocatalysis underlies several key clean energy technologies such as fuel cells and electrolyzers that are expected to be key components of a sustainable energy future. However, continued improvements in electrocatalyst activity, durability, and selectivity must be achieved in order to make this future a reality. In this study, we explore the use of electrocatalyst architectures based on metallic electrocatalysts that have been encapsulated by ultra-thin silicon oxide overlayers. These oxide layers are synthesized with a room temperature process and deposited as continuous layers onto metallic thin films and nanoparticles. The catalytic metal used in these studies is Platinum (Pt), an "old catalyst" material that is one of the most commonly used electrocatalysts for electrolyzers and fuel cells. Inspired by past studies on glass microelectrodes[1] and a recent study from our lab on oxide-coated nanoparticles,[2] we show that well-defined oxide overlayers deposited on smooth metal substrates can be highly permeable to certain electroactive species and thereby enable efficient and selective electrocatalysis at the metal oxide / metal interface. By systematically varying the thickness of the oxide overlayer, we explore its transport properties and influence on the performance of thin film hydrogen evolution electrocatalysts. Our studies show that the properties of an "old catalyst" like Pt can be greatly altered, and perhaps tuned, through the use of the ultra-thin metal oxide overlayers, which can be thought of as nanomembranes that can selectively control transport of reactants and products. These results highlight the potential of this membrane-coated electrocatalyst architecture to serve as a tunable, efficient, and stable platform for a variety of electrochemical reactions.

References

1. J. Velmurugan, D. Zhan, and M. V. Mirkin, *Nature Chemistry*, 2, 498-502, 2010.

2. N. Y. Labrador, X. Li, Y. Liu, J. T. Koberstein, R. Wang, H. Tan, T. P. Moffat, and D. V. Esposito, *Nano Letters*, 16, 6452-6459, 2016.

493787 Using Surface Chemistry to Understand Aqueous-Phase Thermal Catalytic and Electrocatalytic Hydrogenation of Bio-Oil Model Substrates

Abstract:

One of the major challenges of the future is matching the increasing demand for energy using renewable resources. The energy density of hydrocarbons makes them ideal for transportation fuels, especially for aviation, where electric vehicles are not available. As the amount of energy transitions to renewable energy such as solar and wind, which are intermittent and potentially more decentralized, developing ways to store this energy also becomes important. Catalysis can play an integral role in the chemical processes required to produce renewable transportation fuels and store renewable energy. In this talk, I will discuss developments in catalysts used in a method to convert waste biomass to renewable fuels. The feedstock of this process is bio-oil, a product of pyrolysis or hydrothermal liquefaction of biomass. Bio-oil (modeled here using phenol and benzaldehyde) can be hydrogenated and then deoxygenated to yield fuel grade hydrocarbons.[1] Although hydrogen generally comes from steam methane reforming, which emits carbon dioxide, renewable electricity can be used to drive carbon-neutral (i) electrochemical hydrogen production (from water) which is then used for thermal catalytic hydrogenation (TCH), or (ii) direct electrochemical hydrogenation (ECH) of the bio-oil. Electrochemical routes are more economically competitive in decentralized applications and can make use of intermittent renewable electricity. The challenge of both electrochemical hydrogen production coupled with TCH and ECH is developing hydrogenation catalysts with high rates for commercial applications. The target is to achieve comparable rates (at approximately 80 °C) to existing commercial electrochemical processes, on the order of hundreds of mA per square centimeter (geometric current density) or turnover frequencies near 10 s^{-1}. In this work we discuss some advancements in the understanding of the aqueous-phase hydrogenation of phenol and benzaldehyde on different platinum group metals, and how this leads to improved rates for ECH and TCH.

For phenol hydrogenation, ECH and TCH on rhodium or platinum have similar activation barriers due to a shared rate determining step of adsorbed H adatoms reacting with adsorbed phenol. Using a flow cell designed for X-Ray Absorption Spectroscopy (XAFS), we probe Pt nanoparticles under reaction conditions for both near-edge (XANES) and extended X-ray absorption fine structure (EXAFS). This allows us to look at the catalyst state (from XANES) and surface coverages (from Pt-C scattering that we show is due to the adsorbed reactant species) under operating conditions. This spectroscopy, coupled with kinetic measurements, affords us a better understanding of the surface chemistry and its dependence on hydrogen pressure and applied potential. Using this understanding, we demonstrate methods to prevent catalyst deactivation at temperatures at 80 °C.[2] We also show how modifying the adsorption energy of the adsorbed species controls the reaction rate, and discuss how other reaction conditions (ex. pH) affect the surface coverages and rates. These results lead us to propose material properties that may result in more active electrocatalysts for phenol TCH or ECH, and how to reach reaction rates that may be necessary for a commercial ECH electrolyzer system.

For benzaldehyde, on metals such as platinum and rhodium, TCH and ECH have different benzaldehyde rate order dependence, unlike what is observed for phenol. This may be due to a difference in the reaction mechanism for the

ECH and TCH of benzaldehyde on Pt and Rh, or surface coverages under operating conditions. Using XAFS (as well as traditional cyclic voltammetry), we try to understand how the surface coverage of the species may be responsible for the observed reaction rates and orders. We connect this with kinetic measurements and adsorption energies in an attempt to understand how the surface chemistry controls the reaction rates that are observed.

References

(1) Zhao, C.; He, J.; Lemonidou, A. A.; Li, X.; Lercher, J. A. *J. Catal.* **2011**, *280*(1), 8.

(2) Singh, N.; Song, Y.; Gutiérrez, O. Y.; Camaioni, D. M.; Campbell, C. T.; Lercher, J. A. *ACS Catal.* **2016**, *6*, 7466.

498568 Design of Heterogeneous Catalysts for High Temperature Electrochemical Reduction of CO_2

Abstract:

The high levels of carbon dioxide (CO_2) emissions, consequential from the extensive use of fossil fuels, are major contemporary challenges. An approach to alleviate this process is to activate the reverse chemical pathways in which CO_2 is reduced into high-energy molecules. The electrochemical reduction of CO_2 to CO has attracted increasing attention, since CO represents a valuable intermediate to the production of synthetic fuels using established processes, such as Fischer-Tropsch. The use of solid oxide electrolysis cells (SOECs) to electrochemically reduce CO_2 to CO is an attractive technology, given that this solid-state electrochemical systems can, in principle, facilitate the CO_2 reduction with potentially very high rates due to the favorable kinetics at high operating temperatures. While electrochemical reduction of CO_2 using SOECs offers a great deal of promise, these systems are still limited by the high activation overpotential losses induced by the sluggish kinetics of CO_2 reduction at the cathode. In the present work, we combine experimental and theoretical techniques to understand the chemical/electrochemical process that governs CO_2 electrolysis, and develop a structure/performance relation in order to identify the optimum electrocatalyst for this process. Controlled electrochemical experiments show that upon modification of the electrode metal electrocatalyst, significant decrease in the overpotential is achieved.

REFERENCES

1. Ebbesen, S. D., Jensen, S. H., Hauch, A., & Mogensen, M. B. (2014). High temperature electrolysis in alkaline cells, solid proton conducting cells, and solid oxide cells. *Chemical reviews*, 114(21), 10697-10734.

2. Gu, X. K., Carneiro, J., Nikolla, E. (2017). Electrolysis on Carbon Dioxide. In Catalysis (pp. 94-116). The Royal Society of Chemistry.

3. Gu, X. K., & Nikolla, E. (2017). First-Principles Study of High Temperature CO_2 Electrolysis on Transition Metal Electrocatalysts. *Industrial & Engineering Chemistry Research*, under review 2017.

494155 Enhancing Methanol Electrooxidation Activity Using Double Oxide Catalyst Support

Abstract:

Electrooxidation of methanol was investigated on platinum catalyst supported on a double oxide nanocomposite of tin oxide clusters and carbon-doped titanium dioxide nanocoatings on carbon nanotubes. The electrocatalyst demonstrated a higher forward peak current density in methanol eectrooxidation than the electrocatalyst with a single tin oxide or just carbon as support. Formation kinetics of the oxygenated species (-OH) groups was dramatically enhanced in the catalyst with double oxide support. This leads to a much lower onset potential for adsorbed CO oxidation with an enhanced bifunctional effect in tin oxide. Doping with carbon in the titanium oxide enables backward electron donation, making supported platinum more electron negative and stable. Binding energy shifts in oxidation states was observed for the double oxide support catalyst and is attributed to strong metal support interactions. This study demonstrated a synergistic effect of combining bifunctional and electronic mechanisms with double oxide support in the enhancement of methanol electrooxidation.

501794 Study of Impact of Electrode Catalyst Loadings on Thin Flexible Fuel Cell (TFFC) Performance

Abstract:

Thin Flexible Fuel Cell (TFFC) is a flat and flexible device without any rigid support capable of being integrated into portable electronics. TFFC uses hydrogen and oxygen as fuel to provide high volumetric and gravimetric energy densities capable of competing with battery technology as a power source. The group is currently investigating a TFFC with an air-breathing design that maximizes the cathode interface with ambient air. The current challenges of making TFFC commercialized are the cost and performance. Enhanced cell performance is critical for a competitive cost of the device. The group investigated the impact of catalyst loading on the cell performance to increase the overall performance of the device. The catalyst loading of the cathode electrode was varied while keeping the anode catalyst loading constant and vice versa. The effects of the various catalyst loadings were characterized using various analytical techniques. Polarization curves and Electrochemical Impedance Spectroscopy (EIS) were used to separate and quantify the sources of losses for the TFFC. Power curves provided the maximum power density of the device. Cyclic voltammetry (CV) was implemented to characterize the electrochemically active surface area (ECSA) and utilization of the various catalyst layers. Rotating Disc Electrodes (RDE) tests were performed for a direct assessment of the catalyst. We will report the results of these assessments and summarize the resulting guidance for optimizing catalyst loading on similar fuel cell structures.

501059 Transition Metal Modified Graphitic Mesoporous Carbon for PEMFC Application

Abstract:

Proton exchange membrane fuel cell (PEMFC) has drawn considerable attention due to its high power density, zero emission and fast startup features. In PEMFC, platinum metal is the catalyst of choice in anode and cathode electrodes. The cost of Pt is a significant issue in commercialization of PEMFC. Further due to its sensitivity to CO, PEMFC requires high purity hydrogen as fuel. Conventional catalyst support (activated carbon) lacks in adequate porous structure, which results in underutilization of Pt metal particles. To better utilize the Pt-catalyst, we synthesized high surface area metal modified mesoporous carbon using hard template-based (SBA-15) method. Highly porous structure of this carbon allows well-dispersed Pt metal loading ensuring its proper utilization. Transition metals (Ni, Fe and Co) and Ni-Fe alloy were used for catalytic graphitization of the carbon. Resulting carbon showed graphitic structure in X-ray diffraction (XRD) and Raman spectroscopy. Electrocatalysts synthesized using these metal modified carbons were tested on the fuel cell. The structural characteristics of these graphitic mesoporous carbon and their performance in PEMFC will be presented in this paper.

494532 Influence of Bubble Growth at the Catalytic Surface on Heat and Mass Transfer in Gas-Liquid-Solid Microreactors

Abstract:

Influence of bubble growth at the catalytic surface on heat and mass transfer in gas-liquid-solid microreactors

Renée M. Ripken, Jeffery A. Wood, Johannes G.E. Gardeniers and Séverine Le Gac

University of Twente, The Netherlands

Here, we studied the influence of gas bubble growth at the catalytic surface on the heat and mass transfer in a microreactor using a numerical simulation framework built in COMSOL Multiphysics 5.2. Although the gas bubbles are detrimental for the heat transfer, mass transfer is increased as a result of the surface heterogeneity.

Industrial catalytic systems are rarely monophasic, and the majority of the processes are heterogeneous; the catalyst is solid while the reactants and products are in the gas and/or liquid phase. During some catalytic reactions, such as the reforming of alcohols and sugars, the reactants are dissolved in the liquid and typical products such as H_2, CO and CO_2 are formed as gases. When the gaseous products nucleate and grow into bubbles on the catalytic surface, a multiphase system is formed where the interplay of different transport phenomena is relatively complex. To obtain insight in this matter, we have developed a theoretical model to study the heat and mass transfer at the microscale in such a multiphase system.

A 2D microreactor with a channel of 250 Î¼m deep etched into a silicon substrate, covered with a glass top-layer was considered as the model system. A 10 Î¼m thick TiO_2-layer is coated at the bottom of this channel that acts as the catalyst support layer. The system is heated from the silicon side and dissipates heat to air through the glass top-layer, while the rest of the system is thermally insulated. A no-slip boundary condition is assumed on the channel wall and catalytic surface, with the bubble treated as a perfect slip wall. We focus on the external mass transfer limitations, while, including reaction kinetics.

We evaluated the bubble deformation caused by shear force as a function of fluid flow and bubble size. Typically, growing bubbles that are smaller than the channel diameter deform because of the fluid flow, but also relax to their initial state due to the surface tension. This oscillatory behavior is much faster than the reaction kinetics; therefore, bubbles can be approximated as being static. When the bubble size reaches the channel diameter, it detaches from the catalytic surface.

By adjusting the size of the bubbles and the distance between them, higher fluid velocity near the catalytic surface can be achieved. Similarly, mass transfer of the reactants to and products from the catalytic layer is enhanced. However, the heat transfer is negatively affected; gas bubbles act as an insulating layer, increasing the thermal entrance length compared to a system without bubbles. For highly endothermic reactions, a longer residence time is required to reach similar conversion as for a system without bubbles on the surface.

In the future, these simulation results can be potentially useful towards enhancing microreactor performance by giving insight into the role of geometry and operating parameters on relevant transport phenomena and how these affect conversion, yield and selectivity for gas-liquid-solid microsystems in which bubbles are growing at the catalytic surface.

Acknowledgements

This work was supported by the Netherlands Center for Multiscale Catalytic Energy Conversion (MCEC), an NWO Gravitation programme funded by the Ministry of Education, Culture and Science of the government of the Netherlands.

493657 Anomalous Behavior of Liquid-Liquid Two-Phase Reaction in a Slug Flow Microreactor

Abstract:

This work depicts the experimental investigation of a two-phase heterogeneous reaction in the slug flow regime in a micro-reactor. The high surface area to volume ratio and shorter diffusion paths offer by microfluidic devices intensify the performance of these systems. Phase transfer catalyzed benzyl alcohol (BA) oxidation using sodium hypochlorite yielding benzaldehyde is chosen as a test system. BA oxidation is dependent on pH of aqueous phase with a lower pH favoring the reaction. The reaction rate constants at three different pH values 10.4, 8.4 and 7.4 is determined using batch experiments. Slug flow experiments carried out at these pH values are analyzed for benzaldehyde yield. The yield of benzaldehyde increases with an increase in organic phase residence at higher pH i.e. 10.4 and 8.4. However at pH 7.4 the system shows an anomalous behavior i.e. benzaldehyde yield decreases with an increase in organic phase residence time.

The anomalous behavior observed in BA oxidation is explained using hydrodynamics of slug flow and is shown to be due to a change in the rate determining step. The relative importance of three time scales i.e. residence time, mass transfer time and reaction time in two-phase reaction in microfluidic devices is captured using dimensionless Damkohler numbers. Criteria in terms of these dimensionless numbers are obtained to determine where a particular step is rate determining.

493800 Motion Tracking of Liquid-Liquid Segmented Flows in Microfluidics and Application to the Briggs-Rauscher Oscillating Reaction

Abstract:

The junction of computer programming and chemical engineering has provided ubiquitous tools for engineers to study chemical reactions within microfluidic reactors. However, a future with autonomous laboratory-scale reactors that can recommend requires new advancements. In this study, a MATLAB program was written to track individual droplets within steady-state liquid-liquid segmented flows captured in real-time, frame-by-frame, using a camera. Three different residence times (2.0, 5.0, and 10 min) of a liquid-liquid system (*e.g.*, silicone oil-water dosed with UV-active (~260 nm) fluorescene) were investigated using a spiral micoreactor with straight sections. Two algorithms were written for analyses of the resultant aqueous phase in recorded videos (at 30 fps). The system was first calibrated according to the 2D geometry and characteristic length scales (*e.g.*, 300 Î¼m channel width). The first algorithm, the more crude approach, converted each image into a binary (black and white) image and used a connective component analysis (CCA) to split each cluster of pixels into its own structure. This allowed a rigorous frame-by-frame analysis. Distributions of length, velocity, and color were mapped in MATLAB under CCA. Here, the average pixel length along the width of the 300 Î¼m wide channel was found to be 10 pixels, which was axially consistent even throughout the turns. The measured distributions were within reasonable size, confirmed by measurement with a caliper during an experiment. Due to the extensive amount of frames (30 fps), the color distribution mapped was accurate and only limited by the CCD detector resolution. Velocity distributions for each 2.0, 5.0, and 10 min residence times were available *via*a subtraction algorithm that compared adjacent frames of videos of each residence time. The method of CCA also used a reference â€œcentroidâ€, calculated as a weighted average of pixels in a 2D array, that deviated from the actual center of the slug when the slugs traveled through turns.

A second algorithm of analysis along the parameterization of a reactor path was developed. Utilizing each video of the spiral microreactor, any 1D path could be generated. Along the 1D path, color distributions per slug, length, and velocity distributions along the axis of the reactor for each residence time were estimated and found to be more accurate than the first algorithm. The average slug size for each 2.0, 5.0, and 10 min residence times was calculated. The center of the reactor, in the second algorithm, was calculated precisely through a rigorous image-processing algorithm limited solely by the resolution of the camera. This method of skeletonizing a path specified by user parameters calculates the center of the reactor, provided the reactor is continuous and contains no branches. The 1D parameterization of any continuous microreactor geometry using a 5-20 sec video of multiphase flows, provided the residence time is >20 sec, is possible using the algorithm. The same routine was then tested by the Briggs-Rauscher oscillating reaction occurring within the dispersed slugs in order to determine useful kinetics. This non-invasive, visual method of data acquisition could revolutionize the way we discover catalyst and kinetics for certain classes of chemistry.

491435 Highly Controlled Material Transfer into Microfluidic Droplets from an Active Colloidal Continuous Phase

Abstract:

Droplet microfluidics enables the use of monodisperse micro-droplets as reaction vessels with very precise control over the reaction conditions, such as reagent concentration and rapid mixing. However, once the droplets are generated in a microfluidic device, it is usually difficult to add new material into the droplets for subsequent operations. Fusion, via both electrical and hydrodynamic means, of droplets of similar sizes has been the most common strategy to achieve this objective. We have developed a novel and versatile method to transfer materials into the microfluidic droplets using a nano- or mini-emulsion as the continuous phase. We conduct a detailed study on the droplet generation mechanism, and three different strategies for controlled mass transfer: addition of co-surfactant, heat treatment, and electro-coalescence. In particular, electro-coalescence is a non-invasive and highly responsive method for extremely sensitive control over the amount of material to be added and, equally importantly, the rate of addition of these materials. The relationship between the applied electrical field and the coalescence between nano- and micro-droplets is studied. Finally, we demonstrate an application of this method in the controlled nucleation and growth of nanocrystals. Our method expands the toolbox of droplet microfluidics, enabling reaction engineering with precise mass transfer control at the sub-micron level using an active colloidal continuous phase carrying chemical payloads.

493039 Commercial 3D Printed Microreactor for Chemical Analysis

Abstract:

Gas chromatography is a popular technique for determining the composition of chemicals in feed and effluent streams. Yet, this technique still requires the painstaking calibration of individual organic compounds due to the compound-specific response of the detectors used for analysis, including flame ionization detection (FID), mass spectrometry (MS) and thermal conductivity detection (TCD). In this talk, we present a new paradigm for the analysis of compounds using a chemical reactor that converts organic analytes to methane after their chromatographic separation. The microreactor technology is made possible by metal additive manufacturing, and when paired with an FID, yields the worldâ€™s first and only equimolar carbon detector. Here, we present the design and theory behind the PolyarcÂ® reactor, and discuss how an equimolar carbon response improves analysis accuracy, speed and capabilities, especially in the context of high throughput analysis and renewable fuel research. As the first commercial 3D printed metal microreactor, the Polyarc reactor is also an exciting case study on the future of reactor engineering and design using additive manufacturing.

488240 Mechanistic Modeling of Fast Pyrolysis of Hemicellulose

Abstract:

Hemicellulose is one of the major components of lignocellulosic biomass and has potential for the production of renewable drop-in transportation fuels and multiple commodity chemicals.[1] Previously, however, the successful modeling of fast pyrolysis of hemicellulose was hampered by lumped kinetic models, which failed to describe the heterogeneous structure of hemicellulose and cannot predict the bio-oil composition. In this work, a structural model for hemicellulose extracted from corn stover that captured the experimentally measured properties was proposed, and a mechanistic model for hemicellulose pyrolysis was constructed based on the reaction family approach that we used for cellulose pyrolysis[2]. The model described the decomposition of hemicellulose chains, reactions of intermediates, and formation of a range of low molecular weight compounds at the mechanistic level. Rate constants were specified for each elementary reaction with Arrhenius format. Overall, more than 500 reactions of 144 species were included in the mechanistic model for fast pyrolysis of hemicellulose. The mechanistic model was able to closely match experimental yields of all the major products (with yield ≥ 1 wt%) from hemicellulose pyrolysis at 500 °C reported by Shanks and coworkers[3]at Iowa State University. The model predicted about 4 s residence time for complete thermo-conversion of hemicellulose. Model results showed that both degree of polymerization (DP) and polydispersity index (PDI) of hemicellulose had an insignificant effect on the pyrolysis products distribution. The mechanistic model of extracted hemicellulose is extendable to fast pyrolysis of native hemicellulose, which will be also discussed in this work.

[1] X. Zhou, W. Li, R. Mabon, L. J. Broadbelt, *Energy Technol. 2016, 5, 52 –79.*

[2] a) R. Vinu, L. J. Broadbelt, *Energy Environ. Sci.* **2012**, *5*, 9808-9826; b) X. Zhou, M. W. Nolte, H. B. Mayes, B. H. Shanks, L. J. Broadbelt, *Ind. Eng. Chem. Res.* **2014**, *53*, 13274–13289; c) X. Zhou, M. W. Nolte, H. B. Mayes, B. H. Shanks, L. J. Broadbelt, *AIChE J.* **2016**, *62*, 766–777.

[3] J. Zhang, Y. S. Choi, C. G. Yoo, T. H. Kim, R. C. Brown, B. H. Shanks, *ACS Sustainable Chem. Eng.* **2015**, *3*, 293-301.

496698 Fast Pyrolysis Kinetic Study for Ten Ecuadorian Agricultural Residual Biomass Samples

Abstract:

Fast pyrolysis of biomass is a promising technology to produce fuels, commodity chemicals as well as fine chemical from renewable sources. Due to its CO2 neutrality replacing fossil resources by biomass helps to mitigate global warming. The thermochemical route of biomass conversion has the advantage that any type of organic matter can be used. Ecuador's economy is mainly based on the export of primary agricultural products, which creates large amounts of currently unused waste streams. In order to assess the feasibility of these waste materials for thermal conversion to valuable products, laboratory fast pyrolysis data are needed. Although numerous TGA studies are available in the literature, detailed product composition analyses obtained for pyrolysis at *high* heating rates are scarce. A new micro-pyrolysis setup was built to address this gap. The setup consists of a single-stage Frontier micro-pyrolyzer reactor which is connected to a comprehensive GCxGC coupled to a flame ionization detector (FID) and a time-off-flight mass spectrometer (TOF-MS). This configuration allows for the simultaneously identification and quantification of large molecules. In addition to the GCxGC, a customized GC is used to detect permanent gases, light oxygenated hydrocarbons, water and C2 hydrocarbon molecules.

The product spectra of fast pyrolysis of 10 biomass residues selected from the main agricultural processes in Ecuador are reported. These residues are: forestry residues from wood processing, palm oil production wastes (rachis, fiber and kernel shells), cocoa recollection (fresh and rotten cocoa kernel), coffee husks, banana stem, rice husk and sugar cane bagasse. The chosen set of biomass sample covers a wide range of cellulose, hemicellulose and lignin contents. Three pyrolysis temperatures are studied (500 ^{o}C, 550 ^{o}C and 600 ^{o}C) and quantitative products yields are obtained using either calibration factors of pure compounds or the effective carbon number approximation. The obtained experimental product distributions are compared to predictions using a literature biomass pyrolysis mechanism.

Acknowledgements

The SBO proposal "Bioleum" supported by the Institute for promotion of Innovation through Science and Technology in Flanders (IWT) is acknowledged. SG acknowledges the financial support from the Research Board of Ghent University (BOF). This research has been supported by the Belgian Development Cooperation through VLIR-UOS. VLIR-UOS supports partnerships between universities and university colleges in Flanders (Belgium) and the South looking for innovative responses to global and local challenges. Visit www.vliruos.be for more information. The research leading to these results has received funding from the European Research Council under the European Union's Seventh Framework Programme (FP7/2007-2013) / ERC grant agreement n° 290793.

494767 Biomass Pyrolysis: Can a Single Severity Factor Describe the Effect of Pyrolysis Conditions on the Final Biochar Product?

Abstract:

The amendment of soils with biochar has been heralded as a sustainable method for increasing crop yield and preventing pollution problems caused by fertilizer runoff. However, the available experimental results are highly variable. While some studies report significant beneficial effects with up to 60% higher yield after biochar addition, other studies report that biochar amendments either has no net effect or can even lower agricultural productivity by up to 30% [1]. A major reason for this variability is biochar heterogeneity. Numerous biomass feedstocks were used in these studies to produce biochars in multiple types of reactors under varying temperature and oxygen conditions, leading to thousands of biochars with widely varying properties.

Understanding the heterogeneity of biochars is a major challenge for this research. Although the maximum temperature is often used as a metric of the charring intensity or severity of slow pyrolysis conditions, the literature is full of studies reporting that multiple factors control the properties of biochars produced from the same feedstock. These factors include the peak pyrolysis temperature, the treatment time at that temperature, the heating rate, the particle size, the presence or absence of oxygen and others. Since not all these conditions are carefully controlled or even reported, it becomes almost impossible to compare biochars produced under different environmental conditions or in different laboratories. And, the problem becomes even more difficult when different feedstocks are used.

This study presents an approach that combines experimental results and reaction modeling to relate some key properties of biochars to charring intensity metrics that combine (a) the full pyrolysis temperature history and (b) the kinetics of pyrolysis reactions. To achieve very accurate control of the temperature history of our biochars, we used a thermogravimetric analyzer (TGA) to produce 71 biochar samples from slash pine wood under nitrogen and a variety of slow pyrolysis conditions. Two different particle sizes (210-250 and 850-1000 microns) were used and the initial amount of feedstock varied between 130 and 230 mg. Proximate analysis was used to determine the fixed carbon content, volatiles and ash of the produced biochars. Additional TGA experiments were carried out to identify the network of pyrolysis reactions and estimate the kinetic constants. After running a series of step isothermal and/or constant heating rate TGA runs, the weight vs. time data sets were post-processed with an optimization technique to estimate (a) the reaction network (i.e. number of parallel and sequential pyrolysis reactions), and (b) the kinetic constants (i.e. activation energy, pre-exponential factor and order) of each reaction.

As expected, the peak pyrolysis temperature was not a good metric for characterizing total biochar yield, volatiles or fixed carbon content. Fixed C showed a good correlation with peak temperature only for the longest processing times (3 hours) used in our experiments. The charring intensity metric (CITT) defined as the integral of the temperature history $T(t)$ of pyrolysis [2] also did a rather inadequate job in describing total biochar yield, volatiles or fixed carbon content. However, much better correlations were obtained when biochar properties were plotted vs. the charring severity function (CSF) defined as the integral over temperature of the kinetic rate constant of the reaction leading to the formation of char, the highly aromatic and crystalline biochar component that is the final product of the pyrolysis reactions.

Numerical simulations with the pine reaction network revealed that the same value of the CITT metric (the temperature-based charring intensity) can give us multiple concentrations of char and, thus, multiple total biochar yields, volatile contents etc. Clearly, we cannot infer the temperature history of a biochar from the properties of the final biochar product. We reached the same conclusion for the CSF metric. In the case of the CSF metric, however, our simulations revealed much smaller ranges of biochar properties for the same CSF value.

Our experiments also revealed the smaller particle size fraction (210-250 microns) contained a larger proportion of biomass that undergoes pyrolysis at lower temperatures. This was the result of the grinding. The different chemical composition of the two size fractions resulted in significant differences in the properties of the final biochar products, differences that could not have been predicted by the small variations in the effective heating rates resulting from the different particle sizes. The initial mass of the biochar sample also had a rather significant effect on the properties of the final biochar product due to the subtle temperature variations within the pyrolyzing biomass sample we have identified in earlier work.

References:

[1] Crane-Droesch, A; S. Abiven; S. Jeffery; M.S. Torn. "Heterogeneous global crop yield response to biochar: a meta-regression analysis." *Environmental Research Letters* **8** (2013) 044049.

[2] Pyle, L.A; W.C. Hockaday; T. Boutton; K. Zygourakis; C.A. Masiello, "Chemical and Isotopic Thresholds in Charring: Implications for the Interpretation of Charcoal Mass and Isotopic Data," *Environmental Science & Technology*, **49** (24), 14057-14064 (2015).

493304 Real-Time Mass Spectroscopy in Fast Pyrolysis Process

Abstract:

Biomass pyrolysis is inherently dynamic. Each individual particle entering an otherwise steady-state reactor must be heated to reaction conditions. Upon heating, regardless of the rate, the biomass particle will begin to pyrolyze under conditions of changing temperature. Thus, there are no true stead-state pyrolysis experiments from which kinetics can be extracted. In an effort to provide true real-time kinetic information from which reliable kinetic information can be extracted, a novel micro-reactor micro-particle technology was develop and coupled to a fast mass spectrometer. The newly developed micro-sphere micro-reactor (MSMR) was coupled with a fast mass spectrometer capable of sampling pyrolysis vapors at a rate of 100 Hz over a total reaction time of 100 to 10000 ms. A Q Exactive (QE) mass spectrometer at the Pacific Northwester National Laboratory (PNNL) Environmental Molecular Science Laboratory (EMSL) was used. In this study two types of biomass (tall fescue and switchgrass) and microcrystalline cellulose were formed into microspheres of different sizes and pyrolyzed at different temperatures to study the effect of particle size and process conditions on kinetics and chemistry of reaction. The results show clear dynamics even for fast pyrolysis conditions in which particles are complete pyrolyzed in less than 100 ms.

501350 Measuring the Torrefaction Products of Xylan and D-Xylose

Abstract:

Xylan torrefaction plays a crucial role in the pre-treatment of biomass, so this work examines its reactions by measuring the differences between torrefaction products of xylan vs. its monomer, D-xylose at 250° and 300°C. Xylan is a major component of hemicelluloses and one of the least thermally stable polymers in lignocellulosic biomass. While xylan pyrolysis at higher temperatures has received attention [1], its torrefaction kinetics and its differences with the torrefaction kinetics of D-xylose are largely unstudied.

A sample of 0.1 – 0.3 mg of xylan (Sigma Aldrich) or d-xylose (Acros Organics) is dried at 105°C for 15 mins before being flash-torrefied (Pyroprobe Model 5000, CDS) at the selected temperature and held for 30 s. Product analysis is by two-dimensional gas chromatography followed by time-of-flight mass spectrometry (Pegasus 4D, LECO). Products are considered to be identified if they have greater than 80% similarity with the library spectra and their S/N ratio is greater than 100. The products are divided into three categories: common products, products identified only in D-xylose torrefaction, and products identified only in xylan torrefaction. The % peak areas of common products are used to analyze the effects of temperature on the product distribution.

In both raw materials and at both torrefaction temperatures, the % total peak area of water is the largest among all the products. It is above 30% in each case, while the peak areas for the other products are less than 10 %. Interestingly, water % peak area significantly decreases at higher temperature (from 64% to 32 % in d-xylose torrefaction), possibly due to formation of additional products formation at higher temperatures. Water production due to water elimination reactions may play an important role in D-xylose and xylan torrefaction.

D-Xylose torrefaction yields higher % peak areas at 300°C than at 250°C for the most common products (CO_2, glycoladehyde, furfural, furanmethanol, acetol, acetic acid and formic acid), while for xylan, % peak area of CO_2 and formic acid decreased at 300°C. Furfural had the second largest peak area % among the other products of xylan and the highest % peak area for D-xylose (6.56% at 300°C). Another significant result is that methanol, ethanol, propanoic acid, and butyrolactone are identified as torrefaction products only from xylan. On the other hand, furancarboxaldehyde and 4-cylcopentene-1,3-dione are identified in only in D-xylose torrefaction.

[1] X. Zhou, W. Li, R. Mabon, L.J. Broadbelt, L. J., "A critical review on hemicellulose pyrolysis," *Energy Tech.*, 2016, 4, 1 –29.

495111 Supported Transition Metal Carbides As Catalysts for CO_2 Conversion to Syngas Via Reverse Water Gas Shift Reaction

Abstract:

Supported transition metal carbides as catalysts for CO_2 conversion to syngas via Reverse Water Gas Shift reaction

Yichen Zhuang, Faisal Mohamed Khan and David Simakov

Department of Chemical Engineering, University of Waterloo, Waterloo, ON N2L 3G1, Canada

Converting CO_2 into synthetic platform chemicals is an attractive approach to decrease CO_2 emissions and to reduce our dependence on fossil fuels. Thermocatalytic hydrogenation provides advantages of fast reaction rates and high conversion efficiencies (as compared to electro-catalytic and photo-catalytic routes), thus allowing for compact, high-throughput operation. Among possible reaction pathways is the Reverse Water Gas Shift reaction ($CO_2 + H_2 = CO + H_2O$) that has the advantage of low H_2 consumption. The generated CO can be mixed with H_2 to produce syngas, which is a valuable feedstock for chemical industry, i.e., methanol synthesis. Challenges include insufficient activity, incomplete selectivity to CO production, and low stability against coking and sintering.

Copper-based catalysts are highly selective, but their activity is relatively low. Addition of small amounts of noble metals (e.g., Ru) with proper nanostructuring can improve the catalytic activity by as high as 250% (right panel in the figure below). However, it is highly desirable to avoid the use of platinum group metals. Among emerging low-cost catalytic materials, transition metal carbides were recently identified as promising candidates for a number of heterogeneous reactions. In this study, a series of supported transition metal carbides are investigated for their catalytic performance in the Reverse Water Gas Shift reaction. Some catalytic formulations, such as Mo_2C/Al_2O_3, show reasonably high conversions with complete selectivity to CO formation. Molybdenum oxide initially shows similar conversion (left plot in the figure below), but deactivates rapidly by coking. On the contrary, the Mo_2C/Al_2O_3 catalyst exhibits excellent stability maintaining high CO_2 conversions over 100 h of operation at elevated space velocity (GHSV = 15,000 1/h). Results presented include catalytic performance evaluation for several formulations of transition metal carbides, including mixed carbides, as well as characterization of fresh and spent catalytic materials by means of electron microscopy, infrared spectroscopy, thermal gravimetric analysis, physical adsorption, and chemisorption.

494024 Molten Salt Chemical Looping Catalysis for Reactive Separation of HBr in a Halogen Based Natural Gas to Liquids Process

Abstract:

Halogen based conversion of methane to liquid fuels is improved by using molten salts for hydrogen halide reactive separation and regeneration. Molten bromide salts are used in a chemical looping process to selectively convert hydrogen bromide to molecular bromine in the presence of hydrocarbons. The process involves two steps: oxygen reacts with $NiBr_2$ that is dissolved in molten KBr-LiBr and forms bromine and NiO particles; the slurry is then pumped to another reactor where HBr reacts with the NiO to re-form the $NiBr_2$ and water, which closes the salt chemical looping cycle. Sixty-eight metal oxides/bromides were considered, and NiO-$NiBr_2$ was the only material that cycled at 500 °C with reasonable thermodynamics and kinetics.

In contrast to solid-based chemical looping systems, the liquid bromide salt was found to be cycleable without attrition or deactivation. Further, selectivity to hydrogen bromide oxidation over hydrocarbon oxidation was very high when mixtures of olefins and hydrogen bromide were bubbled through the oxide-salt slurry. This was justified by the fact that hydrogen bromide is soluble in the melt whereas hydrocarbons did not dissolve.

In addition, a process model that makes use of reactive separation of hydrogen bromide from hydrocarbons using molten salt-based chemical looping is presented as a potentially lower cost alternative to a process model using conventional separations in bromine-base methane conversion. The heat exchange duties and separations in corrosive and non-corrosive environments were compared using the two models.

AUTHOR INDEX

AUTHOR INDEX

AUTHOR INDEX

AUTHOR INDEX

AUTHOR INDEX

AUTHOR INDEX

AUTHOR INDEX

AUTHOR INDEX

AUTHOR INDEX

AUTHOR INDEX

AUTHOR INDEX

AUTHOR INDEX

AUTHOR INDEX

AUTHOR INDEX

AUTHOR INDEX

AUTHOR INDEX

AIChE
120 Wall Street, FL 23
New York, NY 10005-4020

CURRAN ASSOCIATES INC.
proceedings.com

ISBN 978-1-5108-5806-0

Catalysis and Reaction Engineering Division 2017

Core Programming Area at the 2017 AIChE Annual Meeting

Minneapolis, Minnesota, USA
29 October – 3 November 2017

Volume 2 of 2

AIChE
The Global Home of Chemical Engineers

Catalysis and Reaction Engineering Division 2017

Core Programming Area at the 2017 AIChE Annual Meeting

Minneapolis, Minnesota, USA
29 October – 3 November 2017

Volume 2 of 2

ISBN: 978-1-5108-5806-0

Printed from e-media with permission by:

Curran Associates, Inc.
57 Morehouse Lane
Red Hook, NY 12571

Printed by Curran Associates, Inc. (2018)

For permission requests, please contact AIChE
at the address below.

AIChE
120 Wall Street, FL 23
New York, NY 10005-4020

Phone: (800) 242-4363
Fax: (203) 775-5177

www.aiche.org

Additional copies of this publication are available from:

Curran Associates, Inc.
57 Morehouse Lane
Red Hook, NY 12571 USA
Phone: 845-758-0400
Fax: 845-758-2633
Email: curran@proceedings.com
Web: www.proceedings.com

TABLE OF CONTENTS

Volume 1

Volume 2

Some abstracts in this book refer to images that were not made available by the publisher

500371 Investigation of a New Preparation Method for Cobalt Fischer-Tropsch Catalysts

Abstract:

Investigation of a New Preparation Method for Cobalt Fischer-Tropsch Catalysts

Mahmood Rahmati, Saeed Safdari, Calvin Bartholomew, William C. Hecker, Morris D. Argyle*

Department of Chemical Engineering, Brigham Young University, Provo, UT 84602

Fischer-Tropsch synthesis (FTS) is a catalytic process for production of a wide range of environmentally friendly liquid fuels from synthesis gas. Supported cobalt is used as a commercial catalyst for FTS. Conventional methods for large scale catalyst preparation are time and energy intensive. Research is continuing to develop and improve new preparation methods for cobalt FT catalysts [1]. Recently, we developed a new, simple one-step catalyst preparation method that was successfully applied to produce unsupported iron FT catalysts [2]. This method has been modified to prepare supported cobalt catalysts. This novel preparation method consumes about 75% less energy and 50% less water compared to the incipient wetness impregnation (IWI) method that is one of the most popular methods for cobalt FT catalyst preparation. The goal of this research was to compare supported cobalt FT catalysts prepared by the IWI method, similar to those used industrially, with the same catalyst prepared by the new technique called solvent deficient impregnation (SDI). The macro and nano-scale structures of the catalysts have been compared with various characterization techniques, including TPO, TPR, XRD, TEM, BET, EOR, and CO and H_2 uptake. Catalytic activity, selectivity, and stability were then investigated using a packed bed reactor operated at industrial FTS conditions (220°C and 20 atm) over 240 hour runs. The kinetic data show that catalysts prepared with the SDI method have the same rate, selectivity, and stability compared to catalysts prepared by IWI. Due to low cost and simplicity, this novel one-step preparation method should be preferred for future catalyst development studies.

1. Munnik, P., P.E. de Jongh, and K.P. de Jong, *Recent Developments in the Synthesis of Supported Catalysts.* Chemical Reviews, 2015. **115**(14): p. 6687-6718.

2. Brunner, K.M., et al., *Preparation of an Unsupported Iron Fischerâ€"Tropsch Catalyst by a Simple, Novel, Solvent-Deficient Precipitation (SDP) Method.* Energy & Fuels, 2015. **29**(3): p. 1972-1977.

488873 Short Chain Hydrocarbons Growth Probability in Fischer-Tropsch Synthesis over Cobalt Catalyst

Abstract:

The light hydrocarbon (C_1-C_6) product distributions from a fixed bed Fischer-Tropsch reactor with a 10% Co/TiO_2 catalyst were analysed over an extended period of time as the activity of the catalyst changed. These short chain hydrocarbon product distributions were studied starting at 200 ^oC with syngas for 48 hours and then increasing the temperature to 220 ^oC. The feed was then switched back and forth from syngas to syngas + extra N_2 with increased total pressure in such a way as to keep the reactant partial pressures the same in both cases.

The results showed that at the higher temperature (220 ^oC), more short chain paraffins were obtained than short chain olefin products. The chain growth probability (α) values for the olefins (O_3-O_6) were smaller than those for the paraffins (P_3-P_6) under all the FT reaction conditions studied. The results showed that the co-feeding of N_2 to the syngas during FT runs slightly increased the α values for most of the FT runs. At low reaction conversion the olefin selectivity was higher than that of the selectivity to paraffin. The light paraffin α values were high reaching 1. The data also showed that light olefins α values increased with increasing selectivity to the olefins asymptotically reaching a value of approximately 0.79 whereas the values for the paraffins were not affected by these changes in conditions. This asymptotic behaviour might be indicative of an approach to reaction equilibrium amongst the olefin species.

Overall, these results show that the change operating conditions caused by loss of activity with time affects olefin and paraffin product distributions differently. This suggests that olefin and paraffin product distributions should be considered separately in the development of models for FT processes.

499542 2-D Modeling of Fischer Tropsch Packed Bed Reactor: First Step Towards Scale-up

Abstract:

Fischer Tropsch (FT) synthesis enables conversion of natural gas, shale gas or bio gas to liquid hydrocarbons, and is a boon for gas rich countries like the United States and Qatar to monetize their vast natural gas wealth. This process is an exothermic chemical reaction that converts synthesis gas or â€˜syngasâ€™ (a mixture of hydrogen and carbon monoxide produced by reforming of methane) into a variety of value-added chemicals and ultra clean fuels. Commercial FT reactor designs are the multi tubular fixed bed reactor and the slurry bubble column reactor. The slurry bubble column reactor provides excellent heat transfer, as the reaction takes place in a liquid slurry phase. However, it faces non-trivial process challenges including product recovery and catalyst attrition. On the other hand, the multi tubular reactor comprises of several tubes containing fixed catalyst bed, and enjoys advantages of good product selectivity, recovery and ease of maintenance. However, it faces challenges related to heat and mass transfer resistances that greatly limit its scale up and operation at high capacities. This modeling work aims at providing detailed assessment of the effect of mass and heat transfer in a fixed bed reactor to assess new fixed bed materials aimed at alleviating radial heat transfer limitations.

In order to understand higher resolution picture of such effects, a multi-scale CFD model is developed in COMSOL®, which facilitates observation both at the catalyst pellet level and in bulk fluid conditions. The reactor bed model is developed using different kinetic models which represent FT mechanisms on different cobalt-based catalysts, and thus facilitates comparison under a variety of operational conditions. In particular, a novel Micro Fibrous Entrapped Cobalt Catalyst (MFECC) developed by our collaborator (Sheng *et al*. 2012 , Tatarchuk *et al*. 2013) has been simulated in a bed scale model, which demonstrates orders of magnitude improvement in bed thermal conductivity. Additionally, the model investigates the potentials of scaling up the diametrical geometry of the reactor to improve throughput by utilizing the MFECC bed, as it could help in minimizing the impact of hot spot formations on the product profile and undesired reaction routes that result in formation of methane and light hydrocarbons. Current study results for a catalyst at 20 bar and at a gas hourly space velocity of 5000 1/h in a reactor tube of 0.59 inch ID show hotspot formation of about at the tube center. In contrast to this, the temperature rise in MFECC bed for same operating condition is only , which proves tremendous improvement upon conventional packed bed design. This work is a first stage towards tackling challenges related to reactor scale up and runaway hotspot formation in a fixed bed FT reaction, and is a part of broader project involving numerous experimental and simulation studies.

References:

Sheng, M., *et al.*, High conductivity catalyst structures for applications in exothermic reactions. Applied Catalysis A: General, 2012. 445: p. 143-152.

Tatarchuk, B., *et al.*, Microfibrous media and packing method for optimizing and controlling highly exothermic and highly endothermic reactions/processes. 2013, US Patent: 8420023 B2

Heat Generation and Removal in Fixed-bed Reactors for Fischer-Tropsch Synthesis

Paper 450g

B. Todic[1], *M. Mandic*[1], N. Nikacevic[2], D. B. Bukur[1,3,*]

[1] *Texas A&M Univesity at Qatar, PO Box 23874, Doha, Qatar*

[2] *Faculty of Technology and Metallurgy, University of Belgrade, Karnegijeva 4, Belgrade, Serbia*

[3] *Texas A&M University, 3122 TAMU, College Station, Texas 77843, United States*

1. Introduction

Fischer-Tropsch synthesis (FTS) is one of the main ways utilized commercially to convert natural gas, coal and biomass into liquid hydrocarbon fuels and other value-added products. The most frequently used type of reactor for FTS is a multi-tubular fixed bed reactor (FBR). Due to the high exothermicity of FTS reaction, special care has to be taken of removal of generated heat during the design of such reactors. Understanding the influence of process parameters on heat generation and removal is very important in predicting hot spots and avoiding potential temperature runaways.

Majority of previously published studies on this topic used one-dimensional pseudo-homogeneous models of FBR reactor for FTS [1]. Relatively little attention has been given to the topic of heat management, despite its practical significance. Also a common assumption was that gas-phase correlations for heat transport parameters are applicable, even though a certain amount of liquid is present both accumulated in the FTS catalyst pellets and trickling down the reactor bed.

We will present analysis of heat management characteristics in FTS fixed bed reactor using a cobalt-based catalyst. Results of a two-dimensional model simulations which incorporates the effect of both gas and liquid phase will be presented. Based on these the influence of various relevant process parameters on FTS FBR heat generation and removal will be examined.

2. Fixed-bed reactor model

The reactor model is developed as a two-dimensional pseudo-homogeneous model without axial mixing, where intra-particle transport is taken into account via the component effectiveness factor η_i. Components considered in the system are CO, H_2, H_2O and hydrocarbons products CH_4, C_2, C_3, C_4 … C_{24} and lumped C_{25+}. The main model equations are shown below.

Two-dimensional mass balance is given by:

$$\frac{D_i^{er}}{u_s}\left(\frac{\partial^2(u_s C_i)}{\partial r^2}+\frac{1}{r}\frac{\partial(u_s C_i)}{\partial r}\right)-\frac{\partial(u_s C_i)}{\partial z}+\eta_i\rho_b R_i=0 \qquad (1)$$

where C_i is the molar concentration of component i (i = CO, H_2, H_2O and hydrocarbons CH_4, C_2, … C_{24} and lumped C_{25+}), u_s is superficial velocity, z and r are axial and radial dimension, A_{cs} is cross-section area of a tube with diameter d_t ($A_{cs} = \frac{d_t^2 \pi}{4}$), D_i^{er} is effective radial diffusivity of species i, η_i is catalyst effectiveness for species i, ρ_b is reactor bed density and R_i is rate of species i disappearance or formation. Rates of reactants (CO and H_2) disappearance and products (H_2O, n-paraffin and 1-olefin) formation are calculated at each point along the reactor bed using a hybrid kinetic approach [2-4].

Heat balance equation is defined as:

$$\lambda_{er}\left(\frac{\partial^2 T}{\partial r^2}+\frac{1}{r}\frac{\partial T}{\partial r}\right)-u_s\rho C_p\frac{\partial T}{\partial z}+\eta_{CO}\rho_b\left(-\Delta H_r\right)R_{CO}=0 \quad (2)$$

where T is temperature, $\left(-\Delta H_r\right)$ is reaction enthalpy per mole of CO consumed, , ρ is fluid phase density, λ_{er} is effective radial thermal conductivity, η_{co} is effectiveness factor for CO and C_p is fluid heat capacity. Pressure drop is calculated using Ergun's correlation [5]. Literature correlations considering both gas and liquid phase influence on heat transport parameters were incorporated in the model [6, 7]. The split between the gas and the liquid phase was estimated using Raoult's law. Effectiveness factor values were estimated based on recent work on the single particle modeling done in our group [8].

The boundary conditions needed to solve Eqs. (1) and (2) were defined as:

$$(z=0,\ 0\le r\le R_t) \qquad C_i=C_i^{in},\ u_s=u_s^{in} \quad (3)$$

$$(0<z\le L,\ r=0\ and\ r=R_t) \qquad \frac{\partial(u_s C_i)}{\partial r}=0 \quad (4)$$

$$(z=0,\ 0\le r\le R_t) \qquad T=T_{in} \quad (5)$$

$$(0<z\le L,\ r=0) \qquad \frac{\partial T}{\partial r}=0 \quad (6)$$

$$(0<z\le L,\ r=R_t) \qquad \lambda_{er}\frac{\partial T}{\partial r}=-h_{wall}\left(T-T_{wall}\right) \quad (7)$$

where L is reactor tube length, R_t is reactor tube radius, F_i^{in}, T_{in} and P_{in} are inlet flowrate of species i, inlet temperature and inlet pressure, respectively, h_{wall} is radial heat transfer coefficient at the wall and T_{wall} is wall temperature.

Model equations were discretized and solved using gPROMS software for a number of process conditions and geometric parameters.

3. Results and discussion

The influence of several process parameters (inlet flowrates and temperature, wall temperature, tube and particle diameter) on temperature distribution in radial and axial direction was investigated. At base case conditions used in our study (T_{in} = 473 K, P_{in} = 25 bar, H_2/CO feed ratio = 2, F_{in} = 0.15 mol/s, T_{wall} = T_{in}, d_t = 2.6 cm, d_p = 2.5 mm, L = 10 m) temperature increases in axial direction and reaches a maximum core temperature of 479 K close to the reactor inlet (z ~ 1 m), after which it decreases gradually along the reactor. The radial gradient inside the tube is about 3

K at the maximum temperature position, which highlights the importance of using two-dimensional models to represent this process.

Simulations with varying one process parameter and keeping others constant show temperature runaway occurring for $F_{in} < 0.03$ mol/s, $T_{in} > 497$ K and $d_t > 5.9$ cm. Thermal behavior has a very complex dependency of process conditions and often variation of one parameter can have conflicting influences (both positive and negative effects) on heat management. For example, increasing inlet temperature increases the reaction rate and thus amount of generated heat. However, it also decreases the catalyst effectiveness which results in a lower amount of heat generation.

Varying inlet flowrate shows that the gas phase velocity increase has a very positive effect on the heat transfer coefficient at the wall, however the impact of gas phase composition is noticeable as well (Fig. 1). The highest h_{wall} values are achieved with pure syngas (i.e. inlet conditions), while generation of hydrocarbons along the reactor ($X_{CO} > 0\%$) changes the physical properties of the gas phase and results in h_{wall} decrease (Fig. 2a). Increasing gas phase velocity also has a positive effect on the effective radial thermal conductivity. However, we also see a positive influence of a liquid phase velocity (i.e. CO conversion). The liquid phase is generated along the reactor and thus Reynolds number of liquid increases from 0 to 0.57 for the base case inlet and outlet conditions, respectively, resulting in about 18% increase of λ_{er} (Fig. 2b).

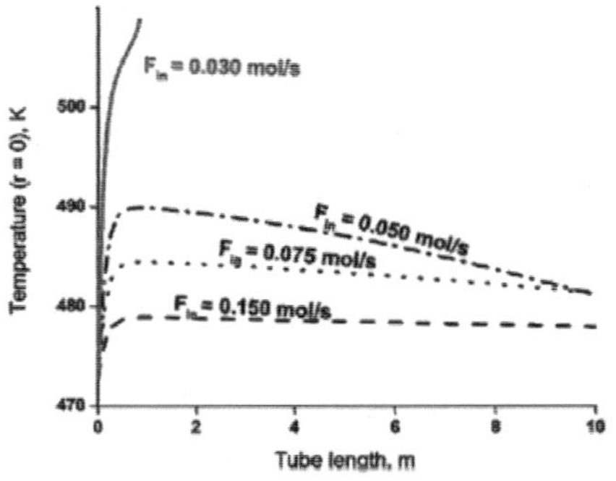

Figure 1. Effect of velocity (flow rate)

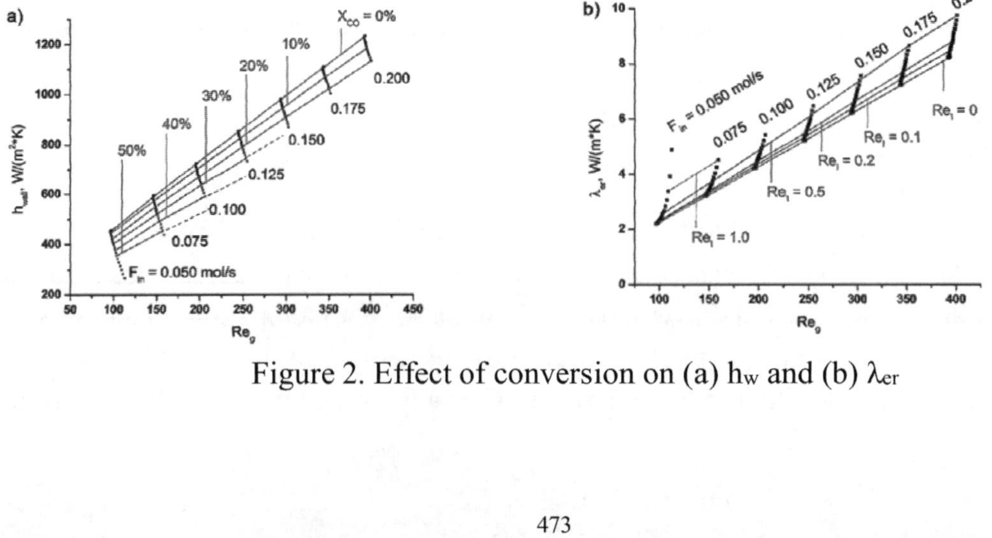

Figure 2. Effect of conversion on (a) h_w and (b) λ_{er}

Increase of particle diameter is beneficial from standpoint of heat management, both from the perspective of decreased heat generation due to lower effectiveness factor and higher heat transport coefficients due to increased Reynolds numbers (Fig. 3). However, this leads to low productivity and poor FTS product selectivity. Increase of tube diameter results in much higher hotspots and more likely runaways, due to higher radial thermal resistance and lower transport coefficients related to lower gas and liquid velocities (Fig. 4). These effects can be somewhat alleviated by increasing the inlet flowrate, but not completely avoided.

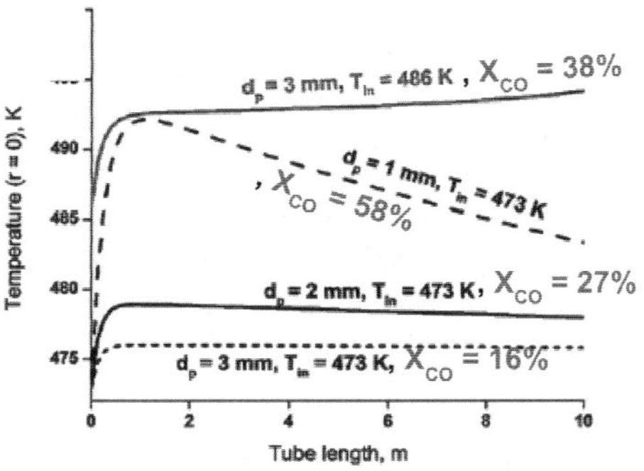

Figure 3. Effect of particle size

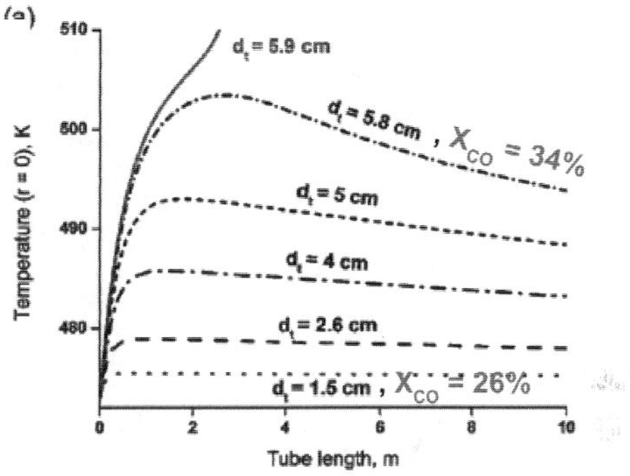

Figure 4. Effect of tube diameter

4. Conclusions

Simulation results with variation of process parameters show that FTS fixed bed reactors are very sensitive in terms of heat management and any large disturbance can lead to potential temperature runaway. Reasons for this behavior of FTS fixed bed reactors are analyzed in detail. Especially interesting and previously unknown in open literature is the contribution which liquid phase has on heat transport in FTS fixed bed reactors.

References

[1] A.P. Steynberg, M.E. Dry, B.H. Davis and B.B. Breman, in S. André and D. Mark (Editors), Studies in Surface Science and Catalysis, Volume 152, Elsevier, 2004, p. 64.

[2] I.C. Yates and C.N. Satterfield, Energy & Fuels, 5 (1991) 168.

[3] W. Ma, G. Jacobs, T.K. Das, C.M. Masuku, J. Kang, V.R.R. Pendyala, B.H. Davis, J.L.S. Klettlinger and C.H. Yen, Industrial & Engineering Chemistry Research, 53 (2014) 2157.

[4] B. Todic, W. Ma, G. Jacobs, B.H. Davis and D.B. Bukur, Catalysis Today, 228 (2014) 32.

[5] S. Ergun, Chemical Engineering Progress, 48 (1952) 89.

[6] A. Matsuura, Y. Hitaka, T. Akehata and T. Shirai, Heat Transfer - Japanse Research, 8 (1979) 44.

[7] V. Specchia, G. Baldi and S. Sicardi, Chemical Engineering Communications, 4 (1980) 361.

[8] M. Mandić, B. Todić, L. Živanić, N. Nikačević and D.B. Bukur, Industrial & Engineering Chemistry Research, 56 (2017) 2733.

499589 A General Adsorption/Reaction Framework for Modular and Multi-Functional Process Design

Abstract:

Modular chemical processes are easily deployable for small scale operations in remote locations to produce value-added chemicals from unconventional resources and distributed feedstocks such as stranded natural gas, CO_2 from flue gas and associated shale gas. The advantages of modular processes include compactness, prevention of inefficiencies of on-site equipment manufacturing, freedom of constructing high-quality equipment in specialized locations, and faster production time. However, modular processes often fail to achieve economies of scale. One way to address this challenge is to design *multifunctional* modular processes using existing single-functional materials. To this end, we first propose a multi-material design approach which integrates multiple phenomena within a single unit and thereby leads to multifunctional modular processes using existing materials. We use a systematic framework to select and arrange materials (e.g., adsorbent and catalyst blocks) with unique tasks (e.g., adsorption, reaction, storage) in a sequential block-by-block arrangement inside a packed column, and configure operational cycles for integrated, periodic and multifunctional processes. The task-integrated column operates in a cyclic or periodic manner, where each cycle consists of several steps. We use a single superstructure that embeds different design alternatives using blocks, where each block represents a specific material performing a specific task for a specific time step. The functional block-based superstructure is general such that it contains numerous cyclic process configurations for separation processes including pressure/temperature swing adsorption (PSA/TSA), as well as for integrated separation-conversion processes including sorption-enhanced reaction processes (SERP).

Next, in order to accurately predict the performance of different processes generated via multi-material design, we develop Generalized Reaction-Adsorption Model (GRAM) â€“ a high-fidelity process model for dynamic adsorption-reaction systems. GRAM is a one-dimensional, pseudo-homogeneous, non-isothermal, non-adiabatic, and non-isobaric model that captures the reaction and adsorption dynamics in hybrid adsorption-reaction systems. The model is validated using different experimental data from literature. Specifically, the model predictions show agreement with experimental observations for sorption-enhanced steam-methane reforming (SMR)[1,3], sorption-enhanced reverse water gas shift (RWGS)[2], and conventional SMR[4,5]. Computationally, the model handles multi-feed, multi-functional and multi-product processes well.

The GRAM model is further incorporated within the multi-material design framework to design several novel multifunctional modular processes for cost-effective carbon capture and utilization using unconventional energy sources such as stranded and distributed natural gas, CO_2-contaminated shale gas, biogas, landfill gas, fuel gas and flare gas. The objective is to produce valuable products such as hydrogen and syngas. Computational studies are performed for several configurations of the column containing (i) pure catalyst, (ii) pure adsorbent, (iii) heterogeneously-compartmentalized adsorbent and catalyst, and (iv) homogeneously-distributed uniform mixture of adsorbent and catalyst. These intensified processes use sections of adsorbent and catalyst for simultaneous capture and

conversion of reactants.

References:

[1] Carvill, B.T., Hufton, J.R., Anand, M., Sircar, S. Sorption-Enhanced Reaction Process. AIChE Journal. 1996.

[2] Jang, H.M., Lee, K.B., Caram, H.S., Sircar, S. High-purity hydrogen production through sorption enhanced water gas shift reaction using K_2CO_3-promoted hydrotalcite. Chemical Engineering Science. 2012.

[3] Hufton, J.R., Mayorga, S., Sircar, S. Sorption-enhanced reaction process for hydrogen production. AIChE Journal. 1999.

[4] Shu, J., Grandjean, B.P.A., Kaliaguine, S. Methane steam reforming in asymmetric Pd- and Pd-Ag/porous SS membrane reactors. Applied Catalysis. 1994.

[5] Gallucci, F., Paturzo, L., Basile, A. A simulation study of the steam reforming of methane in a dense tubular membrane reactor. International Journal of Hydrogen Energy. 2004.

496493 Ultrasonic Tuning of TiO_2 Based Mixed Oxides Structural Properties and Catalytic Activity

Abstract:

Ultrasound (US) has a potential to impact every stage of the preparation of a material in liquid phase. US can produce nanomaterials in amorphous state (due to the very high cooling rates), which is relevant to fields such as catalysis, magnetism and coating processes [1]. US generates shock waves that facilitate coating and insertion/intercalation processes and improves the distribution of the active phase on a support. We pursue ultrasound powder synthesis as a greener method to lower solvent amount and catalyst precursor, as well as to accelerate lengthy preparation method such as sol-gel and impregnation. US-assisted synthesis produces catalysts with better morphology and activity. Simple tuning of operating conditions increases the range of obtainable features for different purposes [2,3].

We report the ultrasound assisted sol-gel and impregnation synthesis of sulphated ZrO_2-TiO_2 and Mn oxides-TiO_2 systems. We characterized the catalysts with N_2 adsoption, XRD, SEM-EDX and TEM. We tested the sulphated ZrO_2-TiO_2 in the free fatty acids (FFA) esterification reaction and the sulphated ZrO_2-TiO_2 and the Mn_xO_y-TiO_2 systems in the ultrasound-assisted photocatalytic degradation of recalcitrant water pollutants (pharmaceutical products). We adopted a ultrasound probe with a tip of 13 or 20 mm diameter. The ultrasound operating frequency was fixed at 20 KHz. The nominal power was 500 W and we varied the amplitude from 20 to 60 % adopting continuous ultrasound or ultrasound pulses (0.5 on and 0.5 off). In the sol-gel synthesis of sulphated ZrO_2-TiO_2 systems, ultrasound pulses increased the surface area up to 220 m^2g^{-1} for the lowest solvent: precursors ratio of 15:1 mol, corresponding to the highest power density. Solvent: precursors molar ratios of 30:1 and 60:1 gave a SSA of 150 m^2g^{-1} and 30 m^2g^{-1}, respectively. The samples obtained with pulses were also the ones exhibiting the highest Brønsted surface acidity. The catalysts with the highest acidity/SSA ratio were the most active in the free fatty acids esterification reaction.

We deposit Mn_xO_y on TiO_2 in one step with the ultrasound-assisted sol-gel synthesis or via ultrasonic impregnation in a second step. We prepared Mn/TiO_2 catalyst in one step with a classical sol–gel method, starting from butyl-titanate, deionized water, nitric acid, anhydrous ethanol, and manganese nitrate [4]. After mixing at room temperature to yield a yellowish transparent sol, we dried the precursor at 80°C for 24h, converting it into a xerogel. The calcination to obtain the Mn/TiO_2 catalyst was standard at 500°C for 5h. We prepared the same catalyst applying US, either maintaining the same calcination parameters or decreasing the calcination temperature to 400°C for 2 hours to trigger the formation of anatase instead of rutile as an active phase.

Mn/TiO_2 two steps synthesis consisted of the US-assisted surface deposition of Mn-oxides on TiO_2 (both nanometric or micrometric), synthesized in a preceding stage [5]. Generally, US promote homogeneous nucleation in a first step, and enhance the crystals growth in a second stage. Acoustic cavitation, in particular for the two-steps Mn deposition, generates high-speed micro jets which impact the TiO_2 support with a cleaning and de-agglomerating effect. Ultrasound increases the dispersion of the Mn species over both supports and has a role in stabilizing Mn_xO_y species that samples synthesized with traditional sol-gel don't normally exhibit (preliminary XRD results) [6]. The modulation of the main sonication operating parameters produces a mixed-size distribution of Mn and Mn-Ox species on TiO_2 rather than a narrow distribution of small crystallites, which often turns out to be better for the photocatalytic activity. US also improves the distribution of Mn_xO_y over micro-metric TiO_2 (SEM-EDX analysis), which is meaningful from a safety point of view since body barriers may not "recognize" nano-powders. The photocatalytic activity of Mn_xO_y over nano and micro-size TiO_2 is comparable. Considering instead the comparison between Mn and other metals, Mn is cheaper and one of a kind, considering the variety of oxides that can form. By means of US we can also modulate and enhance the formation of specific oxidation states [6,7].

References

[1] T.J. Mason, Ultrasound in synthetic organic chemistry, *Chem. Soc. Rev.*, 26 (1997), pp. 443–451.

[2] S.J. Musevi, A. Aslani, H. Motahari, H. Salimi, Offer a novel method for size appraise of NiO Nanoparticles by PL analysis: synthesis by sonochemical method, *J. Saudi Chem. Soc.*, 20 (2016), pp. 245–252.

[3] C. Pholnak, C. Sirisathitkul, D.J. Harding, Characterizations of octahedral zinc oxide synthesized by sonochemical method, *J. Phys. Chem. Solids*, 72 (2011), pp. 817–823.

[4] Qi-lin Chen, Rui-tang Guo, Qing-shan Wang, Wei-guo Pan, Ning-zhi Yang, Chen-zi Lu, Shu-xian Wang, *J. Taiwan Inst. Chem. Eng.*, 64 (2016) 116–123.

[5] M. Stucchi, C.L. Bianchi, C. Pirola, G. Cerrato, S. Morandi, C. Argirusis, G. Sourkouni, A. Naldoni, V. Capucci, *Ultrason. Sonochem.*, 31 (2016) 295–301.

[6] M.I. Díez-García, V. Manzi-Orezzoli, M. Jankulovska, S. Anandan, P. Bonete, R. Gómez, T. Lana-Villarreal, Effects of ultrasound irradiation on the synthesis of metal oxide nanostructures, *Phys. Procedia*, 63 (2015), pp. 85–90.

[7] K.D. Bhatte, D.N. Sawant, D.V. Pinjari, A.B. Pandit, B.M. Bhanage, One pot green synthesis of nano sized zinc oxide by sonochemical method, *Mater. Lett.*, 77 (2012), pp. 93–95.

497703 Multi-Model Operability Approach for Process Design, Intensification and Modularity: Application to Nonlinear and High-Dimensional Membrane Reactors

Abstract:

Membrane reactors (MRs) have arisen as alternatives for conventional reactors, enabling previously economically unfavorable processes. MRs are also intensified processes that are candidates for modular applications. Such reactors provide a built-in membrane separation layer along the entire reactor length, where one or more reaction products can be selectively removed. This removal shifts the reaction equilibrium towards the products, increasing reactants conversion and production rate. However, the process design of MRs is intrinsically challenging. MRs combine chemical reactions with separation transport phenomena, which results in models of complex and nonlinear nature for their representation. MRs are also commonly placed in highly constrained environments, which poses an additional difficulty to the process design task, especially if process integration is needed for modularity. In this presentation, a multi-model operability approach is proposed to address these challenges. Operability was first introduced as a tool for the design and control interface to calculate the inputs that are needed to achieve specified desired outputs in a process [1]. Here, the operability pathway is extended to provide a framework able to tackle nonlinear high-dimensional and constrained systems. This framework attains process intensification, ensuring high efficiency units, suitable for modular systems applications.

In the past decades, process operability has been traditionally applied to low-dimensional nonlinear systems. The implementation of operability for high-dimensional nonlinear systems was successfully performed by employing optimization algorithms based on nonlinear programming (NLP) [2,3]. The novelty of this work derives from the fully utilization of linear programming (LP) tools to handle systems of the same complexity, with the advantage of computational time reduction. In particular, the original nonlinear system is divided into several subsystems and a linear model is calculated for each subsystem. The properties of all linear models are then analyzed, such as the eigenvalues (or singular values), so that the minimum number of model sets that adequately represents the process nonlinearity is obtained. A multi-level optimization algorithm is then formulated based on LP concepts, aiming intensification and control targets, while considering process requirements and constraints.

The developed framework is applied to a membrane reactor for the direct methane aromatization conversion (DMA-MR) to hydrogen and benzene. DMA-MR corresponds to a candidate system for modular natural gas utilization onsite at stranded gas fields without the need of building expensive pipelines [2]. Preliminary results for a low-dimensional system show that benzene production can be optimized and the membrane reactor footprint reduced by 60% in terms of volume (catalyst zone) and 80% in membrane area. These results indicate the potential of the proposed framework to facilitate process intensification towards modularity. In this presentation, a comparison will be established between this approach and previously developed NLP-based approaches [4] to analyze the trade-off between accuracy and

computational expense. The expansion to a higher dimensional system will then be discussed, contemplating membrane parameters, design and operating conditions that are critical for process intensification and modularity.

References:

1. Vinson D. R. and Georgakis C. "A new measure of process output controllability". J. Proc. Cont. 2000, 10(2-3), 185-194.
2. Carrasco J. C. and Lima F.V. "An optimization-based operability framework for process design and intensification of modular natural gas utilization systems". Accepted for publication in Comput. Chem. Eng. 2017; DOI: CACE-5645.
3. Carrasco J. C. and Lima F. V. "Novel operability-based approach for process design and intensification: application to a membrane reactor for direct methane aromatization". AIChE J. 2017; 63(3): 975-983.
4. Carrasco J.C. and Lima F. V. "Operability-based approach for process design, intensification, and control: application to high-dimensional and nonlinear membrane reactors". In proceedings of the FOCAPO/CPC, 2017.

493441 A Modeling Methodology for Predicting Intensified CO_2 removal Efficiency with Rotating Packed Bed

Abstract:

The opportunity in process intensification (PI) of carbon dioxide (CO_2) capture has drawn much attention recently. Since last decade, rotating packed bed (RPB) has been considered as an intensified alternative to conventional packed bed (PB) in CO_2 chemical absorption processes which require excessive absorber volume resulting from mass transfer resistance across vapor-liquid interface during the absorption process [1-3].

Reliable predictions of intensified CO_2 removal efficiency in RPB are essential to design the intensified CO_2 absorption/stripping process. In this study, we present a modeling methodology with sound theoretical basis for predicting the CO_2 removal efficiency in RPB. The methodology assumes the modeling of RPB and that of PB are identical if the gravitational force presented in RPB can be properly taken into account in modeling the mass transfer resistance. The methodology consists of three steps. First, we validate the PB model with PB experimental data [4]. The PB model is based on Aspen rate-based PB model with liquid film discretization (Aspen RateSep) [5]. Second, we predict the CO_2 removal efficiency of RPB experiments [1,4] with the validated PB model. Through this step, we identify the difference in CO_2 removal efficiency with and without the gravitational effect. Third, we account for the gravitational effect in RPB with the adjustable mass transfer factors in the validated PB model. Specifically, the liquid mass transfer coefficient factor and the interfacial area factor are chosen to account for the RPB gravitational effects with the validated PB model. The absolute average deviation (AAD) between the prediction results and the experimental RPB data [1,4] are found to be less than 10% when lean loading of MEA is in the typical operating range of 0.16 ~ 0.44 mol CO_2/mol amine, suggesting a robust predictive capability of our modeling methodology.

This modeling methodology not only provides new insights on RPB, but also avails chemical engineers a practical and theoretically sound modeling capability to estimate the intensified CO_2removal efficiency when replacing PB with RPB. Furthermore, by taking advantages of proven modeling tools in commercial simulators, the methodology could be widely practiced by engineers worldwide to examine the process intensification potential of replacing PB with RPB.

[1] Yu, C.H., Chen, M.T., Chen, H., Tan, C.S., "Effects of Process Configurations for Combination of Rotating Packed Bed and Packed Bed on CO_2 Capture", *Applied Energy*, 175, 269-276 (2016).

[2] Yu, C.H., Lin, Y.X., Tan, C.S., "Effects of Inorganic Salts on Absorption of CO_2 and O_2 for Absorbents Containing Diethylenetriamine and Piperazine", *Int. J. Greenh. Gas Cont.*, 30, 118-124 (2014).

[3] Yu, C.H., Cheng, H.H., Tan, C.S., "CO Capture by Alkanolamine Solutions Containing Diethylenetriamine

2

and Piperazine in a Rotating Packed Bed", *Int. J. Greenh. Gas Cont.*, 9, 136-147 (2012).

[4] Thiels, M., Wong, D.S.H., Yu, C.H., Kang, J.L., Jang, S.S., Tan, C.S. Modeling and Design of Carbon Dioxide Absorption in Rotating Packed Bed. *11th IFAC Symposium on Dynamics and Control of Process Systems*.895-900 (2016).

[5] Zhang, Y., Chen, H., Chen, C.C., Plaza, J.M., Dugas, R., Rochelle, G.T. Rate-Based Process Modeling Study of CO_2 Capture with Aqueous Monoethanolamine Solution. *Ind. Eng. Chem. Res.*48, 9233-9246 (2009).

Keywords: Process Intensification; Rotating Packed Bed; Aspen Plus; CO_2 Chemical Absorption; Mass Transfer Coefficient; Interfacial Area

482122 Carbon Dioxide Capture By Sodium Hydroxide-Glycerol Aqueous Solution in a Rotating Packed Bed

Abstract:

Carbon dioxide capture by chemical absorption has been one of the most popular approaches. However, low solubility of CO_2 in water and high energy demand for absorbent regeneration hinder its feasibility. In this study, glycerol, a biodiesel waste by-product, was added to sodium hydroxide (NaOH) aqueous solution to relax these drawbacks, and to serve as an efficient absorbent in a rotating packed bed (RPB) for CO_2 capture. The experimental results suggested that the addition of glycerol in the NaOH aqueous solution boosted the overall CO_2 absorption efficiency to more than 90% easily. Under the high G (rotating speed) condition, the (chemical) enhancement factor could go up to more than 50. Furthermore, at same NaOH concentration, the mass transfer coefficients (K_Ga) could be doubled even when the viscous glycerol was added because of intensified centrifugal field. It is clear indicated that, in an RPB, glycerol addition in NaOH solution showed very positive effect on CO_2 capture not only for better mass transfer performance but also for the potential utilization of biodiesel by-product.

484778 Graphic Synthesis Method for Multi-Technique Integration Separation Sequences of Multitudinous Refinery Gases

Abstract:

Graphic Synthesis Method for Multi-technique Integration Separation Sequences of Multitudinous Refinery Gases

Xuehua Ruan, Hongyan Xiao, Xiaobin Jiang, Xiaoming Yan, Yan Dai, Gaohong He

School of Petroleum and Chemical Engineering, State Key Laboratory of Fine Chemicals, Dalian University of Technology, Panjin, Liaoning, P.R. China, 124221

Corresponding Author Email: hgaohong@dlut.edu.cn

In petroleum and chemical industries, considerable crude oil, i.e., 3~8 %, is converted into refinery gases and used as furnace fuels. Many valuable species in the refinery gases, e.g., H_2, C_2H_6, C_2H_4, C_3H_8 and C_4H_{10}, are depleted. Currently, this wastage is going on increasing because more and more hydro-processes are built up for refining heavy sour crude oils to high quality products. Accordingly, the effective separation of valuable species from refinery gases has been one of the most potential approach to improve crude oil utilization percentage.

Nowadays, some separation unit operations have been attempted to fractionate and reclaim refinery gases, e.g., pressure swing adsorption, glassy polymer membrane permeation and cryogenic condensation for recovering H_2, shallow condensation, gasoline absorption and rubbery polymer membrane permeation for concentrating light hydrocarbons, and distillation for further separating hydrocarbons. Nevertheless, any simplex technique cannot fractionate all the valuable components efficiently and sufficiently from multitudinous refinery gases. It is necessary to integrate various techniques together as the hybrid processes to achieve the recovery task. In this instance, a shortcut method is established in this work to construct the multi-technique integration processes, in which highly efficient separation unit operations are coupled together within an optimal sequence to improve efficiency.

This shortcut synthesis method is a graphic means based on expressing separation units by vectors in composition space. Three supporting technical measures are summarized:

1. After ranking components with their separation characteristics and belongings in products, refinery gases are reduced to be pseudo-ternary mixtures of H_2, light hydrocarbons and fuel gases, meanwhile the composition space is visualized into triangular coordinate system.

2. Based on the efficiency analysis, dominant feed ranges for adsorption, condensation, glassy and rubbery membrane separation are determined, in which the represented technique could be more efficient than the others for the

composition matched refinery gases.

3. Separation operations, usually acting with the selectivity unique to a certain species group, are estimated quickly by the shortcut calculations and then represented with vectors in the triangular coordinate system.

The shortcut synthesis method and its application for construct multi-technique integration systems are sketched out in Figure 1.

Figure 1. Graphic synthesis of a multi-input cascade coupling separation sequence.

The synthesis is launched by positioning the feedstock in the triangular coordinate system according to the composition. According to the location of the feed stream, the separation unit suitable to deal with this stream can be determined facilely by matching with the dominant ranges. Afterwards, the shortcut calculations are conducted by vector analysis to orientate the residue from a separation unit and select the next separation unit.

Thanks to running the separation unit operations in their preferred ranges and avoiding the stream mixing with great composition difference, the multi-technique systems designed by the expert synthesis approach can reclaim refinery gases sufficiently and efficiently. For multitudinous refinery gases in large-scale petrochemical complex, the overall recovery ratio for valuable species is higher than 95 %, while the specific energy consumption is usually lower than 0.50 kWh/Nm^3 refinery gases. The integration processes designed by this means have been successfully industrialized in Sinopec and PetroChina.

Acknowledgement:

The authors are grateful to the financial support from National Natural Science Foundation of China (No. 21606035, U1663223, 21527812), Education Department of the Liaoning Province of China (LT2015007), and Changjiang Scholars Program (T2012049).

492614 Mechanism for Selective Fructose Etherification on Hierarchical Sn-SPP Zeolite

Abstract:

Isomerization of glucose to fructose is a key reaction for enabling the conversion of sugars into a wide range of biorenewable chemicals through furanic intermediates. The highest glucose to fructose yields on Sn-Beta are 33%[1], but a new catalyst, Sn-SPP can achieve a record 65% yield[2], even in excess of the glucose/fructose equilibrium (50%). Sn-SPP catalyzes a new reaction: fructose etherification (formally a ketalization) to ethyl fructoside, which reverts to fructose after water is added post-reaction (see Scheme 1). After these initial experimental results, many fundamental questions were unanswered, including how the structure of the active site enables this chemistry, why Sn-SPP performs ketalization while Sn-Beta does not, and why fructose ketalization is catalyzed but not glucose acetalization.

Scheme 1 - Reaction scheme for improving fructose yields from glucose using tandem reactions with fructose etherification.

We answered these questions using periodic DFT calculations and Monte Carlo simulations. A screening algorithm identified the most thermodynamically stable active site geometries; Sn-SPP has a "closed" Sn site with three nearby silanol groups, while Sn-Beta has a hydrolyzed "open" Sn site with a single, more distant silanol group. Identification of the most favorable ketalization mechanism revealed that the silanols in Sn-SPP facilitate ketalization through key H-bonding interactions at the transition state, particularly via a Sn-O-Si-OH moiety. However, the silanol in the Sn-Beta active site is not positioned to stabilize the TS, giving a larger barrier and indicating why Sn-Beta is not active for this chemistry. Analyzing glucose acetalization revealed differences in stability of the key oxonium intermediate at the respective transition states, indicating the reason for the remarkable selectivity of this process for fructose ketalization over glucose acetalization.

1. Moliner, M.; Roman-Leshkov, Y.; Davis, M. E. E.; Román-Leshkov, Y.; Davis, M. E. E. *Proc Natl Acad Sci U S A* **2010**, *107* (14), 6164–6168.

2. Ren, L.; Guo, Q.; Kumar, P.; Orazov, M.; Xu, D.; Alhassan, S. M.; Mkhoyan, K. A.; Davis, M. E.; Tsapatsis, M. *Angew. Chemie - Int. Ed.* **2015**, *54* (37), 10848–10851.

493506 Group IV and V Periodic Trends in Olefin Epoxidation: Effects of Local Environment and Electronic Structure

Abstract:

Hydrogen peroxide (H_2O_2) is a green oxidant that can epoxidize a variety of olefins over site-isolated group IV and V metal oxides (e.g., grafted onto mesoporous silica[1] or incorporated into zeolites[2,3]). These group IV and V metals incorporated into the *BEA framework (M-BEA) have been shown to activate H_2O_2 to form epoxidation-active hydroperoxide (M-OOH) or superoxide ($M\text{-}(O)_2^-$) (collectively referred to as $M\text{-}(O_2)^*$) species, respectively. The reactivity and selectivity towards epoxidation is proportional to the functional Lewis acid strength of the catalyst (i.e., adsorption enthalpies of Lewis-acid bound deuterated acetonitrile) or the ligand-to-metal charge transfer (LMCT) energy of the reactive intermediate.[3] In this work, we explore the electronic properties of these activated metal complexes and extend our previous findings to include an analogous series of catalysts grafted onto mesoporous silica ($M\text{-}SiO_2$) via calixarene complexes. The differences in the reactivity towards olefin epoxidation over M-BEA and $M\text{-}SiO_2$ catalysts are attributed to two main effects. First, the microporous structure of the *BEA framework solvates, and thus stabilizes, the interaction between the olefin and $M\text{-}(O_2)^*$ intermediate in the transition state, which decreases the apparent activation enthalpies relative to SiO_2. Second, the coordination environment of the M-BEA or $M\text{-}SiO_2$ affects the electronic structure of the active intermediate (as detected via UV-vis spectroscopy), which results in a systematic decrease in the LMCT band energies for $M\text{-}SiO_2$ catalysts.

Turnover rates for styrene (C_8H_8O) oxide formation through primary reaction pathways as a function of styrene (C_8H_8) and H_2O_2 concentrations over M-BEA and $M\text{-}SiO_2$ (M = Ti, Nb, or Ta) reveal two distinct regimes, which correspond to a change in the most abundant surface intermediate (MASI). The observed dependencies are consistent with a rate expression that describes the irreversible activation of H_2O_2 to form a pool of $M\text{-}(O_2)^*$ intermediates, which then react with C_8H_8 or H_2O_2 via an Eley-Rideal mechanism to form C_8H_8O or H_2O_2 decomposition products, respectively. Time-resolved UV-vis measurements acquired *in situ* during reaction of cyclohexene with H_2O_2-activated M-BEA and $M\text{-}SiO_2$, in conjunction with molecular probe reactions with *cis*-stilbene, support the trend that the reactive intermediate on Ti-based catalysts is Ti-OOH (i.e., hydroperoxide), while Nb- and Ta-materials react through a $M\text{-}(O_2)^-$ (i.e., superoxide) intermediate.

Understanding of the fundamental properties of the catalysts and reactive intermediates that are responsible for the differences in apparent activation barriers for epoxidation require comparisons of these catalysts made under similar conditions (i.e., MASI). Activation enthalpies for C_8H_8O formation ($\Delta H^{\ddagger}_E$) are measured on similar $M\text{-}(O_2)^*$ saturated surfaces on all catalysts (shown by rates that are proportional to $[C_8H_8]$ and independent of $[H_2O_2]$), and correlate linearly with both the measured adsorption enthalpies of pyridine (ΔH_{Py}, determined via van't Hoff analysis of FTIR spectra of pyridine coordinated to the Lewis acidic Ti, Nb, and Ta atoms within the BEA framework or grafted onto SiO_2) and also to the LMCT band energy of the reactive intermediates on these materials. These comparisons reveal two critical differences among these materials. First, $\Delta H^{\ddagger}_E$ and ΔH_{Py} values for $M\text{-}SiO_2$ materials are systematically greater by 19 ± 2 and 11 ± 2 kJ mol^{-1} than M-BEA materials, respectively, which is consistent with the difference in bulk-averaged adsorption enthalpies for C_8H_8 and pyridine in the *BEA framework and in mesoporous SiO_2 (measured in the absence of metal atoms). This difference is attributed to the microporous environment of *BEA solvating the transition state (relative to the initial state) to a larger extent than SiO_2, which results in the systematic decrease of apparent activation and adsorption enthalpies between *BEA and SiO_2. Second, the linear relationship between $\Delta H^{\ddagger}_E$ and the LMCT band energy of the active intermediate possesses the same systematic offset in $\Delta H^{\ddagger}_E$ between *BEA and SiO_2 that matches the differences in adsorption enthalpies of C_8H_8 (*vide supra*). Interestingly, the $M\text{-}SiO_2$ catalysts possess systematically lower energy LMCT bands, which may be attributed to one of two phenomena, or both. First, the hydroperoxidolysis of a M-OSi bond in the zeolite framework leaves a vicinal SiOH group in a tetrahedral position that may contribute a small amount of electron density towards the metal atom, which decreases the Lewis acid strength of the metal atom. Second, the neighboring SiOH group in *BEA likely distorts the geometry of the $M\text{-}(O_2)^*$ intermediates, which will lower the extent of orbital overlap, lengthening the M-O bonds, and thus increase the LMCT band energy. The overarching trend relating $\Delta H^{\ddagger}_E$ and the LMCT band energy suggests that metal centers that are highly electron withdrawing (i.e., possess more highly polarized $M\text{-}(O_2)^*$ bonds) result in more electrophilic and reactive intermediates. To develop an intuition for the electrophilicity of the reactive intermediates, turnover rates for the epoxidation of *para*-substituted styrene ($X\text{-}C_8H_8$, where X = OMe, Me, H, Br, and NO_2) were measured under identical reaction conditions (3 mM $X\text{-}C_8H_8$, 0.01 M H_2O_2, 313 K). The corresponding Hammett plot reveals a reaction constant (ρ) of -0.9 ± 0.05 for all M-β and $M\text{-}SiO_2$, which definitively shows that the reactive intermediates implicated in olefin epoxidation have a strong preference to react with electron-rich olefins. Collectively, these data show that the rational design of next-generation epoxidation catalysts should seek to increase the Lewis acid strength of the metal center to increase the electrophilicity and reactivity of the active intermediates and leverage the decrease in apparent activation barriers that are attributed to the solvation of the transition states in microporous environments.

References:

1) Thornburg, N.E.; Thompson, A.B.; Notestein, J.M.; *ACS Catal.*, **2015**, *5*, 5077-5088.

2) Bregante, D.T.; Priyadarshini, P.; Flaherty, D.W.; *J. Catal.*, **2017**, *348*, 75-89.

3) Bregante, D.T.; Flaherty, D.W.; **2017**, *In Revision*

498047 Sulfur Deactivation Pathways in Cu-SSZ-13 Determined through First-Principle Modeling and X-Ray Spectroscopy

Abstract:

SSZ-13 zeolites with exchanged copper ions are used commercially in diesel exhaust treatment for the selective catalytic reduction (SCR) of nitrogen oxides into nitrogen and water because of their high activity and stability [1]. Exposure to sulfur oxides, which are produced from the combustion of sulfur in diesel fuel, causes a decrease in catalytic activity. The sulfur-containing species will accumulate in the catalyst in the long term and cause worse deactivation over time. Understanding of the poisoning intermediates and reaction pathway at the molecular level is insufficient due to the lack of theoretical studies. In this work we use first-principle density functional theory (DFT) calculations and ab initio molecular dynamics (AIMD) simulations to identify poisoning intermediates and examine molecular configuration at finite temperatures. These simulations provide a molecular interpretation of ex situ and in situ X-ray adsorption spectroscopy (XAS) experiments performed on poisoned Cu-SSZ-13 samples.

The interaction between Cu active sites and sulfur-containing species with variable O/H/N stoichiometry ($SO_xH_yN_z$) was explored by thermodynamic phase diagrams. We used AIMD simulation to explore molecular configurations of the adsorbate and picked lowest energy snapshots for structure optimizations. The potential of mean force method was used to calculate adsorption entropies of $SO_xH_yN_z$ on Cu sites. The phase diagrams show that ammonium sulfate and ammonium bisulfate are thermodynamically favorable below 600 K for a range of experimental conditions. Two Cu-SSZ-13 samples, containing either Z_2Cu or ZCuOH sites, were synthesized in lab [2] and exposed to dry SO_2 and O_2 at variable temperatures to simulate accumulated sulfur poisoning in actual catalytic converters. We performed ex situ and in situ X-ray spectroscopy on the sulfated samples and analyzed the near edge structure (XANES) region of the spectra. Oxidation states elucidated from both Cu K-edge XANES and S K-edge XANES were consistent with the conclusion from first-principle studies. The difference in poisoning species on the two types of Cu sites suggested the possibility of designing a more sulfur-resistant catalyst.

[1] Kumar, A., Smith, M., Kamasamudram, K., Currier, N.W., An, H., Yezerets, A.. *Catal. Today.* 231, 75 (2014).

[2] Paolucci, C., Parekh, A.A., Khurana, I., Di Iorio, J.R., Li, H., Albarracin Caballero, J.D., Shih, A.J., Anggara, T., Delgass, W.N., Miller, J.T., Ribeiro, F.H., *J. Am. Chem. Soc.* 138, 18 (2016).

486158 Catalytic Consequences of Framework Polarity for Ethanol Dehydration on Sn-Beta Zeolites

Abstract:

Catalytic Consequences of Framework Polarity for Ethanol Dehydration on Sn-Beta Zeolites

Jason S. Bates, Brandon C. Bukowski, Jeffrey P. Greeley, Rajamani Gounder*

Charles D. Davidson School of Chemical Engineering, Purdue University, West Lafayette, IN 47907

*rgounder@purdue.edu

Lewis acidic Sn-Beta zeolites containing different silanol defect density and framework metal site coordination (open: (HO)-Sn-$(OSi)_3$, closed: Sn-$(OSi)_4$) have been recognized to behave catalytically different in reactions of oxygenated compounds. In aqueous-phase glucose isomerization, titration and quantification of sites in Sn-Beta zeolites using pyridine and CD_3CN [1], together with theoretical estimates of 1,2-intramolecular hydride shift activation barriers [2], indicate that open sites are the dominant active sites. Furthermore, glucose isomerization *initial* rate constants (373 K, per open Sn) are >10× higher for open sites confined within Sn-Beta-F (fluoride-mediated, low-defect) than Sn-Beta-OH (hydroxide-mediated, high-defect) [1], which has been ascribed to the presence of extended hydrogen-bonded water structures that increase activation free energies in high-defect pores. Here, we employ gas-phase ethanol dehydration as a probe reaction to clarify the kinetic consequences of defect density in the absence of bulk solvent effects, and assess the influence of reaction conditions on active Sn site coordination.

Bimolecular ethanol dehydration rates to form diethyl ether (DEE) were measured over low-defect Sn-Beta-F and high-defect Sn-Beta-OH zeolites at 404 K, ethanol pressures between 0.5–35 kPa, and water pressures between 0.1–50 kPa. Rate data were interpreted using a mechanism-derived rate expression that describes measured reaction orders, supported by density functional theory calculations and microkinetic modeling to discriminate among candidate reaction pathways and surface intermediates. Infrared spectroscopic quantification of open and closed Sn sites using CD_3CN titrants before and after reaction indicated that Sn sites change structure dynamically during catalysis, leading to turnover rates (per total Lewis acidic Sn) that are invariant (within ~2–3× at 404 K) among six Sn-Beta-OH samples with varying Sn content (Si/Sn=30–95). Polar silanol groups located within the pores of Sn-Beta-OH preferentially stabilize, relative to Sn-Beta-F, dimeric intermediates that involve adsorption of a second polar water or ethanol molecule with an ethanol monomer, reflected in more negative apparent water orders at high water pressure, in lower apparent ethanol orders at high ethanol pressure, and in ~4× higher values of the regressed thermodynamic equilibrium constants (at 404 K) associated with ethanol-water dimer formation. These kinetic observations are consistent with single-component adsorption isotherms and two-component adsorption measurements that indicate higher uptakes of water relative to ethanol in high-defect micropores. Hence, differences in DEE formation rates (404 K) between the two materials arise solely from

differences in prevalent surface coverages of inhibiting and reactive dimers rather than the intrinsic rate constants for DEE formation, which are insensitive to framework polarity.

[1] Harris et al., J. Catal. 335 (2016) 141–154.

[2] Bermejo-Deval et al., Proc. Natl. Acad. Sci. 109 (2012) 9727–9732.

497596 Size Selective Oxidation of Alkanes By Microporous Oxides

Abstract:

Introduction

MoVTeNb mixed oxides, with a particular crystal phase called the M1 phase, are among catalysts that provide the highest selectivity to desired C_2H_4 products in C_2H_6 oxidative dehydrogenation (ODH) with the least amounts of combustion products (CO, CO_2).[1] The origin of their selective catalysis is typically attributed to electronic properties of their lattice oxygens.[2] The M1 phase contains network arrangement of the corner and edge shared octahedra that form pentagonal, hexagonal and heptagonal channels. The heptagonal channels have dimensions comparable to kinetic diameters of linear alkanes (~ 0.4 nm).[3] Recently, Ueda et al. reported analogous crystalline materials containing only Mo and V elements and, by varying crystallite sizes, showed that C_2H_6 molecules react within heptagonal channels, while C_3 oxygenates react on external surfaces.[4] Here, we probe the origin of selective catalysis of these materials by examining oxidations ethane (C_2H_6) molecules that can enter the heptagonal channels, and cyclohexane (C_6H_{12}) molecules that are too big to enter these channels. We measure the C_2H_6/C_6H_{12} rate ratios and selectivity trends on MoVTeNb oxides and compare them with silica supported vanadium oxides (VO_x/SiO_2) without micropores. Such comparisons lead us to conclude that confinement within micropores plays a significant role in selective C_2H_6ODH catalysis.

Materials and methods

MoVTeNb mixed oxide samples were prepared by a hydrothermal method,[5] followed by thermal treatment in flowing Helium. The crystal structures were determined using X-ray diffraction (XRD) and microstructural details were examined using electron microscopy. VO_x/SiO_2 (41%wt.) samples were prepared by wet impregnation of ammonium metavanadate and followed by thermal treatments, flowing dry air. Surface areas and elemental compositions of oxides were measured using nitrogen physisorption methods and inductively coupled plasma atomic emission spectra, respectively. Samples were diluted with $SiO_2$to eliminate heat and mass transfer limitations during reactions. Catalytic alkane oxidation experiments were carried out in quartz plug flow reactors and the products were measured via online gas chromatography.

Results and discussion

The XRD pattern of MoVTeNbO is consistent with crystalline oxides with majority M1 phase. Electron micrographs reveal rod-like structures of crystalline oxides typical of the M1 phase (~1 μm length and 0.2-0.4 μm diameter) with a small fraction of much smaller particles. The catalytic sites of VO_x/SiO_2, a catalyst without micropores, are equally accessible to C_2H_6 and C_6H_{12} (Figure 1a) In contrast, the micropores of MoVTeNbO allow access to ethane, leading to rates from pores and external surfaces ($r1_{ext} + r1_{pore}$; Figure 1a), while restricting C_6H_{12} rates to external surfaces ($r2_{ext}$; Figure 1a). Therefore, the C_2H_6/C_6H_{12} ODH rate ratio on MoVTeNbO relative to that on VOx/SiO_2 reflects the contribution of micropores to C_2H_6 conversion. The measured rate ratio is 52 times higher for MoVTeNbO than that of VO_x/SiO_2, suggesting that most of the C_2H_6 ODH reactions on MoVTeNb occur within the pores (Figure 1b). The C_2H_6 oxidation product selectivity to C_2H_4, at similar conversions, is much higher on MoVTeNb than on VOx/SiO (Figure 2a), while both catalysts exhibit similar C_6H_{12} oxidation product selectivity to cycloalkenes and COx (Figure 2b). These trends confirm that only the pores, and not the electronic properties of external surfaces, are responsible to high alkene selectivity. The dependence of rates on reactant pressures are consistent with Mars-van-Krevelen redox cycles. Activation enthalpies for measured rate constants are lower for C_2H_6 than for C_6H_{12}, in spite of the weaker C-H bonds in the latter reactant, which is consistent with the role of van der Waals (vdW) interactions within pores in stabilizing the C_2H_6 activation transition states. We conclude that the high C_2H_4 selectivity arises from steric restriction to secondary reactions in the heptagonal channels. Our preliminary DFT calculations confirm that such vdW stablizations and steric restrictions are prevalent in the micropores.[6]

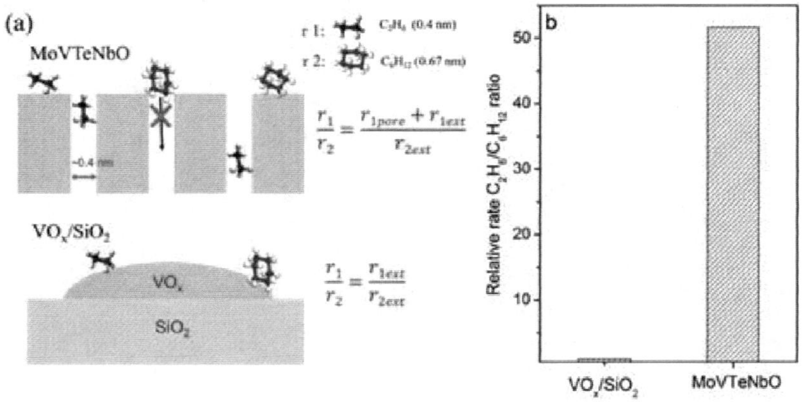

Figure 1. (a) Scheme representing size selectivity of alkane oxidation on MoVTeNb oxides and VO_x/SiO_2 using C_2H_6/C_6H_{12} rate ratios. (b) Measured rate ratios. [Reaction conditions: 30 cc/min; 3 kPa C_2H_6 or C_6H_{12}; 3 kPa O_2; 648 K; Conversions: VO_x/SiO_2 - 5% (C_6H_{12}) and 0.4 % (C_2H_6); MoVTeNbO - 8% (C_6H_{12}) and 1.7% (C_2H_6)]

Figure 2. (a) C_2H_6 and (b) C_6H_{12} product selectivity on the two oxides. [Reaction conditions: 30 cc/min; 3 kPa C_2H_6 or C_6H_{12}; 3 kPa O_2; 648 K; Conversions: VO_x/SiO_2 - 5% (C_6H_{12}) and 0.4 % (C_2H_6); MoVTeNbO - 8% (C_6H_{12}) and 1.7% (C_2H_6)]

Financial support from Tufts University and its Faculty Research Fund are gratefully acknowledged.

References

1. Botella, P.; López Nieto, J. M.; Solsona, B.; Mifsud, A.; Márquez, F. *J. Catal.* **2002,** *209*, 445-55.

2. Getsoian, A. B.; Zhai, Z.; Bell, A. T. *J. Am. Chem. Soc.* **2014,** *136*, 13684-97.

3. Ishikawa, S.; Ueda, W. *Catal. Sci. Technol.* **2016,** *6*, 617-29.

4. Ishikawa, S.; Yi, X.; Murayama, T.; Ueda, W. *Appl. Catal., A* **2014,** *474*, 10-17.

5. Gaffney, A. M.; Song, R, *US 7074956 B2*, B2, **2004**.

6. Cheng, M.-J.; Goddard, W. A. *Top. Catal.* **2016,** *59*, 1506-17.

500609 Impact of Hydrocarbon Trapping on Temperature Programmed Oxidation Â Â over a Pt/Pd/BEA Monolith Catalyst

Abstract:

Impact of hydrocarbon trapping on temperature programmed oxidation over a Pt/Pd/BEA monolith catalyst

Po-Yu Peng, Michael P. Harold, Dan Luss***

Department of Chemical and Biomolecular Engineering, University of Houston, Houston, TX 77204

*mharold@uh.edu, **dluss@uh.edu

The function of the diesel oxidation catalyst (DOC) comprising a washcoat of Platinum (Pt) and Palladium (Pd) (PGM = precious group metals) is to catalyze the oxidation of diesel effluent mixture of CO and hydrocarbons (HCs). A DOC operates at very high CO and HC conversion (>90%) after a sufficiently high operating temperature has been reached. However, during the "cold start" a fraction of the exhaust hydrocarbons may escape unreacted [1, 2]. A large-pore zeolite such as Beta (BEA) is added to the DOC to trap diesel exhaust hydrocarbons during the cold start. High molecular weight hydrocarbons are especially prone to trapping in BEA-modified PGM DOCs. Although extensive research has been conducted on the BEA storage capacity, there is a need to gain insight on how BEA affects the performance of HC light-off (LO) and of the HC oxidation transient performance [3-7]. We conducted experiments to gain a deeper understanding of functioning of the HC trap and to enable improved trap design and operating strategy. The insight gained from this study will help optimize the multi-functional and component catalyst formulation, reactor design and operating strategy for achieving high conversion of multiple hydrocarbon species.

In a typical experiment dodecane (C_{12}) is trapped on a 0.8 in. diameter by 3-in. long mixed washcoated Pt-Pd/BEA/Al_2O_3 400 cpsi cordierite monolith. The catalyst contains 60 g/ft^3 PGM in a Pt:Pd ratio of 2:1 by weight (~1:1 atomic ratio) on a 0.7 g/in^3 load of γ-Al_2O_3 support and incorporated with the BEA content (0 and 0.5 g/in^3) into the washcoat. The C_{12} and CO_2 gas concentrations were measured by spatially resolved mass spectrometry (SpaciMS, QMS-Omnistar GSD 300) and a FTIR. Spatial resolved temperature profiles were measured by a coherent optical frequency domain reflectometer (c-OFDR with a 1.26 m long gold-coated optical fiber placed in the center of monolith channel. The trapping experiments were conducted by flowing 160 ppm C_{12}, 10% O_2 and the balance Ar at 60 ^oC until the catalyst was saturated. This was followed by ~10 ^oC/min inlet temperature ramp. The total gas flowrate was 8 L/min (GHSV= 12,438 h^{-1}). During the dynamic hysteresis analysis, the saturated catalyst was heated up by 1 ^oC/min from 110 ^oC to 170 ^oC inlet temperature and then cooled by -1 ^oC/min from 170 ^oC to 110 ^oC. The total gas flowrate varies from 6 L/min (GHSV= 10,262 h^{-1}) and 8 L/min (GHSV= 12,438 h^{-1}). Adiabatic behavior was confirmed by measuring temperature profiles under non-reaction conditions.

Unusual CO_2 formation spikes were observed during the oxidation of HC that was pre-stored on the catalyst before exposure to the temperature ramp. Fig. 1 shows the effluent C_{12} concentrations along with the effluent generated CO_2 concentration during a linear feed temperature ramp. The C_{12} is completely trapped by C_{12} breakthrough with a slow approach to the feed concentration value at uniform temperature of ~60 °C. However, during the subsequent temperature ramp, in addition to the release of C_{12} at the 48 minute mark, two impressive spikes of CO_2 appeared between 50 and 57 minutes. Corresponding to the first two spikes are measureable spikes in the outlet temperature.

The c-OFDR measurements enabled construction of a 3-D contour plot of temperature versus time and position. (Figure 2) The temperature spikes correlate with the measured CO_2 yield spikes. The feed temperature increased linearly during the ramp. The local temperature sequentially increased and decreased three times at a nearly fixed position in the monolith. During the first and second cooling period, the temperature went back close to the inlet temperature. This behavior suggests that the oxidation stops abrubtly as the generated heat is quickly removed by the flowing gas stream. The drop in the CO_2 concentration between the first and second peaks in Fig. 1 also supports the stoppage of the HC oxidation. A mass balance analysis indicates that the fed C_{12} during this time period is released and re-trapped by the catalyst. The C_{12} continues to accumulate until that the catalyst saturates. This intermittent HC oxidation behavior continues until the HC trapping capacity becomes smaller at high temperature. The rapid propagation of the maximum temperature from the midstream to the downstream during the first peak also supports high heat convection by the gas flow. This rapid temperature peak propagation greatly accelerates the HC oxidation downstream.

A slow heating/cooling rate leads to a significant counter-clockwise hysteresis behavior, underscoring the competition between HC trapping and oxidation. Figure 3 shows the C_{12} and CO_2 concentrations during the hysteresis under the slow heating/cooling rate of ±1 °C/min. In the ramp-up period, the C_{12} (CO_2) gradually decreases (increases), indicating a delicate balance between trapping and oxidation. After ignition at ~159 °C, more impressive peaks occur. The occurrence of multiple peaks suggests that the C_{12} trapping occurs before the C_{12} oxidation. After a certain amount of C_{12} is trapped, the C_{12} oxidation reaction then triggers the release of stored-C_{12}. The HC trapping capability becomes smaller at high temperatures. During the ramp-down period, a high C_{12} conversion remains until the feed temperature is below ~136 °C. The reactor extinguishes at ~23 °C below the ignition, showing a strong ability to reduce HC emission when the engine stalls.

The non-isothermal nature of this adiabatic system suggests that HC trapping and oxidation dominates in the different regions in the catalyst. Since the downstream is much hotter than the upstream, the HC is re-trapped by BEA at the upstream and the HC is drastically oxidizes at the downstream. Both events greatly consume the C_{12}, helping to reduce C_{12} emission, and lead to a much lower extinction temperature. Even at such low heating/cooling rate of ±1 °C/min, a significant counter-clockwise hysteresis behavior still exists. This study is the first to provide insight into the spatiotemporal features of transient behavior of HC trapping and oxidation of BEA/DOC catalysts, and also the first discovery of multiple HC LO phenomena.

Figure 1. Dodecane and CO_2 concentration measured during adsorption followed by TPO with BEA-zeolite added Pt/Pd/Al_2O_3 monolith catalyst. Condition during saturation: 160 ppm dodecane and 10% O_2 balanced with Ar. Condition during desorption and TPO: 10% O_2 balanced with Ar.

Figure 2. The 3-D plot of the spatially-resolved temperature versus time in feed of 10% O_2 balanced with Ar. (Catalyst: D-05)

Figure 3. Experimental C_{12} and CO_2 concentration versus inlet fluid temperature in a dynamic hysteresis analysis. (C_{12} = 160 ppm, 10% O_2; space velocity of 12,438 h^{-1}; temperature ramp = ±1 °C/min).

References

1. Mukai, K., et al., *Adsorption and desorption characteristics of the adsorber to control the HC emission from a gasoline engine*. 2004, SAE Technical Paper.

2. AL-Harbi, M., et al., *Competitive no, co and hydrocarbon oxidation reactions over a diesel oxidation catalyst.* The Canadian Journal of Chemical Engineering, 2012. **90**(6): p. 1527-1538.

3. Li, H.-X., et al., *Application of zeolites as hydrocarbon traps in automotive emission controls.* Studies in Surface Science and Catalysis, 2005. **158**: p. 1375-1382.

4. Chang, H., et al., *Gasoline Cold Start Concept (gCSC™) Technology for Low Temperature Emission Control.* SAE Int. J. Fuels Lubr, 2014. **7**(2): p. 480-488.

5. Murakami, K., et al., *Development of a high performance catalyzed hydrocarbon trap using Ag-zeolite*. 2004, SAE Technical Paper.

6. Elangovan, S., et al., *A comparative study of zeolites SSZ-33 and MCM-68 for hydrocarbon trap applications.* Microporous and mesoporous materials, 2006. **96**(1): p. 210-215.

7. Elangovan, S., et al., *Silicoaluminophosphate molecular sieves as a hydrocarbon trap.* Applied Catalysis B: Environmental, 2005. **57**(1): p. 31-36.

495377 Consequences of Diffusion, Acid Strength, and Confinement on Bifunctional Reactions of Alkanes

Abstract:

Hydrocracking and hydroisomerization reactions convert large-molecular-weight alkane feedstocks to more valuable smaller and more highly branched molecules, respectively, through cascade reactions mediated by alkene intermediates [1]. Such reactions require presence of metal function—sufficient to establish alkane-alkene equilibrium for reactants and to scavenge intermediates and alkene products—that resides within diffusion distances from kinetically relevant acid function. These distances are seldom well-defined and are sensitive to synthetic and formulation protocols but, in part, define concentration gradients of reactant and product alkenes with significant consequences for rates and product selectivities, respectively, especially when acid sites reside within channels and voids of zeolite or zeotype materials (with characteristic sizes ~0.4-1.1 nm [2]). These apertures and voids restrict diffusion of reactants and products—and do so sensitively based on molecule size and shape—and also confer higher reactivity through stabilizing van der Waals and electrostatic interactions with carbocation moieties at ion-pair transition states. Resultant concentration gradients of product alkenes depend on the size of the acid domain and on the density of protons and lead to product selectivities reflecting multiple secondary interconversion events occurring before product alkenes exit acid domains and become hydrogenated at extracrystalline metal functions. Such diffusion-enhanced secondary reactions are rarely treated using rigorous kinetic-transport formalisms, and their effects on measured isomerization and β-scission rates and selectivities are typically attributed, unconvincingly and often inaccurately, to transition state selectivity or to effects of framework structure on acid strength.

Here, we dissect the effects of diffusion, acid strength, and transition state confinement to assess their respective catalytic consequences. Kinetic data, benchmarked against density functional theory, are reported for n-heptane (nH) and 2,4-dimethylpentane (24DMP) isomerization and β-scission on physical mixtures of Pt/SiO_2 and zeolites with diverse confining environments (MFI, MTW, BEA, SFH, FAU) but similar acid strengths and MFI zeotypes with different framework heteroatoms (Al, Fe, Ga, B) which contain H^+ of different acid strength within the same confining environment. To attain intrinsic product formation rates, we assess the effects of product-alkene concentration gradients within acid domains using rigorous reaction-transport formalisms that account for diffusion timescales independently measured from isochoric, isothermal transient uptake of 2,2-dimethylbutane. We additionally develop and report descriptors of confinement as van der Waals interaction energies, E_{vdw}, for transition states based on Lennard-Jones potential force fields [3] and of acid strength as deprotonation energy (DPE), defined as that required to heterolytically cleave the Brønsted O-H bond; these descriptors, together with thermochemical cycles providing the conceptual framework for interpreting activation barriers in terms of separable properties of the acid (E_{vdw} and DPE), the reactant (proton affinity), and the interacting moieties at the confined ion-pair transition state, permit systematic interpretation of the independent effects of acid strength and confinement on intrinsic isomerization and β-scission rates and selectivities.

n-Heptane and 2,4-dimethylpentane isomerization and β-scission turnover rates (per H^+; 548 K; 0.5-25 kPa alkane; 60-101 kPa H_2) on all physical mixtures are proportional to alkane/H_2 molar ratio (a surrogate for the prevalent pressure of heptenes) at low values of this ratio and approach constant values as molar ratio increases. First-order rate constants ($k_{isom,nH}K_{prot,nH}$ for n-heptane isomerization and $(k_{isom,24}+k_{\beta,24})K_{prot,24}$ for combined 2,4-dimethylpentane isomerization and β-scission) reflect the free-energy differences between their respective confined, ion-pair transition states and unconfined, neutral precursors (gas phase alkene & acid sites). Measured ratios of $k_{\beta,24}/k_{isom,24}$ reflect ubiquitous secondary reactions that occur within the acid domain before alkene products leave an acid domain and become alkanes. Similarly, on all acids and for both reactants, measured product selectivities (the ratio of the measured formation rate for a given product to the consumption rate of the reactant) were invariant with reactor residence time or alkene reactant pressure; these selectivities do not reflect the intrinsic values of primary formation events because primary alkene products undergo multiple secondary interconversions before they form their respective less-reactive alkane analogs. The extent of these diffusion-enhanced secondary reactions was assessed for each aluminosilicate by varying the intracrystalline residence time (t_{int}) through monotonic changes in the proton density ([H^+]) during desorption of weakly bound NH_3 titrants from the active protons. Figure 1 shows selectivities measured during NH_3 desorption as a function of t_{int} for (a) β-scission products and (b) isomer products during 2,4-dimethylpentane conversion on FAU-Pt/SiO_2 catalyst mixtures. Product selectivities at low values of t_{int} are invariant with t_{int}, indicating that such β-scission selectivities reflect primary chemical events. As t_{int} increased, β-scission, n-heptane, 2-methylhexane, and 3-methylhexane selectivities increased, while that for to 2,3-dimethylpentane decreases, indicating the prevalence of secondary alkene interconversions of primary 2,4-dimethylpentane products as [H^+] increases.

Figure 1. Measured **(a)** β-scission and **(b)** 2-methylhexane (; left), 3-methylhexane (▲; left), n-heptane (¦; left), and 2,3-dimethylpentane (•; right) selectivities for reactant 24DMP as a function of the intracrystalline residence time (varied by desorption of NH_3) on physical mixtures of Pt/SiO_2 with FAU (548 K). Dashed lines are guides for the eye.

Rigorous reaction-transport treatments reveal that a Thiele parameter for the primary isoalkene products determines the extent of such diffusion-enhanced secondary reactions (, where $k_{i,iso}K_{prot,iso}$ is the first-order rate constant for secondary reaction i of the isoalkenes, D_{iso} is diffusivity of the isoalkenes, and R^2 is zeolite crystallite size). Figure 2 shows β-scission selectivities for nH reactant during NH_3 desorption on FAU, BEA, and MFI (as physical mixtures with Pt/SiO_2; BEA and MFI of different $[H+]R^2/D_{22DMB}$ are indicated, where D_{22DMB}/R^2 was measured using transient uptake of 2,2-dimethylbutane) as a function of the Thiele modulus for n-heptane ( ); the dashed curve is the regressed fit to the theoretical expression describing these secondary reactions (Fig. 2, inset equation, where and) with a single fitting parameter ( ) that has the same value for all aluminosilicates. *β*-Scission selectivities increased monotonically as increases and are a single-valued function of this parameter; selectivities for all aluminosilicates lie along the same curve, despite their very different void sizes, indicating that all frameworks stabilize the two transition states (for *β*-scission and isomerization) to the same extent.

Figure 2. Measured *β*-scission selectivity for reactant nH as a function of the n-heptane Thiele modulus (F_{nH}^2; varied by desorption of NH_3) on physical mixtures of Pt/SiO_2 with FAU, BEA, and MFI (548 K; 100 kPa H_2, 0.5 kPa nH). MFI and BEA of different [H^+] are indicated. Dashed line represents the regressed fit to the inset equation.

These confinement effects, however, lead to preferential stabilization of both *β*-scission and isomerization transition states with respect to their gas phase precursors; the stabilization of transition states by van der Waals interactions with the framework is reflected in E_{vdw}, where a better match between void size/shape and organic carbocation size/shape results in more negative values. First-order nH isomerization rate constants decrease exponentially with increasing E_{vdw}. The effects of E_{vdw} on activation barriers are attenuated by energy penalties brought forth by distortions of the framework (E_{str}) and the carbocation at the transition state in order to establish more effective van der Waals contacts, especially as the organic moiety approaches the size of the confining voids.

Such organic carbocations, together with the conjugate anion of the zeolite or zeotype, constitute ion pair transition states mediating isomerization and *β*-scission reactions. First-order isomerization and *β*-scission rate constants for 2,4-dimethylpentane, measured on MFI zeotypes with H^+ of different acid strength (by varying framework heteroatom, in order of decreasing acid strength: Al, Ga, Fe, B; DPE = 1226-1292 kJ mol^{-1}; in physical mixture with Pt/SiO_2), were corrected for diffusion-enhanced secondary reactions to obtain intrinsic values. These first-order rate constants decrease exponentially with increasing DPE, predominantly a reflection of the less stable conjugate anion at ion-pairs of weaker acids. The ratio of rate constants $k_{\beta,24}/k_{isom,24}$, however, increases with increasing DPE, indicating that *β*-scission occurs preferentially on weaker acids. Cations at transition states mediating *β*-scission resemble "harder" carbenium ions (according to classical acid-base formalisms [4]) compared to those for isomerization. These "harder" carbocations interact more effectively with the conjugate anion, recovering more of the energy required to deprotonate a weaker acid, and consequently are less sensitive to DPE. Thus, for transition state structures differing in their tendency to delocalize charge, acid strength can influence product selectivity through stronger effects of DPE for the "softer" transition state.

This work mechanistically assesses the effects of both confinement and acid strength on intrinsic isomerization and *β*-scission turnover rates and selectivities. Methods of untangling the dramatic and ubiquitous effects of diffusion-enhanced secondary reactions on measured product selectivities are also demonstrated. The insights gained from this work can be applied to design catalytic architectures to obtain desired selectivities and reaction rates.

References

[1] H.L. Coonradt, W.E. Garwood, Ind. Eng. Chem. Process Des. Dev. 3 (1964) 38–45.

[2] C. Baerlocher, L. McCusker, Database of Zeolite Structures, http://www.iza-structure.org/databases/.

[3] S. Herrmann, E. Iglesia, J. Catal. 346 (2017) 134–153.

[4] G. Pearson, J. Am. Chem. Soc. 85 (1963) 3533–3539.

The authors acknowledge technical discussions with Drs. Stacey I. Zones (Chevron) and William Knaeble (UC Berkeley), diffusion timescale measurements by Zhichen (Jane) Shi (UC Berkeley), financial support from Chevron Energy Technology Company, and computational resources from the National Science Foundation's Extreme Science and Engineering Discovery Environment (ACI-1053575).

500019 Hydrophobic Zeolites for Solketal Production from Crude Glycerol

Abstract:

Functionalizing zeolites (i.e., CBV600 with a SiO_2/Al_2O_3 ratio of 5.2) by organosilanes increase the degree of hydrophobicity. These hydrophobic zeolites can catalyze the solketal production from crude glycerol (i.e., glycerol containing moisture, methanol and salt impurities) and show the higher catalytic activity than the pristine zeolites. Glycerol is an under-utilized by-product from biodiesel industries. The global biodiesel consumption was ~34 billion liters in 2016 and is expected to reach 38 billion liters by 2020 from a rapid growth in fuel and power generation sectors. Hence, a large amount of crude glycerol will be available. Herein, crude glycerol and acetone were used for solketal production at low temperatures of 30-50 ËšC. Glycerol conversion and solketal yield were 84% and 80%, respectively. The physico-chemical characteristics of hydrophobic and pristine zeolites were characterized by N_2 physisorption, thermogravimetric analysis (TGA), X-ray diffraction (XRD), and infrared spectroscopy. The obtained characteristics were correlated with glycerol conversion and solketal yield. These results provide an understanding of how these hydrophobic zeolites function and how to tailor their catalytic activities.

495434 Support Effects in Heterogeneous Catalysis Using Au Nanoparticles on Oxides: A DFT Analysis

Abstract:

Supported Au nanoparticles have been found to be active for numerous heterogeneously catalyzed reactions [1]. In addition, these catalysts have been found to exhibit a bifunctional behavior, where the combined Au/oxide system works in a synergistic manner, giving lower reaction barriers than those of either the Au or the oxide in isolation [2]. The identity of the support plays a key role in determining the activity of Au/oxide systems, with Au nanoparticles showing different turnover rates when supported on different oxides [2,3]. Although support effects have been observed for individual systems, a fundamental understanding, especially in terms of a common unifying descriptor for the activity of these systems, is poorly understood. In this work, we have studied model Au nanowire supported on a range of oxides â€“ MgO, ZnO, Al_2O_3, Fe_2O_3, ZrO_2 and TiO_2 - to understand how the catalytic properties at the interface of these systems are influenced by the choice of the oxide support. The geometric effect of strain in these systems was minimized by orienting the Au nanowire in non-epitaxial registries on the oxide supports. Water Gas Shift (WGS) is known to be promoted at the interfaces of supported Au/oxide systems [2,4] and have, therefore, been used as a probe reaction in this study. Rigorous DFT calculations were performed for reactive intermediates usually associated with WGS, at the interfaces of the various model Au/oxide systems. Our results correlate well with the experimental trend in WGS reactivity on these systems. In particular, we found the WGS rates to correlate well with experimental water orders and calculated adsorption energies of dissociated water at the interface of Au/oxide, suggesting dissociated water to be a common intermediate involved in kinetically significant reaction steps on all these systems. This combined theoretical and experimental analysis hints at the possibility of a common mechanistic basis for trends in support effects.

References

[1] Rodriguez, J.A., *Catal. Today*. **2011**, 1, 160

[2] Zhao, Z.-J., Li, Z., Cui, Y., Zhu, H., Schneider, W. F., Delgass, W. N., Ribeiro, F. and Greeley, J., *J. Catal.* **2017**, *345*, 157

[3] Shekhar, M., Wang, J., Lee, W.S., Williams, W.D., Kim, S.M., Stach, E.A., Miller, J.T., Delgass, W.N. and Ribeiro, F.H. JACS 2012, 134, 4700

[4] Williams, W.D., Shekhar, M., Lee, W-S., Kispersky, V., Delgass, W.N., Ribeiro, F.H., Kim, S.M., Stach, E.A., Miller, J.T. and Allard, L.F. JACS, 2010, 132, 14018

501423 Coordination-Based Descriptors for Rational Design of Metal Nanocatalysts

Abstract:

One major challenge for designing metal nanocatalysts is to know a priori appropriate surface structure for a given catalytic reaction. Understanding the effects of particle morphology and composition on the interaction of surface atoms with adsorbates is of pivotal importance for identifying optimal active sites and engineering nanoparticles with maximized fraction of such sites. Many efforts have been made aiming to predict the binding energy of an adsorbate at metal surfaces using the electronic and/or geometric factors of the adsorption site (termed descriptor).[1] While these models have been successfully used as reactivity descriptors for either extended-metal surfaces or simple metal nanoparticles, their extension to complex particle systems with varying strains and/or metal ligands remains elusive due to a formidable computational cost or the lack of an explicit consideration of interatomic interactions.

In this work, we used the previously developed reactivity descriptor, i.e., the orbital-wise coordination number (CN^a), for understanding catalytic effects of size/shape/composition of metal nanocatalysts[2]. The CN^a, quantified by interatomic coupling matrix elements between the site of interest and its all neighbors within a certain cutoff radius, provides a robust description of CO, O_2, and O adsorption energies on Au metal nanoparticles of varying size and shape attributed to its explicit consideration of broken-bond strains. Furthermore, the CN^a shows promise as a general descriptor for predicting adsorption properties of core-shell alloyed structures of Au nanoparticles with d10 metal ligands (e.g., Cu and Ag). We further tailored the metal ligands interacting with surface Pt atoms in a core-shell type alloy structure for a rapid screening of metal nanoparticle sites that have optimized activity toward O_2 reduction. The approach can be readily extended to understand and predict reactivity trends of large and more complex metal catalysts with defects, impurities, transition-metal additions, supports, etc, and thus provides a general basis for rational design of metal nanocatalysts[3]

1. Hammer, B. & NÃ¸rskov, J. K. Electronic factors determining the reactivity of metal surfaces. *Surf. Sci.* **343,** 211â€"220 (1995).

2. Ma, X. & Xin, H. Orbitalwise Coordination Number for Predicting Adsorption Properties of Metal Nanocatalysts. *Phys. Rev. Lett.* **118,** 36101 (2017).

3. Wang, S. & Xin, H. Size Effects of Au Nanocatalysts: The Role of Local Coordination of Surface Atoms. Submitted (2017).

501459 Theoretical Investigation of the Ring Opening Mechanism of Cyclohexanes on Ir Surfaces

Abstract:

Diesel engines have higher fuel economy and produce less greenhouse gas than equivalent gasoline engines. An important requirement in diesel production is to reduce the amount of aromatic compounds, which have poor ignition quality (low cetane numbers, or CN) and form soot during burning. Traditional hydrocracking produces a large fraction of low-molecular weight products that cannot be used in diesel fuel. A better approach would be to convert aromatics to saturated cyclic compounds, followed by selective ring opening (SRO). Although catalyst support, metal dispersion, and reaction conditions all make a difference, iridium compared to other metals possesses much higher activity for ring-opening vs. cracking, but it tends to favor branched isoparaffins, which have lower CN than linear paraffins [1-4]. The dicarbene mechanism, which cleaves secondary-secondary C-C bonds, is postulated to dominate over those that open rings at substituted carbon positions (e.g., the metallocyclobutane mechanism), but neither the details of these mechanisms nor the reason why the dicarbene mechanism is favored by Ir is fully understood.

The focus of this talk will be on the effects that surface morphologyand substitution have on the ring opening reaction on Ir. We are theoretically investigating the ring opening mechanism of cyclohexane and substituted cyclohexanes on the close-packed Ir(111) and stepped Ir(211) surfaces as a basis for understanding the fundamental ability of Ir to promote SRO. We perform detailed density functional theory calculations to identify the more facile reaction channels for each reaction intermediate in order to construct the minimum-energy reaction pathway for the ring opening of a given cycloalkane species. We find that in the low coverage limit, repeated dehydrogenation of cyclohexane at positions closer to the surface is kinetically highly facile, so that neither cyclohexene nor benzene is a major product. Instead, C-C cleavage ensues after 4~5 sequential dehydrogenation steps depending on the surface, producing a metallocycle. This stands in contrast to the decomposition of cyclohexane on Pt, where benzene is a major product [5,6]. The reaction intermediates generally adsorb more strongly on Ir(211) than on Ir(111), but the reaction activation barriers are not always lower on the step edge. Our findings provide clues for enhancing the ability of Ir catalysts to more selectively cleave substituted C-C bonds, thereby increasing the CN of the resultant hydrocarbons.

[1] G. B. McVicker et al., *J. Catal.* **2002**, *210*, 137.

[2] P. T. Do, W. E. Alvarez, D. E. Resasco, *J. Catal.* **2006**, *238*, 477.

[3] H. Shi, X. Li, G. L. Haller, O. Y. Gutiérrez, J. A. Lercher, *J. Catal.* **2013**, *295*, 133; H. Shi, O. Y. Gutiérrez, G. L. Haller, D. Mei, R. Rousseau, J. A. Lercher, *J. Catal.* **2013**, *297*, 70.

[4] D. W. Flaherty, A. Uzun, E. Iglesia, *J. Phys. Chem. C* **2015**, *119*, 2597.

[5] J. A. Rodriguez, C. T. Campbell, *J. Phys. Chem.* **1989**, *93*, 826.

[6] D. H. Parker, C. L. Pettiette-Hall, Y. Li, R. T. McIver, Jr., J. C. Hemminger, *J. Phys. Chem.* **1992**, *96*, 1888.

498309 CO_2 Reduction on the Nickel Surface

Abstract:

Nickel is a typical catalyst for carbon dioxide reforming. In this study, we investigate the mechanism of reducing CO_2 on the Ni(110) surface in the presence of the subsurface or impinging hydrogen using spin-polarized periodic density functional theory and ab initio molecular dynamics simulations. In the case of the subsurface H atoms, they act as both spectator species and reactants of the reactions and hence provide the extra energy for overcoming the reaction barriers, in which the major contribution comes from the energetics of subsurface H emerging to the surface as a reactant. While in the case of the impinging hydrogen, the simulation results show direct theoretical evidence for both associative and redox mechanisms in the reaction of atomic hydrogen with CO_2. Because H_2 is dissociatively chemisorbed on Ni(110) with nearly unit probability, the mechanisms we find are also relevant to the reverse water-gas shift reaction (H_2 with adsorbed CO_2). Furthermore, we provide the first real-time demonstration of both Eley-Rideal and hot atom mechanisms when H impinges on adsorbed CO_2.

497771 Designing Cu-Based Bimetallic Nanoparticles for CO_2 Activation

Abstract:

Copper-based catalysts are good candidates for the conversion of carbon dioxide into useful chemicals. Using Density Functional Theory calculations, we investigate the adsorption and activation of carbon dioxide on CuM (M = Au, Cu, Mn, Mo, Ni, Pd, Rh, Ru, Sc, V, Zn, and Zr) nanoparticles. Candidate heteroatoms (M) for CO_2 adsorption were selected based on the heteroatom preference to occupy a surface site of the nanoparticle. We identified two significant descriptors for carbon dioxide chemisorption on the nanoparticles: (1) the heteroatom local d-band center must be positioned higher in energy than the LUMO of carbon dioxide, and (2) the electropositivity of the heteroatom must be higher compared to Cu. Both descriptors lead to an effective charge transfer from the nanoparticle to CO_2. Using these criteria, bimetallic nanoparticles can be screened rapidly for the ability to chemisorb (and activate) CO_2, which is an important step for the subsequent catalytic conversion of CO_2. Our work highlights the importance of generating binding sites on a nanoparticle surface based on stability and electronic structure properties, which can lead to designing bifunctional catalysts.

497290 Understanding Heterogeneous Catalyst Deactivation By Biogenic Impurities on Ni (111) Surface and Bimetallic Alloy

Abstract:

Interaction of biogenic impurities with the Ni(111) catalyst surface was studied to understand the important role played by them in inhibiting catalysis. Non-bonding interactions of amino acids (Alanine (Ala), Cysteine (Cys), Methionine
(Met), Tryptophan (Trp), Histidine (His), Lysine(Lys), Glutamic acid (Glu) and Threonine (Thr)) in aqueous enivronment were examined using classical molecular dynamics (MD) simulations. The potential of mean force (PMF) profiles of amino acids revealed qualitative differences, resulting into altered orientation and preferential interacting site with the metal surface. Most of the preferential interacting sites were observed to be in direct contact with the surface except Lys which interacted with the backbone nitrogen oriented away from the surface. The strength of nonbonded
interactions varied from -13 to -71 kJ/mol with Ala as the weakest and Trp as the strongest adsorber. The interaction energies scaled linearly with the solvent accessible surface area of the amino acids. DFT calculations showed strong binding for both Cys (-123 kJ/mol) and Met (-102 kJ/mol), undergoing dissociation to form atomic S with an intrinsic activation barrier of 133 kJ/mol and 60 kJ/mol respectively for the C-S bond cleavage on the Ni (111) surface. However, the mechanistic routes for dissociation followed by the two amino acids were a bit different. The mechanistic insights thus obtained lead to an explorative theoretical framework for the design of a bimetallic catalyst which may be used in such an integrated bio and chemo-catalytic process with reduced chances of deactivation. The NiAu alloy surface was calculated to show both reduced Met (-88 kJ/mol) and S binding (132 kJ/mol), as compared to pure Ni surface, indicating promising prospects for the activity of such an alloy catalyst.

492352 Determination of Iridium Alloy Properties for Catalytic Applications

Abstract:

Alloys can have a number of structures, such as ordered, fully mixed, segregated, shell-core, or near surface. The alloy structure in turn determines the electronic and catalytic properties of the alloy. Iridium-based alloys have shown promise in a number of catalytic applications, including ethanol oxidation[1, 2], oxygen reduction[3], and hydrogen generation[4]. Iridium alloys possess high melting point, high corrosion and oxidation tolerance, as well as excellent mechanical properties. Our goal was to provide models and predictions of realistic iridium alloys, including their stability and electronic properties at various temperatures and synthesis conditions. Identifying how structure affects stability and electronic properties is key to predicting catalyst performance. We used density functional theory (DFT) calculations to model these alloys. Statistical methods, such as the cluster expansion approach, were also combined with DFT to model a large number of possible bimetallic configurations. We studied iridium-platinum alloys as a prototypical bimetallic system. We have identified viable structures of these alloys at various composition levels (IrxPt1-x; $0 < x < 1$), and shown how these metals prefer to segregate at low temperature. While identification of the most thermodynamically stable alloy 0 K structures can be straight forward, real working catalysts often have not reached equilibrium. Monte Carlo simulations at finite temperature (600 K+) indicate that disordered iridium-platinum structures occur with larger degree of mixing. We have examined how the electronic structure (e.g. density of states) changes at various compositions and structures (e.g. segregated versus mixed), and potential implications of these changes on catalysis. We have also modeled a large number of other iridium alloys with various metals, including Pd, Rh, Sn, Cr, Fe, etc. This systematic study of iridium alloys provides details on which bimetallic systems are stable, their expected electronic properties, and how to realistically model these catalysts from various synthesis conditions. Our results show how theoretical methods can advance alloy development and lead to potential new catalysts.

1. Cao, L.; Sun, G.; Li, H.; Xin, Q., Carbon-supported IrSn catalysts for direct ethanol fuel cell. Fuel Cells Bulletin 2007, 2007 (11), 12-16.

2. Du, W.; Wang, Q.; Saxner, D.; Deskins, N. A.; Su, D.; Krzanowski, J. E.; Frenkel, A. I.; Teng, X., Highly active iridium/iridiumâ€“tin/tin oxide heterogeneous nanoparticles as alternative electrocatalysts for the ethanol oxidation reaction. J Am Chem Soc 2011, 133 (38), 15172-15183.

3. Ioroi, T.; Yasuda, K., Platinum-iridium alloys as oxygen reduction electrocatalysts for polymer electrolyte fuel cells. Journal of the Electrochemical Society 2005, 152 (10), A1917-A1924.

4. Singh, S. K.; Xu, Q., Bimetallic nickel-iridium nanocatalysts for hydrogen generation by decomposition of hydrous hydrazine. Chemical Communications 2010, 46 (35), 6545-6547.

488974 Are Undercoordinated Sites Poisoned? a Kinetic Monte Carlo Study of NO Direct Decomposition on Pt_{55}

Abstract:

Undercoordinated sites are commonly invoked to explain the high activity of small nanoparticles, as well as the structure sensitivity exhibited by some reactions. However, it is unclear if the higher activity of undercoordinated sites is generalizable to most reactions and reaction conditions. Sabatier reminds us that the ideal catalyst is not the one with exceptionally high or low reactivity, but instead the one that binds all species modestly. Undercoordinated sites are especially active for bond breaking as they bind the products of a reaction strongly, increasing the exothermicity of the reaction.[1,2] However, precisely due to their strong binding affinity, they can be easily saturated and thus poisoned by reactants or decomposition products. Whether these undercoordinated sites can maintain their intrinsic activity under catalytic conditions, or are instead poisoned and rendered inactive, is a highly important and yet rarely studied issue.

To gain insights into this question, we chose as our model system nitric oxide (NO) direct decomposition on a cuboctahedral Pt_{55} nanoparticle. Pt_{55} presents a variety of undercoordinated edge and corner sites which can bind and dissociate NO, a notorious pollutant found in vehicle exhaust. However, the mean-field approximation breaks down when looking to model such a small (~1 nm) nanoparticle due to the strong spatial correlations between its various site types. In order to obtain meaningful reaction kinetics, the use of Kinetic Monte Carlo (KMC) simulations, as opposed to mean-field microkinetic models, is thus necessary. KMC also allows us to accurately take into account coverage effects, which were found to be significant under the studied conditions. This was implemented using a cluster expansion, which parameterizes adsorbate-adsorbate interaction energies terms of point and multibody interactions.[3,4] The Zacros graph-theoretical KMC code[5,6] was used to perform the KMC simulations.

Our results show that undercoordinated sites can protect themselves from being poisoned by exerting strong repulsive interactions on nearby sites. Under our reaction conditions, NO is the most abundant surface intermediate. At high NO coverages, occupation of the sites on Pt_{55} by NO does not follow the Langmuir adsorption isotherm, in which the strongest binding site type is saturated much faster compared to the more weakly binding site types. Instead, we find that site types are occupied in a staggered fashion, such that no one site type is saturated and poisoned, even if certain site types bind NO much more strongly in the dilute limit. The insights provided by our work will provide an interesting dimension to understanding how undercoordinated sites are able to achieve high activity under catalytic conditions.

References

1 B. Hammer and J. K. Nørskov, in *Advances in Catalysis*, 2000, vol. 45, pp. 71–129.

2 R. I. Masel, *Catal. Rev.*, 1986, **28**, 335–369.

3 J. M. Sanchez and D. De Fontaine, in *Structure and Bonding in Crystals*, eds. A. Nsvrotsky and M. O'Keeffe, Academic Press, Inc, New York, 1981, vol. 2, p. 117.

4 D. J. Schmidt, W. Chen, C. Wolverton and W. F. Schneider, *J. Chem. Theory Comput.*, 2012, **8**, 264–273.

5 M. Stamatakis and D. G. Vlachos, *J. Chem. Phys.*, 2011, **134**, 214115.

6 J. Nielsen, M. D'Avezac, J. Hetherington and M. Stamatakis, *J. Chem. Phys.*, 2013, **139**, 224706.

489524 The Contribution of Nickel and Oxygen Vacancies and of Low Valent Dopants in Altering the Surface Reactivity of NiO

Abstract:

Metal oxides are increasingly being used as oxidation catalysts for biomass and hydrocarbon derived feedstocks. Nickel oxide (NiO) is an effective oxidation catalyst and is used in the chemical looping combustion and reforming of methane, in addition to its application in photo and electrochemical systems. Since oxidation reactions involve the direct participation of lattice oxygen, its reactivity on different exposed facets of typical NiO morphologies determine the overall catalytic performance. Using Density functional theory (DFT) calculations incorporating the Hubbard U correction (DFT + U method), we reveal 1) the intrinsic differences in the reactivity of two different crystal facets of NiO, NiO(100) and NiO(110), in C-H activation of methane and 2) the contribution of surface nickel and oxygen vacancies, and of a low valent dopants like Li in altering the mechanism of methane dissociation and the reactivity of the surfaces. Methane dissociation by the synergistic participation of the 5-coordinated Ni-O pairs on the most stable surface of NiO, NiO(100), proceeds with a large activation barrier of 136.6 kJ mol^{-1} while the 4-coordinated Ni-O pairs on the relatively less stable NiO(110) surface can dissociate methane with an activation barrier of 57.1 kJ mol^{-1}. The presence of a surface Ni vacancy on the inactive NiO(100) surface increases its reactivity to enable C-H dissociation by the hydrogen abstraction mechanism by oxygen alone (activation barrier 90 kJ mol^{-1}), in addition to the Ni-O pair mediated mechanism (activation barrier 97 kJ mol^{-1}). This increase in the surface reactivity results from the weakening of the binding strength of lattice oxygen, increasing the surface reducibility, and allowing strong chemisorption of the dissociated H. An equivalent increase in the surface reactivity can be achieved by doping the inactive NiO(100) surface with low valent metals like Li. Low valent metals lower the surface oxygen binding strength, marginally increasing surface reducibility, although not to the extent of surface Ni vacancy. The hydrogen chemisorption energy on the oxygen sites is identified as a descriptor for estimating the reactivity of surfaces, irrespective of the mechanism of dissociation. The experimentally determinable H chemisorption energy as a descriptor, together with the technique of doping, makes a useful tool set to screen metal oxides, their different surfaces, and potential dopants to tune the reactivity of the oxides for desired reactions.

497374 Design of Heterogeneous Catalyst By Stoichiometric Tuning of Intermetallic Gamma-Brass Crystal Structure

Abstract:

DFT Studies of Intermetallic Gamma-Brass Crystal Structures for Selective Hydrogenation

Haoran He, Anish Dasgupta, Gaurav Kumar, Robert M. Rioux, and Michael J. Janik

Department of Chemical Engineering, The Pennsylvania State University, University Park, PA 16802, USA

**mjanik@ psu.edu; (814) 863-9366*

Bimetallic compounds can offer tunable site electronics and ensemble structure for selective hydrogenation catalysis. In this study, we consider the γ-brass phase (Cu_5Zn_8 prototype), in order to expose surfaces with controlled M_x nuclearity to control the selectivity for hydrogenation. The γ-brass structure has a 52 atoms unit cell with 4 distinct symmetry sites —outer tetrahedral (OT), inner tetrahedral (IT) octahedral (OH) cuboctahedral (CO) as shown in Figure 1. In particular, the Pd-Zn γ-brass atomic arrangement as well as the substitution pattern of Zn by Pd in the Pd-Zn γ-brass phase (15.4-24%) has been extensively studied by Edstrom and Westman through x-ray diffraction analysis[1]. Surface energy calculations indicated that the most stable Pd_8Zn_{44} facet is (1 -1 0), which exposes only monomers for Pd_8, but includes Pd trimers for $Pd_{9\text{-}11}$. We vary the number of Pd atoms per isolated active site and investigate its effect on H_2 dissociation and acetylene hydrogenation mechanisms. DFT calculations agreed with experimental results that H_2 activation is faster on trimer sites, substantiating the formation of Pd_3 trimer sites on Pd_9Zn_{43} catalyst surfaces. The activation barrier for H_2 dissociation is nearly identical experimentally on Pd9, Pd10 and Pd11, further substantiating the isolation of the Pd trimer sites. DFT calculations indicate that acetylene binds strongly on both monomer bridge and trimer sites, whereas ethylene binds strongly on monomer atop and trimer side atop sites. At the same time, H_2 dissociation and binding adjacent to ethylene is only possible on the trimer sites. DFT calculations showed the apparent barrier of ethylene hydrogenation is higher than the ethylene desorption barrier, which indicates Pd_8Zn_{44} is superior catalyst in selectively hydrogenating acetylene to ethylene. $Pd_{9\text{-}11}$ on the other hand, contains trimers on the surface, which can lower the ethylene hydrogenation barrier, compared with Pd_8Zn_{44}. The full path of acetylene hydrogenation on these isolated sites, as well as a microkinetic model for acetylene hydrogenation on these intermetallics, will be presented. The gamma-brass intermetallic structures offer isolated active sites with controlled nuclearity, allowing both the design of active and selective catalysts as well as the elucidation of site requirements.

Figure 1. Illustration of the sites in the γ-brass structure: inner tetrahedral (IT, red); outer tetrahedral (OT, blue);

octahedral (OH, green); and cuboctahedral (CO, orange).

References

1. Edstrom, V.A and S. Westman, X-ray Determination of Structure of Cubic Gamma Pd, Zn Phase. Acta Chemica Scandinavica, 1969. 23 (1): p.279-&. *ADDIN EN.REFLIST*

499942 Surface Studies of Functional Alcohol Decomposition on Pt and Pd

Abstract:

Biomass-derived compounds are highly functionalized, making them difficult to selectively upgrade into a desired product. This is because the functional groups interact with the surface, resulting in multiple and often uncontrolled adsorption geometries. Previous studies on Pd indicated that adsorption geometry of these oxygenates dictates product selectivity, where flat lying configurations result in higher production of decarbonylation products, which are undesired, and upright conformations favor Câ€"O bond activation, producing the desired hydrodeoxygenation (HDO) products. It was shown that on Pd(111), high surface coverages of furfuryl alcohol lead towards increased selectivity towards the HDO product, methylfuran. However, the mechanism is poorly understood, hampering efforts to design better catalysts. HDO can proceed through two pathways, 1) a low-barrier direct transfer of H atoms between alcohol adsorbates and 2) a high-barrier H transfer via metal-bound H atoms. Surface science techniques, such as temperature-programmed desorption (TPD) and high-resolution electron energy loss spectroscopy (HREELS), using single crystals of Pd and Pt have provided a fundamental understanding of the low-barrier, direct H transfer mechanism.

Our recent studies have found that there appears to be key differences in the kinetics and mechanism for furfuryl alcohol hydrodeoxygenation. On Pd(111), the lowest-barrier reaction pathway was found to involve direct transfer hydrogenation between neighboring adsorbates, but this mechanism does not appear to be operable on Pt(111). However, on Pt(111), furfuryl alcohol can either proceed through high-barrier transfer to produce methylfuran or decarbonylation to furan, which also occurs on Pd(111). In the HDO pathway, furfuryl alcohol undergoes Câ€"O bond scission to form methylfuran and water. Decarbonylation occurs through the dehydrogenation of furfuryl alcohol to make furfural followed by decarbonylation to produce furan, which undergoes further decomposition on Pt(111) to produce propylene and CO. Most furfuryl alcohol reaction and desorption pathways are similar on Pt(111) and Pd(111) and the relative barriers of the major products is the same. However, on Pt(111) benzene is also produced as a volatile product, and is likely formed via coupling of C_3 surface intermediates.

500719 Spectroscopic Insight into the Reactivity of Oxide Electrocatalysts with Water

Abstract:

The intermittent nature of renewable energy sources requires storing and converting energy in a clean, scalable way. Water electrolysis and hydrogen fuel cells present a promising solution, however often require precious metals to reduce the kinetic overpotential required for the oxygen evolution and reduction reactions (OER and ORR), which limit the overall efficiency. In alkaline environments, transition metal oxides present an alternative to noble metals for oxygen electrocatalysis. However, a lack of fundamental understanding of the reaction mechanism has hindered their rational design.

Investigation of model epitaxial oxide thin films allows spectroscopic examinations of their chemical speciation in an aqueous environment using ambient pressure X-ray photoelectron spectroscopy.[1] By quantifying the formation of hydroxyl groups, we compare the relative affinity of different surfaces for this key reaction intermediate in oxygen electrocatalysis.[2] The coverage of hydroxyl groups measured spectroscopically at a fixed relative humidity trends with the free energy of a hydroxylated surface calculated by density functional theory, providing an experimental handle on the binding strength of this reaction intermediate.[3] Understanding the electronic structure of oxides in an aqueous environment is also critical for promoting charge transfer reactions in both electrocatalytic and photocatalytic reactions. To this end, we investigate changes in the electronic structure as a function of the oxygen and water chemical potential, enabling comparison with the metal redox potential and catalytic activity. The understanding developed from epitaxial surfaces builds molecular insight regarding interactions at oxide/water interfaces, and guides the rational design of high-surface-area oxide catalysts for technical applications.

References

1. K.A. Stoerzinger, W.T. Hong, E.J. Crumlin, H. Bluhm, and Y. Shao-Horn, *Accounts of Chemical Research,* **48**, 2976 (2015).

2. K.A. Stoerzinger, R. Comes, S.R. Spurgeon, S. Thevuthasan, K. Ihm, E.J. Crumlin, S.A. Chambers. *J. Phys. Chem. Lett.* **8**, 1038 (2017).

3. K.A. Stoerzinger, W.T. Hong, G. Azimi, L. Giordano, Y.-L. Lee, E.J. Crumlin, M.D. Biegalski, H. Bluhm, K.K. Varanasi, Y. Shao-Horn. *J. Phys. Chem. C* **119**, 18504 (2015).

500809 Understanding and Tuning the Activation of Unsaturated C-C Bonds over Nickel-Based Intermetallic Compounds

Abstract:

Intermetallic compounds (IMCs) comprised of atomically ordered mixtures of transition and post-transition metal elements exhibit uniquely tunable surface chemistry toward unsaturated Câ€"C bonds, which can be beneficial in many reactions that require gentle or limited activation of the unsaturated Câ€"C bond.Â Specifically, improved selectivity can be achieved in the semi-hydrogenation of alkynes, the production of olefins, and the selective hydrogenation of multiply functionalized unsaturated molecules.Â It is, however, unclear at this time how the surface chemistry of these promising materials changes as a function of element selection and bulk stoichiometry. This study aims to build a systematic understanding of the surface and catalytic chemistry of a suite of nickel-based ordered IMCs through a combined quantum chemical modeling and experimental approach.

Computational studies have isolated clear trends in the activation of unsaturated hydrocarbons as a function of both p-block element selection and the bulk stoichiometry of the IMCs. A volcano-like trend was encountered with respect to surface reactivity with the minimum in reactivity existing near the optimum of d- and p-orbital overlap (NiAl and Ga). The reduced reactivity encountered for NiAl and NiGa was universal for alkynes, alkenes, and aromatics studied as probe adsorbates. The surface reactivity was further a finer function of bulk stoichiometry, which may lend additional degrees of control.

Experimentally, the synthesis of high surface area oxide-supported IMCs (Ni + Ga and Ni + In) has been developed to control stoichiometry, phase segregation, and particle size and morphology. Critical aspects of the synthesis were precursor choice and the rate of its reduction, reduced constituent metal reactivity towards the oxide support and their diffusion, and annealing time and temperature for particle morphology control. A wide range of stoichiometries of Ni + Ga and Ni + In IMCs over different supports have been synthesized using various methods and shown to exhibit moderate activity and high selectivity in olefin production and selective hydrogenation. Reaction results closely followed predicted activity and selectivity from quantum chemical surface reaction modeling. These studies help to build a fundamental understanding of how the surface chemistry of IMCs is manipulated as a function of constituent element selection and bulk stoichiometry as well as the ability to modify the activation of unsaturated hydrocarbons to various degrees such that different catalytic transformations may be promoted. This degree of control is crucial for the development of any PGM-free catalysts associated with the production or functionalization of olefins or aromatics.

503502 Use of Competitive Adsorption with 2,5-Dimethylfuran for the Inhibition of Isomerization and Dimerization Products

Abstract:

The reaction of 2,5-dimethylfuran (DMF) with ethylene to form *p*-xylene has achieved high selectivities (>90%) using an H-BEA (Si/Al=12.5) zeolite acid catalyst.[1,2] As minimal xylene isomerization and transalkylation occurs under these conditions, this phenomenon was investigated and attributed to the surface site coverage of the ring opening product of DMF, 2,5-hexanedione.[3]Using alcohol dehydration as a probe chemistry, the ability of 2,5-hexanedione (always present with the addition of DMF to solution), to inhibit the reactivity of unsaturated bonds was investigated. In the absence of DMF cyclohexene was quickly produced followed by subsequent consumption of the cyclohexene product (Figure 1a). The addition of DMF increases the cyclohexene yield by almost 50% and inhibits its subsequent consumption through secondary pathways. Apparent activation energies of cyclohexanol dehydration appear to be identical in the presence and absence of DMF, indicating that the presence of DMF does not affect the energetics of the dehydration reaction (Figure 1b).

Figure 1. Cyclohexanol dehydration on H-BEA in the presence and absence of 2,5-dimethylfuran. A) Normalized concentration of cyclohexene to initial cyclohexanol concentration over time in the presence of and in the absence of 2,5-dimethylfuran. Reactions Conditions: 0.5 M cyclohexanol, 1.74 mM H-BEA (Si/Al 12.5), 160 °C, 200 psi N_2 added at 22 °C, 82 mL heptane, 2 mL tridecane (internal standard), and 0.74 M 2,5-dimethylfuran, when added. B) Arrhenius plot of apparent rate of cyclohexanol conversion versus inverse temperature for reactions including and excluding 2,5-dimethylfuran. Apparent activation energies are displayed and within error. Reaction Conditions: 135 – 200 °C, 0.5 M cyclohexanol, 1.74 mM H-BEA (Si/Al 12.5), 160 °C, 200 psi N_2added at 22 °C, 82 mL heptane, 2 mL tridecane (internal standard), and 0.74 M 2,5-dimethylfuran, when added.

[1] Williams, C. L.; Chang, C.-C.; Do, P.; Nikbin, N.; Caratzoulas, S.; Vlachos, D. G.; Lobo, R. F.; Fan, W.; Dauenhauer, P. J. *ACS Catal.*, 2012, 2, 935-939.

[2] Chang, C.-C.; Green, S. K.; Williams, C. L.; Dauenhauer, P. J.; Fan, W. *Green Chem.* 2014, 16, 585-588.

[3] Williams, C. L.; Vinter, K. P.; Patet, R. E.; Chang, C.-C.; Nikbin, N.; Feng, S.; Wiatrowski, M.

R.; Caratzoulas, S.; Fan, W.; Vlachos, D. G.; Dauenhauer, P. J. *ACS Catal.*, 2016, 6, 2076-2088.

499675 NO Uptake and Desorption on Fe-Modified Pd/ZSM-5:Â Impact of Pd and Fe Loading

Abstract:

NO Uptake and Desorption on Fe-Modified Pd/ZSM-5:

Impact of Pd and Fe Loading

Kyle Karinshak*, Michael P. Harold*, and Kiran Premchand+

*Dept. of Chemical & Biomolecular Engineering, University of Houston, Houston, Texas

+FCA US LLC Auburn Hills, MI 48236

* corresponding author: M.P. Harold (mpharold@central.uh.edu)

Reducing cold-start emissions, emissions under 200 °C, will play a significant role in meeting future EPA and EU emissions standards. By adsorbing NO and NO_2 at low temperatures and releasing these gases in temperatures regimes where traditional emission control catalysts can operate, Passive NOx Adsorbers (PNAs) augment existing emission control systems and are a promising new field of catalysis. The scant published literature has focused on Pd supported on various materials, including ceria or alumina. In contrast, the patent literature suggests the efficacy of ZSM-5 as a support as well as the addition of Fe to PNA compositions [1-6]. Our preliminary investigations into Palladium-exchanged MFI zeolites show significant NOx uptake capabilities and complex desorption behavior.

Catalyst samples were prepared through a wet impregnation protocol before being deposited onto monolith cores. Palladium(II) nitrate was used as the precursor and dissolved into water equal to the pore volume of the ZSM-5 zeolite (SAR 30). Alumina solution was added to the washcoating solution to improve binding. Four catalysts – 0 wt% Pd, 1wt% Pd, 1 wt% Pd/1 wt% Fe, and 2 wt% Pd, were washcoated onto a 0.25in diameter, 2in length, 600 cspi monolith cores at a fixed loading of 1.5 g/in^3. A bench-scale quartz reactor was utilized to simulate emissions and study uptake and desorption characteristics. DRIFTS (diffuse reflectance infrared Fourier transform spectroscopy) experiments of catalysts samples were performed to examine surface species formation and the mechanism through which NOx uptake occurs on Pd/ZSM catalysts.

Figure 1 shows uptake characteristics of NOx uptake at 50 °C over a five minute duration for the four catalyst. 400 ppm of NO was flowed through the catalyst accompanied by 2% O2 and balance Ar. Similar trends were observed at uptake temperatures of 80 °C and 150 °C with decreasing uptake quantities and breakthrough periods. NO adsorption was accompanied by noticeable NO_2 slip from the catalyst. The addition of Fe to the catalyst composition significantly improved the performance of NOx uptake.

Figure 2 shows the NOx desorption peaks from the same experiments. Over half of desorbed NOx occurred in the form of NO_2. As NO adsorbs as a mono nitrosyl palladium complex, this shows that NO is being converted into NO_2 through a zeolite-surface reaction and likely storing on the Pd-less acid sites of the zeolite and binder. Unlike Murata et al, who reported a single NOx desorption peak during TPD experiments [7], our results reveal multiple desorption peaks. Three distinct peaks are identified in Pd-containing samples whereas only two exist in the Pd-less sample. A preliminary assignment of these peaks include acid sites of the binder, the acid sites of the zeolite, and the Pd nitrosyl complexes.

Figure 3 shows DRIFTS measurements of three samples: 0 wt% Pd, 1 wt% Pd/1 wt% Fe, and 2 wt% Pd. Experiments were conducted in the presence of oxygen. Peak analysis matches that reported by Descorme and Gelin, confirmed the NO adsorption mechanism [8]. The Pd/Fe sample did not show the peaks associated with NO adsorption onto Fe complexes, confirming that the role of Fe in the catalyst composition is ensure a high dispersion of Pd [9]. Experiments conducted without oxygen (not shown) display peaks associated with NO adsorption onto Pd but do not show peaks associated with NO_2 formation and storage.

Additional TPD experiments (not shown) have shown that the presence of water significantly suppresses NOx uptake and desorption. The introduction of water inhibits the surface-reaction formation of NO_2, likely through the suppression of zeolitic acid sites. Ongoing experiments include powdered catalyst TPD experiments as well as additional DRIFTS experiments to better understand and quantify the effects the addition of binder.

Ongoing work will include modifications to the catalyst compositions as well as the addition of a modelling component for both uptake and desorption. Additionally, concurrent experiments are being carried out to understand the effects of adding hydrocarbons to the gas environment.

Figure 3: Evolution of DRIFTS data over time from three samples.

488343 Insights into Influencing Factors on Oxygen Storage Characteristics for Ceria-Zirconia Catalysts

Abstract:

Ceria systems have been investigated extensively in the field of automotive catalysis as oxygen storage materials in order to compensate for deviating stoichiometric operation during exhaust aftertreatment. The ability of ceria to cycle between Ce^{4+} and Ce^{+3} provides an attractive characteristic desirable for reactions such as water-gas shift, hydrocarbon oxidation, and hydrocarbon reforming [1]. Often, an additional dopant is added into the ceria lattice in order to increase oxygen storage capacity (OSC) and oxygen mobility, with ceria zirconia (CZO) being the most common system used [2]. The oxygen release rate (ORR) by these materials can be further enhanced by depositing platinum group metal (PGM) such as Pd or Rh, which provides beneficial metal-support interface in order for rapid oxygen transfer to occur [3]. While these catalytic materials have been in use, additional fundamental insight in the mechanism of OSC and the effect of PGM dispersion, CZO surface area, material pre-treatment, and varying testing conditions on catalyst performance are explored in this study.

A series of Pd/CZO catalysts was prepared by incipient wetness impregnation with different loadings and aged in redox conditions at high temperatures. Characterization by BET, H_2 chemisorption, and XRD will be discussed. OSC measurements were conducted via lambda probes before and after the catalyst as a function of temperature, while ORR were evaluated by the production of CO_2 as measured by FTIR as a function of time. Structure-activity relationships were developed to determine what factors were most important in achieving high OSC and fast ORR, particularly after high temperature aging. In addition, catalysts were evaluated for CO oxidation and water-gas shift to establish the importance of OSC in either reaction. Results from this study will help provide guidelines in the design of future oxygen storage materials in order to meet future emission standards.

[1] R. Gorte, AIChE Journal 56 (2010) 1126-1135

[2] E. Mamontov, et al., J. Phys. Chem. B 104 (2000) 11110-11116

[3] H.W. Jen, et al., Catal. Today 50 (1999) 309-328

Pd doped $CaCo_xZr_{1-x}O_{3-\delta}$ perovskites for automotive emissions control

Nov 01, 2017, AIChE Annual Meeting

Qinghe (Angela) Zheng and **Marty Lail*, PhDs**
Energy Technology Division, RTI International
Research Triangle Park, NC 27709
qizheng@ri.org

Conventional supported PGM-based TWC *vs.* Perovskite-TWC

Conventional PGM-based TWCs
($PdRh/\gamma Al_2O_3$ (CeO_2-ZrO_2))

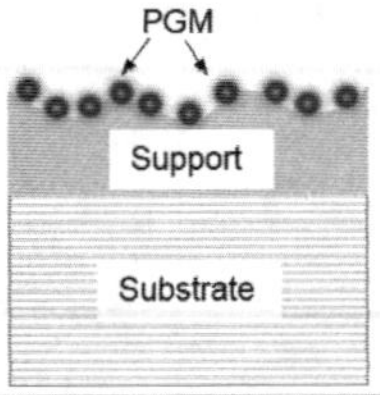

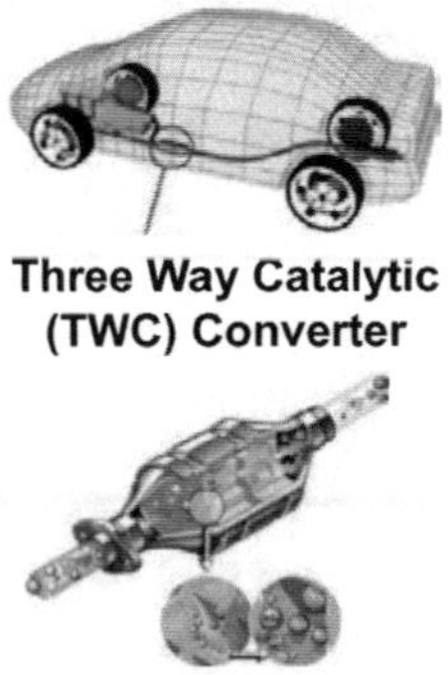

Three Way Catalytic (TWC) Converter

$ABO_{3-\delta}$ Perovskites

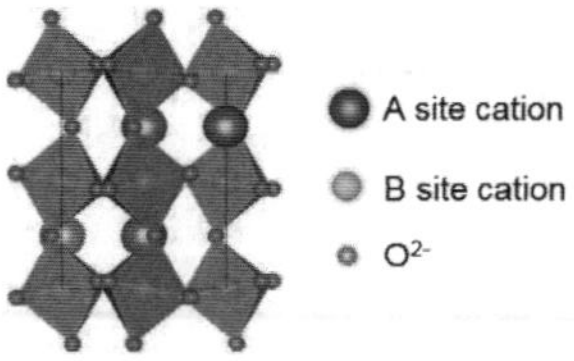

$$ABO_{3-\delta} + \frac{1}{2}\Delta\delta\, CO \rightarrow ABO_{3-\delta-\Delta\delta} + \frac{1}{2}\Delta\delta\, CO_2$$

$$ABO_{3-\delta} + \Delta\delta\, C_xH_y \rightarrow ABO_{3-\delta-\left(2x+\frac{1}{2}y\right)\Delta\delta} + \Delta\delta\,(x\, CO_2 + \frac{1}{2}y\, H_2O)$$

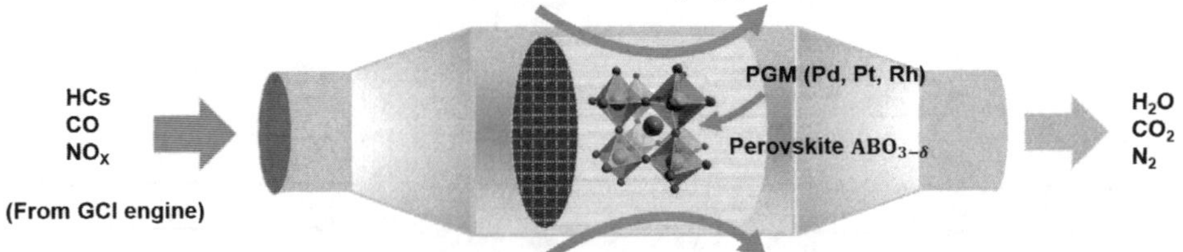

$$ABO_{3-\delta-\Delta\delta} + \frac{1}{2}\Delta\delta\, O_2 \rightarrow ABO_{3-\delta}$$

$$ABO_{3-\delta-\Delta\delta} + \Delta\delta\, NO\left(\text{or}\,\frac{1}{2}\Delta\delta\, NO_2\right) \rightarrow ABO_{3-\delta} + \frac{1}{2}\Delta\delta\, N_2\ (\text{or}\,\frac{1}{4}\Delta\delta\, N_2)$$

Conventional supported PGM-based TWC *vs.* Perovskite-TWC

RTI INTERNATIONAL

Conventional PGM-based TWCs
($PdRh/\gamma Al_2O_3$ (CeO_2-ZrO_2))

- ✓ High activity (@ stoichiometric)
- ✓ Good performance with CeO_2 added
- ✓ High surface area support
- o Low thermal stability (@ fuel cutoff)
- o High cost of PGM
- o Market control of Ce resources

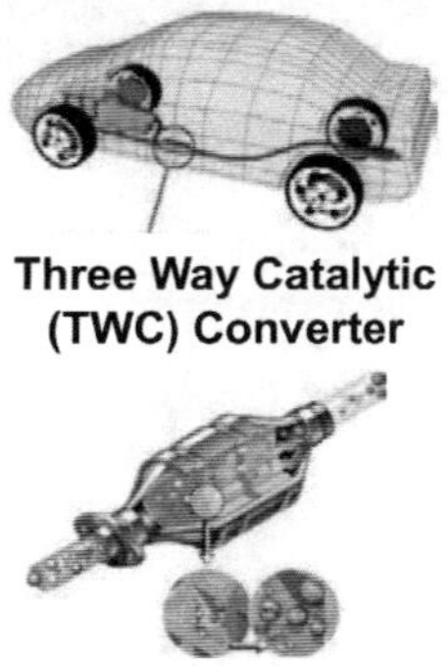

Three Way Catalytic (TWC) Converter

$ABO_{3-\delta}$ Perovskites

- ✓ Improved thermal stability
- ✓ Low cost
- ✓ Tunable compositions
- ✓ Large availability
- o Low surface area
- o Less activity

$$ABO_{3-\delta} + \frac{1}{2}\Delta\delta\, CO \rightarrow ABO_{3-\delta-\Delta\delta} + \frac{1}{2}\Delta\delta\, CO_2$$

$$ABO_{3-\delta} + \Delta\delta\, C_xH_y \rightarrow ABO_{3-\delta-\left(2x+\frac{1}{2}y\right)\Delta\delta} + \Delta\delta\,(x\, CO_2 + \frac{1}{2}y\, H_2O)$$

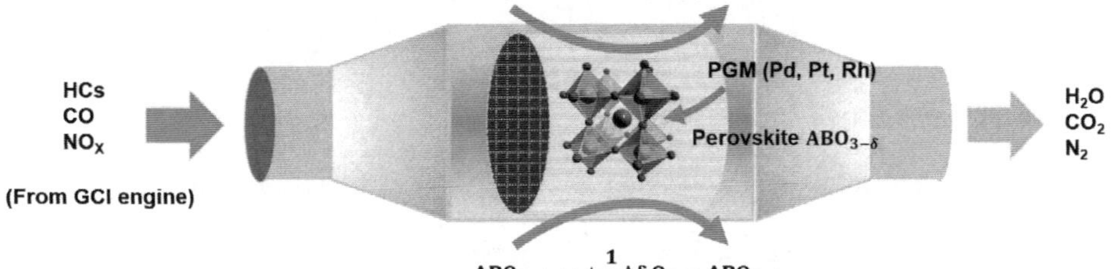

$$ABO_{3-\delta-\Delta\delta} + \frac{1}{2}\Delta\delta\, O_2 \rightarrow ABO_{3-\delta}$$

$$ABO_{3-\delta-\Delta\delta} + \Delta\delta\, NO \left(\text{or} \frac{1}{2}\Delta\delta\, NO_2\right) \rightarrow ABO_{3-\delta} + \frac{1}{2}\Delta\delta\, N_2 \left(\text{or} \frac{1}{4}\Delta\delta\, N_2\right)$$

3

$ABO_{3-\delta}$ perovskite structure

RTI INTERNATIONAL

$ABO_{3-\delta}$ perovskite structure

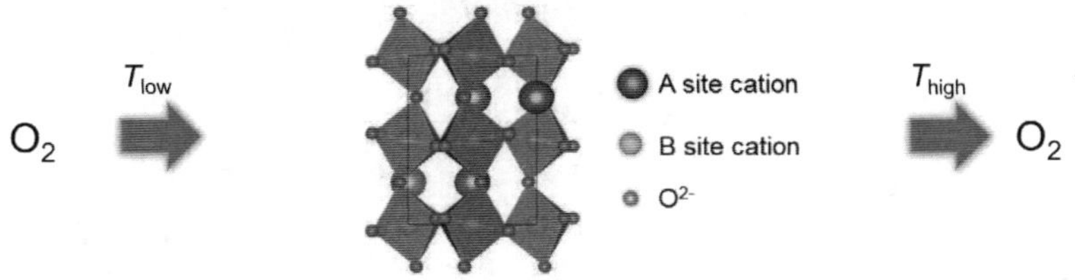

A site cation: alkaline earth, alkaline, and lanthanide ions in a 12-fold coordination (e.g. A= La, Sr, Ca, Ba)

- Important in perovskite structural formation
- Influence oxygen vacancy and structural defects density
- Influence charge-valence balance

B site cation: smaller transition metal ions in a 6-fold coordination with oxygen ions (e.g. B=Co, Mn, Fe, Cu)

- Perovskite oxygen storage capacity depends on the change of oxidation state of the B-site cation
- Perovskites with Co at B site was reported to show best activity in HC and CO oxidation
- Co-based perovskites have low thermal stability due to decomposition at high T
- Improved perovskite properties can be achieved by B-site doping

Oxygen non-stoichiometry *δ*: oxygen anion deficiency or excess depending on the total valence amount of A-site and B-site cations

4

Oxygen mobility within $ABO_{3-\delta}$ perovskite lattice

- $ABO_{3-\delta}$ perovskite can exhibit significant swings in δ through the reaction with gas phase O_2.
- δ can be tailored by doping A- or B-sites with appropriately selected cations
- δ may vary with T and P_{O2}
- Adsorption of superficial [O] at T<600 °C
  Activation of interfacial lattice [O] at T>600 °C } Mars-van-Kreveken mechanism

RTI developed novel $CaCo_xZr_{1-x}O_{3-\delta}$ type perovskites wi or w/o PGM doping

(Patent application RTI 16P027-PROV)

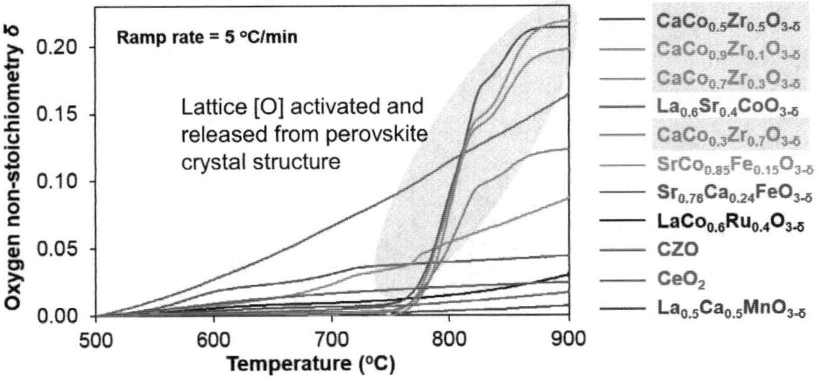

Fig. 1. O_2 non-stoichiometry as a function of sample temperature (500 ~ 900 °C) during TGA temperature programmed study of $ABO_{3-\delta}$ perovskite materials.

Perovskite with high OSC as potential TWC

CeO_2-ZrO_2 as a support for TWC

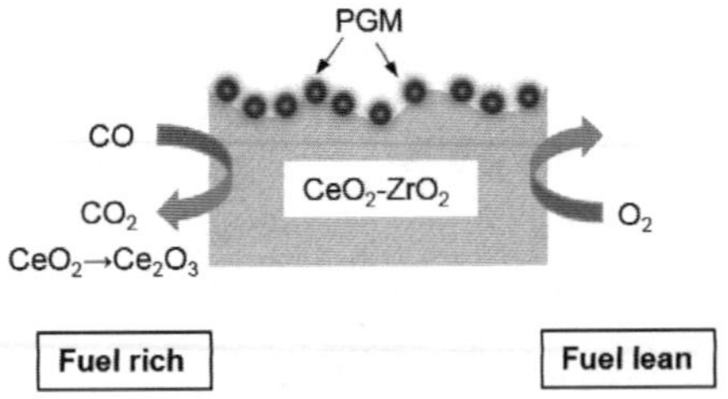

$ABO_{3-\delta}$ perovskite-type TWC

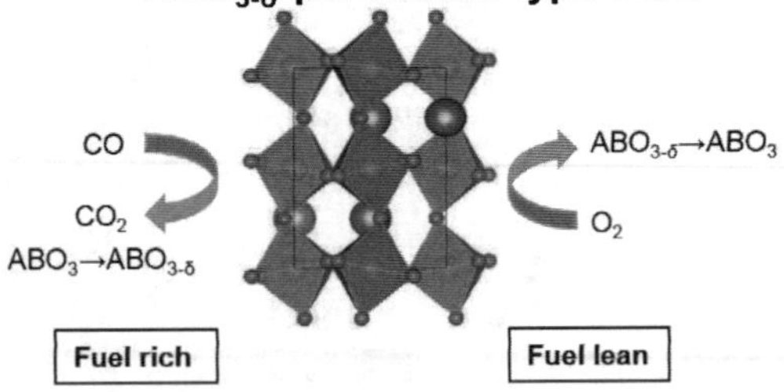

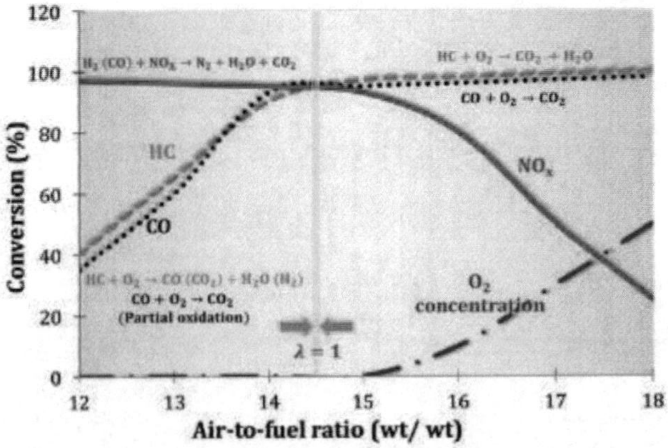

Figure 2. The TWC conversion profile as a function of air-to-fuel ratio (λ).

Experimental methodology

RTI INTERNATIONAL

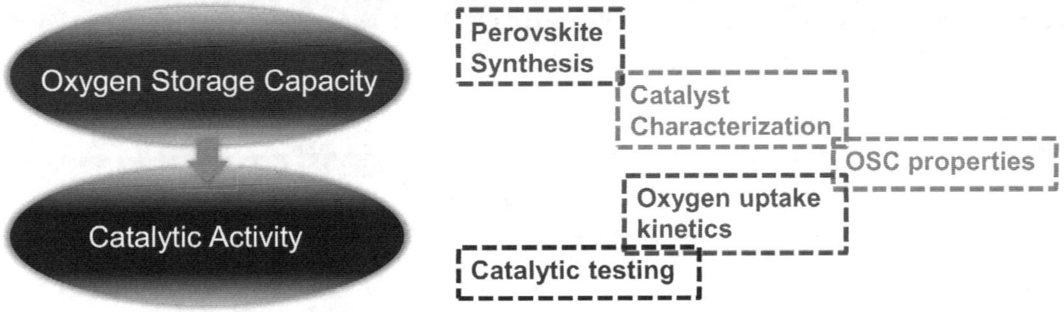

7

Synthesis and characterization of $CaCo_xZr_{1-x}(Pd)O_3$ perovskites

RTI INTERNATIONAL

Table 1. Analytical result of as-synthesized (Pd-doped) $CaCo_xZr_{1-x}O_3$ perovskites.

#	Theoretical composition	Actual composition (XRF)	Tolerance factor, τ
CCZPd2	$CaCo_{0.4}Zr_{0.4}Pd_{0.2}O_3$	$CaCo_{0.4}Zr_{0.3}Pd_{0.2}O_3$	0.81
CCZPd1	$CaCo_{0.5}Zr_{0.4}Pd_{0.1}O_3$	$CaCo_{0.5}Zr_{0.3}Pd_{0.1}O_3$	0.82
CCZPd05	$CaCo_{0.55}Zr_{0.4}Pd_{0.05}O_3$	$CaCo_{0.55}Zr_{0.3}Pd_{0.06}O_3$	0.83
CCZ91	$CaCo_{0.9}Zr_{0.1}O_3$	$CaCo_{1.0}Zr_{0.1}O_3$	0.86
CCZ73	$CaCo_{0.7}Zr_{0.3}O_3$	$CaCo_{0.8}Zr_{0.2}O_3$	0.85
CCZ55	$CaCo_{0.5}Zr_{0.5}O_3$	$CaCo_{0.5}Zr_{0.5}O_3$	0.84
CCZ37	$CaCo_{0.3}Zr_{0.7}O_3$	$CaCo_{0.3}Zr_{0.6}O_3$	0.82
CZ10	$CaZrO_3$	$CaZrO_3$	0.80

- **Perovskite synthesis:** Pechini method, calcination @ 1200 °C;
- **Perovskite target compositions:**
  $CaCo_xZr_{1-x}O_{3-\delta}$ (x=0, 0.3, 0.5, 0.7, and 0.9),
  Pd doped $CaCo_{0.6-y}Zr_{0.4}Pd_yO_{3-\delta}$ (y=0.05, 0.1 and 0.2);

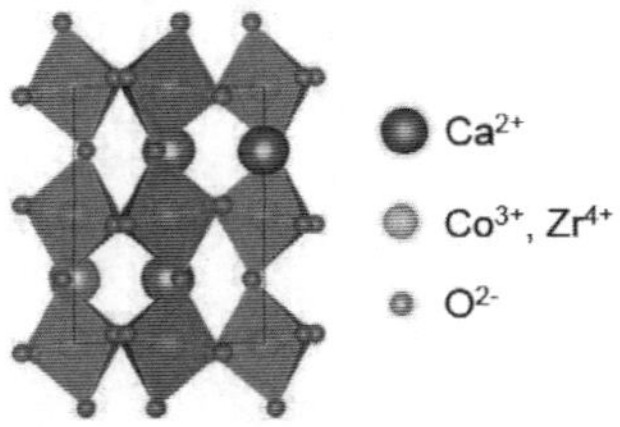

8

Synthesis and characterization of $CaCo_xZr_{1-x}(Pd)O_3$ perovskites

RTI INTERNATIONAL

Table 1. Analytical result of as-synthesized (Pd-doped) $CaCo_xZr_{1-x}O_3$ perovskites.

#	Theoretical composition	Actual composition (XRF)	Tolerance factor, τ
CCZPd2	$CaCo_{0.4}Zr_{0.4}Pd_{0.2}O_3$	$CaCo_{0.4}Zr_{0.3}Pd_{0.2}O_3$	0.81
CCZPd1	$CaCo_{0.5}Zr_{0.4}Pd_{0.1}O_3$	$CaCo_{0.5}Zr_{0.3}Pd_{0.1}O_3$	0.82
CCZPd05	$CaCo_{0.55}Zr_{0.4}Pd_{0.05}O_3$	$CaCo_{0.55}Zr_{0.3}Pd_{0.06}O_3$	0.83
CCZ91	$CaCo_{0.9}Zr_{0.1}O_3$	$CaCo_{1.0}Zr_{0.1}O_3$	0.86
CCZ73	$CaCo_{0.7}Zr_{0.3}O_3$	$CaCo_{0.8}Zr_{0.2}O_3$	0.85
CCZ55	$CaCo_{0.5}Zr_{0.5}O_3$	$CaCo_{0.5}Zr_{0.5}O_3$	0.84
CCZ37	$CaCo_{0.3}Zr_{0.7}O_3$	$CaCo_{0.3}Zr_{0.6}O_3$	0.82
CZ10	$CaZrO_3$	$CaZrO_3$	0.80

- **Perovskite synthesis:** Pechini method, calcination @ 1200 °C;
- **Perovskite target compositions:**
  $CaCo_xZr_{1-x}O_{3-\delta}$ (x=0, 0.3, 0.5, 0.7, and 0.9),
  Pd doped $CaCo_{0.6-y}Zr_{0.4}Pd_yO_{3-\delta}$ (y=0.05, 0.1 and 0.2);
- **Perovskite actual compositions** confirmed by XRF;

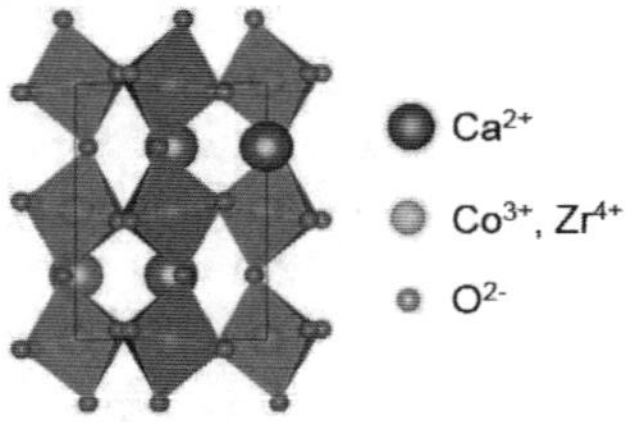

9

Synthesis and characterization of $CaCo_xZr_{1-x}(Pd)O_3$ perovskites

RTI INTERNATIONAL

Table 1. Analytical result of as-synthesized (Pd-doped) $CaCo_xZr_{1-x}O_3$ perovskites.

#	Theoretical composition	Actual composition (XRF)	Tolerance factor, τ
CCZPd2	$CaCo_{0.4}Zr_{0.4}Pd_{0.2}O_3$	$CaCo_{0.4}Zr_{0.3}Pd_{0.2}O_3$	0.81
CCZPd1	$CaCo_{0.5}Zr_{0.4}Pd_{0.1}O_3$	$CaCo_{0.5}Zr_{0.3}Pd_{0.1}O_3$	0.82
CCZPd05	$CaCo_{0.55}Zr_{0.4}Pd_{0.05}O_3$	$CaCo_{0.55}Zr_{0.3}Pd_{0.06}O_3$	0.83
CCZ91	$CaCo_{0.9}Zr_{0.1}O_3$	$CaCo_{1.0}Zr_{0.1}O_3$	0.86
CCZ73	$CaCo_{0.7}Zr_{0.3}O_3$	$CaCo_{0.8}Zr_{0.2}O_3$	0.85
CCZ55	$CaCo_{0.5}Zr_{0.5}O_3$	$CaCo_{0.5}Zr_{0.5}O_3$	0.84
CCZ37	$CaCo_{0.3}Zr_{0.7}O_3$	$CaCo_{0.3}Zr_{0.6}O_3$	0.82
CZ10	$CaZrO_3$	$CaZrO_3$	0.80

- **Perovskite synthesis:** Pechini method, calcination @ 1200 °C;
- **Perovskite target compositions:**
  $CaCo_xZr_{1-x}O_{3-\delta}$ (x=0, 0.3, 0.5, 0.7, and 0.9),
  Pd doped $CaCo_{0.6-y}Zr_{0.4}Pd_yO_{3-\delta}$ (y=0.05, 0.1 and 0.2);
- **Perovskite actual compositions** confirmed by XRF;

- **Goldschmidt tolerance factor:**
  $$\tau = \frac{r_A + r_O}{\sqrt{2}(r_B + r_O)}$$
  (r_A, r_B, and r_O are the ionic radii for the A^{n+}, B^{m+}, and O^{2-});
- All the **τ** values are within the range of 0.8 ~0.9:
  Suggest orthorhombic structures
- **BET surface areas: <10 m²/g**

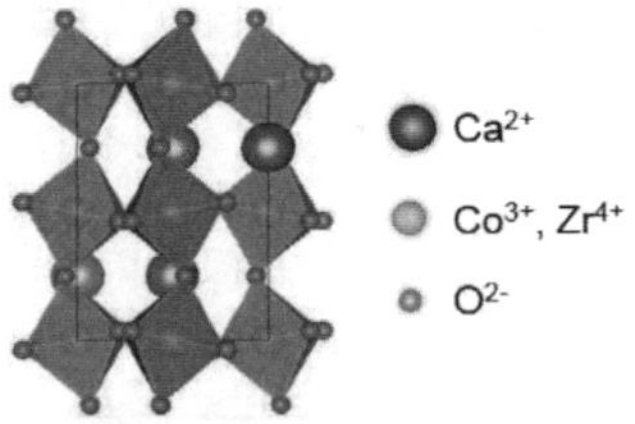

10

XRD: Partial B-site substitution by Zr enhances crystallinity Pd dopant type includes surface Pd^0 and bulk Pd^{2+}

RTI INTERNATIONAL

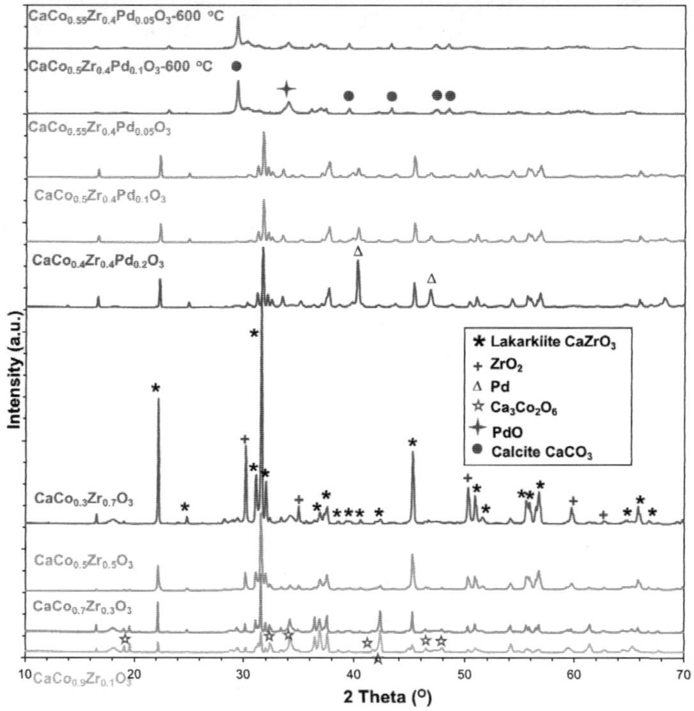

Fig. 3. XRD patterns of the $CaCo_xZr_{1-x}O_3$ (x = 0, 0.3, 0.5, 0.7 and 0.9) and Pd-doped $CaCo_{0.6-y}Zr_{0.4}Pd_yO_3$ (y = 0.05, 0.1 and 0.2) perovskite samples.

- **For $CaCo_{0.6-y}Zr_{0.4}Pd_yO_3$ (calcined at low temperature 600 °C):**
  - Perovskite structure not formed
  - Palladium presents as PdO
- **For $CaCo_{0.6-y}Zr_{0.4}Pd_yO_3$:**
  - Main phase remained $CaCo_{0.5}Zr_{0.5}O_3$ perovskite;
  - Doped Pd includes bulk Pd^{2+} and surface Pd^0.
- **For $CaCo_xZr_{1-x}O_3$:**
  - Retain Lakarkiite $CaZrO_3$ main structure;
  - Partial substitution of Co by Zr enhances the structural crystallinity of $CaCoO_3$;
  - When x=0.5, optimum crystallinity and minimum impurity.

11

H_2-TPR: Enhanced Co reduction by Zr substitution and Pd doping

RTI INTERNATIONAL

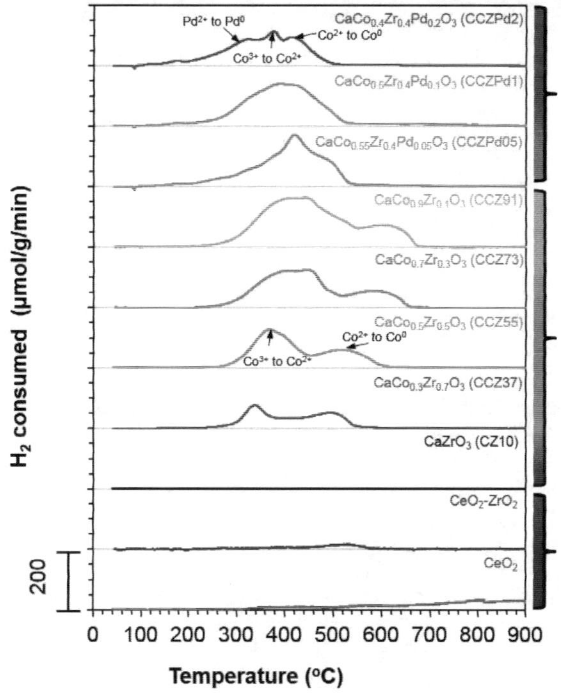

- **For $CaCo_{0.6-y}Zr_{0.4}Pd_yO_3$:**
  - Pd doping also facilitates the reduction of Co species;
  - $CaCo_{0.4}Zr_{0.4}Pd_{0.2}O_3$:
    Bulk $Pd^{2+} \rightarrow Pd^0$ (T_{max}=320 °C)
    $Co^{3+} \rightarrow Co^{2+}$ (T_{max}=390 °C)
    $Co^{2+} \rightarrow Co^0$ (T_{max}=420 °C);
  - Possible H_2 dissociation on surface Pd^0 particles followed by the successive spill-over of dissociated H atoms to the Co species.
- **For $CaCo_xZr_{1-x}O_3$:**
  - $Co^{3+} \rightarrow Co^{2+}$ (280-500 °C)
    $Co^{2+} \rightarrow Co^0$ (450-680 °C);
  - B-site substitution by Zr enhances the reduction of Co species by lowering the reduction T;
  - No reduction peak observed with $CaZrO_3$.
- **Perovskite vs. CeO_2 and CeO_2-ZrO_2 (CZO):**

$CaCo_xZr_{1-x}(Pd)O_3$ perovskites have significantly higher reducibility/OSC than CeO_2 and CZO.

Fig. 4. H_2 consumption as a function of reaction time on stream and temperature of studied samples during H_2-TPR measurements (temperature ramp rate at 5 °C/min).

12

Redox cyclic tests: $CaCo_{0.5}Zr_{0.5}O_3$ showed optimum OSC and stability

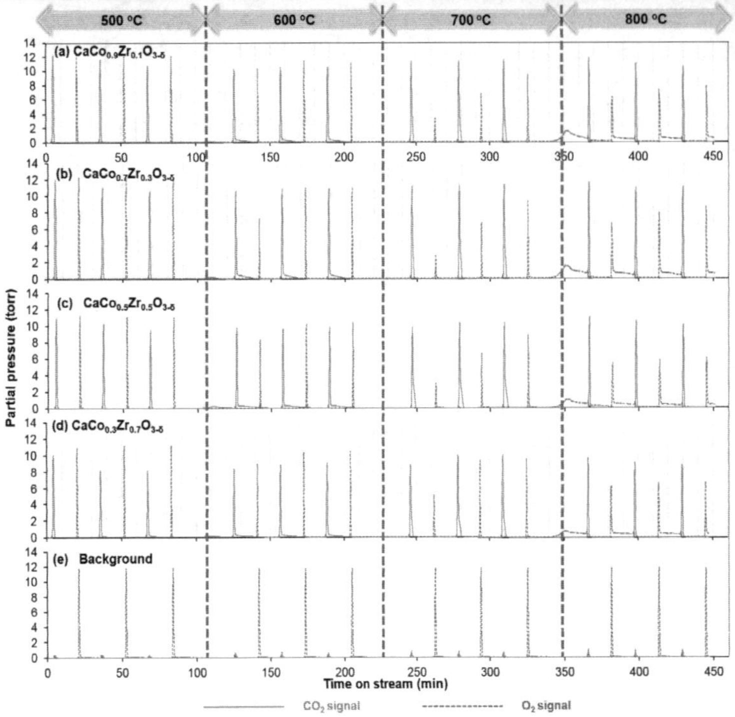

Reduction-Oxidation (3 cycles)

Reduction by 20% CO/He
Oxidation by air
Temperature: 500, 600, 700 and 800 °C

- Perovskite OSC increases with increasing T;
- Perovskite OSC increases with increasing *x* (Co content);
- Stable oxygen storage performance by Zr substitution;
- Optimum perovskite OSC and stability with $CaCo_{0.5}Zr_{0.5}O_3$.

Fig. 5. CO_2 production and O_2 intake vs. reaction time on stream (TOS) of $CaCo_xZr_{1-x}O_{3-\delta}$ (*x*=0.3, 0.5, 0.7, or 0.9) perovskite samples, as represented by MS partial pressure signals during isothermal CO reduction-air oxidation cyclic tests at 500, 600, 700, and 800 °C.

Oxygen intake kinetics measurements

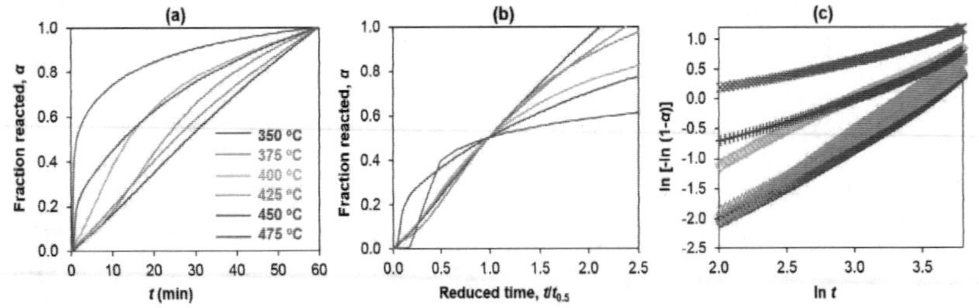

$$\alpha = \frac{m_0 - m_t}{m_0 - m_{final}}$$

$$\alpha \in [0, 1]$$

Fig. 6. Plots for oxygen-intake kinetic calculation of $CaCo_{0.5}Zr_{0.5}O_{3-\delta}$ perovskite showing (a) Fraction reacted α with time variation, (b) Fraction reacted α against reduced time $t/t_{0.5}$, and (c) $\ln[-\ln(1-\alpha)]$ as a function of $\ln t$ (with *t* in min) for the determination of reaction mechanisms at temperatures varied from 350 °C to 475 °C.

Table 2. List of solid-state rate equations with suggested *m* values studied for the oxygen-intake reaction for $CaCo_{0.5}Zr_{0.5}O_{3-\delta}$ and $CaCo_{0.9}Zr_{0.1}O_{3-\delta}$ perovskite samples. *m* values: slopes in Fig.6(c) at each temperature.

Eq. #	Function	Equation	m	Reaction mechanism
1	$D_1(\alpha)$	$\alpha^2 = kt$	0.62	Diffusion controlled
2	$D_2(\alpha)$	$(1-\alpha)\ln(1-\alpha) + \alpha = kt$	0.57	Diffusion controlled
3	$D_3(\alpha)$	$[1-(1-\alpha)^{1/3}]^2 = kt$	0.54	Diffusion controlled
4	$D_4(\alpha)$	$1-2\alpha/3-(1-\alpha)^{2/3} = kt$	0.57	Diffusion controlled
5	$F_1(\alpha)$	$-\ln(1-\alpha) = kt$	1.00	First-order kinetics
6	$R_2(\alpha)$	$1-(1-\alpha)^{1/2} = kt$	1.11	Surface reaction controlled
7	$R_3(\alpha)$	$1-(1-\alpha)^{1/3} = kt$	1.07	Surface reaction controlled
8	Zero order	$\alpha = kt$	1.24	Zero-order kinetics

Oxygen intake kinetics measurements: First order kinetic model

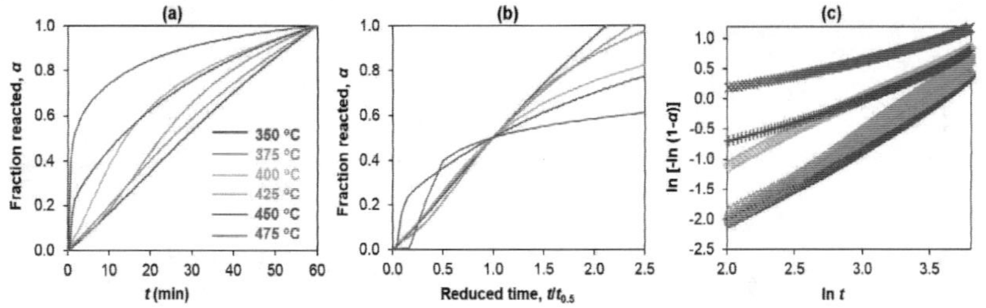

$$\alpha = \frac{m_0 - m_t}{m_0 - m_{final}}$$

$$\alpha \in [0, 1]$$

Fig. 6. Plots for oxygen-intake kinetic calculation of $CaCo_{0.5}Zr_{0.5}O_{3-\delta}$ perovskite showing (a) Fraction reacted α with time variation, (b) Fraction reacted α against reduced time $t/t_{0.5}$, and (c) $\ln[-\ln(1-\alpha)]$ as a function of $\ln t$ (with t in min) for the determination of reaction mechanisms at temperatures varied from 350 °C to 475 °C.

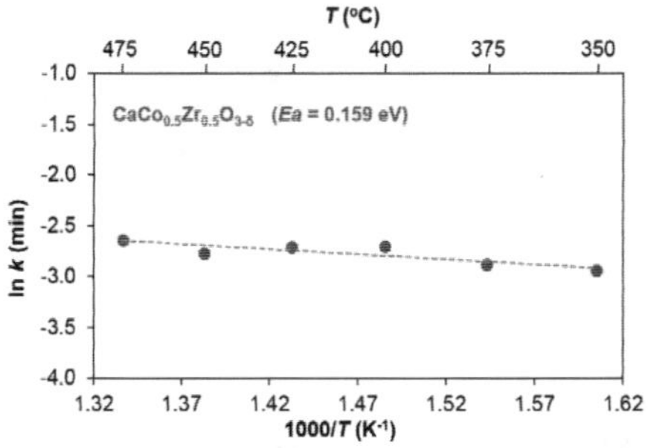

$-\ln(1-\alpha) = kt$ m = 0.6~1.3 First-order kinetics

Fig. 7. Temperature dependent-reaction rate constants k in natural log scale as a function of inverse temperature for the calculation of reaction activation energy for $CaCo_{0.5}Zr_{0.5}O_{3-\delta}$ perovskite. The k values were achieved by fitting reaction data with various kinetic models, and first-order kinetic model was selected.

15

Catalytic oxidation activity of Pd doped $CaCo_{0.6-y}Zr_{0.4}Pd_yO_3$

Simulated dry exhaust feed: 0.060-0.066% C_3H_6, 0.750-0.825% CO, 0.675-0.750% O_2, 19.5% CO_2, N_2 bal.

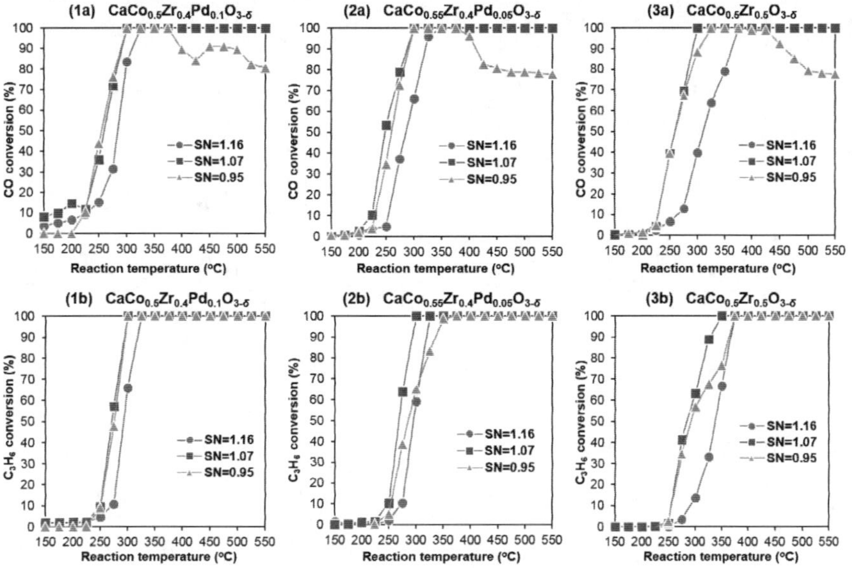

- Complete conversions of CO and C_3H_6 below 350 °C;
- Pd-doped samples showed higher oxidation activities than Pd-free one at all three conditions;
- Higher conversions at richer condition (SN=0.95) for Pd samples.
- "Pd regeneraetion" effect at rich condition (Tanaka *et al*)

Fig. 8. (1) CO and (2) C_3H_6 conversions during catalytic oxidation tests over (a) $CaCo_{0.5}Zr_{0.4}Pd_{0.1}O_{3-\delta}$, (b) $CaCo_{0.55}Zr_{0.4}Pd_{0.05}O_{3-\delta}$ and (c) $CaCo_{0.5}Zr_{0.5}O_{3-\delta}$ perovskite with simulated exhaust feed **at stoichiometric numbers (SNs) of 1.16, 1.07 and 0.95. SN=2[O_2]/([CO]+9[C_3H_6])**

16

Catalytic oxidation activity with addition of H_2O in the feed

RTI INTERNATIONAL

Simulated wet exhaust feed: 0.056% C_3H_6, 0.702% CO, 0.648% O_2, 17.55% CO_2, 10% steam, N_2 bal.

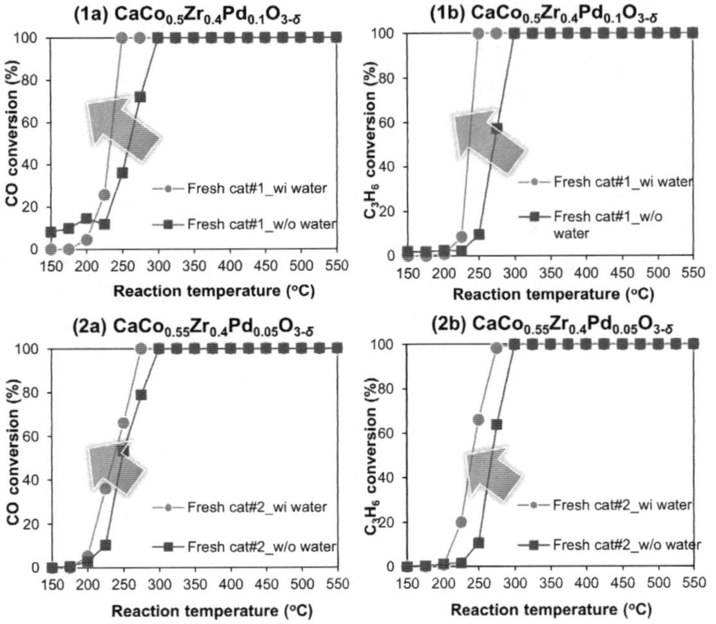

Fig. 9. (1) CO and (2) C_3H_6 conversions during catalytic oxidation tests over fresh and aged (a) $CaCo_{0.5}Zr_{0.4}Pd_{0.1}O_{3-\delta}$, (b) $CaCo_{0.55}Zr_{0.4}Pd_{0.05}O_{3-\delta}$ perovskites **at SN of 1.07, with presence of 10 vol-% H_2O in the feed.**

Enhancements in both CO and C_3H_6 conversions when water (steam) was added to the feed;

17

Catalytic oxidation activity of Pd doped $CaCo_{0.6-y}Zr_{0.4}Pd_yO_3$

RTI INTERNATIONAL

Table 3. Main reaction pathways during catalytic conversions of CO and C_3H_6 over Pd doped $CaCo_xZr_{1-x}O_3$ perovskites by feeding simulated engine exhaust at SN=1.07, when water is present in the feed.

Eq. #	Reaction pathway	Reaction	ΔH^0_{298} (kJ/mol)
(1)	Oxidation of C_3H_6	$C_3H_6 + 4.5\ O_2 \xrightarrow{\text{Perovskite}} 3CO_2 + 3H_2O$	-2058
(2)	Oxidation of CO	$CO + 0.5\ O_2 \xrightarrow{\text{Perovskite}} CO_2$	-283
(3)	Steam reforming of C_3H_6	$C_3H_6 + 6\ H_2O \xrightarrow{\text{Perovskite}} 3CO_2 + 9H_2$	+110
(4)	Water gas shift	$CO + H_2O \xrightarrow{\text{Perovskite}} CO_2 + H_2$	-41
(5)	Combustion of H_2	$H_2 + 0.5\ O_2 \xrightarrow{\text{Perovskite}} H_2O$	-286

When water was added to the simulated exhaust feed:

- Reaction pathways other than oxidation reactions were favored;
- Conversions of C_3H_6 and CO were enhanced respectively by catalytic SR and WGS on the same catalyst ;
- H_2 generated by the two reactions was readily consumed by stoichiometric amount of O_2 in the feed.

18

Catalytic oxidation activity of perovskites after accelerated aging

RTI INTERNATIONAL

Simulated wet exhaust feed: 0.056% C_3H_6, 0.702% CO, 0.648% O_2, 17.55% CO_2, 10% steam, N_2 bal.

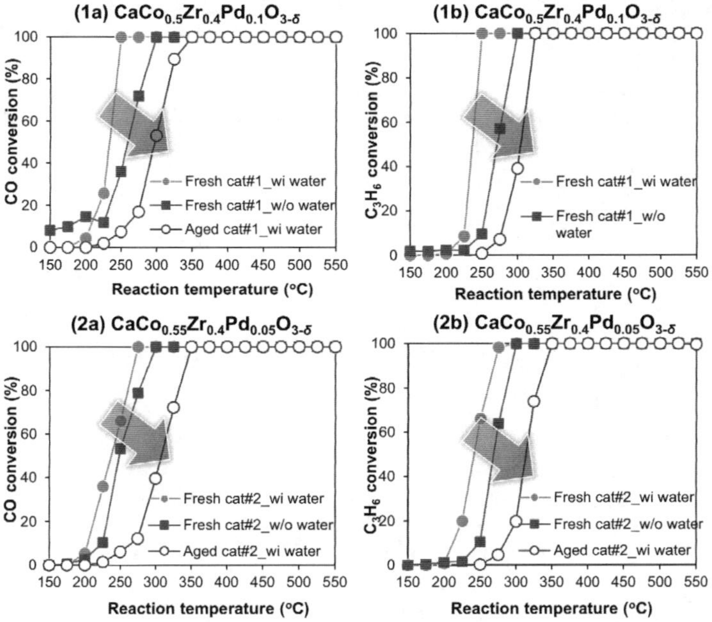

Fig. 10. (1) CO and (2) C_3H_6 conversions during catalytic oxidation tests over fresh and aged (a) $CaCo_{0.5}Zr_{0.4}Pd_{0.1}O_{3-\delta}$, (b) $CaCo_{0.55}Zr_{0.4}Pd_{0.05}O_{3-\delta}$ perovskites with at SN of 1.07, with presence of 10 vol-% H_2O in the feed. **Aging condition: 1000 °C in air for 50 hrs.**

Certain degree of catalyst deactivation observed after accelerated aging in air at 1000 °C for 50 hrs, due to extensive catalyst sintering.

19

Summary

RTI INTERNATIONAL

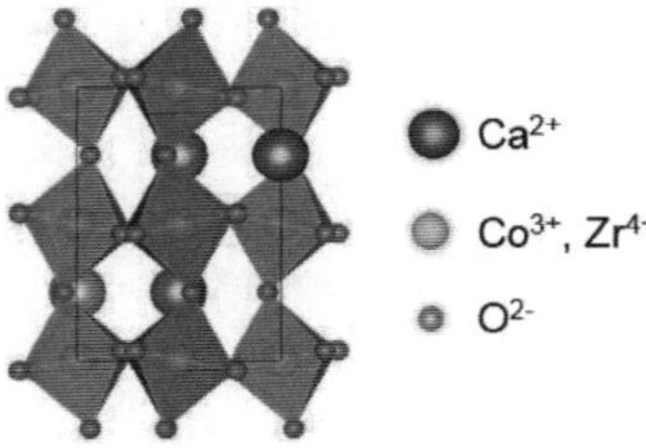

- Novel $CaCo_xZr_{1-x}O_{3-\delta}$ (x = 0, 0.3, 0.5, 0.7 and 0.9) and Pd-doped $CaCo_{0.6-y}Zr_{0.4}Pd_yO_3$ (y = 0.05, 0.1 and 0.2) perovskites were synthesized;
- The studied perovskites retain their main phase of orthorhombic Lakargiite $CaZrO_3$ structure, partial substitution of Co by Zr at B sites enhances the perovskite structural crystallinity;

20

Summary

RTI INTERNATIONAL

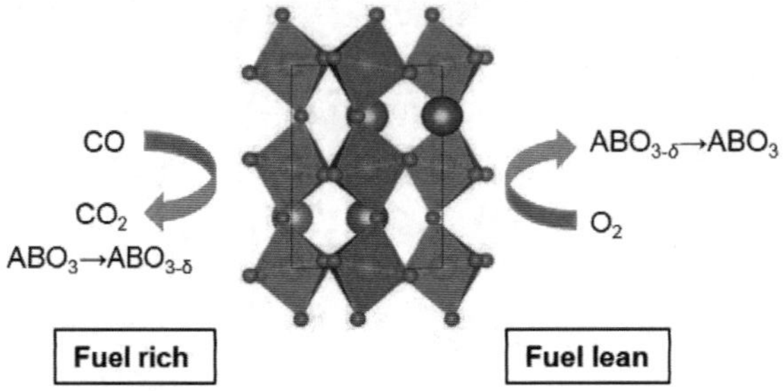

- Novel $CaCo_xZr_{1-x}O_{3-\delta}$ (x = 0, 0.3, 0.5, 0.7 and 0.9) and Pd-doped $CaCo_{0.6-y}Zr_{0.4}Pd_yO_3$ (y = 0.05, 0.1 and 0.2) perovskites were synthesized;
- The studied perovskites retain their main phase of orthorhombic Lakargiite $CaZrO_3$ structure, partial substitution of Co by Zr at B sites enhances the perovskite structural crystallinity;
- The studied perovskites showed improved redox property and OSC compared to the ceria-based oxygen storage materials (OSMs).

21

Summary

RTI INTERNATIONAL

$$ABO_{3-\delta} + \frac{1}{2}\Delta\delta\, CO \rightarrow ABO_{3-\delta-\Delta\delta} + \frac{1}{2}\Delta\delta\, CO_2$$

$$ABO_{3-\delta} + \Delta\delta\, C_xH_y \rightarrow ABO_{3-\delta-\left(2x+\frac{1}{2}y\right)\Delta\delta} + \Delta\delta\, (x\, CO_2 + \frac{1}{2} y\, H_2O)$$

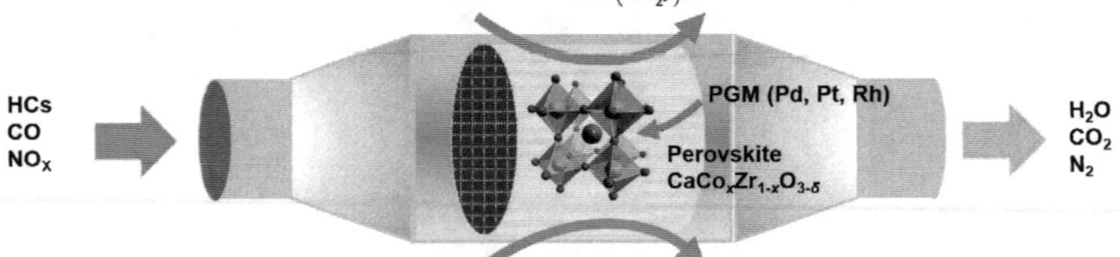

$$ABO_{3-\delta-\Delta\delta} + \frac{1}{2}\Delta\delta\, O_2 \rightarrow ABO_{3-\delta}$$

- Novel $CaCo_xZr_{1-x}O_{3-\delta}$ (x = 0, 0.3, 0.5, 0.7 and 0.9) and Pd-doped $CaCo_{0.6-y}Zr_{0.4}Pd_yO_3$ (y = 0.05, 0.1 and 0.2) perovskites were synthesized;
- The studied perovskites retain their main phase of orthorhombic Lakargiite $CaZrO_3$ structure, partial substitution of Co by Zr at B sites enhances the perovskite structural crystallinity;
- The studied perovskites showed improved redox property and OSC compared to the ceria-based oxygen storage materials (OSMs).
- Pd-doped samples showed promising catalytic activity towards C_3H_6 and CO oxidation under simulated exhaust conditions

22

Selected references

RTI INTERNATIONAL

[1] A. Simson, K. Roark, R. Farrauto, A feasibility study of niobium-containing materials for oxygen storage in three way catalytic converters, Appl. Catal. B Environ. 158–159 (2014) 106–111. doi:10.1016/j.apcatb.2014.04.006.

[2] A. Schön, C. Dujardin, J.-P. Dacquin, P. Granger, Enhancing catalytic activity of perovskite-based catalysts in three-way catalysis by surface composition optimisation, Catal. Today. 258, Part 2 (2015) 543–548. doi:10.1016/j.cattod.2014.11.002.

[3] Q. Zheng, R. Farrauto, M. Deeba, I. Valsamakis, Part I: A Comparative Thermal Aging Study on the Regenerability of Rh/Al2O3 and Rh/CexOy-ZrO2 as Model Catalysts for Automotive Three Way Catalysts, Catalysts. 5 (2015) 1770–1796. doi:10.3390/catal5041770.

[4] R. You, Y. Zhang, D. Liu, M. Meng, L. Zheng, J. Zhang, T. Hu, YCeZrO Ternary Oxide Solid Solution Supported Nonplatinic Lean-Burn NOx Trap Catalysts Using LaCoO3 Perovskite as Active Phase, J. Phys. Chem. C. 118 (2014) 25403–25420. doi:10.1021/jp505601x.

[5] Q. Zheng, R. Farrauto, M. Deeba, Part II: Oxidative Thermal Aging of Pd/Al2O3 and Pd/CexOy-ZrO2 in Automotive Three Way Catalysts: The Effects of Fuel Shutoff and Attempted Fuel Rich Regeneration, Catalysts. 5 (2015) 1797–1814. doi:10.3390/catal5041797.

[6] S. Royer, H. Alamdari, D. Duprez, S. Kaliaguine, Oxygen storage capacity of La1−xBO3 perovskites (with A′ = Sr, Ce; B = Co, Mn)—relation with catalytic activity in the CH4 oxidation reaction, Appl. Catal. B Environ. 58 (2005) 273–288. doi:10.1016/j.apcatb.2004.12.010.

[7] A. Klimkowicz, K. Świerczek, A. Takasaki, B. Dabrowski, Oxygen storage capability in Co- and Fe-containing perovskite-type oxides, Solid State Ion. 257 (2014) 23–28. doi:10.1016/j.ssi.2014.01.018.

[8] A. Mishra, N. Galinsky, F. He, E.E. Santiso, F. Li, Perovskite-structured AMnxB1−xO3 (A = Ca or Ba; B = Fe or Ni) redox catalysts for partial oxidation of methane, Catal. Sci. Technol. 6 (2016) 4535–4544. doi:10.1039/C5CY02186C.

[9] B. Levasseur, S. Kaliaguine, Effects of iron and cerium in La1−yCeyCo1−xFexO3 perovskites as catalysts for VOC oxidation, Appl. Catal. B Environ. 88 (2009) 305–314. doi:10.1016/j.apcatb.2008.11.007.

[10] G. Perin, J. Fabro, M. Guiotto, Q. Xin, M.M. Natile, P. Cool, P. Canu, A. Glisenti, Cu@LaNiO3 based nanocomposites in TWC applications, Appl. Catal. B Environ. 209 (2017) 214–227. doi:10.1016/j.apcatb.2017.02.064.

[11] S. Sartipi, A.A. Khodadadi, Y. Mortazavi, Pd-doped LaCoO3 regenerative catalyst for automotive emissions control, Appl. Catal. B Environ. 83 (2008) 214–220. doi:10.1016/j.apcatb.2008.02.014.

[12] Z. Say, M. Dogac, E.I. Vovk, Y.E. Kalay, C.H. Kim, W. Li, E. Ozensoy, Palladium doped perovskite-based NO oxidation catalysts: The role of Pd and B-sites for NOx adsorption behavior via in-situ spectroscopy, Appl. Catal. B Environ. 154–155 (2014) 51–61. doi:10.1016/j.apcatb.2014.01.038.

23

Acknowledgement

RTI INTERNATIONAL

Material Chemistry Manager: Dr. Marty Lail

Material Group: Dr. Mustapha Soukri
Dr. Jianping Shen
Dr. Ignacio Luz
Dr. John Thompson
Kelly Amato
Timothy Bellamy
Devaughn Body

U.S. DEPARTMENT OF ENERGY

Thank you and please feel free to ask any question!

One patent application (RTI 16P027-PROV) on perovskite material development and applications.

One manuscript just accepted by *Catalysis Today*.

25

500801 Probing the Oxidation and Reduction Halves of the Cu Redox Cycle during Standard NH_3-Selective Catalytic Reduction of NO_x on Cu-SSZ-13

Abstract:

The selective catalytic reduction of NO_x (x=1,2) using NH_3 as a reductant (NH_3-SCR) is a leading strategy for NO_x abatement in oxygen-rich automotive exhaust to meet increasingly stringent emissions constraints. It was shown that low temperature (473 K) NOx SCR on Cu-SSZ-13 occurs via a Cu(II)/Cu(I) redox cycle [1] on two types of isolated Cu sites, both divalent Cu^{2+} cations compensating paired framework Al atoms and monovalent $[CuOH]^+$ complexes at isolated framework Al atoms, which are solvated by NH_3 during reaction [2]. Two different kinetic regimes that are characterized by different apparent orders, activation energies, and steady-state Cu(I)/Cu(II) distributions in the Cu(I)/Cu(II) redox cycle have been identified under standard SCR conditions, in which O_2 is used as an oxidant. The rate-limiting steps during standard SCR depends on the Cu density (determined by the Si/Al and Cu/Al) of the Cu-SSZ-13 catalysts. One regime corresponds to SCR cycles limited by Cu(I)->Cu(II) oxidation with O_2 and the other regime corresponds to cycles limited by Cu(II)->Cu(I) reduction, which we probe in *operando* (varying Cu density, gas pressures and temperature) and in transient XAS experiments. O_2-oxidation during low temperature SCR requires two Cu(I) sites to be oxidized, in contrast to reduction occurring at single, isolated Cu(II) sites. Therefore, standard SCR rates depend on Cu cation density and distribution in Cu-SSZ-13 zeolites when Cu(I) oxidation steps are kinetically-relevant, but are independent of Cu density otherwise. We probe the nature of oxidation and reduction half-cycles independently by measuring their intrinsic kinetics. In-situ monitoring of Cu(I)/Cu(II) fractions under oxidation and reduction-limited conditions showed the existence of two different pools of Cu sites, those that are within diffusion distances of another Cu cation and can be oxidized with O_2 in standard SCR cycles, and others that are stranded and isolated from other Cu.

[1] C. Paolucci, A.A. Verma, S.A. Bates, V.F. Kispersky, J.T. Miller, R. Gounder, et al., Isolation of the Copper Redox Steps in the Standard Selective Catalytic Reduction on Cu-SSZ-13, (2014) 12022â€“12027. doi:10.1002/ange.201407030.

[2] C. Paolucci, A.A. Parekh, I. Khurana, J.R. Di Iorio, J. Albarracin, A. Shih, et al., Catalysis in a Cage: Condition-Dependent Speciation and Dynamics of Exchanged Cu Cations In SSZ-13 Zeolites, J. Am. Chem. Soc. (2016) 1â€“63. doi:10.1021/jacs.6b02651.

493807 Sulfur Oxidation Studies with BimetallicÂ Pd/Pt Catalysts

Abstract:

Bimetallic Pd/Pt catalysts are good automotive-engine oxidation aftertreatment catalyst candidates due to their improved activity and potential sintering resistance relative to the monometallic formulations. Any of these catalyst systems have to endure extended times on stream and tolerate exposure to high temperatures, water, and trace sulfur. Of note, the literature states that the Pd/Pt bimetallic benefits only exist in the absence of sulfur. Previous studies with bimetallic Pd/Pt catalysts show that sulfur sorption and oxidation characteristics are influenced by both Pd:Pt mole ratio and precious metal particle size. Here, studies were conducted with mono- and bimetallic $Pd/Pt/Al_2O_3$ catalysts to investigate more specifically how SO_2 oxidation activity is influenced by the Pd:Pt mole ratio. Diffuse reflectance infrared Fourier transform spectroscopy (DRIFTS), temperature-programmed oxidation, and steady-state oxidation studies were conducted on catalysts with different precious metal composition but similar precious metal crystallite particle sizes. During surface studies, it was found that catalysts with a higher Pd content tended to have greater activity for oxidizing sulfur species at low temperatures and as a result formed more SO_3 and SO_4 species. However, reactor studies showed that catalysts with a lower Pd content appeared to have greater activity for oxidizing sulfur species at a low temperature because more conversion of SO_2 to SO_3 species was detected at the reactor outlet. Due to this obvious disconnect, studies are now focused on investigating how the oxidation products formed on the catalyst surface influence the catalystâ€™s apparent activity for the SO_2 oxidation reaction. This work provides evidence that the SO_2 oxidation characteristics, as well as the potential for sulfur decay, for mono- and bimetallic Pd/Pt catalysts are heavily influenced by the Pd:Pt mole ratio.

495480 Nature of SO_2 Poisoned Cu-SSZ-13 Catalysts Under Ammonia Selective Catalytic Reduction (NH_3-SCR) Conditions

Abstract:

Nature of SO_2 poisoned Cu-SSZ-13 Catalysts under Ammonia Selective Catalytic Reduction (NH_3-SCR) Conditions

Arthur J. Shih[1], Hui Li[2], Ashok Kumar[3], Juan M. Gonzalez[1,4], Ishant Khurana[1], Christopher Paolucci[2], Jonatan D. Albarracin Caballero[1], Atish A. Parekh[1], Aída Luz Villa[4], W. Nicholas Delgass[1], Rajamani Gounder[1], Aleksey Yezerets[3], William F. Schneider[2], Jeffrey T. Miller[1], and Fabio H. Ribeiro[1]

[1]Davidson School of Chemical Engineering, Purdue University, West Lafayette, IN

[2]Department of Chemical and Biomolecular Engineering, University of Notre Dame, IN

[3]Cummins Incorporated, Columbus, IN

[4]Chemical Engineering Department, Universidad de Antioquia, Medellín, Colombia

Reduction of NO_x emissions from diesel engine exhaust is an environmental issue driven by increasingly stringent emission regulations. Cu-SSZ-13 catalysts have been shown to be promising for this application due to their hydrothermal stability and selectivity in reducing NO_x to N_2 using NH_3 compared to other zeolite catalysts. SO_x formed from the combustion of ppm levels of sulfur in diesel fuel deactivates the Cu-SSZ-13 catalyst. However, its effect on Cu^{2+} and CuOH active sites and the SCR mechanism is unclear and under debate.[1-4]

Two Cu-SSZ-13 catalysts, one with only Cu^{2+} (3.8 Cu wt%, 100% Cu^{2+}) and another with primarily $[CuOH]^{1+}$ (1.5 Cu wt%, 80% CuOH) active sites were synthesized, hereby denoted as Z_2Cu and ZCuOH, respectively.[5] Each catalyst was sulfated with dry SO_2 at 200°C and 400°C, resulting in sulfur contents of 0.7 wt% and 1.0 wt%, respectively. Sulfation at 400°C decreased the standard SCR rate by 90% for the ZCuOH catalyst but by only 60% for the Z_2Cu catalyst. In addition, the apparent activation energy for ZCuOH sites decreased from 70 kJ mol^{-1} to 11 kJ mol^{-1} with increasing extents of sulfur poisoning but remained constant at 70 kJ mol^{-1} for the Z_2Cu catalyst. These kinetic results suggest two different deactivation mechanisms for the Z_2Cu and ZCuOH sites. The decrease in apparent activation energy for the ZCuOH catalyst is attributed to changes in the binding energy of NH_3 to the Cu active sites, corroborated by evidence from UV-Visible where sulfur-Cu ligands were observed on ZCuOH but not on Z_2Cu. The unchanged apparent activation energy and reactant orders for the Z Cu catalyst after sulfation are attributed to pore-

2

filling by ammonium sulfate salts; this reduces the total number of accessible Cu, as supported by argon micropore measurements.

Z_2Cu sites are preferred over ZCuOH sites due to the lower impact of sulfur poisoning on the rate and apparent activation energy. These results predict that synthesizing catalysts with higher fractions of Z_2Cu sites will lead to improved emission control catalysts for commercial application.

References:

1. Jangjou, Y.; Wang D.;Kumar, A.; Li J.; Epling, W.S. *ACS Catal.* **2016**, *6*, 10.
2. Shen, M.; Wen, H.; Hao, T.; Yu, T.; Fan, D.; Wang, J.; Li, W.; Wang, J. *Catal. Sci. Technol.* **2015**, *5*, 1741.
3. Zhang, L.; Wang, D.; Liu, Y.; Kamasamudram, K.; Li, J.; Epling, W. *Appl. Catal. B: Environ.* **2014**, *156-157*, 371.
4. Luo, J.; Wang, D.; Kumar, A.; Li, J.; Kamasamudram, K.; Currier, N.; Yezerets, A. *Catal. Today* **2016**, *267*, 3.
5. Paolucci, C.; Parekh, A.A.; Khurana, I.; Di Iorio, J.R.; Li, H.; Albarracin Caballero, J.D.; Shih, A.J.; Anggara, T.; Delgass, W.N.; Miller, J.T.; Ribeiro, F.H.; Gounder, R.; Schneider, W.F. *J. Am. Chem. Soc.* **2016**, *138*, 18.

496535 Measurement of the Oxygen Storage Capacity and Its Role in the Catalytic Oxidation

Abstract:

CeO_2, thanks to its unique redox property of oxygen storage and release under lean and rich conditions, respectively, has been widely used as the catalyst support component. This property of oxygen storage and release can usually be quantitatively measured, called the oxygen storage capacity (OSC), via a reduction method using different techniques. The amount of OSC measured depends on the type of reducing agent and the technique used. Furthermore, the OSC is also influenced by the synthesis method of CeO_2, the modification by other elements like ZrO_2, the physical and geometric properties, and the metal supported on CeO_2-based materials. On the other hand, it is well accepted that the catalytic activity benefits from the unique characteristic of oxygen storage and release for reactions carried out over the CeO_2-based catalysts. Unfortunately, the significant enhancement of catalytic activity is still poorly correlated by the OSC, or the role of OSC in favoring the catalytic reaction is still poorly understood. We establish the measurement of OSC by comparing the use of different reducing agents (CO and H_2) and the use of different techniques (TCD and TGA). The effects of synthesis method, high temperature aging, precious metal loading, and the structure modification by ZrO_2 (i.e. CZO) and other stabilizer elements on CeO_2 have been also investigated. The OSC measured by different methods and the influencing factors are analyzed with emphasis on to the catalytic activity like CO oxidation for the objectives to establish a correlation between OSC and catalytic activity.

495203 Activation of Pd/SSZ-13 in Low Temperature NO Adsorption for Cold Start Application

Abstract:

As the regulation on NOx emission in exhaust gas has been globally issued and enforced, the purification of NOx from diesel engine is regarded as a great challenge. Although selective catalytic reduction (SCR) catalyst effectively reduce NOx in exhaust gas, this system cannot operate properly below 200 Â°C (i.e. during cold start period, initial ca. 100 â €" 200 sec of engine operation). Therefore, during the warm-up of NOx reduction catalyst, the considerable amount of NOx is emitted into atmosphere without any treatment [1]. Recently, cold start NOx trap catalysts attract significant attention with a concept of storing NOx at low temperature (i.e. during cold start) and releasing them at elevated temperature, where SCR operates. In this work, we tried to apply various Pd/SSZ-13 catalysts with different Pd loading method and investigate the effects of hydrothermal aging (HTA) treatment on the NOx adsorption ability and the physicochemical change of Pd/SSZ-13.

501343 Colloidal Pd Nanoparticle Synthesis: The Effect of Ligand-Metal-Solvent Thermodynamics on Kinetics and Final Size

Abstract:

Metal nanoparticles are of great scientific interest due to their unique electronic, catalytic and optical properties. Colloidal metal nanoparticles of different sizes have been successfully synthesized by careful choice of precursors, solvents, and ligands. The roles of ligands and solvents in controlling the size are still not well understood. We have recently reported that surface capping by ligands as well as ligand-precursor coordination can affect the final size of Pd nanoparticles by modulating the surface growth and nucleation rates, respectively. This critical ligand role provides an opportunity for controlling the particle size by tuning the ligand interactions with both the metal precursor and nanoparticle surface. Herein, we report a study on how the thermodynamics of Pd coordination with phosphine and amine ligands in different solvents affect the kinetics of nanoparticles nucleation and growth, and consequently the final particle size. The kinetics were measured by in-situ small-angle X-ray scattering (SAXS) and nuclear magnetic resonance (NMR) was used to characterize the solvent-ligand coordination with the Pd precursors. The possible structures and free energies of ligand-solvent exchange reactions in solution were calculated using Density function theory (DFT). Additionally, we used isothermal titration calorimetry (ITC) to quantitatively compare the thermodynamic interactions between different ligands and palladium precursor and nanoparticle surface atoms by measuring the binding equilibrium constant, heat of binding and stoichiometry of binding. We show how the nucleation and growth rates, and final size are correlated with ligand-Pd(0) and ligand-Pd(II) binding equilibrium. The results give a better understanding of the ligand effect on synthesis kinetics and show the potential for tailoring the solvent-ligand-metal interactions to move closer towards predictive synthesis of metal colloidal nanoparticles.

489858 Switchable Surfactants for the Preparation of Monodisperse, Supported Nanoparticle Catalysts

Abstract:

Synthetic methods of preparing monodisperse, supported nanoparticles remain problematic. Organic surfactants are typically required to passivate the surface of nanoparticles to limit growth and control the size of the nanoparticles. Calcination is typically required following the deposition of the nanoparticles onto a catalyst support to remove the surfactant and activate the catalyst. In general, this calcination causes significant and unpredictable growth of the nanoparticle, undoing the monodispersity maintained during the synthesis. This undesired growth additionally leads to a decline in the catalytic activity of the nanoparticles due to a loss in surface area. We present the use of switchable surfactants (SwiS) for the preparation of monodisperse, supported nanoparticles. SwiS are molecules that undergo reversible, physical changes in response to an external stimulus. The SwiS can be "turned-on" to template nanoparticles during synthesis and then "turned-off" to release bare nanoparticles for deposition onto a catalyst support. The templating synthesis via a hypothesized reverse-micelle process allows for the production of monodisperse nanoparticles controllable by varying the size of the SwiS or the continuous phase. Nanoparticle size has shown to increase as the alkyl chain length of the SwiS is increased. The template is disassembled by reversal of the SwiS by simply heating the solution to drive off the CO_2 in the presence of a catalyst support. The deposition of bare nanoparticles allows for the formation of highly active supported nanoparticles that do not require calcination thus maintaining monodispersity as shown by analysis with transmission electron microscopy (TEM). Supported gold nanoparticles prepared using SwiS are more active in the hydrogenation of 4-nitrophenol than their traditionally prepared counter-parts while requiring less processing thus maintaining the size of the nanoparticle catalysts throughout preparation. Additionally, supported nanoparticles prepared with SwiS are significantly more resistant to sintering than their traditionally prepared analogs.

497875 Colloidal Iron Nanoparticles Provide Size-Control in the Catalytic CO Hydrogenation Reaction

Abstract:

Recent advances in synthetic nanotechnology to produce uniform metal nanoparticle catalysts and to use such "monodisperse" particles in catalytic reactions clearly allowed establishing correlations between the catalyst activity/selectivity patterns and the size and shape of metal catalyst particles. The rationale behind these efforts is the quest for high activity and selectivity in multi-path catalytic reactions, with little to no waste or polluting by-products. Among the various methods employed to prepare well-defined metal nanoparticles, those based on colloidal recipes have gained major importance. While the use of stabilizers ("capping agents", sometimes simply "ligands") to obtain stable colloidal dispersions with narrow particle size distribution is considered mandatory, there is ample evidence that these stabilizers cause alterations in the catalytic properties of the metal nanoparticles. On the other hand, an intriguing reaction for various green feedstock production on an industrial scale is the hydrogenation of carbon monoxide according to Fischer-Tropsch. Fe-, Co- and Ru-based catalysts have been most commonly investigated in the past for their capability to produce paraffins, olefins, and oxygenates with varying hydrocarbon chain length. Tuning these catalysts for high activity and selectivity to specific product classes is therefore of utmost interest. In this research we demonstrated that monodisperse Fe(0) particles in silica could be successfully prepared and tested in the catalytic CO hydrogenation (so-called Fischer-Tropsch process) without ligand effects to play any visible influence.[1] Different from all of the previous studies, which use iron oxide as precursor to prepare catalysts, we applied metallic (0) iron particles. This Fe(0) metal can be easily converted into iron carbide during Fischer-Tropsch synthesis since it is much more active than iron oxide. Palmitic acid and hexadecylamine used as capping agents in suitable ratios, were quantitatively eliminated through the reaction with hydrogen. We further reveal the occurrence of pronounced particle size effects with respect to catalytic activity and selectivity. In particular, an increase in the particle size causes increasing TOF (activity). TOF values of the same magnitude and particle size dependence have been found when relating the geometry-derived TOF to the active Fe(0) metal surface as determined through hydrogen-deuterium exchange. As to the selectivity signatures of Fe/MCF-17, methane and short-chain (terminal) olefin formation significantly increase with decreasing Fe particle size. Quite differently, both the long-chain C_{5+} olefin and (total) oxygenate slates show the opposite behavior. As to the C_{2+} paraffins, it seems that small particles are less efficient in chain lengthening than larger ones. According to XRD results, bulk Fe(0) of the as-prepared catalyst is chemically reconstructed into iron-carbides, e- and χ-Fe_5C_2, after reaction. The occurrence of iron-oxide phases, though not clearly detectable here, cannot be ruled out either. Reaction-induced nanoparticle sintering was limited according to TEM and microtome studies.

Summarizing, the methodological approach to study the catalytic CO hydrogenation over monodisperse Fe nanoparticles intercalated in the mesopores of silica provided new insights into the structure-sensitivity of the reaction and is anticipated to help design catalysts with specific selectivity signatures at high reaction rates.

1. V. Iablokov, Y. Xiang, A. Meffre, P.F. Fazzini, B. Chaudret, N. Kruse *ACS Catalysis* **2016**, *6*, 2496-2500

502479 Selective Dehydrogenation and Oxidation Reactions on Niau Alloys at the Single Atom Limit

Abstract:

Gold, while inert as a bulk metal, shows significant activity and selectivity when in nanoparticle (< 5 nm dia.) form[1]. Despite its high selectivity, monometallic Au is less active than commercial catalysts in reactions of industrial importance, such as the dehydrogenation of ethanol to acetaldehyde and the oxidative methanol-methacrolein esterification to methyl methacrylate. Therefore, the addition of a metal able to coordinate and activate unsaturated organic compounds (e.g. alcohols), such as Ni, could lead to improved catalyst reactivity. Atomically embedded Ni in Au surfaces would form an alloy at the atomic level (Single Atom Alloy), which allows us to reap the benefits of the high selectivity of Au coupled with the remarkable activity of Ni, thus obtaining better catalysts than their monometallic counterparts.

In the present work, we compare various NiAu structures, prepared as Ni-Au nanoparticles (NPs) supported on SiO_2, unsupported Ni-doped nanoporous Au (np-NiAu) and the industrially used Au@Ni core-shell catalyst[2]. The formation of highly dilute Ni-Au alloys is followed by CO-DRIFTS. Temperature Programmed Surface Reaction studies on ethanol dehydrogenation show 100% selectivity and noticeable stability for the NPs and np-NiAu materials. Kinetic measurements find the SAA NiAu to be more active than gold, and further show that the increased activity is induced by Ni, as the apparent activation energy values are lower for the SAA than for the monometallic Au samples. Comparison between Ni-Au NPs and the commercial catalyst was made via *in situ* ATR-IR spectroscopy combined with HPLC for the oxidative methacrolein-methanol coupling reaction in the liquid phase. The catalyst structure and composition was also followed by ICP, STEM/EDS and XAS.

Acknowledgements: This work was supported as part of the Integrated Mesoscale Architectures for Sustainable Catalysis, an Energy Frontier Research Center funded by the U.S. Department of Energy, Office of Science, Basic Energy Sciences, under award #DESC0012573.

References

1. M. Haruta, N. Yamada, T. Kobayashi and S. Iijima, *J. Catal.*, 1989, **115**, 301-309.
2. K. Suzuki, T. Yamaguchi, K. Matsushita, C. Iitsuka, J. Miura, T. Akaogi and H. Ishida, *ACS Catal.*, 2013, **3**, 1845-1849

488371 The Effect Pd Size on TiO_2 for Photocatalytic NO_x Removal

Abstract:

The effect Pd size on TiO_2 for Photocatalytic NOx Removal

Sotiris E. Pratsinis

Particle Technology Laboratory, Institute of Process Engineering,

Department of Mechanical & Process Engineering, ETH Zurich, Switzerland

Reducing the particle size of noble metals on ceramic supports can maximize noble metal performance and minimize its use (Flytzani-Stephanopoulos & Gates, *Annu. Rev. Chem. Biomol. Eng.* 2012, 3:545). Here Pd clusters onto nanostructured TiO_2 particles are prepared in one step by scalable flame aerosol technology while controlling the Pd cluster size from a few nanometers to that of single atoms (Fujiwara et al., *ACS Catalysis*, 2017, 6:1887). Annealing such materials at appropriate temperatures leads to photocatalytic NO_x removal in a standard ISO reactor up to10 times faster than that of commercial TiO_2 (P25, Evonik). Such superior performance can be attained by only 0.1 wt% Pd loading on TiO_2 (Fujiwara & Pratsinis, *AIChE J.*2017, 63:139-146).

Annealing these flame-made powders in air up to 600 °C decreases also the amorphous TiO_2 fraction and increases its crystal and particle sizes as observed by X-ray diffraction (XRD) and N_2 adsorption. The growth of single Pd atoms to Pd clusters on TiO_2 prepared at different Pd loading and annealing conditions was investigated by scanning transmission electron microscopy (STEM) and XRD. The single Pd atoms and clusters on TiO_2 are stable up to, at least, 600 °C for 2 hours in air but at 800 °C they grow into PdO nanoparticles whose fraction is comparable with the nominal Pd loading. Hence, most of Pd atoms are on the TiO_2 surface where at 800 °C they diffuse and coalesce. Diffuse reflectance infrared Fourier transform spectroscopy (DRIFTS) reveals NO adsorption on single, double, 3- and 4-fold coordinated Pd atoms depending on their synthesis and annealing conditions. The peak intensity of NO adsorption sites involving multiple Pd atoms is substantially lower in TiO_2 containing 0.1 wt% than 1 wt% Pd but that intensity from single Pd atoms is comparable. This indicates the dominance of isolated Pd atoms compared to clusters in Pd/TiO_2 containing 0.1 wt% Pd that match or exceed the photocatalytic NO_x removal at much higher Pd contents. Here, the number of isolated Pd sites in different Pd/TiO_2 catalysts is quantitatively compared.

492477 Exploring Fundamental Properties of Fe- and Mn-Promoted Rh Catalysts for Syngas Conversion to Higher Alcohols

Abstract:

As of today, large parts of the chemical industry rely on crude oil as a feedstock. However, in recent years dwindling supplies of this carbon source have led to the exploration of alternative feedstocks. A particularly promising candidate is syngas, a mixture of CO and H_2, which can either be produced from biomass or natural gas. Most commonly, syngas is either converted to methanol or to longer alkanes using the Fischer-Tropsch process, but the direct conversion of syngas to C_{2+} oxygenates would be highly desirable, due to their wide range of applications in chemical, fuel and polymer industries. A catalyst with particular promise in this regard is Rh. While monometallic systems show selectivity towards acetaldehyde, promoters, such as Fe and Mn, significantly influence selectivity. However, the atomic scale nature of the active site and its impact on the reaction network and selectivity has not been explored in detail.

In this contribution, we present our efforts in carefully controlling the synthesis and preparation of Fe- and Mn-promoted Rh catalysts supported on SiO_2. Specific amounts of either one or both promoters are deposited using controlled surface reactions and the impact of systematically varying the amounts of deposited Fe, Mn and Fe/Mn on the product selectivity is explored. This procedure reveals that both, Fe and Mn, increase the selectivity towards ethanol, an effect that is additive and leads to highest selectivity towards ethanol for a ternary system with molar ratios of 1:0.15:0.10 (Rh:Fe:Mn). In a subsequent step, the nature of the active sites is interrogated using a combination of XPS, CO-FTIR spectroscopy and density functional theory calculations. Under reaction conditions, Fe is found to be in a metallic state in a subsurface site. Mn on the other hand is present as an oxide on the 211 step edge. Finally, the implications of the structural changes due to the presence of the promoters on the reaction network and selectivities will be discussed.

The systematic study presented here is a step towards elucidating the effects and interactions of promoters on the selectivity of the studied catalysts. Especially the insights into the nature of the active site are a stepping-stone towards the computational exploration of the impact of various promoters on the activity, which is prerequisite for rational, modeling-based catalyst design.

500296 Controlling Energy Flow in Plasmonic Photocatalysis through the Design of Hybrid Plasmonic Nanostructures

Abstract:

It has been shown that photo-excitation of plasmonic metal nanoparticles (Au, Ag, and Cu) can induce direct photo-chemical reactions on the nanoparticles.[1,2] However, the widespread application of this technology in catalysis has been limited by the relatively poor chemical reactivity of noble metal surfaces. In this presentation, we present our recent work in which we conceptualized and designed hybrid nanostructures in which a plasmonic metal harvests the energy of visible light photons and selectively channels that energy into catalytically active centers on the nanostructure. To accomplish this, we developed a synthetic protocol to deposit a few monolayers of Pt onto plasmonic Ag nanocubes. This model system allows us to conclusively separate the optical and catalytic functions of the hybrid nanomaterial and analyze the interaction between the two functions. Through experimental and theoretical studies of the optical properties of these nanostructures, we show that the flow of energy is strongly biased towards absorption (i.e. excitation of energetic charge carriers) in the thin Pt shell. We demonstrate the utility of these nanostructures for photo-catalytic chemical reactions in the preferential oxidation of CO in excess H_2. The reactor studies conclusively show that photo-excitation of these nanostructures results in photo-chemical reactions on the Pt surface. We describe the fundamental physical reasons for the observed directed flow of energy and discuss how these discoveries impact the field of plasmonic catalysis.

1. Linic, S., Aslam, U., Boerigter, C. & Morabito, M. Photochemical transformations on plasmonic metal nanoparticles. *Nat. Mater.* **14,**567â€“576 (2015).

2. Christopher, P., Xin, H. & Linic, S. Visible-light-enhanced catalytic oxidation reactions on plasmonic silver nanostructures. *Nat. Chem.* **3,** 467â€“472 (2011).

501832 Microwave-Responsive CNT/Polyaniline Core/Shell Structured Catalysts for High Energy Efficiency in Saccharide-HMF Conversion

Abstract:

Microwave absorption and catalytic activity have been synergistically integrated into a core/shell structured CNT/polyaniline(PANI) hybrid catalyst that enables localized heating on the catalyst surface and uplifts energy efficiency in the dehydration reaction of saccharides to 5-hydroxymethylfurfural (HMF). Results reveal that the integration of CNT and polyaniline significantly improves the overall saccharide conversion and HMF yield. The PANI shell thickness and morphology are controlled by both reactant feeding ratio and polymerization reaction rate. It is found that a thinner and rougher PANI shell is favourable to enhance the overall microwave absorption, improve heat transfer efficiency to catalyst surface and thus reaction efficiency. Moreover, the packing of polyaniline shell on CNT core facilitates its separation from reaction mixture and effectively prevents catalyst loss. These catalysts show excellent recyclability that the catalyst activity can be completely recovered by a simple re-doping process in 1.0 M sulfuric acid. This work opens up a new window in microwave responsive catalyst design that enhances energy efficiency in saccharide-HMF conversion reactions and promotes the economic feasibility of such reactions in practice.

493480 Base-Free Aerobic Oxidation of 5-Hydroxymethylfurfural to 2, 5-Furandicarboxylic Acid over Nanoscale Pt Catalysts Prepared By Atomic Layer Deposition

Abstract:

Pt/ZrO_2 catalysts synthesized by atomic layer deposition (ALD), incipient wetness impregnation (IW) and co-precipitation (CP) were employed for aerobic oxidation of 5-hydroxymethylfurfural (5-HMF) to produce 2, 5-furandicarboxylicacid (2, 5-FDCA) in aqueous phase without using alkali or acid. Under a mild reaction condition (T = 100 °C, P_{O2} = 0.4 MPa, time = 12 h), the Pt/ZrO_2 synthesized by ALD of 30 Pt deposition cycles gave a complete 5-HMF conversion and 2, 5-FDCA yield of 97% compared with the Pt/ZrO_2 synthesized by IW method (a 2, 5-FDCA yield of 9.4 %) and CP method (a 2, 5-FDCA yield of 3.9 %) with the same Pt loading of 3.23 wt%. The three catalysts were characterized by ICP, XRD, TEM and CO-TPD, and the results indicated that high Pt dispersion and uniform Pt particle size distribution on the ZrO_2 surface were obtained by ALD and the performance of C=O adsorption was strong. While the growth of Pt particles synthesized by incipient-wetness impregnation was not controllable resulting in the metal particles non-uniformed and the performance of C=O adsorption was weak. The Pt particles synthesized by co-precipitated were wrapped in the ZrO_2 and the catalytic activity of the Pt are completely eliminated due to the difficulty of accessing the Pt active surface for reactant and intermediate products. The ALD produce uniform Pt particles to ensure the unitive active sites for the aerobic oxidation of HMF to FDCA. As the deposition cycles of Pt for Pt/ZrO_2 (ALD) increased, the average size of the Pt nanoparticle increased but the dispersion did not change, and the Pt/ZrO_2 with ALD 30 deposition cycles exhibited the highest TOF. This work has high commercial value for developing efficient heterogeneous catalysts using the cheap and available metals oxide that with low surface area but with high catalytic active which have potential applications in many heterogeneous reactions and maximize the atom efficiency of noble metals.

Mo@Pt Overlayer as Effective Catalysts for Hydrodeoxygenation of Guaiacol and Anisole

Joseph H. Holles
Qinghua Lai

Department of Chemical Engineering
University of Wyoming
Laramie, WY 82071

UNIVERSITY OF WYOMING

1

Hydrodeoxygenation (HDO) of Bio-oils

Biomass can be converted into crude bio-oil via thermochemical treatment such as pyrolysis or liquefaction.

Drawbacks caused by high oxygen content of crude bio-oils:
low heating value, poor stability, poor volatility, high viscosity and corrosiveness.

Upgrading of crude bio-oils is necessary

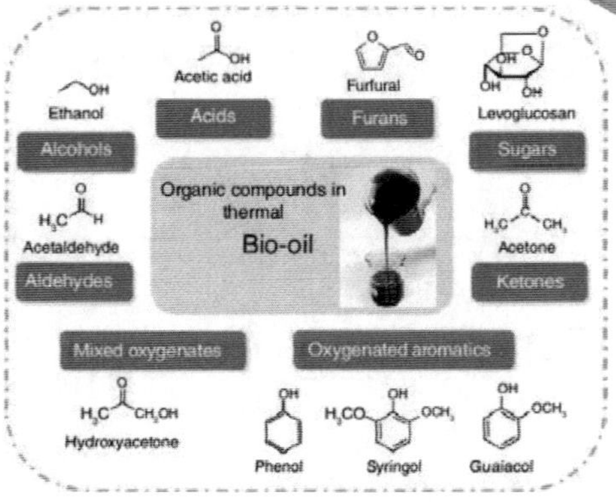

Fermoso J.; Wiley Interdisciplinary Reviews: *Energy Environ.* **2017,** 6:e245

Model compounds for lignin-derived bio-oils:

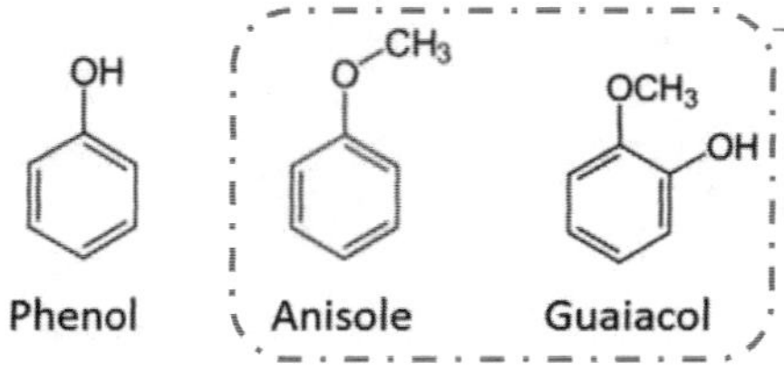

One or two types of C–O bond

Prototype compound to represent lignin-derived bio-oils

Strong surface-reactant binding has been shown to inhibit HDO reaction

Resasco D.; *Energy Fuels* **2011,** 25, 4155-4162

2

Bimetallic Overlayer Catalysts

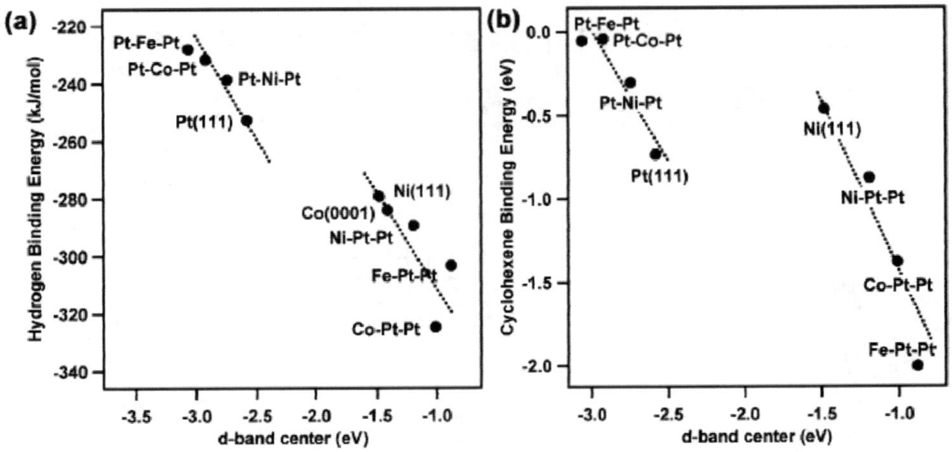

DFT calculations of binding energies of hydrogen and cyclohexene on Pt−3d−Pt(111) and 3d−Pt−Pt(111) surfaces.

- Overlayer structure could change the d-band center energy;
- Changing d-band center energy would alter bonding characteristic to adsorb stronger/weaker as desired;

Yu et al., *Chem. Rev.* **2012**, *112*, 5780−5817

3

Mo@Pt Overlayer Core@Shell Catalyst

- Mo@Pt/SiO_2-Al_2O_3 for Hydrodeoxygenation:
  - Decreased adsorbate binding strength
  - Mo favors the C-O bond cleavage compared to Pt only catalyst
  - Pt facilitates the dissociation of H_2 in comparison to Mo catalyst

- Objective:

  Study how synthesis of Mo@Pt overlayer affects surface properties and activities in HDO of guaiacol and anisole

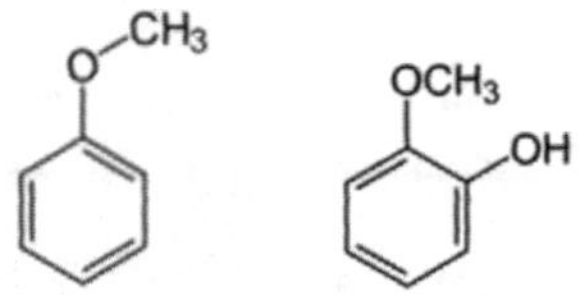

Anisole Guaiacol

4

Catalyst Synthesis

- Silica-Alumina (Alfa Aesar, grade 135) used as support for all synthesized catalysts.
- Mo/SiO_2-Al_2O_3, Mo-Pt/SiO_2-Al_2O_3, and Pt/SiO_2-Al_2O_3 prepared by using incipient wetness impregnation :
  - Pt(acac)$_2$ /toluene slurry.
  - Incipient wetness impregnation with nickel (II) nitrate hexahydrate precursors.
- The Mo@Pt overlayer catalysts used were prepared by using the directed deposition technique.

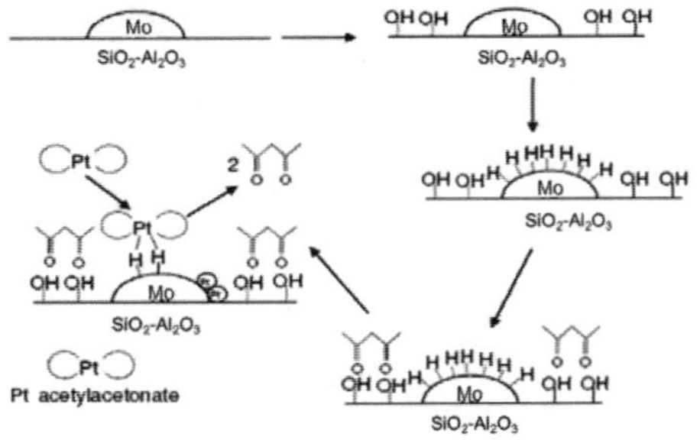

5

Physicochemical Properties

Sample	Metal loading [wt%]		Dispersion [%]	Overlayer coverage [%]	S_{BET} [m^2g^{-1}]	Pore volume [cm^3g^{-1}]
Mo/SiO_2-Al_2O_3	4.98	Mo	5.78		339	0.58
Mo@Pt SD	0.171	Pt	5.45	29	335	0.55
Mo@Pt DD	0.303	Pt	5.75	52	319	0.53
Mo-Pt Bimetallic	4.97	Mo	6.10		304	0.56
	0.285	Pt				
Pt	4.96	Pt	12.9		409	0.64

Typically, additional depositions lead to an increase in Pt loading

With similar S_{BET} and pore volume

6

XRD Characterization

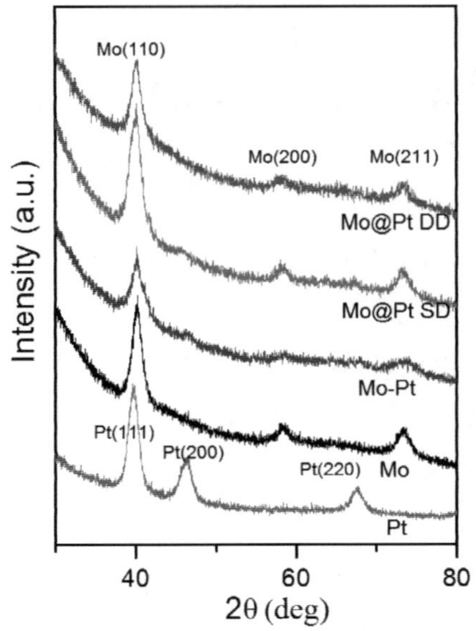

Metallic Pt peak was not observed in overlayer catalysts.

7

H_2 Temperature Programmed Reduction

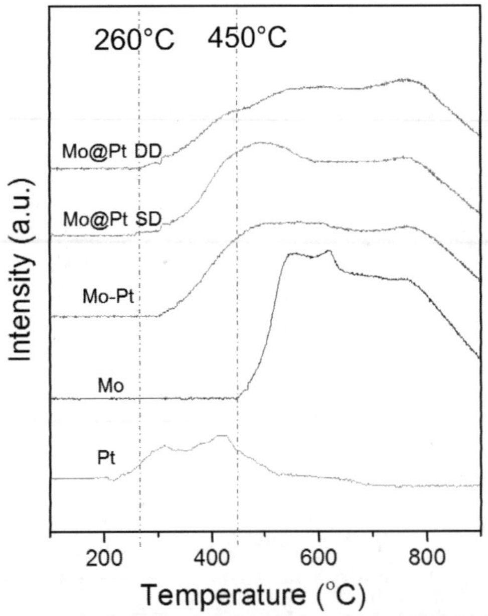

Deposition of Pt on Mo enhances the reducibility of molybdenum oxides

Indicates a close interaction between Pt and Mo species

8

H_2 and CO Heat of Adsorption

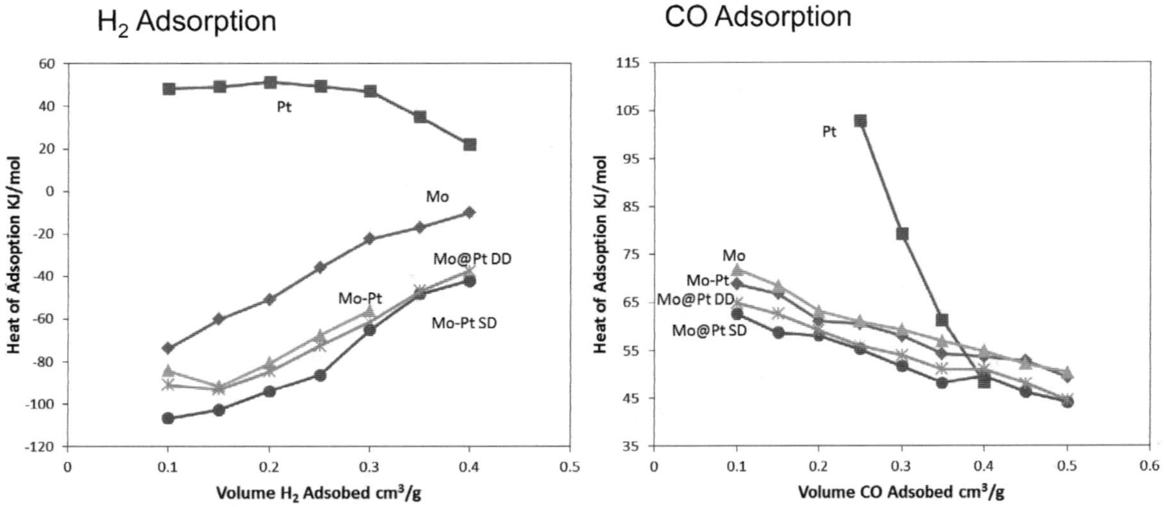

- Overlayer synthesis results in lower heat of adsorption.
  - Lower than Mo and Pt.
  - Similar to computational predictions.

9

Ethylene Hydrogenation Experiments

$$C_2H_4 + H_2 \rightarrow C_2H_6$$

- ~50 mg catalyst (<100 mesh) diluted in 450 mg silica gel (Alfa Aesar, 70-150 mesh).
- Reactivity experiments utilize ethylene hydrogenation in an 8 mm tubular glass reactor.
  - Ceramic furnace fitted around reactor to maintain temperature (30-200°C).
- Electronic mass flow controllers used to control inlet gas concentrations/flow rates:
  - Standard conditions: $3\%C_2H_4/20\%H_2/N_2$ at 750 mL/min
    - Inlet concentrations consistent with literature
- Thermo-Finnigan Trace GC/ISQ GC/MS system.
  - 25 m Varian CP-Al_2O_3/KCl PLOT column.

10

Ethylene Hydrogenation Activity

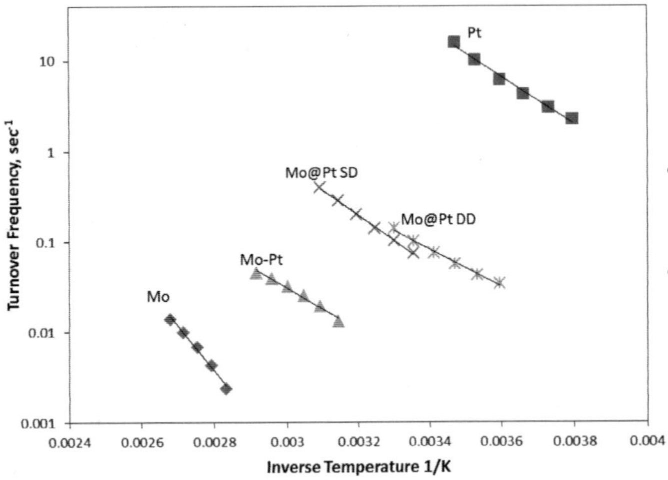

- Ethylene hydrogenation sensitive to hydrogen surface coverage
- Decreased binding strength decreases coverage and reaction rate

- All overlayers show reduced activity compared to pure Pt.

  Overlayer catalysts have lower hydrogen surface coverage and reduced binding strength

Guaiacol HDO Result

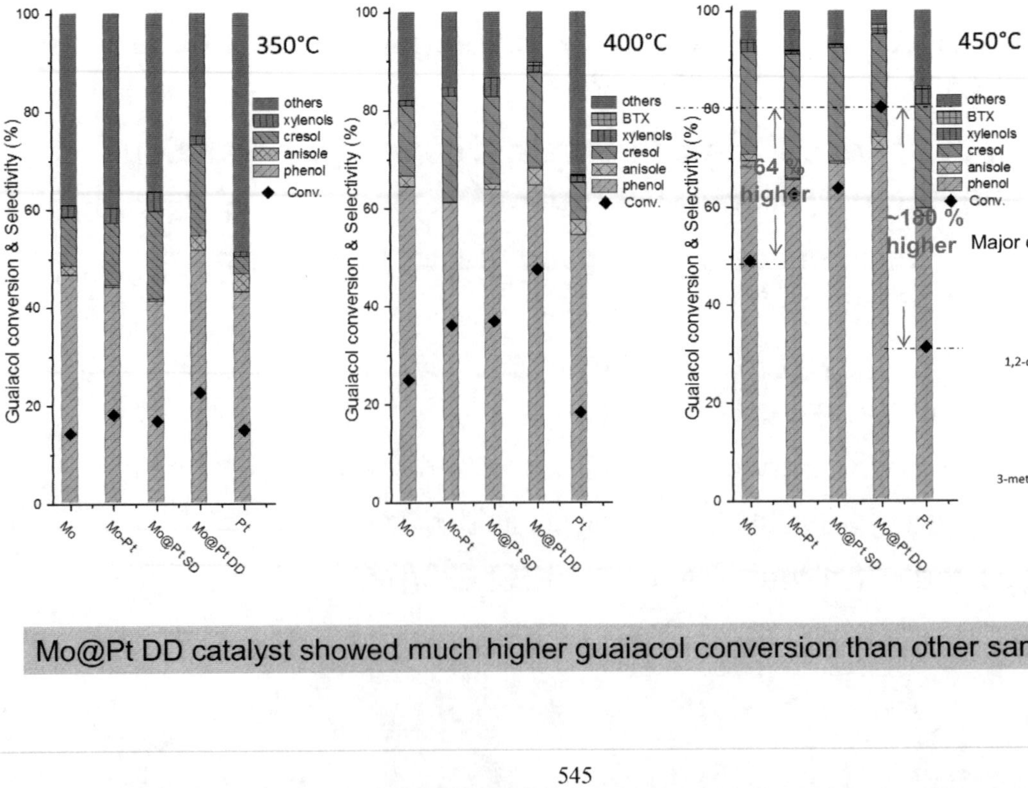

Major others product:

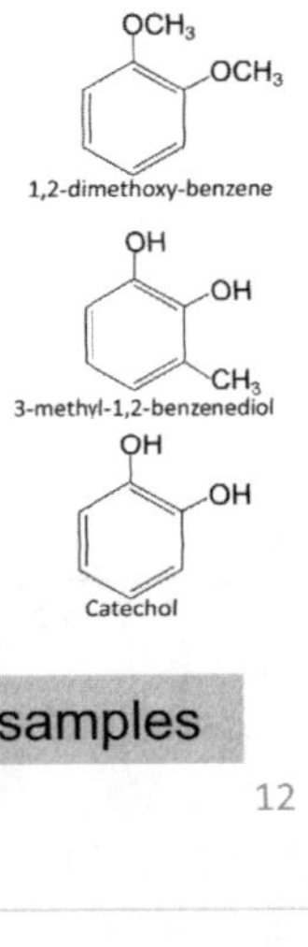

Mo@Pt DD catalyst showed much higher guaiacol conversion than other samples

Guaiacol HDO Result

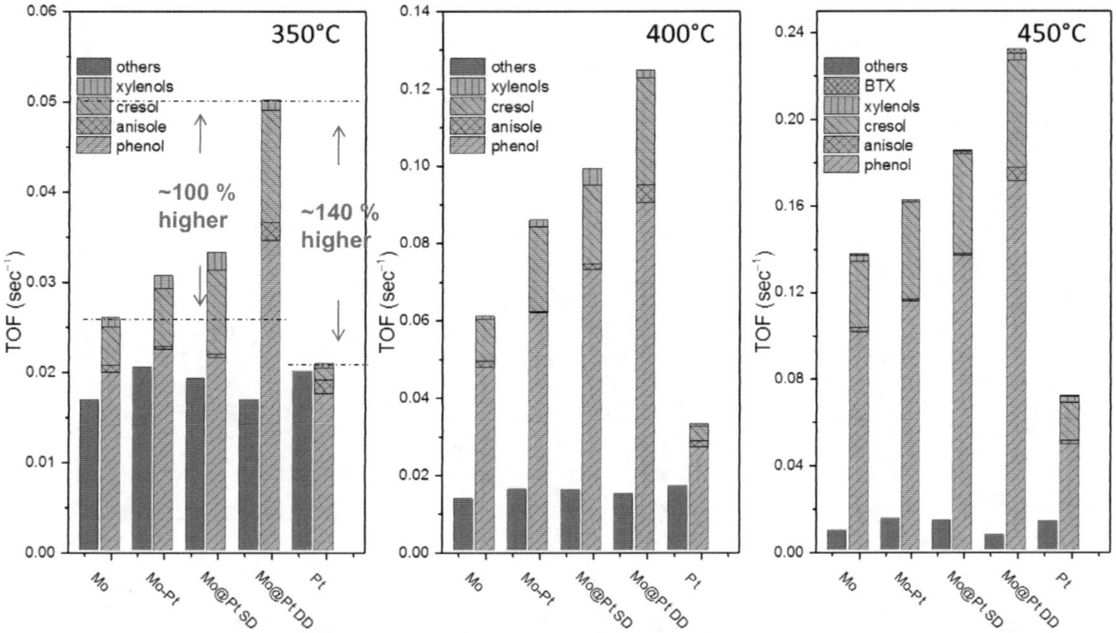

Mo@Pt overlayer catalysts showed much higher activity than other samples

13

Guaiacol HDO Result

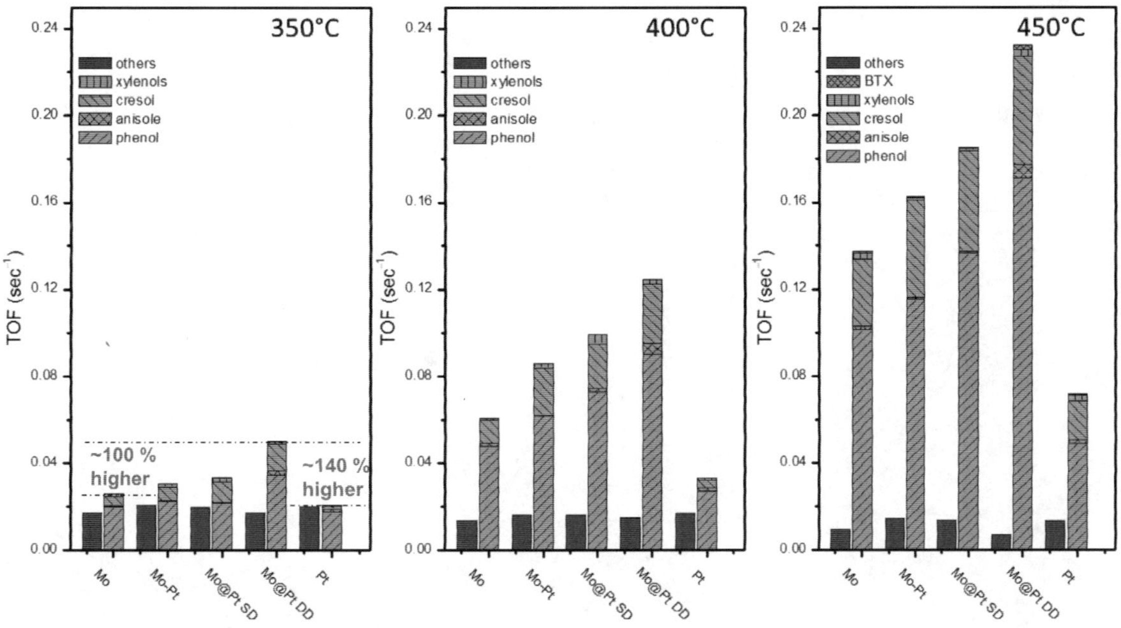

Mo@Pt overlayer catalysts showed much higher activity than other samples

14

Guaiacol HDO with Varying W/F

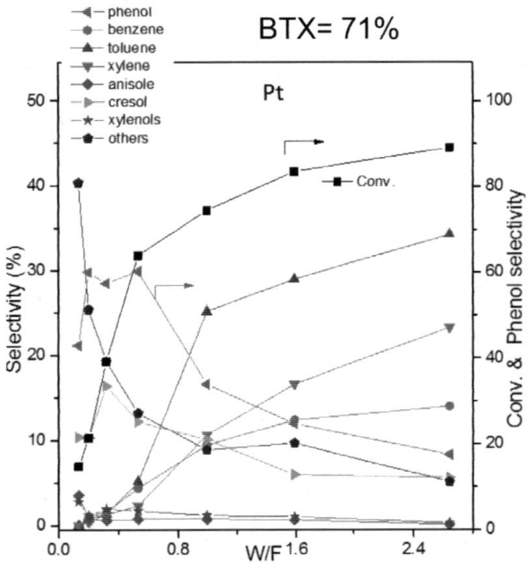

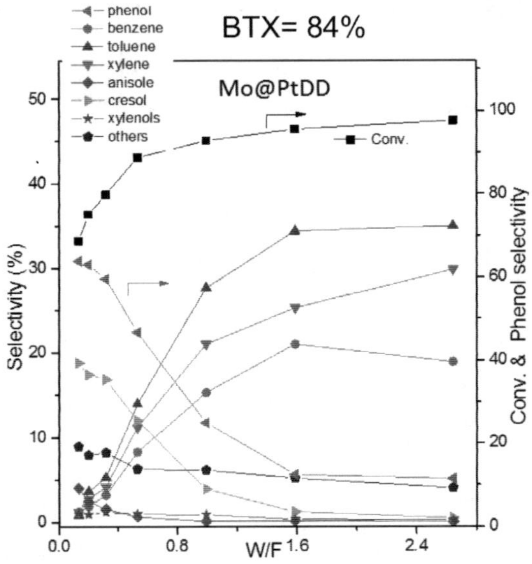

A BTX selectivity up to 84% over Mo@Pt DD at W/F of 2.65 h.

15

Anisole HDO Result

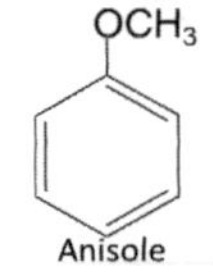

Non-deoxygenation products:

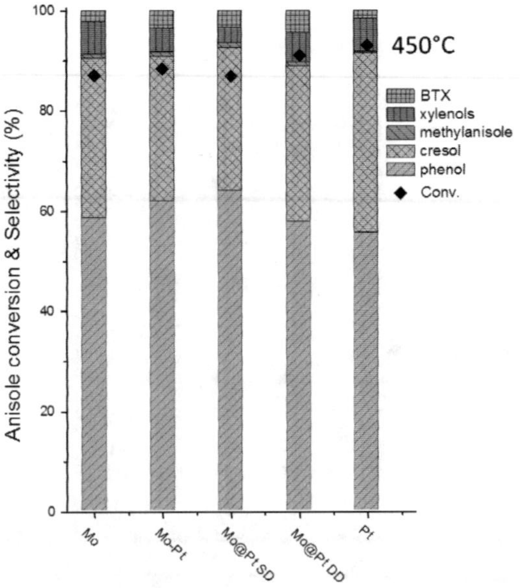

Deoxygenated products:

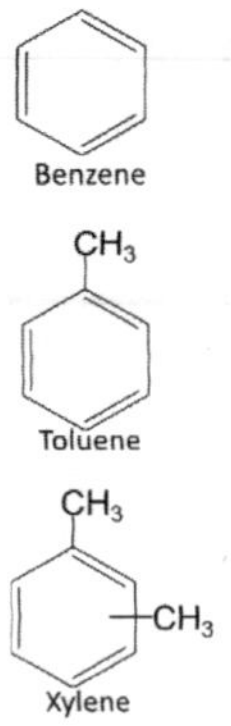

Mo@Pt overlayer catalysts showed a higher selectivity for BTX than Mo and Pt

16

Anisole HDO Result

TOF of Deoxygenation

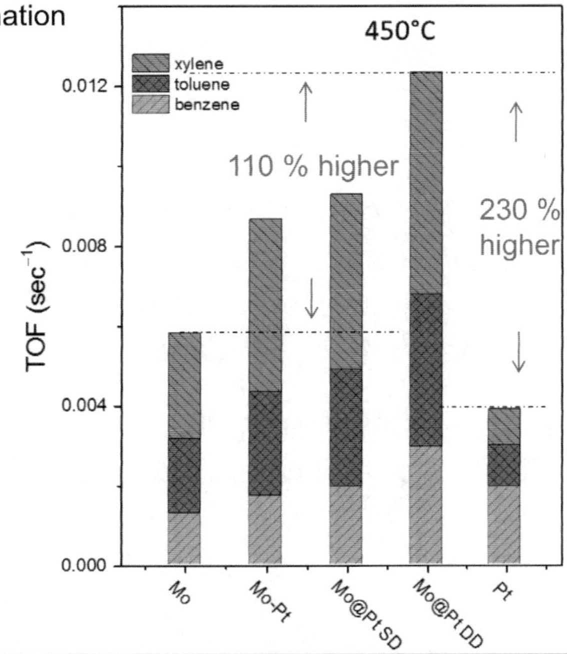

Mo@Pt overlayer catalysts showed higher deoxygenation activity than other samples

17

Anisole HDO with Different W/F

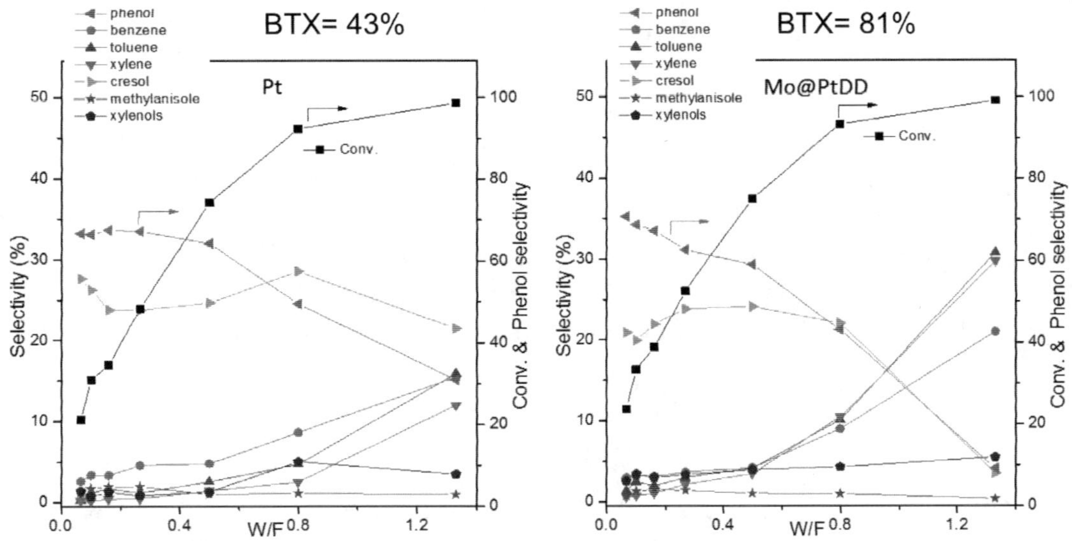

A BTX selectivity of 81% was achieved over Mo@Pt DD at W/F of 1.33 h

18

Conclusion

- Mo@Pt overlayer catalysts have demonstrated weaker H_2/CO adsorption than Pt, which is consistent with computational study.
- Mo@Pt overlayer show significantly increased turnover frequencies for deoxygenation product in both guaiacol and anisole HDO, confirming that the decrease in binding strength increases reactivity in HDO reaction.
- A selectivity up to 80% for BTX could be achieved over Mo@Pt DD for HDO of guaiacol and anisole.

19

Acknowledgements

Funding provided by National Science Foundation,
Chemical, Bioengineering, Environmental and
Transport Systems
[CBET-0933017]

National Science Foundation
WHERE DISCOVERIES BEGIN

Financial support from University of Wyoming
School of Energy Resources
through its Graduate Assistantship program.

School of
Energy Resources
UNIVERSITY OF WYOMING

Research Group Members
Allen R. Morris
Chen Zhang

20

HDO Experimental

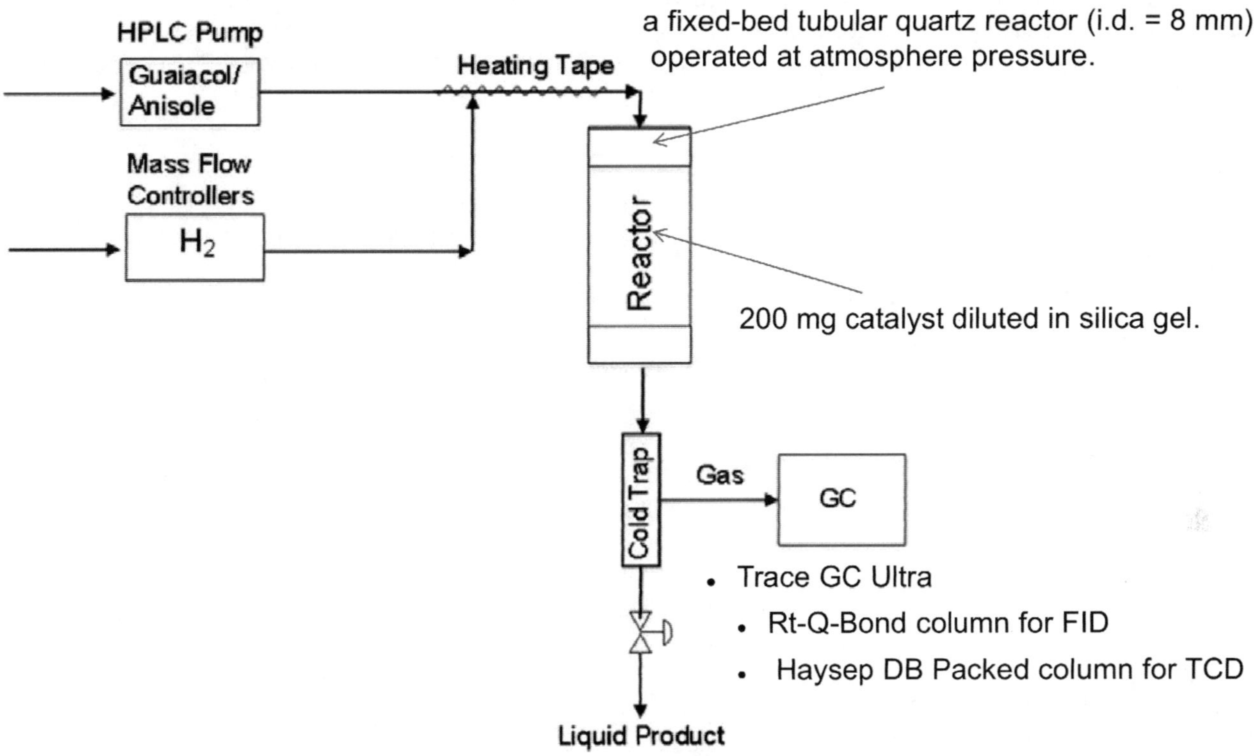

- Trace GC Ultra
  - Rt-Q-Bond column for FID
  - Haysep DB Packed column for TCD

Schematic Diagram of Guaiacol/Anisole HDO Set-up

21

497210 Reaction Engineering for Bio-Renewable Furan-Derived Surfactants

Abstract:

Despite years of technology development, most large-volume surfactants are made from petrochemical sources, while efforts to make renewable surfactants are focused on making existing surfactant structures from renewable sources. In this work, we demonstrate a new surfactant based on the natural structure and chemistry of plant-based oils and sugars with superior function and suitability as a replacement to petrochemicals. Furans obtained from lignocellulosic biomass can be acylated with natural oils-derived fatty acids and anhydrides in the presence of a heterogenous zeolite catalyst. The results obtained for the reaction of lauric anhydride with furan show that different pore sizes, structures and acidity of zeolites result in varying acylation activity. Following acylation, the molecule can be upgraded via several independent and sequential chemistries such as etherification, hydrogenation and aldol condensation and finally subjected to sulfonation to yield surfactant molecules termed as oleo-furan sulfonates (OFS) in high yield. Evaluation of surfactant performance of OFS revealed hundredfold better detergency and stability in hard water conditions in comparison with petroleum-derived counterparts. The synthesis of OFS molecules is highly tunable and selective where the number of carbon atoms in the linear or branched chain can be easily varied without compromising on reaction yields to achieve desired surfactant properties.

References

[1] Joseph, K. E., Park, D. S., Koehle, M., Krumm, C., Ren, L., Damen, J. N., Shete, M. S., Lee, H. S., Xiao, Z., Lee, B., Fan, W., Vlachos, D. G., Lobo, R., Tsapatsis, M., Dauenhauer, P. J. ACS Cent. Sci. **2016**, 2, 820

501750 Upgrading of in-Situ Catalytic Fast Pyrolysis Bio-Oil to Hydrocarbon Fuels

Abstract:

Domestically, the US is challenged with improving the environment, energy security, and energy productivity, thus many efforts are concerted in utilizing biomass as a viable source of renewable liquid transportation fuels and contributing to a robust carbon-neutral energy production. However, achieving an economically competitive process is an ongoing challenge for current biomass upgrading technologies. Specifically, catalyst cost is one of the primary driver of thermochemical conversion cost associated with biomass upgrading to biofuels. The current state of technology for fast pyrolysis route includes a bio-oil stabilization step, which uses catalysts incorporating precious metals such as ruthenium to conduct hydrogenation of bio-oil at low temperatures. However, precious metal catalysts are orders of magnitude more expensive than conventional hydrotreating catalysts and very sensitive to sulfur contamination which limits there lifetime significantly. Catalytic fast pyrolysis (CFP) provides hope for a competitive production of an improved bio-oil. This bio-oil is also stable and higher in quality, and is also more readily hydroprocessed to liquid hydrocarbon fuels. Specifically, In-situ catalytic fast pyrolysis (in-situ CFP) uses a circulating fluidized-bed reactor which offers a straightforward, cost effective application of the technology. This presentation will report our recent research on in situ CFP and produced bio-oil hydrotreating. CFP bio-oils produced by using a low cost red mud catalyst (recovered waste from the Bayer process) and different reaction conditions from different biomass feedstocks were directly hydrotreated without a stabilization step. The impact of CFP parameters and feedstocks on hydrotreating will be elucidated. The overall biomass to fuel yield and cost information based on technoeconomic analysis will be reported. It showed that in situ CFP with an appropriate catalyst, such as red mud, has great potential for an efficient and economic viable route to convert biomass to fuels.

502374 Bio-Terephthalic Acid Synthesis from Cross Metathesis of Unsaturated Carboxylic Acids and Consecutive One-Pot Cycloaddition and Aromatization Reactions

Abstract:

Introduction

Terephthalic acid in its pure form (PTA) is essential for the preparation of a variety of polyester fibers and is largely produced via the catalytic oxidation of petroleum-derived p-xylene [1]. The ongoing exchange of naphtha crackers with shale gas crackers is causing a shortage in some aromatic species. As a consequence, alternative and sustainable routes to produce terephthalic acid from biomass derivatives, are highly desirable [2]. PTA precursors can be prepared via Diels-Alder cycloaddition with ethylene from bio-derived chemicals, like furnaic compounds or carboxylic dienes. These synthetic pathways are, however, limited by the low yields of either the starting chemicals or the final product [3-5]. In the last decade, Ru-catalyzed olefin metathesis, and in particular cross-metathesis, has become an important tool for the formation of carbon–carbon bonds and the formation of valuable organic molecules [6]. Di-ester muconates, in particular, can be synthesized from cross-metathesis of bio-sourced olefins. This route has never been considered before for the synthesis of bio-sourced terephthalates.

Results and discussion

We have investigated the ruthenium-catalyzed cross metathesis of sorbates with acrylates to form di-ester muconates. The effect of alkyl substituents and solvents was investigated over various organometallic ruthenium-based catalysts to optimize the conversion and selectivity towards the muconates. Conversions of up to 60% and 40% yields of the muconates were achieved within the first hour of reaction with low amounts (3 mol.%) of ruthenium-based catalyst containing a N-heterocyclic carbene (NHC) ligand. With the aim to develop a greener process, the acrylate reactant was used in excess with no other solvent added. In a later step, a one pot synthesis of diester terephthalates from the as-prepared di-ester muconates was achieved. First, the Diels-Alder cycloaddition between muconate esters and ethylene forms a cyclohexene dicarboxylic compound in >95% yield, and a consecutive aromatization on the same reaction batch in the presence of Pd/C catalyst affords the formation of terephthalic acid diesters (~ 40% yield). Upon further development, this route can provide a sustainable alternative for the industrial production of biomass derived polyethylene terephthalate (PET).

References

[2] Roger A. Sheldon. Green Chem. 16 (2014) 950.

[3] C. Luke Williams, Chun-Chih Chang, Phuong Do, Nima Nikbin, Stavros Caratzoulas, Dionisios G. Vlachos,

Raul F. Lobo, Wei Fan, and Paul J. Dauenhauer. ACS Catal. 2 (2012) 935.

[4] S. Berard, C. Vallee, D. Delcroix, Ind. Eng. Chem. Res. 54 (2015) 7164-7168.

[5] J.W. Frost, Preparation of trans,trans muconic acid and trans,trans muconates, Draths Corporation, U.S. Patent US2010314243-A1, 2010

[6] Georgios C. Vougioukalakis, and Robert H. Grubbs. Chem. Rev. 110 (2010) 1746

499197 Continuous Flow Nitration: Research and Process Development

Abstract:

Bi-phasic nitration is commonly executed for making nitro intermediates in the pharmaceutical industry. Their scale-up, however, is generally problematic because of the following concerns: high exothermicity, need for proper mixing and liquid-liquid dispersion, and formation of energetic di-nitro by-products. A detailed understanding of these aspects is required to ensure proper scale-up of nitration.

In this talk we will present a case study of a scale-up of 1-methyl-4-(trifluoromethyl)benzene nitration using KNO_3/H_2SO_4 as the nitrating agent. To understand these aspects, targeted experiments were performed to prepare a mass transfer with chemical reaction model (using film theory) to quantify the relation between liquid-liquid dispersion with the conversion and di-nitro impurity levels. Model predictions showed that acceptable conversion required a Hatta number of at least 1.9, with 80% of reaction done in the film. To meet these requirements, a CSTR was chosen as the preferred contacting device, with proper heat transfer protocols in place to handle the exotherm.

The work presented here was sponsored by AbbVie. AbbVie contributed to the design, research, and interpretation of data, writing, reviewing, and approving the presentation. All authors are employees of AbbVie.

494380 On-Demand Medicinal Chemistry and Compound Synthesis in Oscillating Droplets

Abstract:

There is increasing pressure in drug discovery to deliver a steady stream of active compounds for physicochemical profiling and potency testing. In order to accelerate lead optimization, it is imperative to reduce the time required for each iteration of design, synthesis, and screening [1]. With recent advances in automation and high-throughput biological screening, the bottleneck has crept toward the synthesis step [2, 3, 4]. Medicinal chemistry laboratories have also benefitted from this trend in automation [5, 6, 7, 8], where one integrated platform has even been used for the discovery of novel AbI kinase inhibitors [9].

Here, we describe an automated chemistry platform that can efficiently screen a wide range of reactions, including single/multi-phase, single/multi-step, and photochemical reactions at the 20 microliter scale. Individual droplets are prepared by a liquid handler and moved through fluoropolymer tubing by a carrier gas at elevated pressures. Reactions occur in a heated reactor where the liquid droplet is oscillated back and forth to ensure thorough mixing â€" even for biphasic liquid-liquid reactions or liquid-gas reactions â€" without being limited to a finite residence time [10]. Optional inlet and/or outlet injections enable multistep chemistry. A portion of the crude product mixture is sampled and sent directly to an online HPLC/MS for analysis, purification, and product collection, typically of 10-500 Î¼g.

The system offers the enhanced heat and mass transfer characteristics, increased safety, and opportunity for automation associated with flow chemistry while enabling the following key advantages: (a) reduction of material consumption, preparing just 40 microliters for each reaction condition, (b) elimination of residence time dispersion, (c) elimination of the inverse relationship between residence time and mass transfer rate for a flow reactor of fixed length, (d) simplification of continuous and discrete variable screening through the use of a liquid handler to prepare reaction mixtures, and (e) elimination of the time- and material-waste associated with waiting for flow reactors to reach steady state.

Importantly, because the reaction droplet is equivalent to one segment in a continuous segmented flow, reaction conditions identified using the platform can be directly translated to a continuous synthesis. This helps bridge the connection between the early research stages of drug discovery and the later stages, where material demands grow from the Î¼g-mg scale to the g-kg scale.

As a demonstration of this platform, 36 examples of frequently used N-X bond forming reactions using typical small-molecule building blocks are screened. A representative biphasic Suzuki-Miyaura cross coupling reaction, a multi-step synthesis of the drug product diclofenac, and several visible-light catalyzed oxidations are also reported.

[1] K. C. Nicolaou, *Angew Chem Int Edit* **2014**, *53*, 9128-9140.

[2] A. G. Godfrey, T. Masquelin, H. Hemmerle, *Drug Discovery Today* **2013**, *18*, 795-802.

[3] X. Niu, F. Gielen, J. B. Edel, A. J. deMello, *Nat Chem* **2011**, *3*, 437-442.

[4] R. Ostafe, R. Prodanovic, W. L. Ung, D. A. Weitz, R. Fischer, *Biomicrofluidics* **2014**, *8*.

[5] M. Peplow, *Nature* **2014**, *512*, 20.

[6] B. Ahmed-Omer, E. Sliwinski, J. P. Cerroti, S. V. Ley, *Org Process Res Dev* **2016**.

[7] A. Buitrago Santanilla, E. L. Regalado, T. Pereira, M. Shevlin, K. Bateman, L.-C. Campeau, J. Schneeweis, S. Berritt, Z.-C. Shi, P. Nantermet, Y. Liu, R. Helmy, C. J. Welch, P. Vachal, I. W. Davies, T. Cernak, S. D. Dreher, *Science* **2015**, *347*, 49-53.

[8] B. J. Reizman, K. F. Jensen, *Accounts of Chemical Research* **2016**, *49*, 1786-1796.

[9] B. Desai, K. Dixon, E. Farrant, Q. Feng, K. R. Gibson, W. P. van Hoorn, J. Mills, T. Morgan, D. M. Parry, M. K. Ramjee, C. N. Selway, G. J. Tarver, G. Whitlock, A. G. Wright, *J. Med. Chem.* **2013**, *56*, 3033-3047.

[10] M. Abolhasani, N. C. Bruno, K. F. Jensen, *Chem. Commun.* **2015**, *51*, 8916-8919.

501379 Ultrahigh-Throughput Screening of Chemical Synthetic Pathways Using Desorption Electrospray Ionization Mass Spectrometry

Abstract:

We describe a state-of-the-art automated platform for ultrahigh-throughput screening of chemical reactions based on reaction acceleration in charged microdroplets and utilizing Desorption Electrospray Ionization Mass Spectrometry (DESI-MS). The goal is to deliver an extremely rapid yet insightful qualitative yes/no analysis for the discovery of new chemistry in the order of up to 10,000 reactions per hour. The positive results are then used to guide synthesis in a more scalable setting, including in batch and flow. Using this system, particular chemical routes are prepared in parallel using reagents which are robotically pipetted and mixed in chemically-inert well plate. The routes are tested by stamping the reaction mixtures onto an inert reaction substrate, e.g. PTFE membrane, followed by electrospraying of various solvents at an angle relative to the substrate to generate secondary ionized reaction microdroplets. These microdroplets are propelled into the mass spectrometer for mass analysis. The results are then analyzed using an automated MS data analysis system developed in-house, which provides chemical reaction monitoring and intelligence for optimization of future reaction screening conditions to increase the probability of pathway discovery.

494694 Heuristics for Implementing Photoredox Catalysis in Flow Reactors

Abstract:

Heuristics for Implementing Photoredox Catalysis in Flow Reactors Eric G. Moschetta,* Steven M. Richter, and Steven J. Wittenberger AbbVie, Inc. Process R&D, 1401 Sheridan Road North Chicago, IL 60064 *corresponding author: eric.moschetta@abbvie.com Photoredox catalysis is reemerging in organic synthesis as a powerful method for constructing small molecules under mild conditions. Such transformations often employ transition metal catalysts to absorb visible light, exciting the catalysts into energetic states that facilitate single electron transfer events between the catalysts and organic substrates.[1] These reactions frequently occur at mild temperatures, including room temperature, using visible light sources such as LEDs that provide intense light with narrow wavelength emissions. As such, materials for conducting photoredox chemistry at the lab scale are readily available for chemists and engineers. However, significant challenges remain in visible-light photoredox catalysis. The primary challenge in industrial applications is scaling up photoredox reactions from laboratory scale batch reactors to reactors that are capable of multi-kilogram scale production. Proper reactor design relies on rigorous understanding of the inherent kinetics of photoredox reactions, which are largely unexplored. The photon flux, the overall observed rate at which photons are received in the reactor, must be high enough to ensure that the inherent kinetics of the reaction are the operating conditions of the reactor. The lamp-reactor geometry, therefore, is a critical design factor in scalable photoredox flow chemistry. Maximizing the photon flux also likely depends on careful tuning of the catalyst concentration, reactant concentration, and light intensity. Continuous flow reactors can facilitate improved photon fluxes and increased throughput through significantly decreased reaction times compared to batch reactions, which are both necessary for industrial photoredox catalysis.[2] Flexible fluoropolymer tubing is chemically resistant to many materials and highly transparent to visible light, meaning that tubing can be wrapped around light sources such as LEDs for access to customizable tubular flow reactors for photoredox applications. The short tube diameters allow for shorter path lengths, meaning that photons are more evenly distributed throughout the reactor, increasing the apparent rate of reaction. However, general principles and heuristics for designing such flow reactors are not yet available to process chemists and engineers. This work establishes heuristics for employing flow reactors in photoredox chemistry. The effects of reactant and catalyst concentration, tube diameter, and LED intensity on the observed photon fluxes and rates of reaction will be discussed. The model reaction for establishing the heuristics will be the oxidation of 9,10-diphenylanthracene using molecular oxygen and a common ruthenium photoredox catalyst, $Ru(bpy)_3Cl_2$. The quantum yield for this reaction is known,[3] meaning that the photon flux under a given set of reaction conditions can be calculated directly once the rate of reaction is measured. These heuristics will then be examined against a pyrazole synthesis driven by photoredox catalysis, using the same ruthenium catalyst, to see if heuristics accurately predict how the observed rates of reaction change with changes in reaction conditions in flow. **References**

[1] C. K. Prier, D. A. Rankic, D. W. C. MacMillan, *Chem. Rev.* **2013**, *113*, 5322-5363.

[2] Y. Su, N. J. W. Straathof, V. Hessel, T. Noel, *Chem. Eur. J.* **2014**, *20*, 10562-10589.

[3] S. P. Pitre, C. D. McTiernan, W. Vine, R. DiPucchio, M. Grenier, J. C. Scaiano, *Sci. Rep.* **2015**, *5*, 16397.

This abstract presentation was sponsored by AbbVie. AbbVie contributed to the design, research, and interpretation of data, writing, reviewing, and approving the publication. All authors are AbbVie employees.

500424 Operando Raman of the Nitroldol Reaction over Pore-Expanded Fdu-12 â€“ Dendron Hybrids

Abstract:

Many labs have used the Nitroaldol reaction to probe the accessibility and reactivity of supported organic ligands.[1] Our lab has reported prior work on simple ligands and branched melamine-dendrons as catalysts for this reaction. The current work describes our effort on using pore-expanded FDU-12[2] as a novel support for growing melamine dendrons and their use as catalysts for this reaction. Kinetics measurements were obtained using a suite of dendrons with different diamine spacers and peripheral groups to determine how the diamine identity and density influences reactivity. To understand the chemistry at a more basic level operando Raman spectroscopy was used to monitor the various intermediates formed.

A library of samples of first through third generation dendrons supported on pore expanded FDU-12 were synthesized, characterized, and tested for the Nitroaldol reaction. Consistent with prior work the conversion/rate decreases with increasing dendron generation for all samples. We ascribe this finding to mass transfer limitations. We observe that diamines with secondary amines give higher conversions than primary amines. In the case of the G1 samples these samples all have within +/-5% the same number of functional groups.

On a per site basis the dendrons with bulky diamines such as trimethylenedi(piperidien) have a rate a factor 2-3 times higher than the dendrons made with 4-aminomethylpiperidine. Our work shows the effect of primary versus secondary amines, and the effect of the free volume on reactivity, wherein high free volumes are advantageous for reactivity. In all cases we see high (>80%) selectivity to the alcohol product except for the amino-silica.

We have also performed operando Raman[3] measurements to monitor the reaction. A band at 1715 cm^{-1} (C=O) can be seen to systematically decrease with conversion, and the appearance of the C=C double bond of the nitroalkene at 1650 cm^{-1} is clearly observed. Finally a feature at 1350 cm^{-1} assigned to C=N intermediates is observed to increase during reaction, illustrating the utility of Raman as a tool to probe this reaction.

In summary our results show that 1) while we can synthesize more uniform dendrons on pore expanded FDU-12 than SBA-15 the windows between the cages impose severe transport restrictions which are clearly discernible in our results, and 2) operando Raman can be used to easily monitor the formation of the alkene product (C=C bond) as well as to observe the imine (C=N) intermediate. The results show these hybrid materials are interesting catalytic materials for this reaction and that they are amenable to analysis using operando methodologies which have not been widely used prior in this sub-community.

References

1. Bass, J. D., Solovyov, A., Pascall, A. J., and Katz, A. *J. Am. Chem. Soc.* 128 373 (2006).

2. Fan, J, Yu, C., Gao, F., Lei, J., Tian, B., Wang, L., Luo, Q., Tu, B., Zhou, W. and Zhao, D. *Angew. Chem. Int. Ed.* 42 3146 (2003).

3. Grabow, K., and Bentrop, U. *ACS Catal.* 4 2153 (2014).

500195 Scale-up of a Reaction Step from Lab to Commercial Scale, Switching from Batch to Continuous Manufacturing Using a Mechanistic Modelling Approach

Abstract:

This work presents the journey of scaling-up a manufacturing process in a multi-purpose plant that is focused on the reaction step. The scale-up process is based on a mechanistic modelling approach and resorts to both batch and flow data that was generated at lab scale. The modelling applied in this process is used not only to increase process understanding and further optimize the process to highest performance possible, but also to guide the outcome in terms of process safety.

The case-study to be shown is composed of a multi reaction mechanism, comprising one very exothermic reaction, in a gas-liquid system that will be run at commercial scale in a continuous mode. The commercial continuous reactor is composed of several plates, where the first plate is intended to promote mixing of the gas and liquid streams, while the following plates will be use to carry out the reaction. In the last two plates, there will be a quench step, before proceeding to the downstream of the process.

The lab data was generated both in a batch reactor and also in a flow reactor. In the first reactor, it was possible to monitor not only the reaction itself by HPLC sampling of main compounds, but also following the reagent feeding profile, allowing to gathering of knowledge on reaction kinetics. Additional experiments were run at different temperatures in order to improve knowledge on the activation energy of the multiple reaction mechanism. In the continuous reactor, it was possible to gain experience on the reaction outcome in a continuous mode and to apply scale-up engineering rules. On top of that, it was possible to improve the kinetic model using screening experiments around typical process parameters (flowrate, temperature, reagent-to-starting material ratios).

With the accumulated knowledge built upon both reactors, it is possible to simulate the large scale reaction in a commercial scale continuous reactor, based on the kinetic, heat and mass transfer models derived from the lab experiments. The modelling approach developed using Dynochem will allow the optimization of the large scale conditions of the process but also a much safer scaling-up of the process to the commercial scale.

AIChE Annual Meeting 2017, Minneapolis 1

University of Chemistry and Technology, Prague

Microreactors driven by electric field for enzyme catalysis in ATPS

Alexandr Romanov, Lucie Vobecka,
Zdenek Slouka and **Michal Pribyl**

Department of Chemical Engineering,
University of Chemistry and Technology, Prague, Czech Republic

AIChE Annual Meeting 2017, Minneapolis 2

University of Chemistry and Technology, Prague

Content

- Motivation
- Microreactor concepts
  - With droplet flow
  - With parallel flow
- Aqueous two-phase systems (ATPS)
  - Basic properties
  - Droplet addressing by electric field
- Case study
  - Enzyme hydrolysis of penicillin G a cephalexin synthesis
- Conclusions

University of Chemistry and Technology, Prague

Motivation – Why micro?

- Large interfacial area – intensive mass and heat transport
- Short transport distances – reduction of transport resistances
- Laminar (Stokes) flow – predictable behavior
- Fast prototyping, fast debugging
- Possible integration of several unit operation into one chip – mixing, reaction, separation, detection, quantification

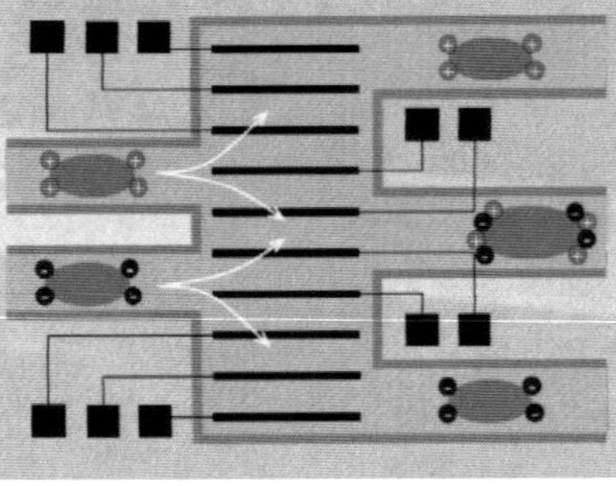

University of Chemistry and Technology, Prague

Motivation – Why electric field?

- Driving force – electric field instead of pressure gradient $\Delta p \approx \frac{32 l v \eta}{d^2}$
- No moving parts and complex mechanical devices are necessary
- Different electrokinetic phenomena can be exploited – electrophoresis, electroosmosis, dielectrophoresis, electrowetting, ...
- Pumping of solutions and multi-phase mixtures
- Differences in ion charge numbers and molecule sizes lead to electrophoretic separation
- Noncharged droplet and particles can be separated by dielectrophoresis
- Droplet addressing into particular parts of microfluidic structures
- Induced electrocoalescence, electrojetting, electrosplitting

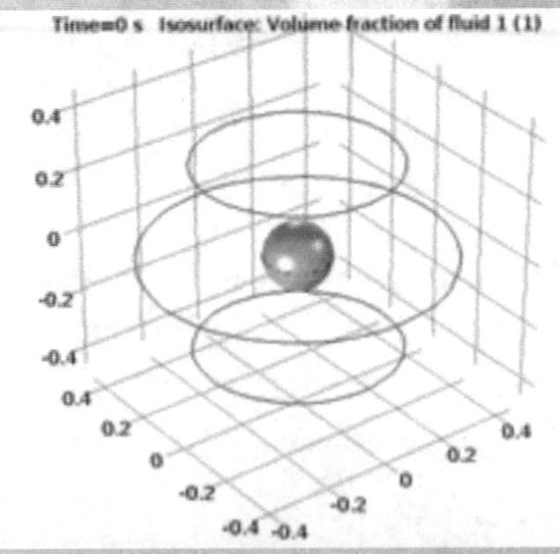

Flittner & Pribyl, J. Fluid Mechanics 822, 31-53, 2017.

Motivation – Properties of micro-bio-reactors?

- Fast fabrication
- Low cost
- Only limited amount of mechanical parts
- Easy control and maintenance
- Continuous regime
- Low operating costs
- Integration of different unit operations into one device
- High safety
- …

Microbioreactor concepts

University of Chemistry and Technology, Prague

Bioreactor with droplet flow

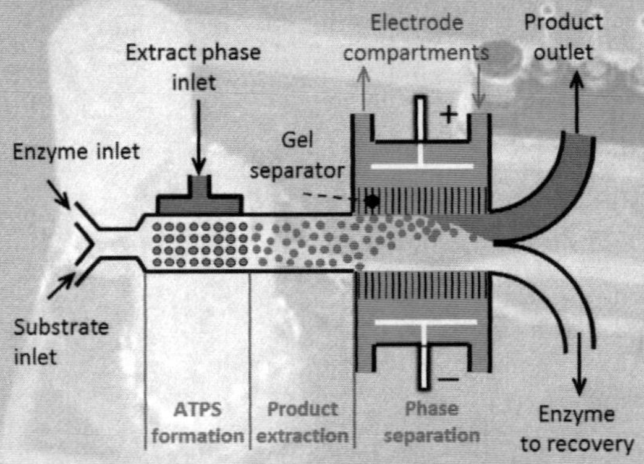

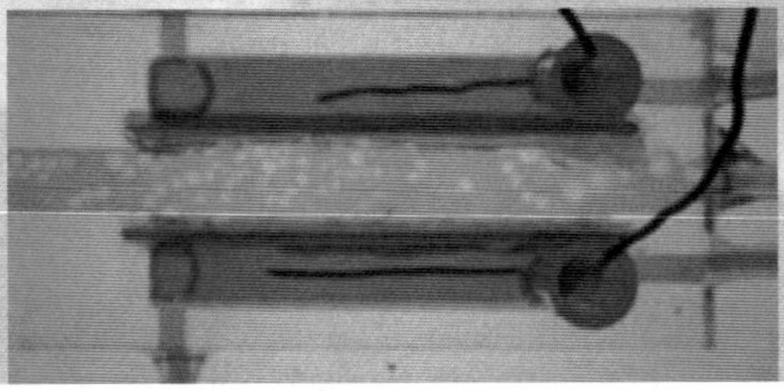

- Integrated reactor-separator
- Droplet flow of ATPS to extract reaction products
- Large interfacial area
- Electric field assisted separation of ATPS
- Possible recycle of enzyme solution or enzyme recovery

University of Chemistry and Technology, Prague

Bioreactor with parallel flow

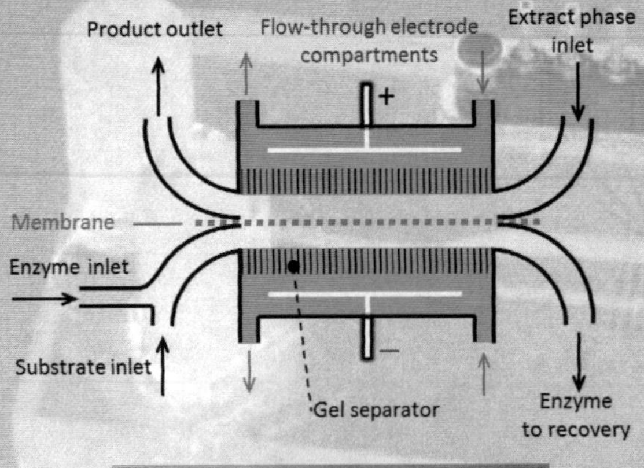

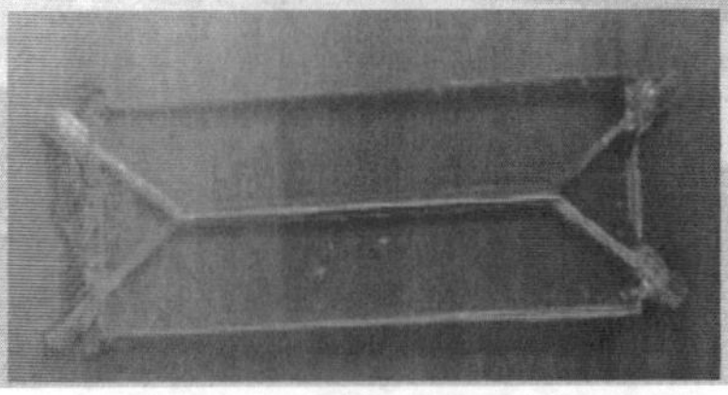

- Integrated reactor-separator
- Parallel flow of ATPS
- Thin membrane for flow stabilization
- Co-current or counter-current arrangement
- Enhancement of concentration of ionic species (products)
- Enzyme recovery

University of Chemistry and Technology, Prague

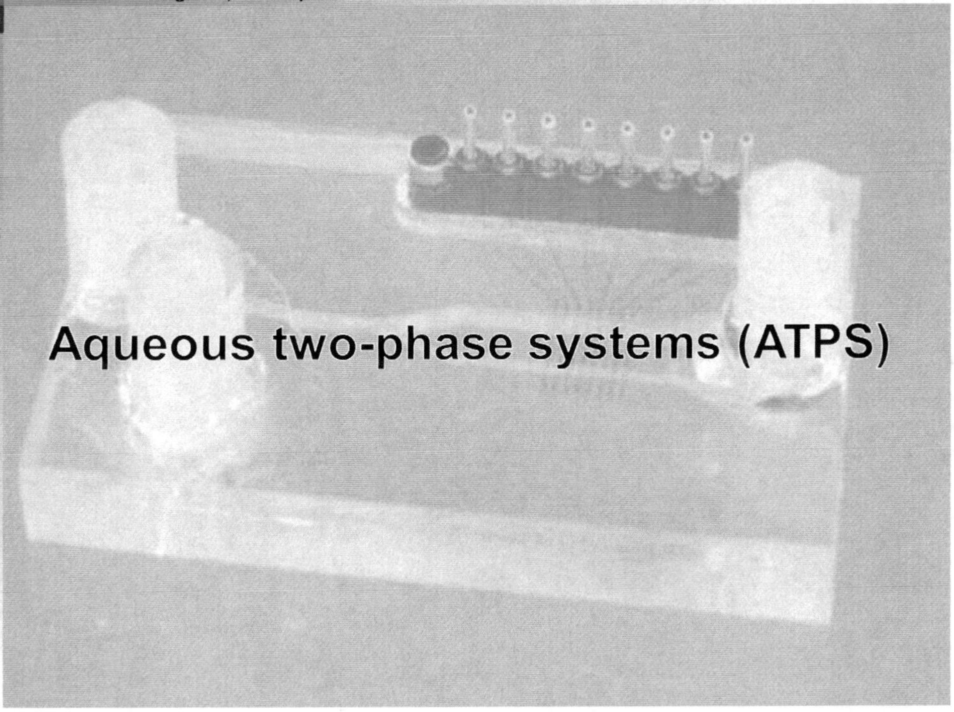

Aqueous two-phase systems (ATPS)

University of Chemistry and Technology, Prague

Aqueous two-phase systems (ATPS)

- Two immiscible aqueous phases
  - Polymer/polymer/water (polyethylene glycol, dextran, water)
  - Polymer/salt/water (polyethylene glycol, phosphates, water)
- Milder conditions than in the extractions by organic solvents
  - Protein, organelles, cells, DNAs separation
- Extractive fermentation
  - Biological reaction in one phase, extraction of products or reactant dosing to/from the other phase
  - Elimination of substrate/ product inhibitions
  - Shift of reaction equilibrium towards the product side

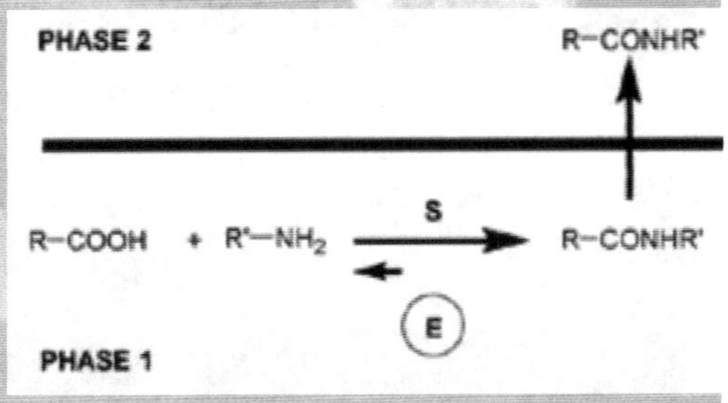

Terreni et al., Enzyme and Microbial Technology 36, 2005, 672–679.

University of Chemistry and Technology, Prague

Aqueous two-phase systems

- Mixture of polyethylene glycol (PEG) 4000, salt, water
- Two immiscible phases
  - Top PEG phase
  - Bottom salt phase
- Both phases conductive!

Low interfacial tension (σ ~ 1 mN/m)

Despite of it, generation of relatively stable slug flow is possible

Extraction in microscale

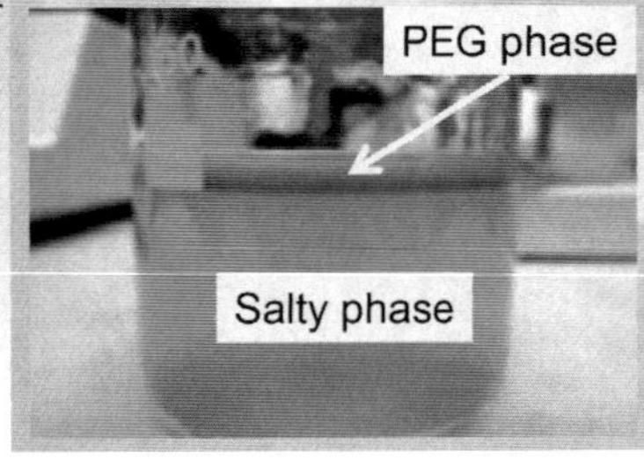

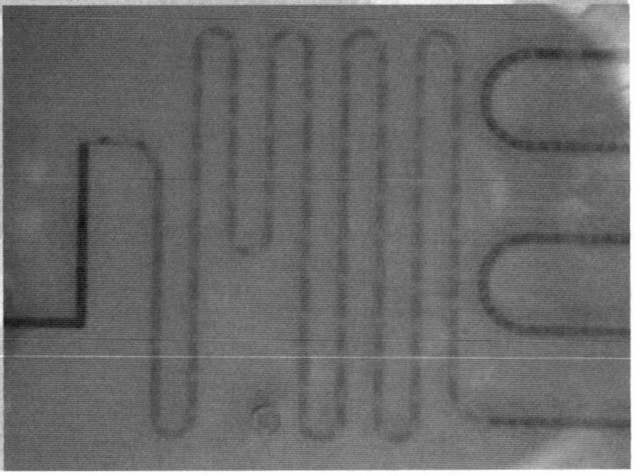

University of Chemistry and Technology, Prague

ATPS manipulation by electric field

Open chip

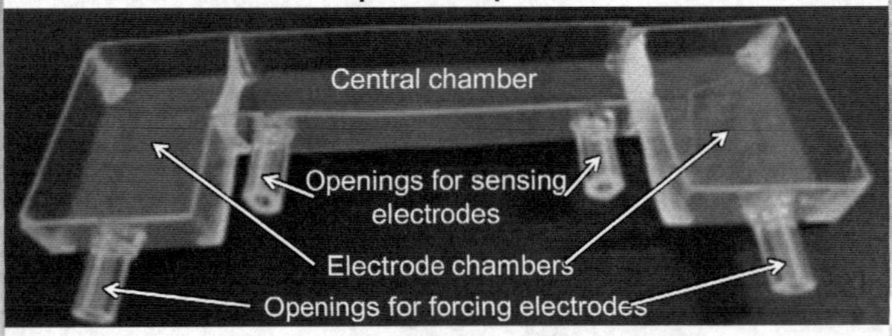

Closed chip

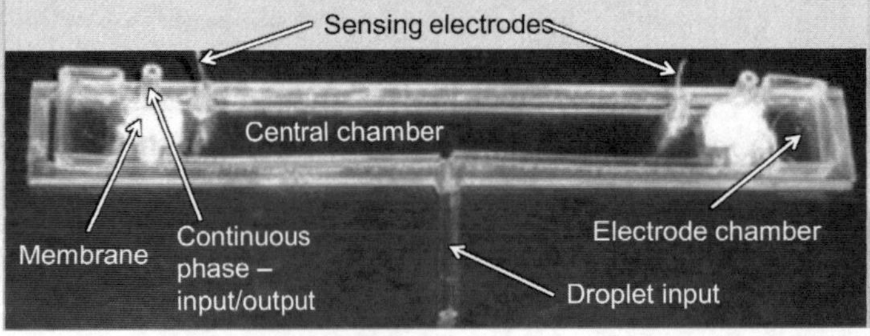

- Droplet of salty (bottom) phase dropped on PEG (top) phase
- Salty phase is heavier but small droplets are kept at the PEG surface
- Both phases are conductive
- Electrode reservoirs are separated membranes treated by agar gel

University of Chemistry and Technology, Prague

Time course of a typical experiment

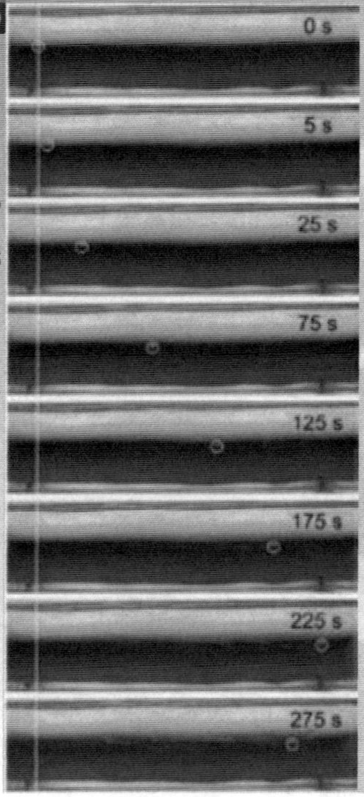

- PEG – continuous phase
- Droplets of the salty phase always move to negatively charged electrode
- Typical electric field strength 10 V cm^{-1}

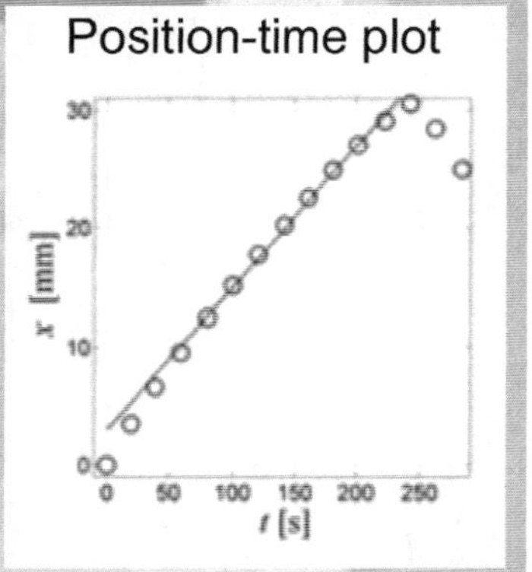

Vobecka et al., Microfluid Nanofluid (2017) 21:51

University of Chemistry and Technology, Prague

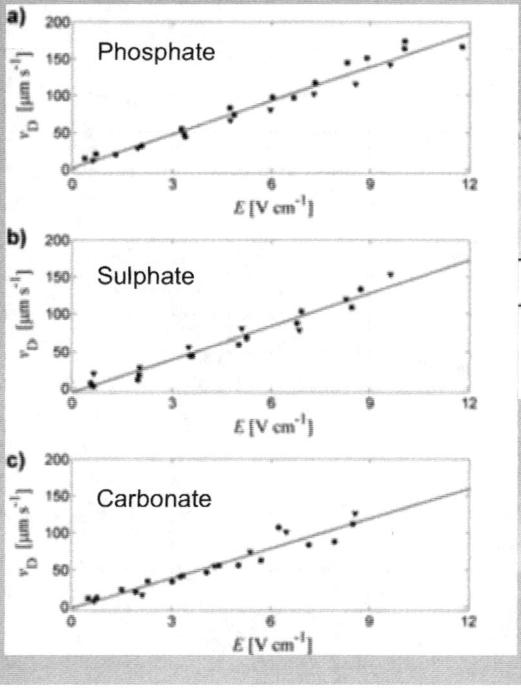

Open system – droplet mobilities

Mobility values

Phosphate>Sulphate>Carbonate

ATPS	$\mu_E/[m^2\ V^{-1}\ s^{-1}]$
PEG/phosphate	$(1.52 \pm 0.12) \times 10^{-7}$
PEG/sulphate	$(1.48 \pm 0.16) \times 10^{-7}$
PEG/carbonate	$(1.34 \pm 0.17) \times 10^{-7}$

- Linear dependences of droplet velocity on applied electric field strength
- Mobilities higher by at least factor 10 if compared to water-oil systems
- Low field is necessary for addressing

Closed system – droplet mobilities

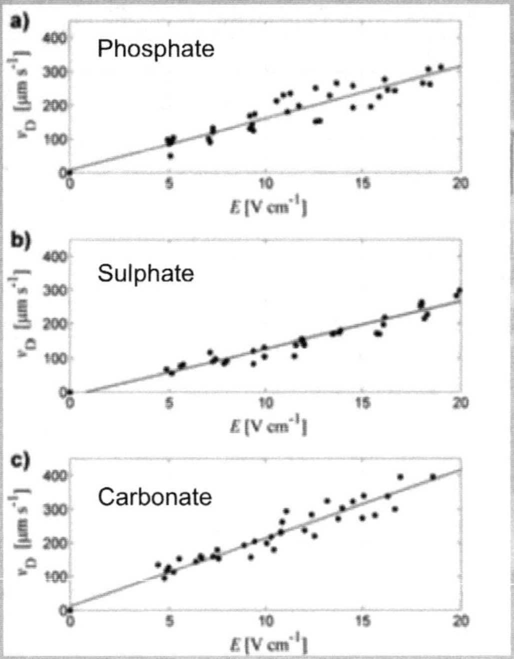

Mobility values

Carbonate>Phosphate>Sulphate

ATPS	$\mu_E/[m^2\ V^{-1}\ s^{-1}]$
PEG/phosphate	$(1.54 \pm 0.16) \times 10^{-7}$
PEG/sulphate	$(1.38 \pm 0.10) \times 10^{-7}$
PEG/carbonate	$(2.02 \pm 0.16) \times 10^{-7}$

- Linear dependences of droplet velocity on applied electric field strength
- In PEG/carbonate significantly higher mobility if compared to open system

Droplet addressing

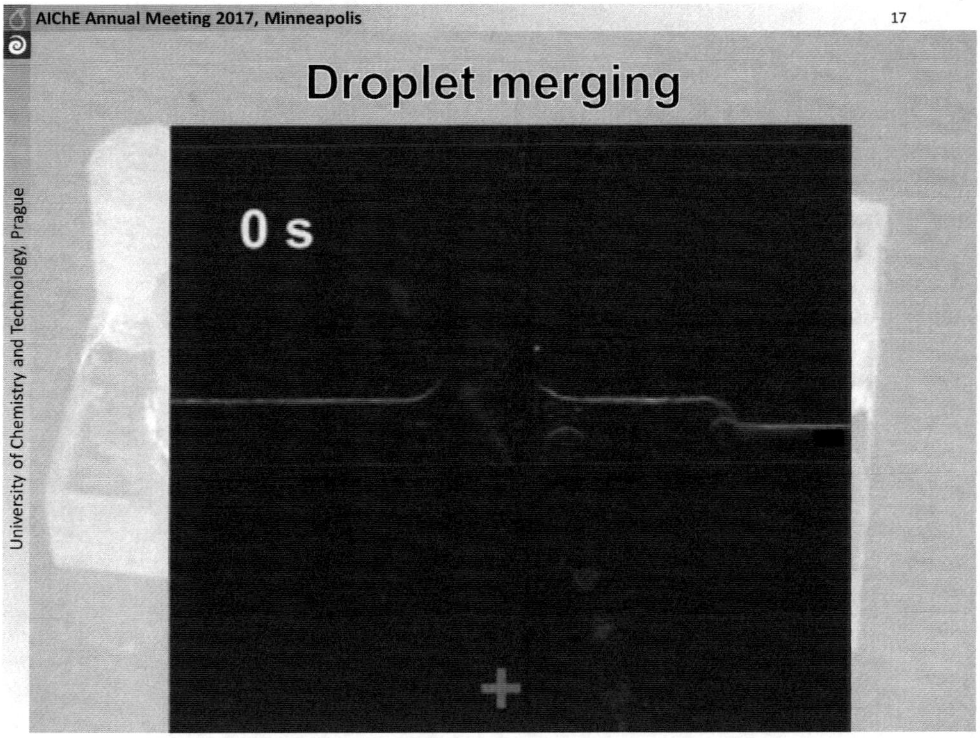
AIChE Annual Meeting 2017, Minneapolis
17
University of Chemistry and Technology, Prague
Droplet merging
0 s

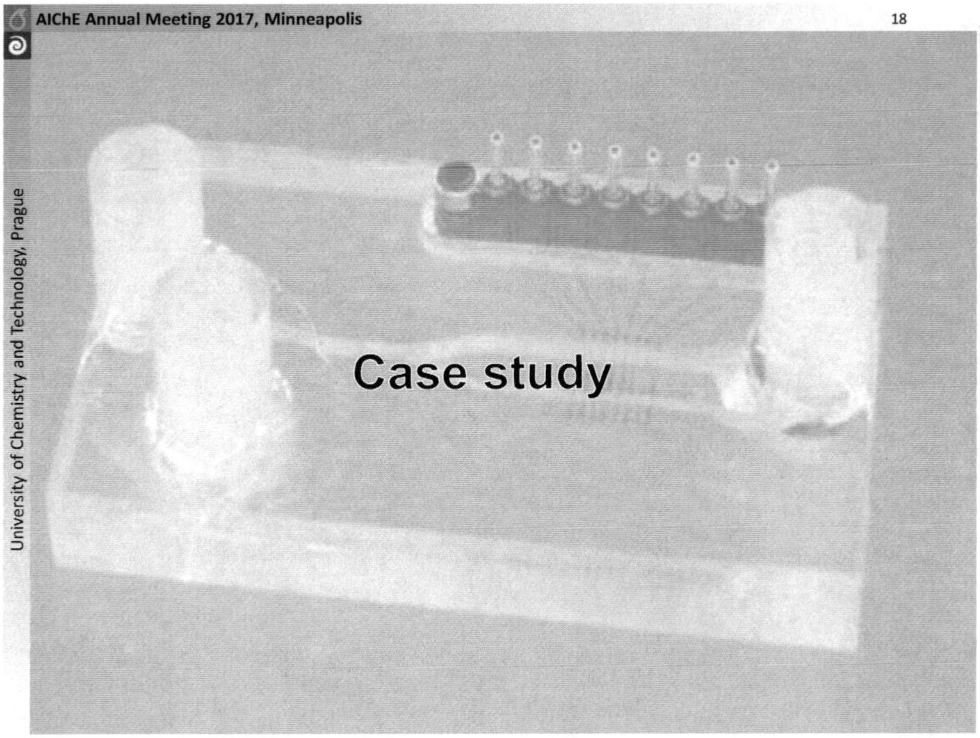
AIChE Annual Meeting 2017, Minneapolis
18
University of Chemistry and Technology, Prague
Case study

Optimization of ATPS composition

$$PenG^- + H_2O \underset{pH \approx 8}{\overset{\text{Penicillin acylase}}{\rightleftharpoons}} 6\text{-}APA^- + PAA^- + H^+$$

$$7\text{-}ADCA + PGME \underset{pH \approx 6.5}{\overset{\text{Penicillin acylase}}{\rightleftharpoons}} \text{Cephalexin} + MeOH$$

ATPS system appropriate for the continuous production and separation of 6-APA has to fulfil the following conditions:

(i) To provide partition coefficients significantly higher than one for the reaction products and significantly lower than one for PA or vice versa.

(ii) To buffer reaction mixture close to the pH optimum of the enzyme.

(iii) To allow for electrophoretic separation of the aqueous phases.

(iv) To avoid high expenses.

Initial ATPS screening

ATPS	Composition
A	PEG 400 - 20 wt %, phosphates - 15 wt%, water - 65 wt%
B	PEG 4000 - 10 wt %, citrates - 12 wt%, water - 78 wt%
C	PEG 4000 - 10 wt %, dextran 40 - 8 wt%, 0.05 M phosphate buffer - 82 wt%
D	PEG 4000 - 10 wt %, dextran 40 - 8 wt%, 0.25 M phosphate buffer - 82 wt%
E	**PEG 4000 - 15 wt %, phosphates - 10 wt%, water - 75 wt%**

Partitioning tests

ATPS	Experiments T2		Experiments T3	
	$L_{6-\mathrm{APA}}$	L_{PAA}	$M_{6-\mathrm{APA}}$	M_{PAA}
A	9.1	25.5	5.6	5.4
B	2.1	3.1	0.36	0.21
C	1.1	1.4	0.23	0.19
D	1.6	2.0	0.22	0.18
E	2.7	6.1	0.13	0.09

The partition coefficient L_i is defined as the ratio of the concentrations of *i*-th component in the top and bottom phases, respectively.

Enzyme partition coefficient M_i

Effect of dextran molecular weight

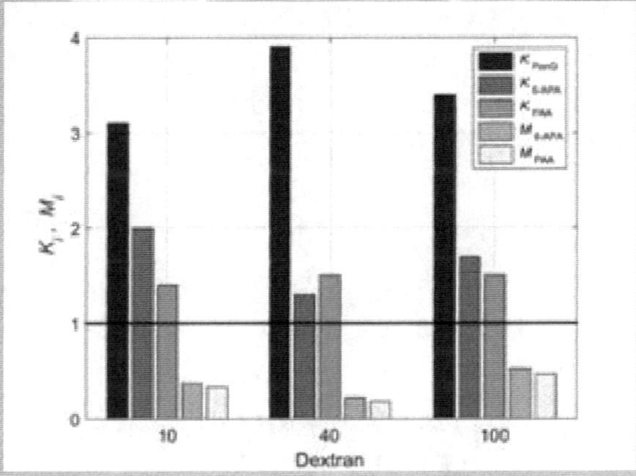

Problems:

- Relatively low values of the partition coefficients for 6-APA
- PEG/dextrans ATPS are more expensive compared to PEG/salt ATPS
- Dextran phase is much more viscous that any salty phase

University of Chemistry and Technology, Prague

Effects of PEG molecular weight and salt content

Enzyme partitioning
Effect of PEG MW

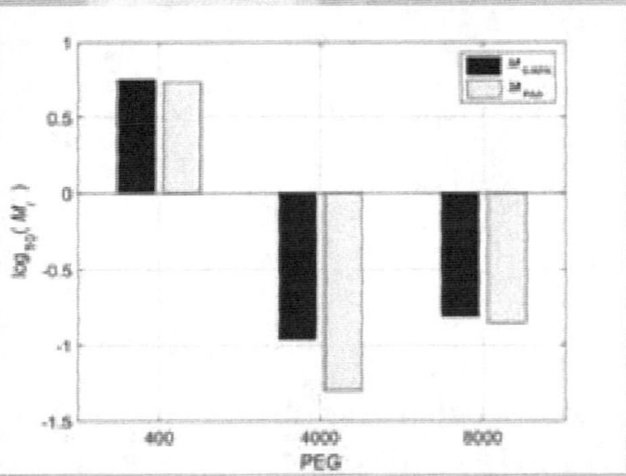

Enzyme and product partitioning
Effect of salt content

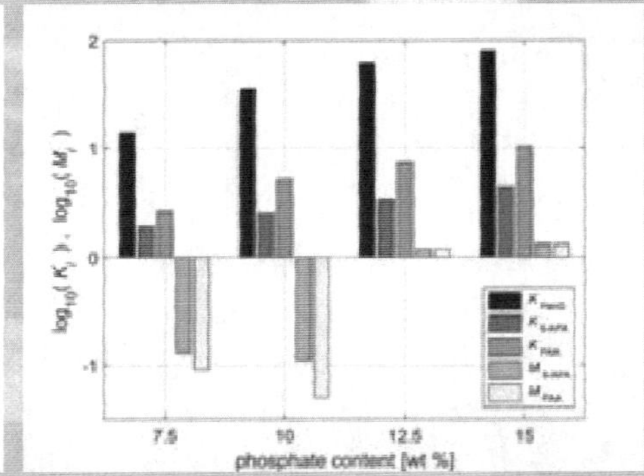

Extraction experiments with ATPS based on PEG 4000, phosphates, water

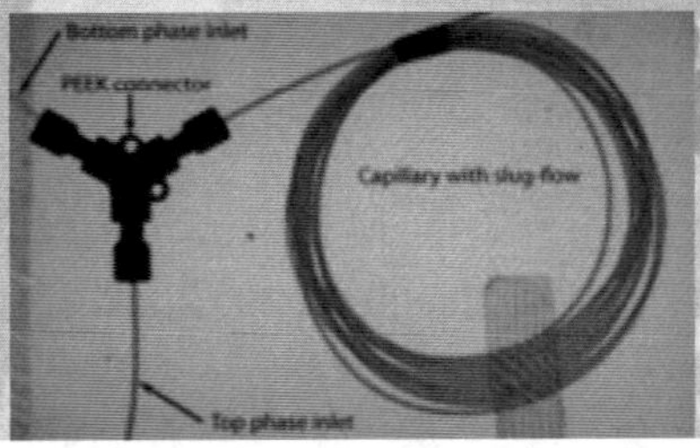

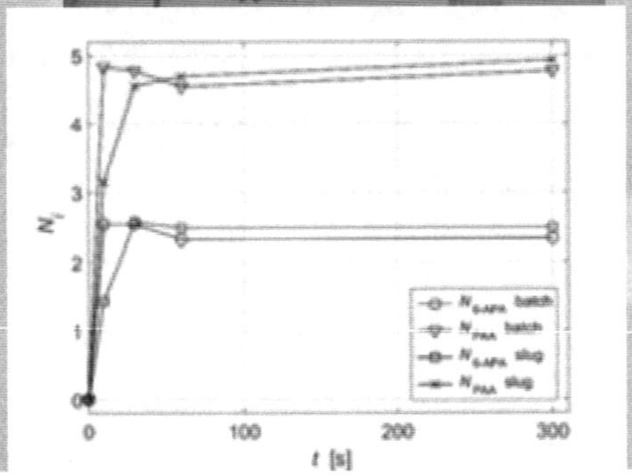

- Extraction of products from the reaction mixture in slug flow and reference batch arrangement
- Equilibrium is attained within 1 min in both arrangements
- Lower energy consumption in slug flow arrangement
- Possible numbering up

Concluding remarks

- The use of ATPS allows for the efficient separation of reaction products (antibiotics or their precursors) from homogeneous reaction mixture
- Enzyme recovery is possible
- Integrated microfluidic reactors-separators are currently under development
  - Droplet flow
  - Parallel flow
- DC electric field can improve their performance by two ways
  - Easy phase separation
  - Enhancement of concentrations of reaction products in outlet streams
- Apparent electrophoretic mobilities of salty droplets in ATPS are at least 10 times higher than those in typical water-oil systems particle-water systems

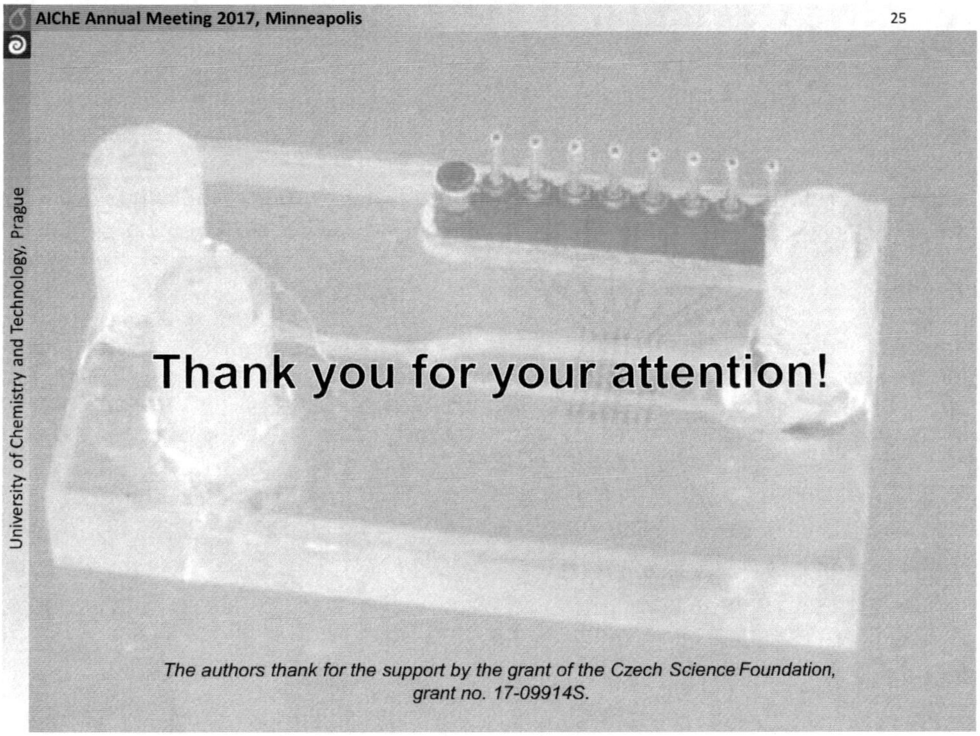
AIChE Annual Meeting 2017, Minneapolis
25
University of Chemistry and Technology, Prague
Thank you for your attention!
The authors thank for the support by the grant of the Czech Science Foundation,
grant no. 17-09914S.

498238 Selective Oxidation of Methane to Methanol in Zeolites: A Window of Opportunity

Abstract:

Despite decades of active research, selective oxidation of methane to methanol remains an unsolved 'grand challenge' of heterogeneous catalysts.[1] Although, previous experimental and computational[2-4] works have studied the mechanism in specific cation-exchanged zeolites; a universal, molecular level understanding of the methane activation and methanol extraction process is not yet available.

In this work, we use periodic DFT calculations to develop a unified framework that describes methane activation reaction and subsequently, enables us to identify strategies for improving the performance of cation-exchanged zeolites. Specifically, we explore a range of active site motifs, transition metal cations, zeolite topologies as well as binding sites for this reaction. Using the activation energy and formation energy of the active site as metrics, we obtain a 'window of opportunity' where we predict active site motifs to be both stable and sufficiently reactive to selectively oxidize methane. Encouragingly, a number of the known active sites (i.e. $[Cu\text{-}OH]^+$, $[Cu\text{-}O\text{-}Cu]^{2+}$, $[Cu_3O_3]^{2+}$, copper-oxide clusters and $[Fe\text{-}O]^{+2}$) lie within this region. More importantly, we also identify $[M\text{-}O]^+$ (M = Ni, Co, Fe), $[M\text{-}OH]^{+2}$ (M = Ni, Co, Fe), and $[bis(\text{-}O)Ni]^{2+}$as alternative target motifs for future catalyst design and synthesis. Finally, the origin of the high methanol selectivity can be explained by the (a) strong binding energy of methanol and (b) formation of surface methoxy species.

The atomistic and mechanistic insight presented in this work will guide future experimental syntheses, and will lead to the design and discovery of highly active cation-exchanged zeolite catalysts for the direct partial oxidation of methane to methanol.

References

1. Olivos-Suarez, A.I.; Szécsényi, À.; Hensen, E.J.M.; Ruiz-Martinez, J.; Pidko, E.A.; Gascon, J., *ACS Catal.* **2016**, 6 (5)

2. Zhao, Z.-J.; Kulkarni, A. R.; Vilella, L.; Nørskov, J. K.; Studt, F., ACS Catal., **2016**, 6 (6)

3. Kulkarni, A.R.; Zhao, Z.-J.; Siahrostami, S.; Norskov, J.K.; Studt, F., *ACS Catal.* **2016**, 6 (10)

4. Latimer, A.A.*; Kulkarni, A.R.*; Yoo, J.S.; Aljama, H.; Tsai, C.; Abild-Pedersen, F.; Studt, F.; Nørskov, J.K., Nat. Mater. **2016** (DOI: 10.1038/nmat4760)

497859 Computational Screening and Characterization of Single-Site Active Porous Surface for Natural Gas to Liquid Feedstock Conversion Process

Abstract:

Catalysts are needed in applications regarding the worldâ€™s fuel and chemical production. To assist these processes, catalysts are used to simplify the chemical reaction by lowering the energy required for reactions to complete. The goal of our project is to identify and characterize specific catalysts through computational research and more specifically the conversion from methane to methanol. Previous work shows that the hydrogen affinity of a structure, the energy needed to move electrons in a reaction, can be directly linked to the reactivity of a catalyst for conversions involving hydrocarbons. By recognizing the correlation between hydrogen affinity, C-H bond activation and transition state activation energy, new optimal catalysts based on a single-site active porous surface can be identified for the target conversion. Also, this work will support the idea of not only decreasing the computational costs for future simulation but also reducing the future experimental efforts by providing the necessary inputs to experiments.

501038 Selective Oxidation of Methane to Methanol or Acetic Acid on Rhodium Single-Site Catalysts at Mild Conditions

Abstract:

The selective activation of the first C-H bond of methane on a metal site at low temperatures and its functionalization to produce oxygenates in a single step using molecular oxygen has been a long-standing goal in catalysis. Recently, the renewed interest in methane utilization has led to extensive studies on several catalytic systems, including low temperature gas phase reactions [1], electrophilic C-H bond activation in acidic media [2] and aqueous- phase conversion using peroxide [3]. However, these systems either require an expensive oxidant, e.g. peroxide or sulfuric acid that poses economic and environmental concerns, or the process is non-catalytic in nature. A complete catalytic cycle of this reaction involves three elementary steps: activation of CH_4 to a M-CH_3 complex, functionalization of the methyl species with OH, halides or CO groups and desorption of the product from the catalyst surface. Fundamental understanding of each of the three steps is necessary to design and further optimize the catalyst system that can activate C-H bond of methane.

In this presentation, we will show that over single-site, isolated Rh/ZSM-5 catalysts, methane can be oxidized directly to acetic acid in the presence of molecular oxygen and carbon monoxide in aqueous solutions under mild pressures and temperatures. The reaction is found to be catalytic and heterogeneous. The Brønsted acidity of the support is crucial for the selective production of acetic acid. On the other hand, methanol formation does not require acid sites and can be realized on Na ion-exchanged ZSM-5, and even on an open oxide support. Thus, anchoring rhodium single ions on the appropriate solid support offers a catalytic system with product tunability between methanol and acetic acid. Single batch production of acetic acid can reach 13,500 μmol/g-cat at more than 4.4% conversion of methane, and a fed-batch process is designed by replenishing limiting reactants to further push the conversion.

Acknowledgment

Financial support by the U.S. Department of Energy (ARPA-e grant# SU0433) is gratefully acknowledged.

References

[1] Alayon, E. M., Nachtegaal, M., Ranocchiari, M., and van Bokhoven, J. A. Chem. Comm. 48, 404 (2012).

[2] Periana, R. A., Taube, D. J., Gamble, S., Taube, H., Satoh, T., and Fuji, H. Science 280, 560 (1998).

[3] Hammond, C., Forde, M. M., and Ab Rahim, M. H., et al. Angew. Chem. Int. Ed. 51, 5129 (2012).

486400 Synergistic Effects between CHA Zeotypes and Basic Metal Oxides in Bifunctional Strategies for Improved Lifetime in Methanol-to-Olefins Catalysis

Abstract:

Methanol to olefins catalysis by CHA zeolites and zeotypes proceeds through a chain carrier mechanism [1,2]. Impurities initiate formation of unsaturated organic moieties within the cavities of the CHA zeolite/zeotype engendering organic-inorganic hybrid chain carriers that serve as scaffolds propagating C-C formation and scission [1,2]. Active chain carriers are comprised of olefinic and monocyclic aromatic organic co-catalysts which undergo C-C scission events via cracking and dealkylation, respectively, to give light olefin products, and C-C formation events via C1 homologation with methanol close the sequence of chain carrier propagation reactions [2,3]. Formaldehyde, formed in bimolecular transfer dehydrogenation of methanol, alkylates active organic co-catalysts instigating their transformation to polycylic aromatic hydrocarbons thereby terminating chain carriers and inducing catalyst deactivation [4]. CHA zeolites and zeotypes confer dynamics of chemical and physical events that ultimately render too large the frequency of formaldehyde-mediated chain carrier termination relative to chain carrier propagation resulting in short catalyst lifetimes.

Here, we describe a bifunctional strategy to impede formaldehyde-mediated chain carrier termination that utilizes rare earth metal oxides to selectively and catalytically decompose formaldehyde, thus increasing lifetime of CHA zeolites and zeotypes for methanol-to-olefins catalysis without disrupting the inherent, high selectivity to ethylene and propylene. We demonstrate that the efficacy of rare earth metal oxides to decompose formaldehyde improves with increasing spatial proximity between the surface of the rare earth metal oxide and the chain carriers confined within the zeolite/zeotype. We evince site requirements for alkanal decomposition on rare earth metal oxide surfaces by combining kinetic, transient chemical probe, and spectroscopic techniques and develop causative and correlative relationships between dynamics of alkanal decomposition and lifetime in methanol to olefins (bifunctional) catalysis using formalisms of reaction-transport phenomena.

[1] J. F. Haw, W. Song, D. M. Marcus, and J. B. Nicholas, Accounts of Chemical Research, 36 (2003) 317-326.
[2] S. Ilias and A. Bhan, ACS Catalysis, 3 (2013) 18-31.
[3] A. Hwang, D. Prieto-Centurion, and A. Bhan, Journal of Catalysis, 337 (2016) 52-56.
[4] A. Hwang, M. Kumar, J. D. Rimer, and A. Bhan, Journal of Catalysis, 346 (2017) 154-160.

482581 Depressed Deactivation of SAPO-34 during Methanol-to-Olefins Procee By MgO

Abstract:

SAPO-34 molecular sieve is the best catalyst for methanol to olefins (MTO) reaction owing to its high ethene and propene selectivity. However, the SAPO-34 molecular sieve is deactivated rapidly due to coke deposition in its small pores, and has relatively short working-life time when used as the MTO catalyst. In order to prolong the working lifetime of the catalyst in MTO reaction, a great deal of efforts, such as optimizing the crystallite size and modifying the pores of zeolite to create hierarchical structure, have been done to enhance mass transfer and to reduce coke formation rate. What's more, modification of SAPO-34 catalyst with different cations by various methods, such as direct hydrothermal synthesis, incipient wet impregnation and ion exchange, can also improve the catalyst working lifetime. However, little attention is paid to prolong SAPO-34 lifetime through physically mixing SAPO-34 with another component.

Reported here in this paper, the rate of SAPO-34 coking was reduced by physical mixing SAPO-34 with MgO, and even the MgO, separated from the SAPO-34 by inert quartz sand layer, was also able to significantly reduce the rate of SAPO-34 coking, and therefore increase the working life time. That is to say, the catalytic performance of SAPO-34 can be significantly affected by MgO which stands "far away" from the SAPO-34. The coke content and coking deposition rate of SAPO-34 were measured, and it was shown that once the SAPO-34 was deactivated, content of coke deposited on the SAPO-34 was the same, regardless with or without the MgO as catalyst promoter. The MgO decreased the rate of SAPO-34 coking and hence increase the working life time. The CO and CO2 were formed when SAPO-34 was promoted with MgO, while no CO and CO2 were detected when pure SAPO-34 was used as the catalyst. MgO by itself shows almost no catalytic reactivity for methanol conversion. But when MgO is sandwiched between layers of SAPO-34 the catalytic activity and production distribution from methanol is significantly altered to favor carbon oxides compared to SAPO-34 by itself. MgO is only catalytically active when sandwiched between layers of SAPO-34, indicating that reaction intermediates are transported from SAPO-34 to MgO and back again. A mechanism for MgO to reduce the coking rate of SAP-34 and the formation of CO, CO2 and H2 was proposed. This research work provides not only a new route for the modification of SAPO-34, but also provided a simple, and convenient way for improving catalytic performance of SAPO-34 used for commercial MTO.

496719 Different Dependency on Copper Oxidation States between H_2 and CO_2 Production during Partial Oxidation of Methanol

Abstract:

The increasing impact of climate change and pollution necessitates the need to replace our current fossil fuel based energy supply with cleaner energy production methods, such as fuel cells. Methanol is of particular interest due to its relatively high energy density and safe handling. Methanol can be directly or inderectly oxidized to power fuel cells. The partial oxidation of methanol (POM, $CH_3OH + 0.5O_2$ â†' $CO_2 + 2H_2$) catalyzed by a copper based catalyst can provide on-board generation of high purity H_2 streams. Although copper-containing catalysts has been invesigated for decades, the reaction mechanism and active sites are still poorly understood. In particular, the chemical nature of the active phase (Cu^0, Cu^+ or Cu^{2+}) is unclear and their impacts on activity and selectivity of POM are unknown.

Results from the study of partial oxidation of methanol (POM) catalyzed by Cu/ZnO were presented here with the objective to identify correlations between POM reactivity and Cu oxidation state. A 30wt% Cu/ZnO nanoparticle catalyst was prepared by a co-precipitation synthesis. Catalyst morphology and elemental distribution on nanoscale were determined by transmission electron microscopy (TEM). The catalytic performance was measured at different O_2 /methanol molar ratios in a home-built micro-reactor. The Cu oxidation state was evaluated at different time-points via ex-situ X-ray photoelectron spectroscopy (XPS). We found that both reactivity for POM and the oxidation state of copper change with reaction time and with O_2 to methanol feed ratio. Most importantly, a strong correlation between H_2 selectivity and (metallic) Cu^0 content of the catalyst was observed. Surprisingly, the CO_2 selectivity was not significantly impacted by the oxidation state of the catalyst, but showed a strong correlation with the O_2 partial pressure. Based on the observed correlations, we propose a mechanism for POM including different bonding configurations of reaction intermediates between metallic Cu and Cu_2O surfaces. We are currently in the process of verifying key reaction steps by firstâ€"principle calculations. The knowledge we gain from this study will benefit the optimization of current Cu-based catalysts which may lead us to a promising methanol based energy economy.

500543 Regioselective Alcohol Ring Opening of Epoxides Using Lewis Acidic Catalysts

Abstract:

Epoxides are versatile chemical intermediates that can be conveniently synthesized from alkenes.[1] Opening of the epoxide ring can be achieved with various nucleophiles like water, alcohols, amines, and thiols yielding synthetically important di-substituted compounds.[2–5] Although heterogeneous acids and bases are known to catalyze this reaction, limitations like low selectivity and/or low stability can limit their scope.[6–8] In this work we have studied silica-based Lewis acidic catalysts with varying pore-sizes and Lewis acid-site strength. We report that these catalysts can selectively open the epoxide ring using alcohols under mild conditions for different systems including epichlorohydrin-methanol. Systematic investigations of the effect of acid site strength, temperature, and pore-confinement of these materials revealed that these factors affect the catalytic activity but not the regioselectivity. In case of straight chain mono-functional epoxides, the regioselectivity increases with an increase in nucleophile size. Overall, we have studied catalytic performance and reusability of different silica-based Lewis acidic catalysts for epoxide ring opening with alcohols.

References:

(1) van der Waal, J. C.; Rigutto, M. S.; van Bekkum, H. *Appl. Catal. A Gen.* **1998**, *167* (2), 331–342.

(2) Tang, B.; Dai, W.; Wu, G.; Guan, N.; Li, L.; Hunger, M. *ACS Catal.* **2014,** *4* (8), 2801-2810.

(3) Kahandal, S. S.; Kale, S. R.; Disale, S. T.; Jayaram, R. V. *Catal. Sci. Technol.* **2012**, *2*, 1493-1499.

(4) Dhakshinamoorthy, A.; Alvaro, M.; Garcia, H. *Chem. - A Eur. J.* **2010**, *16* (28), 8530–8536.

(5) Zhu, X.; Venkatasubbaiah, K.; Weck, M.; Jones, C. W. *J. Mol. Catal. A Chem.* **2010**, *329* (1–2), 1–6.

(6) Brunelli, N. A.; Long, W.; Venkatasubbaiah, K.; Jones, C. W. *Top. Catal.* **2012**, *55*, 432–438.

(7) Takeuchi, H.; Kitajima, K.; Yamamato, Y.; Mizuno, K. *J. Chem. Soc. Perkin Trans. 2* **1993**, 199–203.

(8) Feng, Y.; Lydon, M. E.; Jones, C. W. *ChemCatChem* **2013**, *5* (12), 3636–3643.

BEILSTEIN JOURNAL OF ORGANIC CHEMISTRY

Self-optimisation and model-based design of experiments for developing a C–H activation flow process

Alexander Echtermeyer[1,2], Yehia Amar[2], Jacek Zakrzewski[2] and Alexei Lapkin[*2,§]

Full Research Paper

Open Access

Address:
[1]Aachener Verfahrenstechnik – Process Systems Engineering, RWTH Aachen University, Aachen, Germany and [2]Department of Chemical Engineering and Biotechnology, University of Cambridge, Cambridge, United Kingdom

Email:
Alexei Lapkin* - aal35@cam.ac.uk

* Corresponding author
§ + 44 1223 334796

Keywords:
automated reaction system; C–H activation; design of experiments; flow chemistry; process modelling; self-optimisation

Beilstein J. Org. Chem. **2017,** *13,* 150–163.
doi:10.3762/bjoc.13.18

Received: 31 October 2016
Accepted: 05 January 2017
Published: 24 January 2017

This article is part of the Thematic Series "Automated chemical synthesis".

Guest Editor: I. R. Baxendale

Abstract

A recently described $C(sp^3)$–H activation reaction to synthesise aziridines was used as a model reaction to demonstrate the methodology of developing a process model using model-based design of experiments (MBDoE) and self-optimisation approaches in flow. The two approaches are compared in terms of experimental efficiency. The self-optimisation approach required the least number of experiments to reach the specified objectives of cost and product yield, whereas the MBDoE approach enabled a rapid generation of a process model.

Introduction

The development of manufacturing processes to produce functional molecules, such as pharmaceuticals or fine chemicals, often relies on experience and trial-and-error, rather than on mechanistic process models [1]. The only reason for this is the complexity of chemistry and the duration of time required for the development of good mechanistic models. A game changer in this area is the recently emerged field of automated continuous-flow experiments driven by algorithms for sequential design of experiments (DoE), which significantly reduce the effort in running routine reactions and generating data for optimisation of reaction conditions [2-7]. An illustration of the concept is shown in Figure 1.

Mainly, self-optimisation experimental platforms are used to rapidly obtain optimal reaction conditions using either flow [8-10] or batch experiments [11]. In these cases, the optimisation is driven by the global or target optimisation towards the selected performance criteria. This is rather different from the objectives of model development. In the case of model development, the key criterions are the ability of a model to describe

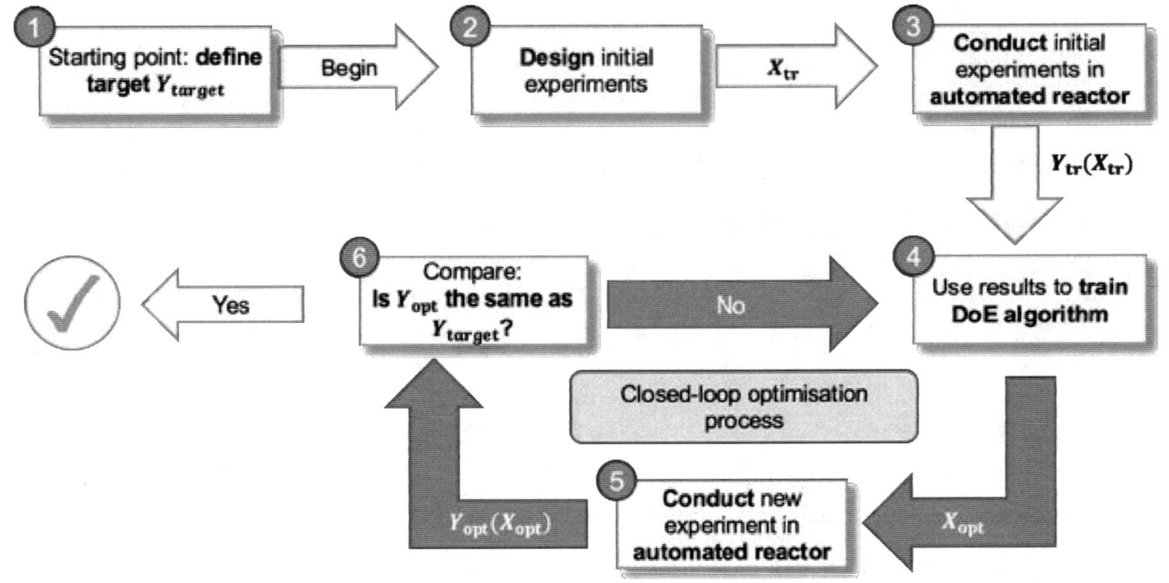

Figure 1: A framework of closed-loop or self-optimisation combining smart DoE algorithms, process analytics, chemoinformatics and automated reactor systems.

the observed experimental data and to predict process performance under unseen conditions. Thus, experiments required for model development are frequently what would be considered as 'bad' experiments in the case of optimisation.

A model-development framework has been demonstrated on the basis of an automated microreactor experimental system for several complex reactions [8,12,13]. The framework uses factorial design of experiments to obtain an initial data set for parameter estimation, followed by an iterative search with online model discrimination and parameter estimation, guided by D-optimal design. In a different approach, transient data from continuous-flow experiments were used to identify parameters of a known mechanistic scheme to discriminate between several alternative model structures and to identify model parameters, but no specific design of experiments method was used [14].

The framework proposed in the present publication is using a model-based design of experiments method (MBDoE) [15-17], which incorporates the model with its parameters, as well as details of the experimental setup, such as measurement accuracy and experimental limitations, to design the most informative experiments. This approach requires some model structures to be known a priori which may restrict the methodology to reactions with known mechanism, or to empirical parametric models. A discussion of how a priori knowledge of chemistry, i.e., reaction mechanisms, is included in self-optimisation and model-development frameworks is not well documented in the literature. Very recently we have shown that a priori knowledge in the form of density functional theory level (DFT) mechanistic calculations can be used to propose process models and to perform in silico design of novel flow processes [18]. In this publication, we present an extension of this methodology, in which an initial process model is developed through a MBDoE methodology coupled with an automated self-optimisation flow system.

This approach was tested on the Pd-catalysed C–H activation reaction of **1** resulting in the formation of an aziridine **2** (Scheme 1) [19]. The reaction was recently discovered [20] and its mechanism studied [21] and later proven [18]. A simplified mechanism is shown in Scheme 2. In the reaction of interest, the starting material **1**, an aliphatic secondary amine, is converted into an intermediate species **B** in a catalytic first step and

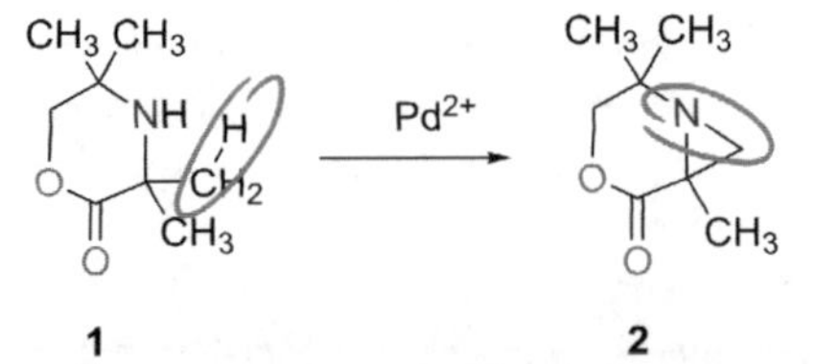

Scheme 1: Catalytic reaction scheme showing C–H activation of an aliphatic secondary amine **1** to form the aziridine product **2** [19,20]. Orange rings show C–H and C–N bonds in the substrate and the product, respectively, indicating the location of the C–H activation.

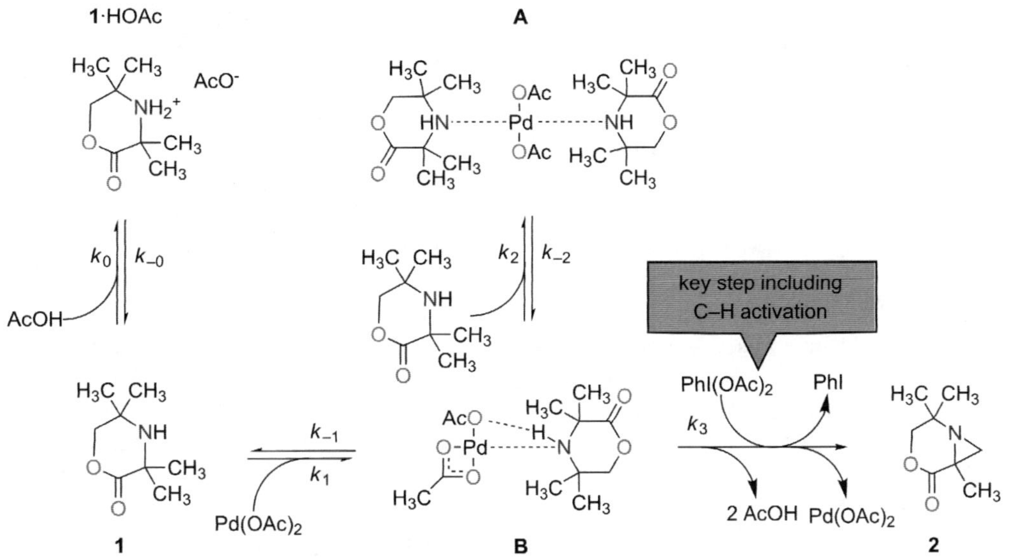

Scheme 2: A simplified reaction mechanism based on literature [21], showing intermediate **B** and the side reaction compounds **1**·HOAc and **A**. The key step includes the C–H activation. **1**: starting material, **1**·HOAc: coordinated starting material, $Pd(OAc)_2$: catalyst, **2**: product, $PhI(OAc)_2$: oxidant.

consecutively transformed to product **2** in the second step, which comprises the C–H activation. In addition to the main reaction pathway, **B** can form the relatively unreactive resting state complex **A**, and compound **1** can also form a coordinated species **1**·HOAc upon protonation with a molecule of acetic acid. This limits the formation of **A** due to reduced concentration of **1**.

Table 1 gives an overview of the a priori knowledge used in this study. Fast reaction steps were lumped into a single one, containing the critical C–H activation, and described by reaction rate constant k_3 in Scheme 2. Empirical information provided constraints of process conditions, such as temperature and concentration ranges, whereas initial values of kinetic parameters were estimated based on a DFT model. Further details can be found in Supporting Information File 1.

Here we demonstrate an MBDoE approach on the basis of the model structure and the initial model parameters from DFT calculations and using automated flow experiments. We then use the obtained process model to develop a surrogate model for optimisation, and compare the different methodologies: classical kinetic modeling approach, MBDoE with automated flow experiments and black-box optimisation in achieving the different objectives of the methods.

Results and Discussion

Experimental system for model development and optimisation in flow

Although a number of experimental systems for self-optimisation were reported in the literature, this number is fairly small and very few examples of using flow experiments for model development are reported [8,9]. In this study a commercial

Table 1: Details of information considered as a priori knowledge in this study and source of this knowledge.

A priori knowledge	Source
reaction mechanism, concentration constraints of species due to degradation of starting material and product.	[21]
Gibbs free energies of reaction, obtained from DFT study.	[18]
target values based on best results from previous experimental study.	[18]
physical constraints (maximum oxidant concentration to prevent crystallisation, maximum temperature to prevent excessive catalyst decomposition).	empirical
technical details of experimental set-up (e.g., variance of gas chromatography (GC) used in variance model for MBDoE, minimum and maximum flow rates).	empirical

Vapourtec R2+/R4 system was used with a standard 10 mL coiled reactor. To save on expensive reagents, reagents and catalyst were injected using 2 mL sample loops, with the solvent being continuously pumped between the reaction slugs. The two employed sample loops were filled with the same reaction mixture (further information on sample preparation is given in Supporting Information File 1) to avoid potential experimental errors due to inaccuracies of generating mixtures with specific concentrations by pumps. Laminar flow through long pipes will necessarily cause dispersion, which dictates the minimum reaction slug length that can be used. This was determined experimentally, which also allowed to develop the method of detection of the reaction slugs (by a flow UV cell) and the protocol for GC sampling. A schematic depiction of the experimental system is shown in Figure 2.

Physical model generation and refinement

The initial model structure and parameters were taken from the earlier published DFT study of the reaction [18]. Performing MBDoE in the process modelling software gPROMS [22] resulted in a design indicating the experimental conditions, the reaction times and the number of samples required in each experiment for the estimation of a particular parameter or the combination of parameters. Table 2 shows the different experiments conducted for estimation of the given parameters. Each experiment refers to a particular composition of the reaction mixture, but with various reaction times for each sample within the experiment. Neither in the MBDoE step for $k_{j,ref}$ nor in the step for $E_{a,j}$ could an experiment be designed for the estimation of all parameters simultaneously. This is likely due to correlations between the parameters, which is common for reaction networks and consecutive reactions. To overcome this problem, sophisticated decoupling techniques and special design criteria considering direct measures of correlation could be used [23,24]. However, as shown by Franceschini and Macchietto, a simple design-by-grouping method can also yield reasonable results [25]. Following this approach experiments were designed for either a single or groups of parameters. Parameters, which showed a maximum in their normalised local sensitivity curves in the same time interval were grouped together. This is reasonable, as a sample taken in this time interval likely yields sensible data for the estimation of the respective parameters. As can be seen from Figure S9 (Supporting Information File 1), all parameters of the same type showed maximum sensitivity in approximately the same time interval. Hence, all possible combinations of single and grouped parameters were tested in the two MBDoE steps and those with the lowest correlation, maximum number of included parameters and a t-value larger than a reference t-value were selected. It is worth noting, that this method overcomes problems with parameter correlations during the experimental design for parameter subgroups and the subsequent estimation. The effects of the neglected parameter correlation may reoccur during the overall parameter estimation, but can be reduced due to the refinement of the parameters in the subgroups.

The best possible design with minimum analytical effort was selected. It can be seen from Table 2 that the t-test is successful for the experiments 1–3. This is not surprising as possible correlations between the parameters are neglected by splitting them into subgroups or even singles. For the experiments 4, 5 and 6 (Table 2) not all parameters pass the t-test. The best possible ex-

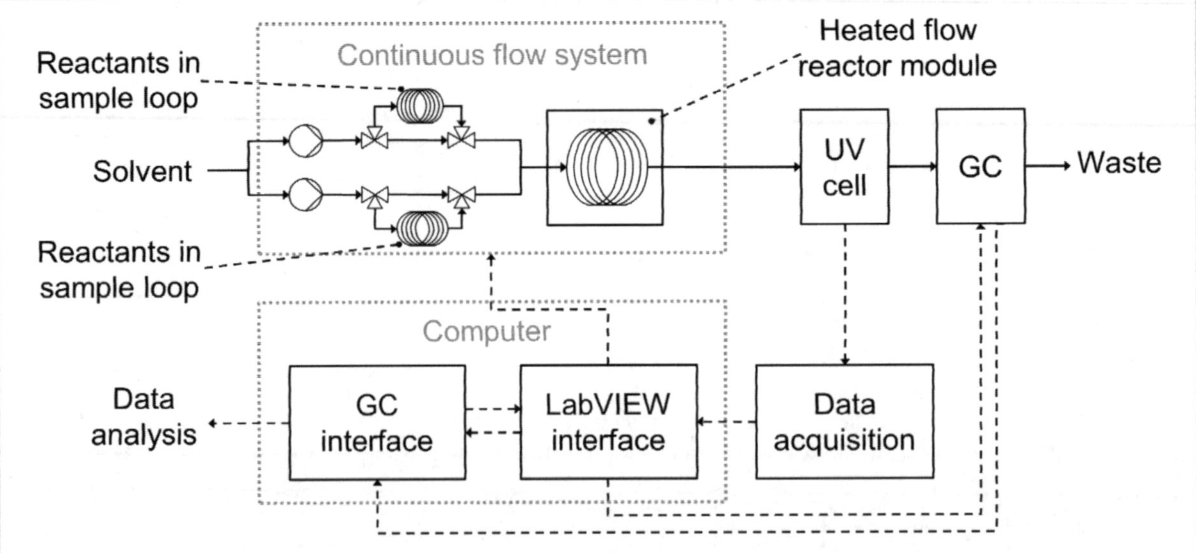

Figure 2: Schematics of the automated continuous-flow system used for model development and 'black-box' sequential optimisation.

Table 2: Results of the MBDoE for kinetic parameters, showing the number of samples needed in the experiment and the statistical t-test results.

Experiments	Parameter(s)	Number of samples	t-value	t_{ref}
1	$k_{0,ref}$	7	76.19	2.92
2	$k_{2,ref}$	6	23.36	2.92
3	$k_{3,ref}$	5	23.36	2.92
4	$k_{0,ref}$, $k_{2,ref}$, $k_{3,ref}$	11	5.34, 0.03, 6.42	1.94
5	$E_{a,0}$, $E_{a,2}$, $E_{a,3}$	11	0.05, 0.04, 2.88	1.94
6	$E_{a,0}$, $E_{a,2}$, $E_{a,3}$	11	0.63, 0.25, 2.33	1.94
7	$E_{a,0}$, $E_{a,2}$	10	2.79, 17.1	2.02
8	$E_{a,0}$, $E_{a,3}$	10	3.99, 46.8	1.94

perimental design was selected. Due to failed estimability analysis, no experiment design included parameters for the reaction $j = 1$. Experimental conditions associated with each experiment sequence are given in Supporting Information File 1, Table S4 and Table S5.

Parameter estimation and comparison of effort

For the investigated reaction in Scheme 2, Zakrzewski et al. generated and validated a kinetic model using a classical kinetic approach [18]. For this they used 38 batch experiments, each comprising approximately 10 sample points at different reaction times, which in total resulted in more than 400 sample points used for the estimation of kinetic model parameters. In contrast to that, we used MBDoE and flow experiments. As Table 2 shows, the MBDoE resulted in 8 experiments with a total of 71 samples required to determine the model parameters. These numbers highlight the benefit of MBDoE for parameter estimation, reducing the consumption of materials, cost and time associated with sample generation. Due to some failed experiments only 64 experimental sample points were used for the model development.

The parameter estimation was employed to obtain estimates of the kinetic parameters $k_{j,ref}$ and $E_{a,j}$, where $j \in \{0, 1, 2, 3\}$ in a two-step procedure using standard solver settings in gPROMS. By applying the initial guesses for the parameters, each experiment was first used to estimate only the parameter for which it was designed, while keeping the others fixed at their current values. Afterwards, all experiments were included in an overall estimation with the parameter values obtained from the previous estimations as new initial guesses to account for possible parameter correlations, which were neglected by grouping the parameters. Even though no experiment design comprised the parameters for reaction $j = 1$ specifically, they were still included in the overall estimation to refine their initial values as much as possible. To avoid stopping the estimation at undesired local optima, several such estimation runs were performed. The final results of the obtained parameter values are shown in Table 3. The final values of parameters $k_{1,ref}$ and $E_{a,1}$ do not differ much from the initial guesses, which is not surprising as the estimability analysis had already predicted a weak influence of $k_{1,ref}$ and $E_{a,1}$ on the model output, i.e., this cannot be estimated with precision. However, this was not necessary, as the parameters do not change the model prediction. Therefore, also the very large 95% confidence interval can be explained. For all other parameters, the difference between initial guess and final value is significant, which might be caused be the simplifications employed for computing the guesses and the uncertainty of the DFT calculation in the exponentially amplified van't Hoff equation. The 95% confidence intervals for the parameters of reaction $j = 0$ are both one magnitude smaller than the final parameter values indicating sufficiently low uncertainty and good significance. The confidence intervals for the parameters of reaction $j \in \{2, 3\}$ are larger than the final parameter value. However, as the values for $k_{j,ref}$ and $E_{a,j}$ can only be positive, this indicates still some level of uncertainty in the parameters estimates. This uncertainty is further revealed by a comparison of the t-values and the reference t-values. For the overall estimation of $k_{j,ref}$ it was impossible to attain t-values exceeding the reference t-value, even though $k_{0,ref}$ comes close. Furthermore, the predicted t-values from the MBDoE, shown in Table 2, could not be reached. In the overall estimation of $E_{a,j}$ three out of four parameters could not be estimated with high statistical significance. Only for $E_{a,0}$ the t-test was satisfied with a t-value close to the predicted one in Table 2. The problem of diminished statistical significance of the estimates is likely due to practical identifiability issues as the measurement data employed for the estimation was affected by experimental errors. Additionally, parameter correlation effects reappeared during the overall estimation making it more difficult to obtain useful results.

Table 3: Results of the parameter estimation showing the final values, initial guesses, 95% confidence intervals (CI) and t-values for each of the 8 parameters.

Parameter	Initial guess	Final value	Units	95% CI	t-values
$k_{0,ref}$	3.019	3.035	L mol^{-1} s^{-1}	±0.396	1.403^{a}
$k_{1,ref}$	2,551,604	2,728,600	L mol^{-1} s^{-1}	±4.817·10^{11}	2.49·10^{-6} ^{a}
$k_{2,ref}$	8,591	16,997	L mol^{-1} s^{-1}	±6.282·10^{5}	0.012^{a}
$k_{3,ref}$	0.001756	0.140378	s^{-1}	±8.175	0.012^{a}
$E_{a,0}$	84,132	128,517	J mol^{-1}	±5.152·10^{4}	2.495^{b}
$E_{a,1}$	45,019	44,941	J mol^{-1}	±1.709·10^{10}	2.63·10^{-6} ^{b}
$E_{a,2}$	59,508	20,995	J mol^{-1}	±3.525·10^{6}	0.006^{b}
$E_{a,3}$	98,831	144,942	J mol^{-1}	±3.517·10^{6}	0.041^{b}

^{a}Refers to t_{ref} = 1.725 and ^{b}refers to t_{ref} =1.688.

Figure 3 shows a comparison of the simulated model response incorporating the final parameter values vs the experimentally observed product concentrations for the experiments 4 and 8 of the MBDoE in Table 2. These show a reasonably good model fit. Only experiments suggested by MBDoE were conducted to generate data for parameter estimation. Thus, as the method did not suggest samples to be taken at reaction times longer than 50 minutes in experiment 4, Figure 3a, or between zero and 24 minutes in experiment 8, Figure 3b, there was no data collected. In total, 8 such experiments were conducted, four for each of the two parameter types (shown in Supporting Information File 1).

Despite the remaining uncertainty in some of the parameters, indicated by the large 95% confidence intervals, the quality of model prediction was considered to be good-enough for the purpose of in silico training of the smart DoE algorithm for target optimisation. Thus, the final parameter values in Table 3 were accepted and used in the model employed for the subsequent in silico target optimisation steps.

Improvement of process conditions using an a priori model and in silico optimisation

Access to automated experimental systems allows to perform black-box sequential optimisation using sequential DoE algorithms. However, if a process model is available, there exist two more options for optimisation: optimisation using the available process model directly, or optimisation using a surrogate model. The latter is frequently used in expensive computer experiments, and in the case of large-scale process simulations, when evaluations of process models is computationally too expensive. In the case of our test reaction the MBDoE approach enabled us to develop a reasonably good process model in a small number of flow experiments. We can use this process model to perform

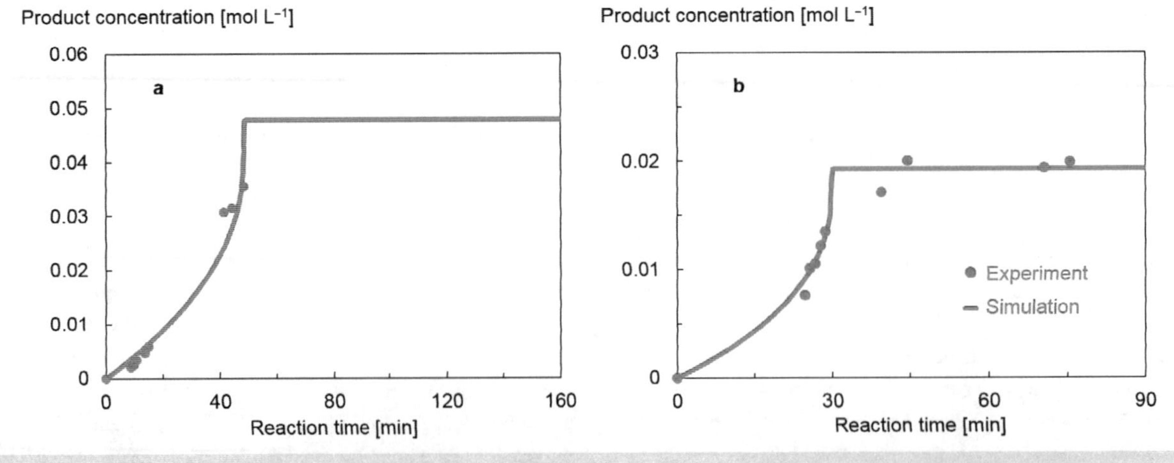

Figure 3: Results of experiments from the MBDoE in Table 2, conducted for parameter estimation, and their corresponding simulated model responses based on the estimated parameters. The only experiments conducted were those calculated by the MBDoE, defined by the sampling times and recipes suggested. (a) Experiment 4 conducted at a reference temperature of 70 °C to estimate rate constants. (b) Experiment 8 used for determining activation energies, at 75 °C.

optimisation. Although this model is not expensive to evaluate we resorted to building a surrogate model, which allowed us to use an efficient target optimisation algorithm we have demonstrated earlier [11,26]. Target optimisation is significantly easier compared to global optimisation as the optimiser is allowed to stop after finding only few conditions that satisfy a target, compared to the problem of finding a global optimal.

The target functions and their corresponding values in the optimisation presented below were the yield, y, of **2** defined in Equation 2, with a desired value of 100%, and a specific cost function given in Equation 3 with a target value of 2,108 £ h kg^{-1}. This cost function was selected to account for the material and energy consumption, and the reaction time with respect to the amount of product **2**. Thus, $cost_{el}$ and $cost_i$ represent the electricity and material costs, whereas W_{el} and $m_{i,0}$ denote the consumed electricity and materials, respectively. The product output $\dot{m}_2$ with

$$\dot{m}_2 = \frac{m_2}{t_{reaction}} \quad (1)$$

combines the amount of product **2** with the necessary reaction time. The cost target value was derived based on a reaction with the shortest reaction time and highest yield, identified from a series of prior experiments (see Table S2 and Figure S5 in Supporting Information File 1 for further details) [18].

$$y = \frac{m_2 - m_{2,0}}{m_{1,0}} \cdot 100\% = 100\% \quad (2)$$

$$cost = \frac{cost_{el} W_{el} + \sum cost_i m_{i,0}}{\dot{m}_2} = 2{,}108 \text{ £h kg}^{-1} \quad (3)$$

The surrogate model was trained on the 64 experimental points obtained for model parameter identification. In case of an unknown mechanism, the experiments for mechanism discovery could also be included in the training set, which leads to a data-efficient approach. However, they might not be the most informative for training the surrogate. The output of the surrogate model is the suggested next experiment to perform, which was used as an input to the process model. Upon reaching the targets in silico after a number of optimisation iterations, the successful input conditions were verified experimentally, to confirm the predictions.

The in silico results for the optimisation target cost and yields are shown in Figure 4 and Table 4. It can be seen that out of 174 iterations, several points were very close to the targets and two optimal sets of conditions satisfy both targets (these iterations are marked with stars). The simulation results of the two identified successful sets of conditions both predict a yield of 98.72%. The experimentally obtained yields in the validation of the two sets of conditions were determined to be >99%, which is caused by the uncertainty of the applied GC method including sample preparation, which lead to ±1% variance in the yield value. The algorithm is not expected to exhibit fast convergence, since it is exploring the reaction space to develop a better statistical model. The physical experiments performed thereafter confirmed prediction of the successfully attained target

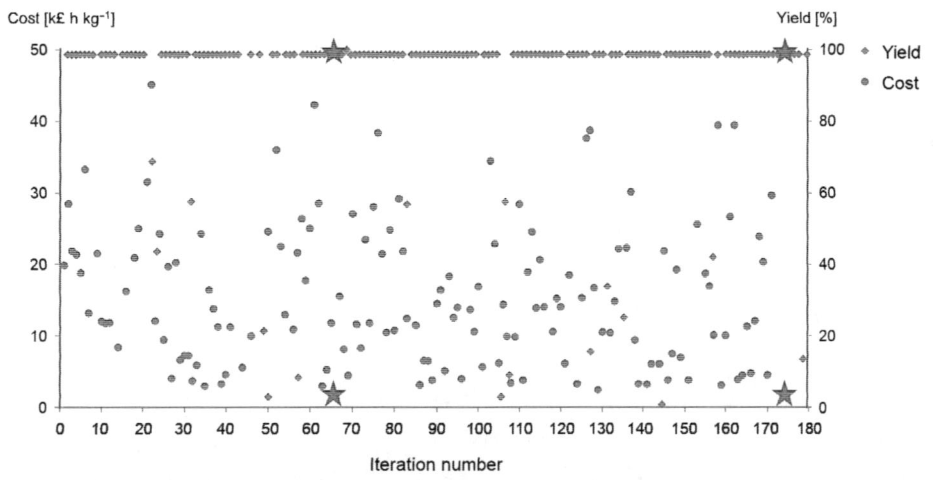

Figure 4: Results of in silico iterations of the multi-objective active learner (MOAL) algorithm [26]. Each iteration produces two resulting values, one for yield and one for the cost function. Targets were 100% for yield and 2,108 k£ h kg^{-1} for the cost function. Green stars signify experiments that satisfy the selected targets.

Table 4: Experimental conditions and results of the experimental validation of the two successful predictions that met the target specifications.[a]

Iteration	T [°C]	$t_{reaction}$ [min]	R_{acid-1}	R_{cat-1}	Yield [%]	Cost [k£ h kg^{-1}]
66	107	9	46.1	0.077	>99 (98.72)	1.79 (1.92)
174	101	10	41.4	0.077	>99 (98.72)	1.79 (1.93)

[a]R_{acid-1}: ratio of the concentrations of acetic acid and compound **1**, R_{cat-1}: ratio of the concentrations of catalyst and compound **1**. Values in brackets are the predicted values by the physical model. Further information regarding the experimental conditions is given in Supporting Information File 1, Table S6.

values. Hence, only the successful predicted experimental conditions X_{opt} were tested in real experiments, which saved time, cost and material, otherwise associated with testing false predicted reactions.

The cost target was more difficult to reach than the yield target, which was already fulfilled after the first iteration and later for most of the proposed experiments. This can be seen by the large fluctuation in the cost values for the proposed experiments over the 174 iterations. One possible reason might be the structure of the cost function with many input variables and strong sensitivity with regards to product amount and reaction time. The reaction conditions shown in Table 4 indicate relatively similar conditions with respect to temperature, reaction time as well as acid and catalyst loading, and do not at this stage demonstrate a case of multimodality.

We have also applied the same target optimisation algorithm for direct improvement of this chemical system as a 'black-box' sequential optimisation. For this approach five experiments were used as a training set, using Latin Hypercube space filling algorithm; the results are shown in Supporting Information File 1, Table S7 (Expt. 1–5). Figure 5 shows results of the initial set of experiments on the left side of the plot. It is noted that two of the five training experiments did incidentally meet the target value for yield at the conditions set. All outputs, regardless whether they reached the desired target values, were included into the training set and the algorithm was re-trained on the updated set once more. This iterative process was conducted six times. After a single iteration, the results of the first suggested set of conditions were already more promising than any of the training points. The target for yield has been met, as perhaps expected as it was optimal for two of the training exam-

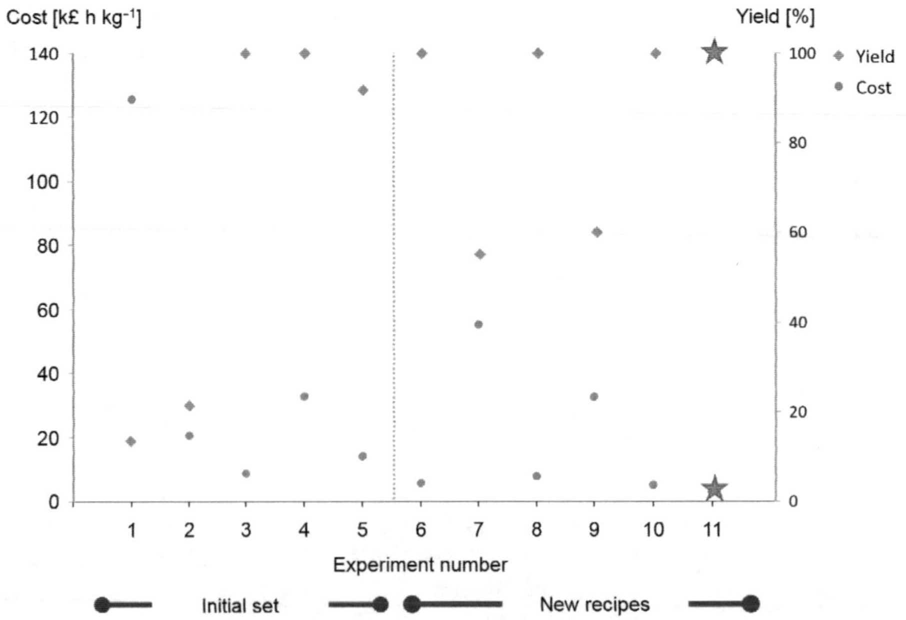

Figure 5: Results of the optimisation driven by a statistical algorithm and in the absence of a physical process model. Results of the training set and of sequential optimisation are shown. Information regarding the experimental conditions is given in Supporting Information File 1, Table S7.

ples as well, and the value for cost has been significantly reduced, getting closer to the pre-defined target. Whilst it is observed that experiments 7 and 9 have a large margin of error with regards to the targets, this is due to the exploration function of the algorithm.

Four of the new suggested experimental conditions achieved high yields with the accepted accuracy and had lower cost scores than even the lowest that was found in the initial set. The recipe at the 6th and final iteration following the training set corresponds to a temperature of 102 °C, reaction time of 15 minutes, acid–substrate ratio of 27.85 and catalyst–substrate ratio of 0.084. With these final conditions, the algorithm converged as both targets were met simultaneously for the first time.

Comparison of the two optimisation approaches

In this work we used the automated flow set-up combined with MBDoE approach to rapidly develop a good-enough process model, which was then used to train a surrogate model and perform a target optimisation. This resulted in two new sets of reaction conditions which both provided better results than the ones obtained previously. In our second approach we used the experimental flow system as a 'black-box' and employed the same statistical target optimisation algorithm to experimentally find the conditions that satisfy the set targets. In this specific case the 'black-box' target optimisation is extremely efficient and found suitable reaction conditions within a very small number of experiments. However, no knowledge about the system was generated. The approach of using automated flow experiments in combination with MBDoE allows to minimise experimental effort compared to classical kinetic studies, but results in a process model that can be directly used in optimisation. This approach is clearly preferred for the cases when a model structure could be identified. There would be many practical cases when due to complexity of chemistry it would be unrealistic to develop a physical model within a reasonable timescale. Then the 'black-box' approach is a viable option.

Conclusion

In conclusion, we present an approach of using model-based design of experiments, based on the first principles model structure, in automated flow experiments, and coupling of the process models with a statistical machine learning based target optimisation. We demonstrate that MBDoE offers a significant potential for efficient and rapid generation of process models in flow experiments. The developed process model enables in silico training of the optimisation surrogate model and cost effective determination of process conditions that satisfy the set performance targets. While this is certainly faster than physical experiments, we also show that the self-optimisation works well when trained on a space-filling method to avoid many necessary experiments for model generation. This results in a set of experiments that reach the pre-defined targets in six iterations, although it does not provide any process knowledge. Hence, a combined approach, leading to generation of a surrogate model and a physical model has unique advantages of rapid optimisation and simultaneous generation of process knowledge.

Experimental

Reaction system and analysis

All reactions were performed in continuous segmented flow using the R2$^+$/R4 system by Vapourtec, see Figure 2. The reaction mixture segments and the solvent were pumped through a 10 mL polytetrafluoroethylene (PTFE) tubular reactor and quenched in an ice bath at the reactor outlet. A minimum segment volume of 2 mL was found to be necessary to avoid dispersion effects in the centre of the segment. The segments were detected using the in-line UV cell, which allowed automatic triggering of the GC (Agilent 6850) to sample the segment at its centre. The flow GC vial was designed by Daniel Geier and Ralf Thelen from the Institut für Technische und Makromolekulare Chemie (ITMC) at RWTH Aachen University and manufactured in-house in Cambridge. GC analysis was performed for product **2** with an accuracy of ±0.0005 mol L^{-1}. Due to the decomposition of **1**·HOAc, **B** and **A** to **1** during sampling following a reaction, the reaction mixture was analysed for species **1** prior to beginning a reaction, with an accuracy of ±0.0003 mol L^{-1}. All communication between instruments was custom-coded in LabVIEW and communication with Vapourtec was via its proprietary Excel interface. Further details of the set-up and the on-line auto-sampling strategy, as well as a protocol for sample preparation and experiment execution are provided in Supporting Information File 1.

Materials

Toluene (Sigma-Aldrich, anhydrous, 99.8%), acetic acid (Sigma-Aldrich, ReagentPlus, ≥99.0%), acetic anhydride (Sigma-Aldrich, ReagentPlus, ≥99.0%), 1,1,2,2-tetrachloroethane (Sigma-Aldrich, reagent grade, ≥98.0%), palladium(II) acetate (Sigma-Aldrich, reagent grade, 98.0%, no further purification steps were applied, the same batch was used for all experiments, stored according to manufacturer's suggestions), (diacetoxyiodo)benzene (Sigma-Aldrich, ≥98.0%) were all used as received. 3,3,5,5-Tetramethylmorpholin-2-one was synthesised as described elsewhere [18].

Model development and analysis

A process model was developed on the basis of the previously discussed reaction mechanism and DFT estimates of the rate constants (see Table S1, Supporting Information File 1) [18].

The kinetic model was developed as a well-stirred tank reactor. The model includes kinetic equations, energy and material balances as well as constitutive equations. A lumped model was created as each reaction segment was assumed to be ideally mixed; thus no excess volume was considered for mixing. As the reaction takes place in a homogeneous liquid phase, and as the tube dimensions are small, there was no need to account for mass transfer effects. For simplicity, the slightly endothermic nature and hence the heat of reaction for the C–H activation was neglected.

The temperature effect in the reaction steps shown in Scheme 2 were expressed using the Arrhenius equation in its re-parametrised form, shown in Equation 4 and Equation 5 [27,28]. This facilitates subsequent parameter estimation by decoupling the kinetic parameters of each reaction. Assuming equilibrium for the three reversible reaction steps in Scheme 2, the 8 kinetic parameters of interest in this reaction sequence were the reference reaction rate constants ($k_{j,ref}$) and activation energies ($E_{a,j}$), where $j \in \{0, 1, 2, 3,\}$ given by Equation 4 and Equation 5).

$$k_j = k_{j,ref} \cdot e^{\frac{E_{a,j}}{R}\left(\frac{1}{T} - \frac{1}{T_{ref}}\right)} \tag{4}$$

$$k_{j,ref} = A \cdot e^{\frac{E_{a,j}}{RT_{ref}}} \tag{5}$$

The temperature-dependent volumetric reaction rates $R^V_{i,j}$ of compound i in the reaction j were modelled by Equation 6, in which $v_{i,j}$ are the stoichiometric coefficients of a compound i in the reaction j, c_i denotes the molar concentration of the compound i, k_j represents the reaction rate constant of the reaction j and $n_{i,j}$ gives the order of the reaction. All reaction steps in Scheme 2 were found to be first order with respect of the participating compounds, except for the oxidant $PhI(OAc)_2$ which is of zero-order dependency [21].

$$R^V_{i,j} = v_{i,j} k_j \prod c_{i,j}^{n_{i,j}} \tag{6}$$

In addition, the overall and the component mole balances, Equation 7 and Equation 8, were written for the process model, where V denotes the reaction volume.

$$\frac{dn}{dt} = V \sum_{i=1}^{n_{comp}} \sum_{j=1}^{n_{reac}} R^V_{i,j} \tag{7}$$

$$\frac{dn_i}{dt} = V \sum_{j=1}^{n_{reac}} R^V_{i,j} \tag{8}$$

The balances were constructed for a single reaction mixture segment, which was assumed to behave as a batch reactor, as samples were taken in the dispersion-free centre of the segment.

For the purpose of calculating the cost associated with heating the system, a steady state energy balance, Equation 9, was established.

$$W_{el} = \frac{Q_{heat}}{\eta_{heat}} = \frac{n_0 \overline{c_{p,0}}(T - T_0)}{\eta_{heat}} \tag{9}$$

were η_{heat} denotes an overall efficiency of conversion of electrical into thermal energy of the reaction mixture segment. This efficiency was determined experimentally for the employed reactor system by measuring the electrical power input to the Vapourtec heating system needed to increase the temperature of a reaction mixture stream with a set flowrate and of known composition, thus with known molar flow and heat capacity, from ambient temperature of approx. 20 °C to a reaction temperature of 70 °C. By inserting these values into Equation 10, the value for the energy-conversion efficiency was calculated and kept constant.

$$\eta_{heat} = \frac{\dot{n}_0 \overline{c_{p,0}}(T - T_0)}{P_{el,Vapourtec}} \tag{10}$$

To complete the process model, simple constitutive equations were applied and initial parameter values were computed. The latter were identified based on Gibbs free energies of reaction for the chemical system, which were obtained from a priori DFT calculations with an accuracy of ±7 kJ mol^{-1} [18]. These values were related to kinetic parameters through the exponential van‘t Hoff equation.

Subsequently, the model was investigated and tested for identifiability to ensure its structural soundness, i.e., that it can be used to uniquely determine its parameters. This was done using an established method, detailed elsewhere [29,30]. It is worth noting that structural identifiability is tested under the assumption of noise-free measurement data and no uncertainty of the model. Thus, it does not necessarily imply practical identifiability as the measurement data for the parameter estimation is usually superimposed by noise and errors [31]. Thereafter, an estimability analysis, based on visual inspection of the local dynamic sensitivity curves [32], confirmed that all parameters except those for reaction with $j = 1$ can be determined with sufficient accuracy. For this reaction step, low sensitivity of the measurable quantity, concentration of compound **2**, was found (see Supporting Information File 1, Figure S9 for further details). The local dynamic sensitivity curves of the remaining parameters were used to identify the time intervals with the maximum sensitivity of the parameters (see Supporting Information File 1, Figure S9), indicating reasonable sample points to obtain sensible measurement data for parameter estimation.

The physical model was implemented in gPROMS. The D-optimality criterion was selected to refine the model using the model-based design of experiments (MBDoE) and parameter estimation suite of gPROMS by employing standard solver settings. Further details on this procedure are given in Supporting Information File 1 together with constraints employed for the experiment control variables in the MBDoE optimisation problem (Equations S8–S21, Supporting Information File 1).

The experimental design and parameter estimation strategy included two steps. In the first step, experiments were designed at a reference temperature $T_{ref} = 70$ °C to eliminate $E_{a,j}$ as a parameter in each reaction j, see Equation 4 and Equation 5). A t-test was used as a statistical method for judging the increase in precision of the predicted parameter and, hence, the statistical significance of the estimates which is attained if the predicted t-value exceeds a reference value t_{ref}. The performed experiments would therefore generate data to enable estimation of the reference reaction rate constants $k_{j,ref}$ independent of $E_{a,j}$. After these parameters have been determined to sufficient accuracy, they were kept constant and experiments were designed at temperatures different from T_{ref} in the second step to obtain data for the subsequent estimation of $E_{a,j}$. The combination of MBDoE and subsequent parameter estimation was repeated twice in both steps to increase parameter precision, whilst keeping the experimental effort low. This was necessary to ensure good parameter improvement in the case of poor initial parameter guesses.

Algorithm for statistical optimisation

One key element of the proposed framework for self-optimisation of reaction conditions is the statistical multi-target optimisation method. For this purpose, the multi-objective active learner (MOAL) algorithm coded in the numerical computing environment MATLAB (v.2015b) was adopted, which combines Gaussian processes as a surrogate model with the concept of mutual information and a genetic algorithm [26]. To apply it to the chemistry under investigation, the algorithm was provided with specified targets Y_{target} for the optimisation and defined experiment design variables $X = [T, t_{reaction}, c_{1,0}, c_{AcOH,0}, c_{cat,0}]$ as the degrees of freedom. The latter were bounded by the corresponding constraints (see Supporting Information File 1 for details). A set of 2,000 randomly generated candidate solutions, uniformly distributed within the allowed design space was employed, because the algorithm works with discrete evaluation techniques for the optimisation. The initial training set $[X_{tr}, Y_{tr}(X_{tr})]$ contained the input variables X_{tr} and measurements of the corresponding target values $Y_{tr}(X_{tr})$, which were adopted from the MBDoE approach. It was updated continuously, so that at each iteration of the algorithm, a new training point was added. Binary Gaussian process classification was included into the algorithm to account for feasible and infeasible solutions in X, hence learning the promising regions of the design space and evolving some internal process knowledge stepwise with each new iteration. An infeasible solution could occur if an experiment fails in the laboratory. Thus, each point in the training was equipped with one more value, providing information on its feasibility (1) or infeasibility (−1). The current limitation of this approach is that it does not automatically distinguish if the experiment failure identifies the region of design space where the specific reaction is not working, or the failure was due to a random fluke and the same experiment, if repeated, would be successful. There is a way of dealing with this problem, which we will implement, when the algorithm will be published.

After classification of the candidate solutions and training samples, the Gaussian process was trained by fitting the so called hyperparameters of its covariance and likelihood functions by maximising the marginal likelihood with a conjugate gradient optimiser. In this way, a statistical surrogate model was created to provide an approximated response surface for the underlying problem of investigation. This response surface was used to evaluate the feasible candidate solutions and subsequently identify a best solution $\tilde{Y}_{opt}(\tilde{X}_{opt})$ with corresponding experimental conditions $\tilde{X}_{opt}$. As only discrete candidate solutions were evaluated by this method, the Non-dominated Sorting Genetic Algorithm-II (NSGA-II) was employed for perturbation of $\tilde{Y}_{opt}(\tilde{X}_{opt})$ to explore the neighbourhood for further refinement of the generated solution. The resulting combination of input and output conditions were subsequently assessed against the targets. If the targets were attained within acceptable tolerance, the results were accepted and the statistical algorithm converged. Otherwise, the results were fed back into the training set and a new iteration was started.

In silico optimisation

The in silico optimisation process was initiated by first training the MOAL algorithm (with the data generated for the purpose of parameter estimation of the physical model described above). This enables the algorithm to construct a statistical surrogate model and suggest a set of experimental conditions which might give results that are closer to the targets. This set is then fed into the physical model to predict what outputs are expected as though the experiment had been conducted. This process is repeated until the required tolerance is reached. A margin of tolerance was included, such that results within 10 and 1.5% of the target values for cost and yield, respectively, were taken as successful, due to the expected difficulty in achieving those targets. Subsequently, the successful reaction conditions found in silico were tested experimentally. In the case of failure, the experimental results were fed back into the algorithm and the target optimisation loop starts again. Otherwise, the algorithm converged and suitable experimental conditions were identified.

In principle, standard optimisation approaches employing the physical model directly to identify optimum operating conditions could be used, but would give poor results in case of uncertainty and restricted validity of the physical model. However, by applying the MOAL algorithm, technical difficulties regarding multi-objective global optimisation can be overcome. Furthermore, the proposed optimisation procedure can deal with potential uncertainties and restricted validity in the physical model. This is achieved by the machine learning functionalities of the MOAL algorithm, which retrain the algorithm not just on the physical model but also on unsuccessful experiments, erroneously predicted as suitable by the physical model. Thus, it obtains information beyond the capabilities of the physical model. An additional point is, that the MOAL algorithm proved to be especially suited for the detection of multiple possible solutions to indicate multimodality, which is challenging for standard optimisation methods, but can yield valuable information in the current case.

Statistical closed-loop optimisation

Statistical target optimisation was performed using the MOAL algorithm. Latin hypercube sampling (LHS) was used to discretise the experimental space initially. An overview of this sampling strategy is laid out for one variable in Figure S10 in Supporting Information File 1. A uniform distribution was taken for the input variables and hence the cumulative distribution function was linear. In this case k is five and N was the number of initial experiments to be conducted, which was decided to be five. This number of initial training experiments was selected as in the previous application of the MOAL algorithm for laboratory optimisation [11], the same number of training samples was applied for an optimisation of two targets, but with a 14 dimensional design space, instead of five dimensions in the current case. Still, the algorithm converged within 17 iterations (including the five training experiments). Hence, we assumed, that for the current work with less design space dimensions, the algorithm would learn the response surface as fast as it was the case for its previous application.

Table 5: Nomenclature.

Symbol	Definition	Units
η_{heat}	heat efficiency	–
$v_{i,j}$	stoichiometric coefficient of component *i* in reaction *j*	–
A	pre-exponential factor in Arrhenius equation	case dependent
$cost$	investigated target value	£ h kg^{-1}
$cost_{el}$, $cost_i$	cost of electricity, cost of material component *i*	case dependent
$c_{i,0}$	initial component concentration	mol L^{-1}
$\overline{c_{p,0}}$	average molar heat capacity	J mol^{-1} K^{-1}
$E_{a,j}$	activation energy of reaction *j*	J mol^{-1}
$k_{j,ref}$	reference rate of reaction in reaction *j*	case dependent
m_i, $m_{i,0}$	mass of component *i*, initial mass of component *i*	kg s^{-1}
n_i, n_0	number of moles of component *i*, total number of moles initially	mol
$P_{el,Vapoutec}$	electrical power uptake of Vapourtec flow system	W
Q_{heat}	heat	J
R	universal gas constant	J mol^{-1} K^{-1}
$R^V_{i,j}$	reaction rate of component *i* in reaction *j*	mol L^{-1} s^{-1}
t, $t_{reaction}$	time, reaction time	s, min
T, T_{ref}, T_0	temperature, reference temperature, initial temperature	°C
V	volume of system	L
W_{el}	electrical work	J
X	experiment design variables	–

Table 5: Nomenclature. (continued)		
$\widetilde{X}_{\text{opt}}$	proposed suboptimal inputs	–
X_{opt_1}	proposed optimal inputs	–
X_{tr}	input training matrix	–
y	yield	%
Y_{target}	target outputs matrix	–
$\widetilde{Y}_{opt}(\widetilde{X}_{\text{opt}})$	proposed suboptimal outputs	–
Y_{opt_l}	proposed optimal outputs	–
Y_{tr}	output training matrix	–

Supporting Information

Supporting Information File 1

Details of experimental set-up and protocols, table of a priori data taken from our previous study, details of model development, MBDoE results, and LHS results.
[http://www.beilstein-journals.org/bjoc/content/supplementary/1860-5397-13-18-S1.pdf]

Acknowledgements

Yehia Amar is grateful to UCB Pharma for his Ph.D. scholarship. Jacek Zakrzewski is grateful to the Department of Chemical Engineering and Biotechnology for funding his Ph.D. scholarship within Catalysis@Cambridge programme.

References

1. Dolas, R. T.; Siddheshwar, S. S.; Somwanshi, S. B.; Merekar, A. N.; Godge, R. K.; Pattan, S. R. *Int. J. Curr. Trend. Pharm. Res.* **2013,** *5,* 137–143.
2. Sans, V.; Cronin, L. *Chem. Soc. Rev.* **2016,** *45,* 2032–2043. doi:10.1039/C5CS00793C
3. Ley, S. V.; Fitzpatrick, D. E.; Myers, R. M.; Battilocchio, C.; Ingham, R. J. *Angew. Chem., Int. Ed.* **2015,** *54,* 2–17. doi:10.1002/anie.201410932
4. Houben, C.; Lapkin, A. A. *Curr. Opin. Chem. Eng.* **2015,** *9,* 1–7. doi:10.1016/j.coche.2015.07.001
5. McMullen, J. P.; Jensen, K. F. *Org. Process Res. Dev.* **2010,** *14,* 1169–1176. doi:10.1021/op100123e
6. McMullen, J. P.; Stone, M. T.; Buchwald, S. L.; Jensen, K. F. *Angew. Chem., Int. Ed.* **2010,** *49,* 7076–7080. doi:10.1002/anie.201002590
7. McMullen, J. P.; Jensen, K. F. *Annu. Rev. Anal. Chem.* **2010,** *3,* 19–42. doi:10.1146/annurev.anchem.111808.073718
8. Reizman, B. J.; Jensen, K. F. *Acc. Chem. Res.* **2016,** *49,* 1786–1796. doi:10.1021/acs.accounts.6b00261
9. McMullen, J. P.; Jensen, K. F. *Org. Process Res. Dev.* **2011,** *15,* 398–407. doi:10.1021/op100300p
10. Parrott, A. J.; Bourne, R. A.; Akien, G. R.; Irvine, D. J.; Poliakoff, M. *Angew. Chem., Int. Ed.* **2011,** *50,* 3788–3792. doi:10.1002/anie.201100412
11. Houben, C.; Peremezhney, N.; Zubov, A.; Kosek, J.; Lapkin, A. A. *Org. Process Res. Dev.* **2015,** *19,* 1049–1053. doi:10.1021/acs.oprd.5b00210
12. Reizman, B. J.; Jensen, K. F. *Org. Process Res. Dev.* **2012,** *16,* 1770–1782. doi:10.1021/op3001838
13. Moore, J. S.; Jensen, K. F. *Org. Process Res. Dev.* **2012,** *16,* 1409–1415. doi:10.1021/op300099x
14. Hone, C. A.; Holmes, N.; Akien, G. R.; Bourne, R. A.; Muller, F. L. *React. Chem. Eng.* **2017**. doi:10.1039/C6RE00109B
15. Walter, E.; Pronzato, L. *Automatica* **1990,** *26,* 195–213. doi:10.1016/0005-1098(90)90116-Y
16. Espie, D.; Macchietto, S. *AIChE J.* **1989,** *35,* 223–229. doi:10.1002/aic.690350206
17. Franceschini, G.; Macchietto, S. *Chem. Eng. Sci.* **2008,** *63,* 4846–4872. doi:10.1016/j.ces.2007.11.034
18. Zakrzewski, J.; Smalley, A. P.; Kabeshov, M. A.; Gaunt, M. J.; Lapkin, A. A. *Angew. Chem., Int. Ed.* **2016,** *55,* 8878–8883. doi:10.1002/anie.201602483
19. Hu, X. E. *Tetrahedron* **2004,** *60,* 2701–2743. doi:10.1016/j.tet.2004.01.042
20. McNally, A.; Haffemayer, B.; Collins, B. S. L.; Gaunt, M. J. *Nature* **2014,** *510,* 129–133. doi:10.1038/nature13389
21. Smalley, A. P.; Gaunt, M. J. *J. Am. Chem. Soc.* **2015,** *137,* 10632–10641. doi:10.1021/jacs.5b05529
22. Process Systems Enterprise. gPROMS. https://www.psenterprise.com/products/gproms. 1997-2005.
23. Franceschini, G.; Macchietto, S. *AIChE J.* **2008,** *54,* 1009–1024. doi:10.1002/aic.11429
24. Franceschini, G.; Macchietto, S. *Ind. Eng. Chem. Res.* **2008,** *47,* 2331–2348. doi:10.1021/ie071053t
25. Franceschini, G.; Macchietto, S. *Ind. Eng. Chem. Res.* **2007,** *46,* 220–232. doi:10.1021/ie060758c
26. Peremezhney, N.; Hines, E.; Lapkin, A.; Connaughton, C. *Eng. Optim.* **2014,** *46,* 1593–1607. doi:10.1080/0305215X.2014.881997
27. Asprey, S. P.; Naka, Y. *J. Chem. Eng. Jpn.* **1999,** *32,* 328–337. doi:10.1252/jcej.32.328
28. Agarwal, A. K.; Brisk, M. L. *Ind. Eng. Chem. Process Des. Dev.* **1985,** *24,* 203–207. doi:10.1021/i200028a034

29. Asprey, S. P.; Macchietto, S. *Comput. Chem. Eng.* **2000,** *24,* 1261–1267. doi:10.1016/S0098-1354(00)00328-8
30. Grewal, M.; Glover, K. *IEEE Trans. Autom. Control* **1976,** *21,* 833–837. doi:10.1109/TAC.1976.1101375
31. Galvanin, F.; Ballan, C. C.; Barolo, M.; Bezzo, F. *J. Pharmacokinet. Pharmacodyn.* **2013,** *40,* 451–467. doi:10.1007/s10928-013-9321-5
32. McLean, K. A. P.; McAuley, K. B. *Can. J. Chem. Eng.* **2012,** *90,* 351–366. doi:10.1002/cjce.20660

498655 Engineering Porous Polymer Hollow Fiber Microfluidic Reactors for SustainableÂ C-H Functionalization

Abstract:

The highly hydrophilic and solvent-stable porous polyamide-imide (PAI) hollow fibers were created by crosslinking of bare PAI hollow fibers with 3-aminopropyl trimethoxy silane (APS). The obtained composite hollow fibers were used as continuous-flow process for heterogeneously catalyzed the aldol and nitroaldol condensations of aromatic aldehydes with acetone and nitomethane at 50 and 40 °C, respectively. Furthermore, the APS-grafted PAI hollow fibers were also functionalized with salicylic aldehyde for binding catalytically active Pd(II) ions through covalent postmodification method. The catalytic activity of the composite hollow fiber microfluidic reactors (Pd(II) immobilized APS-grafted PAI hollow fibers) was tested via heterogeneous Heck coupling reaction of aryl halides under both batch and continuous flow reaction in polar aprotic solvents at high temperature (120 °C) and low operating pressure. XPS and ICP analyses of the startings and recycled composite hollow fibers indicated that the fibers contain very similar loading of Pd(II) implying no degree of catalyst leaching from hollow fibers during reaction. The composite hollow fiber microfluidic reactors showed long-term stability and strong control over the leaching of Pd species.

489997 Comparison of Pd Catalyst Activation Protocols in Suzuki Coupling and Associated Impacts on in-Process Control and Process Performance

Abstract:

Suzuki coupling of an aryl-bromide species to boronic acid forms a pharmaceutical process intermediate. The intermediate contains an atropisomeric stereo-center that affects downstream API. To set the atropisomeric center, a ligated-Pd pre-catalyst was generated in MeTHF or THF prior to being activated with aqueous base; the Suzuki coupling reaction occurs in a biphasic mixture.

Over the course of development, two different processes were developed and implemented on pilot-plant scale: In process [A], the pre-catalyst was activated with aqueous base and subsequently charged to the starting material solution. The resulting active catalyst is a Pd(0) species that is sensitive to oxidation and exhibited degradation over time. In process [B], a small amount of the aryl-bromide starting material (10%) is charged to the pre-catalyst solution. The resulting active catalyst is a Pd(II) species that is not sensitive to oxidation (given its 2+ oxidation state) at room temperature. The two processes generated catalyst species that are different but yet still part of the same catalytic cycle.

The catalyst stability afforded by process [B] had several impacts on in-process control, processing feasibility, and reaction performance. First and foremost, an in-process check was implemented to verify whether sufficient active catalyst was generated to complete the downstream Suzuki coupling; this improvement to the control strategy was not feasible with process [A] given the time instability of the Pd(0) species. Second, the number of processing vessels was reduced from three to two. Oxygen sensitivity during the Suzuki coupling required process [A] to use three separate vessels - one each for the catalyst, base, and starting materials. With process [B], the starting materials could be charged to the catalyst vessel after generating the stable Pd(II) species. Finally, process [B] exhibited faster reaction kinetics at equivalent Pd loading. The improved kinetics allowed for process [B] to be implemented on scale at a lower reaction temperature that resulted in a minor increase in product selectivity.

In summary, the two processes showed marked differences with respect to control, feasibility, and performance. The comparison demonstrates the effects and importance of catalyst stability in a pharmaceutical manufacturing process.

494865 Plasmonic Catalysts As Efficient Catalysts for Cross Coupling Reactions

Abstract:

Plasmonic Catalysts as Efficient catalysts for Cross Coupling Reactions

Andishaeh Dadgar, Farshid Mohammadparast, Mari Andiappan, School of Chemical Engineering, Oklahoma State University, Stillwater, OK, USA

Cross coupling reactions are a diverse set of reactions that are widely used throughout industry. For instance, cross coupling reactions make up 25% of all pharmaceutical reactions. They are also used for many advanced materials such as metal-organic frameworks (MOFs) and OLED materials. The main drawback that limits the reactions though is the wide use of homogenous palladium catalyst at high temperatures.

To overcome this limitation, we propose heterogeneous plasmonic catalysts to drive cross couplings under relatively mild conditions. With plasmonic nanocatalyst, we show improved rate of reaction and stability at a lower temperature than previously seen with conventional Pd catalysts. Furthermore, with in operando measurements we demonstrate the stability of the catalyst as a function of time, solvent, and temperature. The implications of this work are a.) find a suitable replacement for homogenous palladium b.) develop a heterogeneous catalyst capable of long term stability for scale up to a continuous flow system.

494177 Stereoconvergent Tandem Systems Combining Photocatalyzed Isomerization and Enzymatic Reduction for Chiral Building Blocks Synthesis

Abstract:

Abstract

The increasing demand for chiral building blocks, especially those used for the preparation of biologically active compounds, has motivated the development of novel strategies for enantioselective synthesis. Particularly, there is a growing interest in the development of one-pot tandem catalytic systems combining the reactivity of a chemical catalyst with the selectivity engendered by an enzyme. Such tandem systems can achieve levels of chemo-, regio- and stereo-selectivity that are unattainable with a small molecule catalyst. Here, we report the development of a conceptually distinct cooperative chemoenzymatic system by coupling a photosensitizer-catalyzed alkene isomerization in tandem with an enantioselective ene-reductase that selectively reduces one of the isomers for the synthesis of various chiral building blocks of drugs and biologically active compounds. With this system, we can stereoconvergently reduce the isomer mixtures of 2-substitued 1,4-dicarbonyl alkenes, cyano-substituted acrylate esters or ketones, cyano- or ester-substituted acrylamides to corresponding enantiopure reduced products at high yield (>80%) and excellent enantiomeric excess (>95%). These products can be used to easily synthesize *γ*-butyrolactones, *γ*-lactams, *β*-amino acids, *γ*-amino acids, amines and diamines, a structural motif found in many bioactive compounds, such as Phenibut and Baclofen. Importantly, this is the first example of combining photoisomerization with enzymes for enantiopure product synthesis rather than photocatalyst-directed co-factor regeneration in an oxidoreduction reaction. In addition, we show for the first time, that flavin mononucleotide (FMN) can catalyzes two different reactions, alkene isomerization and hydride transfer, in a single system, making our tandem reaction potentially metal-free. We anticipate our system to be a starting point for more synthetic applications when coupled with different enzymes.

A model-based approach for the identification of kinetic models from laboratory reactors

Lu Han, PhD

Pfizer Inc.

Eastern Point Road, Groton, CT

Session: Model Based Integrated Design of Pharmaceutical Drug Substance Processes

Mathematical models of complex chemical reactions in the form of ordinary differential equations provide the basis for systems understanding. In the early phase of modeling, there is a large pool of plausible mechanistic nonlinear models because the underlying mechanism is relatively unknown or uncertain. Thus, building high fidelity and validated kinetic models at limited experimental resource is an important activity. These models in turn can be used to guide process design and decision-making of pursuing batch or continuous manufacturing.

In this work, an application of model-based experimental design to understand the reaction kinetics of a pharmaceutical drug intermediate is presented. The motivation of this model-based guidance is to increase process understanding and control of a reaction impurity, for which purification can require additional post-processing steps downstream. Traditional batch-type experiments were leveraged early on to study the experimental conditions that favored this impurity. Model structures denoting different semi-mechanistic models were postulated and were equally successful at matching the experimental observations at normal operating conditions. Optimal experimental design for model discrimination was utilized to find an experimental condition for the purpose of identifying the most accurate model structure. Eq. 1 (below) outlines the optimization problem,

$$\boldsymbol{\phi}^{MD} = \arg\max_{\mathbf{u}(t)} \int_0^t \sum_{i=1;}^{N_M} \sum_{\substack{i=i+1;\\ i\neq j;}}^{N_M} \left(\mathbf{y}^i(\boldsymbol{\theta}^i, \mathbf{u}(t), t) - \mathbf{y}^j(\boldsymbol{\theta}^j, \mathbf{u}(t), t)\right)^2 dt \qquad (1)$$

subject to:

$$\frac{d\mathbf{z}^i(t)}{dt} = \boldsymbol{f}^i\left(\mathbf{z}^i(t), \boldsymbol{\theta}^i, \mathbf{u}(t), t\right), \qquad (2)$$

$$\mathbf{y}^i(t) = \boldsymbol{g}\left(\mathbf{z}^i(t), \boldsymbol{\theta}^i\right),\ i = 1, \ldots, N_m. \qquad (3)$$

which seeks to maximize the differences in response trajectories between the candidate models with any number of model structures N_M over the duration of the batch experiment. It's worth to note that the response matrix, **y**, is related to the concentration matrix **z** by function **g**. The model parameters, $\boldsymbol{\theta}$, contains the Arrhenius kinetics and absorbance matrix for the species. Parameter estimation was performed only on the kinetic parameters in this work.

At the optimal experimental conditions $\boldsymbol{\phi}^{MD}$, all except for one of the candidate models were rejected on the basis of high lack of fit. The best-fitting model was considered to have the most accurate model structure within the bounds of the design space. This model was subsequently used to design a continuous process in a homogeneous plug flow reactor. The flow process can offer some benefits over the batch process: better residence time control of the reaction, greater control of the process disturbances, real-time control to handle changing inputs, and smaller/more compact of equipment. A new solvent system was designed for this chemistry, in order to solubilize all species. An optimal experimental design procedure was performed to improve the estimation of kinetic parameters. An optimization problem was formulated to minimize a metric of the parameter uncertainty region predicted by the model:

$$\boldsymbol{\phi}^{PE} = \arg\min_{\mathbf{u}(t)}\{\psi[\boldsymbol{V}_{\theta}(\boldsymbol{\theta}, \mathbf{u}(t), t)]\} \qquad (4)$$

where $\mathbf{V}_{\theta}$ is a function of the sensitivity matrix and variance-covariance matrix of the residuals and parameters, subject to the model equations (Eq. 2), response vector (Eq. 3), and experimental bounds. The quality of the optimized experiment design $\boldsymbol{\phi}^{PE}$ was compared to a factorial design. The effect of the different designs on the parameter estimation will be discussed. The applicability of the derived kinetic parameters will be demonstrated on the basis of predictions to a new experiment designed to maximize the discriminability between the two parameter sets.

Through these scenarios, we will show real-life examples of system engineering methodologies applied to pharmaceutical chemistries. Model-based methodologies offer a powerful tool to enhance process understanding of new chemical systems by exploiting the most informative regions of experimentation.

497735 Acid Catalyzed Production of 1,3-Butadiene from Biomass Derived Tetrahydrofuran

Abstract:

Butadiene is an industrially relevant precursor in the production of polymers, where it is typically polymerized with itself or other monomers to produce synthetic rubber. Polymerizing butadiene with styrene and acrylonitrile produces the plastic acrylonitrile butadiene styrene (ABS), while polymers of butadiene and styrene are also used in the production of automobile tires.

Here we present a renewable pathway to produce butadiene from biomass based tetrahydrofuran (THF), proceeding via a dehydrative ring opening of THF (Dehydra-Decyclization). A screening of various solid acid catalysts reveals that an all silica phosphorous containing zeolite (Phosphrous-self pillared pentasil, P-SPP) provides the highgest selectivity to butadiene of ~ 95%.[1] The most significant undesired reaction pathway is the conversion of THF to propene and formaldehyde, which proceeds via a retro-prins condensation reaction. Increased methyl substitution at the 2 and 5 carbon positions of THF leads to significant increase in the rate, without change in selectivity to dienes, while substitution at the 3 carbon position leads to no change in rate. This suggests that a carbenium ion formed at the 2 and 5 carbon positions, is a kinetically relevant surface intermediate. A combination of in-situ titrations of pyridine and a sterically hindered base, 2,6-ditertbutyl pyridine, suggest that the surface chemistry is pre-dominantly Brønsted acid catalyzed.

Figure 1. Dehydra-decyclization in a catalytic packed-bed flow reactor. Reaction of tetrahydrofuran to butadiene at 250-350 °C, WHSV = 1.0 h^{-1} and 5 Torr with catalysts: P-SPP (red), Sn-BEA (blue), amorphous silica-alumina (SiAl, yellow), and ZSM-5 (Si:Al = 140, green). Conversion for each experiment noted numerically within each bar.

[1] Abdelrahman, O. A.; Park, D. S.; Vinter, K. P.; Spanjers, C. S.; Ren, L.; Cho, H. J.; Vlachos, D. G.; Fan, W.; Tsapatsis, M.; Dauenhauer, P. J. *ACS Sustainable Chem. Eng.* **2017**, In Press.

502430 External Surface and Pore Mouth Catalysis in Hydrolysis of Inulin over Zeolites with Different Micropore Topologies and Mesoporosities

Abstract:

Hydrolysis of inulin over zeolite catalysts with various micropore topologies (FER, MFI, MOR, BEA, MWW and FAU) and mesoporosities (pillared MFI (PMFI) and pillared MWW (PMWW)) was studied. The reaction includes cleavage of four types of glycosidic bonds: the terminal glucosyl to fructosyl bond to produce glucose, the terminal sucrosyl to fructosyl bond to form sucrose, and the terminal fructosyl to fructosyl bond and internal fructosyl bonds within the polymer chain to generate fructose. Inulin conversion has shown an initially slow rate followed by pseudo first-order kinetics. Fructose production occurred at a much faster rate than that of sucrose followed by glucose. Rigorous kinetic data analysis showed that the reaction was inclined to proceed on the external surface acid sites of zeolites with cleavage of terminal sucrosyl to fructosyl and terminal fructosyl to fructosyl bonds. The increase in the micropore size in zeolites promoted pore mouth catalysis for the cleavage of the terminal fructosyl to fructosyl bond and terminal glucosyl to fructosyl bond. The mesoporosity in PMFI and PMWW zeolites enhanced external surface and pore mouth catalysis compared to those of their microporous analogues, but did not enable new types of catalytic events. The measured kinetic data were interpreted using a mathematical model based on a network involving parallel and series reactions. Inulin hydrolysis was probed in the transition from external surface to pore mouth catalysis depending on the zeolite topology and mesoporosity in bulky biomass processing. The present study provides guidelines for the utilization of zeolites with variable topologies and porosities for processing inulin and other biomass feedstocks for food and energy applications.

501211 On the Role of Na Cations and Solvent in Glucose Isomerization and Epimerization in Sn-BEA

Abstract:

The efficient conversion of glucose to platform chemicals such as 5-hydroxymethylfurfural (HMF) is crucial in establishing new green chemistry routes to useful products ranging from transportation fuels to chemical intermediates. Since the direct conversion of glucose to HMF is challenging, isomerization of glucose to fructose provides an alternative path that leverages the more favorable dehydration of fructose to HMF. Davis and coworkers[1, 2] have recently demonstrated that the hydrolyzed metal site of the Lewis-acidic zeolite Sn-BEA is particularly selective to glucose-fructose isomerization. However, the selectivity tilts in favor of the epimerization reaction to mannose in Na-exchanged Sn-BEA (Na-Sn-BEA). The reasons behind the selectivity switch have been the source of speculation and debate. Using DFT calculations, we have been able to uncover for the first time the reasons underlying the change in selectivity. Specifically, we shall show that selectivity in Na-Sn-BEA is effectively determined by H_2O molecules vicinal to the active site.

The glucose-fructose isomerization proceeds via a 1,2-hydride shift (HS) from the C2 to the C1 carbon atom, while there are two pathways for mannose formation: a direct pathway via an intramolecular 1,2-carbon shift (CS) from C2 to C1 and an indirect pathway via a second 1,2-HS from C1 to C2 through fructose. We have compared all three reaction pathways both in the absence and presence of water molecules in the vicinity of the active site. In Sn-BEA, the isomerization to fructose is the kinetically dominant pathway, while mannose is formed via the indirect epimerization pathway. In Sn-BEA, water does not influence the selectivity, but it does open the energy gap between the isomerization and epimerization pathways, making the latter even less favorable on account of the extensive solvent re-organization that it entails. In Na-Sn-BEA the lack of water molecules at the active leads to an incorrect prediction that favors isomerization. However, calculations with water at the active site clearly demonstrate that epimerization via the direct 1,2-CS pathway is favored over isomerization, in complete agreement with experiment. Upon coordination to Na^+, the polar H_2O molecules screen the strong electrostatic interaction between the C3 carbon of GLU and Na^+ and thus facilitate the 1,2-CS, which is now the preferred pathway. Our calculations resolve for the first time the role of water and Na cations in the chemistry while rationalizing the experimental data.

References

[1] Bermejo-Deval, R.; Orazov, M.; Gounder, R.; Hwang, S. J.; Davis, M. E., *ACS Catalysis* **2014,** *4* (7), 2288-2297.

[2] Hwang, S. J.; Gounder, R.; Bhawe, Y.; Orazov, M.; Bermejo-Deval, R.; Davis, M. E., *Topics in Catalysis* **2015,** *58* (7-9), 435-440.

Strategies for Improved Design of Nb-Silicates for Propylene Epoxidation

Swarup K. Maiti[1]
Anand Ramanathan[1]
Bala Subramaniam[1, 2]

[1] *Center for Environmental Beneficial Catalysis, Univ. of Kansas (Lawrence, KS)*
[2] *Chemical & Petroleum Engineering, Univ. of Kansas (Lawrence, KS)*

Center for Environmentally Beneficial Catalysis CEBC

THE UNIVERSITY OF KANSAS

AIChE Annual Meeting on "Catalysis with Microporous and Mesoporous Materials", Minneapolis, MN, October 29 - November 3, 2017

Why Propylene Oxide (PO)?

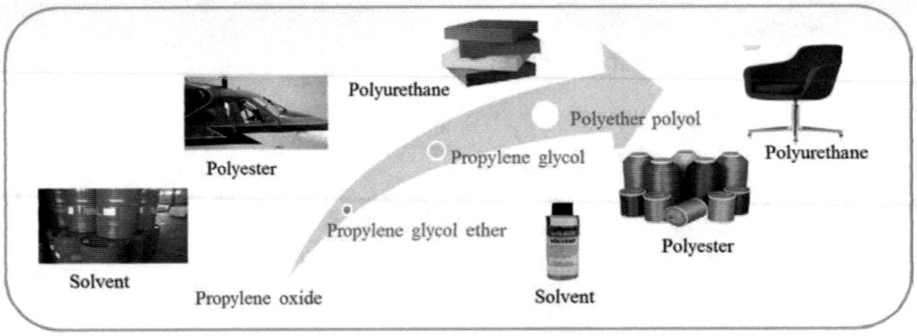

- PO is an important chemical intermediate
- Global PO demand is ~10 million metric tons/y (2013)
- Market is growing @ 3-4 %/y

www.ihs.com., published in March **2016**

The University of Kansas CEBC

THE UNIVERSITY OF KANSAS

Industrial Propylene Epoxidation Processes

Process	Reagents	Conditions	PO Yield (Y) /Selectivity (S)	Disadvantages
Chlorohydrin process	Cl_2, caustic, lime	45-90 °C; 1.1-1.9 bar	S = 88-95 %	• Adverse environmental impact
Hydroperoxide process	Ethylbenzene O_2, W, V, Mo	100-130 °C; 1.1-1.9 bar	Y = 95% S = 98%; 2h	• Dependent on stable byproduct value
DOW/BASF HPPO process	H_2O_2, TS-1	40-60 °C; ~20-25 bar	S = 95%	• PO selectivity ≤95% • TS-1 is expensive • H_2O_2 not fully utilized

Lee et al., *Chem. Eng. Sci.*, **2007**, 62, 7282
Nijhuis et al., *Ind. Eng. Chem. Res.*, **2006**, 45, 3447
Lu et al., *J. Catal.*, **2016**, 342, 173

The University of Kansas CEBC

3

THE UNIVERSITY OF KANSAS

Nb-based Silicates as Epoxidation Catalysts

Catalyst	Substrate	T	t	Conversion	$S_{Epoxide}$
		°C	h	%	%
Nb-MMM-E	Cyclohexene	50	0.5	20	40
#Nb/SiO_2	Cyclohexene	reflux	1	53	94
Nb-β	Cyclohexene	40	0.5	0.12	96
#Nb/SiO_2	Limonene	reflux	1	52	78
Nb-EISA	Cyclohexene	50	6	45.7	77.1
Nb-SBA-16	Cyclohexene	50	2	47.1	61.6
Nb-KIT-5			2	13.2	48.5
Nb-KIT-6			2	26.2	56.4

Solvent in all cases is acetonitrile, # Catalysts prepared by grafting/impregnation

1. Ivanchikova et al., J. Catal., **2015**, 332, 138–148.
2. Tiozzo et al., Catal. Today, **2014**, 235, 49-56.
3. Bregante et al., J. Catal., **2017**, 348, 75-89.
4. Tiozzo et al., Inorg. Chem. Acta, **2015**, 443, 190-196.
5. Ramanathan et al., Ind. Ing. Chem. Res., **2015**, 54, 4236-4242.
6. Ramanathan et al., *Microporous Mesoporous Mater.* **2017**, in press.

The University of Kansas CEBC

4

THE UNIVERSITY OF KANSAS

Goal

- Investigate Nb-based mesoporous silicates for liquid-phase propylene epoxidation under similar operating conditions to those of DOW/BASF HPPO process.
  - Develop Nb-silicate catalyst to maximize PO selectivity and H_2O_2 utilization efficiency.

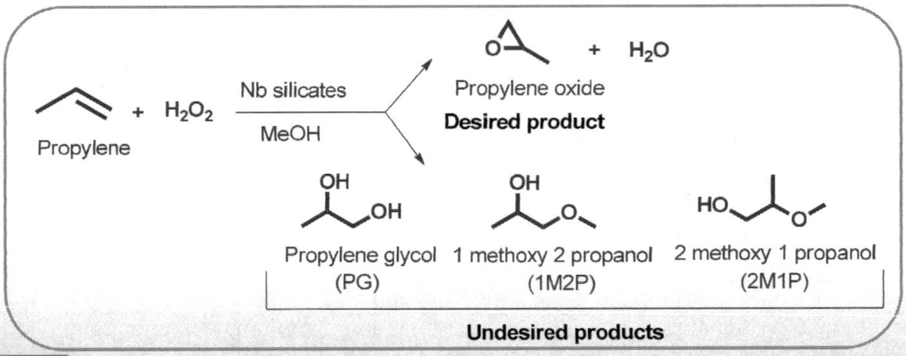

The University of Kansas CEBC

5

KU THE UNIVERSITY OF KANSAS

Catalyst Performance Evaluation

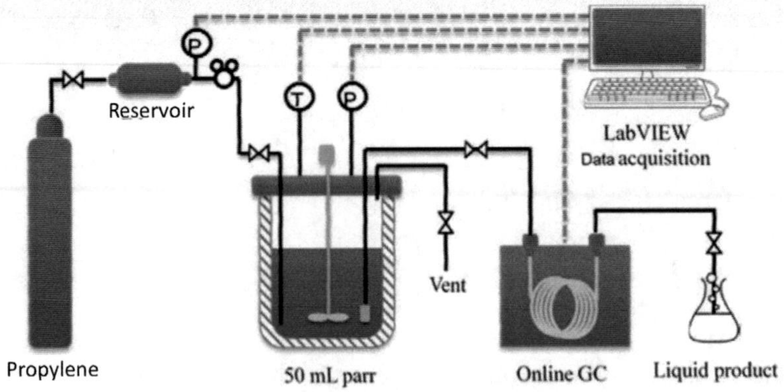

H_2O_2 utilization efficiency

$$U_{H_2O_2} = \frac{Epoxidation\ product\ (moles)}{H_2O_2\ consumed\ (moles)} \times 100\%$$

PO productivity

$$P_{PO} = \frac{m_{PO}}{(time)(m_{metal})}$$

The University of Kansas CEBC

6

KU THE UNIVERSITY OF KANSAS

Compressed Propylene "Expands" Reaction Mixture

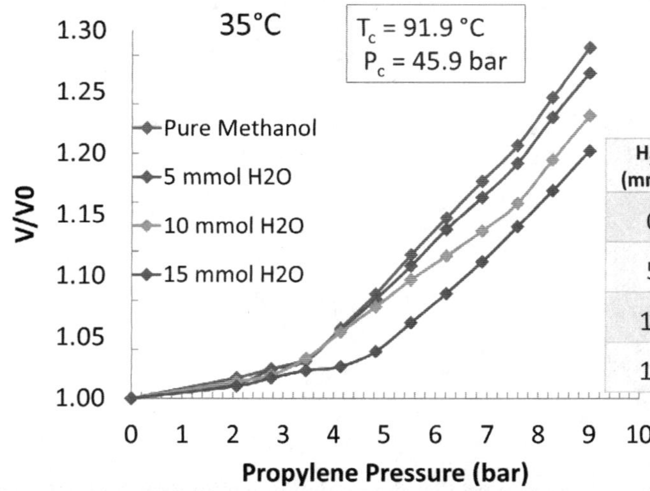

At 9 bar

H_2O (mmol)	Volume expansion
0	29%
5	27%
10	23%
15	20%

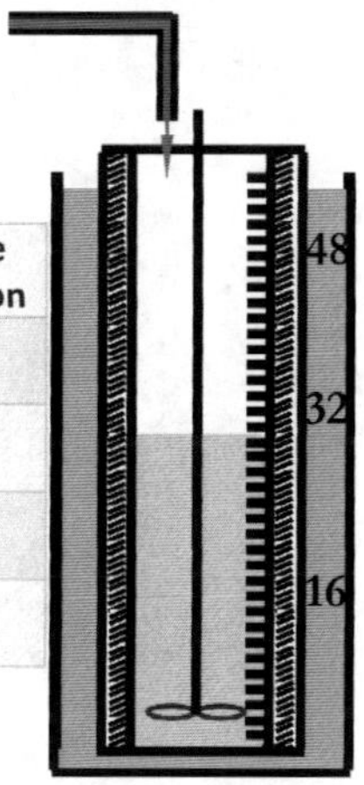

View Cell

Nb-TUD-1 Shows Promise

Entry	Nb wt. %	P_{PO} (mg/h/g)	S_{PO} (%)	U_{H2O2} (%)	X_{H2O2} (%)	Nb leaching %
1	0.8	11680	>99	80.1	17	24.8
2	3.9	4075	>99	89.4	24	20.9

Reaction conditions

- MeOH = 20 g (625 mmol), Acetonitrile = 0.1 g (2.4 mmol),
- H_2O_2 = 10 mmol, catalyst = 0.3 g,
- Propylene pressure = 9.5 bar (maintained constant), T = 35°C, t = 3h,
- Stirrer speed =1400 rpm, catalyst size ~100 μm

Maiti *et al., Ind. Eng. Chem. Res.,* **2017**, 56, 1999-2007
Yan *et al., J. Catal.,* **2016**, 336, 75-84

H_2O_2 Interaction with Simplest Niobyl Silicate

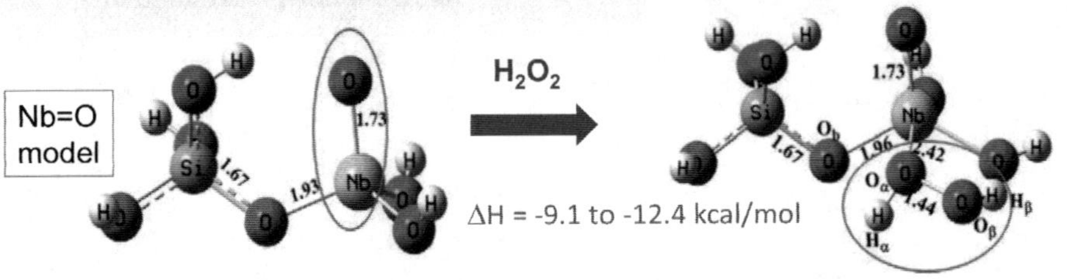

No H_2O_2 Decomposition Over Lewis Niobyl Silicate

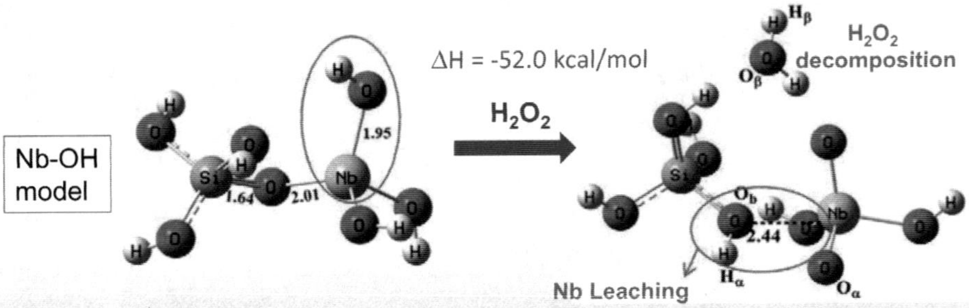

H_2O_2 Decomposition and Nb Leaching Are Tied to Brønsted acidity

The University of Kansas CEBC

Yan *et al., J. Catal.,* **2016**, 336, 75-84

9

KU THE UNIVERSITY OF KANSAS

Desired Catalyst Attributes

- Predominantly Lewis acid sites
- Lower Brønsted acid sites
- Better dispersion of metal species

Strategy

- Prepare and evaluate Nb-Silicates prepared by Evaporation Induced Self Assembly (EISA)

Yan *et al., J. Catal.,* **2016**, 336, 75-84.
Maiti *et al., Ind. Eng. Chem. Res.,* **2017**, 56, 1999-2007.
Wu *et al., J. Catal.,* **2017**, 350, 182-188.

The University of Kansas CEBC

10

KU THE UNIVERSITY OF KANSAS

W-EISA Catalysts Enhance Metathesis Activity

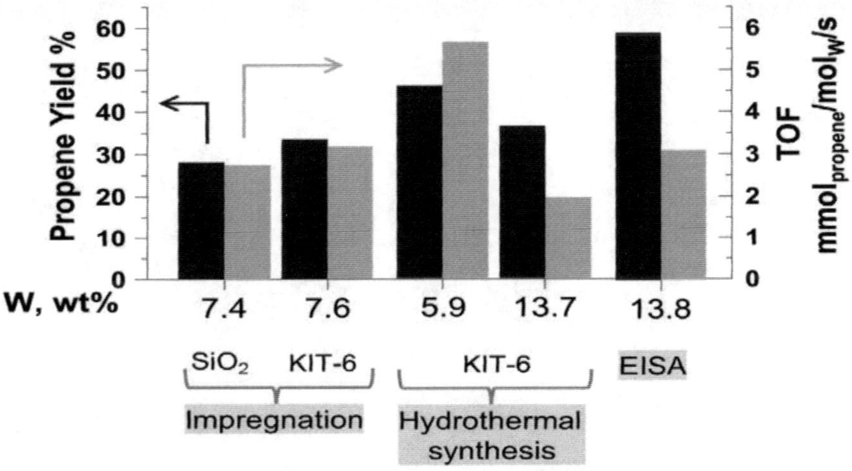

Wu et al., *J. Catal.*, **2017**, 350, 182-188
Yan *et al., J. Catal.,* **2016**, 336, 75-84

The University of Kansas CEBC

11

KU THE UNIVERSITY OF KANSAS

Preparation of Nb-EISA: General Synthesis Procedure

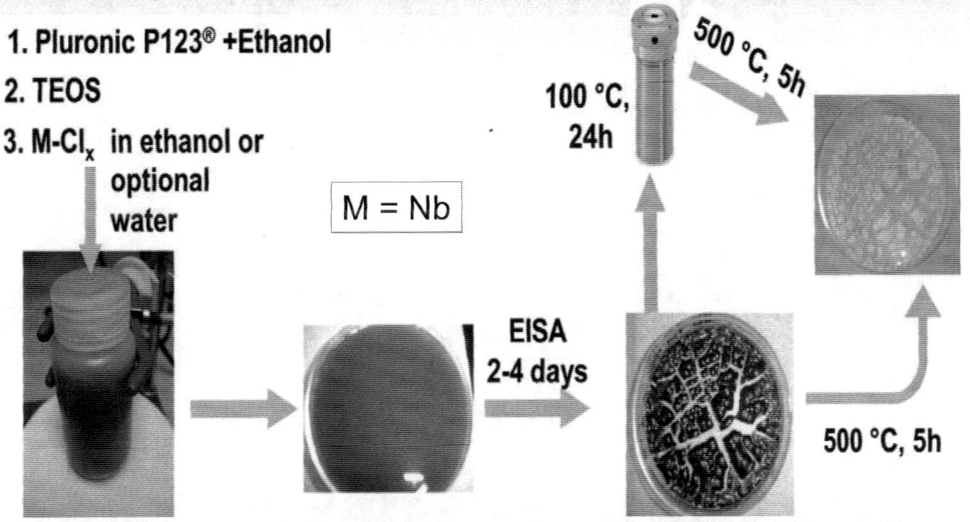

Wu et al., *J. Catal.*, **2017**, 350, 182-188.
Ramanathan et al., *Microporous Mesoporous Mater.*, **2017**, 245, 118-125.
Zhu et al., *Microporous Mesoporous Mater.*, **2016**, 223, 46–52.

The University of Kansas CEBC

12

KU THE UNIVERSITY OF KANSAS

Nb-EISA Materials are Mesoporous with No Detectable Bulk Oxides

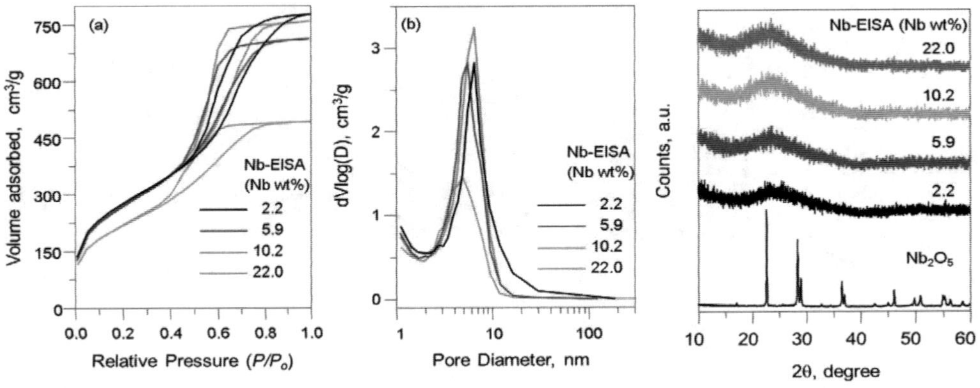

- Type IV isotherm with average pore dia 3.9 - 4.9 nm
- No bulk oxide even at very high loading of 22 wt%

DR-UV-Vis Reveals Isolated Tetrahedral (T) and Polymeric Octahedral (O) Sites

Diffuse Reflectance UV-Vis

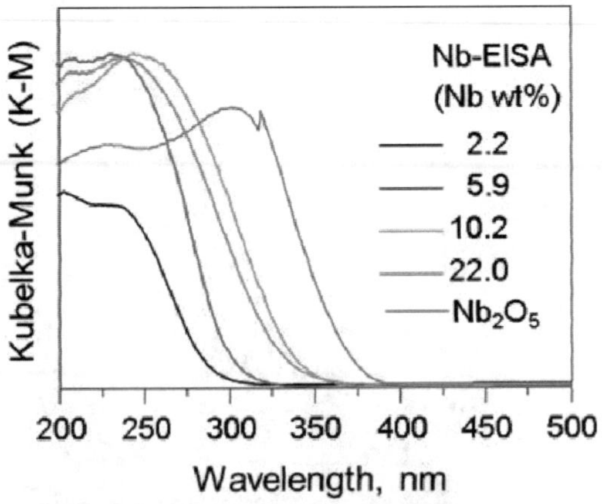

EISA	Isolated	Oligomeric
	Tetrahedral (T, nm)	Octahedral (O, nm)
Nb	220	270

STEM Images Reveal Uniformly Dispersed Nb-O_x Species

Scale bar 50 nm

Scale bar 20 nm

Scale bar 20mm

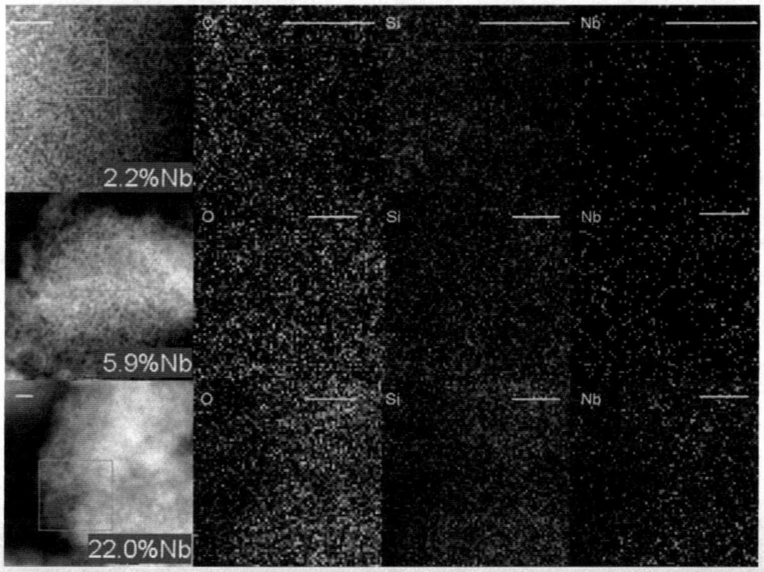

The University of Kansas
CEBC

KU THE UNIVERSITY OF KANSAS

Relatively Low Brønsted Acidity of Nb-EISA Materials

Pyridine FT-IR spectra

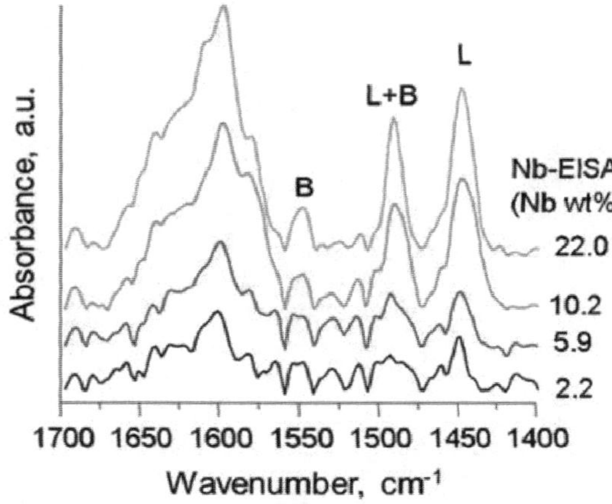

- An increase in Lewis acid sites (1450 cm^{-1}) at higher Nb loading.
- Relatively stable and low Brønsted acid sites (1550 cm^{-1}) at higher Nb loading.

The University of Kansas
CEBC

KU THE UNIVERSITY OF KANSAS

Nb-EISA Performs Better than Nb-TUD-1

Entry	Catalyst	Nb wt.%	S_{PO} (%)	U_{H2O2} (%)	P_{PO} (mg/h/g)	% Nb leaching (mg)
1	Nb-TUD-1	3.9	>99	89	4075	20.9 (2.50)
2	Nb-EISA	1.9	>99	>99	3174	6.3 (0.36)
3	Nb-EISA	4.7	>99	95	2071	5.4 (0.50)
4	Nb-EISA	9.3	>99	>99	1235	3.1 (0.86)

The University of Kansas CEBC

17

KU THE UNIVERSITY OF KANSAS

Parametric Study

[Reaction conditions: T = 35°C, P = 9.5 bar, 10 mmol H_2O_2, 0.3 g Nb-EISA (1.9 wt. % Nb)]

Temperature effect

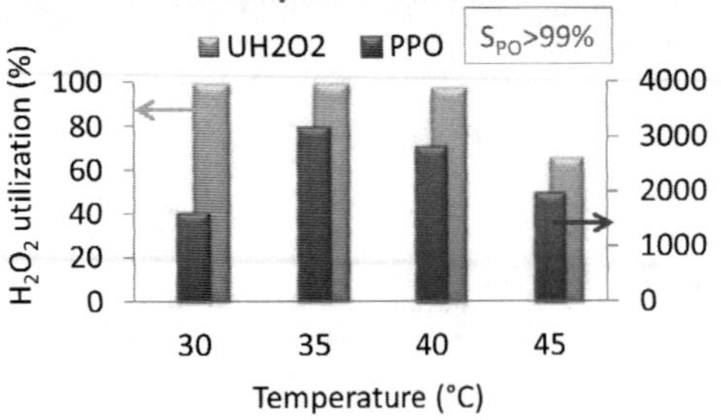

Propylene pressure effect

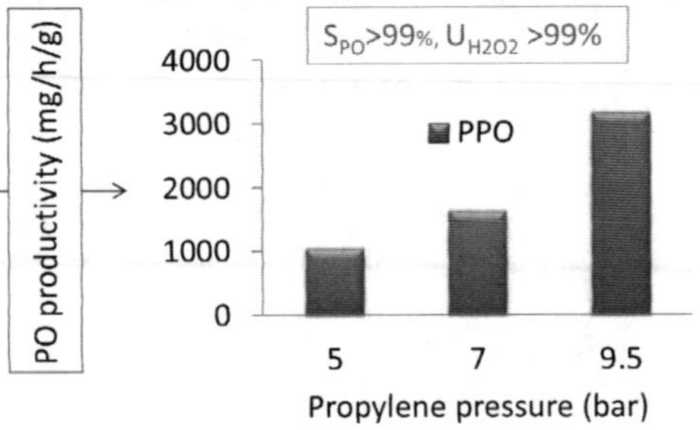

[H_2O_2] effect

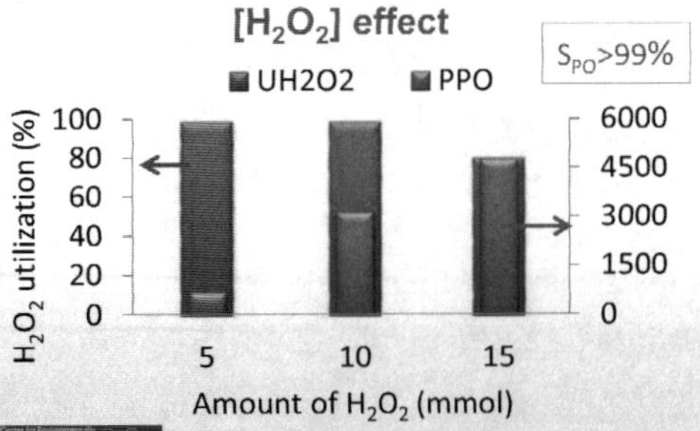

Catalyst loading effect

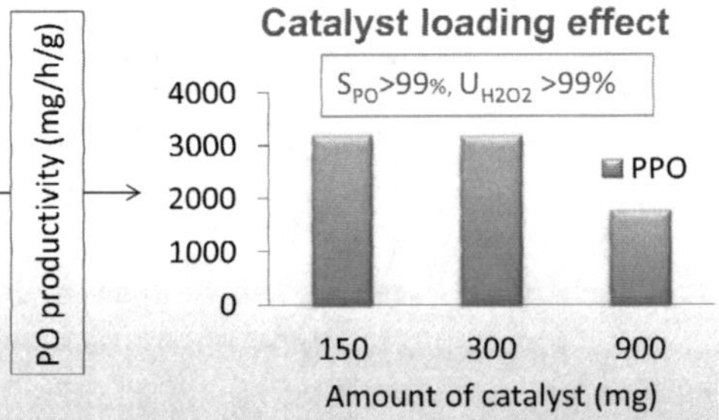

CEBC

18

KU THE UNIVERSITY OF KANSAS

Nb-EISA (1.9 wt.% Nb) Shows Relatively Stable Performance on Recycling

Entry	Y_{PO} (mmol)	P_{PO} (mg/h/g)	S_{PO} (%)	U_{H2O2} (%)	Nb leaching %
Fresh	1.88	3274	>99	>99	6.3
Recycle-1	1.66	4682	>99	98	6.7
Recycle-2	1.55	4960	>99	>99	5.7
Recycle-3	1.46	4525	>99	>99	5.2
Recycle-4	1.38	4279	>99	>99	5.8

CEBC 19 KU THE UNIVERSITY OF KANSAS

Conclusion

- Nb-EISA shows promise as propylene epoxidation catalyst at commercial process conditions
  - >99% epoxide selectivity/H_2O_2 utilization, <5% Nb leaching
- Computational chemistry calculations implicate Brønsted acidity for H_2O_2 decomposition and metal leaching
  - Provide valuable guidance for catalyst design/selection

Path Forward

- Prepare and evaluate NbS-1 catalysts (similar to TS-1 support) to enhance catalyst activity.
- Perform intrinsic kinetics studies with Nb-EISA catalysts for rational reactor design and scale-up.

CEBC 20 KU THE UNIVERSITY OF KANSAS

Acknowledgments

Funding

NSF

EPA
United States
Environmental Protection
Agency

Current CEBC industry partners

ADM

Chevron Phillips Chemical Company LP

GRACE
Talent | Technology | Trust™

INVISTA™

Chemours™

DUPONT®

Reliance
Industries Limited

uop
A Honeywell Company

Thank you!

21

The University of Kansas
CEBC

KU THE UNIVERSITY OF KANSAS

497710 Catalytic Activity of Unique Base Sites Formed Via Nitridation and Subsequent Methylation of Mesoporous Silica SBA-15

Abstract:

Ordered micro and meso porous silica-based materials (e.g., mesoporous silica and zeolite) have attracted wide attention because of their unique structures and attended high surface area, besides their low toxicity, and high chemical and thermal stabilities. Among their applications, their use as heterogeneous base catalysts is one of a challenge. Nitridation of silica materials (i.e., substitution of N-atom with framework O-atom) can create Lewis base sites. Although they have been known to catalyze basic reaction such as Knoevenagel condensation, their basicity is relatively lower than conventional homogeneous catalysts and applicable reactants have been limited.

Here, we demonstrate the preparation of methylated nitrogen substituted mesoporous silica SBA-15 (Me-NSBA-15), and confirmed that the basicity and also nucleophilicity were enhanced to catalyze several reactions.[1),2)] Methylation of N-substituted mesoporous silica SBA-15 was achieved by reacting methyl iodide under reflux condition, and the functionalization was confirmed by elemental analysis and FTIR spectroscopy. Powder XRD showed the ordered porous structure was maintained. The catalytic test was performed with a batch reactor and formed products were analyzed with GC-FID. The Me-NSBA-15 catalyzed the Knoevenagel condensation of benzaldehyde and diethyl malonate (2 mmol C_7H_6O, 2 mmol $C_7H_{12}O_4$, 2.13 mL ethanol, 20 mg catalyst, 353 K for 24 h), which has never been catalyzed with simple N-substitution SBA-15. This suggests that the methylation enhanced the basicity of Lewis base sites, and allowed to deprotonate a higher pK_a reactant (i.e., pK_a of diethyl malonate is 16.4, whereas the pK_a of ethyl cyanoacetate is 13.1). Methylation also enhanced the nucleophilicity. In Morita-Baylis-Hillman reaction of formaldehyde and methyl acrylate (4mmol CH_2O, 2 mmol $C_4H_6O_2$, 2.0 mL 1,4-dioxane, 20 mg catalyst, 423 K for 48 h), the yield of the product (64% at 24 h) was comparable with that of conventional homogeneous catalyst, DABCO (69 % at 24 h), and higher than that of the catalyst without methylation, NSBA-15 (30% at 24 h).

1) Y. Furukawa, M. Ogura, J. Am. Chem. Soc. 136, 2014, 119.

2) K. Sugino, N. Oya, N. Yoshie, and M. Ogura, J. Am. Chem. Soc. 133, 2011, 20030.

486833 The Effect of Water on the Reusability of Aminated Mesoporous Silica Catalysts for Aldol Condensations

Abstract:

Aldol condensations are important reactions which are used in a wide range of processes to produce new carbon-carbon bonds, for example in the pharmaceutical industry for the production of chalcones and in the fine chemical industry for the production of 2-ethylhexanol for PVC plasticizers [1]. A potential novel application in the bio-based industry is the production of liquid fuels from lignocellulosic biomass by upgrading smaller furanic molecules to heavier hydrocarbons [2]. Currently, homogeneous bases such as KOH or NaOH are typically used as aldol condensation catalysts. Despite their excellent activity, these catalysts are dangerous to handle, pose an environmental risk, and are difficult to separate from the final product stream [1]. Hence, innovative heterogeneous catalysts are pursued to establish a more sustainable alternative.

Due to their large pore sizes, large surface areas and the easy incorporation of different types of active sites, mesoporous silicas are promising supports for many types of catalysts. By grafting amine groups on a fraction of the weakly acidic surface silanols, an acid-base aldol condensation catalyst can be obtained. Previously, it has been demonstrated that the residual silanols enhance the catalytic activity of the amine active site by promoting the attack of its nitrogen lone electron pair on the carbonyl group of the reactants [3]. Previous research on acid-base mesoporous silica catalysts has shown that the highest catalytic activity can be achieved when a methyl substituted secondary amine is attached with a propyl linker to the support, and is fully promoted by surface silanols [4]. When the amines are randomly grafted on mesoporous silica, full promotion is achieved when at least 1.7 silanols are available per amine [5].

Although catalyst lifetime is an important factor for the industrial application of a material, no systematic investigation of the deactivation behavior of these cooperative acid-base catalysts has been performed, yet. In this work, a detailed investigation is performed on the factors which induce catalyst deactivation. Using this information, complete catalyst reusability has been achieved.

First, a heterogeneous acid-base cooperative catalyst has been synthesized by grafting of N-methylaminopropyltrimethoxysilane (MAPTMS) on a mesoporous silica gel. The amine density is kept low in order to ensure full promotion by an excess of weakly acidic surface silanol groups. An unpromoted amine catalyst is obtained by, subsequently, end-capping the promoting silanol groups using 1,1,1,3,3,3-hexamethyldisilazane (HMDS). Both materials are characterized by means of nitrogen adsorption-desorption measurements and elemental (CHN) analysis. These catalysts were then employed in the aldol condensation reaction of 4-nitrobenzaldehyde with acetone at 55 °C in a batch-type reactor, using DMSO as a solvent and methyl-4-nitrobenzoate as internal standard. The results of these experiments show that the reaction rate remains constant throughout an experiment, i.e., the slope of the conversion versus batch time graph remains constant. However, subsequent reuse of the spent catalyst yields a lower, albeit also constant, reaction rate. Hence, it seems that catalyst deactivation has occurred upon unloading the catalyst from the reactor. The ratio of the turnover frequency in the second run to that of the first run is used as a measure for the catalyst reusability. For the cooperative acid-base catalyst this ratio amounts to 38 ± 10%, while with the unpromoted amine catalyst a reusability of 77 ± 10% is achieved.

To investigate the nature of the deactivation, the spent catalysts have been characterized via nitrogen adsorption-desorption and CHN analysis. No evidence of pore-blocking was observed. Interestingly, after reaction, the amount of nitrogen in the samples increased with 60% for the cooperative catalyst and with 10% for the unpromoted amine catalyst. This indicates the presence of reaction intermediates on the surface. Raman spectroscopy confirms the presence of aromatic rings, conjugated with double bonds and nitro groups, covalently attached to the active site.

Figure 1: catalytic reaction cycle of the aldol condensation of acetone with 4-nitrobenzaldehyde, catalyzed by a secondary cooperative aminated silica catalyst [3-5]

Possible intermediate species that comply with the organic groups identified using Raman spectroscopy are displayed in Figure 1 as species (VII), (V) and (VI). Species (VII) is the iminium ion of a 4-nitrobenzaldehyde molecule. However, this undesired side-reaction is suppressed by using a large excess of acetone. The other potentially remaining species are the product iminium ion (V) and the product carbinolamine (VI). This iminium ion can undergo hydration to form the carbinolamine and subsequently desorb from the catalytic site. However, when drying the catalyst after a catalytic run, iminium ions that are still present on the surface could dehydrate and form a stable conjugated compound that acts as a site-blocking species, as displayed in Figure 2.

Figure 2: the iminium intermediate species (a) from the catalytic cycle is in equilibrium with its enamine form (b) and can further dehydrate into a stable conjugated species (c) which acts as a site blocker to the amine active site

Adding extra water to the reactor could increase the water-assisted desorption step of the iminium intermediate, as displayed in Figure 1, and thereby reduce its surface coverage upon unloading the catalyst from the reactor. In this work, 0.5 wt% to 3 wt% water is co-fed to the batch reactor. The first and second run turnover frequencies of both the acid-base catalyst, as well as the unpromoted amine catalyst, as a function of the water amount in the reactor, are displayed in Figure 3. When adding at least 0.5 wt% water, the unpromoted amine catalysts are completely reusable, exhibiting a turnover frequency ratio of 92 ± 10%. The maximum reusability achieved with the cooperative acid-base catalysts amounted to 72 ± 10%. This indicates that the promotional ability of the silanol groups irreversibly decreases due to the presence of water, while the activity of the amine sites can be fully recovered.

Figure 3: turnover frequency (s^{-1}) for the aldol condensation of 4-nitrobenzaldehyde with acetone as a function of water in the reactor. Aminated silicas promoted by silanols (red) and unpromoted (blue), empty bars are a first run, full bars are a second run. Error bars indicate a 95% confidence interval. Karl-Fischer titrations have been performed to correct for water in the organic solvents.

Additionally, Figure 3 shows that a small amount of water has a beneficial effect on the reaction rate. In contrast, a high amount of water has a negative effect on the reaction rate. This is directly related to the effect of the water concentration in the catalytic cycle, as shown in Figure 1. Firstly, the enamine (species IV) can react with water leading to the desorption of acetone, which decreases the overall reaction rate. Secondly, by adding water, the rate of the water-assisted desorption step of the iminium ion (V) increases, thereby freeing up catalytic sites and increasing the overall reaction rate.

In this work it has, thus, been shown that the catalyst reusability can be drastically increased with the addition of water, causing a reduction in surface coverage of iminium species that could otherwise dehydrate to a stable site blocking species when the catalyst is dried. For the unpromoted amine catalyst, 0.5 wt% additional water already causes the catalyst to be fully reusable. However the cooperative acid-base catalysts cannot be made fully reusable due to a loss of the promotional ability of the silanol groups. The applicability of aminated mesoporous silica catalysts for continuous aldol condensation processes has, hence, been confirmed in this work by proving that a stable catalyst regime can be obtained by co-feeding a small amount of water. Further testing of these catalysts in a continuous-flow reactor will yield more information about the deactivation behavior of the catalyst with time-on-stream.

[1] G. Kelly, F. King, and M. Kett, "Waste elimination in condensation reactions of industrial importance," *Green Chemistry,* vol. 4, no. 4, pp. 392-399, 2002.

[2] Q. Deng *et al.*, "Efficient synthesis of high-density aviation biofuel via solvent-free aldol condensation of cyclic ketones and furanic aldehydes," *Fuel Processing Technology,* vol. 148, pp. 361-366, 2016.

[3] N. A. Brunelli, K. Venkatasubbaiah, and C. W. Jones, "Cooperative catalysis with acid–base bifunctional mesoporous silica: impact of grafting and co-condensation synthesis methods on material structure and catalytic properties," *Chemistry of Materials,* vol. 24, no. 13, pp. 2433-2442, 2012.

[4] J. Lauwaert, E. De Canck, D. Esquivel, P. Van Der Voort, J. W. Thybaut, and G. B. Marin, "Effects of amine structure and base strength on acid–base cooperative aldol condensation," *Catalysis Today,* vol. 246, pp. 35-45, 2015.

[5] J. Lauwaert, E. De Canck, D. Esquivel, J. W. Thybaut, P. Van Der Voort, and G. B. Marin, "Silanol-Assisted Aldol Condensation on Aminated Silica: Understanding the Arrangement of Functional Groups," *ChemCatChem,* vol. 6, no. 1, pp. 255-264, 2014.

500731 Novel Method for Synthesizing Non-Leaching Functionalized Mesoporous-Silica for Acid Catalyzed-Alkylation Reaction

Abstract:

Novel method for synthesizing non-leaching functionalized mesoporous-silica for acid-catalyzed alkylation reaction

*Santiago J. Umbarila[§], Tuong V. Bui[§], Tawan Sooknoi, Daniel E. Resasco**

[§]*Both authors contributed equally to this work*

School of Chemical, Biological and Materials Engineering, University of Oklahoma, 100 East Boyd Street, Norman, Oklahoma, 73019, United States

**Corresponding Author: resasco@ou.edu*

ABSTRACT

The ordered mesoporous silica materials such as MCM-41 and SBA-15 have attracted special attention because of its chemical and thermal stability[1]. Several studies have been reported modification method of the silica surface through conventional grafting (CG) and co-condensation (CC) functionalization[1–6]. The CG modifies the surface via sylilation of the free silanols with organosilanes. In turn, the CC method uses the TEOS as the a building block for generating the silica structure in the one-pot mixture with organosilanes, which act as functional moieties. The resulting hybrid materials has the characteristic mesoporous hexagonal structure together with organic functional groups anchoring on the surface.

It has been reported that the Si-O bonds obtained from conventional functionalization process are thermally stable, but only under rather mild conditions, which in turn do not cause any loss of active moieties via the leaching or decomposition. There are only a few studies that report the leaching under more severe chemical conditions in either non-polar [5–7] or polar solvents [8]. In this study, the leaching process will be investigated at the molecular level by utilizing a combination of techniques including Fourier-Transform Infrared Spectroscopy (IR), Transmission Electron Microscopy (TEM), Solid-State Nuclear Magnetic Resonance (NMR) of ^{29}Si, ^{1}H and cross polarization ^{1}H-^{29}Si, X-ray Diffraction (XRD).

In addition, a new synthesis method that we identify as "dry-grafting" will be proposed. This method provides a novel and simple way to synthesize stable non-leaching materials useful as catalysts in acid-catalyzed reactions, even under severe conditions (e.g. in polar solvents, at high temperature, etc). The catalytic stability is examined for the alkylation reaction between cyclopentanol (CPOL), acting as the alkylation agent and m-cresol as a substrate. This model compound reaction has practical relevant to biomass conversion and upgrading of pyrolysis vapors into chemicals and drop-in transportation fuels.

Key words: Dry grafting, Leaching, Mesoporous silica, Alkylation and Biomass conversion

(1) Hoffmann, F., et al., *Silica-Based Mesoporous Organic–Inorganic Hybrid Materials.* Angewandte Chemie International Edition, **2006**, 45(20): p. 3216-3251

(2) Huang, Y. Functionalization of Mesoporous Silica Nanoparticles and Their Applications in Organo-, Metallic and Organometallic Catalysis, Iowa University, **2009**.

(3) van der Graaff, W. N. P.; Olvera, K. G.; Pidko, E. A.; Hensen, E. J. M. Stability and Catalytic Properties of Porous Acidic (Organo)silica Materials for Conversion of Carbohydrates. *J. Mol. Catal. A Chem.* **2014**, *388*, 81–89.

(4) Hoffmann, F.; Cornelius, M.; Morell, J.; Fröba, M. Silica-Based Mesoporous Organic-Inorganic Hybrid Materials. *Angew. Chem. Int. Ed. Engl.* **2006**, *45* (20), 3216–3251.

(5) García, N.; Benito, E.; Guzmán, J.; Tiemblo, P.; Morales, V.; García, R. A. Functionalization of SBA-15 by an Acid-Catalyzed Approach: A Surface Characterization Study. *Microporous Mesoporous Mater.* **2007**, *106* (1), 129–139.

(6) Gholami, M.; Talaie, M. R.; Aghamiri, S. F. CO2 Adsorption on Amine Functionalized MCM-41: Effect of Bi-Modal Porous Structure. *J. Taiwan Inst. Chem. Eng.* **2015**.

(7) Testa, M. L.; La Parola, V.; Venezia, A. M. Transesterification of Short Chain Esters Using Sulfonic Acid-Functionalized Hybrid Silicas: Effect of Silica Morphology. *Catal. Today* **2014**, *223*, 115–121.

(8) Canilho, N.; Jacoby, J.; Pasc, A.; Carteret, C.; Dupire, F.; Stébé, M. J.; Blin, J. L. Isocyanate-Mediated Covalent Immobilization of Mucor Miehei Lipase onto SBA-15 for Transesterification Reaction. *Colloids Surfaces B Biointerfaces* **2013**, *112*, 139–145.

500634 Catalytic Conversion of Biomass Derived Ethanol to Advanced Hydrocarbon Fuels and Valuable Chemicals

Abstract:

Global ethanol production is increasing rapidly and anticipated to reach more than 33 billion gallons by 2017. The ethanol market is at a saturation point in transportation sector since its use is limited by the 10-15% blend-wall due to technological and infrastructure constraints. Therefore, it is expected that excess ethanol will be available as a platform molecule for the production of advanced hydrocarbon fuels and value-added chemicals. We will present our findings in exploring conversion of ethanol into hydrocarbon fuels and valuable co-products (e.g., benzene, toluene and xylenes (BTX)). Selective conversion of ethanol over Ga-ZSM-5 will be presented. At 450 °C and 0.4 h^{-1}, Ga-ZSM-5 produces 55% BTX which is two-fold increase over H-ZSM-5 (26% BTX). The BTX yield increases with increase in temperature (300-500 °C) and Ga loading (0.5% to 6.2%). Enhanced hydrogen production and reduced light paraffins (propane and butane) formation are also observed with the increase of Ga loading. The role of Ga in promoting BTX formation and molecular hydrogen production will be discussed. The active gallium species will be further presented from the studies of physical mixture experiments and extensive catalyst characterizations. We will also discuss our findings in alternative zeolite structures for ethanol conversion, and the effect on product selectivity and catalyst stability.

491403 Continuous Hydrogenation of Furfural to Cyclopentanone Using Activated Carbon Monolith Catalysts

Abstract:

Cyclopentanone, defined as a cyclic ketone, plays a key role in synthesis of pharmaceuticals, rubber chemicals, flavors, fragrances, and aviation fuel. The current industrial process in production of cyclopentanone involves a petroleum-based feedstock. During this process, adipic acid undergoes catalytic cyclization at high temperatures to form cyclopentanone. Adipic acid is a petroleum-based chemical that is obtained from oxidization of a cyclohexanol and cyclohexanone mixture in presence of excessive amounts of nitric acid. Generating large amounts of N_2O as by-product is one of the major drawbacks of this adipic acid production process.

Furfural is a green chemical that can be hydrogenated to form cyclopentanone. Acid hydrolysis of hemicellulose-rich biomasses such as corn stover, wheat bran, or corncobs results in xylose. In a second step, furfural can be obtained through catalytic dehydration of xylose. This work focuses on continuous hydrogenation of furfural to cyclopentanone using activated carbon monolith catalysts. Activated carbon monolith catalysts (ACMC), derived from woody biomass, are impregnated with precious metals and used to form cyclopentanone from furfural in the presence of hydrogen. Multiple advantages of using ACMC, in addition to low pressure drop, high mass transfer rates and surface area to volume ratios, are first, the monolith is made from renewable carbon sources, second, high stability in aqueous/acid/base reaction medium and finally, the precious metal can be recovered after combustion of the catalyst. The kinetics of cyclopentanone formation from furfural using ACM impregnated with precious metals will be studied over temperatures of 100 to 180 °C, hydrogen pressures of atmospheric to 300 psig, liquid flow rates of 0.5 to 4 mL/min and hydrogen flow rates of 40 to 100 mL/min. The immediate goal of this work is to develop a kinetic model and rate law, perform comprehensive catalyst characterization correlating structure with function, and determine the longevity of the catalyst.

496742 Soluble Sugars from Cellulose Using Polyvinyl Chloride Derived Hydrochloric Acid As Catalyst

Abstract:

Municipal Solid Waste (MSW) contains a large portion of organic material that has the potential to be converted into valuable fuels through thermochemical conversion. In particular, cellulosic waste-derived sugars are important platform chemicals of biofuels and chemicals. In addition to organic wastes, MSW also contains significant quantities of halogenated plastics like polyvinyl chloride (PVC) that are difficult to recycle using traditional methods. When thermally converted (e.g., pyrolysis, gasification, and incineration), PVC produces corrosive hydrochloric acid (HCl) that can react with organic waste-derived products to form organohalogens and dioxins that create various environmental and health hazards. In the current work, we co-converted cellulose and PVC in tetrahydrofuran (THF) at elevated temperatures and pressures. The reactions were carried out in miniature Swagelok reactors with reaction temperature, time and PVC concentration as the experimental variables. The liquid product was analyzed and quantified with GC/MS/FID, and the solid residue was characterized by pyrolysis using a Frontier micropyrolyzer coupled to a GC/MS/FID. The results showed that HCl released during the thermal decomposition of PVC acts as an effective catalyst in the polar aprotic solvent to promote cellulose decomposition to soluble sugars. Levoglucosan was found to be the most abundant product. The highest monomeric sugar yield was 42% of cellulose at 335 °C and 8 minute reaction time with 2.5% cellulose weight equivalent PVC. The liquid products also contained up to 46% of soluble oligomeric sugars and small amounts of hydroxymethylfurfural and furfural. Pyrolysis of the solid residue also produced a high yield of levoglucosan (~22%, residue weight basis), indicating the residue is unconverted or partly decomposed cellulose. Chlorinated organics were nearly absent in both the liquid product and the pyrolyzates of the solid residue, suggesting the chloride content in PVC is converted to the solvent-dissolved HCl. The present study aims to explore an environmentally friendly approach to convert mixed solid wastes into value added products.

495204 Effect of Temperature and Vapor Residence Time on Product Distribution of High Density Polyethylene Fast Pyrolysis

Abstract:

In search of more sustainable options for plastic waste disposal, thermochemical conversion to produce alternative fuels or chemicals appears as an outstanding alternative due to its environmental benefits such as having a lower carbon footprint than fossil fuel extraction.[1] One method of thermochemical conversion is fast pyrolysis, the thermal degradation of plastic waste in the absence of oxygen. Fast pyrolysis at high temperatures, with reaction times shorter than 10 s, has been used to produce hydrocarbon products from plastic waste that can be converted into fuel or chemical intermediates. This study analyzes the degradation products of high density polyethylene (HDPE), the most abundant plastic waste, with respect to temperature and vapor residence time. Through the use of a novel, high-temperature reactor extension to an existing micro-pyrolysis unit (CDS Analytical 5200 HP Pyroprobe), the residence time at each temperature was varied by adjusting the helium carrier gas flowrate. Residence times were varied from ~0.0 to 5.6 s (25mL/min He to 100mL/min He). The temperature of the micropyrolysis unit and the reactor extension was established at three different temperatures (650°C, 675°C, and 700°C) for each set of residence times. The pyrolysis products were analyzed in a ThermoFisher K8880181 Trace Gas Chromatograph coupled to a DSQ II mass spectrometer. The observed chemical species were alkanes, alkenes, alkadienes and aromatics and the carbon content of the pyrolysis products ranged from 2 to 32 atoms per molecule. At constant reactor temperature of 700°C, the slight variation in residence time from ~0.0 to 1.3 s greatly shifted the product distribution towards lower molecular weight compounds (LMWC, C15 and below) as opposed to higher molecular weight compounds (HMWC, C16 and above). Without extended reaction, 25% of the products were LMWC, while with extended reaction, 96% were LMWC. When the pyrolysis temperature was lowered to 675 °C the amount of LMWC decreased 6% to 90%. These results illustrated the trend that as temperature and residence time increased, there was a greater abundance of lower molecular weight compounds in the product distribution, mostly alkenes with some alkanes but almost lacking dienes. Changes in chemical species distribution were also observed. An increase in the composition of aromatic compounds was exhibited at longer residence times and higher temperatures whilst the opposite trend occurred for high molecular weight alkenes and alkadienes. These results show a promising pathway for HDPE waste as a potential source for hydrocarbon fuel fractions and chemical intermediates.

References:

1. Gracida-Alvarez, U.R.; Keenan, L.M.; Sacramento-Rivero, J.C. & Shonnard, D.R. Resource and Greenhouse Gas Assessments of the Thermochemical Conversion of Municipal Solid Waste in Mexico. *ACS Sustainable Chemistry and Engineering* **4**, 5972-5978 (2016).

497522 Effects of Acid-Pretreatment on Co-Pyrolysis of Biomass and Plastic

Abstract:

Over 521 million pounds of agricultural plastics are annually used in US farm field to protect plants and improve crop yield. Due to the non-biodegradability of the plastics, disposal of the used plastics is problematic. The agricultural waste plastics are mostly burned in farm land, buried or sent to landfill, which cause many environmental problems and health hazards. Since waste plastics and crop residues are often present together in the farm field, co-pyrolyzing the two streams of materials could be an attractive solution to address the environmental issue while promoting energy recovery from waste plastics. In the present study, the methods for improving the co-pyrolysis efficiency of biomass and plastic are explored. It is known that the inorganic impurities presenting in cellulosic biomass favor undesired ring opening and charring reactions to reduce bio-oil yield. Acid washing and acid infusion of biomass prior to pyrolysis are known as effective methods to overcome the detrimental effect of inorganic impurities to increase bio-oil yield. The effect of acid pretreatment on co-pyrolysis of biomass and plastic is studied. Corn stover biomass pretreated with sulfuric acid (both acid washed and acid infused) were co-pyrolyzed with polyethylene in a tandem micropyrolyzer with and without zeolite catalyst. During non-catalytic pyrolysis, levoglucosan yield increased from 0.5wt% with non-treated corn stover to 9.5wt% in acid treated biomass, which is expected. It was found that co-pyrolysis of acid-treated corn stover with PE can further increase the yield of levoglucosan, as well as the yields of phenolic monomers due to synergistic effect. The stronger synergistic effects were also observed when the acid treated corn stover and PE were co-pyrolyzed using HZSM-5 zeolite catalyst in comparison to co-pyrolyzing non-treated corn stover and PE using same catalyst. Due to the synergistic effects, the amount of total aromatic hydrocarbons and the selectivity of mono-ring aromatic hydrocarbons both increased significantly.

497745 Effect of Dielectric Oxides on Microwave Assisted Deconstruction of Lignin Â

Abstract:

Lignin, a natural polymer is a potential source of fuels and platform chemicals due to its aromatic structure. The conventional industrial processes which utilize plant polysaccharides, burn lignin to generate process heat. With the depletion of fossil fuels, cellulosic biorefineries would generate surplus lignin than required to fuel the biomass conversion process [1]. Conventional lignin conversion methods require temperatures greater than 200°C and longer reaction time. A non-conventional source of energy such as microwave reduces the reaction time and gives higher heating rates. The heating of reaction medium by microwaves is through the mechanism of dipolar polarization and ionic conduction. In this study, alkali lignin was solubilized in 1:1 Dimethyl Sulfoxide (DMSO)/water mixture. The DMSO/water mixture has a high polarity (0.7) and solubilizes lignin completely. Metal oxides such as lanthanum ferrate ($LaFeO_3$), iron oxide (Fe_2O_3) and zirconium oxide (ZrO_2) were synthesized by the combustion method. Experiments were carried out in a microwave reactor at different temperatures (80-100°C) and samples were withdrawn at regular intervals between 0-120min. The catalysts were suspended in the reaction media and the solution vessel was continuously stirred for the entire reaction time. The microwave treated samples were analyzed using Gel Permeation Chromatography to study the effect of catalysts on reduction in the average molecular weight of lignin. Liquid chromatography Mass Spectrometry was used for the identification of phenolic monomers produced. The formation of products such as vanillin, acetoguaiacone, homovanillic acid and coniferaldehyde were observed at temperatures as low as 80°C while using $LaFeO_3$ and Fe_2O_3 as catalysts. The effect of different catalyst properties such as structure, dielectric constant and surface area, on the yield of monomers was studied and possible reaction mechanisms were proposed.

References:

1. A.J. Ragauskas *et.al.*, Lignin valorization:improving lignin processing in the biorefinery. *Science*, 344 (2014).

495084 Saddle Point Searches in Electrochemical Reactions

Abstract:

Electrochemical reactions have gained considerable interest as potential solutions for renewable fuels and energy storage. An accurate computational model of electrochemical systems remains challenging, due in part to the presence of a solvent and the need for charged interfaces. Moreover, electrode potentials are governed by an external circuit, which keeps the electrochemical potential of the electrode constant throughout the reaction. Conventional density functional theory (DFT) is carried out within a canonical framework, where the number of electrons is fixed. This constant-charge approach results in a potential bias in the course of an electrochemical reaction since the surface charge of the electrode is depleted [1], and practical models to overcome this constant-electron situation have no obvious energy-referencing scheme.

The nudged elastic band (NEB) method has proven useful in evaluating minimum energy pathways (MEP) in chemical reactions. One can devise methods that continuously vary the number of electrons or the local field in the unit cell to keep the chemical potential of the electrode constant. This presents another difficulty: on-site energies of systems with different numbers of electrons cannot be directly compared since the energy reference varies between images. To circumvent this issue, we introduce a force integration method within the NEB method. This method relies solely on canonical force components parallel to the MEP in each image in the band. As the resolution of images increases, the potential energy surface approaches that of a grand-canonical system, and the path-independence of the potential energy means that the interpolated images need not lie on the minimum energy pathway. We employ this method to study the kinetics of the proton discharge step of the hydrogen evolution reaction.

[1] Skúlason, E.; Tripkovic, V.; Björketun, M. E.; Gudmundsdóttir, S.; Karlberg, G.; Rossmeisl, J.; Bligaard, T.; Jónsson, H.; Nørskov, J. K. J. Phys. Chem. C 2010, 114 (42), 18182–18197.

495535 Enhanced Activity for Oxygen Reduction Reaction By Gold at Step/Edge Sites of Ni@Aupt Core-Shell Nanoparticles: A DFT Investigation

Abstract:

Enhanced Activity for Oxygen Reduction Reaction by Gold at Step/Edge Sites of Ni@AuPt Core-Shell Nanoparticles: A DFT Investigation

Wei An*, Hao Wang

College of Chemistry and Chemical Engineering, Shanghai University of Engineering Science, Songjiang District, Shanghai 201620, China.

*E-mail: weian@sues.edu.cn

ABSTRACT

Modifying Ni@Pt nanoparticle (NP) with Au has been proved to be an efficient strategy to enhance not only the activity of Ni@Pt nanoparticles for oxygen reduction reaction (ORR) demonstrated by previous theoretical study,[1] but also the stability of cathode from experimental observations.[2-3] Still, there is an urgent need for understanding the ORR behaviors at step/edge sites of a NP considering the large portion of surface Pt atoms consumed by step/edge sites for the closure of a NP buildup besides the key roles of step/edge ¡®defect¡¯ sites that play in promoting a catalytic reaction.

In view of the remarkable effect of Au, we investigate ORR at step/edge sites of Ni@AuPt NPs employing periodic spin-polarized density functional theory (DFT) method. We built high-index (221) facets as model catalysts and found that the ORR activity can be greatly enhanced by replacing the surface Pt atoms at edge site with Au atoms. Our results show that ORR activity at *edge* site is much higher than that at *step* site. Inert Au atoms localized at the surface edge sites of Pt-based NPs can effectively promote the elementary steps of ORR. At working potential (U=0.8V), the rate-determining step of ORR at step/edges sites of Ni@AuPt core-shell NPs is likely to be O_2 protonation to *OOH, compared with *O protonation to *OH on Pt NPs. Furthermore, the BEP relations are held for O_2 dissociation in the form of the linear correlation between the transition-state and final-state energies on potential energy surface, and are held for *O protonation in the form of the linear correlation between the activation barrier and reaction energy in which the rate of *O protonation is increased as the reaction becomes more exothermic. We identified that the surface activity of Pt atoms of core-shell NPs can be well explained by either d-band model or coordination-number model, which have the same origin of chemical nature. Our study provides fundamental understanding of functionality of step/edge sites of Pt-based core-shell NPs for ORR that can be used for engineering of practical cathode catalyst with

atomic efficiency and precision by maximizing occupancy of gold atoms at edge sites.

REFERENCES

1. An, W.; Liu, P. Rationalization of Au Concentration and Distribution in AuNi@Pt Core-Shell Nanoparticles for Oxygen Reduction Reaction. *ACS Catal.* **2015,** *5*, 6328-6336.

2. Zhang, J.; Sasaki, K.; Sutter, E.; Adzic, R. R. Stabilization of platinum oxygen-reduction electrocatalysts using gold clusters. *Science* **2007,** *315*, 220-222.

3. Gatalo, M.; Jovanovic, P.; Polymeros, G.; Grote, J. P.; Pavlišic, A.; Ruiz-Zepeda, F.; Šelih, V. S.; Šala, M.; Hocevar, S.; Bele, M.; Mayrhofer, K. J. J.; Hodnik, N.; Gaberšcek, M. Positive Effect of Surface Doping with Au on the Stability of Pt-Based Electrocatalysts. *ACS Catal.* **2016,** *6*, 1630-1634.

492128 Ammonia Synthesis Using Plasma Assisted Catalysis: Understanding Rate Enhancements By Excited Species

Abstract:

The industrial Haber-Bosch process for catalytic ammonia synthesis from nitrogen and hydrogen is carried out at high pressures (100-200 bar) and temperatures (400-500 °C). The robust N-N triple bond is the fundamental barrier for this reaction, and there has been a longstanding desire to develop ways to activate this bond under less extreme conditions. A few reports in the literature [1-4], as well as recent experiments performed by us, have demonstrated that is possible to produce ammonia at atmospheric pressure and temperatures between 100-300°C by coupling the catalyst with a non-thermal plasma. It is speculated that the plasma assists in activating the source gas by generating reactive species, such as vibrationally or electronically excited N2*, ions, and radicals, which interact on the catalyst surface to produce NH3. In this talk, we explore the limits of ammonia synthesis rate enhancements on transition metal surfaces by plasma activated intermediates using a simple microkinetic model based on density functional theory calculations. Our model predicts that plasma-assisted catalysis could lead to low temperature and pressure reaction rates that are competitive with the Haber-Bosch process, but the optimal metal catalysts and active-site motifs are distinct from those predicted to be most active for thermal catalysis. The models are useful for interpreting literature results and guiding new experiments.

References

1. Peng, P., Li, Y., Cheng, Y., Deng, S., Chen, P., & Ruan, R. Plasma Chemistry and Plasma Processing, 36(5), 1201â€“1210 (2016).

2. Xie, D., Sun, Y., Zhu, T., Fan, X., Hong, X., & Yang, W. RSC Advances, 6(107), 105338â€“105346 (2016).

3. Kim, H., Teramoto, Y., Ogata, A., Takagi, H., & Nanba, T. Plasma Processes and Polymers, Published online: DOI:10.1002/ppap.201600157 (2016).

4. Akay, G., and Zhang, K. Industrial & Engineering Chemistry Research, 56(2), 457â€“468 (2017).

501604 Fundamental Investigation of C-C Coupling of Carbonyl Compounds on Ceria

Abstract:

Ceria is an abundant and widely used technological material owing to its ability to readily store and release oxygen and change oxidation state. There is a growing body of evidence indicating that ceria can promote a variety of reactions for organic compounds, including dehydrogenation/hydrogenation, deoxygenation, and C-C coupling, depending on the extent of surface reduction and crystalline orientation of the surface. This suggests that ceria can be a promising heterogeneous catalyst in its own right for the conversion of organic compounds, including the upgrading of pyrolysis and fermentation products, if its reactivity can be understood and tailored at the atomic level. In this talk we will present mechanistic insights generated from surface science experiments and theoretical modeling based on density functional theory calculations into how two representative organic oxygenates, acetaldehyde and acetic acid, interact with the $CeO_2(111)$ surface and undergo self C-C coupling. This is accomplished in part by modeling and predicting the identity of surface reaction intermediates in comparison to infrared vibrational spectroscopy and the thermal decomposition and reaction events observed in temperature programmed desorption experiments. Our work sheds light on the intrinsic reactivity of this ceria surface toward the formation of the most closely related C_4 products, crotonaldehyde and acetone, and highlight the rate-limiting steps in the respective thermal conversion processes.

486079 N_2OÂ Dissociation on Co_3O_4 Based Catalysts: Reducibility of Co_3O_4 and its Catalytic Consequences

Abstract:

N_2O Dissociation on Co_3O_4 Based Catalysts: Reducibility of Co_3O_4 and Its Catalytic Consequences

*Yongchun Hong[1], Andrew Getsoian[2], Christine Lambert[2] and Enrique Iglesia[1]**

[1] *University of California at Berkeley, Berkeley, CA 94720*

[2] *Ford Research and Advanced Engineering, Dearborn, MI 48121*

*iglesia@berkeley.edu

N_2O is a greenhouse gas with up to 300-fold higher global warming potential than CO_2 and it is also one of the major ozone depleting chemicals. [1] The abatement of N_2O emissions by its catalytic decomposition on metal oxides such as Co_3O_4 forms N_2 and O_2, but the latter, a major component in effluent streams (> 10%) inhibits reaction rates. [2] N_2O-CO reaction forms N_2 and CO_2, but requires conditions that balance the residual concentrations of N_2O and CO. Co_3O_4 based catalysts are among the most promising candidates for redox catalysis enabled by Co^{2+}/Co^{3+} interconversion, including NO oxidation [3] and N_2O decomposition [4]. Our study addresses the mechanistic details of N_2O decomposition and N_2O-CO reactions on Co_3O_4 catalysts, including bulk Co_3O_4 powders and Co_3O_4 domains dispersed on insulators (γ-Al_2O_3, SiO_2, ZrO_2) and semiconductors (TiO_2, CeO_2-ZrO_2, CeO_2). The results show that reducibility of Co_3O_4 domains affects the reactivity and the nature of the N_2O dissociation pathways for both N_2O decomposition and N_2O-CO reaction by controlling the number of reduced centers that prevail during catalysis and the ratio of the steady-state concentrations of surface oxygen atoms and O-vacancies ([O*]/[*]). Co_3O_4 domains dispersed on semiconducting materials are more reducible than bulk Co_3O_4 or Co_3O_4 domains dispersed on insulators; therefore, they exhibit much higher reactivities for both N_2O decomposition and N_2O-CO reaction.

The turnover rates for N_2O decomposition and N_2O-CO reaction differ markedly for each Co_3O_4 based catalysts. All catalysts show significant N_2O-CO turnover rates (> 0.001 s^{-1}) below 430 K, while such turnover rates require temperatures above 600 K for N_2O decomposition. These differences reflect the ability of the CO reductant to form vacancies more effectively than via O_2 evolution during N_2O decomposition, thus leading to much higher [*]/[O*] during catalysis and to a much higher concentration of bound N_2O* for a given N_2O pressure. For both reactions, however, turnover rates were similar for Co_3O_4 powders and Co_3O_4 domains dispersed on insulators, while semiconducting supports, such as CeO_2, lead to Co_3O_4 domains with higher reactivity. These trends reflect interfacial heterojunctions between Co_3O_4 and semiconductor, for which band bending allows electron delocalization and a decrease in the energy of the lowest unoccupied molecular orbital (LUMO) in Co_3O_4 domains; such lower energies improve the thermodynamics of reduction of Co_3O_4, thus increasing the density of vacant sites during steady-state catalysis.

N_2O decomposition on Co_3O_4 based catalysts showed first order kinetics on N_2O and strong inhibition by O_2 and rates were accurately described by Langmuir models of surface reactivity. Such treatments show that N_2O dissociates in the kinetically-relevant step via N_2O* reactions with O* (route I), and the O* species formed desorb as O_2 in a quasi-equilibrated step, with O* as the most abundant surface intermediate (MASI). The ratio of N_2O decomposition turnover rates on dispersed Co_3O_4 domains on each support to those on Co_3O_4 powders did not depend on O_2 pressure, indicating that differences in reactivity among the catalysts are independent with the [O*]/[*], presumably because vacancy sites on all Co_3O_4 catalysts have similar oxygen binding energy and therefore the reactivity of all catalysts are inhibited by O_2 to the same extend; and differences therefore only reflect the differences in the number of reducible sites among catalysts.

N_2O-CO turnover rates on all Co_3O_4 based catalysts were proportional to N_2O pressure and independent of CO pressure for [CO]/[N_2O] reactant ratios above unity, but decreased monotonically as [CO]/[N_2O] ratios decrease below unity. The addition of O_2 to CO/N_2O mixtures did not influence N_2O-CO turnover rates at 400 K, because O_2 dissociation does not occur at detectable rates, thus preventing the capping of vacant sites by O* or the required activation of O_2 for CO oxidation catalysis. N_2O-CO turnover rates are accurately described by Langmuir models in which N_2O* reacts with a vicinal * in the kinetically-relevant step (route II), with the relative coverages of * and O* set by the kinetic coupling between the formation of O* via N_2O* reactions with * and the removal of O* via its reactions with CO*. The coverages of oxygen vacancy ([*]) during N_2O-CO reaction on Co_3O_4 catalyst were measured using N_2O titration: [*] increased monotonically with increasing [CO]/[N_2O] ratio and approaching to constants (0.07 and 0.3, at 403 K and 483 K, respectively), which are the portions of reducible sites on the surface.

The findings suggest that the reducibility of Co_3O_4 domains, reflected by the number of oxygen vacancies and the [O*]/[*] ratio, governs the reactivity and reaction pathway for N_2O dissociation: oxygen vacancies can be generated more effectively by CO reduction than by O_2 evolution and therefore Co_3O_4 domains showed higher reactivity in N_2O-CO reaction than in N_2O decomposition; Co_3O_4 domains with semiconducting supports are more reducible than bulk Co_3O_4 powders or Co_3O_4 domains dispersed on insulators, leading to more vacancy site available for catalysis and consequently higher reactivity in both reactions; scavenging of oxygen vacancy (*) by dissociative adsorption of O_2 leads to higher [O*]/[*] ratios in N_2O decomposition and therefore route I is favored; while scavenging of surface oxygen species (O*) by CO leads to lower [O*]/[*] ratios in N_2O-CO reaction and therefore route II is favored. The findings also suggest that N_2O-CO reaction on Co_3O_4 based catalysts represents more promising approach for N_2O abatement than N_2O decomposition, because it can be operated under stoichiometric CO/N_2O ratio at temperature (403 K) that is close to cold start conditions and is not inhibited by O_2, while N_2O decomposition requires much higher temperatures and is inhibited by O_2. The findings shed some lights on the understanding of the catalytic consequences of metal oxide reducibility in catalysis mediated by oxygen vacancy, such as NO oxidation and N_2O decomposition, and provide guidance for the development of better N_2O abatement strategy.

The authors acknowledge financial support from Ford Motor Company.

References

1. Ravishankara, A. R., Daniel, J. S., Portmann, R. W., *Science* 326, 123 (2009).

2. Lambert, C., Dobson, D., Gierczak, C., Guo G., Ura, J., Warner, J. *Int. J. Powertrains*., 3, 4 (2014).

3. Weiss, B.M., Artioli, N., Iglesia, E., *ChemCatChem*, 4, 1937 (2012).

4. Kapteijn, F., Rodriguez-Mirasol, J., Moulijn, J. A., *Appl. Catal. B*, 9, 25 (1996) ADDIN EN.REFLIST

498286 Computational Study of Selective Oxidation of Ethane on Oligomeric VO_x/SiO_2 and Bulk Mixed Oxide Catalysts

Abstract:

Computational Study of Selective Oxidation of Ethane on Oligomeric VO_x/SiO_2 and Bulk Mixed Oxide Catalysts

Yilang Liu[1] and Prashant Deshlahra[1]*

[1]Tufts University, Medford, MA 02155 (USA)

*prashant.deshlahra@tufts.edu

Introduction

Selective oxidations of light alkanes are promising routes to produce more valuable chemicals.[1] However, selectivity to the desired products are limited by parallel and secondary reactions leading to undesired fully oxidized products. Primary alkane activations are typically limited by C-H activation, but mechanisms of parallel and secondary reactions are still unclear, while they are important for selectivity limitation. Here, to understand reaction steps limiting selectivity, we use hybrid-DFT calculations to probe all plausible paths of C_2H_6 oxidation reactions on different sites of VO_x/SiO_2catalyst.

Materials and Methods

To model oligomeric VO_x/SiO_2, silsesquioxane clusters ($H_8Si_8O_{12}$) with four of its SiO_x moieties replaced by VO_x are used. (Figure 1)[2,3]DFT calculations were performed using Gaussian09 program with B3LYP functional and 6-31g(d,p) basis sets. Transition states were calculated using quadratic synchronous transit method. The optimized structures are characterized as minima points for intermediates and saddle points for transition states by frequency calculations. Intrinsic reaction coordinates (IRC) were calculated to confirm that transition states connect reactants and products. Activation energy barriers are calculated as differences between electronic energies of relevant intermediates or transition states to gaseous reactants and bare sites, and will be converted to free energies to determine reactivity trends at realistic temperatures and pressures and the effect of van der Waals interactions will be probed.

Results and Discussion

C_2H_6 oxidative dehydrogenation (ODH) on metal oxides occurs via Mars-van Krevelen redox cycles in which the C-H activation at lattice oxygens of oxides (O*) leads to reduced catalytic centers in the form of surface hydroxyls (OH*) or oxygen vacancies (*) (Figure1a). Re-oxidation of isolated reduced centers by O_2can lead to peroxo species (OO*). Therefore, plausible reaction paths for primary, parallel and secondary reactions at both types of species and steps accounting for the formation and consumption of these species were considered (Figure 1).

Figure 1. (a) Lattice oxygens (O*), vacancies (*), and peroxo species (OO*) involved in reduction-oxidation cycles of metal oxides during ethane-O_2 reactions. DFT derived structures and the metal-oxygen bond distances for (b) oxo (O*) and (c) peroxo (OO*) sites on VO_x/SiO_2.

On oxo site, the primary C_2H_6 ODH to form C_2H_4 is accomplished by two C-H activations on terminal O-atoms, with the highest barrier of 186 kJ/mol. C_2H_6 can also react in parallel to form C_2H_5OH via bonding between C_2H_5 radical and HO* after first C-H activation, with an activation barrier of 180 kJ/mol. C_2H_5OH can then be oxidized to CH_3CHO with activation barrier of 125 kJ/mol. The first step in secondary reactions after formation of C_2H_4 involves O-insertion to form C_2H_4O, which has a much lower activation barrier of 132 kJ/mol. Thus, both parallel and secondary paths have comparable or lower barriers and can contribute to C_2H_4 selectivity limitation. The removal of bound OH* or products of O-insertion is assisted by O_2 to form peroxo species in rapid steps with low barriers. On peroxo site, first C-H activation to form C_2H_4 or C_2H_5OH has a barrier of 122 kJ/mol while C_2H_4 can undergo O-insertion even more preferentially, with a 55 kJ/mol barrier. The calculated activation energies suggest that peorxo sites are more selective to undesired O-insertions than oxo sites, consistent with known tendency of peroxo sites to favor combustion reactions.[4]These steps lead to oxidized products and oxo sites to complete the catalytic turnovers. The overall reaction barriers for highest points along the reaction coordinate for primary, parallel and secondary reactions are listed in Table 1.

Table 1. Electronic activation energies relative to bare sites and gaseous reactants for ethane and ethylene oxidation at oxo and peroxo sites.

We further studied subsequent reactions of C_2H_4O on oxo sites. The most favorable reaction paths involve isomerization of C_2H_4O to form CH_3CHO, with an activation barrier of 103 kJ/mol. The CH_3CHO molecules bind strongly on oxo sites to form stable and relatively unreactive acetate species with 101 kJ/mol energy barrier and a -117 kJ/mol intermediate energy relative to gas phase CH_3CHO. On acetate covered surfaces, bridge O-atoms are more reactive than terminal O-atoms for C-H activation and oxygen insertion. Activation barriers of C_2H_4 formation and secondary oxidation of C_2H_4 to form C_2H_4O are 218 kJ/mol and 217 kJ/mol, respectively, shifted toward favoring C-H activation relative to oxo site.

These results lead us to conclude that selectivity to primary products is determined by O-insertion reactions parallel and sequential to desired C-H activations on oxo, peroxo and acetate covered surfaces. Preliminary computations of such reaction steps on microporous MoVTeNb mixed metal oxides suggest that undesired O-insertion reactions can be suppressed by steric hindrance in their small micropores while stabilizing the C-H activations by van der Waals interactions. Detailed comparisons of reaction steps in the two oxides help understand the role of electronic properties and size restrictions in determining selectivity limitations in oxidative conversion of small molecules.

Acknowledgements

Financial support from Tufts University and access to computing resources from Tufts Research Computing Cluster and the Extreme Science and Engineering Discovery Environment (XSEDE; NSF ACI-1053575; Proposal number TG-CTS150005) are gratefully acknowledged.

References

1. Labinger, J. A. and Bercaw J. E. *Nature*, **2002**, *417*, 507

2. Döbler, J., Pritzsche, M., and Sauer, J. *J. Am. Chem. Soc.*, **2005**, *127*, 10861

3. Goodrow, A., and Bell, A. T. *J. Phys. Chem. C*, **2007**, *111*, 14753

4. Liu, J., Mohamed, F., Sauer, J. *J. Catal.,* **2014**, *317,* 75

Computationally-Enhanced Spectroscopic Studies of Supported Vanadium Oxide

Abstract: 499534

Nicholas R Jaegers,[1,2] Chuan Wan,[1] Mary Y Hu,[1] Monica Vasiliu,[3] David Dixon,[3] Eric Walter,[1] Israel Wachs,[4] Yong Wang,[1,2] and Jian Zhi Hu[1*]

[1]*Institute for Integrated Catalysis and Earth and Biological Science Directorate, Pacific Northwest National Laboratory, Richland, Washington 99354 (United States)*
[2]*Voiland School of Chemical Engineering and Bioengineering, Washington State University, Pullman, Washington 99163 (United States)*
[3]*Department of Chemistry, University of Alabama, Tuscaloosa, Alabama 35487 (United States)*
[4]*Operando Molecular Spectroscopy & Catalysis Laboratory, Department of Chemical Engineering, Lehigh University, Bethlehem, Pennsylvania 18015 (United States)*
*Jianzhi.Hu@pnnl.gov

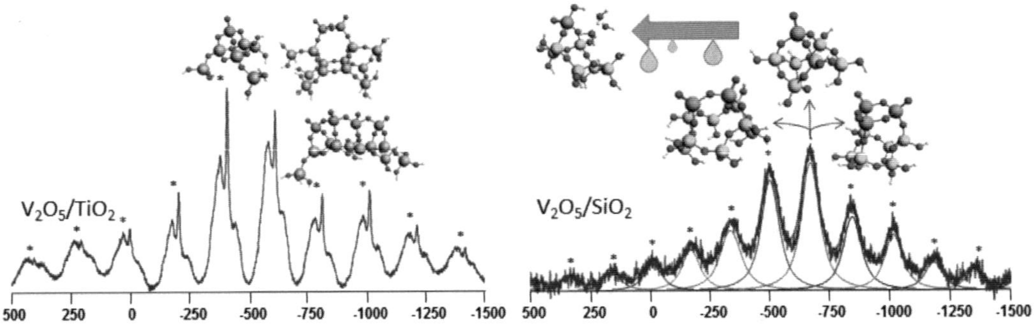

Introduction

Supported vanadium oxide catalysts are important materials due to their wide utilization for industrial chemical conversions. Their ability to promote an extensive array of oxidation reactions and functionality for selective NO_X reduction have enabled their application in the pharmaceutical, petrochemical, textile, plastic, agricultural, and emissions controls sectors. Due to their broad applicability, thorough characterization of their properties is invaluable for a clear description of the catalytic process. The structural arrangement of these materials provides the atomic scale depiction of these properties, promoting an improved understanding of the structure-function relationships that may be useful in the rational design of new materials. Despite the strong motivation and efforts to better understand the surface structure of these materials, questions remain regarding the arrangement of vanadium sites on different supporting materials, particularly under different chemical environments.

Previous studies have suggested the exclusive presence of VO_4 monomers at low surface coverages and progressive oligomerization up to the monolayer limit on most oxide supports. Less decisive evidence on this matter is present for silica –supported catalysts due to their inability to accommodate large quantities of vanadium (low surface reactivity) and their high surface area.[1] Further ambiguity exists when considering the environmental effects on the catalyst's structure. Hydration has been shown to alter the surface structure where contrasting claims of decavanadate and vanadia gel formation exist.[2-3] In this study, solid-state nuclear magnetic resonance (NMR) spectroscopy and quantum chemistry simulations of the NMR spectra are used as aides in understanding the nature of V_2O_5/SiO_2 and SCR $V_2O_5(-WO_3)/TiO_2$ catalysts exposed to different hydration environments.[4] Surface coverages of 22% and 63% on silica were explored to assess the potential for oligomerization where low loadings of vanadium on titania were used for comparison.

Methods

Solid-state ^{51}V and ^{1}H MAS NMR experiments were conducted on two instruments: a Varian-Inova 850 MHz and a Bruker WB600 NMR spectrometer. The Larmor frequencies for vanadium were 223.367 and 157.778 MHz, respectively. A $\pi/2$ pulse width (ca. 2 μs) was used with a recycle delay of 0.2 s. Magic-angle spinning rates up to 38 kHz were used for the 1.6 mm pencil-type rotors on the 850 MHz spectrometer and 35 kHz for the 2.5 mm rotors used on the 600 MHz spectrometer. Chemical shifts were externally referenced to the center-band of bulk V_2O_5.

Electronic structure simulations accompanied spectroscopic measurements to aide in interpretation of the results. These NMR chemical shift calculations were carried out with the Amsterdam Density Functional (ADF) package.[5] Cluster model geometries were optimized using the generalized gradient approximation with the dispersion-corrected Becke-Lee-Yang-Parr functional (GGA: BLYP-D). The all-electron triple-ζ, two-polarization (TZ2P) basis set with Slater-type orbitals and 'very good' numerical accuracy was used for all models. The models also accounted for relativistic effects using the zero-order regular approximation (ZORA). NMR calculations were conducted using the gauge independent atomic orbital (GIAO) approach.

To relate the calculated nuclear shielding of the DFT-based chemical shift calculations to the chemical shift from the spectroscopic experiments, $VOCl_3$ was used as a computational standard. Its calculated shielding was used to interpret the models using the following equation: $\delta_{obs}(\text{ppm}) = \sigma_{ref}(VOCl_3, \text{calc}) - \sigma_{calc}$. To gauge the suitability of this method, benchmarking references were simulated to compare their calculated chemical shifts to known experimental values. Table 1 compares these calculations to their experimental analogs. The bulk vanadium oxide and silsesquioxane models are in reasonable agreement with their experimental analogs, only showing deviations up to 14 ppm.

Table 1. Calculated NMR reference chemical shifts

Compound	Experimental	Theoretical
V_2O_5	-610 ppm	-622 ppm
$[(c\text{-}C_6H_{11})_7(Si_7O_{12})V{=}O]_2$	-714 ppm	-700, -715 ppm

Results & Discussion

Silica-supported vanadium oxide under both hydrated and dehydrated conditions can be seen in Figure 1. Both dehydrated V_2O_5/SiO_2 catalysts exhibited one broad feature at -675 ppm while the hydrated forms offered a majority peak (86%) at -566 ppm and a minority one at -610 ppm. No significant differences or new features were visible comparing the two loadings in either the hydrated or dehydrated conditions, suggesting similar surface structures. To better assign these resonances, NMR chemical shift calculations were conducted.

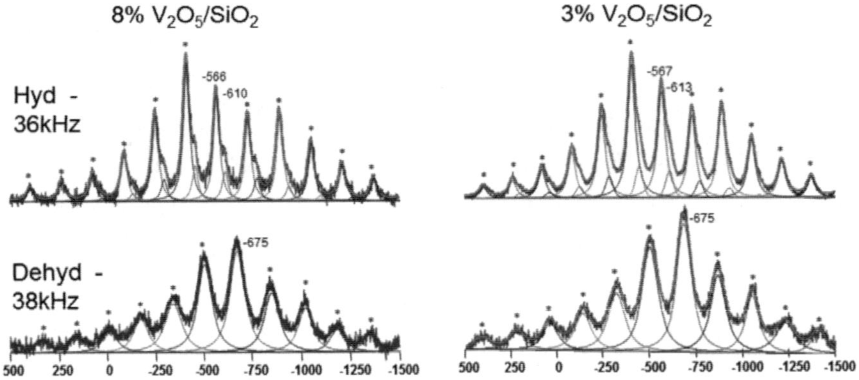

Figure 1. ^{51}V **MAS NMR of hydrated (top) and dehydrated (bottom)** V_2O_5/SiO_2 **catalysts at 8% and 3% loading.**

Computational modeling of the ^{51}V NMR chemical shifts of dehydrated surface species (Figure 2) indicated that anchoring of vanadium oxide does not occur on three-ringed silica (-459 ppm) and requires larger silica anchoring clusters such as 6-ringed support anchoring positions (-676 ppm). The results also show that, unlike titania-supported vanadium oxide, the chemical shifts for monomeric and dimeric vanadium species are similar (-676 and -671 ppm). Cyclic trimer species may also resonate near the observed chemical shift, but linear trimers, tetramers, and V_2O_5 crystallites are unlikely candidates due to either the expectation of two unique signals or the prediction of signals too far deshielded to represent those observed.

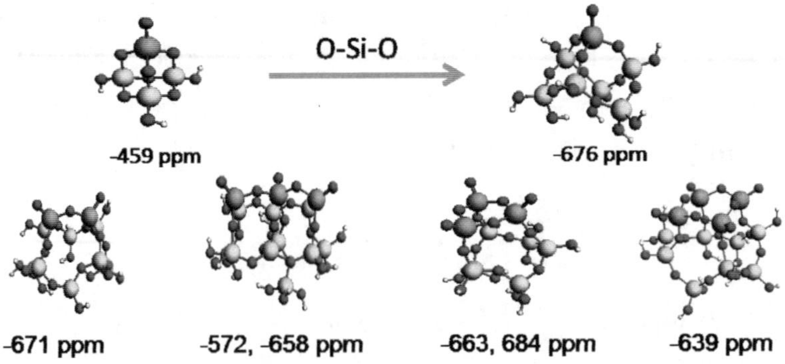

Figure 2. DFT cluster models and calculated chemical shifts for dehydrated surface species.

Hydrated structures were considered on the basis of water addition and previously proposed olation mechanisms as seen in Figures 3 and 4. Coordination of 2-4 water molecules resulted in deshielded vanadium nuclei, but was insufficient in the simulation of the spectroscopic results, the nearest being the hydrated dimer at -621 ppm or select tetramer nuclei. No claim to any relation to the major feature at -566 ppm could be made. The olation mechanism was evaluated by optimizing structures along the course of support bond hydrolysis and vanadium agglomeration as both an anchored and mobile species over SiO_2. Among the structures considered, both di-grafted, isolated VO_4 units and di-grafted dimer species demonstrated reasonable agreement with the main peak at -566 ppm observed in the spectra. More mobile species were too deshielded for assignment to the species (ca. -488 – -525). Digrafted monomers have previously been described as stable species, which strengthens this assignment.[6]

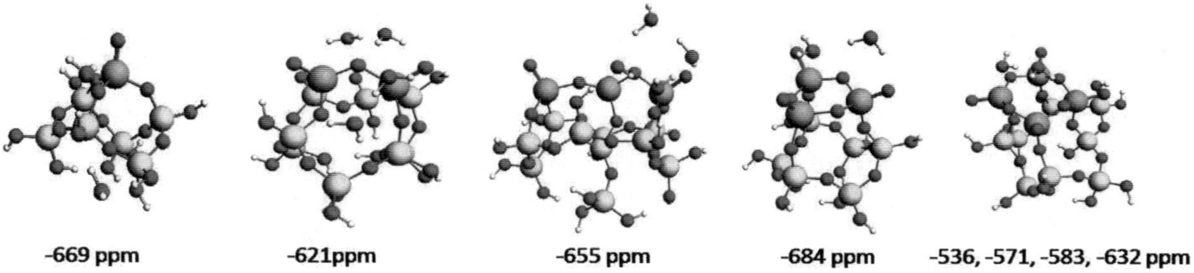

Figure 3. DFT cluster models of water-coordinated surface species.

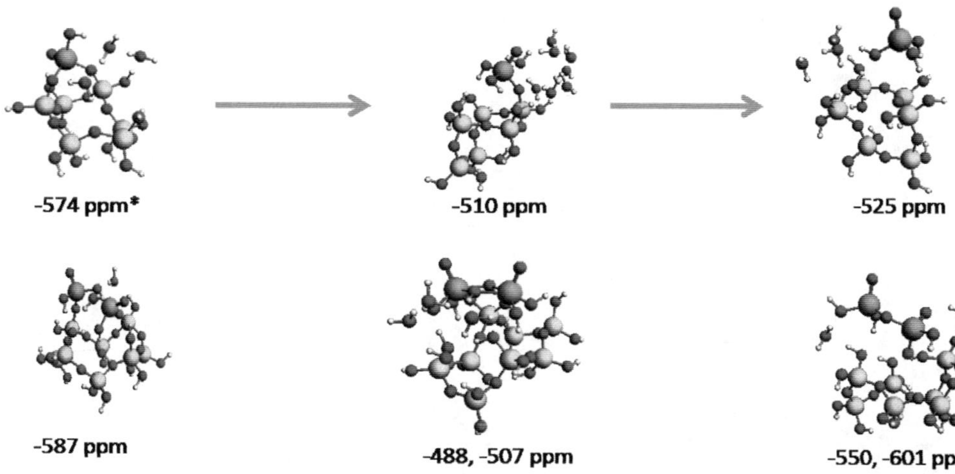

Figure 4. DFT cluster models of previously proposed olation structures. *Digrafted monomer has been reported as a stable surface species on silica.

Olation proceeding to decavanadate or vanadia gels was considered as well. Reported chemical shifts and our DFT calculations for decavanadate suggest that this species is not present in either of the silica-supported samples. Vanadia gels are more difficult to generalize due to their ambiguous structure and wide range of chemical shifts reported in the literature. On the basis of previous report comparison, no evidence for their formation upon hydration can be supported with these NMR spectra. As such, the remaining unassigned peak at -610 ppm is ascribed to bulk-like vanadium oxide due to its proximity to the compound's resonance.

While V_2O_5/SiO_2 catalysts demonstrate only a single resonance regardless of loading that splits into two upon hydration, V_2O_5/TiO_2 catalysts show a much wider range of unique vanadium nuclei in a given sample. As can be seen in Figure 5, the signals that arise are indeed dependent upon vanadium loading, presence of tungsten, and even some minute differences between catalysts synthesized by impregnation (I) and co-precipitation (C). Spectral deconvolution of the lines show largely the same behavior mid-field in the V_2O_5/TiO_2 spectra with nearly identical peaks at -533, -562, and -583 ppm, mirroring some similarity to NMR spectra collected at a similar vanadium loading.[7] Differences are present at low fields where the impregnated sample shows deshielding of the -516 ppm peak to -507 ppm and the co-precipitated hosts a feature at -459 ppm. Furthermore, the co-precipitated sample provides a broad feature at -636 ppm and a small peak at -665 ppm where the impregnated sample shows features at -613 and -648 ppm. Upon application of greater quantities of vanadium or the inclusion of tungsten onto the surface, a notable difference in the signals is observed. In these instances, vanadium signals move upfield where a majority (>75%) of the vanadium resonates at or above -568 ppm. Due to the variety of signals observed, assignment of these peaks requires DFT interpretation.

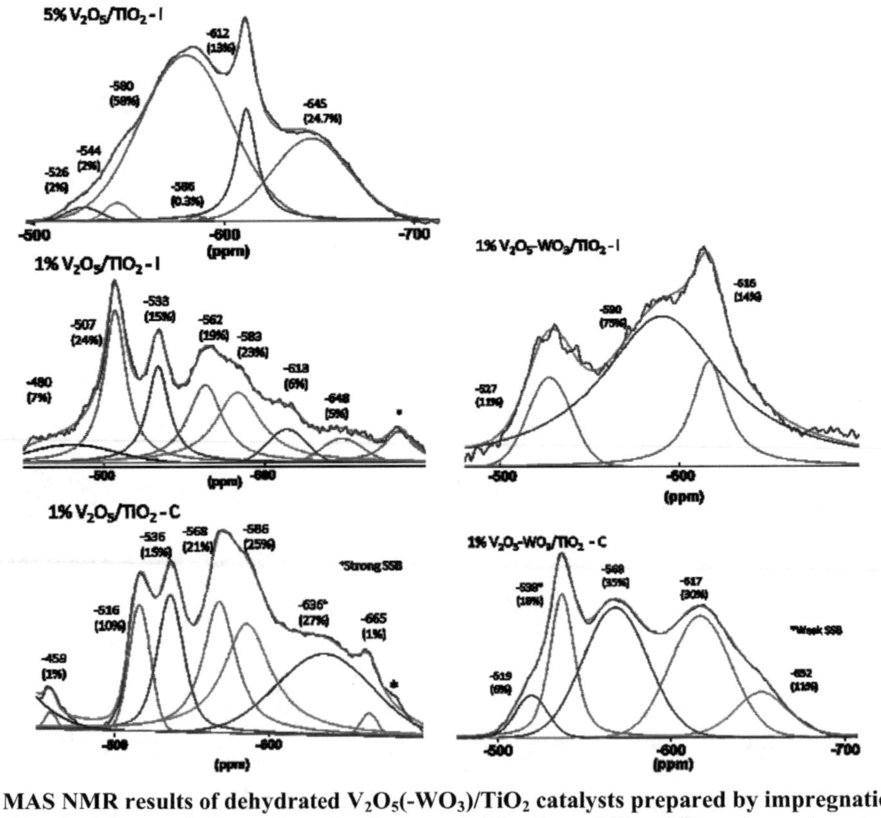

Figure 5. **^{51}V MAS NMR results of dehydrated $V_2O_5(-WO_3)/TiO_2$ catalysts prepared by impregnation (I) and co-precipitation (C). Spectral deconvolution summation is presented by the red line. All spectra were externally referenced to V_2O_5 at -613.8 ppm. Sidebands displayed are indicated by (*).**

Cluster models of titania-supported vanadium oxide have been evaluated for chemical shift assignment in previous ^{51}V NMR investigations.[7] It was found that monomers on a relatively amorphous surface resonate around -500 ppm, dimers around -550 or -640 ppm, and tetramers around -640 ppm. While these calculations are useful in understanding the general surface structure of these materials, they are inadequate in describing each observed signal. As such, additional large cluster models were generated to include the surface effect on geometry. Since both titania supports are comprised of primarily anatase (X-ray diffraction), models of around 190 atoms were removed from the bulk structure model described by Howard *et al.* and cut along the {101} and {001} facets.[8] The bottom segment was frozen to retain the anatase properties and terminal oxygen atoms were charge balanced with protons. The resulting cluster models for monomeric species show that the various observed species between -460 and -536 ppm are likely related to monomeric species on different titania facets or with different bonding configuration such as a square pyramidal structure on {001} at -529 ppm and a tetrahedral configuration on {101} at -506 ppm. Larger vanadia domains do indeed show upfield shifts on {001} to around -600 to -660 ppm, supporting the previous results on a less defined surface.

Exposure to a hydrated environment incites changes in the vanadium structure of titania-supported catalysts as reported for silica. However, a large number of distinct signals are retained. The most apparent transformation occurs with the relatively high loaded sample of 5% V_2O_5/TiO_2. The primary feat (81%) resonates at -614 ppm, exactly as reported for the updated V_2O_5 reference chemical shift of -613.8 ppm. Additionally, this peak is accompanied by a strong side band pattern and displays a comparable line width to the reference of 1600 Hz. The other peak at -628 ppm is the minority species and is present without a side band pattern, suggesting high species mobility and confirming a dramatic surface rearrangement upon hydration. Lower loaded samples show an elimination of the downfield peaks above -500 ppm, leaving monomeric species around -510 unaltered. Hydration also resulted in an enhancement of the peak at -536 ppm, a reduction in the peak around -580, and the impregnated sample hosts a broad signal at -630 ppm that dominates the upfield region. By comparison, tungsten-promoted catalysts showed mild changes to higher shielding and broader signals, but were relatively less impacted by hydration.

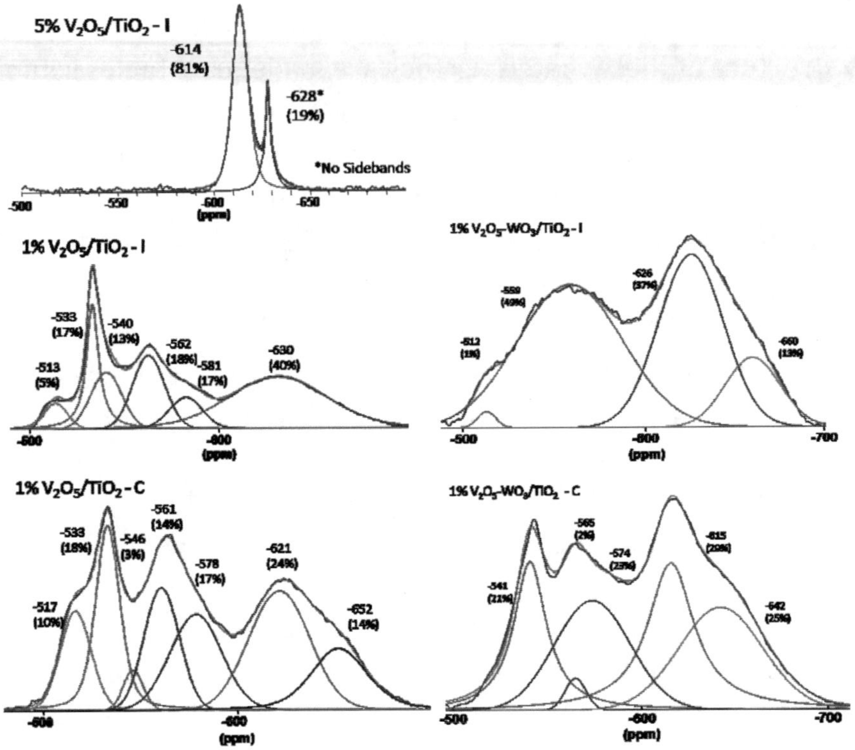

Figure 6. ^{51}V **MAS NMR results of hydrated** $V_2O_5(-WO_3)/TiO_2$ **catalysts prepared by impregnation (I) and co-precipitation (C). Spectral deconvolution summation is presented by the red line. All spectra were externally referenced to** V_2O_5 **at -613.8 ppm.**

Conclusion

The results suggest the possible presence of monomers, dimers, and cyclic trimers on the dehydrated V_2O_5/SiO_2 samples. Upon exposure to ambient moisture, the V-O-Support bridge bonds partially hydrolyze to form di-grated species and a bulk-like species. No support for decavanadate or vanadia gel formation is available from this study. These results are compared with those obtained from SCR V_2O_5/TiO_2 catalysts where a variety of surface vanadium species are clearly observed by ^{51}V MAS NMR and clear changes are observed upon both a change in surface coverage and environmental conditions. Higher loadings appear to lead to larger vanadium domains. Additionally, the inclusion of tungsten on the surface appears to promote the formation of large vanadium domains. Hydration also has a notable effect on the structures, favoring the formation of larger vanadium domains, including bulk V_2O_5. Understanding the structure of these materials and how they respond to changes in chemical environment is essential for rational design of novel materials with predictable behavior.

Acknowledgements

This work is supported by the US Department of Energy, Offices of Basic Energy Sciences, Division of Chemical Sciences, Biosciences, and Geosciences, the Center for Understanding & Control of Acid Gas-Induced Evolution of Materials for Energy, the ARCS Foundation, and used NMR and supercomputing resources in the William R. Wiley Environmental Molecular Sciences

Laboratory, a national scientific user facility sponsored by the Department of Energy's Office of Biological and Environmental research at Pacific Northwest National Laboratory.

References

1. Roozeboom, F.; Mittelmeijer-Hazeleger, M. C.; Moulijn, J. A.; Medema, J.; De Beer, V. H. J.; Gellings, P. J., Vanadium Oxide Monolayer Catalysts. 3. A Raman Spectroscopic and Temperature-Programmed Reduction Study of Monolayer and Crystal-Type Vanadia on Various Supports. *The Journal of Physical Chemistry* **1980**, *84*, 2783-2791.

2. Wachs, I. E., Catalysis Science of Supported Vanadium Oxide Catalysts. *Dalton transactions* **2013**, *42*, 11762-11769.

3. Xie, S.; Iglesia, E.; Bell, A. T., Effects of Hydration and Dehydration on the Structure of Silica-Supported Vanadia Species. *Langmuir* **2000**, *16*, 7162-7167.

4. Jaegers, N. R.; Wan, C.; Hu, M. Y.; Vasiliu, M.; Dixon, D. A.; Walter, E.; Wachs, I. E.; Wang, Y.; Hu, J. Z., Investigation of Silica-Supported Vanadium Oxide Catalysts by High-Field ^{51}V Magic-Angle Spinning NMR. *The Journal of Physical Chemistry C* **2017**, *121*, 6246-6254.

5. te Velde, G.; Bickelhaupt, F. M.; Baerends, E. J.; Fonseca Guerra, C.; van Gisbergen, S. J. A.; Snijders, J. G.; Ziegler, T., Chemistry with ADF. *Journal of Computational Chemistry* **2001**, *22*, 931-967.

6. Islam, M. M.; Costa, D.; Calatayud, M.; Tielens, F., Characterization of Supported Vanadium Oxide Species on Silica: A Periodic DFT Investigation. *The Journal of Physical Chemistry C* **2009**, *113*, 10740-10746.

7. Hu, J. Z., et al., Investigation of the Structure and Active Sites of Tio2 Nanorod Supported Vox Catalysts by High-Field and Fast-Spinning ^{51}V MAS NMR. *ACS Catalysis* **2015**, *5*, 3945-3952.

8. Howard, C. J.; Sabine, T. M.; Dickson, F., Structural and Thermal Parameters for Rutile and Anatase. *Acta Crystallographica Section B* **1991**, *47*, 462-468.

496396 Mechanistic Origins of Unselective Products in Acrolein Synthesis on Mixed Metal Oxide Catalysts

Abstract:

Mechanistic origins of unselective oxidation products during acrolein synthesis on commercial mixed metal oxide catalyst

Linh Bui, Aditya Bhan

University of Minnesota â€“ Twin Cities, Minneapolis, MN 55455

The selective oxidation of propylene to acrolein is the first step in the production of high value acrylate-based chemicals. The formation of aldehydic and acidic organic byproducts, even in small quantities (less than 10% yield), poses significant challenges to the down-stream processing of acrolein and its derivatives. We describe in this talk the mechanistic origins of over-oxidation and C-C bond scission products in the partial oxidation of propylene to acrolein on mixed metal oxide catalysts. We combine transient kinetic studies, co-feed experiments of aldehydes and carboxylic acids formed as byproducts in propylene oxidation, and isotopic-labeling studies to elucidate the reaction mechanisms, identify the existence and the involvement of relevant surface intermediates, and develop an extensive reaction network describing the formation of all C_2 â€“ C_6 products (> 20 C_2-C_6 products are formed in this chemistry at carbon selectivity as low as 0.001%), and illustrate the underlying mechanisms for C-C bond cleavage and formation reactions. The carbon backbone of propylene is preserved in sequential oxidation of propylene to allyl alcohol, acrolein, and acrylic acid, as well as propylene oxidation to acetone and propanal via water-mediated pathways. Transient kinetic measurements in conjunction with co-feed experiments of C_2 and C_3 aldehydes and carboxylic acids show that decarbonylation and decarboxylation reactions, reactions of organic compounds with surface-adsorbed oxygen species, and total combustion reactions are the three mechanisms for C-C bond cleavage. C-C bond formation reactions that result in C_4 â€“ C_6 byproducts is shown to occur via: (i) the addition reaction of a propylene-derived surface allyl (C_3H_5) species with formaldehyde to form C_4 products and with propylene and/or allyl alcohol to form C_6 products, or (ii) the addition reaction of an acrolein (acrylic acid)-derived surface ethenyl (C_2H_3) intermediate with propylene to form pentadiene and with acrolein to form C_5 cyclic oxygenates. These mechanistic insights provide guidance for process conditions and catalyst development to minimize the formation of undesired products.

497019 Silica As a Redox Support for Enhanced CO_2 to CO By Perovskite Oxides in the Reverse Water Gas Shift Chemical Looping Process

Abstract:

Conversion of CO_2 is crucial for realigning the carbon cycle and essentially providing the means for renewable fuel production. Perovskite oxides, more notably $La_{0.75}Sr_{0.25}FeO_3$ (LSF), have exhibited promising CO_2 to CO conversion capabilities through high temperature redox applications such as the reverse water gas shift chemical looping process (RWGS-CL). In this work, LSF was combined with SiO_2 (25% w/w) thus resulting in CO production of 2850 μmoles/g LSF, 150% greater than the unsupported control. Silica also increased the reduction rate of LSF in isothermal RWGS-CL which has been the limiting reaction step. X-ray diffraction patterns revealed the minor presence of inevitable secondary phases $FeSiO_3$ and La_2SiO_5 along with a minute tridymite phase. These were all proven have no direct contribution to the activity enhancement through energy of vacancy formation calculations done by DFT accompanied by experiments with different LSF loading (25, 50, and 75% w/w). Crystallite size estimations by Scherrer analyses of post experimental samples showed that silica is capable of reversing the coalescence of perovskite grains, from high temperature syntheses, throughout redox applications. The increased surface area alleviates the transport resistance of bulk oxygen to establish a greater degree of vacancy formation for an improved CO formation yield.

490066 Computational Study of a MOF-Supported Single Site Ni Catalyst for Ethylene Dimerization

Abstract:

Metal-organic frameworks (MOFs) with ultrahigh porosity and large surface area have recently emerged as excellent heterogeneous supports for transition metal catalysts, either through functionalization of the linker or direct modification of the nodes. The mesoporous MOF, NU-1000, is of particular interest due to its chemical and thermal stability as well as its large pore size (~30 Å), which facilitates the diffusion of gaseous molecules. More importantly, the coordinatively unsaturated metal sites at the inorganic nodes provide specific sites for metal oxide species to anchor, enabling the synthesis of a highly dispersed and isolated single-site metal catalysts on the internal surface of NU-1000. Through atomic layer deposition, a highly dispersed Ni catalyst (Ni-NU-1000) was successfully synthesized; it shows good activity and stability for ethylene hydrogenation and was predicted to be a potential catalyst for ethylene oligomerization.(1,2)

In this work, we aim to study the structure of the Ni-NU-1000 catalyst and from there to identify the catalytic active sites for ethylene dimerization. In addition, the effect of the pore size on the selectity for 1-butene production have been studied. Furthermore, we have screened the first-row transition metals for their activity toward the ethylene dimerization. The goal is to provide physical insight into properties controlling the activity and selectivity for ethylene dimerization and then guide experimentalists to design a better catalyst for ethylene dimerization.

The current study provides physical insight into the properties of MOF supported metal clusters in controlling the activity and selectivity for dimerization of ethylene to produce 1-butene, which is a basic petrochemical building block and largely used as a co-monomer for the production of light weight polyethylene. As compared to homogeneous catalysts, the novel heterogeneous catalyst has many industrial advantages such as ease of product separation and catalyst recycling.

References
1. Li, Z.; Schweitzer, N. M.; League, A. B.; Bernales, V.; Peters, A. W.; Getsoian, A. B.; Wang, T. C.; Miller, J. T.; Vjunov, A.; Fulton, J. L. *J. Am. Chem. Soc., 138*, 1977 (2016).
2. Mondloch, J. E.; Bury, W.; Fairen-Jimenez, D.; Kwon, S.; DeMarco, E. J.; Weston, M. H.; Sarjeant, A. A.; Nguyen, S. T.; Stair, P. C.; Snurr, R. Q. *J. Am. Chem. Soc., 135*, 10294 (2013).

490094 Two-Dimensional Nanomaterials As Supports for Heterogenous Catalysis

Abstract:

Two-Dimensional (2D) nanomaterials have garnered remarkable interest recently due to their promising potential in versatile applications such as energy storage devices including lithium ion batteries (LIB) and capacitors. However, catalytic properties of 2D nanomaterials, especially for thermo-catalytic reactions, are rarely studied. The 2D structures may facilitate the characterizations of the supported nanoparticles and help determine the chemistry of active sites.

Water gas shift (WGS) reaction is widely used in industry to produce clean hydrogen for plenty of processes such as ammonia synthesis, methanol synthesis and hydrocarbon processing. Moreover, WGS is a useful probe reaction that can shed light on fundamental catalytic processes over considerable catalyst systems. In this presentation, we studied the effects of 2D nanomaterials as supports of noble metals for WGS reaction. Kinetics such as orders of reactants cast light on absorption and coverage of the reactants. Scanning transmission electron microscopy (STEM) as well as X-ray absorption fine structure (XAFS) were employed to identify the nature of active sites.

496895 Mitigating the Oxidative Deactivation on a Multi-Faceted Fe Catalyst: A Multi-Scale Model from First Principles

Abstract:

One promising method for addressing the growing need for diverse, sustainable energy sources is bio-oil produced via the fast-pyrolysis of biomass. However, these bio-oils have high concentrations of oxygenated compounds, resulting in fuel quality issues[1] that require catalytic hydrodeoxygenation (HDO), thereby reducing the overall oxygen content and improving fuel quality. Bimetallic, doped iron, catalysts have demonstrated synergistic behavior that contributes to a more cost-effective and longer-lasting HDO catalyst[2-6]. In order to improve the performance of such bimetallic iron catalysts, the oxidative deactivation mechanism that limits these catalysts must be determined.

While the addition of dopants to Fe catalysts exhibits an overall synergistic interaction, further investigation into the behavior of oxygen on a doped catalytic Fe grain is required to minimize the use of precious metals while maximizing its synergistic properties[2,4,7]. Here, we model the oxidative deactivation of a full catalytic Fe grain by exposure to O_2 as well as taking into account both coverage and facet effects on the kinetic behavior. Additionally, as the application of external electric fields has demonstrated favorable effects on other catalytic processes, such as methane steam reforming, the effect of a range of electric fields were also investigated[8-11]. Such a realistic, multi-faceted model of a doped Fe grain is an effective tool for optimizing the use of precious metal dopants and taking advantage of the effect of an electric field. These realistic, predictive models help to bridge the gap between experiment and theory, streamlining research efforts and accelerating progress.

In order to build the multi-faceted catalytic Fe grain model, the adsorption of oxygen was calculated on both clean and doped Fe(100), Fe(110), and Fe(111) over a range of surface coverages. The effect of an electric field on oxygen adsorption on these three Fe facets was studied over a range of ±1.0 V/Å, examining each available adsorption site over a range of coverages. Surprisingly, we found that the undoped Fe(111) facet favors a high oxygen coverage even under low O_2 exposure and all electric field strengths (see Figure 1). Results indicate that the undoped Fe(110) surface is the most strongly affected by the electric field over the coverage range, while the undoped Fe(100) surface is the least effected. As seen in Figure 1, the undoped Fe(100) and Fe(111) surfaces have the strongest adsorption energy, resulting in high oxygen coverage upon exposure to O_2. Preliminary work on Pd doped Fe(100) with varying Pd concentration has shown that increasing the surface concentration of Pd significantly decreased the adsorption strength of oxygen, which is consistent with other studies on the adsorption of benzene and phenol on Fe doped surfaces[12,13]. These results suggest that the primary function of Pd is the disruption of oxide formation due to the aversion that oxygen has for Pd on the Pd/Fe(100) surfaces. Consequently, the degree of catalyst deactivation due to the formation of a surface oxide likely decreases with the addition of Pd to an Fe surface, thereby promoting HDO with minimal deactivation of the active Fe surface.

The three Fe facets studied here were then used to construct a model of a multi-faceted grain surface, incorporating the anisotropic effects of the surface orientation on kinetic characteristics[14]. Such a model accounts for the effect of pressure, temperature, electric field, dopants, and surface orientation on the adsorption/desorption and oxidative deactivation behavior. This doped Fe catalytic grain model can be used as an experimentally predictive, useful tool for optimizing methods in which HDO catalysts are designed to reduce oxidative deactivation while minimizing the use of costly precious metals.

Figure 1: Equilibrium oxygen coverage on a multi-faceted catalytic iron grain at 500 K,1.5 Pa, and an electric field of -1.0 V/Å.

References:

1. Wang, H., Male, J. & Wang, Y. Recent Advances in Hydrotreating of Pyrolysis Bio-Oil and Its Oxygen-Containing Model Compounds. *ACS Catal.* **3,** 1047–1070 (2013).

2. Hong, Y., Zhang, H., Sun, J., Ayman, K. M. & Hensley, A. Synergistic catalysis between Pd and Fe in gas phase hydrodeoxygenation of m-cresol. *ACS Catalysis* (2014). doi:10.1021/cs500578g

3. Nie, L., de Souza, P. M., Noronha, F. B. & An, W. Selective conversion of m-cresol to toluene over bimetallic Ni–Fe catalysts. *Journal of Molecular …* (2014). doi:10.1016/j.molcata.2013.09.029

4. Sun, J. *et al.* Carbon-supported bimetallic Pd–Fe catalysts for vapor-phase hydrodeoxygenation of guaiacol. *Journal of Catalysis* **306,** 47–57 (2013).

5. Sitthisa, S., An, W. & Resasco, D. E. Selective conversion of furfural to methylfuran over silica-supported Ni Fe bimetallic catalysts. *Journal of Catalysis* (2011). doi:10.1016/j.jcat.2011.09.005

6. González-Borja, M. Á. & Resasco, D. E. Anisole and guaiacol hydrodeoxygenation over monolithic Pt–Sn catalysts. *Energy & Fuels* (2011). doi:10.1021/ef200728r

7. Hensley, A. J. R. *et al.* Enhanced Fe 2O 3Reducibility via Surface Modification with Pd: Characterizing the Synergy within Pd/Fe Catalysts for Hydrodeoxygenation Reactions. *ACS Catal.* **4,** 3381–3392 (2014).

8. Che, F., Ha, S., McEwen, J. S., Hensley, A. J. & Zhang, R. Elucidating the field influence on the energetics of the methane steam reforming reaction: A density functional theory study. *'Applied Catalysis B, Environmental'* (2016). doi:10.1016/j.apcatb.2016.04.026

9. Che, F., Gray, J. T., Ha, S. & McEwen, J.-S. Improving Ni Catalysts Using Electric Fields: A DFT and Experimental Study of the Methane Steam Reforming Reaction. *ACS Catal.* (2016). doi:10.1021/acscatal.6b02318

10. Che, F., Ha, S. & McEwen, J. S. Hydrogen Oxidation and Water Dissociation over an Oxygen-Enriched Ni/YSZ Electrode in the Presence of an Electric Field: A First Principles-Based Microkinetic …. *Industrial & Engineering Chemistry …* (2017). doi:10.1021/acs.iecr.6b04028

11. Che, F., Gray, J. T., Ha, S. & McEwen, J. S. Catalytic water dehydrogenation and formation on nickel: Dual path mechanism in high electric fields. *Journal of Catalysis* (2015). doi:10.1016/j.jcat.2015.09.010

12. Hensley, A. J. R., Zhang, R., Wang, Y. & McEwen, J.-S. Tailoring the Adsorption of Benzene on PdFe Surfaces: A Density Functional Theory Study. *J. Phys. Chem. C* **117,** 24317–24328 (2013).

13. Hensley, A., Schneider, S., Wang, Y. & McEwen, J. S. Adsorption of aromatics on the (111) surface of PtM and PtM 3 (M= Fe, Ni) alloys. *RSC Adv.* (2015). doi:10.1039/C5RA13578H

14. McEwen, J. S., Gaspard, P., de Bocarmé, T. V. & Kruse, N. Oscillations and Bistability in the Catalytic Formation of Water on Rhodium in High Electric Fields. *J. Phys. Chem. C* **113,** 17045–17058 (2009).

485927 Heterogeneous Dephosphorylation Using Ceria Nanocatalysts: Identifying the Active Site and the Rate-Determining Step

Abstract:

As an essential element in living systems, phosphorus is crucial for cell division and growth, energy storage and conversion, respiration, photosynthesis, and other biological processes. Phosphorus is found naturally in soil, but concentrations can fall to low levels depending on geographic conditions, and artificial phosphorus fertilizers are demanded to sustain the growth of crops. Before the early 1900s, most of the worldâ€™s phosphorus was derived from animal wastes. Today the vast majority (nearly 80%) of phosphorus in fertilizers comes from phosphate rocks, which are primarily harvested in the remote Western Sahara region. It is predicted that the production of phosphate rocks will reach its peak before 2040 and the reserves will be completely depleted by the end of this century. Therefore it is imperative to develop innovative and sustainable methods for production of phosphorus from renewable sources.

One solution toward this sustainability challenge is to extract phosphorus from phosphorylated biomolecules (e.g., phospholipids and nucleic acids) and recycle it for fertilizer production. This is conventionally done via fermentation, which produces struvite, a precipitate that can be processed to make phosphorus fertilizers. Alternatively, a more robust approach is catalytic dephosphorylation. By hydrolytic cleavage of the phosphate ester bond, free phosphate anions can be released from biomass and agriculture wastes, which can then be captured and regenerated as chemical streams for further applications. Previously studies of cerium oxide (CeO_2) nanocrystals for biological applications (e.g., cancer therapy, pharmacology and toxin mitigation) have shown that these nanomaterials can function as artificial phosphatases and catalyze the dephosphorylation of nucleic acids, peptides, DNA and RNA under ambient conditions. Thereby CeO_2 is believed to be a promising catalyst for dephosphorylation and recovery of phosphorus from biomass and organic wastes

Here we report on catalytic dephosphorylation for recovery of phosphates from organic and biological molecules. Ceria (CeO_2) nanocrystals were synthesized with shape control and applied as â€œartificial phosphatasesâ€ to cleave the phosphate ester bond in para-nitrophenyl phosphate (p-NPP) and release free phosphate anions in aqueous solutions. After a fundamental study on this model reactant, CeO_2 nanocrystals were also applied to feedstocks to phospholipids and nucleic acids. The dephosphorylation reaction was studied on the CeO_2 nanocrystals at various temperatures to evaluate the dependences of rate constant, activation energy and recyclability on the particle shape. The structure-property relationship established in these studies suggests that the oxygen vacancies on the surface of CeO_2 are the active sites for dephosphorylation. Adsorption studies of poisoning species allow for the proposal of a potential rate-determing step in this reaction.

487048 Surface Reconstruction of Pd Concave Icosahedra: Atomic-Scale Mechanisms from First-Principles and Experiments

Abstract:

Concave surfaces exhibit a large density of high-index facets, such as steps, edges, and kinks. These structural features offer sites of high surface energy, and low coordination number[1]. Consequently, they have shown improved catalytic activities toward a number of reactions[2], including formic acid electro-oxidation[3], and Suzuki coupling reaction[4]. A common aspect of this diverse set of reactions is that they take place at low temperatures (<100 °C), which does not compromise the structural integrity of the as-synthesized concave structures. However, a myriad of structure-sensitive catalytic processes take place at much higher temperatures. Therefore, acquiring an atomic-level understanding of the mechanisms by which the concave surface structures transform into more common surface facets at elevated temperatures is of paramount interest. Here, we focus on the atomic surface reconstruction of Pd concave icosahedra to regular icosahedra. Applying self-consistent periodic density functional theory (PW91-GGA) to an atomistic model of a concave icosahedron edge, we quantify activation energy barriers for key steps responsible for the surface reconstruction, including step deformation and adatom diffusion. Our calculated energetics are surmountable at the temperatures at which the reconstruction experimentally takes place, as observed by *in situ* high-resolution transmission electron microscopy (HRTEM). Combining experimental and theoretical methods, we offer insight into the mechanism of surface reconstruction of Pd concave icosahedra to regular icosahedra, which offers an opportunity to rationally design thermally-durable concave nanoparticles.

References

1. Ruditskiy A, Peng HC, Xia YN. Shape-Controlled Metal Nanocrystals for Heterogeneous Catalysis. *Annual Review of Chemical and Biomolecular Engineering, Vol 7.* 2016;7:327-348.

2. Zhang JA, Langille MR, Personick ML, Zhang K, Li SY, Mirkin CA. Concave Cubic Gold Nanocrystals with High-Index Facets. *Journal of the American Chemical Society.* Oct 2010;132(40):14012-14014.

3. Huang XQ, Zhao ZP, Fan JM, Tan YM, Zheng NF. Amine-Assisted Synthesis of Concave Polyhedral Platinum Nanocrystals Having {411} High-Index Facets. *Journal of the American Chemical Society.* Apr 2011;133(13):4718-4721.

4. Jin MS, Zhang H, Xie ZX, Xia YN. Palladium Concave Nanocubes with High-Index Facets and Their Enhanced Catalytic Properties. *Angewandte Chemie-International Edition.* 2011;50(34):7850-7854.

496857 Enhanced Catalysis By Optical Nanoantenna Reduced on Monolayer Transition Metal Dichalcogenide

Abstract:

Enhanced Catalysis by Optical Nanoantenna Reduced on Monolayer Transition Metal Dichalcogenide

D. Keith Roper[a,b*], Jeremy R. Dunklin[c], Gregory T. Forcherio[d], Alexander O'Brien[a]

[a]*Department of Chemical Engineering, University of Arkansas, Fayetteville, AR 72701*

[b]*Microelectronics and Photonics Graduate Program, University of Arkansas, Fayetteville, AR 72701*

[c]*National Renewable Energy Laboratory, Golden, CO 80401*

[d]*Army Research Laboratory, Adelphia, MD 20783*

Heterostructures of two-dimensional (2D) transition metal dichalcogenide (TMD) and optical nanoantenna (NA) offer optoelectronic inducibility with superior electron mobility and gate tunability. This enables enhanced photocatalytic activity of interest for chemical and biological catalysis, energy, sensing, desalination, and nanoelectromechanical systems. However, characterization of interrelated electro-optical and thermal effects at TMD-NA heterointerfaces is predominantly empirical. Simulation and its integration with microscopic and spectroscopic analysis of TMD-NA has been limited to date by complexity and convolution of effects, particularly for characterizing dynamic interactions at nanometer (nm) scales. Compact, multi-scale, integrated analysis of optical, electronic and catalytic effects could identify extraordinary features to guide design and implementation.

This work compared simulated vs. measured optical, electronic, and spectrocopic properties of TMD-NA for photoreductive catalysis. Heterostructures of monolayer (1L) TMD and NA were self-assembled via exfoliation and redox chemistry (Fig 1). Nanometer- and femtosecond-resolved electron energy loss spectroscopy (EELS) was used to simulate and measure low-energy NA plasmon modes, damping and electric near fields at heterointerfaces (Fig 2). Discrete dipole approximation (DDA) was used to analyze transmission ultraviolet-visible (TUV-vis) extinction spectra (Fig 3). Heterostructures of 1LTMD-NA were tested as catalyst electrodes for hydrogen evolution reaction using linear sweep voltammography.

Transmission electron microscopy (TEM; Fig 1) and x-ray photoelectron spectroscopy (XPS) showed NA were reduced directly onto 1LTMD flake edges forming covalent Au-sulfur bonds, the first such reported result. EELS (Fig 2) showed enhanced plasmon damping attributable to direct electron transfer at the catalytic 1LTMD-NA heterointerface. DDA (Fig 3) revealed plasmon and exciton features enhanced in far field extinction spectra of 1LTMD-NA heterostructures. Photocatalytic reaction rates measured for these novel 1LTMD-NA were over 10-fold higher than those reported for other TMD-based electrodes.

This integration of novel experimental, computational, microscopic and spectroscopic results leads to a new paradigm for improved photocatalysis at 1LTMD-NA interfaces. These new tools and data support improved design of 1LTMD-NA photocatalysts and their implementation in a variety of chemical, biological, energy and water systems.

[1] J.R. Dunklin, D.K. Roper et al., in preparation. [**2**] G. T. Forcherio, J.R. Dunklin, D.K. Roper et al., in submission. **[3]** J.R. Dunklin, D.K. Roper et al., in preparation.

502033 Kinetics of the thermal decomposition of sulfur compounds

Abstract:

Abstract

In this contribution the kinetics of the thermal decomposition of sulfur compounds will be discussed both experimentally and theoretically under typical steam cracking and pyrolysis conditions. The experimental investigation has been conducted in a new 85 cm long annular quartz flow reactor, having an outer and inner tube diameters of 2.1 and 1.5 cm respectively. Both pure sulfur compounds pyrolysis in nitrogen (0.04 mol% S in N2) as well as in the presence of hydrocarbons (0.04 mol% S and 13.7 mol% of HC in N2) were studied covering a wide temperature range of 600 to 850 °C, at a fixed pressure of 1.5 bara and residence time of ~0.5 s. The reaction effluent is sampled on-line at 300 °C. The unique on-line sampling system consists of valve-based sampling manifold and uniformly heated transfer lines connected to a comprehensive two-dimensional gas chromatograph (GC × GC) equipped with a flame ionization detector (FID) and a sulfur chemiluminescence detector (SCD). This dedicated set-up allows to detect and quantify both pure (with the FID) and sulfur-containing hydrocarbons (with the SCD) using heptane and 3-chlorothiophene as internal standards, respectively. The GC × GC-FID/-SCD settings are optimized for trace analysis of sulfur compounds in olefin-rich (ethylene- and propylene-) hydrocarbon matrices produced during steam cracking.

To understand the underlying decomposition mechanism of sulfur compounds in the presence of a complex hydrocarbon matrix formed by its decomposition, a group additive kinetic model was constructed via an in-house automated reaction mechanism generation software called Genesys [1]. It relies on ab-initio CBS-QB3 based kinetic coefficients for the reference/ parent reactions for each elementary reaction family, and for the others, contribution of groups to the single event Arrhenius parameters. No adjustment of the Arrhenius parameters to the experimental data is performed. The kinetic model is validated by comparison with the set of experimental data gathered in the annular quartz flow reactor by providing a satisfactory representation of the latter.

[1] Nick M. Vandewiele, Kevin M. Van Geem, Marie-Françoise Reyniers, Guy B. Marin, Genesys: Kinetic model construction using chemo-informatics, Chemical Engineering Journal, Volumes 207-208, 2012, Pages 526-538.

500300 A Cylindrical Formulation of the One Dimensional Turbulence (ODT) Model for Turbulent Jet Flames

Abstract:

The One Dimensional Turbulence (ODT) model has been successfully applied to a wide range of turbulent flows including homogeneous turbulence, jets, wakes, mixing layers, channel flows, Rayleigh-Taylor, buoyant wall flows, and others. ODT solves a flow by temporally-evolving one-dimensional transport equations for mass, momentum, and (optionally) reacting scalars. Advection is modeled through stochastic eddy events that are implemented as triplet maps. The size, location, and rate of these maps are drawn from an eddy rate distribution in a manner that this consistent with turbulent inertial range advection. The model is an extension to the Linear Eddy Model (LEM), and differs from LEM in that, in ODT, momentum components are solved in addition to scalar quantities, and these momentum components are used to define eddy rate distribution function. Hence, ODT is a dynamic turbulence model and can accurately capture velocity and mixing profiles in nonuniform turbulence. Because the model is one dimensional, it is limited to boundary-layer like flows, but such flows are important and widely studied, with a wealth of experimental data available. ODT is unique in that it is able to resolve a full spectrum of turbulent time and length scales (in one dimension). Only direct numerical simulation is able to resolve all scales of a turbulent flow, but this is done at a computational cost that is prohibitive for engineering flows at realistic Reynolds numbers. Large Eddy Simulation directly captures large eddies, but models small subgrid scale motions. Conversely, ODT resolves fine scale structures, allowing diffusion processes in the physical space coordinate, but models the large-scale advection. Often the fine scales are more difficult to model since nonlinear multi-component diffusion and reaction processes occur at dissipation scales. Large-scale advection is often not too difficult to model in boundary layer flows.

ODT has been recently validated quantitatively against several DNS in compatible configurations, namely, temporal, planar jets. This has been done for soot formation, and flame extinction and reignition using ethylene and syngas fuels. Premixed flames have also been studied. As the model has been extended to more and more complex flows, there has been a desire to use ODT as a surrogate DNS, due to its low cost compared to DNS. Such validation studies comparing ODT and DNS are promising in this area. However, DNS is limited to relatively few simulations and relatively low Reynolds number. Application of ODT to experimental flows is limited by the configuration. Experimental jets are usually cylindrical and evolve spatially, whereas existing ODT simulations are planar and (normally) evolve temporally. (There is a spatial formulation of the model in which a one-dimensional flow evolves spatially in a down-stream direction like a parabolic boundary layer formulation.)

We extend the ODT model to a cylindrical formulation. The formulation is presented, along with challenges and interpretations. We present comparisons of the cylindrical ODT formulation to round pipe flow, round nonreacting jets at high Reynolds number, and to reacting nonpremixed jet flames. The new formulation is expected to improve previous predictions of jet flows, including, for instance, particle flows. We discuss such expected improvements.

AUTOMATED TRANSITION STATE THEORY CALCULATION OF HYDROGEN ABSTRACTION FROM NOVEL BIOFUELS

Nathan D. Harms, Richard H. West
Northeastern University, Boston MA 02115

Abstract

Understanding and predicting the combustion properties of novel fuels allows for development of greener and more efficient fuels and engines. To predict the detailed combustion kinetics of a single novel fuel requires the construction of a reaction mechanism consisting of thousands of elementary reactions and their kinetics. The reaction rates can be determined by performing transition state theory (TST) calculations on each elementary reaction. However, with dozens of novel biofuels being considered, and each fuel having its own reaction mechanism, the TST calculations must be automated. We present a fully-automated method to determine reaction rates via automated Transition State Theory (AutoTST) calculations, coupled to the Reaction Mechanism Generator (RMG) [1, 2]. We focus here on the abstraction of hydrogen by six common radicals present in combustion ($HO\cdot$, $H\cdot$, $HOO\cdot$, $O\cdot$, $H_3C\cdot$, and O_2) from several biofuels being considered by the U.S. Department of Energy "Co-Optima" program. Hydrogen abstraction reactions are one of the primary reaction families that facilitate combustion and is arguably the most important class of reactions in the combustion process; therefore, accurately determining kinetic parameters for these reactions is crucial. Our approach uses RMG to determine the possible hydrogen abstraction reactions for each fuel, and the reaction kinetics are determined using our AutoTST method. The AutoTST method estimates the transition state geometry using a group contribution and distance geometry method, performs DFT geometry optimization, performs intrinsic reaction coordinate calculations to verify the transition state, and uses CanTherm [3], a canonical transition state theory calculator, to determine kinetic parameters. This method has previously been applied to hydrogen abstractions and results have been compared to published values and benchmark calculations [4] The application of AutoTST is broad and can be applied to all families of combustion reactions given proper resources.

Introduction

Long term energy, environmental, and economic needs have motivated the study of renewable energy sources to replace fossil fuels. Currently, the United States is responsible for 25 % of the world's CO_2 production despite the size of its population. [5] In order to limit the United State's production of CO_2 from motor vehicle transport, the Department of Energy (DOE) developed the "Co-Optima" program to simultaneously optimize the selection of novel biofuels and the design of efficient engines, hoping co-optimization of both will maximize performance, efficiency, and adoption. Designing an engine for a fuel that has not yet been made in significant quantities, however, requires a *predictive* kinetic model, as does evaluating a fuel's performance in an engine that does not yet exist.

From this investigation, 23 biofuels were identified as potential replacements for fossil fuels or blend-stocks (Figure 1). Few of these fuels have accurate kinetic models.

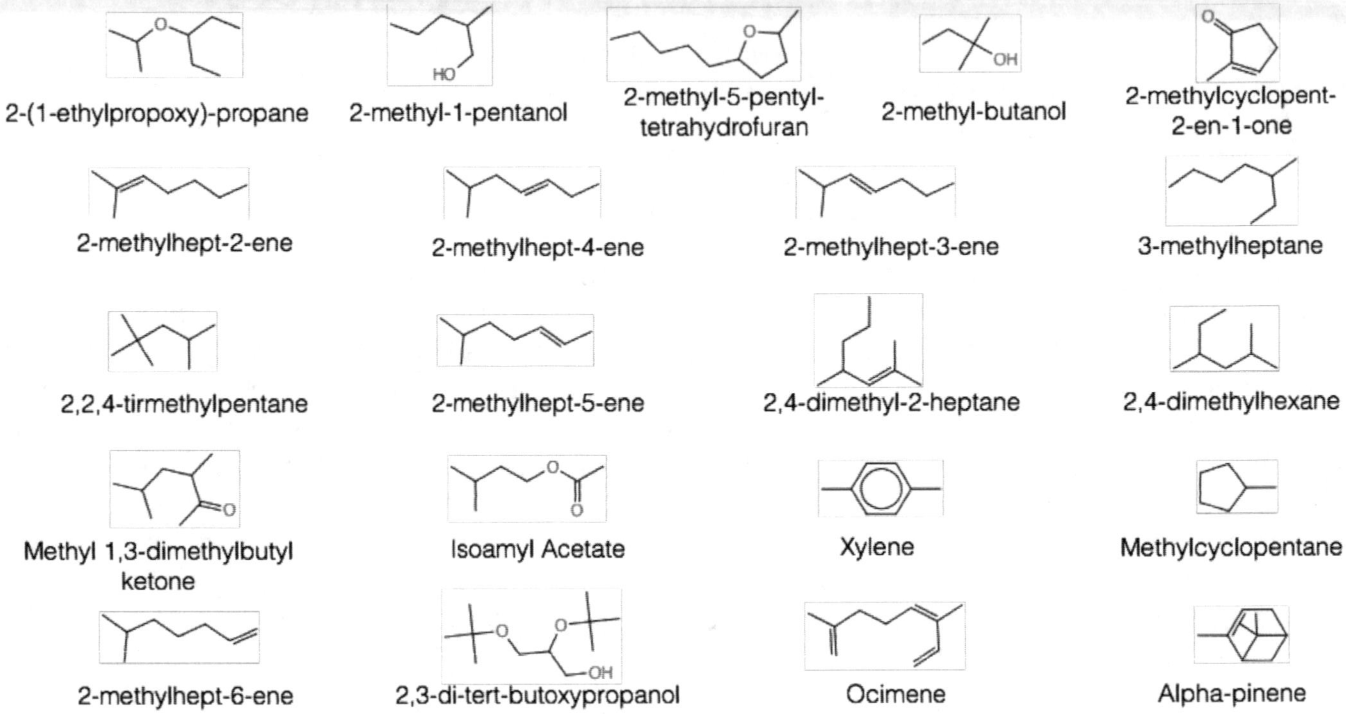

Figure 1: *The 23 novel biofuels being investigated in this study.*

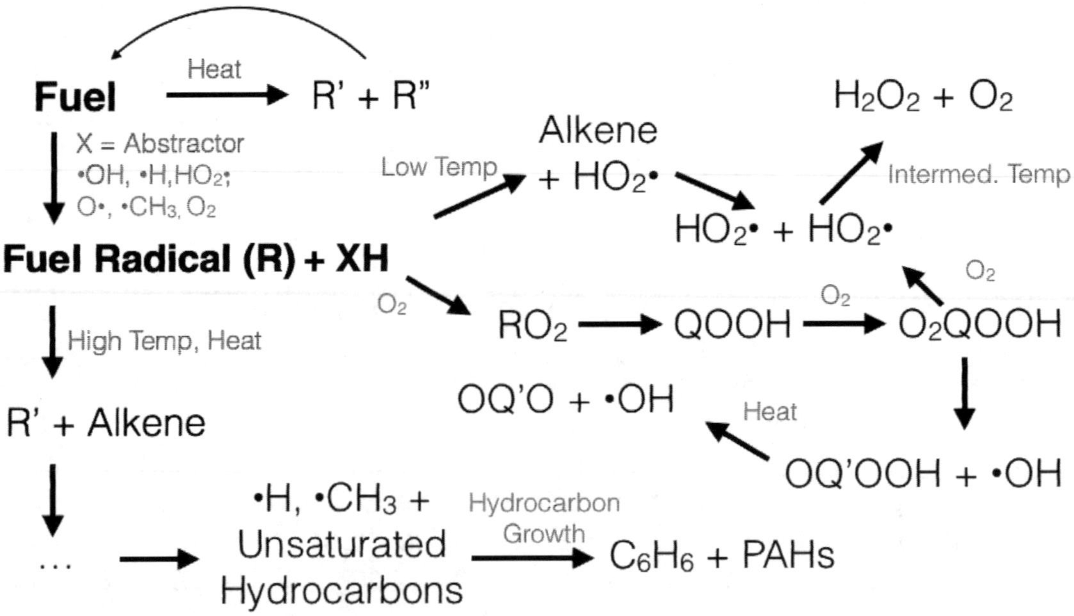

Figure 2: *Overview of a generic combustion process*

Reaction mechanisms often follow a typical road map outlined in Figure 2. For this work, we focus on the abstraction of hydrogen by six common radicals present in combustion: $HO\cdot$, $H\cdot$, $HOO\cdot$, $O\cdot$, $H_3C\cdot$, and O_2. Hydrogen abstraction reactions are some of the most important reactions in the combustion of fuel, and abstractions from the fuel molecule are the first in a set of thousands of reactions that allow for the rest of combustion to occur.

When constructing a reaction mechanism, rates are usually generated in three ways: estimates and intuition, rate rules, or transition state theory on the bases of quantum chemistry calculations [6]. Decent estimation require decades of prior knowledge in the combustion field and, even then, these estimates will be just that – estimates. Rate rules are an efficient method to determine kinetics for reactions, essentially codifying the estimates an experienced scientist might make. These methods are orders of magnitude faster than quantum calculations but their accuracy suffers when applied outside of the range of "chemical space" where the rules were developed and trained: These rate rules can be misapplied (i.e. applied to a reaction that is too dissimilar to the reactions from which it was derived), or the chemistry of a reaction might not follow the rules. Because of this, quantum calculations are often used to develop more accurate reaction rates.

Although experiments are essential to test and validate an overall kinetic model, it is often challenging to isolate an individual elementary reaction in an experimental study, and with the latest computational techniques some reactions can be calculated *ab initio* with an accuracy exceeding experiments, and at a fraction of the cost. Because of this, theoretical calculations are often utilized to determine kinetics in combustion mechanisms.

Utilization of quantum packages like Gaussian [7] allows one to optimize geometries of reactants and products by walking down the potential energy surface (PES) until a minimum energy has been found. This has been automated for on-the-fly calculation of thermochemical properties of intermediate species in RMG [8]. For transition state geometries, an initial guess of the transition state will undergo a similar search, but the program will search for a saddle point on the PES instead: a stationary point with negative curvature in one direction (the normal mode corresponding to the reaction coordinate) and positive curvature in all other directions. The search algorithm is usually only successful if the initial guess is close enough to the correct saddle point geometry that the direction of negative curvature is correct and pointing towards the saddle point. In other words, it needs a very good initial guess, which often requires significant input from the user. In order to automate the search for saddle points, Bhoorasingh et al. developed a procedure to provide a guess for the transition state geometry based on the reactants, products, and the type of reaction [9]. The algorithm used group-additivity to determine inter-atomic distances for the atoms participating in the reaction, and distance geometry to convert these into a 3D structure. Successful optimization of the TS geometry is validated using Intrinsic Reaction Coordinate (IRC) calculations. These calculations use the TS as a starting point and walk down the PES until a local minimum energy is attained. If the reactants and products of these local minima match the actual reactants and products of the reaction, the TS is validated.

Successful optimization of these geometries allows one to determine the the canonical partition functions from the results of the DFT calculations and subsequently kinetics using transition state theory (equation 1).

$$k(T) = \frac{k_B T}{h} \frac{Q^{TS}}{\Pi_i^n Q_i} \exp\left(\frac{-\Delta E_0}{k_B T}\right) \tag{1}$$

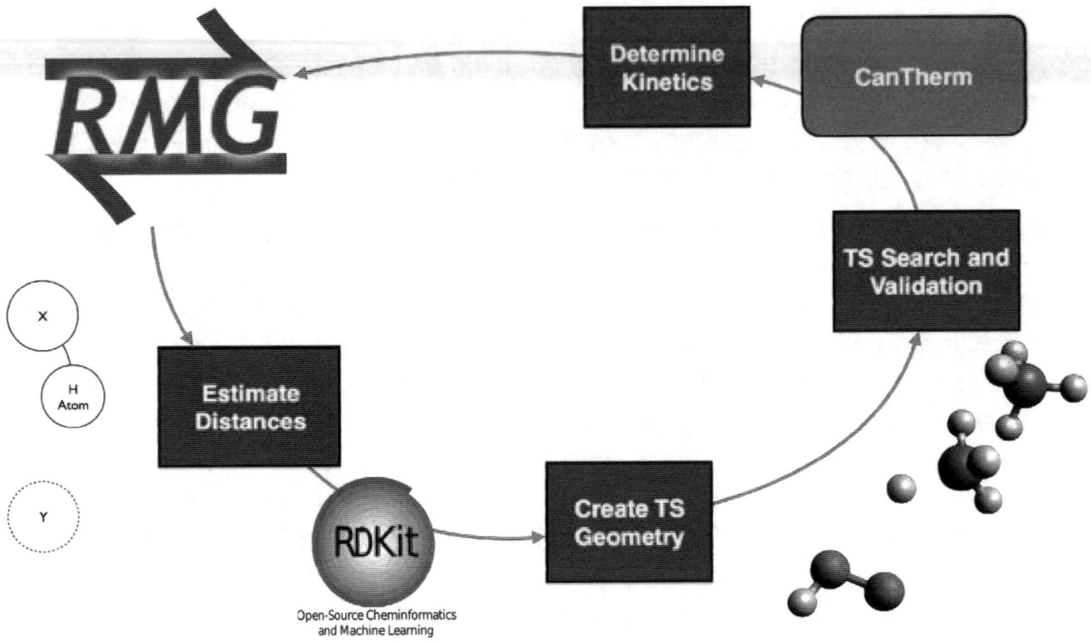

Figure 3: *Overview of the AutoTST process developed by [10]*

$k(T)$ represents the reaction rate, k_B is the Boltzmann constant, T is temperature, h is Plank's constant, Q^{TS} is the canonical partition function for the TS, $\Pi_i^n Q_i$ represents the product of the canonical partition functions for the reactants and ΔE_0 is the activation energy. This highlights that the temperature-dependent reaction rate can be determined by the partition functions and activation energy, which can all be determined from the electronic structure or DFT calculations of the reactants and transition state.

This geometry-producing algorithm was coupled to a canonical TST calculator to provide a fully automated AutoTST workflow (procedure outline in Figure 3) [4, 10].

Methods

Identification of possible hydrogen abstraction reactions

Identification of the hydrogen abstraction reactions of each of the proposed biofuels was accomplished by using Reaction Mechanism Generator (RMG) [1]. RMG identified 692 hydrogen abstraction reactions that were subsequently run using AutoTST.

Perform automated transition state theory calculations

The overall workflow is described in the recent paper by Bhoorasingh et al. [4] but a brief overview is given here. Reactant and product geometries were created using algorithms in RMG developed by Magoon and Green [8]. Transition state geometries were estimated using a group contribution method developed by Bhoorasingh and West [9]. This group contribution method identifies functional groups in each of the reactants and products, which are used to estimate the distances between atoms at the reacting center using previously identified transition states as training data. The transition state geometry estimate was optimized to an energy minimum with the reaction

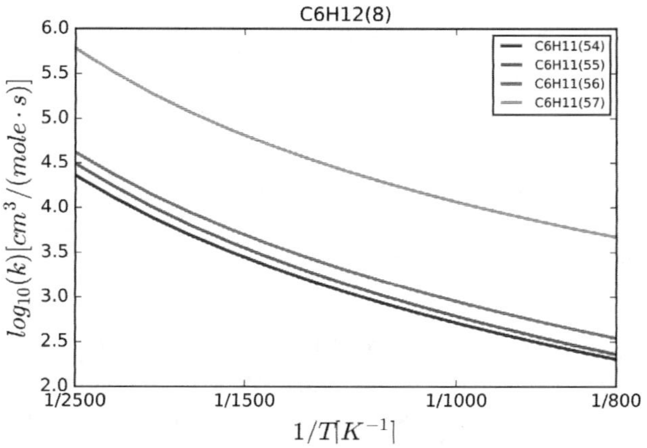

(a) *Arrhenius plot of HOO· abstracting a hydrogen from methylcyclopentane. Colors correspond to subfigure (b)*

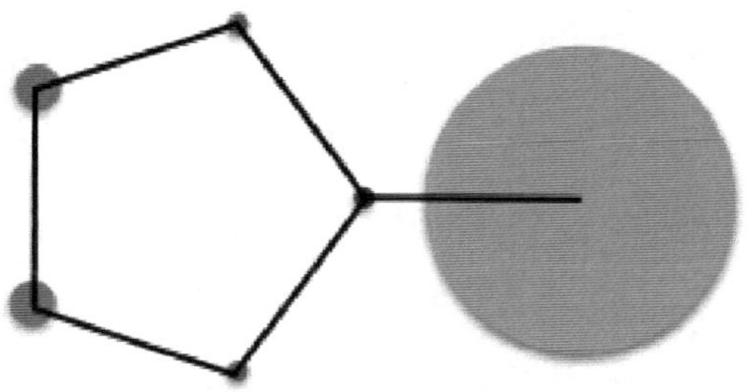

(b) *Depiction of methylcyclopentane. Colors correspond to subfigure (a). Area of each circle proportional to H abstraction rate at that site by HOO·*

Figure 4: *An Arrhenius plot and heat map of the hydrogen abstraction reactions from methylcyclopentane by HOO·.*

center distances frozen followed by a saddle point search with all distances frozen except for the reaction center. The resulting geometry is then fully optimized to a saddle point. If a saddle point is found, an intrinsic reaction coordinate (IRC) calculation is performed to verify the reactants and products. If the reactants and the products from the IRC calculations, determined from the Cartesian coordinates using a connect-the-dots algorithm, match the reactants and products from the initial reaction, the transition state validation is complete. All the geometries were optimized and validated with the M06-2X DFT functional with the MG3S basis set in the Gaussian 09 quantum chemistry package.

Following this validation, the CanTherm software package [3] was used to determine kinetic parameters using TST. Symmetry numbers for the rate calculations were determined using the SYMMETRY software [11]. SYMMETRY takes the optimized transition state geometry, allows for small deviations, and calculates the point group. Product properties were also calculated so that tunneling corrections could be applied via the Eckart model.

Results

Through our presented work flow, we have identified 692 hydrogen abstraction reactions on 23 novel biofuels, started calculations on 540 of these reactions, and successfully calculated kinetics for 106 of them. At time of preparation of this manuscript, the completion rate is 19.6%, which is substantially lower than the success rate previously reported for this method of $\sim$70% [4].

Although the current success rate of this work flow is low, we have successful identified the reaction rates of HOO· abstracting a hydrogen from methylcyclopentane at all unique sites, for example. In figure 4, we see an Arrhenius plot of the hydrogen abstraction rate by HOO· and a corresponding "heat map" of methylcyclopentane. This has told us that the methyl group of this proposed biofuel is most likely to lose a hydrogen.

Total Reactions Identified	692
Reaction Calculations Started	540
Overall Successes	106
Success Rate	19.6%

Table 1: *Success rates at time of manuscript preparation.*

Conclusions

We are still investigating the currently low success rate. Some failure modes include: not finding the lowest energy conformer for a reactant so that the barrier height is negative; failed TS optimizations in Gaussian; converging upon a saddle point that is not the required transition state (eg. a conformational change); and long computational time associated with IRC calculations leading to interrupted calculations. Many abstraction reactions will succeed in one direction but not the other, which is not reflected in these results.

In order to obtain the lowest energy conformers, we intend to implement Fafoom, a conformer searcher that utilizes a genetic algorithm technique [12]. We have tested this approach on reactant and product molecules, and plan to extend the approach to transition states.

A number of our calculations resulted in a non-existent barrier height. This could have occurred because tolerances on calculated energies of reactants, products, and TSs can overlap if the optimization is not exact enough. We propose running a stricter calculation on these geometries to minimize their tolerances and obtain barrier heights.

Some of our TS optimizations failed to converge as well. This could be because the guess provided by AutoTST was too far from the actual TS. One method to try and correct this is to retrain the data which the distance geometry estimates in AutoTST are based on. This will result in more accurate reaction coordinates for the TS guess and should give a higher success rate. Another avenue to explore is performing a looser optimization on the geometry in order to get convergence, then performing a tighter optimization on the loosely optimization geometry.

Finally, we would like to find a faster method of validation for AutoTST calculations. Currently, the IRC calculations are the most computationally expensive calculation. We have identified Quick Reaction Coordinate calculations [13] and nodal analysis as possible avenues to explore.

References

[1] Gao CW, Allen JW, Green WH, West RH. Reaction Mechanism Generator: Automatic construction of chemical kinetic mechanisms. *Comput Phys Commun*. 2016;203:212–225. https://doi.org/10.1016/j.cpc.2016.02.013.

[2] Green WH, Allen JW, Barlow J, Bhoorasingh PL, Buesser BA, Class CA, Gao CW, Han K, Khanshan FS, Lambert V, Merchant SS, Slakman BL, Troiano S, Vandeputte A, Vandewiele NM, West RH, NathanYee. Reaction Mechanism Generator;.

[3] Allen JW, Green WH. CanTherm: Open-source software for thermodynamics and kinetics. Included in: Reaction Mechanism Generator, v2.0.0. 2016.
URL http://reaction.mechanismgenerator.github.io

[4] Bhoorasingh PL, Slakman BL, Seyedzadeh Khanshan F, Cain JY, West RH. Automated transition state theory calculations for high-throughput kinetics. *J Phys Chem A*. 2017; https://doi.org/10.1021/acs.jpca.7b07361.

[5] U.S. Energy Information Administration - EIA - Independent Statistics and Analysis. URL https://www.eia.gov/tools/faqs/faq.php?id=32&t=6

[6] Van de Vijver R, Vandewiele NM, Bhoorasingh PL, Slakman BL, Seyedzadeh Khanshan F, Carstensen HH, Reyniers MF, Marin GB, West RH, Van Geem KM. Automatic Mechanism and Kinetic Model Generation for Gas- and Solution-Phase Processes: A Perspective on Best Practices, Recent Advances, and Future Challenges. *Int J Chem Kinet*. 2015;47(4):199–231. https://doi.org/10.1002/kin.20902.

[7] Frisch MJ, et al. Gaussian 09. Gaussian, Inc., Wallingford, CT. 2009.

[8] Magoon GR, Green WH. Design and implementation of a next-generation software interface for on-the-fly quantum and force field calculations in automated reaction mechanism generation. *Comput Chem Eng*. 2012;52:35–45. https://doi.org/10.1016/j.compchemeng.2012.11.009.

[9] Bhoorasingh PL, West RH. Transition state geometry prediction using molecular group contributions. *Phys Chem Chem Phys*. 2015;17(48):32173–32182. https://doi.org/10.1039/C5CP04706D.

[10] Bhoorasingh P, Slakman B, Khanshan FS, Cain J, West R. Kinetic data for manuscript describing the AutoTST algorithm for automated Transition State Theory calculations of chemical reaction rates. *figshare*. 2016;https://doi.org/10.6084/m9.figshare.4234160.v1.

[11] Patchkovskii S. SYMMETRY software, University of Calgary (2003). 2003. URL http://www.cobalt.chem.ucalgary.ca/ps/symmetry/

[12] Supady A, Blum V, Baldauf C. First-Principles Molecular Structure Search with a Genetic Algorithm. *J Chem Inf Model*. 2015;55(11):2338–2348. https://doi.org/10.1021/acs.jcim.5b00243.

[13] Goodman JM, Silva MA. QRC: a rapid method for connecting transition structures to reactants in the computational analysis of organic reactivity. *Tetrahedron Letters*. 2003; 44(45):8233 – 8236. https://doi.org/10.1016/j.tetlet.2003.09.074.

490269 First Principles Based Automated Kinetic Model Generation Using on-the-Fly *Ab Initio* Calculations

Abstract:

Introduction

Many kinetic models for combustion and pyrolysis processes have been published in the past decade, and a common challenge in most work is often being stressed, i.e. the shortage of accurate rate coefficients for a large number of elementary steps. Currently, many rate coefficients are calculated based on rough approximation methods, and their accuracy can be questioned when these approximations are used outside of the original application domain of the applied method. Furthermore, building these approximation methods requires a complete training and test set of reactions of which the rate coefficients are known from either experiments or theory. Manually obtaining all these rate coefficients is time consuming, and this work is aiming at automatically calculating a large number of rate coefficients using quantum chemistry, minimizing the manual work. The developed methods can furthermore be used after the model generation, i.e. when not all the kinetics are sufficiently accurate.

This work has been performed in the framework of Genesys[1], an automatic kinetic model generation tool developed at the Laboratory for Chemical Technology which employs both the rule-based and the rate-based termination criteria. The latter can in particular benefit from the automatic *ab initio* calculations in the search for more complete, more accurate and predictive kinetic models.

Methodology

Genesys offers a representation of chemical reactions, supported by the Chemistry Development Kit[2]. Molecules, radicals and transition states are treated as mathematical graphs, in which the atoms are nodes and the bonds are edges of the graph. This is a very powerful molecular representation to convert reactants into products, to find sub-molecular patterns, to search in databases for appropriate data, etc. However, it is insufficient to perform *ab initio* calculations, where the coordinates of the atoms in the 3D space are required. The conversion of connectivity information to 3D coordinates is already well established for molecules and radicals, for which rule-based heuristic methods and numerical methods exist. "Distance geometry" is a method belonging to the latter group in which the coordinates are estimated by minimizing an error function describing all the interatomic distances compared to expected distances, which are tabulated. Additional optimization steps, among others using the Merck Molecular Force Field (MMFF)[3], allow the generate a meaningful set of 3D coordinates, directly applicable for *ab initio* calculations. This has been implemented in Genesys.

To obtain accurate rate coefficients, quantum chemical calculations of transition states are also required. The 3D coordinates of atoms in transition states are more complex to deduce and cannot be summarized in the same simple rules used for molecules and radicals. Since Genesys has been programmed to offer the end-user flexibility in terms of which reaction families to use and how to constrain them, this philosophy is further employed for the generation of 3D coordinates for transition states. The end-user has to possibility to guess the interatomic distances of the atoms in the transition state that change in connectivity throughout the reaction, called the reactive atoms. This information, the so-called "template" is used in Genesys to obtain 3D coordinates of all the atoms via "distance geometry", i.e. similar to stable molecules and radicals. The template can be built from a small set of manually calculated transition states or from the vast literature data on quantum chemical calculations. Refining the "distance geometry" output to a more meaningful set of coordinates is also done. For this, the MMFF is used in which equilibrium distances for the reactive atoms originate from the template, and the force constants are taken from analogous stable molecules and radicals.

Once initial 3D coordinates are available for molecules, radicals and transition states, quantum chemical calculations are performed using the Gaussian09 suite of programs[4] in several steps. First, the geometry is energy-optimized using fast semi-empirical calculations such as PM3. Using the resulting geometry, the conformational space is scanned by rotating around single bonds in a systematic and exhaustive manner. Each of these conformers are optimized using DFT methods, and the geometries belonging to the lowest 5 kJ mol^{-1} are used for high-level optimization, i.e. CBS-QB3 calculations. Finally, hindered rotor potentials are obtained to correct for the harmonic oscillator approximation.

For transition states, it is necessary to verify if the imaginary frequency indeed corresponds to the expected reaction. This is done in two steps. First, Genesys controls if the final geometry is close to the user-provided template. Second, the normal mode corresponding to the imaginary frequency is analyzed and it is evaluated whether the bonds that are broken or formed have a large contribution to this normal mode. For this, the normal mode in Cartesian coordinates is transformed into a normal mode in internal coordinates.

Via statistical thermodynamics, standard enthalpies of formation, standard entropies and heat capacities of all the molecules, radicals and transition states are calculated. The frequencies that resemble internal rotations are automatically searched for via an analysis of the normal mode. Their contributions to the vibrational partition function are automatically replaced by the hindered rotor partition functions. Also, the total rotational symmetry number of the species is automatically calculated and used to obtain correct entropy values. In the case where stereocenters are present, several approaches are possible. In Genesys, it is possible to explicitly track each stereoisomer as a separate species and to calculate separate thermodynamic data. However, when this is not necessary, they can also be lumped together, for which the entropy is then corrected by the number of optical isomers.

The thermodynamic data of reactants and the transition state of a reactions are subsequently used to obtain high-pressure limit rate coefficients. Tunneling corrections are automatically calculated using Eckart potentials, and Arrhenius parameters are determined by regression of the rate coefficients.

Results

The new methods have been illustrated by calculating rate coefficients for 8 reactions, each belonging to a different reaction family, i.e. inter- and intramolecular hydrogen abstractions, inter- and intramolecular radical additions, a substitution, a dehydration, a dehydrogenation and a retro-ene reaction. For each reaction family, the template was created based on known geometries of transition states from the literature or from manual calculations. The resulting Arrhenius parameters and rate coefficients at 1000K are given in Table 1.

These rate coefficients have been compared to experimental and theoretical values in literature and an excellent agreement is found, which is the same agreement found when calculating rate coefficients from CBS-QB3 calculations "manually". Most of the rate coefficients are within a factor of 2 of experimental and theoretical literature rate coefficients.

Table 1: Arrhenius parameters and rate coefficients at 1000K of the eight reactions automatically calculated using

Genesys. The first line of each reaction corresponds to the forward reactions, the second line to the reverse reaction. The pre-exponential factor and rate coefficients are expressed in s^{-1} for monomolecular reactions and m^3 mol^{-1} s^{-1} for bimolecular reactions. The activation energies are expressed in kJ mol^{-1}.

Conclusions

New methods have been developed to automatically calculate rate coefficients from *ab initio* calculations. For this, Genesys is used, which provides a connectivity based representation of the species, transition states and reactions. This information is used to create 3D coordinates of the atoms in the species and transition states. These coordinates are inputted to quantum chemistry calculations, yielding high level thermodynamic and high pressure limit kinetic data for gas-phase reactions.

The new methods are of great value to calculate a large number of rate coefficients from *ab initio* calculations without the need for manual interventions or user knowledge. This is especially of great interest for (1) building approximation methods for rate coefficient calculations such as group additivity models and (2) increasing the accuracy of important rate coefficients in a kinetic model.

1. Vandewiele NM, Van Geem KM, Reyniers MF, Marin GB. Genesys: Kinetic model construction using chemo-informatics. *Chem Eng J.* 2012;207:526-538.

2. Steinbeck C, Han Y, Kuhn S, Horlacher O, Luttmann E, Willighagen E. The Chemistry Development Kit (CDK): An Open-Source Java Library for Chemo- and Bioinformatics. *Journal of Chemical Information and Computer Sciences.* 2003;43(2):493-500.

3. Halgren TA. Merck molecular force field .1. Basis, form, scope, parameterization, and performance of MMFF94. *Journal of Computational Chemistry.* 1996;17(5-6):490-519.

4. *Gaussian 09* [computer program]. Wallingford, CT, USA: Gaussian, Inc.; 2009.

492655 Quantifying the Catalytic Effect of Calcium on Initiation during Cellulose Pyrolysis

Abstract:

During high temperature biomass pyrolysis, biopolymers undergo rapid decomposition to form hundreds of organic compounds which condense to form bio-oil and can be subsequently upgraded to fuel and chemicals. Inorganic metals naturally present in lignocellulosic biomass such as calcium and magnesium have been shown to catalyze pyrolysis chemistry[1]. However, little information is available that quantifies the catalytic behavior of these metals on pyrolysis chemistry. This work focuses on identifying the fundamental chemical mechanisms and kinetics involved in this catalytic chemistry. Using Î±-cyclodextrin as a cellulose surrogate[2] and a novel reactor system PHASR (Pulsed Heated Analysis of Solid Reactions)[3-4], time-resolved conversion data is obtained between 370 and 430 ^{0}C. To study the effect of calcium catalysis, varying concentrations of $CaNO_3$ (0.4 to 2.0 wt%) were mixed with Î±-cyclodextrin to simulate calcium concentrations in lignocellulosic biomass[1]. Cellulose conversion exhibits first order like kinetic behavior among all temperatures and calcium concentrations. A catalyzed kinetic rate constant was extracted and exhibits similar energetics compared to a high-temperature non-catalyzed intra-chain scission regime determined in an earlier study[4].

References:

1. Zhu, C.; Maduskar, S.; Paulsen A. D.; Dauenhauer, P. J., *Alkaline-earth-metal-catalyzed thin film pyrolysis of cellulose*. ChemCatChem, **2016**, 8, 818-829.
2. Mettler, M. S.; Mushrif, S. H.; Paulsen, A. D.; Javadekar, A. D.; Vlachos, D. G.; Dauenhauer, P. J., Revealing pyrolysis chemistry for biofuels production: Conversion of cellulose to furans and small oxygenates. *Energy & Environmental Science* **2012,** 5, (1), 5414-5424.
3. Krumm, C.; Pfaendtner, J.; Dauenhauer, P.J., Millisecond pulsed films unify the mechanisms of cellulose fragmentation. *Chemistry of Materials* **2016**, 28, 3108-3114.
4. Zhu, C.; Krumm, C.; Facas, G. G.; Neurock, M.; Dauenhauer, P.J., Energetics of cellulose glycosidic bond cleavage. *Reaction Chemistry and Engineering* **2017**, 2, 201-214.

486729 Artificial Neural Networks for Flare Modeling and Set Point Determination

Abstract:

The refineries and chemical industries are mandated to meet the EPAâ€™s regulations for flare performance, i.e., smokeless flaring while maintaining a minimum combustion zone heating value of 270 BTU/scf and NHV_{dil} >22 Btu/ft^2 for all steam/air/non-assisted flares. While field tests of the many flare operation parameters are expensive, it is highly efficient to use mathematical models to analyse data and develop the right tools to make decisions in the real plant scenario.Â Modelling flare data using artificial neural networks to achieve the desired combustion efficiency (CE) and eliminate smoke can be done robustly. Selected flare data obtained from various flare tests having a wide range of exit velocity, heating value, and fuel composition have been analysed and the most influencing independent variables are chosen. Both steam and air assisted flares were modelled using the neural network toolbox in MATLAB with high correlation coefficients of greater than 0.95. Determining the set point (amount of steam/air/make-up fuel required) at the Incipient Smoke Point (ISP) and for smokeless flaring has been performed as a part of this study. Desirable operating inputs can be set for the ISP (NHV_{dil} >22 BTU/ft^2 & Opacity<$Opacity_{ISP}$) and for smokeless flaring with a high CE (>96.5).

489000 Hydrolysis of Corncob Using a Modified Carbon-Based Solid Acid Catalyst

Abstract:

Lignocellulosic biomass is a very important renewable resource that can be used to produce biofuels and various platform chemicals. The conversion of biomass to reducing sugars is an important intermediate step that can impact the economic benefits of the process. Modified carbon-based solid acid is a novel catalyst used in the hydrolysis process of lignocellulosic materials. The catalyst material was prepared through the incomplete carbonization and subsequent sulfonation of a mixture of glucose and p-toluenesulfonic acid, followed by oxidation using 30% H_2O_2 to promote the acidity. The hydrolysis efficiency of the solid acid catalyst thus prepared was considerably increased. The effects of the carbonization temperature, time, and the glucose/p-toluene-sulfonic acid ratio on catalytic efficiency were investigated in this study. Under the appropriate preparation conditions, the catalyst could hydrolyze corncob to produce 78% xylose. The activity of the catalyst slightly declined after five cycles of application.

499188 Utilizing Solvent-Effects within Heterogeneous Catalysts for Selective Production of 5-Hydroxymethylfurfural

Abstract:

Biomass, such as lignocellulose, is a highly abundant carbon source giving it great potential as a green alternative to fossil resources as a renewable source for production of chemicals and fuels. Current methods to valorize biomass break lignocellulose down into its monomeric components before converting it to useful platform chemicals such as 5-hydroxymethylfurfural (HMF). Previous research has shown that high selectivity can be achieved using organic solvents, which stabilize HMF preventing it from being converted to byproducts. However, the current challenge for HMF production is achieving high selectivity in water alone to avoid the difficult separation of HMF from the bulk organic solvent. This study involves implementing organic-like solvent effects within a mesoporous silica catalyst to promote fructose dehydration to HMF selectively in water. Testing these materials, we find that the catalytic material design can impact the overall selectivity, resulting in high selectivities. Ongoing work is investigating the ability to reuse the catalyst. Overall, this work provides a new direction to tune catalytic activity and selectivity.

499555 Understanding the Effect of Alloying Pd and Sn on Direct Synthesis of H_2O_2

Abstract:

Direct synthesis (DS) of H_2O_2 has the potential to replace the current method of H_2O_2 production, anthraquinone auto-oxidation (AO) process, which is energy- and cost-intensive. But Pd, the most commonly used catalyst for DS, is not very selective towards H_2O_2 formation (> 90% by AO vs. < 60% using DS) in the absence of additives like acids and halides.[1] Alloying metals (such as Au, Zn, and Sn) with Pd has been effective in increasing H_2O_2 selectivity, however, the reasons for these changes remain unknown and likely depends on the identity of the second metal.[2,3] Here, we compare activation enthalpies for H_2O_2 and H_2O formations and the mechanisms of these reactions among Pd and $PdSn_x$ catalysts (where $0.5 \le x \le 10$) to understand the underlying cause of improved selectivity towards H_2O_2 production that result from the addition of Sn to Pd and different post synthesis oxidative and reductive heat treatments.

H_2O_2 formation rates on Pd and Pd_2Sn are significant in a protic solvent (e.g., methanol), whereas those in aprotic solvents (e.g., acetonitrile) are immeasurable. These results are in agreement with a proton-electron transfer (PET) mechanism on Pd in which the presence of a proton is necessary for DS to occur.[4] Steady state H_2O_2 and H_2O formation rates over silica supported $PdSn_x$ nanoparticles (7-15 nm in diameter) synthesized by colloidal techniques depend on the number of oxidative and reductive treatments as well as the final treatment used prior to DS, however, Pd nanoparticles (~10 nm in diameter) do not. Catalysts which were subjected to a second oxidation treatment after an oxidation-reduction cycle were less selective than catalysts which were just oxidized and reduced. Activation enthalpies ($\Delta H^{\ddagger}$) were calculated from rates measured as a function of temperature (281-305 K) and show that $PdSn_2$ catalysts have greater $\Delta H^{\ddagger}$ for H_2O_2 (9 kJ mol^{-1}) and H_2O (15 kJ mol^{-1}) formation than Pd nanoparticles (-6 kJ mol^{-1}, 11 kJ mol^{-1}). These differences in $\Delta H^{\ddagger}$ values indicate that the addition of Sn raises the barriers of H_2O_2 formation much more than H_2O formation, however, H_2O_2 selectivities on $PdSn_2$ (45%, oxidized and reduced) exceed those on Pd (23%, reduced). Spectra of adsorbed CO on $PdSn_x$ obtained using Fourier transform infrared spectroscopy (FTIR) show that the ratio of the peak area for the bridge bound CO to that of the atop bound CO decreases from 17 on Pd to 2 on $PdSn_2$, indicating that Sn breaks the ensembles of surface Pd atoms that are proposed to be necessary for O-O bond rupture leading to H_2O formation. Together, the $\Delta H^{\ddagger}$ values and spectroscopic data suggest that the addition of Sn to Pd increases the selectivity towards H_2O_2 production predominantly through ensemble effects, which is significantly different from the electronic modification of Pd that are mostly responsible for the high H_2O_2 selectivities observed on $PdAu_x$ catalysts.[3]

Factors such as the presence of additives in the reaction medium (i.e., acids, halides) and the type of reactor used (e.g., PFR vs. BSTR) also affect H_2O_2 selectivities and rates of DS apart from alloying metals.[1] Such results indicate that the active form(s) of the catalyst may include soluble forms of Pd in addition to supported nanoparticles. Homogeneous Pd complexes may form by dissolution of Pd from surfaces and complexation with anionic species added to act as "promoters." The contributions of both homogeneous and heterogeneous Pd can be distinguished by adding specific titrants into the reaction medium that deactivate either heterogeneous or homogeneous catalysts (e.g., Hg, CS_2, C_4H_4S).[5] These results will show that comparisons between DS catalysts operating in continuous flow or batch modes require corrections to account for more than just the simple difference between the design equations of the systems.

(1) Wilson, N. M.; Bregante, D. T.; Priyadarshini, P.; Flaherty, D. W., In *Catalysis*; Spivey, J., Han, Y., Eds.; Royal Society of Chemistry: 2017; Vol. 29, p 122.

(2) Freakley, S. J.; He, Q.; Harrhy, J. H.; Lu, L.; Crole, D. A.; Morgan, D. J.; Ntainjua, E. N.; EDwards, J. K.; Carley, A. F.; Borisevich, A. Y.; Kiely, C. J.; Hutchings, G. J., *Science* **2016**, *351*, 965.

(3) Edwards, J. K.; Thomas, A.; Carley, A. F.; Herzing, A. A.; Kiely, C. J.; Hutchings, G. J., *Green Chem.* **2008**, *10*, 388.

(4) Wilson, N. M.; Flaherty, D. W., *J Am Chem Soc* **2016**, *138*, 574.

(5) Widegren, J. A.; Finke, R. G., *J Mol Catal A: Chemical* **2003**, *198*, 317.

501798 Optimal Control of a Fluid Catalytic Cracking Unit

Abstract:

Fluid Catalytic Cracking (FCC) is the most important refinery process in petroleum industry. Feedstock of low economic value due to high molecular weight is transformed to high value gasoline, naphtha, LPG and other products. In the fluid catalytic cracking process heavy long chain hydrocarbons are decomposed into shorter chain molecules in a zeolite used as catalyst.

Oil from the reduced viscosity unit, vacuum fractionators or a mixture is heated and steam sprayed into the reactor which is a vertical straight tubular reactor called the riser, small sized catalyst at high temperature is introduced also in the reactor. Almost all reactions take place in the first length section of the reactor lowering its temperature as most reactions are endothermic. In the last part of the reactor, the reactor have a 180° turn, entering into the disengager where deactivated by coke catalyst and gas products are separated in a cyclone system.

Hydrocarbon reactions form coke in the surface of the catalyst and when separated is fluidized into a valve which connects the disengager with the regenerator in which the coke, composed mainly by carbon is burned to CO and CO_2 thus recovering its activity and feed into the riser again to form a continuous cycle.

Gasoline and naphtha have a better economic value while LPG and heavy naphtha must still be processed. Then obtaining the good process conditions and feedstock composition are very important to improve the economic profit of the process.

In this work the process is simulated inspired in the work of Santos, 2000 which is a set of ordinary differential equation including a 10 lump kinetic scheme, mass and energy balances on the riser, disengager and regenerator , non linear valves and added equations for air blowers and wet gas compressor.

The set comprised 21 differential algebraic equations and 64 algebraic equations.

In the operation of an FCC the process variables easier to change by operator are the feed temperature and the feed mass flow.

Optimal control is used to determine trajectories of control for a system expressed as a set of ordinary differential equations over a period of time to minimize a performance index. In the FCC problem it is intended to determine the trajectories of temperature and pressure that must be followed so the composition of gasoline is improved.

502518 Enzymatic Cascade Reactions for Synthesis of High Value Products in a Multiphase System

Abstract:

The production of high value products is still a growing market in the field of white biotechnology. Due to the demand on high selectivity and purity of products, enzymatic cascade reactions have a quite high potential. Usually *in-vivo* systems are used, although they have a low time-space yield, are quite complex and hard to describe and optimize mathematically. In addition, a lot of by-products are formed. *In-vitro* systems with isolated enzymes are contrary: less by-products, less complex, easy to optimize. However, a big drawback is the lack of co-factor regeneration. In this study we describe an enzymatic cascade reaction system with integrated co-factor regeneration within the total synthesis pathway.

By the example of high value products derived from natural educts a cascade reactor system is presented and described. Because the co-factor depending dehydrogenases are only active in water and the natural educts are organic, a multiphase system has to be set up. In addition an in-situ product removal is strived to enhance the total conversion and realize a continuous process. The individual reactions and unit operations are modeled and aggregated to a complex system.

An alcohol dehydrogenase and an aldehyde dehydrogenase are used. While one enzyme needs NAD as co-factor and NADH is produced, the other enzyme needs NADH producing NAD. In addition a lipase is used within the reaction pathway to synthesize high value products. As model reactants organic acids, alcohols, aldehydes and esters are used which are relevant as flavors and fragrances.

This presentation outlines the whole production process and gives first experimental results of single unit operations. This decomposition offers the opportunity for a structured analysis of the complete process.

483040 Hydrogenation of Phenol to Cyclohexanone Via Tubular Nanofiber Supported Catalyst

Abstract:

Hydrogenation of phenol to cyclohexanone via tubular nanofiber supported catalyst

Lin Pan and G. G Chase

Abstract

Cyclohexanone is the key intermediate in the manufacture of nylon-6 and nylon-66. The hydrogenation of phenol process is commonly used in industry due to the lower temperature requirement and less byproducts generation compared with the oxidation of cyclohexane[1]. The hydrogenation process could happen through two path ways: one step or two step reaction. The one step reaction (Figure 1) is applied here. The hydrogenation can be conducted either in liquid phase (low temperature) or gas phase (high temperature). The temperature needed for the one-step reaction is lower than that needed for the two step reaction[2].

The liquid phase reaction is preferred in our work because the operation conditions are easier to establish and control in a laboratory environment. Researchers have evaluated many catalysts for use in liquid phase phenol hydrogenation. The catalysts have been applied as dispersed particles in the liquid as a pseudo homogeneous reaction and the catalyst particles have been immobilized on monolithic support structures for heterogeneous reaction. Each approach has its advantages and disadvantages. In general, hydrogen bubbles must transport through the liquid phase to the catalyst particles for the reaction to occur. The transport of individual bubbles to a catalyst particle is somewhat random and challenging to predict. To our knowledge no previous research has evaluated the performance of catalyst supported on a gas-liquid barrier membrane. The membrane is in the form of a hollow tube. Hydrogen gas flows through the inside of the tube and aqueous phenol flows on the outside of the tube. The objective of this work is to the feasibility and to evaluate the reaction kinetics of a prototype tubular membrane reactor shown in Figure 2. The membrane is fabricated using electrospinning techniques.

Figure 1. The one step reaction pathway of phenol hydrogenation

Figure 2. Designed reactor of tubular hydrogenation reactor

Reference

1. Liu, Huizhen, et al. "Selective phenol hydrogenation to cyclohexanone over a dual supported Pd–Lewis acid catalyst." Science 326.5957 (2009): 1250-1252.

2. Shore, Sheldon G., et al. "Vapor phase hydrogenation of phenol over silica supported Pd and Pd-Yb catalysts." Catalysis Communications 3.2 (2002): 77-84.

482388 Catalytic Aqueous-Phase Reforming of Methanol to Produce Hydrogen

Abstract:

Catalytic aqueous-phase reforming of methanol to produce hydrogen

I. Coronado[a,*], M. Stekrova[b], L. García Moreno[c], M. Reinikainen[a], P. Simell[a], R. Karinen[b], J. Lehtonen[a]

[a] *VTT Technical Research Centre of Finland Ltd., Espoo, FI-02044 VTT, Finland*

[b] *Department of Chemical and Metallurgical Engineering, Aalto University, Espoo, 02150, Finland*

[c] *Department of Chemical and Energy Technology, Rey Juan Carlos University, c/Tulipan s/n, Móstoles, 28933, Spain*

**Corresponding author: irene.coronado@vtt.fi*

Aqueous-phase reforming (APR) is an energetically efficient reaction to process water fractions derived from biofuel production in processes including pyrolysis and Fischer-Tropsch (FT) reaction. The organic compounds commonly found in those fractions can be upgraded to different added value molecules such as hydrogen. H_2 can be for instance further utilized in the hydrotreatment of the pyrolytic bio-oil and to adjust the CO:H_2 ratio in the FT reaction. Moreover, applying APR reduces the environmental impact caused by organic-water disposal [1].

The operation conditions of APR, low temperature and moderate pressure, introduce kinetic and mass transfer limitations that negatively affect the activity and selectivity. Furthermore, the feedstock composition and hydrothermal conditions compromise the stability of the catalysts. Platinum-based catalysts are widely applied in APR because of their high H_2 selectivity [2]. The selectivity to hydrogen decreases over less expensive Ni-based catalysts [3]. Nevertheless, they exhibit high activity in APR and the hydrogen production can be increased by metal promotion and selection of suitable supports [4]. Accordingly, this study presents unpublished, highlighted results of the methanol aqueous-phase reforming over γ-alumina supported nickel doped with copper and cerium.

1. Scope

Aiming at high activity and hydrogen selectivity in APR, the catalysts screened for this work were nickel and nickel promoted with copper and cerium. The metals were supported on β-SiC, α- and γ-Al_2O3, and ZrO_2. Additionally, platinum on alumina was considered as a bench mark. Except the α-alumina supported, which was a commercial catalyst, the rest of catalysts tested for this work were prepared by impregnation of metal precursors on the indicated supports. The catalysts were characterized by atomic absorption spectroscopy (AAS) or X-ray fluorescence (XRF), nitrogen physisorption, and X-ray diffraction (XRD) to estimate the amount of active metals, surface properties, and surface species and particle size respectively. These features were regarded to assess the performance of the catalysts.

Methanol was selected as a model compound for catalytic APR because it is a major compound, representative of pyrolysis and FT reaction waters. The experiments were conducted in a continuously operated tubular reactor at 230°C and 32 bar. The gaseous and liquid products were analysed by online and offline gas chromatograph respectively. The results were evaluated considering the carbon yield to carbon containing gases and the product yield.

2. Results and discussion

Nickel on β-silicon carbide and α-alumina exhibited inferior results compared to the rest of catalyst. The carbon yield to carbon containing gases (CtG) was under 1% for both catalysts. The low values can be attributed to the lower surface area of these catalysts that suggest poor metal dispersion. Although the CtG was similar for Ni/ β-SiC and Ni/ α-Al_2O_3, the later catalyst yielded more methane. In contrast, the inert character of β-SiC might have allowed the more selective and abundant production of hydrogen by lower participation of this in secondary reactions.

The CtG of Ni/ γ-Al_2O_3 was 5 times higher than the value obtained with the platinum-based catalyst. In contrast, the hydrogen selectivity was higher over Pt / γ-Al_2O_3 than over the supported nickel. The experiments revealed higher methane selectivity also over zirconia supported nickel catalyst. Higher alkane yields and lower CtG was attributed to the lower surface area and larger nickel crystallites that characterized the zirconia catalyst.

To improve the activity and selectivity of the Ni/ γ-Al_2O_3, this catalyst was promoted with copper and cerium. Hydrogen yield and selectivity over the Cu-doped catalyst were significantly improved. Moreover, the addition of cerium increased the CtG and resulted in higher hydrogen yield and lower alkane production. The enhancements were attributed to a synergic effect of Cu and Ce with nickel.

3. Conclusions

The goal of this study was to find a nickel-based catalyst that exhibited high activity and hydrogen selectivity in the APR of methanol. Nickel on γ-alumina showed considerable activity to convert methanol. However, the hydrogen selectivity was lower compared to the values commonly obtained with platinum-based catalysts. Promotion of Ni/ γ-Al_2O_3 by copper and cerium increased the hydrogen production and selectivity indicating a synergy of metals for the studied reaction. The catalysts proved their suitability to convert methanol into hydrogen by APR. Therefore, the authors encourage further research to optimize the metal loadings, and prove the suitability of the catalysts for APR of real water fractions of refinery.

References

1. Coronado I, Stekrova M, Reinikainen M, Simell P, Lefferts L, Lehtonen J, *Int J Hydrogen Energy* **2016**, 41, 11003-11032.

2. Shabaker JW, Davda RR, Huber GW, Cortright RD, Dumesic J, *J Catal* **2003**, 215, 344–52.

3. Huber GW, Dumesic J, *Catal Today* **2006**, 111, 119–32.

4. Davda RR, Shabaker JW, Huber GW, Cortright RD, Dumesic J, *Appl Catal B Environ* **2005,** 56, 171–86.

Acknowledgement

This work was funded by the Academy of Finland (Project no. 285398).

493530 Glycerol Valorization to Oligomers of Glycerol By Etherification over Supported Sr and Ca Catalysts

Abstract:

Increasing demands for alternative fuels, such as biofuel, has led to the large surplus of glycerol, generated as a byproduct in conventional biodiesel production. The glycerol valorization could help the cost reduction of biofuel and, consequently, make the production process of biofuel more economically favorable.

In this work, glycerol oligomers synthesized from the catalyzed etherification of glycerol were studied using supported Sr and Ca catalysts. These catalysts were obtained with impregnation of strontium onto γ-alumina (γ-Al_2O_3) and zeolite LTA substrates in 2M $Sr(NO_3)_2$ or 1M $Ca(NO_3)_2$ solutions. Effect of calcination temperatures of these supported catalysts, from 500°C to 700°C, on the resultant conversion efficiency of glycerol to oligomers and the selectivity of linear and branched oligomer to cyclic ones were investigated. Various instruments including BET, TPD, XRD, FT-IR and SEM were employed to characterize the obtained solid catalysts. With Sr supported on γ-alumina catalyst (Sr/γ-Al_2O_3), the catalyzed etherification reactions were proceed optimally at 260°C with 2.5 wt% Sr/γ-alumina calcined at 500°C under the inert atmosphere. Under optimal reaction conditions, a conversion efficiency of glycerol near 82% glycerol cold be obtained after 4-h reaction. Moreover, a selectivity near 90% of di- and tri-glycerol in linear and branched forms over the cyclic form could be achieved. In contrast, calcination of Sr on zeolite LTA (Sr/LTA) catalysts at 600°C would be ideal for the production of glycerol oligomers. The optimal reaction condition would fall at (1) catalyst loading at 2 wt% of glycerol, (2) reaction temperature at 240°C, and (2) 6 h for reaction. Under these conditions, a conversion efficiency about 90% for glycerol and the 95% selectivity of di- and tri-glycerol could be achieved. In contrast, the Ca-supported zeolite LTA gave only about 60% conversion of glycerol.

Keywords: glycerol, diglycerol, triglycerol, γ-alumina, zeolite LTA, etherification, strontium, biodiesel.

498799 Positive Synergy in Bimetallic Wzr Mesoporous Silicates for Ethanol Conversion Reactions

Abstract:

We previously reported that metal-incorporated mesoporous silicates (Zr-KIT-6, W-KIT-6 and Nb-KIT-6[1]) are active in catalyzing ethanol conversion. While Zr-KIT-6 materials are highly selective for olefin production, they are much less active than catalysts that exhibit stronger Brønsted acidity, such as H-ZSM-5.[2] Therefore, we explored bimetallic metal-incorporated silicates by simultaneously incorporating W and Zr into the KIT-6 framework to better tune the activity of the hybrid catalyst (denoted as W/Zr-KIT-6). Through systematic studies with various W and Zr loadings, we discovered that the dehydration activity of the W/Zr-KIT-6 catalysts exceeded that of either Zr-KIT-6 or W-KIT-6 monometallic catalysts. The performance of WZr/KIT-6 and other bimetallic catalyst formulations along with detailed catalyst characterization will be presented.

1. Zhu H, Ramanathan A, Wu J-F, Chaudhari RV, Subramaniam B. Effects of tunable acidity and basicity of Nb-KIT-6 catalysts on ethanol conversion: Experiments and kinetic modeling. AIChE Journal. 2017.
2. Nash CP, Ramanathan A, Ruddy DA, et al. Mixed alcohol dehydration over Brønsted and Lewis acidic catalysts. Applied Catalysis A: General. January 25, 2016 2016;510:110-124.

490163 The Role of External Acidity of Hierarchical ZSM-5 Zeolites in n-Heptane Catalytic Cracking

Abstract:

The role of external acidity of hierarchical ZSM-5 zeolites in n-heptane catalytic cracking

Xiaoxiao Zhang, Dang-guo Cheng*, Fengqiu Chen and Xiaoli Zhan

College of Chemical and Biological Engineering, Zhejiang University, 38 Zheda Road, Hangzhou 310027, China

Abstract: The hierarchical ZSM-5 zeolites, which contain meso-/or macroporosity in addition to the intrinsic microporosity, is expected to enhance the bulky molecular diffusion and accessibility of acid sites. These sites, distributed in the internal surface (micropore) or external one (including external surface and mesopore), play different roles in the chemical reaction. To explore the effect of internal/external acidity on the reaction, in this study, a series of hierarchical ZSM-5 zeolites with similar pore structures and different Si/Al ratios were synthesized. Tartaric acid was employed to poison the acidic sites in external surface selectively and extract aluminum from the framework. And catalytic cracking of n-heptane (the main compound of naphtha) was chosen as a model reaction. The results show that active sites located on the external surface tend to provoke the secondary reactions. This will lead to the coke formation and the selectivity decrease of ethylene and propylene. The lower Si/Al ratio of hierarchical zeolite is, the more external acid sites can be selectively eliminated. Therefore, the better catalytic performance can be achieved because of suppressing the secondary reactions of ethylene and propylene, as illustrated in Fig. 1.

Fig. 1 The effect of external acid sites on catalytic cracking of n-heptane

* *Corresponding author*. Tel.: +86 0571 87953382; Fax: +86 0571 87951227

E-mail address: dgcheng@zju.edu.cn (D.G. Cheng) fqchen@zju.edu.cn (F.Q. Chen)

xlzhan@zju.edu.cn (X.L. Zhan) xxzhang@zju.edu.cn (X.X. Zhang)

488915 Catalytic Cracking of Light and Heavy Crude Oil Blends to Maximise Light Olefins: Impact of Conversion and Product Yields

Abstract:

Resid fluid catalytic cracking of two crude oils have been evaluated over an equilibrated RFCC catalyst (E-Cat) blended with olefin additive zeolite catalyst in a fixed fluidised bed reactor test unit at 540 to 550â—¦C, different metal content between 3000 to 13000 ppm, olefin additive from 5 to 15 wt% and catalyst/oil ratio between 3 to 8. At 79 wt% conversion, the Murban (MUR) crude oil yielded about 24 wt% C2â€"C4 olefins and 28 wt% gasoline (C-190â—¦C cut) over E-Cat, Upper Zakkum (UZM) crude oil yielded 23 wt% light olefins and 25 wt% gasoline cut, while the mixed crude at 75/25 and 50/50, light olefins and gasoline produced were26 and 25 and 29 and 26 wt.%, respectively at 10 wt% olefin additive. The addition of olefin additive to E-Cat increased the yield of light olefins with a maximum of 10 wt%. GC X GC analysis of naphtha produced showed an increase in aromatics content over the range of additive content. The conversion of Light naphtha, heavy naphtha, light light cycle oil and light cycle oil portion has been cracked individually to gases and liquid products indicates that olefin additive facilitated the increased cracking of the light and heavy cut species to light olefins.

494350 Kinetics of Resid Fluid Catalytic Cracking Using a Fixed Fluidized Bed Reactor

Abstract:

Kinetics of resid fluid catalytic cracking using a fixed fluidized bed reactor Â

Abdulmajed Alkatheeri[1], Gnanapragasam Singaravel*, 1, Mohammad A Rakib[1], Mustafa Karakaya[1], Â Anood Taher[1], Syed Bashir Umar[1], Mohamed Al-Musharfy[1], and Stepan Spatenka[2]

[1]TAKREER Research Centre, Abu Dhabi Oil Refining Co. (TAKREER)

P.O. Box 3593, Abu Dhabi â€“ United Arab Emirates

[2]Process Systems Enterprise Limited

5th Floor East, 26-28 Hammersmith Grove

London W6 7HA - United Kingdom

Abstract

Kinetics of Resid Fluid Catalytic Cracking process was determined utilizing a batch fluidized bed system. A commercial atmospheric residue feed was used in a fluidized bed of commercial catalyst. Parameters varied during the experimental runs included reactor temperature (530°C-560°C), equilibrium metal content on catalyst (3000-13000 ppm) feed injection times (64-150 s), catalyst-to-oil ratio (3-7), and injection of main feed (atmospheric residue), and intermediate liquids (light cycle oil, light and heavy cracked naphtha), in order to capture several scenarios likely to incur during commercial operations. A 2-phase bubbling bed reactor model was utilized to develop the reactor model, which was utilized to extract the kinetics of the cracking reactions. A 10-lump model was utilized for the kinetics, and intermediate lumps were also used as a feed in the experiments in order to improve the accuracy of the kinetics of intermediate steps in the reaction network.

Keywords

Kinetics, RFCC, cracking reactions, 9-lump model

500992 Assembly of Kaolin and SAPO-34 Catalyst Using Silica Sol As Binder By Spray Drying for Conversion of Methanol to Light Olefins in Fluidized Bed Reactor: Effect of Ion Exchange Method

Abstract:

Light olefins are of most important intermediates in organic chemical process and are essential feedstocks in petrochemical industry, which are generally produced through thermal and catalytic cracking. High reaction temperature, low yield and dependency to oil sources are of drawbacks of those methods. Catalytic methanol-to-olefins (MTO) can be an alternative. The availability of huge quantities of natural gas and other non-oil sources for methanol production such as coal, biomass and industrial waste has led to an increased interest in the MTO process. SAPO-34 as a novel molecular sieve with CHA structure has the best catalytic performance in methanol-to-olefin (MTO) reaction due to the contribution of small pore, medium acidity, shape selectivity, high surface area and good thermal/hydrothermal stability. However, rapid deactivation and short lifetime are challenges for SAPO-34 catalysts. To overcome these drawbacks, fluidized bed reactor was suggested by consideration of reactor design principles. It provides the possibility of catalyst regeneration. In this study, the successful synthesis of kaolin-bound SAPO-34 catalyst by using silica sol was achieved. The used silica sol during the slurry preparation was derived from sodium silicate. The Na^+ ion in the binder lead to the formation of catalyst with almost no activity in MTO reaction. Thus, ion exchange stage should be considered when sodium silicate is the desired Si sol precursor due to its low price. Slurries with constant ratios of SAPO-34 catalysts, kaolin (as matrix) and silica sol (as binder) were subjected to the spray dryer at constant operational conditions. Two different methods of ion exchange, using ammonium nitrate and resin, were applied after spray dring for Na^+ ions exchange with H^+. Different physiochemical characterization methods including XRD, FESEM, BET, NH_3-TPD and FTIR techniques were applied. At first, the catalysts activities were examined in a fixed bed reactor and then the catalyst with the best performance in MTO reaction was used in fluidized bed reactor. Using acid cation exchanger caused the successful removal of Na^+ ions. The result of XRD analysis demonstrates appearing of corresponding peaks of SAPO-34 with the most intensity for the sample in which ion exchange resin was applied. Using ammonium nitrate to exchange Na^+ ions could not completely lead to H^+ ions substitution. Catalyst life times increased when sprayed spherical particles were subjected to ion exchange resin bed.

503141 Functionalized Metal–Organic Framework As a Biomimetic Heterogeneous Catalyst for Transfer Hydrogenation of Imines

Abstract:

Functionalized Metal–Organic Framework as a Biomimetic Heterogeneous Catalyst for Transfer Hydrogenation of Imines

Jingwen Chen†, Zhiguo Zhang†,* Zongbi Bao†, Huabin Xing†, Qiwei Yang†, and Qilong Ren†

Key Laboratory of Biomass Chemical Engineering of Ministry of Education, College of Chemical and Biological Engineering, Zhejiang University, Zheda Road 38, Hangzhou 310027, P. R. China

Abstract

Nature enzymes have evolved to carry out diverse chemical transformations with enormous proficiencies and exquisite specificities. Mimicking the action modes of enzymes by artificial materials, whose features are akin to enzymatic active sites, has become a fertile pursuit. Reduced nicotinamide adenine dinucleotide (NADH) is the most abundant electron carrier in cell metabolism, and its capacity as a redox agent has been exploited by numerous enzymes for the production of amines, alcohols, and so on. Inspired by nature occurring hydrogenations, a biomimetic approach employing enzyme-like organocatalysts and Hantzsch ester (HEH) as the NADH mimic for the reduction of C=N has received increasing interest. In this context, Brønsted acids such as diphenyl phosphate and trifluoromethanesulfonic acid have been used as efficient catalysts in the reduction of ketimines with Hantzsch ester as the hydrogen source. However, from a practical point of view, use of the homogeneous Brønsted acids faced detrimental problems, such as severe equipment corrosion, tedious separation process, and serious environmental pollution. As a consequence, the development of a biomimetic solid acid catalyst is desirable to alleviate or eliminate those problems. Metal-organic frameworks (MOFs), a class of hybrid materials constructed via the linkage of metal ions/clusters with various organic ligands, have received much attention in catalysis, adsorption, and so on. In view of their ordered porous structure, hydrophobic confined environment, and fruitful functionalizations, MOFs have been envisioned as one of the most appropriate platforms for the design of biomimetic catalyst.

In the present work, we employed MIL-101(Cr) as a biomimetic platform, and customized a sulfonic group (SO_3H) into its hierarchical pores to generate a heterogeneous catalyst (MIL-101(Cr)-SO_3H) for the transfer hydrogenation of imines with Hantzsch ester as the reductant (Scheme 1). Both aldimines and ketimines were efficiently converted to their hydrogenated counterparts in moderate to high yields. Experimental investigations revealed that the superior catalytic activity of MIL-101(Cr)-SO_3H to MIL-101(Cr) or its homogeneous counterparts may be resulted from the synergistic interaction of Lewis acid sites and Brønsted acid sites within the hydrophobic pores. This protocol much resembles the action mode of metalloenzyme in biology and demonstrates an attractive choice of using MOFs to mimic enzymatic

catalysis. Furthermore, MIL-101(Cr)-SO_3H could be recovered and reused several times without considerable loss in reactivity.

Scheme 1. MIL-101(Cr)-SO_3H catalyzed biomimetic transfer hydrogenation of imines

References

1. Marchetti L, Levine M. Biomimetic catalysis. ACS Catal. 2011; 1(9), 1090-1118.
2. Zheng C, You SL. Transfer hydrogenation with Hantzsch esters and related organic hydride donors. Chem. Soc. Rev. 2012; 41(6), 2498-2518.
3. Nath I, Chakraborty J, Verpoort F. Metal organic frameworks mimicking natural enzymes: a structural and functional analogy. Chem. Soc. Rev. 2016; 45, 4127-4170.
4. Chen Y, Ma SQ. Biomimetic catalysis of metal-organic frameworks. Dalton Trans. 2016; 45(24), 9744-9753.
5. Ferey G, Mellot-Draznieks C, Serre C, et al. A chromium terephthalate-based solid with unusually large pore volumes and surface area. Science 2005; 309(5743): 2040-2042.
6. Akiyama G, Matsuda R, Sato H, Takata M, Kitagawa S. Cellulose hydrolysis by a new porous coordination polymer decorated with sulfonic acid functional groups. Adv. Mater. 2011; 23, 3294-3297.

495304 Study on the Influence of Particle Size Distribution on the Solid-Liquid Reaction to Produce Sucrose Ester

Abstract:

Biobased surfactants are chemicals produced from renewable raw materials, which reduces environmental impact compared with traditional petroleum based surfactants. Fatty acid sucrose esters (emulsifier food ingredient E473) is a value added entirely biobased surfactant used in niche applications such as food, cosmetics and pharmaceuticals. These are produced from sucrose and fatty acid methyl esters, and they have interesting properties as green chemical such as rapid biodegradability, biocompatibility, and biocide potential for certain microorganisms [1]

Sucrose esters are produced by basic transesterification of sucrose and fatty acid methyl esters (FAMEs) at temperatures between 90 and 180°C, and under vacuum conditions to remove the produced methanol. A main challenge of this reaction is to deal with reactants incompatibility: sucrose is highly polar solid, and FAMEs are highly non-polar liquids. Consequently, solubility of sucrose in FAMEs is very low, nearly 0.1 g of sucrose per 100g of FAME at 140°C [2]. This reduced solubility limits mass transfer to the reactive media, reducing reaction rate and conversion.

One of the processing alternatives proposed to overcome reactants incompatibility is the use of emulsifiers to provide contact and enhance solubility of sucrose in FAME [3–6]. Main emulsifiers used are fatty acid soaps (mono- and divalent), sucrose esters (product itself), mono- and diacylglycerols. However, in this heterogeneous reaction, solid-solid-liquid interfaces are present between catalyst (usually K_2CO_3), sucrose and FAMEs, respectively. In this kind of reactions, particle size distribution of solid species become a major variable o the process. Reaction to produce sucrose esters should be operated at sufficient small particles to ensure that the reaction is not mass transfer limited by diffusion effects [7].

In this work, transesterification reaction to produce sucrose ester was performed using different particle size of sucrose and catalyst (between 30 and 1000 μm) at 120 and 140°C. Sucrose and catalyst were grinded in a ball mill and then sieved to determine their particle size distribution. Reactions were carried out in batch reactors of 100ml of volume using potassium palmitate as emulsifier (5%wt) and K_2CO_3 as catalyst (5%wt). Initial sucrose/FAMEs molar ratio was 2.5. For quantification of FAME consumed and sucrose esters produced, high performance liquid chromatography (HPLC) was used. The effect of particle size distribution on the reaction productivity and the reaction rate were evaluated. Results showed that below a certain particle size of sucrose and catalyst, productivity and reaction rates achieves an optimal value. Experimental data were used to construct a kinetic model for sucroester production including the effect of particle size distribution sucrose ester production.

References

[1] N. Otomo, Basic properties of sucrose fatty acid esters and their applications, in: D.G. Hayes, D. Kitamoto, D.K.Y. Solaiman, R.D. Ashby (Eds.), Biobased Surfactants Deterg., AOCS Press, Urbana, Illinois, 2009.

[2] R. Zhao, Z. Chang, Q. Jin, W. Li, B. Dong, X. Miao, Heterogeneous base catalytic transesterification synthesis of sucrose ester and parallel reaction control, Int. J. Food Sci. Technol. 49 (2014) 854–860. doi:10.1111/ijfs.12376.

[3] K.J. Parker, K. James, J. Hurford, Sucrose Ester Surfactants — A Solventless Process and the Products Thereof, in: J.L. Hickson (Ed.), Sucrochemistry, American Chemical Society, Washington, DC, 1977: pp. 97–114.

[4] J. Fitremann, Y. Queneau, J. Maitre, A. Bouchu, Co-melting of solid sucrose and multivalent cation soaps for solvent-free synthesis of sucrose esters, Tetrahedron Lett. 48 (2007) 4111–4114. doi:10.1016/j.tetlet.2007.04.015.

[5] H.R. Galleymore, K. James, H.F. Jones, C.L. Bhardwaj, Process for the production of a surfactant containing sucrose esters, 4298730, 1981.

[6] G.P. Rizzi, H.M. Taylor, Synthesis of higher polyol fatty acid polyesters, 3963699, 1976.

[7] H.S. Fogler, Elements of Chemical Reaction Engineering, 4th. ed., Pearson Education Inc., 2006.

498176 Kinetics Studies on Redox Flow Battery Electrolytes

Abstract:

Global energy demand continues to grow in the face of decreasing total supply of fossil fuels. Not only is the supply of fossil fuels limited, but the technologies required produce the final products pollute the environment by releasing greenhouse gases. Thus, there is a need to increase availability and decrease the cost of renewable energy technologies, as they are environmental friendly and available in abundance. However, the primary challenge in using renewable energy is storage. For instance, solar energy is only available during the day and unavailable at night. Hence, storage plays an important role for using renewable energy technologies continuously. The redox flow battery (RFB) is an exciting option to harness and store energy for use when it is needed.

In this poster presentation, we discuss recent studies of RFB electrolyte materials aimed at developing a systematic approach for performing kinetics studies that can be used for characterization of new electrolytes that are currently being developed, and to resolve some of the conflicting data in the literature regarding RFB kinetics. Rotating disk electrode voltammetry was used for these measurements, owing to its effective use in RFB and fuel cell kinetics studies which have been demonstrated in literature. Analytical measurements were carried out under mutually similar conditions to find kinetic and diffusional properties of a set of electrolytes comprising both well-established and nascent flow battery chemistries. We compared kinetics data for a range of model electrode surfaces including platinum, gold and glassy carbon with systematically controlled surface oxidation. We performed Levich analysis to find the diffusivity and Koutecky-Levich analysis to find the reaction rate constant and Tafel slope for each combination of electrolyte and electrode. We further extended this study towards practical measurements using RFB electrolytes at high concentrations to demonstrate the validity of analytical measurements for predicting real battery performance.

500571 Direct Phtocatalytic Reduction of Bicarbonate to Formate on Plasmonic Metallic Nanoparticles

Abstract:

A remarkably enhanced yield of formic acid (~270 mM formic acid/g cat-hr) is observed by photoreduction of bicarbonate under solar light using ~15 nm gold nanoparticles in the presence of glycerol as a hole scavenger. Interestingly, the productivity of formic acid using gold nanoparticle catalysts was found to be even higher than conventional TiO_2 nanocatalysts. In addition, we studied the Au nanoparticle catalyzed photoreduction of resazurin in parallel. For both of the photoreduction reactions 15 nm gold nanoparticles were more efficient than 50 nm Ag nanoparticles. The enhanced efficiency of gold nanoparticle is attributed to their higher light absorption efficiency.

501168 Engineering Metal/SnO_x Interfaces for Electrochemical CO_2 Reduction

Abstract:

With dwindling fossil fuels and rising concern about their impacts on climate, electrochemical reduction of CO_2 has attracted researchers' attention as it has the potential to utilize the abundant greenhouse gas in the Earth's atmosphere and store intermittent energy from solar panels and wind turbines in chemical bonds[1,2]. A commercially viable catalyst for CO_2 electroreduction should meet the cost-effective requirement while possessing high efficiency and selectivity. Experimental studies showed that tin oxides (SnOx) have the ability to selectively convert CO_2 to formate ($HCOO^-$), although the reaction mechanism is not clear with respect to the nature of active sites and the process occurs at prohibitively high overpotentials (~1 V) [3].

The goal of this study is to understand the reaction mechanism of CO_2 electroreduction on SnOx and leverage that knowledge to design metal/SnOx composite catalysts with superior activity. In this poster, we will discuss the reaction mechanism of CO_2 reduction on SnOx obtained from first-principles density functional theory calculations, focusing on the competing pathways leading to the production of CO, H_2, and $HCOO^-$. We found that the outcome of CO_2 electroreduction on SnOx is highly dependent on its surface structure, which can explain a wide range of faradaic efficiencies for formate production on SnOx reported in different studies. By engineering the lattice strain of SnOx, we illustrated that the electrocatalytic reactivity toward formate can be further optimized, while the CO formation pathway is barely affected[4,5].

1. Hori, Y. in *Modern Aspects of Electrochemistry* (eds. Vayenas, C. G., White, R. E. & Gamboa-Aldeco, M. E.) 89–189 (Springer New York, 2008).

2. Ma, X., Li, Z., Achenie, L. E. K. & Xin, H. Machine-Learning-Augmented Chemisorption Model for CO2 Electroreduction Catalyst Screening. *J. Phys. Chem. Lett.* **6,** 3528–3533 (2015).

3. Chen, Y. & Kanan, M. W. Tin Oxide Dependence of the CO2 Reduction Efficiency on Tin Electrodes and Enhanced Activity for Tin/Tin Oxide Thin-Film Catalysts. *J. Am. Chem. Soc.* **134,** 1986–1989 (2012).

4. Wang, S., Wang, J. & Xin, H. Insights into electrochemical CO2 reduction on tin oxides from first-principles calculations. *Green Energy Environ.* doi:10.1016/j.gee.2017.02.005

5. Luc, W. *et al.* Ag–Sn Bimetallic Catalyst with a Core–Shell Structure for CO2 Reduction. *J. Am. Chem. Soc.* **139,** 1885–1893 (2017).

493351 Mechanistic Insights into the Effect of Electrolyte Composition on the Electroreduction of Carbon Dioxide (CO_2) to C_1-C_2 Chemicals Using a Flow Electrolyzer

Abstract:

The electroreduction of carbon dioxide (CO_2) to value added C_1-C_2 chemical feedstocks (such as formic acid, carbon monoxide (CO), ethylene, ethanol) has been proposed as a potential strategy for utilizing excess anthropogenic CO_2 emissions.[1, 2] Starting with the seminal work of Hori *et al.*,[3] a variety of catalysts and electrolytes have been investigated for this reaction.[4] Specifically, for the electroreduction of CO_2 to CO on silver nanoparticle based electrodes, the electrolyte composition (especially pH) has been shown to play an important role with high concentrations of alkaline electrolyte improving the activity and lowering the overpotential required for this reaction.[5, 6]. However, the exact mechanism behind such a behavior is poorly understood.

In this poster presentation, we will attempt to provide mechanistic insights into the effect of electrolyte concentration and pH on the electroreduction of CO_2 to CO, by analyzing the onset potentials and the Tafel slope data on two different cathode catalysts *i.e.*, silver nanoparticles, and carbon nanotube supported gold nanoparticles. All experiments were performed in a gas diffusion electrode based flow electrolyzer,[7] to avoid mass transfer limitations associated with the low solubility of CO_2 in aqueous solutions (~35 mM under standard conditions). Our results indicate that the rate determining step for the electroreduction of CO_2 to CO is a pH independent single electron transfer step and increasing the electrolyte pH can be a simple strategy for reducing the overpotential required for the reaction. In addition, this presentation will also focus on our recent results with regards to the mechanistic pathway involved in the electroreduction of CO_2 to C_2 products (such as ethylene and ethanol) on copper nanoparticle based catalysts.

References:

[1] A.M. Appel, J.E. Bercaw, A.B. Bocarsly, H. Dobbek, D.L. DuBois, M. Dupuis, J.G. Ferry, E. Fujita, R. Hille, P.J.A. Kenis, C.A. Kerfeld, R.H. Morris, C.H.F. Peden, A.R. Portis, S.W. Ragsdale, T.B. Rauchfuss, J.N.H. Reek, L.C. Seefeldt, R.K. Thauer, G.L. Waldrop, *Chem. Rev.* 2013, 113, 6621-6658.

[2] D.T. Whipple, P.J.A. Kenis, *J. Phys. Chem. Lett.* 2010, 1, 3451-3458.

[3] Y. Hori, H. Wakebe, T. Tsukamoto, O. Koga, *Electrochim. Acta* 1994, 39, 1833-1839.

[4] H.R.M. Jhong, S. Ma, P.J.A. Kenis, *Curr. Opin. Chem. Eng.* 2013, 2, 191-199.

[5] B. Kim, S. Ma, H.R.M. Jhong, P.J.A. Kenis, *Electrochim. Acta* 2015, 166, 271-276.

[6] S. Verma, X. Lu, S. Ma, R.I. Masel, P.J.A. Kenis, *Phys. Chem. Chem. Phys.* 2016, 18, 7075-7084.

[7] D.T. Whipple, E.C. Finke, P.J.A. Kenis, *Electrochem. Solid-State Lett.* 2010, 13, B109-B111.

499070 Gold Nanoparticle Clusters As Fenton Reaction Photocatalysts

Abstract:

The formation of inter-particle junctions (hotspots) increases the electron density and near-field intensity between assemblies of plasmonic nanoparticles in the presence of light. The influence of these hotspots on the rate of heterogeneous photocatalysis is not well understood. In this talk, we will discuss experiments designed to understand the influence of hotspots on heterogeneous photocatalysis. The degradation of methylene blue using gold nanoparticles in the presence of light will be studied as the basis system. These experiment use the Fenton reaction mechanism where hydroxyl radicals (OH·) will be generated for the oxidation of organic molecules. Previous work by Navalon et.al demonstrated that gold nanoparticles enhanced the degradation of an organic compound using the Fenton reaction mechanism in the presence of a monochromatic light of 532 nm and using broad-spectrum solar light[1]. In the previous study, gold nanoparticles enhanced the formation of solvated electrons that aided in the formation of hydroxyl radical and increased the oxidation rate of the basis compound, phenol[1]. In this presentation, we discuss experimental results that were designed to study to what extent clusters of gold nanoparticles demonstrate an enhanced photocatalytic effect compared to gold monomers based on the rate of degradation of methylene blue. Nanoparticles with enhanced hotspots due to electromagnetic radiation and directed self-assemble have a number of applications, including but not limited to wastewater treatment and commodity chemical catalysis.

Reference:

1. Navalon, S.; Miguel, M.; Martin, R.; Alvaro, M.; Garcia, H.; *J. Am. Chem. Soc.* **2011,** *133,* 2218-2226.

498095 Interconnection of Spillover and Electrochromism at Tungsten Oxide

Abstract:

Our research group is working to improve the efficiency of energy conversion technologies like fuel cells by developing multifunctional catalysts for electrochemical processes. In this context, we are interested in studying the related phenomena of hydrogen spillover and electrochromism in tungsten oxide. Spillover is defined as a process of transferring a species formed or adsorbed on one surface to another surface that cannot form or adsorb the same species under the same conditions [1]. Spillover can take place from a metal to metal, from oxide to oxide, from metal to oxide, or from oxide to metal, and it can occur over relatively long distances [2]. Electrochromism is a phenomenon of reversibly changing the optical properties of a material through electrochemical oxidation or reduction. Materials with cathodic coloration show color change at negative potentials while those with anodic coloration exhibit color change at positive potentials [3].

Platinum (Pt) catalyst supported on tungsten oxide (WO_3) in the presence of hydrogen gas undergoes hydrogen spillover, as first reported by Khoobiar in 1964 [4]. The product of this reaction is hydrogen tungsten bronze (H_xWO_3) which can transport protons and electrons over distances of at least a few nanometers in short timescales [5]. Electrochromism of uncatalyzed WO_3 was first reported by Deb in 1969 [6]. Applying a negative potential to WO_3 yields a blue coloration due to formation of H_xWO_3. Thus, hydrogen spillover and electrochromism at WO_3 are clearly analogous processes in that they lead to the same product. Moreover, both processes are known to be accelerated in the presence of water. Nevertheless, the mechanistic details of hydrogen spillover and electrochromism at WO_3 have not been studied under mutually similar conditions.

In this poster presentation, we highlight our recent work to understand the similarities and differences between the mechanisms of hydrogen spillover and electrochromism at WO_3. We have combined electrochemical and optical measurements to correlate color changes, corresponding to H_xWO_3 formation, with current-voltage data under conditions intended to induce or suppress hydrogen spillover or electrochromism. Based on prior work and our own studies to date, we conclude that both processes occur by essentially the same mechanism comprising discrete proton and electron-transfer steps. We have extended these mechanistic studies to practical systems targeting multifunctional reactivity where hydrogen intermediates (i.e., protons and electrons) migrate across the interface between H_xWO_3 and another catalyst component.

[1] Prins, R. *Chem. Rev.* **2012**, *112* (5), 2714–2738.

[2] Conner, W. C.; Falconer, J. L. *Chem. Rev.* **1995**, *95* (3), 759–788.

[3] Velevska, J.; Stojanov, N.; Pecovska-Gjorgjevich, M.; Najdoski, M. *J. Electrochem. Sci. Eng.* **2017**, *7* (1), 27–37.

[4] Khoobiar, S. *J. Phys. Chem.* **1964**, *68* (2), 411–412.

[5] Bond, G. C. *Studies Surf. Sci. Cat.* **1983**, *17*, 1–16.

[6] Deb, S. K. *Appl. Opt., AO* **1969**, *8* (101), 192–195.

497527 An Integrated Workflow for Numerical Generation and Meshing of Packed Beds of Non-Spherical Particles: Applications in Chemical Reaction Engineering

Abstract:

An integrated workflow for numerical generation and meshing of packed beds of non-spherical particles: Applications in chemical reaction engineering

Behnam Partopour, Anthony G. Dixon

Department of Chemical Engineering, Worcester Polytechnic Institute,

Worcester, MA, USA, 01609

Packed beds are widely used in different areas of the chemical industry such as reaction engineering.In particular, packed bed models are needed where different particle shapes are used to fill low tube-to-particle diameter ratio (N) beds. Therefore, along with experimental methods, the numerical generation of these beds is extensively investigated. Unlike packed beds of spherical particles, the contacts between non-spherical particles could include lines and surfaces. Therefore, collision detection and subsequent meshing in those areas can computationally be challenging.

In this work we introduce the automated packed bed generator (PBG) package for spherical and non-spherical particles based on the Bullet physics library. The library includes a robust collision detection module which uses the impulse based collision detection method[1]. The package gets the user input for parameters such as shape, dimensions and number of particles, as well as friction and restitution factors, etc., and returns an STL file of the constructed geometry along with the bed properties (e.g. void fraction and angle distributions, see first figure).

The constructed packed bed STL file then is imported to meshing software and a 3-dimensional mesh is generated based on the shrink-wrap method. The bed properties for packed beds of different shapes of particles are validated against existing experimental and computational data in the literature. The meshed geometries are imported to CFD software, and momentum and heat transfer simulations are carried out. Velocity and temperature profiles are validated against the experimental data, and previous CFD simulations, respectively. Finally, for the first time the generated and meshed geometries are successfully used for resolved particle fixed bed CFD simulations. An ethylene epoxidation reaction mechanism[2] is coupled to three-dimensional CFD simulations of randomly packed bed of spheres, Raschig rings, and quadrilobes. Ethylene mass fraction gradients on the surface of the quadrilobe particles as well as velocity contours are shown in the second figure.

References

1- Pytlos M, Gilbert M, Smith CC. Modelling granular soil behaviour using a physics engine. Géotechnique Letters 2015; 5:243–249

2- Linic S, Barteau MA. Construction of a reaction coordinate and a microkinetic model for ethylene epoxidation on silver from DT calculations and surface science experiments. J. Catal. 2003; 214:200-2012

498257 First-Principles Studies of CO Oxidation on $MgAl_2O_4$ supported Iridium Single Atoms

Abstract:

Single atom catalysts (SACs) can maximize the efficiency of metal atom. Previously, a number of supported noble metals (e.g., Pt, Au) have been identified as active SACs for CO oxidation [1]. In this poster, we will discuss our recent studies of CO oxidation on supported Ir single atoms using density-functional theory (DFT) calculations and in-situ/operando experimental measurements.

[1] Liang, Jin-Xia, Jian Lin, Xiao-Feng Yang, Ai-Qin Wang, Bo-Tao Qiao, Jingyue Liu, Tao Zhang, and Jun Li. The Journal of Physical Chemistry C 118, no. 38 (September 25, 2014): 21945â€"51. doi:10.1021/jp503769d.

490168 Preparation of Biodiesel from Waste Cooking Oil Via Lithium Metasilicate Catalyzed Transesterification Reaction

Abstract:

In this study, biodiesel was produced from transesterification of waste soybean oil in excess methanol with lithium metasilicate (Li_2SiO_3) catalysts prepared from diatomite Lithium metasilicate was synthesized via hydrothermal processes by using diatomite in starting gel material with an addition of $LiOH_{(aq)}$ at 150°C for one day. After the hydrothermal reaction, the solids were rinsed by deionized water and calcined at 500°C. The properties of solid catalysts were characterized with XRD, FTIR, solid-state NMR, SEM, BET and TPD. The acid values of the waste soybean oil collected ranged from *ca.* 0.5 mg KOH/g oil to 1.5 mg KOH/g oil. Hence, oleic acid was employed to spike soybean oil to obtain an acid value of 0.54 mg KOH/g oil and 1.5 mg KOH/g oil to simulate the waste cooking oils (WCO). Variables in transesterification process were studied to optimize the reaction condition. Consequently, the yields of biodiesel were above 95% in an hour by using lithium metasilicate catalysts under the optimal condition. The activation energy of the catalyzed transesterification using lithium metasilicate is near 61.65 kJ/mol. Furthermore, the spent catalysts of could be recycled and reused for at least for 19th cycles without significant deactivation observed in catalysis. With a decreasing loading of catalyst-to-oil to 25wt%, the yield of biodiesel could still reach *ca.* 85% in four hours at 60°C.

501160 Electrode-Electrolyte Interfaces Probed By Quantum-Chemical Simulations and Machine Learning for Lithium-Ion Batteries

Abstract:

Understanding electrode-electrolyte interactions and predicting their consequence on structural properties of the interface is crucial for optimizing performance of various energy capture and storage devices, such as electrolyzers, fuel cells, and batteries [1]. For example, the oxidation of organic electrolytes at the cathode surfaces of rechargeable lithium-ion batteries leads to the formation of the electrode-electrolyte interface (EEI) that has high impedance and is partially responsible for the degradation of lithium-ion batteries [2]. Due to the complexity of interfacial processes involving charge transfer and structural transformations, the composition and formation mechanism of such surface layers remains elusive, rendering the rational design of cathode materials rather challenging. Quantum chemical simulations of materials properties have significantly improved our understanding of electrochemical processes, however, due to the lack of physical and reliable descriptors, the community has not been able to understand the interfacial interactions and their relevance to the essential characteristics of lithium-ion batteries, such as safety, energy, and power densities.

[1] Gauthier, Magali, Thomas J. Carney, Alexis Grimaud, Livia Giordano, Nir Pour, Hao-Hsun Chang, David P. Fenning, et al. 2015. "Electrode–Electrolyte Interface in Li-Ion Batteries: Current Understanding and New Insights." *The Journal of Physical Chemistry Letters* 6 (22): 4653–72.

[2] Lin, Feng, Isaac M. Markus, Dennis Nordlund, Tsu-Chien Weng, Mark D. Asta, Huolin L. Xin, and Marca M. Doeff. 2014. "Surface Reconstruction and Chemical Evolution of Stoichiometric Layered Cathode Materials for Lithium-Ion Batteries." *Nature Communications* 5 (March): 3529.

496743 A DFT Study of CO Adsorption and Coverage on Co_7Pd_6 13 Atom Cluster

Abstract:

With the new shale gas revolution in United States methane is found in abundance. This abundance of methane is demanding for new technological developments for converting methane to other forms of energy. Methane can be converted to syngas which is a mixture of CO and H_2. Syngas can be further converted to liquid fuel alternative Ethanol. In the current research we are interested in catalytic conversion of syngas to ethanol. First and foremost, step in the ethanol conversion process is CO adsorption on Catalytic surface.

In this study, CO adsorption and coverage is studied on 13 atom bimetallic clusters of cobalt and palladium as catalyst. Atomic orbital approach is used to study CO adsorption on the cluster. Atomic simulation software Jaguar is used. Two hybrid exchange correlation functionals B3LYP and MO6 are considered. Number of CO molecules on the surface is varied from 1 to 6. Three adsorption sites atop, bridge and threefold are considered.

CO adsorbs molecularly on both cobalt and palladium further CO preferentially binds on palladium than cobalt. On the palladium surface initially when only one molecule of CO is adsorbed CO prefers to adsorb on bridge site and as number of molecules of CO on surface increases beyond 3 adsorption site preference shifts from bridge to atop. This can further be explained by shift in HOMO from palladium surface to Cobalt surface. Adsorption energy of CO is calculated as -1.38 ev using M06 and -1.32 using B3LYP. The difference in the values can be attributed to B3LYP not including dispersion effects in this case CO back donation of electrons to metal surface. Vibrational frequency of CO is also determined. When CO adsorbs on atop site vibrational frequency is in between 2015-2065, on bridge values are between 1920-2000, and on threefold site it is in the range 2040-2100. Further as surface coverage of CO increases adsorption energy decreases, also vibrational frequency shifts to a higher value. Bond lengths between palladium metal and carbon atom are underestimated by B3LYP whereas bond length between carbon and oxygen is overestimated. This further explains that B3LYP does not include dispersion effects in CO bonding.

Calculated values for adsorption energy, bond lengths and vibrational frequency are in very good agreement with experimental values.

501828 Microkinetic Modeling of Hydrogen Oxidation on Transition Metal Surfaces for SOFC Anode

Abstract:

Microkinetic modeling was used to understand the mechanism of hydrogen oxidation for the {111} and {211} surfaces of transition metals, on SOFC anode and the effects of temperature were studied to understand the change in trends. TOF were calculated and Cu{111} was obtained as the best catalyst for low-temperature SOFC and RH {111} for high temperature. For the {211} case Pt was best at 673K whereas Cu had the highest reactivity at 1273K. MKM was successfully implemented for the analysis of hydrogen oxidation for SOFC anode

497485 Bifurcation Analysis of a Two-Dimensional Homogeneous Reactor Model

Abstract:

Abstract:

We present a comprehensive bifurcation analysis of a full two-dimensional, steady-state homogeneous reactor model with a focus on the dependence of ignition-extinction and hysteresis loci on the reactor design and operation parameters. The mathematical model consists of a pair of partial differential equqations with Danckwerts type boundary conditions and with parabolic/flat velocity profile and heat transfer at the wall.

We show that the hysteresis behavior of the full two-dimensional model is bounded by three limiting models: the two-dimension convection model (which ignores axial heat and mass diffusion), the one-dimensional axial dispersion model (which ignores radial gradient) and the short tube model (which ignores axial gradients), when taking limit values of transvers and axial Thiele moduli or Peclet numbers.

For adiabatic case with parabolic velocity profile, we show that multiple ignition and extinction points can exist even for a single step exothermic reaction, and the types of bifurcation diagrams are classified by hysteresis and double limit varieties. When multiple ignitions exist, the first ignition takes place near the reactor wall due to the boundary layer generated by slow local convection, while the second ignition occurs at the center of the tube, depending on the relative contribution of radial thermal and mass diffusion.Â For adiabatic case with a flat velocity profile, only axial gradient is important thus thereâ€™s only one ignition. The system can exhibit complex hysteresis behaviors when heat transfer occurs at the wall and when the heat and mass transfer rates in the radial direction are not equal. When the residence time is taken as the bifurcation variable, there exist bifurcation diagrams containing isolated solution branches (isolas) even in the adiabatic case when the Lewis number is less than unity.

493536 Studies on Redox Properties of Dual Metal Substituted Ceria

Abstract:

CO and NOx are piosonous gases emitted from automobile exhaust posing health and environmental issues. For the past three decades, noble-metal doped ceria has been used as the most prominent catalysts for converting CO and NOx to less toxic gasses. Substitution of different metals into single a crystalline phase has been shown to exhibit better catalytic properties than single metal substituted ceria. One of the issues with these materials is the identity of metal used for substitution and how material property are affected by substitution of these metals. In the present study, catalytic activities of these dual metal substituted ceria are explained with the help of DFT+U calculations. Both singlly doped and dually doped ceria were compared for their vacancy formation energies, structural characteristics and charge distribution for explaining redox properties of ceria-based catalytic materials. Study of all of above three properties is expected to provide through understanding of dual substituted ceria and help in develpoing optimal composition of catalytic materials.

Simulation of adiabatic trickle bed reactor for liquid phase catalytic exchange of hydrogen isotope

Ran Wang, Feng Xin
School of Chemical Engineering and Technology, Tianjin University, Tianjin 300350, China

Introduction

Liquid phase catalyzed isotopic exchange (LPCE) between hydrogen and water is promising in heavy water upgrading for fission reactor and waste water treatment for fusion reactor, the main facility of which is a trickle bed reactor referred as catalytic exchange reactor or column. Most experimental or pilot plant catalytic exchange reactor is operated under non-adiabatic condition with a temperature control jacket, which results in extra construction and operation cost on industrial scale with uncertain effect on isotopic exchange. Thus an adiabatic exchange reactor is simulated based on modified model in comparison with reactor under non-adiabatic condition.

- **Catalyzed isotopic exchange between hydrogen and water**

$$HD_{(g)} + H_2O_{(l)} \rightleftharpoons H_{2(g)} + HDO_{(l)} \begin{cases} HDO_{(g)} + H_2O_{(l)} \rightleftharpoons H_2O_{(g)} + HDO_{(l)} \\ HD_{(g)} + H_2O_{(g)} \rightleftharpoons H_{2(g)} + HDO_{(g)} \end{cases}$$

Modeling

- **Spherical Pt/SDB pellet and catalyst bed**

2.5mm (outer diameter) ×2.5mm (height) Dixon ring of phosphor bronze is used as hydrophilic packing

Item	Property	Value
Catalyst pellet	Mean pore diameter (nm)	4.9
	Pore volume (cm^3/g)	0.702
	Porosity	0.31
	BET specific surface area (m^2/g)	549.2
Catalyst bed	Catalyst diameter (mm)	0.9
	Bed density of catalyst ($g\cdot cm^{-3}$)	0.265
	Bed voidage of catalyst,	0.4
	Bed voidage of Dixon ring	0.91
	Volume fraction of catalyst in the bed	0.2

- **Model**

Ran et al's model[1] modified to adiabatic condition.

Adiabatic LPCE facility

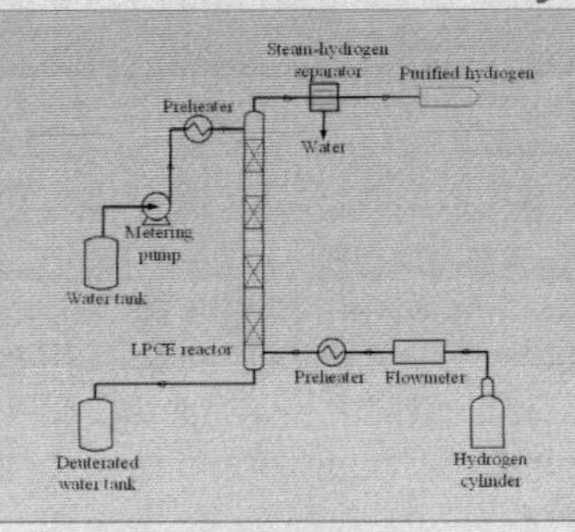

Humidification

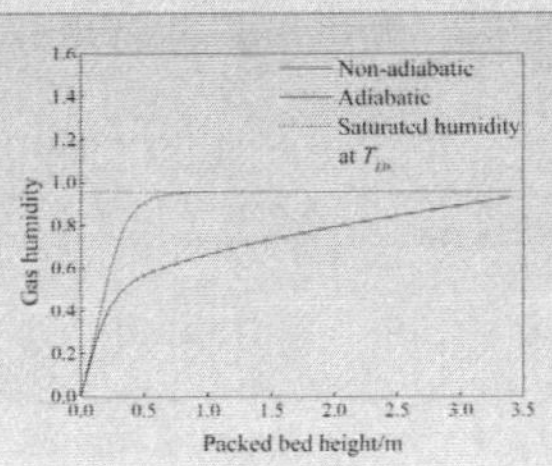

Adiabatic operation leads to less sufficient humidification, which affects gas-solid catalytic exchange and gas-liquid mass transfer of HDO.

Isotopic exchange

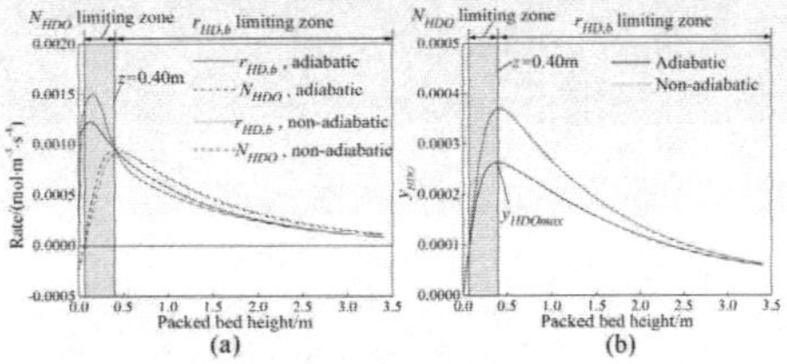

- **Gas-liquid mass transfer rate of HDO slightly rises while gas-solid catalytic exchange rate decreases at adiabatic operation.**
- **A lower HDO accumulation in gas phase**

Operation comparison

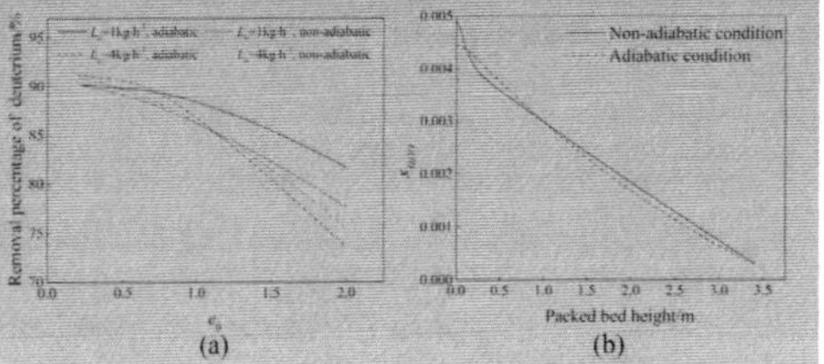

- **Adiabatic operation promotes overall isotopic exchange in the reactor at L0=4kg·h-1,e0<1.2 or L0=1kg·h-1**
- **At L0=4kg·h-1,e0>1.2, non-adiabatic operation leads to higher deuterium removal percentage.**
- **Lower HDO rise in liquid during adiabatic operation.**

Sensitivity study

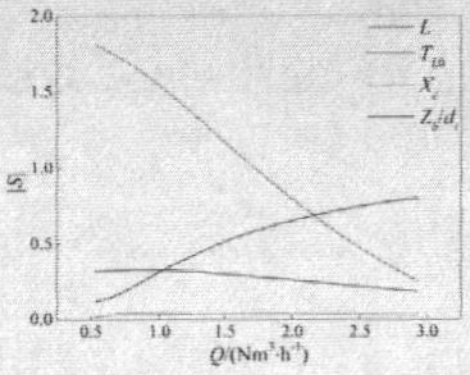

$$S = \frac{(\Delta Y_{HDO,in} / Y_{HDO,in})}{(\Delta B / B)}, B = L, X_c, Z_b/d_r, T_{L0}$$

Conclusion

- **Adiabatic operation improves gas-liquid mass transfer of HDO while it deteriorates gas-solid catalytic exchange because of less sufficient gas humidification.**
- **Adiabatic operation promotes overall isotopic exchange in the reactor at relatively low liquid flow rate and molar flow rate ratio between gas and liquid.**
- **Main operation condition effects isotopic exchange is searched**

Reference

1. Wang R, Xin F, Chen X, Xia X. Simulation of the catalyzed isotopic exchange between hydrogen and water in a trickle bed reactor. Chem. Eng. Sci.. 2017;171:179-188.

Funding Source: International Science & Technology Cooperation Program of China (2015DFR60380)

492687 Spectroscopic and Kinetic Assessment of Sn Sites Incorporated into Chabazite Frameworks at Intracrystalline and Extracrystalline Locations

Abstract:

Chabazite molecular sieves with isomorphously substituted framework tin heteroatoms (Sn-CHA) were synthesized via fluoride-mediated hydrothermal routes, and constitute model stannosilicate zeolites containing a single crystallographically unique tetrahedral-site (T-site). Lewis acid sites in Sn-CHA were titrated by ammonia and deuterated acetonitrile and quantified from temperature programmed desorption profiles and infrared (IR) spectra, respectively, giving values identical to the total Sn content. Thus, integrated molar extinction coefficients for CD_3CN IR vibrations measured on Sn-Beta [1] can be used for Sn-CHA. The predominance of framework Sn centers is also consistent with UV-visible and X-ray absorption spectra that detect tetracoordinate Sn centers under dehydrated conditions. IR spectra of Sn-CHA zeolites titrated by pyridine indicate that ~20% of the framework Sn sites are located at extracrystalline surfaces, consistent with ^{15}N DNP-NMR spectra of ^{15}N-pyridine saturated Sn-CHA that detect pyridine bound to Lewis acidic Sn sites. Despite the presence of a single T-site in CHA, ^{119}Sn dynamic nuclear polarization NMR (DNP-NMR) detected two Sn sites with the same isotropic chemical shift but with different chemical shift anisotropy (CSA) parameters in hydrated Sn-CHA zeolites, reflecting two Sn sites of the same bond-connectivity but with slight differences in the symmetry of bound oxygen atoms. Comparison of experimentally measured ^{119}Sn NMR chemical shifts and spans with those estimated using density functional theory (DFT) supports the presence of framework Sn sites with ("defect open") and without ("closed") neighboring framework Si vacancy defects. Aqueous-phase glucose isomerization turnover rates (1% (w/w) glucose, 398 K) on Sn-CHA zeolites, normalized by the number of sites titrated by pyridine, were similar to rates on an amorphous stannosilicate xerogel (Sn-xerogel), consistent with the presence of framework Sn sites at extracrystalline surfaces. Intermolecular reactions involving Meerwein-Pondorff-Verley reduction of propionaldehyde with Oppenauer oxidation (MPVO) of ethanol confirm that Lewis acidic Sn sites within microporous voids of Sn-CHA are capable of catalyzing intermolecular hydride shift steps. The approach used here, which combines the synthesis of model catalysts with theoretical and experimental probes sensitive to local structure, enables more precise descriptions of the structures and locations of metal heteroatoms substituted within zeolite frameworks.

[1] J. W. Harris et al. J. Catal., 335 (2016) 141-154.

493748 Zeolite Catalyst Design and Optimization: Impact of Synthesis Parameters on Crystal Properties

Abstract:

Zeolite Catalyst Design and Optimization: Impact of Synthesis Parameters on Crystal Properties

James Sutjianto, Rui Li, and Jeffrey D. Rimer*

Department of Chemical and Biomolecular Engineering, University of Houston, Houston, TX, 77204

*jrimer@central.uh.edu

Zeolites possess well-defined networks of pores that give rise to shape selectivity as catalysts for chemical reactions and provide an essential advantage for their use as sieves and sorbents in commercial applications. The performance of zeolite catalysts (i.e., their activity and lifetime) is closely aligned with their physicochemical properties, which include (but are not limited to) shape, size, and composition. The design of zeolite catalysts *a priori* is challenging owing to the complexity of their synthesis that is derived from the multiple growth units and pathways of crystallization. Previous studies have demonstrated that zeolites, such as zeolite L (LTL type), grow via nonclassical mechanisms involving the assembly of so-called *worm-like particles* (WLPs), which are bulk amorphous precursors that serve as putative growth units during crystallization.[1] Evidence of crystallization by particle attachment (or CPA) is mounting for a range of materials that include biominerals[2], metal oxides,[3] and zeolites.[4] In this relatively new area of research, there are many knowledge gaps between the design of synthesis conditions, the resultant precursors, and the final physicochemical properties of the zeolite products. Here, we will discuss how the judicious selection of synthesis parameters, such as reagent sources, alkalinity, and supersaturation, alter the morphology and structure of crystalline products. Our findings reveal the involvement of diverse precursors and multifaceted pathways of crystal growth. The results presented here provide a generalized platform for rational design towards the development of optimal zeolite catalysts through facile, commercially-viable synthesis approaches that bridge fundamental research with practical applications.

(1) Kumar, M.; Li, R.; Rimer, J. D.; *Chem. Mater.* 28 **(2016)** 1714-1727

(2) Politi, Y.; Arad, T.; Klein, E.; Weiner, S.; Addadi, L.; *Science.* 306 **(2004)** 1161-1164

(3) Penn, R. L.; Banfield, J. F.; *Science.* 281 **(1998)** 969-971

(4) Lupulescu, A. I.; Rimer, J. D.; *Science.* 344 **(2014)** 729-732

494919 Elucidating Zeolite Crystal Growth Mechanisms By Atomic Force Microscopy

Abstract:

Zeolites are widely used in commercial processes spanning from ion exchange in detergents to catalysis in the (petro)chemical industry. Understanding zeolite growth mechanisms at the molecular level aids the *a priori* selection of synthesis parameters to tailor the physicochemical properties of zeolites. Despite tremendous effort to elucidate the mechanisms of nucleation and crystal growth, these pathways in zeolite synthesis are not well understood. This is due in large part to the inherent complexity of zeolite crystallization and the synthesis conditions (i.e., high pH, high temperature, etc.) that render *in situ* characterization challenging. Recently, our group developed a system to carry out solvothermal *in situ* atomic force microscopy (AFM) wherein we can observe zeolite surfaces at near molecular resolution under realistic growth conditions.[1]

Here, we will present *in situ* AFM measurements of silicalite-1 (MFI) crystallization using zeolite growth modifiers (ZGMs) to modulate the shape of zeolite crystals. ZGMs are molecules or macromolecules that selectively bind to zeolite crystal surfaces and mediate the anisotropic rate(s) of growth to achieve desired crystal size and morphology. We have reported the efficacy of ZGMs in silicalite-1 bulk crystallization experiments.[2] Here, we will discuss *in situ* AFM studies of ZGM effects on silicalite-1 crystals wherein we observe differences in the relative rates of growth by two distinct pathways: classical processes involving molecule addition and nonclassical pathways involving the attachment of amorphous nanoparticle precursors. We show that these pathways can be influenced by the presence of ZGMs.

We will also present *in situ* AFM results of industrially relevant aluminum-containing zeolite LTA where we observe distinct growth regimes as a function of supersaturation. At high supersaturation, we observe the three-dimensional assembly and structural evolution of gel-like islands on LTA crystals leading to a mechanism that differs from kinetic roughening in classical theory. We also report a distinct switch in the growth mode at low supersaturation marked by two-dimensional nucleation of single layers with step heights corresponding to the composite building units of the crystal structure.

(1) Lupulescu, A. I.; Rimer, J. D.; *Science* 344 (**2014**) 729-732.
(2) Lupulescu, A. I.; Rimer, J. D.; *Angew. Chemie Int. Ed.* 51 (**2012**) 3345â€“3349.

498031 Crystallization of One-Dimensional Zeolites By Nonclassical Pathways: Perspectives on Nucleation and Crystal Growth

Abstract:

Zeolites are microporous materials that are heavily utilized in commercial applications such as catalysis, separations, ion exchange, and adsorption, among others. Their utility is based largely on their unique characteristics that include excellent stability, well-defined networks of pores, and tunable physicochemical properties (i.e., shape, size, and composition). Despite significant effort to understand the formation of zeolites,[1,2] their pathways of growth are not well understood owing to complex nonclassical mechanisms involving the assembly of precursors. For example, syntheses commonly lead to the formation of *worm-like particles* (WLPs), which are bulk amorphous precursors that serve as putative growth units during crystallization[3]. Evidence of crystallization by particle attachment (or CPA) is mounting for a range of materials that include biominerals[4], metal oxides[5], and zeolites[6]. Here, we will discuss the pathways of formation for two one-dimensional zeolites (LTL and TON types). We will discuss the results of standard synthesis protocols that incorporate crystal seeds and/or zeolite growth modifiers (ZGMs), which are molecules that alter the anisotropic growth rate(s) of zeolite crystallization. The results presented here focus on fundamental aspects of zeolite nucleation and growth, as well as describing how we can capitalize on this knowledge to rationally design materials with superior performance than those derived from conventional synthesis methods.

(1) Lupulescu, A. I.; Kumar, M.; Rimer, J. D.; *JACS.* 135 **(2013)**6608-6617

(2) Choi, M.; Na, K.; Kim, J.; Sakamoto, Y.; Terasaki, O., et al.; *Nature.* 461 **(2009)**246-249

(3) Kumar, M.; Li, R.; Rimer, J. D.; *Chem. Mater.* 28 **(2016)**1714-1727

(4) Politi, Y.; Arad, T.; Klein, E.; Weiner, S.; Addadi, L.; *Science.* 306 **(2004)**1161-1164

(5) Penn, R. L.; Banfield, J. F.; *Science.* 281 **(1998)**969-971

(6) Lupulescu, A. I.; Rimer, J. D.; *Science.* 344 **(2014)** 729-732

502226 Tailoring the Morphology and Active Site Distribution of ZSM-5 Catalysts

Abstract:

The micropores of large zeolite crystals often impose severe mass transport limitations that adversely affect catalyst activity.[1] One of the challenges in zeolite design is to predictively control crystal size and habit to improve molecular diffusion. Here, we present a novel approach to tailor ZSM-5 crystallization using zeolite growth modifiers (ZGMs), which are molecules that bind to specific crystal surfaces and alter the anisotropic growth rates as a means of tailoring crystal habit.[2,3] We have observed that subtle changes in synthesis conditions (e.g., pH) can alter ZGM binding specificity and produce crystals with different morphology. This is due to the fact that modifier speciation can be controlled by adjusting alkalinity, which could prove to be a facile and highly effective method of tailoring zeolite growth.[4]

The unique properties of ZSM-5 catalysts, such as hydrothermal stability and shape-selectivity, are ideal for many reactions. Catalyst performance is dependent upon its chemical composition. A phenomenon known as *aluminum zoning* in ZSM-5 crystals has been reported in many studies; however, its origin remains elusive.[5] To address this topic, we have performed parametric investigations to assess the role of synthesis conditions on Al siting. We show that ZSM-5 crystals can be prepared with either Si- or Al-rich exteriors, which impacts their hydrothermal stability and product selectivity in catalytic reactions.[6]

The ability to design ZSM-5 catalysts with tunable morphology and Al distribution opens new avenues for tailoring catalyst performance, and understanding property-performance relationships. ZSM-5 properties can be selectively optimized for a wide range of applications. To this end, this study provides a general platform for zeolite design and optimization that could potentially be applied to other microporous framework types.

(1) Pérez-Ramirez, J. et al.; *Chem. Soc. Rev.* 37 **(2008)** 2530-2542

(2) Lupulescu, A. I.; Rimer, J. D.; *Angew. Chem. Int. Ed.* 51 **(2012)** 3345-3349

(3) Lupulescu, A. I.; Qin, W.; Rimer, J. D.; *Langmuir* 32 **(2016)** 11888-11898

(4) Qin, W.; Clark, R.J.; Palmer, J.C.; Rimer, J.D. (in preparation)

(5) Danillina, N. et al.; *J. Phys. Chem. C* 114 **(2010)** 6640-6645

(6) Qin, W.; Patton, M.D.; Rimer, J.D. (in preparation)

497793 Knoevenagel Condensation over Ion-Exchanged Low-Silica Beta Zeolites: Their Catalytic Properties and Kinetic Analysis

Abstract:

Aluminosilicate zeolites are well known heterogeneous acid catalysts possessing, in many cases, Brønsted acid sites, which is originated by protons compensating the framework negative charges. On the other hand, the Lewis basicity also expresses on framework oxygen when the counter cations are alkaline metal or organic cations.[1] The presence of counter cation, however, sometimes limits the access of reactant to the catalytic sites through the crowded micropore. Thus, a large 12-membered ring pore zeolite, FAU-type, have been widely used as a model basic zeolite. Besides other zeolitic properties (e.g., Si/Al ratio), framework structure has been known as an important parameter in the catalytic reaction. There are, however, not enough dissection between the basicity and zeolite framework, because it is sometimes difficult to prepare high Al content large pore zeolties.

Here, in this study, we prepared the alkaline metal or alkylammonium ion-exchanged form of another large pore zeolite, *BEA-type, and applied to Knoevenagel condensation reactions. The synthesis of zeolite beta with a higher Al content (Si/Al ratio of around 6) has been recently achieved by using seed-assisted structure-directing agent free approach.[2],[3] The catalytic activity is compared among the type of counter cations and with conventional FAU-type zeolites. The kinetic analysis clarified the reaction mechanism of Knoevenagel condensation over this heterogeneous basic catalyst.

References

1) R. J. Davis, J. Catal. 2003, 216, 396.

2) B. Xie, J. Song, L. Ren, Y. Ji, J. Li, F.-S. Xiao, Chem. Mater. 2008, 20, 4533.

3) Y. Kamimura, W. Chaikittisilp, K. Itabashi, A. Shimojima, T. Okubo, Chem. Asian J. 2010, 5, 2182.

How pH affects the Metal Dispersion on Silica HMS, MCM-41 and SBA-15 Supports

Dilip Kumar Mondal, Chandona Mondal, Shyamal Roy,*

Department of Chemical Engineering, Jadavpur University, Kolkata-70032, West Bengal, India

Abstract: This work investigated the possibility of incorporation of platinum and palladium using impregnation method into several meso-structured silica supports (HMS, SBA-15, MCM-41) prepared by using sol-gel method followed by reduction process and to obtain supported Pt, Pd catalysts with highest Pt, Pd metal loading. Supports and catalysts were characterized by XRD, SEM, TEM analyzer and the metal dispersions at varied pH on silica mesoporous support were carried out by CO chemisorption technique and a range of metal dispersion from 64 to 94% was obtained and it was demonstrated that there exists pH optimum at about 11 for HMS, 12 for MCM-41 and 10 for SBA-15, which result in the highest metal dispersions, 95% for SBA-15, 94% for HMS and 82% for MCM-41. The pH of the impregnation solution influences the surface charge of the mesoporous support and therefore determines the strength of interaction between the metal precursor and the mesoporous support.

Keywords: HMS, SBA-15, MCM-41, pH, pH, point zero charge, Pt-Pd catalyst

1. Introduction

Supported platinum and palladium catalysts as heterogeneous catalysts have attracted research attentions because of their potential application in many important petrochemical industries such as hydrogenation, deoxygenation, methanation, reforming, hydrocracking and catalytic wet air oxidation. Besides good metal particle dispersion in the catalyst support, pore size is a crucial variable affecting the catalyst performance as the activity usually relies on the presence of accessible active centres located in the internal pore of the catalysts [1]. Larger-pore sizes of the catalyst provide better diffusion of reactants and products during the course of reaction[2].Therefore, high dispersion of small particles of metal precursor and high porosity of the catalyst are always desirable. Catalyst support may play a more active role in increasing the dispersion and stability of metal particles [3]. The main function of catalyst support is to achieve a fine dispersion of metal nanoparticles and to prevent the nanoparticles from aggregating and the latter relies on confining nanosized environment of the catalyst [4].

Most metal reagents are adsorbed to varying degrees on most supports, but the characteristics of the process are complicated since various types of adsorptions are possible. Silicas and zeolites are acidic and adsorb cations. Alumina is amphoteric. Titania is also amphoteric, but more acidic than alumina. For these adsorbents the relative adsorption of cations and anions can be controlled by adjusting pH. Metal ions may be cation-exchanged with a surface containing hydroxyl groups or containing alkali or alkaline earth-metal ions, or they may be held by coordination. The surface structure of the carrier may also be altered by the impregnation procedure, thus changing its adsorption characteristics.

The ultimate degree of dispersion of metal through the catalyst pellet is also determined by the interplay of a large number of factors whose relative importance varies with circumstances. These include the method of impregnation, the strength of adsorption, the extent to which the metal compound is present as occluded solute (that in the bulk liquid in the pores) in contrast to adsorbed species on pore walls, and the chemical reactions that occur upon heating and drying. The situation may be further complicated by attack on the support. Silica gel is attacked at high pH, and

Email: shyamalroy.iitd@gmail.com

alumina, which is amphoteric, is attacked at a pH that is either too or too low. Indeed, during the impregnation of an alumina support with an acidic liquid, some solution of alumina may first occur, followed by precipitation as the pH increases. It may desirable to control this effect by using a buffer.

Mesoporous silica materials such as MCM-41, SBA-15 and HMS with high porosity (pore size of up to 100 Å) have been widely studied as catalyst supports for incorporation of sulphated metal oxides [5, 6], platinum nanoparticles [7] and propyl sulfonic acid [8-10]. These mesoporous silica materials have also been extensively used as supports for incorporation of nickel particles. Nickel functionalized mesoporous silicas have been successfully applied for hydrochlorination of chlorobenzene [11] and catalytic reforming of methane with carbondioxide to produce synthesis gas (syngas) [12, 13]. By using direct synthesis [12], post synthesis-grafting and impregnation methods [13] for incorporation of platinum and palladium mesoporous silica materials (MCM-41, SBA-15, HMS) would form catalysts with good dispersion of metal particles. However, higher metal loadings could lead to structural collapse and a significant drop in the well-defined framework mesoporosity because of local blockage of pore cannel and agglomeration of metal nanoparticles. This could result in diffusion limitation of reactants as well as the products to consequently reduce the activity of the catalyst. Recently, MCM-41 was incorporated with nickel particles using deposition-precipitation method followed by a reduction process that resulted in catalyst with good nickel particles dispersion at high nickel loading of 12.8 wt% [14]. However, the supported MCM-41 nickel catalyst had low porosity (average pore diameter of 39 Å) that would not be suitable for reactions involving bulky molecules such as hydrogenation of edible or non edible oil and deoxygenation of fatty acid. These reactions require catalysts with pore diameters of above 50 Å to diminish diffusion limitation of reactants and products [15]. Besides that, catalysts derived from MCM-41 support faced serious drawback of low hydrothermal stability [16-18].

Mesostructured silica is a very interesting new mesoporous silica material. It is a class of three-dimensional (3D) materials with ultra-large mesopores (up to 500 Å) that are hydrothermally robust [19-22]. In terms of the textural and framework structures, mesostructured silica materials are composed of uniform spherical cells interconnected by window pores with a narrow size distribution [19]. In the present study, meso porous silica materials have been prepared and used as supports for platinum and palladium incorporation. The incorporation of metal particle in the mesoporous silica has been carried out using excess solution impregnation method followed by a reduction process. The aims of this study are to investigate the influence of pH on metal loading on mesoporous silica supports. Thus, the best mesoporous silica support can be identified in order to obtain a catalyst with highest content of metal nanoparticles together with high porosity.

2.Experimental

2.1. Preparation of silica HMS, SBA-15, MCM-41

Hexagonal molecular sieve (HMS) material was prepared via a neutral (S^0I^0) templating route proposed by Tanev and Pinnavaia [23]. The primary amines-dodecylamine was used as neutral structure directors in order to control a pore size of the resulting mesoporus silicas. In a typical preparation, dodecylamine (DDA) 4.15 g was dissolved in 55 ml of ethanol, and then 75 ml of H_2O were added to the solution. The synthesis was carried out at 55°C. Tetraethoxysilane (TEOS) 20 ml was added to the surfactant solution resulting in a reaction mixture with a ratio TEOS : DDA : EtOH : H_2O of 4 : 1 : 41.5 : 170. The mixture was vigorously stirred in sealed flask at 55 °C for 20 h.

The resulting white precipitate was filtered, washed with water and dried at room temperature. Template removal was performed by calcination at 550°C for 4 h (heating rate: 2 °C/min).The final product was air dried over night. The preparation of mesoporous silica SBA-15 was carried out according to the method of Zhao et al.[24]. In the synthesis procedure, 16 g amphiphilic triblock copolymer, poly (ethylene glycol)–poly (propylene glycol)–poly (ethylene glycol) was first dissolved in 120 ml of water and 480 ml of 2(N)HCl (Merck, 37wt%) solution at 40°C for 2 h under continuous stirring. After addition of 36.4 ml of TEOS the reaction mixture was stirred first at 40°C for 24 h and then at 80°C for 48 h in a Teflon-lined steel autoclave. The resulting solid was filtered, washed with deionized water, dried at 100°C for over night and calcined in air at 550°C for 6 h at a heating rate of 2°C/min to decompose the triblock copolymer.

A typical synthesis MCM-41 was carried out as follows: the appropriate amount of cetyltrimethylammonium bromide (CTAB), template was dissolved in solvent glycerol, glycol, water, respectively. The pH of this solution was adjusted to 12 with ethylenediamine (EDA), followed by the addition of tetraethyl orthosilicate (TEOS) with vigorous stirring for about 30 min at room temperature. The molar composition of the reaction mixture was 1 TEOS: 0.7 CTAB: 130 solvent (glycerol or glycol or water): 13 EDA. The mixture was stirred for 8 h at room temperature. Then the mixture was aged at room temperature for 24 h. The as-synthesized product was filtered and washed with water to pH=7, dried at 100°C for 24 h, then calcined at 550°C in air for 5 h to remove the organic templates.

2.2. Catalyst preparation

The catalyst was prepared by excess solution impregnation (ESI) method. Catalysts supported on HMS , MCM-41 and SBA-15 with different Pt, Pd loadings were prepared using prepared HMS, SBA-15 and MCM-41 supports and tetra amine platinum nitrate (ALDRICH) and tetra amine palladium (II) nitrate (ALDRICH) as metal precursors. Double distilled water was used in all these experiments. The quantities of platinum and palladium salts required for the specific catalyst with pre-determined % Pt and % Pd content were calculated from the stoichiometry. The amount of Pt and Pd salts of desired concentration was used for 15 g of support and stirred continuously for 6 h at pH ca.11 for HMS, pH ca. 12 for MCM-41 and at pH ca.10 for SBA-15 supports, respectively. The excess water from the slurry was removed in a rotary vacuum evaporator at 80°C. The residue was then dried at 115°C for 24 h in an oven followed by calcination at 420°C in controlled flow of oxygen for 6 h for complete decomposition of Pt and Pd oxide salts and deposition of the metals on the support structure. The prepared catalyst was further reduced using H_2. The reduction process was carried out under controlled temperature condition to avoid rapid reduction. The temperature was regulated by increasing the temperature of H_2 from ambient temperature to 280°C at a ramp rate of 1°C/min rise and was maintained at 280°C for 6 h to reduce the precursor's salts and to reduce the platinum and palladium from oxide state to metal form. The reduced catalysts were used for the chemisorptions process.

3. Results and Discussion

3.1. Small-angle X-ray scattering analysis

The XRD measurements were carried out to study the mesoporous structure of the supports and catalysts. The crystalline phases of the calcined catalysts can also be obtained by this technique. The low angle XRD pattern of supports and catalysts supported are shown in Fig. 1 in the interval between the 2θ values of 0.5-10°. The small-angle X-ray scattering patterns of the calcined MCM-41, HMS and SBA-15 exhibit well-resolved peaks, characteristic of mesoporous MCM-41, HMS and SBA-15 materials.

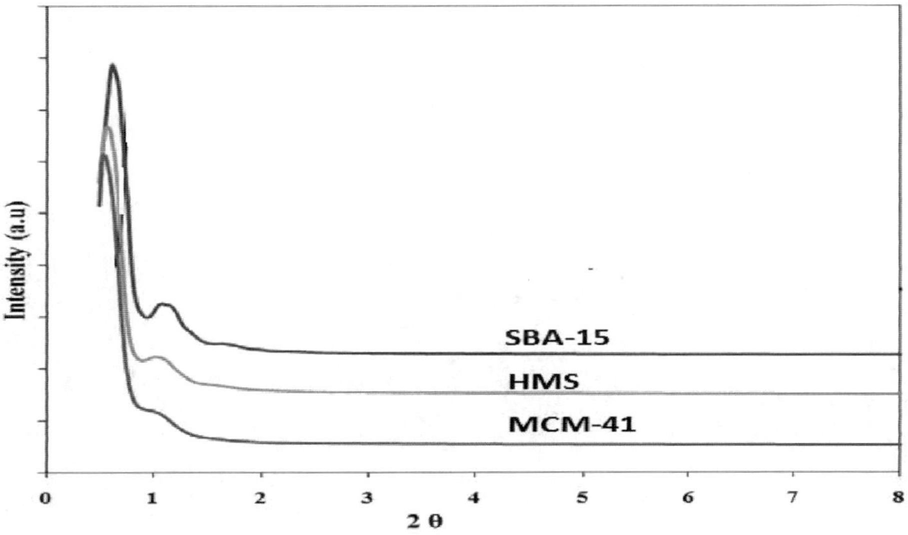

Figure 1. Small-angle X-ray scattering analysis for silica HMS, MCM-41and SBA-15 supports

3.2. SEM

Scanning electron microscopy of the sample was carried out on Jeol JSM-5610-SLV scanning electron microscope. SEM pictures for MCM-41, HMS and SBA-15 are shown in Fig. 2a, Fig. 2b, 2c. SEM image of MCM-41, HMS and SBA-15 exhibit irregular particle size and reveal that MCM-41, HMS and SBA-15 are made up of submicrometer-sized free-standing or aggregated sphere shaped particles.

Fig.2a. SEM image of MCM-41

Fig.2b. SEM image of HMS

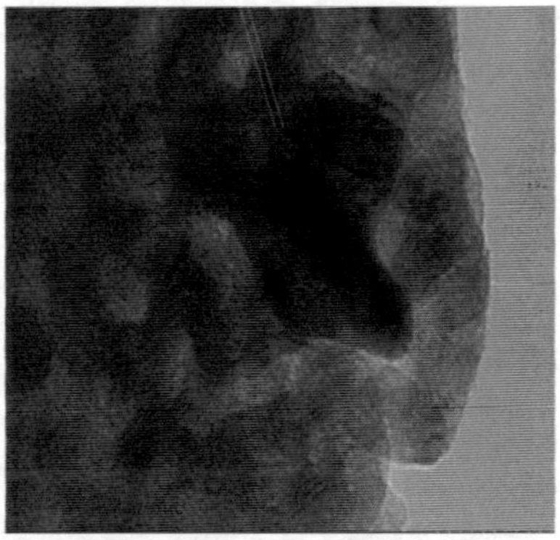

Fig.2c. SEM image of SBA-15

3.3. TEM

The morphological features of the support and catalysts were studied from transmission electron micrographs obtained with a Tecnai F20 at 200 kV. In this technique, the powder samples were grounded softly in an agate mortar and dispersed in heptane in an ultrasonic bath for several minutes. A few drops were then deposited on 200 mesh copper grids covered with a holey carbon film. The electron micrographs were recorded in electron negative films and in a digital PC system attached to the electron microscope.

The TEM images of calcined Pt-Pd/MCM-41, Pt-Pd/HMS and Pt-Pd/SBA-15 catalysts are shown in Fig. 3a, 3b, 3c. The TEM technique evidenced well-ordered cylindrical channel-like pore morphology with a 2D hexagonal mesostructure. The TEM images provide a pictorial evidence of an ordered mesostructured MCM-41 silicate material. These images confirmed that the mesoporous and well-ordered structures of the MCM-41, SBA-15 and HMS were preserved even after metals loading and that the metals are well dispersed in the pore channels since no patches of metal aggregates or agglomerates were found for samples analyzed. The average distance estimated between two consecutive centers of hexagonal pores estimated from the TEM images are in the 2.5-4.9 nm range, which are consistent with N_2 physisorption results.

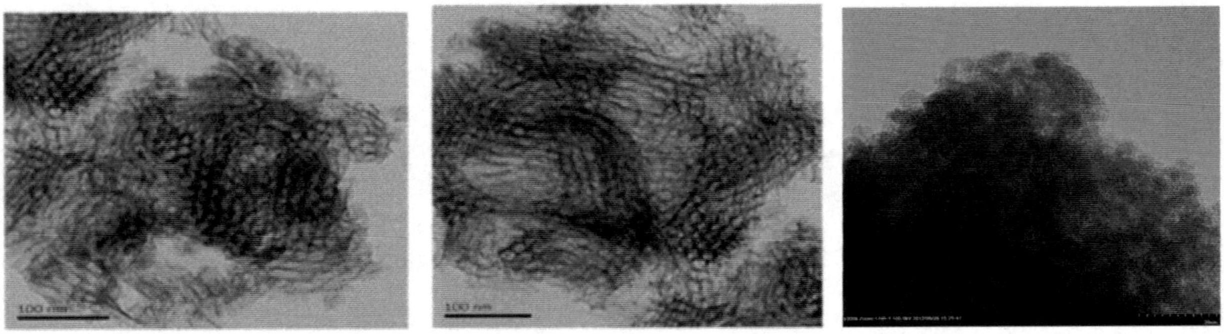

Fig.3a.TEM image of Pt-Pd/MCM-41 **Fig.3b.**TEM image of Pt-Pd/HMS **Fig.3c.**TEM image of Pt-Pd/SBA-15

3.4. CO chemisorptions

Volumetric CO chemisorptions isotherms at 298 K were recorded separately in order to estimate metal dispersion using an Autosorb-1C instrument. Canton et al. [25] reported that hydrogen chemisorption gives many problems, due to the formation of Pd hydride and of subsurface hydrogen. Satterfield [26] summarized that H_2 chemisorption of supported Pd gives erroneous result as hydrogen dissolves in Pd metal. Sheu et al.[27] and Zilm et al.[28] brought attention to the phenomenon by showing that CO chemisorbs on supported Pd as both linear and bridge-bonded species, whose ratio may depend both on the nature of the support and on the metal dispersion. For CO chemisorption catalyst was pretreated as for the catalytic experiments and then out gassed at 9×10^{-6} Torr. After cooling the sample to 303 K, CO was admitted at pressures below 20 Torr. Since in the region of low pressures, the CO isotherm can be described by the Langmuir equation, only one isotherm was recorded. The extrapolation of its linear part to zero pressure gives the amount of CO adsorbed in monolayer. Metal particle size was estimated assuming an CO:Pt(Pd) = 1:1 surface stoichiometry. In the bimetallic catalysts, platinum was assumed to behave like palladium.

3.5. Effect of pH on metal dispersion

The metal dispersions at various pH on HMS, MCM-41 and SBA-15 supports were carried out by CO chemisorptions method and plotted against corresponding pH value are given in Fig. 4. A range of metal dispersion from 64-95% was obtained and it was demonstrated that there exists a pH optimum at about 11 for HMS, 12 for MCM-41 and 10 for SBA-15 respectively, which results in the highest metal dispersions (95 % for SBA-15, 94 % for HMS and 82% for MCM-41). In order to get different metal dispersion on different silica mesoporous supports, the pH of the impregnating solution was systematically varied.

The pH was adjusted through the addition of dilute NH_4OH to initial values of 3-12. One observation was that the solution which was clear became cloudy, perhaps due to the formation of a micro emulsion upon the addition of NH_4OH. However at a higher pH, obtained by the addition of NH_4OH, the surface charge on the support becomes more negative [29-30], and attraction between the negatively charged surface and the cationic precursor increases with increasing pH.

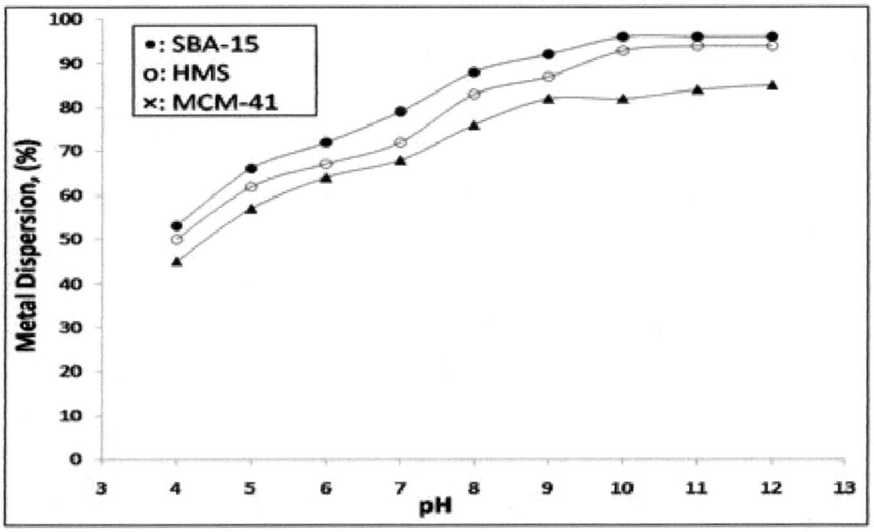

Fig. 4. Plot for different pH versus metal dispersion (%) on silica HMS, MCM-41and SBA-15 supports

Giudo and Jacob [31] brought attention to the phenomenon by showing that soluble catalyst precursors are fixed to the support either by reaction, exchange with surface OH groups, and /or by adsorption. In the former case, the concentration (density) of the surface OH groups, which depends on the pretreatment of the support, is crucial. In the latter case, the surface charge plays an important role. At a pH value of the so-called Point Zero Charge (PZC) of the surface is electrically neutral. At a pH values above PZC, the surface is negatively charged, while at pH values below PZC the surface is positively charged (Table 1).

For silica, this can be illustrated as follows. At pH = 3, the surface is neutral. In a mildly basic environment, H^+ ions are removed, and, as a result, the surface is negatively charged. In an acidic environment, the surface will become protonated. If it is intended to deposit anions onto the carrier surface, the preparation should proceed values below the PZC, whereas if cation are to be deposited, a pH value above that of the PZC is preferred. Table 2 gives PZC values for alumina, silica and a mixture of alumina and silica.

It should be mentioned that the exact PZC values not only depend on the chemical nature of the carrier, but also on to its history and the method by which it was prepared. Of course, for the solid support, a window of applicable pH exists. At improper pH values (e.g. pH >12 for alumina), the carrier itself may be dissolved.

Satterfield [26] reported that most reagents are adsorbed to varying degrees on most supports, but the characteristics of the process are complicated since various types of adsorptions are possible. Silicas and zeolites are acidic and adsorb cations. Metal ions may be cation-exchanged with a surface containing hydroxyl groups or containing alkali or alkaline earth-metal ions, or they may be held by coordination.

It is also possible to control the deposition by competitive adsorption, e.g., by adding a citrate to the impregnating solution. This procedure has been used to embed a catalytically active layer slightly inside a catalyst particle [32]. Such a structure may be desired for prolongation of catalyst life in an application in which poisons are deposited on the outside surface of porous catalyst support. An example is supported platinum catalysts for oxidation of pollutants in automobile engine exhaust.

4. Conclusions

To obtain a more uniform dispersion, the adsorptivity of $PtCl_6^{2-}$ ions may be reduced by adjusting pH. Alternatively, platinum could be applied to alumina as $Pt(NH_3)_4^+$, in which case platinum is in the form of a cation. It is then less readily adsorbed on alumina, but more strongly adsorbed on silica HMS, SBA-15 and MCM-41. If a halogen-free catalyst preparation is desired, a compound such as platinum diaminodinitrite, $Pt(NH_3)_2(NO_2)$, may be used. The low metal dispersion observed is, in all likelihood, due to the weak interaction between the metal precursor and the support. In addition to this, metal aggregation due to the formation of an emulsion and the mobility of $[Pt](NH_3)_4^+$ and $[Pd](NH_3)^{4+}$ on the silica support surfaces during reductive pretreatment may be partially responsible for this low metal dispersion. The interaction between the positively adsorbed cations, $Pt(NH_3)_4^+$, $[Pd](NH_3)^{4+}$ and the negatively charged surface is strong. Under these conditions, dispersions are much higher. The use of $Pt(NH_3)_4(NO_3)_2$, $Pd(NH_3)_4(NO_3)_2$ results in a high metal dispersion (95 %). This is interpreted in terms of strong metal-support interaction because of the positive nature of the cation in solution and the negatively charged surface.

Acknowledgements

Author thanks Prof. A. K. Biswas (Former Professor and Head, Chemical Engineering Department, Indian Institute of Technology, Kharagpur, India) and Dr. Chancal Samanta (Deputy Manager, Corporate Research & Development Centre, Bharat Petroleum Corporation Ltd., India) for numerous stimulating discussions. Financial support was provided by the Chemical Engineering Department, Jadvpur University, Kolkata, India.

References

[1] T. Maschmeyer, Curr. Opin. Solid St. M. 3 (1998) 71-78.

[2] T. M. Lancaster, S. S. Lee, J. Y. Ying, Chem. Commun. 28 (2005) 3577-3577.

[3] W. F. Taylor, D. J. C.Yates, J. H. Sinfelt, J. Phys. Chem. 68 (1964) 2962-2966.

[4] K. Niu, D. Shi, W. Dong, M. Chen, N. Zhongbin, J. Colloid Interf. Sci. 362 (2011), 74-80.

[5] L. Hermida, A. Z. Abdullah, A. R. Mohamed, J. Appl. Sci. 10 (2010) 3199-3206.

[6] V. Degirmenci, A. Yilmaz, D. Uner, Catal. Today 142 (2009) 30-33.

[7] T. Huang, W. Tu, Appl. Surf. Sci. 255 (2009) 7672-7678.

[8] L. Hermida, A. Z. Abdullah, A. R. Mohamed, Chem. Eng. J. 174 (2011) 668-676.

[9] R. I. Kureshy, I. Ahmad, N. H. Pathak, S. H. R Khan, R. V. Abdi, Jasra, Catal. Commun. 10 (2009) 572-575.

[10] L. Hermida, A. Z. Abdullah, A. R. Mohamed, J. Porous Mat. 19 (2012) 835-846.

[11] J. Chen, J. Zhou, R. Wang, J. Zhang, Ind. Eng. Chem. Res. 48 (2009) 3802-3811.

[12] D. Liu, R. Lau, A. Borgna, Y. Yang, Appl. Catal. A: Gen. 358 (2009) 110-118.

[13] D. Liu, X. Y. Quek, H. H. A. Waha, G. Zeng, Y. Li, Y. Yang, Catal. Today 148 (2009) 243-250.

[14] R. Nares, J. Ramirez, A. Gutierrez-Alejandre, R. Cuevas, Ind. Eng.Chem. Res. 48 (2009) 1154-1162.

[15] M. Nele, A. Vidal, D. I. L. Bhering, J. V. Pinto, V. M. M. Salim, Appl. Catal. A: Gen. 178 (1999) 177-189.

[16] R. Ryoo, S. Jun, J. Phys. Chem. B 101 (1997) 317-320.

[17] S. Inagaki, Y. Sakamoto, Y. Fukushima, O. Terasaki, Chem. Mater. 8 (1996) 2089-2095.

[18] K. A. Koyano, T. Tatsumi, Microporous Mater. 10 (1997) 259-271.

[19] P. Schmidt-Winkel, et al. J. Am. Chem. Soc. 121 (1999) 254-255.

[20] J. S. Lettow, et al. Langmuir 16 (2000) 8291-8295.

[21] D. T. On, S. Kaliaguine, J. Am. Chem. Soc. 125 (2003) 618-619.

[22] Q. Li, Z. Wu, D. F. Eng, B. Tu, D. J. Zhao, J. Phys. Chem. C 114 (2010) 5012-5019.

[23] P.T. Tanev, T.J. Pinnavaia, Science 267 (1995) 865-867.

[24] D. Zhao, J.Feng, Q.Huo, N. Melosh, G.H. Fredrickson, B.F. Chmelka, G.D. Stucky, Science 279 (1998) 548-552.

[25] P. Canton, G. Fagherazzi, M. Battagliarin, F. Menegazzo, F. Pinna, N. Pernicone, Langmuir 18 (2002) 6530-6535.

[26] C. N. Satterfield, Heterogeneous catalysis in industrial practice, McGraw-Hill, Inc., New York, Second Edition, 108-109.

[27] L.L. Sheu, Z. Karpinski, W.M.H. Sachtler, J. Phys. Chem. 93 (1989) 4890-4894.

[28] W. K. Zilm, L. Bonneviot, G.L. Haller, O.H. Han, M. Kermarec, J. Phys. Chem. 94 (1990) 8495-8498.

[29] T.A. Zepeda, B. Pawelec, J.L.G. Fierro, T. Halachev, J. Catal. 242 (2006) 254-269.

[30] W. Zou, R.D. Gonzalez, Catal. Lett. 12 (1992) 73-86.

[31] M. Guido, A.M. Jacob, Preparation of supported metal catalysts. Wiley-VCH Verlag Gmbh & Co. 2005.

[32] J.A. Hoekstra, Catalyst for treatment of combustible waste products. U.S. Patent 3.388, 077 (to U.O.P.), 968.

500872 Nature and Consequences of Al - Al Interactions in SSZ-13 Zeolite

Abstract:

Cu-exchanged SSZ-13 zeolite is now widely used in diesel vehicles for on-board selective catalytic reduction (SCR) of NO_x to nitrogen. It is now generally accepted that catalytic activity is associated with isolated, exchanged Cu ions within the zeolite pores, and that those ions can be present either charge-compensated by two framework Al (Z_2Cu) or by a single Al (ZCuOH) [1]. Under some synthesis conditions those Al appear to avoid first-nearest-neighbor locations but to be otherwise randomly distributed [2], and synthesis conditions can be adjusted to bias towards one type of site or the other [3]. The origins and consequences of these biases remain to be fully elucidated, both for the catalytic activity and durability of the catalyst. Dealumination is a common mode of zeolite catalyst failure, and the number and proximity of Al may have an influence on the tendency towards or resistance against dealumination in addition to its influence on Cu speciation. Thus, it is important to understand how the Al environment influences the stability of Al within the framework. In this work, we use density functional theory (DFT) methods to compute the energetics of Al siting in SSZ-13 as a function of Al proximity.

We calculated the total energies of two protons each associated with an O adjacent to Al, and reported these energies relative to the isolated BrÃ¸nsted site energy. At a given Al separation there is a range of energies corresponding to the proton associated with each of the four symmetry-distinct O atoms.Â At third-nearest-neighbor separation, energies begin toÂ decrease relative to the infinite limit. Energies are minimized at first-nearest-neighbor separation, suggesting that Al sites experience an attractive potential in the acid-form, and in this form the Al will tend to aggregate. As laboratory-prepared zeolites generally follow LÃ¶wensteinâ€™s rule, the DFT results show that this rule is not a consequence of the underlying energy landscape experienced by the Al sites. Further, at conditions at which Al become mobile on the SSZ-13 lattice, e.g. high temperature steaming, the Al experience an attractive interaction that may lead to Al aggregation and ultimately provide a route to dealumination. The results provide a baseline for considering the influence of other extra-lattice ions.

[1] Paolucci, C., Di Iorio, J. R., Ribeiro, F. H., Gounder, R, and Schneider, W. F. Adv. Catal. 57 (2016).

[2] Paolucci, C., Parekh, A. A., Khurana, I., Di Iorio, J. R., Li, H., Albarracin Caballero, J. D., Shih, A. J., Anggara, T., Delgass, W. N., Miller, J. T., Ribeiro, F. H., Gounder, R., and Schneider, W. F. J Am. Chem. Soc. 138, 6028 (2016).

[3] Di Iorio, J. R., and Gounder, R. Chem. Mater. 28 2236 (2016).

500979 Aqueous One-Pot Synthesis of Pd-Based Core@Shell Catalysts with Tunable Core and Shell Sizes

Abstract:

Encapulating an active metal core such as palladium (Pd) in a porous oxide shell material can lead to improved catalytic activity, selectivity, and thermal stability compared to conventional supported catalysts. However, many reported synthetic methods include some drawbacks which increase the difficulty of synthesis and limits the scalability of the method. For example, many methods require the use of organic solvents and surfactants in large quantities to form microemulsions, which are used as templates for core@shell particle formation. However, such a method is not very versatile in terms of tuning the catalyst composition, since microemulsions require very specific chemical conditions to form. Such a requirement also limits the scalability. Therefore, prior to a larger-scale implementation of core@shell catalysts, which has been hindered, a new synthesis method needs to be developed.

Here, we are reporting a versatile, water-based and therefore environmentally benign synthesis method to facilitate the use of core@shell catalysts. The described method can be used to prepare core@shell nanoparticles with Pd cores and oxide shells (e.g., SiO_2, CeO_2) with independently tunable core and shell sizes. By using water as the synthesis medium, the new method effectively overcomes the limitations of previously reported synthetic approaches that require the use of toxic organic solvents. In addition, not only the sizes of the cores and the shell, but also the compositions of the resultant core@shell nanoparticles could be adjusted by making simple modifications to the synthesis procedure. The prepared catalysts were characterized by transmission electron microscopy, X-ray diffraction, and nitrogen-physisorption to confirm the morphology with different compositions. Following the characterization, the accessibility and the catalytic activity of the core was tested using simple probe reactions such as the oxidation of CO. Some of these novel core@shelll structures exhibited remarkable thermal stability, maintaining the particle size and pore structure at very high temperatures (800-900 °C), close to those one may encounter in automotive exhaust applications.

495906 Application of Alginate Film Supported Nano-Silver Catalyst

Abstract:

Silver nanoparticles (AgNP) as promising catalysts have attracted key attention in various chemical reactions due to their high catalytic efficiency. Though homogeneous catalysts owe efficiency for their uniform molecular dispersion in the reaction medium, the disadvantage associated with this dispersion is that, the recovery of the catalysts is challenging and their reuse is questionable. The later one is an essential factor in terms of cost. The strategy of immobilising the active catalyst on a suitable support achieves a harmonious combination of the efficiency of as well as its reusability like that of heterogeneous ones. Metal nanoparticles progress one step farther by showing molecular level activity, which when framed in a suitable support, such as polymer, can act as heterogeneous catalyst. The important factor that attracts the attention for metal nanoparticle is their catalytic property, which is a dramatic contrast between the material in bulk and in nano form. Keeping this in mind, noble metals, Ag which are very active in their nano level, are thought to be synthesized and suitable polymers will be chosen as both stabilizing agent and support for the metal nanoparticles. In this work silver nanoparticle encapsulated with alginate film has been prepared by in-situ chemical reduction method. However alginate acts both as reducing agent as well as stabilizing agent. The successful formation of silver nanoparticles in alginate matrix has been validated by transmission electron microscope (TEM), UVâ€"Vis spectroscopy techniques. The effectiveness of the highly stabilized silver nanoparicles in alginate film as a heterogeneous catalyst has been estimated on the well-known p-nitrophenol (p-NP) reduction to p-aminophenol (p-AP) in the presence of excess sodium borohydride.

501056 Reduction Kinetics of Hercynite Materials Using Isoconversional Methods for Solar Thermochemical H_2O Splitting

Abstract:

The solar thermochemical water splitting process (STWS) is a promising technology for producing renewable H_2. A metal oxide is first reduced with high temperature concentrated sunlight to produce O_2. Then, the reduced material is exposed to steam to reoxidize it and produce H_2. Un-doped hercynite ($FeAl_2O_4$) is shown to be a promising and low-cost iron and aluminum based solar thermochemical water splitting material [1]. The reduction kinetics of $FeAl_2O_4$ was studied using a thermogravimetric analysis (TGA) technique. Isoconversional methods and XRD analysis show that hercynite is first formed by reacting aluminum oxide with iron oxide as shown in R1. Then, hercynite is further reduced, forming O-vacancies, while maintaining the spinel structure as shown in R2. Apparent activation energies for both reactions have been determined. This is the first kinetics study to be reported for the total reduction process to form herycnite and then to form the spinel, and to identify H_2 production from oxidizing the reduced hercynite with H_2O as shown in R3.

$Fe_2O_3 + 2Al_2O_3 \rightarrow 2FeAl_2O_4 + 0.5\ O_2$ R1

$FeAl_2O_4 \rightarrow FeAl_2O_{4-\delta} + \delta/2\ O_2$ R2

$FeAl_2O_{4-\delta} + \delta H_2O \rightarrow FeAl_2O_4 + \delta H_2$ R3

1. Muhich, C.L., et al., *Predicting the solar thermochemical water splitting ability and reaction mechanism of metal oxides: a case study of the hercynite family of water splitting cycles.* Energy & Environmental Science, 2015. **8**(12): p. 3687-3699.

495144 Hydrothermal Co-Carbonization (HTCC) of Coal-Biomass Blend

Abstract:

A 7.5 billion current population of the world is expected reach nearly 10.3 billion in 2040 while energy consumption for electric power is projected to reach 42 quadrillion BTU in 2040. Meeting these energy demands requires high output from our existing energy technologies as well as introduction of emerging technologies. In the US, coal-fired plants accounted for approximately 40% of electricity generation in 2015 and coal consumption has been expected to increase by 17.6% quadrillion BTU by 2040. To meet the projected energy demand without contributing additional greenhouse gas emission, co-combustion of biomass with coal is a viable solution. However, very low bulk density, energy content, and wet nature implements severe risks on co-firing biomass with coal.

Hydrothermal carbonization is one of the thermochemical processes, where biomass is treated with a hot compressed water. As the reaction medium is water, there is no need to dry the feedstock. Therefore, the expensive drying process can be eliminated by applying this technology. Now, sub-and supercritical water, unlike water at ambient condition, are very reactive due to its high ionic product, low density, and low dielectric constant. As a result, hot compressed water reacts with biomass components and release oxygen and ash from the biomass structures. The resulted solid product, often called hydrochar, is hydrophobic and energy dense, just as bituminous coal. Moreover, hydrochar can be further densified by palletization. Energy density via pelletization of hydrochar can be reached much higher than energy density of raw coal.

In this project, hydrothermal co-carbonization (HTCC) of coal-biomass blend was studied at various water temperatures (200, 230, and 260 Â°C) at water saturation pressures. For coal samples, Ohioâ€™s Clarion #4a was used, where, Ohio grown miscanthus (*miscanthus x gigantus*), a perennial grass was used as biomass source. The main objective of this study was to produce a homogeneous hydrochar product in terms of density, fuel properties, and elemental composition. Resulted hydrochar produced from blend (50:50 wt %) was compared with control miscanthus and coal hydrochars. Furthermore, hydrochar pellets were produced using bench-scale single press pellet-press. Pellets mass and energy densities were measured and compared with control biomass or coal pellets.

Hydrochar produced from HTCC reveals that, coalâ€™s elemental carbon content was not effected by water temperatures in the tested temperature range. However, sulfur and chloride content are reduced significantly. Hydrochar produced from biomass is located at the surface of the coal, which ensures the homogenization in the micro-level. The resulted hydrochar was hydrophobic, contains high energy (30-33 MJ kg^{-1}), lower ash, sulfur, chloride, and nitrogen than raw coal. The blend hydrochar pellets have very high mass and energy densities compared to their raw feedstocks. For instance, the blend hydrochar pellets can reach as high as 38 GJ m^{-3} energy density, when biomass-coal blend was treated at 260 °C.

Therefore, HTCC technology can be applied as an upstream processing to feasible co-combustion of coal-biomass mixture. This technology produces homogeneous feedstock with superior fuel quality and low pollutants.

500913 Versatile Surface Modifications for Functionalization of Fibers

Abstract:

Chemical modification of fibers and textiles makes possible both functional consumer products and novel industrial applications with a range of properties. This work focuses on the chemistry and kinetics of covalent attachment of functional groups that impart super-hydrophobicity to cotton fibers. One synthesis method involves cyanuric chloride (2,4,6-trichloro-1,3,5-triazine), which has been employed as an anchor for attachment of various compounds and dyes to the fabric surfaces for several decades. The action of cyanuric chloride on the surface of cellulose materials allows for facile, high yield attachments of compounds by nucleophilic substitution with amines; the functionality of the fabric can be varied by the nature of the amine. In this study, applications of functionalized cellulose fibers have been evaluated. A super-hydrophobic fabric (contact angle > 150°) has been created by attaching functionalized alkyl amines to the surface of cotton fabric. Notably, this process creates a super-hydrophobic surface without the use of perfluorocarbons, the precursors of which are hazardous and highly pervasive environmentally. In addition to the hydrophobic functionalization, other surface properties have been achieved with a range of precursors, including acidic and basic behavior. The examined syntheses have attained high yields, and the materials synthesized show thermal and chemical stability. A second method for covalent attachment of hydrophobic functional groups to cotton fibers employs readily available hydrocarbons and mild reagents and conditions. The novel chemistry provides a straightforward means to modify fibers using inexpensive starting materials. The effects of process variables including temperature and concentrations on the resulting surface coverage and properties will be discussed. These syntheses offer versatile, efficient, and scalable methods for permanent attachment of diverse functional groups not only to fabrics and cellulosic materials, but also to a variety of other surfaces.

Massachusetts Institute of Technology

Extinction Strain Rate (ESR) Calculation for Large Kinetic Models

Alan E. Long, Paul I. Barton, & William H. Green [aelong@mit.edu, whgreen@mit.edu]
Department of Chemical Engineering, Cambridge MA 02139

Motivation

Fig. 1: *Int. Comb. Engine*

In the transition to more sustainable energy production, the ability to theoretically vet and analyze potential new fuels and fuel mixtures has become increasingly important. In internal combustion engines and turbines where these new fuels will be used, the combusting mixtures are exposed to high deformation (strain) rates. Therefore, understanding and predicting combustion behavior under these conditions is important.

Flame extinction is the limit phenomena that occurs under the most severe strains, but is currently difficult to calculate using large detailed chemistry models. We aim to remove this limitation to allow for greater study of extinction.

Fig. 2: *Gas Turbine*

Introduction

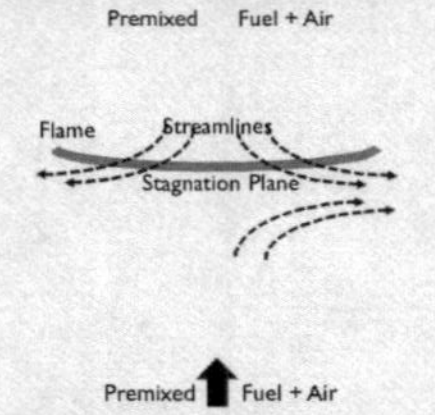

Fig. 3: *Diagram of experimental setup for studying extinction. Twin flame arrangement*

Numerical Simulation

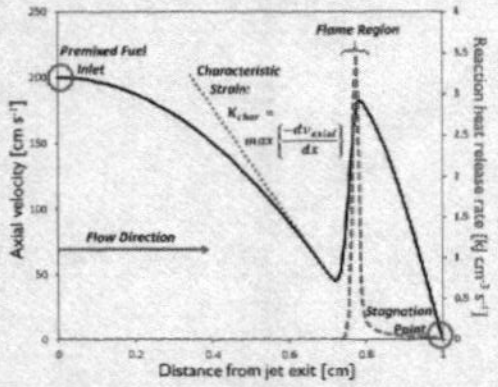

Fig. 4: *Detailed example profile from numerical simulation of experiment*

Continuation to Extinction

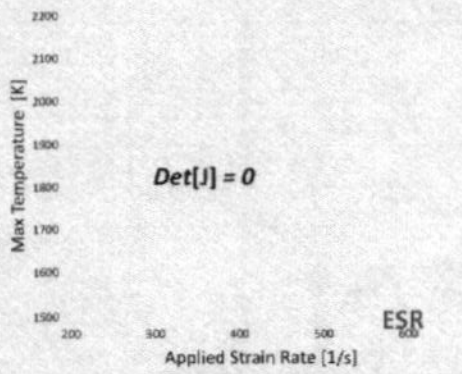

Fig. 5: *Turning point curve showing how extinction strain rate (ESR) is determined from numerical simulations*

Experimental investigation of extinction behavior is most commonly achieved through study of axisymmetric opposed flow [1]. Under these conditions, the strain (deformation) rate on the flame is well defined and the governing equations can be reduced to one dimension. This allows for computation using smaller (~100 species) kinetic models [2], but current solution methods still fail or take prohibitively long for larger models. Extinction strain rate (ESR), or the maximal deformation rate that the flame can sustain, is obtained experimentally (Fig. 3) by increasing the jet velocity until the flame(s) extinguish. Computationally (Fig. 4), arc-length continuation is used to traverse the turning point where the Jacobian needed for the numerical solver becomes singular (Fig. 5). This progression is depicted in Fig. 3-5 for the scenario of twin flames. However, extinction is also often investigated for diffusion flames [3] where the fuel and oxidizer/air mixtures oppose one another from opposite jets and as single flames [4] where the premixed fuel/air mixture opposes an inert gas such as nitrogen. For simplicity and consistency, the remainder of the figures on this poster have used the twin flame arrangement.

[1] CK Law, et al. Symposium (International) on Combustion. Vol. 21. No. 1. (1988)
[2] RJ Kee, et al. Symposium (International) on Combustion. Vol. 22. No. 1. (1989)
[3] O Park, et al. Combustion and Flame. 1073-1094. (2014)
[4] C Ji, et al. Combustion and Flame 157.2, 277-287. (2010)

Proposed Solution

Software developed at MIT, EMBER [5], for the primary purpose of evaluating dynamically strained, curved and planar flamelets shows promise as an alternative to the incumbent methods of ESR computation provided in the CHEMKIN program. Orders of magnitude improvement in computational performance for large, detailed chemical models is observed when EMBER is extended to calculation of extinction (Fig. 6). The increased performance is predominantly attributed to use of Rebalanced Strang Splitting [6]. However, for our purpose of ESR computation, EMBER requires the following improvements:

1. Axisymmetric jet (experimental) formulation
2. Sensitivity analysis for kinetic model refinement
3. Methods for finite boundary condition (experimental) ESR. Potential flow (infinite boundary) formulation currently implemented.

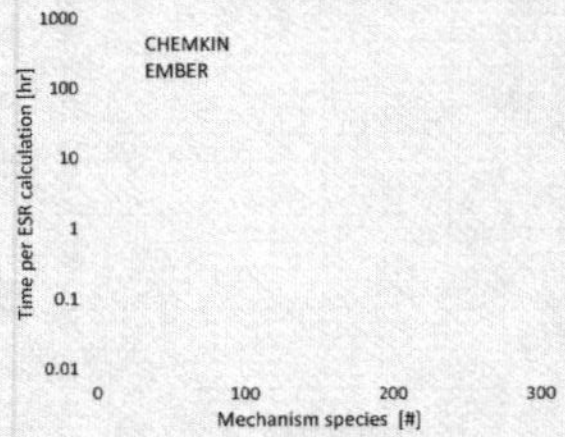

Fig. 6: EMBER vs. CHEMKIN computational cost in calculating planar ESR (Twin flames, mixture averaged diffusion, no Soret effect)

[5] RL Speth. EMBER. https://github.com/speth/ember
[6] RL Speth, et al. *SIAM Journal on Numerical Analysis* (2013).

Solution Progress

1. EMBER governing equation reformulation: Addition of β

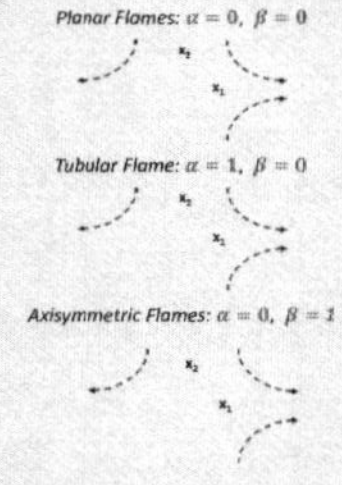

Momentum:

$$\rho\frac{\partial v_{x_1}}{\partial t}+\rho v_{x_1}\frac{\partial v_{x_1}}{\partial x_1}+\rho v_{x_2}\frac{\partial v_{x_1}}{\partial x_2}=\frac{\rho_\infty x_1}{2^\beta}\frac{da}{dt}+\frac{\rho_\infty a^2}{2^{2\beta}}x_1+\frac{1}{x_2^\alpha c_p}\frac{\partial}{\partial x_2}x_2^\alpha\mu\frac{\partial v_{x_1}}{\partial x_2}$$

Continuity:

$$\frac{\partial\rho}{\partial t}+\frac{1}{x_2^\alpha}\frac{\partial}{\partial x_2}x_2^\alpha\rho v_{x_2}+\frac{\rho}{x_1^\beta}\frac{\partial}{\partial x_1}(x_1^\beta v_{x_1})=0$$

Enthalpy:

$$\rho\frac{\partial T}{\partial t}+\rho v_{x_2}\frac{\partial T}{\partial x_2}+\frac{1}{c_p}\sum_{k=1}^{K}\hat{h}_k w_k+\sum_{k=1}^{K} j_k c_{p,k}\frac{\partial T}{\partial x_2}=\frac{1}{x_2^\alpha c_p}\frac{\partial}{\partial x_2}x_2^\alpha\lambda\frac{\partial T}{\partial x_2}$$

Species:

$$\rho\frac{\partial Y_k}{\partial t}+\rho v_{x_2}\frac{\partial Y_k}{\partial x_2}=-\frac{1}{x_2^\alpha}\frac{\partial}{\partial x_2}x_2^\alpha j_k+w_k W_k$$

EMBER was initially developed for flamelet calculations to be used in turbulent combustion simulations, and currently only has capabilities for solving planar flames and cylindrical/curved flames. Fortunately, through comparison with a derivation described in [6] we are able to show that the axisymmetric jet formulation can be concisely added to the current governing equations as shown above with the introduction of a β parameter.

[6] RJ Kee. *Chemically Reacting Flow: Theory and Practice.* (2003)

2. ESR sensitivity analysis possible through internal temperature constraint on solution

Fig. 7 and Fig. 8 show the profile results of brute force perturbation of typically the most sensitive reaction in combustion chemistry: $H+O_2=O+OH$. Here it is observed that while the strain rate at extinction has changed significantly, the max temperature at extinction remains predominantly fixed. This means that by fixing T_{max} and allowing jet velocity to become a dependent parameter, ESR sensitivities can be recovered. This is demonstrated in Fig. 9 where the brute force sensitivity is compared with that obtained by Jacobian forward sensitivity analysis of the fixed temperature formulation implemented in CHEMKIN.

ESR results for brute force rate perturbation of: $H + O_2 = O + OH$

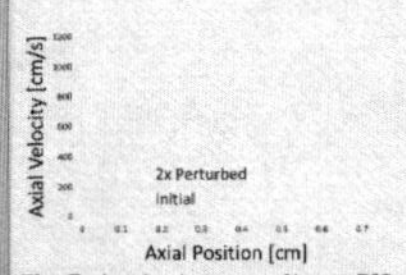

Fig. 7: Axial velocity profiles at ESR before and after rate perturbation

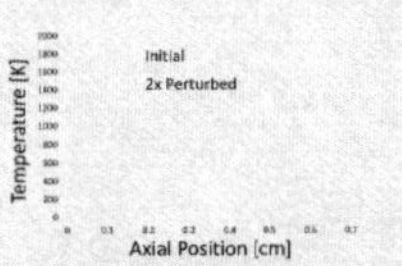

Fig. 8: Temperature profiles corresponding to Fig. 7

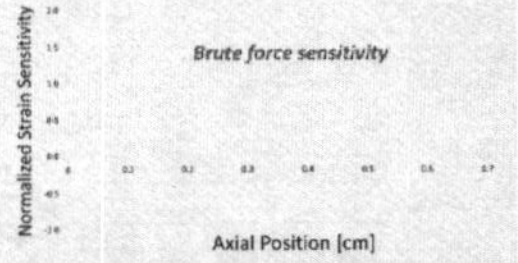

Fig. 9: Jacobian based forward sensitivity for fixed temp formulation vs brute force result

Solution Progress Cont.

3. Finite boundary conditions handled through analytical solution to non-reacting cold flow regime

Observing that the 'hot' regime at extinction is the same for a variety of boundary conditions, we propose that rather than modifying EMBER for the finite BVP, a more effective method for obtaining results relevant to experimentally measured values is to use an analytical solution for the cold, non-reacting flow regime. This is demonstrated in Fig. 10 and Fig. 11. The cold flow governing equation and its analytical solution are:

Governing Eq:

$$2U\frac{\partial U}{\partial x_2}-\left(\frac{\partial U}{\partial x_2}\right)^2+\frac{\rho_\infty^2\,ESR_\infty^2}{4}=0$$

Analytical Solution:

$$U=\frac{4K_0^2-ESR_\infty^2\rho^2}{16\,C_1}z^2+K_0 z+C_1$$

$U=\frac{v_{x_2}\rho}{2}$ = re-scaled axial velocity
ESR_∞ = ESR from pot. flow formulation
ρ_∞ = initial gas density
K_0 = initial strain rate (typically = 0)
C_1 = fitting constant

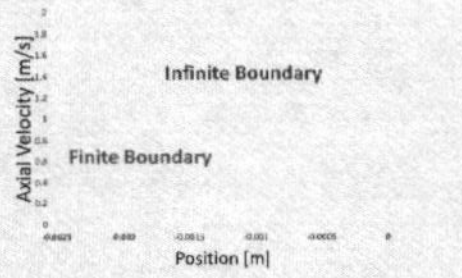

Fig. 10: Infinite vs finite boundary solutions at extinction

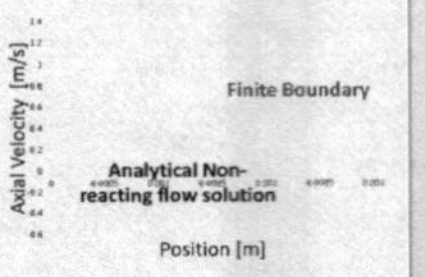

Fig. 11: Demonstration of analytical solution to approximate finite boundary non-reacting flow

Conclusions

EMBER should soon be able to replace conventional methods of calculating extinction strain rate (ESR) for detailed chemical models. This improved computational performance will be attained principally through addition of axisymmetric jet formulation of the governing equations as well as sensitivity analysis through internal temperature specification. This capability will stimulate use of ESR as a validation parameter in kinetic model generation, and will also allow for better understanding of the extinction behavior of fuels whose kinetic models were previously too detailed for ESR calculations.

Future Work

Next efforts will focus primarily on:

- Implementation of necessary changes in EMBER
- More detailed theoretical examination of extinction behavior. Why does T_{max} at extinction stay constant? Is ESR as currently measured really the principal quantity of interest for extinction phenomena?

Acknowledgements

The authors thank ExxonMobil for financial support throughout the duration of this project. In addition, Alan would like to thank Ray Speth, the lead developer of EMBER, for many helpful discussions.

488639 Catalytic Performance and Regeneration of Gallium and Platinum Promoted ZSM-5 Zeolite Catalysts in Ethane Aromatization

Abstract:

Ethane is separated from shale gas to minimize condensation in natural gas pipeline. Direct non-oxidative conversion of ethane to aromatics has become an effective way of upgrading shale gas. Shape selective zeolite catalysts are often used for aromatization. Although the coking issue of the catalysts in ethane aromatization has been widely noticed, the performance of regenerated gallium and platinum contained catalysts needs to be further investigated. The objective of this study is to elucidate coking mechanism over metal-promoted zeolite and investigate feasibility of regenerating deactivated catalysts for commercial viability. In this study, four catalysts will be analyzed: 3 wt% Mo/HZSM-5; 3 wt% Ga/HZSM-5, 0.5 wt% Pt/HZSM-5 and 2.5 wt% Ga-0.5 wt% Pt/HZSM-5. The metal fraction was calculated based on elemental content. For each aromatization reaction, 45 minutes of reaction will be performed for each catalyst with a total flow rate of 50 cm^3/min. To regenerate the catalysts used for ethane aromatization, diluted air (10 vol% of O_2) will be used to oxidized the coke deposit on the catalyst under 500 degree Celsius. The regeneration time is set to be 60 minutes. Eight cycles of reaction and regeneration will be performed in a micro-reactor for each catalyst; and the reactivity will be evaluated and compared by mass spectrometry.

489976 Catalytic Decomposition of Methane into CO_x Free Hydrogen and Carbon Nanotubes over Mono and Bimetallic Ni, Fe, Co Catalysts

Abstract:

Methane is the major component and the most stable hydrocarbon of natural gas. We have investigated the catalytic cracking of methane as an alternative route for the production of carbon nanofibers (CNFs) and COx-free H_2 (Scheme 1). Mono and bimetallic Ni, Fe, Co catalysts were prepared on SiO_2 support by incipient impregnation technique. The effect of doping of metals (Ni-Fe, Ni-Co and Fe-Co) with different mole ratios on the activity and stability of the reaction is investigated here. The performance of the catalyst is evaluated at a constant temperature (T = 650 °C) with a flow rate 70ml/min. (30% CH_4/N_2). After the catalyst screening, catalyst regeneration study was performed on the most stable catalyst. The physico-chemical characteristics are evaluated by XRD analysis, TPR, H_2-chemisorption, HR-TEM and BET surface areas of fresh and used forms of catalysts. XRD and TPR analysis of the fresh catalysts confirmed the alloy formation in the bimetallic catalysts. TEM characterization of the spent catalyst indicated that carbon deposits on the catalyst in the form of hollow cylindrical filaments or carbon nanotubes, with the metal particle located on the tip of each filament. It was also observed that bimetallic catalysts produce a carbon nanotube with a smaller diameter compared to the monometallic ones. A detailed study on the structural characteristics and the mechanism of carbon nanotube formation is in progress.

Scheme 1: Catalytic decomposition of CH4 into COx free hydrogen and carbon nanotubes

490176 Novel ApproachÂ of NO_x Removal from Exhaust Gas

Abstract:

Abstract: As a result of harmful environmental and health effects brought by NO_x emission, it is necessary to minimize NO_x pollutant released into the atmosphere. Conventional SCR systems involve anhydrous or aqueous ammonia (NH_3) or urea ($CO(NH_2)_2$) as a reducing reagent to reduce NO_x to harmless N_2 and H_2O at high temperature 300-400 °C.[1] It is still a big challenge to develop novel low-temperature catalyst (LTC) for SCR of NO_x.[2,3] This work focused on investigation on novel supported ionic liquid phase (SILP) catalysts and application of hydrocarbon (C_xH_y) or CO presented in exhaust gas from marine engine as reducing agent for SCR instead of using urea or ammonia.

Acknowledgements: Funding support for the research from Singapore Maritime Institute Maritime Sustainability (MSA) R&D Programme is gratefully acknowledged.

References

[1] M. F. Fu, C. T. Li, P. Lu, L. Q. M. Y. Zhang, Y. Zhou, M. Y, Y. Fang, *Catal. Sci. Technol*. **2014**, *4*, 14-25.

[2] P. Wang, H. Zhao, H. Sun, H. Yu, X. Quan, *RSC Adv*. **2014**, *4*, 48912-48919.

[3] L. J. Wang, H. X. Deng, H. Furukawa, F. Gándara, K. E. Cordova, D. Peri, O. M. Yaghi, *Inorg. Chem*. **2014**, *53(12)*, 5881-5883.

492115 CO_2-Utilizing Chemical Looping Reforming with the Phase Merge of Fe_2O_3â€“NiO to $NiFe_2O_4$ in a Perovskite Shell

Abstract:

Chemical Looping Reforming (CLR) converts methane into syngas by using metal oxide as an oxygen carrier. Reduced metals by CH_4 are then completely regenerated by air in another reactor. In this research, NiO-Fe_2O_3/$La_{0.8}Sr_{0.2}FeO_3$ was synthesized as an oxygen carrier and CO_2 was utilized for combined oxidizer for reduced metal with coke gasification. CH_4 was further injected to the reduced oxygen carrier and hydrogen was additionally produced through the decomposition of CH_4. Through continuous redox reactions, bimetallic phases of NiO-Fe_2O_3 were merged to singular $NiFe_2O_4$ phase. H_2-Temperature Programmed Reduction (H_2-TPR) measured the reduction kinetics of fresh and 10-cycle spent particle. The result demonstrates that through cyclic redox, the maximum reduction extent of the oxygen carrier increase with the slower reduction kinetics. This can inhibit the total combustion of CH_4 and coke formation on the surface, thus both syngas selectivity and reaction stability can be enhanced.

496289 The Effect of CO_2 on FTS over FBR System for Applications in Gtl-FPSO Process

Abstract:

The crude oil price on fluctuation, together with world energy demand is continuously increasing. The Gas to Liquid (GTL) process is spotlighted in energy industry and one of the promising alternative methods of petroleum fuels production technologies. In addition to the emergence of shale gas, large quantity of stranded gas has been discovered in the offshore. For better use of offshore platform, our group suggests the GTL-FPSO process based on a new concept of combining Compact GTL with Floating Production Storage and Offloading (FPSO).

In offshore stranded gas field, lots of gas fields have CO_2 content more than 30%. Steam carbon dioxide reforming (SCR) of methane which can produce synthesis gases is one of the processes where CO_2 can be utilized as a feed stock.

The synthesis gases from SCR process were adjusted for applications in GTL-FPSO system. The Co based catalyst was prepared by impregnation method and characterized by XRD, TGA, TPR and FT-IR technologies. The effect of CO_2 on Fischer-Tropsch synthesis (FTS) reaction over FBR system was investigated over the Co based catalyst and analyzed by product distributions and temperature profiles.

496354 Hydrogen Production By Steam Reforming of Methane over Nickel Based Catalysts Supported on the Alumina Mixed with SiC

Abstract:

Steam reforming of methane is an extremely important process for the hydrogen and syngas(H_2, CO) production. Hydrogen has a long tradition as an energy carrier as well as an important feedstock in petrochemical and refinery industries. In these days, hydrogen has attracted even more attention as a transportation fuel. Steam reforming of methane at a lower temperature range is one such promising approach. The mild operating conditions lead to lower operation and materials costs.

In this study, Ni based catalysts supported on the alumina mixed with SiC were prepared by physical mixing and impregnation methods. The prepared catalysts were characterized by XRD, TPR and N_2-phyisoprtion techniques. The catalytic performance for steam reforming of methane was investigated at 650~850 °C, 1~20 bar and GHSV of 5,000~30,000 h^{-1}. The effect of SiC contents on the catalytic performance for steam reforming of methane was investigated and compared with the catalytic performance for steam CO_2 reforming of methane.

497274 Calcium and Manganese-Doped Lanthanum Iron Perovskite Oxides As Candidate Redox Materials for CO_2 Reduction to CO

Abstract:

Carbon dioxide is the sole source for renewable carbon-based energy on Earth. CO_2 processing into fuel production intermediates such as CO, a major component of synthesis gas, is thermodynamically challenging and requires catalytic materials capable of redox cycling. Perovskite-type oxides in particular have shown promise in efficiently converting CO_2 to CO in the reverse water gas shift chemical looping process (RWGS-CL) by retaining nonstoichiometric lattices as previously demonstrated by our group. Metals of greater Earth abundancy are required to decrease both conversion temperature and material cost to make large scale CO_2 to fuels processes more economically feasible for commercial industry. For instance, calcium is capable of replacing previously used elements such as strontium or barium while improving catalytic activity. Metals such as Ca and Mn may be tailored into the perovskite structure of $LaFeO_3$ to simultaneously improve CO production rates and decrease rare metal content. Density functional theory calculations also conducted by our group suggest that compositional optimization of $La_XCa_{1-X}Fe_YMn_{1-Y}O_3$ (LCFM) will further increase CO_2 convertibility of the alloy material. Currently, $La_{0.6}Ca_{0.4}MnO_3$ demonstrated a total CO yield of 1240 Î¼moles/g cat while that of $La_{0.6}Ca_{0.4}Fe_{0.4}Mn_{0.6}O_3$ was 970. However, a peak CO production rate of 160 Î¼moles/g cat/min was achieved with the latter species at 500 Â°C as opposed to 60 Î¼moles CO/g cat/min at 550 Â°C by the conventional $La_{0.75}Sr_{0.25}FeO_3$ perovskite. These LCFM materials will be tested in temperature programmed experiments monitored by mass spectrometry to observe the effects of doping on ideal reaction temperatures and characterized via several techniques including X-Ray diffraction and BET analysis for crystalline structure and surface area.

498872 Fast Cycling to Achieve High NO_x Conversion in Exhaust: Role of Ceria

Abstract:

Fast Cycling to Achieve High NOx Conversion in Exhaust: Role of Ceria

Zhiyu Zhou, Michael P. Harold*, Dan Luss**

Department of Chemical and Biomolecular Engineering, University of Houston, Houston, TX 77204

* *mharold@uh.edu*, ** *dluss@central.uh.edu*

The abatement of NOx (NO+NO_2) remains a challenge under the continuously more stringent emission standards. Commercialized deNOx systems, NOx storage and reduction (NSR) and selective catalytic reduction (SCR), function well from ~250 °C to ~400 °C, but have unsatisfactory performance outside that temperature range. Toyota researchers recently invented a new deNOx system, Di-Air (diesel NOx aftertreatment by adsorbed intermediate reductants), which involves fast injection of fuel into the exhaust feed to the NSR converter [1]. Previous studies show that ceria increases the deNOx performance of the Di-Air system in both low [2] and high [3] temperature regions. Data suggest that NOx conversion occurs via the reaction of an intermediate produced through the reaction with O_2 and NO [1, 2]. However, a recent study [3] suggests that catalysts containing ceria may involve an alternative pathway of NOx abatement via NO decomposition at high temperatures. We studied in a bench-scale flow reactor system the NO decomposition and reduction on ceria under different frequencies with CO as sole reductant.

In our study, a lean to rich feed time ratio 6 to 1 of was controlled to keep the same fuel penalty (fuel to oxidant ratio). The lean/rich cycling conditions included lean/rich times (in seconds) of 90/15, 60/10, 30/5 and 6/1. Different reaction conditions were applied to study NO conversion behavior under favorable and adverse conditions. 500 ppm NO was added to both rich phase and lean phase. 2.5% CO was only added to rich phase and O_2 with varied concentration was only added to lean phase. In some cases, CO_2 and H_2O were introduced continuously along with NO, O_2 and CO to simulate lean exhaust.

When CO is used as the sole reductant and no H_2O is introduced in the feed, NO is mainly converted to N_2 with only trace amounts of N_2O and NO_2 detected. Figure 1 shows the cycle-averaged NO conversion with four cycle times from 450 °C to 625 °C in two feed conditions. In the first case, when 1400 ppm O_2 is introduced in lean feed, Figure 1(a) shows that increase of lean/rich switching frequency enhances NO conversion. The optimized lean/rich switching frequency is always 6/1 above 500 ^{o}C. In the second case, when 2100ppm O_2 is introduced in the lean feed, Figure 1(b) shows that a different optimal lean/rich switching frequency exists for different temperatures. The optimal lean/rich switching frequency is 30/5 at intermediate temperatures (500 ^{o}C and 550 ^{o}C) and 6/1 at high temperatures (600 ^{o}C and 625 ^{o}C) respectively. Figure 1 shows that the optimal lean/rich switching frequency changes with reaction conditions (e.g., temperature and feed condition).

Figure 1. (a) NO conversion with different lean/rich switching frequency; (a) rich: 2.5%CO, 500ppm NO; lean: 500ppm NO, 1400ppm NO; (b) rich: 2.5%CO, 500ppm NO, lean: 500ppm NO, 2100ppm O_2

Figure 2 shows the cycle-averaged NO conversion with the same four cycle times at 600 ^{o}C with fixed ratio of lean and rich feeds. The short cycle limit represents NO conversion obtained with total mixed lean and rich feeds. The long cycle limit represents weight averaged NO conversion in lean phase and rich phase. The optimal lean/rich switching frequency of 30/5 exceeds those of both short and long cycle time limit. This shows that the lean/rich cycling operation is beneficial to NO conversion.

Figure 2. NO conversion at 600C, with a fixed switching frequency as 90s lean/15s rich; rich: 2.5%CO, 500ppm NO, lean: 500ppm NO, 3600ppm O_2

Figure 3 shows that NO conversion changes when CO_2 and H_2O are fed. By comparing the blue line (reference) with either red line (with H_2O only) or purple line (with CO_2 only), it is obvious that both H_2O and CO_2 inhibit NO decomposition on ceria but H_2O has a smaller detrimental effect than CO_2. The green line (with both CO_2 and H_2O) lies between the red line (with H_2O only) and the purple line (with CO_2 only), which indicates that H_2O can mitigate the detrimental effect of CO_2.

Figure 3. NO conversion at 550C, with a fixed switching frequency as 90s lean/15s rich; (rich: 2.5%CO, 500ppm NO, lean: 500ppm NO, varied O_2 concentration, CO_2 and H_2O in both lean and rich if introduced)

This study on NO decomposition on ceria during lean and rich cycling helps explain the beneficial function of ceria on NOx reduction at high temperatures, which also provide guidance for optimization of catalyst formulation and operation strategies.

Reference

1. Y. Bisaiji et al., *SAE Int. J. Fuels Lubr.* 5 (2012) 380-388
2. Y. Zheng et al., *Catal. Today*, 276 (2016), 192-201.
3. Y. Wang et al., *Top. Catal.*, 59 (2016), 854-860

499248 Cobalt Supported on Hydrothermally Synthesised Carbon Spheres for Fischer-Tropsch Synthesis

Abstract:

Carbon spheres (CS_{hydr}) were synthesised by a low temperature (190 °C) hydrothermal synthesis method using sucrose as a substrate. The as-synthesised CS_{hydr} were then annealed at different temperatures ranging from 250 °C to 800°C under N_2 gas. The effect of annealing changed the physical properties of the CS_{hydr}. BET studies showed that annealing of CS_{hydr} increased their surface area from 2.1m^2/g to 431m^2/g. CS_{hydr} were characterised using TEM, XRD, TGA and Raman spectroscopy. The Raman spectral ratios I_D/I_G, I_{D2}/I_G, I_{D3}/I_G, I_{D4}/I_G were analysed and correlated with the graphitic nature of CS_{hydr}. Co/CS_{hydr} were evaluated for Fischer-Tropsch synthesis and the graphitic nature of the support correlated with the FT activity.

499463 A User-Friendly Set up for Undergraduate Research: Combining Thermogravimetric Analysis with Micro Gas Chromatography

Abstract:

Thermogravimetric analysis (TGA) is a common, easy to use analytical tool to provide data on solid/gas reactions or to characterize materials such as catalysts or polymers. Because the technique provides limited information on the perceived weight change, the coupling of thermogravimetric analysis with mass spectrometry (MS) for gas evolution data is becoming more common. In this study, we couple the TA Q600 TGA with the INFICON Fusion micro gas chromatograph (Micro GC) to conduct several standard catalyst characterization tests (temperature programmed oxidation/reduction/desorption) and to investigate several gas/solid reactions. We looked at reactions where expected products were light gases and were therefore able to use GC cycle times less than 90 seconds; getting a resolution of gas composition data that could be correlated with the TGA signal. During temperature programmed reduction of several common catalysts were able to correlate weight loss on the TGA with water production and hydrogen consumption on the GC. Additional data was collected for CO_2 methanation, biomass oxidation (with O_2 or CO_2) reactions demonstrating the usefulness of the coupling of these instruments. Compared to coupling a TGA-MS, this Micro GC approach allows for easier quantification of gas species, is less expensive, and use-friendly enough for undergraduate research.

Transmission FT-IR Spectroscopy

Abstract:

A great deal of work has been devoted to study the surface properties of zirconia over many years, mostly due to its excellent catalytic behavior in dehydration, hydrogenation, and hydrogen exchange reactions as both a catalyst and a support. In this study the interactions of CO with amorphous zirconia(am-ZrO_2), monoclinic zirconia(ZrO_2), and tetragonal zirconia(t-ZrO_2) were investigated by diffuse reflectance FT-IR spectroscopy and transmission FT-IR to determine the role of hydroxyl on CO adsorption and conversion. The CO bands and surface intermediates detected by DRIFT were compared with transmission FT-IR at different temperature and dehydroxylation degree. Moreover, the effect of different zirconia polymorphs on CO adsorption was discussed.

The results showed various CO bands and surface intermediates were observed on three zirconia polymorphs. In DRIFT spectra, formate and/or bicarbonate species formed by CO reacting with hydroxyls at high temperature were detected on am-ZrO_2, m-ZrO_2 and t-ZrO_2 and then the active sites for CO adsorption were produced by consumption of hydroxyls. While in the case of transmission FT-IR spectra, three different CO bands, bicarbonate and carbonate species were observed on am-ZrO_2 and m-ZrO_2 at room temperature. Only carbonate was formed on t-ZrO_2 and no CO band was detected. Note that active sites for CO adsorption could be produced by thermal dehydroxylation in vacuum or CO reaction with hydroxyls at high temperature for am-ZrO_2 and m-ZrO_2. While in the case of t-ZrO_2, active site for CO adsorption was only created by the formation of bicarbonate species at high temperature.

A: DRIFT Spectra

B:Transmission infrared spectra

Fig. 1 FT-IR of CO adsorption on ZrO_2 polymorphs.

References

1. Kouva S K Honkala, L Lefferts J Kanervoa (2015) Review: monoclinic zirconia, its surface sites and their interaction with carbon monoxide. Catal. Sci. Technol. 5: 3473-3490

2. Köck EM, Kogler M, Bielz T, Klötzer B Penner S (2013) In Situ FT-IR Spectroscopic Study of CO_2 and CO Adsorption on Y_2O_3, ZrO_2, and Yttria-Stabilized ZrO_2. J. Phys. Chem. C, 117: 17666-17673

3. Pokrovski K, Jung K T, Bell A T (2001) Investigation of CO and CO_2 Adsorption on Tetragonal and Monoclinic Zirconia. Langmuir 17:4297-4303

4. Kondo J, Abe H, Sakata Y, Maruya K I, Domen K Onishi T (1995) Infrared studies of adsorbed species of H_2, CO and CO_2 over ZrO_2, Chem. Soc. Faraday Trans. 91: 125-132

5. Jung K T, Bell A T (2000) The effects of synthesis and pretreatment conditions on the bulk structure and surface properties of zirconia. J. Mol. Catal. 163: 27-42

499215 Measurement of Raman Spectra during Thermal Oxidation of Hydrocarbon Fuels

Abstract:

Spectral measurements of Raman scattering in fluids at elevated temperatures and pressures can be challenging because of fluorescence and thermal radiation. These limitations severely hinder the ability to track small changes to the chemical composition in reacting environments, such as during the thermal stressing of hydrocarbon fuels. To aid in these measurements, a fiber-coupled, in-line, Raman flow cell was developed that is capable of withstanding 300 °C (572 °F) and 3.5 MPa (510 psi) without active cooling. The flow cell was used to collect high quality spectra of different fuels and model compounds from room temperature up to 300 °C, and the signal-to-noise (S/N) reduction at temperature was quantified. Chemometric multivariate calibration models were created that were able to remove the fluorescence background from the samples. The system was used to examine thermal degradation in multi-component fuel mixtures over a 24 hour period and quantify deposition of degradation products on the vessel walls. Flow cell development, measurement techniques, model development, and application cases will be presented.

502420 Effects of Controlled Crystalline Plane of Hydroxyapatite on Methane Conversion Reactions

Abstract:

Hydroxyapatite (HAP, $Ca_{10}(PO_4)_6(OH)_2$) is a type of calcium phosphate with a hexagonal crystal structure, commonly studied in biomaterials, electrolyte membranes and catalysts. The orientation of HAP, for example, along the crystallographic c-axis direction with the (002) plane exposed as end face, significantly enhances the ion conductivity of HAP electrolyte membranes and wear resistance of HAP-containing teeth and bones. The effect of HAP orientation on catalytic reactions over HAP-based catalysts, however, has not been explored. This work reports the influences of controlled HAP crystal orientation (along c-axis direction) and exposed crystalline plane ((002) as end face) on methane conversion reactions over HAP and lead-substituted HAP (Pb-HAP) catalysts. The alignment of HAP-based crystals in the catalysts was achieved via an electrochemical deposition of HAP seeds on a titanium substrate, the secondary hydrothermal crystallization to promote the c-axis growth normal to the substrate, and peeling off of the HAP-based crystals from the substrate. Two catalytic reactions, methane combustion and oxidative coupling of methane to C_2 (ethylene and ethane) were tested over the oriented HAP-based catalysts. The results show that oriented HAP-based catalysts with exposed (002) crystalline facet promoted oxygen adsorption which facilitated the reaction between methyl radicals and adsorbed oxygen species to form CO_2 product when compared to randomly oriented HAP-based catalysts. The present analysis explored for the first time the effects of HAP crystal orientation and controlled crystalline plane exposure in methane conversion reactions, which may also be applied to other apatite materials and catalytic reactions.

498069 Catalysis Research Gas Analysis Using Micro GC Fusion

Abstract:

Greater demand for alternative energy sources has led to an increase in research involving sustainable energy conversion systems. Micro GC Fusion is an ideal tool for catalysis gas analysis. A 2 module configuration can be used as a solution to several gas analysis applications including general purpose energy gas analysis such as landfill gas and syngas analysis. It can also be used in Catalysis Research gas analysis or thermal and catalytic conversion of carbon based material to desired products. Some examples of these alternative energy applications include the conversion of solid waste and biomass to synthetic fuels, liquid fuels to hydrogen, and greenhouse gases to fuels.

A micro gas chromatograph (GC) uses microelectromechanical systems (MEMS) components to miniaturize injection and detection subsystems, greatly reducing size and weight. Micro GC Fusion offers significant throughput gain through rapid temperature programming and modular architecture. The modular GC platform allows for parallel analysis of fixed gases and hydrocarbons. In addition, Micro GC Fusionâ€™s rapid temperature ramping through resistive column heating technology extends the application range on each column reducing downtime by providing the option to build a bakeout into the method.

Analytical results in less than 3 minutes ensures that ongoing projects and deadlines can be effortlessly met.Â An integrated design and a web-based user interface to add to ease of use.Â Micro GC Fusion offers exceptional precision and a simple, cost effective means to meet the needs of catalysis gas analysis.

492333 Investigation of the Effect of Reducing Agents (Syngas, H_2 and CO) on Catalyst Deactivation during Low Pressure Fischer-Tropsch Synthesis

Abstract:

An understanding of catalyst deactivation during Fischer-Tropsch Synthesis (FTS) is an important piece of knowledge. In the literature, research has been done experimentally and theoretically and models have been developed on the deactivation of FTS catalyst. In the present study, experimental deactivation rates of three differently reduced catalyst in FTS are presented. Three fixed bed reactors were set in parallel and each was loaded with 1 gram of the same iron based catalyst. The catalyst was then simultaneously reduced with syngas, H_2 and CO, respectively, for 48 hours at 250 ^oC, under atmospheric pressure using a flow-rate of 60 (NTP)ml/min. The reactors were then subjected to the same FT reaction procedures for 5000 hours at a pressure of 2 bar(abs). The results obtained indicate that the use of different reducing agent does have an impact on the catalyst deactivation rates. The average activity loss due to catalyst deactivation was found to be 0.0025%/day for syngas reduced, 0.0016%/day for hydrogen reduced and 0.0027%/day for carbon monoxide reduced catalyst. The experimental data could imply that for long time FTS operations, the hydrogen reduced catalyst tend to take more time to deactivate.

501513 Integration of Random Pore Model & Langmuir-Hinshelwood Kinetics to Study High Temperature Coal Gasification

Abstract:

Random pore model (Bhatia and Perlmutter, 1980; 1981; 1983) has been applied extensively to both char oxidation systems (Su and Perlmutter, 1985) and char gasification systems with CO_2 (Bhatia, 1981) and H_2O (Chi and Perlmutter, 1989). The model solutions developed in these works assume the reaction to be first order with respect to the reactant gas. But it is well known that coal char systems are highly porous and the gasification reactions follow the Langmuir-Hinshelwood (LH) kinetics (Ergun, 1955).

There have been attempts to combine random pore model with LH kinetics for competing CO_2 and H_2O gasification reactions (Umemoto et. al., 2013). But these efforts are restricted to the low temperature kinetically controlled regime. There is a need to extend the combined model to systems at higher temperature that are pore diffusion controlled to better understand the industrial systems of coal gasification.

In this study, we extend the random pore model with transport and diffusion effects to incorporate LH kinetics first to CO_2 gasification systems with varying amounts of CO using multi-species reaction diffusion modeling. We then study the H2O gasification systems in presence of H_2O, CO and H_2. Finally we combine both these efforts to study the complete industrial gasification systems that involve both CO_2 and H_2O in the presence of all four gases: CO_2, CO, H_2O and H2. Using this model we gain important insights into optimizing the feed compositions to obtain the final product: syngas, with the desired specifications.

References

Bhatia, S. K., and D. D. Perlmutter, "**A Random Pore Model for Fluid-Solid Reactions: I. Isothermal, Kinetic Control**,"AIChEJ., 26, 379 (1980).

Bhatia, S. K., and D. D. Perlmutter, "**A Random Pore Model for Fluid-Solid Reactions: II., Diffusion & Transport Effects**,"AIChEJ., 27, 247 (1981).

Bhatia, S. K., and D. D. Perlmutter, "**Unified Structural Effects in Fluid-Solid Reactions**,"AIChEJ., 29, 281 (1983).

Bhatia, S. K., "**The Effect of Pore Structure on the Kinetics of Fluid-Solid Reactions**," PhD Thesis, University of Pennsylvania (1981).

Su, J. L., and D. D. Perlmutter, "**The Effect of Pore Structure on Char Oxidation Kinetics**,"*AIChEJ.*, 31, 973 (1985).

Chi, W. K., and D. D. Perlmutter, "**The Effect of Pore Structure on the Char-Steam Reaction**,"*AIChEJ.*, 35, 1791 (1989).

Umemoto, S., S. Kajitani and S. Hara, "Modeling of Coal Char Gasification in Coexistence of CO2 and H2O Considering Sharing of Active Sites" *Fuel.*, 103, 14 (2013).

Ergun, S., "Kinetics of the Reaction of Carbon Dioxide with Carbon,"J. of Phys. Chem., 60, 480 (1955).

Synthesis of Pd nanocatalysts on the multilayered polyelectrolyte film in a gas-liquid-solid microreactor for nitrobenzene hydrogenation

Jian Liu [a, b], Xun Zhu [a, b *], Qiang Liao [a, b], Rong Chen [a, b], Dingding Ye [a, b], Biao Zhang [a,b]

[a] *Chongqing University, China;* [b] *Key laboratory of Low-grade Energy Utilization Technologies and Systems, China*

Introduction

Polyelectrolyte multilayers (PEMs) thin film can be easily prepared via the layer-by-layer (LbL) electrostatic assembly technique.

LbL assembly is a simple, versatile and significantly inexpensive approach and it is widely applicable to various substrates of different shapes and sizes as long as the substrate surface is charged.

The PEMs thin film can serve as regenerable surface coating to bind metal ions via a simple aqueous ion exchange procedure and these metal ions can subsequently be reduced to nanoparticles .

these nanoparticles supported by PEMs thin film, catalysis provide a natural application because their large surface area-to-volume ratio allows effective utilization of expensive metals.

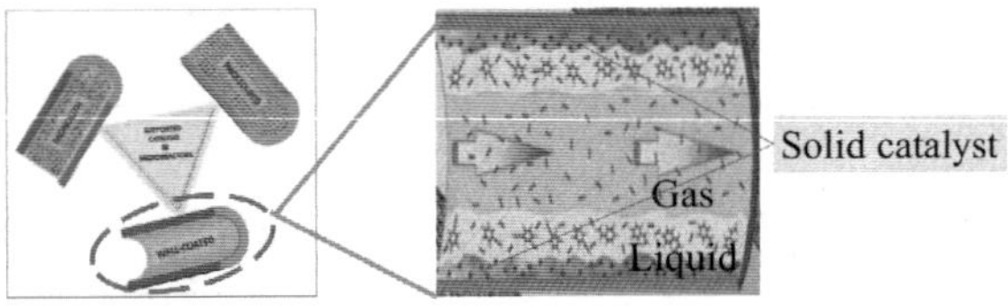

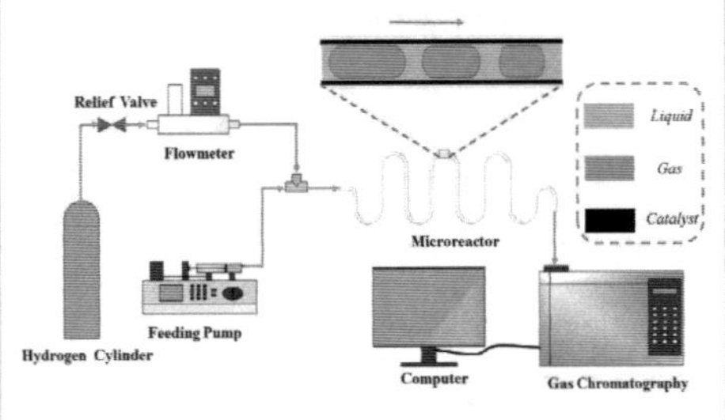

Materials and Methods

Preparation of Pd nanocatalysts supported on inner surface of microreactor involves three steps. Namely, (1) the PEMs thin film deposition on the PDA modified surface, (2) loading of the Pd precursors, (3) generate within the PEMs thin film of the Pd nanocatalysts.

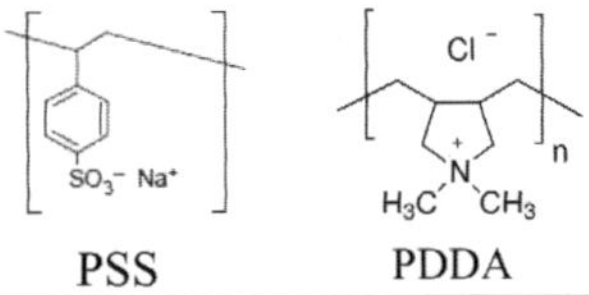

Inspired by mussel-adhesion phenomena in nature, polydopamine (PDA) was employed for the inner surface modification of microreactor.

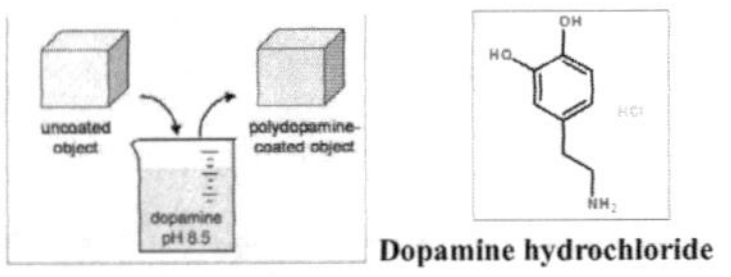

Schematic illustration of the surface modification of substrate by dopamine hydrochloride solution

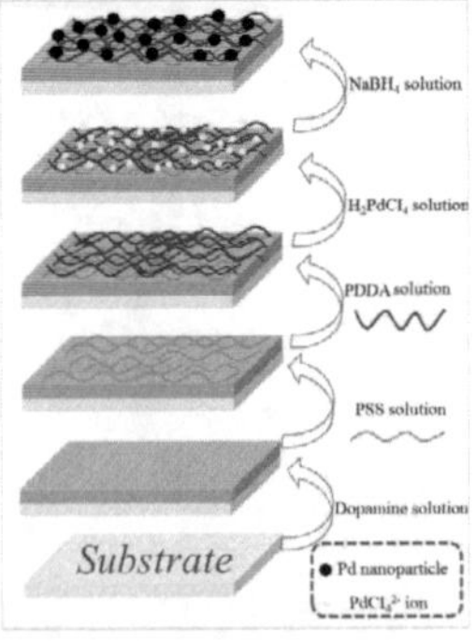

Schematic illustration of nanocatalysts-PEMs composite thin film prepared by LbL self-assembly

Results and Discussion

Materials characterization

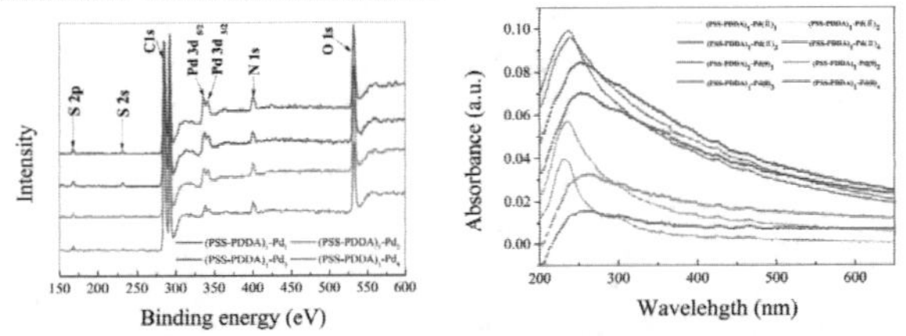

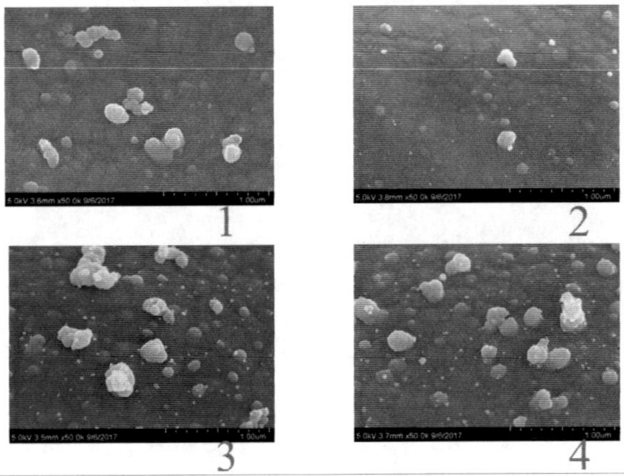

Evaluation of the developed microreactor.

c_{NB} = 60 mM. u_G =0.15 sccm, u_L = 15 μL/min

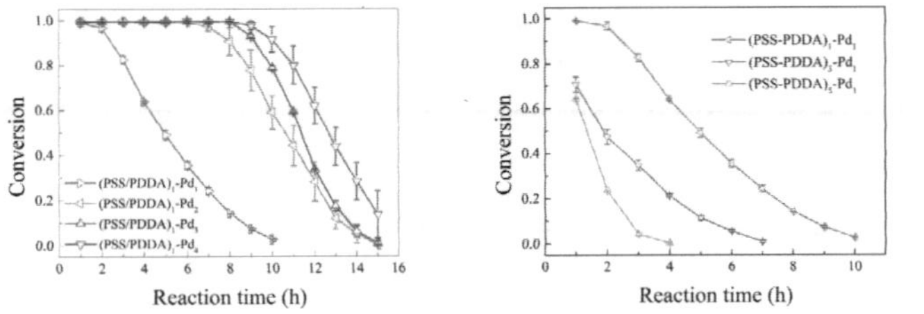

Conclusion

1. Incorporation of Pd nanocatalysts into polyelectrolyte multilayers via counterion exchange and in situ reduction is a facile approach.
2. In the case of low salt concentration, Pd nanocatalysts are only present at the surface of the PEM , the content of which should be dependent of the number of bilayers.

Acknowledgments

The authors gratefully acknowledge the financial supports of the National Natural Science Foundation of China (No.51325602, No.51276208, No.51576021).

重庆大学
CHONGQING UNIVERSITY
工程热物理研究所
Institute of Engineering Thermophysics

508951 Evaluating the Surface Science of Photocatalytic Nitrogen Fixation

Abstract:

Photocatalysis is the ideal solution to sustainable ammonia synthesis, as it will *enable the distributed production of fertilizers and fuels from reactants that are readily available at ambient conditions: air, water and photons.* Given the massive amount of energy currently expended to produce and transport fertilizer, such a process would have an enormous impact on energy consumption and carbon emissions. Photocatalytic nitrogen reduction has been demonstrated over titania catalysts, although the production rates are too low to be practical. Despite the enormous promise of this reaction, first demonstrated 40 years ago, there is no understanding of the fundamental chemical and physical processes that enable it.

Previous investigations of photocatalytic fixation of nitrogen over titania have produced a myriad of contradictory results, including reduction to ammonia and hydrazine, oxidation to nitrate ions, and absence of nitrogen fixation activity. The reaction mechanisms have therefore been intensely debated for decades[3, 6, 7], with no consensus reached. It has been pointed out that the single-photon reduction of nitrogen and water to ammonia is thermodynamically unfeasible, yet ammonia production has been consistently observed by several independent groups, suggesting a complex multi-photon mechanism. Furthermore, it has been shown that photocatalytic ammonia production can be enhanced by the addition of transition-metal co-catalysts. Since these reports, tremendous progress has been made towards understanding the molecular-level processes involved in titania photocatalysis and thermochemical nitrogen fixation; however, a modern surface-science perspective has not been applied to photocatalytic nitrogen fixation. Here we will detail initial experiments which were conducted to evaluate surface properties of titania under dark and conditions with various reactants.

497372 Tunable Oleo-Furan Surfactants Via Acylation of Biomass-Derived Furans

Abstract:

Surfactants are widely used in household detergents, cleaners, emulsifiers, foaming agents, and personal care products. Limited selectivity and tunability of petroleum-derived, commercial surfactants such as linear alkylbenzene sulfonates result in poor hard water performance, thereby demanding the use of expensive and toxic chelants in detergent formulations. We propose a new, renewable pathway to produce surfactants made from biomass-derived feedstocks, such as furan obtained from the sugar, xylose, in biomass and lauric acid obtained from the hydrolysis of triglycerides found in natural oils such as coconut and soybean oil. The key chemistry enables acylation of furan with lauric acid or anhydride in the presence of a heterogenous catalyst such as a zeolite or a homogenous catalyst such as trifluoroacetic anhydride. The synthesis of these oleo-furan molecules is highly tunable, where surfactant properties can be selected by using different sources of triglycerides and by coupling various chemistries to obtain a wide range of surfactants incorporating linear or branched structure, or different chemical functionalities. Oleo-furan surfactants (OFS) demonstrates a hundredfold improvement in hard water stability when compared to conventional surfactants, thereby bypassing the need for chelants. The ability to tailor furans with the functionality desired in the final surfactant molecule enables the expansion of this chemistry to make a wide variety of OFS and opens the doors to explore the potential of furan-based surfactants for other applications.

497904 Catalytic Upgradation of Biomass-Derived Bio-Oil By C-C Coupling of Phenolic Compounds with Light Oxygenates

Abstract:

Catalytic upgradation of biomass-derived bio-oil by C-C coupling of phenolic compounds with light oxygenates

Gul Afreen, Tanmoy Patra, Ratan Mohan, Sreedevi Upadhyayula*

Department of Chemical Engineering, Indian Institute of Technology Delhi, New Delhi-110016, India

**corresponding author: sreedevi@chemical.iitd.ac.in*

With the rapid depletion of crude petroleum resources, the demand for renewable energy sources has increased immensely. Extensive research in utilization of biomass indicates that it is a promising feedstock for the sustainable production of fuels and chemicals. Pyrolysis of lignocellulosic biomass results in liquid bio-oil which contains a high percentage of light oxygenates like acetic acid, ketones, etc as well as lignin-derived phenolic compounds. Large amount of oxygen present in the bio-oil is removed by performing hydrodeoxygenation, to make it compatible with the crude petroleum. However, during this process small oxygenates are lost in the form of light gases. So to retain the carbon of such small oxygenates as well as to bring the phenolic compound in the fuel range (C10-C13), C-C coupling by alkylation of the phenolic compounds with small oxygenates is thought to be beneficial for the biofuel upgradation, before the hydrodeoxygenation step.[1] The resulting compounds not only fall in the desirable carbon-chain fuel range but also are widely used as chemical intermediates in the synthesis of antioxidants, fragrances, agrochemicals, and fuel additives.[2-4] In this context, to represent the phenolic compounds and light oxygenates, the model compounds chosen are *m*-cresol and *iso*propanol, respectively. The main product of alkylation of *m*-cresol with *iso*propanol is 2-*iso*propyl-5-methylphenol (IMP), also known as thymol, a precursor of commercially useful pharmaceutical product called menthol. A number of different types of catalysts like mineral acids, Lewis acids, ionic liquids, mesoporous solid superacidic materials, modified zirconia, HY zeolites, Pt metal loaded on Hβ zeolite, iron catalyst containing Cr, Si and K oxides, zinc aluminate spinel, silica-alumina clays, metals loaded on mesoporous materials, γ-alumina, etc. have been reported earlier for the production of thymol by alkylation of *m*-cresol with *iso*propanol.[1,2,5-9] However, the drawbacks of these catalysts are that the use of mineral acids and Lewis acids result in equipment corrosion, environmental pollution, low reactant to catalyst ratio and loss of selectivity to the desired product. Homogeneous ionic liquid catalysts have difficulties in product purification and catalyst recovery. Reported mesoporous materials and other solid acidic catalysts have the advantages like easy separation of the products, excellent reusability, etc. However, the use of solid catalysts results in disadvantages like rapid deactivation due to leaching of active metals, pore blocking due to coke formation, etc. Among the reported heterogeneous catalysts, zeolites contain Bronsted acid sites and have high shape selectivity and thermal/hydrothermal stability which can give high conversion of *m*-cresol to form thymol.[10] Hence, for the large scale production of thymol, the zeolite-based catalysts with good thermal stability, least coking issues, and controlled Bronsted and Lewis acidity is to be developed.

In this study, *iso*propylation of *m*-cresol was performed over HY, ZnY, Hbeta, HZSM5 and HMCM22 zeolites as catalysts to ascertain the effect of the nature and strength of superficial acid sites as well as their pore size on the catalytic activity and product distribution. The reaction was conducted using a fixed-bed continuous flow reactor with length to diameter (L/D) ratio of 30. The liquid products were analyzed using a NUCON-GC supplied by AIMIL India Ltd. equipped with a CHROMSORB-WHP (2 m × 3.175 mm × 2 mm) column and flame ionization detector (FID). Conversion of *m*-cresol at temperatures between 200 ^{0}C and 300 ^{0}C is shown in Figure 1. The maximum conversion of *m*-cresol (86%) was achieved at 275 ^{0}C for ZnY catalyst and selectivity to thymol was 41% . The use of HY as a catalyst resulted in 69% conversion of *m*-cresol at 250 ^{0}C, while the conversion of *m*-cresol was very low for HZSM5, HMCM22 and Hbeta catalysts. The higher conversion of *m*-cresol over ZnY catalyst can be attributed to the higher Lewis as well as Bronsted acidity of the active metal containing catalyst. This concludes that a strong Lewis plus Bronsted acidity is required for this reaction. The low conversion of *m*-cresol on HZSM5, HMCM22 and Hbeta catalysts can also be explained due to diffusional constraint. The selectivity to thymol was also analyzed by using these catalysts.

Figure 1: Conversion of *m*-cresol using different catalysts. (WHSV=2.8 h^{-1}, *m*-cresol:IPA mole ratio=1:2, Catalyst loading=1.03 g, P= 1 bar)

The maximum conversion for each catalyst, except ZnY, was obtained at 250 ^{0}C, thereafter, it decreases with further increase in temperature due to the deactivation of catalysts at higher temperatures. However, for ZnY the maximum conversion was achieved at 275 ^{0}C, which shows that it deactivates comparatively slower than other catalysts. The deactivation studies were performed using characterization tools like NH_3-TPO. The influence of reaction parameters such as reaction temperature, WHSV, mole ratio of reactants and time-on-stream has also been investigated for the best catalyst i.e ZnY. Thus, development of a continuous, environmentally benign, energy efficient and economical process route for the production of alkylated *m*-cresol for bio-fuel upgradation can be made using metal-containing solid acid catalysts.

References

1. P.A. Zapata, J. Faria, M.P. Ruiz, R.E Jentoft, D.E Resasco, *J Am Chem Soc.* **2012**, *134*, 8570–8578.
2. H. Grabowska, L. Syper, M. Zawadzki, *Appl Catal A.* **2004**, *277*, 91–97.
3. V. Umamaheswari, M. Palanichamy, V. Murugesan, *J Catal.* **2002**, *210*, 367–374.
4. M.C. Samolada, E. Grigoriadou, Z. Kiparissides, I.A. Vasalos, *J Catal.* **1995**, *152*, 52–62.
5. G. D. Yadav, G. S. Pathre, *Microporous Mesoporous Mater.*, **2006**, *89*, 16–24.
6. A. C. Cole, J. L. Jensen, I. Ntai, K. J. Weaver, D. C. Forbes, J. H. Davis, K. L. T. Tran, *Communication*, **2002**, 5962–5963.

7. L. S. Ferreira, J. O. Trierweiler, *IFAC Proc. Vol.*, **2009**, 7, 405–410.

8. H. Grabowska, J. Wrzyszcz, *Res. Chemcial Intermed.*, **2001**, 27, 281–285.

9. X. Liu, M. Liu, X. Guo, J. Zhou, *Catal. Commun.*, **2008**, 9, 1–7.

10. M. D. Acevedo, G. A. Bedogni, *Catal Lett,2014*, 144,1946–1954.

501826 Development of High Performance Heterogeneous Catalysts for Valorization of Biogenic Chemicals

Abstract:

Development of High Performance Heterogeneous Catalysts for Valorization of Biogenic Chemicals

Tomoo Mizugaki,[1] Kohei Uesugi,[1] Kodai Nitta,1 Zen Maeno,[1] Takato Mitsudome,[1] Koichiro Jitsukawa,[1] and Kiyotomi Kaneda*[1,2]

1. Graduate School of Engineering Science, Department of Materials Engineering Science, Osaka University

2. Research Center for Solar Energy Chemistry, Osaka University

kaneda@cheng.es.osaka-u.ac.jp

Biomass has attracted significant attention as a carbon-neutral resource with the potential to reduce the vast quantity of CO_2 emissions. Recently, the biomass-derived compounds such as glycerol, furfural, and levulinic acid are identified as the best suited alternatives for petroleum-derived chemicals. From the stand point of green sustainable chemistry, development of highly selective and efficient catalytic transformation of the above compounds has been desired.

We have developed novel nanostructured heterogeneous catalysts capable of cooperative activation by several active species using inorganic crystallites as unique macro ligands. We report here the precise design of heterogeneous metal catalysts for highly selective transformation of glycerol,[1-5] furfural,[6] and levulinic acid [7,8] into value added chemicals. For example, highly selective hydrogenolysis of glycerol to 1,3-propanediol could be catalyzed by Pt-WO_x supported on aluminum oxide. The combination catalyst system using the lanthanum-exchanged montmorillonite and the aluminum oxide-embedded Cu nanoparticles enabled one-pot transformation of glycerol to 1,3-diacetoxyacetone via the acetylation, oxidation, isomerization, and hydrogenation. Furthermore, the Pt nanoparticles supported on hydrotalcite and Pt-MoO_x on hydroxyapatite selectively converted furfural and levulinic acid into 1,2-pentanediol and 1,4-pentandiol, respectively. The unique activities of the above nanostructured heterogeneous catalysts will be also discussed.

References

[1] T. Mizugaki, R. Arundhathi, T. Mitsudome, K. Jitsukawa, K. Kaneda, *ACS Sustainable Chem. Eng.* **2014**, *2*, 574.

[2] R. Arundhathi, T. Mizugaki, T. Mitsudome, K. Jitsukawa, K. Kaneda, *ChemSusChem* **2013**, *6*, 1345.

[3] T. Mizugaki, R. Arundhathi, T. Mitsudome, K. Jitsukawa, K. Kaneda, *Chem. Lett.* **2013**, *42*, 729.

[4] T. Mizugaki, T. Yamakawa, R. Arundhathi, T. Mitsudome, K. Jitsukawa, K. Kaneda, *Chem. Lett.* **2012**, *41*, 1720.

[5] T. Mitsudome, T. Matsuno, S. Sueoka, T. Mizugaki, K. Jitsukawa, K. Kaneda, *Heterocycles* **2012,** *84*, 371.

[6] T. Mizugaki, T. Yamakawa, Y. Nagatsu, Z. Maeno, T. Mitsudome, K. Jitsukawa, K. Kaneda, *ACS Sustainable Chem. Eng.* **2014**, *2*, 2243.

[7] T. Mizugaki, Y. Nagatsu, K. Togo, Z. Maeno, T. Mitsudome, K. Jitsukawa, K. Kaneda, *Green Chem.* **2015**, *17*, 5136–5139.

[8] T. Mizugaki, K. Togo, Z. Maeno, T. Mitsudome, K. Jitsukawa, K. Kaneda, *ACS Sustainable Chem. Eng.* **2016**, *4*, 682–685.

492126 Metal H-Beta Zeolite Catalytic Upgrading of Bio-Crude Oil Derived from Hydrothermal Liquefaction of Algae

Abstract:

Hydrothermal liquefaction (HTL) conversion of algae is a potential method for producing high-yield bio-crude oils with high heating values for liquid fuel applications. The quality of the bio-crude oils produced from HTL has not yet reached the levels needed for transportation applications because of the presence of undesirable heteroatom-containing aromatic compounds and the related high N/C and O/C ratios. Upgrading is needed to improve HTL bio-crude oil properties. Here, we describe synthesis of bi-functional zeolite catalysts (Pd-Fe/H-Beta and Ru-Fe/H-Beta) to promote bio-crude oil upgrading reactions (i.e., hydrodeoxygenation, hydrodenitrogenation, cracking, alkylation). H-beta zeolites have shown high carbon conversions as support materials for upgrading bio-oils derived from lignocellulosic biomass like wood and grasses. Nobel metals, like Ru and Pd, effectively promote hydrodeoxygenation reactions under milder conditions than other metals. Fe can activate hydrogen to react with oxygen atoms in heteroaromatic compounds leading to the cleavage of C-O bonds. The goal of this research is to investigate the relationship between catalyst synthesis conditions (i.e., calcination temperature and time, metal content ratios) and upgrading effectiveness. The physicochemical properties of the catalysts will be studied by gas adsorption analysis, X-ray photoelectron spectroscopy (XPS), and transmission electron microscopy (TEM). The catalytic upgrading experiments will be performed using a Pyroprobe micropyrolyzer-GC/MS system at high hydrogen (400 psi) pressures under varying temperature and catalyst loading conditions. While small, the Pyroprobe-GC/MS system provides a fast and easy way to screen catalysts and reaction conditions through immediate characterization of the upgraded bio-crude oils. Results from this study will be used to design follow-on bench scale catalyst synthesis and testing studies in a continuous flow reactor system.

495221 Catalytic Depolymerization of Lignin into Value-Added Chemicals over CuO(X)MgAlO$_y$ Catalysts in Supercritical Ethanol

Abstract:

The depletion of fossil fuels leads to pay more attention to finding alternative energy or chemicals resources. Lignin has been regarded as one of the most promising natural feedstock despite of its recalcitrant structure. Therefore, it is essential to develop effective route to convert lignin into value-added chemicals. In this work, CuO(X)MgAlOy catalysts were examined to depolymerize lignin into monomeric compounds in supercritical ethanol without supplying additional hydrogen. The products were identified and quantified by using GC/MS and GC-FID. The monomers mainly consist of oxygen-containing benzenes resulting from the cleavage of Î^{2}-O-4 linkage. The maximum yield of monomeric compounds reaches up to 18.4 wt.% over CuO(30)MgAlOy catalyst.

Analysis over both products and the catalysts by means of various analytical techniques such as FT-IR, GPC, NMR, XRD and NH_3-TPD allows us to understand the characteristics of catalytic depolymerization of lignin under supercritical ethanol condition.

498242 Thermodynamic and Kinetic Analysis of γ-Valerolactone Ring Opening in Multiphase Reactors

Abstract:

Title: Thermodynamic and kinetic analysis of γ-valerolactone ring opening in multiphase reactors

Author: Xinlei Huang, Zijian Wang, and Jesse Q. Bond

Lignocellulosic biomass is an alternative source of industrial carbon. In general, biomass upgrading is predicated upon intermediate formation of functional platform molecules, which can be converted as desired into a range of commodity products. γ-valerolactone (GVL) is an example of one such platform as it provides a spectrum of interesting opportunities for the production of gasoline, jet, and diesel fuels; polymer precursors; and commodity and specialty chemicals. Of interest here is the acid-catalyzed ring opening of GVL to form pentenoic acid (PEA) isomers, which are a family of bifunctional alkene-acids that are of potential interest in the production of nylon.

GVL ring opening is kinetically facile, and it occurs readily in the gas phase over solid acids; however, thermodynamic limitations and secondary reactions make it generally difficult to recover pentenoic acids in good yield. Although GVL ring opening is entropically favorable, it is substantially endothermic, such that PEA yields are equilibrium limited below ≈ 20% at low temperatures. Raising operating temperatures will theoretically address this issue; unfortunately, increasing temperatures to the point where ring opening becomes thermodynamically favorable also makes undesired secondary reactions kinetically accessible. In particular, PEA isomers will undergo decarboxylation over solid acids, irreversibly forming butenes and carbon dioxide. Thus, there exists no gas-phase operating window where PEA isomers can be prepared in good single pass yield.

Instead, we propose that homogeneous mineral acids can be used in biphasic (aqueous-organic) reactors to navigate thermodynamic and kinetic constraints and deliver high PEA yields. Doing so rationally requires a detailed understanding of kinetic and thermodynamic aspects of ring opening in solution. Furthermore, it requires an understanding of how GVL, protons, and PEA isomers partition between water and various organic phases. This poster focuses on recent progress in understanding both reaction kinetics and thermochemistry of ring opening in condensed media and discusses the complex solution thermodynamics that are important in dictating the effectiveness of biphasic reactors.

Key Words: GVL ring opening, solution thermodynamics, liquid-phase kinetics

495541 Integrating Hydride Donor Regeneration with Size-Selective Capsules for Efficient and Sustainable Biohydrogenation

Abstract:

Integrating Hydride Donor Regeneration with Size-Selective Capsules for Efficient and Sustainable Biohydrogenation

Shaohua Zhang†, , Zhongyi Jiang†, , Jiafu Shi*,‡

†Key Laboratory for Green Chemical Technology of Ministry of Education, School of Chemical Engineering and Technology, Tianjin University, Tianjin 300072, China;

‡School of Environmental Science and Engineering, Tianjin University, Tianjin 300072, China;

Collaborative Innovation Center of Chemical Science and Engineering (Tianjin), Tianjin 300072, China.

ABSTRACT:

Enzymatic asymmetric reductions driven by the high selective hydride transfer from enzyme cofactor play a vital role in the manufacture of pharmaceutical intermediates and fine chemicals. The hydride donor (cofactor) can be *in situ* regenerated by introducing a second reaction using cheap reducing agents. However, the unidirectional diffusion of cofactors between the consuming and regenerating sites would dilute its concentration and restrict the performance of biohydrogenation. Here, we show that an enclosed cycle of NAD^+/NADH based on size-selective MOF capsules is used to integrate NADH regeneration and aldehydes/ketones hydrogenation. Specifically, alcohol dehydrogenase (ADH) is positioned in the capsule wall, while glucose dehydrogenase (GDH) and NADH are encapsulated in the lumen. The NADH regenerated by GDH is restricted in the capsule lumen and can be directly consumed by ADH without dilution by the bulk solution. The self-sufficient heterogeneous biocatalyst can continuously catalyze the hydrogenation of formaldehyde to methanol for over 72 hours without the exogenous addition of NADH and can be recycled by centrifugation for 10 times with ~80% of the initial activity retained. The integration of reactions by manipulating the diffusion of intermediates paves the way for the design of catalysts and reactors.

489653 Selective Dehydration of Polyols on Copper Modified BrÃ¸Nsted Acid Support

Abstract:

Selective Dehydration of Polyols on Copper Modified Brønsted Acid Support

Yan Cheng[1,2] and Brent H. Shanks[1,2]

1. Department of Chemical and Biological Engineering, Iowa State University, Ames, IA
2. NSF Engineering Research Center for Biorenewable Chemicals (CBiRC)

Abstract

Polyols, derived from carbohydrates in biomass, are rich in hydroxyl groups, which typically need to be at least partially removed to produce commodity chemicals. In order to maintain the carbon from the starting material, dehydration is an important reaction. However, selective dehydration in the presence of multiple hydroxyl groups has received limited attention. While dehydration can be accomplished at relatively mild reaction conditions, the reaction systems tend to be challenged by rapid catalyst deactivation and low selectivity towards linear unsaturated structures compared with ether ring structures. Interesting catalysts to potentially overcome these issues are copper modified silica-aluminas, which are examined in the dehydration of model compounds. By comparing the conversion of 1,2-propanediol over a Brønsted acid, copper on silica and a copper supported Brønsted acid catalyst, we propose a reaction network for the overall dehydration reaction. The addition of copper on silica-alumina not only affects the reaction kinetics by manipulating catalyst acidity, but also introduces a new pathway for 1,2-propanediol conversion. In competition to the initial reaction step on Brønsted acid sites, this new pathway on copper reduces the interactions between hydroxyl groups, so as to inhibit the formation of ethers.

490084 Kinetic Study of Enzyme-Catalyzed Phenol and Its DerivativesÂ Oxidative Coupling Products

Abstract:

Human derived pharmaceuticals in water have attracted great attention over the past decade. The employment of enzymes for biotransformation of pharmaceutical contaminants has become an increasingly attractive alternative approach. A large class of pharmaceuticals contained the moieties of phenol derivatives, which can be removed by the laccase-catalyzed oxidative coupling, *i.e.*, the phenol derivatives are oxidized into radicals which are further polymerized into biologically inactivity polymers that precipitate from the water. The kinetics of enzyme-catalyzed oxidative coupling process was studied to remove products effectively based on higher enzyme activity. The higher enzyme activity was acquired by adjustment of the concentration of enzyme and the microenvironment (*e.g.* pH) for enzyme. The enhancement of radicals formation rate catalyzed by laccase with higher activity accelerated the generation of products with higher degree polymerization and further improved the removal performance of phenol derivatives. The change of UV absorbance over time was compared using phenol, 2,2-azino-bis-(3-ethylbenzthiazoline-6-sulfonic acid) (ABTS), p-aminophenol or acetaminophen as different substrate. The contents of oxidative coupling products of varied coupling degree were analyzed by LC-MS. The mechanism of coupling reaction process was conducive to the further study of kinetics. The precipitation caused by coupling reaction was analyzed by dynamic light scattering (DLS) measurements. The results showed that the formation time of polymerized products which could naturally precipitated became earlier and precipitation rate was also faster under higher enzyme activity. The micron-grade products were increasing with time increasing. The generation of micron-grade products (precipitation easily) could explain the decline of absorbance over time well. The kinetics is helpful for understanding the reaction process of enzyme-catalyzed coupling of phenol and its derivatives in depth and finding the way to removal of them from water with high performance.

490085 Effect of Dehydrating Agent in Cu/Y-Zeolite Catalyst on Oxidative Carbonylation of Methanol for Dimethyl Carbonate Synthesis

Abstract:

Cu/Y-zeolite catalyst was prepared by impregnation of $CuCl_2 \cdot 2H_2O$ and commercial Y-zeolite in its NH_4 form and characterized by XRF, BET, XRD and NH_3-TPD to check Cu, Cl content, surface area and acidity. The influence of various dehydrating agent was examined with aim of increasing conversion of methanol (X_{MeOH}) in oxidative carbonylation. The conversion of methanol was increased with the addition of metal oxide in the order CaO > MgO > ZnO, but with selectivity (S_{DMC}) in order of MgO > CaO > ZnO, respectively. After oxidative carbonylation of methanol, solid was analyzed by XRD, TG/DTG to confirm that the dehydrating reaction ($MeO + H_2O \rightarrow Me(OH)_2$) was occurred by the metal oxide. The conversion of methanol was increased by 67.8 % and the DMC selectivity was increased by 36.1 % with applying dehydrating agent, CaO. Reaction tests were carried out using autoclave (batch reactor) for 18 hours at 403 K and 5.5 Mpa.

490103 CFD Simulation about Isobutane Alkylation Catalyzed By Ionic Liquid in Cyclone Reactor

Abstract:

CFD simulation about isobutane alkylation catalyzed by ionic liquid in cyclone reactor

Shuiqiang Duan, State Key laboratory of Heavy Oil Processing, China University of Petroleum-Beijing

Mengzijing Chen, State Key laboratory of Heavy Oil Processing, China University of Petroleum-Beijing

Rui Zhang, State Key laboratory of Heavy Oil Processing, China University of Petroleum-Beijing

Xianghai Meng, State Key laboratory of Heavy Oil Processing, China University of Petroleum-Beijing

Haiyan Liu, State Key laboratory of Heavy Oil Processing, China University of Petroleum-Beijing

Zhenbo Wang, State Key laboratory of Heavy Oil Processing, China University of Petroleum-East China

Zhichang Liu*, State Key laboratory of Heavy Oil Processing, China University of Petroleum-Beijing

Key Words: CFD, cyclone reactor, ionic liquid, isobutane alkylation

Cyclone reactor[1] is a new and promising device which is suitable for the fast chemical reaction and separation for liquid-liquid system. Ionic liquid catalyzed isobutane alkylation is one of the typical liquid-liquid reaction and the products need separating quickly to decrease byproducts. In this study, a computational fluid dynamics (CFD) simulation of liquid-liquid two phase flow was carried out about cyclone reactor. The multiphase model and the turbulence model were described by the volume of fluid model and Reynolds stress model, respectively. The vertical velocity distribution, tangential velocity distribution, velocity vector plot and pressure distribution were obtained according to the simulation. The locus of zero vertical velocity (LZVV) and the representative streamline of the ionic liquid and organic were also attained. The location and morphology of the ionic liquid/organic interface were acquired and a pseudo-steady organic phase core was observed after the organic phase had entered into the cyclone reactor about 0.8s.

(a) (b) (c) (d)

Figure 1. (a) the organic phase core at 0.8s(y=0,z/x=5); (b) the locus of zero vertical velocity(y=0,z/x=5); (c)(d) the streamline of the organic phase and ionic liquid(z/x=5)

Acknowledgement

Financial support was provided by the Natural Science Foundation of China (Nos. 21425626, 21436001 and 21276275), the National Science and Technology Support Program of China (2014BAE13B01), and the Science Foundation of China University of Petroleum, Beijing (No. 2462015QZDX03).

References

[1] Zhenbo Wang; Zhichang Liu; Chunming Xu; Aqiang Chen; Rui Zhang. C.N. Patent 201 410 137 641.1

491456 Microreactor Techniques for Analysis of Complex Reactions

Abstract:

Microreactor Techniques for Analysis of Complex Reactions

Saurabh Maduskar, Paul J. Dauenhauer

Department of Chemical Engineering and Materials Science, University of Minnesota, 421 Washington Ave. SE, 484 Amundson Hall, Minneapolis, MN 55455, USA.

Limitations of analytical systems is the major hurdle in understanding complex chemical reactions such as biomass pyrolysis. Cellulose is a major constituent of biomass and the most abundant biopolymer in the world. During high temperature pyrolysis, it transforms into a liquid intermediate before depolymerizing and volatilizing into condensable bio-oil products for upgrading into fuels and chemicals.[1] The underlying pyrolysis chemistry has been studied for decades, and numerous conflicting mechanisms and kinetics models have been proposed. Lack of the experimental verification of such kinetic models is primarily due to the complexity of the analytical problem. Limitations of conventional analytical systems to measure kinetics of cellulose pyrolysis were identified in terms of five experimental criteria, such as sample length scale, temperature ramp during heating and cooling, sweeping gas flowrate, temperature measurement and control, and product quantification.

To overcome these challenges, Pulse-Heated Analysis of Solid Reactions (PHASR), an experimental microreactor system was developed. The PHASR reactor rapidly heats and cools (~10^6 ^{0}C/min) thin film solid samples (10-50 μm thick) at millisecond time scales to control reaction progression, allowing for quantification of complex mixture of vapor, gas, and intermediate products as a function of reaction time.[3] Further, quantification of each carbon-containing analyte in such complex mixtures by existing methods (flame ionization detection) requires extensive identification and calibration. An integrated microreactor system called the Quantitative Carbon Detector (QCD) for use with current gas chromatography techniques for calibration-free quantitation of analyte mixtures was developed.[5] Combined heating, catalytic combustion, methanation and gas co-reactant mixing within a single modular reactor fully converts all analytes to methane (>99.9%) within a thermodynamic operable regime. Residence time distribution of the QCD reveals negligible loss in chromatographic resolution consistent with fine separation of complex mixtures including cellulose pyrolysis products.

PHASR/GC-QCD reactor system was used to measure millisecond-resolved evolution of cellulose and its volatile fragmentation products over a wide range of temperature (400-525 ^{0}C). The results obtained clearly demonstrates the effect of experimental conditions such as sample length scale on product distribution. Energetics of glycosidic bond cleavage was measured by analyzing cyclodextrin, a demonstrated surrogate of cellulose for pyrolysis reactions.[1,4] Combining the rates of glycosidic bond cleavage with that of product formations, apparent kinetics of product formation was evaluated. Using these next-generation analytical techniques, mechanisms of complex chemical

reactions such as pyrolysis of biomass can be elucidated.

(1) Mettler, M. S.; Mushrif, S. H.; Paulsen, A. D.; Javadekar, A. D.; Vlachos, D. G.; Dauenhauer, P. J. *Energy Environ. Sci.* **2012**, *5*, 5414.

(2) Mettler, M. S.; Vlachos, D. G.; Dauenhauer, P. J. *Energy Environ. Sci.* **2012**, *5*, 7797.

(3) Krumm, C.; Pfaendtner, J.; Dauenhauer, P. J. *Chem. Mater.* **2016.**

(4) Zhu, C.; Krumm, C.; Facas, G.; Neurock, M.; Dauenhauer, P.J. *Reaction Chemistry & Engineering* **2017.**

(5) S. Maduskar, A.R. Teixeira, A.D. Paulsen, C. Krumm, T.J. Mountziaris, W. Fan, P.J. Dauenhauer, *Lab on a Chip*, 2015, 15, 440-447.

494464 Membrane Reformers: Optimization of Catalysts and Membranes for Production of Ultra-Pure Hydrogen through Steam Reforming of Methanol

Abstract:

Membrane reformers couple reaction and separation in a single unit which makes them a promising tool to increase efficiency of many chemical processes. Selective removal of a reaction product through membrane in these devices enhances the per-pass conversion for equilibrium limited reactions such as dehydrogenation compared to conventional fixed bed reactors. Therefore, in order to generate and separate high purity hydrogen, on-board membrane reformer is considered a promising technology. Further, as the name suggests, reforming catalysts and membranes are the two intertwined constituents which control the performance of this system. A catalyst with low CO selectivity will enable higher hydrogen partial pressures near membrane. With higher hydrogen partial pressure difference across the membrane, permeation is also enhanced. Further, permeation of hydrogen through the membrane also maximizes the mass transfer rate of hydrogen from the catalyst by maintaining a concentration gradient. This work is therefore focused on the primary optimization of varying catalysts and membrane performance prior to their integration in a single unit.

A light weight and high energy density liquid fuel such as methanol is easy to store and can be reformed to produce hydrogen rich gas at temperatures as low as 250°C. Catalysts for stable methanol reforming activity were synthesized with Cu as an essential element supported on various mixed oxides. Catalyst testing was performed with varying operational parameters such as weight hour space velocity (WHSV, $kg_{cat}/mol_{methanol}.s$)), temperature and steam-to-methanol (S/M) ratio. Further, active metal% and bi-metal compositions were varied to study the product selectivity for the optimal reaction. The obtained data was also fitted using different rate equations to identify the most probable reaction mechanism.

In addition to catalysts, separation studies were performed with membranes synthesized with a series of modified electroless deposition-hydrogen heat treatment steps. This was continued until a non-porous palladium-based film was achieved on asymmetric tubular support. A non-porous Pd-film morphology allows only hydrogen to permeate through it according to solution-diffusion mechanism. With support total surface area of 28.3 cm^2, membrane deposition was carried out for three combinations a) pure Pd, b) 90%Pd-10%Ag and c) 90%Pd-8%Ag-2%Au. Surface characterizations were performed with FESEM-EDX and AFM. Further, the prepared membranes were tested with simulated compositions of reformate to determine membrane perm-selectivity.

The studies illustrated in this work will thereby present a complete insight to individual membrane and catalyst performances. Lastly, with reference to the prior art, this work will also present the feasibility of the optimal catalyst and membrane integration using a dimensionless DaPe number. DaPe is the product of Damkohler and Peclet number which provides the ratio of maximum reaction rate per unit volume over maximum permeation rate per unit volume, a

defining feature of a membrane reactor.

495252 Fe and Zn Promoted Mo/ZSM-5 Catalyst for the Conversion of Ethane into Aromatic Products

Abstract:

Ethane can be extracted from conventional natural gas and shale gas resources, then converted to petrochemicals such as olefins and aromatic compounds by the non-oxidative dehydro-aromatization reaction. Benzene and toluene are the targeted products in this study which can be further processed into larger components for polymers, textiles, and plastics, thus providing a stable market value for benzene and toluene. The goal of this study is to investigate catalytic performance and reaction mechanism over Mo/ZSM5 catalyst promoted with Fe and/or Zn for the conversion of ethane into aromatic compounds. The catalysts used in this study include 4%Mo-ZSM-5, 4%Mo-0.3%Fe, 4%Mo-0.3%Fe-0.3%Zn-ZSM-5, 0.3%Fe –ZSM-5 and 0.3%Zn-ZSM-5 which were prepared by incipient wetness technique and calcined at 550oC in the air. The fresh and spent catalysts were characterized by a number of analytical techniques including XRD, NH3-TPD, pyridine-FTIR, TEM, SEM, and surface area micropore analysis. Elemental surface composition was obtained from energy dispersive spectroscopic (EDS). The reaction was carried out in a fixed bed reactor where gas chromatography monitored the formation of ethylene, benzene and toluene. Thermogravimetric Analysis (TGA) and Temperature Programed Oxidation (TPO) coupled with mass spectroscopy analysis was performed on the spent samples for coke characterization. The effects of Fe and Zn promoters on the aromatics production and coke formation are elucidated. The NH3-TPD profiles [Figure 1] of the fresh catalysts revealed the correlation between promoters and surface acidity, further discussion will be made. The influence of surface acidity, especially the ratio of Lewis/ Brønsted acid sites, on the catalytic activity and deactivation behavior will also be discussed.

Figure 1: NH_3-TPD profiles of fresh catalyst

495794 Reforming of Methanol Aqueous Solution Use of Superheated Liquid-Film Concept

Abstract:

Hydrogen is produced by water electrolysis using renewable energy, and is regarded as an ideal energy carrier. Organic hydride is adequate for hydrogen to store, transport and utilize because of its advantages on safety and easiness-to-treat as well as of its effectiveness on energy and cost. Recently, methanol is also considered as a storable and transportable hydrogen carrier by reforming reaction of methanol aqueous solution. Moreover, it is reported that methanol is produced using carbon dioxide. Therefore, hydrogen energy utilization system using reforming reaction of methanol aqueous solution is efficient system from the view point of carbon capture and utilization. In our previous study, nanosized metal particles supported on carbon, immersed with liquid substrate, were heated externally under boiling conditions. Superheated liquid could contact with the catalyst active sites immediately after the evolution of bubbles, into which the reaction products had been enforced to desorb under boiling conditions with sharp temperature gradient. Gaseous hydrogen was disturbed by liquid film to meet the catalyst again. In this study, reforming reaction of methanol aqueous solution using superheated liquid-film concept was carried out, and the effects of the molar ratio of water to methanol (S/C ratio), and species of metal catalyst on reforming reaction of methanol aqueous solution were investigated. The effects of S/C ratio on reforming of methanol aqueous solution was investigated using Ni-Ru/C catalyst, and the optimal conditions of S/C ratio was determined 1.0. The generation rate of gas containing hydrogen, carbon oxide, and carbon dioxide increased with increasing reaction temperature. The generation rate of gas was also influenced by the species of metal catalyst, and Ni-Ru/C catalyst was the best metal from the view point of generation of gas. On the other hand, there is no generation of carbon oxide as a byproduction using Cu-Mn/C catalyst. However, the reaction rate was low. It is considered that hydrogen energy utilization system using reforming reaction of methanol aqueous solution is efficient system.

496129 Methanol to Propylene Conversion: Recent Development Trends

Abstract:

The conversion of methanol to propylene (MTP) is a value-added process and has gained extreme significance because of high demand of propylene in the production of petrochemicals. The demand for propylene is increasing due to increasing usage of polypropylene. The catalytic materials used in MTP conversion include SAPO-34 (small-pore molecular sieves), ZSM-5 (medium-pore zeolites) and its modified forms. Limited research has also been found using large pore zeolites such as mordenite and EU-1. MTP research has been carried using structured catalysts including ceramic based honeycomb or monolith and silicon carbide (SiC) foam. The major difference in process design between SAPO-34 and H-ZSM-5 is that the SAPO-34 is used in fluidized bed process while H-ZSM-5 catalyst is used in fixed bed process. SAPO-34 is a selective catalyst for olefins but deactivates fast and thus requires frequent regeneration. The H-ZSM-5 is less selective for olefins but shows less deactivation and thus quite stable. A number of structured supports such as monolith, foam and mesh are being research for coating the active zeolite based catalysts. The structured catalysts are used to reduce the diffusional limitations of pellet catalyst system and have exhibited excellent results in terms of activity and selectivity.

496378 Resin Based Solid Acid Catalyst for Dehydration of Fructose to HMF

Abstract:

Sustainable chemical production can be achieved through using biomass. A key biomass conversion strategy produces 5-hydroxymethylfurfural (HMF), an important platform chemical that has applications for manufacturing solvents, polymer precursors, and fuel additives. HMF can be produced by dehydrating fructose under acidic conditions, thus allowing us to exploit the renewable carbon source i.e. lignocellulosic biomass. The studies to understand the fructose dehydration suggest that the HMF production is more favored in organic solvents. However, separating HMF from bulk organic solvents is an energy intensive process given the high boiling point of the bulk solvent. To make this process more sustainable work is being done to design catalysts that can selectively dehydrate fructose to HMF while using water as the bulk solvent. In this work, we present a bifunctional resin based catalyst with polystyrene backbone for selective dehydration of fructose to HMF in water. The effect of individual functional groups and loading of functional groups on the conversion and selectivity are studied. Overall, in this work we present a bifunctional catalyst design to modify the catalytic site to improve the conversion and selectivity of fructose dehydration to HMF.

496859 Optimizing ZSM-11 Catalysts for Methanol-to-Hydrocarbon Reactions

Abstract:

ZSM-11 (MEL) is an emerging catalyst for a wide variety of applications. The MEL framework is very similar to ZSM-5 (MFI), which is one of the most widely used zeolite catalysts in industry. Recent studies have shown that ZSM-11 has better catalytic performance in several different reactions compared to ZSM-5 [1-3], possibly due to improved intra-crystalline diffusion within the less-tortuous straight channels of the MEL topology; however, there are relatively few systematic studies of ZSM-11 synthesis, specifically with respect to tailored physicochemical properties such as size, shape, and Si/Al ratio (SAR). Here, we will present a systematic study of ZSM-11 catalyst preparation and testing for C_1 upgrading reactions. Through the judicious selection of synthesis parameters, we successfully developed a facile protocol to generate a library of ZSM-11 catalysts with tunable size, shape, and SAR. Optimization of the synthesis procedure focused on reducing the use of organic structure-directing agent (OSDA), thereby reducing the cost of synthesis. These methods for tailoring zeolite crystallization will be discussed along with catalytic studies of methanol-to-hydrocarbon (MTH) reactions. Collectively, our findings are part of a broader initiative to establish quantitative structure-performance relationships as a platform for designing zeolite catalysts.

[1] Kustova, M. Y.; Rasmussen, S. B.; Kustov, A. L.; Christensen, C. H., *Applied Catalysis B-Environmental* **2006**, *67* (1-2), 60-67.

[2] Zhang, L.; Liu, H.; Li, X.; Xie, S.; Wang, Y.; Xin, W.; Liu, S.; Xu, L., *Fuel Processing Technology* **2010**, *91* (5), 449-455.

[3] Bleken, F.; Skistad, W.; Barbera, K.; Kustova, M.; Bordiga, S.; Beato, P.; Lillerud, K. P.; Svelle, S.; Olsbye, U., *Physical Chemistry Chemical Physics* **2011**, *13* (7), 2539-2549.

498441 Selective Glucose Isomerization to Fructose Using Heterogeneous Amine Catalysts

Abstract:

Upgrading biomass is an important step toward a more sustainable chemical industry. Selective glucose isomerization to fructose is a bottleneck in upgrading cellulosic biomass to high valued products like high fructose corn syrup and useful intermediates like 5-hydroxymethylfurfural. The Lewis acidic heterogeneous catalyst, Sn-beta has been extensively researched for this reaction in the past few years.[1] With recent reports demonstrating the ability of homogeneous amines to selectively convert glucose to fructose, there is a growing interest in developing means to translate these properties of homogeneous amines to heterogeneous materials.[2] Herein, we report heterogeneous amine catalysts developed using organosilanes immobilized onto porous supports for selective conversion of glucose to fructose. We investigated catalyst design parameters including the organosilane structure and loading, finding that tuning both these parameters results in the best catalytic performance.

References:

(1) Moliner, M.; RomÃ¡n-Leshkov, Y.; Davis, M. E. *Proc. Natl. Acad. Sci. U. S. A.* **2010**, *107* (14), 6164â€“6168.

(2) Liu, C.; Carraher, J. M.; Swedberg, J. L.; Herndon, C. R.; Fleitman, C. N.; Tessonnier, J. P. *ACS Catal.* **2014**, *4* (12), 4295â€“4298.

498478 Group IV and V Periodic Trends in Olefin Epoxidation: Effects of Electronic Structure and Local Environment

Abstract:

Hydrogen peroxide (H_2O_2) is a green oxidant that can epoxidize a variety of olefins over site-isolated group IV and V metal oxides.[1-3] Group IV and V metals incorporated into the *BEA framework (M-BEA) have been shown to activate H_2O_2 to form active hydroperoxide (M-OOH; group IV) or superoxide ($M-(O)_2^-$; group V) (collectively referred to as $M-(O_2)^*$) species. The observed reactivity and selectivity towards epoxidation is proportional to the functional Lewis acid strength of the catalyst (determined by the adsorption enthalpies of Lewis-acid bound deuterated acetonitrile) or the ligand-to-metal charge transfer (LMCT) energy of the reactive intermediate.[3] The electronic properties of these activated metal complexes are extended past our previous findings to include an analogous series of catalysts grafted onto mesoporous silica ($M-SiO_2$) via calixarene complexes. The differences in the reactivity towards olefin epoxidation over M-BEA and $M-SiO_2$ catalysts are due to two main effects. First, the micropores of the *BEA framework solvates, and thus stabilizes, the epoxidation transition state, which decreases the apparent activation enthalpies relative to SiO_2. Second, the coordination environment of the M-BEA or $M-SiO_2$ have slightly different electronic structures of the active intermediate (as detected via UV-vis spectroscopy), evidenced by the systematic decrease in the LMCT band energies for $M-SiO_2$ catalysts.

Turnover rates for styrene (C_8H_8O) oxide formation through primary reaction pathways as a function of styrene (C_8H_8) and H_2O_2 concentrations over M-BEA and $M-SiO_2$ (M = Ti, Nb, or Ta) reveal two distinct regimes, which correspond to a change in the most abundant reactive intermediate (MARI). The observed dependencies are consistent with a mechanism that describes the irreversible activation of H_2O_2 to form a pool of $M-(O_2)^*$ intermediates, which then react with C_8H_8 or H_2O_2 via an Eley-Rideal mechanism to form C_8H_8O or H_2O_2 decomposition products, respectively. Time-resolved *in situ* UV-vis measurements acquired during reaction of cyclohexene with H_2O_2-activated M-BEA and $M-SiO_2$, in conjunction with probe reactions with *cis*-stilbene, support the trend that the reactive intermediate on Ti-based catalysts is the hydroperoxide (i.e., M-OOH), while Nb- and Ta-materials react through a superoxide (i.e., $M-(O_2)^-$) intermediate.

Understanding of the fundamental properties of the catalysts and reactive intermediates that are responsible for the differences in rates for epoxidation require comparisons of these catalysts made under similar surface coverages (i.e., MARI). Activation enthalpies for C_8H_8O formation ($\Delta H^{\ddagger}_E$) are measured on similar $M-(O_2)^*$ saturated surfaces on all catalysts (shown by rates that are proportional to $[C_8H_8]$ and independent of $[H_2O_2]$), and correlate linearly with both the measured adsorption enthalpies of Lewis-acid bound pyridine (ΔH_{Py}, determined via van't Hoff analysis of FTIR spectra of pyridine coordinated to the Ti, Nb, and Ta atoms within the BEA framework or grafted onto SiO_2) and also to the LMCT band energy of the reactive intermediates on these materials. These comparisons reveal two critical differences among these materials. First, $\Delta H^{\ddagger}_E$ and ΔH_{Py} values for $M-SiO_2$ materials are systematically greater by 19 ± 2 and 11 ± 2 kJ mol^{-1} than M-BEA materials, respectively, which is consistent with the difference in bulk-averaged adsorption enthalpies for C_8H_8 and pyridine in the Si-BEA framework and in mesoporous SiO_2 (i.e., measured in the absence of metal atoms). This difference is attributed to the microporous environment of *BEA solvating the transition state (relative to the initial state) to a larger extent than SiO_2, which results in the systematic decrease of apparent activation and adsorption enthalpies. Second, the linear relationship between $\Delta H^{\ddagger}_E$ and the LMCT band energy of the active intermediate possesses the same systematic offset in $\Delta H^{\ddagger}_E$ between *BEA and SiO_2 that matches the differences in adsorption enthalpies of C_8H_8 (see above). Interestingly, the $M-SiO_2$ catalysts possess systematically lower energy LMCT bands, which may be attributed to one of two phenomena, or both. First, the cleavage of a M-OSi bond in the zeolite framework (upon H_2O_2 activation) leaves a vicinal SiOH group in a tetrahedral position that may contribute a small amount of electron density towards the metal atom, which, in turn, decreases the Lewis acid strength of the metal atom. Second, the adjacent SiOH group in *BEA (after H_2O_2 activation) likely distorts the geometry of the $M-(O_2)^*$ intermediates, which will lower the extent of orbital overlap, lengthening the M-O bonds, and thus increase the LMCT band energy.

The overarching trend relating $\Delta H^{\ddagger}_E$ and the LMCT band energy suggests that metal centers that are highly electron withdrawing (i.e., possess more highly polarized $M-(O_2)^*$ bonds) result in electrophilic intermediates. To develop an intuition for the electrophilicity of the reactive intermediates, turnover rates for the epoxidation of *p*-substituted styrene ($X-C_8H_8$, where X = OMe, Me, H, Br, and NO_2) were measured under identical reaction conditions (3 mM $X-C_8H_8$, 0.01 M H_2O_2, 313 K). The corresponding Hammett plot reveals a reaction constant (ρ) of -0.9 ± 0.05 for all M-β and $M-SiO_2$, which strongly suggests that the reactive intermediates implicated in olefin epoxidation have a strong preference to react with electron-rich olefins. Collectively, these data show that the rational design of increasingly active and selective epoxidation catalysts should seek to increase the Lewis acid strength of the metal center, thus increasing the electrophilicity and reactivity of the active intermediates, and leverage the decrease in apparent activation barriers that are attributed to the stabilization of the transition states in microporous environments.

References:

1) Thornburg, N.E.; Thompson, A.B.; Notestein, J.M.; *ACS Catal.*, **2015**, *5*, 5077-5088.

2) Bregante, D.T.; Priyadarshini, P.; Flaherty, D.W.; *J. Catal.*, **2017**, *348*, 75-89.

3) Bregante, D.T.; Flaherty, D.W.; **2017**, *In Revision*

498975 Mechanistic and Spectroscopic Evidence for Reactive Intermediate Structures during C-O Bond Rupture in Oxygenates over Metal Phosphide Clusters

Abstract:

Selective hydrogenolysis of C-O bonds in small oxygenates is an important step for the conversion of biomass-pyrolysis oils into fuels and valuable platform chemicals, such as α,ω-diols from furanic species.[1] Metal phosphides (MPs) are promising catalysts for C-O bond rupture as they are inexpensive, chemically and thermally stable, and selectively cleave C-X bonds (where X=S, N, or O) over C-C bonds.[2,3]

Here, we combine kinetic analysis of reaction networks, modulation excitation *in situ* infrared spectroscopy (MES), and density functional theory (DFT) calculations to describe the mechanism and reactive intermediates for C-O bond rupture over MPs. Rate measurements of C-O bond rupture within 2-methyltetrahydrofuran (2-MTHF; 1-50 kPa 2-MTHF; 0.1-6 MPa H_2; 543 K) and DFT calculations show adsorption and dehydrogenation steps of 2-MTHF are quasi-equilibrated and precede kinetically relevant C-O bond rupture.[4] The reactive species have lost 1-2 H-atoms from the fully hydrogenated reactant, and ratios of sterically-hindered C-O (^{3}C-O) bond rupture rates to those for unhindered C-O (^{2}C-O) bonds increase with $[H_2]^{1/2}$, which suggests H-content of the reactive intermediates for ^{3}C-O and ^{2}C-O rupture differ by the removal of one H-atom. Sites that bind the reactive intermediates for both ^{3}C-O and ^{2}C-O bond rupture resemble those that bind CO*, NH_3*, and H* on all Ni_xP (where x = 2-2.4) catalysts, which combined with predicted adsorption enthalpies suggest the reactive intermediates are coordinated to Ni ensemble active sites within Ni-rich terminating planes of Ni_xP surfaces. Selectivities for ^{3}C-O bonds are higher than for ^{2}C-O bonds over Ni_xP and much greater than the ^{3}C-O bond selectivities on Ni. Measured and DFT-predicted apparent activation enthalpies ($\Delta H^{\ddagger}$, 473-583 K) for C-O bond rupture indicate that the phosphorus content decreases the $\Delta H^{\ddagger}$ for ^{3}C-O bond rupture relative to that of ^{2}C-O bond rupture (i.e., $\Delta H^{\ddagger}_{2CO}$ - $\Delta H^{\ddagger}_{3CO}$ increases with phosphorus content). Selectivity differences between specific C-O bonds within 2-MTHF reflect differences in the H-content of reactive intermediates, activation enthalpy barriers, and phosphorus content of Ni and Ni_xP clusters.

Direct experimental evidence for the structure of reactive intermediates would distinguish clear connections between the relationship between C-O bond rupture selectivity, surface coordination, and enthalpic barriers over Ni_xP clusters. MES experiments coupled with phase sensitive detection reveal spectral features of reactive intermediates by suppressing those of inactive surface species.[5] Multivariate curve resolution-alternating least squares algorithm[6] extract pure-component spectra and surface coverages that indicate the presence of distinct reactive surface intermediates during C-O bond rupture of 2-MTHF on Ni_xP, whose

composition and surface coordination are consistent with those implied by kinetic measurements and predicted by DFT calculations. Modulations in H_2 pressure < 2 ks^{-1} (i.e., the measured turnover rate for C-O bond rupture in 2-MTHF) allow all intermediates to form and oscillate at the applied frequency; however, experiments performed with modulation frequencies > 2 ks^{-1} (e.g., 2.9 ks^{-1} – 0.13 s^{-1}) suppress the creation of reactive intermediates whose barriers for formation correspond to characteristic timescales of formation that are greater than the period of the modulation. Such experiments show that the compositions of the most abundant reactive intermediate (MARI) on Ni and Ni_xP clusters during C-O bond rupture of 2-MTHF are identical; however, the MARI changes from $Ni_3(\mu^3\text{-}C_5H_{10}O)$ to $Ni_3(\eta^5\text{-}C_5H_{10}O)$ with increasing phosphorus content. The shift in binding configuration with phosphorus content is likely the fundamental cause for increased selectivity towards ^{3}C-O bond rupture. The combination of these kinetic, spectroscopic, and theoretical techniques provides insight into underlying causes of selectivity for reactions of small oxygenates across MP clusters.

References

(1) Chia, M.; Pagan-Torres, Y. J.; Hibbitts, D.; Tan, Q. H.; Pham, H. N.; Datye, A. K.; Neurock, M.; Davis, R. J.; Dumesic, J. A. *J. Am. Chem. Soc.* **2011**, 133, 12675.

(2) Oyama, S. T.; Wang, X.; Lee, Y. K.; Chun, W. J. *J. Catal.* **2004**, 221, 263.

(3) Rensel, D. J.; Rouvimov, S.; Gin, M. E.; Hicks, J. C. *J. Catal.* **2013**, 305, 256.

(4) Witzke, M. E., Almithn, A.; Coonrod, C. L., Hibbitts, D. D.; Flaherty, D. W. *J. Catal.* to be submitted.

(5) Urakawa, A.; Burgi, T.; Baiker, A. *Chem. Eng. Sci.* **2008**, 63, 4902.

(6) Jaumot, J.; de Juan, A.; Tauler, R. *Chemometr. Intell. Lab.* **2015**, 140, 1.

493330 Experimental and Simulation Studies of High-Temperature Ethane Dehydrogenation in Microporous Zeolite Membrane Reactor

Abstract:

Experimental and Simulation Studies of High-Temperature Ethane Dehydrogenation in Microporous Zeolite Membrane Reactor

*Shailesh Dangwal, Ruochen Liu, Seok-Jhin Kim**

School of Chemical Engineering, Oklahoma State University,

420 Engineering North, Stillwater, OK 74078, USA

ABSTRACT

Ethylene is an important chemical engineering ingredient and used in polymerization, oxidation, and alkylation. The increasing demand for ethylene has stimulated substantial research into the development of new processes to reduce energy consumption. Catalytic ethane dehydrogenation using a membrane reactor is an attractive solution because the cracking equilibrium can be shifted in favor of ethylene by selective removal of hydrogen. We reported the intensification of ethane dehydrogenation reaction in MFI zeolite membrane reactors operating with a Pt/Al_2O_3 catalyst. The effects of the reaction conditions and the membrane separation properties on the ethane conversion in the membrane reactor were investigated experimentally and by simulations using a simple one-dimensional model which was validated by the experimental data. It was demonstrated that the rate of H_2 membrane permeation significantly enhances the ethane conversion. MFI membrane reactors allow the equilibrium limit of ethane conversion to be surpassed at high temperatures. The model calculations have shown that near-completion ethane conversion (>99.5%) may be achieved under realistic operating temperature, pressure, space velocity and catalyst load even for membranes with moderate H_2 selectivity, ($aH_2/C_2H_4<10$) and H_2 permeance ($<1.8\times10^{-7}$ mol.s.m^2.Pa).

*Corresponding author: E-mail: seokjhin.kim@okstate.edu

501253 Experimental and Numerical Study of an Intensified Water-Gas Shift (WGS) Reaction Process Using a Membrane Reactor (MR)/Adsorptive Reactor (AR) Sequence

Abstract:

Reactive separation processes, such as membrane reactors (MRs) and adsorptive reactors (ARs), have been attracting increased interest for industrial applications. Such a hybrid system combining a membrane reactor (MR) and an adsorptive reactor (AR) in tandem (with the AR following the MR, and the MRâ€™s reject stream serving as the ARâ €™s feed) can produce an ultra-pure H_2 product (without the need for using a post-processing step), and can deliver WGS reaction conversion exceeding equlibrium. The proposal configuration (AR following MR) provides significant flexibility in addition to high-purity hydrogen and carbon dioxide production.

The primary objective of this work is to first develop and simulate an advanced, and detailed mathematical model for the MR/AR sequence and then validate this model using experimental data obtained at the bench-scale. The MR system is composed of a reaction zone packed with catalyst and quartz pellets, and a water/hydrogen permeation zone. In the AR, both catalyst and adsorbent materials are used for simultaneous reaction and separation. The AR is a dynamically operated process. Both steady-state and dynamic experiment are carried out in a laboratory setting to capture the individual MR/AR performances. Experiments are carried out for a broad range of temperatures (up to 300 ^{o}C) pressures (up to 25 bar) relevant to realistic application conditions. The velocity, temperature and species concentration profiles in the MR reaction and permeation zones are captured by momentum/species transport models accounting for convection/reaction /diffusion/conduction mechanisms. The rigorous Maxwell-Stefan and dusty gas models are applied to describe mass diffusion fluxes. Finally, the performances of the combined system (the membrane reactor (MR) followed by an adsorptive reactor (AR) is simulated for a broad range of the operating conditions and design parameters. The simulation results are compared with the experimental findings.

[HC1]and pressures
[HC2]experimental

493066 Synthesis & Performance Study of Pd-Au/PSS Membrane Reactor to Produce Highly Pure Hydrogen Via Natural Gas Steam Reforming

Abstract:

One possible use of hydrogen, without direct greenhouse gas emissions, is as feed for a fuel cell (FC), with the most readily available technology being a proton exchange membrane FC (PEMFC). In order to avoid poisoning of PEMFCâ€™s Pt-based catalyst due to the presence of ppm levels of CO, the hydrogen feed needs to be ultra-pure. The industrial process for hydrogen production, which is a multi-step energy-intensive process followed by further separation/purification, can be a potential source [1]. However, as an alternative method, a Pd-based membrane reactor (MR) may be used due to its ability to provide high-purity hydrogen directly. Moreover, the MR works at milder operating conditions compared to a traditional system.

In recent years, Pd-based composite membranes, i.e., thin metallic layers supported on porous materials such as ceramics or stainless steel, have been considered due to their lower cost (thin Pd layer) and higher mechanical resistant (porous support) compared to dense Pd-based membranes [2,3]. Moreover, the use of Pd-alloys, in particular, Ag, Au, Pt, has grown in the recent years given their ability to resist sulfur poisoning in addition to their enhanced hydrogen permeability [4]. Therefore, the aim of this study was to fabricate a Pd-Au membranes supported on porous stainless steel (PSS) with the intent to produce high-purity hydrogen from natural gas steam reforming. The membrane used in this study was fabricated via an electroless plating technique for depositing Pd and electro-plating for the deposition of Au.

The initial characterization of the membrane by way of ideal selectivity took place at 400 °C with H_2, He and N_2, and with DP in the range of 1.5 - 3.0 bar. After ideal selectivity characterization of the Pd-Au/PSS membrane, the natural gas steam reforming reaction was carried out and the MR performance in terms of CH_4 conversion, hydrogen production and hydrogen permeate purity has been analyzed and will be discussed.

References

[1] Rostrup-Nielsen, J.R., Catalytic steam reforming. 1984: Springer.

[2] S. Liguori, A. Iulianelli, F. Dalena, P. Pinacci, F. Drago , M. Broglia, Y. Huang, A Basile, "Performance and long-term stability of Pd/PSS and Pd/Al_2O_3 membranes for hydrogen separation", *Membranes*, 4 (2014) 143-162.

[3] B. Anzelmo, J. Wilcox, S. Liguori, "Natural gas steam reforming reaction at low temperature and pressure conditions for hydrogen production via Pd/PSS membrane reactor", *Journal of Membrane Science*, 522 (2017) 343-350.

[4] Chen, C.-H. and Y.H. Ma, "The effect of H_2S on the performance of Pd and Pd/Au composite membrane". *Journal of Membrane Science*, 362 (2010) 535-544.

497163 Non-Isothermal CFD Study of Ethanol Steam Reforming in a Catalytic Membrane Reactor

Abstract:

Non-isothermal CFD Study of Ethanol Steam Reforming in a Catalytic Membrane Reactor

Rui Ma[1], Bernardo Castro-Dominguez[2], Anthony G. Dixon[1], Yi Hua Ma[1]

[1.]Center for Inorganic Membrane Studies, Department of Chemical Engineering, Worcester Polytechnic Institute, 100 Institute Road, Worcester, MA 01609, USA.

[2.] University of Limerick, Sreelane, Castletroy, Co. Limerick, Ireland

Single unit hydrogen production and separation achieved by conducting ethanol steam reforming (ESR) in a catalytic membrane reactor (CMR) has attracted more interest due to the carbon neutral property of bio-ethanol, and the high reactant conversion due to process intensification.[1] In ESR process, temperature control helps to achieve higher conversions. Compared with traditional packed bed reactors (PBR), heat supply for CMR needs to compensate for the enthalpy that hydrogen carried across the membrane during the permeation process, in addition to the heat loss from endothermic reactions. Furthermore, hydrogen flux through the membrane is temperature dependent as the hydrogen permeability follows Arrhenius law. In this work, a comprehensive heat and mass transfer study of an ESR process in a membrane reactor is carried out using an experimentally-validated non-isothermal CFD model. Retentate and permeate domains were modeled by introducing a negative and positive hydrogen flux term respectively, using Sieverts' law. In order to increase the hydrogen permeation driving force, a purge gas was applied to the permeate side. Effective thermal conductivity and diffusivity were introduced to the simulation in order to take into account the catalyst bed.

A "cold spot" adjacent to the membrane was observed (Fig 1) since this highly endothermic process benefits from hydrogen removal adjacent to the membrane. From the reaction rate study of the ESR process, the process takes place very rapidly, and a slight reverse reaction for methane steam reforming was observed due to the decrease in equilibrium conversion at lower temperature. This phenomenon is most noticeable within the "cold spot". A clear hydrogen generation and depletion along the reactor on the retentate side was observed. In addition, the reactions reach equilibrium shortly after reactants enter the reactor which leads to an inefficient use of the reactor and the membrane surface area; thus, counter-current flow of the purge gas is studied, which helps with shifting the reaction equilibrium throughout the entire reactor. Using this model, the effect of temperature and pressure on hydrogen recovery and yield are studied in both co-current and counter-current cases, and a process optimization study was carried out.

Fig 1. Temperature distribution in the reactor at LHSV=3.77h-1, Pret =6 bar, S/E=5 and Tin=723K on both permeate and retentate side.

1) Ma, R., Castro-Dominguez, B., Mardilovich, I.P., Dixon, A.G. and Ma, Y.H., "Experimental and Simulation Studies of the Production of Renewable Hydrogen through Ethanol Steam Reforming in a Pilot-Scale Catalytic Membrane Reactor," Chem. Eng. J., 303, 302-313 (2016).

511513 Heat Integrated Steam Reforming Membrane Reactor

Abstract:

A heat integrated steam reforming membrane reactor was investigated for development of a gasoline fuel processor for fuel cell vehicles. This reactor utilizes Le Chatelierâ€™s Principle for driving thermodynamically equilibrium limited steam reforming reactions at higher pressures. Steam reforming at higher pressures also reduces the palladium alloy membrane area required. Coupling steam reforming with a hydrogen selective membrane provides essentially pure hydrogen for optimal fuel cell performance.

Engineering models were used extensively to help define and refine the fuel processor design. The catalytic heat exchange reactor was coupled to a separate palladium-alloy membrane and used recycle of hydrogen depleted reformate to achieve high efficiencies at moderate operating temperatures and pressures (650Â°C, 20 Bar). Recycle was achieved by using a steam ejector. Catalytic combustion of the hydrogen depleted reformate with lean air fuel cell cathode exhaust (11% O2) supplied the heat required for steam reforming. Our modeling suggested that our steam reforming approach could have thermodynamic efficiencies of up to 40%.

Screening studies were undertaken to identify high activity noble metal steam reforming catalysts. A bimetallic Ru-Ir catalyst was found to have exceptionally high activity and stability. Order-of-magnitude, improvements in catalyst performance and reactor geometry were made.

In laboratory studies excellent heat transfer was observed when conducting catalytic combustion and steam reforming co-currently on opposing walls of the wash coated reactor tubes. Several membrane geometries were used. Hydrogen permeate was obtained from a variety of feeds, including isooctane, naphtha and methane, at reaction temperatures of 600-700°C. The experimental results were in close agreement with the modeling predictions.

While considerable research and development would be required to advance this technology, these results demonstrate the potential of the heat integrated steam reforming membrane reactor system.

486015 Improving Dehydrogenation Conversion and Selectivity By Utilizing Thermally Stable Membranes

Abstract:

In an equilibrium-limited alkane dehydrogenation reaction, due to its endothermic nature, high temperature is demanded to reach high conversion. Removal of product H_2in a dehydrogenation reaction shifts the equilibrium forward to realize a higher conversion at mild conditions. Thermodynamic computation demonstrates that with the assistance of a membrane separation unit, equilibrium limit can be exceeded in alkane dehydrogenation by selective removal of hydrogen. By operating at a lower temperature, high catalyst performance can be achieved; while catalyst coking is less likely to occur. In this research, dehydrogenation of methylcyclohexane to toluene is studied as a model reaction.

To study the impact of membrane utilization on the dehydrogenation process, blend polymer materials with high thermal stability, chemical resistance and transport properties has been selected as membrane materials. The permeabilities of dense films of Polybenzimidazole (PBI) and Matrimid® blends were evaluated from room temperature to 300 C. Permeabilities increased exponentially with temperature and the selectivities decreased. All blend materials were stable over the multi-day testing.

The potential impact of selective hydrogen removal via thermally-stable polymer blends was calculated from finite element modelling. The model incorporated the measured permeation properties, measured reaction equilibrium information, and catalyst performance (kinetics and coking). As the process can be operated at a range of temperatures, the influence of system temperature on conversion, permeation rate, and catalyst deactivation is analysed. The use of a single membrane unit inter-stage between two packed beds has the potential to increase reactant conversion by 35% compared to operation without inter-stage conversion.

494458 Integration of Synthesized Catalysts and Membrane in Packed and Fluidized Bed Membrane Reformer for Production of Hydrogen through Steam Reforming

Abstract:

Hydrogen is the only energy carrier which can cater to the rising power demands across the world without compromising with the environmental impact. Its high efficiency to generate power comes in conjunction with fuel cells that can produce a desired output with absolutely no carbon emissions. However, the hydrogen fed to fuel cells must be of high purity to ensure stable performance of Pt-based catalysts of fuel cell anode. Therefore, production of hydrogen with high purity is considered crucial. In the current work, hydrogen generation is targeted with methanol steam reforming using synthesized Cu-based catalysts. To achieve high hydrogen purity, Pd-based dense supported membranes were employed after their synthesis, characterization and testing. The target of this work is to suitably integrate the catalysts with the membranes in one single unit. Such integration will cause continuous permeation of hydrogen through a H_2 selective Pd membrane. Disturbed reaction equilibrium will therefore cause an increase in the forward reaction rate according to Le Chatelierâ€™s rule and hence improve methanol conversion. This system of simultaneous generation and separation of hydrogen in one single unit is termed as a membrane reformer (MR). Catalyst placing relative to the membrane determines the reaction residence time in a membrane reformer and thereby plays a key role. Further, this factor is inter-related with reaction kinetics, hydrogen mass transfer rate from reaction zone to membrane, and lastly membrane permeation flux and perm-selectivity. In order to observe the effect of these parameters on total system performance, the membrane reformer study was paralleled between a packed bed and a fluidized bed MR. Packed bed consists of a catalyst bed (~bed ht. 2 cm) in the shell of the reactor with membrane in its lumen separated by a distance of approximately 15 cm. In contrast to that, fluidized bed gives the advantage of catalyst being fluidized in the same shell that brings it in close contact with the membrane.

The current study would determine performance of packed bed MR with varying catalyst positions and parameters such as temperature, flow rate and weight hour space velocity (WHSV). Based on these operating conditions, fluidization regime was characterized by u/u_{mf} in the range of 1.4â€“3.6 (where u_{mf} corresponds to the minimum fluidization velocity). All testing was performed in an in-house designed test rig. Dense Pd-based membranes were prepared using modified electroless deposition approach on porous SS (PSS) with yttria stabilized zirconia (YSZ) as the intermediate layer. YSZ minimizes the inter-metallic diffusion between Pd and SS. Further, catalysts with varying bi-metallic compositions were synthesized by incipient wetness impregnation. Optimization of standalone membrane as well as catalyst performance was performed prior to the integration.

Once integrated, the packed bed membrane reformer (MR) showed >95% methanol conversion and 45% H_2 recovery at 723 K and 3 bar. Further, a comparison of packed bed MR was made with a fluidized bed MR. It was observed that with increasing u/u_{mf} greater contact between the catalyst and the membrane enhances the hydrogen yield and thereby leads to better separation in comparison to packed bed. Moreover, a complete description of these results will be

elaborated in terms of effectiveness factor, reaction kinetics and hydrogen permeation flux through the membrane.

498421 Highly Selective FT Synthesis for Production of JP-8 Jet Fuel from Coal and Coal/Biomass Mixtures

Abstract:

An indirect liquefaction process employing advanced process intensification approaches is being developed to reduce costs and accelerate commercialization of a coal to liquid (CTL) process for converting coal and coal-biomass mixtures to JP-8 jet fuel. In this Air Force and DOE/NETL sponsored cooperative agreement, Southern Research is leading a team consisting of IntraMicron, Inc., Southwest Research Institute, and Nexant, Inc., along with Chevron as selective Fischer-Tropsch (FT) catalyst supplier and the National Carbon Capture Center (NCCC) as the host site. The process reduces the number of parasitic syngas cleaning steps and eliminates costly upgrading of wax produced in conventional indirect liquefaction processes, resulting in improved economics and lower lifecycle greenhouse gas (GHG) emissions for CTL-based jet fuel production. These improvements directly reduce capital costs of indirect CTL processes, resulting in significantly improved economics and cost competitiveness of CTL when compared to petroleum refining. Specific objectives for FT experiments are to demonstrate a selective and active second generation cobalt-zeolite hybrid Fischer-Tropsch (FT) catalyst from Chevron that predominantly produces liquid range C5-C20 hydrocarbons with high productivity (5X conventional FT catalysts) and jet fuel selectivity while eliminating the production and associated handling of solid C21+ wax. In fall of 2015, Southern operated a bench-scale FT reactor with a 2" diameter heat exchange reactor tube using IntraMicron technology in a nominal 5 gallons per day skid-mounted fixed-bed installed at NCCC. In those experiments, both productivity and selectivity goals were achieved during a 300 hour experiment. Another experiment has just been completed in spring 2017 with the reactor diameter increased to 4". This work will compare the results from the two long-term tests of the hybrid FT jet-selective catalyst using the skid-mounted fixed-bed FT reactor system at NCCC.

501942 Microfibrous Entrapped Catalyst Structure for Highly Exothermic Reaction: Fischer-Tropsch Synthesis

Abstract:

An Abstract Submission to 2017 AICHE Meeting

Session: Alternative Fuels

Topical Area: Catalysis and Reaction Engineering Division

Microfibrous Entrapped Catalyst Structure for Highly Exothermic Reaction: Fischer-Tropsch Synthesis

Xinquan Cheng and Bruce J. Tatarchuk*

Center for Microfibrous Materials Manufacturing, Department of Chemical Engineering, 212 Ross Hall, Auburn University, AL 36849

**Corresponding author. Tel.: +1 334 844 2023; fax: +1 334 844 2065; E-mail address: Tatarbj@auburn.edu*

Abstract

Fischer-Tropsch synthesis (FTS) is a well-known technique for converting gas to liquid fuels. However, FTS is a highly exothermic reaction, which producing extremely amount of heat during the reaction process. More important, the product distribution of FTS is strongly dependent on the reaction temperature. High temperatures favor the formation of undesirable light products like methane. Therefore, a reactor system with an excellent heat transfer property is necessary for highly exothermic reaction like FTS. Our group developed a novel catalyst structure called Microfibrous Entrapped Catalyst (MFEC) structure. MFEC is a catalyst network supported by micron-sized metal fibers with high thermal conductivity. Previous efforts on this unique catalyst structure have shown significant enhancement in intra-bed heat transfer and maintains a stable reaction temperature.

In this investigation, both iron-based and cobalt-based FTS catalysts have been tested for FTS reactions in a large tubular reactor (34.0 mm I.D.) packed with Cu MFEC. Typically, the MFEC reactor can be prepared in two steps: 1) the prepared catalyst particles were entrapped into a sintered copper microfiber media to form a Cu MFEC structure; 2) circular disks with a diameter of 36.0 mm were punched out from the Cu MFEC structure to pack into the tubular reactor. The disk size (36.0 mm) were designed 6 % larger than the inside diameter of the tubular reactor (34.0 mm) in order to increase the contact points between the copper fiber media and the reactor wall. By doing this, the inside wall heat transfer coefficient increased greatly when compared with the traditional packed bed. A four-point thermocouple

was sticked in from the side when the reactor was loaded half way, and then loading the other half. In that way, the radial temperature distribution profiles can be measured precisely. Take iron-based catalyst as an example, the reaction demonstrated a radial temperature gradient around 5 ºC. While with the same test conditions, the comparative packed bed reached a radial temperature gradient around 139 ºC. It was also found that the Cu MFEC based FTS reaction required negligible amount of time to reach its steady state compared with the traditional packed bed. Considering these attributes and advantages, Cu MFEC structure is a promising reactor system for highly exothermic reactions such as the FTS reaction.

488458 Hydrogen Production from Thermal and Catalytic Gasification of Biomass in a Fluidized CREC-Riser Simulator: Thermodynamic Modeling and Experimental Results

Abstract:

Biomass gasification is an alternative path of critical future value. It produces synthesis gas, a mixture of hydrogen and carbon monoxide, the first one being an energetic vector. [1, 2]

Gasification of several biomasses was performed in a CREC Riser Simulator reactor. Thermal experiments were performed at 550 °C and 600 °C with broza, bark and CANMET Pellets, under helium or CO_2. On the other hand, catalytic experiments were developed using glucose and metoxy-phenol. Temperature, steam/biomass ratio, catalyst loading and reaction time were changed in a wide range of operating conditions. Two fluidizable catalysts were used: 20%Ni/5%La_2O_3/g-Al_2O_3 and 15%Ni-5%CeO_2/γ-Al_2O_3. They were synthesized via incipient wetness impregnation and characterized using PSD, DA, TPD, TPR and BET. [2]

Thermal conversion showed that hydrogen increases with temperature, with this being consistent for the three feedstocks employed (Broza, Canmet Pellet and Bark). Furthermore, the gasifying agent influences hydrogen production, with maximum yields obtained for hydrogen at 600 C, under inert atmospheres (helium). As well, higher temperatures led to a decrease in the H_2 /CO ratio for "Broza". For the "Canmet Pellet", it showed a slight H_2 /CO increase, while for "bark" remained almost constant. The maximum value of the ratio H_2 / CO was obtained at 550 °C, being 5.88.

Regarding the catalytic experiments with fluidizable methoxy-phenol (representing lignin) were used. It was shown hydrogen production increased with temperature. The change with temperature was more notorious from 600 °C to 650 °C than from 650 °C a 700 °C. Hydrogen production augmented as well, with reaction time, with a major change observed in the 10 and 20 second range.

Concerning reaction equilibrium eight chemical reactions were first considered[1]. The Gauss- Jordan procedure was applied to identify five independent equations. On this basis a thermodynamic model, with five algebraic nonlinear equations and five unknowns was considered. The equations were solved with Newton-Raphson method. Experimental hydrogen production was systematically compared with chemical equilibrium values and this for H_2, CO, CO_2 and CH_4. An experimental maximum hydrogen molar fraction of 0.54 for S/B = 2.5 was observed. A similar situation was noticed with CO_2 produced with however, CO and methane being reduced at those conditions.

(1) Salaices E., Serrano B. *de Lasa H.* "Steam Gasification of a Cellulose Surrogate Over Fluidizable Ni/alpha alumina: A Kinetics Model" *AIChEJ* Vol.58, Issue 5, 1588–1599 (2012), *IF: 2.03*

(2) Mazumder, A., *de Lasa H.* Fluidizable La_2O_3 Promoted Ni/ γ-Al_2O_3 Catalyst for Steam Gasification of Biomass: Effect of Catalyst Preparation Conditions. *Applied Catalysis B,* Vol.169-169, 250-265 (2015). *IF:3.42*

500597 A Generalized Kinetic Model for Transesterification and Saponification

Abstract:

Owing to the depletion of fossil fuel resources and increasing environmental concerns, the search for alternative fuels such as biodiesel has gained enormous impetus in the past two decades. Plant and vegetable oils are so far the most prominent feedstocks for these alternative fuels. Rudolf Diesel was the first to test peanut oil as an alternative fuel for his engine about a century ago (Shay, 1993). During 1930s and 1940s, vegetable oils were used as alternative fuels in case of emergency (Singh and Singh, 2010). However, vegetable oils cannot directly substitute diesel due to their high viscosity, low volatility, and poor cold flow properties. Hence, they are processed to obtain biodiesels with properties similar to those of the petroleum-based diesel. This has attracted much attention in recent years (Verma and Sharma, 2016). Biodiesel is a non-toxic, renewable, and biodegradable fuel composed mainly of fatty acid esters. It is industrially produced via transesterification of vegetable oils with alcohol in the presence of a catalyst.

Though biodiesel holds promise as an alternative fuel, its production offers a significant challenge to researchers due to higher production costs compared to the production of petroleum based diesel. Hence, it becomes crucial to model and optimize the biodiesel production process with the aim to achieve lower costs along with meeting the standard property specifications. There are numerous research articles and literature review in modeling, simulation and optimization of biodiesel production process (Aransiola et al., 2014). However, in all of the studies, modeling the reaction kinetics is the foremost important step. The studies regarding kinetics of biodiesel production can be divided into two categories. In the first category (Vicente et al., 2006; Eze et al., 2014), the kinetic models are oversimplified and strongly dependent on the parent oil used for that study. Hence, there are a lot of inconsistencies in the reported values of the kinetic parameters. In the second category (Likozar and Levec, 2014a,b), the effect of the saponification reactions is not accounted for in the model which restricts their application to the case of water-free chemicals only. Therefore, in this article, we address these issues by proposing a kinetic parameter ranking and estimation framework that has the following key components,

1. A generalized kinetic model that considers both both transesterification and saponification reactions.
2. The model is suitable for any biodiesel feed containing the glycerides formed using the five most common fatty acids *viz.*, Oleic (O), Linoleic (L), Linolenic (Ln), Palmitic (P) and Stearic (S).
3. A detailed sensitivity analysis of all model parameters.
4. Application of a Bayesian approach to estimate the most important six parameters.

The purpose of this work is to develop a generalized kinetic model for both transesterification and saponification reactions. It consists of a differential mass balance for the process of transesterification and saponification of each glyceride and ester formed by the major fatty acids present in oil. Eq. (1) shows the general form of differential equations in the model.

$dC_X/dt = f_X(C,\Theta)$ (1)

where x represents the different species such as glycerides, esters, methanol, glycerol, soaps, and free fatty acids; C_x represents the concentration of species x; and Θ represents the set of kinetic rate constants. Here, heterogeneity in the reaction system is neglected and the model is assumed to be in pseudo-homogeneous phase. This is a sound assumption for the industrial-scale process as they employ high mixing intensity. Our model consists of 71 species and 257 reactions. User-defined operational inputs for the model are temperature of the reactor, molar ratio of alcohol to oil, catalyst concentration (wt.% based on oil), and the composition of oil. The model response/output is the yield of fatty acid methyl ester (biodiesel) at different time points. The model parameters include the pre-exponential (Arrhenius) factors and the activation energies with respect to all the reactions. This sums up to 2 x 257 = 514 parameters. However, it is not practically feasible to estimate all the parameters of a reaction mechanism simultaneously. Thus, we employ finite difference approach based sensitivity analysis to identify the most significant parameters for estimation. Typically, the number of parameters estimated is limited by the available experimental observations. Here, we estimate the 6 most sensitive parameters using Bayesian approach of parameter estimation and 32 experimental observations. As the direct application of the Bayesian method is challenging over wide ranges in parameter space, it is prudent to employ a preliminary step of optimization to narrow down the ranges using sum of squared errors as an objective function. This is performed in 2 stages *viz.* a quasi-random global search as the first stage followed by an optimization using the Hooke-Jeeves algorithm. We then employ a Bayesian approach described in (Mosbach et al., 2012; Beers, 2006) over the optimized parameter ranges to study the effects of uncertainties in parameters on the model response. Due to the large number of evaluations of the model required for sampling from the distributions, we employed a surrogate model instead of directly evaluating the original model. The surrogate model was developed using the High Dimensional Model Representation (HDMR) technique.

Resulting distributions showed that we obtain a single optimal value obtained for each of the parameters respectively. We validate our model by conducting parametric studies over varied ranges of temperature, molar ratio of oil to alcohol and catalyst concentration. Our analysis indicates that the model could predict the cetane number and higher heating value of biodiesel formed using various parent oils with an average accuracy of around 3%. Finally, we performed mechanism reduction in order to find the minimum set of reactions that can represent the kinetics of 4 major oils *viz.*, Cottonseed oil, Palm oil, Rapeseed oil and Soybean oil without significant loss of accuracy. We observe that the deviation in the dynamic concentration profiles increased from less than 1% to around 10% as we vary the size of the mechanism from 211 reactions to 175 reactions.

References

1. Aransiola, E., Ojumu, T., Oyekola, O., Madzimbamuto, T., Ikhu-Omoregbe, D., 2014. A review of current technology for biodiesel production: State of the art. Biomass and bioenergy 61, 276–297.
2. Beers, K. J., 2006. Numerical methods for chemical engineering: applications in Matlab. Cambridge Univer-sity Press.
3. Eze, V. C., Harvey, A. P., Phan, A. N., 2015. Determination of the kinetics of biodiesel saponification in alcoholic hydroxide solutions. Fuel 140, 724–730.
4. Eze, V. C., Phan, A. N., Harvey, A. P., 2014. A more robust model of the biodiesel reaction, allowing identifica-tion of process conditions for significantly enhanced rate and water tolerance. Bioresource Technology 156, 222–231.
5. Keera, S., El Sabagh, S., Taman, A., 2011. Transesterification of vegetable oil to biodiesel fuel using alkaline catalyst. Fuel 90 (1), 42–47.
6. Likozar, B., Levec, J., 2014a. Effect of process conditions on equilibrium, reaction kinetics and mass transfer for triglyceride transesterification to biodiesel: experimental and modeling based on fatty acid compo-sition. Fuel Processing Technology 122, 30–41.
7. Likozar, B., Levec, J., 2014b. Transesterification of canola, palm, peanut, soybean and sunflower oil with methanol, ethanol, isopropanol, butanol and tert-butanol to biodiesel: Modelling of chemical equilib-rium, reaction kinetics and mass transfer based on fatty acid composition. Applied Energy 123, 108–120.
8. Marjanovic,´ A. V., Stamenkovic,´ O. S., Todorovic,´ Z. B., Lazic,´ M. L., Veljkovic,´ V. B., 2010. Kinetics of the base-catalyzed sunflower oil ethanolysis. Fuel 89 (3), 665–671.
9. Mendow, G., Veizaga, N., Querini, C., 2011. Ethyl ester production by homogeneous alkaline

transesterifica-tion: influence of the catalyst. Bioresource technology 102 (11), 6385–6391.
10. Mosbach, S., Braumann, A., Man, P. L., Kastner, C. A., Brownbridge, G. P., Kraft, M., 2012. Iterative improvement of bayesian parameter estimates for an engine model by means of experimental design. Combustion and Flame 159 (3), 1303–1313.
11. Richard, R., Thiebaud-Roux, S., Prat, L., 2013. Modelling the kinetics of transesterification reaction of sun-flower oil with ethanol in microreactors. Chemical engineering science 87, 258–269.
12. Shay, E. G., 1993. Diesel fuel from vegetable oils: status and opportunities. Biomass and bioenergy 4 (4), 227–242.
13. Singh, S., Singh, D., 2010. Biodiesel production through the use of different sources and characterization of oils and their esters as the substitute of diesel: a review. Renewable and Sustainable Energy Reviews 14 (1), 200–216.
14. Verma, P., Sharma, M., 2016. Review of process parameters for biodiesel production from different feedstocks. Renewable and Sustainable Energy Reviews 62, 1063–1071.
15. Vicente, G., Martinez, M., Aracil, J., 2006. Kinetics of brassica c arinata oil methanolysis. Energy & fuels 20 (4), 1722–1726.
16. Vicente, G., Martínez, M., Aracil, J., Esteban, A., 2005. Kinetics of sunflower oil methanolysis. Industrial & Engineering Chemistry Research 44 (15), 5447–5454.

495814 Microwave Assisted Co-Pyrolysis of High Ash Indian Coal and Rice Husk and Detailed Product Characterization

Abstract:

With a share of over 60% in the commercial energy sector, coal has become Indiaâ€™s energy security. The relative abundance of coal in India compared to other fossil fuels makes it a natural choice as the primary source of fuel. The reserves to production ratio at the current rate of production for coal is around 230 years as against 15 and 23 years, respectively, for indigenous oil and natural gas. This makes domestic coal a very secure energy source for the country. India has got 192 billion tons of coal reserves but the majority contain high amount of ash [1]. Coal is a complex heterogeneous material which consists of a number of distinct organic (maceral) and inorganic (mineral) compounds. Although regained interest in the production of transportation fuels from coal is positive sign, it is not sufficient as liquefaction of coal yields relatively low liquid yield compared with biomass liquefaction.

Coal and biomass to liquid hydrocarbon fuels offer great hope in commercialization than other alternatives as it not only reduces the carbon footprint in the environment but also facilitates the utilization of locally available low-rank coals [2]. To improve the utilization of coal and to increase the carbon and energy density, co-pyrolysis, co-gasification and co-combustion processes have been developed. Among these, co-pyrolysis of coal and biomass is a promising strategy to produce liquid bio-oil that can be either combusted for power/heat or can be further catalytically upgraded to liquid transportation fuels [3]. Co-processing of coal and lignocellulosic biomass can provide a solution for the use of biomass on a commercial scale. Conventional pyrolysis involves heat transfer in the form of conduction, convection and radiation, while in microwave heating electromagnetic energy is converted to thermal energy from inside of the sample to the surface. Due to its uniform, volumetric, rapid, selective and efficient heating mechanism, microwave heating reduces the processing time and enables the selective production of chemicals [4]. Moreover, secondary cracking of primary pyrolysates can be controlled in microwave heating using different microwave absorbing materials/susceptors.

In this study, microwave assisted co-pyrolysis of high ash Indian coals and rice husk is investigated to evaluate the liquid fuel potential and to understand the interaction effect in the liquid yields and its composition. Microwave co-pyrolysis experiments were conducted in a bench scale unit, and the effects of various parameters such as microwave power, coal: rice husk ratio, feed: susceptor ratio on product yields and oil composition were thoroughly evaluated. The bio-oil composition was semi-quantitatively analyzed using gas chromatograph-mass spectrometry. Gas composition was analyzed using an online gas analyzer. The oil yield was not significantly affected by the microwave power and feed: susceptor ratio for a particular coal: rice husk ratio. Additive effect was observed with increase in rice husk ratio. The bio-oil yields were 11±2 wt.%, 15± 2 wt.%, 18 ± 2 wt.%, 21 ± 2 wt.% and 26 ± 2 wt.% for 100:0, 75:25, 50:50, 25:75 and 0:100 mass ratios of coal to rice husk. The use of graphite as susceptor improved the heating rate significantly owing to its high dielectric tangent value. High microwave powers increased the second cracking of condensables, which resulted in reduction of oil yield as compared to oil yield from low microwave power. Coal oil is a mixture of aromatics, aliphatic HC and phenolic compounds, while pure rice husk oil was a mixture of furans,

aromatics, phenolic compounds, guaiacols, syringols and cyclooxygenates. The addition of coal to rice husk demoted the yield of phenolic compounds and promoted the aromatic content. HHV of bio-oil was 32-35 MJ kg^{-1}, while HHV of char was 11-15 MJ kg^{-1}. Although char was highly porous, its HHV was low due to the high amount of ash that was present in both feedstocks (20% ash in rice husk and 33% ash in Indian coal). Hydrogen gas constituted a major fraction in the pyrolysis gases. Addition of rice husk promoted CO_2 evolution.

References

[1] Haque E. Indian Coal: Production and Ways to Increase Coal Supplies. Intl J Sci Res Publ. 2(2013), 1-3.

[2] Arvelakis S, Folkedahl B, Dam-Johansen K, Hurley J. Studying the melting behavior of coal, biomass, and coal/biomass ash using viscosity and heated stage XRD data. Energy Fuel 20(2006), 1329-1340.

[3] Guan Y, Ma Y, Zhang K, Chen H, Xu G, Liu W, Yang Y. Co-pyrolysis behaviors of energy grass and lignite. Energy Convers Manage. 93(2015), 132-140.

[4] Shraâ€™ah A A, Helleur R. Microwave pyrolysis of cellulose at low temperature. J Anal Appl Pyrol. 105(2014), 91-99.

496717 Highly Selective Catalytic Conversion of Furfural to Î³-Butyrolactone

Abstract:

Highly selective catalytic conversion of furfural to ¦Ã-butyrolactone

Tiefeng Wang* **and Xiaodan Li, Yafei Li**

Beijing Key Laboratory of Green Reaction Engineering and Technology

Department of Chemical Engineering, Tsinghua University, Beijing 100084, China

**Corresponding author: wangtf@tsinghua.edu.cn*

The depletion of fossil fuel reserves and growing demand for energy are drawing efforts to find alternative sources of energy and organic carbon. Biomass-derived molecules can serve as renewable substitutes for petroleum-based building blocks to produce fuels and chemicals.[1] Furfural, a renewable platform compound produced by hydrolysis and dehydration of xylan contained in lignocellulose, has already been used to produce fine chemicals such as furfuryl alcohol, tetrahydrofurfuryl alcohol, furan, and furfural resin in industry.[2-4] During the past decades, novel technologies and processes have been developed for furfural production. This will necessarily promote the further utilization of furfural to produce more value-added chemicals.

One of the most promising products is ¦Ã-butyrolactone (GBL), which has wide applications in the fields of solvents, medicine, spice and fine organic synthesis.[5] Currently, GBL is mainly produced by dehydrocyclization of 1,4-butanediol (BDO) and hydrogenation of maleic anhydride, but they usually have problems of harsh operating conditions, complicated process, and environmental pollution.[6] Therefore, it is attractive to sustainably produce GBL from a renewable biomass-derived molecule with high atom economy. This work reports a two-step process to produce GBL from furfural with 2(5*H*)-furanone as the intermediate. In the first step, furfural was oxidized to 2(5*H*)-furanone with a promising yield over 60%. In the second step, the purified 2(5*H*)-furanone was further hydrogenated to GBL using SiO_2 supported metal catalysts.

The catalytic oxidation of furfural to 2(5*H*)-furanone was studied using hydrogen peroxide as the oxidant and formic acid as the catalyst. A 60-62% yield of 2(5*H*)-furanone, with a total yield of diacids (*i.e.*, maleic acid and succinic acid) in the range of 15-20% as by-products, was obtained in a bi-phasic system using 1,2-dichloroethane or ethyl acetate as the solvent. The good solubility in the organic phase and the strong oxidizing nature of performic acid, which was the real active oxidant generated from formic acid and hydrogen peroxide, contributed to the promising yield of 2(5*H*)-furanone.[7]

Having obtained 2(5*H*)-furanone by furfural oxidation, we next turn to the hydrogenation of 2(5*H*)-furanone to GBL. With only one more additional C=C bond than GBL, 2(5*H*)-furanone is an ideal intermediate for converting furfural to GBL. The selective hydrogenation of the C=C bond in 2(5*H*)-furanone was first studied in a liquid phase over a series of SiO_2 supported monometallic catalysts (M = Pt, Pd, Rh, Ru, Ni, Co and Cu) prepared by incipient wetness impregnation. The catalyst activity followed the trend of $Pd0.5/SiO_2 \sim Rh0.5/SiO_2 > Ru0.5/SiO_2 > Pt0.5/SiO_2$, and $Ni4.5/SiO_2 > Co4.5/SiO_2 > Cu4.5/SiO_2$. Furthermore, the initial TOF for 2(5*H*)-furanone hydrogenation over these monometallic catalysts showed a volcano-type dependence on their metal *d*-band center values, with Pd having the highest activity.[8] This correlation further guided our design of non-noble metal-based bimetallic catalysts to reduce cost.

Ni was shown to be active and selective for the hydrogenation of the C=C bond in 2(5*H*)-furanone. The activity and selectivity of a metal catalyst can be significantly altered by incorporating a second metal due to electronic and geometric effects.[9] Thus, a series of Ni-based bimetallic catalysts, Ni-M/SiO_2 (M= Fe, Co, Cu and Zn), were prepared and evaluated for the selective hydrogenation of 2(5*H*)-furanone to GBL. Among the prepared bimetallic catalysts, Ni-Fe/SiO_2 showed the best performance. The effect of addition of Fe on the catalyst performance was further studied with different Ni/Fe ratios. It was found that the introduction of a suitable amount of Fe significantly enhanced the activity while maintaining the selectivity of the Ni/SiO_2 catalyst. Characterization measurements by XRD, BET, TEM and chemisorption were used to investigate the promotional effect of Fe, showing that the formation of NiFe alloy, optimized reducibility and metal particle size distribution contributed to the enhanced catalytic performance of the Ni-Fe/SiO_2 bimetallic catalyst.

In summary, a two-step process was developed for producing GBL from furfural, providing a new strategy for converting biomass-derived platform compound to high value-added fine chemicals. In the oxidation step, the highest yield of the intermediate, 2(5*H*)-furanone, was obtained in a bi-phasic system with formic acid as the catalyst, ethyl acetate as the solvent, and hydrogen peroxide as the oxidant. In the subsequent hydrogenation step, a series of Group VIII monometallic catalysts were first investigated and their catalytic performances were correlated with intrinsic properties of the metals, based on which an efficient non-noble bimetallic catalyst, Ni-Fe/SiO_2, was further designed.

References

[1] P. Gallezot, *Chem. Soc. Rev.*, **2012**, 41, 1538.

[2] D. M. Alonso, J. Q. Bond and J. A. Dumesic, *Green Chem.*, **2010**, 12, 1493.

[3] J. P. Lange, E. van der Heide, J. van Buijtenen and R. Price, *ChemSusChem*, **2012**, 5, 150.

[4] S. Dutta, S. De, B. Saha and Md. I. Alam, *Catal. Sci. Technol.*, **2012**, 2, 2025.

[5] X. Li, Y. Y. Cui, X. L. Yang, W. L. Dai and K. N. Fan, *Appl. Catal. A*, **2013**, 458, 63.

[6] J. R. Ebner and V. M. Francetti, *Chem. Rev.*, **1988**, 88, 55.

[7] X. D. Li, X. C. Lan and T. F. Wang, *Catal. Today*, **2016**, 276, 97.

[8] X. D. Li, X. C. Lan and T. F. Wang, *Green Chem.*, **2016**, 18, 638.

[9] S. Zafeiratos, S. Piccinin and D. Teschner, *Catal. Sci. Technol.*, **2012**, 1787.

497585 In-Situ Drifts Studies over Cobalt Catalyst for Hydrogen Production from Ethanol

Abstract:

Ethanol dehydrogenation mechanism was investigated over cobalt catalyst prepared by novel cellulose assisted combustion synthesis techniques [1-4]. Nanoparticles synthesis mechanism was studied using TGA/DTA analysis. The effect of synthesis precursor ratio was also studied and correlated with microstructural properties of the product. The amount of fuel content in the combustion solution was found to greatly affect the phase and microstructure of the synthesized catalysts. Along with the microstructure, changes in surface cobalt oxidation states as a function of fuel ratio was also studied for various fuel to oxidizer ratios using XPS. The cobalt catalyst was found to be highly active for ethanol dehydrogenation reaction over a temperature of 50°C– 400°C. In situ DRIFTS studies were conducted to understand the ethanol-metal interaction and deduce a reaction pathway by monitoring the gas phase and surface adsorbed species. The reaction mechanism was found to proceed via ethoxy formation and subsequent generation of aldehyde species at 50°C – 200°C and formation of carbon dioxide at 400°C. Carbon formation was also observed as shown in the TEM image (Fig. 1) of the spent catalyst, which supposedly comes from sequential removal of H from adsorbed $*CH_3$ to produce hydrogen and carbon.

Figure 1: TEM image of spent Co catalyst after ethanol dehydrogenation reaction

-

References

[1] A. S. Mukasyan, P. Dinka, Advanced engineering materials 9 (2007) 653-657.

[2] A. Kumar, A.S. Mukasyan and E.E. Wolf, Applied catalysis.A, General 372 (2010) 175-183.

[3] A. Kumar A.S. Mukasyan, E.E. Wolf, Industrial engineering chemistry research 49 (2010) 11001-11008.

[4] A. Ashok, A. Kumar, R.R. Bhosale, M.A.H. Saleh and Leo J.P. van den Broeke, RSC Advances 5 (2015) 28703-28712.

500197 Towards the Design of Sulfur-Tolerant CO_2-Reforming Catalysts

Abstract:

Mixed metal oxide catalysts, combining late transition metals (TMs) doped into earth oxides (REOs), offer hydrocarbon conversion (e.g., tar reforming) activities and selectivities that meet or exceed supported metal systems,[1] while also offering continued catalytic performance in the presence of sulfur poisons and sintering agents such as water.[2] These materials also have potential as sulfur tolerant CO_2-reforming or tri-reforming catalysis. The objectives of the proposed work are to acquire a "proof of principle" data set towards the development of sulfur- and water-tolerant (especially the former) TM-doped rare-earth oxysulfide catalysts for high temperature reforming, and 2) develop a mechanistic rationale for how surface sulfidation alters reforming and coke formation kinetics. Integrated experimental and computational efforts will provide a mechanistic rationale for designing sulfur- and water-tolerant catalysts. Density functional theory (DFT) methods examine C-H and C-C bond activation and coke formation over partially sulfided metal oxide surfaces, identifying materials promising for retaining, or possibly enhancing, conversion in the presence of sulfur. Synthesis, testing, and characterization evaluate the computationally predicted catalyst performance and provide feedback and further hypothesis development to aid the computational efforts.

To date, we have found that Ce-Zr supports at the 1:1 or higher elemental ratio are superior to the Ce-La supports at similar elemental ratios. The latter are relatively inactive. We have also found that coking rates in the absence of water are strongly dependent upon stoichiometric equivalence between CO_2 and CH_4. Structurally, both the Ce-Zr and Ce-La families of supports show no post-mortem evidence of separate TM (Ni, Mn, Fe) or TM oxide phases, at least for TM:REO ratios kept below thresholds determined in past work on tar reforming. Characterizations of the used catalysts have included XANES, XAFS, Raman, XRD and chemisorption techniques.

Extended coking rate measurements by DSC-TGA show that for the TM-doped REO catalysts a coking plateau is attained, while for simpler Co- and Ni-supported Al_2O_3 catalysts the coking rates are nonzero and almost constant. This suggests that regardless of their short-term performance, the latter family of catalysts supported on Al_2O_3 will ultimately deactivate more rapidly.

1. R. Li, A. Roy, J. Bridges and K.M. Dooley, "Tar Reforming in Model Gasifier Effluents: Transition Metal/Rare Earth Oxide Catalysts," Ind. Eng. Chem. Res. 53, 7999–8011 (2014)
2. M. Krcha, K. Dooley, and M. Janik, "Alkane Reforming on Partially Sulfided CeO_2 (1 1 1) Surfaces," J. Catal., 330, 167-176 (2015)

500717 Development of Catalytic Process for CO_2 Utilization

Abstract:

The utilization of CO2 as a raw material in the synthesis of chemicals and liquid energy carriers offers a direct way to mitigate the increasing CO2 build-up in the atmosphere. However, high-energy reagents and harsh reaction conditions are typically required for transforming CO2 into other chemicals. Renewable hydrogen is a high-energy material that can be stored by hydrogenating CO2 to valuable secondary energy carriers, such as methane, methanol, and formic acid. Indeed, formic acid/formate has been widely used in agriculture and industry and is the potential hydrogen storage. Herein, we developed a process of hydrogenating liquid amine captured CO2, the intermediates in the Bioenergy with Carbon Capture and Storage (BECCS) process, over supported palladium nano-catalysts, to produce formates. We further developed the rechargeable â€œhydrogen batteryâ€ based on the redox equilibrium between amine captured CO2 / formate amine adducts. The stored hydrogen can be catalytically released under mild conditions as the feed to fuel cells for regenerating electricity. The strategies of the process development has been focused on the efficiency (rate of CO2 hydrogenation and dehydrogenation reactions) and energy use (the amount of energy required to utilize CO2 in existing products). This development also includes gaining insight into the relationships among the various structures of amines, solvent effect, distribution of intermediates, catalytic performance, and reaction mechanism.

487340 Catalytic Methanation of Carbon Dioxide with Ni/MgO Catalyst in Dual-Step Reductive Calcination

Abstract:

The heterogeneously catalyzed gas-phase conversion of carbon dioxide with hydrogen to produce methane has gained major interest in recent years [1]. One of the reasons for this increased interest is the quest for hydrogen storage systems. Renewably produced hydrogen is stored via methanation of the greenhouse gas carbon dioxide. Methane can partially substitute natural gas in an established gas grid. Our research focuses on process integration in this context and addresses potential carbon dioxide sources. Calcination is a widely used preparatory step in mineral carbonate beneficiation. During calcination carbon dioxide is released in high concentration into the off-gas. If calcination is performed in hydrogen atmosphere (= reductive calcination), a highly concentrated (CO_2/H_2) process gas stream suitable for direct gas phase methanation can be prepared. Our research [2,3] revealed that reductive calcination is a low temperature process compared to calcination in oxidizing atmosphere, and carbon dioxide is partially converted into methane and/or carbon monoxide in a single-step process. The yield of methane and carbon monoxide can be partially controlled by reaction temperature and pressure. Selective methanation, however, is not achieved in the calcining step. Therefore, methanation of the process gas from reductive calcination by catalytic methanation at ambient pressure with a Ni/MgO catalyst was investigated [4]. Although methanation of CO_2 according to the Sabatier reaction has been investigated since decades, little information about process performance at ambient pressure is available. At moderate temperature (534 K) and ambient pressure CO_2-conversion close to equilibrium conversion beyond 90% is possible. Adsorption of reactants and products does not affect kinteics. At moderate temperature conversion does draw advantages from interaction of carbon dioxide with MgO. Latter effect was confirmed by direct methanation of $MgCO_3$ with hydrogen without catalytic support.

1. Bailera, M., Lisbona, P., Romeo, L.M., and Espatolero, S. (2017) Power to Gas projects review: Lab, pilot and demo plants for storing renewable energy and CO2. Renew. Sustain. Energy Rev., 69 (January 2016), 292–312.

2. Baldauf-Sommerbauer, G., Lux, S., Aniser, W., and Siebenhofer, M. (2017) Synthesis of carbon monoxide from hydrogen and magnesite/dolomite. Chemie Ing. Tech., 89 (1–2), 172–179.

3. Baldauf-Sommerbauer, G., Lux, S., Aniser, W., and Siebenhofer, M. (2016) Reductive calcination of mineral magnesite: hydrogenation of carbon dioxide without catalysts. Chem. Eng. Technol., 39 (11), 2035–2041.

4. Baldauf-Sommerbauer, G., Lux, S., and Siebenhofer, M. (2016) Methanation of carbon dioxide with a nickel/magnesium oxide catalyst. in: AIChE Annual Meeting 2016, San Francisco, USA.

501005 Glycerol Transfer Hydrogenation of CO_2 Using Ir and Ru Carbene Organometallics Immobilized on Active Hydrotalcites in a Packed Bed Flow Reactor

Abstract:

Utilization of CO_2 has become a focal point for reducing carbon emissions. A long-standing method involves direct hydrogenation and transfer hydrogenation (TH) using isopropanol as a hydrogen donor to convert CO_2 to value-added commodity chemicals such as formic acid. Less explored is the use of glycerol, a low value byproduct of biodiesel synthesis made from trans-esterification of triglycerides, as a substitute for isopropanol due to thermodynamic and kinetic challenges. Using CO_2 as a substrate and glycerol as a hydrogen donor for TH yields formate and lactate under basic condition following a Cannizzaro reaction. Herein we explored the use of ruthenium and iridium N-heterocyclic carbenes (NHC) homogeneous catalysts immobilized on a tunable hydrotalcite (HT) supports for CO_2 TH, converting CO_2 to potassium formate and glycerol to potassium lactate in a packed bed flow reactor. HT was synthesized using 3:1 ratio of Al:Mg in flow, and composition was varied by incorporating Cu or Fe in place of Al up to a total of 15 wt. %. Comparison of the heterogenous catalysts to their respective homogenous catalysts indicated that the HT is non-innocent during the reaction. The use of a flow reactor allowed for continuous data collection of kinetic parameters as well as catalytic efficiencies within a single reaction setup and demonstrated that immobilized catalysts are stable for over 300 hours on stream with no sign of deactivation in the presence of 0.25 - 2.0 M potassium hydroxide (KOH) at 150-225ËšC. Precise control over the gas-to-liquid ratio within the reactor provided insight into the efficiency of catalysts at utilizing CO_2, giving scale-up details on required gas stream purity and operating pressure requirements. Reactions were initially carried out in 45 bars of CO_2 in a batch reactor and then scaled to a flow reactor using less than 35 bar CO_2. Upon depletion of accessible carbonate in the solution, the formate yield stabilizes while lactate/lactic acid yield continues to form under acid conditions at a reduced rate. Due to similar turn-over-frequencies (TOF) in 0.25-2.0 M KOH exceeding 4.4 K and 1.5 K per hour for lactate and formate, respectively.

489777 Combining CO_2 Reduction with Propane Oxidative Dehydrogenation over Bimetallic Catalysts

Abstract:

Combining CO_2 reduction with propane oxidative dehydrogenation over bimetallic catalysts

Elaine Gomez and Jingguang G. Chen*

Department of Chemical Engineering, Columbia University, New York City, NY 10027, USA

*jgchen@columbia.edu

Abstract: Propylene is one of the most diverse petrochemical building blocks used for the production of many chemicals (e.g. polypropylene, propylene oxide, and acrylonitrile). The inherent variability and insufficiencies in the co-production of propylene from steam crackers has raised concerns regarding the global propylene production gap and has directed industry to more on-purpose propylene technologies[1]. The oxidative dehydrogenation of propane by CO_2 (CO_2-ODHP) can potentially fill this gap while consuming a greenhouse gas. Utilizing propane and CO_2 presents several advantages over the conventional processes of direct or O_2 assisted propane dehydrogenation, including lower operating temperatures, less coke formation, and higher equilibrium conversions.[2] The reactions of CO_2 with propane may occur through two distinct pathways, oxidative dehydrogenation ($CO_2 + C_3H_8 \rightarrow C_3H_6 + CO + H_2O$) and dry reforming ($3CO_2 + C_3H_8 \rightarrow 6CO + 4H_2$). The two reactions should occur simultaneously at temperatures around 823 K and above with considerable conversions, allowing the formation of both dehydrogenation products (propylene) and reforming products (synthesis gas). Thus, it is of great interest to identify catalytic systems that can selectively break either the C-H bond to produce propylene or the C-C bonds to produce synthesis gas ($CO + H_2$).

Non-precious Fe_3Ni and precious Ni_3Pt supported on CeO_2 have been identified as promising catalysts for CO_2-ODHP and dry reforming, respectively, in flow reactor studies conducted at 823 K. For oxidative dehydrogenation, nonprecious metal based Fe_3Ni had higher stability and propylene production rates. Fe_3Ni showed distinct C_3H_6 yield compared to its parent metals and the Ni_3Pt catalyst showed both greater activity and stability than the corresponding monometallics. In-situ X-ray absorption spectroscopy measurements revealed the oxidation states of metals under reaction conditions and density functional theory was utilized to identify the most favorable reaction pathways over the two types of catalysts. Additional flow reactor studies have been performed on the Fe_3Ni and Ni_3Pt bimetallic systems with other oxide supports (TiO_2, ZrO_2, SiO_2, γ-Al_2O_3, and V_2O_5) to examine the effects of strong support metal interactions, surface defects, inert surfaces, irreducibility, as well as alkane activation on both the CO_2-

ODHP and DRP pathways. Overall, two types of bimetallic catalysts have been identified for the $CO_2 + C_3H_8$ system; one is selective toward reforming (C-C bond cleavage) and the other favors dehydrogenation (C-H bond scission).

References

(1) Plotkin, J. S. *ACS News Industrial Chemistry & Engineering*. 2016, pp 1–4.

(2) Wang, S.; Zhu, Z. H. *Energy and Fuels* **2004**, *18* (4), 1126–1139.

494828 CO_2 Splitting Using Miec Membranes

Abstract:

In this presentation, we discuss fuel production using membrane-supported CO_2 splitting. A mixed ionic-electronic conductive (MIEC) $La_{0.9}Ca_{0.1}FeO_{3-\delta}$ (LCF-91) membrane is utilized. At intermediate temperatures (850 – 990°C), CO_2 undergoes thermochemical reduction on the feed-side producing CO, while oxygen incorporates into the perovskite as lattice oxygen. Oxygen diffuses to the sweep side, driven by the chemical potential gradient across the membrane generated in this case by fuel (either H_2 or CO in this study) oxidation on that side.

LCF-91 membranes have been examined in our laboratory before for air separation [1-3] and water splitting [4, 5]. In this work, we report a detailed study on CO_2 reduction on the same membrane. We examine the dependence of CO_2 splitting rate on various operating conditions such as CO_2 concentration on the feed-side, fuel (H_2 or CO) concentration on the sweep side and temperatures. CO_2 splitting rate is proportional to the oxygen flux across, and the measured maximum flux is 0.19 μmol/cm^2•s on a 1.3 mm membrane at 990°C, when 9.5% H_2 is added to the sweep side (balance with argon gas). Long term study shows almost a constant flux, i.e., stable performances for 106 hours. Following the experiment, morphology changes, carbonate formation and calcium/iron enrichment are observed on the feed-side surface, while no obvious changes are found on the sweep-side surface as revealed by SEM, EDS and XRD. A resistance-network model is developed to describe the oxygen flux, in which CO_2 direct-incorporation mechanism is used on the feed side, and the Mars-van Krevelen (MvK) mechanism for the oxidation reaction on the sweep side (for both H_2 and CO) [6]. Using the flux measurements, surface kinetics parameters are derived, which shows that H_2 has larger oxidation rate constant than CO. The model also shows that in CO_2 splitting, the rate limiting step changes from the CO_2 splitting reaction on the feed-side to the fuel (H_2 or CO) oxidation reaction on the sweep-side at higher temperatures. This is different from the H_2O splitting cases, where fuel oxidation is always the limiting step within the same temperature range tested. This is supported by the change in the slope of the Arrhenius plot of the oxygen flux. Moreover, the transition temperature for H_2 is higher than that of CO (both used as fuels).

Porous layers made from the same materials are added on both side of the membrane to enhance the oxygen flux. Results shows that the flux is raised by about 0.02 μmol/cm^2•s when CO is used as a fuel. And the same transition temperature is observed in the Arrhenius plot of oxygen flux with or without porous layers.

Reference

[1] Hunt, A., Dimitrakopoulos, G., and Ghoniem, A. F., 2015, "Surface oxygen vacancy and oxygen

permeation flux limits of perovskite ion transport membranes," J. Membr. Sci., 489, pp. 248-257.

[2] Dimitrakopoulos, G., and Ghoniem, A. F., 2016, "A two-step surface exchange mechanism and detailed defect transport to model oxygen permeation through the $La_{0.9}Ca_{0.1}FeO_{3-\delta}$ mixed-conductor," J. Membr. Sci., 510, pp. 209-219.

[3] Dimitrakopoulos, G., and Ghoniem, A. F., 2017, "Developing a multistep surface reaction mechanism to model the impact of H_2 and CO on the performance and defect chemistry of mixed-conductors," J. Membr. Sci., 529, pp. 114-132.

[4] Wu, X. Y., Chang, L., Uddi, M., Kirchen, P., and Ghoniem, A. F., 2015, "Toward enhanced hydrogen generation from water using oxygen permeating LCF membranes," PCCP, 17(15), pp. 10093-10107.

[5] Wu, X. Y., Ghoniem, A. F., and Uddi, M., 2016, "Enhancing co-production of H_2 and syngas via water splitting and POM on surface-modified oxygen permeable membranes," AIChE J., 62(12), pp. 4427-4435.

[6] McFarland, E. W., and Metiu, H., 2013, "Catalysis by Doped Oxides," Chem. Rev., 113(6), pp. 4391-4427.

494005 Molten Metal Catalysts for Methane Pyrolysis

Abstract:

Molten metals were investigated as catalysts for the thermochemical pyrolysis of methane. Conversion of methane to hydrogen and solid carbon is demonstrated in bubble column reactors where the solid carbon can be readily separated and removed enabling continuous processes. Twenty-one molten metal systems consisting of elemental metals and binary alloys are compared for methane pyrolysis activity. Nickel-bismuth was found to be the most active catalyst, and 95% methane conversion was observed at 1060 °C in a 1.1 meter bubble column. Initial characterization of the carbon product indicated predominately sp^2 graphitic carbon as the solid product. Although the hydrogen produced from partial oxidation of methane by pyrolysis retains only approximately half the combustion energy potential as complete methane combustion, pyrolysis produces CO_2-free hydrogen, and catalytic molten metals may provide a means for performing reactive separation of the hydrogen from the potentially valuable carbon co-product.

500486 Increasing Activity, Reducing Coking, and Promoting Unexpected Reaction Pathways during Methane Steam Reforming By Applying Uniform Electric Fields in a Scalable Reactor

Abstract:

It has been shown both computationally and in small-scale studies using STM tips for single-molecule analysis that very high electric fields can change both the kinetics and thermodynamics of chemical reactions. This could be an exciting and unexplored new area for reaction engineering, but generating uniform electric fields in a larger scale continuous flow reactor has proven to be a significant challenge. We have designed a system capable of generating uniform electric fields over Ni catalysts for methane steam reforming, which has been compared to density functional theory computations by our collaborating group in the past. This talk focuses on the experimental observations, with special focus on the properties of the catalyst itself after methane steam reforming in the presence of electric fields. We report both expected findings and a new mystery â€" and our attempts to unravel said mystery. On the expected side, our DFT calculations correctly predicted that, in a positive electric field, reaction rates would increase consistent with the lowering of the energy barrier of the first methane C-H bond scission; H_2O would bind more readily to the surface; H_2O would have a more difficult time decomposing; and that coke would have a more difficult time forming. Unexpectedly, the negative field apparently promotes alternate reaction pathways which may lead to C-C coupling reactions to form larger unsaturated hydrocarbons without coking. Both post-reaction catalysts were scrutinized using visual observations including macroscopic photographs and SEM imaging, x-ray photoelectron spectra, temperature-programmed oxidation, and x-ray diffraction patterns.

487516 Oxidative Dehydrogenation of Propane Using CO_2 over Molybdenum Oxycarbidic Catalysts

Abstract:

Transition metal carbides have been shown to catalyze oxidative dehydrogenation of alkanes using CO_2 as a soft oxidant. In this work, we investigate reactive pathways of propane and CO_2 including oxidative dehydrogenation to form propylene, reforming to form CO, and hydrogenolysis to form C_2 and C_1 hydrocarbons over molybdenum oxycarbidic formulations at ambient pressure and 803 â€“ 843 K. Co-feeds of CO_2 and H_2 can result in an order of magnitude change in steady-state propane dehydrogenation rates at constant propane pressure, and these rate changes are reversible. Product carbon selectivities vary between >95% dehydrogenation, >80% reforming, or >95% hydrogenolysis depending upon reaction conditions and the introduction of co-feeds. CO_2 co-feed can produce alkane-derived CO via a reforming pathway, and excess H_2 can result in methane produced via propane hydrogenolysis. Product selectivities and rates can be rationalized and quantitatively fit using a proposed kinetic model predicated upon surface site interconversion between oxidized or reduced active sites as functions of H_2/CO_2 co-feed partial pressures.

496959 Alkali Promoted Mg_6MnO_8 Redox-Catalysts for Chemical Looping-Oxidative Dehydrogenation of Ethane: Mechanistic Investigations

Abstract:

Steam cracking of ethane to ethylene is one of the most energy and carbon intensive processes in the chemical industry. While several decades of process improvements have pushed cracking technologies to near 95% thermal recovery, high reaction endothermicity, rapid temperature swings, and significant downstream separation costs, result in significant exergy loss (lost work). Oxidative dehydrogenation (ODH), which partially oxidizes ethane to ethylene and water, has been suggested as a pathway to process intensification. Conventional ODH, which uses gaseous oxygen, has significant challenges to overcome due to the difficulty of achieving high per-pass yields in a controllable manner and the high capital and energy costs imposed by air separation. We have introduced a chemical looping approach to ODH (CL-ODH), which uses the lattice oxygen of a redox catalyst as an oxidant. The redox catalyst lattice oxygen is subsequently regenerated in air in a separate reaction step, producing heat to drive the process. Such a cyclic redox scheme eliminates the needs for O_2 co-feeding and hence cryogenic air separation. Redox catalyst performance is vital to this process; the redox catalyst must be highly selective towards water vs. CO_x formation. It is also important to keep hydrocarbon residence times low to prevent the formation of undesired hydrocarbon byproducts such as benzene. As such, a detailed mechanistic understanding of redox catalyst in ODH is highly desirable to allow rational catalyst design and optimization.

We have reported a model Mg_6MnO_8 catalyst system with excellent CL-ODH properties. This system is interesting due to its facile lattice oxygen (O^{2-}) transport ability, and the very strong effects of alkali promoters on product selectivity. Without promotion Mg_6MnO_8 is highly active for deep oxidation, but the use of alkali salts as promoters significantly shifts the product selectivity towards ethylene even at high ethane conversions (>70% single pass). In this work, we describe the mechanistic aspects of promoted and unpromoted Mg_6MnO_8 redox catalysts. Extensive characterizations, including surface characterization, temperature programed reactions, and *in-situ* DRIFTS were performed. We also explored the lattice oxygen behavior with TGA, O^{18} labeling experiments, and *in-situ* XRD. These experimental data indicate that alkali salt promoters suppress ethane and ethylene surface activation while the bulk structure maintains fast lattice oxygen transport during the redox reactions.

487726 Propane Dehydrogenation on MOF-Derived Iron Carbide Catalysts

Abstract:

The dehydrogenation of light alkanes, sourced from cracking of traditional petrochemical feedstocks and from emerging shale gas, represents an attractive route to alkenes (ethene, propene, butenes), which are the building blocks for valuable polymers and chemicals. Catalysts for alkane dehydrogenation must be active, but also selective, because several undesired side reactions, such as hydrogenolysis, isomerization, and coke formation. Recent studies in our group have found that iron/phosphorous catalysts supported on alumina are active for propane dehydrogenation (PDH) and are reasonably selective to propene [1]. In situ X-ray absorption spectroscopy indicated the precatalyst ($FePO_4/Fe_2O_3$) is initially reduced under $H_2$and then proceeds through an induction period under propane flow during which iron carbide is formed as the active and selective phase of the catalyst.

Here, we aim to generate the iron-carbide phase directly. Carbon-supported iron carbide from pyrolysis of Fe-containing metal organic frameworks (MOFs; mainly, Fe-BTC) have been reported as active and stable catalysts for Fischer-Tropsch synthesis at 613 K [2â€“4]; however, the temperature of pyrolysis was found to influence the iron-carbide phase and the size of the nanoparticles formed. The state of these catalysts under the higher temperature reaction conditions needed for PDH (773-873 K) is investigated here. We have found that the reactivity and selectivity of these MOF derived iron-carbide catalysts are dependent on the temperature of pyrolysis and reaction and correlate these results to differences in phase or structure as determined from various characterization techniques. This work demonstrates how using MOFs as precursors may allow the synthesis of unique catalyst structures that can be designed to obtain desired rates and product selectivities of hydrocarbon processes.

[1] S. Tan, B. Hu, W.-G. Kim, S.H. Pang, J.S. Moore, Y. Liu, R.S. Dixit, J.G. Pendergast, D.S. Sholl, S. Nair, C.W. Jones, ACS Catal. 6 (2016).

[2] V.P. Santos, T.A. Wezendonk, J.J.D. Jaén, A.I. Dugulan, M.A. Nasalevich, H.-U. Islam, A. Chojecki, S. Sartipi, X. Sun, A.A. Hakeem, A.C.J. Koeken, M. Ruitenbeek, T. Davidian, G.R. Meima, G. Sankar, F. Kapteijn, M. Makkee, J. Gascon, Nat Commun 6 (2015).

[3] T.A. Wezendonk, V.P. Santos, M.A. Nasalevich, Q.S.E. Warringa, A.I. Dugulan, A. Chojecki, A.C.J. Koeken, M. Ruitenbeek, G. Meima, H.-U. Islam, G. Sankar, M. Makkee, F. Kapteijn, J. Gascon, ACS Catal. 6 (2016).

[4] T.A. Wezendonk, Q.S.E. Warringa, V.P. Santos, A. Chojecki, M. Ruitenbeek, G. Meima, M. Makkee, F. Kapteijn, J. Gascon, Faraday Discuss. (2017).

497242 Selective Oxidative Dehydrogenation of Propane to Propylene Using Boron Nitride Catalysts

Abstract:

The oxidative dehydrogenation of propane (ODHP) as a method of "on-purpose" propylene production has the potential to be a game-changing technology in the chemical industry.[1] However, even after decades of research establishing supported vanadium oxide (V/SiO_2) as the state-of-the-art catalyst for ODHP,[2,3] selectivity to the olefin product remains too low to be commercially attractive because of overoxidation of propylene into CO and CO_2 (CO_x).

Here, we report that hexagonal boron nitride (h-BN) and boron nitride nanotubes (BNNTs) are highly selective catalysts for the ODHP reaction.[4,5] Despite structurally similar materials (graphene and carbon nanotubes) showing catalytic activity for partial oxidations in recent years, BN materials have yet to be explored for their own catalytic activity. Both h-BN and BNNT are among the most-selective catalysts reported for ODHP (Figure 1A). Aside from only propylene, it is equally interesting to compare the complete product distribution between BN and V/SiO_2 catalysts (Figure 1B). While the main byproducts of V/SiO_2 are overoxidized products CO_x, the main byproduct of h-BN and BNNT is ethylene, another important small olefin. For reasons that are still under investigation, BN materials are able to largely avoid the overoxidation reaction pathway.

We have kinetic evidence for an Eley-Rideal reaction mechanism when using h-BN, showing the rate of propane consumption dependent on the adsorption of oxygen and second order with respect to P_{C3H8}. Our spectroscopic investigations tell us the surface of h-BN and BNNT is functionalized by oxygen only when exposed to ODHP reaction. Considering these insights, as well as previous DFT studies investigating stable O-terminated BN armchair edges, we calculated reaction intermediates suggesting the reaction active site may be a peroxo-like >B-O-O-N< armchair edge.

Figure 1. (**A**) Selectivity to propylene plotted against propane conversion for ODHP, comparing h-BN (green), BNNT (blue) and V/SiO_2 (black) to previously reported catalysts (open shapes); (**B**) Comparisons of product selectivity between V/SiO_2 (X_{C3H8}=5.8%), h-BN (X_{C3H8})=5.4%) and BNNTs (X_{C3H8}=6.5%). Product selectivities are represented by colored bars.

References

1. T. Degnan, *Focus on Catalysts,* 2 (2016) 1

2. C. Carrero, R. Schlogl, I. Wachs, R. Schomaecker, *ACS Catal.*, 4 (2014) 3357

3. J. Grant, C. Carrero, A. Love, R. Verel, I. Hermans, *ACS Catal.*, 5 (2015) 5787

4. J. Grant, C. Carrero, F. Goeltl, J. Venegas, P. Mueller, S. Burt, S. Specht, W. McDermott, A. Chieregato, I. Hermans, *Science*, 354 (2016) 1570

5. J. Grant, C. Carrero, J. Venegas, A. Chieregato, I. Hermans, US Patent Application 15/260,649; Filed September 9, 2016.

490322 Pt-Fe Intermetallic Alloy Nanoparticles As Selective Propane Dehydrogenation Catalysts

Abstract:

Bimetallic Pt-Fe catalysts were synthesized with nominal Fe:Pt atomic ratios of 0.7, 2.9, and 4.4 and compared to monometallic Pt and Fe catalysts for propane dehydrogenation at 550 °C. The three bimetallic catalysts were greater than 95% selective towards propylene. The Pt catalyst was only 70% selective and the Fe catalyst produced almost exclusively methane.
intermetallic alloys and that the phase varied with the Fe:Pt ratio. As the atomic ratio was increased, first the Pt_3Fe phase was formed, then PtFe, and lastly $PtFe_3$. The increase in selectivity is proposed to arise from the elimination of Pt ensembles responsible for hydrogenolysis, the primary competing reaction. Changes in the energy and shape of the XANES of each phase indicate differences in the electronic structure of Pt in each alloy.

491303 Investigating the Role of Molybdenum during the Direct Conversion of Methane to Liquids Under Non-Oxidative Conditions

Abstract:

Methane dehydroaromatization (MDA) over Mo/HZSM-5 is still being researched due to the fact that the mechanism is still not well-deciphered. In an attempt to understand the contribution of molybdenum in the MDA process, this research has studied the catalytic conversion of methane over MoO_3/SiO_2 catalyst. Silica was chosen as the inert support so that the role of molybdenum could be better understood. A recirculating batch reactor was used for that purpose. Catalysts of different metal loadings were prepared by incipient wetness impregnation and sol gel method. The prepared catalysts were characterized by Brunauer-Emmett-Teller (BET) surface area, inductively coupled plasma-mass spectroscopy (ICP-MS), scanning electron microscopy (SEM), thermogravimetric analysis (TGA) and X-ray diffraction (XRD). XRD analysis showed that at low Mo loadings, molybdenum was uniformly dispersed on the silica support but at higher Mo loadings crystalline Î±-MoO_3 phase was present. Sintering and agglomeration of metal take place at higher loadings which is evident from SEM.Â The catalytic activity of both catalysts were studied at two different temperatures (973 K and 1023 K). It was found that with increase in temperature the conversion of methane increases. Selectivity of ethylene increases with increase in reaction time and temperature. Benzene was the major product formed with trace amounts of ethylene and ethane. A maximum of 1.2% conversion of methane was achieved at 1023 K using 2 wt % Mo/SiO_2 prepared by sol gel method. The role of Mo and a mechanism of the reaction was proposed. The XRD analysis of used catalysts revealed the presence of Î²-Mo_2C phase, obtained during the interaction of molybdenum with methane. These molybdenum carbide sites act as active sites for C-H bond activation and ethylene formation. The ethylene undergoes subsequent oligomerization to form aromatics like benzene. It is interesting to note that benzene was formed even in absence of acidic zeolite sites. Thus, acidic zeolite sites are not the primary sites for methane aromatization. It can be postulated that the free H^+ radicals formed on C-H bond activation would help in ethylene oligomerization on active metal sites.

494718 Catalytic Consequences of Isolated and Paired Acid Sites in CHA Zeolites on Monomolecular Propane Cracking

Abstract:

Catalytic Consequences of Isolated and Paired Acid Sites in CHA Zeolites on Monomolecular Propane Cracking

Philip M. Kester, John R. Di Iorio, and Rajamani Gounder*

Charles D. Davidson School of Chemical Engineering, Purdue University, West Lafayette, IN 47907-2100 (USA)

*rgounder@purdue.edu

Turnover rates for monomolecular cracking and dehydrogenation of alkanes in zeolites are sensitive to local void environments [1] and to the local density of Brønsted acid sites as inferred from changes in bulk Si/Al ratio [2]. Here, we utilize single tetrahedral-site chabazite (CHA) zeolites of fixed bulk composition (Si/Al = 15) with controlled arrangements of framework Al atoms (Al-O(-Si-O)$_x$-Al) in paired (x = 1, 2) and isolated (x = 3) configurations [3] to decouple the effects of void environment and active site proximity. Turnover rates of monomolecular propane cracking (per H^+, 748 K) increase systematically in CHA zeolites as the fraction of Al pairs increases from 0 to 44%. Apparent rate constants (per H^+) on CHA with all isolated Al are constant, suggesting that all isolated Brønsted acid sites are equivalently reactive. Apparent rate constants (per H^+) are an order of magnitude larger for paired Al sites (4.0×10^{-3} mol (mol H^+)$^{-1}$ s^{-1} bar^{-1}) than for isolated Al sites (3.3×10^{-4} mol (mol H^+)$^{-1}$ s^{-1} bar^{-1}). Furthermore, rate constants decrease upon partial Na^+ poisoning of H^+ sites in CHA zeolites containing paired Al sites, consistent with the different reactivity of isolated and paired Brønsted acid sites. Infrared spectra are used to detect and quantify paired proton sites in CHA zeolites using molar extinction coefficients estimated from a suite of model samples of varying paired acid site density, allowing for the quantification of the distribution of residual protons in partially ion-exchanged CHA zeolites. Apparent activation energies are independent of acid site proximity, while apparent activation entropies are less negative at paired acid sites, suggesting that transition states are later and more loosely bound at such sites. This work illustrates that the catalytic behavior of zeolites for alkane activation can be tuned by synthetically manipulating the local arrangement framework aluminum atoms and, in turn, of active Brønsted acid sites.

[1] R. Gounder, E. Iglesia. *J. Am. Chem. Soc.* 131 (2009) 1958-1971.

[2] C. Song, et al. *J. Catal.* 349 (2017) 163-174.

[3] J. R. Di Iorio, R. Gounder. *Chem. Mater.* 28 (2016) 2236-2247.

495752 Developing Multi-Scale Models of Bimetallic Catalysts for the Hydrodeoxygenation of Bio-Oil Compounds

Abstract:

With the ever-increasing need to find a sustainable source of renewable energy, bio-oil produced via the fast pyrolysis of biomass is a promising source of liquid fuel; however, the resulting bio-oil contains oxygenated products that contribute to poor fuel quality [1]. Hydrodeoxygenation (HDO) is used to refine the bio-oil by reducing the oxygen content, ideally using a minimal amount of H_2 in order to form H_2O. Bimetallic catalysts such as Pd on Fe have demonstrated synergistic behavior that contributes to a more cost-effective and longer-lasting catalyst [2-4]. Similarly, Pt/Sn catalysts have also exhibited traits favorable for the HDO of phenolic compounds [5], namely a propensity to promote lower temperature desorption of aromatic compounds like benzene [6]. Identifying the cause of the different synergetic effects of these systems can provide insight into the synthesis of a superior HDO bimetallic catalyst. However, the behavior of bimetallic catalysts is complex and necessitates a thorough understanding of the nanoscale behavior of these systems in order to develop a truly predictive model on relevant time scales. In particular, we present a DFT study that examines the importance of including coverage effects into theoretical models in order to accurately predict experimental behavior.

Quantifying the lateral interactions between adspecies is necessary in the development of truly predictive, theoretical models for bimetallic catalysts, as they will significantly affect the coverage and diffusion of adspecies and therefore the heterogeneous catalytic environment. However, such a characterization of the lateral interactions between adspecies for the highly complex HDO reaction has not yet been attempted. To that end, we have characterized the benzene-benzene lateral interactions on a Pt (111) surface and a Pt_3Sn (111) surface alloy. This coverage-dependent adsorption behavior was incorporated into a kinetic model for each surface and used to simulate the temperature-programmed desorption (TPD) spectra. We benchmarked our simulated spectra by comparing it to experimental TPD [6]. We found that it is imperative to incorporate the effect of surface coverage in these theoretical models if the desorption behavior is to be accurately predicted (see Figure 1). Furthermore, we found that a mean-field model is sufficient in describing the benzene-benzene lateral interactions. Using our models, we were able to interpret the desorption behavior at an atomistic level and provide a deeper level of insight into these systems. While it was previously speculated that the broad desorption peak for benzene on Pt (111) was the result of desorption from two adsorption sites of different binding strengths [6], our results indicate that the behavior is due to benzene desorbing from a single site but experiencing strong lateral interactions on the surface. We also found that the adsorption of benzene was significantly weaker on the Pt_3Sn (111) than on Pt (111), and as such, we were able to assign the monolayer desorption of benzene to a TPD peak previously attributed entirely to multilayer benzene desorption [6]. In this regard, we were also able to attribute the higher temperature desorption peaks observed on Pt_3Sn (111) to benzene desorbing from defects in the alloy surface, where Pt atoms were not substituted for Sn atoms in the alloying process.

Our investigation into modeling the lateral interactions of an aromatic molecule on both the Pt (111) and the Pt_3Sn (111) surfaces warrants the application of our approach onto systems less experimentally characterized, such as Fe surfaces doped with different noble metals.

Figure 1. TPD spectra for benzene on Pt (111) simulated with coverage effects (blue) and without coverage effects (light green) compared to experiment (black) [6].

References

[1] H. Wang, J. Male, Y. Wang, ACS Catal. 3 (2013) 1047-1070.

[2] Y. Hong, H. Zhang, J. Sun, K.M. Ayman, A.J.R. Hensley, M. Gu, M.H. Engelhard, J.-S. McEwen, Y. Wang, ACS Catal. 4 (2014) 3335-3345.

[3] J. Sun, A.M. Karim, H. Zhang, L. Kovarik, X.S. Li, A.J. Hensley, J.-S. McEwen, Y. Wang, J. Catal. 306 (2013) 47-57.

[4] A.J.R. Hensley, Y. Hong, R. Zhang, H. Zhang, J. Sun, Y. Wang, J.-S. McEwen, ACS Catal. 4 (2014) 3381-3392.

[6] C. Xu, Y.L. Tsai, B.E. Koel, J. Phys. Chem. 98 (1994) 585-593.

496761 Mechanistic Insights into Hydroeoxygenation of Phenol on Bimetallic Phosphide Catalyst

Abstract:

Bimetallic transition metal catalysts provide immense opportunity to tailor catalytic function via composition modulation to promote desired chemical transformation. Addition of non-metal components such as oxygen, nitrogen, and phosphorous can provide for better charge density distribution to enhance Lewis acidic character. Recent work by Rensel et al.[1] has shown that FeMoP catalyst is highly selective for Câ€"O bond cleavage for aryl ethers. Furthermore, the catalytic function of FeMoP can be tuned by altering the Fe/Mo ratio to improve benzene selectivity upon hydrodeoxygenation of phenol.[2] In this presentation, we will share density functional theory calculations that provide mechanistic insights into hydrodeoxygenation of phenol on $Fe_XMo_{2-X}P$ catalyst. We find that optimal Fe/Mo composition has smallest activation energy barrier for Câ€"O bond cleavage which correlates well with experimental TOFs. Bader charges and density of states analysis provide pointers to the change in the electronic structure of the catalyst upon changing Fe/Mo composition.

495987 Catalytic Conversion of Furfural to Methylfuran: Investigating Reaction Mechanisms on Ni and the Effect of Boron Doping on the Activity and Selectivity of the Catalyst

Abstract:

Fast pyrolysis of lignocellulosic biomass is emerging as a promising route for the production of liquid transportation fuels. However, the bio-oil produced from pyrolysis has a high oxygen content (~45%) which results in a poor heating value, thus limiting its direct usability as a fuel. The catalytic hydrodeoxygenation (HDO) of bio-oil in the presence of hydrogen is an effective technique for the removal of oxygen groups from bio-oil. Furfural is one of the major bio-oil components and is a potential starting material for the production of compounds like 2-methylfuran (MF), which can be blended into transportation fuels. The challenge is to develop catalysts which can selectively break C – O bonds, while preserving the C – C bonds. In this context, hydrogenolysis of furfural (C_{formyl} – O cleavage) to produce MF is desirable, as the high octane number of MF allows easy blending into gasoline/diesel pools. Vapor phase HDO of furfural over Ni catalysts leads to furfuryl alcohol (via hydrogenation of formyl group) as the dominant product at low temperatures, while furans (via decarbonylation by C – C cleavage) and C4 products (via furan ring opening) dominate at higher temperatures[1-2]. We investigated the potential competitive pathways for the catalytic conversion of furfural on model Ni(111) surfaces using Density Functional Theory. Our calculations show that decarbonylation (C – C cleavage) and ring opening (C_{ring} – O cleavage) of furfural to form furan and ring opened C4 products are kinetically favored pathways on Ni(111) surface, in agreement with experimental studies. Recent studies have suggested that the activity and selectivity of catalyst can be improved by doping boron[3-6]. On Ni(111) surfaces boron binds strongly at the subsurface positions, resulting in a corrugated Ni surface, similar to a stepped surface. The presence of the subsurface, interstitial boron expands the lattice of Ni and also alters its electronic properties by donating electrons. The incorporation of B in the Ni surface was found to increase the C – C cleavage and ring opening barriers and drastically lower the C_{formyl} – O cleavage barriers, relative to the Ni(111) surfaces. We propose the hydrogenolysis of furfuryl alcohol to form MF to be kinetically favorable over C – C cleavage (furan formation) and ring opening (C4 products), leading to significantly enhanced MF selectivity in the vapor phase hydrogenation of furfural on boron doped Ni surfaces.

References

1. S. Sitthisa, D. E. Resasco, Catalysis Letters 2011, 141, 784-791.
2. S. Sitthisa, W. An, D. E. Resasco, Journal of Catalysis 2011, 284, 90-101.
3. J. Xu, M. Saeys, Journal of Physical Chemistry C 2009, 113, 4099-4106.
4. T. Q. Thang, A. V. Nguyen, D. C. Huynh, T. H. Pham, S. H. Mushrif, Catalysis Science and Technology 2016, 6, 5871-5883.
5. B. Yang, R. Burch, C. Hardacre, P. Hu and P. Hughes, Journal of Physical Chemistry C, 2014, 118, 3664-3671.

6. J. Xu, L. Chen, K. F. Tan, A. Borgna and M. Saeys, Journal of Catalysis, 2009, 261, 158-165.

494725 First-Principles Insights into the Mechanisms and Sites for Base Catalyzed Aldol Condensation and Esterification over Copper

Abstract:

Base-catalyzed aldol condensation and esterification are important reactions in the synthesis of large organic molecules from biomass-derived precursors.[1] These molecules can then subsequently be converted to fuels and useful chemicals. Such reactions are typically catalyzed on homogenous catalysts in the presence of a basic promoter or on basic metal oxides, that catalyze C-C and C-O bond formation.[1-3] Recent studies have shown that non-basic materials such as Cu supported on SiO_2 can actively catalyze both condensation and esterification, when equilibrated mixtures of propanol, propanal and H_2 are reacted.[4] These supported Cu systems result in the formation of 47% 3-pentanone (C-C formation via condensation) as well as 34% propyl propionate (C-O formation via esterification).[4] The results suggest that mixtures of propanol, propanal and H_2 readily equilibrated via the formation of surface propoxide intermediates, that can act as a base, thus, catalyzing both condensation and esterification. Ab-initio density functional theoretical calculations were carried along with detailed kinetic analyses to elucidate elementary steps and sites involved in propanal conversion to propanol and the subsequent C-C and C-O bond formation steps. Detailed charge analyses show that alkoxides which readily form on Cu as well as other group 11 metals readily extract electron density from the metal surface and behave as basic intermediates. For the aldol condensation path, the surface bound propoxide abstracts the weakly acidic hydrogen from a bound propanal intermediate to generate an enolate species in a rate-determining step having a Gibbs free energy barrier of 69 kJ/mol. The enolate subsequently undergoes rapid C-C coupling with a second vicinally-bound propanal to generate a Î²-alkoxide intemediate. The Î²-alkoxide can either eliminate hydrogen to the Cu or react with a surface propanal to form a 2-formyl-3-pentanone intermediate that subsequently decarbonylate to form 3-pentanone. The same alkoxide intermediates can also catalyze esterification. The calculations show that the bound propoxide can readily carry out a nucleophilic attack on the carbonyl group of a surface bound propanal in a rate-determining step to form a hemiacetalate intermediate with a Gibbs free energy barrier of 66 kJ/mol. The hemiacetalate further reacts on Cu to form the propyl propionate. The theoretical results indicate that the rate limiting steps for both C-C and C-O formation essentially involve the same propoxide and propanal surface intermediates, which is fully consistent with experimental findings. We predict similar activation barriers for both C-C and C-O formation with C-O formation being more favorable by 3 kJ/mol, while experiments report that C-C formation is slightly more favorable than C-O formation. In addition, we show that esterification and condensation preferentially proceed over coordinatively-saturated Cu 111 terrace sites. The alkoxide as well as other critical intermediates bind too strongly to the edge and corner sites and as such are blocked and much less active than the terrace sites. The active alkoxide intermediates bind much more weakly to the terrace sites and thus catalyze condensation and esterification. This is is fully consistent with the experimental results which show that larger particles are more active than the smaller Cu particles.

References

[1]Chheda, J.N., and Dumesic, J. A. Catal. Today (2007), 123, 59.

[2]Hamilton, C. A., Jackson, S.D., and Kelly, G.J. Appl. Catal. A (2004), 263, 63.

[3]West, R.M., Liu, Z.Y., Peter, M., Gartner, C.A., and Dumesic, J.A. J. Mol. Catal. A: Chem. (2008), 296, 18.

[4]Sad, M.E., Neurock, M., and Iglesia, E. J.Am.Chem.Soc. (2011), 133, 20384.

500107 Advancing the Selective Oxidation of Ethylene Glycol Via Combining Novel Catalyst Design and Density Functional Theory (DFT) Calculations

Abstract:

Ethylene glycol (EG) is the simplest diol that can be produced from hydration of ethylene oxide or hydrogenolysis of biomass (cellulose) [1]. In recent work, significant research work is focused on EG electroxidation for fuel cells [2]. This status is attributed to the fact that EG has higher boiling point (198 ^{o}C) making it a safer liquid fuel and has higher energy density compared to ethanol and methanol. However, in order to harvest the total amount of energy during this process, EG is fully converted into CO_2 that aggravates global warming. In this context, conversion of EG into value-added products chemically could be more attractive. A number of papers are available on EG oxidation in aqueous alkali medium using a non-hazardous oxidizing agent – oxygen gas [3-6]. However, the data reported on this reaction system are mainly on EG conversion and little is known on the reaction mechanism or the viability of applying different catalysts. Moreover, glycolic acid is usually the main product, while formic acid and CO_2 are believed as the side products of over oxidation rather than desirable oxalic acid (dicarboxylic acid) [6]. At the same time, current oxalic acid production suffers from the problems of using corrosive and stoichiometric reagents (nitric acid and sulfuric acid), with low yield and selectivity [7,8]. Suitable catalysts for a single-step conversion of EG to oxalic acid are highly desirable, that would be important to both industrial application and fundamental understanding of alcoholic group oxidation in polyols.

In this study, a series of monometallic catalysts (including Ag, Pd, Ru, Rd, Fe, Pt) supported on CeO_2 were synthesized by using a novel solvothermal method. The oxidation tests were conducted at a mild condition (70 ^{o}C, ambient pressure O_2) in aqueous alkali solution. The catalysts were characterized by H_2 chemisorption, inductively coupled plasma (ICP), transmission electron microscopy (TEM) for the metal content, dispersion and morphology information. Volcano plots of normalized TOF values vs. density functional theory (DFT) calculated Gibbs Free energy was established for EG oxidation. Furthermore, a bimetallic Pt-Fe catalyst showed a significant rate enhancement and selectivity in one-step conversion of EG-to-oxalic acid. The reaction mechanism for EG catalytic oxidation by O_2 on (100) and (111) surfaces was studied using DFT calculations which rationalizes the experimental observations that the bimetallic Pt-Fe catalyst has a significantly higher activity (TOF) compared to the monometallic catalysts (Pt and Fe). Fundamental concepts in the computational chemistry and its relation to experimentally observed performance of bimetallic catalysts will be presented in this work.

References

1. Yue, H., Zhao, Y., Ma, X., & Gong, J. (2012). Ethylene glycol: properties, synthesis, and applications. *Chemical Society Reviews*, *41*(11), 4218-4244.

2. Serov, A., & Kwak, C. (2009). Review of non-platinum anode catalysts for DMFC and PEMFC application. *Applied Catalysis B: Environmental*, *90*(3), 313-320.

3. Sakharov, A. M., Sakharov, P. A., & Zaikov, G. E. (2012). Catalytic Oxidation of Ethylene Glycol by Dioxygen in Alkaline Medium. The New Example of One-Stage Oxidative Cleavage of C–C Bond. *Molecular Crystals and Liquid Crystals*, *555*(1), 168-176.

4. Berndt, H., Pitsch, I., Evert, S., Struve, K., Pohl, M. M., Radnik, J., & Martin, A. (2003). Oxygen adsorption on Au/Al 2 O 3 catalysts and relation to the catalytic oxidation of ethylene glycol to glycolic acid. *Applied Catalysis A: General*, *244*(1), 169-179.

5. Case, J. M. (2009). *Gold catalysts prepared by ion exchange for use in ethylene glycol oxidation: An exploratory study* (Doctoral dissertation, University of Cape Town).

6. Kopeykina, L. Y., Vodyankina, O. V., Kurina, L. N., & Golovko, A. K. Selective ethylene glycol oxidation on the oxide catalysts. In *Science and Technology, 2001. KORUS'01. Proceedings. The Fifth Russian-Korean International Symposium on* (Vol. 2, pp. 177-179). IEEE.

7. Isshiki, T., Suzuki, T., Yashima, Y., & Yonemitsu, E. (1972). *U.S. Patent No. 3,678,107.* Washington, DC: U.S. Patent and Trademark Office.

8. Knepper, W., & Rohl, H. (1975). *U.S. Patent No. 3,864,393.* Washington, DC: U.S. Patent and Trademark Office.

488083 Quantum Chemical Characterization of Catalytic Ester Decarbonylation: Olefins from Biomass

Abstract:

Quantum Chemical Characterization of Catalytic Ester Decarbonylation: Olefins from Biomass

Synthetic transformations of carboxylic acids into olefins are of paramount importance since bio-mass derived carboxylic acids could then be a renewable source for polyolefin syntheses. An important goal, then, is to develop catalytic transformations that sustainably transform these structurally diverse acid educts into olefin products with high selectivity. Production of olefins via transition-metal-catalyzed decarbonylation of bio-derived carboxylic acids is one approach that has been explored and demonstrated to be possible with low toxicity, and tolerant of diversity.[1-2]. As a subclass, linear a-olefins (LAOs) are important commodity chemicals that find utility in a wide range of industrial applications. LAOs are usually generated from petroleum, which is a nonrenewable and depleting resource. A green and sustainable approach would be to produce LAOs by decarbonylative dehydration of fatty acids, which are inexpensive and renewable.

Direct routes for the conversion of carboxylic acids to olefins typically proceed with high activation energies. However, carboxylic acid derivatives such as esters are more reactive and can serve as key precursors in carboxylic acid transformations, and in particular transition-metal-based catalysts have been shown capable of transforming them to high-value olefins. However, controlling selectivity for terminal alkene formation is a major challenge. A highly selective, palladium-catalyzed decarbonylation of *p*-nitrophenyl esters of fatty acids to their corresponding LAOs has been very recently developed.[3] A cooperative ligand system including phosphines and N-heterocyclic carbenes is reported to exhibit enhanced selectivity to terminal olefins. Herein, we investigate the decarbonylation mechanism of *p*-nitrophenylbutyrate to rationalize the switching reactivity and selectivity trends based on either individual or cooperative roles of phosphine and heterocyclic carbene ligands. The primary goal of the project is to elucidate the origin of terminal alkene **A** selectivity vs the thermodynamically more stable internal alkene **B** which can be generated as an undesirable alternative product (Scheme 1).

Scheme 1. Pd-catalyzed decarbonylation of *p*-nitrophenylester to a-olefins. XantPhos and IPr ligands comprise the dual ligand system.

To gain mechanistic insight at the atomic level of detail, we performed density functional theory (DFT) modeling. We focused on the Pd-catalyzed decarbonylation of *p*-nitrophenylbutyrate in order to fully characterize reaction coordinates for the formation of terminal **A** and internal **B** alkene products and to explain the roles of the ligands in promoting a-selectivity. To assess selectivity, we have thoroughly compared the competing reaction mechanisms using three ligand systems: IPr, XantPhos, and a mixture of IPr and XantPhos. Calculations suggest that the bulkiness of IPr combined with the bidentate nature of XantPhos prompt a rapid decoordination of the terminal alkene, thus precluding the alkene isomerization pathway.

References

1. Rodriguez, N.; Goossen, L. J., Decarboxylative coupling reactions: a modern strategy for C-C-bond formation. *Chemical Society Reviews* 2011, *40* (10), 5030-5048.
2. Le Nôtre, J.; Scott, E. L.; Franssen, M. C. R.; Sanders, J. P. M., Selective preparation of terminal alkenes from aliphatic carboxylic acids by a palladium-catalyzed decarbonylation–elimination reaction. *Tetrahedron Letters* 2010, *51* (29), 3712-3715.
3. Alex John, Levi T. Hogan, Marc A. Hillmyer and William B. Tolman, Olefins from biomass feedstocks: catalytic ester decarbonylation and tandem Heck-type coupling, Chem. Commun., 2015, 51, 2731.

Elucidating the Effect of External Electric Fields and Surface Dopants

Abstract:

Fe catalysts are of great interest as they have been shown to be highly selective for the catalytic upgrading, via hydrodeoxygenation (HDO), of bio-oil compounds. Unfortunately, these catalysts alone are unacceptably prone to oxidative deactivation. One method of improving the oxidation resistance of Fe catalysts has been to dope the Fe with a second metal (e.g. Pd, Pt, etc.). However, the addition of the second metal increases the catalyst cost and complexity, thereby making it harder to experimentally optimize such systems to both increase activity and prevent oxidative deactivation. Previous work has shown that externally applied electric fields can dramatically improve the coke resistance of a Ni catalyst used in methane steam reforming (MSR).[6-9] These results suggest that external electric fields are another adjustable parameter that can be used to increase activity, tune selectivity, or prevent catalyst deactivation. Due to the highly polar nature of oxygen-containing compounds, it is expected that electric fields will have a large effect on HDO and could potentially be used in conjunction with doping to create a highly stable and selective Fe-based HDO catalyst. To this end, we must develop multi-scale models that can guide the selection of dopants in a rational manner—with both surface coverage and electric fields explicitly accounted for.

Here, we develop density functional theory (DFT) parameterized lattice gas models of oxygen on both undoped and doped Fe(100) surfaces to show the direct effect of an applied electric field on oxygen adsorption. Typical DFT models consider a uniform coverage distribution of adsorbates on the surface, when in reality, this is not the case.[2-4] Our cluster expansions of oxygen on Fe(100) surfaces both with and without an externally applied electric field accounts for the more realistic non-uniform adsorbate distributions and characterizes the lateral interactions between oxygen adatoms.[5] The apparent sensitivity of the lateral interactions on an applied external electric field reveals their influence on the oxidative deactivation of Fe catalysts.

Our preliminary results for O/Fe(100) indicate that an applied electric field has a significant effect on the attractive/repulsive interactions between adsorbed oxygen atoms. As seen in Figure 1, the clusters that are most important to the predictiveness of the model also change after the application of an electric field. These changes suggests that electric fields will have major effects on the Fe oxidation process as the electric field has a significant effect on the dynamic surface behavior of oxygen. Using such a lattice gas model for oxygen on both undoped and doped Fe(100), we will characterize the synergy between surface dopants and external electric fields in order to minimize the oxidation potential of the Fe surface. Our preliminary results suggest that one can expect significant deviations in catalytic activity after the application of external electric fields, but also, they suggest that mean-field models will not be sufficiently predictive for the work needed moving forward. Overall, the work presented here provides the first ingredients necessary for the selection of potential dopants for optimizing Fe-based HDO catalysts in the presence of external electric fields.

Figure 1: Effect of an electric field on oxygen cluster lateral interactions on the Fe(100) surface.

References:

1. Wang, H., Male, J. & Wang, Y. Recent Advances in Hydrotreating of Pyrolysis Bio-Oil and Its Oxygen-Containing Model Compounds. *ACS Catal.* **3,** 1047–1070 (2013).

2. Govender, A., Curulla Ferré, D. & Niemantsverdriet, J. W. The Surface Chemistry of Water on Fe(100): A Density Functional Theory Study. *ChemPhysChem* **13,** 1583–1590 (2012).

3. Blonski, P., Kiejna, A. & Hafner, J. Theoretical study of oxygen adsorption at the Fe (110) and (100) surfaces. *Surface Science* **590,** 88–100 (2005).

4. Freitas, R. R. Q., Rivelino, R., de Brito Mota, F. & de Castilho, C. M. C. Dissociative Adsorption and Aggregation of Water on the Fe(100) Surface: A DFT Study. *J. Phys. Chem. C* **116,** 20306–20314 (2012).

5. Frey, K., Schmidt, D. J. & Wolverton, C. Implications of coverage-dependent O adsorption for catalytic NO oxidation on the late transition metals. *Catalysis Science & ...* (2014). doi:10.1039/C4CY00763H

6. Che, F., Gray, J. T., Ha, S. & McEwen, J.-S. Improving Ni Catalysts Using Electric Fields: A DFT and Experimental Study of the Methane Steam Reforming Reaction. *ACS Catal.* (2016). doi:10.1021/acscatal.6b02318

7. Che, F., Ha, S., McEwen, J. S., Hensley, A. J. & Zhang, R. Elucidating the Role of the Electric Field at the Ni/YSZ Electrode: A DFT Study. *J. Phys. Chem. C* (2016). doi:10.1021/acs.jpcc.6b01292

8. Che, F., Ha, S. & McEwen, J. S. Hydrogen Oxidation and Water Dissociation over an Oxygen-Enriched Ni/YSZ Electrode in the Presence of an Electric Field: A First Principles-Based Microkinetic .... *Industrial & Engineering Chemistry ...* (2017). doi:10.1021/acs.iecr.6b04028

9. Che, F., Gray, J. T., Ha, S. & McEwen, J. S. Catalytic water dehydrogenation and formation on nickel: Dual path mechanism in high electric fields. *Journal of Catalysis* (2015). doi:10.1016/j.jcat.2015.09.010

Methane Conversion to Ethylene and Aromatics on PtSn Catalysts

Duygu Gerceker, Ali Hussain Motagamwala, Keishla R. Rivera-Dones, James B. Miller, George W. Huber, Manos Mavrikakis, James A. Dumesic

University of Wisconsin-Madison

2017 AIChE Annual Meeting, Minneapolis
11/02/17

Motivation

→ Methane is the most abundant component of natural gas, an important alternative to depleting petroleum for liquid fuel and chemicals production[1,2,3]

→ Non-oxidative conversion route : Methane → Ethane, Ethylene, Benzene, Naphthalene using transition metal catalysts supported on H-ZSM-5[2,4,5]

→ Thermodynamics dictates (T > 700°C) for feasible operation

↓

Favors coke formation + severe deactivation of the catalyst

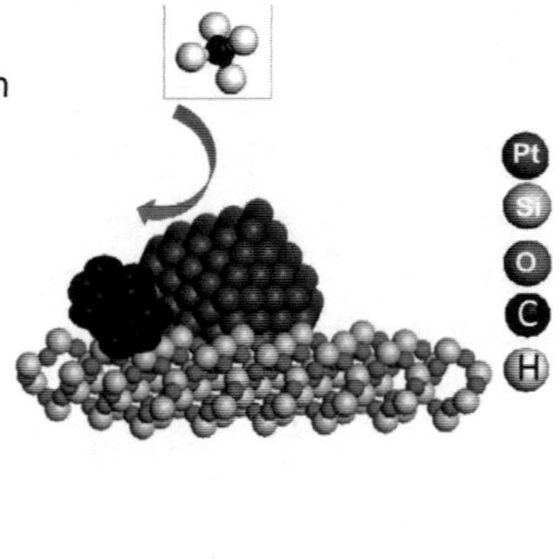

[1]Ma, S.; Guo, X.; Zhao, L.; Scott, S.; Bao, X. *J. Energy Chem.* **2013**, *22* (1), 1–20.
[2]Shu, Y.; Ichikawa, M. *Catal. Today* **2001**, 71, 55– 67
[3]Karakaya, C.; Morejudo, S. H.; Zhu, H.; Kee, R.J. *Ind. Eng. Chem. Res.* **2016**, *55*, 9895-9906
[4]Wang, L.; Ohnishi, R.; Ichikawa, M. *J. Catal.* **2000**, 190, 276– 283
[5]Weckhuysen, B. M.; Wang, D.; Rosynek, M. P.; Lunsford, J. H. *J. Catal.* **1998**, 175, 338– 346

2

Methodology

Target: Obtain an active, selective and stable Pt-based catalyst for non-oxidative methane conversion to ethylene and aromatics

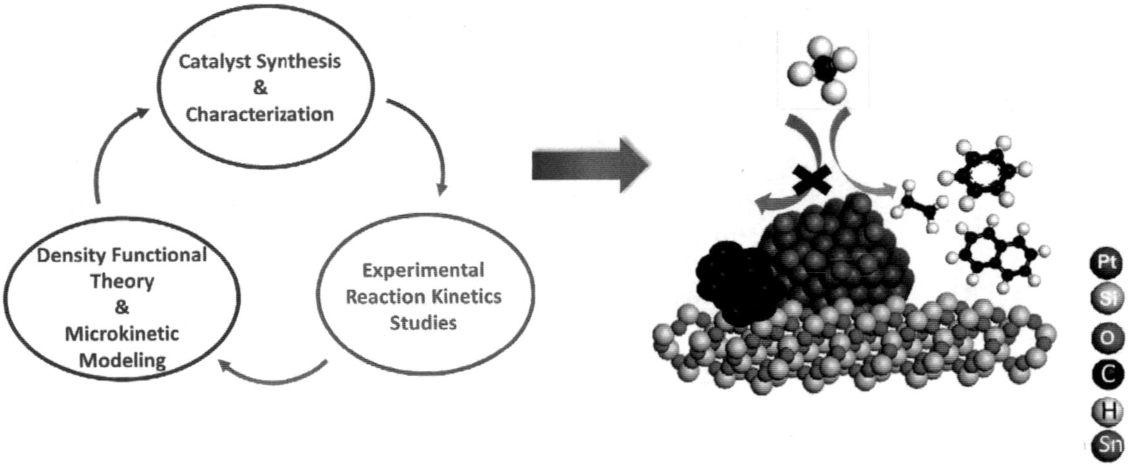

3

Catalysts & Experimental Conditions

→ Pt and PtSn catalysts supported on SiO_2 and H-ZSM-5 with SiO_2:Al_2O_3 ratio of 23, 50, 80 and 280 were prepared by impregnation technique.

Labeled as Pt-Sn/Z-XX

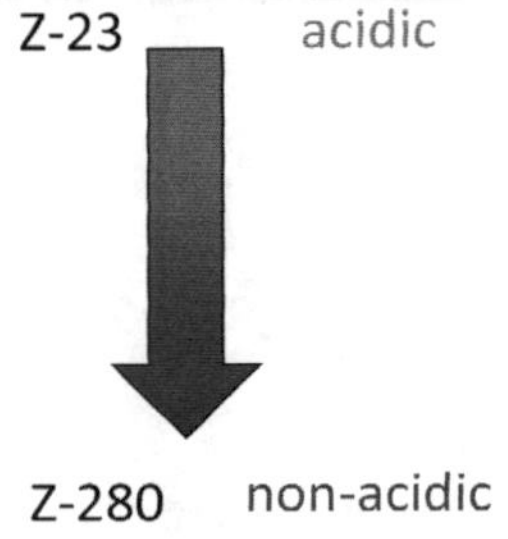

→ Fresh and spent catalysts were characterized by
N_2 Physisorption
CO Chemisorption
NH_3-TPD (Temperature Programmed Desorption)
TGA (Thermogravimetric Analysis)
STEM (Scanning Transmission Electron Microscopy)

→ Reaction conditions : 700°C, 1 atm
42 cm^3(STP)/min CH_4 + (5% H_2 for co-feeding experiments)
Reactor effluent analyzed by online Gas Chromatography

4

Challenges

→ How to **decrease coke deposition** on SiO_2-supported catalysts?

→ How to **increase activity** and **selectivity** of the catalysts?

→What is the **effect of particle size** on the activity?

→How does **acidity of the zeolite influence** the activity?

→What is the **reaction mechanism** on the zeolite supported catalysts?

Coking problem on SiO_2 catalysts

→Pt/SiO_2 and $PtSn/SiO_2$ catalysts pretreated at 500°C yielded poor carbon balance

→To decrease the coke formation, catalysts were pretreated at higher temperature (850°C).

Carbon deposition by TGA

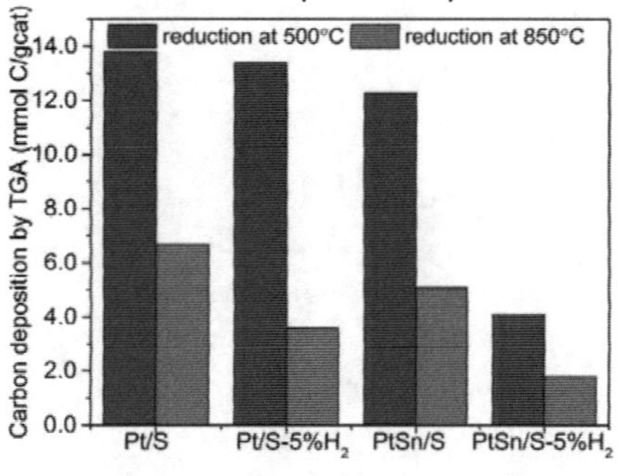

Carbon loss during the reaction

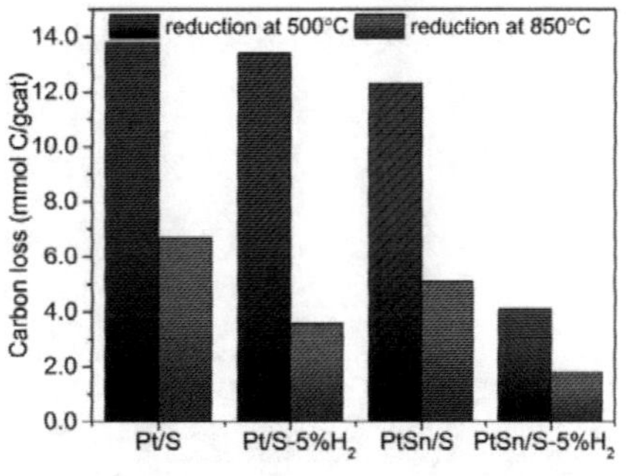

Pretreatment at 850°C decreased surface silanol groups and provided catalysts with less coke deposition: improved carbon balances

Gerceker, D., A. H. Motagamwala et al. (2017). "Methane Conversion to Ethylene and Aromatics on PtSn Catalysts." ACS Catalysis **7**(3): 2088-2100.

How to improve ethylene formation?

Effect of Sn addition

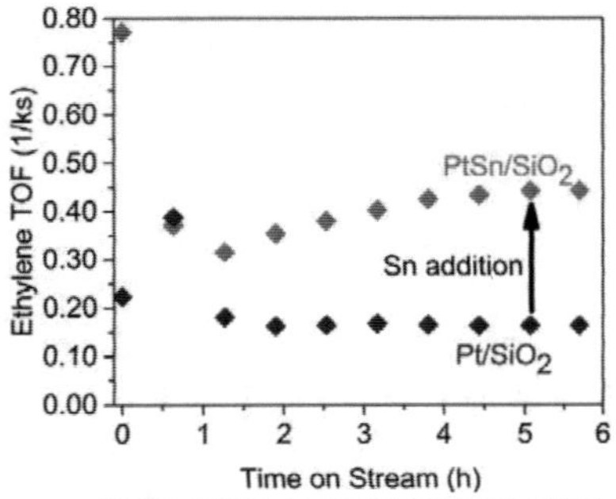

Effect of H_2 co-feeding*

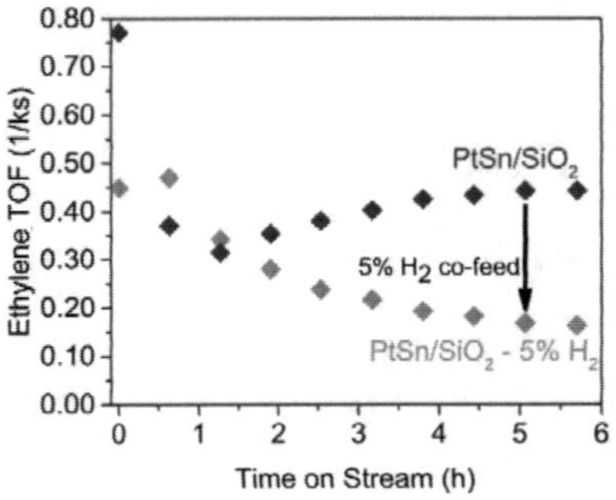

Changing support from SiO_2 to Z-280 (SiO_2:Al_2O_3 =280)

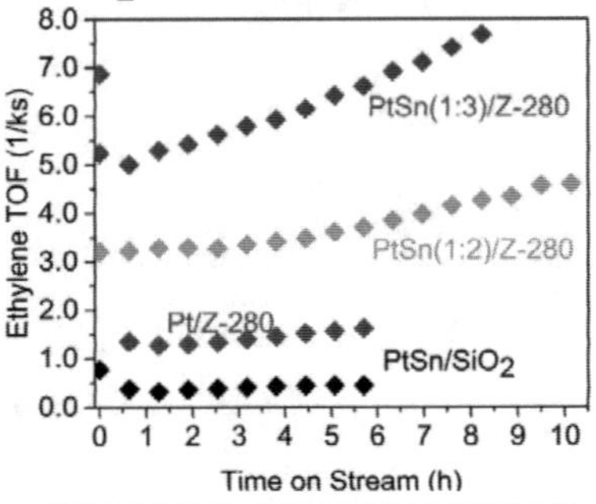

Adding Sn to the Pt/SiO_2 enhanced ethylene TOF by a factor of 5 times

H_2 co-feeding decreased ethylene TOF significantly

H-ZSM-5 support enhanced ethylene TOF on PtSn catalysts ~20 times

Conditions : 42 cm^3(STP)/min CH_4 (*5% H_2 co-feeding) , T = 700°C, **2 g** (*1 g*) for **SiO_2** (*H-ZSM-5*) supported catalysts 7

How to improve benzene formation?

Effect of Sn addition

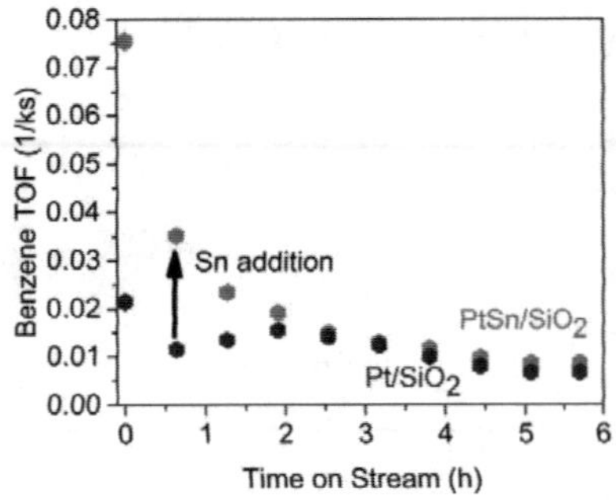

Effect of H_2 co-feeding*

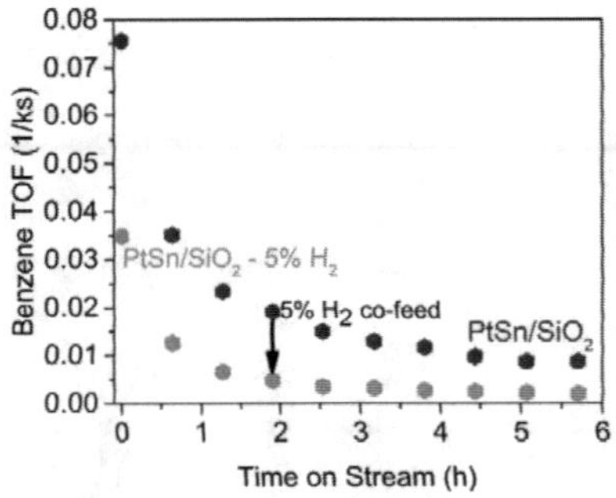

Changing support from SiO_2 to Z-280 (SiO_2:Al_2O_3 =280)

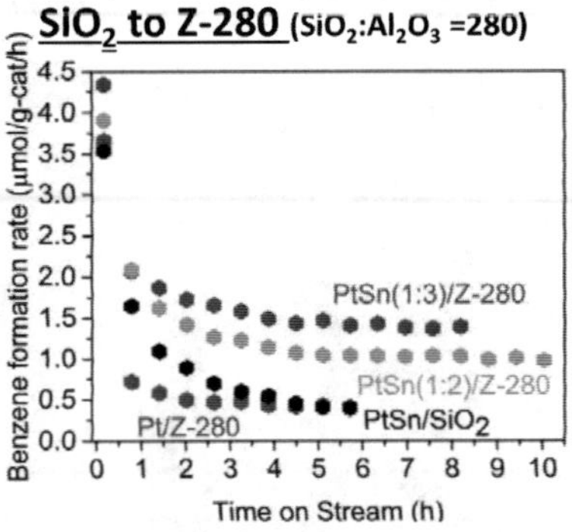

Sn addition enhanced initial activity, H_2 co-feeding decreased activity, zeolite support promoted benzene formation

Conditions : 42 cm^3(STP)/min CH_4 (*5% H_2 co-feeding) , T = 700°C, **2 g** (*1 g*) for **SiO_2** (*H-ZSM-5*) supported catalysts 8

Catalysts are highly selective for ethylene

PtSn/SiO₂

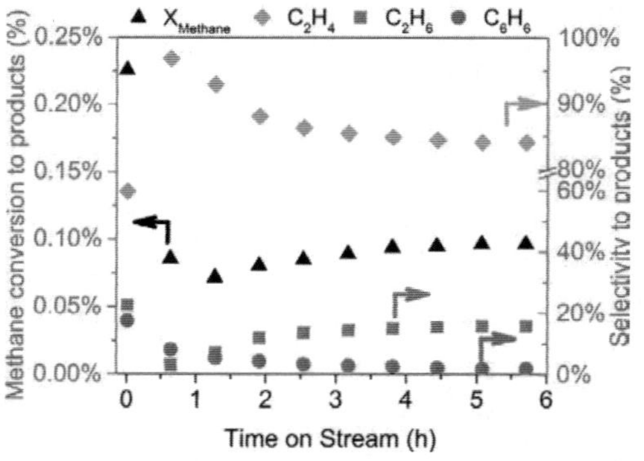

PtSn(1:3)/Z-280

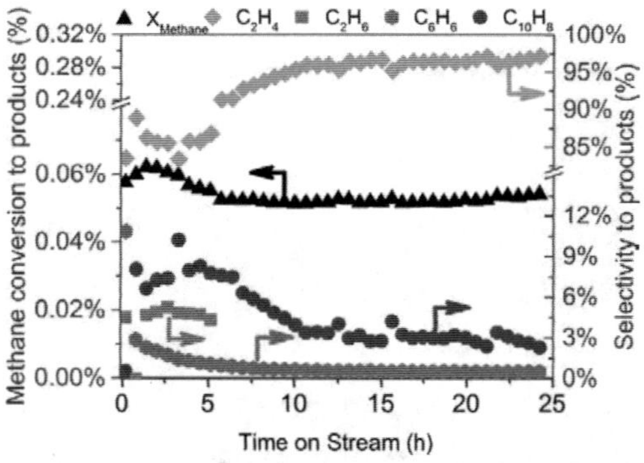

Both SiO_2 and H-ZSM-5 supported catalysts showed ~90% selectivity towards ethylene
Zeolite support promoted higher selectivity for aromatics

Conditions : 42 cm³(STP)/min CH_4, T = 700°C, **2 g** (*1 g*) for **SiO_2** (*H-ZSM-5*) supported catalysts

9

Why H-ZSM-5-supported catalysts are more active?

STEM studies to investigate particle size

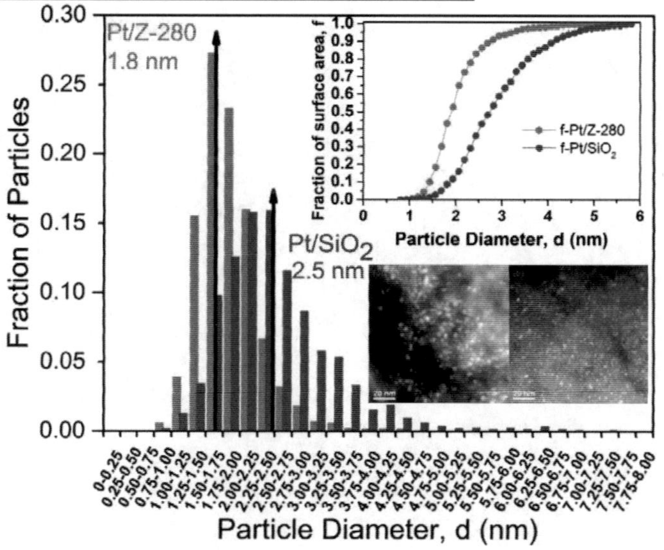

Pt/H-ZSM-5 catalyst has smaller particle size
In the particle diameter range (1-2 nm) it exposes 8 times higher values of surface area (inset).

Abundance of small PtSn nanoparticles might be the source of superior activity of zeolite-supported catalysts.

10

Why H-ZSM-5-supported catalysts are more active?

A microkinetic model is built using DFT data to predict reaction rates and surface coverages under reaction conditions

Surfaces studied with the model

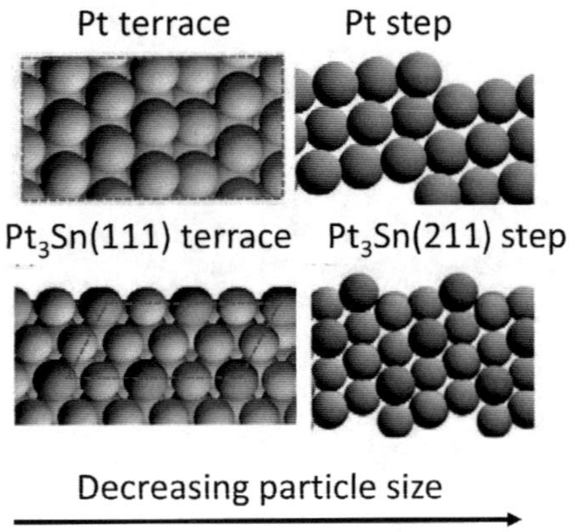

Reaction network

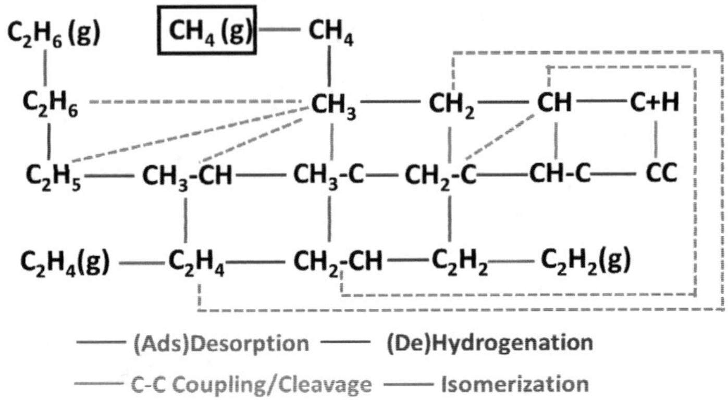

11

Recap of experimental findings

→ Adding Sn to Pt **promoted** ethylene production

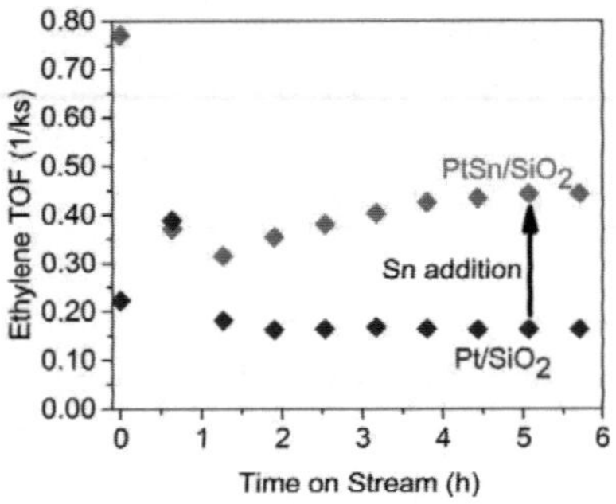

→ H_2 co-feeding **decreased** the catalyst activity

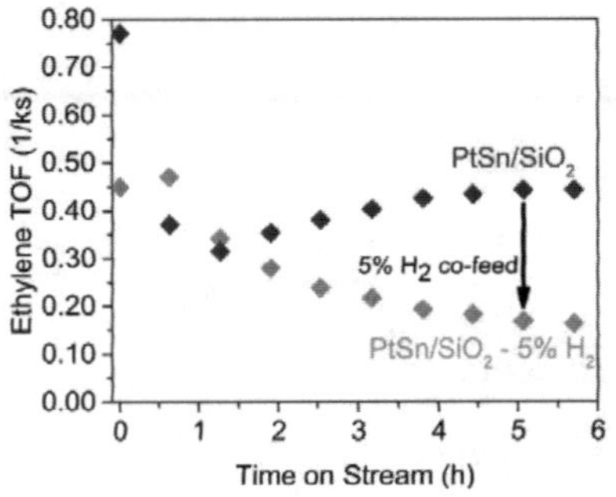

→ More active zeolites catalysts have smaller particle size

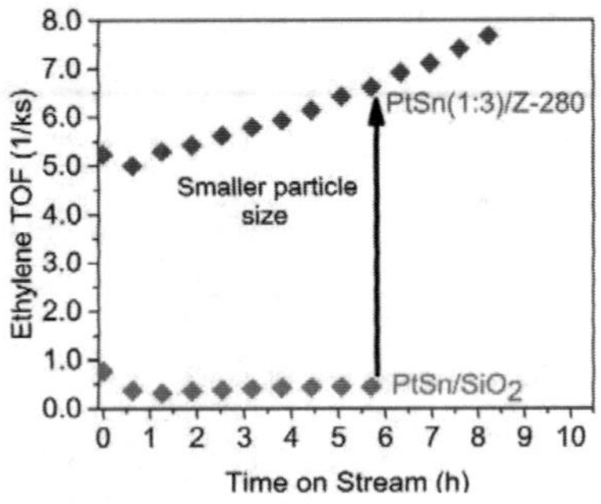

12

Surface coverages influence activity

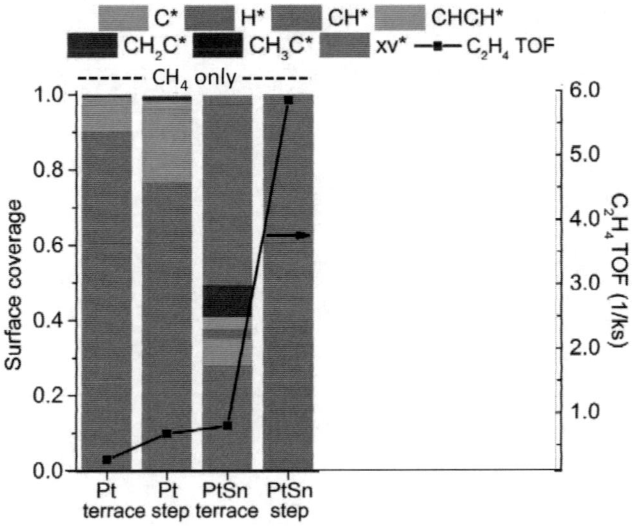

Under only CH_4 flow, Pt sites are covered with C* and CH*
Pt_3Sn have more than 50% vacancy and H*

13

Surface coverages influence activity

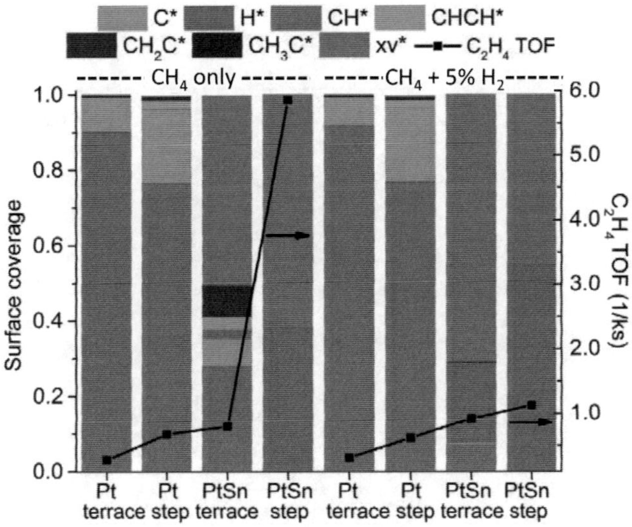

Under only CH_4 flow, Pt sites are covered with C* and CH*
Pt_3Sn have more than 50% vacancy and H*

5% H_2 co-feeding increased H* coverage on Pt_3Sn sites, but on the step site decreased the C_2H_4 TOF ~4 times

Steps of Pt_3Sn has the highest C_2H_4 TOF value, implying smaller particles are more active

14

Effect of support acidity

PtSn(1:2) catalysts on H-ZSM-5 support with varying SiO_2: Al_2O_3 ratio

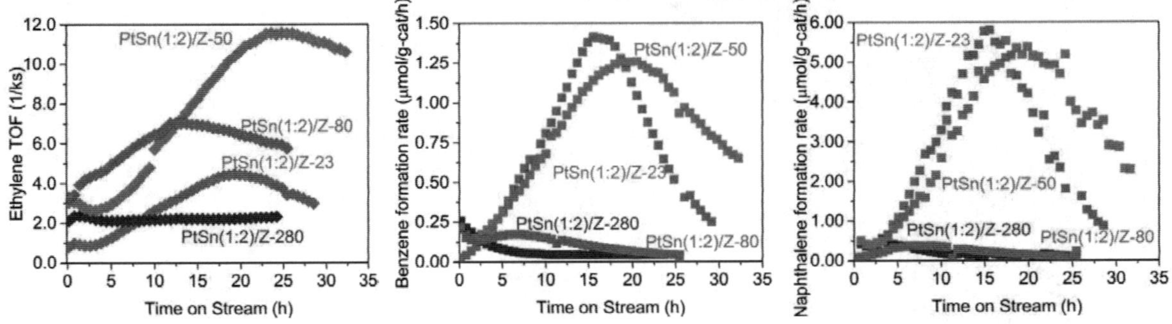

Initially increasing reactivity profile : formation of a reactive hydrocarbon pool

Ethylene is formed on the metal sites, oligomerized to aromatics on the acid sites

Aromatics reach maximum earlier than ethylene: faster oligomerization than desorption

Conditions : 10.5 cm^3(STP)/min CH_4, T = 700°C , 0.25 g catalyst

Effect of support acidity

PtSn(1:2)/Z-280

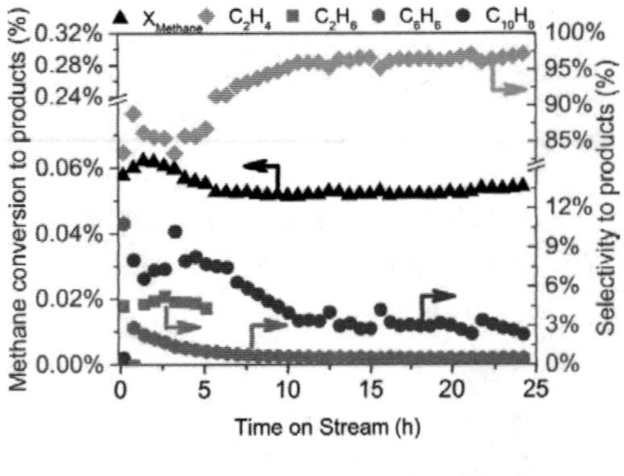

PtSn(1:2)/Z-50

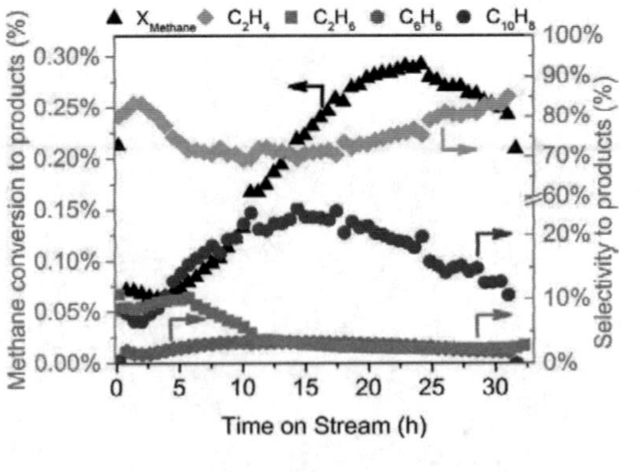

More acidic zeolite support **Z-50** significantly increased methane conversion and naphthalene selectivity at the expense of ethylene production

Conditions : 10.5 cm^3(STP)/min CH_4, T = 700°C , 0.25 g catalyst

Pt-Sn catalysts are comparable with Mo/H-ZSM-5

Activity comparison of PtSn and Mo systems

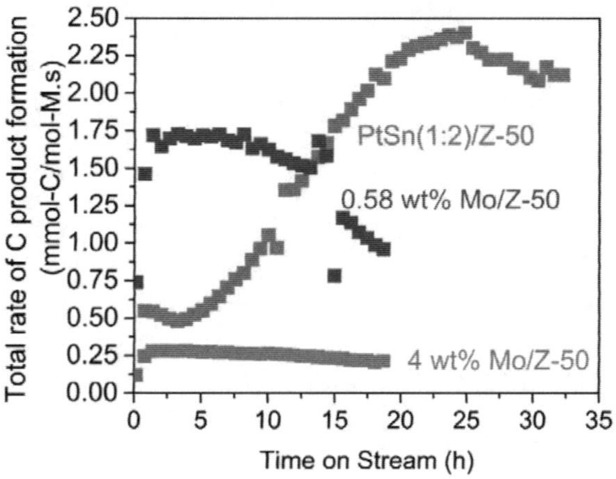

To compare PtSn/Z-50 with the well-studied Mo/H-ZSM-5 system, two Mo/Z-50 catalysts with high and low Mo loadings were prepared by impregnation technique

Total carbon production formation rate per moles of metal sites of PtSn/Z-50 is comparable with Mo/Z-50 catalysts

Summary

For non-oxidative methane conversion:

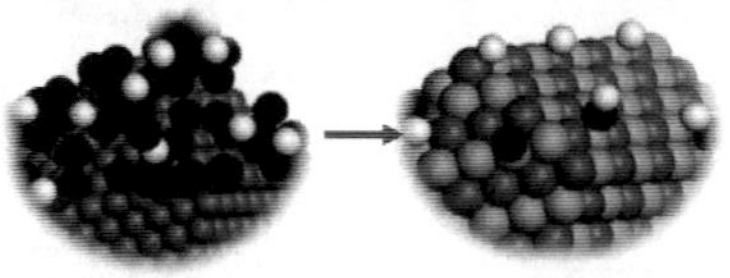

Pt PtSn

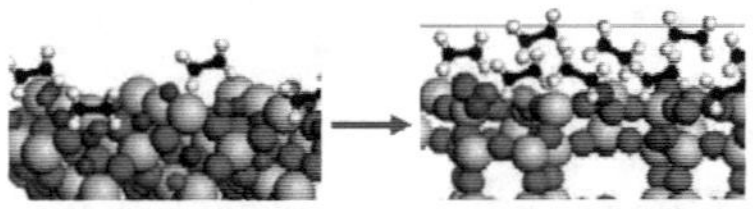

$PtSn/SiO_2$ PtSn/H-ZSM-5

Low acidity H-ZSM-5 → High acidity H-ZSM-5

Obtained a PtSn/H-ZSM-5 catalys

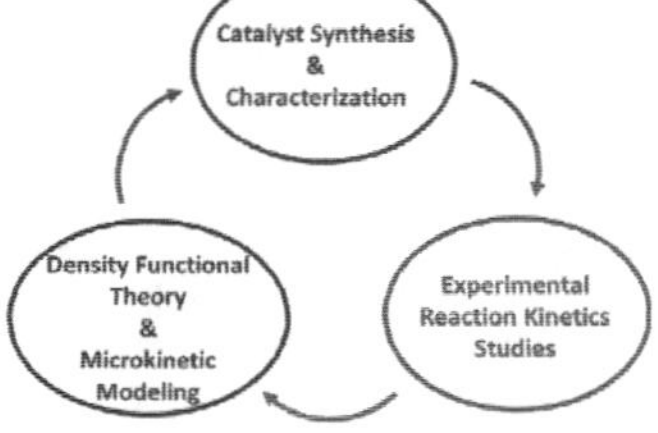

- ✓ **Active**
- ✓ **Stable**
- ✓ **Selective**
- ✓ **Bifunctional**

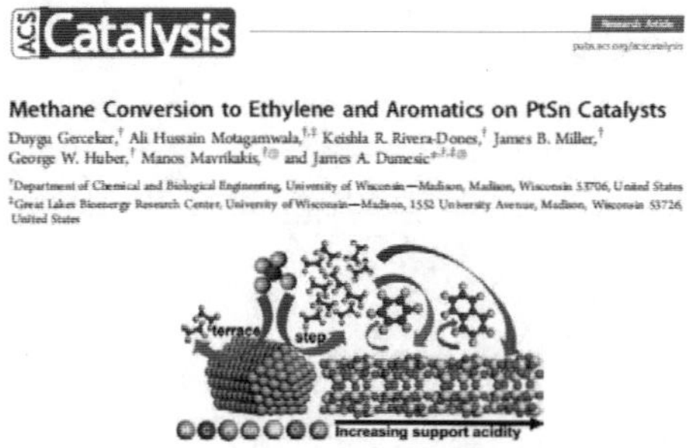
ACS Catalysis

Methane Conversion to Ethylene and Aromatics on PtSn Catalysts

Duygu Gerceker, Ali Hussain Motagamwala, Keishla R. Rivera-Dones, James B. Miller, George W. Huber, Manos Mavrikakis, and James A. Dumesic

Gerceker et al., ACS Catalysis **7**(3): 2088-2100.

Acknowledgements

Funding : Air Force Office of Scientific Research

Computational Resources : EMSL at PNNL
NERSC
CHTC at UW-Madison

19

500278 Geometric and Electronic Effects of Zn Promotion on Pt for Ethane Dehydrogenation

Abstract:

Geometric and Electronic Effects of Zn Promotion on Pt for Ethane Dehydrogenation

Viktor J. Cybulskis[1], Brandon C. Bukowski[1], Han-Ting Tseng[1], James R. Gallagher[2], Zhenwei Wu[1], Evan C. Wegener[1], A. Jeremy Kropf[2], Bruce Ravel[3], Fabio H. Ribeiro[1], Jeffrey Greeley[1], Jeffrey T. Miller[1]

[1]Davidson School of Chemical Engineering, Purdue University, West Lafayette, IN 47907

[2]Chemical Sciences and Engineering Division, Argonne National Laboratory, Argonne, IL 60439

[3]Materials Measurement Laboratory, National Institute of Standards and Technology, Gaithersburg, MD 20899

The study of hydrocarbon C-H bond activation has become more important because of recent shale gas boom in the United States. Supported Pt catalysts have been shown to selectively promote C-H bond activation for light alkanes when alloyed with other inactive metals [1][2]. We tried to correlate the structure of a $PtZn/SiO_2$catalyst with its catalytic performance for ethane dehydrogenation.

In the present work, 100% C_2H_4 selectivity and six-fold turnover rate (TOR) enhancement compared to Pt were achieved during ethane dehydrogenation (EDH) at 600 °C after the addition of Zn to Pt/SiO_2. The apparent activation energy (E_{app}) was determined to be 72±4 kJ mol^{-1} for Pt/SiO_2 and 99±5 kJ mol^{-1} for $PtZn/SiO_2$. High-angle annular dark-field scanning transmission electron microscopy (HAADF-STEM) showed that Pt and PtZn metal particle sizes were 3.3±1.9 nm and 2.0±0.6 nm, respectively. The Pt_1Zn_1 nanoparticle alloy with geometrically isolated Pt sites was identified by synchrotron X-ray diffraction (XRD) and X-ray absorption spectroscopy (XAS) over $PtZn/SiO_2$. XAS also revealed that the Pt edge energy increased by 0.9 eV for $PtZn/SiO_2$. Resonant inelastic X-ray scattering (RIXS) showed that the energy difference between the unoccupied and occupied Pt 5*d* valence orbitals increased from 4.0 eV for Pt/SiO_2 to 6.0 eV for $PtZn/SiO_2$, which agrees with the results from density functional theory (DFT) calculations. It is suggested that the geometrically isolated Pt sites by Zn are responsible for the enhanced C_2H_4 selectivity as the structure-sensitive hydrogenolysis is suppressed by breaking Pt ensembles. The change in the energy of Pt 5*d* electrons caused the change in the relative energy difference between Pt 5*d* orbitals and the ones from the adsorbate. We propose that this difference leads to the increased TOR and E_{app} for $PtZn/SiO_2$.

References:

[1] Galvita, Vladimir, et al. "Ethane dehydrogenation on Pt/Mg (Al) O and PtSn/Mg (Al) O catalysts." *Journal of Catalysis* 271.2 (2010): 209-219.

[2] Siddiqi, Georges, et al. "Catalyst performance of novel Pt/Mg (Ga)(Al) O catalysts for alkane dehydrogenation." *Journal of Catalysis* 274.2 (2010): 200-206.

496856 CO, C_2H_2, and C_3H_6 Oxidation on Pd/Ceria-Zirconia/Al_2O_3 Three-Way Catalysts

Abstract:

CO, C2H2, and C3H6 oxidation on Pd/Ceria-Zirconia/Al2O3 Three-way Catalysts

Wendy Lang[1], Michael P. Harold[1*], Yisun Cheng[2], Carolyn Hubbard[2], and Paul Laing[2]

[1]*Department of Chemical and Biomolecular Engineering, University of Houston, Houston, TX*

[2]*Research and Innovation Center, Ford Motor Company, Dearborn, MI*

*mharold@uh.edu

Three-way catalyst (TWC) light-off (LO) performance is critical in cost-effectively meeting emission standards. Improvements in catalyst LO should include catalyst modifications that minimize inhibition effects of exhaust species such as CO and hydrocarbons (HCs), lower precious group metal (PGM) loading, or use a lower cost PGM such as Pd. Alumina (Al_2O_3) and zirconia (ZrO_2) supports provide high surface area and stability. Furthermore, ceria (CeO_2) provides the key benefit of oxygen storage, compensating for deviations from stoichiometric TWC operating conditions by supplying oxygen for CO and HC oxidation under rich transient conditions. Inclusion of ceria in PGM catalysts enhances CO and HC conversion and minimizes self-inhibition; this improved activity is mainly attributed to ceria containing more efficient oxygen activation sites at the PGM-support interface relative to other support materials such as Al_2O_3 [1-4].

Monolith catalysts having washcoats with 1 wt% Pd, 3 g/in^3 loading, 400 cpsi and formulations Pd/Al_2O_3, Pd/CZO, and Pd/CZO/Al_2O_3 (17 wt% CZO, balance Al_2O_3) were synthesized and compared. Catalyst performance experiments were conducted to study the role of Pd and support compositions and loadings on inhibition during oxidation of CO, acetylene (C_2H_2), and propylene (C_3H_6) under near-stoichiometric conditions ($\lambda = 1.01$). Transient and steady-state bench scale reactor studies with simulated exhaust gas mixtures were conducted to extract oxidation kinetics, compare LO behavior, and develop a predictive model for understanding and optimizing catalyst performance.

Catalyst activities were compared via temperature-programmed oxidation experiments. Figure 1 compares CO LO curves using feed concentrations of 0.5% CO and 1% CO. The Pd/CZO catalyst was tested when "fresh" (initial tests after synthesis) and 10+ months later ("deactivated"). Using all three catalysts, LO temperatures increase with feed CO concentration, showing that CO oxidation is self-inhibiting. With increasing CZO content, LO temperatures decrease, demonstrating the promotional effects of using ceria.

Figure 2 shows C_2H_2 LO (feed concentrations of 375 ppm C_2H_2 and 750 ppm C_2H_2) and C_3H_6 LO (feed concentrations of 250 ppm C_3H_6 and 500 ppm C_3H_6) using the deactivated Pd/CZO catalyst. Like CO, C_2H_2 and C_3H_6 are self-inhibiting. For the same concentrations of carbon in the feed mixture (750 ppm C_1 or 1500 ppm C_1), C_2H_2 oxidation occurs at higher temperatures than C_3H_6 oxidation.

Exhaust hydrocarbons propylene and acetylene reveals some interesting differences during co-oxidation with CO. The individual LO features of 1% CO and 750 ppm C_2H_2 in Figures 1 and 2 are compared to their co-oxidation (1% CO + 750 ppm C_2H_2) in Figure 3. CO lights off at higher temperatures in the mixture, i.e. C_2H_2 inhibits CO oxidation. During co-oxidation, C_2H_2 oxidation is inhibited by the presence of CO below ~15% conversion, but is otherwise improved compared to oxidation of C_2H_2 alone. Whereas 1% CO lights off before 750 ppm of C_2H_2 during individual oxidation, the reverse is true during co-oxidation. This is in contrast to earlier findings of CO + C_3H_6 trends in which was found that CO inhibits C_3H_6; i.e. propylene conversion follows CO light off (see below).It is instructive to compare the behavior of the two hydrocarbons at the same C1 concentration. Though the same feed concentration of 1% CO + 1500 ppm C_1 (from HCs) is used, C_3H_6 and C_2H_2 co-oxidation with CO exhibits different trends in terms of LO temperatures and order of species oxidation. As shown in Figure 4, using the deactivated Pd/CZO catalyst, 1% CO lights off before 500 ppm C_3H_6 both during individual oxidation and mixture LO. During co-oxidation, 500 ppm C_3H_6 is 50% converted at ~ 246 °C, compared to ~ 266 °C for 750 ppm C_2H_2.

Differential kinetics studies were conducted to quantify reaction orders and activation energies. The reaction order with respect to CO was found to be approximately -1 using Pd/Al_2O_3 and Pd/CZO/Al_2O_3 and 0 to slightly negative using Pd/CZO, consistent with literature values [5-7]. The activation energies for CO oxidation were found to be 134-213 kJ/mol using Pd/Al_2O_3 and Pd/CZO/Al_2O_3, and 25-70 kJ/mol using Pd/CZO. Using all three catalysts, the reaction order with respect to C_3H_6 was found to be approximately -0.89 to -0.75 and the activation energy in the range of 72-105 kJ/mol, approximately half of that observed for CO oxidation and similar to literature values [5]. Using the Pd/CZO catalyst, the reaction order with respect to C_2H_2 was found to be approximately -0.52 to -0.96 and the activation energy of C_2H_2 oxidation approximately 73 kJ/mol.

Steady-state and transient CO, C_2H_2, and C_3H_6 oxidation experiments using fresh Pd/Al_2O_3, Pd/CZO, and Pd/CZO/Al_2O_3 monolith catalysts were conducted to compare LO behavior and kinetic parameters (i.e. reaction orders and activation energies). CO LO temperatures decreased with increasing proportion of ceria used in the TWC, and C_2H_2 and C_3H_6 LO studies using Pd/CZO demonstrated self- and mutual-inhibition during species oxidation. Kinetics studies demonstrated further the self-inhibiting oxidation behavior.

Ongoing work involves the use of a mechanistic-based kinetic model incorporated into a low-dimensional monolith model [8-9]. The global reactor models are used with CO, C_2H_2, and C_3H_6 oxidation on Pd with steps involving ceria. The 1+1 dimensional model formulation comprises coupled species balances with transverse-average equations having external and internal mass transfer coefficients to verify experimental LO behavior during individual and co-oxidation using catalysts and feed mixtures of different formulations. Finally, experiments involving C_2H_2 oxidation on Pd/Al_2O_3 and Pd/CZO/Al_2O_3 catalysts and other hydrocarbons including toluene are being conducted to investigate inhibition during LO, extract kinetic parameters for oxidation reactions, and provide a basis for validating modeling simulations.

Figure 1: CO conversion (%) versus monolith temperature using Pd/Al_2O_3, Pd/CZO/Al_2O_3, and Pd/CZO catalysts for feed concentrations of 0.5% and 1% CO.

Figure 2: C_2H_2, C_3H_6 conversion (%) versus monolith temperature using deactivated Pd/CZO for feed concentrations of 375 ppm C_2H_2, 750 ppm C_2H_2, 250 ppm C_3H_6, and 500 ppm C_3H_6.

Figure 3: CO, C_2H_2 conversion (%) versus monolith temperature using deactivated Pd/CZO for feed concentrations of 750 ppm C_2H_2 and/or 1% CO.

Figure 4: CO, C_3H_6 conversion (%) versus monolith temperature using deactivated Pd/CZO for feed concentrations of 500 ppm C_2H_2 and/or 1% CO.

References

[1] Fernandez-Garcia, M.; Martinez-Arias, A.; Iglesias-Juez, A.; Hungria, A.B.; Anderson, J.A.; Conesa, J.C.; and Soria, J. *App. Cat. B* 2001, **31**, 39-50.

[2] Martinez-Arias, A., Fernandez-Garcia, M.; Iglesias-Juez, A.; Hungria, A.B.; Anderson, J.A.; Conesa, J.C.; and Soria, J.. *App. Cat. B* 2001, **31**, 51-60.

[3] Sharma, S.; Hegde, M.S.; Das, R.N.; and Pandey, M. *App. Cat. A* 2008, **337**, 130-137.

[4] Harmsen, J.M.A.; Hoebink, H.B.J.; and Shouten, J.C. *Chem. Eng. Sci.* 2001, **56**, 2019-2035.

[5] Yao, Y-.F. Y. *J. Catal.*, 1984, **87**, 152-162.

[6] Cant, N.W.; Hicks, P.C.; and Lennon, B.S. *J. Catal.* 1978, **54**, 372-383.

[7] Rainer, D.R.; Koranne, M.; Vesecky, M.; and Goodman, D.W. *J. Phys. Chem. B* 1997, **101**, 10769-10774.

[8] Joshi, S.; Harold, M.P.; Balakotaiah, V. *AIChE J.*, 2009, **55,** 1771-1783.

[9] Raj, R.; Harold, M.P.; Balakotaiah, V. *Chem. Eng. J.*, 2015, **281**, 322-333.

487884 Sintering and Deactivation Mechanism of Platinum/Palladium Two-Phase Catalysts

Abstract:

Despite efforts to create novel catalysts with inexpensive materials, precious metal catalysts remain widespread in many applications due to their high activity. Maintaining the long-term activity of these catalysts by preventing degradation is therefore of substantial economic importance. One model system for studying deactivation is a well-controlled platinum/palladium catalyst for methane oxidation, in which platinum represents a necessary catalytic component but its sintering causes deactivation of the palladium catalyst toward fuel oxidation.[1-2] We seek to understand the mechanism by which this deactivation occurs, toward the goal of designing new catalysts with improved resistance to sintering.

In this presentation, we use density functional theory calculations to elucidate the mechanism by which an oxidized palladium catalyst (PdO) on an Al_2O support deactivates toward methane combustion in the presence of platinum. We show that gas-phase PtO_2 acts as the primary carrier of Pt to PdO, on which it nucleates in surface oxygen vacancies generated by the methane combustion cycle. The migration is thermodynamically driven, and leads to the formation of large Pt-Pd bimetallic particles with uniform particle size and severely reduced methane oxidation activity. Experimental evidence allows differentiation of this O_2-assisted gas-phase migration from whole-particle surface migration, as catalyst deactivation is severely reduced in an argon atmosphere (which cannot form PtO_2, in the absence of O_2). This work demonstrates the power of integrating theory and experiments on well-defined materials to understand and elucidate interaction mechanisms in complex catalysts composed of multiple phases.

[1]C. F. Cullis, B. M. Willatt, *J Catal* **83**, 267 (1983)

[2]R. Strobel, J.-D. Grunwaldt, A. Camenzind, S. E. Pratsinis, A. Baiker, *Catal Lett* **104**, 9 (2005)

498817 Functional Descriptors, Active Intermediates, and the Influence of the Porous Environment for Epoxidations at Lewis Acidic Metal Atoms in Zeolite BEA

Abstract:

The epoxidation of olefins is a critical reaction to produce building blocks for the production of polyurethanes and pharmaceuticals. Group IV and V metal containing zeolites have been used for olefin epoxidation reactions for decades, yet the underlying properties of these catalysts that determine the turnover rates and the selectivities with which they utilize the H_2O_2 as a terminal oxidants have not yet been shown. Isolated early transition metal atoms stabilized in zeolite frameworks (e.g., BEA) exhibit large difference in epoxidation that depend on metal identity and which have been postulated to depend on the energy of the lowest unoccupied molecular orbital of the metal center among based on computational comparisons of Ti-BEA and Zr-BEA [1]. Here we show that cyclohexene epoxidation rates and selectivities vary by a factor of 10^5 and 10^2, respectively, across a series of group IV and V substituted into BEA [2,3].

Turnover rates for cyclohexene oxide ($C_6H_{10}O$) production exhibit two distinct sets of dependencies on the concentrations of cyclohexene (C_6H_{10}), hydrogen peroxide (H_2O_2), and $C_6H_{10}O$ on all M-BEA (M = Nb, Ta, Ti, Zr, and Hf) materials. When operating at high $[C_6H_{10}]$ to $[H_2O_2]$ ratios, epoxidation rates do not depend on $[C_6H_{10}]$ but increase in proportion to $[H_2O_2]$ and depend inversely on $[C_6H_{10}O]$. At low $[C_6H_{10}]$ to $[H_2O_2]$ ratios, epoxidation rates are proportional to $[C_6H_{10}]$ but do not depend on $[H_2O_2]$. These observations reflect a reaction mechanism that proceeds via irreversible activation of H_2O_2 over M-β to form active H_2O_2-derived intermediates (M-(O_2)), which then reacts with weakly coordinated C_6H_{10} on sites that are either predominantly covered by $C_6H_{10}O$ or M-O_2 at high and low ratios of $[C_6H_{10}]$ to $[H_2O_2]$, respectively. *In* situ UV-vis spectroscopy, attenuated total reflectance infrared spectroscopy, and X-ray photoelectron spectroscopy show that these group IV and V materials activate H_2O_2 to form pools of hydroperoxide, peroxide, and superoxide intermediates. Time-resolved UV-vis spectra and the isomeric distributions of stilbene epoxidation products demonstrate, however, that the active species for epoxidations on group IV and V transition metals are only M-OOH/-$(O_2)^{2-}$ and M-$(O_2)^-$ species, respectively.

Epoxidation rates and selectivities vary by a factor of 10^5 and 10^2, respectively, at a given set of conditions and on similarly M-(O_2) saturated surfaces of M-BEA catalysts. These differences reflect a linear correlation between the activation enthalpies ($\Delta H^{\ddagger}$) for each of the reaction pathways (i.e., epoxidation and H_2O_2 decomposition) and both the energy for ligand-to-metal charge transfer (LMCT; i.e., from -O_2^- or –OOH to the transition metal center) and also the functional Lewis acid strength of the metal centers (quantified by the heat of adsorption of base titrants on these sites). These differences in electronic properties lead to differences in $\Delta H^{\ddagger}$ for epoxidation of 50 kJ mol^{-1} and for H_2O_2 decomposition of 30 kJ mol^{-1} across the series of M-BEA materials with largely hydrophilic pores (produced by substituting M-atoms into 20% of the framework defects created by dealumination of BEA with Si/Al of 12). H_2O_2 decomposition (the undesirable reaction pathways) possesses a weaker dependence on Lewis acidity than epoxidation, which suggests that the design of catalysts with increased Lewis acidity will simultaneously increase the reactivity and selectivity of olefin. These results will be compared to similar trends obtained on an analogous series of M-BEA materials containing predominantly hydrophobic pores.

References:

1) Boronat, M.; Corma, A.; Renz, M.; Viruela, P. M.; *Chem. Eur. J.,* **2006**, *12,* 7067-7077.

2) Bregante, D.T.; Priyadarshini, P.; Flaherty, D.W.; *J. Catal.*, **2017**, *348*, 75-89.

3) Bregante, D.T.; Flaherty, D.W.; **2017**, *submitted.*

499262 A Comparative Study of ZSM-5 and BEA Zeolites for Low Temperature Passive Adsorption

Abstract:

Regulations on vehicle emissions are becoming more stringent worldwide due to increasing concerns of the impact of air pollution on environment and public health. Significant attention has been paid to develop methods to treat automobile exhausts for the regulated emissions such as hydrocarbons especially during engine cold starts. An effective solution is to employ suitable porous materials which can trap and retain hydrocarbons (HCs) temporarily until emission control catalysts are lit off (i.e., exhaust gas temperature reaches high enough for catalysts to become active) [1,2]. For this purpose, zeolites have been found to be a preferred type of materials due to their affinity to HCs and relative thermal stability [3]. The goal of this work was to understand key factors controlling hydrocarbon trapping and release performance of zeolites. Gained insights can, in turn, help to design HC traps with enhanced performance suitable for addressing future emission control challenges related to low temperature exhausts expected from more efficient engines. Zeolites were evaluated using propylene (C_3H_6) as a model HC compound in simulated exhaust conditions relevant to vehicle cold starts. The specific objective was to investigate in detail trends observed between C_3H_6 trapping-release performance, zeolite framework structure (ZSM-5 vs. BEA), acidity (high vs low), exchanged cation (e.g. H^+ vs. Ag^+ vs Pd^+), synthetic procedure (incipient wetness vs. ion exchange) and gas composition. Our results show that the amount of C_3H_6 adsorbed on H-zeolites significantly decreased in the presence of water, likely due to competitive adsorption and inhibition of adsorption sites. The results demonstrated that ion-exchanged metal is necessary to achieve increased storage and release of C_3H_6 in the presence of H_2O. Increased metal loading led to increased C_3H_6 adsorption ability and decreased desorption temperature.Â Most of stored C_3H_6 was not released as C_3H_6, but as various products of oxidation and cracking catalyzed by zeolites.Â We will discuss potential strategies to design enhanced HC traps based on the gained insights.

Acknowledgements. Research sponsored by the U.S. Department of Energy, Vehicle Technologies Office.

References

[1] A. Iliyas, M.H. Zahedi-Niaki, M. EiÄ‡, S. Kaliaguine, Micropor. Mesopor. Mat. 103 (2007) 171-177.

[2] J.-H. Park, S.J. Park, I.-S. Nam, G.K. Yeo, J.K. Kil, Y.K. Youn, Micropor. Mesopor. Mat. 101 (2007) 264-270.

[3] T. Kanazawa, Catal. Today 96 (2004) 171-177.

489239 Kinetic Monte Carlo Simulation of Propylene Epoxidation on Supported Gold Nanoparticles

Abstract:

Propylene oxide (C_3H_6O, PO) is an important chemical intermediate for the production of a number of commodity chemicals, including polyol, propylene glycol, and glycol ethers.[1] However, current industrial methods that produce propylene oxide from propylene pose environmental risks because of the production of chlorinated or peroxycarboxylic waste.[2, 3] Using supported gold-based catalysts to produce propylene oxide directly from propylene (C_3H_6) and molecular H_2 and O_2 provides an alternative, clean, and potentially more efficient route.

The direct propylene epoxidation reaction has been previously investigated by several different groups, and gold-based catalysts tend to provide high selectivity for propylene oxide. It has been found that Au supported by titanium silicalite (TS-1) is particularly effective at catalyzing the propylene epoxidation reaction. The isolated Ti active sites are necessary for obtaining a high selectivity towards propylene oxide,[4-6] since propylene oxide molecules that adsorb on adjacent Ti sites lead to catalyst deactivation and the formation of unwanted byproducts. Although the selectivity is very high (>90%), even the best catalysts found to date still suffer from multiple challenges, including low propylene conversion (<10%), poor stability and inefficient usage of H_2 (<50%). Therefore, significant improvement with regard to these issues is necessary, and this requires a more thorough understanding of the underlying chemical reaction network.

In this work, KMC simulations are used to model the direct epoxidation of propylene to propylene oxide on an $Au/TiO_2/SiO_2$ catalyst. Although the model is a simple two-dimensional representation of the actual catalyst surface, the basic catalyst features are preserved (Ti concentration, Au loading, Au particle sizes), and the KMC results are benchmarked against relevant experimental data. In addition, the composition in the bulk gas phase is synchronized with the dynamic reaction events occurring on the surface, and this coupling allows us to relax the typical assumption of a constant gas-phase composition. By acquiring mechanistic information from various DFT studies, and compiling this information into a KMC model, several aspects of the experimental catalytic behavior can be captured. We find close agreement between the KMC simulations and the experimental data in several aspects, but this is only achieved by considering the re-adsorption of trace amounts of the oxidant (H_2O_2) from the gas phase, versus merely assuming that the desorbed products are swept away in the gas stream.

A shortcoming in the KMC model is the H_2 sensitivity, which is higher than in most previous experimental studies. This may be an artifact that arises from our neglect of other competing reaction channels on the catalyst surface which may serve as hydrogen sinks. Also, several of the individual event rates used in the model deserve additional scrutiny. If some of these secondary effects can be resolved, this approach can provide an excellent modeling tool for connecting the atomistic-level features of such a catalytic system to the predicted experimental behavior. In particular,

bimetallic catalyst and optimally-designed supports will likely be needed for advancing the direct synthesis of propylene oxide beyond current limits, while maximizing the availability of the oxidizing species near the Ti sites is critical for accelerating the production of PO.

References

1. K. Weissermel, H.-J. A., *Industrial Organic Chemistry*. Fourth ed.; WILEY-VCH: Weinheim, Germany, 2003.

2. Monnier, J. R., *Applied Catalysis a-General* **2001,** *221* (1-2), 73-91.

3. Nijhuis, T. A.; Makkee, M.; Moulijn, J. A.; Weckhuysen, B. M., *Industrial & Engineering Chemistry Research* **2006,** *45* (10), 3447-3459.

4. Stangland, E. E.; Stavens, K. B.; Andres, R. P.; Delgass, W. N., *Journal of Catalysis* **2000,** *191* (2), 332-347.

5. Nijhuis, T. A.; Huizinga, B. J.; Makkee, M.; Moulijn, J. A., *Industrial & Engineering Chemistry Research* **1999,** *38* (3), 884-891.

6. Chen, J.; Halin, S. J. A.; Pidko, E. A.; Verhoeven, M. W. G. M.; Perez Ferrandez, D. M.; Hensen, E. J. M.; Schouten, J. C.; Nijhuis, T. A., *Chemcatchem* **2013,** *5* (2), 467-478.

494088 A Computational Mechanism Study of Ethylene Dimerization and Hydrogenation on Iridium Loaded Nu-1000 and Uio-66

Abstract:

Metal-organic frameworks (MOFs) are ordered, nanoporous materials built from metal-oxides and organic linkers. Due to their high porosity, surface area, chemical versatility and functionalization, they possess good potential for applications in catalysis, gas storage/separation, sensing etc. Recently, two synthetic methods, atomic layer deposition (ALD) and solvothermal deposition, have been used to load MOFs with metal complexes to enhance their catalytic potentials.

Zr-based MOFs have been a target for experimentalists since they are known to be stable under various experimental conditions which is mostly attributed to strong Zr-O bonding. In this computational study, mechanisms of ethylene dimerization and hydrogenation are investigated for $Ir(C_2H_4)_2H_2$ loaded NU-1000 and UiO-66 MOFs using density functional theory (DFT) employing M06-L functional. The cluster models of these two MOFs are obtained from their respective DFT optimized periodic unit cells by truncating organic linkers to acetate groups. To characterize the reaction mechanisms, activation barriers are compared for different mechanisms and they are contrasted with experimental selectivities. Moreover, CO stretching frequencies are computed for $Ir(CO)_2$ deposited structures to compare with experimental IR frequencies. Lastly, the interatomic distances in the structures are compared between optimized DFT models and EXAFS measurements.

497553 Design, Fabrication, Testing, and Operation of a Continuous Reactor for Hydrothermal Carbonization

Abstract:

Hydrothermal carbonization (HTC, thermal hydrolysis, or wet torrefaction) is a treatment process which converts moist feedstocks into homogenized, carbon rich, and energy dense solid fuel, called hydrochar. One of the main advantages of HTC compared to other thermochemical treatment processes is the use of moisture as reaction medium and reactant, thereby precluding the need for drying prior to treatment. Thermodynamic properties of water change greatly in the subcritical region from 180-280 °C, and as a result, subcritical water behaves as a non-polar solvent and mild acid and base catalyst simultaneously. When subjected to HTC biomass undergoes rapid hydrolysis and other reactions, releasing oxygen-containing volatiles and producing a highly hydrophobic hydrochar.

Although HTC offers a relatively simple and straightforward solution to process diverse biomass feedstocks, the requirements of high pressure and high temperature make the process complex and potentially costly to design and operate. The lab-scale batch process has already been demonstrated in various laboratories around the world, but batch process is not cost-effective for industrial-scale deployment. A continuous process offers a relatively small footprint, energy recovery and efficiency, and notably, economies of scale.

In this study, a bench-scale continuous HTC reactor system has been designed, constructed, commissioned, and operated with cellulose and with whole biomass feedstocks. The maximum temperature and pressure of the reactor are 250 °C and 40 bar. The throughput of the reactor system was maintained at 0.3 L/min with solids loading of ~5%, while the reaction time was maintained at 5 min. The basic reactor configuration is a section of straight pipe, and numerous engineering challenges were encountered during design and commissioning. The primary challenges are related to instrumentation durability, temperature control, and flow control.

Results are presented for hydrothermal conversion of microcrystalline cellulose. Both solid and liquid products were tested for their physico-chemical properties and compared with the corresponding products from batch process generated in a 2L batch Parr reactor. It was found that temperature and pressure were stable during operation and products are relatively similar to that of batch process. Similar results were found for HTC of cow manure, when comparing solid yield, char fuel value and total organic content in the aqueous phase, when corrected for water volume.

495138 Fate of Nitrogen during Hydrothermal Treatment of Septage

Abstract:

Septic tanks are a popular method in rural areas to treat human waste. Waste enters the septic tank and settles so that solids collect in the bottom of the tank. Here organic matter is aerobically digested. Digestion only occurs partially, and is variable with each design of tank and the conditions of the tank (such as temperature and extent of cleaning). Septic tanks offer many advantages, such as the containment of the digestate and the containment of human waste. However, the resulting solid and liquid products must be pumped and disposed of, or further treated. One concern with the use of septic tank is the long term effects of untreated pollutants that are only present in the microgram/liter range, which have been shown to have serious impacts on aquatic ecology of surrounding water systems [1],[2]. Failure of septic tanks can also result in contamination of surrounding water sources with the various macronutrients present in the digestate and effluent of the tank.

Hydrothermal treatment has been investigated as a possible method for the treatment of municipal solid waste and sewage sludge [3]. Septage, or the combination of both septic sludge and effluent, can also be treated by hydrothermal means. Hydrothermal treat is an attractive option for additional septage treatment due to its high water content; there is no need for any drying or separation for solid treatment. More so, the high nitrogen content of septage can be recovered and reused as fertilizer. In this study septage was collected from local health department and hydrothermally treated. A full nitrogen balance was performed between soluble nitrogen species, as well as solid nitrogen content determined by CHNS. Experimental trials for this treatment used a batch reactor with a reaction time of 30 minutes. It was found that hydrothermal treatment at temperatures between 180-260 °C results in increased nitrogen in the process liquid compared to the untreated septage. In addition, treatment results in increased ammonia concentration in the liquid phase and a pH between 6 and 8. This generation of soluble nitrogen species is a result of the solid phase decomposing as it is heated.

References

[1]. Schwarzenbach, René P., et al. "The challenge of micropollutants in aquatic systems." *Science* 313.5790 (2006): 1072-1077.

[2]. Withers, Paul JA, et al. "Do septic tank systems pose a hidden threat to water quality?." *Frontiers in Ecology and the Environment* 12.2 (2014): 123-130.

[3]. Zhao, Peitao, et al. "Clean solid biofuel production from high moisture content waste biomass employing hydrothermal treatment." *Applied Energy* 131 (2014): 345-367.

491306 Hydrothermal Treatment of Paper Mill Sludge

Abstract:

Each year the pulp and paper industry generates large amounts of effluent treatment sludge. The sludge is made up of three different fractions namely, primary, secondary, and tertiary sludge. Depending on the layout of the sludge handling in the individual mill, these three sludges can either be extracted separately or mixed into a stream called mixed sludge. Usually secondary and tertiary sludge are mixed together. The primary sludge, often rich in fiber material and easy to dewater, can easily be used as fuel source. However, water is difficult to remove from mixed sludge.[1-2] Moreover, a low fuel heating value of mixed sludge results against economical combustion. Therefore, a new technology that converts mixed sludge to viable solid fuel may be required for mixed sludge valorization.

Hydrothermal carbonization (HTC) has received considerable attention during the last several years as a prominent alternative for upgrading biomass feedstock ranging from wood and algae to agricultural and forest residues and industrial or municipal wastes. One of the main advantages of HTC is the use of aqueous solutions under high temperature (180-260 °C) and autogenously generated pressure (5-50 bar).[3] As a result, moist and wet feedstocks are favorable for HTC.[4]

This goal of this study was to determine the effects of HTC conditions on paper mill sludge dewaterbility and fuel value. A 100 ml Parr batch reactor was be used for HTC. Reaction temperature was varied from 180-260 °C, while reaction time was varied from 30 min-1h. The solid hydrochars were analyzed for energy content, ultimate analysis, FTIR, and NMR. Meanwhile, the process liquid is characterized with GCMS, HPLC, UV-VIS, NMR, and pH. The energy content of the hydrochar was increased with reaction temperature and longer reaction time as a result of the decrease in oxygen content. The resulted hydrochar is more hydrophobic than the feedstock. Therefore, dewaterbility is enhanced with the hydrothermal treatment of paper mill sludge.

References

1. Makela, M.; Benavente, V.; Fullana, A., Hydrothermal carbonization of industrial mixed sludge from a pulp and paper mill. *Bioresource Technol* **2016,** *200*, 444-450.

2. Makela, M.; Yoshikawa, K., Simulating hydrothermal treatment of sludge within a pulp and paper mill. *Appl Energ* **2016,** *173*, 177-183.

3. Reza, M. T.; Andert, J.; Wirth, B.; Busch, D.; Pielert, J.; Lynam, J. G.; Mumme, J., Hydrothermal Carbonization of Biomass for Energy and Crop Production. *Applied Bioenergy* **2014,** *1* (1).

4. Reza, M. T.; Coronella, C.; Holtman, K. M.; Franqui-Villanueva, D.; Poulson, S. R., Hydrothermal Carbonization of Autoclaved Municipal Solid Waste Pulp and Anaerobically Treated Pulp Digestate. *Acs Sustain Chem Eng* **2016,** *4* (7), 3649-3658.

486662 Structural Analysis of Humins Formed in the BrÃ¸nsted-Catalyzed Dehydration of Fructose

Abstract:

Humins are carbonaceous and polymeric by-products formed during acid-catalyzed hydrothermal processing of sugars to 5-hydroxymethylfurfural (HMF). These are formed via cross-polymerization of HMF with sugars in acidic solutions, and cause significant carbon loss and unfavorable process economics of biorefineries. The deposition of these solid particles on the reactors and processing equipment also increases the maintenance costs. Currently humins are combusted to supply heat in biorefineries. Minimization of humins formation or their valorization is essential for improving the process economics of the biorefineries. In this work, we developed methods to understand microstructure of humins through solubility experiments. First, we have conducted dissolution experiments in various solvents and correlated the solubility data to molecular properties of solvents to develop suitable descriptors. By doing this, we have identified solvents for high solubility of humins at room temperature and provided insights into the mechanism of humins dissolution. Next, Liquid Chromatography- Mass Spectrometry (LC-MS) was used to determine and compare the masses of components in the humins solubilized in different solvents. Finally, a six stage dissolution experiment was done to investigate the structural inhomogeneity of humins. Both Fourier Transform Infrared Spectroscopy (FT-IR) and LC-MS were used to characterize the solubilized and insoluble humins obtained at each stage. Based on our results, an inhomogeneous structure where macromolecular and molecular components are connected through weak forces, was proposed.

Catalysis Center for Energy Innovation UNIVERSITY OF DELAWARE

Super Oleophilic and Water Repellent Graphene Oxide Sponge for Catalytic Coupling of Furans and Oil Absorption

Saikat Dutta

Advisers: Basu Saha, Dion G. Vlachos

Catalysis Center for Energy Innovation

University of Delaware

Newark, Delaware

Session: Hydrothermal Carbonization

November 2, 2017

AIChE Annual Meeting 2017, Minneapolis, MN

U.S. DEPARTMENT OF ENERGY Office of Science

CCEI is an Energy Frontier Research Center funded by the U.S. Department of Energy, Office of Basic Energy Sciences

Catalysis Center for Energy Innovation UNIVERSITY OF DELAWARE

Why Graphene Sponge

- Graphene is generally hydrophobic, can be tuned to hydrophilic

Hydrophobic properties

Graphene@SiC

CA 92°

SiC CA 69°

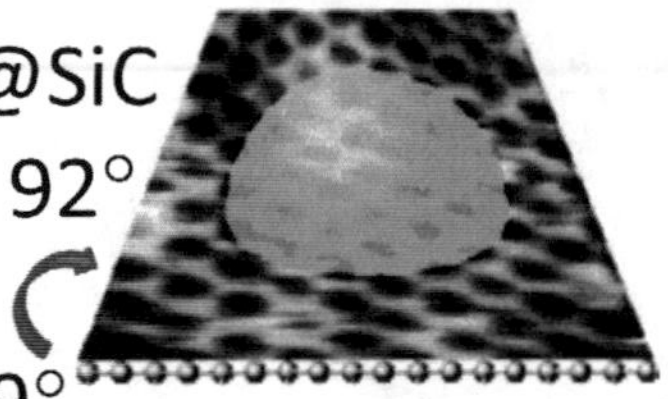

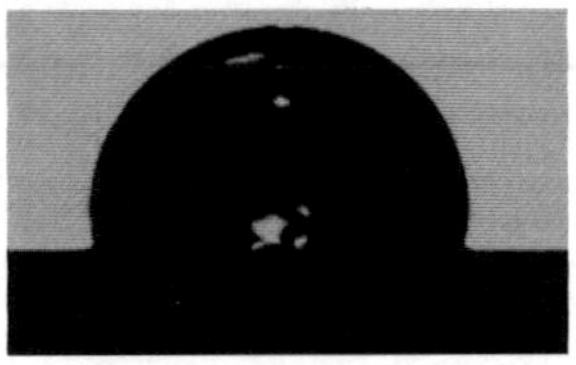

O_2 plasma

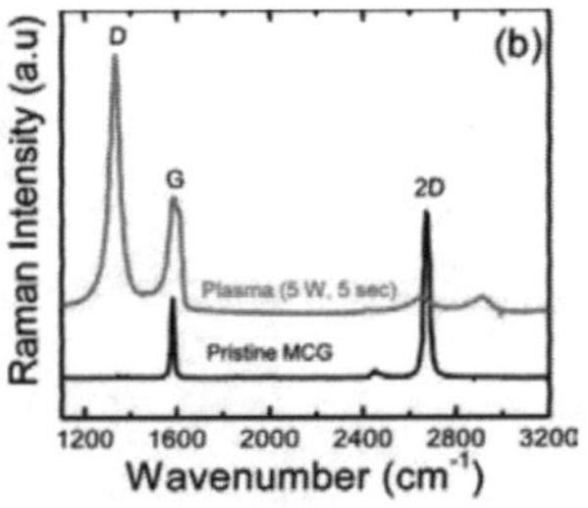

- Thickness Independent Hydrophobicity

Energy Environ. Sci. **2012**, *5*, 7908

Langmuir **2010,** *26*, 3798

- More defects, more hydrophilic
- GO sponges are not explored extensively

2

Sponge Materials: Why Is It Important

Marine Oil Spill

Threats to the planet

- High frequency
- Long-lasting adverse environmental impacts
- Significant economic impact

Current Methods

- Skimming on calm water
- Chemical dispersants spread

Sponge Materials

carbon-based aerogels

Adv. Mater. **2013**, 25, 2554

porous boron-nitride

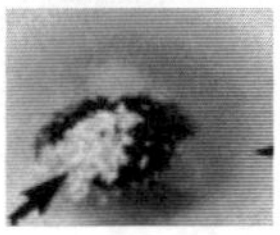

Nat. Commun. **2013**, 4, 1777

modified commercial sponges

Energy Environ. Sci. **2012**, 5, 7908

GO-wrapped sponge

Nat. Nanotechnol. **2017**, 12, 434

3

Catalysis Center for Energy Innovation UNIVERSITY OF DELAWARE

Graphene Oxide Sponge (GOS)

- Explore hydrophobic and hydrophilic behaviors of GOS

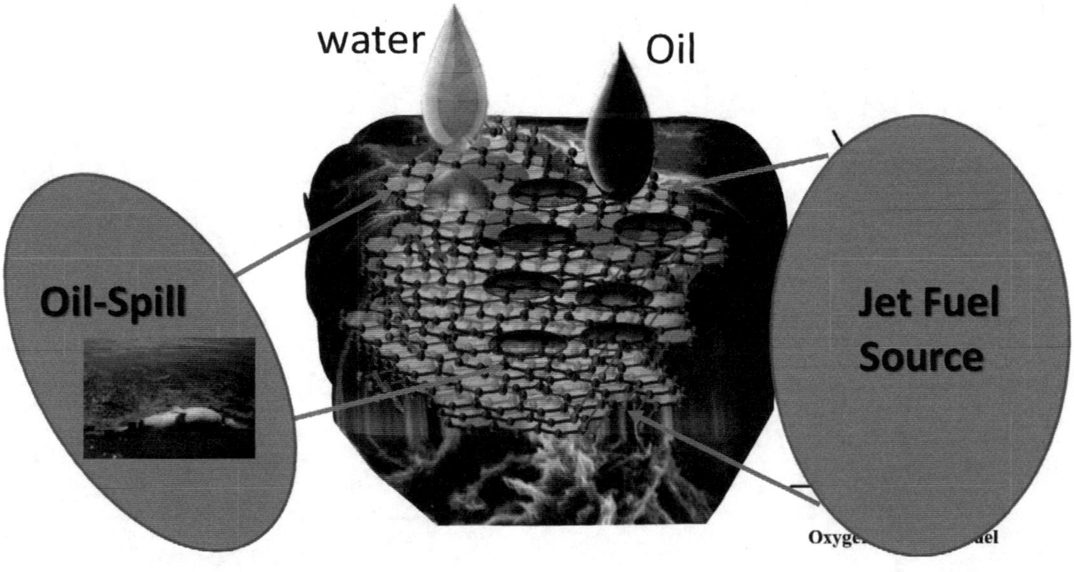

4

Catalysis Center for Energy Innovation

UNIVERSITY OF DELAWARE

Hydrophilic IGO to GOS

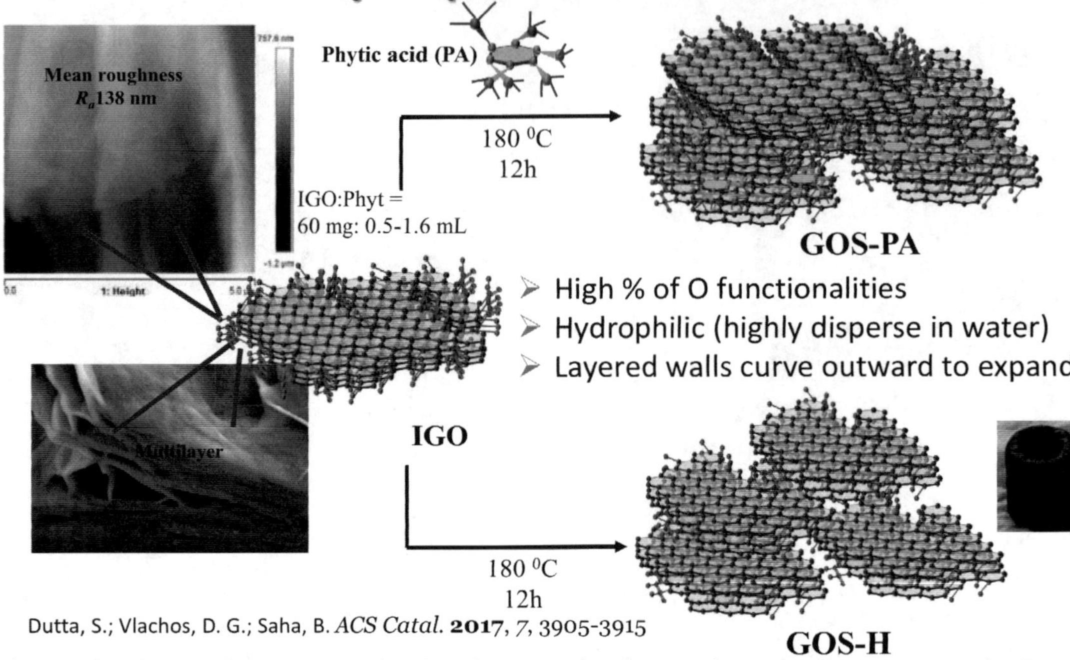

- High % of O functionalities
- Hydrophilic (highly disperse in water)
- Layered walls curve outward to expand

Dutta, S.; Vlachos, D. G.; Saha, B. *ACS Catal.* **2017**, *7*, 3905-3915

- Hydrothermal treatment is simplest method to reduce surface oxygen (gain hydrophobicity) compared to cross-linking/freeze-drying

5

Catalysis Center for Energy Innovation

UNIVERSITY OF DELAWARE

Hydrophobicity of GOS from XPS C1s Peak

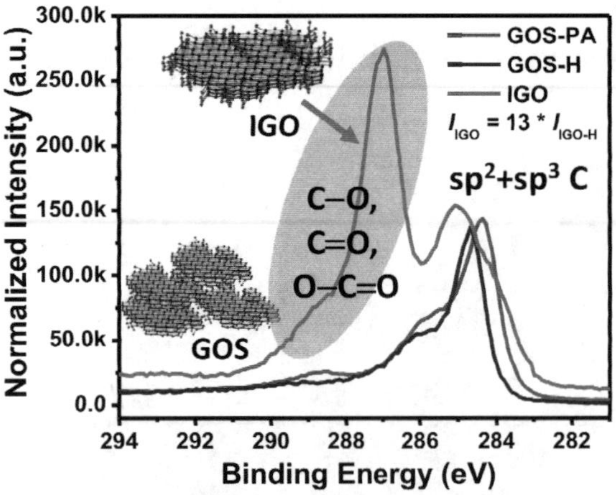

More surface oxygen (hydrophilic)

Less surface oxygen (hydrophobic)

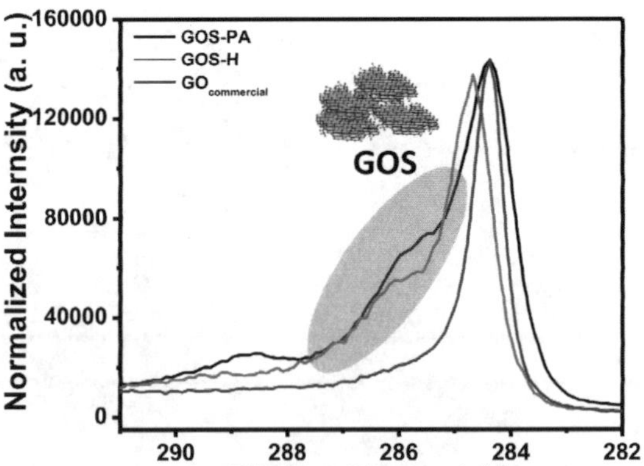

- Increase in hydrophobicity: selective for oil/organic absorption

- Surface oxygen GOS > $GO_{commercial}$
- GOS more reactive

6

Catalysis Center for Energy Innovation

Defect Analysis: Raman Spectra

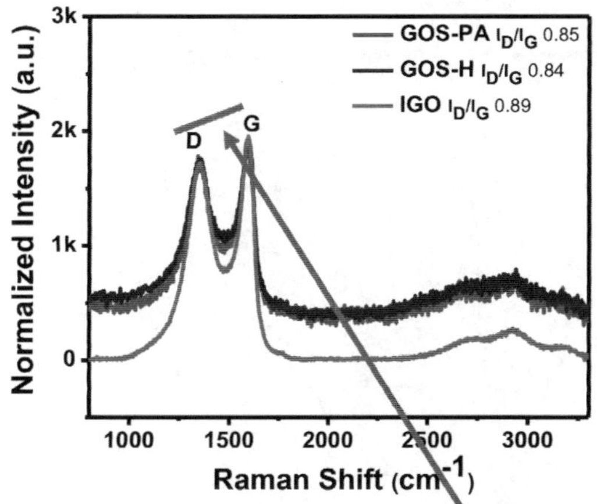

O_2 plasma (introduction of defect)
G@SiC

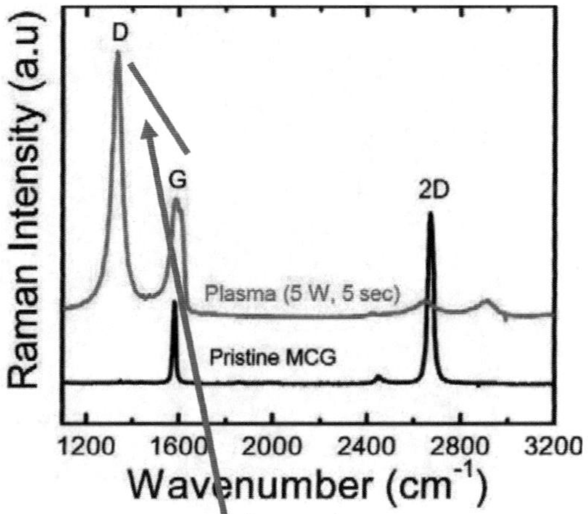

Langmuir **2010,** *26*, 3798

- GOS and IGO comparable crystalline defect
- G contains less crystalline defect before plasma treatment

7

Catalysis Center for Energy Innovation

Raman Imaging of D and G Bands

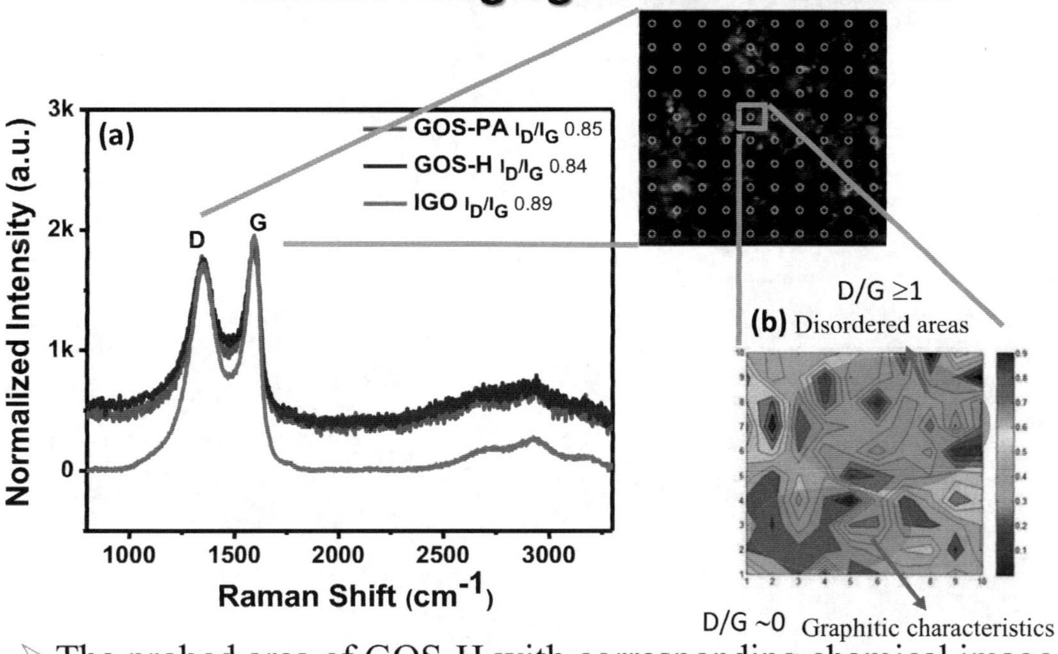

- The probed area of GOS-H with corresponding chemical image displaying the ratio of D-band and G-band intensity (b)

8

Catalysis Center for Energy Innovation

UNIVERSITY OF DELAWARE

SEM Characterization of GOS-PA and GOS-H

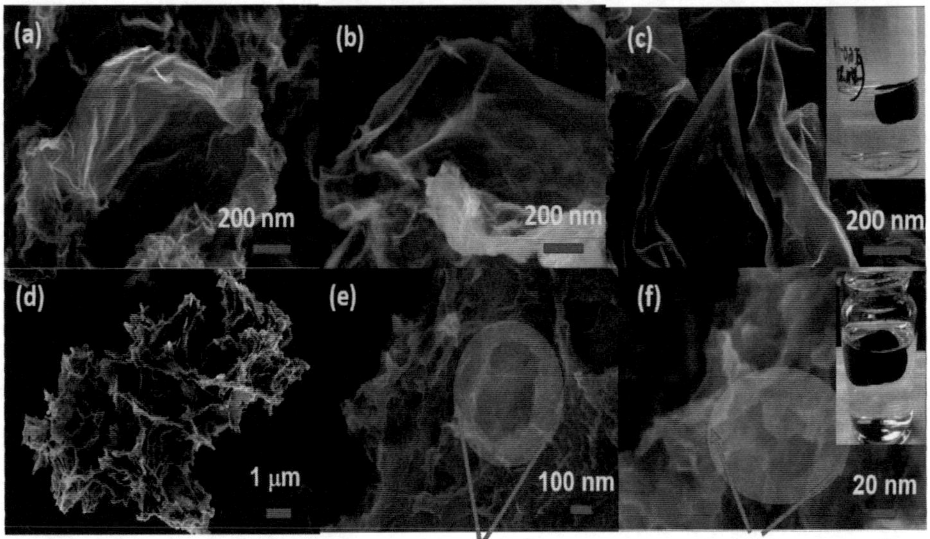

➢ GOS-H contains more rough surface as compared to GOS-PA

9

Catalysis Center for Energy Innovation

UNIVERSITY OF DELAWARE

Water-Contact Angle of GOS

Sample	Contact Angle (Degrees)	Error (%)
HOPG (std)	67.3	3.3%
GOS-H	86.5	0.7%
GOS-PA	84.2	1.8%

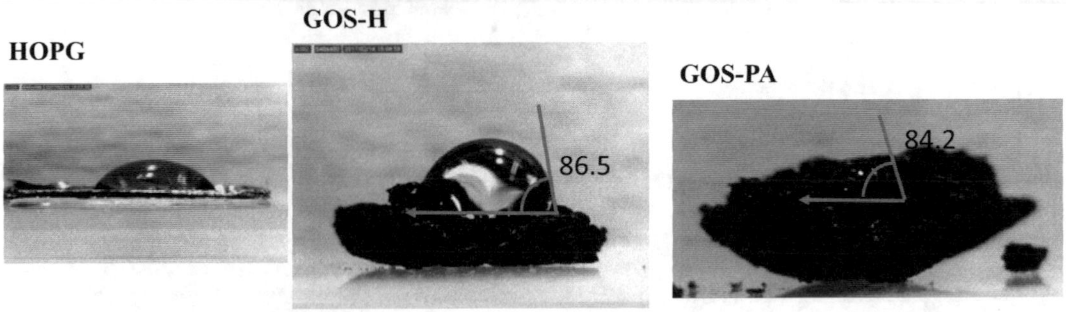

➢ GOS hydrophobic with contact angle ~90°
➢ Enhancement of contact angle is countered by surface oxygen

10

Oil Absorption by GOS-H

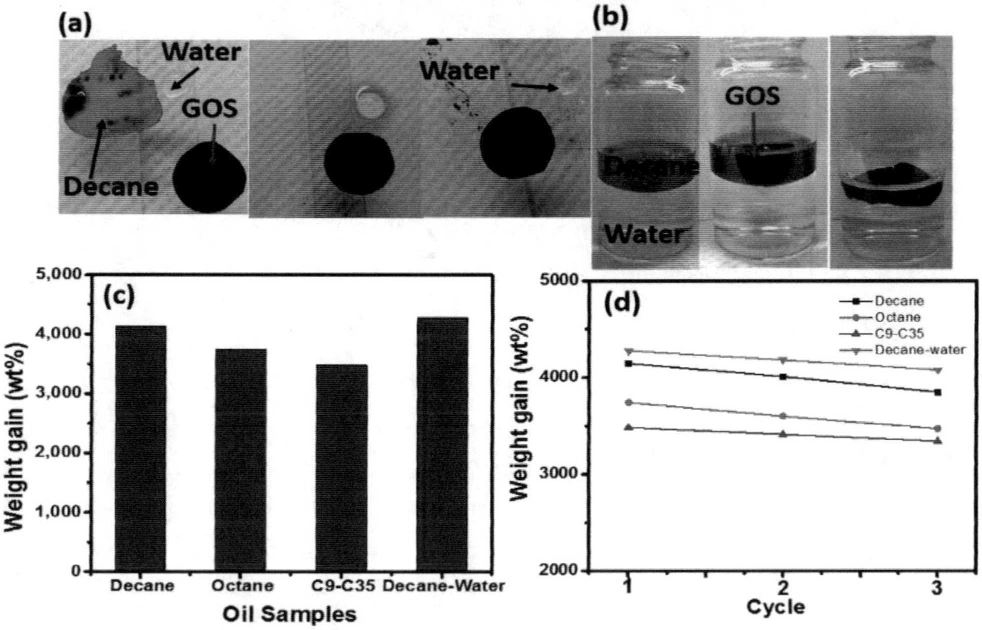

- ➢ Selective oil absorption by GOS in 15 min
- ➢ Exhibit recyclability for oil absorption

11

Origin of Hydroxyalkylation-Alkylation (HAA) Catalysis

- ➢ Surface-oxygen active sites of IGO enables tailor the C-numbers of lignocellulosic fuel precursors

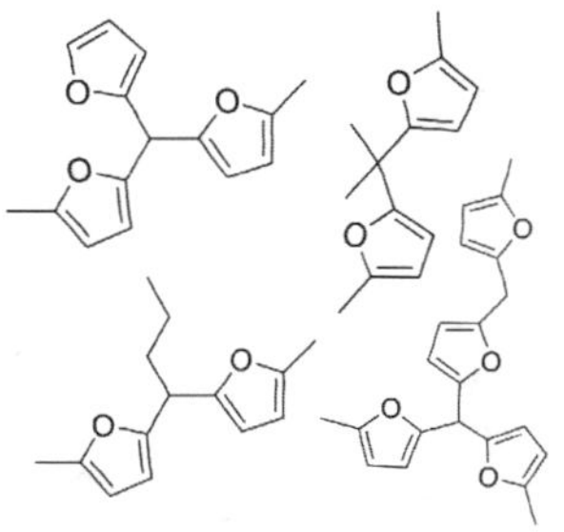

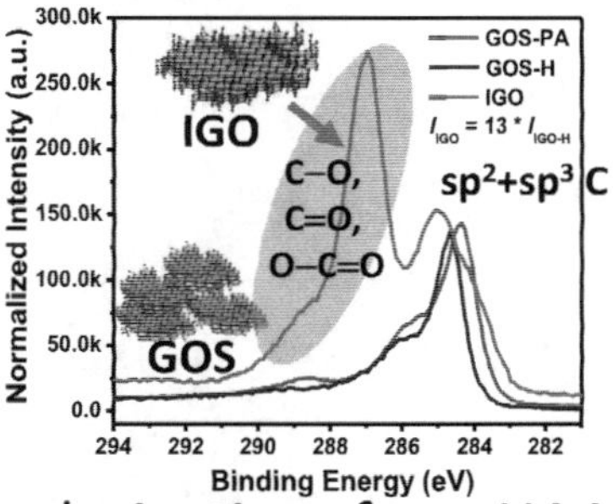

Entry	C (%)	O(%)	P(%)	S(%)
IGO-H	73.4	25.1		1.4
IGO-PA	72.5	24.2	2.1	1.2
IGO	59.2	36.6		3.1
HOPG	98.2	0.41		

- ➢ C13-C21 furan-derivatives from HAA with IGO

Dutta, S.; Saha, B. *ACS Catal.* **2017**, *7*, 5491-5499

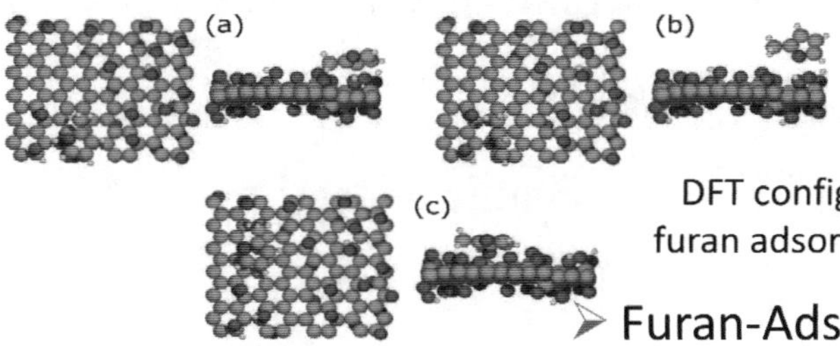

DFT configuration of furan adsorption on IGO

- ➢ Furan-Adsorption on a Oxygenated Surface

Dutta, S.; Vlachos, D. G.; Saha, B. *ACS Catal.* **2017**, *7*, 3905-3915

12

HydroxyAlkylation-Alkylation (HAA) of Furans

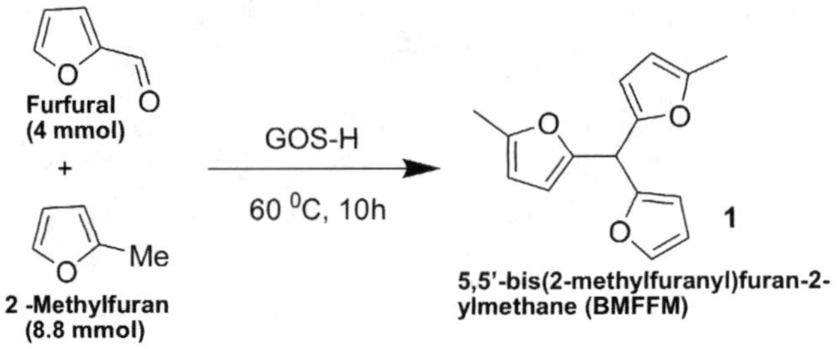

GOS-H recycle

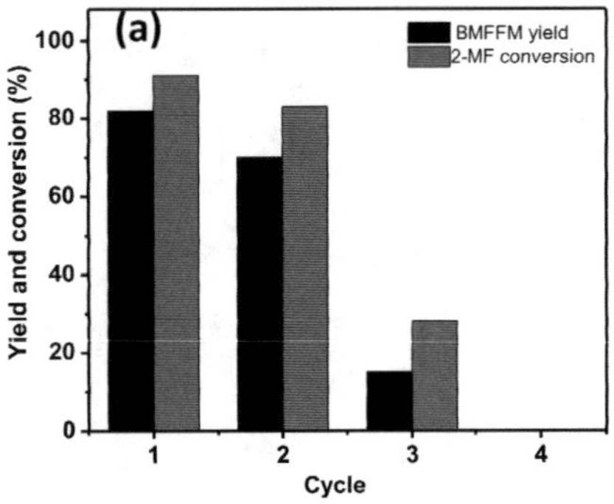

- Carbon extension of furans to desired fuel precursors
- Intermediate surface oxygen % with hydrophilic behavior GOS-H for HAA
- Loss of activity due to further reduction of surface oxygen (XPS)

13

Catalysis Center for Energy Innovation
UNIVERSITY OF DELAWARE

Summary

- GOS derived from aqueous dispersion of IGO.
- GOS contains surface roughness responsible for wettability
- Raman imaging, XPS analysis, and SEM were used for characterization
- Percentage of surface oxygen in GOS is higher than graphene or rGO sponge
- GOS-H is effective for HAA reactions for furans
- GOS-H offers selective oil absorption from oil-water mixture

Moving Forward

- Technique for cross-linked GOS for mechanical stability

14

Acknowledgements

Joseph P. Smith (Raman Spectroscopy)
Prof. Karl S. Booksh (Raman Spectroscopy)
Dr. Weiqing Zheng (SEM,TEM imaging)
Rachel Pupillo (XPS)
Dion G. Vlachos (insightful discussions)
Jeff Everhart (GC analysis)

U.S. Department of Energy, Office of Science, Office of Basic Energy Sciences, under Award No. DE-SC0001004

U.S. DEPARTMENT OF ENERGY | Office of Science

15

502391 Simultaneous Isomerization and Reactive Extraction Followed By Back Extraction of Sugars from Biomass Hydrolysate for High Purity and Yield of Ketose Sugars

Abstract:

Biomass hydrolysates are often low in sugar concentration and contain contaminants associated with the pretreatment method and sugar and lignin degradation. Cost-effective and high-yield downstream processing requires sugar streams at higher concentration, free of contaminants generated during the pretreatment process. Instead of removing contaminants from the hydrolysate, we have taken the approach of removing sugars from the hydrolysate, leaving contaminants behind. In the process of removing sugars, we enzymatically isomerize them from their aldose (glucose, xylose) to their ketose (fructose, xylulose) forms. These ketose sugars are ideally suited for production of furans â€“ HMF from fructose and furfural from xylulose.

Unfortunately, the isomerization reactions do not have favorable equilibria with respect to ketose formation. To overcome this limitation, we have previously developed a method of simultaneous-isomerization-and-reactive-extraction (SIRE) to produce the ketose isomer in high yield. We have demonstrated that with a xylose/xylulose mixture, SIRE followed by 2-stage back-extraction (BE) allows high yield recovery of xylulose in nearly pure form.

For mixtures of xylose and glucose found in biomass hydrolysate, separation and purification of the ketose sugars (xylulose and fructose) from the hydrolysate will be more complex. The extraction affinities of glucose and fructose are lower than xylulose, but higher than xylose. SIRE-BE has not been previously evaluated for biomass hydrolysates with high sugar concentrations (up to 90 g/L) for recovery of a high purity, mixed C5 and C6 ketose product stream. We have developed strategies for high purity recovery of both xylulose and fructose from mixed sugar streams using SIRE-BE. We will present these strategies as well as xylulose and fructose yields and purity obtained under different experimental configurations.

487883 A Simple Coordination-Based Model for Bimetallic Nanoparticles

Abstract:

The stability of nanoparticle catalysts to phenomena such as sintering is of fundamental importance to heterogeneous catalysis.[1] Great effort has been dedicated to understanding the mechanisms of nanoparticle sintering, which range from single-atom Ostwald ripening to whole-particle migration depending on reaction conditions and the nature of the catalyst support.[2-3] In recent work, we presented a kinetic Monte Carlo (kMC) algorithm to describe the surface migration of supported platinum nanoparticles.[4] We seek to develop a more general understanding of nanoparticle structural evolution under reaction conditions to design catalysts with enhanced stability.

In this presentation, we demonstrate a new model for predicting the energies of monometallic and bimetallic configurations of *fcc* metal atoms based only on the identity of the metals and the coordination number of individual atoms. Using parameters obtained from a small set of density functional theory calculations for monometallic surface adsorption, we reliably predict energies of bimetallic nanoparticles with varying size, shape, composition, and atomic configuration. In combination with a kMC algorithm, this model enables a deeper understanding of the dynamic nature of supported nanoparticle catalysts under reaction conditions and subsequent design of catalysts with improved resistance to sintering.

[1]C.H. Bartholomew, *Appl Catal A-Gen* **212**, 17 (2001)

[2]E. Ruckenstein, B. Pulvermacher, *J Catal* **245**, 224 (1973)

[3]P. Voorhees, *J Stat Phys* **38**, 231 (1985)

[4]L. Li, P. N. Plessow, M. Rieger, S. Sauer, R. S. Sánchez-Carrera, A. Schaefer, F. Abild-Pedersen, *J Phys Chem C* **121**, 4261 (2017)

496944 Design of Metallic Surface Nanostructures Using Mathematical Optimization

Abstract:

Technological advances in the area of nanofabrication of transition metal crystalline surfaces are enabling new opportunities for synthesizing nanostructured catalysts at ever-increasing levels of structural control. The developed transition metal materials are particularly exciting for their high-performing, unique catalytic properties that often differ dramatically from bulk catalysts [1]. More specifically, rational design of catalysts is being enabled by advances in electrochemical deposition [2], formation of near-surface alloys [3], and photochemical lithography [4], among other synthesis techniques. These advances result in a large space of possible catalyst designs, and as time goes on, the complexity of realizable designs will only continue to grow. This growing design space represents a great opportunity for identifying novel, high-performing materials, but also poses a significant challenge to rigorously identify the best designs in terms of reactivity and stability. We propose formulating this materials design question as a mathematical optimization model [5], which leverages the use of well-established approaches for solving highly combinatorial problems.

In this work, we identify the nanostructured surfaces that maximize the reactivity of catalysts by solving a mathematical optimization model that represents the relevant choices using an appropriate set of decision variables, an objective function, and algebraic constraints. We consider the key decision variables to be the choice of placement of atoms at the surface of a periodic crystal lattice, thereby forming a particular surface nanostructure that exhibits complex active sites. To link the structure to catalytic functionality, we utilize demonstrated correlations from density functional theory (DFT) that link surface geometric descriptors to chemical properties such as adsorption energy or activation energy [6,7]. In many cases of interest, the emergence of volcano plot correlations to describe reactivity on a catalyst site allows us to model this reactivity through the identification of "ideal sites"; that is, active sites that correspond to the tip of the volcano plots and that tend to dominate the exhibited turnover on the catalyst surface [8]. This binary nature of our decision variables leads to mixed integer linear program (MILP) models that can be solved with well-developed optimization software with rigorous guarantees on the optimality of identified solutions.

The identified surface structures serve as theoretical bounds on the activity that can be achieved by a catalyst surface and may also serve as targets for experimental synthesis, as new nanofabrication methods are developed. However, it is possible that the identified solutions of the aforementioned optimization model are in fact unstable under reaction conditions and would therefore rearrange to unproductive surfaces. Therefore, we have developed several approaches to account for the stability of designs as a critical element of the process of materials design via mathematical optimization. The first approach collects a solution pool of highly reactive designs, which are offline filtered for structures that are determined to be stable. This evaluation of stability can be carried out using computational methods such as DFT or the embedded atom method (EAM) [9]. As a second approach, the EAM calculations can be embedded directly into the optimization model, resulting in algebraic constraints that directly restrict the search space to only include designs that are considered stable.

Using our mathematical optimization paradigm, we have proposed optimal nanostructured surface designs for a variety of crystallographic surfaces and across an array of descriptions of ideal reactive sites. The results demonstrate patterns which may be predictable in some cases, but also reveal that the optimal reactive surfaces are often non-intuitive. We show that the optimal surface is strongly dependent on the definition of the ideal reactive site and that the characterization of a "stable" surface can have dramatic impact on the resulting design. A feature of this framework is the flexibility to encode a multitude of materials design considerations in a generic fashion, thereby decoupling the complications of modeling a material design problem from solving the resulting optimization problem.

References

[1] Narayanan R., El-Sayed M. A., "Shape-dependent catalytic activity of platinum nanoparticles in colloidal solution," *Nano Letters*, 4(7):1343–1348, 2004.

[2] Calle-Vallejo F., Koper M. T. M., Bandarenka A. S., "Tailoring the catalytic activity of electrodes with monolayer amounts of foreign metals," *Chemical Society Reviews*, 42(12):5210–30, 2013.

[3] Stamenkovic V. R., Mun B. S., Mayrhofer K. J. J., Ross P. N., Markovic N. M., "Effect of surface composition on electronic structure, stability, and electrocatalytic properties of Pt-transition metal alloys: Pt-skin versus Pt-skeleton surfaces," *Journal of the American Chemical Society*, 128(27):8813–8819, 2006.

[4] Geissler M., Xia Y., "Patterning: Principles and some new developments," *Advanced Materials*, 16(15):1249–1269, 2004.

[5] Hanselman C. L., Gounaris C. E., "A mathematical optimization framework for the design of nanopatterned surfaces," *AIChE Journal*, 62(9):3250–3263, 2016.

[6] Nørskov J. K., Bligaard T., Logadottir A., Bahn S., Hansen L. B., Bollinger M., Bengaard H., Hammer B., Sljivancanin Z., Mavrikakis M., Xu Y., Dahl S., Jacobsen C. J. H., "Universality in heterogeneous catalysis," *Journal of Catalysis*, 209(2):275–278, 2002.

[7] Calle-Vallejo F., Martínez J. I., García-Lastra J. M., Sautet P., Loffreda D., "Fast prediction of adsorption properties for platinum nanocatalysts with generalized coordination numbers," *Angewandte Chemie - International Edition*, 53(32):8316–8319, 2014.

[8] Calle-Vallejo F., Loffreda D., Koper M. T. M., Sautet P., "Introducing structural sensitivity into adsorption-energy scaling relations by means of coordination numbers," *Nature Chemistry*, 7(5):403–410, 2015.

[9] Daw M. S., Baskes M. I., "Embedded-atom method: Derivation and applications to impurities, surfaces, and other defects in metals," *Physical Review B*, 29(12):6443–6453, 1984.

492669 Accelerated Catalyst Screening Using Computational Alchemy

Abstract:

Large number of screening studies identifying new catalysts for different reactions have been reported over the past decade. Almost all of them employ Kohn-Sham density functional theory (KS-DFT) and thermodynamic descriptors (see for example ref. [1]) to screen for new catalysts. Though usually considered reliable for descriptor-based analyses, KS-DFT calculations are computationally expensive and intractable for use when screening across the full chemical space of all possible alloy materials. In order to accelerate screening of catalysts, we employ a model Hamiltonian method, â€˜computational alchemyâ€™ [2-4] to approximate KS-DFT energies at a fraction of the computational cost. We will discuss about computational alchemy and how it can be used to reliably screen for oxygen reduction reaction (ORR) catalysts using Volcano plot descriptors.

Reference:

1. Greeley, J. *et al.* Alloys of platinum and early transition metals as oxygen reduction electrocatalysts. *Nat Chem* **1,** 552â€“556 (2009).

2. Lilienfeld, O. A. von, Lins, R. D. & Rothlisberger, U. Variational Particle Number Approach for Rational Compound Design. Phys. Rev. Lett. 95, 153002 (2005).

3. Lilienfeld, O. A. von & Tuckerman, M. E. Molecular grand-canonical ensemble density functional theory and exploration of chemical space. The Journal of Chemical Physics 125, 154104 (2006).

4. Sheppard, D., Henkelman, G. & Lilienfeld, O. A. von. Alchemical derivatives of reaction energetics. The Journal of Chemical Physics 133, 084104 (2010).

494816 DFT Study of Trends in Reactivity at Bifunctional Interfaces: A Case Study of Water Gas Shift on Doped Au/MgO

Abstract:

Metal nanoparticles supported on oxides are ubiquitous in their usage as heterogeneous catalysts. Such materials often have bifunctional properties whereby the interface in the supported system allows for higher turnovers than either the oxide or the metal alone. Catalysis using supported Au nanoparticle is an example where this effect is often observed [1]. While bifunctional effects have been shown for specific cases both theoretically and experimentally [1-4], a systematic theoretical design paradigm for understanding trends in reactivity due to changes in electronic properties at bifunctional interfaces is lacking. In this work, we used Density Functional Theory (DFT) to study model Au nanowire supported on MgO, wherein the electronic properties at the Au/MgO interface was tuned through substitutional doping of MgO. The effect of dopants on reactivity at the perimeter sites of the supported nanowire was studied using Water Gas Shift (WGS) as a probe reaction, which is known to depend on support properties [1]. Using extensive DFT calculations, it was found that scaling relations for atomic as well as bifunctionally bound species exist at the doped Au/MgO interface, with slopes different from those identified for extended surfaces [5]. The scaling properties originate from electrostatic interactions between adsorbate and redistributed charges at the interface of doped Au/MgO. In addition, Bronsted-Evans-Polanyi relations were also identified for reaction steps that involved both Au and MgO sites. These trends were then combined with a dual site microkinetic model to develop a volcano plot, relating WGS turnover rates with descriptor for adsorption energies at the interface. Our results show that doping the oxide can change turnover rates at the interface even when the reactive intermediates are not in direct contact with the dopant. In addition, there exists an optimum descriptor value at which the WGS rate of the doped Au/MgO is maximized, with the volcano behavior likely arising from different rate limiting steps at different ranges of descriptor values. These results lay the groundwork for constructing a systematic design framework for bifunctional interfaces, and suggest a possible means to enhance activity of Au/oxide catalysts through doping, if they can be experimentally synthesized.

References

[1] Shekhar, M., Wans, J., Lee, W.-S., Williams, W.D., Kim, S.M., Stach, E.A., Miller, J.T., Delgass, W.D. and Ribeiro, F.H. *JACS* **2012**, 134, 4700

[2]Rodriguez, J.A., Ma, S., Liu, P., Hrbek, J., Evans, J., Perez, M. *Science* 318, 1757 (2007)

[3] Duan, Z. and Henkelman, G. *Phys. Chem. Chem. Phys.* **2016**, 18, 5486

[4] Zhao, Z.-J., Li, Z., Cui, Y., Zhu, H., Schneider, W. F., Delgass, W. N., Ribeiro, F. and Greeley, J., *J. Catal.* **2017**, *345*, 157

[5] Abild-Pedersen, F., Greeley, J., Studt, F., Rossmeisl, J., Munter, T. R., Moses, P. G., Skúlason, E.; Bligaard, T. and Norskov, J. K. *Phys. Rev. Lett.* **2007**, *99,*16105

498390 Perovskite Oxides for Low Temperature Carbon Dioxide Conversion Towards Hydrocarbon Generation

Abstract:

Carbon dioxide is the primary cause towards global warming and climate change. Carbon dioxide reduction thus, demands acute attention. Though carbon capture and sequestration (CCS) routes for CO_2 reduction attracted sufficient investment in the recent years, its scale of operation (about 35 Mtpa, in early 2017) is far below the CO_2 emissions (about 35 Gtpa, in 2015). Thus, it is imperative to spend efforts on repurposing CO_2 for hydrocarbon generation. Amongst the different routes of CO_2 conversion, solar energy based options like solar photocatalytic or solar thermochemical cycles (STC) garnered a lot of interest. Photocatalysis can be a low temperature process, however extreme low rates of conversion makes it unsuitable for industrial application. STC can achieve superior rates of CO_2 conversion, but at the cost of high temperatures of operation (more than 1000 °C). Reverse water gas shift – chemical looping (RWGS-CL) presents the benefits of these processes without facing their limitations. It can operate at low temperatures of about 500 °C with even better CO_2 conversion rates.

Mixed metal oxides like perovskite oxides (ABO_3) are the main vehicles of conversion of CO_2 to CO via RWGS-CL process. RWGS-CL is essentially a two-step process in which the perovskite oxide is first reduced under hydrogen to its oxygen deficient form ($ABO_{3-\delta}$). In the second step, CO_2 is converted over these oxygen vacant materials while the materials regain their stoichiometric forms. Since oxygen vacancy formation is the most crucial factor that drives this process, DFT-calculated oxygen vacancy formation energy (E_{vac}) presents an appropriate descriptor for the process. This E_{vac} is largely dependent on the material composition and perovskite oxides are capable of accommodating different elements in their crystal structure. Hence, perovskite oxides present the perfect platform for tuning the oxygen vacancy formation characteristics of a material. We hence, undertook a screening process to predict different perovskite oxide with CO_2 conversion capabilities. Based on our DFT calculation we predicted several earth abundant materials which can prove worthy of our purpose. CO_2 conversion capability of these predicted materials was tested via temperature programmed experiments. Lanthanum and calcium based materials revealed unprecedented CO formation yields and rates. Detailed investigation of the effect of composition tuning of lanthanum and calcium based materials has been done to optimize the CO_2 conversion performance of these materials.

501232 Design of Ruddlesden-Popper Oxides with Optimal Activity for Surface Oxygen Exchange and Electrochemical Oxygen Reduction

Abstract:

Implementation of efficient energy conversion systems (i.e., solid oxide fuel cells) are of high importance due to the rapid depletion of nonrenewable energy sources. Oxygen reduction reaction (ORR) plays an important role in the performance of these systems. Ruddlesden-Popper (R-P) oxides have shown promise as electrocatalysts for this reaction due to their excellent mixed ionic electronic conductivities and high electrocatalytic activity for surface oxygen exchange (the process that governs oxygen reduction in these materials).[1-3] In this study, we combine density functional theory calculations along with controlled kinetic experimental studies on well-defined nanostructures of first-order R-P oxides of varying compositions to understand the underlying factors that govern their surface oxygen exchange kinetics and develop a structure-function relation that can guide the development of optimal R-P oxides for ORR. A series of R-P oxides with different compositions have been synthesized using a controlled reverse microemulsion approach. Isotopic labeled oxygen ($^{18}O_2$) exchange and electrochemical oxygen reduction studies were used to investigate the effect of composition on the activity of these materials. The experimental results were compared to the computational findings. A structure-activity relation was developed providing insights into the optimization of complex mixed ionic-electronic oxides for oxygen reduction.

References

1. Ma, X.; Carneiro, J. S. A.; Gu, X.-K.; Qin, H.; Xin, H.; Sun, K.; Nikolla, E., *ACS Catal.* **2015**, *5*, 4013-4019.
2. Das, A.; Xhafa, E.; Nikolla, E., *Catal. Today* **2016**, *277*, 214-226
3. Tarancon, A.; Burriel, M.; Santiso, J.; Skinner, S. J.; Kilner, J. A., *J. Mater. Chem.* **2010**, *20*, 3799-3813.

500048 Identification of Highly Active Catalytic Sites for Oxygen Reduction Reaction in Carbon Nanostructures from First-Principles Investigation

Abstract:

Proton-exchange membrane (PEM) fuel cells are considered to be a promising technology for use as alternative energy sources for transport, stationary and portable applications. They use hydrogen and oxygen to generate electricity through an electrochemical process which includes the oxygen reduction reaction (ORR) at the cathode. Currently, platinum (Pt)-based catalysts are commonly used to speed up the sluggish ORR that limits the efficiency of low temperature PEM fuel cells. However, Pt is expensive and scarce, drawing much interest in developing non-Pt catalysts.

Carbon nanomaterials, including fullerenes, carbon nanotubes (CNTs), graphene, and their composites, have been actively studied as a potential candidate to replace Pt-based ORR catalysts. These sp^2-bonded carbon nanostructures are inherently constituted by largely extended -conjugation systems, making them essentially unreactive toward molecular oxygen (O_2) in their pristine forms. However, the conjugated -systems can be severely disrupted with localized electrons through dopant incorporation, defect creation, and/or local lattice distortions. The doped, defective and distorted carbon structures are currently thought to be responsible for O_2 adsorption/activation and thus ORR, but the active sites and the exact molecular mechanisms are still uncertain.

We have theoretically explored the mechanisms by which modification can activate graphene for the adsorption and reduction of oxygen, with an aim to define design guidelines for carbon ORR catalysts. Our effort has been focused on identification of active sites that allow favorable O_2 adsorption and subsequent reduction based on extensive first-principles quantum-mechanical calculations. In this talk, we will present some of our recent findings.

Using density functional theory (DFT) calculations, we have identified a novel method for the activation of graphene. This modification creates quasi-localized electron states in the graphene n-bonding system which are vulnerable to form bonds with O_2. The result is favorable O_2 adsorption, which despite being necessary for ORR activity is not predicted for previously proposed active sites. In this work, we applied an explicitly solvated system to model the ORR at the solid-liquid interface with aqueous solvent. Many previous DFT studies were modeled in the gas phase where O_2 adsorption is consistently predicted to be endothermic on carbon. Our study clearly demonstrates that a more accurate description of interactions at the graphene-solvent interface is necessary to properly describe O_2 adsorption, which appears strongly influenced by solvation. Ab-initio MD simulations were performed with the explicit solvent model to outline a full reaction mechanism including novel elementary reaction steps. The rate limiting step of the ORR in this system is predicted to be desorption of final reduction products. DFT allows us to study this desorption separately from the accompanying electron injection. We predict that there is a small but surmountable activation barrier that must be overcome for the reduction product to be removed. Excitingly, the predicted maximum overpotential for any elementary reaction step of the modified carbon structures that we have found is approximately 0.6 eV, which can be competitive when compared to standard carbon supported Pt (Pt/C) catalysts. Our study provides insight into molecular mechanisms underlying the superior ORR activity of nanostructured carbon materials and also guidelines for the rational design and development of high-efficiency, cost-effective carbon-based ORR catalysts for fuel cell applications.

489608 Confined Ni Nanoparticles in Mesoporous Silica Via a Polyethyleneimine-Assisted Route and Their Catalytic Performance for Methane Dry Reforming

Abstract:

The major greenhouse gases of CH_4 and CO_2 were converted into CO and H_2 synthesis gas via dry reforming of methane with the minimization of the carbon coking on the nickel confined mesoporous silica catalyst. To immobilize nickel nanoparticles in the ordered channels of the mesoporous silica, polyethyleneimine (PEI) was used as a stabilizer, reducing, and capping agent of nanoparticles. The nickel-PEI chelated complexes were formed by dissolving PEI and the nickel precursor into an aqueous solution. Then, the positively charged NH_2 terminal groups of the complexes were strongly interacted with the negatively charged surface silanol groups of the mesoporous silica. From this strong interaction, the nickel-PEI chelates were anchored in the porous framework of their parent support during the impregnation steps. A long hydrocarbon chain of PEI and its steric hindrance prevented the aggregation of the nickel-PEI complexes during the drying and calcination steps. Therefore, nickel nanoparticles were highly dispersed inside the porous framework even after the reduction step, suggesting the successful synthesis of the nickel well-dispersed mesoporous silica catalyst via a PEI-assisted route. Whereas the carbon coking was severely generated on the catalyst without PEI (177 g_{carbon}/g_{cat} for 40hr), the PEI-assisted catalyst showed the superior catalytic stability for dry reforming of methane with the minimization of the carbon coking (3.63 g_{carbon}/g_{cat} for 40hr). The synthesized nickel nanoparticles without PEI were placed outside of the porous structure and then, easily aggregated more than their parent support with the severe carbon coking. On the other hands, the sintering of nickel nanoparticles was more effectively hindered and the generation of carbon coking was suppressed on the confined nickel nanoparticles, since the prepared nickel particles via a PEI-assisted route were well-confined inside the pore. Therefore, combined with the simple procedure and low cost, the PEI-assisted route is one of the most promising methods for embedding nickel nanoparticles in mesoporous framework.

495924 Effects of Morphology and Site Proximity on Sorption-Enhanced Steam Methane Reforming Using Hybrid Ni-CaO Based Nanofibers

Abstract:

Hydrogen production is an essential component of the refining and petrochemical industries. Hydrogen also represents a critical energy carrier as energy portfolios shift from carbon-based to renewable technology. Currently 95% of the hydrogen in the US is produced via steam-methane reforming (SMR), with the byproducts being CO and CO_2. Sorption enhanced steam-methane reforming (SE-SMR) is an attractive adaptation of SMR technology. SE-SMR can produce a much higher hydrogen purity stream (>85%) at lower operating temperatures (500-600°C) and pressures (1-10 atm) compared to conventional steam reforming. The primary drawbacks of SE-SMR materials are the high temperatures required for sorbent regeneration and loss of sorption capacity and catalytic activity over repeated sorption-regeneration cycles. This work focuses on developing hybrid catalyst-sorbent materials that improve reaction and sorption rates by reducing the size and increasing the dispersion of metal and metal oxide domains.

Site proximity and morphology changes were achieved with nanofibrous metal oxides produced via electrospinning. Calcium oxide, nickel oxide, and magnesium oxide nanofibers were synthesized via electrospinning using calcium acetate, nickel acetate, and magnesium acetate precursors. Electrospinning solutions consisted of the salt precursor and polyvinylpyrrolidone (PVP) dissolved in an ethanol and water mixture. Electrospinning processing was conducted at 12.5-30 kV, 0.25-2 mL/hr extrusion rate, and 3-20 inch separation between the collector plate and the tip of the extrusion needle. Salt-PVP nanofibers were finally calcined at 600°C (2°C/min ramp rate) for 2 hours to decompose PVP and produce oxide nanofibers. MgO-supported Ni catalysts were also synthesized via incipient wetness impregnation of MgO powder with nickel nitrate solutions and calcined at 650°C (5°C/min ramp rate) for 5 hours.

Materials were characterized via scanning electron microscopy (SEM), x-ray diffraction (XRD), and N_2 physisorption. XRD spectra confirmed the formation of the desired metal oxide phases. SEM images indicated nanofiber diameters of approximately 80-200 nm, 65-95 nm, and 100-300 nm for calcined calcium oxide, nickel oxide, and magnesium oxide, respectively. The nanofibrous materials exhibited higher porosity, smaller crystallites, and smaller metal oxide domain sizes than their supported particle analogs.

The reactivity and catalytic activity of the nanofibrous materials were tested using CO_2 thermogravimetry, temperature-programmed reduction, and steam-methane reforming experiments. Calcium oxide nanofibers exhibited higher rates and higher saturation CO_2 capacities (78% vs 31%) compared to bulk calcium oxide powder. The increased rate and conversion of the calcium oxide nanofibers result form their smaller more reactive domains. Temperature-programmed reduction experiments were used to characterize Ni-based materials. Commercial Ni-based reforming catalyst exhibited reduction peaks at 500°C and 720°C, however, nickel oxide nanofibers exhibited a single reduction peak centered at 350°C. The much lower reduction temperature results from smaller NiO domains that are more evenly dispersed throughout the nanofiber compared to the particle analogs. The performance of absorber materials for SE-SMR was tested in a plug flow reactor at 500-650°C, 12-80 kPa of H_2O, 12-80 kPa of CH_4, and 1-5 atm total pressure. Catalyst-sorbent beds comprised of nanofibrous materials operated for longer periods of time and could be regenerated at milder conditions compared to beds consisting of supported Ni-catalysts and bulk CaO powder.

The results of this work indicate a promising opportunity for the application of nanofibrous materials in SE-SMR. Smaller, more reactive metal oxide domains allow for these materials to achieve stoichiometric CO_2 capacities without mass transfer restricted sorption rates. These hybrid catalyst-sorbent materials also retain their maximum capacity and catalytic activity over multiple cycles because they are less susceptible to particle agglomeration and morphology changes during thermal regeneration. These improvements in SE-SMR catalyst-sorbent materials has the potential to reduce costs associated with reforming and potentially lead to economically competitive distributed H_2 generation by (i) replacing expensive alloys required for high temperatures and pressures with less expensive materials of construction, (ii) eliminating water-gas shift reactors and downstream hydrogen purification, (iii) reducing carbon deposition in the reactor, and (iv) reducing the size of heat exchange equipment.

497081 Methane Partial Oxidation and Dry Reforming to Syngas Using the $La_{0.9}Ca_{0.1}FeO_{3-}$Î´Â Mixed-Conductor

Abstract:

One of the most promising technologies for carbon dioxide (CO_2) Capture and Sequestration is oxyfuel combustion, i.e. the use of pure oxygen in the combustion of fossil fuels. The major energy penalty in oxyfuel combustion comes from oxygen production. The current technology uses cryogenic methods and this process is energy intensive; it can lead to approximately 25% reduction in the net efficiency of a power plant.

Mixed ionic-electronic conducting (MIEC) membranes can serve as a better substitute. These materials have the potential of decreasing the energy consumption for oxygen separation significantly given that they operate under an oxygen chemical potential difference between the two sides. In addition, due to both oxygen separation and reaction (process intensification), chemical processes become simpler and this leads to a lower operating cost. MIEC materials are becoming popular in the energy sector, especially for chemical conversion processes.

A stable oxygen-conducting membrane under reducing conditions is $La_{0.9}Ca_{0.1}FeO_{3-}$Î´ (LCF). Our study investigates the performance of LCF in applications combining oxyfuel combustion, CO_2 reuse and syngas production under a reactive environment using methane-carbon dioxide mixtures in the oxygen lean side. In the presence of fuel, experimental measurements show that the oxygen flux through the membrane increases by one order of magnitude compared to the non-reactive case due to heterogeneous reactions of hydrogen and carbon monoxide with lattice oxygen ions. These surface reactions produce additional steam and carbon dioxide which further increase the syngas yield by reforming methane in the gas-phase. The aforementioned performance enhancement is accompanied by significant reuse of CO_2. Our results suggest that the performance of LCF is limited by the surface reactions on the fuel side and hence, we show that addition of nickel nanoparticles on the fuel side increase the oxygen permeation and the methane performance significantly by reducing the surface polarization resistance.

494091 Probing the Catalytically Active Phase of Cobalt Fischer-Tropsch Catalysts from First Principles: Predicting High Coverage Surface Hydroxyl Conformations Under Reaction Conditions

Abstract:

The reaction mechanism of Fischer-Tropsch (FT) synthesis is a continual source of intrigue and debate. While two classes of mechanisms have been proposed (namely, carbide vs. CO-insertion), the composition and configuration of the catalytically active phase has received a comparatively small amount of attention. Chemical transient kinetic (CTK) experiments performed in our group on model Co FT catalysts have yielded evidence for a CO-insertion mechanism, but just as importantly, these experiments have also provided evidence for a "crowded" reactive surface, where oxygen, hydrogen, and carbon bind to the surface of the catalyst in quantities exceeding the monolayer limit [1]. Because of the considerable delay in water formation in CTK experiments it is considered likely that the catalytically active phase contains substantial amounts of surface hydroxyl (OH). Such OH may therefore be considered a candidate precursor for constructing the "most abundant surface intermediate" ("masi") necessary for inducing hydrocarbon chain lengthening. Irrespective of what the mechanistic pathway for constructing the "masi" looks like, we present here a DFT-based investigation of an OH-rich surface, anticipating that such information will provide guidance for more complete mechanistic studies addressing chain lengthening initiation and growth.

By utilizing the Alloy Theoretic Automated Toolkit (ATAT) [2,3] to generate a large variety of extended Co (111) surfaces with different coverages and configurations of OH, we constructed a large library of OH/Co(111) structures. The face centered cubic facet is chosen because we wish to model smaller particle size Co, which typically transitions from its native hcp phase to the fcc phase. From this library we find that proximal OH have a large tendency to hydrogen bond as seen in Figure 1. These structures are quite common in the calculations, which suggests they are likely relevant to their stable surface phases. However, to actually test this notion, a corresponding lattice gas (LG) model using these structures will be constructed and subsequent grand canonical Monte Carlo simulations will be performed. In this way, we will elucidate the nature of OH patches on Co catalysts under FT synthesis conditions and their susceptibility to an interaction with CO molecules. A prevalence of surface OH was previously considered key in the CO dissociation on Co catalysts during FT [4], but may also react with CO molecules to provide an oxygen- and hydrogen containing C1 "masi". Perhaps more importantly, however, this result, combined with our previous CTK evidence [1], suggests that modeling the FT reaction over purely metallic Co—a common practice—is ill-advised if one wishes to properly test the mechanism of the reaction

Figure 1. Example of predicted ground state structures of OH on Co(111) at 0.5 ML (left panel) and 0.75 ML (right panel). All ground state structures below 0.5 ML do not exhibit hydrogen bonding and compression like that seen above. However, all ground state structures above 0.5 ML do. Blue spheres are Co, red spheres are O, and white spheres are H.

CITATIONS

1. Schweicher, J., A. Bundhoo, and N. Kruse, Hydrocarbon Chain Lengthening in Catalytic CO Hydrogenation: Evidence for a CO-Insertion Mechanism. Journal of the American Chemical Society, 2012. 134(39): p. 16135-16138.

2. van de Walle, A., M. Asta, and G. Ceder, The Alloy Theoretic Automated Toolkit: A User Guide. Calphad - Computer Coupling of Phase Diagrams and Thermochemistry, 2002. 26: p. 539-553.

3. van de Walle, A. and G. Ceder, Automating First-Principles Phase Diagram Calculations. Journal of Phase Equilibria, 2002. 23: p. 348.

4. Gunasooriya, G.T.K.K., et al., Key Role of Surface Hydroxyl Groups in C–O Activation during Fischer–Tropsch Synthesis. ACS Catalysis, 2016. 6: p. 3660-3664.

496731 Promoted Mixed Oxides for "Low-Temperature" Methane Partial Oxidation in Absences of Gaseous Oxidants

Abstract:

Promoted Mixed Oxides for "Low-Temperature" Methane Partial Oxidation in Absences of Gaseous Oxidants

Reforming of methane to syngas is an important but energy intensive operation in the synthesis of chemical and liquid fuels from natural gas. Steam reforming does not give an ideal H_2 to CO ratio for liquid fuel or chemical synthesis and conventional methane partial oxidation (POx) approaches require costly air separation. Chemical looping reforming (CLR), which partially oxidizes methane into H_2 and CO in the absence of steam or gaseous oxygen, offers a simpler and potentially more efficient route for syngas generation. This is achieved by cyclic removal and replenishment of active lattice oxygen in oxygen carrier particles also known as redox catalysts. Selecting a redox catalyst particle with high selectivity and facile oxygen transport is very important for CLR. Several redox catalysts have been reported in literature, but their activities toward methane POx, at low temperatures (<800 °C), are often limited due to the high activation energy for the migration and removal of lattice oxygen. Moreover, syngas selectivity is often low, as metal oxide surface species that catalytically activate methane at these temperatures tend to be non-selective. To address these limitations, we investigated the effects of promoting catalytic activity of oxide surfaces for a number of mixed metal oxide based redox catalysts. Our findings indicate that surface promotion can lower the onset temperature of methane POx by as much as 300 °C while achieving >90% syngas selectivity.

While the highly dynamic nature of the system makes mechanistic studies challenging, understanding the mechanism is important to aid rational catalyst design. We report pulse reaction and isotope studies indicating that the selectivity of the redox catalysts is related to the relative rates of lattice oxygen (O^{2-}) conduction to the surface and the surface oxygen removal by the gas-solids reactions. This is particularly the case at the early stage of the reaction. Surface promotion can increase the oxygen removal rate from the surface, thereby changing the type oxygen species present on the surface. This increase in oxygen removal is caused by the enhancement of surface methane activation. It is also shown that the rate of bulk lattice oxygen diffusion to the surface can be notably affected by surface promotion of the redox catalysts. This results from the change in driving force for O^{2-} conduction.

498172 A Highly Selective Route from Syngas to Ethanol: Tandem Catalysis Unconstrained By Anderson-Schulz-Flory Distribution

Abstract:

Selective higher alcohols synthesis (HAS) from $CO/CO_2/H_2$ mixtures (syngas) has been motivated by the demand of such products in the chemical and energy sectors, as well as the versatility and economics of syngas production. Research efforts in catalyst development for HAS have spanned decades, and in many cases drew inspiration from methanol synthesis and Fischer-Tropsch technologies. However, unlike the latter processes, insufficient selectivity and activity in HAS from syngas have prevented commercialization of chemocatalytic routes. Thus, further improvement in both metrics is required.

In many cases of HAS catalyzed by single catalysts, the reaction is believed to proceed by concerted hydrogenation and chain growth (C-C coupling) steps, resulting in co-production of aldehydes, alkenes, and alkanes, with chain-length approximately following the Anderson-Schulz-Flory (ASF) distribution. Such a wide spread in products increases separation costs for targeted HAS, and results in lower-value products (such as methane) competing for feed utilization. Advances in the engineering of the catalyst surface composition and structure have led to significant improvements in selectivities, but, to date, the aforementioned limitations have not been completely eliminated. One approach to break away from the constraints of the ASF distribution in HAS is to spatially separate the hydrogenation and chain-growth steps of the process by pairing catalysts that, individually, selectively catalyze only one of the steps, but, in a multi-bed tandem catalysis scheme, enable both.

In this work, we developed a combination of precious-metal-free, inexpensive, heterogeneous catalysts that, under differential conversions of CO/H_2 feed, are able to achieve upwards of 80% selectivity towards ethanol, on a CO_2-free carbon basis, with <1% methane selectivity. We will discuss the response of this system to process variables, and its stability. Additionally, some of the thermodynamic and kinetic considerations that are important in the design of such tandem-catalytic schemes for HAS will be highlighted.

494763 Control of Metal-Support Interaction of Fe@CNTs By Surface Modification of CNT and Its Application to Direct Olefin Synthesis from Syngas

Abstract:

Rapid economic development in China arouses the intense interest to produce increasing amounts of light olefins (e.g., ethylene, propylene) in both academia and industry. Traditionally, steam cracking of naphtha with dwindling and non-renewable crude oil is the main route for olefin synthesis [1,2]. Most recently, direct synthesis of light olefins from coal-/biomass-based syngas is regarded as the most promising non-petroleum route due to low-energy consumption, low-cost, high-efficiency, etc [1,3,4]. Generally, the catalyst design strategy for direct syngas route is mainly from the aspect of support modification, promoter doping, metal precursor synthesis and construction of multifunctional active sites [1,3,4]. Regardless of multifunctional catalyst, Fe is considered as the most promising active metal for FTO with _-Fe_5C_2 commonly recognized as active phase [1,5,6]. The key to high olefin selectivity is to modulate the Fe catalyst with appropriate capability for CO dissociation and hydrogenation.

Carbon materials possess high surface area, chemical inertness, mechanical strength, fruitful surface chemistry and electronic properties via covalent modifications. The type and quantities of surface chemical groups determined the interactions between carbon support and metal species [7,8]. Previous work demonstrated that iron-carbon interactions played an important role in Fischer-Tropsch process. Comparing with other conventional supports, the iron species on carbon materials are generally more reducible and can be transformed to iron carbides more facilely in the FTS reaction. In this work, the interaction between Fe and CNT is regulated by different surface modifications on CNT. The effect of the chemical and physical properties of the different CNT supports on the FTO performance of the Fe/CNTs was discussed.

Table 1 FTO performance of various carbon materials supported Fe catalysts

The CNTs were prepared by surface modification under different conditions. Fe supported catalysts were prepared by incipient wetness impregnation of the modified CNTs after drying and calcination. Structural and electronic properties of Fe/CNTs were characterized by BET, FTIR, Raman, XPS, XRD, TEM, TPR, TPD, XAFS, etc.

Figure 1 Catalytic performance of Fe catalysts supported on CNTs with different modifications

We evaluated first the FTO performance on multiple carbon materials supported Fe catalysts and found that CNT is the most favorable carbon support for olefin synthesis (Table 1). Moreover, we found that there existed large difference in performance of Fe catalysts supported on CNT with different modification methods (Figure 1). In similar conversion level, defect-rich HTCNT supported Fe showed highest initial conversion and olefin selectivity. Besides, OCNT with large amounts of O-functional groups supported Fe catalyst exhibited the lowest olefin selectivity but most stable activity. Based on the performance results, we tried to carry on various characterizations to elucidate the intrinsic effect of CNT modification on Fe-CNT interaction. BET, zeta potential and point of zero charge suggested that defect-rich HTCNT possessed highest surface area and less negative charge, while OCNT is the opposite. Raman and FTIR spectra evidenced the large amounts of O-contained groups (-C=O, -COOH) and rich defect on OCNT and HTCNT, respectively. In addition, there existed obvious electron transfer between Fe and modified CNTs, which is also demonstrated by XPS, in situ XRD, TPR, XAFS, etc.

Figure 2 Possible structure-performance relationship for FTO over different kinds of Fe NPs immobilized on CNT

The presence/absence of surface functionalities can directly affect the catalytic behavior of the active phases. The defects on CNT benefits to the initial activity and selectivity to light olefins. High concentration of O groups enhanced the stability of Fe during FTO. Moreover, the performance strongly depends on the degree of intimacy between CNT and Fe NPs. The surface modification caused the difference in Fe-CNT interface property, which significantly affected the electronic environment of Fe, ultimately altering the FTO performance. An optimal surface chemistry also facilitates catalyst conversion into its catalytically active state.

References

[1] H.M. Torres Galvis, J.H. Bitter, C.B. Khare, M. Ruitenbeek, A. I. Dugulan, K. P. de Jong, Science, 335 (2012) 835-838.

[2] L. Zhong, F. Yu, Y. An, Y. Zhao, Y. Sun, Z. Li, T. Lin, Y. Lin, X. Qi, Y. Dai, L. Gu, J. Hu, S. Jin, Q. Shen, H. Wang, Nature, 538 (2016) 84-87

[3] F. Jiao, J. Li, X. Pan, J. Xiao, H. Li, H. Ma, M. Wei, Y. Pan, Z. Zhou, M. Li, S. Miao, J. Li, Y. Zhu, D. Xiao, T. He, J. Yang, F. Qi, Q. Fu, X. Bao, Science, 351 (2016) 1065-1068

[4] K. Cheng, B. Gu, X. Liu, J. Kang, Q. Zhang, Y. Wang, Angew. Chem. Intern. Ed., 128 (2016) 4803-4806

[5] Y. Cheng, J. Lin, K. Xu, H. Wang, X. Yao, Y. Pei, S. Yan, M. Qiao, B. Zong, ACS Catal., 6 (2016) 389-399

[6] D. Fu, W. Dai, X. Xu, W. Mao, J. Su, Z. Zhang, B. Shi, J. Smith, P. Li, J. Xu, Y.-F. Han ChemCatChem, 7 (2015) 752-756

[7] T.O. Eschemann, W.S. Lamme, R.L. Manchester, T. E. Parmentier, A. Cognigni, M. R¿nning, K. P. de Jong, J Catal., 328 (2015) 130-138

[8] M. Oschatz, J.P. Hofmann, T. W. van Deelen, W. S. Lamme, N. A. Krans, E. J. M. Hensen, K. P. de Jong, ChemCatChem, 9 (2017) 620-628

491814 Heterogeneous, Carbon-Supported Cobalt Oxide Catalysts for the Oligomerization of Light Olefins (C_2-C_4) into Linear Olefins

Abstract:

Olefin oligomerization is a useful reaction to upgrade abundant, light hydrocarbons stemming from natural gas and petroleum manufacturing to higher-valued, longer-chain derivatives. Despite its commercial relevance, olefin oligomerization is one of the few industrial processes still dominated by homogeneous catalyst systems, which require an activator such as MAO. An early publication by Shultz et al. identified carbon-supported cobalt oxide as a particularly selective heterogeneous catalyst for linear oligomerization of C2-C6 alpha-olefins into linear olefins. [1] Recent work from our research group has expanded upon the original work of Shultz et al. to provide rational for the exceptional selectivity of this system and highlight the sensitivity of catalytic activity which can be improved by making variations to the synthesis procedures. [2] Here, we will report our most recent efforts to optimize these carbon-supported, cobalt oxide-based systems for the oligomerization of ethylene to 1-butene with selectivities as high as 82%. The influence of the carbon support on the final state of the active catalyst will be presented.Â We will suggest reasons for the inactivity of similarly synthesized materials on SiO2 and Al2O3 supports. Finally, we will demonstrate the potential industrial application of this system by showing that the less-valuable linear internal oligomers can be fed directly to a hydroformylation catalyst to form linear aldehydes.

[1] (a) Schultz, R. G.; Schuck, J.; Wildi, B. J. Catal. 1966, 6, 385â^'396. (b) Schultz, R. G.; Engelbrecht, R. M.; Moore, R. N.; Wolford, L. T. J. Catal. 1966, 6, 419â^'424.

[2] Zhuoran Xu, Joseph P. Chada, Dongting Zhao, Carlos A. Carrero, Yong Tae Kim, Devon C. Rosenfeld, Jessica L. Rogers, Steven J. Rozeveld, Ive Hermans, and George W. Huber. ACS Catalysis 2016 6 (6), 3815-3825

496347 Understanding Catalytic Impacts on Vacuum Gas Oil (VGO) Hydrotreating Using Electrospray Ionization-Ion Mobility-Mass Spectrometry (ESI-IMMS)

Abstract:

Vacuum gas oil (VGO) is a complex and challenging petroleum feedstock containing high double bond equivalent (DBE) and polycyclic aromatic species with heteroatoms like nitrogen and sulfur. The presence of these heteroatoms in hydrotreated VGO has great impact including high fouling rate to downstream hydroprocessing processes for making high-value nitrogen and sulfur-free products. Analyzing the composition of VGO is currently a challenge due to the complexity of petroleum-derived hydrocarbons. Analytical methods are increasingly more important as refining utilizes more diverse petroleum feedstocks. In this work, ion-mobility mass spectrometry (IMMS) is used to understand the effect of different hydroprocessing catalysts on the efficiency of nitrogen removal from VGO feedstocks. Analysis of different VGO samples before and after hydrotreating was done using high-resolution positive ion electrospray ionization IMMS. IMMS measurements demonstrated that catalytic hydrotreating altered the overall composition of the VGO samples, removing undesired heteroatom containing compounds. IMMS measurement also revealed different nitrogen residues in hydrotreated VGO products over varying catalysts, indicating that different catalysts had complementary outcomes for removing nitrogen containing compounds. Overall, the IMMS results demonstrate that the efficiency of nitrogen removal can be improved by modifying hydrotreating catalyst systems for diverse and changing VGO feedstocks.

498925 Comprehensive Mathematical Modeling of Fischer-Tropsch Synthesis within a Microreactor

Abstract:

The Fischer-Tropsch Synthesis (FTS) is a set of chemical reactions that involves the catalytic hydrogenation of carbon monoxide to various liquid hydrocarbons as well as water. It has received considerable interest of late, primarily driven by recent natural gas booms in the United States and elsewhere. Natural gas, which is chiefly methane, can be steam reformed to synthesis gas (a mix of CO and H_2) before being fed to a FTS reactor for conversion to more valuable liquid hydrocarbons such as diesel and naphtha. The goal of this work is to construct a mathematical model that successfully incorporates microkinetics, diffusion, and convection to simulate the FTS process within a microreactor.

This computational model will use COMSOL Multiphysics to simulate the FTS reactions on the surface of a cobalt-based catalysis using the CH_2 insertion mechanism. The aim will be to compare the theoretical product distribution determined from this model to experimental results obtained from an already existing microreactor platform. In this contribution, 2-D models of the microreactor will be presented along with steady-state calculations of the final products as well as analysis of surface species concentrations.

499497 Investigation into the Superior Coking Resistance of ALD-Coated, Nickel-Based Reforming Catalysts

Abstract:

Traditional supported-Nickel catalysts, although common, are known to be inadequate for harsh natural gas reforming conditions, especially in the Dry Reforming of Methane (DRM). This is primarily due to sintering and carbon deposition, or coking, resulting in rapid catalyst deactivation. Several approaches to improving Ni-based catalyst stability have been pursued over the years with varying degrees of success. This paper evaluates a novel catalyst prepared by the Atomic Layer Deposition (ALD) technique. A nanometer-thick over-layer of amorphous alumina is deposited via ALD on conventional wet impregnation-prepared 4% (nominal) Ni/Al2O3 catalyst with the purpose of protecting the underlying active Nickel sites from sintering, as well as inhibiting coke formation tendencies at an atomic level. Several types of reference Nickel catalysts are prepared with different supports (alumina, silica, titania) and compared to the ALD coated catalyst. Methane decomposition is widely accepted as the main mechanism of carbon precursor formation in DRM at higher temperatures. Therefore, the catalysts are evaluated by performing methane cracking reaction to facilitate rapid coking. Catalyst testing is performed in a custom TGA-DSC system with the evolved gases being monitored via an on-line MS. The data generated is correlated with data from DRIFTS studies to elucidate possible coke formation pathways. Preliminary results of carbon deposition studies showcase the superior coking resistance of the ALD catalyst as opposed to conventional ones. Investigations into the removal of coke using CO2 as an oxidant are also carried out. Detailed SEM, HR-TEM, XPS and XRD analysis provides further insight into the mechanism of coke formation and the effect of the ALD overcoat on catalyst performance. Since DRM is being promoted globally as an effective CO2 utilization pathway in gas processing (via ultra-clean fuels (Fischer-Tropsch) and value-added chemicals), there is significant merit in the application of ALD to synthesize reforming catalysts with tunable activities and stabilities.

499759 Hydrothermal Stability of ZSM-5 Zeolite

Abstract:

Upgrading of heavy oils into fuels and chemical precursors is a major technical challenge. Recent advances to improve process intensity and energy efficiency motivate interest in zeolite catalysis in the presence of liquid water or water-rich liquid reaction mixtures. Zeolites are crystalline microporous structures composed of silica and alumina with high surface area, commonly used for catalysis in environmental, fuel upgrading and biomass industries. Zeolites have the advantage of being extremely tunable based on the selection of zeolite framework, Si/Al ratio as well as a variety of post-synthesis modification techniques. However, many zeolites are prone to framework degradation under aqueous phase conditions. Larger pore zeolites such as H-Î² and H-Y will fully break down at 200 Â°C within hours when placed in hot liquid water. While many zeolites are unsuitable for hot liquid water conditions, ZSM-5 zeolite has been shown to retain crystallinity at for at least 6 hours when placed in liquid water at 200 °C. Understanding the relative stability of ZSM-5 in hot liquid water may provide the insight needed for further development of zeolite catalysis in hot liquid water phases.

In our study, we exposed ZSM-5 (silica/alumina ratio of 38:1) to liquid water for three hours at temperatures ranging from 200 °C to 450 °C to determine its hydrothermal stability. Post-run, the catalyst was characterized with a battery of complementary techniques including XRD, IR, TEM, XPS and 27Al MQMS NMR. While other zeolites completely degrade when exposed to liquid water at temperatures greater than 250 °C, ZSM-5 retained approximately 80% of its original crystallinity after 3 hours. ZSM-5 degradation occurs due to a combination of acid site loss at intermediate temperatures followed by surface degradation at higher temperatures. Interestingly, ZSM-5 exhibits improved crystallinity stability in supercritical water conditions as the familiar properties of liquid water (i.e., high dielectric and auto-ionization constant) shift dramatically near the critical point, contributing to decreased rates of ionic reactions.

Modification methods of both Brønsted acid sites and the zeolite surface were also investigated to improve ZSM-5 stability under hot liquid water conditions. To understand the effects of proton species desorption on the rate of degradation, ZSM-5 was ion exchanged with sodium before being treated in hot liquid water conditions. The lack of protons in solution led to an increase in zeolite stability at 300 °C under hot liquid water conditions. Improvements of zeolite crystallinity were also shown when altering ZSM-5 with organo-silane groups, an observation which is consistent with surface defects playing a key role in ZSM-5 degradation. Lastly, we measured the activity of liquid-water treated ZSM-5, finding that it retained some of its original activity for ethanol dehydration.

500834 Carbon Monoxide Oxidation on Nitrogen and Copper Doped TiO_2

Abstract:

Titanium dioxide (TiO_2), is a well-known support for photocatalysis, where it shows its efficiency in the removal of unwanted impurities from water and air, as well as generating chemicals, particularly hydrogen from water splitting [1]. Meanwhile, various TiO_2-supported catalysts including Cu-TiO_2 have been previously developed for numerous industrial applications [2], such as the water gas shift (WGS) reaction, the selective catalytic reduction of NO_x, and carbon monoxide oxidation. Among those diverse catalytic processes, carbon monoxide oxidation is attractive especially for preventing the poisoning of Pt catalysts by CO in proton exchange membrane fuel cell.

We developed one efficient approach to improve the interaction between copper and titania support with the aid of nitrogen addition, and elucidated the important role that copper cluster plays in improved activity and stability in CO oxidation. The nitrogen concentration on the surface was optimized by tuning the initial weight ratio of urea and TiO_2 and calcination temperatures. The XPS analysis revealed the maximum nitrogen concentration on the surface was achieved once calcination temperature reached 450 °C. Compared with undoped TiO_2, the addition of nitrogen facilitated the copper bound on the surface as well as the increase of actual copper loading. TEM characterization also confirmed that copper formed small clusters on nitrogen modified TiO_2. Kinetic studies focused on the measurement of the reaction rate regarding the carbon monoxide oxidation reaction. Three-fold increases were observed over copper modified nitrogen-titania. Stability results from time-on-stream test and start-shutdown also favored the nitrogen modified Cu-TiO_2. The enhanced activity and stability could be attributed to nitrogen facilitating the formation of copper clusters on TiO_2 surface.

References:

1. Fujishima, A., Zhang, X.T., Tryk, D. A. *Sur.Sci.Rep.* 63, 515 (2008).
2. Bartholomew, C. H., Farrauto, R. J. in â€œIndustrial Catalytic Processâ€ p.67. Wiley-Interscience, New Jersey, 2006.Â

496975 Nano-Engineered Ni Catalyst for Tri-Reforming of Methane

Abstract:

Tri-reforming of methane has been proposed to utilize the components from a direct flue gas without CO_2 separation step with desirable H_2/CO product ratio of 2. The activity of a tri-reforming catalyst increases with an active site depending on metal content and/or size. However, the catalyst can lead to sintering of metal particles during the highly endothermic reaction such as dry or steam reforming of methane reactions. In this study, we prepared the core-shell structure of Ni@CeO_2 catalyst, and applied to the try-reforming of methane. CeO_2 exhibits high oxygen vacancies with reversible valence change property (Ce^{4+}/Ce^{3+}), which is an active site in reaction accompanying CO_2. The prepared catalysts were characterized with transmission electron microscopy (TEM), X-ray diffraction (XRD), Fourier transform infrared spectroscopy (FTIR), and H_2-temperature-programmed reduction (TPR). Preliminary results showed that the core-shell structure prevents the activation of methane onto the Ni surface. Therefore, diffusion of methane through the CeO_2 shell should be increased to obtain higher activity in tri-reforming.

498248 Density Functional Theory Study of Dry Reforming of Methane on Pure Nickel As Well As Transition Metal Overlayer Deposited Nickel Surfaces

Abstract:

Dry reforming of methane (DRM) reaction utilizes CO_2, a major greenhouse gas to convert natural gas (mainly composed of methane) to synthesis gas, an important feedstock which could further be converted into valuable chemicals and cleaner fuel. This reaction presents a clear incentive in terms of its positive impact on the global environment and thus it has gained increasing attention from the scientific community lately. The superior catalytic activity of Nickel based catalysts and also their relatively lower costs make them the most promising catalyst for this reaction. However, these catalysts also deactivate rapidly owing to their high susceptibility to coke formation and filamentous carbon deposition. This severe catalyst deactivation is one of the major drawbacks that is obstructing the widespread commercialization of DRM. Several alternative catalysts have been explored for this reaction, including noble metals such as Rh and Ru. Even though these metals are found to be more reactive, as well as more resistant to carbon deposition, their high cost generally hinders their use [1]. One of the techniques that has been proposed to suppress the coke formation on the nickel surface is the substitution of single foreign transition metal atom which could modify the electronic structure [2].

In the current work, we present our exhaustive work on the solid state density functional theory (DFT) model results to study theoretically the wide network of elementary reactions comprising the DRM reaction on various facets of pure nickel catalyst, such as Ni(111), Ni(100) and Ni(110) surfaces. Calculations were performed using rev-PBE as exchange-correlation functional within the generalized gradient approximation (GGA) as implemented in the software VASP. Long term interactions such as the van der Waals forces are quite critical for the accurate description of the adsorption energies. Thus the DFT-D3 method of Grimme is used to account for dispersion corrections. Adsorption energies were calculated for all the DRM reaction intermediate species and then subsequently the activation barriers were calculated for all the elementary reactions in the DRM cycle. The catalytic activity of these pure nickel surfaces in terms of DRM reaction rate are then compared to the rates obtained on the single overlayer deposited nickel (X/Ni) surfaces where X is a transition metal such as Cu, Rh and Pd. Electronic structure analysis of various pure as well as overlayer modified surfaces is performed in terms of the d-band theory of catalysis. Our results indicate that the catalyst stability is greatly improved by a transition metal overlayer deposition on nickel surface.

References:

1. Pakhare, D. and J. Spivey, *A review of dry (CO2) reforming of methane over noble metal catalysts.* Chemical Society Reviews, 2014. **43**(22): p. 7813-7837

2. Fan, C.; Zhu, Y.; Xu, Y.; Zhou, Y.; Zhou, X. and Chen, D; *Origin of synergistic effect over Ni-based bimetallic surfaces: A density functional theory study.* The Journal of Chemical Physics, 2012. 137: 014703

500455 A Feasibility Study of Biogas Reforming to Improve Energy Efficiency and to Reduce Nitrogen Oxide Emissions

Abstract:

The feasibility of employing the catalytic steam reforming of biogas to increase its energy content was determined by converting the methane content of it into a hydrogen-rich syngas mixture; and using this reformate product intermixed with raw biogas in an Internal Combustion (IC) engine to enhance combustion stability and to reduce NO_X emissions. The field-testing component of the project involved catalytically reforming a side stream of biomethane from a landfill gas collection system at a California landfill via a waste-energy chemical recuperation process, in which waste heat from a gas engine was used to promote the reforming reaction of biogas. In the study, the total flow of raw biogas diverted to the engine remained constant. A fraction of that biogas was, however, separated and directed to a catalytic reactor packed with a Pt/Al_2O_3 catalyst. The reformer exit stream was then dehydrated (dried), blended with the remaining portion of biogas, and burned in the engine to produce electricity. When operating on the blended biogas mixture, combustion stability was enhanced and the engine ran smoothly at the full speed of 3600 rpm with a 60 Hz output frequency. On the other hand, when burning raw biogas without any reformate gas in the fuel mixture, the test engine ran poorly, sputtering and never reaching 3600 rpm. It was determined that reforming increased the energy value of the biogas by nominally 24% and estimated that a 50/50 blended reformed gas with biogas would have nominally 12% higher energy content. It was also demonstrated that the engine operating on mixtures of reformer gas had significantly lower NO_x emissions than when operating on pure propane under the same load conditions. Similar comparative testing could not be performed on raw landfill gas alone because the engine would not operate at full speed, as noted above. The conducted research has validated the technical and economic feasibility of using reforming products during biogas combustion in an internal combustion engine to reduce NO_X emissions and to enhance combustion stability. This, in our opinion is an important contribution to the scientific/technical literature and a significant advance for the NO_x emissions reduction and the field of biogas and engine waste heat utilization.

498372 Confined Nickel Nanoparticles Supported on Silica with and without Ceria Promoter for the Partial Oxidation of Isooctane

Abstract:

A confined Ni-based catalyst was synthesized using ceria promoters over silica supports via a wet impregnation method in ethylene glycol solvent, tested for its catalytic performance toward a partial oxidation reaction (POX) of isooctane at high space velocities, and compared with Rh- based catalyst that is commonly used for the same reaction. The catalytic performance of the catalysts was investigated at 750 ^{o}C, atmospheric pressure, GHSV of 43,681 h^{-1} and O_2/C ratio of 0.5. Various characterizations of prepared catalytic samples (including XRD and TEM analysis) demonstrated that the addition of ceria promoter and utilization of ethylene glycol solvent lead to the reduced Ni particle size and the improved Ni metal dispersion. Our reforming test results showed that isooctane conversion, syngas yields, and long-term stability of Ni supported catalyst samples significantly improve as the Ni metal dispersion increases. Thus, the Ni supported catalyst sample that is prepared with ethylene glycol solvent and ceria promoter showed the best reforming performances toward the POX of isooctane, which even outperformed the supported Rh catalyst.

501690 Aqueous Phase Reforming of Ethanol over Cobalt Doped Bismuth Vanadate

Abstract:

Cation doped bismuth vanadate ($BiMeVO_x$) has generated interest as a heterogeneous catalyst support due to its high oxygen conductivity. This characteristic is due to the unusual morphology of the material, in which crystalline perovskite sheets of $(Bi_2O_2)^{2+}$ alternate with $(VO_4)^{2-}$. This anisotropy is responsible for the material's high ionic conductivity, while limiting diffusion to a plane axis. In order to stabilize the desirable tetragonal phase, which ordinarily occurs only above 570 °C, ~15% of the vanadium is replaced with a dopant (Fe^{3+}, Cu^{2+}, Ti^{4+}, Co^{2+}). This creates defects in the structure which increases the number of oxygen vacancies, further improving the ionic conductivity.

In this study, BiCoVOx has been investigated as a support for nickel and platinum catalysts for the aqueous phase reforming of ethanol. This process is of interest because it allows dilute fermentation broths to be directly and inexpensively converted into syngas, an important precursor for synthetic natural gas and specialty chemicals. Testing has shown BiCoVOx supported catalysts perform favorably against an alumina support. In order to best exploit the material's unique properties, supports were prepared both by conventional combustion synthesis, yielding a highly textured, porous substrate; and by molten salt synthesis, producing micro-platelets of tunable dimensions. This highly ordered product has the highest oxygen conductivity along its axis. Thus, the best activity can be achieved by selective deposition of the catalyst on the platelet edges, which has been accomplished via strong electrostatic adsorption (SEA). To cheaply investigate this, a protocol for selective fluorescent dye uptake has been developed. By using this, a proposed SEA study can be quickly analyzed by fluorescent microscopy, to determine spatial deposition on materials that exhibit heterogeneous surface charging.

491253 Nickel-Based Catalysts for Aqueous-Phase Reforming of Short-Chain Alcohols

Abstract:

Aqueous-phase reforming (APR) is a technology to convert organic compounds in aqueous medium primarily to hydrogen and carbon dioxide. Because of the energetic efficiency of the process, APR has been considered as a promising route to upgrade the organic residue found in biorefinery water fractions to value added products. APR of oxygenated hydrocarbons to produce hydrogen requires catalysts that promote C-C bond cleavage and water-gas-shift (WGS) reaction. The highest activity and hydrogen selectivity in reforming reactions have typically been achieved over Pt-based catalysts. However, comparable activities were achieved over cheaper Ni catalysts[1]. However, the most tested nickel-based catalysts have been observed to deactivate rapidly[2].

The target of this work was to prepare and test suitable nickel modified catalysts using zirconium and cerium mixed oxides as supports. These supports have been chosen because of their good hydrothermal stability which is the main issue to be considered in APR studies.

Nickel was introduced to CeO_2-ZrO_2 supports by incipient wetness impregnation method. Reduction of the catalyst was done in-situ in APR reactor before the reaction. Successful modification of all tested oxide supports by nickel was proven by various characterization methods (AAS, N_2 physisorption, TEM and XRD).

APR experiments were performed in a continuously operated stainless steel tubular reactor using downward flow. Reaction conditions of experiments were 230 °C, 32 bars and 5 wt% aq. solutions of short chain alcohols, namely methanol, ethanol, propan-1-ol and butan-1-ol, were used as a feedstock.

Deep characterization of prepared catalysts revealed differences in pore size, surface areas, and metal particle size.

The prepared catalysts showed very high activities in APR of short-chain alcohols and conversions over 50 % were achieved. In the case of methanol reforming, hydrogen, methane, CO and CO_2 were detected as the main gaseous products. In case of other alcohols APR, also significant formation of light alkanes and alkenes was detected in gaseous phase and of aldehydes in liquid phase products. Comparable conversions of longer alcohols were determined but lower hydrogen efficiency was achieved due to the significant formation of side products.

Furthermore, the possibility of the catalyst reuse (after their re-reduction) in the APR of methanol was tested and proven. However, some loss of activity, which was addressed to the Ni particles sintering, was observed.

References:

[1] Wikberg, H.; Grönberg, V.; Jermakka, J.; Kemppainen, K.; Kleen, M.; Laine, C.; et al. Tappi J. **2015**, 14, 195-207.

[2] Coronado, I.; Stekrova, M.; Reinikainen, M.; Simell, P.; Lefferts, L.; Lehtonen, J. Int J Hydrogen Energy **2016**, 41, 11003-11032.

489614 Production of Hydrogen-Rich Syngas from Steam Reforming of Acetone over Ni-Co-Mg-Al Hydrotalcite Catalysts

Abstract:

Production of Hydrogen-rich SynGas from Steam Reforming of Acetone over Ni-Co-Mg-Al Hydrotalcite Catalysts

Sanchari Basu and Narayan C. Pradhan

Department of Chemical Engineering, Indian Institute of Technology,

Kharagpur - 721302, India.

E-mail:88sanchari@gmail.com, ncp@che.iitkgp.ernet.in

Abstract

In present time, the world is witnessing gradual exhaustion of conventional fossil fuels.This is driving us to look for alternate sources of energy. Hydrogen is a potential source of green energy. Today the most common process for hydrogen generation is the catalytic steam reforming of smaller hydrocarbons and alcohols such as methane, ethane, methanol, ethanol, glycerol, etc. The prominent catalysts for steam reforming are nickel supported on various oxides, hydrotalcites, etc. There are very few work reported on steam reforming of acetone, which is a low-cost by product of the cumene route for phenol. Also, acetone is an important intermediate in ethanol and acetic acid steam reforming. So, acetone steam reforming process can provide a better understanding of ethanol and acetic acid steam reforming processes.

In the present work, the steam reforming of acetone has been investigated over Ni-Co-Mg-Al hydrotalcite catalysts. The catalysts were made by varying the Ni:Co mole ratio (0.33, 1 and 3). Co-precipitation method was used to prepare the catalysts. The synthesized catalysts were characterized by XRD, N_2-BET, FESEM-EDS, FTIR and H_2-TPR techniques. CHNS analysis was performed to estimate the extent of carbon deposition on the spent catalyst. The steam reforming activity was evaluated in a continuous packed bed reactor at atmospheric pressure. The operating parameters were temperature, 450 °C - 550 °C; water/acetone molar ratio, 6 and space-time, 12.10 kg cat h/kmol acetone. The reactant and product compositions were determined by gas chromatography. Initially, the experiments were conducted at 550°C to compare the activities of different catalysts as shown in Figure 1. All the catalysts exhibited around 99% conversion. The methane selectivity was very less for all the catalysts. However, a maximum hydrogen selectivity of ~ 80% was obtained over Ni-Co-Mg-Al (0.33). Therefore, all further experiments were conducted over Ni-Co-Mg-Al (0.33) catalyst. The effect of temperature is shown in Figure 2. The conversion of acetone was >90% throughout the temperature range. The selectivity of hydrogen increased with temperature up to 500 °C and remained constant beyond this temperature. The selectivity of CO_2 and CO was high initially and decreased at high temperature. The catalyst did not show any considerable selectivity for CH_4. The stability test was carried out for 4 h at 500 ^oC and water/acetone mole ratio of 6. The time-on-stream behavior of the catalyst is shown in Figure 3. As can be seen from Figure 3, there is very little change in conversion over a period of 4h. Therefore, the catalyst is quite stable at the reforming conditions.

Figure 1: Activity of different catalysts at 550 ^oC.

Water:Acetone = 6:1 (molar), Space-time = 12.1 kg cat h/kmol acetone;

A: Ni-Co-Mg-Al (0.33), B: Ni-Co-Mg-Al(1) &C: Ni-Co-Mg-Al(3)

Figure 2: Effect of temperature on acetone conversion and product selectivity.

Water:Acetone = 6:1 (molar), Space-time = 12.1 kg cat h/kmol acetone;

Catalyst: Ni-Co-Mg-Al (0.33)

Figure 3: Time-on-stream behavior of the Catalyst A

References

1. Braga A. H., Sodre E. R., Santos J. B. O., Marques C. M. D. P, Bueno J. M. C., Steam reforming of acetone over Ni- and Co-based catalysts: Effect of the composition of reactants and catalysts on reaction pathways. *Applied Catalysis B: Environmental*, 195 (2016)16-28.

2. Guil-Lopez R., Navarro R.M., Ismail A.A., Al-Sayari S.A., Fierro J.L.G., Influence of Ni environment on the reactivity of Ni-Mg-Al catalysts for the acetone steam reforming reaction. *International Journal of Hydrogen energy*, 40 (2015) 5289-5296.

3. Hu X., Zhang L., Lu G., Pruning of the surface species on Ni/Al_2O_3 catalyst to selective production of hydrogen via acetone and acetic acid steam reforming. *Applied Catalysis A: General*, 427(2012) 49-57.

4. Navarro R. M., Guil-Lopez R., Ismail A. A., Al-Sayari S. A., Fierro J. L. G., Ni- and PtNi-catalysts supported on Al_2O_3 for acetone steam reforming: Effect of the modification of support with Ce, La and Mg. *Catalysis Today*, 242 (2015) 60-70.

5. Sun J., Mei D., Karim A.M., Datye A.K., Wang Y., Minimizing the formation of coke and methane on Co nanoparticles in steam reforming of biomass-derived oxygenates. *ChemCatChem*, 5 (2013) 1-5.

489377 Probing Topology and Reactant Effects on Hydride Transfer in Various Zeolites from First Principles

Abstract:

Theoretical Insights into the Effect of the Zeolite Environment and Reactant Identity on Alkane-Alkoxide Hydride Transfer

Thomas Chen and Matthew Neurock, Deparment of the Chemical Engineering and Materials Science, University of Minnesota, Minneapolis, MN, 55455.

The thermal stability and tunability of the pore topology of zeolites make them attactive solid acid catalysts for carrying out the elementary transformations involved in hydrocarbon chain growth processes important in the production of chemicals and fuels. Hydride transfer is ubiquitous to hydrocarbon transformation processes. However, detailed mechanistic studies of hydride transfer are limited as it is a secondary reaction that reactive intermediates can undergo, and direct experimental observations of hydride transfer are difficult to obtain. Herein, first principle density functional theory calculations are used to examine light alkane and olefin hydride transfer in TON, MFI, MOR, and BEA zeolites. We examine the effects of the alkane, the alkoxide, and the zeolite and show how transition state selectivity is determined by the zeolite topology. The transition state for hydride transfer are carbenium ions whose stability increase with reactant substitution thus increasing reactivity. However, calculations with TON show the existence of pore size limitations where hydride transfer reactivity decreases when either alkane or alkoxide exceeds a certain size threshold. Geometric and charge analyses are used to describe the competing effects of destabilizing steric interactions and stabilizing solvation effects. These interactions cause the pore size limitation where repulsive interactions outweigh zeolite solvation that delocalizes the positive charge of the carbenium ion transition state. Zeolite comparisons show the relative reactivity of alkane and alkoxide pairs for hydride transfer shifts to bulkier reactants as the pore size and dimensionality increase. Detailed comparison between TON and BEA show differences in energetics and transition state structure. Due to the larger pore size of BEA, no size threshold is observed for the molecules studied indicating a size threshold which is shifted towards bulkier species.

499776 Simulations of Ammonia Adsorption for the Characterization ofÂ Acid Sites in Metal-Doped Amorphous Silicates

Abstract:

The acidity of amorphous mesoporous silicates is directly related to the catalytic activity of these materials, and metal doping increases the acidic behavior. However, the local structure of the active sites is not well known as it is difficult to probe it experimentally. Additionally, it is not clear what drives the formation of specific acid sites for a specific metal dopant, i.e. Brønsted and/or Lewis, and their relative strength. All of this inhibits the systematic improvement of the catalytic activity of these materials. We use quantum mechanical simulations based on density functional theory to characterize the local structure of Zr-, W-, and Nb-doped amorphous silicates, for which experimental data of acidity is available. We consider a different number of hydroxyl groups directly bonded to the metal centers and of nearby silanol groups, and their effect on the Lewis and Brønsted acidity of the active site. The acidity is studied via adsorption of one or two ammonia molecules on the metal, M-OH, and Si-OH sites, respectively, to model measurements with the NH_3 temperature controlled desorption (NH_3-TCD) technique. Our calculations reproduce the experimental trends of acidity strength across metals as seen by Ramanathan (2012, 2013, 2014): Zr > W > Nb, where Zr mostly shows Lewis acidity, while Nb and W showing both Brønsted and Lewis sites. Metal atoms that are more grafted into the silica (have more M-O-Si interactions) exhibit the strongest Lewis acidity. Furthermore, silyl oxoniums stabilized by a nearby metal atom exhibit the strongest affinity for ammonia, and represent the most likely source of Brønsted acidity in these materials.

501323 First-Principles Grand-Canonical Simulations of Water Adsorption in Proton-Exchanged Zeolites Using a Highly Parallelizable Algorithm

Abstract:

Water is a ubiquitous solvent or an unavoidable impurity in many important engineering and biological processes. For reactions catalyzed by microporous materials, quantifying the amount or the structure of adsorbed water clusters at reaction conditions is prerequisite to understanding the relevant reaction mechanisms. Computational studies, such as those based on static electronic structure methods or first-principles molecular dynamics in the canonical ensemble, relies on preloading the framework with an often arbitrary amount of water molecules in order to examine their effects. We present here first-principles calculations in the grand-canonical ensemble of water adsorption onto proton-exchanged zeolites with different Si/Al ratios (Si/Al = 31--95). These simulations use pre-sampling with inexpensive force fields and a new parallelization algorithm to improve simulation efficiency and to bypass the scaling limit of the underlying electronic-structure method. These developments enable us to predict water adsorption under specified thermophysical conditions. Under almost all conditions examined (T = 373 -- 573 K, p = 10 -- 100 kPa), we observe an appreciable amount of water adsorption, and the zeolitic protons are delocalized by participating in the hydrogen-bonding chains formed by adsorbed water molecules.

495946 The Role of ABC-6 Zeolite Cavity in Methanol to Olefin Conversion

Abstract:

A density functional study (DFT) study is reported to investigate methanol to olefin (MTO) conversion in four ABC-6 zeolite cavities (*cha, avl, aft* and *hl*) differing in shape and size. The reaction barriers and energies of methylations, deprotonations and eliminations based on the side-chain of hydrocarbon pool (HP) mechanism are calculated in these cavities. By comparing their catalytic performance, we find that *cha* with low barriers in methylations and eliminations exhibits a stronger confinement effect due to its elliptical shape and small size. *Avl* and *aft* have low barriers in the first and second methylations, and the elimination barriers similar to *cha* for producing olefins. With a slightly wider dimension, *aft* shows high stability for HP species, especially the charged HP species due to long-range interaction with zeolite framework. These results demonstrate that the activity and selectivity of MTO conversion depend strongly on the size and shape of zeolite cavity, which is highly controlled by the confinement eï¬€ect of zeolite framework. Our calculations also predict a candidate zeolite with *aft* cavity might perform well in MTO conversion. This theoretical study reveals the relationship between zeolite structure and activity performance, and provides an efficient approach to design and screen ideal zeolites for MTO conversion.

496463 Understanding the C-H Activation and Dehydrogenation Mechanisms of Alkanes on Metal Oxides

Abstract:

Olefins are important chemical building blocks for the production of a wide range of valuable chemicals and plastics. A promising route to produce olefins is the non-oxidative dehydrogenation of alkanes on metal oxides. The Lewis acid-base functionalities of metal oxides play a key role in their catalytic behavior. However, how these functionalities affect the alkane dehydrogenation behavior and how they can be rationalized on multisite surfaces of metal oxides, is still elusive. In this work, we provide fundamental insights into the various mechanisms of alkane dehydrogenation on Î³-Al_2O_3 and identify chemically intuitive, structure activity relationships, by using Density Functional Theory calculations. These relationships are able to screen different sites of metal oxides towards their C-H activation and alkane dehydrogenation behavior. The obtained relationships can be utilized to accelerate the discovery of active dehydrogenation metal-oxide catalysts.

486831 Prediction and Screening of Product Distribution in Nanoporous Material-Catalyzed Propene Dimerization Via Molecular Simulations

Abstract:

The ability to predict the product distribution of a nanoporous catalyst *in silico* would enable better understanding of catalyst structure-function relationships and advance discovery of more selective catalysts for industrially important reactions. One such reaction is the propene dimerization reaction, which yields a mixture of branched and linear C_6 olefin isomers, of which the linear isomers are valued as industrial feedstocks for fuels and lubricants[1]. We have developed a conceptual framework based on shape selectivity theory[2] to explain and predict changes in the product distribution of the propene dimerization reaction resulting from the use of different zeolites and metal-organic frameworks (MOFs) as catalysts.

Using our model, experimentally observed changes in the product distribution from the literature[1] could be explained with quantitative agreement in terms of the contribution of the pores to the free energies of formation of the product isomers, which were computed using molecular simulations. We applied our model to screen a database of 118 existing and hypothetical MOF and zeolite structures to study how product distribution could be tuned by changing pore size, shape, and composition of porous materials. The screening identified a few not-yet-synthesized materials with improved linear isomer selectivity compared to experimentally identified catalysts, and determined catalyst properties that enhanced selectivity toward the valued linear olefin isomers, such as presence of open metal sites. This molecular picture of the mechanisms influencing product distribution provides insights for prediction and rational design of improved catalyst materials.

[1] Mlinar, A. N.; Keitz, B. K.; Gygi, D.; Bloch, E. D.; Long, J. R.; Bell, A. T. ACS Catal. 2014, 4, 717â€“721.

[2] Smit, B.; Maesen, T. L. M. Nature 2008, 451, 671â€“678.

493406 Catalytic Hydrogenation of Carbon Dioxide in Functionalized Metal Organic Frameworks

Abstract:

Metal Organic frameworks (MOFs) are a class of porous materials having potential for both gas adsorption and catalysis. Functionalization of MOFs with catalytic moieties results in heterogeneous catalysts with active sites similar to those of molecular catalysts. In this work, we will present how the capture and conversion of CO_2 to C1 products can be enhanced in MOFs by incorporating Lewis pairs (LPs) and chelated metal centers. In contrast to CO_2 reduction on metal surfaces, the pathways here involve concerted transfer of a hydride and proton to the adsorbed or co-adsorbed CO_2. The density functional theory results presented here were calculated using the CP2K code, with the PBE functional to describe the exchange correlation energy. Girmmeâ€™s D3 dispersion correction was used to account for van der Waals interactions. We computed reaction pathways for H_2 dissociation, CO_2 chemisorption, and CO_2 hydrogenation.

We have shown that inclusion of Lewis pair (LP) moieties into UiO-66 and UiO-67 gives effective catalysts for hydrogenation of CO_2 to formic acid and methanol [1, 2]. The reaction pathway in this system has a lower barrier than the barriers reported for the reduction of CO_2 on traditional heterogeneous catalysts [3]. MOFs have a similar structure topology to zeolites, where the effects of confinement on the transition states and intermediates are well documented [4]. Here, we focus on how the effects of pore size could be applied to improving catalysis in MOFs. We do so by comparing and contrasting results from UiO-66 (largest pore), MIL-140C, and MIL-140B (smallest pore).

We have also investigated the reduction of CO_2 to formic acid via a formate intermediate on chelated metal centers. The pathway for Cu active centers are formic acid desorption limited, with low (<0.2 eV) barriers for the bond breaking/formation steps. This system allows for further tuning of the activity by changing the identity of the metal atom, as well as the electron affinities of the chelating ligands.

1. Ye, J.; Johnson, J. K. *ACS Catal.* 5, 6219-6229 (2015)
2. Ye, J.; Johnson, J. K. *Catal. Sci. Technol.* 6, 8392-8405 (2016)
3. Ye, J.; Johnson, J. K. *ACS Catal.* 5, 2921-2928 (2015)
4. Jones, A. J.; Zones, S. I.; Iglesia, E. *J. Phys. Chem. C* 118, 17787-17800 (2014)

491022 Inorganometallic Catalyst Design: Alkane Metathesis Catalysis in Nu-1000 MOFs Functionalized with Transition Metals

Abstract:

Mid-weight (C_3-C_8) *n*-alkanes are petrochemical raw materials that are available in large quantities from oil and gas reserves. They are also produced in Fischer-Tropsch catalysis, which is the major route for the conversion of natural gas and coal to liquid fuels. It is desirable to convert mid-weight *n*-alkanes into higher-value molecules. Alkane metathesis, in which alkane molecules are transformed into higher and lower homologs, is a potential method for achieving such conversion.

The majority of single-site alkane metathesis catalysts are metal hydrides on amorphous SiO_2 and Al_2O_3. In each case, the metal plays dual roles: first, dehydrogenation of alkanes to alkenes, and second, perform alkene metathesis to form C–C bonds. The present work considers using a metal–organic framework (MOF), in particular NU-1000 (molecular formula $[Zr_6(\mu_3\text{–}O)_4(\mu_3\text{–}OH)_4(OH)_4(OH_2)_4](TBAPy)_2$, H_4TBAPy = 1,3,6,8-tetrakis(p-benzoic-acid)pyrene), as a support for new catalyst development. Because of their tunable surface structures, MOFs are ideal for rational heterogeneous catalyst design based on structure–reactivity relationships.

We used Kohn–Sham density functional calculations with the M06-L exchange-correlation functional for computational screening of NU-1000 functionalized with various transition metals, in particular screening for propane metathesis. We considered eight transition metals (V, Nb, Ta, Cr, Mo, W, Mn, Re) along with three variations of the NU-1000 framework: fully hydrated NU-1000, partially dehydrated NU-1000, and NU-1000 decorated with yttrium oxide. The activation energies for C–H bond activation and [2+2] cycloaddition were identified as descriptors of catalytic activity of different catalysts. Using the Sabatier principle, we constructed volcano plots to evaluate the expected efficacy of the trial catalysts. This identified vanadium-functionalized, yttrium oxide-decorated NU-1000 (VYO_x@NU-1000) as the most suitable catalyst among the trial catalysts for the propane metathesis. The complete reaction cycle for VYO_x@NU-1000 catalyst was computed and the free energy of activation is predicted to be 37.8 kcal/mol. This relatively high activation energy obtained by using a dual-role single-site catalyst indicates that the overall performance of the catalyst is compromised in order to achieve a balance between different reactions. We conclude that future developments in alkane metathesis catalysis should focus on dual-site catalysts and/or tandem systems.

490323 Hydrogenolysis of Propionic Acid to 1-Propanol Using a Bimetallic Pd-Re/SiO_2 Catalyst

Abstract:

The hydrogenolysis of carboxylic acids to more reduced compounds like alcohols and aldehydes is an important transformation in the production of biomass derived chemicals and fuels. Fatty alcohols, for example, can be derived from the hydrogenolysis of biologically produced fatty acids for use in plasticizers, lubricants, detergents, and other bulk chemicals. In this work, the structure of a SiO_2 supported 1 wt% Pd, 8 wt% Re catalyst was investigated using *in-situ* X-ray absorption spectroscopy (XAS), electron microscopy, steady-state reaction kinetics, and steady-state isotopic transient kinetic analysis (SSITKA) for the gas-phase hydrogenolysis of propionic acid. Reactions were carried out at 413-453 K under 0-3 atm H_2 and 0-0.05 atm propionic acid in a packed-bed flow reactor. The fresh catalyst consisted of highly-dispersed Re particles ~1nm in size, while the Pd particles were somewhat larger, 3-4 nm in size. Reaction conditions did not significantly affect the particle size. The catalyst demonstrated 80-90% selectivity to 1-propanol, with minor amounts of propionaldehyde, ethane, and propane. The XAS near-edge spectra demonstrated that Re was partially reduced to an intermediate oxide between Re^{+3}-Re^{+5} during pretreatment and gas-phase reaction. Analysis of the transient kinetic response suggested that a majority of Re on the catalyst is active for the hydrogenolysis reaction, and that intermediate Re oxides are active for the hydrogenolysis of carboxylic acids to alcohols. The hydrogenolysis reaction demonstrated a first-order rate dependence on H_2, and a zeroth-order dependence on propionic acid. An inverse kinetic isotope effect of 0.7 was observed when D_2 was substituted for H_2 during the reaction, suggesting the rate limiting step is unlikely to be a hydrogen addition step.

494127 First Principles Insight into the High Activity and Selectivity of Pd3Cu Alloys for the Guerbet Reaction

Abstract:

The Guerbet and related condensation reactions can be used to convert biomass-derived short chain alcohols into higher branched alcohols.[1] A variety of transition metal-based heterogeneous catalysts (e.g. Pd-based catalyst) possess good activity for this reaction, but the suppression of undesired side reactions must be further improved.[2] A newly discovered Pd-Cu alloy catalyst, however, was experimentally found to combine the activity of Pd with the selectivity of less active Cu, resulting in a unique catalyst with exceptional reaction rate and low tendency for the undesired decarbonylation reaction.[3] In this work, we present results from density functional theory simulations and micro-kinetic modeling to explain the origin of the remarkable properties of this bimetallic alloy catalyst.

The computational model was chosen to match the comprehensive experimental characterization showing a Cu-rich surface with a Pd-rich bulk. In agreement with experimental observations our microkinetic model predicts that the PdCu alloy with a Cu-rich surface has comparable activity with monometallic Pd catalysts for the desired (de)hydrogenation steps that covert alcohols to aldehyde and form saturated higher alcohol products. The high activity of the alloy can be attributed to the rise of the d-band center of the surface Cu atoms, making them nearly as active as surface Pd atoms.

While (de)hydrogenation steps are facile on both Pd sites and exposed Cu sites, the undesired decarbonylation reaction, however, requires a pair of adjacent Pd sites. The formation of a Cu-rich surface on the PdCu bulk alloy decreases the availability of these Pd ensembles, thus suppressing decarbonylation. Moreover, any produced CO will gradually block the unselective Pd sites under reaction conditions, but it does not affect the activity of surface Cu sites. This interpretation is consistent with experimental observations, showing that the ratio of decarbonylation to dehydrogenation decreases with increasing conversion and CO formation.

Finally, the effect of particle size was investigated by comparing the key reactions on terrace sites with those on stepped surfaces.[4] Congruent with experimental data the (de)hydrogenation steps are favored on larger particles, while decarbonylation is favored on smaller particles. In addition, we propose that CO can induce re-segregation of Pd atoms to the undercoordinated step sites, which further enhances decarbonylation and amplifies the geometric effect.

In summary, we have used ab-initio methods to elucidate the fundamental characteristics of PdCu alloys that lead to their exceptional catalytic performance for the Guerbet reaction. The overall catalyst properties stem from synergistic alloying, segregation, coverage and particle size effects. We anticipate that other bimetallic alloys with tailored performance can be discovered based on the fundamental understanding contributed by this study.

Reference:

(1) Kozlowski, J. T.; Davis, R. J. *ACS Catal.* **2013**, *3*, 1588–1600.

(2) Sitthisa, S.; Resasco, D. E. *Catal. Letters* **2011**, *141*, 784–791.

(3) Goulas, K. A.; Sreekumar, S.; Song, Y.; Kharidehal, P.; Gunbas, G.; Dietrich, P. J.; Johnson, G. R.; Wang, Y. C.; Grippo, A. M.; Grabow, L. C.; Gokhale, A. A.; Toste, F. D. *J. Am. Chem. Soc.* **2016**, *138*, 6805–6812.

(4) Goulas, K. A.; Song, Y.; Johnson, G. R.; Chen, J. P.; Gokhale, A. A.; Grabow, L. C.; Toste, F. D. *submitted.*

495145 Tunable Mixed Metal Oxides for Selective Hydrogenation and Ring-Opening of Furfuryl Alcohol

Abstract:

Lignocellulosic biomass can be broken down to furfural and furfuryl alcohol, major platform chemicals, which can be further catalytically converted into fuel additives (2-methylfuran, 2-MF) and polymer precursors (1,5-pentanediol, 1,5-PD) under hydrogen atmosphere. A shift towards catalysts based on multi-metal systems has shown promise in developing more selective catalysts for biomass conversion. Prior work for vapor phase hydrogenolysis to 2-methylfuran conducted over Ni-Fe/SiO_2[1], Mo_2C[2], and Co-Al[3] demonstrated yields of 80% (250 °C), 55% (180 °C), and 42% (155 °C), respectively. In liquid phase batch systems other bifunctional catalysts such as Ir-ReO_x[4] and Pt/Co-Al[5] are active for ring-opening of furanics. However, there is still a desire to increase yields to these value-added products, reduce the need for expensive precious metals, and further our understanding of the electronic environment of bifunctional catalysts.

The work presented demonstrates the versatility of mixed metal oxides (MMO) derived from layered double hydroxides (LDH) to conduct furanic chemistry after undergoing a reduction pretreatment. By either doping small amounts of Fe or Cu into a Co-Al oxide matrix improvements towards selectivity and activity are observed for side chain hydrogenolysis and ring-opening, respectively. The resulting multi-metal catalysts can achieve yields of 82% towards 2-MF (Co-Fe-Al) [6] and 42% towards 1,5-PD (Cu-Co-Al) under optimizing conditions, which is comparable or better than prior catalysts with similar mechanisms. To better understand the electronic environment, various spectroscopic experiments were used to develop a model of how each oxide species was changed under reducing conditions.

This work provides a comprehensive account of the versatility of non-precious metal MMO catalysts for biomass conversion. By tuning the metal addition, various chemistries can be targeted in furanic conversion.

References

1. Sitthisa, S., An, W., and Resasco, D.E., *J. Catal.*284, 90 (2011).

2. Lee, W.-S., Wang, Z., Zheng, W., Vlachos, D. G., and Bhan, A., *Catal. Sci. Technol*. 4, 2340 (2014).

3. Sulmonetti, T. P., Pang, S. H., Claure, M. T., Lee, S., Cullen, D.A., Agrawal, P. K., and Jones, C. W., *Appl. Catal., A* 517, 187 (2016).

4. Koso, S., Furikado, I., Shimao, A., Miyazawa, T., Kunimori, K., and Tomishige, K. *Chem. Commun.* 2035 (2009).

5. Xu, W., Wang, H., Liu, X., Ren, J., Wang, Y., and Lu, G. *Chem. Commun.*47, 3924 (2011).

6. Sulmonetti, T. P., Hu, B., Ifkovits Z., Lee, S., Agrawal, P. K., and Jones, C. W., *ChemCatChem* (2017) Accepted.

497568 Support Effects on Ethanol Dehydrogenation over Cu Catalysts

Abstract:

AICHE 2017 Fall Minneapolis Abstract

Title:

Support Effects on Ethanol Dehydrogenation over Cu Catalysts

Abstract:

Ethanol (EtOH) dehydrogenation to acetaldehyde (Acet-H) has been widely studied over oxide supported Cu catalysts as a reaction pathway that can provide a sustainable alternative to the current industrial process that uses ethylene as a feedstock for fuel. The rates and selectivity for Acet-H formation have been shown to be dependent on Cu particle size, reaction conditions, and the catalyst support. Specifically, it has been argued that Acet-H production occurs exclusively over Cu^0 sites, whereas other products such as diethyl ether, ethylene, and ethyl acetate form over the support or metal-support interface. Here, we will discuss how the nature of the oxide support influences Acet-H formation on Cu^0 sites. A combination of N_2O titration experiments, CO probe molecule IR, *in-situ* IR, and reactivity measurements under strict kinetic control are used to provide mechanistic insights into the support effects on Acet-H production with a specific focus on comparative turnover frequency measurements, mechanistic interrogation, and stability analysis.

498017 A Combined Theoretical and Kinetic Assessment of Câ€“O Bond Rupture Pathways within 2-Methyltetrahydrofuran over Nickel Phosphide Catalyst

Abstract:

The conversion of oxygenates in biomass-derived feedstocks into more valuable fuels and chemicals requires selective C–O bond cleavage while leaving other C–O and C–C bonds intact. Previous studies have shown that metal phosphides (MPs) are promising catalysts for selective C–O, C–S, and C–N bond rupture over C–C bonds [1]. MPs also exhibit remarkably higher selectivity towards breaking sterically-hindered C–O (^{3}C–O) bonds compared to pure transition metals. Our measurements indicate that Ni_2P cleaves ^{3}C–O bonds at a rate three times faster than ^{2}C–O bond cleavage, and has 50 times higher selectivity for ^{3}C–O bond cleavage than pure Ni. The specific mechanism behind C–O bond cleavage on Ni_xP_y, as well as the identity of the reactive intermediate and active site, are not fully understood despite recent studies. Here, we combine kinetic studies and density functional theory (DFT) calculations to investigate the mechanisms of hindered and unhindered C–O bond cleavage in 2-methyltetrahydrofuran (2-MTHF) on Ni_xP_y and Ni surfaces.

Previous studies showed that C–C and C–O hydrogenolysis in alkanes and alkanols occur via unsaturated species formed by sequential quasi-equilibrated dehydrogenation steps [2,3]; the removal of H atoms weakens the C–C and C–O bonds by replacing C–H bonds with C–metal bonds. The measured ratio of products to reactants pressures during the dehydrogenation of 2-methyltetrahydrofuran (2-MTHF) to 2-methylfuran (2-MF) was constant at low conversions (< 1%), suggesting that C–H activation steps are quasi-equilibrated and leaving the C–O bond rupture as the kinetically relevant step. The measured ratio of ^{3}C–O to ^{2}C–O bond cleavage rates (χ) has a half order dependence on hydrogen pressure $[H_2]^{1/2}$, indicating that the reactive intermediate for ^{2}C–O bond rupture must lose an additional hydrogen atom compared with the reactive intermediate for ^{3}C–O bond rupture; the identity of those intermediates that undergo C–O bond cleavage can be assessed using DFT calculations.

Periodic plane-wave DFT calculations were performed using the Vienna *ab initio* Simulation package (VASP) with the RPBE exchange correlation functional and PAW potentials. The (001) and (111) surfaces were chosen to model Ni_xP_y and Ni catalysts, respectively, because they exhibit the lowest surface formation energies. Vibrational frequency calculations were performed to estimate enthalpies and free energies of each state at temperatures relevant to the catalytic reactions. Results show that the effective enthalpy barrier ($\Delta H\neq$) for ^{3}C–O bond rupture is 34 kJ mol^{-1} lower than for ^{2}C–O on Ni_2P(001) and 21 kJ mol^{-1} on $Ni_{12}P_5$(001) at 543 K. These DFT-derived enthalpy barriers are consistent with the measured values on 12 nm Ni_2P (23 ± 9 kJ mol^{-1}) and on 19 nm $Ni_{12}P_5$ (25 ± 5 kJ mol^{-1}). ^{2}C–O bond rupture on Ni(111), however, is more favorable by 73 kJ mol^{-1}. Although C–O bonds activations through

partially dehydrogenated intermediates (i.e. $^2CH_2$–O, ^{2}CH–O, and ^{3}CH–O) have lower enthalpy barriers, the removal of H atoms from the carbon atom vicinal to the C–O bond being broken results in an increase in entropy, rendering $\Delta G\neq$ barrier the lowest for the fully dehydrogenated C–O bond. The contribution of each transition states to the measured hydrogenolysis rate depends on their $\Delta G\neq$ values, and thus partially dehydrogenated intermediates are unlikely to contribute to the measured rate. These findings confirm the essential role of entropy in addition to enthalpy in determining the rate constants. DFT-predicted ratio of ^{3}C–O to ^{2}C–O bond cleavage rates χ is equal to 2.7 on Ni_2P(001) and 0.53 on $Ni_{12}P_5$(001) at 543 K and 1 MPa H_2, in agreement with the measured values (~3 on 12 nm Ni_2P and ~0.4 on 19 nm $Ni_{12}P_5$) at low conversions.

References

[1] Oyama, S.; Wang, X.; Lee, Y.; Chun, W., Active phase of Ni2P/SiO2 in hydroprocessing reactions. *J. Catal.* **2004**, 221.2, 263-273.

[2] Flaherty, D.; Hibbitts, D.; Iglesia, E., Theoretical and kinetic assessment of the mechanism of ethane hydrogenolysis on metal surfaces saturated with chemisorbed hydrogen. *J. Am. Chem. Soc.* **2014**, 136, 9664.

[3] , E.; Hibbitts, D.; Iglesia, E., Kinetic and Mechanistic Assessment of Alkanol/Alkanal Decarbonylation and Deoxygenation Pathways on Metal Catalysts. *J. Am. Chem. Soc.* **2015**, 137(37), 11984-11995.

498394 Enhanced Deoxygenation on Bifunctional Pd/Al_2O_3 Modified with Phosphonate Self-Assembled Monolayers

Abstract:

The modification of Pd/Al_2O_3 with self-assembled monolayers (SAMs) deposited from phosphonic acids for improving deoxygenation was investigated. In this study the approach was to control the near-surface environment of the metal-support interface by applying SAMs to the catalyst support. More specifically, the goal was to produce a bifunctional deoxygenation catalyst via the introduction of carboxylic acid functionality in the SAM tails. Phosphonic acids containing either alkyl or carboxylic acid tails were deposited onto Pd/Al_2O_3 from solution, then annealed to covalently bind the SAMs to Al_2O_3 before washing to remove excess phosphonic acid. Fourier-transform infrared spectroscopy (FT-IR) was used to confirm the presence and identity of the different SAMs. The uncoated and SAM-coated catalysts were then tested in the gas-phase deoxygenation of different pyrolysis-oil components, such as furfuryl alcohol, under hydrogenation conditions. At the conditions tested, uncoated Pd/Al_2O_3 exhibited low selectivity (<5%) toward the deoxygenation product, 2-methylfuran, and primarily performed ring hydrogenation. When modified with phosphonate SAMs, a significant improvement in deoxygenation selectivity can be observed, regardless of the tail group tested. Though, when comparing SAMs of similar size and coverage, those having carboxylic acid tails show greater rates of deoxygenation than those having alkyl tails. Modification with phosphonate SAMs typically resulted in an increase of the overall catalyst activity as well, so that the observed rate of deoxygenation was improved by more than an order of magnitude with particular SAMs. FT-IR following the adsorption of pyridine was used to confirm the introduction of Brønsted acidity by phosphonate SAMs, and to compare the relative acid strengths. The results of this work suggest that preparation of bifunctional catalysts with organic monolayers may be a useful way to tune catalyst activity and selectivity, particularly given the ease of preparation and the high stability of the SAMs.

Chemical and Biological Engineering Department

University of British Columbia

The activity and stability of Mo_2C supported on activated petroleum coke in hydrotreating reactions

Haiyan Wang, Shida Liu and Kevin J. Smith

2017AICHE
Nov 2nd, 2017

Outline

- Background
- Objectives
- Results and Discussion
  - Catalyst characterization
  - Catalyst transformation/deactivation study
- Summary

Background

- R.B.Levy and M. Boudart
  - Early transitional metal carbides
  - Platinum-like behavior
  - H_2 efficiency

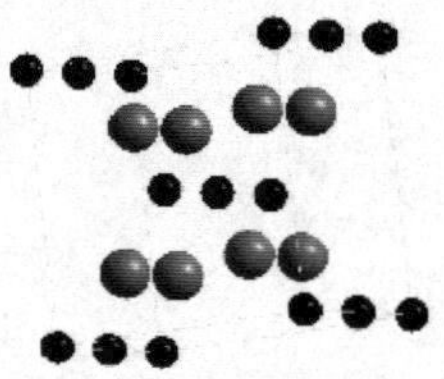

Mo_2C

- Crude oil and bio-oil upgrading

Mo_2C performance:

- HDO[1]

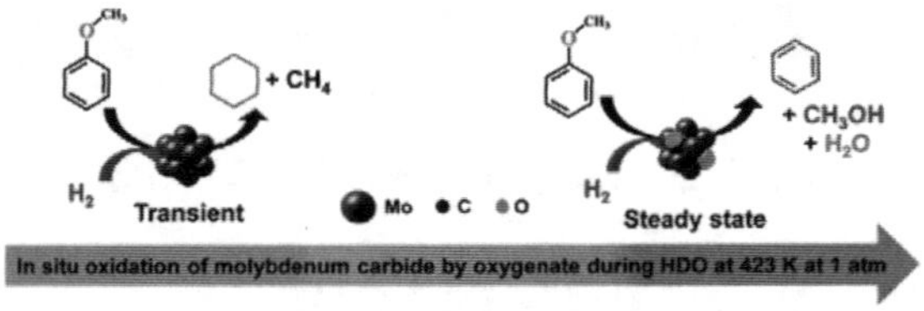

- HDS[2], HDN

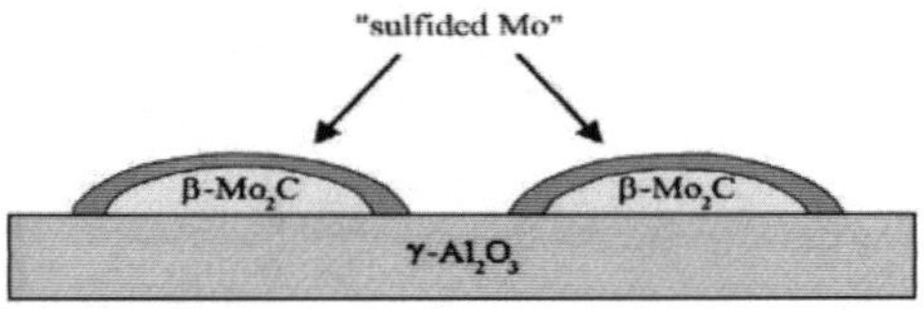

Graphs [1, 2] were adopted from Wen-Sheng Lee et al. (ACS Catalysis, 2015) and K.R.McCrea et al. (1997, J.C.)

Background

Catalysts synthesis

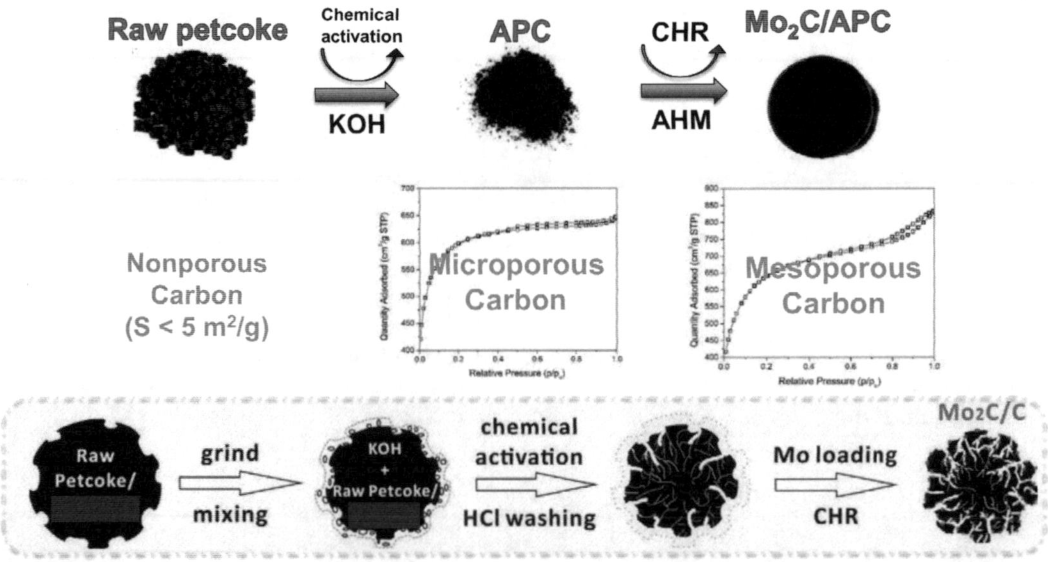

Objectives

- **To evaluate the activity and selectivity of the catalysts in hydrotreating (HDO, HDS, HDN)**
- **To understand the effect of heteroatoms (O, S and N) on Mo_2C catalysts**

2017 AICHE

5

Results and Discussion

Characterization-XPS

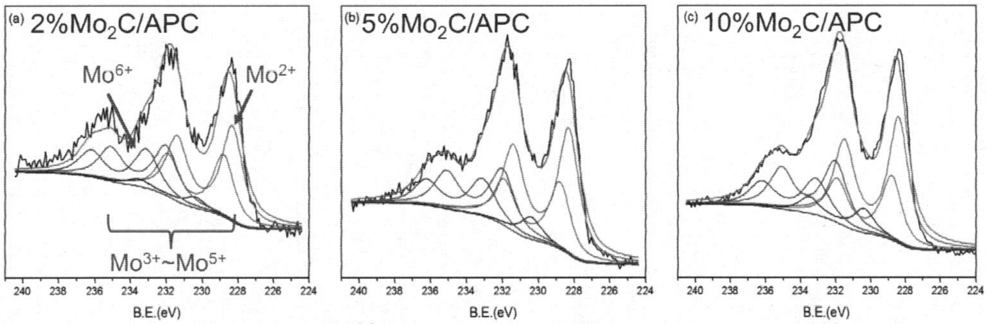

Fresh catalysts	Mo^{2+} (B.E.=228.3 eV)	Mo^{3+} (B.E.=228.8 eV)	Mo^{4+} (B.E.=230.4 eV)	Mo^{5+} (B.E.=232.0 eV)	Mo^{6+} (B.E.=233.1 eV)
	~40%	~45%			~15%
$2\%Mo_2C/APC$	35	23	4	21	16
$5\%Mo_2C/APC$	40	21	4	20	14
$10\%Mo_2C/APC$	39	20	7	21	13
	⇩ Mo_2C	⇩ Mo oxycarbide			⇩ Mo oxide

2017 AICHE

6

Fresh 2%Mo_2C/APC

□ Characterization-TEM

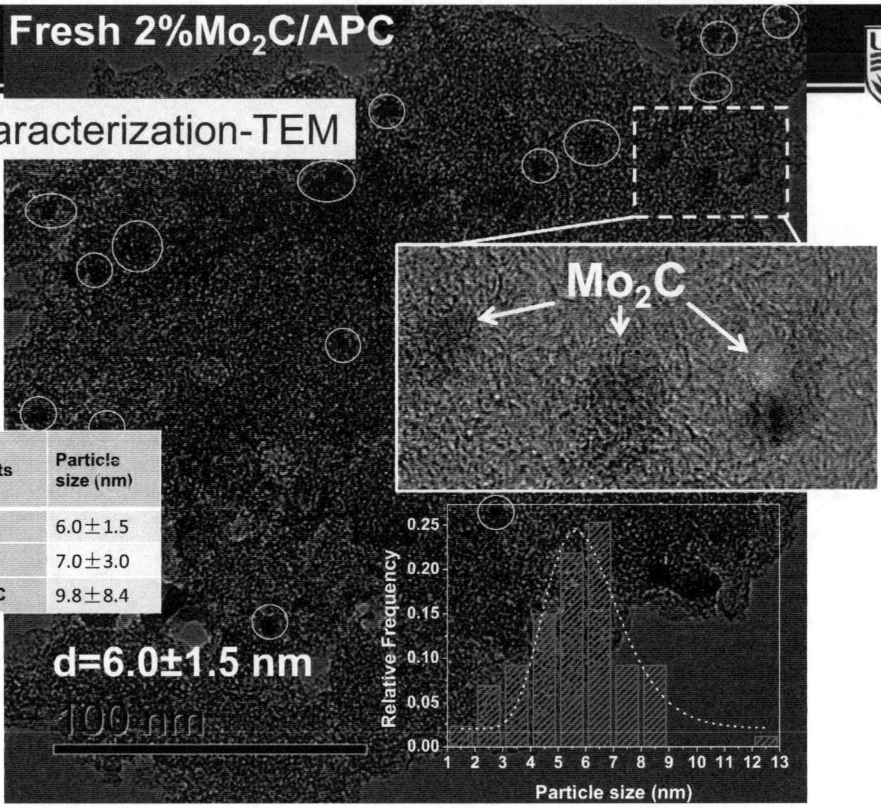

Fresh catalysts	Particle size (nm)
2%Mo_2C/APC	6.0±1.5
5%Mo_2C/APC	7.0±3.0
10%Mo_2C/APC	9.8±8.4

2017 AICHE

7

Results and Discussion-HDS

Reactor: Fixed-bed reactor with down flow

Reaction condition: 350 °C, LHSV=4 h^{-1}, H_2/liquid flow=600 (gas phase reaction).

HDS reactant: DBT (~0.2 wt%)

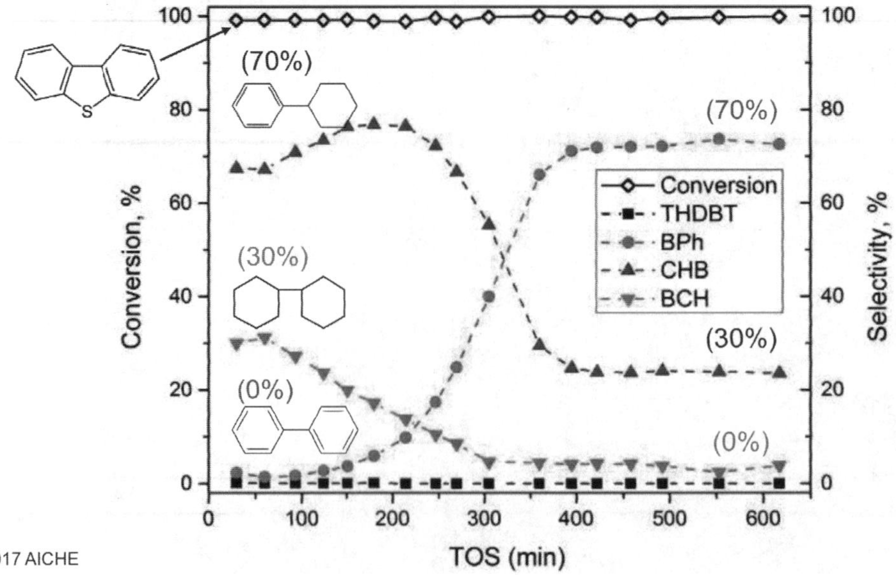

2017 AICHE

8

HDS of DBT on Mo_2C/APC

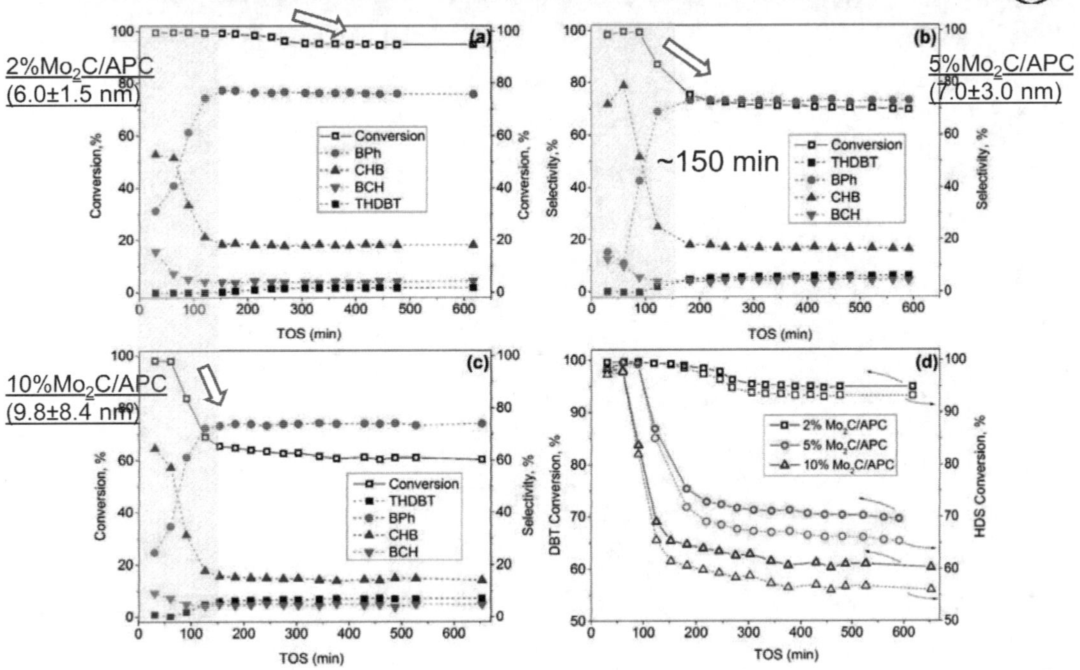

2017 AICHE

HDS of DBT on Mo_2C/APC

Exponential decay rate law

$$-\frac{da}{dt} = k_d a$$

Catalyst	k_d (h^{-1})	R^2
10%Mo_2C/APC	1.96E-2 ± 4.48E-3	0.94
5%Mo_2C/APC	9.65E-3 ± 1.71E-3	0.90
2%Mo_2C/APC	2.60E-3 ± 1.28E-4	0.98

2%Mo_2C/APC

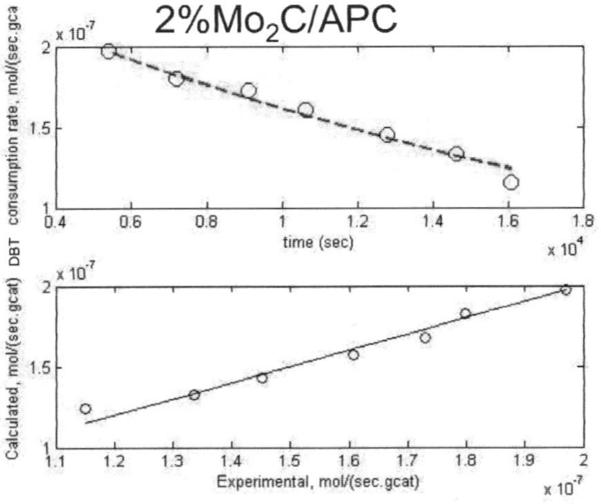

- Deactivation type: Poisoning
- Large particle → large k_d.

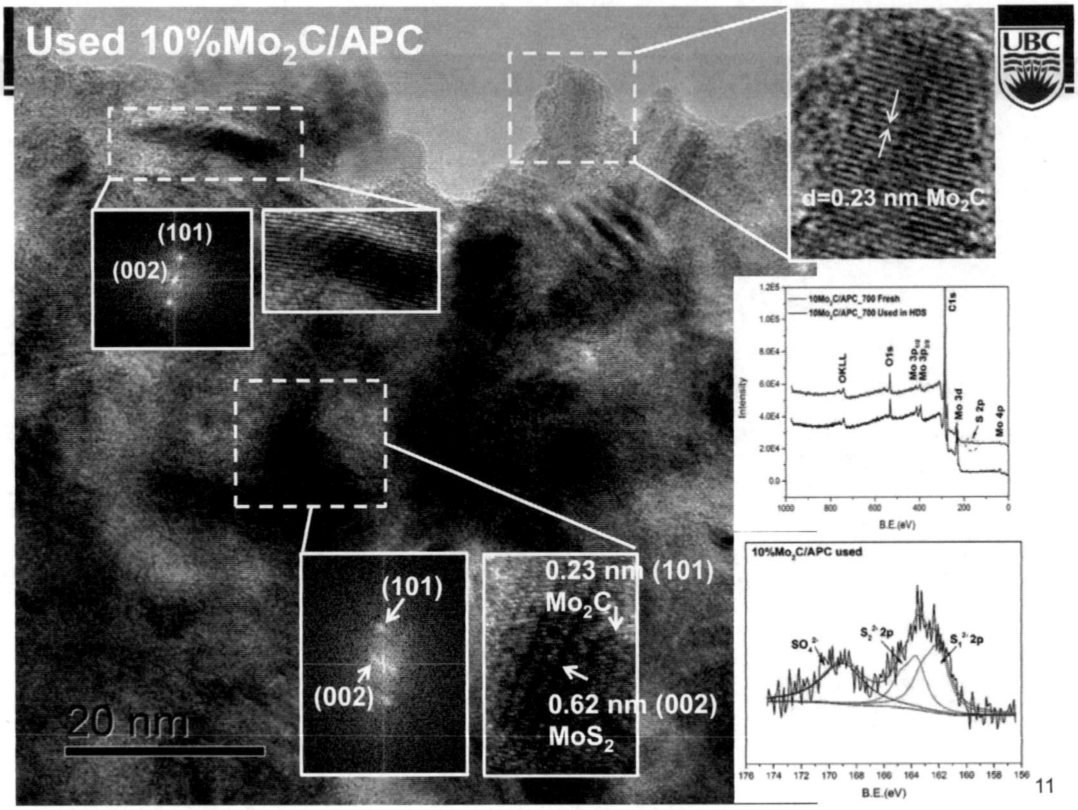

11

Results and Discussion-HDN

Reaction Condition: 350 °C, SV=4 h^{-1}, H_2/liquid flow=600 (gas phase reaction).

HDN Reactant: CBZ (~0.2 wt%)

Classified groups	Identified products
N-containing products	THCZ; CHBA; DHCZ
DDN product	CHB
HYD products	CHCHE; BCH
Isomerization products	CPCHX; HCH
Cracking products	BEN; PhMe; CH

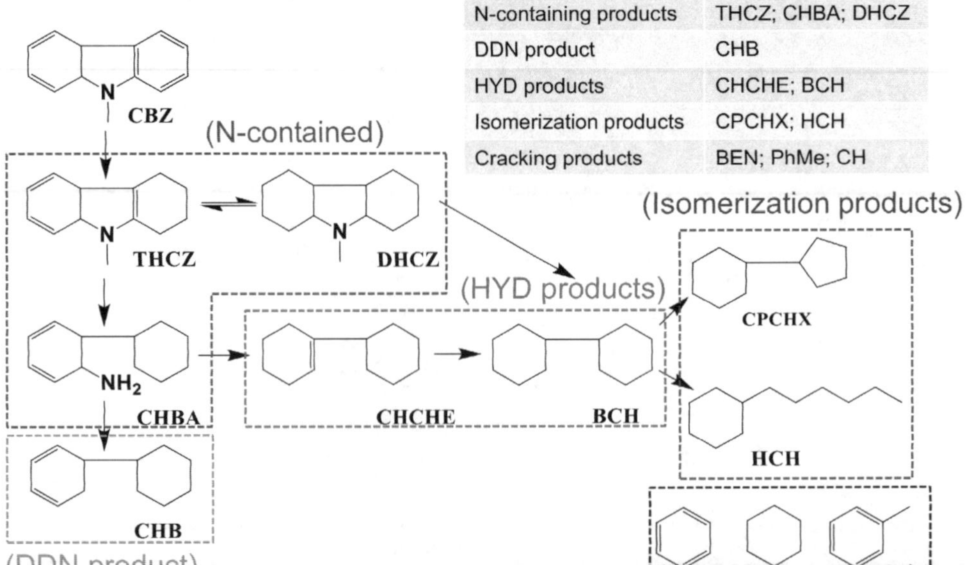

2017 AICHE

12

HDN of CBZ on Mo_2C/APC

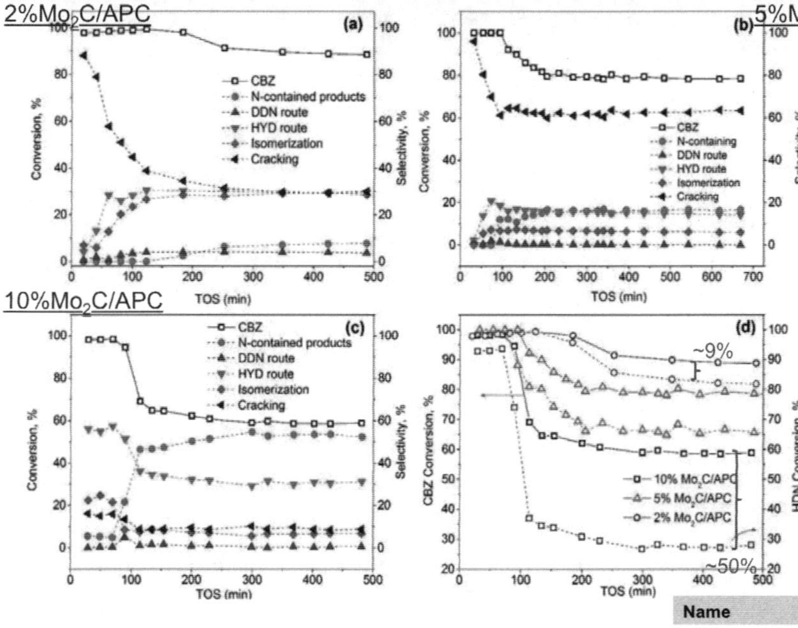

Fresh catalysts	Particle size (nm)
2%Mo_2C/APC	6.0±1.5
5%Mo_2C/APC	7.0±3.0
10%Mo_2C/APC	9.8±8.4

Name	N/Mo
(a) 2%Mo_2C/APC used in HDN	2.05
(c) 10%Mo_2C/APC used in HDN	0.21

2017 AICHE

13

HDN of CBZ on Mo_2C/APC

Deactivation type:
Poisoning
Exponential decay

$$-\frac{da}{dt} = k_d a$$

HDN

Catalyst	k_d (h^{-1})	R^2
10%Mo_2C/APC	2.26E-2 ± 2.72E-3	0.97
5%Mo_2C/APC	1.03E-2 ± 1.74E-3	0.93
2%Mo_2C/APC	3.70E-3 ± 5.83E-4	0.90

HDS

Catalyst	k_d (h^{-1})	R^2
10%Mo_2C/APC	1.96E-2 ± 4.48E-3	0.94
5%Mo_2C/APC	9.65E-3 ± 1.71E-3	0.90
2%Mo_2C/APC	2.60E-3 ± 1.28E-4	0.98

2017 AICHE

14

Results and Discussion-HDO

Reaction Condition: 350 °C, SV=4 h^{-1}, H_2/liquid flow=600 (gas phase reaction).

HDO Reactant: DBF (~0.2 wt%)

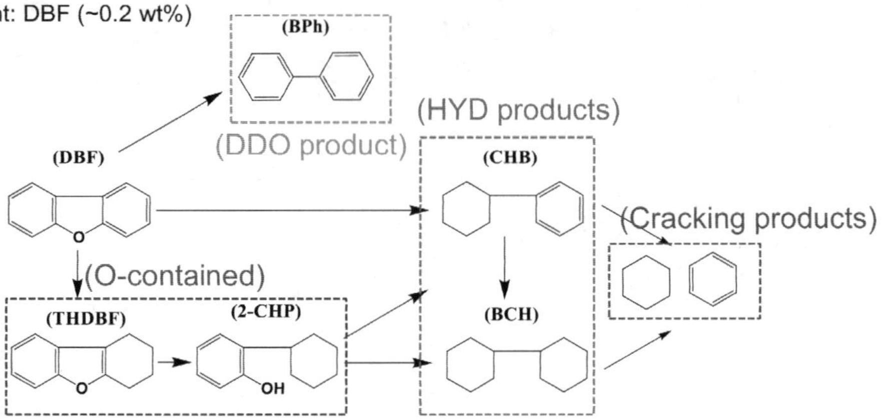

Classified groups	Identified products
O-containing products	THDBF; 2-CHP
DDO product	BPh
HYD products	CHB; BCH
Cracking products	Cyclohexane

Note:
1. O-contained products were not detected under this conditions.
2. If the product amount was < 1%, it won't be considered.

HDO of DBF on Mo_2C/APC

2%Mo_2C/APC

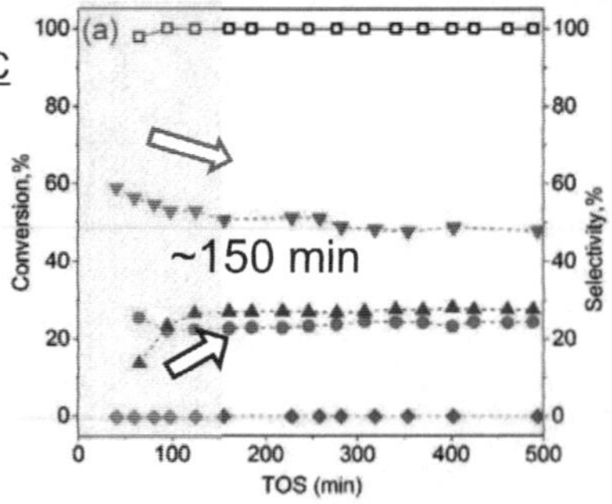

5%Mo_2C/APC

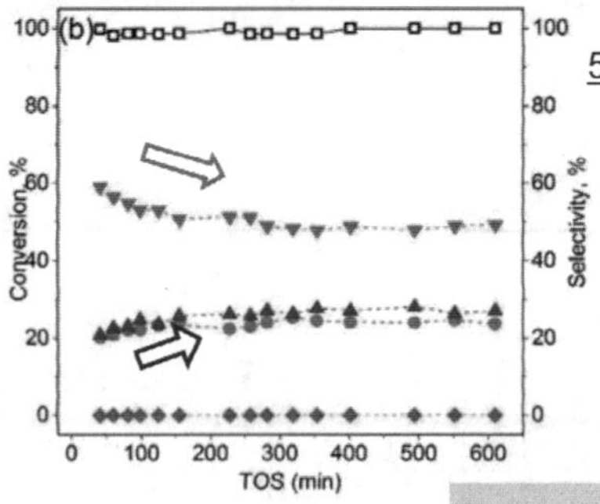

10%Mo_2C/APC

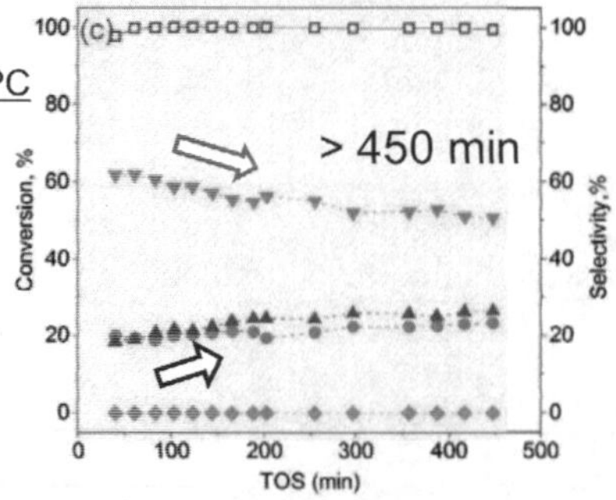

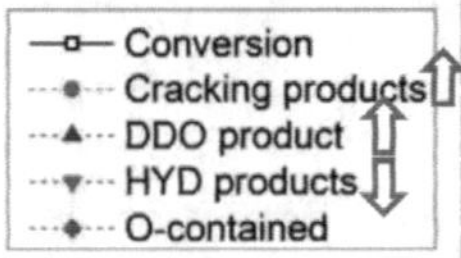

Fresh catalysts	Particle size (nm)
2%Mo_2C/APC	6.0±1.5
5%Mo_2C/APC	7.0±3.0
10%Mo_2C/APC	9.8±8.4

~0.05 mol_{DBF}/g_{Mo}

Catalytic effect comparation of Mo_2C in HDS, HDN and HDO

UBC

$2\%Mo_2C/APC$

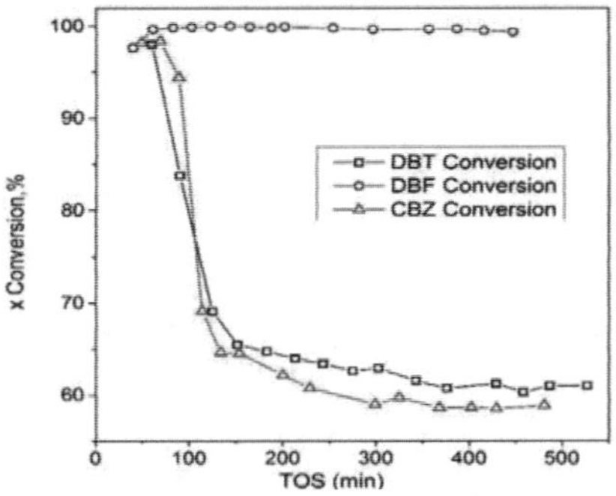

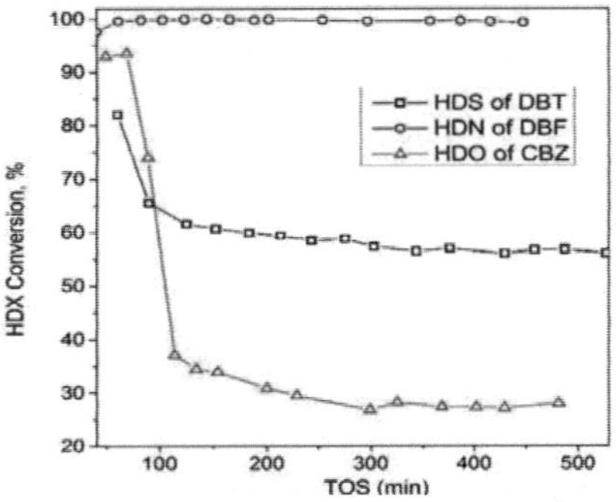

$$F_{Ao}\frac{dX}{dW} = -r_A' = k'C_A = k'C_{Ao}(1-x)$$

$$k' = \frac{F_{AO}}{C_{AO}W}[-\ln(1-x)]$$

Catalyst	k' ($mol_{reactant}/g_{Mo}.s$), x10^{-6}		
	HDS	HDN	HDO
$10\%Mo_2C/APC$	1.308	1.370	4814
$5\%Mo_2C/APC$	1.725	2.421	—
$2\%Mo_2C/APC$	4.220	3.480	7214

Note: this rate constant k'(T=350 °C) was calculated by numbers obtained at the stabilized status.

Summary

UBC

- The deactivation of Mo_2C/APC catalysts in HDS and HDN reactions is caused by poisoning.
- The stabilization of activity and selectivity in HDS and HDN reactions indicates the formation of new Mo species.
- MoS_2 is formed during HDS reaction, and directly covered above Mo_2C confirmed by TEM images.
- Deactivation in HDN is more severe in terms of the heteroatom removal ability.
- Deactivation in HDO reactions is not as serious as in HDS and HDN; Mo_2C catalyst maintained a very high activity.

Acknowledgement

- Catalysis group
- NSERC and Shell Canada for research funding

NSERC
CRSNG

2017 AICHE 19

Thank You for your attention !

?

2017 AICHE 20

Supporting Information

Regeneration of Mo_2C in HDO

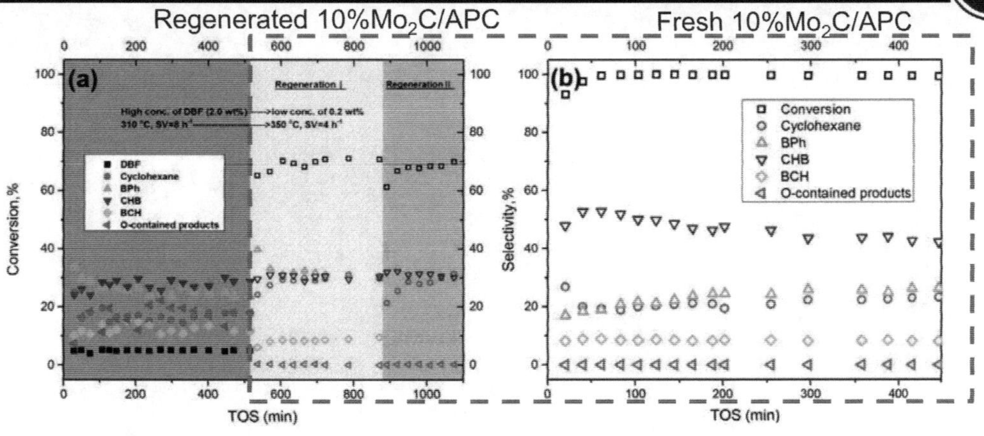

Results: The activity could partially recovered after regeneration; However, the selectivity has changed.

Name	HDO conv.%	BPh, %	CHB,%	BCH,%	Cyclohexane,%	O-contained,%
(b)	99.55	25.79	43.32	8.25	22.63	0.00
(a) Regenerated I and II regions	67.98	31.30	31.11	8.77	28.58	0.24

488587 "Core–Shell" Nanostructured Supported Size-Selective Catalysts Prepared By Molecular Layer Deposition

Abstract:

Heterogeneous catalysts are widely used due to many advantages, including the ease of separating the catalysts by filtration after reaction. However, heterogeneous catalysts usually cannot selectively convert specific molecules in a reactant mixture to only desired products. Size-selective catalysis is an important concept for improving the selectivity of heterogeneous catalysts. In most previous studies, unsupported and nonporous substrates supported metallic nanoparticles were encapsulated in relatively thick porous structures to realize the size-selectivity of heterogeneous catalysts by the size discrimination effect of the porous shell. However, the catalytic activity of the catalysts greatly decreased due to mass transfer limitations brought by relatively thick shells and the contact areas between the porous shells and active sites. In our study, we deposited an ultra-thin porous oxide shell on high surface area substrate-supported Pt nanoparticle catalysts, using molecular layer deposition (MLD), followed by oxidation to remove the organic components in hybrid organic/inorganic MLD films. The encapsulated catalyst showed very high selectivity in catalyzing hydrogenation reactions of n-hexene over cis-cyclooctene. However, there was a decrease of the catalytic activity due to contact areas between the porous shell and metallic nanoparticles. To reduce the contact areas, we introduced gaps between the Pt nanoparticles and porous shells using a sacrifical layer of self-assembled monolayers (SAMs) on the Pt nanoparticles, before the deposition of a MLD layer. This novel nanostructured catalyst showed much higher activity, as compared to the catalyst that had been directly coated with the MLD layer.

500149 Novel ALD-Formed Cobalt/Alumina Nanostructures Active for Fischer-Tropsch Synthesis

Abstract:

Cobalt metal was deposited onto alumina (Al_2O_3) supports using atomic layer deposition (ALD) to create dispersed catalysts for Fischer-Tropsch Synthesis (FTS). We show that the phase of the Al_2O_3 support greatly influences the FTS activity of the resulting catalyst. ALD literature predicts that Co deposited on Al_2O_3 with low cycle numbers should only produce Co nanoparticles less than 2 nm in diameter, which fall into the well-characterized non-active region of nanoparticle sizes for FTS. Our FTS experiments confirm that this inactivity is indeed the case for Co deposited on amorphous Al_2O_3. However, Co deposited on calcined Î³-Al_2O_3 is in fact active, suggesting an alternative ALD deposition regime not previously recognized. Aberration corrected STEM/EELS imaging shows that in addition to nanoparticle deposition, cobalt has also deposited on Î³-Al_2O_3 as semi-continuous planar layers likely less than 3 monolayers thick, which must be responsible for the catalystâ€™s FTS activity. Further, it was found that the number of ALD cycles greatly alters the activity of these catalysts, with 4 and 8 cycle catalysts being much more active than the 1 cycle catalyst. These data suggest Co requires the presence of crystalline Al_2O_3 surfaces in order to achieve a thin semi-continuous planar growth regime. Finally, density functional theory (DFT) calculations were carried out to understand the thermodynamics of Co small particle vs. planar layer (monolayer and bilayer) growth on Al_2O_3 and explain the activity shown by these catalysts.

490326 Controlled Production of Nanostructured Noble-Metal Catalysts Using Atomic Layer Deposition

Abstract:

Controlled production of nanostructured noble-metal catalysts using atomic layer deposition

J. Ruud van Ommen, Fabio Grillo, Hao Van Bui, Jacob A Moulijn, Michiel T Kreutzer

Already an established technique for the growth of conformal thin films, atomic layer deposition (ALD) is also an attractive technique for the controlled growth of nanoparticles (NPs). As such, it can play an important role in the synthesis of nanostructured catalysts. Compared to conventional synthesis routes, ALD boasts a number of advantages: atomic-level control over the amount of material being deposited, conformality, scalability to high-surface-area substrates, and its being solvent-free. Yet, bringing the unparalleled precision of ALD of thin films to ALD of NPs is not straightforward. The precision of ALD relies, in fact, on cyclic repetitions of self-saturating surface reactions that lead to the deposition of less than a monolayer per cycle. Hence, if the growth proceeds in a layer-by-layer fashion, as is the case with ALD of thin films, varying the number of cycles translates into tuning the film thickness with atomic-level precision. However, if the as-deposited atoms form into NPs upon deposition, a growth per cycle of less than a monolayer does not necessarily enable atomic-level control over the NP size. This is because the NP morphology (e.g., size, shape and number density) is dictated by atomistic processes other than "ALD reactions" such as surface diffusion and aggregation of atoms and NPs, and atom attachment to and from NPs. Understanding the role of such kinetic processes during the ALD of NPs is therefore crucial to the development of ALD routes for the synthesis of NPs with a well-defined morphology and thus functionality.

Here, we present an atomistic model of thermal ALD of Pt and Pd NPs on oxides nanoparticles and graphene nanoplatelets. In particular, we study the effect of temperature, number of cycles, coreactant partial pressure and exposure time on the evolution of the NP size distribution and metal loading. Atomistic modelling of our experiments shows that the NPs grow mostly due to Smoluchowski aggregation, that is, NP diffusion and coalescence, rather than through single atom processes such as precursor chemisorption, single atom diffusion and attachment, and Ostwald ripening [1]. While the metal loading can be precisely controlled over a wide range of temperatures, atomic-level precision over the NP size is retained only at low deposition temperatures ($T \leq 100$ °C) when growth by atom attachment/deposition becomes the dominant process. Furthermore, we show that the coreactant partial pressure and exposure time are far more important parameters for the NP size than the number of cycles. We will demonstrate the favourable effect of optimizing the synthesis conditions on the catalytic performance of the obtained material.

[1] Grillo, F., et al. "Understanding and Controlling the Aggregative Growth of Platinum Nanoparticles in Atomic Layer Deposition: An Avenue to Size Selection." The Journal of Physical Chemistry Letters 8.5 (2017): 975.

492271 Controlled Deposition of Metal Oxide Layers By Non-Hydrolytic Sol-Gel for Tailored Acid Sites Generation and Beyond

Abstract:

Bulk materials are omnipresent in catalysis due to their ease of production, but they often suffer from low selectivity and stability due to the lack of precise design of their surface functionalities. We present a new technique for the controlled deposition of metal oxides layers, where precise tuning of material surface functionalities yields optimal performances for various applications.

The method is based on non-hydrolytic sol-gel chemistry using mixtures of metal alkoxides and metal halides. For instance, addition of a 1/3 mol/mol mixture of $Ti(iPrO)_4$ and $TiCl_4$ to a toluene suspension of Cu_2O spheres led to the formation of a thin conformal TiO_2 overcoat. Thickness can be adjusted by the number of deposited layers with a growth rate of 0.5 nm/cycle. Furthermore, this method allows deposition of TiO_2 layers inside the pores of SBA-15 while maintaining 2D hexagonal structure and high porosity. The mixed oxide created at the interface generates medium strength Lewis acid sites which catalyze 1-phenylethanol dehydration to styrene. High Lewis acid sites density with medium acid strength was obtained with 6 TiO_2 layers which lead to the highest yields of styrene and the highest catalyst stability. In contrast, a zeolitic catalyst often used for this reaction (HZSM-5) deactivates due to polystyrene formation on strong Lewis acid sites. The technique has also been employed to overcoat supported metal particle yielding metal-acid bifunctional catalysts, where overcoat porosity can be adjusted by ageing in basic media. Finally, an analogous method was developed for deposition of alumina, yielding strong acid sites.

The easy implementation and tunable porosity of the present technique contrasts with the heavy instrumentation and limited structural control associated with gas phase methods such as atomic layer deposition or chemical vapor deposition on powder substrates. We anticipate that the method reported can be used for the rational design of catalytic materials used in a wide range of applications.

498798 Enhanced Hydrothermal Stability of Phosphonate-Coated Al_2O_3 Supported Catalysts

Abstract:

Gamma phase alumina (γ-Al_2O_3) is a common catalyst support because its high abundance, mechanical and chemical stability, and high surface area, however, γ-Al_2O_3 is unstable in the presence of high temperature/high pressure steam used in aqueous phase reforming. Under these conditions, the support undergoes a phase transition to boehmite (AlO(OH)) accompanied with a significant loss in the support's surface area as the pore structure deteriorates. Consequently, catalytic performance is greatly diminished as active metal clusters become less accessible or detach from support.

To prevent catalyst degradation, hydrophobic coatings can be deposited onto the Al_2O
increased hydrothermal stability. Specifically, we chose to coat catalysts with phosphonic acids (PAs) since these capping agents selectively bind to the oxide support and form robust bonds with the surface. Additionally, the physical properties of catalysts can be further tuned by changing the phosphonate's tail group.

In this work, we discuss recent results that have shown the deposition of alkyl phosphonic acids on alumina and supported catalysts drastically improves their hydrothermal stability. We examine how the alkyl tail length (or hydrophobicity) affects the suppression of boehmite formation and improves the retention of both metal and support surface areas. The beneficial roles of both phosphonate head and tail groups will be assessed as well as their impact on catalytic performance.

498271 Catalytic Activity of Porphyrin-Supported Iron Oxide Clusters for Methane Oxidation

Abstract:

During the last decade, natural gas production in the United States has significantly increased every year, and it is predicted to increase even more rapidly during the coming years. This has provided new incentives for selectively transforming natural gas components and their derivatives into more valuable chemicals through energy efficient and environmentally friendly processes. However, the current route for converting methane into more useful chemicals is through a complex, multi-step industrial process that requires multiple reactors, separations, and high temperatures and pressures[1]. For this reason, there is a great interest in designing catalysts that accomplish this transformation directly while avoiding deeper oxidation to carbon dioxide.

Methane monooxygenase (MMO), an enzyme found in biological systems, is capable of oxidizing methane to methanol using O2 at mild conditions. MMO has two different forms: (1) particulate MMO (pMMO), which is postulated to have a dicopper-oxo active site, and (2) soluble MMO (sMMO), which contains a diiron site.[2] The identification of different iron and copper sites in these enzymes has inspired the synthesis of catalysts that mimic their activity towards the partial oxidation of methane.[3]

Using density functional theory, we have studied the growth and catalytic activity of iron oxide nanoclusters that can be grown via atomic layer deposition (ALD) on a porphyrinic substrate and mimic the structure of the active site in the enzyme for the oxidation of methane to methanol. The porphyrin-supported iron oxide active site consists of an $Fe^{IV}_2(\mu\text{-}O)_2(OH)_2$ diamond core supported by a bridging $Al^{III}(\mu\text{-}O)_2(OH)$ moiety. After a series of ALD cycles to deposit the Al atom and the Fe^{II} sites on the porphyrin substrate, the bridging oxygen atoms that initiate the catalytic cycle can be incorporated via O_2 or N_2O activation. Preliminary results show that N_2O activation is more favorable, but may lead to the formation of sites that contain only one bridging oxygen and two Fe^{III} sites. Two mechanisms for the oxidation of methane using the $Fe^{IV}_2(\mu\text{-}O)_2(OH)_2$ active site and N_2O as the oxidant are proposed: (1) a concerted mechanism in which the C-H bond is cleaved heterolytically and (2) a rebound mechanism in which a bridging oxygen or a terminal oxo group abstracts a hydrogen atom from methane. Our preliminary computational results show that the rebound mechanisms are more likely to occur due to lower activation barriers of about 70 kJ/mol. We also find activation barriers between 100-120 kJ/mol for N_2O activation and CH_3OH desorption. To further understand the effect of cluster composition on the activity for the various oxygen sites in the di-iron cluster, we replaced the aluminum atom and one of the iron atoms with other transition metals, calculated the main reaction barriers, and performed population analysis studies. Using these computational tools, we aim to understand the ways in which we can tune the activity of the proposed catalysts for the selective oxidation of methane.

References

1. Crabtree, R. H. *Chem. Rev.* 95, pp 987–1007 (1995).

2. Lippard, S. J. and Feig, A. L. *Chem. Rev.* 94, pp 759–805 (1994).

3. Wulfers, M. J., Teketel, S., Ipek, B., and Lobo, R. F. *Chem. Commun.* 51, pp 4447-4450 (2015).

488147 Trends in Formic Acid Electro-Oxidation on Pt and Pd Monolayers on Transition Metal Surfaces: A Theoretical and Experimental Study

Abstract:

Formic Acid (FA) is a viable carbon-neutral fuel that can be produced from biomass1 or via the reduction of CO22. Formic acid electro-oxidation (FAO) in low-temperature direct formic acid fuel cells (DFAFCs) has attracted considerable interest due to its numerous advantages compared to other similar fuel cells3. DFAFCs show an equilibrium cell voltage (1.40 V) that is higher than that shown by hydrogen (1.23 V) and methanol (1.21 V) fuel cells among others4. Additionally, FA is liquid (unlike hydrogen) and, compared to methanol, is less toxic4,5 and shows less crossover through Nafion® membranes in fuel cells6. Platinum (Pt) and Palladium (Pd) are the best monometallic catalysts for FAO, however, they suffer from stability issues in acidic media, are poisoned by CO during the reaction, and are too expensive to make DFAFCs economically viable 3. One way to mitigate these issues is to deposit pseudomorphic monolayers of Pt or Pd atop other less expensive metals 7. Such structures would reduce the amount of Pt or Pd in the catalyst (and thereby the cost), and could improve the stability and activity of Pt and Pd8-10. To this end, we present a first-principles, self-consistent periodic density functional theory (PW91-GGA) study of FAO on model (111) and (100) facets of Pt and Pd monolayers atop six fcc metals (Au, Ag, Pt, Pd, Ir, and Rh), and (0001) facets of Pt and Pd monolayers atop three hcp metals (Os, Ru, and Re). On all surfaces, we calculate the free energies of the adsorbed intermediates: carbon monoxide (CO), hydroxyl (OH), carboxyl (COOH), and formate (HCOO). Using these data, we develop thermochemical free energy diagrams and, thereby, calculate onset potentials of three mechanisms: direct oxidation via carboxyl, direct oxidation via formate, and the indirect mechanism leading to the formation of CO. Comparing these onset potentials on the surfaces studied reveals trends in the preference to either direct mechanisms, as well as trends governing their propensity towards CO poisoning. For a subset of the close-packed surfaces, we compare the calculated theoretical trends to experimental trends, as determined by electrochemical measurements. Finally, having studied both close-packed and open facets of fcc systems, we discuss the structure sensitivity of FAO on Pt and Pd monolayers on transition metals.

References

1. Bozell JJ, Petersen GR. Technology development for the production of biobased products from biorefinery carbohydrates-the US Department of Energy's "Top 10" revisited. Green Chem. Apr 2010;12(4):539-554.
2. Moret S, Dyson PJ, Laurenczy G. Direct synthesis of formic acid from carbon dioxide by hydrogenation in acidic media. Nature Commun. Jun 2014;5.
3. Jiang K, Zhang H-X, Zou S, Cai W-B. Electrocatalysis of formic acid on palladium and platinum surfaces: from fundamental mechanisms to fuel cell applications. Phys. Chem. Chem. Phys. 2014 2014;16(38):20360-20376.
4. Demirci UB. Direct liquid-feed fuel cells: Thermodynamic and environmental concerns. J. Power Sources. Jun 20 2007;169(2):239-246.
5. Tedsree K, Li T, Jones S, et al. Hydrogen production from formic acid decomposition at room temperature using a Ag-Pd core-shell nanocatalyst. Nature Nanotechnology. May 2011;6(5):302-307.
6. Rhee YW, Ha SY, Masel RI. Crossover of formic acid through Nafion((R)) membranes. J. Power Sources. May 15 2003;117(1-2):35-38.
7. Kibler LA, El-Aziz AM, Hoyer R, Kolb DM. Tuning reaction rates by lateral strain in a palladium monolayer. Angew. Chem. Int. Ed. 2005 2005;44(14):2080-2084.
8. Bligaard T, Nørskov JK. Ligand effects in heterogeneous catalysis and electrochemistry. Electrochim. Acta. May 10 2007;52(18):5512-5516.
9. Mavrikakis M, Hammer B, Nørskov JK. Effect of strain on the reactivity of metal surfaces. Phys. Rev. Lett. Sep 28 1998;81(13):2819-2822.
10. Sasaki K, Naohara H, Cai Y, et al. Core-Protected Platinum Monolayer Shell High-Stability Electrocatalysts for Fuel-Cell Cathodes. Angew. Chem. Int. Ed. 2010;49(46):8602-8607.

500794 Activity and Thermal Stability in Well-Defined Platinum/Palladium Bimetallic Catalysts for Hydrocarbon Combustion

Abstract:

Emmett Goodman[1], Luke T. Roling[1], Sheng Dai[2], Adam S. Hoffman[3], Simon R. Bare[3], George W. Graham[2], Xiaoqing Pan[2,4], Frank Abild-Pedersen[3], and Matteo Cargnello[1]

Proposed Section: Rational Catalyst Design I

(1)Chemical Engineering, Stanford University, Stanford, CA (2)Department of Chemical Engineering and Material Science, University of California-Irvine (3)SLAC National Accelerator Laboratory, Menlo Park, CA (4)Department of Physics and Astronomy, University of California-Irvine

Activity and Thermal Stability in Well-Defined Platinum/Palladium Bimetallic Catalysts for Hydrocarbon Combustion

Although there is an increasing effort to take advantage of inexpensive base metals in the development of modern materials, rare and expensive metals continue to permeate the fields of thermal and electrocatalysis. Noble metal catalysts, especially when composed of two or more precious metals, often show remarkable activity, making the use of these more expensive materials economically viable. Nevertheless, it remains challenging to establish clear structure-activity-stability relationships using traditional co-impregnation techniques, due to the coexistence of multiple active nanoparticles of different sizes, shapes, and compositions. In this work, we synthesize model bimetallic catalysts with well-defined compositions and properties in order to systematically investigate activity and stability properties in bimetallic catalysts.

We demonstrate a procedure to synthesize small (2.2-3.4 nm), monodisperse bimetallic nanocrystals catalysts, of tunable composition, in order to understand which properties confer advantages in terms of activity for lean methane combustion, with and without the presence of steam. At low temperatures, all bimetallic catalysts show activity similar to Pt/Al_2O_3. Although such bimetallic catalysts are uniformly much less active than pure Pd/Al_2O_3catalyst under dry conditions, they are more active at low temperatures in the presence of steam. In terms of stability, the bimetallic nanocrystal catalysts possess unique, composition-dependent stability properties. Despite lower initial combustion rates, hydrothermal aging of the Pd-rich bimetallic catalyst induces both sintering and segregation of a PdO phase in close contact to a Pd/Pt alloy phase, forming more active and highly stable sites for methane combustion.

Approaching this stability problem from the other direction, we investigate fundamental sintering processes in well-defined two-phase catalysts formed by colloidally synthesizing each active phase independently. We carefully distinguish between two fundamental sintering modes in platinum catalysts: low-temperature, surface-mediated diffusion, and high-temperature vapor-phase transport of platinum species. By using highly active palladium catalysts

as thermodynamic â€˜platinum-trapsâ€™, we can indirectly probe the trajectory of platinum species. Understanding the atomic mechanisms of platinum sintering as a function of temperature and reaction environment can allow us to develop approaches to mitigate these detrimental aging effects. This work correlates activity and stability properties of Pt/Pd bimetallic systems with specific compositions and structures, which provides important guidelines for the preparation of highly active materials with optimal utilization of expensive noble metals.

490160 Rational Design of Supported Ni Inverse Catalysts for Hydrogenolysis

Abstract:

Hydrogenolysis and hydrodeoxygenation reactions are heavily investigated as they have useful applications in biomass and coal processing. Increased insight into factors that control catalytic activity and selectivity could be gained by comparing a series of catalyst with different structural properties. Thus, catalyst preparation methods where the metal particle size as well as the nature of the metal oxide overcoating the metal particle can be tuned are advantageous for both fundamental studies as well as practical applications.

The group of Hartwig first reported low temperature hydrogenolysis of aryl alkyl ethers using homogeneous Ni catalysts. Their observation of small 2 nm Ni colloids at the termination of the reaction suggested possible role of heterogeneous catalysis [1]. Their initial study using homogeneous catalysts was followed with heterogeneous catalysts prepared by supporting Ni on activated carbon using $Ni(COD)_2$ as the Ni precursor [2]. The average Ni particle size after decomposition of the ligand from the precursor was about 2-6 nm. The catalyst was active for selective hydrogenolysis of a broad range of aryl ethers in the presence of a *stochiometric amount of base*. They also observed that commercial $Ni/Al_2O_3/SiO_2$ catalysts can accomplish the reaction under the same condition but at much higher Ni loading. They inferred from their studies that large Ni particles were not as active as smaller Ni particles.

Recent interesting reports from the group of Zhang et al. [3] showed that supported Au nanoparticles did not alter the catalytic activity and selectivity using supports such as ZrO_2, Al_2O_3, SiO_2, activated carbon, and rutile TiO_2 for guaiacol hydrodeoxygenation. However remarkable changes in guaiacol hydrodeoxygenation activity were observed when Au nanoparticles were deposited on anatase titania. The activity was affect by both the Au and anatase particle size, although the selectivity appeared not to be affected.

In view of these examples of the importance of controlled synthesis of supported metal particles for hydrodeoxygenation reaction, we investigated supported Ni inverse catalysts for the dependence on the presence of stochiometric amount of added base by tuning the basicity of the metal oxide overcoating the Ni particles. The Ni inverse catalysts were synthesized by depositing $Ni(COD)_2$ onto silica presilylated with (3-aminopropyl)triethoxysilane. The amine was introduced to deter surface mobility of the Ni(0) atoms once their protective COD ligands were removed upon heating. The Ni particle size could be varied with different degree of silylation. Deposition of the metal oxide overcoat was conducted after desorption of COD ligands from $Ni(COD)_2$ but before oxidative removal of the propylamine groups from the silica surface so as to enhance metal oxide-metal interaction. Thus, starting from the same parent supported Ni catalysts, different metal oxides could be deposited without changing the Ni particle size distribution. Effective deposition of titania on Ni/SiO_2 was evidenced from XPS that showed highly attenuated Ni signal after small amounts of titania deposition (Ti:Ni =2 mole ratio). The Ni signal reference to Ti was as expected for samples without the coating and also quite intense when the coating was removed by ion bombardment, and the order was: Ni/SiO_2 > $Ti\text{-}Ni/SiO_2$ with ion beam bombarded > $Ti\text{-}Ni/SiO_2$. The catalysts with and without titania deposition also showed different TPR profiles with the titania decorated catalyst exhibiting a low temperature reduction peak absent for Ni/SiO_2. Different reactions were conducted to investigate the catalytic activities of the various catalysts. The probe reaction of 1-phenylethanol hydrogenolysis, conducted at 140 °C and ambient pressure by bubbling a stream of N_2/H_2 (50:50) gas through a disperser into the reactor, showed surprised formation of ethylbenzene and acetophenone. The presence of acetophenone, an oxidized product, in the presence of H_2 gas indicated that its formation proceeded via hydrogen transfer from one molecule of the alcohol to another to yield acetophenone (the oxidized product) and ethylbenzene (the reduced product). The result suggests that this catalyst is a good candidate for reactions involving hydrogen transfer.

References

[1] A. G. Sergeev, and A.F. Hartwig, Science, 2011, 332, 439-443.

[2] F. Gao, J.D. Webb and J. F. Hartwig, Angew. Chemie, Int., 2016, 55, 1474-1478.

[3] J. Mao, J. Zhou, Z. Xia, Z. Wang, Z. Xu, W. Xu, P. Yang, K. Liu, X. Guo and Z.C. Zhang, ACS Catal, 2017, 7, 695-705.

501260 Tuning Catalyst Activity Using Self-Assembled Monolayers

Abstract:

One of the biggest problems in heterogeneous catalysis is that of poor selectivity. Often dopants are utilized to tune the properties of various catalysts to improve selectivity or activity. Over the last decade self-assembled monolayers (SAMs) have been developed as a means to control the selectivity of reactions on supported metal catalysts, but little focus has been made in the ability to tune activity. Here, we demonstrate how self-assembled monolayers can be modified in a rational design strategy to tune the surface electronic properties. In doing, so we are able to control the rate of some reactions. Similar strategies have been performed with metal-ligand complexes and semiconductors materials, but have not been applied to heterogeneous catalysts. Two such approaches will be presented 1) functionalizing metal oxide catalysts with phosphonate monolayers and 2) using palladium catalysts with thiolate monolayers. Firstly, when 1-propanol is reacted over titania three main pathways are possible: dehydration, dehydrogenation, and condensation. At low temperatures (~250°C), dehydrogenation is favored on the native catalyst, closely followed by condensation. However, upon functionalization of titania with phosphonic acids with various organic tail functionalities (methyl, octadecyl, benzyl, etc), dehydration becomes the dominant pathway. More importantly, depending on the substituents of the benzylic tail (e.g. 4-aminobenzylphosphonic vs 3-fluorobenzylphosphonic acid), the activity of dehydration can be tuned over a wide range, allowing far higher dehydration activities than on the uncoated catalysts. The applicability of this approach to other catalysts will be demonstrated with results from noble metals functionalized with thiolate monolayers.

494261 Structure Sensitive Phenol Hydrogenation on Pd Nanostructures

Abstract:

Structure Sensitive Phenol Hydrogenation on Pd Nanostructures

Shelaka Gupta[a], Sreedhala.S[b], Tuhin Suvra Khan[a], C.P.Vinod[b] and M. Ali Haider[a]

[b]*Catalysis Division, CSIR-National Chemical Laboratory, Dr. Homi Bhabha Road, Pune - 411 008*

[a]*Renewable Energy and Chemicals Laboratory, Department of Chemical Engineering,*

Indian Institute of Technology Delhi, Hauz Khas, New Delhi-110016

Tel: +91-11-26591016, Fax: +91-11-2658-2037
haider@iitd.ac.in

Phenol hydrogenation on unsupported Pd nanostructures showed structure sensitivity with respect to particle shape and insensitivity with respect to particle size. The hydrogenation reaction was performed over Pd octadehra (~27 nm, Figure 1(a)), cubes (~25 nm, Figure 1(b)), and spheres (~6nm). While Pd cubes (exposing only the (100) facets) were observed to selectively (100%) produce cyclohexanone, octahedra (exposing only the (111) facets) were 100% selective towards cyclohexanol. Interestingly, Pd spheres of varying size (<5 nm tp 6 nm) and exposing both (111) and (100) facets were observed to produce cyclohexanone only. Density functional theory (DFT) calculations showed that the difference in the binding modes of phenol on Pd(111) (Eb = -102 kJ/mol) and Pd(100) (Eb = -135 kJ/mol), primarily lead to the difference in the selectivity (Figure 1(a) and 1(b)). On Pd(111), cyclohexanol was produced by phenol dissociation to phenoxy and subsequent hydrogenation (Figure 1(a)), whereas on Pd(100) cyclohexanone was produced by direct phenol hydrogenation (Figure 1(b)). Further hydrogenation of cyclohexanone to produce cyclohexanol on Pd(100) was observed to be prohibitive (E a = 125 kJ/mol) at reaction conditions. Interestingly, a mixture of cube and octahedra with 40:60 ratio, resembling a spherical supported Pd nanoparticle was observed to yield cyclohexanone with 100% selectivity even on increasing the phenol weight in the reaction from 50 mg to 450 mg. However, a small fraction (8% selectivity) of cyclohexanol was detected when phenol concentration was raised to 600 mg. The shift in selectivity was attributed to the reduction in the adsorption energy of phenol on the (100) facets (Eb = -97 kJ/mol at 0.6 ML coverage) which became comparable to the (111) facet due to increased lateral interactions at high phenol coverage (Figure 1(c)). The reaction however remained structural insensitivity on varying the size of the Pd sphere

Figure 1: Reaction energy diagram for phenol hydrogenation on a) Pd (111) and b) Pd (100) surface. c) Trend in calculated binding energy (E_b) of phenol and d-band centre (ε_d) of Pd (100) versus phenol coverage (shaded area shows the region where selectivity is likely to shift towards cyclohexanol).

500492 Fabrication of Supported Size-Controlled Cobalt Nanoparticles over Porous Silicon Carbide for Superior Catalytic Performance in the Fischer-Tropsch Process

Abstract:

Heterogeneous catalysis by metals frequently involves nanosized particles inside nano- or mesoporous host structures. The high surface area and the easy accessibility of the porous matrix for molecules make them ideal materials in various applications. However, due to the exothermic nature of many catalytic reactions, local overheating and sintering of hosted metal nanoparticles (NPs) may be encountered. As a result, the catalyst activity decreases and the selectivity patterns change. To cope with the need for stable catalyst performance, the heat produced during the reaction has to be efficiently removed. To provide a remedy, materials with high thermal conductivity such as silicon carbide (SiC) may be added to the catalyst. In fact, SiC has outstanding thermal, mechanical and chemical stability. Moreover, rather than diluting the catalyst bed by adding conductive SiC powder, it would appear attractive to texturize it as a mesoporous material amenable to accommodate metal NPs and to exert molecular traffic control. Our group has recently shown that both a "bottom-up" approach of SiC synthesis through nanocasting and a "top-down" approach by partial dissolution of bulk silicon carbide can lead to mesoporosity.[1] In the latter case, highly crystalline samples with large surface areas and well-ordered pore network were obtained. In the present work, such samples were used as a host of metallic nanoparticles for the CO hydrogenation according to Fischer-Tropsch.

The Fischer-Tropsch (FT) process is key to the "gas-to-liquid" technology, converting syngas (a mixture of H_2 and CO) into liquid hydrocarbons and waxes. The highly exothermic nature of the reaction dictates further improvements in the rational design of the catalysts to mitigate the local hot spot formation and sintering of the active phase. Usually Co and Fe-based catalysts are being employed on support materials of low conductivity, such as silica and alumina. Herein, colloidal Co particles were synthesized, and subsequently intercalated into porous SiC using an ultrasonic treatment. After removal of the solvent by centrifugation, the particles were readily dispersed over the support and then directly used in studies of the Fischer-Tropsch reaction. By combining the colloidal cobalt particles supported into highly crystalline and well-ordered porous silicon carbide, we demonstrate here superior catalytic properties of cobalt-based catalysts in terms of activity, selectivity and stability as compared to traditional catalysts in the Fischer-Tropsch process.

Briefly, porous SiC samples were prepared by electrochemical dissolution of highly-doped (1 mO·cm, n-type) polycrystalline 3C-SiC in HF/H_2O/ethanol solution. Electrochemically-derived SiC materials were crystalline demonstrating the occurrence of a mesoporous structure (pore size ranging from 30 to 20 nm) with relatively high pore volume (0.85 to 0.70 mL g-1) and surface areas of about 100 - 120 $m^2 g^{-1}$.[1] The pore framework indicated free access of the pores and absence of bottle-necked pores. The samples were stable up to 500 °C in gas flows of either 20% O_2 in Ar or H_2. To impregnate porous silicon carbides with active metal, colloids of uniform cobalt particles were synthesized applying a "hot injection" technique.[2] Accordingly, dicobalt octacarbonyl was used as a precursor and injected into the dichlorobenzene solution (174 °C). The presence of oleic acid as a capping agent avoided the coalescence of colloidal particles. Typically, 9 ± 0.5 nm cobalt particles were used for impregnation into pSiC. The interest in such particles is associated with a particle size effect according to which the methane selectivity at small particle sizes is high. The impregnation was dominated by van der Waals forces in dry (water-free) chloroform. The as-prepared Co/SiC samples were activated in hydrogen using temperature-programmed reduction up to 460 °C.

Catalytic results in the Fischer-Tropsch reaction (220 °C, 20 bar, H_2/CO = 2/1) showed a superior activity with the observed reaction rate above 350 $umol_{CO}\ g_{Co-1}\ s^{-1}$ over the Co/SiC samples. Co NPs supported on commercially available silica (Co/Davisil) and foam-like mesoporous silica (Co/MCF-17) demonstrated much lower activity of about 4.9 and 27.3 $umol_{CO}\ g_{Co-1}\ s^{-1}$, respectively. TEM and SEM studies highlighted a significant sintering of cobalt particles for Davisil while the Co/MCF-17 and Co/SiC samples were seen relatively unchanged. However, chemisorption studies using in-situ H_2/D_2 exchange after catalysis showed a decrease in the metallic surface area for the Co/MCF-17, probably due to the reaction with the support or a hindered access to the active centers. A considerable increase in the C_{5+} selectivity up to 53% was observed over Co/SiC catalyst while a C_{5+} yield of max. 35-40% was obtained over silica supported samples as well as bulk cobalt. Moreover, the Co/MCF-17 sample showed an increase in the methane formation up to 40% in the long-run testing while Co/SiC catalysts demonstrated stable performance after at least 76 h. This effect seems to match the conceptional idea that methane is formed by hydrogenation of surface carbon after CO dissociation while the chain lengthening involves a different mechanism (possibly via CO insertion, as reported by Schweicher et al.[3]). For a poorly conducting support the heat removal due to the exothermic CO hydrogenation is inhibited, so it may be anticipated that larger amounts of carbon are formed leading to an increase in the methane formation compared to samples using a SiC support.

In conclusion, we report here a rational catalyst design by preparing highly crystalline and well-ordered silicon carbide with optimal morphological properties to impregnate colloidal particles of cobalt metal. The use of uniform cobalt particles ensures high activity and selectivity in the Fischer-Tropsch reaction while the porous silicon carbide provides superior stability of obtained catalysts due to efficient heat release and due to its chemical inertness.

References:

1. a) D. Korytko, S. Gryn, S. Alekseev, V. Iablokov, O. Khaynakov, V. Zaitsev, I. Bezverkhyy, N. Kruse *RSC Adv.* **2016**, *6*, 108828-108839; b) S. Alekseev, D. Korytko, S. Gryn, V. Iablokov, O. Khaynakova, S. Garcia-Granda, N. Kruse *J. Phys. Chem. C* **2014**, *118*, 23745-23750

2. V. Iablokov, S.K. Beaumont, S. Alayoglu, V.V. Pushkarev, C. Specht, J. Gao, A.P. Alivisatos, N. Kruse, G.A. Somorjai *Nano Lett.* **2012**, *12*, 3091-3096

3. J. Schweicher, A. Bundhoo, N. Kruse *J. Am. Chem. Soc.* **2012**, *134*, 16135-16138

501197 Exploring Ruthenium Metal-Support Dynamics for the Low-Temperature Partial Oxidation of Methane

Abstract:

Emmett Goodman[1], An-Chih Yang[1], Allegra Latimer[1], Liheng Wu[2], Frank Abild-Pedersen[2], and Matteo Cargnello[1], (1)Chemical Engineering, Stanford University, Stanford, CA, (2)SLAC National Accelerator Laboratory, Menlo Park, CA

Exploring Ruthenium Metal-Support Dynamics for the Low-Temperature Partial Oxidation of Methane

Department of Chemical Engineering and SUNCAT Center for Interface Science and Catalysis, Stanford University, Stanford, CA 94305, USA

Industrially, the transformation of methane to syngas stands as an invaluable reaction, the latter products being converted downstream into valuable bulk chemicals such as methanol, ammonia, and synthetic fuels. Steam or bireforming are common routes for methane conversion to syngas, but work at high temperatures because of their endothermic nature. Methane partial oxidation ($CH_4 + ½ O_2 = CO + 2 H_2$) is exothermic and benefits from more favorable thermodynamics. Thus, there are significant incentives to explore catalysts that can partially oxidize methane at low temperatures, with high selectivity and yield, into an industrially relevant feedstock.

Methane partial oxidation (MPO) is a complex reaction, in which cycling catalyst oxidation states dramatically influences catalyst structure, activity, and selectivity. In order to rationally design a catalyst for MPO, it is important to understand how interactions between catalyst active phases and underlying support affect catalyst structure and subsequent behavior. However, in non-uniform systems with active phases of a variety of sizes and shapes, it is often difficult to back out such structure-property relationships. By colloidally synthesizing ruthenium nanoparticles prior to support impregnation, we may make a fair comparison of activity, selectivity, and stability across supports and understand how catalyst support affects the morphological and catalytic evolution of ruthenium active phase(s). We report highly active 0.5 wt. % ruthenium catalysts prepared from monodisperse nanocrystals, which show >90% CH_4 conversion and >98% CO selectivity at ~650Â °C.

Mechanistically, at low temperatures, RuO_2 slowly catalyzes complete combustion; upon oxygen depletion in the reactor, and reduction to metallic Ru, we ignite partial oxidation, which tends toward equilibrium values. This study thus identifies two design rules for igniting low-temperature formation of synthesis gas: oxygen depletion in the reaction feed, and the effect of metal-support interaction on oxidation state. We find that selectivity and yield are strongly dependent on the support type, with a size dependence on certain â€˜activeâ€™ supports. These effects are related to how certain supports preferentially stabilize different oxidation states of ruthenium. Using this information, we can gain fundamental mechanistic insight regarding methane partial oxidation on supported ruthenium systems, and work to design more effective catalysts. Taking advantage of these design rules, we suggest the development of highly

active bifunctional catalysts, which take advantage of a palladium phase to deplete feed oxygen, and a reduced ruthenium phase to form partial oxidation products.

500783 Direct Methane Conversion to Ethylene and Ethane By Oxidative Coupling in Packed Bed and Membrane Reactors

Abstract:

The abundance of methane makes it an attractive raw material for chemical production. However, current industrial routes for converting methane to valuable chemicals typically involve a synthesis gas (CO and H_2) production step which is complex and expensive. Direct methane conversion can reduce the need for syngas production and is potentially more suitable for converting methane released from small scale sources, which is typically flared resulting in greenhouse gas emissions. Oxidative coupling of methane (OCM) is a direct route for converting methane into ethylene and ethane (C_2). Ethylene, one of the target products of OCM, is a widely-used feedstock in the petrochemical industry. However, OCM is yet to be commercialized due to the low C_2yields obtained, particularly in conventional packed bed reactors (PBRs).

It has been demonstrated that a reactor with distributed oxygen feed (i.e. an O^{2-} conducting membrane reactor) can give much higher C_2 selectivity and yield compared to a PBR. However, the relatively few membrane reactors that have been studied in OCM literature give C_2 yields that are still below the commercial viability targets[1,2]. The low performance of OCM membrane reactors is partly due to the limited oxygen flux through the membranes and low membrane surface area. Another reason for the low performance of some of these membrane reactors is the absence of a selective OCM catalyst. The integration of such catalyst in the reactor should facilitate methane activation, leading to improvements in C_2 yields. Ideally, the integrated catalyst should be ionically conductive to accept O^{2-} from the membrane, and compatible with the membrane material at the high temperatures used in fabrication and testing[1,3]

In this work, we use PBR experiments to identify O^{2-} conducting catalysts that are selective for OCM. These catalysts include lanthanum and barium containing perovskite oxides. We test these catalysts in membrane reactors that consist of a dense O^{2-} conducting membrane and a porous O^{2-}conducting layer to aid methane activation. We present our results and discuss our efforts to further improve the OCM performance of these membrane systems by increasing the membrane surface area and oxygen flux.

(1) Farrell, B. L.; Igenegbai, V. O.; Linic, S. *ACS Catal.* **2016**, *6,* 4340–4346.

(2) Wang, B.; Albarracín-Suazo, S.; Pagán-Torres, Y.; Nikolla, E. *Catal. Today* **2017**, 1–12.

(3) Farrell, B.; Linic, S. *Catal. Sci. Technol.* **2016**, *6*, 4370–4376.

493526 Sulfur As a Selective Soft Oxidant in the Catalytic Conversion of Methane to Ethylene over Metal Chalcogenides

Abstract:

Oxidative coupling of methane (OCM) in presence of an oxidant over metal oxide catalysts provides an attractive route for direct conversion of methane to ethylene.[1] The tremendous wealth of previous studies using oxygen as the oxidant have provided important insights into the mechanisms for OCM.[2] Oxygen, however, is a very strong oxidant and ultimately limits the selectivity in OCM. New coupling processes that use abundant, thermodynamically mild oxidants together with selective catalysts that reduce the thermodynamic driving force for over-oxidation and result in higher ethylene yields would offer exceptional opportunities. We thus report herein on the use of elemental sulfur as a promising soft oxidant for selective conversion of methane to ethylene. We explored the oxidative coupling of methane via sulfur (SOCM) over a range of different metal sulfide catalysts including MoS_2, RuS_2, TiS_2, PdS and Pd/ZrO_2. Experimental results and density functional theory (DFT) calculations reveal that methane conversion is directly correlated with surface metal–sulfur bond strengths. Surfaces with weakly bound sulfur are more basic and activate the C–H bonds of methane more readily. In contrast, ethylene selectivities scale inversely with surface metal–sulfur bond strengths, and surfaces with the strongest metal–sulfur bonds afford the highest ethylene selectivities. The PdS system was found to be unique in that it showed the highest rates and selectivities. More detailed characterization of the catalyst showed that the catalyst was partially reduced under reaction conditions to form $Pd_{16}S_7$. Theoretical results discussed herein show that this surface exposes both Pd as well as S sites which can more readily activate methane but also promote the C-C bond formation and the production of ethylene. We extend these results to the reactivity of oxides together with sulfur and report C_2 selectivities > 33% for Fe_3O_4. XRD and XPS analyses of Fe_3O_4 indicate that it readily converts to FeS_2. The primary products formed over these FeS_2 catalysts are ethylene, ethane and CS_2. DFT calculations show that ethylene predominantly forms in path similar to that for PdS via CH_2 coupling. The exposed Fe-S site pairs that form on the sulfided Fe_3O_4 catalyst aid in the activation of the C-H bond of methane as well as the production of ethylene. CS_2formation, on the other hand, was found likely to occur over the adsorbed sulfur sites via facile C-H bond activation steps. Ethane formation predominantly occurs via the formation of methyl fragments, their desorption from the sulfide and their recombination in the gas phase.

References:

1) Zavyalova, U., Holena, M., Schlögl, R. & Baerns, M. Statistical analysis of past catalytic data on oxidative methane coupling for new insights into the composition of high performance catalysts. ChemCatChem 3, 1935–1947 (2011).

2) Lunsford, J. H. The catalytic oxidative coupling of methane. Angew. Chem. 35, 970-980 (1995).

500461 Methane Coupling in Atmospheric Pressure Corona Discharge Microreactor

Abstract:

Methane Coupling in Atmospheric Pressure Corona Discharge Microreactor

Adam Shareghi, Ian Reddick, Justin Pommerenck, Scott Harpool, Yu Miao, Yousef Alanazi, Alexandre Yokochi, Annette von Jouanne, Nick AuYeung, Goran Jovanovic

Methane-rich landfill biogas, agricultural gas and stranded natural gas from fracking are vast energy resources that remain largely untapped. Bringing this energy to market is uneconomical with conventional technology. Converting a percentage of this unused resource into an easily transportable form represents an economic opportunity.

Methane coupling using nonthermal plasmas in microscale reactors has been demonstrated. The high temperature electrons disrupt the stable carbon-hydrogen bonds of the methane and allow the fragmented species to polymerize into higher-order hydrocarbons that can be used as liquid fuels. Nonthermal plasmas enable this chemistry to occur at ambient temperatures and pressures thus making this technology economically attractive.

Developing nonthermal plasma reactors on the microfluidic scale reduces the required materials of construction and energy input. Maximal active volume can be achieved at these smaller scales as more reactant gas is forced to move in the vicinity of the high voltage electrodes. Development and optimization of this technology to produce liquid fuels is currently the goal of the study. This presentation will summarize recent developments in this area.

497774 Non-Oxidative Methane Coupling Using Metal Hydrides As Hydrogen Storage Materials

Abstract:

Methane has experienced a meteoric rise due to its vast availability and its prominence in the constitution of natural gas, which is increasingly becoming the dominant energy source in the United States. Although highly desirable, the direct conversion of methane into useful chemicals is limited due to the molecule's high symmetry and strong C-H bonds. In this work, we seek to upgrade methane to higher hydrocarbons. Historically, the most studied method for the upgrade of methane is the oxidative coupling of methane (OCM), $2CH_4 + ½O_2 \rightarrow C_2H_6 + H_2O$ [1]. However, this method has failed to yield an economically viable process due to poor carbon selectivities, which are a result of the over-oxidation of the reactive CH_3 intermediate into CO and CO_2 [2].

To avoid the over-oxidation of methane, we employ hydride-forming metals as hydrogen storage materials (HSMs) in a chemical looping approach. In the first step, hydrogen is catalytically abstracted from methane and stored within the HSM while the methyl groups are combined to form ethane. This first step follows the reaction $2CH_4 + M \rightarrow C_2H_6 + MH_2$, where M represents the hydride-forming metal. In the second step, the HSM is regenerated through thermal release, forming H_2. With the regenerated HSM, the first step can be repeated, thus creating a chemical loop.

Through both density functional theory simulations and experimental work, we have identified five Group III and IV metals (scandium, titanium, yttrium, zirconium, and hafnium) that have strong hydrogen storage capabilities and also exhibit the ability to cleave C-H bonds. In particular, titanium and yttrium have shown promise due to experimentally demonstrated rapid hydride formation kinetics and a theoretical activation energy barrier of less than 0.6 eV for the initial C-H bond scission. In addition, titanium's conversion and selectivity towards higher hydrocarbons can be tuned and improved by alloying titanium with nickel and aluminum, respectively. We have found that the use of metal hydrides and their alloys can be propitious not only for the upgrade of methane, but also for other catalytic dehydrogenation reactions.

References:

[1] Hammond, C., Conrad, S. & Hermans, *ChemSusChem* 5, (2012) 1668–86.

[2] Zaera, F, *Catal. Letters* 91, (2003) 1-10.

499404 Oxidative Coupling of Methane over Mixed Metal Oxides

Abstract:

Oxidative coupling of methane (OCM) is one of the promising routes to convert natural gas to useful chemical building blocks like ethylene. The reactions between methane and oxygen leading to C_2+ products are known to be highly exothermic and some researchers have pointed out that the hot spot temperatures within the catalyst bed in lab scale reactors could exceed the feed temperature by as much as 300°C.

We present experimental and modeling studies on the bifurcation behavior of OCM in fixed bed reactors. Feed temperature and space time are studied as bifurcation variables at various feed CH_4/O_2 ratios. Further, the bifurcation behavior (or lack of) is explained with various reactor configurations, showing the impact of heat transfer characteristics. For systems exhibiting multiple steady-states and bifurcation behavior, this work demonstrates the path dependency of the observed states and provides new insight for systems exhibiting such behavior.

497526 Experimental and Theoretical Investigation on Methane Dehydroaromatization Under Non-Oxidative Conditions

Abstract:

Advances in hydraulic fracturing technology has generated huge industrial attention towards catalytic conversion of low cost natural gas to valuable hydrocarbons. Developing molecular level understanding of catalyst active sites and its deactivation mechanism will help in the process of developing an effective commercial non-oxidative methane dehydroaromatization (MDA) catalyst. Mo_xC_y/H-ZSM-5 is one of the suitable catalyst for the MDA process. Mo/HZSM-5 contains both metal site (Mo species) and acid site (H-ZSM-5) hence act as a bi-functional catalyst to efficiently activate methane on Mo species to produce C_2H_4 which than oligomerize and aromatize to benzene and its derivatives on acidic sites of shape selective zeolite[1]-[2] . Nature of the active site for methane is suggested to be carbide or oxy-carbide but active phase structure is strongly debated. Lezcano-Gonzàlez and co-workers[3] have carried our operando X-ray method and asserted MoC_3 as possible active Mo speciation responsible for methane activation and formation of aromatics. Density functional theory calculations have been employed to study the effect of charge on catalytically active molybdenum carbide nano-cluster and its stability which will help in designing better MDA catalyst with suitable promoter species. In the study we have presented a detail investigation of effect of residual charge, Figure 1(a) showing a linear correlation between charges on MoC_3 cluster on methane activation underlying the effect of reducibility of Mo_xC_y cluster. Further, reaction mechanism insights have been studied for the methane activation and formation of ethylene, through formation of C_2H_6 and further C-H activation. Potential energy diagram for methane following scheme, Figure 1(b) shows ethylene will be formed easily on the catalyst surface, which eventually lead to formation for benzene with an intrinsic methane activation barrier of ~1.23 eV. The two resultant CH_3 groups attached to the Mo_2C_6 cluster undergo C-C coupling reaction with relatively lower barrier of +0.35eV. Following these theoretical insights, H-ZSM-5 impregnated with Mo were tested experimentally for MDA under non-oxidative conditions. In order to minimize coke deposition two different approach a) carbon dioxide (CO_2) as co-feed with methane[4] and b) in-situ topotactic transformation of MoC form hexagonally close packing to face centered cubic[5] by pre-treating catalyst with inflow of H2 at 2400ml/hr-gm GHSV for 6 hours at 350°C and then increasing the temperature to desired reaction condition with constant inflow of 30ml/min during carburization period carried out on a fixed bed reactor. Addition of CO_2 to methane feed reduces not only inert coke but also reactive coke as elucidated by TGA analysis, Figure 1(c). In the current study, the effect of variation in CO_2 concentration was studied. Results depict that addition of CO_2 enhances benzene percentage upto 35% on mole basis, Figure 1(d) in product stream as higher oxygen donor ability of CO_2 eliminate surface carbon by forming CO. Phase transformation of MoC hinders the migration of external surface Mo species to BAS present in zeolite channels resulting in retardation of inert coke deposition at pore entrance which results in enormous catalyst active with high selectivity towards aromatics, Figure 1(e) and less coke deposition resulting in high catalyst stability which is validated through coke gasification kinetics using Kissinger plot.

Figure 1. (a) Effect of residual charges on methane C-H bond activation, (b)Reaction diagram for methane activation and formation of ethylene on Mo_2C_6 nanocluster, (c) TGA analysis for Mo/HZSM-5, (d)-(e) Product selectivity during MDA process for Mo/HZSM-5.

Acknowledgement

Financial support provided by Gas Authority of India limited (GAIL) for carrying out this research work is gratefully acknowledge.

References

1. Majhi, S., Mohanty, P., Wang, H. & Pant, K. K. Direct conversion of natural gas to higher hydrocarbons: A review. *J. Energy Chem.* **22,** 543–554 (2013).

2. Gao, J. *et al.* Identification of molybdenum oxide nanostructures on zeolites for natural gas conversion. **2534,** (2015).

3. Lezcano-González, I. *et al.* Molybdenum Speciation and its Impact on Catalytic Activity during Methane Dehydroaromatization in Zeolite ZSM-5 as Revealed by Operando X-Ray Methods. *Angew. Chemie - Int. Ed.* **55,** 5215–5219 (2016).

4. Ohnishi, R. Catalytic Dehydrocondensation of Methane with CO and CO2toward Benzene and Naphthalene on Mo/HZSM-5 and Fe/Co-Modified Mo/HZSM-5. *J. Catal.* **182,** 92–103 (1999).

5. Liu, H., Bao, X. & Xu, Y. Methane dehydroaromatization under nonoxidative conditions over Mo/HZSM-5 catalysts: Identification and preparation of the Mo active species. *J. Catal.* **239,** 441–450 (2006).

499369 Photocatalytic Hydrogen Generation Onto Mesoporous Titania Doped with Pt

Abstract:

In this research mesoporous TiO_2 photocatlysts were synthesized using the sol-gel technique. A Pluronic F127 triblock-copolymer, a structure-directing agent, was incorporated as a soft template into the sol-gel. In addition, and during a separate synthesis, the sol-gel was doped with a Pt precursor. Photocatalysts were prepared with 1.00 wt. %, 2.50 wt. %, 5.00 wt. % Pt nominal loadings, respectively. They were calcined at 500°C and 550°C following synthesis. Morphological and structural properties were studied by: a) X-ray diffraction, b) UV-Vis spectrophotometry, c) N_2 adsorption-desorption (BET, BJH), and d) X-ray photoelectron spectroscopy (XPS). Optical band gap values for mesoporous TiO_2 and modification with Pt were calculated by Kubelka-Munk (K-M) function coupled with Tauc plot methodology. It was observed that the prepared semiconductors displayed pore sizes in the 10 to 40 nm range with bimodal distributions. Their photocatalytic activity for hydrogen production via water splitting was established in a Photo-CREC Water-II reactor under near-UV light irradiation. The aqueous solution contained 2% v/v ethanol, employed as a renewable organic scavenger. The prepared materials exhibited that the mesoporous 2.50 wt. % Pt-TiO_2 has the highest photoactivity for hydrogen generation. This suggests the key role played by the loading of platinum as a TiO_2 dopant, reducing the optical band gap, increasing electron storage and diminishing, as a result, electron-hole recombination. The measured Quantum Yield (QY), obtained using a rigorous approach, was established for the mesoporous 2.50 wt. % Pt-TiO_2 at a promising level of 22.6 %.

495916 TiO_2 Nanoparticles By Flame Synthesis for Photocatalytic Hydrogen Evolution

Abstract:

Energy shortage and environment deterioration have been increasingly urgent and severe over the past decades. The limited non-renewable fossil fuels can hardly meet the growing energy demand. Thus, the exploitation of new green energy resources has become a global concern in recent years. Sunlight and water are abundant natural resources, which can be utilized to produce renewable energy in the form of hydrogen through water splitting. Thus, our work has been focused on this promising area. Despite the great progress achieved by now, the research in this field is still in an early age. The efficiencies of photocatalytic water splitting for the production of hydrogen are still relatively low and remain to be improved. Our flame-made TiO_2 has many advantages over the traditional TiO_2 which is prepared by wet chemistry. The small uniform particle size (~15nm) and the large specific surface area (~150 m^2/g) make our TiO_2 a good candidate for this reaction and it can improve the hydrogen generation rate largely. The simple procedure and stable mass production (~100mg/h) can largely save the cost of time and money, compared to the wet chemistry method. Besides, the excellent durability and environmental-friendly property can make our TiO_2 nanoparticles as a promising substitution for the toxic heavy metal catalysts, like CdS, CdO, MoP. The highest hydrogen generation rate that we have achieved by now is 24178 Î¼mol $g^{-1}h^{-1}$. So far, this performance has been proved to be better than most results in other works.

UCLA

$Ni(OH)_2$ Decorated Jagged Pt Nanowires Have High Activities for Hydrogen Evolution Reaction

Mufan Li, Kaining Duanmu, Zipeng Zhao, Peng Li, Huilong Fei, Wenxin Chen, Yuanming Zhu, Rong Yu, Jun Luo, Ketao Zang, Zhaoyang Lin, Mengning Ding, Chengzhang Wan, Jin Huang, Hongtao Sun, Yu Huang, Philippe Sautet, Xiangfeng Duan

Nov. 2, 2017

Synthesis of J-PtNWs/$Ni(OH)_2$

UCLA

NWs = Nanowires
CV = Cyclic Voltammetry

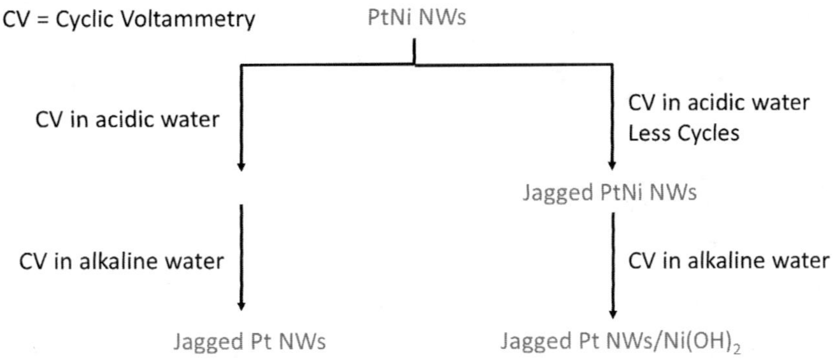

Characterization of J-PtNWs/Ni(OH)$_2$

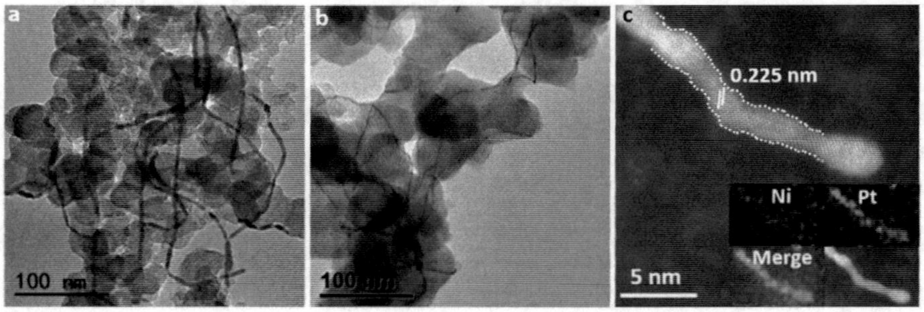

TEM of PtNi NWs   TEM of J-PtNWs/Ni(OH)$_2$ STEM of J-PtNWs/Ni(OH)$_2$

Electrocatalytic activity for HER

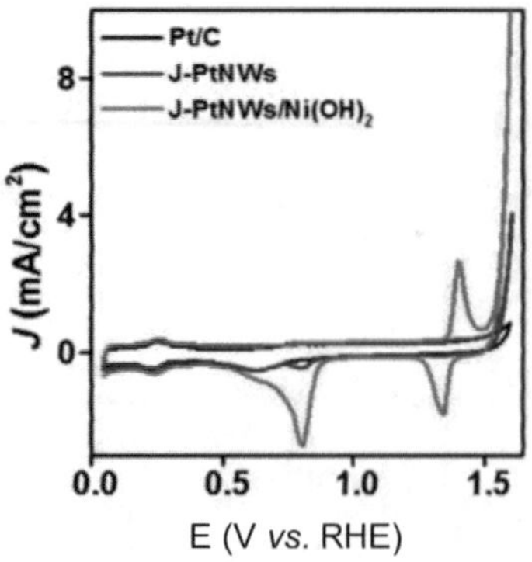

Cyclic voltammetry (CV) diagram

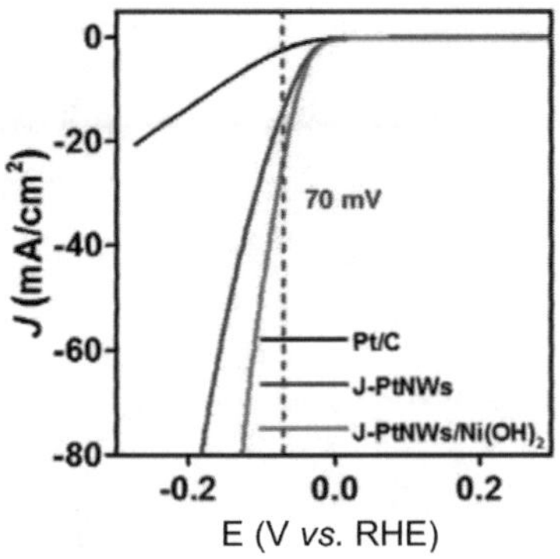

Linear sweep voltammetry diagram

Electrocatalytic activity for HER

UCLA

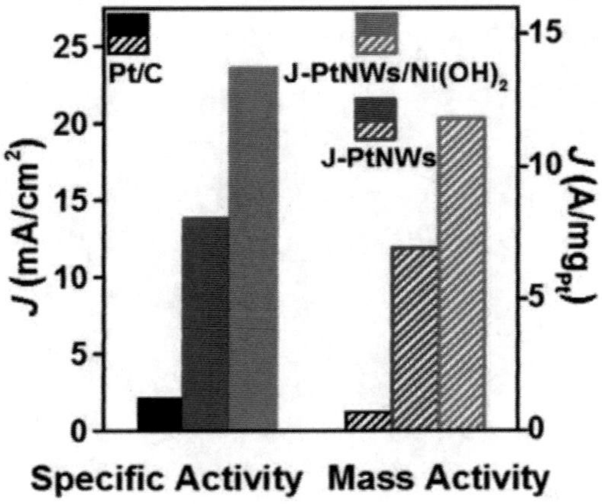

Volcano curve

UCLA

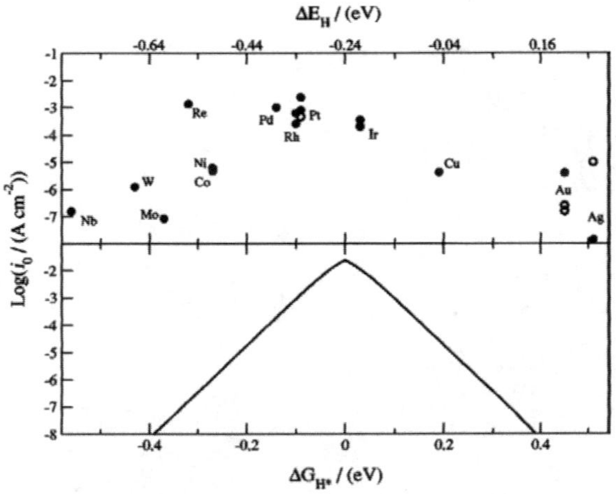

Nørskov, Bligaard, Logadottir, Kitchin, Chen, Pandelov, Stimming *J. Electrochem. Soc.* **2005**, *152*, J23.

Cavity model 1

$\Delta\Delta G_H(\text{site}) = \Delta G_H(\text{site}) - \Delta G_H(\text{Pt111})$

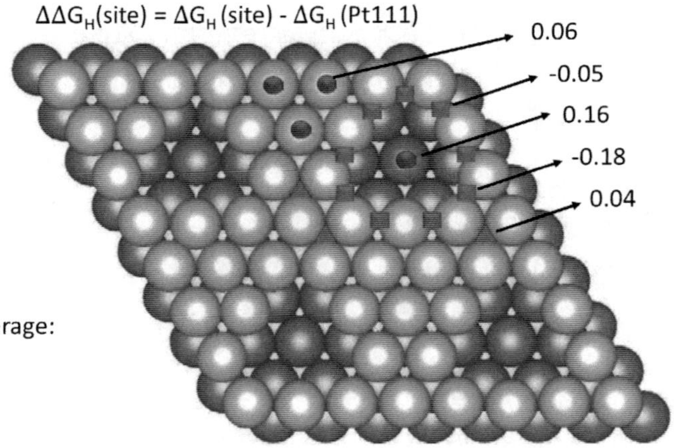

Full coverage:
16/16

Cavity model 2

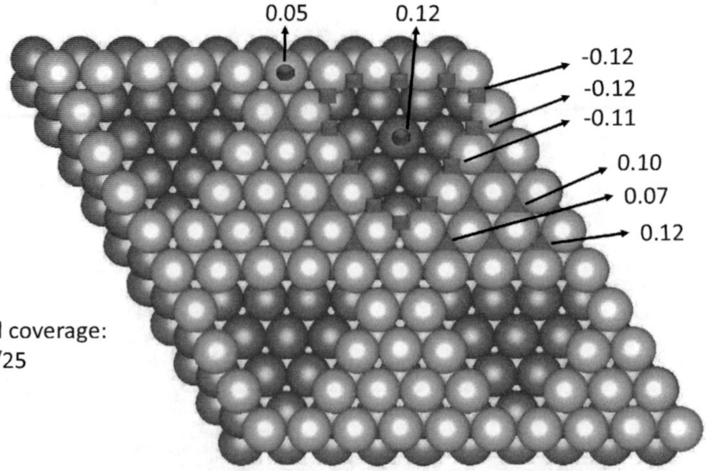

Full coverage:
24/25

Step model 3

UCLA

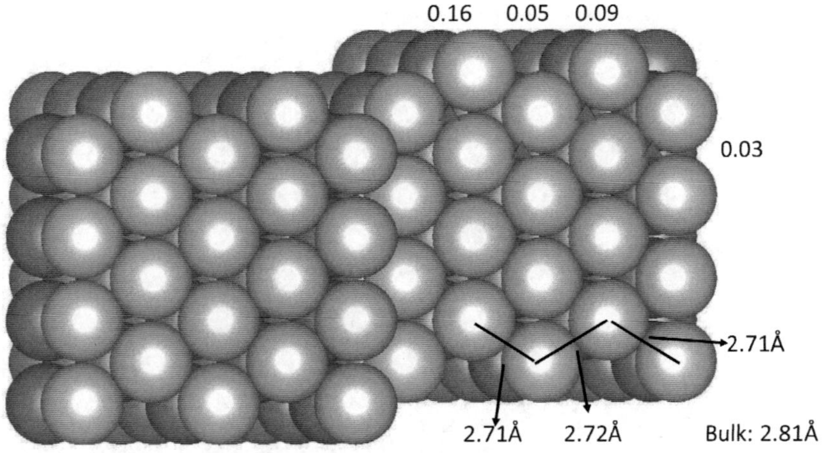

HER for J-PtNWs

UCLA

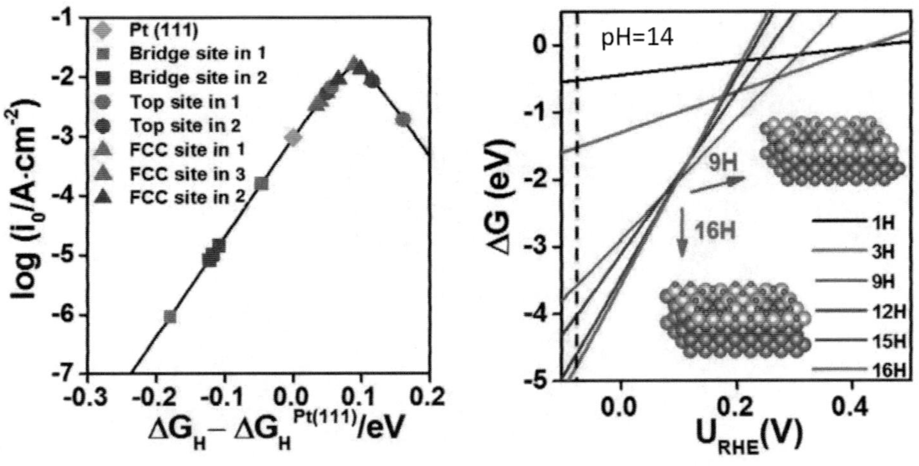

Model J-PtNWs/Ni(OH)$_2$ – low coverage

UCLA

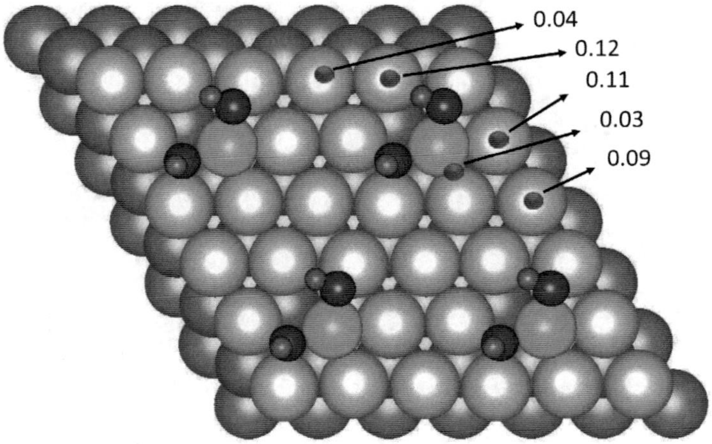

Model J-PtNWs/Ni(OH)$_2$ – high coverage

UCLA

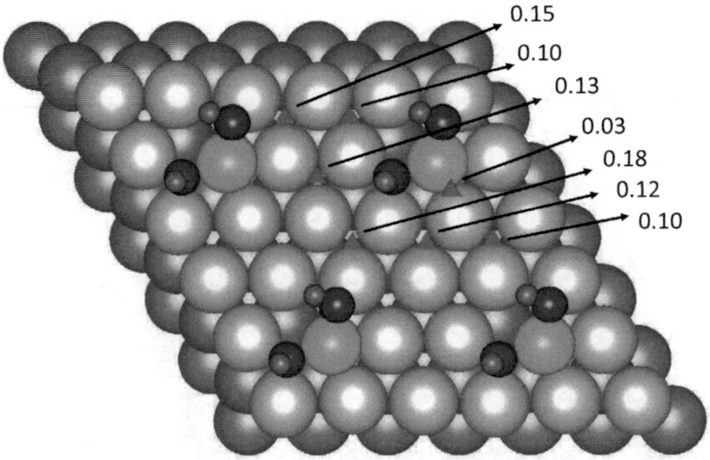

HER for model J-PtNWs/Ni(OH)$_2$

UCLA

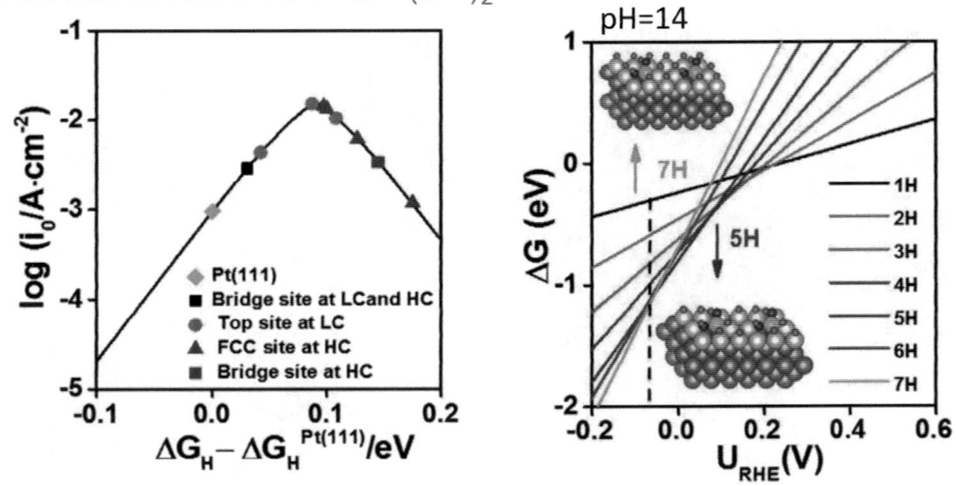

Conclusions

UCLA

- Synthesis of J-PtNWs and J-PtNWs/Ni(OH)$_2$
- HER activities for Pt/C, J-PtNWs and J-PtNWs/Ni(OH)$_2$ are 1: 10: 17
- The defective surface of J-PtNWs creates both active sites and non-active sites. In the experimental condition, hydrogen can adsorb on both of them.
- The decoration of Ni(OH)$_2$ can increase the activity of Pt, and all the sites around Ni(OH)$_2$ have higher activities.

Acknowledgments

UCLA

Sautet Group

Prof. Xiangfeng Duan

Mufan Li

XSEDE
Extreme Science and Engineering
Discovery Environment

NSF

UCLA

Thank You

500531 Selective and Highly Stable H_2 Production from Formic Acid Decomposition over Graphite-Supported Moc

Abstract:

Formic acid (FA) is a good â€œgreenâ€ energy carrier candidate since it can be produced from CO_2, stored as a liquid, and converted to H_2 and CO_2 on demand to power H_2 fuel cells. Several interesting catalytic approaches have been investigated on both ends of this cycle, but the final reforming reaction still needs more work to become a viable commercial fuel source. Unlike the synthesis reaction, which will be conducted at a production plant and can use any materials or processes within the cost and safety limits, the on-demand reforming reaction must meet a number of criteria. First, FA reforming cannot be conducted over expensive, difficult-to-prepare, or unsafe catalysts â€“ for both economic and safety reasons. Any catalyst chosen for a commercial FA reformer must also maintain very good long-term stability, which oftentimes means a loss in activity. We have synthesized MoC nanoparticles (~2 nm) supported on graphite which demonstrate higher FA reforming activity, better CO_2/CO selectivity, and greater long-term stability without H_2 pretreatment than unsupported Mo_2C. We also propose a simple reaction mechanism for both reaction pathways based on previous observations and DFT studies on other materials as well as our own observations, and thoughts on how to further improve the reaction without using expensive or difficult-to-prepare catalysts, including by additive co-feed and with electric fields.

494871 Catalytic Trends for Nanoscale Catalysts from First-Principles Modeling and Tuning of Catalytic Reactivity for Fuel Production

Abstract:

It is still a rather open area regarding our fundamental understanding for effective manipulation of the activity and selectivity of catalyst for chemicals and fuel production. In this study, a first-principles-based kinetic model has been developed in order to understand the catalytic trends of reactivity and selectivity for hydrogen production on nanoscale single crystal nickel particles. The earth abundant nickel is a widely utilized industrial catalyst for hydrogen production through reforming, which is often accompanied by high temperature water-gas shift reaction (WGSR) that further boosts the hydrogen yield. However, nickel also promotes a number of side reactions, such as methanation and hydrogenolysis, both of which would lower hydrogen selectivity, and thus making the effort to tune the activity and selectivity of nickel catalyst an urgent fundamental research topic.

The mechanistic guidelines were first established using periodic density functional theory (DFT) calculations on the Ni(111), Ni(211) and Ni(100) single crystal facets, followed by the establishment of the most thermodynamically and kinetically relevant hydrogen production pathways. Furthermore, the reaction routes involving CO methanation consuming hydrogen will also be considered. In this investigation, the adsorbate-adsorbate interactions - between the key WGSR and methanationÂ reaction intermediate pairs, such as CO-CO, CO-H_2O, CO-OH, CO-O, and CO-CH â €“ have been extensively considered and shown significant impact on the kinetic trends on nickel single crystal surfaces. It has also been noted that, although most of the lateral interactions are repulsive in nature, catalyst lattice structures and intermolecular hydrogen bonding will create exceptions and complicate the trend analysis. Mean-field microkinetic modelings based on the molecular-level mechanism showed that the catalyst reactivity varies at different temperature regimes. At temperature below 570 K, the H_2 production rate is higher on the open flat surface, such as Ni(100), than on the stepped surface, e.g., Ni(211). Above 570 K, this trend has been reversed. Detailed analysis of the hydrogen production route also revealed that the contribution from the redox and carboxyl pathways also vary at different facets. This study will enrich our understanding of the hydrogen chemistry at the molecular level, and would also provide theoretical insights into tuning the properties of nanoscale transition metal catalysts by manipulating the surface structure and morphology.

495800 Sorption-Enhanced Water Gas Shift (SE-WGS) Reaction Processes for the Production of High-Purity Hydrogen

Abstract:

The interest of hydrogen, regarded as an alternative energy carrier, has been increasing with expectation of the upcoming hydrogen economy. As one of methods for the bulk hydrogen production, water gas shift (WGS) reaction processes using synthesis gas feeds produced by the gasification of carbonaceous feedstocks has been highlighted. The performance of WGS reaction is significantly increased by applying the sorption-enhanced reaction (SER) concept. In the reaction of sorption-enhanced WGS (SE-WGS) reactions, the WGS reaction and CO_2 removal by sorption are carried out simultaneously in a single reactor. For the successful fulfillment of SE-WGS reaction, adopting the novel materials that can capture high-temperature CO_2, and managing homogenous packing of catalyst and sorbent are crucial. In this study, the SE-WGS reactions were experimentally demonstrated using $Cu/ZnO/Al_2O_3$ catalyst and double salt-based sorbent pellets. High-purity hydrogen could be directly from the SE-WGS reaction as a result of applying double salt-based CO_2 sorbent, and the effect of the degree of homogeneity for the catalyst and sorbent in a column was further investigated by applying divided section packing concept. The productivity of high-purity hydrogen could be enhanced when more catalyst was packed in the former sections and more sorbent in the later sections.

487615 Aldehyde Water Shift Reaction: Comparative Investigation of Supported Metal Catalysts

Abstract:

There has been significant interest in the use of H_2O as a reactant for biomass conversion given its high concentration in typical feed streams. Steam reforming and 5-hydroxymethylfurfural oxidation are two important examples.[1], [2] The aim of research described in this paper was to investigate the use of H_2O as a terminal oxidant for the partial oxidation of aldehydes. The aldehyde water shift (AWS) reaction oxidizes the aldehyde and produces H_2 (Equation 1).

$CH_3CHO + H_2O \rightarrow CH_3COOH + H_2$ (Eq. 1)

Prior investigations of the AWS reaction have focused on homogeneous catalysts.[3] The commercial implementation of these types of catalysts can be cost-prohibitive. To our knowledge, no heterogeneous catalyst has been reported for this reaction. We explored the use of Mo_2C-, CeO_2-, and Al_2O_3-supported metals (Cu, Pt, Au), the design of which was inspired by that for highly active water gas shift (WGS) catalysts. In particular, the most active heterogeneous WGS catalysts are bifunctional and possess distinct sites for H_2O dissociation and CO oxidation.[4] The target metal loadings were equivalent to 0.1 monolayer. The catalysts were loaded into a quartz flow-through reactor and the reactions were then carried out at 200-280°C.

The AWS activities (Figure. 1) and product distributions (Figure. 2) were strong functions of the support. The observed AWS activities and activation energies suggest that the rate-limiting step, which is believed to be H_2O dissociation, occurred on the support surface. Among all the catalysts studied, the Mo_2C-supported catalysts possessed the highest AWS activities, with activities at least 5 times higher than those for the other materials. For the Mo_2C-supported catalysts, the admetal did not significantly affect the activity. The Cu/CeO_2 showed the highest AWS activity among the CeO_2-supported materials. Like Mo_2C, CeO_2 is active for H_2O dissociation. The high activity of the Cu/CeO_2 catalyst clearly illustrates the importance of Cu particles for this material. In addition to the AWS reaction, the major side reactions were the Cannizzaro reaction, which produces ethanol and acetic acid, and aldol condensation, which forms crotonaldehyde. The Mo_2C-supported catalysts and Cu/CeO_2 were more than 60% selective to the AWS reaction. Our results suggest that the cooperation between the sites for H_2O dissociation and aldehyde oxidation may have contributed to the high AWS activities and selectivities. In contrast, the other CeO_2- and Al_2O_3-supported catalysts produced primarily crotonaldehyde via the aldol condensation reaction, which appeared to be catalyzed by weak acid sites. These and other results will be presented in the paper.

References

[1] Davda, R. R.; Shabaker, J. W.; Huber, G. W.; Cortright, R. D.; Dumesic, J. A. *Appl. Catal. B Environ.* **2005**, *56*, 171-186

[2] Zhang, Z.; Deng, K. *ACS Catal.* **2015**, *5*, 6529-6544

[3] Brewster, T. P.; Goldberg, J. M.; Tran, J. C.; Heinekey, D. M.; Goldberg, K. I. *ACS Catal.* **2016**, *6*, 6302-6305

[4] Schweitzer, N. M.; Schaidle, J. A. ; Ezekoye, O. K.; Pan, X. ; Linic, S.; Thompson, L. T. *J. Am. Chem. Soc.* **2011**, *133*, 2378-2381

491970 2-Propanol Decomposition on $TiO_x/Au/SiO_2$: Probing Properties of the Interfacial Perimeter Sites.

Abstract:

Interfacial perimeter sites of Au and TiO_2 has been shown to be essential in a number of oxidation reactions. One of them is the selective formation of acetone in the oxidation of propane, where the interfacial perimeter site between the metal and $(Ti,Si)O_2$ was shown to be necessary for acetone formation over a Au/SiO_2 catalyst. Since 2-propoxide is a likely intermediate in this reaction and it can be generated in the decomposition of 2-propanol, this reaction was studied as a probe for the properties of the interfacial perimeter sites.

The decomposition of 2-propanol was studied over SiO_2, SiO_2 with an overlayer of TiO_2 (Ti/SiO_2), Au/SiO_2, and Au/SiO_2 with an overlayer of TiO_2 ($Ti/[Au/SiO_2]$) at 170-190 °C. The SiO_2 support was inactive. Without Au in the Ti/SiO_2 catalysts, propene was the only product and its rate of formation increased proportionally with the Ti content. Acetone was the major product (selectivity >65%) on all Au-containing catalysts, and its rate of formation also increased with Ti loading. The presence of Ti in the $Ti/[Au/SiO_2]$ catalysts also resulted in the formation of small amounts of propene and the rate of which increased with Ti loading. Characterization of the catalysts with N_2 adsorption, STEM, DR-UV-vis spectroscopy, XPS, XANES and EXAFS suggested that the Ti formed a porous, amorphous TiO_2 overlayer on the catalyst. Au was present as 1-2 nm nanoparticles which did not change after reaction. Only Lewis acid sites were detected based on results from pyridine adsorption, and the quantities increased with Ti loading. From the characterization and kinetic data, especially the reaction rates and the apparent activation energies, it was concluded that the propene formation was on the TiO_2 overlayer, whereas acetone was formed at the $Au-TiO_2$ interfacial perimeter sites. Breaking of the Î±-C-H bond in 2-propoxide preceding acetone formation occurred on Au at the interfacial site.

489232 Determining How Support pH and Hydrophilicity and Metal Particle Size Influence Activity and Product Distributions in Aqueous Phase Reforming of Glycerol

Abstract:

The catalytic conversion of biomass derivatives, such as glycerol, to platform chemicals or energy products could lessen dependence on fossil fuels, thus providing a sustainable source of energy for the worldâ€™s growing demand. One method for performing these conversions is aqueous phase reforming (APR), which is of interest because it utilizes low operating temperatures.Â Presently, catalysts synthesized to carry out APR are comprised of noble metals catalysts supported on metal oxides or other minerals. One of our goals is to learn how these materials function so that we can design less expensive catalysts for APR and other similar reactions. One of the challenges in understanding how APR catalysts function is the aqueous environment itself: H_2O molecules not only influence catalytic chemistry and physics in numerous ways, they also impede our ability to observe these influences.Â Our group and others have recently reported several ways in which H_2O influences the catalytic chemistry of glycerol APR and related reactions over *transition metal catalysts*. However, how the support participates, and especially how the catalytic chemistry is influenced by the *H_2O/catalyst/support interface* is still largely unknown. In this research, we identify how multiple support properties influence catalytic activity and product distributions in glycerol APR. Specifically, we examine the effects of acidity/basicity, hydrophobicity/hydrophilicity, and catalyst particle size on Pt-catalyzed glycerol APR. These properties are expected to have the following influences on catalytic activity and performance. Acidity/basicity of the support is caused by the presence of Brønsted and Lewis acid/base sites on the support surface, which promote some of the reaction steps in APR. Additionally, they influence the pH of the catalyst due to the presence of H+ and OH- ions, which also alter the hydrophobicity/hydrophilicity of the catalyst. These effects also influence the structure of liquid H_2O at the interface, which in turn influences the thermodynamics and kinetics of catalytic reactions. Catalyst particle sizes are dictated by metal/support interactions and influence catalytic activity and selectivity. Our goal is to identify the support properties that lead to optimal performance but also that have the least influence on performance, as these materials can be best compared with simulation work from our group. Synthesized catalysts undergo various characterization methods, including scanning electron microscopy (SEM), X-ray diffraction (XRD), energy dispersive X-ray (EDX), nuclear magnetic resonance (NMR), thermogravimetric analysis (TGA), temperature-programmed-desorption (TPD), and various infrared (IR) spectroscopy methods. Results from these characterization methods as well as product distributions from catalytic experiments will be presented, and comparisons with simulations will be made where possible.

493074 Measurement of the Concentration and Intrinsic Catalytic Activity of Monometallic and Interfacial Sites

Abstract:

Title: Measurement of the concentration and intrinsic catalytic activity of monometallic and interfacial sites

Insoo Ro[a], Yifei Liu[a], Isaias B. Aragao[b], Madelyn R. Ball[a], Joseph P. Chada[a], Canan Sener[a], Daniela Zanchet[b], James A. Dumesic[a], and George W. Huber[a]

[a] Department of Chemical and Biological Engineering, University of Wisconsin-Madison, Madison, WI

[b] Institute of Chemistry, University of Campinas, Campinas, SP, Brazil

Many previous studies have proposed that several types of catalytic reactions occur at the interface between a metal and metal oxide or an â€œinterfacial catalytic siteâ€. However, the measurement of the intrinsic activity of an interfacial catalytic site has not previously been possible. This is primarily because the synthesis of catalysts by traditional methods produces a wide distribution of metal particle sizes and compositions which results in a distribution of interfacial sites. We have developed synthetic routes to prepare a series of metal-metal oxide catalysts (including Au-MoO_x, Pt-MoO_x, Cu-ZrO_2, and Pt-FeO_x all supported on SiO_2) with well-defined concentrations of interfacial sites using an approach called controlled surface reactions (CSR). We were able to quantify the concentrations of monometallic and interfacial sites using a combination of sub-ambient CO Fourier transform infrared spectroscopy (FT-IR), CO chemisorption, reactive N_2O chemisorption, and scanning transmission electron microcopy (STEM)/energy dispersive X-ray spectroscopy (EDS). We then measured the intrinsic activity in the reverse water gas shift reaction (RWGS), methanol synthesis reaction, hydrogenation of carbonyls, CO oxidation, and ethanol conversion into ethyl acetate [1-4]. The intrinsic reaction rate at these interfacial sites was 8 to 700 times higher than the intrinsic reaction rate on monometallic sites. In this presentation we will attempt to develop a relationship between the structure of the interfacial site and the catalytic activity in different reactions. We will discuss how we can control the catalyst activity by changing the concentration and types of interfacial sites. We will illustrate how this approach can be used to design improved catalysts by maximizing the concentration of interfacial sites.

Reference

[1] Ro et al. Role of the Cu-ZrO_2 Interfacial Sites for Conversion of Ethanol to Ethyl Acetate and Synthesis of Methanol from CO_2 and H_2, *ACS Catalysis*, 6 (2016) 7040

[2] Ro et al. Measurement of intrinsic catalytic activity of Pt monometallic and Pt-MoO_x interfacial sites over visible light enhanced $PtMoO_x/SiO_2$ catalyst in reverse water gas shift reaction, *Journal of Catalysis*, 344 (2016) 784

[3] Carrasquillo-Flores and Ro et al. Reverse water-gas shift on interfacial sites formed by deposition of oxidized molybdenum moieties onto gold nanoparticles, *Journal of the American Chemical Society* 137 (2015) 10317

[4] Ro et al. Intrinsic kinetics of plasmon-enhanced reverse water gas shift on Au and Auâ€"Mo interfacial sites supported on silica, *Applied Catalysis A: General*, 521 (2016) 182

497747 Furfuryl Alcohol Dehydration over Metal Oxides (TiO_2, Al_2O_3 and Nb_2O_5) supported WO_3 Catalysts

Abstract:

The conversion of lignocellulosic biomass into fuels has been investigated as a result of increasing demand for petroleum oil and environmental concerns. Furfuryl alcohol (FA), which is produced from furfural derived from a selective dehydration of xylose, is an attractive intermediate chemical for the production of template chemicals, such as levulinic acid and alkyl levulinate and FA polymer. Under acid condition, FA can be converted to FA oligomers through intermolecular dehydration. Previously we found the use of bulk tungsten oxide was very similar to strong homogeneous acid, such as sulfuric acid. It was known that the interaction of a catalytic active component, such as WO_3, with a metal oxide support can dramatically alter the molecular and electronic structural and catalytic properties of the supported catalytic active component. In the present study, a series of supported WO_3 catalysts were synthesized on TiO_2, Al_2O_3, and Nb_2O_5 supports by incipient wetness impregnation method. The monolayer coverage and the molecular structures of the supported tungsten oxide were determined with in-situ Raman. Two significant parameters, the tungsten oxide coverage (W/nm^2) (loading effect) and the specific metal oxides support (support effect), were carefully examined during liquid phase FA dehydration reaction under mild experimental conditions, 100°C and ambient pressure, for 3 hours. The specific metal oxide support was found to significantly affect the relative catalytic activity of the surface WO_x species to that of the crystalline WO_3 particles.

497234 Support-Induced Control of Surface Composition in Bimetallic Catalytic Particles

Abstract:

Supported bimetallic nanoparticles have been identified as an important class of materials with demonstrated tunable reactivity for a variety of industrial catalytic properties. The design of bimetallic surface structures with controlled geometries and compositions via density functional theory calculations has proven successful for catalyst optimization. However, the actual structure and composition of exposed catalytic surfaces on synthesized structures under reaction conditions is a complex function of interactions between the two metals, the metals and reactive environments, and the metals and support. While reaction environment induced reconstruction of bimetallic particles has been thoroughly explored, relatively little is known about how specific metal-support interactions can be used to tune and control exposed surface structures. In this talk I will describe recent work where we have identified how metal-support interactions can be used to control the surface composition of bimetallic Cu-Ni nanostructures, and how this segregation influences reactivity, selectivity, and stability of the catalysts under harsh biomass-derived compound hydrogenolysis conditions. We found that for irreducible oxide (e.g. Al_2O_3) supported Cu-Ni bimetallic particles, both Cu and Ni domains are exposed at the reactive surface, where for reducible oxide (e.g. TiO_2) supported bimetallic particles, Cu is enriched at the surface while Ni seggregated to the metal-oxide interface. As a result of the seggregation observed for Cu-Ni/TiO_2 catalysts, excellent reactivity, selectivity, and stability were observed in catalytic furfural and hydroxymethylfurfural hydrogenolysis. We propose a mechanism where we envision the support acting cooperatively with the bimetallic structure to control the exposure of desired catalytic surfaces.

500212 Nature of Molecular Interactions of Bio-Oil Derivatives on Thiolate-Coated Pd Nanoparticles

Abstract:

Biomass-derived compounds are highly functionalized, making them difficult to selectively upgrade into a desired product. This is because the functional groups interact with the surface, resulting in multiple and often uncontrolled adsorption geometries. Previous experimental studies on Pd indicated that adsorption geometry of these oxygenates dictates product selectivity, where flat-lying configurations result in higher production of undesired decarbonylation products and upright conformations favor Câ€"O bond activation, producing the desired hydrodeoxygenation (HDO) products. Previously, our group has shown that thiolate self-assembled monolayers (SAMs) are able to block specific active sites on Pd nanoparticles during the adsorption and reaction of furfural and furfuryl alcohol, thereby reducing selectivity towards decarbonylation without affecting HDO to methylfuran. However, the effect of the thiolates is not understood fundamentally, hampering the efforts to design better catalysts. Molecular dynamics (MD) has provided insight into the molecular effects of the SAMs on the reactions and possible mechanisms. Simulations with CHARMM-Interface were conducted using the Nanoscale Molecular Dynamics (NAMD) program and also showed the same accuracy as DFT calculations for smaller systems without surfactants in vacuum.

These studies have found that many aromatic oxygenates, such as furfuryl alcohol, furan, and methylfuran adopt an upright adsorption configuration, with the aromatic ring repelled away from the surface. The results indicate that the thiol ligands affect both binding energies and configurations of these aromatic oxygenates compared to uncoated Pd. These changes impact product selectivity. For example, furfural on a SAMs covered nanoparticle adopts an upright conformation, as well as diffuses to the edge sites, which correlates to the reduced selectivity towards the decarbonylation product, furan. These aromatic oxygenates on SAMs-coated nanoparticles also desorb from the surface and can interact with the SAMs for some residence time before desorbing into vacuum.

493322 Surface Characterization of 2D Metal-Supported Bilayer Silica and Aluminosilicates As Model Zeolites

Abstract:

Zeolites have been a cornerstone of industrial catalysis for many decades and are essential for a wide range of reactions. While three-dimensional silica and aluminosilicates have been studied widely, their typical micron particle sizes coupled with internal 5-10 Å pores containing active sites have limited the application of direct surface microscopic and spectroscopic techniques for measurement of reactions in these materials. Two-dimensional analogs have only recently been discovered and synthesized, with applications as model zeolite surfaces and potentially as atomically-thin size-selective membrane materials. In the silica bilayers, corner-shared SiO_4 tetrahedra arrange on metal substrate surfaces to form two mirror-image layers with all bonds saturated, ultimately consisting of all six-membered rings in the crystalline phase and a range of four- through nine-membered rings in the amorphous (vitreous) phase. Through the substitution of Al into the silica bilayers to form $Al_xSi_{1-x}O_2$, Brønsted acid sites analogous to those in zeolites can be generated, therefore the $Al_xSi_{1-x}O_2$ bilayer can be considered as an acid zeolite model system. In this work, we have focused on the synthesis and fundamental understanding of these two-dimensional silica and aluminosilicate bilayer materials. To this end, we have synthesized a range of bilayers over a continuously tunable Ni-Pd(111) alloy system in ultra-high vacuum to achieve the desired strain match to the bilayer composition. Whether the film becomes crystalline or amorphous depends on the atomic spacing of the underlying metal substrate and the growth conditions, and by careful substrate selection the resulting observed bilayer may be carefully controlled even though the bilayer sheets interact with the substrate through van der Waals forces alone. Surface diffraction, spectroscopy, and scanning tunneling microscopy reveal how the bilayers accommodate the underlying strain and form coherent, atomically flat surfaces for model study. Additionally, chemical and catalytic properties of the aluminosilicate bilayers have been explored through probe reactions (e.g. alcohols and amines) by conducting thermal desorption spectroscopy to demonstrate the applicability of these materials for direct study of zeolite catalysis.

500049 Effects of Electrolyte Ions on Electrochemical CO_2 Reduction

Abstract:

The identity of both metal cations and anionic species in the electrolyte is known to influence the electrochemical reduction of CO_2; however, detailed understanding of these effects is still lacking. Here we present a combination of experimental and theoretical studies of the effects of electrolyte cation size on the activity and selectivity of metal catalysts for the reduction of CO_2. Further, we explore how the identity of anionic and buffering species in the electrolyte influence catalytic activity. Experiments were conducted under conditions where the influence of electrolyte concentration polarization is minimal in order to separate intrinsic kinetic and mass transport effects. Interpretation of the findings for CO_2 reduction was aided by studies of the reduction of gyloxal and CO, key intermediates along the reaction pathway to final products. Density functional theory calculations show that the effect of the presence of alkali metal cations can be understood in terms of electrostatic interactions between solvated cations and adsorbed species with large dipole moments, while trends in activity with cation size arise from additional chemical interactions for larger cations. The effect of anions on CO2 reduction rates can be understood in terms of the hydrogen donation ability of the most willing donor in solution.

497754 Multifunctional Homogeneous-Heterogeneous Polymer Catalysts for the Synthesis of Hydroxymethylfurfural from Glucose

Abstract:

Homogenous catalysts are generally more active than their heterogeneous counterparts because the solubility in the reaction medium contributes to access of reactants to all active sites. However, heterogeneous catalysts are commonly preferred since catalysts can be easily recovered from the medium and be reused. Poly(styrenesulfonic acid) (PSSA) combines the advantages of both homogeneous and heterogeneous catalysis. PSSA is soluble in polar solvents; therefore, all acidic sulfonic groups are readily accessible. In addition, the catalyst cannot be deactivated through coking because there is no surface for the carbonaceous species to be deposited. At the same time, PSSA, due to its high molecular weight, can be easily recovered by ultrafiltration for further utilization.

PSSA can be obtained by sulfonation of polystyrene waste (e.g., yogurt packaging or expanded polystyrene), which is an additional advantage from an environmental point of view [1]. This polymer catalyst demonstrated high activity in several biomass conversion reactions that require Brønsted acid sites: synthesis of biodiesel from vegetable oil [1], dehydration of xylose to furfural, and furfural oxidation to maleic and succinic acids [2]. Here we present the addition of Lewis acid functionality to this polymer for a one-pot synthesis of hydroxymethylfurfural (HMF) from glucose.

1. Alonso-Fagúndez, N.; Laserna, V.; Alba-Rubio, A.; Mengibar, M.; Heras, A.; Mariscal, R.; Granados, M. L., Poly-(styrene sulphonic acid): An acid catalyst from polystyrene waste for reactions of interest in biomass valorization. *Catalysis Today* **2014,** *234*, 285-294.

2. Granados, M. L.; Alba-Rubio, A. C.; Sádaba, I.; Mariscal, R.; Mateos-Aparicio, I.; Heras, Á., Poly(styrenesulphonic) acid: an active and reusable acid catalyst soluble in polar solvents. *Green chemistry* **2011,** *13*(11), 3203-3212.

486961 Nanobowls: A Platform for Selective Acid Catalysis on External Oxide Surfaces

Abstract:

Metal oxide catalysts are workhorse materials in the production of petroleum products and fine chemicals. Crystalline, microporous materials (zeolites) have long dominated catalysis in these industries since being introduced in 1960. Zeolites act as shape selective solid acid catalysts and often possess both strong Brønsted and Lewis acidity. As the chemical industry looks to switch over to biomass and tar sands feedstocks, reactant diffusion and catalyst stability issues have arisen. These problems could be addressed by running these same reactions over (*somehow*) selective, non-porous metal oxides. In addition, using bulk oxides for zeolite chemistry would open the door to tunable materials development covering a wide array of elemental compositions. Therefore, we propose the Nanobowls approach as an alternative, highly tunable synthesis method to create selective acid catalysts. In this approach, we construct 1-2 nm diameter cavities on the surface of nonporous oxides. The Nanobowls approach has been applied to both major types of acid catalysts: those with Brønsted sites and those with Lewis sites.

In the Brønsted acidic case, spherical Al_2O_3 particles were overcoated with SiO_2 in the liquid phase using TEOS as the Si precursor. These overcoated materials were calcined at 650 °C with a high ramp rate, 20 °C/min, that lead to the formation of strong Brønsted acid sites contained within a thin, accessible microporous overcoat. Their performance in a Brønsted acidity probe reaction, the vapor phase catalytic cracking of Triisopropylbenzene (TIPB) was on par with industry standard amorphous SiO_2-Al_2O_3 and slightly less active than zeolite H-Y on a surface area basis. In the Lewis acidic case, shape selective Ti sites were synthesized on SiO_2 spheres by grafting Cp*$TiCl_3$, calcining off the ligands at 550 °C, overcoating with SiO_2 using the same liquid phase deposition technique, and recalcining the material at 550 °C with a high ramp rate of 10 °C/min. This lead to the formation of strong Lewis acid sites contained within a thin, accessible microporous overcoat. Their shape selectivity performance is being tested in several Lewis acidity probe reactions including competitive Tetrahydropyranylation of Benzyl alcohol and 2,4,6-Trimethylbenzyl alcohol, competitive MPV reduction of Acetophenone and 2,4,6-Trimethylacetophenone, and Limonene Epoxidation.

Overall, we have demonstrated this approach as a highly tunable synthetic pathway to alternative selective Brønsted and Lewis acidic oxide materials. Future materials development will include application to bifunctional catalysis such as biomass conversion reactions and the synthesis of shape selective Brønsted sites via sulfation of Al_2O_3 or TiO_2 within 1-2 nm SiO_2 cavities.

493953 Controlled Tin Insertion into Zeolite Framework Vacancy Defects and Catalytic Consequences for Sugar Isomerization

Abstract:

Alternative synthesis routes to hydrothermal crystallization of Sn-Beta zeolites have focused on the post-synthetic incorporation of stannic precursors within framework vacancy defects in dealuminated supports via vapor-phase deposition [1], solid-state ion exchange [2], and liquid-phase grafting under isopropanol reflux [3], each with different consequences for the density and coordination of framework Sn atoms and density of residual silanol defects. We report that grafting Sn heteroatoms in dichloromethane solvent (333 K) enables preparing Sn-Beta zeolites comprising a wide range of Sn content (Si/Sn = 30-144; 1.4-6.1 wt% Sn) [4] and higher framework Sn densities than reported for isopropanol reflux (Si/Sn > 120) [3], because Sn incorporation is limited during isopropanol reflux (383 K) by competitive adsorption of isopropanol at vacancy defects, consistent with IR and TPD evidence [4]. As a result, dichloromethane-assisted grafting of stannic chloride into vacancy defects provides a route to prepare Sn-Beta zeolites with controlled densities of framework Sn and residual vacancy defects. IR spectra of Sn-Beta samples titrated with CD_3CN (303 K) was used to quantify the distribution of Sn heteroatoms in an open ((HO)-Sn-$(OSi)_3$, 2316 cm^{-1}) and closed ($Sn(OSi)_4$, 2308 cm^{-1}) configuration [5], providing evidence that open Sn sites are formed preferentially at low Sn framework densities. Thus, first-order glucose-fructose isomerization rate constants (373 K) decreased systematically with increasing Sn content when normalized to the total number of Sn atoms. Isomerization rate constants (373 K) normalized to open Sn sites, however, were invariant (within 2x) among twelve Sn-Beta samples prepared via liquid-phase grafting under dichloromethane reflux (333 K), highlighting the single-site behavior of open framework Sn sites in post-synthetically prepared Sn-Beta for glucose isomerization [4]. Vapor-phase CH_3OH adsorption isotherms (293 K) of Sn-Beta zeolites of varied Sn content and residual density of vacancy defects suggest that removing internal hydrophilic binding sites may increase glucose-to-fructose isomerization rate constants (per open Sn, 373 K) for post-synthetically prepared solid Lewis acids. These findings provide insight into how reflux solvents participate in grafting procedures used to incorporate heteroatoms within vacancy defects, and highlight the importance of quantifying active sites to normalize rate data prior to kinetic interpretation.

[1] P. Li et al., J. Phys. Chem. C 115 (2011) 3663-3670.

[2] C. Hammond et al., Angew. Chem. Int. Ed. 51 (2012) 11736-11739.

[3] J. Dijkmans et al., J. Catal. 330 (2015) 545-557.

[4] J.C. Vega-Vila et al., J. Catal. 344 (2016) 108-120.

[5] J.W. Harris et al., J. Catal. 335 (2016) 141-154.

499603 Structure Function Relationship between Catalyst Hydrophobicity and Water Tolerance of Alkyl-Modified SBA-15-Supported Propylsulfonic Acid Catalysts

Abstract:

Upgrading of fast pyrolysis-derived bio-oils via acid-catalyzed esterification is hindered by high water content of the bio-oil. This study focused on understanding the relationship between catalyst hydrophobicity and water tolerance using SBA-15-supported propylsulfonic acid catalysts on a model esterification reaction. In addition to the propylsulfonic acid surface moiety, alkyl modifiers of varying lengths were introduced to the surface to change the hydrophobicity of the catalyst active site. The catalysts were tested under dry and wet reaction conditions on the esterification of acetic acid with methanol in the liquid phase. A structure function relationship between hydrophobicity and water tolerance was determined via calorimetry and reaction studies. The degree of hydrophobicity of the catalyst was probed using solution calorimetry. The hydrophobically modified catalysts were immersed in water and the heat evolved was used to quantify the level of hydrophobicity of the catalyst surface. Water tolerance was quantified by measuring reaction rates at varying initial concentrations of water up to 5 M and calculating the water reaction order. Apparent activation energies were determined at increasing water concentrations between 30 °C and 70 °C. The importance of considering non-ideal thermodynamics (i.e. activity coefficients) in solutions was addressed in this system.

501805 Promoted MoS_2 Edge Atoms for Highly Efficient CO_2 Conversion to Syngas

Abstract:

Development of advanced materials that catalyze electrochemical reduction of CO
renewable electricity source has been proven as a promising route to close the anthropogenic carbon cycle.[1,2] Recently, transition metal dichalcogenides (TMDCs) have shown a remarkable potential to be used as a catalyst for CO_2 reduction reaction. It has been shown that the unique electro catalytic properties of this class of materials are due to the low work function and significant overlap of the d-band partial density of states with the Fermi energy level.[5,6]However, unlike noble metals the rate limiting reaction step in TMDCs for conversion of CO_2 to syngas (CO and H_2) is CO* desorption rather than COOH* formation.[4] Here we investigate the effect of stimulating the electronic property of MoS_2, as a model structure of TMDCs, to see if the reaction mechanism can be tailored toward higher CO formation turnover frequencies (TOF).

For this purpose, Niobium (Nb) and Tantalum (Ta) doped MoS_2 samples were synthesized using chemical vapor deposition (CVD) method with different concentrations of dopants, and the electro catalytic performance of doped samples were compared with pristine MoS_2 for CO_2 reduction. Different experimental methods such as in-situ differential electrochemical mass spectrometry (DEMS), scanning transmission electron microscopy (STEM), X-ray photoelectron spectroscopy (XPS) as well as Density functional theory (DFT) were utilized to fully study the role of doping on the electrocatalytic performance of MoS_2. Our results indicate a volcano trend for electrocatalytic performance of Nb doped MoS_2 samples in which $Mo_{0.95}Nb_{0.05}S_2$ sample shows one order of magnitude higher CO formation turnover frequency (TOF) compared to pristine MoS_2, however higher concentrations of Nb doping show negative effect on the catalytic activity of MoS_2. The results also show negative effect of doping in Ta doped MoS_2 samples for all concentrations of dopant. Our DFT results show that introducing optimum concentrations of proper dopants near the edge structure would lead to a Sabatier effect by decreasing the binding strength between the Mo edge and CO, giving rise to the CO formation TOF. However, this is not the case for Nb doping with higher concentrations or Ta doping in which higher work function of doped structures lead to poorer electron transfer properties.

References:

(1) Haszeldine, R. S. Carbon Capture and Storage: How Green Can Black Be? *Science (80).* **2009**, *325*, 1647–1652.

(2) Blanchard, L. a; Hancu, D. Green Processing Using Ionic Liquids and CO2. *Nature* **1999**, *399*, 28–29.

(3) Abbasi, P.; Asadi, M.; Liu, C.; Sharifi-Asl, S.; Sayahpour, B.; Behranginia, A.; Zapol, P.; Shahbazian-Yassar, R.; Curtiss, L. A.; Salehi-Khojin, A. Tailoring the Edge Structure of Molybdenum Disulfide toward Electrocatalytic Reduction of Carbon Dioxide. *ACS Nano* **2017**, *11*, 453–460.

(4) Asadi, M.; Kim, K.; Liu, C.; Addepalli, A. V.; Abbasi, P.; Yasaei, P.; Phillips, P.; Behranginia, A.; Cerrato, J. M.; Haasch, R.; *et al.* Nanostructured Transition Metal Dichalcogenide Electrocatalysts for CO2 Reduction in Ionic Liquid. *Science (80).* **2016**, *353*, 467–470.

(5) Nørskov, J. K.; Bligaard, T.; Hvolbaek, B.; Abild-Pedersen, F.; Chorkendorff, I.; Christensen, C. H.; Norskov, J. K.; Bligaard, T.; Hvolbaek, B.; Abild-Pedersen, F.; *et al.* The Nature of the Active Site in Heterogeneous Metal Catalysis. *Chem Soc Rev* **2008**, *37*, 2163–2171.

(6) Rosen, B. A.; Salehi-Khojin, A.; Thorson, M. R.; Zhu, W.; Whipple, D. T.; Kenis, P. J.; Masel, R. I. Ionic Liquid-Mediated Selective Conversion of CO(2) to CO at Low Overpotentials. *Science (80).* **2011**, *334*, 643–644.

500294 Framework-Topology-Dependent Catalytic Activity of Zirconium-Based, Porphyrinic Metal-Organic Frameworks

Abstract:

One of the goals of rational catalyst design is to increase the rates at which reactions occur in catalytic systems. For suitable reactions, a possible strategy to achieve such rate increase is to design catalysts with active sites that are both abundant and spatially distributed in such fashion that concerted binding of reactants to neighboring sites facilitates the formation of a transition state or intermediate. Metal-organic frameworks (MOFs) are porous crystalline materials constituted by interconnected organic and inorganic building blocks, which can provide well-defined active sites. Because the spatial distribution of building blocks is a function of the underlying framework topology, the topological diversity of MOFs can be exploited to alter the concentration, position and orientation of catalytic sites to enhance catalytic performance. Metalated porphyrins are attractive catalytic sites due to their known ability to catalyze small molecule transformations.

To study to what extent the topological arrangement of immobilized porphyrin metal centers affects catalytic activity of an AB + C -> A + BC reaction, in this work we synthesized three MOFs interconnecting metalated porphyrins and zirconium oxoclusters—which are known to provide high MOF stability—into frameworks of three distinct topologies (*ftw*, *csq*, and *scu*) and tested their catalytic activity for the acyl transfer from N-acetylimidazole (NAI) to three different pyridylcarbinol (PC) regioisomers (2-PC, 3-PC, and 4-PC). The synthesized and tested catalytic MOFs correspond to the well-known **MOF-525** and **PCN-222** materials and the newly introduced **NU-902**. Collected kinetic data reveals rate dependence on both framework topology and PC regioisomerism, which is discussed in the light of molecular modeling results.

Initial reaction rates for 3-PC show that the immobilization of metalated porphyrins into a heterogeneous MOF catalyst increases the reaction rate 40- to 140-fold with respect to the uncatalyzed reaction and 20- to 70-fold with respect to the homogeneous catalyst. The enhancement with respect to the homogenous catalysts is due to pre-concentration (i.e. binding of NAI or PC to porphyrin centers) and orientation effects. Initial reaction rates for all PC/MOF combinations show a dependence on the MOF (topology) and the PC regioisomer. The differences for PC/MOF combinations arise from the different spatial arrangement of porphyrins in the three MOFs. Pre-concentration effects are found to be proportional to the density of Zn sites in each MOF and orientation effects to be proportional to the density of "suitable" Zn pairs. Suitable pairs for each MOF were determined based on MM calculations determining the strain of anchored intermediates. DFT calculations show that contrary to 3-PC and 4-PC, Zn-bound 2-PC cannot orient properly relative to Zn-bound NAI on suitable pairs as for its reaction rate to benefit from orientation effects. Thus, we demonstrated how a high density of catalytic sites with rationally selected spatial orientations can be provided with MOFs to facilitate the formation of relevant transition states and enhance reaction rates. We hope this will motivate exploiting the topological diversity of MOFs for multisite-catalyzed chemical reactions.

References

1. Deria, P., Gomez-Gualdron, D.A., Hod, I. Snurr., R.Q., Hupp, J.T., Farha, O.K. *J. Am. Chem. Soc.*138, 14449 (2016)

1.

501904 Constrained Geometry Single Site Catalysts for Olefin Polymerization

Abstract:

Constrained Geometry Single Site Catalysts for Olefin Polymerization

ABSTRACT

Polyolefins are commercially important plastics. These are available in compounds of widely different structure as copolymers, in polymer mixtures as filled compositions and as cross-linkable products etc. Single site polymerization catalysts, with the generic name of metallocenes, are currently used industrially to produce various polyolefins with more precise molecular architecture than those can be produced by any conventional polymerization methods like Zeigler-Natta type catalysts. Metallocenes are known for the stereoselective polymerization of a range of polymers. It is now a well-accepted fact that metallocene offers several advantages over the traditional Zeigler-Natta catalysts. Many metallocenes are up to hundred-fold more active because the initiators are homogeneous and nearly each atom of transition metal is active. Stereoreselectivity can be controlled and varied to produce different stereoregular products by appropriate choice of the metallocene ligands and reaction conditions. In single site polymerization, there are different kinetic processes, including catalyst activation and deactivation, chain propagation, chain transfer by beta-hydrogen elimination etc. These reactions are mediated by the metal/ligand/counter-ion complex, where the key challenge is to first understand, and then rationally design, the ligands and the counterion to produce a polymer with a desired molecular architecture. The major impact of Metallocene catalysts on the polyolefin industry is related to the fact that these catalysts can produce polymers with controlled microstructure. Presently polymer manufacturing companies the world over are examining the possibility of using the metallocene catalyst to produce polyolefins. In this paper, the development of the metallocene catalysts and co-catalysts from their discovery to their present state of the art is presented. This paper also focuses on various mechanistic aspects of applications of single site catalyst systems in polymer industry.

495312 Limitations of Top-Down Synthesis and Chloride Based Bifunctional Polymer Solid Acid Catalysts for Cellulose Hydrolysis

Abstract:

Limitations of Top-Down Synthesis and Chloride Based Bifunctional Polymer Solid Acid Catalysts for Cellulose Hydrolysis

Maksim Tyufekchiev, Michael T. Timko, Marion Emmert,

Worcester Polytechnic Institute, Worcester, MA

Pu Duan and Klaus Schmidt-Rohr

Brandeis University, Waltham, MA

Sergio Granados-Focil

Clark University, Worcester, MA

Solid acid catalysts have been proposed as a viable alternative to liquid acids and enzymes for cellulose hydrolysis for the recovery of glucose and other valuable compounds. The low catalytic activity of materials possessing only acidic groups in their chemical structure has motivated a new approach. A dominant hypothesis in the literature suggests that introduction of specific binding groups to the catalyst structure in addition to strong acidic groups increases the interaction time between cellulose and the catalyst resulting in increase of hydrolysis rate. A typical procedure for synthesis of such bifunctional is top-down modification of insoluble polymer precursors either commercially available or polymerized prior to modification. Materials that possess chloride as a binding group, thought to interact with cellulose by forming a hydrogen bond with a hydroxyl group, have shown remarkable activity towards cellulose hydrolysis to glucose and levulinic acid. However, the mechanism of catalysis involving solid acid and cellulose solid particles has not been elucidated.

In this work, we investigated the literature hypothesis of structure-related catalytic activity by detailed catalyst structure characterization and a series of hydrolysis tests. Commercially available cross-linked chloromethyl polystyrene beads were functionalized by a well-known synthesis procedure to introduce sulfonic acid groups to the structure, while retaining part of the chlorine groups. The catalyst was characterized using NMR, FTIR, Raman microscopy, and EDS. Interestingly, we discovered that part of the chloromethyl groups in the polymer was converted to hydroxymethyl groups, which could also form hydrogen bonds. Hydrolysis in water of cellobiose (soluble) and cellulose (insoluble) at 150 °C and 175 °C was carried out to establish the catalytic activity. The bifunctional catalyst was able to completely convert cellobiose to glucose and cellulose to glucose, levulinic acid, and formic acid. For comparison, a completely sulfonated catalyst, lacking chloromethyl groups that could potentially bind, showed low catalytic activity towards cellulose hydrolysis.

To elucidate the structure of the catalysts, we performed cross-sectional analysis with Raman microscopy and EDS, each with spatial resolution of 1-2 microns. The analysis revealed the chlorine was almost completely lacking from the structure of the bifunctional catalyst, while the sulfonic acid groups were predominantly on the surface. NMR analysis of the catalyst after hydrolysis reaction revealed that the chloromethyl groups were converted to hydroxyl groups in the hydrothermal environment. Such a degradation would result in generation of hydrochloric acid. To confirm this we tested the hydrothermal stability of the bifcuntional catalyst and ion chromatography and pH measurements confirmed chloride leaching to form acid.Based on the results from the catalyst characterization and hydrolysis reactions, we conclude two things: 1) top-down modification of solid material for catalyst preparation does not result in uniform chemical structure and 2) catalysts utilizing chloride as binding groups are not stable in hydrothermal conditions and leach hydrochloric acid. We suggest that benzylic alcohol groups may instead contribute to cellulose binding. This work points to the disadvantages of top-down bifunctional catalyst synthesis and the hydrothermal stability of chlorides, suggesting some new explanations to account for their performance in cellulose hydrolysis.

501842 Advanced Catalyst for Energy Conversion and Storage Systems

Abstract:

Energy conversion and storage systems have received much more attentions as fossil fuel alternatives due to their great potential to store energy into the chemical bonds[1–5]. In these systems, energy can be converted or stored into chemical bonds as a form of fuels or electricity using renewable energy sources e.g. solar and wind energy[6–8]. However, the performance of these systems are hindered due to the high cost and relatively poor performance i.e. low reaction rates and high overpotentials of state-of-the-art catalysts (e.g., noble metals)[9–11]. Although many physical and chemical approaches have been employed to enhance the existing materials catalytic performance, it appears to be a fundamental limit, connected to the electronic structure of these catalysts.

Here, I will present my recent effort to design, synthesis and characterization of transition metal dichalcogenides (TMDCs) class of catalysts with unique electronic properties suitable for energy conversion and storage systems. We have tested the performance of this class of catalysts for the carbon dioxide (CO_2) reduction reaction, oxygen reduction reaction and oxygen evolution reactions. Our results include 100 times higher catalytic activity for the CO_2 reduction reaction, exceeding the performance of state-of-the-art catalysts and highly efficient bi-functional catalysts for oxygen reduction and evolution reactions compared to expensive noble metal catalysts such as platinum and gold in the aprotic media. We also tested the performance of molybdenum disulfide nanoflakes (MoS_2 NFs) -a member of TMDCs- in lithium-oxygen batteries known as potential alternative of lithium-ion batteries. I will discuss these and other results including the potential of our recent discovery to open a new route towards energy efficient, highly active and cost effective energy storage and conversion systems to replace fossil fuels.

References:

1. Asadi, M. *et al.* Robust carbon dioxide reduction on molybdenum disulphide edges. *Nat. Commun.* **5,** 4470 (2014).

2. Behranginia, A. *et al.* Highly Efficient Hydrogen Evolution Reaction Using Crystalline Layered Three-Dimensional Molybdenum Disulfides Grown on Graphene Film. *Chem. Mater.* **28,** 549–555 (2016).

3. Asadi, M. *et al.* Cathode Based on Molybdenum Disulfide Nanoflakes for Lithium-Oxygen Batteries. *ACS Nano* **10,** 2167–2175 (2016).

4. Lu, J. *et al.* A lithium–oxygen battery based on lithium superoxide. *Nature* **529,** 377–382 (2016).

5. Abbasi, P. *et al.* Tailoring the Edge Structure of Molybdenum Disulfide towards Electrocatalytic Reduction of Carbon Dioxide. *ACS Nano* acsnano.6b06392 (2016). doi:10.1021/acsnano.6b06392

6. May, M. M., Lewerenz, H.-J., Lackner, D., Dimroth, F. & Hannappel, T. Efficient direct solar-to-hydrogen conversion by in situ interface transformation of a tandem structure. *Nat. Commun.* **6,** 8286 (2015).

7. Jingshan, L. *et al.* Water photolysis at 12.3% efficiency via perovskite photovoltaics and Earth-abundant catalysts. *Science (80-.).* **345,** 1593–1596 (2014).

8. Chang, X., Wang, T. & Gong, J. CO2 Photo-reduction: Insights into CO2 Activation and Reaction on Surfaces of Photocatalysts. *Energy Environ. Sci.* (2016). doi:10.1039/C6EE00383D

9. Chen, Y., Li, C. W. & Kanan, M. W. Aqueous CO 2 Reduction at Very Low Overpotential on Oxide-Derived Au Nanoparticles. *J. Am. Chem. Soc.* **134,** 19969–19972 (2012).

10. Li, C. W. & Kanan, M. W. CO 2 Reduction at Low Overpotential on Cu Electrodes Resulting from the Reduction of Thick Cu 2 O Films. *J. Am. Chem. Soc.* **134,** 7231–7234 (2012).

11. DiMeglio, J. L. & Rosenthal, J. Selective Conversion of CO 2 to CO with High Efficiency Using an Inexpensive Bismuth-Based Electrocatalyst. *J. Am. Chem. Soc.* **135,** 8798–8801 (2013).

487678 Study on Mechanism of Clinker Formation in Bamboo Combustion Process

Abstract:

In this study, we burn the bamboo powder packed layer placed on stainless steel mesh to know how the clinker is made and also we examined the possibility of clinker reduction by adding bark powder to bamboo powder. We carried out the mass change of the mesh after burning, mesh surface observation and also measured the adsorption amount and size of clinker. The following results were obtained.As a result, In the process of clinker formation, two stages were confirmed. At stage 1, about 10 Î¼m of fine white particles adsorbed onto the stainless steel mesh at the volatile component combustion of about 1000 K. In step 2, it was suggested that during combustion of char having a surface temperature of about 1200 K, these small grains melted and turned into large chunks of about 200 Î¼m.Furthermore, it was found that clinker clogging on the surface of the stainless steel mesh is reduced by mixing bamboo powder with bark and burning it. This proved that bark contains CaO with high melting point.

493264 Modeling Pyrolysis-Induced Microstructural Changes in Biomass: A Cellular Automata Approach

Abstract:

Lignocellulosic biomass is a chemically and morphologically heterogeneous material. Furthermore, this heterogeneity is in-part responsible for the vast number of thermal decomposition products seen in pyrolysis events. While modeling of biomass pyrolysis has been a subject of much research in past years at length-scales from the macro to molecular, the majority of these works have focused on a range of continuum-based approaches. Though effective at capturing global outcomes, these approaches are less tractable as frameworks for capturing microstructural effects and upscaling molecular information. This work demonstrates the use of kinetic cellular automata (kCA) as an alternative platform for the modeling and simulation of biomass pyrolysis. Asides from being effective at capturing transport and chemical processes in highly heterogeneous system, kCA is capable of modeling microstructural changes that occur as a result of chemical and physical transformations. A number of earlier benchmark trials demonstrated the convergence of the kCA to global continuum outcomes. Application of the kCA to actual biomass microstructures show promise for this platform as an intermediate length-scale tool capable of predicting char morphologies that mimic experimental outcomes at meso-scale resolution.

487672 Shrinking and Heat and Mass Transfer during Pyrolysis of Woody Biomass

Abstract:

Heat and mass transfer during pyrolysis of major components bed have been investigated numerically. For Cellulose powder, the calculation temperature and velocity vector distributions had a quasi-steady state during pyrolysis because the gas volume increased near the side wall of the packed bed and then the effective thermal conductivity decreased. The calculation temperature with the effect of volume shrinking of the packed bed agreed qualitatively with the experimental ones for all kinds of powder when the temperature plateau appeared.

493123 Plasma Gasification of Wood Pellets Using a Laboratory Scale Plasma Reactor to Produce Synthesis Gas: Effect of Increasing Temperature and Addition of Oxygen to the Process

Abstract:

A laboratory scale plasma reactor (max 15Kw power) has been used to convert wood pellets (biomass) into synthesis gas that can be used for downstream processes. A series of experiments have been conducted to see the effects of increasing temperature and adding oxygen to the process on conversion of hydrogen and carbon monoxide. The results show that increase in temperature from 400°C to 1000°C results in overall increase in biomass conversion (mass %) from 30% to 82%. In the case where no additional oxygen is supplied, the carbon and hydrogen conversion increases with increase in temperature up to 82% and 98% respectively but the molar ratio of carbon monoxide and hydrogen is unchanged at 1:1 for the temperatures studied. Addition of controlled oxygen increases the carbon monoxide yield and also results in the CO/H2 ratio of 3:2 for temperature range 800°C -1200°C. Simultaneous increase in reactor temperature and oxygen amounts to the process resulted in increased biomass conversion and a reduction of residues in the reactor.

497756 Multistage Fluidized Bed Reactor for Gasification

Abstract:

An ideal gasification process for a solid organic feedstock entails direct conversion of the feedstock into a synthesis gas, primarily CO and H_2, without any volatile or char byproducts. These byproducts coat the internal surfaces of the gasifier, rapidly degrading the performance of the gasifier and decreasing the yield of synthesis gas.

A three stage fluidized bed reactor has been developed to produce a clean synthesis gas in high yield. High quality synthesis gas can be used as a fuel for a combustion turbine or as feedstock for a chemical refinery. The first stage is the fluidized bed reactor where both the feedstock and heat transfer particles are fluidized in order to quickly heat the feedstock to a high temperature to achieve rapid pyrolysis and gasification. The second stage has a larger cross-sectional area than the first stage. This increase in area minimizes entrainment of solids in the second stage and increases the residence time to gasify char and decompose volatiles into low molecular weight gases. Typically the second stage operates at a high temperature in order to minimize condensation. The diameter of the third stage is greater than the diameter of the second stage in order to permit more char gasification and volatiles decomposition. If most of the volatiles have decomposed before entering stage three then the gases can be partially cooled in stage three. Following stage three the product stream passes through a cyclone to remove some of the entrained solids. Then the gases pass through a condenser to recover residual volatiles. Following the condenser gas samples are captured periodically for chemical analysis.

In the initial trials using this reactor, pine powder was used as the feedstock and alumina particles were used as the heat transfer medium. The temperature profile was set at 600 °C for stage 1, 800 °C for stage 2 and 300 °C in stage 3. Steady state operation was not achieved due to tar deposits on the alumina particles and the reactor walls. A close approach to steady state has been achieved by switching from alumina to zirconia particles and setting each stage at 800 °C. CO_2 was used as the fluidizing gas. Steady state was maintained for about 100 minutes and then the reactor was shut down. At this high temperature less fouling was apparent and the corresponding carbon yield in the gas fraction was about 74 %.

495901 Overall View of Lignocellulosic Biomass Decomposition in Supercritical Water

Abstract:

Supercritical water gasification is a promising technology for gasifying wet biomass. However, detail reaction kinetics of each component, cellulose, hemicellulose, and lignin have not been elucidated, and the interaction between the components has to be elucidated. In this study, mixture of cellulose, hemicellulose, and/or lignin was delivered into a continous reactor for supercritical water gasification and its reaction characteristics were determined. Reaction temperature was changed around critical point of water so that both subcritical and supercritical conditions were considered while the reaction pressure was kept at 25 MPa. Intermediate compounds were analyzed by HPLC, and comparison between avaraged single copound result and mixture was made. The interaction was clearly observed, especially in terms of hydrogen consumption. Reaction network was develped, and reaction rate constants were deteremined for each of the reaction by least square method to fit the experimental data with the calculation by the model. The Arrhenisc plot for these reactions classified ionic reaction and radical reaction. For the case of radical reacitons, Arrhenius plot showed a straight line as usual, but for the case of ionic reactions, reaction rate kinetics deviated from the straight line at high temperature, indicating ions get unstable under supercritical condition. The overall reaction kinetics was also determined, and discussed.

AUTHOR INDEX

AUTHOR INDEX

AUTHOR INDEX

AUTHOR INDEX

AUTHOR INDEX

AUTHOR INDEX

AUTHOR INDEX

AUTHOR INDEX

AUTHOR INDEX

AUTHOR INDEX

AUTHOR INDEX

AUTHOR INDEX

AUTHOR INDEX

AUTHOR INDEX

AUTHOR INDEX

AUTHOR INDEX

AIChE
120 Wall Street, FL 23
New York, NY 10005-4020

ISBN 978-1-5108-5806-0